P9-DTE-826

STRUCTURAL SAFETY & RELIABILITY
VOLUME 3

PROCEEDINGS OF ICOSSAR'93 – THE 6TH INTERNATIONAL CONFERENCE ON STRUCTURAL SAFETY AND RELIABILITY / INNSBRUCK / AUSTRIA / 9-13 AUGUST 1993

Structural Safety & Reliability

Edited by

G.I.SCHUËLLER
University of Innsbruck, Austria

M.SHINOZUKA
Princeton University, New Jersey, USA

J.T.P.YAO
Texas A & M University, College Station, Texas, USA

VOLUME 3

A.A.BALKEMA / ROTTERDAM / BROOKFIELD / 1994

The texts of the various papers in this volume were set individually by typists under the supervision of each of the authors concerned.

Published by
A.A. Balkema, P.O. Box 1675, 3000 BR Rotterdam, Netherlands
A.A. Balkema Publishers, Old Post Road, Brookfield, VT 05036, USA

For the complete set of three volumes, ISBN 90 5410 357 4
For Volume 1, ISBN 90 5410 377 9
For Volume 2, ISBN 90 5410 378 7
For Volume 3, ISBN 90 5410 379 5

Printed in the Netherlands

Structural Safety & Reliability, Schuëller, Shinozuka & Yao (eds) © 1994 Balkema, Rotterdam, ISBN 90 5410 357 4

Table of contents

Importance sampling

Simulation and fuzzy sets

Structural control

Wind engineering

Wind engineering (ongoing research)

Computational stochastic dynamics

Damage assessment and limit states

Risk assessment

Risk assessment (ongoing research)

Geotechnical engineering

Geotechnical engineering (ongoing research)

Earthquake engineering

Earthquake engineering (ongoing research)

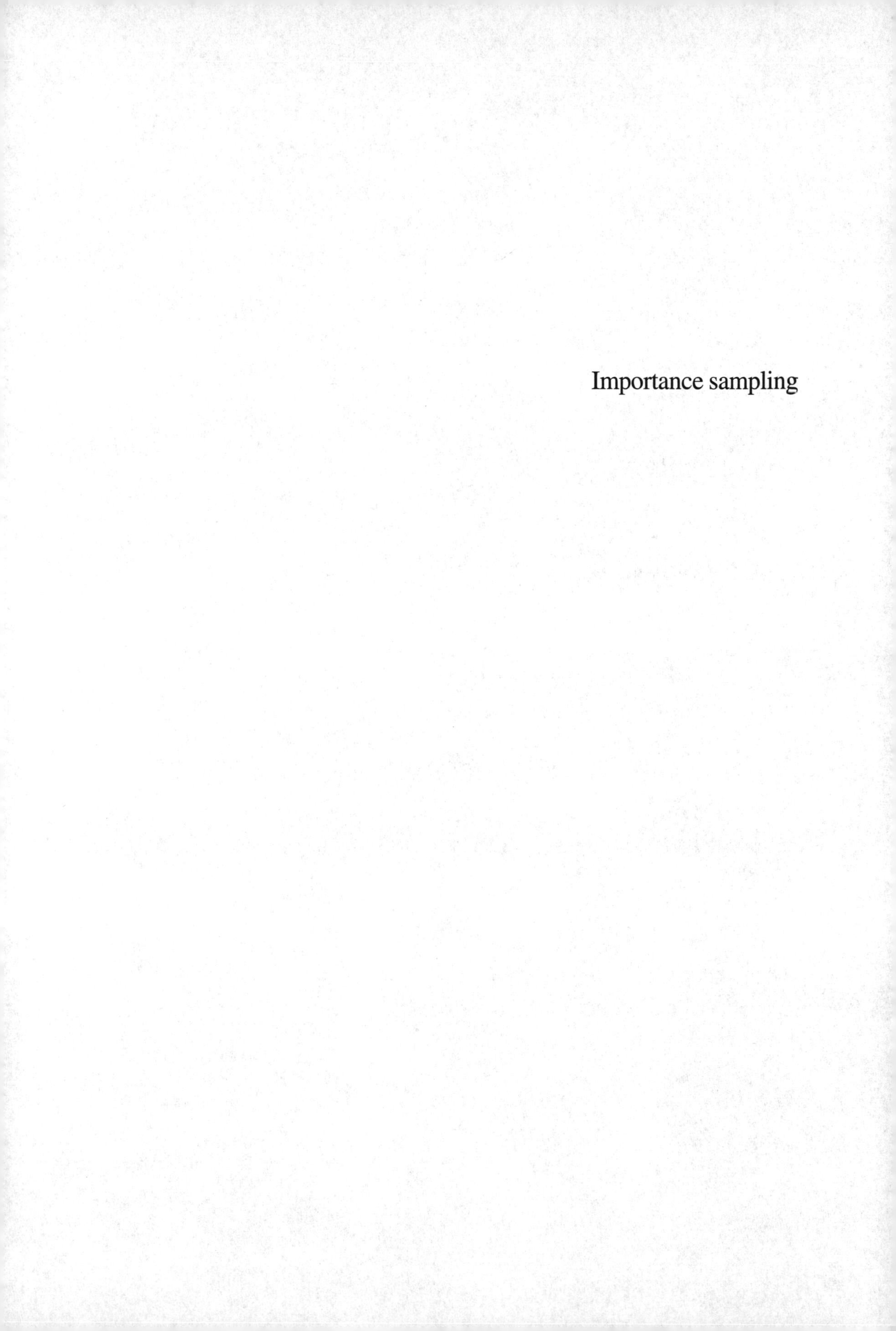

Importance sampling

Structural Safety & Reliability, Schuëller, Shinozuka & Yao (eds) © *1994 Balkema, Rotterdam, ISBN 90 5410 357 4*

Variance reduction in simulation for first order problems

Gongkang Fu
New York State Department of Transportation, Albany, N.Y., USA

ABSTRACT: This paper presents an importance sampling method for the first order problem of structural system reliability analysis, which has a failure domain defined by linear or linearized functions. Truncated multimodal simulation is suggested, providing an advantage of locating all samples in the failure domain. Its variance is evaluated by an analytically derived upper bound. The ratio of this maximum variance to that of the conventional Monte Carlo method is used as a variance change factor for conservative estimation of variance reduction. The upper bound of variance can be used for a priori determination of required sample size, given an acceptable maximum error associated with a confidence level. Various application examples of both series and parallel systems are included for illustration.

1. INTRODUCTION

A structural system reliability problem can be formulated as assessment for system failure probability P_f:

$$P_f = \int_x G(x)f(x)\,dx \qquad (1)$$

where f(x) is the normal probability density function of random variable vector **X** with mean vector $\underline{\mathbf{X}}$ and symmetric covariance matrix **C** that is positive definite; $\mathbf{x} \in \mathbb{R}^n$ is realization of **X** where n is the space dimension of the random variable vector; and G(x) is an indicator for the structure's limit state defined as

$$G(x) = \begin{cases} 0 & \text{any } g_m(x) > 0 \\ 1 & \text{all } g_m(x) \leq 0 \end{cases} \qquad (m=1,2,\ldots,M) \qquad (2)$$

The normal distribution assumption for **X** imposes no restriction on generality of the problem, since random variables can be transformed to normal variables in general [Hohenbichler and Rackwitz 1981]. Note that a parallel system is treated in this paper as a special case of series one with M=1 (with one design point). The first order problems are defined here as those with linear or appropriately linearized failure surfaces, i.e.

$$g_m(x) = \mathbf{a}_m^t(x - \mathbf{x}_m^*) \qquad (m=1,2,\ldots,M) \qquad (3a)$$

where $\mathbf{x}_m^*$ is the maximum likelihood point (or design point) for mode m, and $\mathbf{a}_m$ is the gradient vector of failure function $z_m(x)$ at $\mathbf{x}=\mathbf{x}_m^*$:

$$\mathbf{a}_m = \nabla z_m(x)\big|_{\mathbf{x}=\mathbf{x}_m^*} \qquad (3b)$$

This type of problem is commonly of practical interest. Further, a problem with nonlinear failure surface(s) can be always approximated by a series of piecewise linear hyperplanes, and formulated as a first order problem defined here. On the other hand, attention is needed to assure approximation with desired accuracy.

An importance sampling method is introduced for the first order problems. A major advantage of this method is to locate all samples in the failure region. A theoretical analysis of its variance, error, and confidence level is also provided, based on an analytically derived upper bound for the variance. This information can be used in a priori determination of required sample size, given an acceptable maximum error associated with a desirable confidence level.

2. TRUNCATED MULTIMODAL IMPORTANCE SAMPLING

A multimodal simulation scheme is suggested:

$$p(\mathbf{x}) = \sum_{m=1,M} w_m \, p_m(\mathbf{x}) \tag{4}$$

where $p_m(\mathbf{x})$ is the normal probability density function with mean vector $\mathbf{x}_m^*$ and covariance matrix $\mathbf{C}$, and truncated by the hyperplane $\mathbf{a}_m^t(\mathbf{x}-\mathbf{x}_m^*)=0$:

$$p_m(\mathbf{x}) = \begin{cases} \dfrac{2}{(2\pi)^{n/2}|\mathbf{C}|^{1/2}} \exp\{-0.5(\mathbf{x}-\mathbf{x}_m^*)^t\mathbf{C}^{-1}(\mathbf{x}-\mathbf{x}_m^*)\} & \mathbf{a}_m^t(\mathbf{x}-\mathbf{x}_m^*)\le 0 \\ 0 & \mathbf{a}_m^t(\mathbf{x}-\mathbf{x}_m^*)>0 \end{cases} \tag{5}$$

where $|\mathbf{C}|$ is determinant of the covariance matrix $\mathbf{C}$. The weight coefficients w_m are to be determined by solving the following equations:

$$f(\mathbf{x}_1^*)/f(\mathbf{x}_m^*)=p(\mathbf{x}_1^*)/p(\mathbf{x}_m^*) \qquad (m=2,3,\ldots,M)$$

$$\sum_{m=1,M} w_m = 1; \; w_m \ge 0 \;\; (m=1,2,..,M) \tag{6}$$

Computation of Eq.1 using this sampling scheme is conducted in M groups of simulation. Each group generates correlated normal variates $\mathbf{x}_{k\_m}$ ($k\_m=1,2,\ldots,N_m$; $m=1,2,\ldots,M$) from $p_m(\mathbf{x})$, i.e.

$$P_f = \sum_{m=1,M} w_m \left\{\int_{\mathbf{x}} \frac{G(\mathbf{x})f(\mathbf{x})}{p(\mathbf{x})} p_m(\mathbf{x})d\mathbf{x}\right\}$$

$$\approx \sum_{m=1,M} \frac{w_m}{N_m} \left\{\sum_{k\_m=1,Nm} \frac{G(\mathbf{x}_{k\_m})f(\mathbf{x}_{k\_m})}{p(\mathbf{x}_{k\_m})}\right\}$$

$$= P_f' \tag{7}$$

where P_f' denotes the estimator for P_f. The computation in parentheses is one of the M groups of simulation corresponding to mode m with samples $\mathbf{x}_{k\_m}$ ($k\_m=1,2,\ldots,N_m$) from subdensity $p_m(\mathbf{x})$. N_m, the number of samples for mode m, is determined by

$$N_m \approx w_m N \quad (m=1,2,\ldots,M) \text{ and}$$

$$\sum_{m=1,M} N_m \approx N \tag{8}$$

to reflect relative importance of respective failure modes, with N_m ($m=1,2,\ldots,M$) being integers.

Note that the effort of solving the M simultaneous linear equations in Eq.6 is insignificant, as M is usually small when only significant failure modes are included. Further note that solution to Eq.6 may not exist, when some of the maximum likelihood points are too close to one another in the space of **x** (for example, w_m may be found to be negative). These close points (and corresponding failure modes) should then be covered by a single subdensity. In that case, the problem is treated as if it had correspondingly fewer failure modes in order to determine the weights for the respective modes.

3. VARIANCE REDUCTION

The mean and variance of estimator P_f' are respectively

$$E[P_f'] = P_f \tag{9a}$$

$$Var[P_f'] = \frac{1}{N}\left\{\int_{\mathbf{x}} \frac{G^2(\mathbf{x})f^2(\mathbf{x})}{p^2(\mathbf{x})} p(\mathbf{x})d\mathbf{x} - P_f^2\right\} \tag{9b}$$

and the variance is bounded by

$$Var[P_f'] \le Var[P_f']_{max}$$

$$= \frac{1}{N}\left\{P_f\left[\frac{f(\mathbf{x})}{p(\mathbf{x})}\right]_{max} - P_f^2\right\}$$

$$= \frac{1}{N}\left\{P_f \frac{f(\mathbf{x}^*)}{p(\mathbf{x}^*)} - P_f^2\right\} \tag{10}$$

using

$$[f(x)/p(x)]_{max}=f(x^*)/p(x^*) \qquad (11)$$

where x^* stands for any one of the design points x_m^* (m=1,2,...,M). Eq.11 can be proved by demonstrating that f(x)/p(x) can have extreme values only on boundaries defined by $a_m^t(x-x_m^*)=0$.

The ratio of variances by the importance sampling method and the conventional Monte Carlo method is defined here as variance change factor (VCF), assuming the same sample size N:

$$VCF = \frac{\frac{1}{P_f}\int_x \frac{G^2(x)f(x)}{p(x)} f(x)dx - P_f}{1-P_f}$$

$$\leq VCF_{max} = \frac{f(x^*)/p(x^*) - P_f}{1-P_f} \qquad (12)$$

which provides a direct means of evaluating the importance sampling method with respect to variance reduction. Using the central limit theorem, one can define

$$COV_{max}=\{\frac{1}{N}[\frac{f(x^*)}{p(x^*)P_f} -1]\}^{1/2} \text{ and}$$

$$\epsilon_{max} = k\, COV_{max} \qquad (13)$$

where ϵ_{max} is an upper bound for ϵ, acceptable error; $\epsilon'=|P_f'/P_f-1|$ is error; k is the confidence index associated with the normal probability function; and COV is the coefficient of variation of the estimator. The ratios of error level, confidence index, and sample size by the importance sampling method and the conventional Monte Carlo method are also bounded by [Fu 1988]:

$$\epsilon_{IS}/\epsilon_{MC} \leq VCF_{max}^{1/2};$$

$$k_{IS}/k_{MC} \geq VCF_{max}^{-1/2};$$

$$N_{IS}/N_{MC} \leq VCF_{max} \qquad (14)$$

where subscripts IS and MC stand for importance sampling and Monte Carlo methods.

4. APPLICATION EXAMPLES

Simulation approaches generally use pseudo-random number generators to produce random samples, whose quality is critically important. It has been found that some congruential generators produce random numbers on certain hyperplanes [Marsaglia 1968]. Thus they are not suitable for multidimensional integrals. The Generalized Feedback Shift Register generator was developed to avoid this problem [Ripley 1987]. A such generator [Fushimi and Tezuka 1983] is suggested for high dimensional problems, and it has been implemented on an IBM PS/2 Model 80 for computation of such cases below.

Example 1

This example is given by S.Engelund and R.Rackwitz at Technical University Munich, FRG, in their ongoing benchmark study on the importance sampling method. The limit state function z_1 is defined as

$$z_1(x)=\beta n^{1/2}-\sum_{i=1,n} x_i \qquad (15)$$

where X_i are independent standard normal variables ($\underline{\mathbf{X}}=\mathbf{0}$, $\mathbf{C}=\mathbf{I}$), and n and β are constants. This problem can be used to test the random number generator's quality, especially in higher dimensional cases. It is also a relatively simple problem with a single failure mode (M=1), as a first step of illustrating the present method.

Fig.1 shows the results by the present method for β=1,2,...,10 and n=1,5,10,..., 30, respectively, using N=4,000 samples. Fig.1a provides an overview comparing the estimates and exact values. Fig.1b exhibits the involved errors in more detail, contrasted with ϵ_{max} by Eq.13 that is associated with a confidence level of 95% (k=1.96). It shows that the actual errors do not exceed this maximum level in all but a few cases. They increase insignificantly with β (or equivalently, with decrease of P_f). Fig.1c contains similar comparison for COV vs.

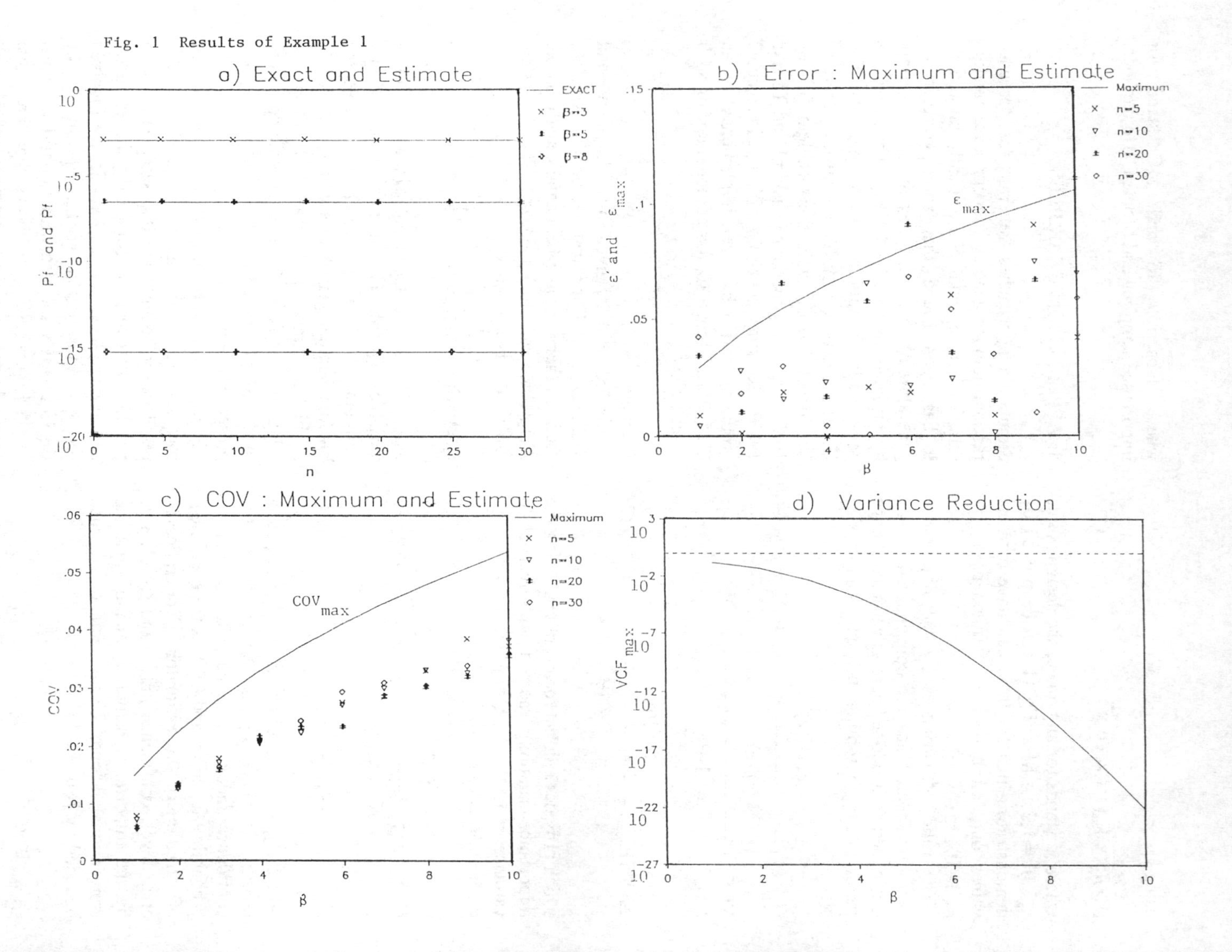

Fig. 1 Results of Example 1

Fig. 2 Results of Example 2

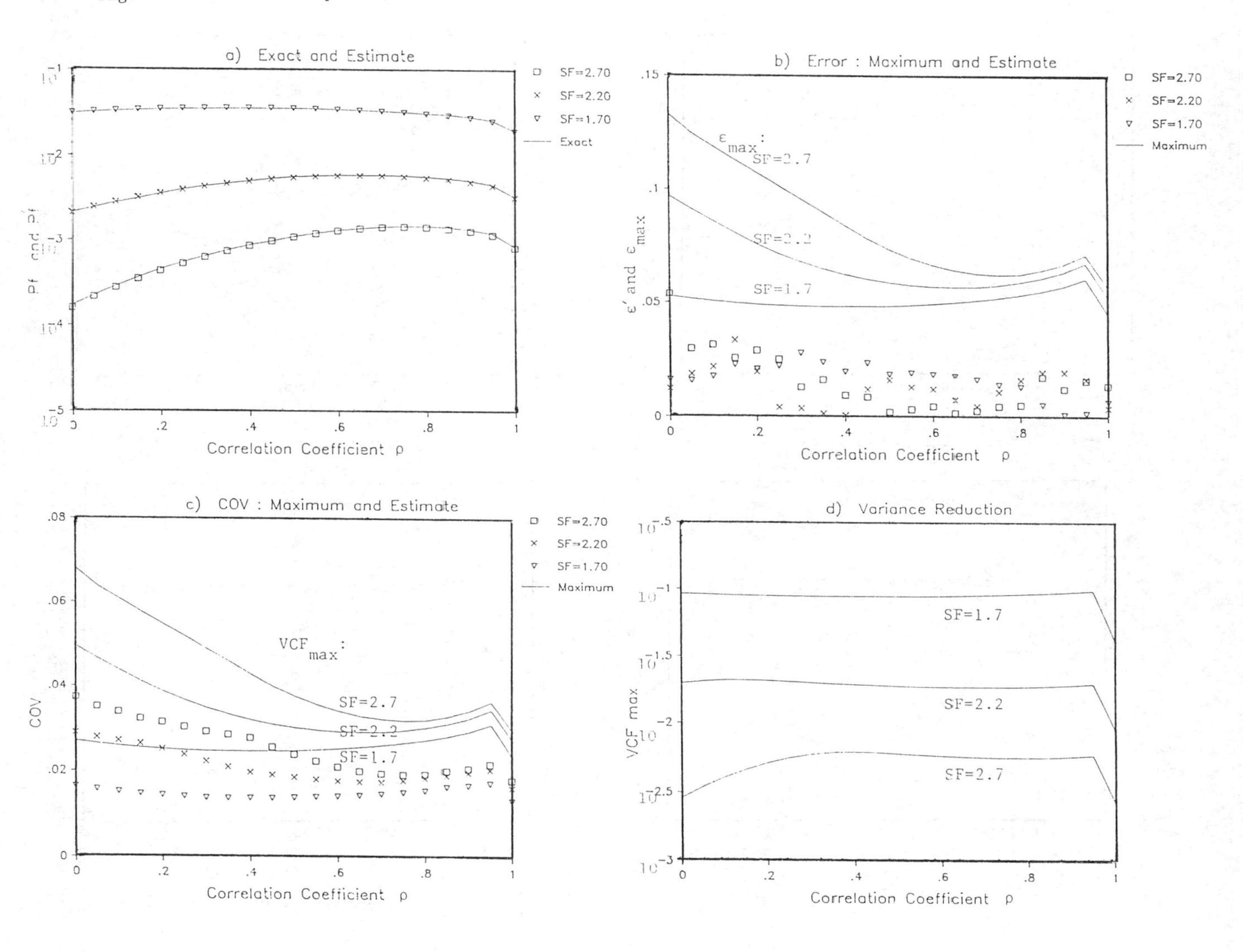

Fig. 3 Results of Example 4

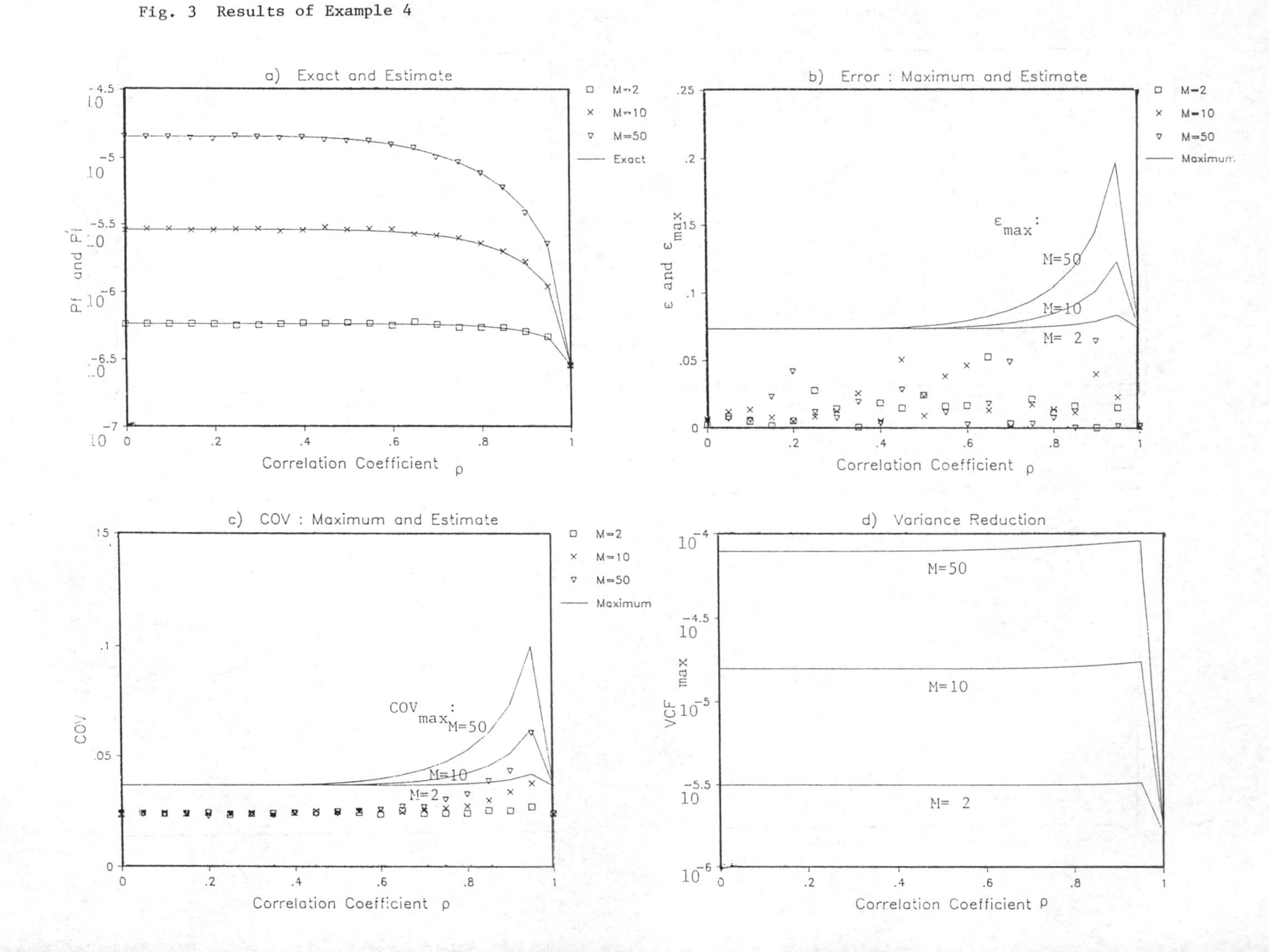

COV_{max}. The solid line is obtained by Eq.13. Estimated COV are marked points and they are bounded by COV_{max}. This also indicates the satisfactory quality of the pseudo-random number generator employed. Fig.1d shows decrease of VCF_{max} with β over a range of practical applications. This demonstrates a much desired merit that simulation efficiency and accuracy do not decrease with P_f to be estimated, as they would if the conventional Monte Carlo method were used. As a matter of fact, Fig.1c shows a virtually small variation range of COV (0 to 6%) over a wide range of β (1 to 10), using the same sample size that is practically affordable. These cases also show that the upper bounds for variance, VCF, and COV are fairly close to the estimated ones by simulation. Thus these bounds can be used for sample size determination without significant overestimation. This will be further demonstrated by the rest of the examples.

Example 2

This is a problem with combination of series and parallel systems. It is treated as a series of parallel systems as defined earlier. The structural system consists of a deterministic load S and two parallel bars made of brittle materials [Moses 1982]. The two nondifferentiable failure surfaces represent two symmetrical sequences of component failures that lead to system failure. They are equally important and thus equally weighted: $w_1=w_2=0.5$. Axial strengths of bars R_1 and R_2 are assumed normally distributed with a correlation coefficient ρ and equal coefficients of variation of 20%. Their mean values are set equal and given by a safety factor (SF): $\underline{R}_1=\underline{R}_2=0.5*S*SF$. For SF=1.7, 2.2, and 2.7, the results by the present method using N=4,000 samples are plotted in Fig.2. The exact P_f used for comparison was obtained using an integration table [National Bureau of Standards 1959]. A good agreement between the two methods is observed in Fig.2a. Error ϵ' of these cases is shown in Fig.2b, most being within the range of 2.5% with a maximum of 5.4%. ϵ_{max} shown here is based on a confidence level of 95% (k=1.96) by Eq.13. Fig.2c shows COV that are all within 3.7%. Note that they are not significantly dependent on SF (or equivalently, P_f) as the conventional Monte Carlo method would be, with the same sample size for all the cases. This indicates that the standard deviation of the estimates is reduced almost proportionally to P_f. VCF_{max} is shown in Fig.2d, which demonstrates again its decrease with P_f.

Example 3

This example shows an application to a nonlinear problem with linearized failure surfaces. The original failure function is $z(x)=6+x_1-0.622x_2^2=0$ with $\mathbf{X}^t=(X_1,X_2)$ and independent $X_1=X_2=N(0,1)$. Linearized failure functions are formed according to Eq.3, using x_1^{*t} = (-0.804,2.890) and x_2^{*t} = (-0.804,-2.890) and $\mathbf{a}_1^t=(1,-3.596)$ and $\mathbf{a}_2^t=(1,3.596)$. The symmetric failure modes give $w_1=w_2=0.5$. Exact $P_f=0.002816$ is computed by a one-dimensional integration using the Gamma distribution [Schuëller and Stix 1987].

Listed below are simulation results using N=4,000: VCF_{max}=.008316, P_f'=.002761, ϵ'=2.0% vs. ϵ_{max}=5.3%, and COV=1.7% vs. COV_{max}=2.7%. Apparently this linearization represents a good approximation. P_f' by the present method based on the linearization is reasonably close to the exact one P_f. The estimated COV and error ϵ' are bounded by the analytically derived upper bounds, respectively. Maximum error ϵ_{max} is based on a confidence level of 95% (k=1.96).

Example 4

This example exhibits applications to series systems of higher dimensions. A system of M components in series is considered, with the components designed to be equally reliable with a component reliability index $\beta=5$ (equivalently a component failure probability of $.287\text{x}10^{-6}$). Given a deterministic load S and normally distributed component strengths R_m (m=1,2,...,M) with equal correlation coefficient ρ among them, P_f' was obtained. Using N=4,000 for M=2, 10, and 50, the results are plotted vs. correlation coefficient ρ in Fig.3. Exact P_f was calculated for comparison by a one-dimensional integration [Grigoriu and Turkstra 1979] and shown in Fig.3a. Error ϵ'

is displayed in Fig.3b, most being within the range of 5% with a maximum of 6.4%. COV of the estimates is shown in Fig.3c, being around 2.5% and again almost independent of P_f. VCF_{max} shown in Fig.3d has a similar behavior to that shown in Fig.2d for Example 2, within the range of 6.1%.

5. CONCLUSIONS

A general importance sampling method is introduced. Its sampling distribution is proportional to the ideally optimal one at the maximum likelihood points and able to locate all samples in the failure domain. This method can be employed in integrals for extremely small probability when assessing system reliability, expected damage, etc. Upper bounds for variance, error, and COV of estimator are derived analytically, which have been used to evaluate the present method and can be used in general applications to determine required sample size given an acceptable maximum error associated with a confidence level. Example applications of the suggested method show the assured improvement of efficiency and accuracy for quite general cases, compared to the conventional Monte Carlo method. These include problems of both series and parallel systems and of higher dimensions. They also show that the required sample size is not much dependent on P_f to be estimated, but by the desired COV of estimator.

6. ACKNOWLEDGEMENTS

The US Federal Highway Administration provided partial support. Dr.J.Tang, Mr.E.W.Dillon, and Mr.I.A.Aziz with the New York State Department of Transportation assisted in partial preparation of the examples and figures.

7. REFERENCES

[1] G.Fu and F.Moses "Multimodal Simulation Method for System Reliability Analysis", ASCE J. Eng. Mech., Vol.119, No.6, June 1993, p.1173

[2] G.Fu, Error Analysis for Importance Sampling Method, in D.M.Frangopol (Ed.) New Directions in Structural System Reliability, (Proc. of An NSF Workshop on Research Needs for Applications of System Reliability Concepts and Techniques in Structural Analysis, Design and Optimization, Boulder, CO, Sept.12-14, 1988), pp.132-146

[3] G.Fu and F.Moses, Application of Lifetime System Reliability, Preprint No.52-1, ASCE Structures Congress'86, New Orleans, LA, Sept.15-18, (1986)

[4] G.Fu and F.Moses, Importance Sampling in Structural System Reliability, in: P.D.Spanos (Ed.), Probabilistic Methods in Civil Engineering, (5th ASCE Specialty Conference, Blacksburg, VA, May 1988), p.340-343

[5] M.Fushimi and S.Tezuka, The k-Distribution of Generalized Feedback Shift Register Pseudorandom Numbers, Communications of the ACM, Vol.26, No.7, (July 1983), pp.516-523

[6] M.Grigoriu and C.Turkstra, Safety of Structural Systems with Correlated Resistances, Appl. Math. Modelg., Vol.3, (1979), pp.130-136

[7] M.Hohenbichler and R.Rackwitz, Non-Normal Dependent Vectors in Structural Safety, ASCE J.Eng.Mech., Vol.107, No.6, (Dec.1981), pp.1227-1238

[8] G.Marsaglia, Random Numbers Fall Mainly in the Planes, Proc. Nat. Acad. Sci. USA 61, (1968), pp.25-28

[9] F.Moses, System Reliability Developments in Structural Engineering", Structural Safety, Vol.1, (1982), pp.3-13

[10] National Bureau of Standards, Tables of the Bivariate Normal Distribution Function and Related Functions, Appl.Math.Series 50, June 1959

[11] G.I.Schuëller and R.Stix, A Critical Appraisal of Methods to Determine Failure Probabilities, Structural Safety, Vol.4, (1987), pp.293-309

Structural Safety & Reliability, Schuëller, Shinozuka & Yao (eds) © 1994 Balkema, Rotterdam, ISBN 90 5410 357 4

Adaptive Kernel method for evaluating structural system reliability

G.S.Wang & A.H-S.Ang
University of California, Irvine, Calif., USA

ABSTRACT: Importance sampling method has been developed with the aim of reducing the computational cost inherent in Monte Carlo methods. This study proposes a new algorithm called the adaptive Kernel method which combines and modifies some of the concepts from adaptive sampling and the simple Kernel method to evaluate the structural reliability of time variant problems. The essence of the resulting algorithm is to select an appropriate "design point" from which the importance sampling density can be generated efficiently. Numerical results show that the method is unbiased and substantially increases the efficiency over other methods.

1 INTRODUCTION

The basic step of simulation generally consists of simulating the behavior of a system using specific realization of element random variables and determining whether the system failure occurs in each simulation. The system failure probability is estimated as the ratio of the number of failures to the total number of simulations. Because a low failure probability is usually expected in structural systems, the total number of simulations required to obtain a failure probability with low variance can be extremely large. To reduce the number of simulations and the statistical error of failure probability in the Monte Carlo method, several variance reduction techniques have been proposed including importance sampling, antithetic variates, and latin hypercube sampling [Harbitz 1986; Schueller, Stix 1987; Rubinstein 1981; Schueller et al 1989]. Importance sampling is to bias the generated realization of the random variables by using an importance sampling density function to increase the number of samples representing system failures. In this regard, a simple Kernel method was proposed to develop an importance sampling density function [Ang 1990]. The main drawback in the simple Kernel method is the need to perform Monte Carlo simulations in order to construct the Kernel density function. If the probability of failure is very small, the required Monte Carlo simulations can still be costly.

The objective of this paper is to employ the technique of adaptive sampling to improve the simple Kernel method so that an effective simulation method, called adaptive Kernel method, can be developed to calculate the system reliability. In the evaluation of structural reliability under stochastic dynamic loadings, the structural parameters such as stiffness, damping and strength, are seldom perfectly known. The effect of these and other uncertainties may be important in the overall reliability of the structural system. In order to investigate this effect and demonstrate the ability of the proposed adaptive Kernel method in solving time variant problems, two stochastic structural systems with single degree-of-freedom and multiple degree-of-freedom, respectively, are presented. A bilinear hysteretic model for the restoring force of these dynamic systems are assumed.

2 MATHEMATICAL BACKGROUND

The general equation for structural failure probability can be written as

$$P_F = \int_{D_f} f_X(x)dx \tag{1}$$

which is the volume integral in the x space over the failure domain D_f, and $f_X(x)$ is the joint density function (PDF) of the random variables x.

Eq. (1) can be written as

$$P_F = \int_{all\,x} I[g(x)]f_X(x)dx \tag{2}$$

where:

$$I[g(x)] = \begin{Bmatrix} 1 & if & g(x) \le 0 \\ 0 & if & g(x) > 0 \end{Bmatrix}$$

in which g(x) is the performance function. Then, by

Monte Carlo simulation, the probability of failure is estimated as

$$P_F = \frac{1}{N}\sum_{i=1}^{N} I[g(x_i)] \tag{3}$$

where N is the number of Monte Carlo samples.

If P_F is small, N has to be very large to obtain a sufficient number of failure samples satisfying $g(x) \leq 0$. To increase the number of failure samples for given N, the sampling can be biased toward the failure domain as follows

$$P_F = \int_{all\,x} I[g(x)] \frac{f_X(x)}{h_X(x)} h_X(x) dx \tag{4}$$

where $h_X(x)$ = importance sampling PDF. Accordingly, the Monte Carlo estimate of the failure probability becomes

$$P_F = \frac{1}{N}\sum_{i=1}^{N} I[g(x_i)] \frac{f_X(x_i)}{h_X(x_i)} \tag{5}$$

Clearly, the appropriate choice of the importance sampling PDF, $h_X(x)$, is crucial for the proper Monte Carlo estimate of P_F with Eq. (5).

With the simple Kernel method, Ang [1990] suggested the following for constructing the required importance sampling PDF:

$$h_X(x) = \frac{1}{M}\sum_{i=1}^{M} \frac{1}{(\lambda_i \omega)^d} K\left(\frac{x - y_i}{\lambda_i \omega}\right) \tag{6}$$

where y_i = failure sample generated from the original joint PDF, $f_X(x)$; $K(\cdot)$ = Kernel function satisfying $\int_{all\,y} K(y)dy = 1$; ω = window width; and λ_i = a scale parameter. Criteria for selecting ω and λ_i are suggested by Ang [1990] including a Gaussian PDF for $K(\cdot)$.

The above simple Kernel method suffers from having to generate the M failure samples from $f_X(x)$ through basic Monte Carlo. To overcome this weakness, the simple Kernel method can be modified by introducing an adaptive sampling scheme [Bucher 1988] for generating the M failure samples.

Observe that the statistical error of Eq. (5) reduces to zero if $h_X(x)$ is the original PDF, $f_X(x)$, conditional on the failure domain D_f; i.e.,

$$h_X(x) = f_X(x \mid x \in D_f) \tag{7}$$

It is, however, difficult to find this function. As an alternative, the condition may be imposed on the first and second moments only; namely,

$$E_h[X] = E_f[X \mid X \in D_f] \tag{8}$$

$$E_h[XX^T] = E_f[XX^T \mid X \in D_f] \tag{9}$$

where the subscripts h and f refer to the expected values with respect to the respective PDF's. From an initial simulation run, the right hand sides of Eqs. (8) and (9) can be estimated. Because knowledge of the mean and covariance uniquely determines a joint normal density, these values are used to adapt $\hat{h}_X(x)$ for the next run. To examine a starting vector for the conditional moments mentioned above, perform a simulation for one variable while keeping all other variables at the mean values. The individual adaptive mean $\bar{X}_{hci}$ and the associated conditional failure probability P_{Fi} are obtained. An approximation for the design point $\bar{X}_h = (\bar{X}_{hi}, \bar{X}_{h2}, \ldots, \bar{X}_{hd})$ is

$$\bar{X}_{h_i} = \frac{P_{F_i}\bar{X}_{h_{c_i}} + \mu_{X_i}\sum_{j=i,\, j\neq i}^{d} P_{F_j}}{\sum_{j=1}^{d} P_{F_j}} \tag{10}$$

where d is the number of basic variables and μ_{Xi} is the mean of X_i. In the next step, a simulation around $\bar{X}_h$ is carried out, again varying one variable X_i at a time. Finally, from these simulations, the conditional standard deviations σ_{Xhi} are estimated. The importance sampling density can then be constructed as a normal density function with the design point and its associated standard deviation serve as the mean and standard deviation. Finally, the importance sampling is carried out. Besides using the standard procedure of adaptive sampling, use the moments obtained above to generate the M failure points and then produce $\hat{h}_X(x)$ on the basis of the Kernel function.

The major difference between the present adaptive Kernel method and the previous simple Kernel method is as follows. To construct $\hat{h}_X(x)$, the present method generates the failure points $y_1, y_2, \ldots, y_M$ from conditional moments without the necessity of a Monte Carlo simulation, whereas the simple Kernel method requires a Monte Carlo simulation. This is a significant improvement in computations, particularly when the failure probability is small.

3 ADAPTIVE KERNEL METHOD

The essence of the proposed adaptive Kernel method is to select an appropriate design point from which the Kernel importance sampling density can be constructed efficiently. Instead of using the mean of each variate as the starting point, it is proposed that the starting point be moved to a point which is close to the failure domain.

The algorithm for the above purpose may be described as follows:

(i) Choose a starting point $\mu'_X = (\mu'_{X1}, \mu'_{X2}, \ldots, \mu'_{Xd})$. The main principle to determine this point is to insure that it is close to the failure domain. With this in mind, one may assume $\mu'_{Xi}=\mu'_{Xi} - \gamma_i\sigma_{Xi}$, where γ_i is a deviation factor that can be determined by satisfying the limit state function $g(x)=0$.

(ii) Let $X_j=\mu'_{Xj}$, where $j=1,2,...,d$, but $j\neq i$ and determine a critical value, a_{Xi} for X_i, from the limit state function $g(\mu'_{X1}, \mu'_{X2}, ..., \mu'_{Xi-1}, a_{Xi}, \mu'_{Xi+1}, ..., \mu'_{Xd})=0$. The conditional failure probability $P_{Fi}=P(X_i<a_{Xi})$ or $P(X_i>a_{Xi})$ can be obtained accordingly. If X_i is a normal variate, i.e., $N(\mu_{Xi},\sigma_{Xi})$, one of the following two equations is used to obtain the required adaptive mean:

$$\bar{X}_{h_{c_i}} = E[X_i|X_i<a_{X_i}] = \mu_{X_i} - \sigma_{X_i} \cdot \frac{e^{-\frac{1}{2}\left(\frac{a_{X_i}-\mu_{X_i}}{\sigma_{X_i}}\right)^2}}{\sqrt{2\pi}\,\Phi\left(\frac{a_{X_i}-\mu_{X_i}}{\sigma_{X_i}}\right)} \quad (11)$$

$$\bar{X}_{h_{c_i}} = E[X_i|X_i>a_{X_i}] = \mu_{X_i} + \sigma_{X_i} \cdot \frac{e^{-\frac{1}{2}\left(\frac{a_{X_i}-\mu_{X_i}}{\sigma_{X_i}}\right)^2}}{\sqrt{2\pi}\,\Phi\left(-\frac{a_{X_i}-\mu_{X_i}}{\sigma_{X_i}}\right)} \quad (12)$$

where μ_{Xi} and σ_{Xi} are the original mean and standard deviation of x_i.

(iii) Repeat step (ii) for all $i=1,2,...,d$ to obtain $\bar{X}_{hc}=(\bar{X}_{hc1},\bar{X}_{h2c},...,\bar{X}_{hcd})$.

(iv) Obtain the design point $\bar{X}_h=(\bar{X}_{h1},\bar{X}_{h2},...,\bar{X}_{hd})$, in which

$$\bar{X}_{h_i} = \frac{P_{F_i}\bar{X}_{h_{c_i}} + \mu'_{X_i}\sum_{\substack{j=i\\ j\neq i}}^{d} P_{F_j}}{\sum_{j=1}^{d} P_{F_j}} \quad (13)$$

Once the point $\bar{X}_h$ is determined, the conditional standard deviation, σ_{Xh}, can be obtained using the following two steps:

(i) Let $X_j=\bar{X}_{hj}$, where $j=1,2,...,d$, but $j\neq i$ and determine a critical value for X_i, which is denoted by β_{Xi}, from the limit state function $g(\bar{X}_{h1},\bar{X}_{h2},...,\bar{X}_{hi-1},\beta_{Xi},\bar{X}_{hi+1},...,\bar{X}_{hd})=0$. The corresponding conditional standard deviation, σ_{Xh}, is calculated by the following equation

$$\sigma_{X_{h_i}} = \sigma_{X_i}\sqrt{1+\left(\frac{\beta_{X_i}-\mu_{X_i}}{\sigma_{X_i}}\right)\left(\frac{\bar{X}_{h_i}-\mu_{X_i}}{\sigma_{X_i}}\right)-\left(\frac{\bar{X}_{h_i}-\mu_{X_i}}{\sigma_{X_i}}\right)^2} \quad (14)$$

(ii) Repeat procedure (ii) for all $i=1,2,...,d$ to obtain $\sigma_{Xh}=(\sigma_{Xh1},\sigma_{Xh2},...,\sigma_{Xhd})$.

With $\bar{X}_h$ and σ_{Xh}, M failure points $y_1,y_2,...,y_M$ can be generated efficiently. The Kernel importance sampling density $\hat{h}_X$ is then constructed with Eq. (6). On the basis of the Kernel density function $\hat{h}_x$, N samples are generated and the failure probability is then calculated through Eq. (5).

4 NUMERICAL EXAMPLES

Consider a single degree-of-freedom structure whose restoring force is governed by a bilinear hysteresis model. The equation of motion is

$$\frac{d^2u}{dt^2} + 2\zeta(2\pi n_0)\frac{du}{dt} + q(u(t),a_1,a_2,a_3,t) = S(t) \quad (15)$$

where n_0 is the natural frequency (cps), ζ is the damping ratio, u is the relative displacement of the structure with respect to the ground, S(t) is a Gaussian white noise with spectral density S_0 and duration T, and q is the nonlinear restoring force which is a bilinear hysteresis model as shown in Fig. 1. The constants in the bilinear hysteresis model are as follows:

a_1 = yielding strength = $0.0495a_2$
a_2 = initial stiffness = $(2\pi n_0)^2$
a_3 = yielding stiffness = $0.1a_2$

The above ordinary differential equation is first transformed into a set of first-order nonlinear differential equations

$$\frac{dy_1}{dt} = y_2$$
$$\frac{dy_2}{dt} = -2\zeta(2\pi n_0)y_2 - q(u,a_1,a_2,a_3,t) = S(t) \quad (16)$$

in which

$$y_1 = u$$
$$y_2 = \frac{du}{dt} \quad (17)$$

The set of first-order differential equations in Eq. (16) is then solved by using Runge-Kutta method with zero initial conditions.

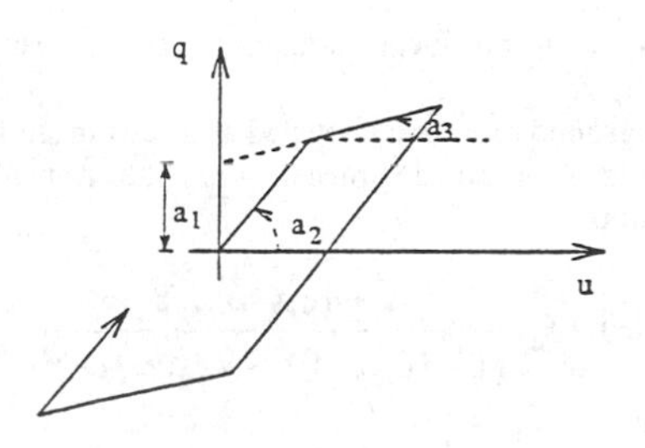

Fig. 1: Bilinear hysteresis model

Failure is assumed to occur when the structure experiences excessive relative displacement, e.g. u > 6in. The distribution of S_0, T, n_0 and ζ are given as follows.

N_0 = Normal, $N(\mu_{N0}=2cps, \delta_{N0}=0.10)$
Z = Lognormal, $LN(\mu_Z=0.02, \delta_Z=0.40)$
S_0 = Type II Extreme Value
$(\mu_{S0}=36in^2/sec^3, \delta_{S0}=0.60)$
T = Lognormal, $LN(\mu_T=10sec, \delta_T=0.30)$

The construction of the Kernel sampling density function and estimation of P_F follow the procedure in the previous section. The failure probability computed by using the adaptive Kernel method and the simple Kernel method are shown in Figs. 2 and 3, respectively. The number of simulations, I1, to generate M failure samples in the adaptive Kernel method is 16. It is far smaller than the number required in the simple Kernel method (I1=1754). The efficiency of the adaptive Kernel method is even more promising when the failure probability of a system is small.

The second example is the 5-degree-of-freedom system shown in Fig. 4, subjected to earthquake ground excitation. It is assumed that the restoring force follows a hysteretic model [Baber, Wen 1981]. The equations of motion of the system may be written as

$$\ddot{u}_i - (1-\delta_{i1})\frac{q_i - 1}{m_i - 1} + \frac{q_i}{m_i}[1 + (1-\delta_{i1})\frac{m_i}{m_{i-1}}] - (1-\delta_{i5})\frac{q_{i+1}}{m_{i+1}}\frac{m_{i+1}}{m_i} = -\ddot{u}_g, \; i = 1,2,...,5 \tag{18}$$

where $\ddot{u}_g$ is the ground acceleration, m_i = mass of the ith floor, δ_{i1}, δ_{i5} = Kronecker deltas, u_i = interstory deformation of the ith floor, and the restoring force is

$$q_1 = C_i\dot{u}_i + \alpha_i K_i u_i + (1-\alpha_i)K_i z_i, \; i = 1,2,...,5 \tag{19}$$

in which K_i = initial stiffness of the ith floor, C_i = viscous damping of the ith floor, z_i = the hysteretic component of the deformation u_i, and

$$\dot{z}_i = \dot{u}_i - (\beta_i|\dot{u}_i||z_i|z_i + \gamma_i\dot{u}_i|z_i|^2), \; i = 1,2,...,5 \tag{20}$$

Values of the pertinent parameters are summarized in Table 2.

The ground motion is modeled as a zero-mean filtered Gaussian shot noise process with the Kanai-Tajimi spectrum

$$S_{aa}(\omega) = S_0 \frac{1 + 4\xi_g^2(\omega/\omega_g)^2}{[1-(\omega/\omega_g)^2]^2 + 4\xi_g^2(\omega/\omega_g)^2} \tag{21}$$

where S_0 is the power spectrum ordinate of the stationary unfiltered shot noise. In this case, ω_g, ξ_g and

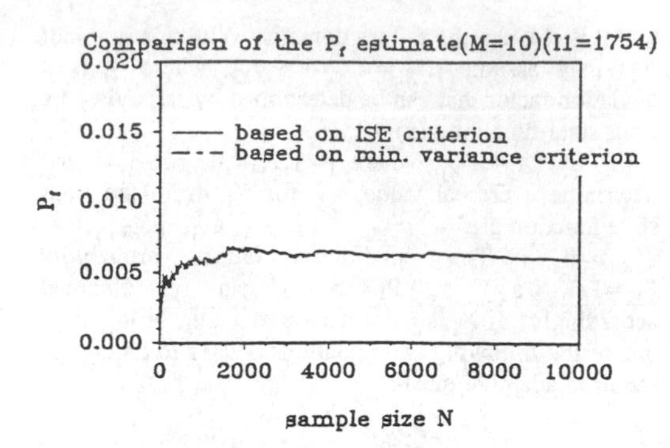

Fig. 2: Estimation of P_F for SDOF bilinear system using simple Kernel method

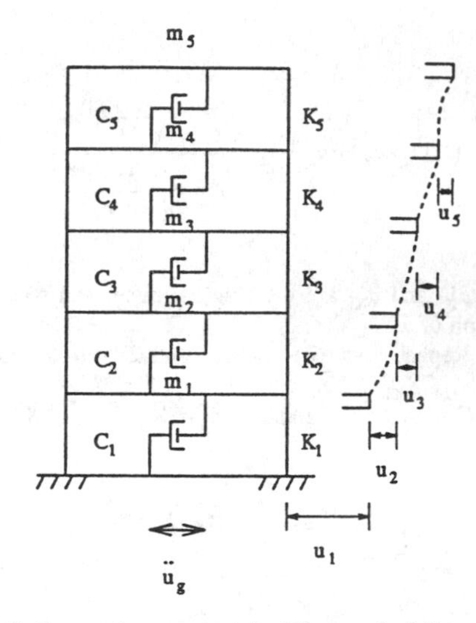

Fig. 3: Lumped mass model of 5-story building

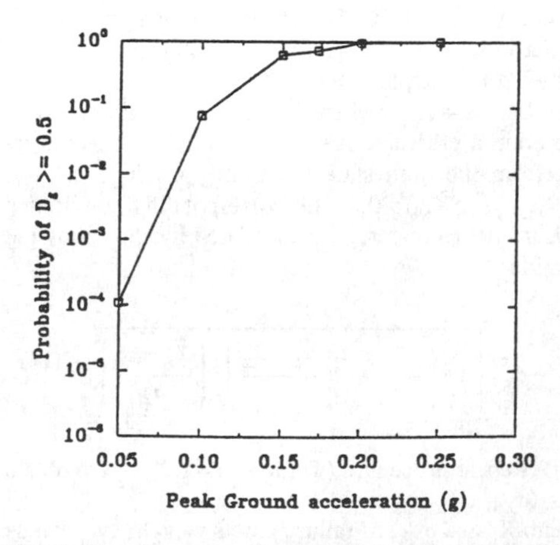

Fig. 4: Fragility curve for $D_g \geq 0.5$ of 5-story building

S_0 are calibrated as 3.2rad/sec, 0.3, and 21.84in²/sec³, respectively, representing the ground motion in Mexico City during the 1985 earthquake.

To obtain an even more representative process for strong ground motion, the nonstationary characteristics of actual accelerograms can be considered. This appearance suggests using a nonstationary process, namely a process $\ddot{u}_g$ given by

$$\ddot{u}_g = \psi(t)a_1(t) \tag{22}$$

where ψ(t) is an intensity function having an appropriate form based on statistical analyses of real accelerograms. In the case of the Mexico City earthquake, this intensity is modulated by a time function given by

$$\psi(t) = \begin{Bmatrix} (t/t_1)^2 \\ 1.0 \\ e^{-2(t-t_2)/t_d} \end{Bmatrix} \tag{23}$$

where

t_1 = 15 sec

t_2 = 45 sec

and

t_d = the strong motion duration = 30 sec

The structure described above suffered substantial damage in the first floor during the Mexico City earthquake of 1985.

According to the Park-Ang model [1985], the damage index for a structural component can be expressed in terms of the maximum interstory displacement δ_{max} and the dissipated hysteretic energy E as follows,

$$D = \frac{\delta_{max}}{\delta_u} + \frac{\beta_0 \int dE}{Q_y \delta_u} \tag{24}$$

where D is the damage index, δ_u is the ultimate deformation, Q_y is the yield force capacity, and β_0 is a constant. The above parameters for each component of the building are also tabulated in Table 1.

The hysteretic energy associated with each degree-of-freedom is described by the following differential equation,

$$\dot{E} = (1-\alpha_i)K_i \dot{u} z \tag{25}$$

The global damage index for a structure is clearly a function of the damage indices of the components, and may be defined as follows:

$$P(D_g \geq d) = P[\cup(D_i \geq d)] \tag{26}$$

where D_g is the global damage index and D_i is the damage index of component i.

From a sensitivity analysis, the failure probability was found to be not sensitive to changes in the means and standard deviations of β_i and γ_i of Eq. (20). According-

Table 1: Structural properties of 5 DOF system

story i	$m_i(k/ins^2)$	$K_i(k/in)$	$C_i(k/ins)$	α_i
1	2.8	758	4.607	0.02
2	2.6	3677	9.778	0.02
3	2.5	3677	9.588	0.02
4	2.4	3749	9.486	0.02
5	2.2	2564	7.51	0.02

story i	β_i	γ_i	$\delta_u(in)$	β_0	$Q_u(k)$
1	0.75	-0.25	6	0.27	1476
2	26.4	-8.8	1	0.24	1163
3	17.1	-5.7	1.1	0.24	1427
4	20.4	-6.8	1.6	0.24	1340
5	6.9	-2.3	1.4	0.24	1156

ly, these variables are treated as constants. The random variables involved in this system are then m_i, C_i, K_i, α_i and S_0, involving 21 variables for this building. The coefficients of variation of these variables are 0.12, 0.5, 0.32, 0.1, and 0.5, respectively.

The interstory deformation and hysteretic energy of each floor are obtained by solving Eqs. (18) and (25) simultaneously using the Runge-Kutta method. After obtaining the maximum deformation and hysteretic energy, the damage index for each floor is obtained through Eq. (24) and the probability that the global damage index D_g will exceed a tolerable value is evaluated through Eq. (26) by the adaptive Kernel method. The fragility curve for the global damage index exceeding 0.5, which is the tolerable damage, is obtained and presented in Fig. 4. The number of failure samples used to construct the Kernel sampling density function, M, is 40 and the number of samples generated from the Kernel sampling density function, N, is 1000 for all the points on this fragility curve. The collapse probability, $P(D_g \geq 1.0)$ is also evaluated by the adaptive Kernel method through the same procedure; the corresponding fragility curve is illustrated in Fig. 5. The same numbers of samples are used for all the points on this fragility curve. Observe that the collapse

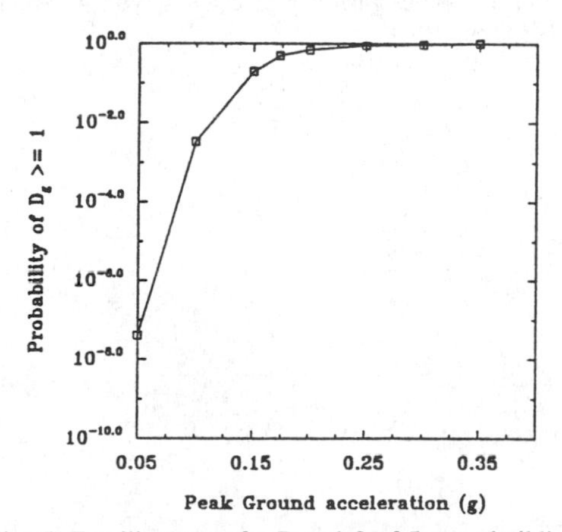

Fig. 5: Fragility curve for $D_g \geq 1.0$ of 5-story building

probability drops to $4.08 \cdot 10^{-8}$ under a peak ground acceleration of 0.05g. This shows that the adaptive Kernel method is equally effective even for problems involving very low failure probability. It would be difficult or expensive to apply any other method, including the simple Kernel method for problems involving such small probabilities.

5 CONCLUSIONS

The proposed adaptive Kernel method is an effective importance sampling technique for evaluating the reliability of structural systems, including those involving very small probabilities of failure (e.g. $\leq 10^{-6}$) which are often required in structural problems. It is equally effective for evaluating systems reliability formulated either with the failure mode approach or the stable configuration approach.

REFERENCES

Ang, G.L., Kernel method in monte carlo importance sampling, Ph.D. thesis, University of Illinois, Urbana, 1990.

Baber, T.T., and Wen, Y.K., Random vibration of hysteretic degrading systems, Jour. of Engineering Mechanics, ASCE, Vol. 107, Dec. 1981.

Bucher, C.G., Adaptive sampling - an iterative fast monte carlo procedure, Structural Safety, Vol. 5, 1988, pp. 119-126.

Harbitz, A., An efficient sampling method for probability of failure calculation, Structural Safety, Vol. 3, 1986, pp. 109-115.

Rubinstein, R.Y., Simulation and the monte carlo method, John Wiley & Sons, New York, 1981.

Schueller, G.I. and Stix, R., A critical appraisal of methods to determine failure probabilities, Structural Safety, Vol. 4, 1987, pp. 293-309.

Schueller, G.I., Bucher, C.G., Bourgund, U., and Quypornprasert, W., On efficient computational schemes to calculate structural failure probabilities, Probabilistic Engineering Mechanics, Vol. 4, No. 1, 1989, pp. 10-18.

Structural Safety & Reliability, Schuëller, Shinozuka & Yao (eds) © 1994 Balkema, Rotterdam, ISBN 90 5410 357 4

An improved importance sampling density estimation for structural reliability assessment

M.Yonezawa
Department of Industrial Engineering, Kinki University, Higashi-Osaka, Japan

S.Okuda
Kumano Technical College, Kinki University, Japan

ABSTRACT: This paper is concerned with the reliability analysis of structural systems based on a new method of the importance sampling simulation. The basic random variables are assumed to be stochastically independent and normally distributed and the reliability index ß to be known beforehand. The first step is to execute the simulation to determine a frequency distribution of the importance sampling density estimator as a function of radius in polar coordinates. The second step is to estimate the structural failure probability through the importance sampling simulation, samples of which are generated within the limit of the failure region from the estimator of the importance sampling density. The proposed method is compared in numerical examples with the Monte Carlo simulation combined with partition of the region technique, ISPUD for multi mode failure etc. It can be said that the proposed method is effective for the estimation of structural failure probability.

1 INTRODUCTION

The probability of structural failure is defined, for time-independent case, by the following multi-dimensional integral:

$$P_f = \int_{g(\mathbf{x})\leqq 0} f(\mathbf{x})d\mathbf{x} \tag{1}$$

where $\mathbf{X} = (X_1, X_2,..., X_k)$ is the vector of the basic random variables in the k-dimensional space, $f(\mathbf{x})$ is the joint p.d.f of the basic random variables and $g(\mathbf{x}) = g(x_1, x_2,..., x_k)$ is the limit state function which describes the state of the structural system. For any realization $\mathbf{x}$ of the basic random variable $\mathbf{X}$, the set of $\{\mathbf{x} \mid g(\mathbf{x})\leqq 0\}$ implies the failure domain.

The integral in Eq.(1) is generally difficult to calculate because the domain of integration has often irregular shape and $f(\mathbf{x})$ is generally complex. Therefore, the simulation approaches are widely adopted to estimate the integral.

Using an indicator function $I[(\cdot)]$ defined by

$$I[(\cdot)] = 1 \text{ if } (\cdot)\leqq 0, \quad I[(\cdot)] = 0 \text{ if } (\cdot)>0 \tag{2}$$

Eq.(1) can be rewritten as follows:

$$P_f = \int_{\text{all } \mathbf{x}} I[g(\mathbf{x})]f(\mathbf{x})d\mathbf{x} = E\{I[g(\mathbf{X})]\} \tag{3}$$

where $E\{I[g(\mathbf{X})]\}$ is an expected value of $I[g(\mathbf{X})]$ with respect to $f(\mathbf{x})$.

Generating an i-th sample vector $\mathbf{x}^{(i)}$ (i=1, 2,..., N) from $f(\mathbf{x})$, the estimate P_{fe} of Eq.(3) and its variance are evaluated by

$$P_{fe} = \frac{1}{N}\sum_{i=1}^{N} P_f^{(i)} \tag{4}$$

$$\mathrm{Var}\{P_{fe}\} = \frac{1}{N(N-1)}\sum_{i=1}^{N}[P_f^{(i)} - P_{fe}]^2 \tag{5}$$

$$P_f^{(i)} = I[g(\mathbf{x}^{(i)})] \quad ,(i=1,2,...,N) \tag{6}$$

A large number of samples are generally required to estimate the integral of Eq. (3) for the case with low structural failure probability. Then various variance reduction techniques such as the importance sampling, method of partition of the region have been introduced to estimate the structural failure probability. It is well known that the importance sampling simulation could be an efficient method to estimate the probability of failure, if a suitable importance sampling p.d.f. is determined.

Various approaches have been proposed to determine the importance sampling p.d.f. such as ISPUD by Schuëller and et al. (1986), adaptive sampling by Bucher (1988), kernel method by Ang and Tang (1989) etc.

In this paper, a new method of an estimation of the importance sampling density is proposed for the multi-dimensional reliability problems in which a frequency distribution of the importance sampling density is determined within the limit of the failure region.

2 IMPORTANCE SAMPLING

2.1 *Importance sampling p.d.f.*

An importance sampling probability density function $h(\mathbf{y})$ can be introduced into Eq.(3) to obtain the well-known equation,

$$P_f = \int_{\text{all}\,\mathbf{y}} I[g(\mathbf{y})] \frac{f(\mathbf{y})}{h(\mathbf{y})} h(\mathbf{y})d\mathbf{y} = E\{I[g(\mathbf{Y})] \frac{f(\mathbf{Y})}{h(\mathbf{Y})}\} \quad (7)$$

If $I[g(\mathbf{y})]h(\mathbf{y})$ is regarded as a new importance sampling p.d.f. on the condition that it satisfies the following relation,

$$\int_{\text{all}} I[g(\mathbf{y})]\, h(\mathbf{y})d\mathbf{y} = 1 \quad (8)$$

then, P_f can be expressed as an expected value of $f(\mathbf{y})/h(\mathbf{y})$ with respect to $I[g(\mathbf{y})]h(\mathbf{y})$,

$$P_f = E\{\frac{f(\mathbf{Y})}{h(\mathbf{Y})}\} \quad (9)$$

Through the simulation, the estimate P_{fe} of Eq.(9) is evaluated by Eqs.(4), (5) and Eq.(10) in place of Eq.(6),

$$P_f^{(i)} = \frac{f(\mathbf{y}^{(i)})}{h(\mathbf{y}^{(i)})}, \quad (i=1, 2,..., N) \quad (10)$$

where $\mathbf{y}^{(i)}$ is an i-th sample vector generated from the importance sampling p.d.f. $I[g(\mathbf{y})]h(\mathbf{y})$. As $I[g(\mathbf{y}^{(i)})]h(\mathbf{y}^{(i)}) = h(\mathbf{y}^{(i)})$ for the samples set of $\{\mathbf{y}^{(i)} \mid g(\mathbf{y}^{(i)}) < 0\}$, therefore samples are most effectively generated, if $h(\mathbf{y})$ is defined within the limit of the failure region. In the next section, how to construct $h(\mathbf{y})$ within the limit of the failure region is proposed.

2.2 *Importance sampling p.d.f. defined in polar coordinates*

Consider that the coordinates of the basic random variables are transformed to polar coordinates $(R, \Theta_1, \Theta_2, \ldots, \Theta_{k-1})$, Kendall (1961), the probability of failure Eq.(3) can be rewritten as follows:

$$P_f = \int_{\text{all}\,r} r^{k-1} I[g(\mathbf{y})] f(\cdot)\, dr \int_{-\pi/2}^{\pi/2} \cos^{k-2}\theta_1 d\theta_1 \cdots \int_{-\pi/2}^{\pi/2} \cos^1\theta_{k-2}\, d\theta_{k-2} \int_0^{2\pi} d\theta_{k-1} \quad (11)$$

$$f(\cdot) = f(r\cos\theta_1, \ldots, \cos\theta_{k-1}) \cdots f(r\sin\theta_1) \quad (12)$$

where $f(\cdot)$ is the multi-dimensional standard normal p.d.f., which is rewritten as a function of radius r in polar coordinates and denoted by

$$f_r(r) = \frac{1}{(2\pi)^{k/2}} \exp(-\frac{1}{2} r^2) \quad (13)$$

As the value of $I[g(\mathbf{y})]$ takes 1 or 0, and it does not affect the result of the integral, so the expression $I[g(\mathbf{y})]$ is left unchanged.

Integrating Eq.(11) with respect to all the angles, it is expressed as an integral of radius r in polar coordinates, that is,

$$P_f = \int_{\text{all}\,r} f_r(r)\, I[g(\mathbf{y})] \frac{2r^{k-1}\pi^{k/2}}{\Gamma(k/2)} dr \quad (14)$$

into which $h_r(r)$, a function of radius r as an importance sampling p.d.f., is introduced, then we obtain a following integral,

$$P_f = \int_{\text{all } r} \frac{f_r(r)}{h_r(r)} h_r(r) I[g(y)] \frac{2 r^{k-1} \pi^{k/2}}{\Gamma(k/2)} dr \tag{15}$$

If $h_r(r)$ is determined so as to satisfy a following relation,

$$\int_{\text{all } r} h_r(r) I[g(y)] \frac{2 r^{k-1} \pi^{k/2}}{\Gamma(k/2)} dr = 1 \tag{16}$$

then P_f can be expressed as an expected value of

$$P_f = E\{\frac{f_r(R)}{h_r(R)}\} \tag{17}$$

We consider here to determine $h_r(r)$ on the condition that it satisfies the relation Eq.(16) through the simulation, and to get it as a frequency distribution $h_r(r_j)$ for a radial interval of $[r_j, r_{j+1}]$ in polar coordinates.

2.3 *Determination of frequency distribution of importance sampling p.d.f.*

As Eq.(16) can be rewitten as a summation of the integral over a small radial interval $[r_j, r_{j+1}]$ as follows:

$$\sum_{j=1}^{C} h_r(r_j) I[g(y)] \frac{2\pi^{k/2}}{k\Gamma(k/2)} (r_{j+1}^k - r_j^k) = 1 \tag{18}$$

then an importance sampling density $h_r(r_j)$ should be determined through the simulation so as to satisfy Eq. (18) and the number of classes C is also selected to hold Eq. (18). The procedure to determine the estimator of the importance sampling density is as follows:

First, the sampling region outside the ß-sphere is classified into C classes bounded by C concentric spheres, which have their center at the origin of polar coordinates as shown in Fig. 1. The class interval $r_{j+1} - r_j = w$ (j=1, 2,...,C) is a constant.

Then execute the simulation, referred as the initial simulation, in order to construct a frequency histogram of the importance

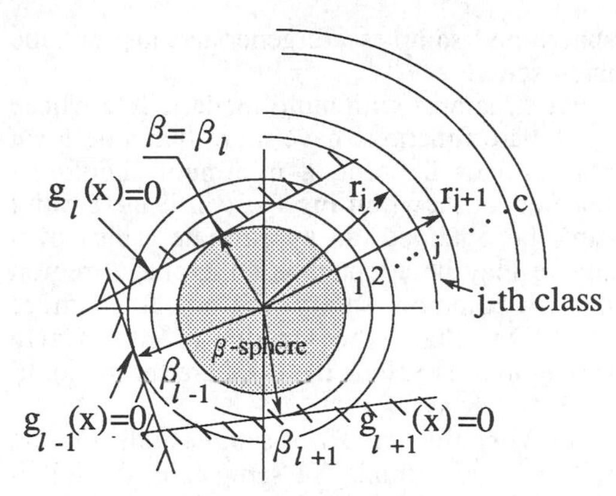

Fig. 1 Classification of sampling region outside the ß-sphere in polar coordinates.

sampling p.d.f.. In the initial simulation, random samples $y^{(i)}$ (i=1, 2,..., N1), N1 is the sample size of it, are generated,

$$y^{(i)} = r^{(i)} \times a^{(i)} \tag{19}$$

$$a^{(i)} = \frac{(v_1^{(i)}, v_2^{(i)}, \ldots, v_k^{(i)})}{\{\sum_{l=1}^{k} (v_l^{(i)})^2\}^{1/2}} \tag{20}$$

where a random sample of direction vector $a^{(i)}$ is generated from the standard normal variates, $v_l^{(i)}$, (l=1, 2,..., k) and a random radius sample $r^{(i)}$ is generated from the truncated chi-square p.d.f. defined for the region outside the ß-sphere, Yonezawa and Okuda (1993).

For the case with law failure probability, almost no sample may be generated in the failure region through the crude Monte Carlo simulation, so the initial simulation is based on the Monte Carlo simulation combined with partition of the region technique. Since the basic variables are assumed to be independent normal variates, a square of the magnitude of the vector of them follows a chi-square distribution with k degrees of freedom, then probability P of the ß-sphere can be obtained by evaluating the chi-square distribution, Harbitz (1986). Therefore, the truncated chi-square p.d.f. is easily defined for outside the ß-sphere and samples are generated just outside the ß-sphere.

For structures with multi mode failure whose limit state functions have more than one local optimum of ß point, a minimum should be selected as shown in Fig.1. Even if the ß point can't be obtained, an appropriate radius of a sphere may be adopted as an excluded region in the common safety region. It is more effective than the crude Monte Carlo simulation to exclude the ß-sphere for the multi mode failure system.

Let M be the number of samples fallen in the failure region among N1 samples of the initial simulation and M_j be the number of samples fallen in the failure region of j-th class. The relative frequency M_j/M is taken as an importance sampling probability for j-th class, Ang and Tang (1989).
Then the estimator of the importance sampling density $h_r(r_j)$ and the cumulative distribution function of the importance sampling probability, $H_r(r_j)$ are determined by

$$h_r(r_j)\,\Delta_j = \frac{M_j}{M}, \quad (j = 1, 2, \cdots, C) \tag{21}$$

$$H_r(r_j) = \sum_{l=1}^{j} h_r(r_l)\Delta_l = \sum_{l=1}^{j} \frac{M_l}{M} \tag{22}$$

$$\Delta_j = I\,[g(y)]\frac{2\pi^{k/2}}{k\Gamma(k/2)}(r_{j+1}^k - r_j^k)$$
$$= I\,[g(y)]\;A_j \tag{23}$$

$$A_j = \frac{2\pi^{k/2}}{k\Gamma(k/2)}(r_{j+1}^k - r_j^k) \tag{24}$$

where Δ_j is a k-dimensional volume of the failure region of j-th class and A_j is a k-dimensional volume of j-th class. Δ_j of Eq.(23) is evaluated by $(L_j / L)*A_j$, where (L_j / L) is a volume ratio of the failure region of j-th class to the whole region of j-th class. This volume ratio must be evaluated through a supplemental simulation for each class, in which L_j is the number of samples fallen in the failure region of

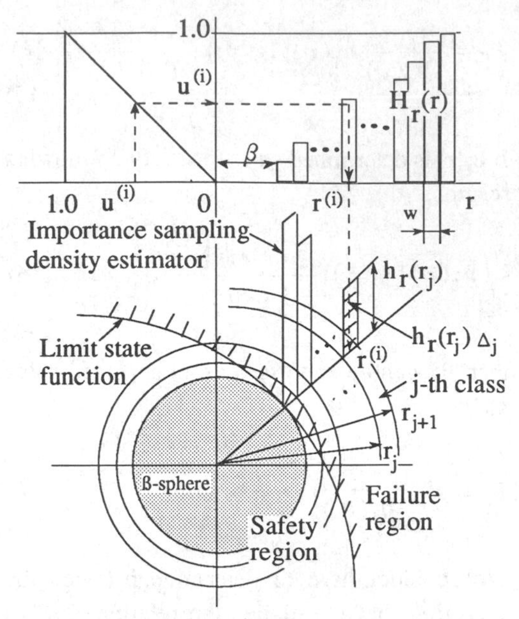

Fig. 2 Sample generation from the estimator of the importance sampling density

j-th class and L is a total number of samples generated for j-th class. $h_r(r_j)\Delta_j$ and $H_r(r_j)$ determined above are illustrated in Fig.2. It should be noted that N1(for the initial simulation)+N2(for the supplemental simulation, C*L) samples are required to construct the estimator of the importance sampling density.

2.4 *Estimation of probability of failure*

The estimate of the probability of failure is obtained by the importance sampling simulation, in which N samples are generated from the estimator of the importance sampling density constructed above. For each simulation cycle, an inverse function value $r^{(i)}$ of the cumulative frequency function $H_r(r^{(i)})$ corresponding to an uniform random variate $u^{(i)}$ is calculated as shown in Fig.2. The estimate P_{fe} and its variance are evaluated by using Eqs.(4), (5) and Eq.(25) in place of Eq.(6).

$$P_f^{(i)} = [\frac{f_r(r^{(i)})}{h_r(r^{(i)})}] , \ (i=1, 2,..., N) \qquad (25)$$

Suppose the importance sampling p.d.f. is given as a function of radius r by

$$h_r(r) = \frac{f_r(r)}{P_f} \qquad (26)$$

it is easily shown that $P_{fe}=P_f$, $Var[P_{fe}]=0$, then Eq.(26) is an ideal importamce sampling p.d.f. for the case P_f is known. Although it is really impossible, as P_f is unknown.

3 NUMERICAL EXAMPLES

In this section, some numerical examples are presented to show some aspects of the proposed method and discuss the effectiveness.

Consider the structures with the limit state functions given by

No.1:

$$g(\mathbf{x})=(1 - x_2 - x_1^2) \cup (1 + x_2 - x_1^2) \qquad (27)$$

No.2:

$$g(\mathbf{x}) = - 0.125(\sum_{i=1}^{3} x_i^2) - x_4 + 4 \qquad (28)$$

Table 1 Statistical data of basic random variables.

Limit state function	No.1	No.2
Basic variable	x_1, x_2	x_i
Mean value	0.1, 0.05	0
Standard deviation	0.1, 0.2	1

Table 2 Probability P of the ß-sphere.

Limit state function	No.1	No.2
P	$1-1.925\times10^{-5}$	$1-3.02\times10^{-3}$

Table 3 Number of samples M generated within the limit of the failure region.

Limit state function	No.1	No.2
M	116	143

(N1=1000)

The basic random variables are assumed to be stochastically independent normal variates, their statistical data for the analysis are given in Table 1.

First, the initial simulation with N1 samples are executed for outside the ß-sphere region to construct the estimator of the importance sampling density.

For the limit state function of Eq.(27), there are two ß-points, that is, $\beta_1=4.66$ and $\beta_2=5.16$,

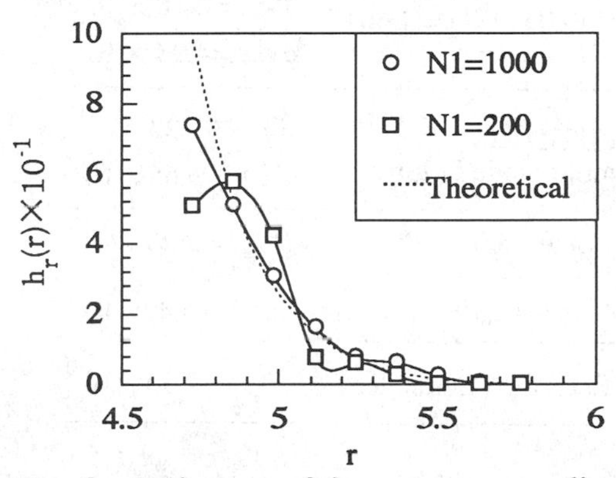

Fig. 3 Estimator of importance sampling density for the limit state function of Eq. (27).

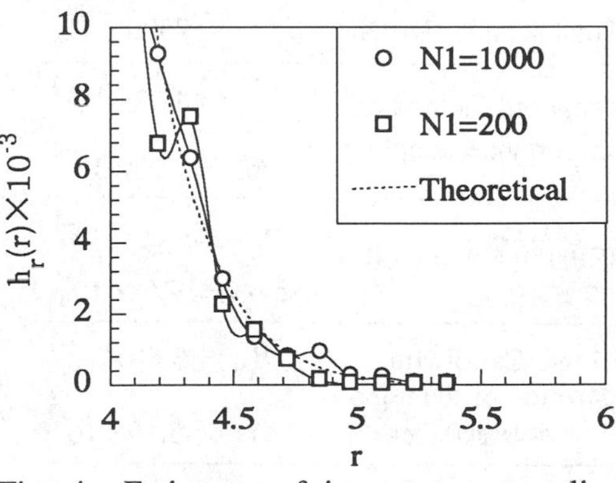

Fig. 4 Estimator of importance sampling density for the limit state function of Eq. (28)

then ß=4.66 is taken as the radius of the ß-sphere, which is excluded from sampling. For the case of Eq. (28), it has only one ß-point, so ß=4.00 is adopted as the radius of the ß-sphere. The probability P of the ß-sphere for each limit state function is given in Table 2. The initial simulation excludes the ß-sphere from the sampling to improve the sampling efficiency as stated above. It can be said that a large amount of samples which might be lost are applied effectively to generate samples outside the ß-sphere. In the initial simulation with N1 samples, there is no sample generated in the range of r > ß+1.3 for both cases of Eqs.(27) and (28). Therefore, the frequency distribution of samples fallen in the failure region is determined by dividing the range of ß ≤ r ≤ ß + 1.3 into 10 classes, that is, the number of classes, C=10 and the class interval, w=0.13. The number of samples, M generated within the failure region among N1=1000 samples are listed in Table 3.

The volume ratio L_j / L (j=1,2,...,C) of j-th class is determined through the supplemental simulation with L=100 samples, then N2= C*L =1000 samples are used in this process.

The estimator $h_r(r_j)$ of the importance sampling density determined for each limit state function using N1 + N2 samples are shown in Figs. 3 and 4. The ideal importance sampling density evaluated by Eq.(26) is also plotted on these figures as a theoretical estimate, in which the value of P_f is taken from works of Schuëller and Stix (1987) for Eq.(27) and those of Harbitz (1986) for Eq.(28), respectively. It can be said that the estimator of the importance sampling density determined by using N1=1000 samples almost coincides well over the range of r with the theoretical importance sampling p.d.f. given by Eq.(26).

Using the estimator of the importance sampling density determined above, the importance sampling simulation with N samples is executed to estimate the failure probability and its coefficient of variation (c.o.v.).

The results obtained by total sample size Nt=2500 are given in Tables 4 and 5 compared with other methods such as ISPUD, ISPUD for multi mode failure denoted by I.S.M. and the Monte Carlo simulation combined with partition of the region denoted by M.C.P., the estimate P_{fe} of which is evaluated by (1-P)*M/N1.

As the proposed method uses Nt= N1+N2+N = 1000+1000+500 samples totally, then other simulation methods are compared with by using the same 2500 samples. The proposed method gives reasonable estimates of failure probability and remarkable lower coefficient of variations compared with other methods.

The estimate of the failure probability and its c.o.v. vs. total sample size Nt with N1=1000 and N2=1000 are illustrated in Figs. 5, 6, 7 and 8 compared with other methods. It can be said

Table 4 Comparision of the probability of failure estimates for the case of No.1

Total sample size Nt	2500
Proposed method of Importance sampling	$P_{fe} = 2.30\times10^{-6}$ c.o.v.= 1.10×10^{-2}
ISPUD (β_1=4.66)	$P_{fe} = 2.18\times10^{-6}$ c.o.v.= 4.93×10^{-2}
ISPUD for multi mode failure	$P_{fe} = 2.23\times10^{-6}$ c.o.v.= 6.65×10^{-2}
Monte Carlo with the region (Excluded region: $\lvert x \rvert < \beta$)	$P_{fe} = 2.23\times10^{-6}$ c.o.v.= 5.49×10^{-2}
Estimate (Schuëller)	$P_f = 2.30\times10^{-6}$

Table 5 Comparision of the probability of failure estimates for the case of No.2.

Total sample size Nt	2500
Proposed method of Importance sampling	$P_{fe} = 4.03\times10^{-4}$ c.o.v.= 1.00×10^{-2}
ISPUD (β = 4.00)	$P_{fe} = 4.54\times10^{-4}$ c.o.v.= 17.5×10^{-2}
Monte Carlo with partition of the region (Excluded region: $\lvert x \rvert < \beta$)	$P_{fe} = 3.80\times10^{-4}$ c.o.v.= 5.19×10^{-2}
Estimate (Harbitz)	$P_f = 4.00\times10^{-4}$

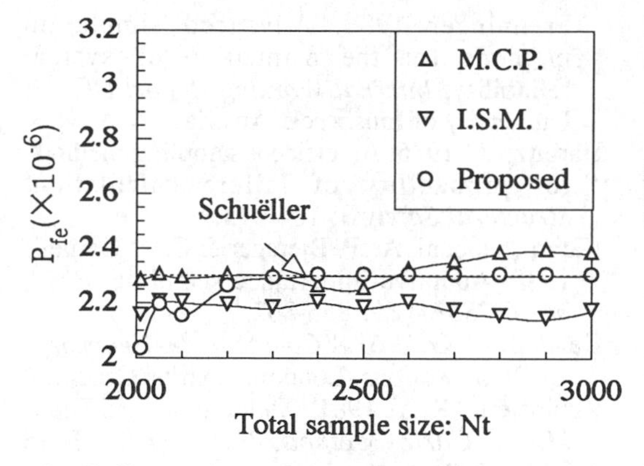

Fig. 5 Comparison of the probability of failure estimates for the case of No.1.

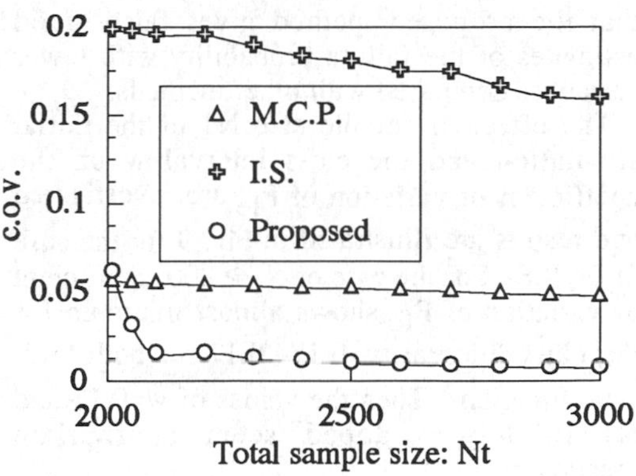

Fig. 8 Coefficient of variations of probability of failure estimates for the case of No.2.

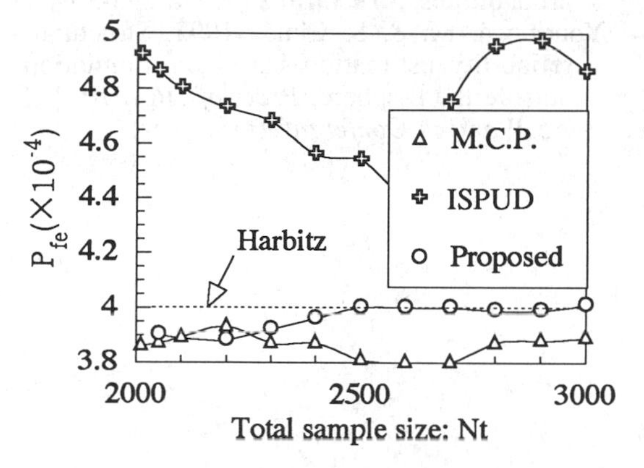

Fig. 6 Comparison of the probability of failure estimates for the case of No.2.

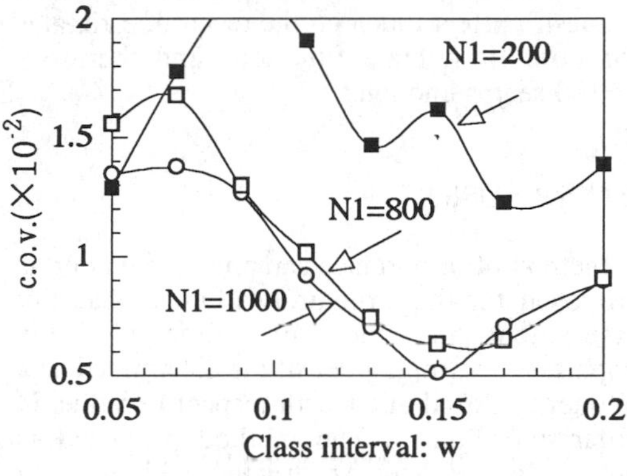

Fig. 9 Effect of class interval on coefficient of variation of P_{fe} for the case of No.2

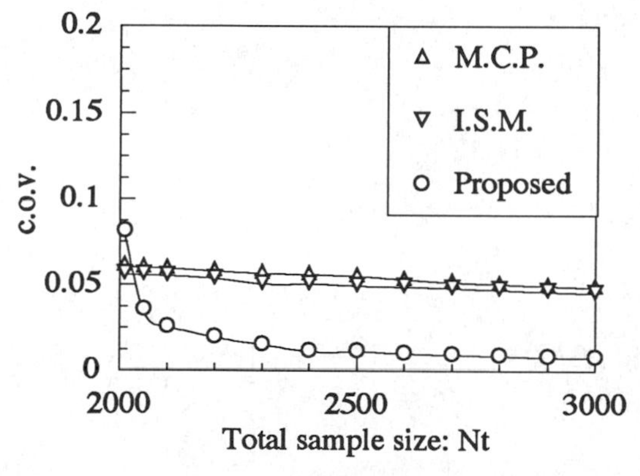

Fig. 7 Coefficient of variations of probability of failure estimates for the case of No.1.

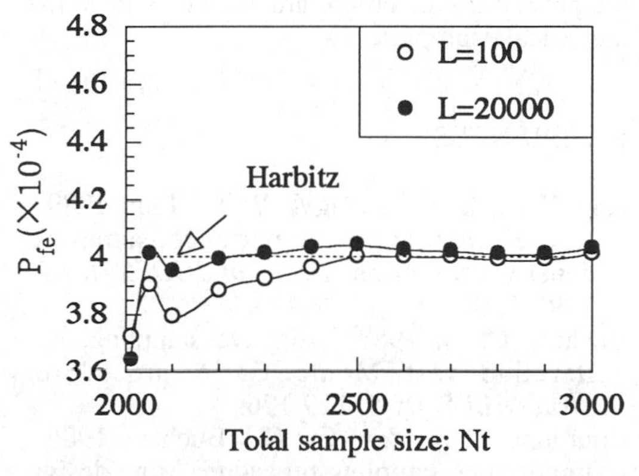

Fig.10 Effect of sample size L on estimate P_{fe} for the case of No.2.

that the proposed method gives fairly good estimates of the failure probability with lower variances compared with other methods.

The effect of sample size N1 of the initial simulation and the class interval w on the coefficient of variation of P_{fe} are investigated and results are illustrated in Fig. 9 for the case of Eq.(28). For the case of $N1 \geq 800$, coefficient of variation of P_{fe} shows almost minimun for the class interval $w \approx 0.13 \sim 0.17$ for both limit state functions. Then the values of w=0.13 and N1=1000 we adoped seem appropriate practically.

The effect of the sample size L of the supplemental simulation on the estimate P_{fe} is investigated and illustrated in Fig.10 for the case of Eq.(28). It can be said that sample size L doesn't affect much on the result of estimates for both limit state functions and therefore L=100 seems enough.

4 CONCLUSION

A method of importance sampling simulation is proposed for the structural failure probability assessment, in which an estimator of the importance sampling density is determined as a frequency distribution with respect to radius in polar coordinates. The method proposed is especially useful for the structure with the limit state functions of highly nonlinear curvatures and structures with multi mode failure. Remarkable lower variances in the estimates of the probability of failure are attained with some numerical examples.

REFERENCES

Ang, G. L., A. H-S. Ang & W. H. Tang 1989. Kernel method in importance sampling density estimation. *Proc. of ICOSSAR'89*: 1193-1200.

Bucher, C. G. 1988. Adaptive sampling, an iterative fast Monte Carlo procedure. *Structural Safty* 5: 119-126.

Bourgund, U. & C. G. Bucher 1986. Importance sampling procedure using design point-ISPUD. *Report* 8-86: Institute of Engineering Mechanics, University of Innsbruck, Austria.

Bourgund, U., W. Ouypornprasert & P. H. W. Prenninger 1986. Advanced simulation methods for the estimation of system reliability. *Internal Working Report* NO.19: University of Innsbruck, Austria.

Harbitz, A. 1986. An efficent sampling method for probability of failure calculation. *Structural Safety* 3: 104-115.

Karamchandani, A., P. Bjerager & C. A. Cornell 1989. Adaptive importance sampling. *Proc. of ICOSSAR'89*: 855-862.

Kendall, M. G. 1961.*A Course in the Geometry of N dimensions*. London: Charles Griffin.

Rubinstein, R. Y. 1981. *Simulation and The Monte Carlo Method*. New York: John Wiley & Sons.

Schuëller, G. H. & R. Stix 1987. A critical appraisal of methods to determine failure probabilities. *Structural safety* 4: 293-309.

Yonezawa, M. & S. Okuda 1993. Structural reliability estimation based on simulation outside the ß-sphere. *Proc. of 5th IFIP WG 7.5 Working Conference*(to appear).

Simulation and fuzzy sets

Structural Safety & Reliability, Schuëller, Shinozuka & Yao (eds) © *1994 Balkema, Rotterdam, ISBN 90 5410 357 4*

Fuzzy-based safety control of construction activities: A case study

M. H. M. Hassan
Higher Technological Institute, Cairo, Egypt

B. M. Ayyub
University of Maryland, College Park, Md., USA

ABSTRACT: In this study, an actual construction failure case study is analyzed. The analysis is performed utilizing a fuzzy-based controller that was developed earlier for the control of construction activities. The referenced system builds a model for a controlled construction activity that emphasizes critical and important properties. The model comprises a condition assessment unit that is responsible for transferring the inputs into outputs. Fuzzy-based control was found to be the most suitable strategy for controlling such highly complex and uncertain systems. The fuzzy controller comprises a rule-base, a self learning unit and a conflict resolution unit for controlling multiple attributes at once.

1 INTRODUCTION

Any structure passes through several stages during its lifetime from the point of conception until it reaches its last stage, i.e., demolition, when it is considered to have served its function or life. The construction stage is identified as one of the most critical stages a structure goes through during its life cycle. During this stage, the structure needs external elements to provide temporary support until the structure can either develop its full strength, e.g., concrete structures, or until it is all connected and tied up, e.g., steel structures. The presence of these temporary support elements in addition to the incremental nature of the construction operation raises several safety concerns that should be investigated. Thus, safety monitoring and control during the construction stage represents an important objective that should be achieved.

Construction activities are highly uncertain and complex systems, and they involve a large number of subjectively defined variables and descriptors. In addition, they are almost impossible to model mathematically. Thus, a fuzzy-based control strategy seems to be the best suitable approach to control such systems. In order to develop such a control system, a model of the controlled system should be built as a first step. In this model, an image of the actual system is projected using a set of critical factors. A condition assessment procedure is developed and incorporated as an integral component of the system's model. The condition assessment procedure transfers the states of the input variables into some attribute measures. These assessments represent the output of the controlled system. Utilizing a fuzzy control strategy, these outputs could be monitored and controlled within a given acceptable range. An actual case of construction failure is considered and modeled throughout this paper. The construction of the concrete deck of an overpass bridge is considered. The collapsed bridge was one of four bridges under construction at the Maryland 198 crossing of the Baltimore-Washington parkway (FHWA 1989). The bridge was designed as a simple span post-tensioned concrete box girder bridge with the east abutment providing a fixed support and the west abutment providing a roller support. The superstructure was supported by timber formwork on steel longitudinal support beams which are supported by metal shoring as schematically shown in Figure 1. The case study is used to demonstrate the applicability and validity of the control system. Several failure initiators are selected to test whether they could have initiated the failure of the modeled system. This analysis is performed in an effort to identify several potential scenarios of events that could have triggered the failure mechanism.

2 SYSTEM DEFINITION

In this paper, the safety of the concrete placement activity was selected as the only attribute of interest. Thus, the concrete placement activity represents the controlled system with the safety attribute defined as the control attribute of interest. The controlled activity comprises two main processes. Each process inherits the same attributes of the higher level, i.e., the concrete placement activity. For each process, a set of critical variables is defined. These variables represent the factors affecting the safety of the process under consideration. Each variable is defined as an unobservable label that consists of a set of observable sub-factors. By observing the sub-factors, an overall state estimate could be made for each variable. Figure 2 shows the defined model with the observable and unobservable factors. The activity as defined comprises a falsework construction process and a concrete pouring process. Each process has its own safety level which when evaluated could be combined together in order to evaluate the activity's safety level. For each process, its safety attribute level is projected given the states of a set of critical factors. For example, Figure 2 shows that the first process, i.e., falsework construction, has four critical factors defined as stability, alignment, condition and tower arrangement. The developed model is built in a hierarchical framework in order to facilitate the control process and to identify better feedback decisions. The developed model identifies the components of the activity together with the factors affecting the attribute of interest, i.e., safety. An integral component of the system is the condition assessment procedure. Any model should be capable of identifying the input and output variables in addition to a function or a procedure that could be utilized in estimating the output values given a set of input values. A safety assessment procedure is utilized in this study in order to evaluate the safety level given a set of states of the input variables. These assessments should be evaluated at both hierarchical levels, i.e., the process level and the activity level.

3 SAFETY ASSESSMENT

The safety assessment procedure is applied at the process as well as the activity levels. At the process level, for each potential state of a given variable, a failure likelihood impact is assigned. The states of the variables as well as the failure likelihood levels are defined using linguistic measures which are quantified using Fuzzy set theory. The assigned likelihood level depends on the state of the variable rather than the variable itself. Thus, a sensitivity factor that reflects the importance of each variable to the attribute of interest, is applied to the state fuzzy set of each variable. The adjusted state fuzzy set is defined as

$$\mu_A{}^a(x) = m\,\mu_A(x) \tag{1}$$

where $\mu_A{}^a(x)$ = adjusted membership function of element x in fuzzy set A; m = importance factor; and $\mu_A(x)$ = original membership function of element x in fuzzy set A. The relationship between each individual potential state and its failure likelihood impact is defined by the Cartesian product of fuzzy relations. Figure 3 shows a block diagram that summarizes the safety assessment procedure at the process level. Effects of potential combinations of states are then evaluated using the union of fuzzy relations as shown in Figure 3 and defined as follows:

$$\mu_{R_1 \cup R_2}(x) = \mathrm{MAX}\left\{\mu_{R_1}(x), \mu_{R_2}(x)\right\} \tag{2}$$

where $\mu_{R_1 \cup R_2}(x)$ = membership function for the union; MAX = maximum operator; $\mu_{R_1}(x)$ = membership function of element x in fuzzy relation R_1; and $\mu_{R_2}(x)$ = membership function of the same element x in fuzzy relation R_2. An aggregation and defuzzyfication procedures are applied to the resulting combined relation matrix in order to reduce the matrix into a single point estimate.

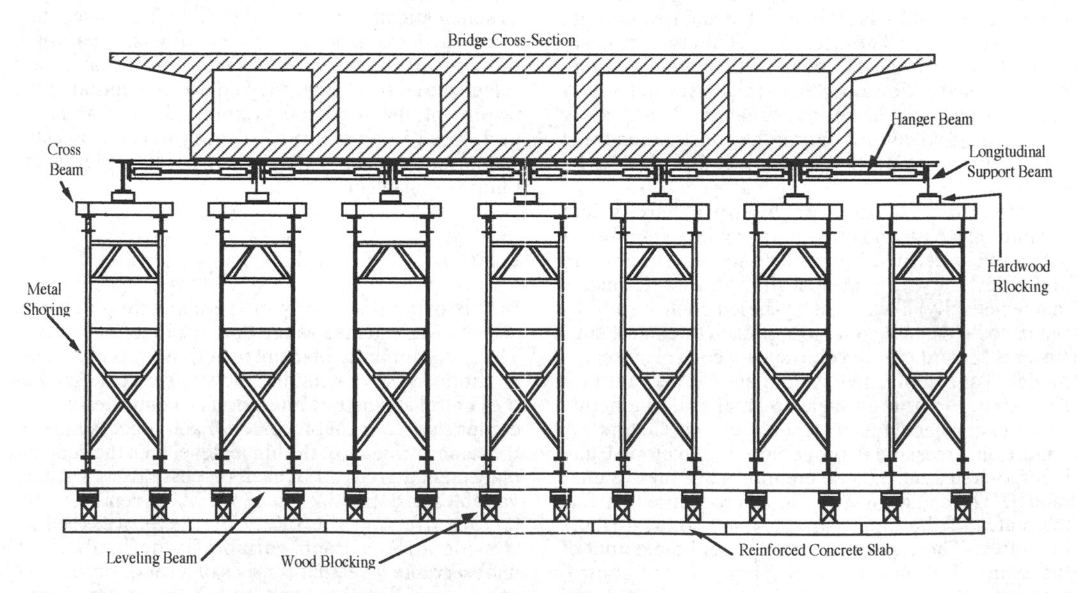

Fig. 1. Falsework system of case study

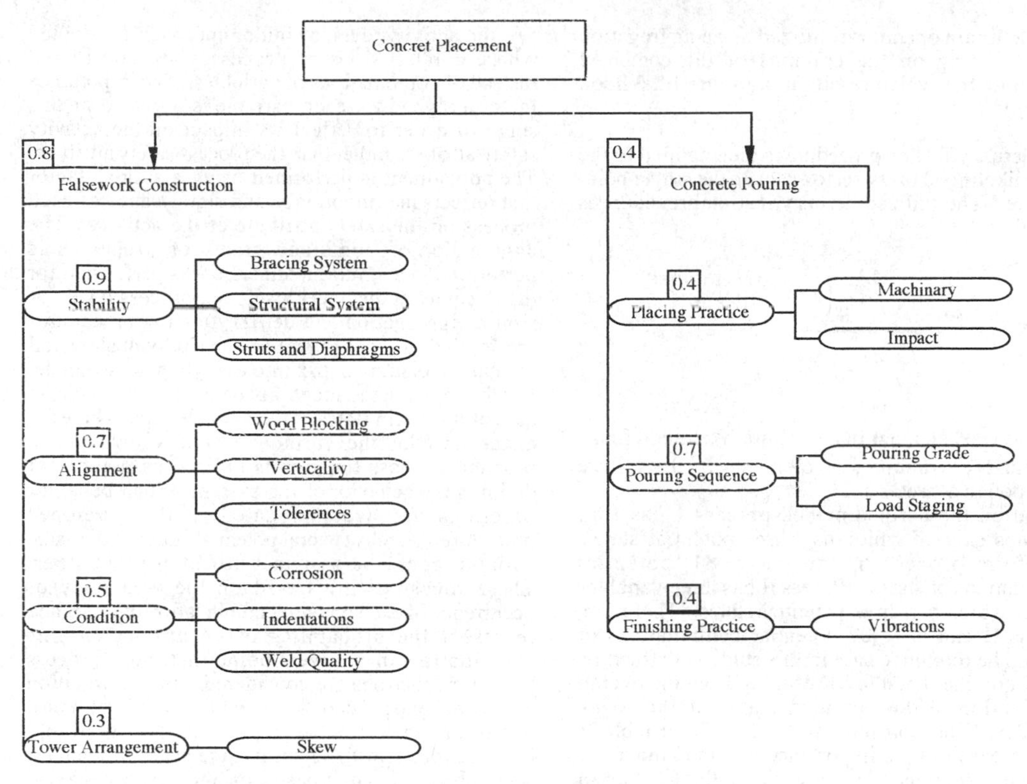

Fig. 2. System's model of case study

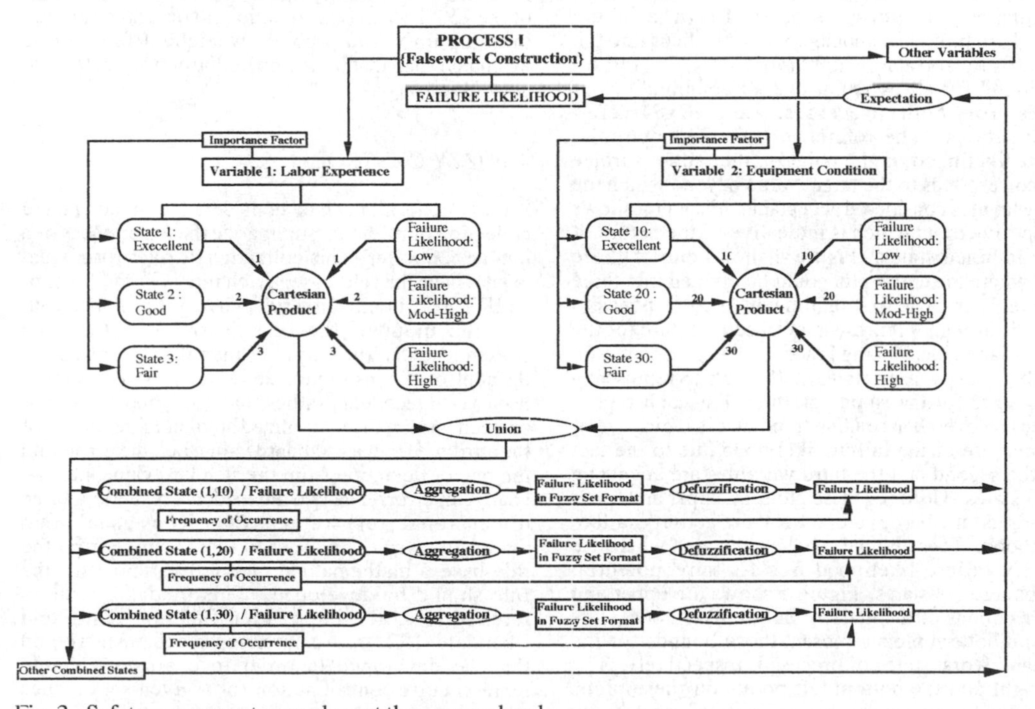

Fig. 3. Safety assessment procedure at the process level

The maximum operator is utilized as an aggregation tool operating on the columns of the combined relation matrix which results in a failure likelihood fuzzy set.

The defuzzyfication procedure is then applied to the failure likelihood fuzzy set to evaluate the single point estimate. The utilized defuzzyfication procedure is defined as

$$\log_{10}\left(P_{f_j}\right) = \frac{\sum_{i=1}^{N_p} \mu\left(P_{f_i}\right)\log_{10}\left(P_{f_i}\right)}{\sum_{i=1}^{N_p} \mu\left(P_{f_i}\right)} \quad \text{for } j = 1, 2, \ldots, 9 \quad (3)$$

where P_{f_i}= i^{th} element in the failure likelihood fuzzy set; and N_p = number of elements in the failure likelihood fuzzy set.

Based on the defined model, process I has four variables each of which has three potential states. Therefore, process I can have 81 potential combinations of states. Process II has three variables each of which has three potential states. Therefore, process II can have 27 potential combinations of states. The reference case in this study is defined by the failure likelihood level resulting from the overall combination of the optimum states of the seven variables. The optimum state of each variable is defined based on its importance and the amount of acceptable tolerance. For each combined state of process I, the failure likelihood level resulting from all potential combinations of process II form a failure plot. This plot corresponds to a vertical cut through the failure surface which represents the failure likelihood for all potential overall combinations of states for both processes, i.e., 27x81=2187 combinations. The reference case, i.e., optimum failure likelihood, is the point on the failure surface that corresponds to the target value beyond which the safety level is considered acceptable. Each plot shows a repetitive trend which is indicative of the nature of the combined states. Figure 4 shows one of these plots where the entire plot could be divided into three main regions. The first region which corresponds to states 1 through 9 represent an Excellent state for the first variable, i.e., Placing Practice. The second region which corresponds to states 10 through 18 represent a Good state for the same variable. For each region there is an overall ascending trend that is indicative of the increase of the failure likelihood due to the fact that the second and the third variables are acquiring lesser states. Utilizing these plots, an upper and lower bounds could be developed for the activity failure likelihood. These bounds enclose the area where the activity failure likelihood lies for any potential combination of states. Figure 5 shows the upper and lower bounds of the activity failure likelihood. The top and bottom plots represent those bounds for the best and worst states of process I, respectively. The top right and the bottom left points on these plots, represent the upper and lower limits of the activity failure likelihood.

At the activity level, a similar approach is adopted where combinations of processes are considered instead of combinations of variables. The importance factor assigned for each variable is adjusted at this stage in order to reflect its impact on the activity safety attribute rather than the process safety attribute. The adjustment is performed using a similar factor that reflects the importance and significance of each process on the safety attribute of the activity. The aggregation of combined states of processes is performed in a similar manner as was performed for the variables at the lower level, i.e., process level. A similar aggregation and defuzzyfication procedures are applied in order to reduce each individual overall combined relation matrix into a single point estimate. For the safety assessment procedure at both levels, in case of uncertain states of the variables, i.e., where the exact state of the variable is not known yet, a probability mass function is utilized as a tool for defining the behavior of the system. Such behavior functions are accommodated in the developed procedure whereby several potential states of the same variable could be assigned probability measures. These measures are based on the frequency of occurrence of each potential combination of states and represent the probability of occurrence of this combination in a given point in time. Process behavior functions are considered input information where an expert can fit one of several standard behavior functions to represent the available information. On the contrary, the activity behavior function is evaluated based on the given behavior functions of its underlying processes. The evaluation of the activity behavior function involves the solution of an optimization problem whereby a function is evaluated such that it maximizes the entropy subject to all additional constraints.

4 FUZZY CONTROL

A fuzzy controller can be considered as an abstracted collection of rules that summarize the experiences of a human controller. This collection of rules form what is known as the rule-base. Each rule is in the form of an IF THEN implication rule where if the antecedent, i.e., the input of the rule, is satisfied then the consequent, i.e., the output of the rule, is implied. In the control of construction activities, it is necessary to use two antecedents rather than just one. The two antecedents are the normalized amount of deviation of the attribute from its standard value, i.e., the error, and the rate of deviation from the standard value, i.e., the change in error. The two antecedents result in a three dimensional rule rather than the usual two dimensional rule. In order to infer actions from the rule-base, a mathematical model that represents the rule should be developed. This model is usually referred to as the implication function (King and Mamdani 1977). An inferring mechanism should then be developed in order to evaluate a single representative control action for a given set of rules and input values.

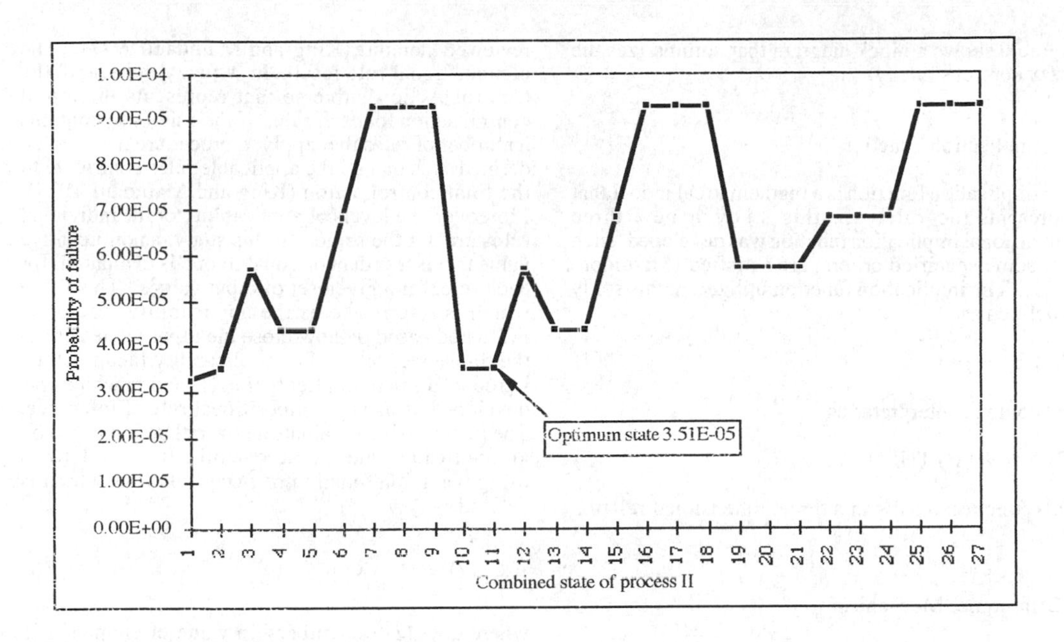

Fig. 4. Impact of state of process II

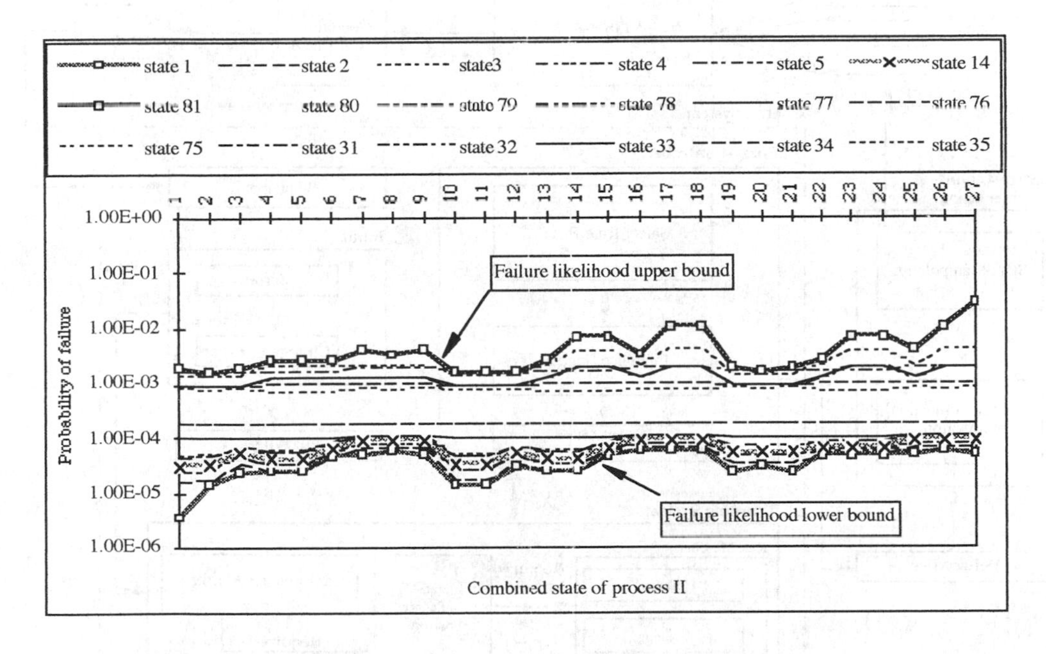

Fig. 5. Failure likelihood bounds

Figure 6 shows a block diagram that summarizes the fuzzy control strategy.

4.1 Implication Function

The implication function is a mathematical model that represents the rule. In this study a new three dimensional implication function was developed such that some specified criteria are satisfied (Mizumoto 1981). The implication function utilized in this study is defined as

$$(A \times B) \longrightarrow C \tag{4}$$

and could be interpreted as

$$\text{IF } (A \text{ AND } B) \text{ THEN } C \tag{5}$$

This function results in a three dimensional relation matrix.

4.2 Inference Mechanism

An inference mechanism is a procedure by which a control action could be inferred given a rule-base and a set of input values. For a single rule, the compositional rule of inference developed by Zadeh (1965) was found satisfactory according to the reviewed literature (King and Mamdani 1977). The compositional rule results in a fuzzy subset of the Control Action universe that represents the initial control action for each rule. If the rule-base contains a number of rules that apply, a procedure needs to be defined such that all the applicable rules contribute to the final control action (King and Mamdani 1977). Moreover, the levels of applicability of the individual rules are not the same. In this study, a non-negative value that is less than or equal to one is evaluated, for each rule, for a given set of input values. This value which is referred to as the applicability factor is evaluated based on how close the input values are to the rule antecedents. The applicability factor is then introduced as a multiplier to the rule action which was developed using the compositional rule of inference. The final step is to evaluate an overall action based on all applicable rules. The overall action is defined using the union function of fuzzy relations which is defined as

$$\mu_{(OA)}(z) = \underset{i=1}{\overset{n}{\text{MAX}}} \; \mu_{(FA)_i}(z) \tag{6}$$

where $\mu_{(OA)}(z)$ = membership value of element z in the overall action fuzzy set; $\mu_{(FA)_i}(z)$ = membership function of element z in the overall action of the ith rule; and MAX = maximum operator applied over all

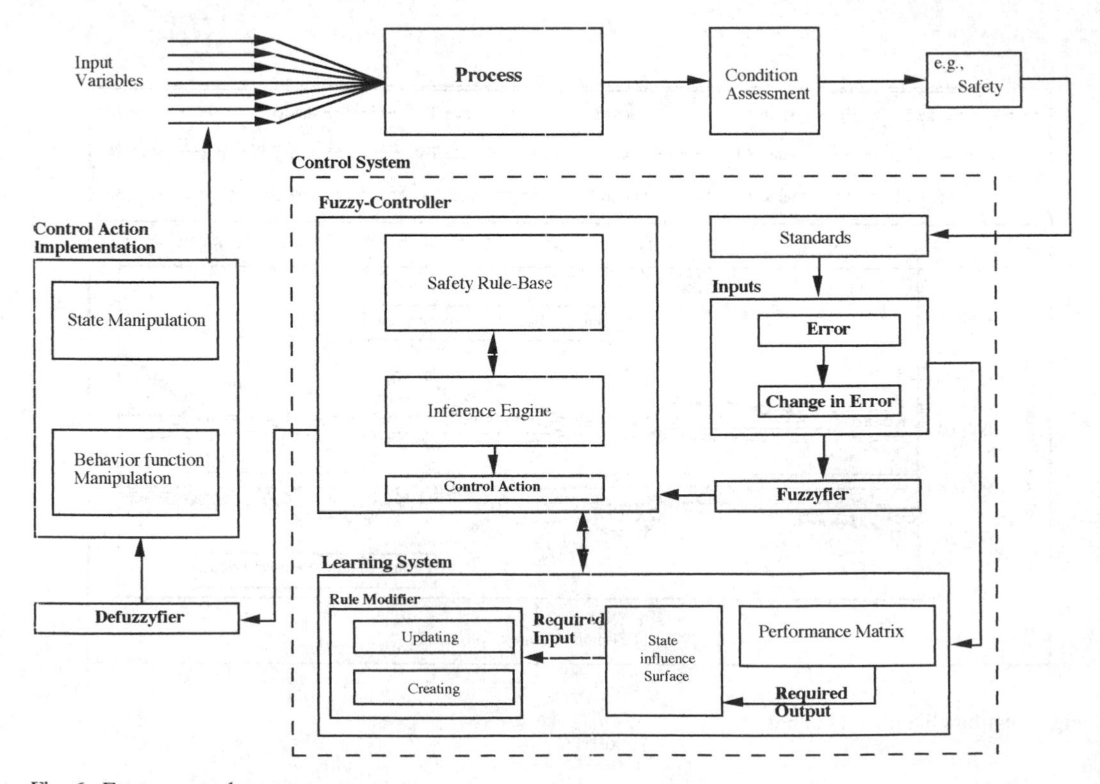

Fig. 6. Fuzzy control strategy

the applicable rules, where i ranges from 1 to the number of applicable rules n. The resulting overall action is a fuzzy subset of the universe of discourse of the fuzzy variable Control Action. In order to be able to apply the control action, it should be in the form of a crisp number. A crisp control action could be defined by defuzzfying the resulting overall action fuzzy set.

5. SELF LEARNING SYSTEM

A fuzzy controller depends on a rule-base where all previous experiences and information are stored. However, it is difficult to construct a complete rule-base that is capable of handling all potential situations. Thus, it is necessary to include a self learning unit within the control system that is responsible for expanding and updating the current rule-base. The self learning unit should identify situations or cases that are not covered in the current rule-base. It should also extract necessary information and construct new rules to handle these situations. In general, the self learning unit monitors the performance of the control system. It compares the performance with an ideal performance that is built in the learning unit and stored in what is defined as the performance matrix (Procyk and Mamdani 1979). This ideal performance should be a reflection of how the control system should react to each potential combination of Error and Change in Error. If the control system deviates from this ideal track, the learning system should be able to identify the underlying causes and improve the control system's performance. Unsatisfactory performances may be due to one of two main causes. The first is due to missing rules, i.e., the current rule-base cannot handle the current situation because none of its rules is applicable. Accordingly, the control system cannot suggest any corrective action. The second is due to untuned rules. In other words, the rules that apply to the current case needs to be adjusted in order to result in the expected ideal performance.

6. SCENARIO ANALYSIS

In this study, the control system is applied to the defined case study. Five failure scenarios were tested where each scenario starts with a given combination of states that is less than the reference case. In each scenario, specific variables were identified as difficult or impractical to adjust once the process/activity has started. Then, the critical control variables were successively adjusted in each control cycle based on the requested control action. This scenario analysis furnishes the required insight necessary to test several potential combined states that would eventually develop a failure mode. Each scenario would terminate when all other factors except for the selected factor(s) reach their ultimate states. Five combined states were then identified as failure initiators. These failure initiators are defined in Table 1.

The failure likelihood of process I and II as well as the activity were evaluated for each combined state. A critical limit which defines the failure zone was assumed based on previous experiences to be 10^{-3}. Figure 7 shows the failure likelihoods for both processes as well as the activity resulting from each combined state. Three zones were identified in the figure, namely, the failure zone, the critical zone and the safe zone. If the failure likelihood lies within the failure zone, this indicates that the responsible state would cause failure. On the other hand, if the failure likelihood lies within the safe zone, the responsible state does not initiate or develop a failure mode. However, if the failure likelihood lies within the critical zone, this indicates that although this combination does not, by itself, develop a failure mode, any slight variation in the states of any of the variables would move the point into the failure zone resulting in a failure. According to Figure 7, the most critical failure initiator for the whole activity is state 4. The modeled system, i.e., the concrete placement of the bridge deck, could have failed given any of the potential combined states developed during the placement of the top deck. In other words, insufficient longitudinal bracing combined with large tolerances combined with a large skew would initiate a failure mode for the bridge deck during construction. The importance of the developed system arises from its real-time monitoring of the failure likelihood and all the involved variables. For some variables, a severe deviation from their optimum states could result in adverse effects which could not be corrected by the controller. However, the feed forward loop integrated within the control system looks at the controllability of the system given a set of uncontrollable variables. This loop has been deactivated during the course of this analysis in order to be able to proceed with all different situations that might face the controller. This loop essentially terminates the activity/process in its early stages or even before it actually starts, if it detects an uncontrollable situation. An uncontrollable situation would result if the control system cannot

Table 1. The failure initiators utilized in the scenario analysis

State	Stability	Alignment	Condition	Arrangement	Placing	Sequence	Finishing
1	Fair	Fair	Moderate	Fair	Good	Excellent	Good
2	Fair	Fair	Fair	Moderate	Good	Excellent	Good
3	Fair	Fair	Moderate	Moderate	Good	Fair	Good
4	Fair	Fair	Fair	Moderate	Good	Fair	Good
5	Fair	Fair	Fair	Excellent	Excellent	Fair	Excellent

improve the failure likelihood to reach the target value after adjusting all other variables to their ultimate states. This inability to control the system might be either because of practical limitations on controllable variables, the presence of several uncontrollable variables or problems with the developed system model. If such a control system is operating on the activity during construction, it could detect a failure potential and adjust those variables that can be responsible for initiating a failure. If these variables were uncontrollable it could at least send an alarming signal and suggest the termination of the activity until any of the involved variables is changed or the practical limitations are lifted.

7 SUMMARY AND CONCLUSIONS

In this paper, a case study of an actual construction failure was analyzed utilizing a fuzzy-based controller for construction activities developed earlier. The system built a model of the construction activity which emphasized critical properties. A scenario analysis was performed in an effort to identify several potential set of events that could have initiated the failure mode. The importance of the developed system arises from its real-time monitoring of the failure likelihood and all the involved variables. If such a control system is operating on the activity during construction, it could detect a failure potential and adjust those variables that can be responsible for initiating a failure. If these variables were uncontrollable it could at least send an alarming signal and suggest the termination of the activity until any of the involved variables is changed or the practical limitations are lifted. During this exercise, the system was essentially simulating the situation that prevailed during the construction of the bridge deck. This was accomplished in an effort to identify the reasons for such a failure. However, during its real-time operation, the system will detect a critical situation based on the resulting failure likelihood. It signals the operator which variable should be adjusted and how much. If the activity is uncontrollable as mentioned above the system will recommend to terminate operation.

8 REFERENCES

Ayyub, B.M. and Hassan, M.H.M. 1991. Control of construction activities: III. Fuzzy controller. Civil Engineering Systems, scheduled for publication.

FHWA Board or review 1989. Report of the investigation into the collapse of the route 198 Baltimore-Washington parkway bridge. U.S. Department of Transportation, FHWA Publication No. PR-90-001.

King, P.J. and Mamdani, E.H. 1977. The application of fuzzy control systems to industrial processes. Automatica, 13, 235-242.

Mizumoto, M. 1981. Note on the arithmetic rule by Zadeh for fuzzy conditional inference. Cybernetics and Systems, 12, 247-306.

Procyk, T.J. and Mamdani, E.H. 1979. A linguistic self-organizing process controller." Automatica, 15, 15-30.

Zadeh, L. A. 1965. Fuzzy sets. Information and control, 8, 338-353.

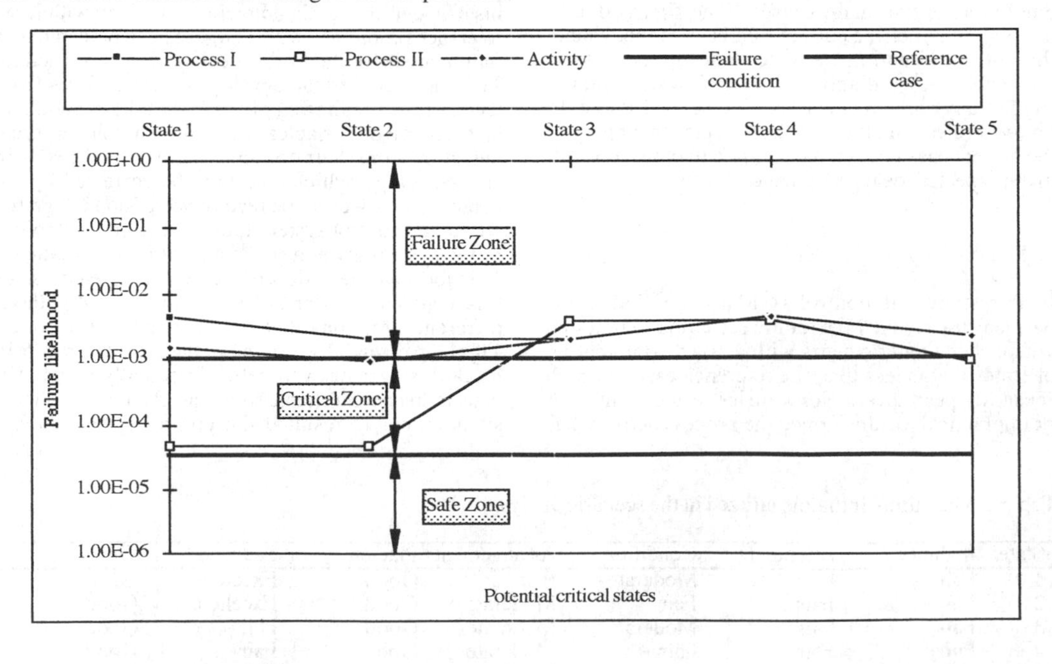

Fig. 7. Potential failure initiators

Structural Safety & Reliability, Schuëller, Shinozuka & Yao (eds) © 1994 Balkema, Rotterdam, ISBN 90 5410 357 4

A combined probability-possibility evaluation theory for structural reliability

Hiroshi Kawamura & Yasuhiko Kuwamoto
Department of Architecture and Civil Engineering, Faculty of Engineering, Kobe University, Rokkodai, Nada, Japan

ABSTRACT: In structural identification and planning, it is very important to evaluate the uncertainties of future loads and structural responses (Yao and Natke 1991). In this paper, probability functions in which additive law holds are combined with possibility functions in which maximal law holds, so that their evaluation measures belong to a kind of fuzzy measures instead of probability measures. By using their joint evaluation distributions and measures, one can evaluate the uncertain strength of parallel or serial structural systems and the survival or fracture possibility of structures. This paper presents several numerical examples. Results show that the proposed combined probability-possibility evaluation theory is effective to evaluate the combined objective and subjective reliability of structures from practical and theoretical points of view.

KEYWORDS: Structural Reliability, Possibility Measure, Probability Measure, Evaluation Measure, Fuzzy Measure, Parallel System, Serial system, Survival Possibility, Fracture Possibility

1.INTRODUCTION

The correct probability of a certain future event can be obtained, only when the sufficient data of its past events are given and their repetition is assured in future. For the evaluation of the probabilities of uncertain future events which do not satisfy the above conditions, subjective probability (Ang and Tang 1984), fuzzy probability of fuzzy events (Zadeh 1968) have been proposed instead of the usual probability theory. In cases where the additive law is not valid in probability measures, Dempster-Shafer's theory (Shafer 1976) and Zadeh's possibility theory (Zadeh 1978) were proposed.

In this paper, for the subjective evaluation of the safety of future structures under planning, a combined probability-possibility theory is proposed. This new theory is based on the relative frequency histograms derived from the numbers of "yes" answers from structural experts to questionnaires about structural strength and loading, to which each expert is allowed to give "yes" answer once or more times. (In the usual probability theory, only one event is allowed.) Therefore, probability-possibility measure can be defined as the total number of "yes" answerers about concerned items, and it belongs to a kind of fuzzy measures (Sugeno 1972).

Consider a future, unique and complex structure under planning. No one can estimate the objective reliability of the structure. This structure has to be designed and planned based on subjective judgments of many kinds of experts (strictly saying, based on many kinds of pseudo-objective theories argued by different experts). The purpose of this paper is to present a theoretical and practical method for the evaluation of the safety of such a future structure.

2.THEORY

2.1 Definitions of Evaluation Distribution and Measure

Suppose that experts are asked to answer "yes" or "no" to a question (e.g., is it safe?) about a given structure under given conditions xi (i=1,...,n). The number of the experts is assumed to be m, and it is assumed that they are allowed to give one or more answers "yes" within the total conditions.

Let the number of experts who answer "yes" to a question under the condition x_i be l_i, and a relative frequency μ_i can be given by

$$\mu_i = l_i / m. \qquad (1)$$

μ_i is plotted as shown in Fig.1 which is called the evaluation mass function. X is the set of x_i. Generally, $\Sigma\mu_i$ is larger than the unity, so a new theoretical treatment is needed.

Now let's define evaluation measure M as the total number of the experts who answer "yes" under a subset of conditions x_i. Then the following equations can be derived:

$$M(\phi)=0,\ M(X)=1, \qquad (2)$$

$$X_1 \subset X_2 \rightarrow M(X_1) \leqq M(X_2), \qquad (3)$$

where ϕ is empty set, X_1 and X_2 are subsets of X. Eqs.2 and 3 show that M belongs to a kind of fuzzy measure (Sugeno 1972).

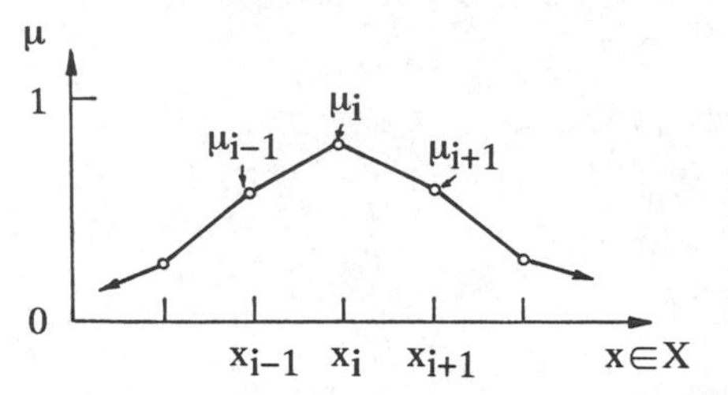

Fig.1 Evaluation distribution as normalized frequencies in which multiple events are countable

Here, every set is assumed to be finite and discrete.

2.2 Compositional Assumption

In this paper, it is assumed that the evaluation mass function μ_i (i=1,...,n) is expressed by the sum of possibility mass function π_i and probability mass function f_i as follows:

$$\mu_i = \pi_i + f_i. \tag{4}$$

A distribution of π_i can be drawn as shown in Fig.2 where a horizontal bar corresponds to an expert who answers "yes" regarding the same subset x_i as the width of the bar. A distribution of f_i can be drawn as shown in Fig.3 where a point corresponds to an expert who answers "yes" only one time.

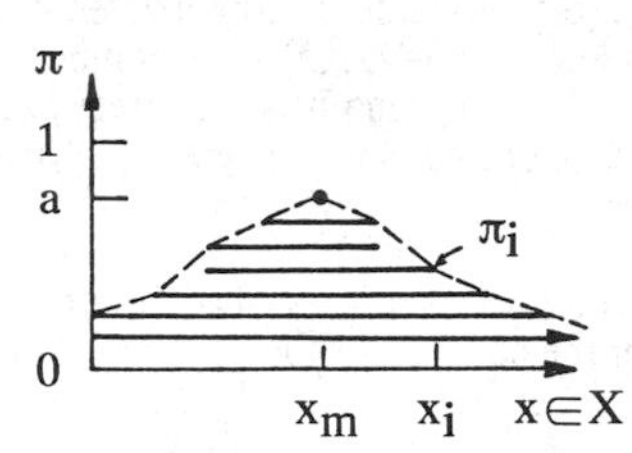

FIg.2 Possibility mass function π_i

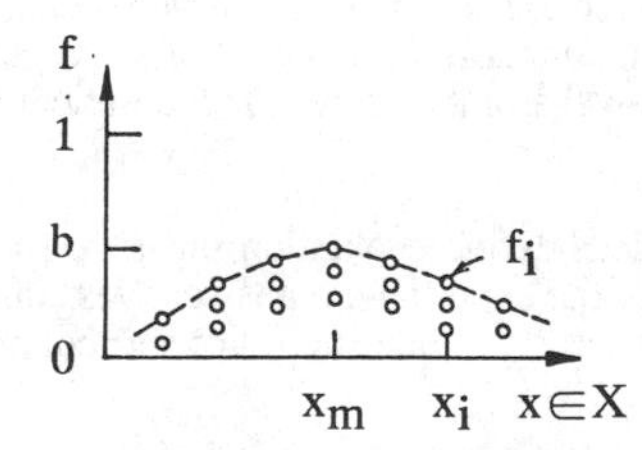

Fig.3 Probability mass function f_i

According to the above assumption, evaluation measure $M(x_i \in X_0)$ can be reduced to

$$M(x_i \in X_0) = \max_{x_i \in X_0}(\pi_i) + \sum_{x_i \in X_0} f_i, \tag{5}$$

which also satisfies the conditions of fuzzy measure, i.e., Eqs.2 and 3. When a, α, b, β are given by

$$\max_{x_i \in X}(\pi_i) = a, \quad \sum_{x_i \in X} \pi_i = \alpha, \tag{6}$$

$$\max_{x_i \in X}(f_i) = b, \quad \sum_{x_i \in X} f_i = \beta, \tag{7}$$

the following relations can be derived:

$$a + \beta = 1, \quad a + b \leqq 1, \tag{8}$$

the former of which can be given by the substitution of X for X_0 in Eq.5.

2.3 Joint Evaluation Mass Function

As for the above evaluation distribution, joint evaluation mass function is needed as well as joint probability mass function. Fig.4 shows that μ_{3ij} under the condition x_{3ij} (=$g(x_{1i}, x_{2j})$), is joint evaluation of μ_{1i} under the condition x_{1i} and μ_{2j} under the condition x_{2j}. According to Eq.4, μ_{1i}, μ_{2j} and μ_{3ij} are expressed by

$$\mu_{1i} = \pi_{1i} + f_{1i},$$
$$\mu_{2j} = \pi_{2j} + f_{2j}, \tag{9}$$
$$\mu_{3ij} = \pi_{3ij} + f_{3ij},$$

where i=1,...,n, and j=1,...,q.

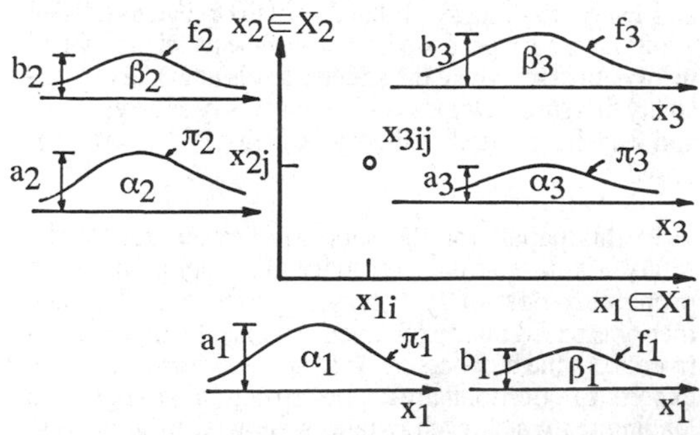

Fig.4 Joint evaluation mass function

In this paper, it is assumed that μ_3 is given by

$$\mu_3 = \mu_1 \mu_2, \tag{10}$$

and that π_3 and f_3 are given by

$$\pi_3 = \pi_1 \pi_2 + \gamma(\pi_1 f_2 + \pi_2 f_1),$$
$$f_3 = f_1 f_2 + (1-\gamma)(\pi_1 f_2 + \pi_2 f_1), \tag{11}$$

where an interpolation coefficient γ is introduced to satisfy Eq.9.

As shown by Eqs.6, 7 and 8, a_3, β_3 are given by

$$a_3=\max_{x_i \in X_0}(\pi_3),\ \beta_3=\sum_{x_i \in X_0} f_3, \tag{12}$$

and have the following relation:

$$a_3+\beta_3=1. \tag{13}$$

If each of the pair of π_1 and f_1, π_2 and f_2 and π_3 and f_3 have single extreme values with respect to the same x as shown in Figs.2 and 3, according to Eq.12, a_3 and β_3 are given by

$$\begin{aligned} a_3 &= a_1a_2+\gamma(a_1b_2+a_2b_1), \\ \beta_3 &= \beta_1\beta_2+(1-\gamma)(\alpha_1\beta_2+\alpha_2\beta_1), \end{aligned} \tag{14}$$

where a_1, α_1, b_1, β_1 and a_2, α_2, b_2, β_2 are given regarding π_1, f_1 and π_2, f_2, respectively, by Eqs.6 and 7.

Using Eqs.13 and 14, γ is given by

$$\gamma=\frac{1-a_1a_2-\alpha_1\beta_2-\alpha_2\beta_1-\beta_1\beta_2}{a_1b_2+a_2b_1-\alpha_1\beta_2-\alpha_2\beta_1}. \tag{15}$$

3. APPLICATION

3.1 Objective Structural Systems

In this paper, as a case study, the whole strengths x_s and x_p of serial and parallel systems composed of two elements with strengths x_1 and x_2 as shown in Figs.5 and 6 are supposed to be evaluated. As another example, let's consider the survival and fracture possibilities p_s and p_f of a structure with strength x_1 subjected to a load x_2 as shown in Fig.7. These evaluations can be performed as evaluation measures which belong to fuzzy measures.

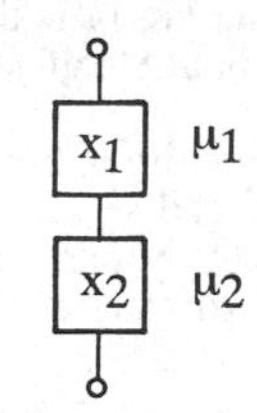

Fig.5 Serial structural system

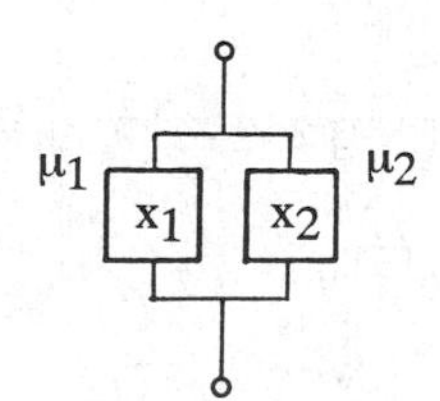

Fig.6 Parallel structural system

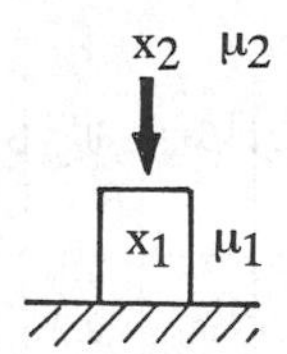

Fig.7 Load-resistance structural system

3.2 Joint Evaluation Mass Functions

Firstly, let's assume that the evaluation mass functions, μ_1 and μ_2, are given as shown in Figs.8 and 9, where μ_1 and μ_2 mean that x_1 is about 50kgf and that x_2 is about 30kgf, respectively. Of course, μ_1 and μ_2 satisfy the above equations 4 to 8, then the joint evaluation mass function μ_3 can be calculated according to the equations 9 to 14, where γ is given as 0.660 by using Eq.15, and a_3 and β_3 are reduced to 0.3720 and 0.6726, respectively, by using the above γ and Eq.14, which satisfies Eq.13, i.e., $a_3+\beta_3=0.3720+0.6276=0.9996 \doteq 1$.

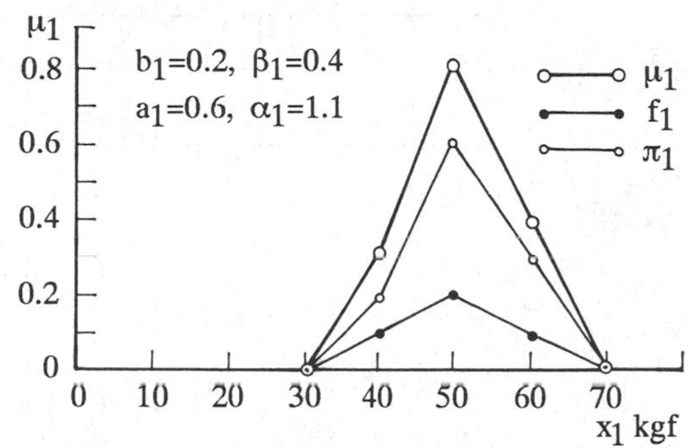

Fig.8 Assumed evaluation mass function μ_1 of x_1

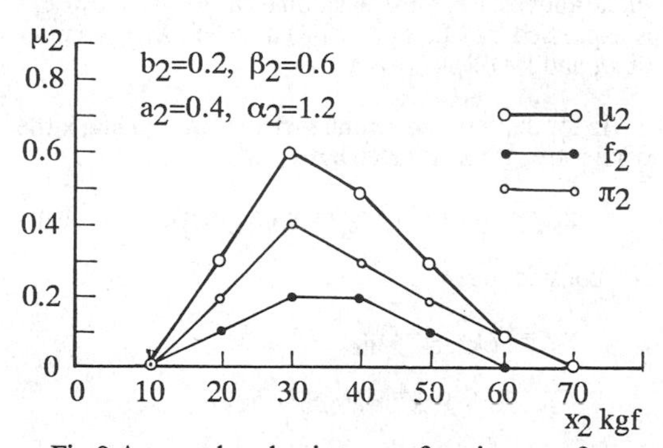

Fig.9 Assumed evaluation mass function μ_2 of x_2

The relation among x_1, μ_1, f_1, π_1, x_2, μ_2, f_2, π_2, x_3, μ_3, f_3 and π_3 are shown in Table 1 in which $g(x_{1i},x_{2j})$ means x_{3ij}, and i=1,...5, j=1,...,7.

Table 1 Joint evaluation mass function

π_2	f_2	μ_2	x_2 (kgf) \ x_1 (kgf)	30	40	50	60	70	
0	0	0	70	0 0 0 g(30, 70)	0 0 0 g(40, 70)	0 0 0 g(50, 70)	0 0 0 g(60, 70)	0 0 0 g(70, 70)	μ_3 f_3 π_3
0.1	0	0.1	60	0 0 0 g(30, 60)	0.03 0.0034 0.0266 g(40, 60)	0.08 0.0068 0.0732 g(50, 60)	0.04 0.0034 0.0366 g(60, 60)	0 0 0 g(70, 60)	
0.2	0.1	0.3	50	0 0 0 g(30, 50)	0.09 0.0236 0.0664 g(40, 50)	0.24 0.0540 0.1860 g(50, 50)	0.12 0.0270 0.0930 g(60, 50)	0 0 0 g(70, 50)	
0.3	0.2	0.5	40	0 0 0 g(30, 40)	0.15 0.0438 0.1062 g(40, 40)	0.40 0.1012 0.2988 g(50, 40)	0.20 0.0506 0.1494 g(60, 40)	0 0 0 g(70, 40)	
0.4	0.2	0.6	30	0 0 0 g(30, 30)	0.18 0.0472 0.1328 g(40, 30)	0.48 0.1080 0.3720 g(50, 30)	0.24 0.0540 0.1860 g(60, 30)	0 0 0 g(70, 30)	
0.2	0.1	0.3	20	0 0 0 g(30, 20)	0.09 0.0236 0.0664 g(40, 20)	0.24 0.0540 0.1860 g(50, 20)	0.12 0.0270 0.0930 g(60, 20)	0 0 0 g(70, 20)	
0	0	0	10	0 0 0 g(30, 10)	0 0 0 g(40, 10)	0 0 0 g(50, 10)	0 0 0 g(60, 10)	0 0 0 g(70, 10)	
			μ_1	0	0.3	0.8	0.4	0	
			f_1	0	0.1	0.2	0.1	0	
			π_1	0	0.2	0.6	0.3	0	

3.3 Evaluation Measures

In the above case, the evaluation measure given by Eq.5 is expressed by $M_k(x_{3ij} \in X_{0k})$ in which X_{0k} is subset of x_3 and $k=1,2,... \leqq n \times q$.

As for the serial structural system shown in Fig.5, the whole strength x_s is given by

$$x_{sk}=x_{3ij}=g_s(x_{1i},x_{2j})=\min(x_{1i},x_{2j}), \quad (16)$$

and considering

$$x_{sk}=X_{0k},\ f_{sk}=f_{3ij}, \qquad \pi_{sk}=\pi_{3ij},\ \mu_{sk}=M_k(x_{3ij} \in X_{0k}), \quad (17)$$

f_s, π_s and μ_s can be plotted with respect to x_s as shown in Table 2 and Fig.10, where f_s, π_s, μ_s present the probability, possibility and evaluation distributions of x_s, respectively, which shows x_s is about 30kgf.

As for the parallel structural system in Fig.6, if its elelemts have sufficient ductility, the whole strength x_p is given by

$$x_{pk}=x_{3ij}=g_p(x_{1i},x_{2j})=x_{1i}+x_{2j}. \quad (18)$$

Theprobability, possibility and evaluation distributions, f_p, π_p and μ_p, of x_p can be shown with respect to x_p as shown in Table 3 and Fig.11 in the same manner as x_s, which means x_s is about 80kgf.

Regarding the structure shown in Fig.7, suppose that strength x_1 and the load x_2 are evaluated by means of Figs.8 and 9, respectively, and the survival and fracture evaluations μ_s and μ_f can be calculated as shown in Table 4, where

$$M_s=M(x_{3ij} \in X_{0s}), \quad (19)$$

in which

$$X_{0s}=X_{0s}\{x_{3ij} \mid x_{3ij}=g_s(x_{1i},x_{2j})=s \mid x_{1i} \geqq x_{2j}\}, \quad (20)$$

and

$$M_f=M(x_{3ij} \in X_{0f}), \quad (21)$$

in which

$$X_{0f}=X_{0f}\{x_{3ij} \mid x_{3ij}=g_f(x_{1i},x_{2j})=f \mid x_{1i} \leqq x_{2j}\}. \quad (22)$$

However, as for X_0 where $x_{1i}=x_{2j}$, M defined as Eq.5 has to be replaced by

$$M(x_i \in X0)=\max(\pi_i)+\Sigma f_i/2. \quad (23)$$

Table 2 Calculated evaluation mass function μ_s of x_s as measures (strength of serial structural system)

x_s kgf	$\Sigma f_3=f_s$	$\max(\pi_3)=\pi_s$	$M=\mu_s$
10	0	0	0
20	0.1046	0.1860	0.2906
30	0.2092	0.3720	0.5812
40	0.2226	0.2988	0.5214
50	0.0878	0.1860	0.2738
60	0.0034	0.0366	0.0400
70	0	0	0

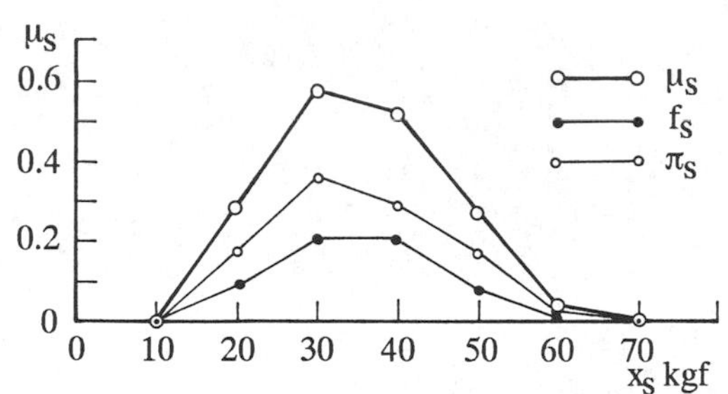

Fig.10 Calculated evaluation mass function μ_s of x_s as measures(strength of serial structural system)

Table 3 Calculated evaluation mass function μ_p of x_p as measures (strength of parallel structural system)

x_p kgf	$\Sigma f_3=fp$	$\max(\pi_3)=\pi_p$	$M=\mu_p$
50	0	0	0
60	0.0236	0.0664	0.0900
70	0.1012	0.1860	0.2872
80	0.1788	0.3720	0.5508
90	0.1788	0.2988	0.4776
100	0.1080	0.1860	0.2968
110	0.0338	0.0930	0.1268
120	0.0034	0.0366	0.0400
130	0	0	0

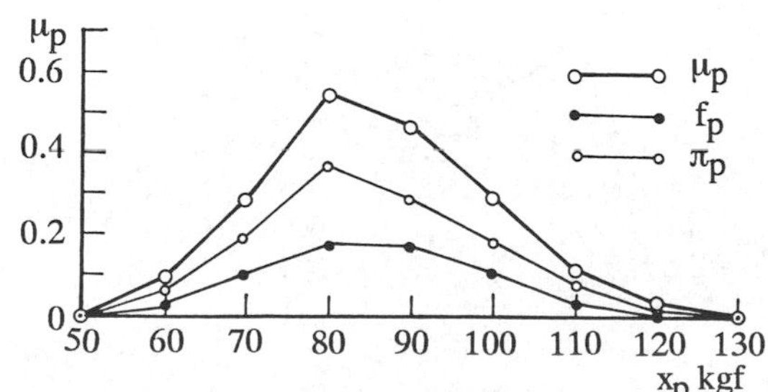

Fig.11 Calculated evaluation mass function μ_p of x_p as measures (strength of parallel structural system)

Table 4 Survival and Fracture evaluation μ_s, μ_f as measures

S or F	Σf_3	$\max(\pi_3)$	μ_s or μ_f
S	0.5432	0.3720	0.9152
F	0.0844	0.1860	0.2704

4. CONCLUSIONS

By introducing the compositional assumption that evaluation distribution is given by the sum of possibility and probability distributions, a very flexible evaluation theory can be presented. This theory called a combined probability-possibility evaluation theory has the following features.

(1) The numerical foundation is based on normalized frequencies in which multiple events are countable under the same conditions.

(2) The sum-compositional assumption holds not only in evaluation distributions but also in their measures.

(3) Therefore, by using joint evaluation distributions, one can analyze the reliabilities not only of a unit structure but also of serial and parallel structural systems.

However, there remain the following some problems which should be solved in future (Kawamura 1992):

(1) Development into infinite continuous sets of variables.

(2) Extension into joint evaluation distributions of three or more marginal evaluation distributions.

REFERENCES

Ang, A.H-S. and Tang, W.H. 1984. Probability Concepts in Engineering Planning and Design. Vol.II-Decision, Risk, and Reliability. John Wiley & Sons.

Kawamura, H. 1992. "Definition and Application of Fuzzy Sets and Fuzzy Measures for Structural Reliability Evaluation". Proc. of 8th Fuzzy System Symposium. Hiroshima. Japan. pp.17-20. May. (in Japanese)

Shafer, G. 1976. Mathematical Theory of Evidence. Princeton University Press.

Sugeno, M. 1972. "Fuzzy Measure and Fuzzy Integral". Trans. of the Society of Instrument and Control Engineers. Vol.8. No.2. pp.94-102. (in Japanese)

Yao, J.T.P and Natke, H.S. 1991. "Uncertainties in Structural Identification and Control". Proc. of International Fuzzy Engineering Symposium on Fuzzy Engineering toward Human Friendly Systems. Vol.2 . Yokohama. Japan. Nov. pp.844-849.

Zadeh, L.A. 1968. "Probability Measures of Fuzzy Events", J.Math.Analysis and Appl., 10, pp.421-427.

Zadeh, L.A. 1978. "Fuzzy Sets as a Basis for a Theory of Possibility". Fuzzy Sets and Systems. 1. pp.3-28.

Structural Safety & Reliability, Schuëller, Shinozuka & Yao (eds) © 1994 Balkema, Rotterdam, ISBN 90 5410 357 4

System reliability analysis using response surface and Monte Carlo approaches

Joo-Sung Lee
University of Ulsan, Republic of Korea
Mikhail B. Krakovski
Research Institute for Concrete and Reinforced Concrete, Moscow, Russia

ABSTRACT: A method to calculate reliability of complex structural systems including offshore structures is presented. Algorithms to build a response surface and to compute reliability on the basis of Monte Carlo simulation are described. The proposed approach is illustrated by a numerical example.

1 STATEMENT OF THE PROBLEM

Consider a complex structural system, e.g., an offshore structure. Basic load and resistance variables are assumed to be random values with known probability density functions. If a basic variable is time varying and described as a stochastic process X(t), then the stochastic process is replaced by a random variable for which the probability distribution function $F_X(x)$ is defined as:

$$F_X(x) = P(X(t') \leq x) \quad (1)$$

where t' is any randomly selected time (Thoft-Christensen & Baker 1982).

Load basic variables $L_1, L_2, \ldots, L_t$ can act in s combinations $Q_1(L_1, L_2, \ldots, L_t)$, $Q_2(L_1, L_2, \ldots, L_t)$, ..., $Q_s(L_1, L_2, \ldots, L_t)$. The probabilities $P_1(Q_1)$, ..., $P_s(Q_s)$ of occurrence of each combination are known. It is clear that

$$P_1(Q_1) + P_2(Q_2) + \ldots + P_s(Q_s) = 1 \quad (2)$$

For each load combination Q_i (i = 1, ..., s) a set of realizations of load basic variables $l = (l_1, l_2, \ldots, l_t)$ represents a load pattern.

The resistance basic variables are denoted by R_1, R_2, ..., R_n and an arbitrary set of realizations of n resistance basic variables is denoted by $r = (r_1, r_2, \ldots, r_n)$.

It is assumed that for any l and r the load bearing capacity of the structure is determined by the load factor u, i.e., one of the failure modes occurs when the load pattern is $u\,l = (u\,l_1, u\,l_2, \ldots, u\,l_t)$ and a set of resistance variables r is fixed. Load factor u can be determined on the basis of any deterministic design method: a load pattern l is being increased proportionally until a failure mode occurs.

The probability of failure P_f is denoted as:

$$P_f = P(u < 1) \quad (3)$$

and the reliability is $1 - P_f$. The task of the system reliability analysis is to determine P_f.

2 RESPONSE SURFACE

To calculate P_f first a response surface in the form of a polynomial for the load factor u should be built. For this purpose a number of deterministic designs are carried out at different points of (n+t)-dimensional space of basic variables and load factor u is determined in each design. The response surface is built in (n+t+1)-dimensional space and passes through the values of u calculated at the above points. These points can be fixed in various ways.

Bucher and Bourgund (1990) have suggested to use a response surface in the form of the following polynomial:

$$u(l_1, \ldots, l_t, r_1, \ldots, r_n) = k_0 + \sum_{j=1}^{2}\sum_{i=1}^{t} a_{ij}\, l_i^j + \sum_{j=1}^{2}\sum_{i=1}^{n} b_{ij}\, r_i^j \quad (4)$$

To obtain the values of unknown factors k_0, a_{ij}, b_{ij} deterministic designs are carried out at the points $(\mu_{l_1}, \ldots, \mu_{l_t}, \mu_{r_1}, \ldots, \mu_{r_n})$, $(\mu_{l_1} \pm \delta_{l_1}\sigma_{l_1}, \mu_{l_2}, \ldots, \mu_{l_t}, \mu_{r_1}, \ldots, \mu_{r_n})$, $(\mu_{l_1}, \mu_{l_2} \pm \delta_{l_2}\sigma_{l_2}, \ldots, \mu_{l_t}, \mu_{r_1}, \ldots, \mu_{r_n})$, ..., $(\mu_{l_1}, \mu_{l_2}, \ldots, \mu_{l_t}, \mu_{r_1}, \ldots,$

$\mu_{r_n} \pm \delta_{r_n}\sigma_{r_n}$) where, μ_{l_i}, μ_{r_i}, σ_{l_i}, σ_{r_i} are respectively mean values (μ) and standard deviations (σ) for i-th load (l_i) or resistance(r_i) basic variable and δ_{l_i}, δ_{r_i} are the number of standard deviations from the mean value μ to the design point $\mu \pm \delta\sigma$. Factors k_0, a_{ij}, b_{ij} are determined as the solution to a system of (t+n+1) linear equations. The right hand sides of the equations are the values of u calculated at the above points.

The investigation has shown that sometimes the accuracy of the approximation by polynomial (4) should be improved. For this purpose other polynomials can be used. First of all the largest power of polynomial (4) can be increased:

$$u(l_1, \ldots, l_t, r_1, \ldots, r_n) = k_0 + \sum_{j=1}^{2h}\sum_{i=1}^{t} a_{ij} l_i^j + \sum_{j=1}^{2h}\sum_{i=1}^{n} b_{ij} r_i^j \tag{5}$$

where h is an integer positive number. In this case the number of deterministic designs (the total number of polynomial terms) is $N = 2h(t+n)+1$.

Sometimes the following polynomial appears to be useful ($N=2^{t+n}$):

$$u(l_1, \ldots, l_t, r_1, \ldots, r_n) = \sum_{i_{l_1}=0}^{1}\cdots\sum_{i_{l_t}=0}^{1}\sum_{i_{r_1}=0}^{1}\cdots\sum_{i_{r_n}=0}^{1} k_{i_{l1}\ldots i_{lt}, i_{r1}\ldots i_{rn}} l_1^{i_{l1}}\ldots l_t^{i_{lt}} r_1^{i_{r1}}\ldots r_n^{i_{rt}} \tag{6}$$

If the accuracy of approximation by polynomial (6) is to be increased then the terms of type (5) can be added to polynomial (6):

$$u(l_1, \ldots, l_t, r_1, \ldots, r_n) = \sum_{i_{l_1}=0}^{1}\cdots\sum_{i_{l_t}=0}^{1}\sum_{i_{r_1}=0}^{1}\cdots\sum_{i_{r_n}=0}^{1} k_{i_{l1}\ldots i_{lt}, i_{r1}\ldots i_{rn}} l_1^{i_{l1}}\ldots l_t^{i_{lt}} r_1^{i_{r1}}\ldots r_n^{i_{rt}} + \sum_{j=2}^{2h+1}\sum_{i=1}^{t} a_{ij} l_i^j + \sum_{j=2}^{2h+1}\sum_{i=1}^{r} b_{ij} r_i^j \quad (h,t,r \geq 1) \tag{7}$$

In this case $N=2^{t+n} + 2h(t+n)$.

To increase the accuracy of approximation the (t+n)-dimensional space of basic variables can be divided into several parts and different polynomials (4) - (7) can be used for different parts.

3 MONTE CARLO SIMULATION

After the response surface has been built the Monte Carlo simulation for each load combination is carried out in the following order.

1. Knowing the probability density functions for all basic variables and using a random number generator obtain m random sets (l_i, r_i) (i=1,.., m) of basic load and resistance variables.
2. Using equation(s) of the response surface determine the values of u_i (i= 1, ..., m) for each random set of basic variables.
3. Assume the values of u_i thus obtained to be m realizations of the random variable U, i.e., the sample size is m. Calculate the first four statistical moments of the sample.
4. Fit appropriate probability density function $f_i(u)$ from the family of Pearson's curves (the first four statistical moments of the distribution coincide with those of the sample).
5. Using numerical integration calculate the probability of failure for i-th load combination:

$$P_{fi} = \int_{-\infty}^{1} f_i(u)\, du \tag{8}$$

Taking into account all load combinations and using the total probability theorem calculate the probability of failure:

$$P_f = \sum_{i=1}^{S} P_i(Q_i)\, P_{fi} \tag{9}$$

The algorithm was implemented in the form of a computer program. Accuracy of the program results was proved by many test examples (Krakovski 1982). A possibility of using statistics of extreme values was demonstrated (Krakovski & Stolipina 1983).

4 NUMERICAL EXAMPLE

A numerical example presented below checks the above approach and illustrates its application. Consider a portal frame with a vertical load applied as shown in Fig.1a (Thoft-Christensen & Murotsu 1986). Four failure modes can occur under the load (Fig.1b). Their safety margins are (Thoft-Christensen & Murotsu 1986):

$$M_1 = R_2 + 2R_5 + R_6 - hL$$
$$M_2 = R_2 + 2R_5 + R_7 - hL$$
$$M_3 = R_3 + 2R_5 + R_6 - hL$$
$$M_4 = R_3 + 2R_5 + R_7 - hL \tag{10}$$

where R_i is the limit moment at the i-th plastic hinge (i = 2, 3, 5, 6, 7)

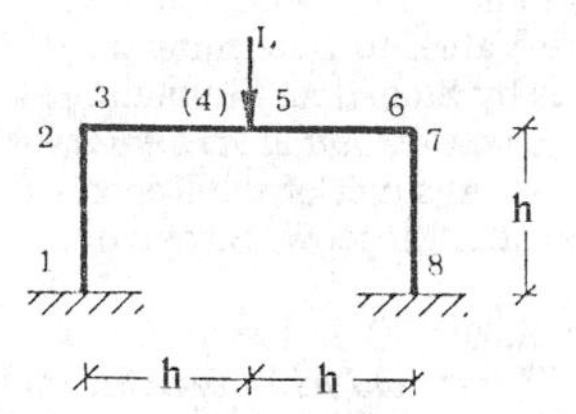

(a) portal frame structure

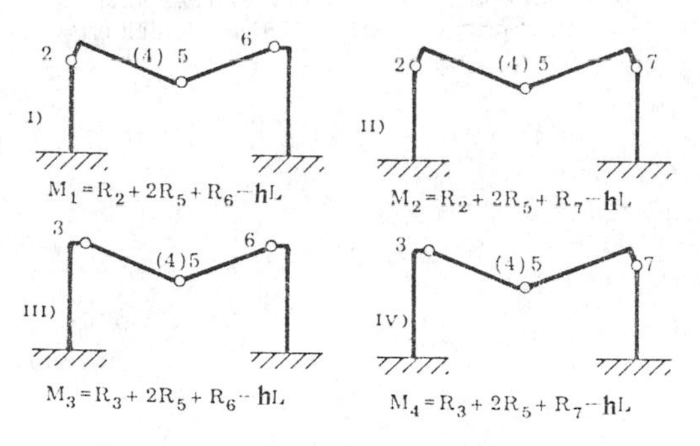

(b) failure modes

Fig.1 Portal frame model and its failure modes

If the probability of the occurrence of each failure mode P_{fi} is known, then the following inequalities must be satisfied for the exact probability of failure of the portal frame P_f (Thoft-Christensen & Murotsu 1986):

$$\max P_{f_i} \leq P_f \leq 1 - \prod_{i=1}^{4} (1 - P_{f_i}) \qquad (11)$$

The procedure for a check of the suggested approach is as follows. First the lower and upper bounds for P_f according to Eq.(11) are calculated. In so doing Eqs.(10) are used. Then the probability of failure P_f is calculated with the use of the above approach and compared with the lower and upper bounds in Eq.(11).

Below are given the numerical data used in calculations. All limit moments as well as the load were assumed to be normally distributed statistically independent random variables with the following mean values and standard deviations:

$$\mu_1 = 50\text{KN}$$
$$\sigma_1 = 2.5 \text{ KN}$$
$$\mu_{r2} = \mu_{r7} = 50 \text{ KN-m}$$
$$\sigma_{r2} = \sigma_{r7} = 2.5 \text{ KN-m};$$
$$\mu_{r3} = \mu_{r5} = \mu_{r6} = 55 \text{ KN-m}$$
$$\sigma_{r3} = \sigma_{r5} = \sigma_{r6} = 2.75 \text{ KN-m};$$

The height of the portal frame is h = 3.6 m.

The probabilities of the occurrence of the failure modes are:

$$P_{f1} = P_{f4} = 0.9676 \times 10^{-3}$$
$$P_{f2} = 0.3573 \times 10^{-2}$$
$$P_{f3} = 0.1854 \times 10^{-3}$$

The lower and upper bounds for P_f calculated in accordance with Eq.(11) are:

$$\max P_{f_i} = 0.3573 \times 10^{-2} \qquad (12)$$

$$1 - \prod_{i=1}^{4} (1 - P_{f_i}) = 0.5685 \times 10^{-2} \qquad (13)$$

To apply the suggested approach first a polynomial (4) was built. The deterministic designs were carried out at the points (μ_1, μ_{r2}, μ_{r3}, μ_{r5},μ_{r6} ,μ_{r7}); ($\mu_1 \pm 3\sigma_1$, μ_{r2}, μ_{r3}, μ_{r5},μ_{r6} ,μ_{r7}) ; (μ_1, $\mu_{r2} \pm 3\sigma_{r2}$, μ_{r3}, μ_{r5},μ_{r6} ,μ_{r7}) ;...; (μ_1, μ_{r2}, μ_{r3}, μ_{r5},μ_{r6}, $\mu_{r7} \pm 3\sigma_{r7}$). At each point all four failure modes were examined and load factors u_i (i=1, ..., 4) were determined so as to obtain M_i =0. The result of this deterministic design is the minimal u_i . For example, at the point (μ_1, μ_{r2}, μ_{r3}, μ_{r5},μ_{r6} ,μ_{r7}) from Eq.(10) one can get:

$$M_1 = 50 + 2 \times 55 + 55 - u_1 \times 3.6 \times 50 = 0$$
$$u_1 = 1.194$$
$$M_2 = 50 + 2 \times 55 + 50 - u_2 \times 3.6 \times 50 = 0$$
$$u_2 = 1.167$$
$$M_3 = 55 + 2 \times 55 + 55 - u_3 \times 3.6 \times 50 = 0$$
$$u_3 = 1.222$$
$$M_4 = 55 + 2 \times 50 + 55 - u_4 \times 3.6 \times 50 = 0$$
$$u_4 = 1.194$$
$$\min u_i = 1.167 \qquad (14)$$

The solution to the system of linear equations gives the factors a_{ij}, b_{ij} in Eq.(4). In this particular case 5 separate systems with two equations in each system were solved. As a result the load factor is approximated by the following polynomial:

$$
\begin{aligned}
u = {} & 1.167 - 0.0597\, l + 0.0115\, r_2 + 0.003\, r_3 \\
& + 0.0305\, r_5 + 0.03\, r_6 + 0.0115\, r_7 \\
& + 2.88 \times 10^{-3}\, l^2 - 8.33 \times 10^{-4}\, r_2^2 \\
& - 0.001\, r_3^2 - 0.001\, r_6^2 - 8.33 \times 10^{-4}\, r_7^2
\end{aligned} \quad (15)
$$

where

$$
l = \frac{l - \mu_l}{\sigma_l}\ ; \quad r_i = \frac{r_i - \mu_{ri}}{\sigma_{ri}}
$$

$$
(i = 2, 3, 5, 6, 7) \quad (16)
$$

Monte Carlo simulation with subsequent approximation of the results by Pearson's curves has shown that two curves can be used for approximation, namely, Pearson's curve of type 2 and normal curve. Numerical integration performed in accordance with Eq.(8) showed that

$$P_f = 0.4044 \times 10^{-2}$$

for the Pearson's curve of type 2 and

$$P_f = 0.4117 \times 10^{-2}$$

for the normal curve. Both these values fall within the interval

$$(0.3573 \times 10^{-2},\ 0.5685 \times 10^{-2})$$

between the lower and the upper bounds, Eqs.(12) and (13).

5 CONCLUSION

In authors' opinion the suggested approach is rather simple and efficient. It permits from one and the same point of view to estimate reliability of different complex structural systems taking into account the requirements of practical design.

REFERENCE

Bucher, C.G. & Bourgund, U. 1990. A Fast and Efficient Response Surface Approach for Structural Reliability Problems. *Structural Safety* 7(1)

Krakovski, M.B. 1982. Determination of the Reliability of Structures by Statistical Simulation Methods. *Stroitelnaya mekhanika e raschet sooruzheniy* 2 (in Russian)

Krakovski, M.B. & Stolipina, L.I. 1983. The Use of Statistics of Extreme Values to Determine the Reliability of Structures by Statistical Simulation Methods. *Problems of Optimization & Reliability in Structural Mechanics*, Abstract of the Reports to the National Conference, Moscow, Stroyizdat, (in Russian)

Thoft-Christensen, P. & Baker, M.J. 1982. *Structural Reliability Theory and Its Applications*, Springer-Verlag, Berlin, Heidelberg

Thoft-Christensen, P. & Murotsu, Y. 1986. *Application of Structural Systems Reliability Theory*. Springer-Verlag, Berlin, Heidelberg, New-York, Tokyo

Structural Safety & Reliability, Schuëller, Shinozuka & Yao (eds) © 1994 Balkema, Rotterdam, ISBN 90 5410 357 4

Reliability of complex structural systems using an efficient directional simulation

Shaowen Shao & Yoshisada Murotsu
University of Osaka Prefecture, Sakai, Japan

ABSTRACT: The investigation for the limit state surfaces of complex structural systems is needed in the evaluation of structural reliabilities. For this purpose, a directional simulation technique is applied in this paper. The direction sampling density is modified based on the information from simulation to efficiently estimate the failure probability of a structural system and in the meanwhile, to provide an outline of the limit state surface. A sampling procedure is proposed to transform the uniform density on the unit sphere to the modified density. Numerical examples are provided to demonstrate the validity of the proposed method.

1 Introduction

For structural systems with multiple limit state functions or multi-modal limit state function, knowledge of the actual limit state surface, i.e., the boundary between failure and safety regions, is needed in applying various methods to perform reliability evaluation. For instance, searching for most likely points (one or multiple) on limit state surface[6, 7] could help to apply FORM/SORM[5], to define a proper importance sampling density[1, 2, 4], or to make multiple checking[7]. This searching work, however, involves repeated structural analyses and is time-consuming. The present paper combines the search with a directional simulation to reduce the total computational effort in reliability evaluation.

A general directional simulation executed in standard normal space uses a uniform density on the unit sphere around the origin to generate random direction vectors. And for each direction represented by such a vector, the distance from the origin to the limit state surface is searched to estimate a conditional failure probability. Here arises an idea that, if the information got in the simulation for the limit state surface is kept, one can gradually find the outline of the limit state surface. But it is also computationally expensive to use a uniform density in generating direction vectors. In this paper, a procedure is presented to modify the direction-generating density based on the information obtained from the simulation. Consider the case where starting from a uniform density distribution on the unit sphere, the limit state surface in some sampled directions is found to be much more close to the origin than other directions. Then, the density distribution is adjusted on the unit sphere. Some densities on the places where the limit state surface is far from the origin are moved to the region where the limit state surface is near the origin. On the other hand, other parts of the unit sphere where there has been no sample generated are kept the same as before. This is repeated until a good direction-generating density is obtained for efficiently estimating the failure probability by a directional simulation. In the meantime, the knowledge of the limit state surface from the simulation is also useful in other reliability evaluation methods.

2 Modification of Direction-Generating Density

Consider a structural reliability problem including n basic random variables which can be transformed and expressed as an n-dimensional standard normal vector $\boldsymbol{U} = (U_1, U_2, \cdots, U_n)^T$. Let

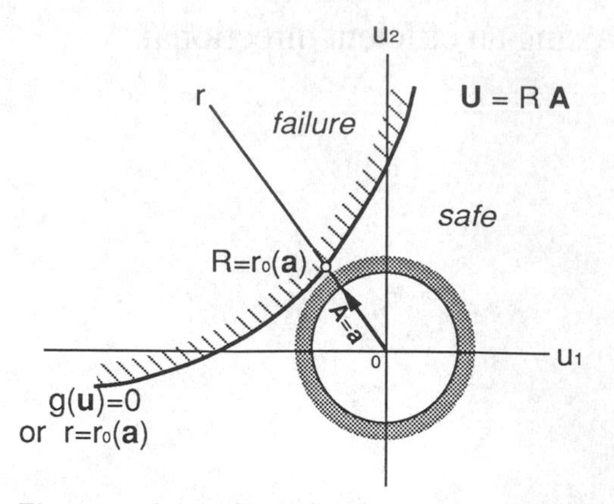

Figure 1: Limit State Surface in Euclidean and Polar Coordinates

$g(\boldsymbol{U})$ denote the safety margin, i.e., $g(\boldsymbol{U}) > 0$ corresponds to a safe state, and $g(\boldsymbol{U}) \leq 0$ represents a failure state. $g(\boldsymbol{u}) = 0$ defines a limit state surface in $\boldsymbol{u}$-space as shown in Fig. 1. Then, the failure probability P_f is given by a multi-dimensional integral

$$P_f = \int_{g(\boldsymbol{u})\leq 0} f_{\boldsymbol{U}}(\boldsymbol{u})d\boldsymbol{u} \tag{1}$$

where $f_{\boldsymbol{U}}(\boldsymbol{u})$ is the joint probability density function of $\boldsymbol{U}$. Eq. (1) is formulated in Euclidean coordinates. It can also be formulated in polar coordinates. Let $\boldsymbol{U} = R\boldsymbol{A}$ where R and $\boldsymbol{A}$ indicate the length and the direction of the vector $\boldsymbol{U}$, respectively, which are also shown in Fig. 1. $\boldsymbol{A}$ is a random unit vector, which is uniformly distributed on the n-dimensional unit sphere Ω_n centered around the origin. R is a random variable independent of $\boldsymbol{A}$ and its square, R^2, is chi-square distributed with n-degrees of freedom. Then, Eq. (1) can be rewritten as[1, 3]

$$\begin{aligned} P_f &= \int_{\boldsymbol{a}\in\Omega_n} \int_{g(r\boldsymbol{a})\leq 0} f_R(r)dr f_{\boldsymbol{A}}(\boldsymbol{a})d\boldsymbol{a} \\ &= \int_{\boldsymbol{a}\in\Omega_n} \int_{r_0(\boldsymbol{a})}^{\infty} f_R(r)dr f_{\boldsymbol{A}}(\boldsymbol{a})d\boldsymbol{a} \\ &= \int_{\boldsymbol{a}\in\Omega_n} \int_{r_0^2(\boldsymbol{a})}^{\infty} f_{R^2}(r^2)dr^2 f_{\boldsymbol{A}}(\boldsymbol{a})d\boldsymbol{a} \\ &= \int_{\boldsymbol{a}\in\Omega_n} \left[1 - \chi_n^2\left(r_0^2(\boldsymbol{a})\right)\right] f_{\boldsymbol{A}}(\boldsymbol{a})d\boldsymbol{a} \end{aligned} \tag{2}$$

where $r_0(\boldsymbol{a})$ denotes the distance from the origin to the limit state surface for $\boldsymbol{A} = \boldsymbol{a}$, and $r = r_0(\boldsymbol{a})$ represents the limit state surface which is identical to $g(r\boldsymbol{a}) = 0$ in polar coordinates. $\chi_n^2()$ denotes a chi-square distribution function. Eq. (2) implies that the conditional failure probability for $\boldsymbol{A} = \boldsymbol{a}$ can be estimated by $1 - \chi_n^2(r_0^2(\boldsymbol{a}))$, which completely depends on $r_0(\boldsymbol{a})$.

By using Monte Carlo simulation technique to solve the integral in Eq. (2), an unbiased estimator of P_f is obtained from N direction samples $\boldsymbol{a}_i$ ($i = 1, 2, \cdots, N$):

$$\hat{P}_f = \frac{1}{N}\sum_{i=1}^{N} p_i = \frac{1}{N}\sum_{i=1}^{N}\left[1 - \chi_n^2\left(r_0^2(\boldsymbol{a}_i)\right)\right] \tag{3}$$

where $p_i = 1 - \chi_n^2(r_0^2(\boldsymbol{a}_i))$ is a sample value for $\boldsymbol{A} = \boldsymbol{a}_i$. An estimate of variance of $\hat{P}_f$ can also be obtained from samples $\boldsymbol{a}_i$ ($i = 1, 2, \cdots, N$):

$$V\hat{a}r(\hat{P}_f) = \frac{1}{N(N-1)}\sum_{i=1}^{N}\left\{p_i - \hat{P}_f\right\}^2 \tag{4}$$

The sampling density to generate $\boldsymbol{a}_i$ used in Eqs. (3) and (4) is $f_{\boldsymbol{A}}(\boldsymbol{a})$, i.e., the uniform density on the unit sphere Ω_n. If there is no idea about the shape of the limit state surface $r = r_0(\boldsymbol{a})$, this density is good. However, as simulation is repeated, the distances from the origin to the limit state surface in various directions, $r_i = r_0(\boldsymbol{a}_i)$ ($i = 1, 2, \cdots$), are evaluated. This information can be fed back, and it can be utilized in constructing a more efficient sampling density. By using a sampling density $h_{\boldsymbol{B}}(\boldsymbol{b})$ other than $f_{\boldsymbol{A}}(\boldsymbol{a})$, Eq. (2) is rewritten as

$$P_f = \int_{\boldsymbol{b}\in\Omega_n}\left[1 - \chi_n^2\left(r_0^2(\boldsymbol{b})\right)\right]\frac{f_{\boldsymbol{A}}(\boldsymbol{b})}{h_{\boldsymbol{B}}(\boldsymbol{b})}h_{\boldsymbol{B}}(\boldsymbol{b})d\boldsymbol{b} \tag{5}$$

For a direction sample $\boldsymbol{b}_i$ generated by $h_{\boldsymbol{B}}(\boldsymbol{b})$, the sample value p_i becomes

$$p_i = \left[1 - \chi_n^2\left(r_0^2(\boldsymbol{b}_i)\right)\right]\frac{f_{\boldsymbol{A}}(\boldsymbol{b}_i)}{h_{\boldsymbol{B}}(\boldsymbol{b}_i)} \tag{6}$$

An ideal sampling density $h_{\boldsymbol{B}}(\boldsymbol{b})$ yielding $p_i = P_f$ can be derived from Eq. (6):

$$h_{\boldsymbol{B}}(\boldsymbol{b}) = \left[1 - \chi_n^2\left(r_0^2(\boldsymbol{b})\right)\right]\frac{f_{\boldsymbol{A}}(\boldsymbol{b})}{P_f} \tag{7}$$

As P_f is unknown, this ideal sampling density, of course, can not be defined a priori. However, Eq. (7) shows that $h_{\boldsymbol{B}}(\boldsymbol{b})$ should be directly proportional to $1 - \chi_n^2(r_0^2(\boldsymbol{b}))$ since $f_{\boldsymbol{A}}(\,)$ and P_f are constants.

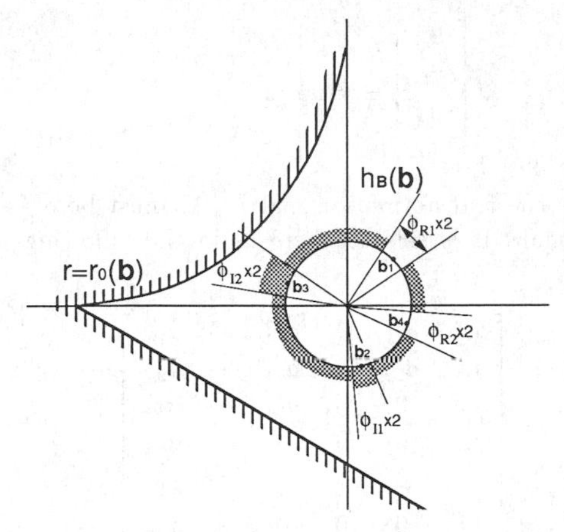

Figure 2: Moving Densities on Unit Sphere Ω_n

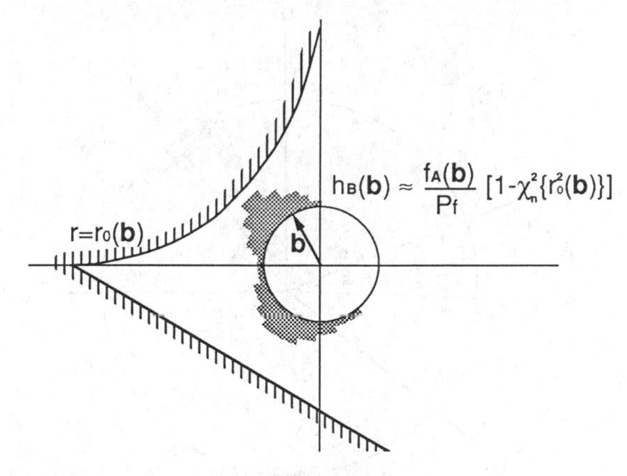

Figure 3: Modified Sampling Density

In this study, the sampling density $h_{\boldsymbol{B}}(\boldsymbol{b})$ is selected to be equal to $f_{\boldsymbol{A}}(\boldsymbol{b})$ at the beginning of a simulation, without knowledge of the limit state surface $r = r_0(\boldsymbol{b})$. For each direction sample $\boldsymbol{b}_i$ generated from $h_{\boldsymbol{B}}(\boldsymbol{b})$, the distance from the origin to the limit state surface, $r_i = r_0(\boldsymbol{b}_i)$, is searched by some relevant methods. Then, $h_{\boldsymbol{B}}(\boldsymbol{b})$ is modified based on $r_i = r_0(\boldsymbol{b}_i)$ $(i = 1, 2, \cdots)$. The basic idea is schematically illustrated in Fig. 2. If $r_i = r_0(\boldsymbol{b}_i)$ is found infinitely far from the origin in a sampled direction, e.g., $\boldsymbol{b}_1$ or $\boldsymbol{b}_4$, this means that $1 - \chi^2_n(r^2_0(\boldsymbol{b}_i))$ is 0 and therefore the sampling density $h_{\boldsymbol{B}}(\boldsymbol{b})$ should be set to 0 in this point. On the contrary, $r_i = r_0(\boldsymbol{b}_i)$ may be relatively close to the origin in some other directions, e.g., $\boldsymbol{b}_2$ and $\boldsymbol{b}_3$. Then, the sampling density in these places should be increased in proportion to $1 - \chi^2_n(r^2_0(\boldsymbol{b}_i))$. Further, the limit state surface near a sample direction $\boldsymbol{b}_i$ is assumed to be continuous, and so the density in $\boldsymbol{b}_i$ can represent that of an area near $\boldsymbol{b}_i$. Here, an angle around $\boldsymbol{b}_i$ is used in which the density is set to the same as in $\boldsymbol{b}_i$. ϕ_{Ri} and ϕ_{Ii} denote the two kinds of angles used respectively for the samples where the density should be reduced from the uniform density $f_{\boldsymbol{A}}(\boldsymbol{b})$ and where the density should be increased. Fig. 2 shows a 2-dimensional case where the densities on some arcs of the unit circle are moved to other arcs, while in a higher dimensional case as shown in Fig. 4, an angle around a direction sample, e.g., $\boldsymbol{x}^0$, corresponds to a circular cap on the unit sphere Ω_n.

The sampling density $h_{\boldsymbol{B}}(\boldsymbol{b})$ is changed step by step until it is close to the ideal density which includes the information of the limit state surface. It must be noted that, in the beginning, only part of the densities on Ω_n is adjusted where some samples have been generated and the distances from the origin to the limit state surface in those directions have been searched. The densities of other places are kept as same as before until some samples are obtained. This can avoid missing other check points on the limit state surface which also have relatively short distances to the origin. Meanwhile, the densities increased in the parts on Ω_n could help searching real check points in those areas. The sampling density $h_{\boldsymbol{B}}(\boldsymbol{b})$ is modified based on the increasing information got from the simulation, and is gradually close to the ideal sampling density function given in Eq. (7), as shown in Fig. 3.

3 Sampling Procedure

An outcome $\boldsymbol{a}_i$ of the random unit vector $\boldsymbol{A}$ with the uniform density $f_{\boldsymbol{A}}(\boldsymbol{a})$ can be obtained by generating an outcome $\boldsymbol{u}_i$ of the standard normal vector $\boldsymbol{U}$ and then using $\boldsymbol{a}_i = \boldsymbol{u}_i/|\boldsymbol{u}_i|$. The sampling density $h_{\boldsymbol{B}}(\boldsymbol{b})$ proposed in section 2 is defined based on the moving of the densities on Ω_n from some places to others. In the following, a sampling procedure is discussed which consists of two steps:

- Generate a sample $\boldsymbol{a}_i$ according to the uniform density $f_{\boldsymbol{A}}(\boldsymbol{a})$ on Ω_n.

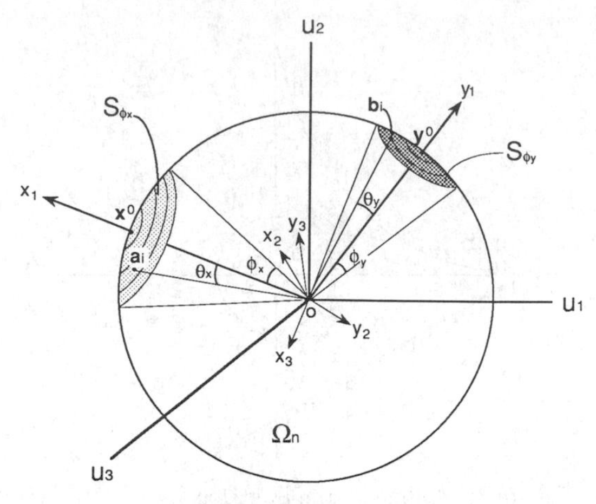

Figure 4: Transformation of a Sample from a Density-Reduced Area to a Density-Increased Area

- Check if the sample $\boldsymbol{a}_i$ is in an area where the density has been reduced from $f_{\boldsymbol{A}}$. If it is, the sample is transformed to another area where the density has been increased. Otherwise, the sample is directly used as $\boldsymbol{b}_i$. This step can be expressed as a transformation $\boldsymbol{b}_i = \boldsymbol{T}(\boldsymbol{a}_i)$.

Fig. 4 schematically shows the transformation from a sample $\boldsymbol{a}_i$ in area S_{ϕ_x} to a sample $\boldsymbol{b}_i$ in area S_{ϕ_y}. S_{ϕ_x} is assumed to be a density-reduced area with its center at $\boldsymbol{x}^0$ and S_{ϕ_y} is a density-increased area with center $\boldsymbol{y}^0$. $\boldsymbol{x}^0$, $\boldsymbol{y}^0$, $\boldsymbol{a}_i$, and $\boldsymbol{b}_i$ are on the n-dimensional unit sphere Ω_n, in which $\boldsymbol{b}_i$ must be determined from others.

Since the simulation is executed in $\boldsymbol{u}$-space, the $\boldsymbol{u}$-coordinates are used as the normal coordinates. For example, the coordinates of $\boldsymbol{x}^0$ in $\boldsymbol{u}$-space is expressed as $\boldsymbol{u}^{x^0} = (u_1^{x^0}, u_2^{x^0}, \cdots, u_n^{x^0})^T$. In Fig. 4, two supplementary coordinate systems $\boldsymbol{x}$-coordinates and $\boldsymbol{y}$-coordinates are introduced corresponding to the areas S_{ϕ_x} and S_{ϕ_y} for the purpose of the transformation of direction samples. $\boldsymbol{x}$-coordinates are selected such that the origin is the same as that of the $\boldsymbol{u}$-coordinates and x_1 axis is in the direction of $\boldsymbol{x}^0$. The other axes $x_2 \cdots x_n$ are selected to be orthogonal to x_1 and each other. Let the transformation between $\boldsymbol{u}$- and $\boldsymbol{x}$-coordinates be expressed as

$$\begin{bmatrix} u_1 \\ u_2 \\ \vdots \\ u_n \end{bmatrix} = \boldsymbol{A}_x \begin{bmatrix} x_1 \\ x_2 \\ \vdots \\ x_n \end{bmatrix} \tag{8}$$

where the transformation matrix $\boldsymbol{A}_x$ must be orthogonal. It is selected here as in the following form:

$$\boldsymbol{A}_x = \begin{bmatrix} u_1^{x^0} & a_{12} & a_{13} & a_{14} & \cdots & a_{1n} \\ u_2^{x^0} & a_{22} & a_{23} & a_{24} & \cdots & a_{2n} \\ u_3^{x^0} & 0 & a_{33} & a_{34} & \cdots & a_{3n} \\ u_4^{x^0} & 0 & 0 & a_{44} & \cdots & a_{4n} \\ \vdots & \vdots & \vdots & \vdots & \vdots & \vdots \\ u_n^{x^0} & 0 & 0 & 0 & \cdots & a_{nn} \end{bmatrix} \tag{9}$$

where the first column is the $\boldsymbol{u}$-coordinates of the point $\boldsymbol{x}^0$ from which all the elements in other columns are also uniquely determined.

$\boldsymbol{y}$-coordinates are determined in the same way. Let the transformation between $\boldsymbol{u}$ and $\boldsymbol{y}$ be expressed by

$$\boldsymbol{u} = \boldsymbol{A}_y \boldsymbol{y} \tag{10}$$

When a sample $\boldsymbol{a}_i$ falls into the area S_{ϕ_x}, i.e., $\theta_x \le \phi_x$, and it is needed to be moved to the area S_{ϕ_y}, the coordinates of sample $\boldsymbol{b}_i$ are determined in the following way. Given the $\boldsymbol{u}$-coordinates $\boldsymbol{u}^{a_i}$ of $\boldsymbol{a}_i$, $\boldsymbol{x}^{a_i}$, $\boldsymbol{y}^{b_i}$, and $\boldsymbol{u}^{b_i}$ which represent the $\boldsymbol{x}$-coordinates of $\boldsymbol{a}_i$, the $\boldsymbol{y}$- and $\boldsymbol{u}$-coordinates of $\boldsymbol{b}_i$ are respectively given as follows,

$$\boldsymbol{x}^{a_i} = \boldsymbol{A}_x^T \boldsymbol{u}^{a_i} \tag{11}$$

$$y_1^{b_i} = \cos\theta_y \tag{12}$$

$$\frac{y_j^{b_i}}{x_j^{a_i}} = \sqrt{\frac{1-\left(y_1^{b_i}\right)^2}{1-(x_1^{a_i})^2}} \qquad j = 2, \cdots, n \tag{13}$$

$$\boldsymbol{u}^{b_i} = \boldsymbol{A}_y \boldsymbol{y}^{b_i} \tag{14}$$

Eq. (13) makes $y_j^{b_i}$ proportional to $x_j^{a_i}$ ($j = 2, \cdots, n$), which means that, except for the first dimension, the vectors $\boldsymbol{y}^{b_i}$ and $\boldsymbol{x}^{a_i}$ are in the same direction in their own $n-1$–dimension spaces.

θ_y in Eq. (12) is calculated from the modified sampling density $h_{\boldsymbol{B}}(\boldsymbol{b})$. Consider the simple case as shown in Fig. 4 where only one density-reduced area and one density-increased area exist. It is assumed that the density in S_{ϕ_x} is 0, for simplicity. In this case, all the probability which had been in S_{ϕ_x} should be moved to S_{ϕ_y}.

The probability P_{ϕ_x} first uniformly distributed on S_{ϕ_x} is given by the ratio of the area S_{ϕ_x} to the

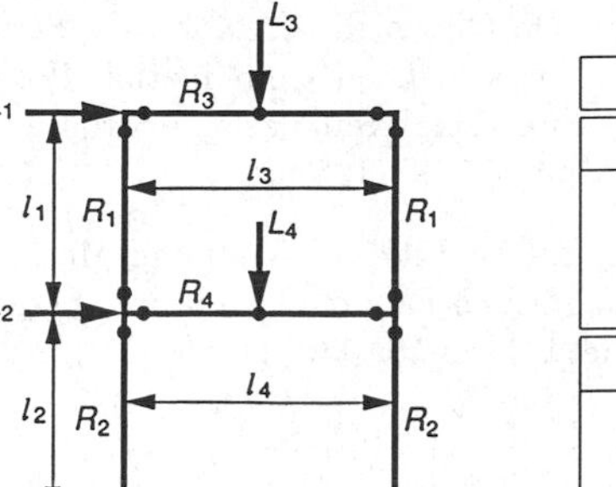
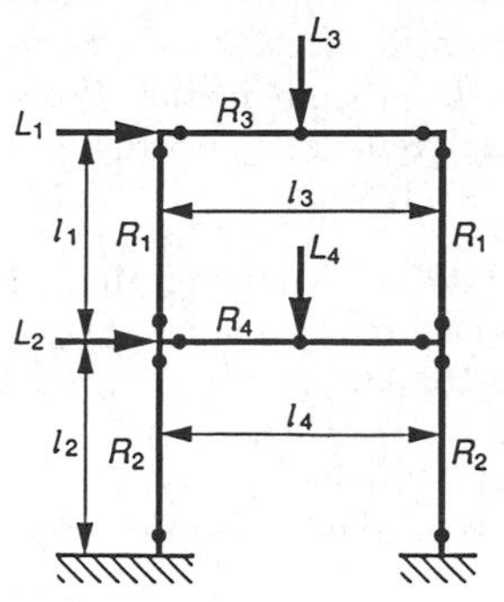

Figure 5: A Two-Storied Frame

Table 2: Simulation Results

	Directional Simulation		ISMMUL	
N	$\hat{P}_f$	$\hat{COV}$	$\hat{P}_f$	$\hat{COV}$
10^2	2.710×10^{-4}	4.973×10^{-1}	2.308×10^{-4}	2.841×10^{-1}
10^3	2.800×10^{-4}	2.694×10^{-1}	3.931×10^{-4}	7.150×10^{-2}
10^4	3.835×10^{-4}	5.735×10^{-2}	4.219×10^{-4}	2.422×10^{-2}
β-point	$\beta = 3.511$		$\beta = 3.500$	
u_1^* (R_1)	-4.008×10^{-1}		-3.613×10^{-1}	
u_2^* (R_2)	-7.626×10^{-1}		-6.085×10^{-1}	
u_3^* (R_3)	-1.325×10^{-1}		-2.766×10^{-9}	
u_4^* (R_4)	-7.647×10^{-1}		-7.635×10^{-1}	
u_5^* (L_1)	2.929×10^{0}		2.989×10^{0}	
u_6^* (L_2)	1.540×10^{0}		1.494×10^{0}	
u_7^* (L_3)	-1.601×10^{-1}		1.482×10^{-9}	
u_8^* (L_4)	7.387×10^{-2}		-3.277×10^{-9}	
Time(sec.)	254.2		274.9	(678.1)*

*: time for searching β-point

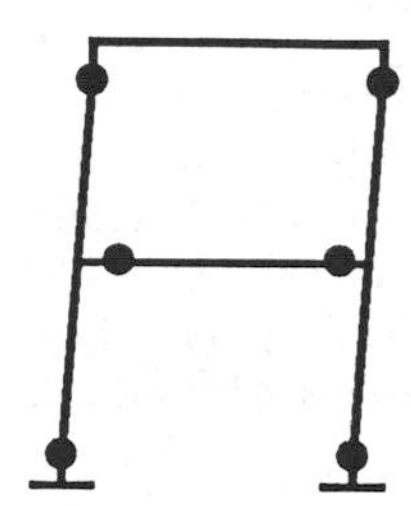

Figure 6: The Most Probable Failure Mechanism

Table 1: Numerical Data of the Frame

Random variable	Normal Distribution Mean	COV	Parameter
R_1	326.4 (kN-m)	0.05	l_1 = 15.0 (m)
R_2	549.7 (kN-m)	0.05	l_2 = 15.0 (m)
R_3	351.6 (kN-m)	0.05	l_3 = 20.0 (m)
R_4	689.7 (kN-m)	0.05	l_4 = 20.0 (m)
L_1	45.0 (kN)	0.2	
L_2	45.0 (kN)	0.2	
L_3	60.0 (kN)	0.2	
L_4	60.0 (kN)	0.2	

total surface area S_n of Ω_n which are given as follows:

$$S_n = \frac{2\pi^{\frac{n}{2}}}{\Gamma(\frac{n}{2})} \tag{15}$$

$$S_{\phi_x} = \frac{2\pi^{\frac{n-1}{2}}}{\Gamma(\frac{n-1}{2})} \int_0^{\phi_x} \sin^{n-2}\theta d\theta \tag{16}$$

$$P_{\phi_x} = \frac{S_{\phi_x}}{S_n} \tag{17}$$

where $\Gamma(\ )$ denotes the Γ-function. When P_{ϕ_x} is moved to S_{ϕ_y}, the density $h_{\boldsymbol{B}}(\boldsymbol{b})$ on S_{ϕ_y} becomes:

$$h_{\boldsymbol{B}}(\boldsymbol{b}) = \frac{1}{S_n} + \frac{P_{\phi_x}}{S_{\phi_y}} = \frac{1}{S_n}\left(1 + \frac{S_{\phi_x}}{S_{\phi_y}}\right) \tag{18}$$

Corresponding to the density $h_{\boldsymbol{B}}(\boldsymbol{b})$ given in Eq. (18), θ_y is determined from the following equation:

$$\frac{S_{\theta_x}}{S_{\theta_y}} = \frac{S_{\phi_x}}{S_{\phi_y}}$$

$$or \quad \int_0^{\theta_y} \sin^{n-2}\theta d\theta = \frac{S_{\phi_y}}{S_{\phi_x}} \int_0^{\theta_x} \sin^{n-2}\theta d\theta \tag{19}$$

Generally, there may be several density-reduced areas and density-increased areas on Ω_n. Movement of the probability between them should follow the principle of keeping the densities proportional to $1 - \chi_n^2(r_0^2(\boldsymbol{b}_i))$.

4 Numerical Results

Consider a two-storied frame structure as shown in Fig. 5. The numerical data concerning the strengths, loads and dimensions are given in Table 1. The potential plastic hinges in the frame are assumed at joints or the places where loads are applied as shown in Fig. 5. The structure has 170 failure mechanisms constituted by different plastic hinges, which correspond to 170 limit state functions. The structural failure event is a union of the formation of the 170 failure mechanisms.

The simulation results using the proposed method are given in Table 2, where the results by a multiple-point importance sampling method (ISMMUL)[7] are also listed. The latter uses an algorithm to systematically find multiple check points on the limit state surface at first and then performs importance sampling based on those points. It is seen from Table 2 that, the directional simulation using the modified sampling density yields a β-point very close to that from ISMMUL. The investigation for the complex limit state surface has been combined with the simulation and the total computation time is reduced. The failure mechanism corresponding to the β-point is shown in Fig. 6.

5 Conclusions

This paper provides an efficient directional simulation procedure for the reliability evaluation of complex structural systems. The simulation starts from the state without any knowledge of the limit state surface, while the direction sampling density is modified step by step, utilizing the information obtained from the simulation to finally yield a good efficiency of sampling. Furthermore, the modified sampling density produces accurate information about the most probable failure area, which is useful in the reliability evaluation and the practical design of structures.

6 Acknowledgement

This research has been supported by Research Aid of Inoue Foundation for Science which is gratefully acknowledged.

References

[1] Bjerager, P., Probability Integration by Directional Simulation, J. of Engineering Mechanics, ASCE, 114, pp.1285-1302, (1988).

[2] Bourgund, U. & Bucher, C. G., Importance Sampling Procedure Using Design Point-ISPUD, Institute of Engineering Mechanics, University of Innsbruck, Austria, Report 9-86 (1986).

[3] Ditlevsen, O., Olesen, R., & Mohr, G., Solution of a Class of Load Combination Problems by directional Simulation, Structural Safety, 4, pp.95-109, (1987).

[4] Harbitz, A., An Efficient Sampling Method for Probability of Failure Calculation, Structural Safety, 3, pp.109-115, (1986).

[5] Madsen, H. O., Krenk, S. & Lind, N. C., Methods of Structural Safety, Prentice Hall (1986).

[6] Rackwitz, R., & Fiessler, B., Structural Reliability Under Combined Random Load Sequences, Computer and Structures, Vol. 9, pp. 489-494, (1978).

[7] Shao, S., & Murotsu, Y., Reliability Evaluation Methods for Systems with Complex Limit States, in: Rackwitz, R., & Thoft-Christensen, P.(eds.), Reliability and Optimization of Structural Systems '91, Springer Verlag, pp. 325-338, (1991).

Structural Safety & Reliability, Schuëller, Shinozuka & Yao (eds) © 1994 Balkema, Rotterdam, ISBN 90 5410 357 4

A fuzzy neural expert system for damage assessment of RC bridge deck

Eiichi Watanabe, Hitoshi Furuta & Jianhong He
Kyoto University, Japan

Motohide Umano
Osaka University, Japan

ABSTRACT: This paper aims to propose a method that helps maintenance engineers to evaluate the damage states of reinforced concrete (RC) bridge deck and to select its appropriate repairing methods by using a fuzzy neural expert system. The evaluation measures include damage cause, damage degree and damage propagation speed. One of important items in the maintenance program is that the repairing methods are automatically selected. Using a neural network as a subsystem, the present system can provide several appropriate repairing methods by taking account of the past experience through its learning ability. Since the system is built on a *32 bit* engineering workstation and is written with *Common Lisp* and *C language*, anyone can use this system without difficulty.

1 INTRODUCTION

In Japan, there are many existing bridges to be repaired or altered (Shiraishi, Furuta & Sugimoto 1985) However, it is impossible to rebuild all the damage bridges because of financial limitation (Komai, Kimata & Kobori 1991). Evaluation of durability becomes necessary for appropriate maintenance and repair of bridge structures. Future progress of the damage state of bridge structures should be estimated based on the damage cause, damage degree and damage propagation speed. Then, whether or not a bridge needs to be repaired or rebuilt, and what method should be adopted if repaired, are decided based on the estimation results.

It is difficult to maintain and repair all the bridges because the number of experts engaging in damage assessment cannot satisfy its demand. Therefore, it is very meaningful to build an expert system to help engineers without sufficient experience to make various judgments on maintenance and repair as the expert does (Shiraishi, Furuta, Umano & Kawakami 1991).

This paper attempts to develop a fuzzy neural expert system for assessing the damage states of RC bridge deck. The system consists of two subsystems: one is a fuzzy production system for making inferences and the other is a neural network system for deriving solutions. The fuzzy production system is used to evaluate damage causes, damage degree and damage propagation speed with appropriate knowledge; and the neural network system is used to select appropriate repairing methods taking into account of many factors comprehensively. It is possible to construct a more practical and useful system by using these two methods with different characteristics.

2 AN EXPERT SYSTEM FOR DAMAGE ASSESSMENT OF RC BRIDGE DECK

2.1 Architecture of expert system

2.1.1 Fuzzy production system

The fuzzy production system consists of inference engine including forward and backward inferences, rule base and working memory.

Both forward and backward inferences can be used in the fuzzy production system by introducing backward inference into the previous system (Furuta, Shiraishi, Umano & Kawakami 1991).

The advantages of using forward rules in the antecedent part of backward rules are as follows (Umano 1987, Nakajima 1991):

1. Rules can be described in the form of module because forward rules can be executed in the antecedent part of backward rules.
2. Knowledge can be easily added, modified and refined by describing rules in modules.
3. When someone wants to know if a conclusion can be valid, only the rules related to such a conclusion can be investigated. Namely, since unnecessary search can be avoided, the system becomes more efficient.

2.1.2 Neural network (Hayashi & Imura 1990)

In the system, a subsystem for choosing appropriate repairing methods is constructed by utilizing the learning and pattern recognition abilities of neural network. Fig.1 shows the structure of a neural network.

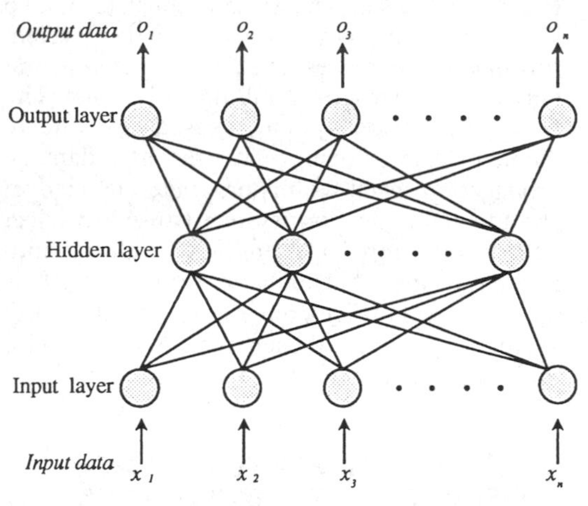

Fig.1. Multi-layer neural network

Learning method: Back-propagation
Network type: Multi-layer neural network
Input and output form: Events
Parameters: Synapse weights, Threshold values
Input and output function of cell: Sigmoid function

$$f(x)=1/((1+\exp(-x)) \quad (1)$$

Introducing the neural network into the expert system, it is possible to consider not only structural factors of RC bridge deck but also economics, construction and environmental factors in choosing the repairing methods.

2.2 Construction of expert system for damage Assessment of RC Bridge Deck

2.2.1 Evaluation of damage cause, damage degree, and damage propagation speed by fuzzy production system

This system estimates the damage cause from the design condition, environmental condition and inspection data of a bridge, and evaluates the damage degree and damage propagation speed for each cause. The flow chart of the inference process is shown in Fig.2.

Firstly, design and environmental conditions of a bridge are surveyed in advance. Secondly, inspection is carried out at site. Inspection items include crack, pavement, concrete, steel etc. Based on the above survey and inspection results, damage causes yielding the bridge damages are estimated among the causes shown in Table 1. For example, the following rules are used.

```
(rules "Damage causes-1-2-2"
 very-true
  if     (structural-type  girder-plate)
         (crack-configuration  width-direction)
         (crack-location  center-of-deck-span)
         (wheel-load-location  center-of-deck-
           span)
  then   (deposit ("Damage cause"  Extreme-
           wheel-load))
         (change-rb "Damage causes-1-3"))
```

Next, damage modes consisting of only damage types with the same damage causes are established.

Furthermore, damage degree and damage propagation speed associated with the damage mode, environmental condition and age of the bridge are considered.

2.2.2 Automatic selection of repairing methods by neural network

As shown in Fig.2, the system first evaluates damage cause, damage degree and damage propagation speed by fuzzy production system. Then, a suitable repairing method is chosen by utilizing the learning ability of neural network taking into account of the effects of importance, economy and construction condition of bridges.

The architecture of the system for choosing automatically repairing methods by neural network is illustrated below.

1.As stated in Section 2.2.1, inference results

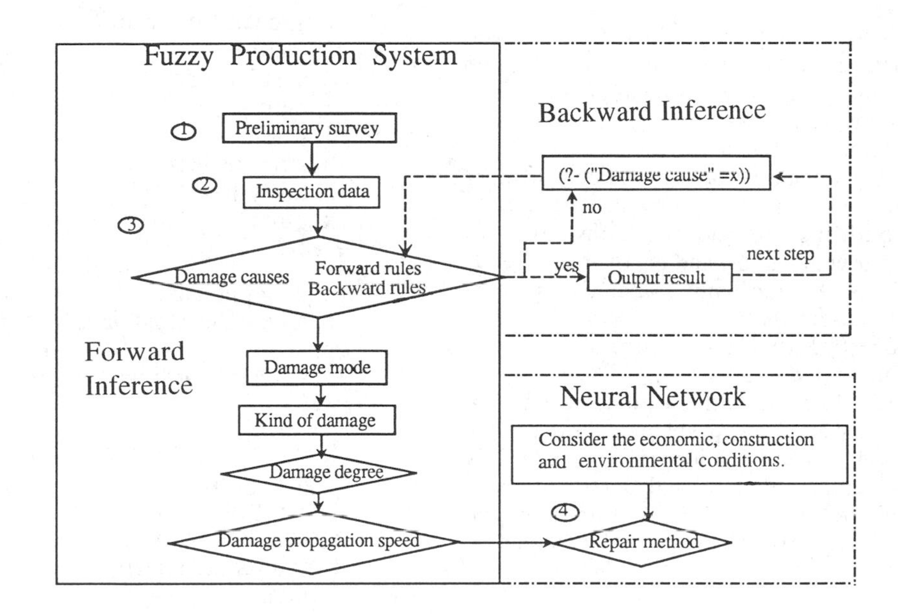

Fig.2. Architecture of the present system

Table 1. Damage causes

Load	g-1: Extreme wheel load g-2: Impact effect g-3: Inadequacy of girder arrangement
Design and structural factors	g-4: Short of deck depth g-5: Lack of main reinforcing steel bar g-6: Lack of distribution reinforcing bar g-7: Inadequacy of cross beam allocation g-8: Additional bending moment due to nonuniform settlement
Construction conditions	g-9: Poor quality of cement g-10: Poor compaction g-11: Inadequate curing of construction joint g-12: Lack of covering
Other factors	g-13: Salt g-14: Poor drainage g-15: Movement of substructure g-16: Action of alkali material

obtained from a fuzzy production system, namely, damage cause, damage degree, damage propagation speed of RC bridge deck, are used as input data. In addition, structural property, economic, construction and social effects etc. are also taken into account together. Here the results obtained from the fuzzy production system are represented in the form of fuzzy set.

2.A four-layered neural network is used (see Fig.1).

First layer (input layer): In the input layer, fuzzy data are represented as a fuzzy unit group and non-fuzzy data as a crisp unit group.

Second layer: 20 unit

Third layer: 20 unit

Forth layer (output layer): Adding stringer method, retrofitting method by steel plate deck and rebuilding method are considered as representative methods, and denoted as A, B, C respectively. As an example, when the output is B, it is represented as:

(A, B, C)=(0, 1, 0)

3.The input data are learned.

4.In order to evaluate the learning accuracy, it is necessary to achieve some criteria for the correctness of identification. In this study, it is thought that a highly precise diagnosis has been made if the result obtained from a neural network is nearly consistent with an actual example.

3 NUMERICAL EXAMPLES

The applicability and usefulness of the present system is demonstrated by using actual data in the durability assessment of RC bridge deck. Numerical experiments on the selection of repairing methods are carried out using both the fuzzy production system and neural network. The data are collected from the actual repairing records made past for 13 bridge in Osaka City.

3.1 Reasoning by fuzzy production system

The input of data such as structural type, inspection results, environmental condition, etc. are implemented interactively as shown in Fig.3. Here, the following rules are used for the durability assessment.

```
> <Input of Data>
  1. Construction of data file
  2. Load existing data file
   => 1

  Curve bridge?
  y: yes
  n: no
   => y

  Degree of truth value?
  1. truth
  2. absolute-true
  3. more-true
  4. true
  5. more-or-less-true
  6. fairly-true
  x. other
   =>1

a. Add input data?
 re: renew the input data from first
 sp: stop
 except keyboard: next input
   =>3

Structural type of bridge?
 1. plate girder
 2. box girder
 3. arch (below road)
 4. arch (upper road)
 x. other
   =>3
......
Write above data into file?
 y: yes
 n: no
   => y
Input file name => file.1

Make another file?
 1. Make another file
 2. Implement the inference
   => 2

Inference results:
......
```

Fig.3. Interactive input process

1. Rules for damage reasoning 31
2. Rules for estimating damage causes 223
3. Rules for damage mode 5
4. Rules for evaluation of crack damage degree 55
5. Rules for evaluation of road pavement damage degree 37
6. Rules for evaluation of reinforcing steel damage degree 4
7. Rules for evaluation of concrete damage degree 20
8. Rules for evaluation of structural damage degree 20
9. Rules for judgment of comprehensive damage degree 187
10. Rules for judgment of damage propagation speed 28
11. Rules for selection of repairing method 31
12. Rules for output of results 51

First, the data obtained from preliminary survey and inspection are input. For example, Table 2 and Table 3 present the design conditions, environmental conditions and inspection data of a RC bridge deck. Using these data, working memory is rewritten as shown in Fig.4. For instance, concerning the age of the RC bridge deck, is rewritten as follows:

very-true/(deck age medium)

Next, forward inference is commenced using inference rules and the rewritten working memory. The results shown in Table 4 can be obtained after about two minutes inference. Here, truth value is expressed in linguistic forms.

If one wants to know whether some damage cause exists, forward inference can be used in the antecedent part of backward-rules when judging damage causes of RC bridge deck in the system. In this case, matching is implemented between inquiry content and consequent part of backward-rules to investigate if the antecedent condition of rules is satisfied. As an example, provided that an inquiry whether or not the damage cause is [g-1: Extreme wheel load], the following inference result is obtained:

```
>(?- ("Damage causes" g-1))
  [Inference Result]
  g-1: Extreme wheel load (Truth value:
       more-large)
```

3.2 Selection of repairing methods by neural network

Since the above inference results are given by fuzzy sets to deal with the inference results and the data shown in Table 2 in a unified manner, the data (a, b, c) containing fuzziness are expressed by fuzzy unit groups and the data (0, space) not containing fuzziness are expressed by crisp unit groups. For example, it is defined that [crack damage is small]={1/0, 0.66/1,

Table 2. Design and environmental conditions

Kind	Factor	Data	Truth value
Design conditions	Structural type	Plate girder(straight)	1
	Design specification	Before 1967	1
	Construction year	Old	very-true
	Deck thick	20cm	1
	Bridge length	69.00m	1
	Bridge width	12.95m	1
	Lanes	3 lanes	1
	Foot way	One side	1
Environment conditions	Erection location	Near city	1
	Road rank	Main road	1
	Ratio of heavy vehicle	Many	very-true
	Traffic flow	Many	very-true
	Wheel load location	Center of deck span	absolute-true

Table 3. Inspection results

Kind	Factor	State	Truth value
Crack	Configuration	Width direction	absolute-true
	Location	Center of deck span	very-true
	Density	6.63	1
	Space	Large	very-true
	Width	Large	very-true
Road surface	Pavement	Large uneven	very-true
		Small log	true
Concrete	Color change	Medium	true
	Oldness	Medium	true
	Alienation lime	Center of deck span	absolute-true
Steel	State of rust	Rusting	absolute-true
Other	water around	Medium	true
	Leaking water	Center of deck span	very-true

```
(working-memory
 {              (structural-form  girder-plate),
                (design-specification  1967),
    very-true/(construction-year old),
                (deck  thick  20),
                (bridge-length 69),
                (bridge-width 12.95),
                (lanes 3),
                (foot-way  one-side),
                (erection-location near-city),
                (road-rank  main-road),
 absolute-true/(location-of-wheel-load center-of-deck-span),
     very-true/(rate-of-heavy-vehicle many),
     very-true/(triffic-flow many),
 absolute-true/(crack-configuration  width-direction),
    very-true/(crack-location  center-of-deck-span),
                (crack-density  6.63),
    very-true/(crack-space large),
         true/(crack-width  w-large),
    very-true/(road-pavement  uneven),
         true/(road-pavement  small-log),
         true/(concrete color-change medium),
         true/(concrete  oldness  medium),
 absolute-true/(location-of-alienation-lime  center-of-deck-span),
 absolute-true/(steel  rusting),
         true/(around-water  medium),
     very-true/(leak-water  center-of-deck-span),
 })
```

Fig.4. An example of working memory

Table 4. Inference results

Damage causes (Truth value)	g-1:	Extreme wheel load	(more-large)
	g-6:	Lack of distribution bar	(small)
	g-15:	Poor drainage	(large)
Damage degree (Truth value)	g-1:	more-large	(small)
	g-6:	more-large	(small)
	g-15:	large	(more-large)
Damage propagation speed (Truth value)	g-1:	more-large	(more-large)
	g-6:	medium	(small)
	g-15:	more-large	(large)

0.33/2, 0/3, 0/4, 0/5, 0/6, 0/7, 0/8, 0/9, 0/10}. As a fuzzy unit group, [crack damage is small]={1, 0.66, 0.33, 0, 0, 0, 0, 0, 0, 0, 0} are input to 11 units of the input layer in the neural network, whereas the data not containing fuzziness are represented by binary values as a crisp group. For example, it is 1 when steel reinforcing bars are exposed and 0 otherwise. The output of results is A or B or C, where A denotes the repairing method by additional stringers, B the retrofitting method by steel plate deck and C rebuilding method. For example, when the output is B, it can be expressed as (A, B, C)=(0, 1, 0). Then, through the neural computation, it is possible to judge which repairing method is suitable for this case. The output values for bridge No.1 are A=0.94, B=0.05, C=0.04. Thus A should be chosen as repairing method. It can be seen that the outputs agree completely with the past record, and hence the learning has been carried out precisely.

4 CONCLUSIONS

Appropriate damage assessment is important in maintenance program of bridge structures. This research has developed a practical expert system for damage assessment of bridge structures, based on the knowledge of experts who make various judgments efficiently in daily maintenance and management work. The durability of RC bridge deck is considered as a main evaluation object in the system. Based on the damage cause, damage degree and damage propagation speed, repairing method can be chosen automatically. By introducing the fuzzy logic into the system, it becomes possible to deal with linguistic data given by the subjective judgments of engineers (Furuta 1990). Practical and useful solutions can also be obtained even from incomplete data. Furthermore, repairing methods can be chosen automatically, while considering a lot of factors by introducing a neural network into the expert system.

The characteristics of the present expert system for damage assessment of RC bridge deck can be summarized as follows:

1.Since the system was built on a 32 bit engineering workstation NEWS (made by SONY) and written in Common Lisp and C language, anyone can use it without difficulty, at any time and place.

2.Using this system, it is possible to evaluate a lot of bridges with relatively short time.

3.Inference time can be reduced by module of rules. With the former system takes 15-20 minutes from beginning of inference to output of results, the present system reduces up to 1/4.

4.Data are input in a dialogue form, everyone can use it without difficulty.

5.Because backward inference can be implemented in the system, both forward and backward inference can be used in the fuzzy production system. Implementing forward-rules in the antecedent of backward-rules, it is possible to reduce the inference time.

6.It becomes clear from this study that, in order to solve efficiently a practical problem, appropriate results can be obtained by using fuzzy production system when knowledge is easily attainable, or relationships between events are clear. However, when it is difficult to make out rules, meaningful results can be obtained by utilizing the learning ability of neural network (Furuta, Umano, Kawakami &

Shiraishi 1990, Miyamoto, Kimura & Nishimura 1989, Mikami, Tanaka, Kurachi & Yoneda 1992, Miyamoto, Morikaw & Kushida 1990).

REFERENCES

Furuta, H. 1990. Fuzzy expert system, *System/Control/Information*, .34: 288-294, (in Japanese).
Furuta, H., M. Umano, K.Kawakami, H. Ohtani & N. Shiraishi 1990. A fuzzy expert system for durability assessment of bridge decks, *Proc. of ISUMA*, 522-527.
Furuta, H., N. Shiraishi, M. Umano & K. Kawakami 1991. Knowledge-based expert system for damage assessment based on fuzzy reasoning, *Computer and Structures*, 40-1: 137-142.
Hayashi, I. & A. Imura 1990. A neural expert Ssystem using fuzzy teaching input, *Proc. of 6th Fuzzy system Symposium*, 49-55, (in Japanese).
Komai, K., N. Kimata & T. Kobori 1991. A fundamental study on evaluating system of bridges for maintenance and control planning, *Proc. of Japan Society of Civil Engineers*, .428/I-15: 137-146 (in Japanese).
Mikami, I., S. Tanaka, A. Kurachi & S. Yoneda 1992. Inference by analogy and negative learning for selecting retrofitting method of steel bridge fatigue damage, *Journal of Structural Engineering*, 38A: 557-570, (in Japanese).
Miyamoto, A., H. Kimura and A. Nishimura 1989. Expert system for maintenance and rehabilitation of concrete bridges, *Proc. of IABSE Colloquium on Expert System in Civil Engineering* ,207-217.
Miyamoto, A., H. Morikawa & M. Kushida 1990, Engineering dealing of subjective uncertainty in bridge rating, *Proc. of ISUMA*,.539-545.
Nakajima, T. 1991. Realization of backward inference and its combination with fuzzy data based on fuzzy production system, *Master Thesis, Kansai University* (in Japanese).
Shiraishi, N., H. Furuta & M. Sugimoto 1985.:Integrity assessment of bridge structures based on extended multi-criteria analysis, *Proc. of ICOSSAR*, 1:505-509.
Shiraishi, N., H. Furuta, M. Umano & K. Kawakami 1991. An expert system for damage assessment of a reinforced concrete bridge deck, *Fuzzy Sets and Systems*, 44: 449-457.
Umano, M. 1987 Fuzzy set manipulation system in lisp, *Proc. of 3rd Fuzzy System Symposium*, 167-172, (in Japanese).

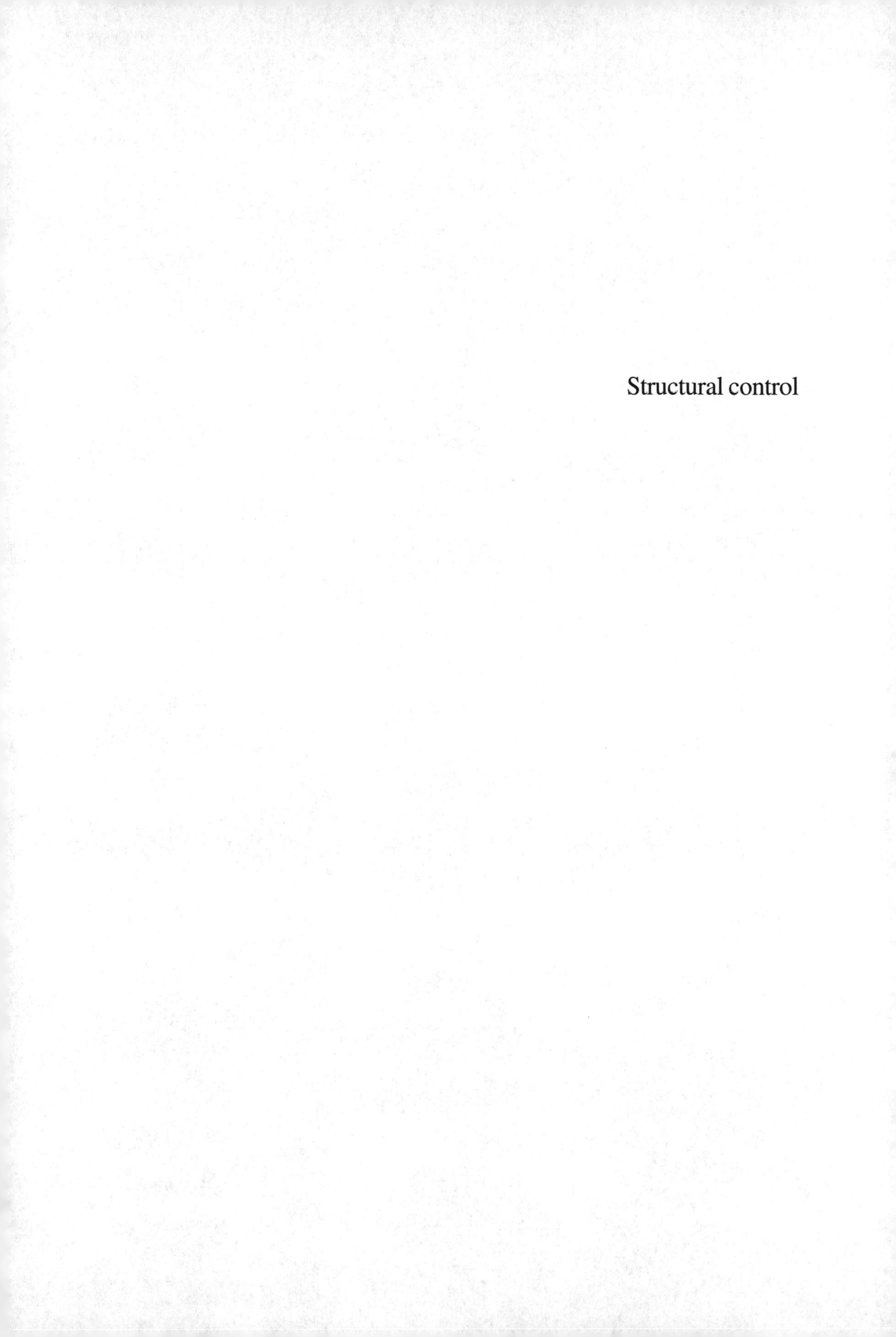

Structural control

Structural Safety & Reliability, Schuëller, Shinozuka & Yao (eds) © 1994 Balkema, Rotterdam, ISBN 90 5410 357 4

Reliability based retrofitting of RC-frames with hysteretic devices

U.E.Dorka
University of Kaiserslautern, Germany

H.J.Pradlwarter
University of Innsbruck, Austria

ABSTRACT: Studies have shown that hysteretic device systems (HYDE-systems) are very efficient for retrofitting of rc-frames. A practical method for the verification of such systems based on stochastic mechanics is presented and applied to two examples: An 8-story rc-frame with eccentric stiffening systems and the Allstate Building in Seattle, the first shear-panel retrofit in the US.

1 INTRODUCTION

Recently, hysteretic devices (Hydes), such as shear-panel dampers (SPDs, Seki et al. 1988) or friction based devices (Pall et al. 1991) have been used successfully for the upgrading of rc-frames against earthquake effects with savings of up to 40% compared to conventional retrofitting techniques. This is due to the fact that such systems can limit the story shear and drift below the quasi-elastic limit sway of the rc-frame which makes upgrading of columns or beams un-necessary. To achieve this goal, the structural layout has to concentrate the story drifts in the Hydes which in turn limit the maximum story shear physically by their maximum yield or frictional forces. In a well designed system, the input energy due to strong earthquakes is largely dissipated during several inelastic cycles and does not cause excessive stresses in the remaining structure which stays elastic (Roik et al. 1988).

The safety of such a system can be checked by evaluating its maximum story drift under earthquake events and compare this to the quasi-elastic limit of the existing rc-frame. Unfortunately, current code procedures like the response spectrum method do not allow for such a check since they do not provide a close enough estimate of the story drift for non-linear systems. Time domain analysis is therefore used to perform this task. The current state-of-the-art in engineering practice calls for three to five "representative" earthquakes (natural or generated ones) in such a case, usually performed on plane models of the lateral force resisting system that contains the Hydes. From non-linear stochastic mechanics, it is well known that such a small number of simulations does not provide the accuracy needed to allow a reasonable estimate of the safety against a defined limit state. Furthermore, plane structural models often over-simplify the system, especially existing ones which are often quite difficult to even identify.

With system identification techniques being another class of problems, this paper is concerned with a practical method of estimating the safety of existing (and already identified) rc-frames retrofitted with Hydes against a defined quasi-elastic drift limit. It is this problem that is most frequently encountered in such structures with important economical and design implications: namely how many Hydes are needed to make expensive upgrading of the rc-columns and beams in the building obsolete.

2 A BASIC MODEL FOR SIMULATING BUILDINGS WITH HYDES

The first step in this process is to find a 3D model of the system that is accurate enough, yet simple and efficient to be used in extensive simulations. For rc-frame structures retrofitted with Hydes, a model like the one given in Fig. 1 will be sufficient, if the following assumptions apply: (1) - masses can be lumped into prismatic bodies with centers in the slabs, (2) - the slabs are rigid in plane, and (3) - the non-linear behavior is concentrated in the Hy-

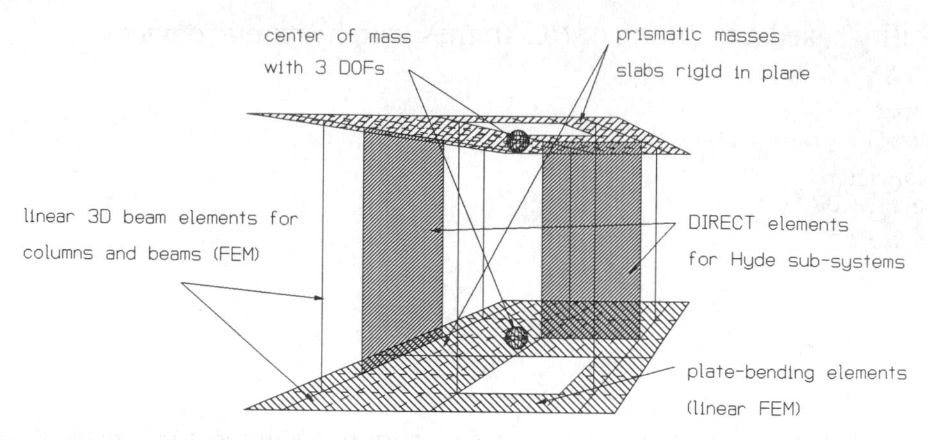

Fig.1 Basic 3D-model of a rc-frame structure retrofitted with Hydes.

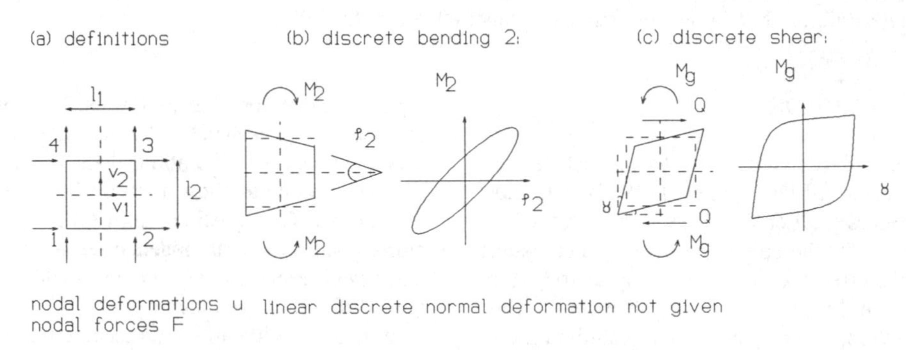

Fig.2 Discrete rectangular element (DIRECT element) to model the stiffening systems containing Hydes (a). Independent deformation states (discrete strains) and their hysteresis laws appropriate for modeling Hyde systems: discrete bending (b) and discrete shear (c).

des. (1) and (2) are assumptions often made in the analysis of multistory buildings and have proven time and again their validity for a large variety of buildings. (3) is an assumption specific to Hyde systems: it actually reflects the design goal to protect the structure from damage by limiting its response to quasi-elastic behavior.

With these assumptions, 3D linear beam elements (they should include shear deformation) can be used to model columns and beams, linear plate bending elements to model the slabs and discrete rectangular or DIRECT elements (Dorka 1991, Fig.2) to model the stiffening system containing the Hydes. These elements model the mechanical behavior directly by specifying appropriate hysteresis laws. After their parameters are found by the simulation process, the actual system is selected as a combination of Hyde and stiffening system to provide the required hysteresis loop. Thus, a Hyde system is appropriately based on the design of its response.

The assumption of lumped masses and damping provides diagonal mass and damping matrices with greatly reduced numbers of DOFs: with vertical mass movements neglected, only 3 dynamic DOFs remain per story. All elastic effects can now be statically condensed on the dynamic DOFs prior to any simulation. Furthermore, the assumption of the slabs being rigid in plane allows a direct geometric coupling of the non-linear discrete shear loops to the dynamic DOFs. Therefore, only the dynamic DOFs and the non-linear M_g-γ loops (Fig.2) remain for the simulation. This will not be altered by including the P-Δ effect (which should be done always and presents no difficulty) or a model for the soil, as long as it does not contain masses or non-linear effects.

For the simulation process of such systems, the

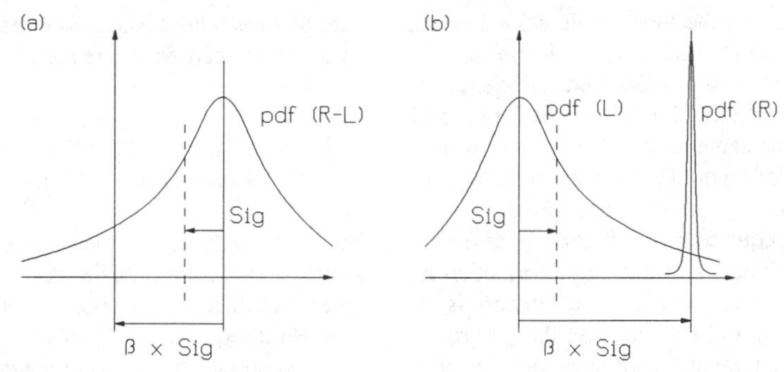

Fig.3 Definition of safety index β for general systems (a) and systems with quasi-deterministic properties (b) (L - load effect, R - resistance, SIG - standard deviation (Std)).

Newmark-*beta* method has proven its applicability. An iteration performed during each time step will provide the response within a specified bound for its accuracy. Using 1/100 sec time steps for earthquake simulations, more than 2 iterations per time step are hardly necessary. Therefore, a 10 sec earthquake simulation of a 60-story building requires not more than $20*10^8$ floating point operations (FLOPs) when the basic 3D model of Fig.1 can be used. On a workstation with RISC-processor (15 Mflops), this can be performed in about 2 minutes. Smaller systems of up to about 15 stories can be accommodated quite well on PC's. A 3D earthquake simulation of an rc-frame retrofitted with Hydes therefore is a viable and accurate tool for the practical response assessment of such systems.

3 A SAFETY INDEX PROCEDURE FOR THE VERIFACTION OF HYDE SYSTEMS

Based on such simulations, a method has to be found that provides a good safety estimate for the building with a reasonable number of simulations. Using the safety index approach of modern codes, the safety of a Hyde system can be expressed in terms of a β-index vector, with each β-value representing a multiple of a standard deviation (Std) of a critical variable with respect to its defined limit state (see Fig.3). Since Hyde systems are designed to behave in a certain manner, the spread of their resistance pdf is usually small compared to the load effect pdf and the β-index definition of Fig.3(b) can be applied. The most critical system parameters responsible for the spread of pdf(R) are the ones describing the hysteresis loops of the DIRECT elements, notably their shear stiffness and maximum shear force. The Stds of these parameters have to be small, if the β-index definition of Fig.3(b) is applied.

For rc-frames retrofitted with Hydes, these β-values can be expressed as ratios of the quasi-linear drift limits to the drift Stds at the most critical locations in a story. The task is therefore to determine these Stds. For this, basically two methods are available for complex non-linear systems: the equivalent stochastic linearization (EQL) and the Monte-Carlo simulation (MCS) (Schuëller (Ed.) 1991).

EQL has the advantage of being fast but is known to underestimate the Stds, especially in highly non-linear systems. Using the correct non-linear hysteresis law also was considered a problem until recently, when EQL-2 was introduced (Pradlwarter, Schuëller 1991). Here, the linearization coefficients for any constitutive law are found numerically using the least-square criterion to minimize the error between the linearized and true representations of the hysteresis laws. By randomly generating permissible sets of state vectors including the truncation of variables to their physical limits (e.g. Hyde forces), the least-square problem is reduced to a set of linear equations that can be solved for the linearization coefficients. Using this approach, an improvement in predicting the shape of the Std vector by EQL is reported but the underestimation effect is not mitigated to acceptable levels.

It is therefore only MCS that can provide results that are accurate enough but at the cost of extensive simulations. Their number can be estimated using the central limit theorem: Assuming a desired confidence of 92% for example, about 500 simulations are needed to estimate the Stds within a 5% interval of the calculated value (Gauss-type behavior assumed). This number can be deemed sufficient for the purpose of estimating the safety index for a real structure and at the same time will not cause prohibitive amounts of computer time when the above mentioned modeling principles are applied.

Checking the required safety index by MCS is therefore recommended and can be performed in a practical environment, if appropriate software is available. The design of the response though, requires methods that are less computer time consuming: a pre-design method is needed.

For Hyde systems, this means basically an estimate for the level of maximum Hyde forces and their distribution over the building. The latter should ensure as an optimum that every Hyde participates equally in the dissipation process. For rc-frames, the following distribution over the height is recommended, which is a slightly modified version of the function suggested by (Dechent 1989):

$$r_i = R\, M_e^{-1}\, [1+2.84*(h_i/H)^4] \sum_{j=N}^{i} m_j \qquad (1)$$

$$M_e = \sum_{i=1}^{N} i m_i + 2.84*H^{-4} \sum_{i=1}^{N} (h_i^4 \sum_{j=N}^{i} m_j) \qquad (2)$$

with H - total height of the building, m_j - story mass, h_i - story height above ground N - number of stories, r_i - maximum story shear (Hyde limit force) and R - sum of all story shears (Hydes) in one principal direction.

Next, the value for R must be estimated. This can be accomplished by calculating the average power of the linear structure without the new Hyde system (prior to retrofitting) under the same site-specific stochastic earthquake loading that is used later for the verification by MCS. The request that all Hydes together must be able to dissipate in one limit cycle the energy thus brought into the system during one second provides, together with Eq.1, a first estimate for the hysteresis loop parameters of all Hydes. This *one-second-request* has provided good estimates that needed only slight modifications later and is therefore recommended by the authors.

An advantage of this pre-design method is that it uses the old 3D structure and same stochastic load model that is needed for the final verification by MCS. This saves considerable time during the design process where changes in the position of the hyde system can occur frequently.

4 EXAMPLES: 8-STORY RC-FRAME AND THE ALLSTATE BUILDING IN SEATTLE

To illustrate the safety index procedure for Hyde systems, first a fictitious 8-story rc-frame retrofit is given and then the shear-panel retrofit of the Allstate Building in Seattle is discussed.

Fig.4 shows the plan and elevation detail of the 8-story frame with the position of the new stiffening systems containing Hydes. Without the new stiffening systems, this can be considered a type of rc-frame of which many have been built during the last decades. A large percentage of these were built exactly according to code provisions that until very recently did not require enough ties in the columns to prevent sudden and brittle failure under side sway.

Assuming such brittle behavior and an identified elastic limit sway of 20 mm for the columns, elastic story drift Stds of the old system (Fig.5) indicate β-values that must be considered dangerous. As stochastic ground loading, a modified Kanai-Tajimi spectrum was used with parameters $\Omega_1 = 15.6$ rad/sec, $\xi_1 = 0.6$, $\Omega_2 = 1.0$ rad/sec and $\xi_2 = 0.995$, reflecting conditions found in many southern European locations.

The positions of the new stiffening systems were considered given by the architecture. Regardless of their eccentricity to the centers of mass, the Hydes (which can be any suitable device), were distributed evenly in one floor and placed in *seismic links* between solid reinforced concrete walls and the slabs (Fig.4). The walls provide a large horizontal stiffness so the Hydes are already activated for very small story drifts.

The pre-design yielded a combined limit force of R = 580.00E+02 KN of all Hydes acting in one direction. Using Eq.1, the local Hyde forces were then selected. As hysteretic model for the shear loops of the DIRECT elements, a formulation was adopted that was used for shear-panel dampers (Pradlwarter, Schuëller 1991):

$$M_g(t) = \alpha K_g \gamma(t) + (1-\alpha) K_g\, q(t) \qquad (3)$$

with K_g - initial stiffness of M_g-γ loop and

$$q(t) = \gamma(t)\, h(q,\gamma) \qquad (4)$$

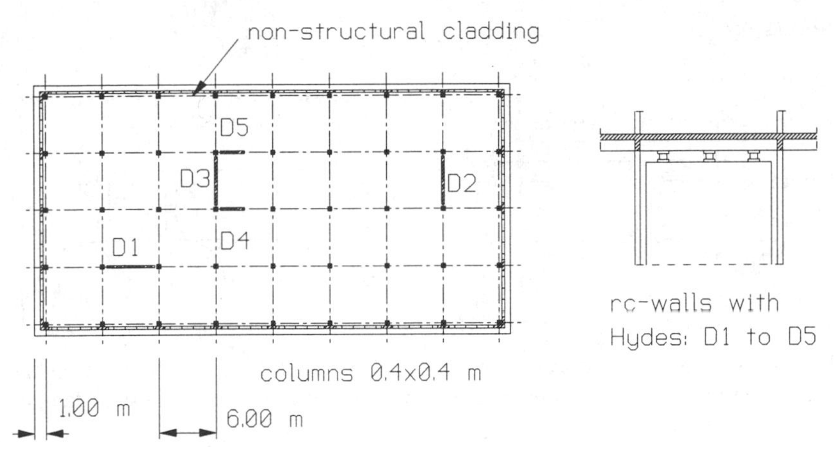

Fig.4 Plan and elevation detail of an 8-story rc-frame studied for retrofitting with Hydes.

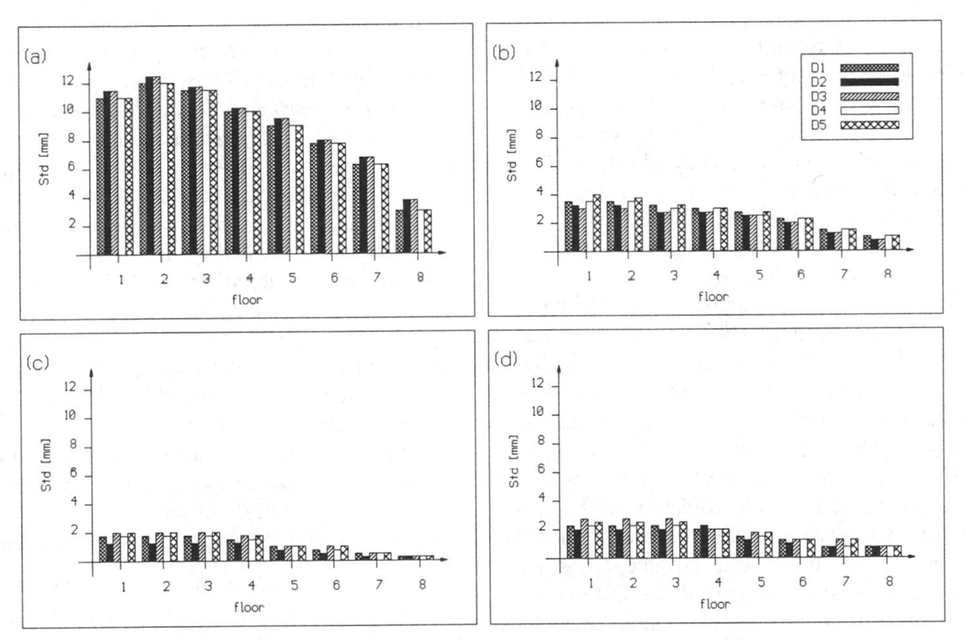

Fig.5 Standard deviations of story drifts for the 8-story rc-frame in locations and direction of D1 to D5 in all stories. Linear system before retrofitting (a), linear system retrofitted with shear walls only (b), retrofitted system calculated with EQL-2 (c) and retrofitted system with MCS (d).

$h(q,\gamma) = 1$ for abs $q <= q_y$ or $\dot{\gamma} <= 0$
$h(q,\gamma) = (q_p-q)/(q_p-q_y)$ for abs $q > q_y$ and $\dot{\gamma} < 0$
and $0 <= \alpha <= 1$ is the ratio of post-yield stiffness to initial stiffness K_g.

The model provides a smooth transition between elastic and non-linear region. This does not only stabilize the time stepping scheme but is also a desirable property for Hydes.

Fig.5 shows the story drift Stds for locations D1 to D5 in all stories for various system configurations and computed by EQL-2 and MCS. As is seen from this Figure, the new design greatly reduces the story drift of the building. A sufficient β value in all stories not smaller than 6.2 is observed. It also verified that all Hydes participate sufficiently in the dissipation. As for EQL-2, it is veri-

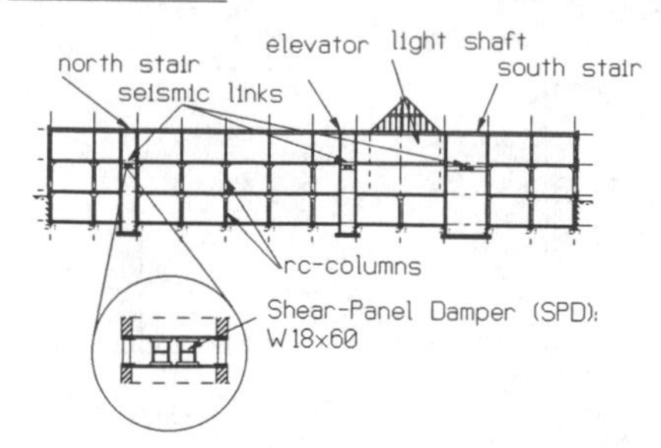

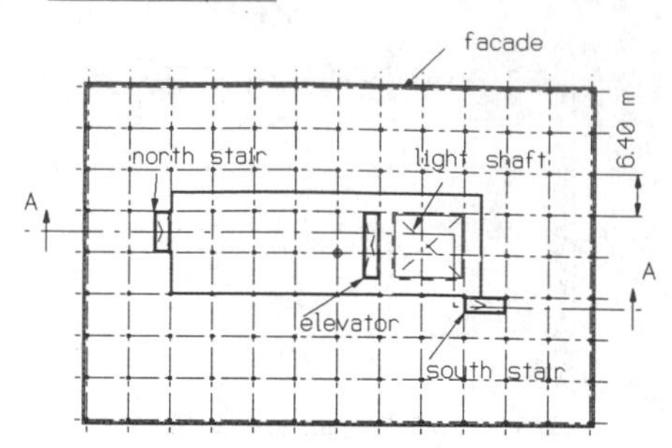

Fig.6 East-West (E-W) elevation and 2nd floor plan of Allstate Building in Seattle, WA including the new cores (after retrofitting): The first building in the US retrofitted with shear-panel dampers (SPDs).

fied that it captures the form of the Std vector but underestimates its values in the order of 30% in this example.

The shear-wall retrofit without Hydes (Fig.5(b)) has Stds in the order of magnitude of the full retrofit, but this design causes stresses far exceeding any possible elastic limit of the rc-walls. This example therefore shows again the ability of Hyde systems to limit displacements to levels that can only be achieved by very stiff systems and at the same time limit the forces physically to reasonable levels (Roik et al. 1988).

The other example discussed here is the first shear-panel retrofit in the US: The Allstate Building in Seattle (Fig.6). At a first glance, the building appears to be quite regular but each story differs considerably with respect to its horizontal load resisting system. The 1st story has solid perimeter rc-walls on the three sides where it is embedded in the ground. Therefore, no considerable differential movement with respect to the ground actually takes place here. The bi-directional rc-frame in this story contributes little to the horizontal stiffness but provides rotational restraints to the column bases of the 2nd floor.

The 2nd story consists of a rc-frame formed by the columns and the slab. The columns are continuous from the foundation upwards. Columns and slab-to-column connections are thought to exhibit basically brittle failure modes that occur at about 20 mm of story drift. This was identified as the crucial limit state by D'Amato Conversano Inc. (DCI) of Kirkland, WA, the engineering firm in charge of the retrofit.

The 3rd story consists of a light perimeter steel frame on all four sides made of truss beams with columns hinged to the concrete floor. Excessive drift was to be expected in this "soft top story". Additionally, the truss beams do not provide adequate ductility.

Since the architecture called for three new cores, shear-panel dampers (SPDs) were implemented in seismic links between the cores and the 2nd story ceiling (Fig.6). The idea was to provide a *seismic link* that effectively limits the forces transmitted to the upper level by the maximum possible forces in the SPDs and to dissipate large amounts of energy by concentrating the deformations in the SPDs. This is a classical approach for Hyde systems.

Verification of the system was done by DCI using the UBC provisions for eccentrically braced frames (EBFs) that also contain shear panels as their main dissipative device (UBC 1988).

For the safety index study presented here, bi-directional earthquake records were generated from a modified Kanai-Tajimi spectrum with the following parameters: $\Omega_1 = 30.0$ rad/sec, $\xi_1 = 0.5$, $\Omega_2 = 5.0$ rad/sec, $\xi_2 = 0.4$. This spectrum has a standard deviation of $\sigma = 0.9786$ m/sec^2 and emphasizes the low frequency range which is thought to contribute the most to plastic deformation. It is assumed that this load model represents the local conditions at the site reasonably well.

The results of an EQL-2 study and MCS with 500 10-sec simulations is given in terms of the Std vector of the story displacements in Fig.7. For MCS, less than 2 hours were needed on an INTEL-80486 processor based machine. The results show again the ability of EQL-2 to capture the correct form of the Std-vector but also its considerable underestimation of its magnitude. In MCS, the Stds stabilized after about 2.5 sec making the response essentially independent of the earthquake's duration.

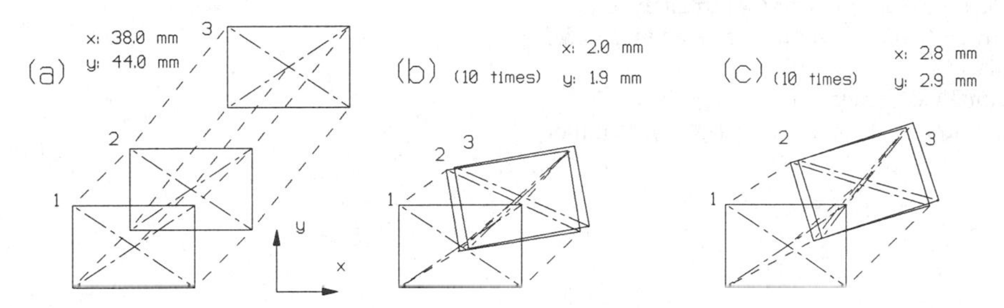

Fig.7 Standard deviation vector of story displacements for the Allstate Building as seen from the top (numbers correspond to floor levels). Linear system before retrofitting with shear panel dampers (a), retrofitted system computed with EQL-2 (b) and retrofitted system computed with MCS (c).

A β-value of about 10 was estimated by MCS with respect to the crucial limit state of 20 mm story drift which is more than sufficient. This is a clear indication that the code design procedures used in the verification of the Allstate Building do not provide a good estimate of safety for such systems: if not unsafe, they might lead to uneconomical designs with too many Hydes.

5 SUMMARY AND CONCLUSIONS

A procedure based on the safety index β as defined in most modern codes is presented to verify rc-frames retrofitted with hysteretic devices (Hydes) against earthquake loading. After using a pre-design method based on linear stochastic mechanics, verification of the designed response is provided by calculating the standard deviations (Stds) of the story drifts using a dynamically reduced 3D nonlinear model and Monte-Carlo simulation. The vector of b-values is computed as the ratio of the identified quasi-linear story drift limits to the story drift Stds.

For the purpose of comparison, the new stochastic equivalent linearization technique EQL-2 was also performed on the two examples studied here. Although it captures the form of the Std vector quite well unlike conventional EQL, it does underestimate its magnitude considerably. It therefore is able to provide a quick estimate of the order of magnitude during the pre-design stage.

The procedure has proven its applicability for practical applications provided the necessary software is available. It yields a much higher degree of accuracy as current code procedures that cannot estimate the safety of such systems. Often, this accuracy is needed in retrofitting complex systems with brittle failure modes that have crucial drift limits, e.g. many rc-frames.

In retrofitting rc-frames with Hydes, the procedure greatly enhances the economy of the design since the number of Hydes is an important economical factor, as is the question whether existing columns and beams have to be upgraded.

6 REFERENCES

Seki, M., Katsumata, H., Uchida, H., Takeda, T. 1988. Study on earthquake response of two-storied steel frame with y-shaped braces. *Proc. 9th world conf. earthq. eng.*, Tokyo-Kyoto, Japan, Vol.4: 4/65-4/70.

Pall, A.S., Ghorayeb, F., Pall, R. 1991. Friction dampers for rehabilitation of Ecole Polyvalente at Sorél, Quebec. *Proc. 6th Canadian conf. on earthq. eng.*, Toronto, Canada: 191-200.

Roik, K., Dorka, U.E., Dechent, P. 1988. Vibration control of structures under earthquake loading by three-stage friction-grip elements. *Earthq. eng. struc. dyn.*, Vol. 16: 501-524.

Dorka, U.E. 1991. System analysis by discrete elements. *Proc. Asian Pacific conf. comp. mech., Hong Kong*: 1257-1262.

Schuëller, G.I. (Ed.) 1991. *Structural dynamics - recent advances*. Springer Verl. Berlin.

Dechent, P. 1989. Berechnung und Bemessung reibgedämpfter Bauwerke unter Erdbebenbeanspruchung. *TWM-Rep. Nr. 89-7*, Inst. f. Konstr. Ing. Bau, Ruhr-Universität Bochum (in German).

Pradlwarter, H., Schuëller, G.I. 1991. Equivalent linearization - an efficient tool to analyze MDOF-systems. *Winter annual meeting, ASME*, Atlanta.
International Conference of Building Officials 1988. *Uniform Building Code* (UBC). Whittier, Cal.

Structural Safety & Reliability, Schuëller, Shinozuka & Yao (eds) © 1994 Balkema, Rotterdam, ISBN 90 5410 357 4

Coupling tall buildings for control of response to wind

Kurtis Gurley & Ahsan Kareem
University of Notre Dame, Ind., USA
Lawrence A. Bergman, Erik A. Johnson & Richard E. Klein
University of Illinois, Urbana, Ill., USA

ABSTRACT: Tall buildings are frequently clustered, and the concept of linking two compatible structures together for response control using one or more passive, semi-active, or active force generating devices has been examined off and on for nearly two decades. However, the problem has not been studied exhaustively, particularly with respect to link-structure interaction and the overall performance of the joined system. Herein, a system consisting of two adjacent and dissimilar buildings, modeled by shear beams, is coupled through a single force link. The structures are placed in a tandem arrangement with the wind approaching parallel to the axis on which they are situated. The structures are represented by their exact open loop transfer functions, and the closed loop transfer function of the system, incorporating the effects of the link, is derived. The spectra of the structural responses are examined, and the mean square response of the primary structure is minimized.

1. INTRODUCTION

Tall buildings are frequently clustered, and the concept of linking two compatible structures together for response control using one or more passive, semi-active or active force generating devices has been suggested. In work that has spanned nearly two decades, Klein and coworkers have studied dissipative links, either viscous elements or skyways, as well as semi-active devices, for example, aerodynamic appendages or a cable providing one-way restraint, to control the response to wind excitation (Klein, Cusano, and Stukel, 1973; Klein and Healey, 1985; Klein, 1990). However, the problem has not been studied exhaustively, particularly with respect to methods of actuation, system modeling, the interaction of the link/controller with the inherent properties of the structure and excitation, and the stability and robustness of the closed loop system.

In this paper, we propose to study a system consisting of two adjacent and dissimilar buildings, modeled by shear beams, coupled through a single force link at height H, shown in Figure 1. The buildings will be exposed to wind induced loads in the alongwind direction.

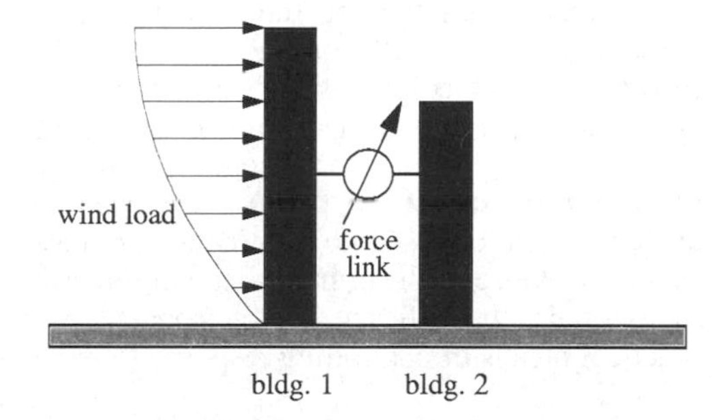

Figure 1. Problem Description

Typically, the complexity of the flow field around an isolated building precludes theoretical evaluation of the loading spectra which are generally established using scale models in boundary layer wind tunnels. In the case of more than one building, the problem of determining aerodynamic loads is further complicated by interference and interaction. These effects can be identified as changes in local pressure fluctuations as well as in static and dynamic components of the loading and can be either adverse or beneficial in nature. Also, in certain cases, buffeting due to impingement of upstream vorticity and wake resonance due to coherent vortical structure in the wake may become important for the downstream building. However, in order to facilitate the study of various controller concepts, a simple building layout is assumed in order to avoid modeling complex flow interferences.

Thus, the buildings are placed in a tandem arrangement with the wind approaching parallel to the axis on which these buildings are located. The alongwind loading is based on quasi-steady and strip theories which lead to analytical expressions for the mode-generalized alongwind spectra (Kareem, 1987). The acrosswind spectrum is based on typical measurements in the wind tunnel and is given in closed form as a function of approach flow conditions (Kareem, 1985). For the initial phase of this study, it will be assumed that no modification to these spectra is required despite the proximity of the two buildings, recognizing that the methodology will remain the same when enhanced spectra for an advanced layout becomes available from wind tunnel tests.

The structures will be modeled using their exact transfer functions, and the exact closed loop transfer function of the system will be derived (Pang, Tsao, and Bergman, 1992), incorporating the effects of the link. The performance of the system will be determined from closed-loop transfer functions as well as from mean square responses. Parametric analyses will be used to minimize the mean square response of the primary structure in order to ascertain the efficacy of the concept as a practical means of controlling response to wind.

2. SYSTEM ANALYSIS

The building layout for this analysis is shown in Figure 2 and consists of two dissimilar buildings joined by a damped elastic link. The following system properties are assumed.

1. Each structure is periodic.
2. Floors are rigid, leading to a deformation pattern which is a shearing type.
3. Each structure is modeled as a shear beam with fixed-free boundary conditions.
4. The properties of each shear beam are given by (Clough and Penzien, 1975)

$$\frac{12\,(EI)}{h^2} \approx GA \qquad (1)$$

$$\frac{m}{h} \approx \rho A \qquad (2)$$

5. Forces $P_n(t)$ act at height $X_n = nh$ on building 1. No external forces act on building 2.
6. The coupling acts at height $X_m = mh$.

The equations of motion of the system are

$$\begin{aligned} G_1A_1w''_1(x,t) - \rho_1A_1\ddot{w}_1(x,t) &= P_n(t)\,\delta(x-x_n) + \\ \{K[w_2(x_m,t) - w_1(x_m,t)] &+ \\ C[\dot{w}_2(x_m,t) - \dot{w}_1(x_m,t)]\,\}\,&\delta(x-x_m) \end{aligned} \qquad (3)$$

and

$$\begin{aligned} G_2A_2w''_2(x,t) - \rho_2A_2\ddot{w}_2(x,t) &= \{K[w_1(x_m,t) - w_2(x_m,t)] + \\ C[\dot{w}_1(x_m,t) - \dot{w}_2(x_m,t)]\,\}\,&\delta(x-x_m) \end{aligned} \qquad (4)$$

We seek the steady-state response and wish to compute the transfer function between input force and building response. Fourier transforming the equations of motion gives

$$\begin{aligned} W''_1(x,\omega) + \frac{\rho_1}{G_1}\omega^2W_1(x,\omega) &= \frac{1}{G_1A_1} \\ \{P_n(\omega)\,\delta(x-x_n) &+ (K+j\omega C) \\ [W_2(x_m,\omega) - W_1(x_m,\omega)]\,&\delta(x-x_m)\,\} \end{aligned} \qquad (5)$$

and

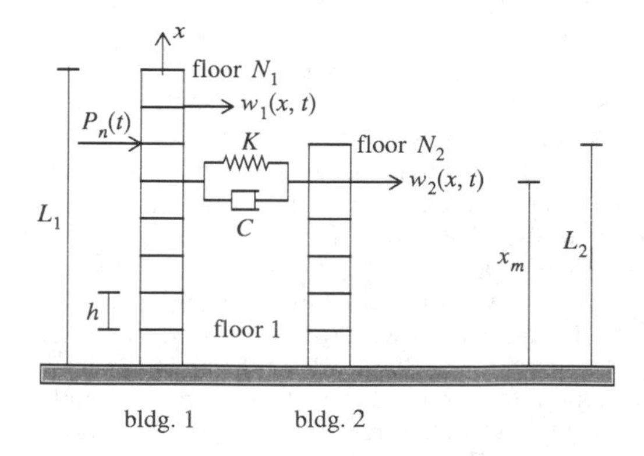

Figure 2. Building Layout

$$W''_2(x,\omega) + \frac{\rho_2}{G_2}\omega^2 W_2(x,\omega) = -\frac{1}{G_2 A_2} \quad (6)$$

$$(K+j\omega C)\,[W_2(x_m,\omega) - W_1(x_m,\omega)]\,\delta(x-x_m)$$

In order to simplify these expressions, examine the equation

$$g_i''(x,\xi;\alpha_i) + \alpha_i^2 g_i(x,\xi;\alpha_i) = \frac{\delta(x-\xi)}{G_i A_i} \quad (7)$$

where

$$\alpha_i^2 = \frac{\rho_i}{G_i}\omega^2 \quad (8)$$

subject to fixed-free boundary conditions. The solution is the Green's function $g_i(x,\xi;\alpha_i)$ which is given later. Then, the solution of each equation can be written as

$$W_1(x,\omega) = P_n(\omega) g_1(x,x_n;\alpha_1) + (K+j\omega C)\,[W_2(x_m,\omega) - W_1(x_m,\omega)]\,g_1(x,x_m;\alpha_1)$$

$$(9)$$

$$W_2(x,\omega) = -(K+j\omega C)\,[W_2(x_m,\omega) - W_1(x_m,\omega)]\,g_2(x,x_m;\alpha_2)$$

Evaluating the system at $x = x_m$ and letting $K+j\omega C = K^*$ and $g_i(x_m,x_m;\alpha_i) = \bar{g}_i(\omega)$ gives

$$\begin{bmatrix} 1+K^*\bar{g}_1(\omega) & -K^*\bar{g}_1(\omega) \\ -K^*\bar{g}_2(\omega) & 1+K^*\bar{g}_2(\omega) \end{bmatrix} \begin{Bmatrix} W_1(x_m,\omega) \\ W_2(x_m,\omega) \end{Bmatrix} = \begin{Bmatrix} P_n(\omega) g_1(x_m,x_n;\alpha_1) \\ 0 \end{Bmatrix} \quad (10)$$

Inversion leads to

$$W_1(x_m,\omega) = \frac{[1+K^*\bar{g}_2(\omega)]\,P_n(\omega) g_1(x_m,x_n;\alpha_1)}{1+K^*[\bar{g}_1(\omega)+\bar{g}_2(\omega)]}$$

$$(11)$$

$$W_2(x_m,\omega) = \frac{K^*\bar{g}_2(\omega) P_n(\omega) g_1(x_m,x_n;\alpha_1)}{1+K^*[\bar{g}_1(\omega)+\bar{g}_2(\omega)]}$$

Substitution into equations (9) gives

$$W_1(x,\omega) = P_n(\omega)\left[g_1(x,x_n;\alpha_1) + \frac{K^* g_1(x,x_m;\alpha_1) g_1(x_m,x_n;\alpha_1)}{1+K^*[\bar{g}_1(\omega)+\bar{g}_2(\omega)]}\right]$$

$$(12)$$

$$W_2(x,\omega) = -P_n(\omega)\frac{K^* g_2(x,x_m;\alpha_2) g_1(x_m,x_n;\alpha_1)}{1+K^*[\bar{g}_1(\omega)+\bar{g}_2(\omega)]}$$

Thus, the transfer function between the displacement of building 1 at floor k to the input force of floor n is

$$H_{kn}^{(1)}(\omega) = \frac{W_1(x_k,\omega)}{P_n(\omega)} = g_1(x_k,x_n;\alpha_1) + \frac{K^* g_1(x_k,x_m;\alpha_1) g_1(x_m,x_n;\alpha_1)}{1+K^*[\bar{g}_1(\omega)+\bar{g}_2(\omega)]} \quad (13)$$

and for building 2 the displacement at floor k due to the input force on building 1 at floor n is

$$H_{kn}^{(2)}(\omega) = \frac{W_2(x_k,\omega)}{P_n(\omega)} = -\frac{K^* g_2(x_k,x_m;\alpha_2) g_1(x_m,x_n;\alpha_1)}{1+K^*[\bar{g}_1(\omega)+\bar{g}_2(\omega)]} \quad (14)$$

The input/output relations can also be expressed as

$$W_1(x_i,\omega) = \sum_{j=1}^{N_1} H_{ij}^{(1)}(\omega)\,P_j(\omega) \quad (i = 1,\ldots,N_1)$$

$$(15)$$

$$W_2(x_i,\omega) = \sum_{j=1}^{N_1} H_{ij}^{(2)}(\omega)\,P_j(\omega) \quad (i = 1,\ldots,N_2)$$

Rewriting in matrix form, multiplying by the respective complex conjugates, and taking expectations gives

$$[S_{W_1}(\omega)] = [H^{(1)}(\omega)]\,[S_p(\omega)]\,[H^{(1)}(\omega)]^{*T}$$

$$(16)$$

$$[S_{W_2}(\omega)] = [H^{(2)}(\omega)]\,[S_p(\omega)]\,[H^{(2)}(\omega)]^{*T}$$

Here $S_{W_1}(\omega)$ and $S_{W_2}(\omega)$ are the response power spectral density matrices for buildings 1 and 2, and $S_p(\omega)$ is the wind force power spectral density matrix. Modeling the wind force distributed along the face of building 1 as point loads (Figure 2), the building face is divided into i segments where i is the number of levels, and a wind point load is applied at each level. The wind force spectral matrix is expressed as

$$[S_p(\omega)] = \qquad (17)$$

$$\left[(\rho A_i C_{D_i} \bar{U}_i)^2 S_u(\omega)\, J_F(\omega)\, \mathrm{coh}(\Delta y, \Delta z, f) \right]$$

Here ρ is the air density, C_{D_i} is the ith component drag coefficient, $\bar{U}_i$ is the mean wind velocity at the center of the ith building face segment, and A_i is the area of the ith building face segment. $S_u(\omega)$ is the wind velocity spectrum, and $J_F(\omega)$ is the aerodynamic admittance function which accounts for the partial correlation of wind velocity fluctuation over the ith segment of the structure. The coherence function

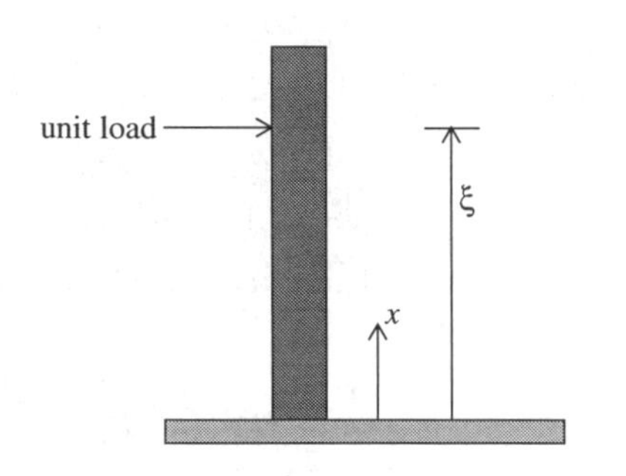

Figure 3. Response at x to point load at ξ.

$\mathrm{coh}(\Delta y, \Delta z, f)$ accounts for partial wind correlation between different segments and results in the off-diagonal terms in the load matrix. The closed form expression for $J_F(\omega)$ is given as

$$J_F(\omega) = \frac{4}{\varepsilon_y^2 \varepsilon_z^2}\left[e^{-\varepsilon_y} + \varepsilon_y - 1\right]\left[e^{-\varepsilon_z} + \varepsilon_z - 1\right] \qquad (18)$$

where

$$\varepsilon_y = \omega\theta C_h W/\bar{U},\ \varepsilon_z = \omega\theta C_v D/\bar{U} \qquad (19)$$

$$\theta = \sqrt{\frac{1+r^2}{1+r}},\ r = \frac{C_h W}{C_v D} \qquad (20)$$

W and D are the width and height of the building face, respectively. The coherence function given by Davenport, 1961 is

$$\mathrm{coh}(\Delta y, \Delta z, f) = \qquad (21)$$

$$\exp\left\{\frac{-f\sqrt{[C_v \Delta z]^2 + [C_h \Delta y]^2}}{\bar{U}_{12}}\right\}$$

where C_h and C_v are experimentally determined decay constants which usually vary between 10 and 16, Δy and Δz are separation distances in the y and z directions, and $\bar{U}_{12}$ is the average velocity between the two locations (Kareem,

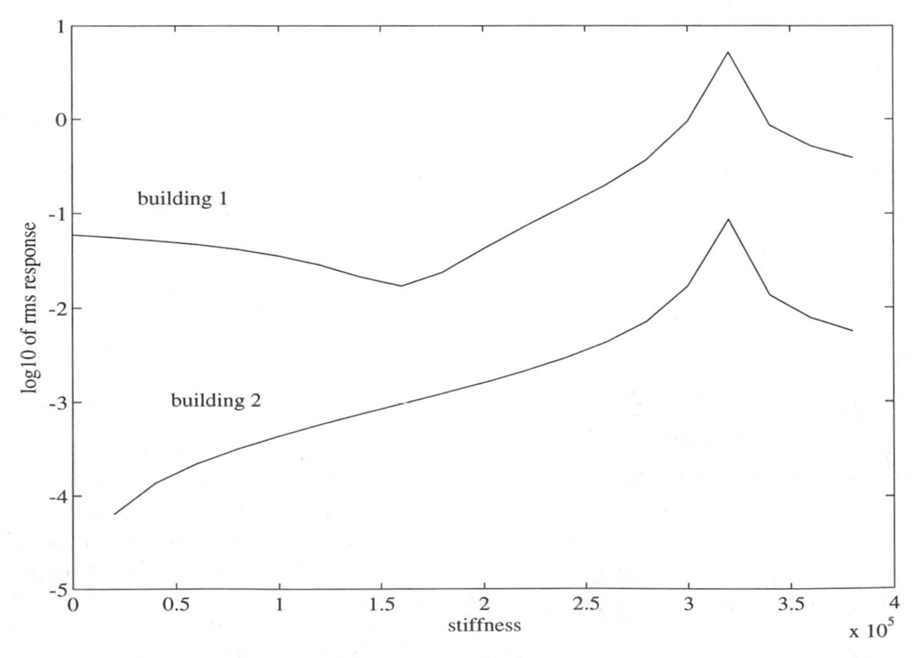

Figure 4. R.m.s. Displacement of Top Level of Both Buildings vs. Stiffness, C=0 (dissimilar buildings)

1987). The wind velocity spectrum $S_u(\omega)$ has been described in many previous works (*e.g.*, Davenport, 1961 and Harris, 1971). The Harris spectrum is used in the results section of this paper.

3. DYNAMIC GREEN'S FUNCTION

In order to determine the response spectral densities of the composite structure, the Green's functions $g_1(x, \xi;\alpha_1)$ and $g_2(x, \xi;\alpha_2)$ are needed for the computation of the transfer functions $H^{(1)}(\omega)$ and $H^{(2)}(\omega)$. The dynamic Green's function for the fixed-free shear beam $g(x, \xi;\alpha)$ is the response at x to a point load at ξ (Figure 3) and has been previously derived (see, for example, Marek and Bergman, 1985). It is given by

$$g(x, \xi;\alpha) = -\begin{cases} \dfrac{\sin\alpha x \cos\alpha(L-\xi)}{GA\alpha\cos\alpha L}, & 0 \le x \le \xi \\ \dfrac{\sin\alpha\xi \cos\alpha(L-x)}{GA\alpha\cos\alpha L}, & \xi \le x \le L \end{cases} \tag{22}$$

4. RESULTS AND CONCLUSIONS

This study considers the connection of both dissimilar and similar buildings. For the former, a fifty story building (building 1) is connected to a forty story building (building 2), with the wind acting directly on building 1. Both buildings are divided into one level every ten stories. The bending stiffness, lumped mass per level, and level heights, identical for both structures, are given as

$$\frac{12EI}{h^3} = 1.28\times10^8\ \frac{\text{N}}{\text{m}}, \tag{23}$$

$$m = 6.567\times10^6\ \text{kg},\quad h = 36.576\ \text{m}. \tag{24}$$

The objective of this study is the minimization of the root mean square displacement of building 1. The energy distribution of the wind spectrum results in a building response almost entirely in the first mode. The reduction of displacement at the top level is therefore the goal of the study. The response of building 1 to the wind spectrum independent of building 2 is first computed. This results in an rms deflection of 0.059 meters at the top level.

The two buildings are linked at the third level.

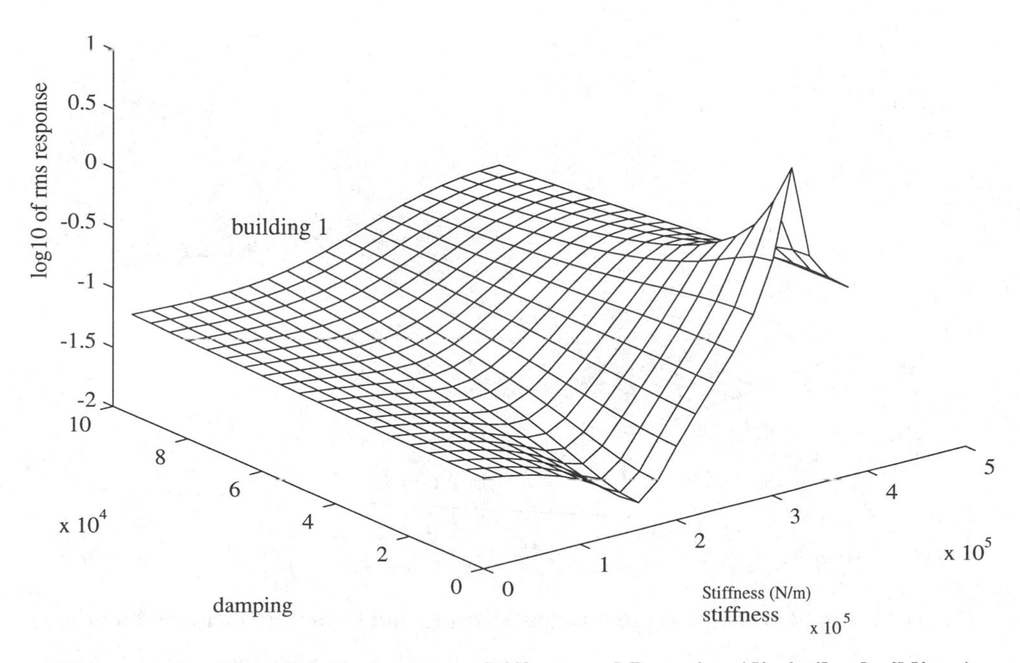

Figure 5. Top Floor Displacement vs. Stiffness and Damping (dissimilar buildings)

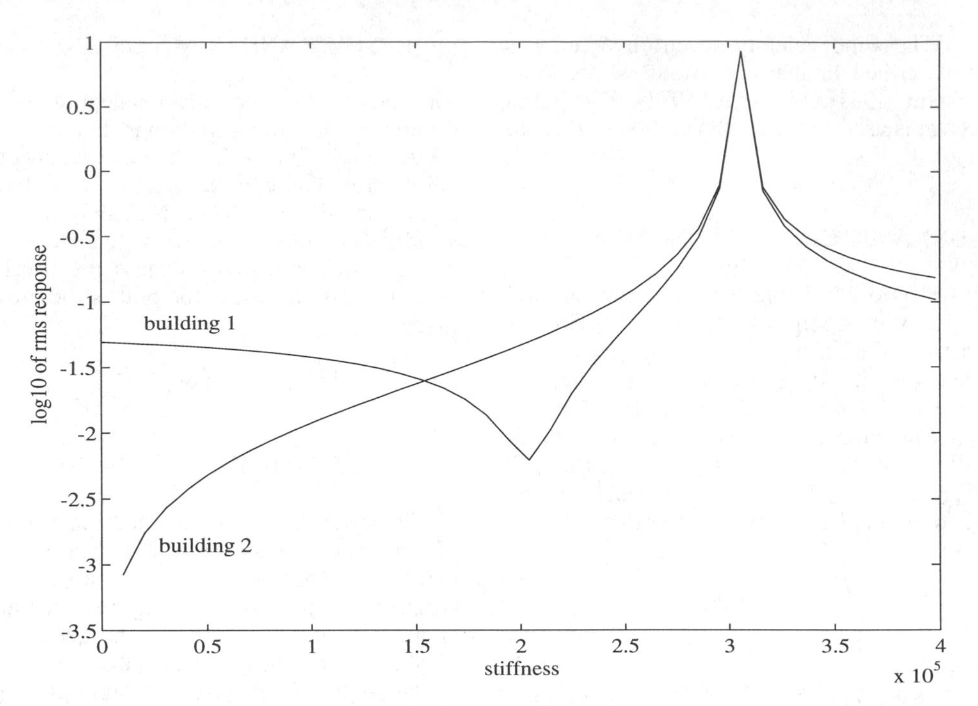

Figure 6. R.m.s. Displacement of Top Level of Both Buildings versus Stiffness, C=0 (identical buildings)

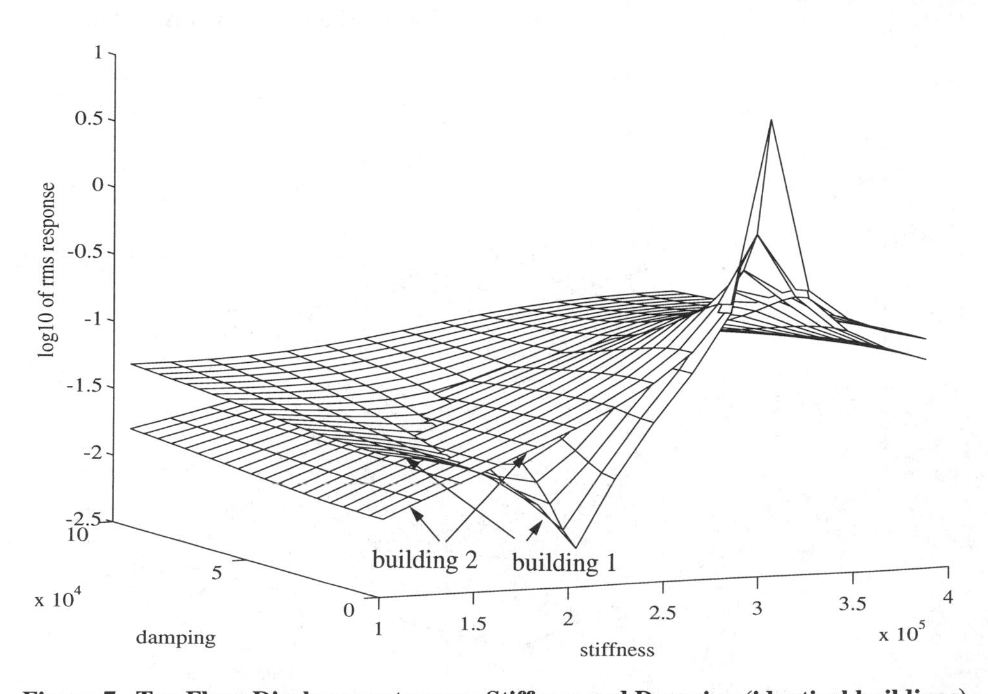

Figure 7. Top Floor Displacement versus Stiffness and Damping (identical buildings)

With link damping held at zero, the stiffness which results in minimum rms response is found to be 160,000 N/m. Figure 4 shows the rms response of the top level of both buildings versus stiffness with damping held to zero. The rms displacement at the top of building 1 is reduced to 0.017 meters, and the response of building 2 is negligible.

When damping is varied while holding stiffness to zero, it was found that damping alone does not mitigate the response of building 1 but, rather, increases it until a plateau is reached where the damper connection with building 2 is essentially rigid.

Damping in the coupling by itself does not decrease the response of building 1, while stiffness in the coupling reduces response for a range of stiffness values. The question to be resolved is whether damping in combination with stiffness can further reduce response. Stiffness is iterated about the ideal stiffness for zero damping. At each stiffness value the damping is iterated. A three dimensional plot of the $\log_{10}$ of rms response at the top level versus stiffness and damping is shown in Figure 5. As can be seen, the addition of damping in the connector does not reduce response. In fact, the more damping, the higher the response. On the positive side, the peak in Figure 4, which indicates a large increase in building 1 response as compared to that of the uncoupled building 1, is attenuated by the addition of damping.

Building 1 from the previous example is now linked to an identical 50 story building. The connection is at the second floor. Figure 6 shows the top level rms displacement of both buildings. The use of identical buildings adds a dimension to the optimization of building response. As seen in Figure 6, the response of building 2 is now of the same order of magnitude as that of building 1 and needs to be considered in the optimization. Choosing the minimum response of building 1 results in a building 2 response larger than that of the uncoupled building. The optimum response occurs when the two building responses are equal, at $k \cong 155000$ N/m. The resulting top level rms response is 0.025 m.

The location of the connector was varied in both examples. The effect of lowering the connector is to increase the optimum stiffness. Raising the connector level lowers the ideal stiffness. The inability of damping to improve the response is not altered.

The conclusion of this study, then, is that the response of an uncoupled building may be improved by coupling it by a spring connection with a second building. The addition of damping hinders the response reduction, so the connection should be designed for minimum damping. The ideal link stiffness is a function of the characteristics of both buildings as well as the location of the link. When linking two dissimilar buildings, the response of the forced building is optimized and the response of the second building is orders of magnitude smaller. When linking two identical buildings, the response of the second building will approach that of the forced building and should be weighted accordingly in the optimization.

Clearly, the reduction of rms displacement from 0.059 meters to 0.0171 and 0.025 meters represents a significant improvement in performance and provides sufficient impetus to continue further development in this area. Several concepts, including semi-active actuators which have variable stiffness and damping characteristics, are currently under investigation by the authors.

5. ACKNOWLEDGMENTS

The support for this study was provided in part by NSF grant BCS-9096274.

6. REFERENCES

Clough, R. N. and J. Penzien 1975. Dynamics of Structures, McGraw Hill, New York.

Davenport, A.G. 1961. The spectrum of horizontal gustiness near the ground in high winds. Journal of the Royal Meteorological Society 87:194-211.

Harris, R.I. 1971. The nature of wind. The Modern Design of Wind Sensitive Structures, Construction Industry Research and Information Association, London.

Kareem, A. 1985. Lateral-torsional motion of tall buildings to wind loads. Journal of the Structural Division (ASCE) 111:2479-2496.

Kareem, A. 1987. Wind effects on structures: a probabilistic viewpoint. Probabilistic Engineering Mechanics 2:166-200.

Klein, R.E., C. Cusano, and J.J. Stukel 1973. Investigation of a method to stabilize wind induced oscillations in large structures. Proc.

Winter Annual Meeting of the ASME, New York.

Klein, R.E. and M.D. Healey 1985. Semi-active control of wind induced oscillations in structures. Proc. Second International Conference on Structural Control, University of Waterloo, Ontario, Canada.

Klein, R.E. 1990. Teaching linear systems theory using Cramer's rule. IEEE Transactions on Education 11:258-267.

Marek, E.L., L.A. Bergman, and J.W. Nicholson 1985. Solution of the nonlinear eigenproblem for the free vibration of linear combined dynamical systems. Technical Report UILU-ENG-85-0507, University of Illinois, Urbana, Illinois.

Pang, S.T., T-C. Tsao, and L.A. Bergman 1992. Active and passive damping of Euler-Bernoulli beams and their interactions. Proc. American Control Conference, Chicago: 2144-2149. Accepted for publication in the Journal of Dynamics, Measurement, and Control (ASME).

Structural Safety & Reliability, Schuëller, Shinozuka & Yao (eds) © 1994 Balkema, Rotterdam, ISBN 90 5410 357 4

Active base isolation of earthquake loaded structures by means of adaptive controlled columns

Gerhard H. Hirsch
RWTH Aachen, University of Technology, Institut für Leichtbau, Germany
Maria R. Józsa
KABE Engineering, Science Division, Aachen (RWTH), Germany

ABSTRACT: The paper opens with a comparison of several methods of active control concepts from engineering point of view. Consequently it introduces intelligent columns, adaptively controlling substructures in a sense of base isolation. Concerning the verification of this concept relating to hardware and software a collaborative research project between Aachen University and the University of Southern California is proposed. Here a wide-band spectrum of different loading tests will be prepared. The program schedule consisting of the experimental model, analytical and experimental studies is presented in detail.

1. INTRODUCTION

As revealed in /1/ the actual seismic strain of a structure results from the inertia force arising from base acceleration. Therefore the first law concerning earthquake resistant design of structures should be the development of lightweight structures.

In general engineers regard earthquake excitation as a nonstationary random process when calculating structure reactions. From the on-line predictive control point of view, however, one has to pay attention to the time history of response which means treating earthquake excitation as a transient event.

The concept of active control has been introduced in recent years in order to reduce the seismic response of civil engineering structures. In general most investigations dealt with the development of control theory and algorithms.

The only exististing full scale examples described in literature can be found in Japan (partially in cooperation with US-scientists /2/)

Within the scope of this paper a possibility of active earthquake control is reported. In this new concept the filtering of seismic time history resulting in a sine-beat is taken into consideration. This fact allows a predictive control of 1-2 filtered modes of interest. From an economic point of view only the substructural movement relative to the main structure is to be reduced. Finally a test model for universal verification of the new method in combination with other concepts will be described.

2. REMARKS ON SOME ACTIVE CONTROL CONCEPTS

2.1 Fundamental remarks

Active feedback control of structures has been used frequently for aircraft (gust response reduction and flutter suppression) and spacecraft, but active control of large and more or less massive civil engineering structures is a relative new technique. In structural engineering, one of the constant challenges is to find new and better means of protecting structures from the damaging effects of destructive enviromental forces. One way open to the researchers and designers is to introduce more conservative designs so that structures such as buildings, towers and bridges are better able to cope with large external loads. This approach, however, can be untenable both technologically and economically.

Another possible approach is to make structures behave more like machines, aircrafts, or human beings in the sense, that they are made adaptive or responsive to external forces. Structural muscles, so to speak, can be flexed when warranted, or appropriate adjustments can be made within the structure as enviromental conditions change. This latter approach has led to active structural control research and has opened up a new field of investigations, which began more as an intellectual curiosity in the early 1970s, but now is at the stage where large-scale experimentation is underway /2, 3/.

2.2 Active control devices

Most common control concepts recently proposed are active tendon and bracing control /2/, active base isolation /4/, active pulse control /5/ and active mass driver /3/, as shown in Fig. 1 - 6.

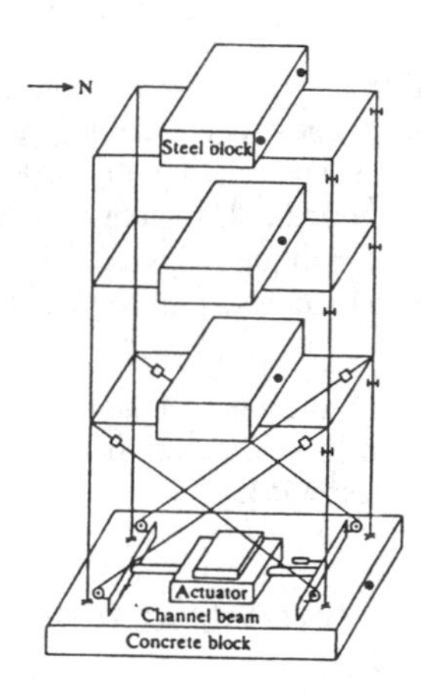

Figure 1: Active tendon control

Active control using structural tendons, proposed as early as 1960 has been one of the most studied mechanisms both on paper and in the laboratory. However, tendons cannot be regarded as stiff actuators /6/. Their dynamic behaviour requires further research work as well as the behaviour of active bracing systems.

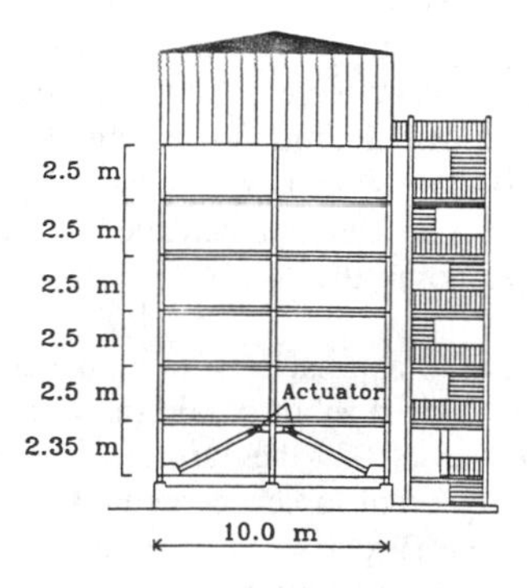

Figure 2: Active bracing control

This concept of active structural control includes as well the philosophy of active tendon control as of control with intelligent columns, the scope of this paper.

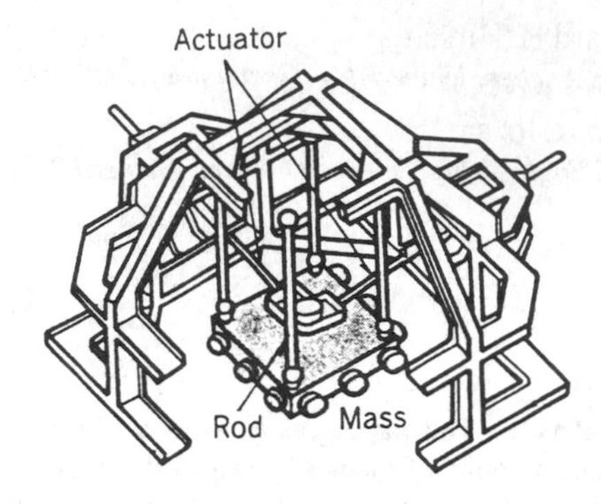

Figure 3: Active mass driver

Kobori reported on the first active mass driver ever verified. The advantage in comparison to passive control is evident because a lower mass is needed. Nevertheless the energy required to move it is high. Therefore Kobori developed a proposal relating to a composite TMD /7/. Figure 4 shows the system, consisting of a prime structure (generalized values of mass m, stiffness k and structural damping c, the passive TMD (m_d, k_d and c_d) and the passive and active coupled satelite system (m_s, k_s, c_s, and u(t)). The exiting force is $f_w(t)$.

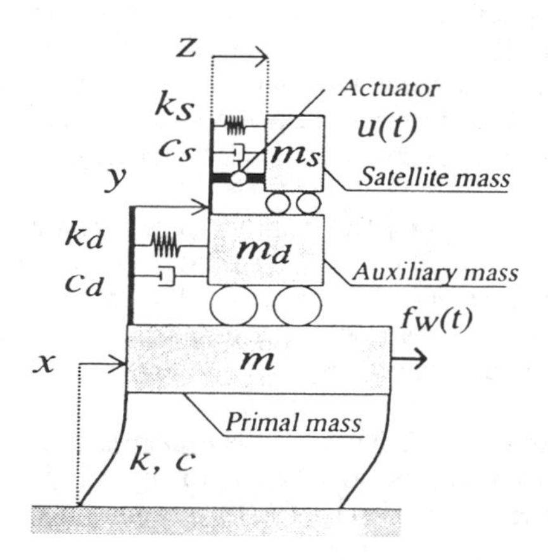

Figure 4: Composite TMD

The size of the active controller is significantly reduced if compared with other conventional active tuned mass damper devices which is the key factor for installing the actual device into large civil engineering structures such as high rise buildings.

The necessary control force and power for activating the device are significantly small as compared with the expected vibration control performance.

Harmonic and random excitation tests were successfully conducted to verify the control effect of this control method. In our opinion this concept appears promising for an effective semi-active reduction of structure response by means of phase shifting in the case of transient loading.

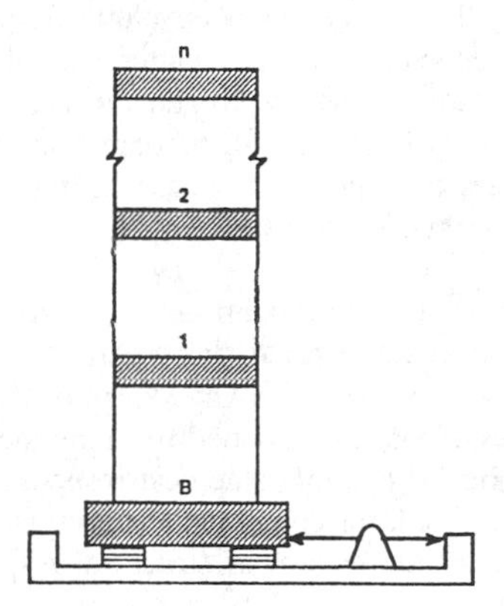

Figure 5: Active base isolation

Base isolation is usually proposed according to Yang. Here the big fundamental mass and its large inertia force are included in the active control. That´s why control forces compared to those from other control mechanisms increase inadequatley. In contrast column control as proposed herewith avoids this disadvantage by isolating only substructures of interest.

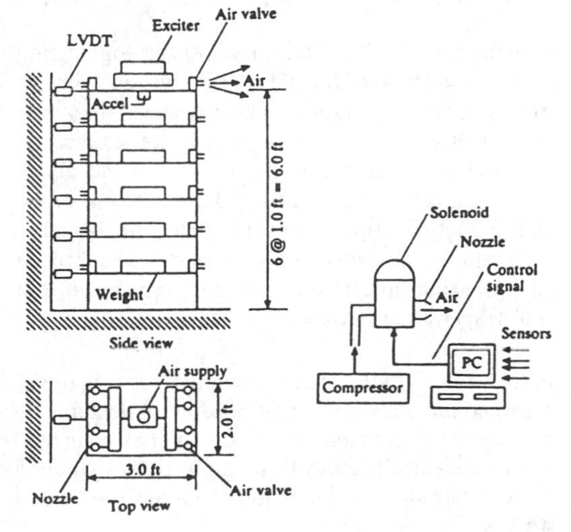

Figure 6: Active pulse control, principle of thruster

In 1983 Masri developed and produced two pulse genererating systems based on pressurized gas to a state of operational readiness suited for earthquake testing of small buildings, large industrial equipment and frame structures. An efficient pulse train algorithm was developed for use in programming the gas pulse system for motion simulation. A simple anti-earthquake algorithm was also developed. When used in conjunction with various types of pulse units it showed promise in the reduction of structural motions at the damage or life threatening thresholds. Nevertheless in its form the anti-earthquake algorithm is limited to preselected fixed amplitude pulses so that further development is needed to improve it

In conclusion all active control concepts require high control forces and thus energy if applied to control whole large civil structures. Therefore control strategy should move towards the reduction of the relative movement of substructures as part of earthquake-excited main structures. Movement of the primal structure might be acceptable.

With this background and based on our own experiences the use of intelligent columns for base isolation of arbitrary substructures is proposed and outlined in the following.

3. CONTROL STRATEGY

3.1 New supporting tools

Generally, structures can be represented by distributed-parameter systems. When they are spacially discretized, the discrete models tend to have a large number of degrees of freedom. However, it is recognized that vibrations of most tall buildings, towers etc. under transient excitation are dominated only by a few low vibrational modes which can be described numerically with sufficient accuracy. Then the vibration of the controlled structure can be described by a system of coupled n second-order differential equations having the matrix form (like single degree of freedom system). In the case of earthquakes, the transient exciting forces induce the relative structure movement (inertia force = mass · base acceleration)

The basic matrix equation of motion of the controlled structural system can be written as

$$M \cdot \ddot{x}(t) + C \cdot \dot{x}(t) + K \cdot x(t) = D \cdot u(t) + E \cdot f(t) \qquad (1)$$

x(t) is the n-dimensional displacement vector, f(t) is an r-vector representing applied load or external excitation and u(t) is the m-dimensional control force vector. The n x m matrix D and n x r matrix E are location matrices which define locations of the control force and the excitation, respectively /2/.

In the development of an active structural concept, based on this equation, one of the first tasks at hand is to develop suitable control laws. The adaptive control moreover can be used in relation to system identification and monitoring the dynamic behaviour of the controlled structure.

Time delay and the constraint of feedback amplitude are the most important factors that affect the efficiency of controllers. Some more important facts in this field of engineering are: modelling errors and spill-over effects (coupling of different vibration modes by control forces, like the non-proportional damping in structures), structural nonlinearities, uncertainties in structural parameters, limited number of sensors and controllers, discrete time control, reliability.

The theory is not developed here because many approaches and detailed treatments can be observed from actual literature /2, 4, 8, 9/. With their help and being supported by advanced available control and mathematics software /10, 11/, quick estimation of an existing system concerning stability and control parameter variation and optimization is possible.

3.2 Control concept

Fig. 7 shows the response of a building + typical earthquake input.

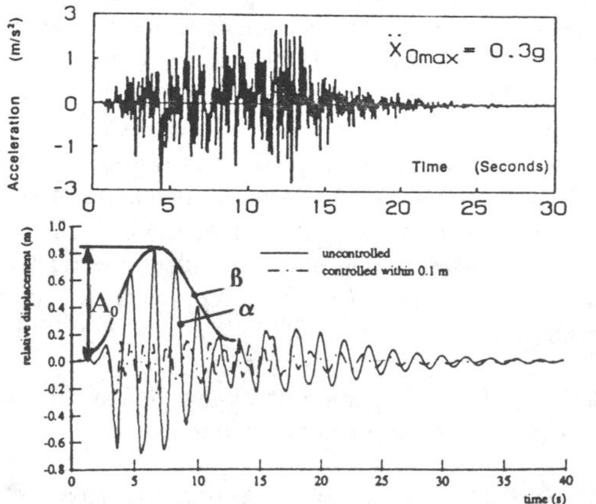

Figure 7: Typical earthquake input and response of structure

The structure acts as a frequency filter, responding with its natural frequency to the corresponding frequency in the earthquake spectrum approximately in the form of a sine beat, exactly described with equation (2).

$$F = A_0 \sin \alpha t \quad \sin \beta t \qquad (2)$$

where A_0 = (unknown) maximum response, α = (known from system identification) system´s natural frequency, ß = (unknown) envelope frequency of response, depending on the endurance of seismic input of natural frequency. In general, eqation (2) is of interest only in the case of fundamental vibration. Nevertheless of course a response in higher modes can be described in the same way.

The above mentioned circumstances open the possibility to realize active predictive control of structures /2/ in a new way: With the knowledge of the natural frequencies of interest obtained from measurement a large number of possible sine beat responses can be stored in the control computer. During registration of the actual time history of structural response the best corresponding sine beat can be used for optimum control in a predictive way.

For verification of the described control strategy research work is necessary as introduced in the following chapter.

4. RESEARCH PROPOSAL

4.1 Interdisciplinary work

Experiences /12, 13/ taken in aerospace engineering (Fig. 8 and 9) were transferred to civil engineering structures to develop this control strategy. It´s base consists of an interface decoupling between substructure and main structure. In contrast to the above mentioned active base isolation the movement of the main structure remains uncontrolled. On the other hand relative movement between main- and substructure (dynamic magnification factor) is reduced to a non-dangerous stress level.

In the case of civil engineering structures it is possible to compensate horizontal bending deflections by vertical force components in the column elements. The piezoceramic actuators are replaced by hydraulic devices because of the required larger forces and strokes.

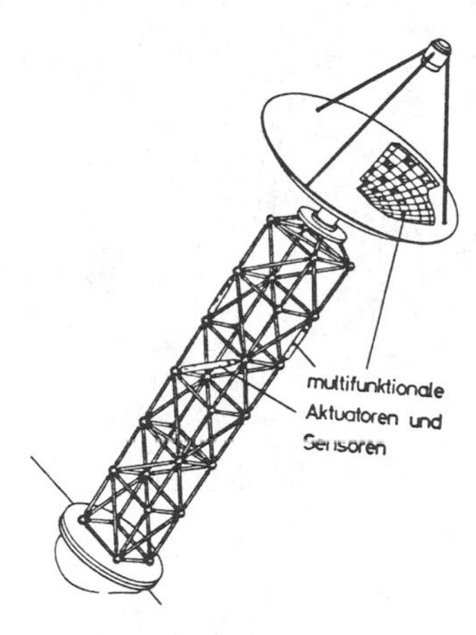

Figure 8: Space structure model

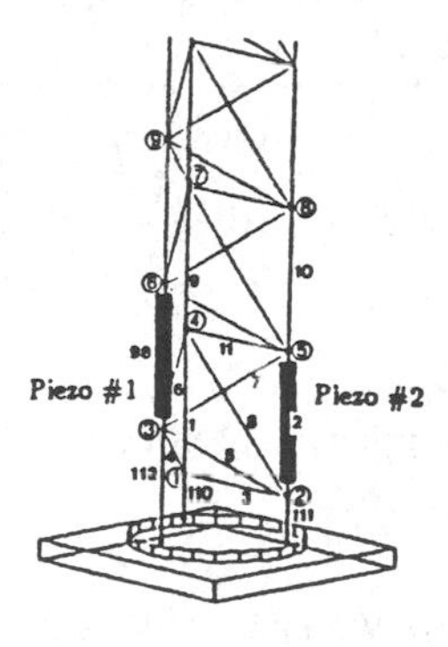

Figure 9: Model with interface decoupling /12/

4.2 Test procedure

As a conclusive concept of all experiences, a full scale test model was developed. It will be used to make experiences and improvements with several active and passive control concepts, alone or in combination. It is the basis of a research cooperation between the University of Aachen and the University of Southern California.

The following tasks apply to each of the proposed actuator classes: gas thrusters, active hydraulic columns, semi-active nonlinear auxiliary mass dampers, etc.

1. Experimental Model

   Design/Construction of Test Fixture
   Instrumentation Setup
   Calibration of Instrumentation
   Actuator Design/Specifications
   Data Acquisition Network Installation
   Microprocessor/Controller Calibration Tests

2. Analytical Studies

   Control Energy Requirements
   Instrumentation Performance Bounds
   Optimum Controller Locations
   Optimum Sensor Locations
   Stability Considerations
   Communication Network Specifications

3. Experimental Studies

   Component Design/Construction/Calibration
   Component Tests
   Active Control Experiments
   Performance Characteristics of Different Approaches
   Quantify Hardware Problems and Limitations
   Correlate Analytical/Experimental Results
   Hybrid Control Experiments
   Determine Technology Impediments

Active, semi-active and passive control items are:

- linear and nonlinear behaviour of seismic loaded structures
- passive control by means of base isolation (linear and nonlinear) and tuned mass dampers (KABE)
- active control with hydraulic actuators and pulse control by means of pneumatic nozzles
- hybrid control observed by any possible combination of active and passive devices

Figure 10 shows the test set-up, consisting of a cantilever steel tube (R) of 5m height (i), 6mm wall thickness (d), 700mm diameter (e), fitted to the test platform of a shaking table (planned on MAVIS I, DLR Jülich /10/) An additional mass at the tube top decreases natural frequency from 28 Hz to 14 Hz (A). An elastic support has the same function, decreasing natural frequency to approximately 2 Hz (C). These measures are necessary for obtaining realistic structural frequencies. Four intelligent columns (D) composed by a hydraulic actuator (I) and a cover tube (column) (II) supplied with a sensor (III) are installed circumferentially around the base of the structure. On top we find a TMD /5/ (B) for

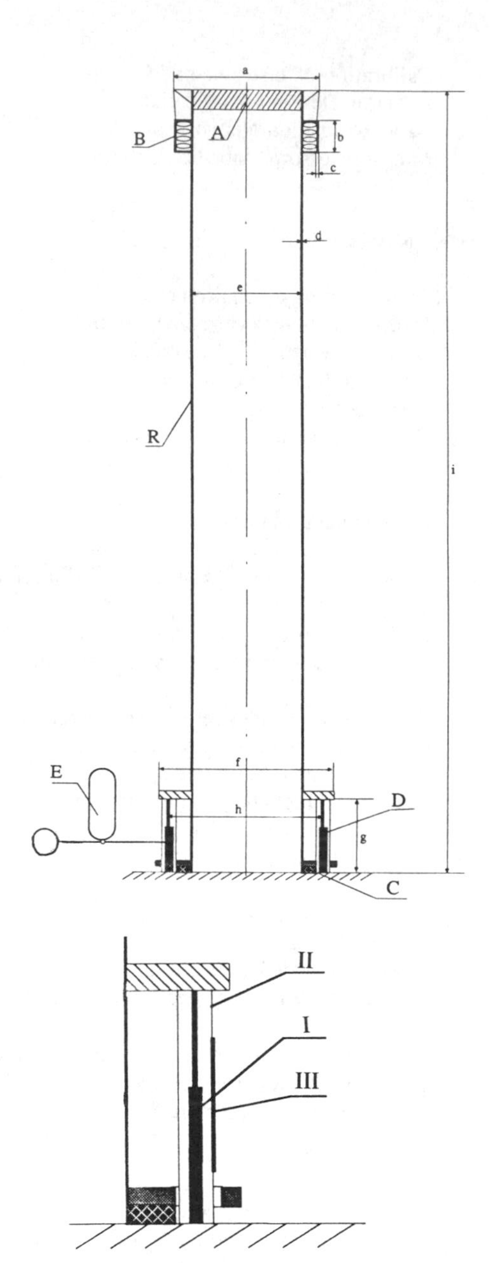

Figure 10: Test Setup

passive and semi-active control research work. A pneumatic accumulator (E) allows immediate counteraction at the beginning of an earthquake excitation.

Figure 11 shows how the intelligent columns work against an arbitrary sinusoidal moment M: Columns A and B build up an alternating countermoment with

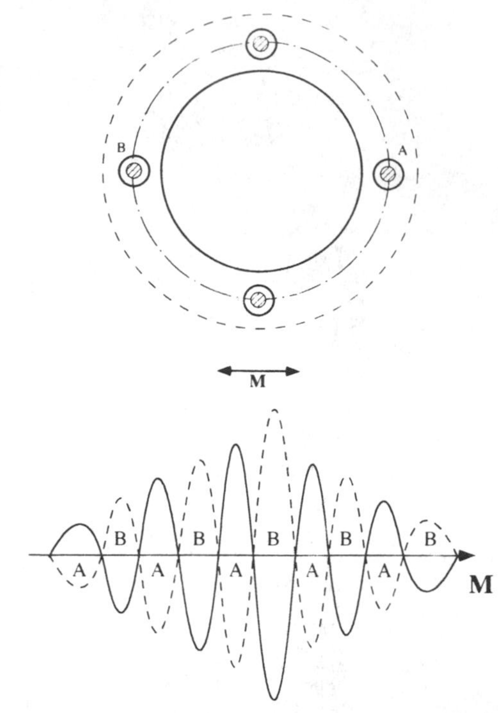

Figure 11: Arrangement of intelligent columns and principle of effect

one column always working in the same direction (pressing).

The main purpose of all tests is the feasibility verification for implementation of the most important control hardware in original structures, economic aspects included. Moreover one goal is the improvement of todays available optimum control algorithms.

The book "Instrumentation and Control" edited from Chester L. Nachtigal /14/ serves as an extremly helpful guide in the realization of the project

5. CONCLUDING REMARKS

Unfortunately the realization of active control in civil engineering structures and the full use of their remarkable advantages is not yet possible owing to the actual lack of official national regulations. Therefore this paper has briefly described the results of our work in structural control research and has outlined the need for international cooperation in implementing the valuable recent findings in a fruitful way. The collaborative activities between USC and Aachen University will demonstrate an example of this.

6. ACKNOWLEDGEMENTS

The authors would like to thank Prof. Sami F. Masri (USC, Los Angeles, California) for taking part in the preparation of cooperative research-work. Furthermore the authors had encouraging and motivating discussions with Prof. Claus (VPI, Blacksburg, Virginia), Prof. Soong and Prof. Reinhorn (SUNY, Buffalo, New York). Also Dr. A. Kleine-Tebbe (DLR) has to be gratefully acknowledged for proposing a test procedure on the shaking table MAVIS 1.

7. REFERENCES

/1/ Hirsch, G.; Kleine-Tebbe, A.; Winkler, A.: "Safety Improvement of Earthquake Loaded Structures..." Proceedings of ICOSSAR ´89, ASCE, 1990, pp 733 - 740

/2/ Soong, T.T.: Active structural control, Theory and Practice, Longman Scientific & Technical, 1990

/3/ Kobori, T.: State-of-the-Art of seismic response control, Structural Control Research, 1990, USC-Publ. No: CE-9013, p1-p21

/4/ Yang, J. N. et al.: Optimal Hybrid Control of seismic-excited nonlinear and inelastic Structures.,Intelligent Structures-2, Monitoring and Control, Elsevier Applied Science, 1992, pp. 293 - 307

/5/ Soong, T. T.: State-of-the-art of structural control in USA, Structural Control Research, 1990, USC-Publ. No: CE-9013, p1-p21

/6/ Józsa, Maria: Optimum Hybrid Control of Transient Loaded Structures... (in German), Master Degree Thesis, RWTH Aachen, 1992

/7/ Nishimura, I. et.al.: An Intelligent Tuned Mass Damper, AIAA-93-1709-CP, 1993, 3561-3569

/8/ Ogata, K.: Modern Control Engineering, Prentice Hall, Englewood Cliffs, New Jersey, 1990

/9/ Dorf, R.C.: Modern Control Systems (Disk included), 6th Edition, Addison-Wesley Publ. Co.Inc., 1992

/10/ MATLAB, The Student Edition (Disk included), Prentice Hall, Englewood Cliffs, NJ, 1992

/11/ Anderson, R.B.: MathCAD, Version 3.1, Addison-Wesley Publ. Company, Inc. 1993

/12/ Wimmel, R.: Adaptive digitale Echtzeitfilterung in der Strukturdynamik (ARES). Report, Inst. für Aeroelastik, DLR, 1990

/13/ Preumont A., Dufour, J.P. und Malekian, C.: Decentralized integral force feedback with piezoelectric actuators for active damping of space structures. Paper 91-141, Proc.Workshop on Smart Material Systems and Structures. Aachen, 1991, S. 23-29.

/14/ Nachtigal, C. L.: Instrumentation and Control, John Wiley & Sons, Inc., 1990

Structural Safety & Reliability, Schuëller, Shinozuka & Yao (eds) © 1994 Balkema, Rotterdam, ISBN 90 5410 357 4

A direct optimization algorithm for stochastic control and its applications

Zhikun Hou
Mechanical Engineering Department, Worcester Polytechnic Institute, Mass., USA
Gongkang Fu
Structures Research, New York State Department of Transportation, Albany, N.Y., USA

ABSTRACT: A new algorithm is presented to optimize the quadratic performance index of linear control systems without loosing information of the external excitation. In the algorithm, the performance index is directly calculated from the moment response of structures and the optimal control gains are determined by the simplex method. The algorithm leads to actual optimal performance index since the effects of the excitation, previously ignored in the conventional Riccati control, are now taken into account. The approach is also accurate and efficient since the algorithm only needs to evaluate the performance index itself which can be explicitly calculated for most existing stochastic models of earthquake ground motion. Numerical results are provided to compare the difference between the conventional Riccati control and the new algorithm proposed. An application to hybrid control of highway bridges under nonstationariy earthquake excitation is included.

1 INTRODUCTION

Structural control has recently received considerable attention for its potential to improve safety and/or functionality of structures under severe environmental loadings, such as earthquakes and wind loads (Kobori 1988, Soong 1988, and Yao 1972). These environmental loads are often modeled as stochastic processes to account for their nature of uncertainty. Among many available control strategies and algorithms, stochastic optimal control appears to be an appropriate approach to general solutions for problems with similar stochastic properties.

In conventional stochastic optimal control, the required control forces are determined in terms of so-called Riccati matrix, which relates the control forces to the measured displacement and velocity of the structural response (Soong 1988, and Yang et. al. 1987). However, as noted by many investigators including Yang (Yang et. al. 1987), the Riccati matrix needs to be solved backwards, which brings some limitations to its application since the environmental loads are usually not known in advance. Therefore, the Riccati matrix is approximately solved in the absence of external excitation. As a result, the performance index may not actually achieve its optimum as expected. Furthermore, using the feedback control force with Riccati control gains in fact modifies the damping and stiffness properties of the structure. If the natural frequency of the modified structure becomes closer to the dominant frequency band of the excitation, the system response may even increases. In that critical case, the structure control will fail.

This paper presents a new stochastic control algorithm which optimize the performance index without loosing information of the external excitation. In this algorithm, the performance index is directly calculated from the moment response of structures under actual stochastic loading, and the optimal control gains are determined by the simplex method.

The algorithm will lead to actual optimal performance index since the effects of the excitation are included. The approach is also accurate and efficient since the algorithm only needs to evaluate the performance index itself which can be explicitly calculated for most existing stochastic models of earthquake ground motion. Numerical results are presented to compare the difference between the conventional Riccati control and the proposed algorithm for a hybrid control problem of highway bridge problem under nonstationary earthquake excitation.

2 FORMULATION

The equation of motion of a single-degree-of-freedom control system is given by

$$m\ddot{x}(t)+c\dot{x}(t)+kx(t)+f_c(t) = f_e(t) \tag{1}$$

where x(t) is the displacement response of the system relative to ground, m,c,k represent the inertia, damping and stiffness properties of the system respectively; $f_c(t)$ is the control force expressed by

$$f_c(t) = \beta\dot{x}(t)+\gamma x(t) \tag{2}$$

where β and γ are control gains determined by minimizing a quadratic performance index defined as

$$J = E\left[\int_0^{t_f}[\frac{R_r}{2}(\omega^2x^2(t)+\dot{x}^2(t))+R_f f_c^2(t)]dt\right] \tag{3}$$

in which t_f is the duration of the external excitation; R_r and R_f are weighing coefficients to indicate the relative importance between safety and economy. E[·] is the operator of mathematical expectation. Though many other performance indices, such as the maximum of the absolute structural response, can be defined, the Eq.3 is employed throughout this investigation.

The external excitation $f_e(t)$ is modeled herein as a nonstationary modulated white noise process. That is

$$f_e(t) = -\eta(t)n(t) \tag{4}$$

where $\eta(t)$ is a deterministic envelope function to describe nonstationarity of the excitation, and n(t) is a stationary white noise process with properties

$$\begin{aligned} E[n(t)] &= 0 \\ E[n(t_1)n(t_2)] &= S_0\delta(t_1-t_2) \end{aligned} \tag{5}$$

in which S_0 is the intensity of the white noise. It is noted that $\eta(t)$ is normalized such that its maximum absolute value is one. For narrow-band excitation, a filtered modulated white noise process may be employed. The dynamics of the filter is governed by the following equation:

$$\ddot{y}(t)+2\zeta_g\omega_g\dot{y}(t)+\omega_g^2y(t) = -\eta(t)n(t) \tag{6}$$

where y(t) is the ground motion relative to the bedrock where the earthquake wave arrives, ζ_g and ω_g describe energy dissipation property and dominant frequency of the site.

The above formulation may be rewritten in terms of state variables as

$$\dot{Z}(t) = AZ(t)+GZ(t)+F_e(t) \tag{7}$$

where

$$Z^T(t) = [x(t),\dot{x}(t),y(t),\dot{y}(t)]$$

is a vector of state variables; A is a 4x4 system matrix obtained from the mass, damping, and stiffness of the system; G is a 4x4 gain matrix relating the feedback control force to state variables; and $F_e(t)$ is a 4x1 vector specifying the external excitation.

3 DETERMINATION OF RICCATI CONTROL GRAINS

Consider a general optimal control problem of linear control systems governed by

$$\dot{Z}(t) = AZ(t)+BF_c(t)+F_e(t) \tag{8}$$

where B is the location matrix indicating the location of actuators; the control force $F_c(t)$ is determined such that the performance index

$$J = E[\int_0^{t_f} (Z^TQZ+F_c^TRF_c)dt] \tag{9}$$

is minimized; Q and R in Eq.(9) are weighing matrices. Note that Eq. (3) is a special case of Eq. (9). Traditional control algorithms neglect the effects of external excitation and lead to the following Riccati matrix cquation for P(t) (Yang et. al. 1987).

$$\begin{aligned} &\dot{P}(t)+P(t)A-\frac{1}{2}P(t)BR^{-1}B^TP(t) \\ &\qquad +A^TP(t)+2Q = 0 \\ &P(t_f) = 0 \end{aligned} \tag{10}$$

After solving P(t), the closed-loop feedback control force can be obtained by

$$F_c(t) = -\frac{1}{2}R^{-1}B^TP(t)Z(t) \tag{11}$$

As noted by many investigators (Yang et. al. 1987), the Riccati matrix equation needs to be solved backwards, which brings some limitations to its application since the environmental loads are usually not known in advance. Therefore, the Riccati matrix is approximately solved in the absence of external excitation. As a result, the performance index may not actually achieve its optimum as expected.

It is found that the Riccati matrix can be approximately treated as constant matrix which can be solved by the steady-state version of Eq. 10 (Yang et. al. 1987). For a linear single-degree-of-freedom system, the constant control gains β and γ can be solved explicitly as

$$\begin{aligned} \gamma &= \omega(\sqrt{\omega^2+\frac{1}{R}}-\omega) \\ \beta &= \sqrt{2\gamma+\frac{1}{R}+4\zeta^2\omega^2}-2\zeta\omega \end{aligned} \tag{12}$$

where the relative weighing coefficient $R = R_f/R_r$. The above result can be reduced to the expressions given by Meirovitch and Silverberg (Meirovitch and Silverberg 1983) for undamped systems.

It should be reemphasized that the Riccati matrix P(t) does not guarantee an optimal active control, since it is derived in the absence of external excitation, which is certainly an unrealistic assumption. For deterministic loading, an instantaneous optimal control algorithm was proposed by Yang (Yang et. al. 1987) to account for effects of the external excitation. This paper presents an alternative algorithm for stochastic loading.

4 A NEW STOCHASTIC OPTIMAL CONTROL ALGORITHM

In the present algorithm, the optimal control gains are searched in a two-dimensional parameter space i.e. (β,γ) space, such that the optimal of the performance index defined in Eq.9 is achieved. Starting with given initial guess $\beta^{(0)}$ and $\gamma^{(0)}$, the proposed algorithm includes the following two elements: 1) for known $\beta^{(i)}$ and $\gamma^{(i)}$, calculate the performance index of the control system under external excitation by the simplified state-variable approach (Hou 1990); and 2) determine new $\beta^{(i+1)}$ and $\gamma^{(i+1)}$ from the previous $\beta^{(i)}$ and $\gamma^{(i)}$ and the corresponding performance index by the simplex method (Himmeblan 1972). The simplex method is effectively used in this study. In the simplex method, no calculation of the gradient of the objective function is required. The only calculation needed is to evaluate the objective function itself, which involves an integral of the covariance matrix of the response. Fundamentals of the simplex method may be found in textbooks on linear and nonlinear programming including Himmeblan (Himmeblan 1972).

By the simplified state-variable approach (Hou 1990, Iwan and Hou 1989), Q(t), the nonstationary covariance matrix of lincar MDOF systems subjected to modulated white noise, can be expressed in a compact matrix form as

$$Q(t)=S_0L\int_0^t \eta^2(\tau)P(t-\tau)P^T(t-\tau)d\tau L^T \tag{13}$$

where S_0 is the intensity of the white noise; L is a constant matrix independent of stochastic excitation; and

$$P(t) = \begin{bmatrix} P_1(t) \\ P_2(t) \\ \cdots \\ P_N(t) \end{bmatrix} \tag{14}$$

$$P_k(t) = e^{-\zeta_k \omega_k t} \begin{bmatrix} \cos\omega_{dk}t \\ \sin\omega_{dk}t \end{bmatrix}$$

in which ζ_k, ω_k, and ω_{dk} are the damping ratio, natural frequency, and damped natural frequency of the kth mode, respectively. For most envelopes used in earthquake engineering, Q(t) as well as its time integral can be explicitly integrated, which makes the proposed algorithm accurate and efficient.

5 NUMERICAL RESULTS

Figure 1 presents the results for equivalent natural frequency and critical damping ratio for linear SDOF systems controlled by Riccati control force based on Eq.12. Figure 1(a) gives ζ^*/ζ, the ratio of the equivalent damping ratio of the modified system to that of the original system, versus ω_n, the natural frequency of the original system. Three different values of ζ are used, i.e. ζ=0.05, 0.10,0.30. R is assumed to be 0.1. Figure 1(b) gives ω_n^*/ω_n, the ratio of the equivalent natural frequency of the modified system to that of the original system, versus ω_n. It is noted that ω_n^*/ω is independent of ζ. Three different values of R are used, that is R=0.1,1,10. Significant modification of stiffness and damping properties is observed for long-period systems. As the natural frequency of the system increases, the Riccati control force becomes ineffective.

Figure 2 presents a comparison between responses of the original uncontrolled system, the system controlled by the Riccati force, and the present algorithm. The system is subjected to a narrow-band excitation which is modeled as an filtered modulated white noise. The system parameters are assumed that ζ=0.05 and ω=4 rad/sec. A unit step envelope is

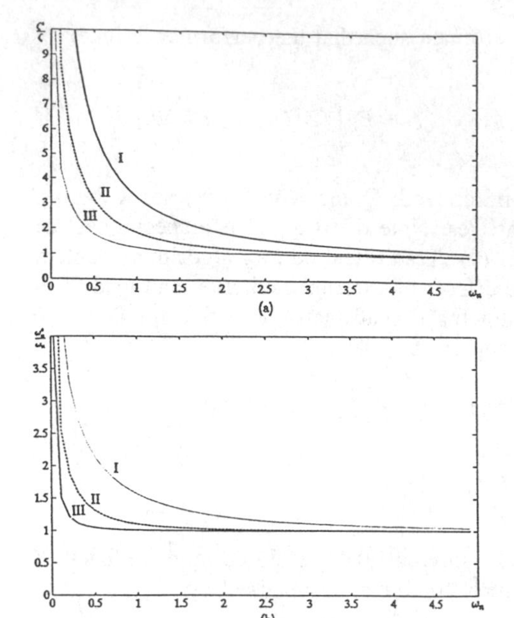

Figure 1 Equivalent natural frequency and damping ratio of controlled system by the Riccati control force. (a) Equivalent damping ratio; R=0.1; I-ζ=0.05, II-ζ=0.10, and III-ζ=0.20. (b) Equivalent natural frequency; I-R=0.1, II-R=1.0, and III-R=10.0.

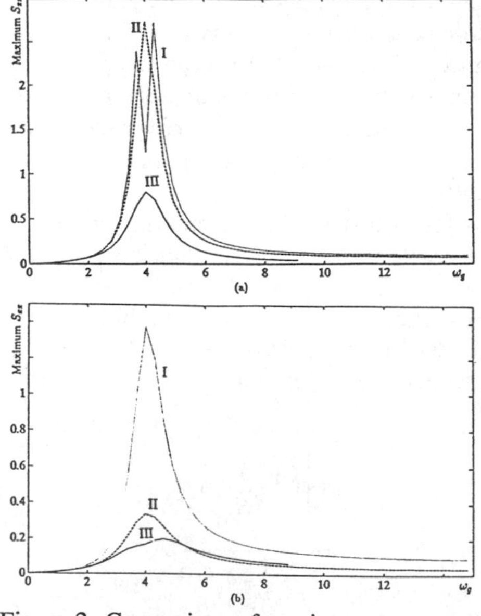

Figure 2 Comparison of maximum mean-square responses for I - uncontrolled system, II - controlled system by Riccati control force, and III - controlled system by the present algorithm. (a)ζ_g=0.05. (b) ζ_g=0.10.

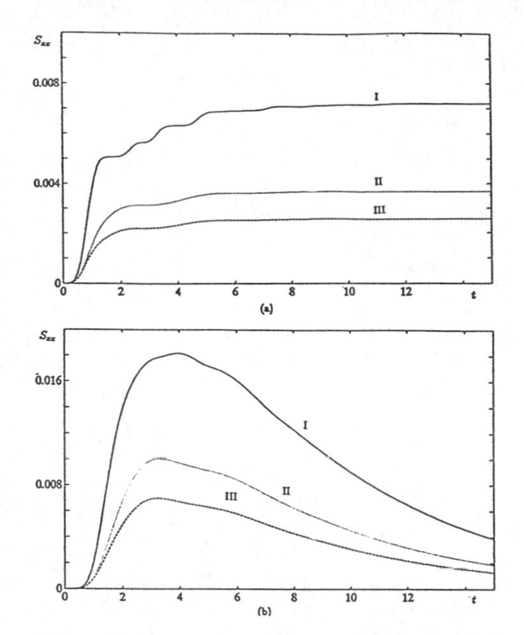

Figure 3 Comparison of nonstationary mean-square responses of a base-isolated highway bridge subjected to filtered modulated white noise. I - uncontrolled system, II - controlled system by Riccati control force, and III - controlled system by the present algorithm. (a) Unit step envelope. (b) Shinozuka-Sato envelope.

used. The intensity of the white noise is assumed to be one. ζ=0.05 and 0.10 in (a) and (b) respectively. In both (a) and (b), results are presented for maximum mean-square response S_{xx} of a SDOF system versus ω_g ranging from 0.1 to 15 rad/sec. Generally speaking, the Riccati control significantly reduces vibration level but may not achieve the optimum; and the present control algorithm gives better result. In Figure 2(a), response by the Riccati control is greater than that of uncontrolled system when ω_g is close to ω_n, indicating that the Riccati control fails. In this critical case, the proposed algorithm still works well. In 2(b), slightly larger response is observed for the present program than that from the Riccati control for large ω_g. However, the overall performance index for the present algorithm is still better.

Figure 3 shows results from an application of the present algorithm to hybrid control problem of highway bridges subjected to nonstationary ground motion. A 30.5M single span bridge of 5 steel girders supporting a reinforced concrete deck slab is modeled as an SDOF system with m=271,500kg, original natural period T=0.7sec, and viscous damping ratio ζ=0.05. A base isolation system is employed to lengthen the original period to T=1.7sec and to increase the damping ratio to an effective value ζ=0.15 (AASHTO 1991). The filter frequency and damping ratio are set ζ_g=0.60 and $\omega_g=5\pi$ rad/s, respectively (Clough and Penzien 1975). Results are presented for the nonstationary mean- square response of the bridge for two types of envelope: the unit step envelope and Shinozuka-Sato envelope; the latter is expressed as

$$\eta(t) = a(e^{-bt}-e^{-ct}) \tag{15}$$

and a=2.32, b=0.09, and c=1.49 are assumed (Corotis and Marshall 1977). In both cases considerably further suppression is gained by the proposed algorithm compared to the Riccati control.

6 SUMMARY

A stochastic optimal control algorithm is proposed to take into account the external excitation which is neglected in the conventional Riccati control. In this algorithm, the performance index is directly calculated from the moment response of structures under actual stochastic loading, and the optimal control gains are determined by the simplex method. The algorithm leads to actual optimal performance index since the effects of the excitation are included. The approach is also accurate and efficient since the algorithm only needs to evaluate the performance index itself which can be explicitly calculated for most existing stochastic models of earthquake ground motion by an simplified state-variable approach. The algorithm works well in the critical case where the Riccati control fails; and provides further reduction of vibration level in general. Numerical results are presented to compare the difference between the conventional Riccati control and the proposed algorithm for a hybrid control problem of highway bridges under nonstationary earthquake excitation. The algorithm can be extended to general linear multi-degree-of-freedom systems.

7 REFERENCES

AASHTO 1990. *AASHTO guide specifications for seismic isolation design.* 1990.

Clough, R.W. and Penzien, J. 1975. *Dynamics of structures*. McGraw-Hill Book Company, 1975.

Corotis, R.B. and Marshall, T.A. 1977. Oscillator response to modulated random excitation. *ASCE Journal of Engrg. Mech. Dic.* 103:EM4: 501-513.

Himmeblan, D.M. 1972. *Applied nonlinear programming*, McGraw-Hill, Inc.

Hou, Z.K. 1990. *Nonstationary response of structures and its application to earthquake engineering*. California Institute of Technology. EERL 90-01.

Iwan, W.D. and Hou, Z.K. 1989. Explicit solutions for the response of simple systems subjected to nonstationary excitations. *Structural Safety* 6: 77-86.

Kobori, T. 1988. State-of-the-art report, active seismic response control. *Proceedings of Ninth World Conference on Earthquake Engineering*. Tokyo-Kyoto, Japan.(Vol. VIII).

Meirovitch, L. and Silverberg, L.M. 1983. Control of structures subjected to seismic excitations. *ASCE, Journal of Engineering Mechanics* 109: 604-618.

Soong, T.T. 1988. State-of-the-art review: active control in civil engineering. *Engineering Structures* 10:74-84.

Yang, J.N., Akbarpour, A., and Ghaemmaghami, P. 1987. New optimal control algorithms for structural control. *ASCE J. Engrg. Mech.* 113:EM9:1369-1386.

Yao, J.T.P. 1972. Concept of structural control. *ASCE, J. Stru, Div.* 98:1567-1574.

Structural Safety & Reliability, Schuëller, Shinozuka & Yao (eds) © 1994 Balkema, Rotterdam, ISBN 90 5410 357 4

Structural safety of a Hubble Space Telescope science instrument

M.C. Lou & D.N. Brent
Jet Propulsion Laboratory, California Institute of Technology, Pasadena, Calif., USA

ABSTRACT: This paper gives an overview of safety requirements related to structural design and verification of payloads to be launched and/or retrieved by the Space Shuttle. To demonstrate the general approach used to implement these requirements in the development of a typical Shuttle payload, the Wide Field/Planetary Camera II, a second generation science instrument currently being developed by the Jet Propulsion Laboratory (JPL) for the Hubble Space Telescope is used as an example. In addition to verification of strength and dynamic characteristics, special emphasis is placed upon the fracture control implementation process, including parts classification and fracture control acceptability.

INTRODUCTION

For a flight hardware system to be launched and/or retrieved by the Space Shuttle, the development of its structures must address both personnel safety and safety of the mission. Safety of personnel and the Shuttle has been a paramount concern for the National Space Transportation System (NSTS) since the first Shuttle flight in 1980. This safety concern covers all aspects of the Shuttle operations, including development of Shuttle payloads. Payload structural components are classified in accordance with their likelihood of creating hazards threatening the safety of the Shuttle and its flight and ground crews. Payload developers are required to pay special attention to those components of which the failure could result in catastrophic safety hazards. Because numerous foreign and domestic agencies, private companies, and universities are developing hundreds of Shuttle payloads, the National Administration of Aeronautics and Aerospace (NASA), as the operator of the Shuttle, has established a set of uniform safety policies and requirements for payload structural development (NASA 1989). These requirements, as well as the methodologies for their implementation, were continuously revised and updated through the past decade. Safety of a Shuttle payload mission is measured by the level of reliability of the payload system. NASA does not impose agency-wide, uniform requirements for mission reliability. Mission reliability is considered a sole responsibility of the payload developer and, in general, is achieved by mission-specific structural design and verification requirements.

This paper discusses the Shuttle payload structural design and verification requirements and the general approach used to meet these requirements. Greater emphasis will be placed upon personnel safety. Also, as an example, structural development of a typical science payload, the Wide-field/Planetary, will be described to illustrate implementation of Shuttle safety requirements.

SAFETY REQUIREMENTS FOR SHUTTLE PAYLOADS

To launch and/or retrieve space flight systems (the payloads), the Space Shuttle provides many services and interfaces during the ground preparation, launch, and flight phases. In order to ensure personnel and Shuttle safety, NASA has established a uniform set of safety requirements for verifying the flight-worthiness of the payload structures (JSC 1982). These structural safety requirements can be divided into three categories: 1) strength design and verification; 2) dynamic characteristics and verification; and 3) fracture control. The requirements in each of these categories will be briefly discussed below.

Strength Design and Verification:

NASA requires that the strength of a payload structure must be demonstrated by analysis and/or testing. Strength requirements are expressed in terms of limit loads. For Shuttle safety, the limit loads are the maximum loads to be experienced by a payload while it is in the Shuttle Cargo bay. This includes all the launch, flight, and normal and emergency landing events. All payload structures are required to be designed to withstand the ultimate loads defined by multiplying the limit loads by an ultimate factor of 1.4. Compliance of this strength design requirement should be demonstrated by qualification-level static testing. Depending on whether the strength demonstration is done on a development article or on the flight article, one of the following two options can be taken:

Option 1- Static test a developmental (i.e., the prototype) article to 1.4 times limit load.

Option 2- Static test the flight (i.e, the protoflight) article to 1.2 times limit load.

For the cases in which the adequacy of the structural design has been demonstrated by previous space applications, the protoflight static test factor of 1.2 may be reduced to 1.1.

Under some circumstances, it may be permissible to verify the compliance of strength design by analysis alone, usually using an ultimate factor of safety higher than the required value of 1.4. Several NASA field centers have selected ultimate factors of safety between 2.0 and 3.0 for the analysis-only verification approach (JPL 1989, MSFC 1981, GSFC 1990). Due to the favorable cost and schedule considerations, as well as the desire to eliminate the risk to flight hardware and personnel imposed by static tests, This analysis-only, or commonly known as the "no-test," verification approach has become increasingly popular among the Shuttle payload developers. It should be emphasized that increasing the factor of safety for the design of a payload structure does not by itself justify the omission of static test verification. Sound engineering rationale must be developed to support the use of the no-test option for any payload development program. Some example rationale accepted by NASA/JSC include: 1) the structural design is simple with well-defined load paths, and has been thoroughly analyzed for all critical load cases; 2) the structural design has been successfully test-verified for previous Shuttle payload applications, and good correlation of test results to analytical prediction have been achieved; and 3) all safety-critical components of the payload have been identified and those that are difficult to analyze have been test verified.

Dynamic Characteristics and Verification:

The vibro-acoustic loads encountered by a payload during Shuttle launch and landing should be determined on the basis of coupled loads analysis results. The coupled load analyses are based on imposing the Shuttle launch and landing forcing functions on a synthesized mathematical model which couple the dynamic model of the payload with that of the Shuttle. The payload dynamic model used in the coupled loads analyses must capture the essential dynamic characteristics of the payload system in the frequency range up to 100 Hz. Test verification by modal survey (or equivalent tests) of the payload model is required except for payload designs whose fundamental frequency, when assuming a fixed interface with the Shuttle Cargo Bay, is higher than 35 Hz.

As for structural damping, it is required that all damping values higher than one percent critical to be used for flight control interaction studies must be test verified.

Fracture Control Requirements:

For cyclically stressed structures containing crack-like flaws, the traditional design approach based on materials yield and ultimate strengths may not be adequate and fatigue and fracture should also be important design considerations for these structures.

Fracture mechanics analysis has been a part of the design process of aircraft structures for many decades. However, except for pressure vessels, fracture is not a major design factor for payloads launched by the expendable launch vehicles. Fracture control is the rigorous application of fracture mechanics analysis and/or testing to the prevention of crack propagation leading to catastrophic failure that may endanger the Shuttle and its flight crew. The application of fracture control to Shuttle payloads is supported by many engineering disciplines, including structural and dynamic analyses, material selection and characterization, fabrication and processes, life testing, non-destructive examination, and quality assurance.

In the early development phase of the Space Shuttle program, NASA decided that fracture control should be imposed on all payloads to assure that the presence of crack-like defects in payload components do not endanger the Shuttle and flight personnel (NASA 1989). The underlying rationale for this requirement is that no matter how carefully a payload part is made, undetected flaws can exist and, under cyclic loading, these flaws may propagate, reach unstable growth, and cause catastrophic failures. Detailed requirements for Shuttle payload fracture control are provided by NASA (NASA 1988).

Prior to a payload begin approved for integration into the Shuttle Cargo Bay, compliance of the above-listed safety requirements must be reviewed and accepted by the NASA/JSC Shuttle Payloads Safety Review Panel. JSC provides submittal requirement and safety review procedures (JSC 1989).

To improve cost effectiveness and to take advantage of recent progress of technology, NASA is constantly reviewing and updating Shuttle payload safety requirements. It is important for a Shuttle payload developer to keep current of safety requirements and to define an acceptable approach to meet the requirements at the very beginning of a payload development process. The Phase 0 Safety Review meeting with NASA/JSC (JSC 1989) provides the best opportunity to achieve this goal.

WF/PC INSTRUMENT

The first generation of the Wide-Field/Planetary Camera (WF/PC I) is the principal science instrument on the Hubble Space Telescope (HST) which was launched into a low Earth orbit by the Space Shuttle Discovery on April 24, 1990. The complement of HST instruments includes: two cameras (WF/PC I and Faint Object Camera), two spectrographs (Faint Object Spectrograph and High Resolution Spectrograph) and one photometer. The WF/PC I and three guidance sensors are mounted radially and the rest are axial modules in the aft of the telescope. The HST configuration is shown in Figure 1.

Due primarily to the constraints on volume, mass, and power, the WF/PC I was built as a single-string instrument with only limited redundancy and a mission life requirement of 2.5 years on-orbit. A second generation of WF/PC, the WF/PC II, was intended to serve as a replacement instrument for WF/PC II in

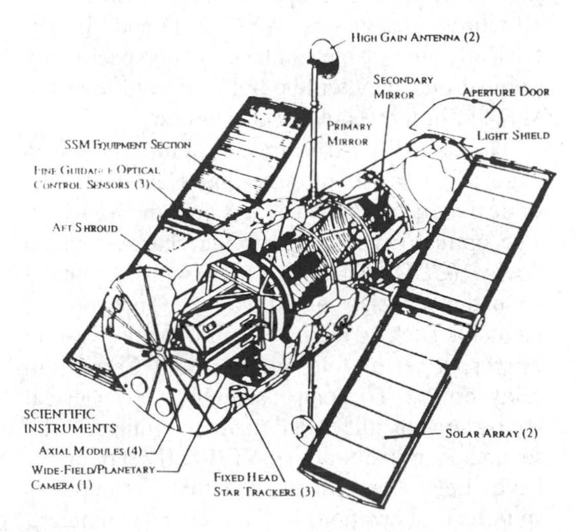

Figure 1 The Hubble Space Telescope

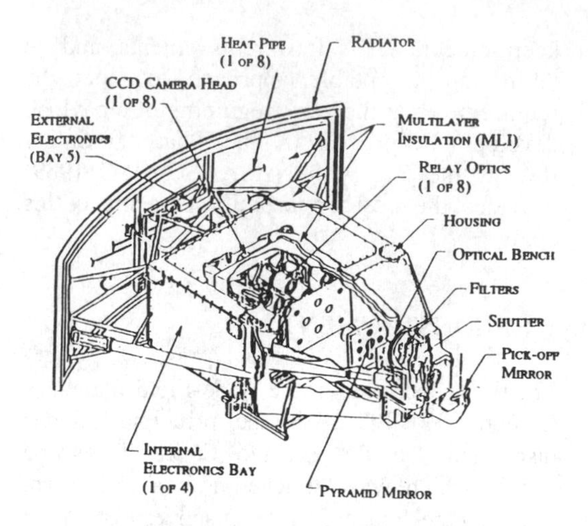

Figure 2 The WF/PC Structure

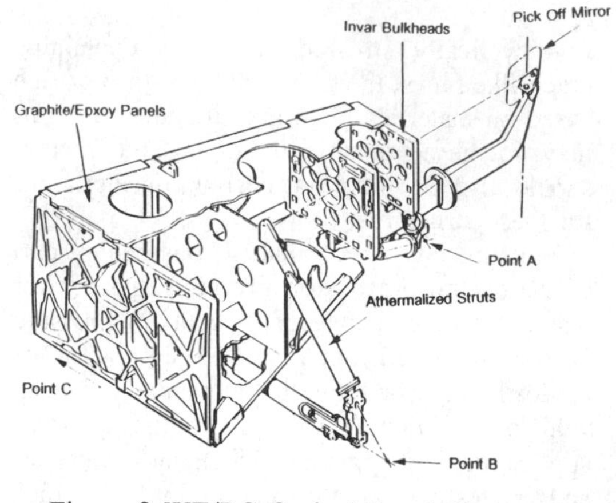

Figure 3 WF/PC Optical Bench Structure

case of an instrument failure and is designed for on-orbit replacement by shuttle astronauts.

The construction of WF/PC II was initiated prior to the launch of WF/PC I. A few months after WF/PC I launch, it was discovered that the HST was unable to meet its intended optical performance due to spherical aberration on the primary mirror. As a consequence, most of the expected "breakthrough" science observations of very faint objects and crowded fields could not be performed. It was then decided to retro-fit the already existing design of the WF/PC II with the required optical fix to compensate for the aberrated telescope mirror. Since the structural design of WF/PC I and II are basically the same, unless it is specifically pointed out, they will be both referred to as the WF/PC in the following discussion.

The WF/PC structural system, shown in Figure 2, consists of three major elements: the optical bench, the housing, and the radiator. The optical bench supports the charge-coupled device (CCD) detectors along with an optical train that consists of critically aligned optical elements such as the pickoff mirror, a pyramid mirror, a set of fold mirrors, and Cassegrain relay optics. To compensate for the spherical aberration of the HST primary mirror, the secondary mirrors of the WF/PC II relay optics have been re-configured with an opposite spherical aberration. This change required extremely precise alignment of the HST primary mirror pupil image on the secondary mirror of the relay optics. To accomplish this alignment, adjustment mechanisms were added to the pickoff mirror and to three of the four fold mirrors.

The optical bench structure, shown in Figure 3, consists of four bulkheads bonded to graphite/epoxy panels. The bench is supported in a determinate manner at the three interface points via sets of athermalized struts. The fold mirrors, pyramid mirror, and relay optics are all supported on invar bulkheads. The pickoff mirror is supported at the end of a graphite/epoxy beam cantilevered off the optical bench bulkheads. The housing structure, shown in Figure 4, shields the optical bench from contamination from the outside HST Aft Shroud Environment. Aside from providing mounting surfaces for the internal electronics, the housing also supports the radiator with the use of a boron/epoxy truss structure at the end of the instrument. The housing is constructed from aluminum sheet and machined sections (6061-T6, T651).

WF/PC STRENGTH DESIGN AND VERIFICATION

Following the traditional structural development practices of JPL flight instruments, preliminary design of WF/PC structures was based on load factors given by a Mass Acceleration Curve

(MAC). The MAC was developed in a semi-emprical manner (JPL 1989), and the use of which greatly simplifies preliminary sizing of flight structural members. It has been repeatedly proven by flight experience that the MAC loads are conservative and envelop the coupled loads analysis results that are used to perform final verifications of the safety margins of the structures.

For strength design and analysis of WF/PC structures, the ultimate factor of safety was selected to be 2.0 minimum. This safety factor exceeds the minimum requirements for Shuttle payloads and forms, (MSFC 1981), the basis for exempting WF/PC structures from static test qualification. The safety margins, M.S., of a WF/PC structure is defined as:

$$M.S. = \frac{Materials\ Allowable}{2.0\ x\ Applied\ stress} - 1.0$$

Safety margins for WF/PC structural components were determined based on results of component-level analyses, using hand stress calculations and computer modeling methods. The minimum safety margins and corresponding load conditions for the WF/PC instrument are summarized in Table 1. Under ground handling conditions where the WF/PC will be supported at Bay 5 and the housing and optical bench are supported by the radiator truss tubes, the minimum margin of safety is +0.02. Under launch loads, the minimum margin of safety for the housing structure is buckling of the top cover at +1.11. The minimum margin of safety for the optical bench is +0.62 for the bolts that attach the optical bench struts to the housing at the A latch.

To verify structural adequacy and workmanship of the WF/PC, environmental tests were performed both at the sub-assembly and system level. Random vibration tests to protoflight levels were conducted on mechanisms and optical assemblies to verify their structural integrity.

Following the assembly and integration of all component parts, WF/PC system random vibration and acoustic tests to protoflight levels were conducted on the system to verify workmanship and the structural integrity of the

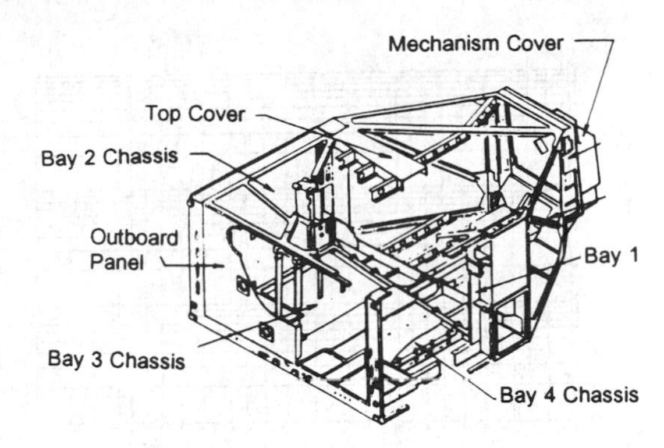

Figure 4 WF/PC Housing Structure

Table 1: WF/PC SAFETY MARGINS

Component	Load Condition	Safety Margin	Failure Mode
Radiator Truss Tubes	Ground Handling	+0.02	Compression
Top Cover	Launch	+1.11	Buckling
Optical Bench Bolts	Launch	+0.62	Tension

electronics assemblies. The system random vibration tests were immediately preceded and followed by low-level sinusoidal vibration tests from 5 to 2000 Hz. These low-level sine tests were used as signature test to ensure that changes of the structural characteristics caused by the random vibration tests were noticed and identified. Figures 5 and 6 are typical responses of WF/PC structures as measured by accelerometers during vibration tests. The system level vibration tests were also followed by optical alignment tests to verify that the critical alignment of the optical elements stayed within acceptable tolerances.

WF/PC DYNAMIC CHARACTERISTICS AND VERIFICATION

The WF/PC dynamic characteristics were determined using finite element analysis. A system finite-element model (FEM) was assembled and run both to determine the instrument mode shapes and frequencies and to be used by the launch integrator for coupled

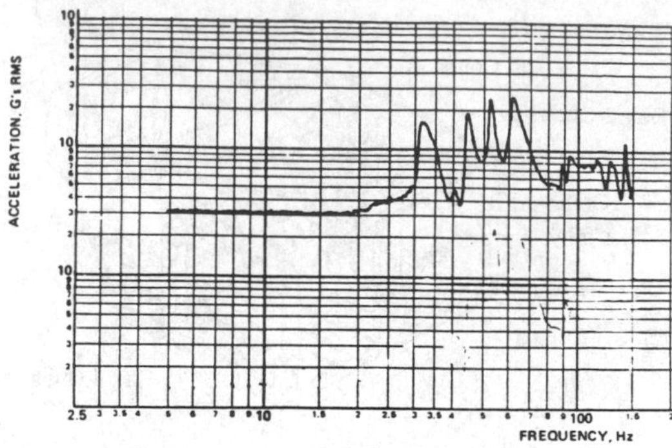

Figure 5: Typical Sine Vibration Response

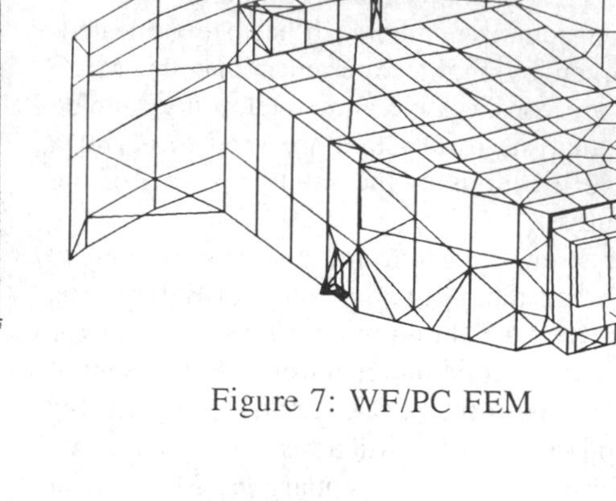

Figure 7: WF/PC FEM

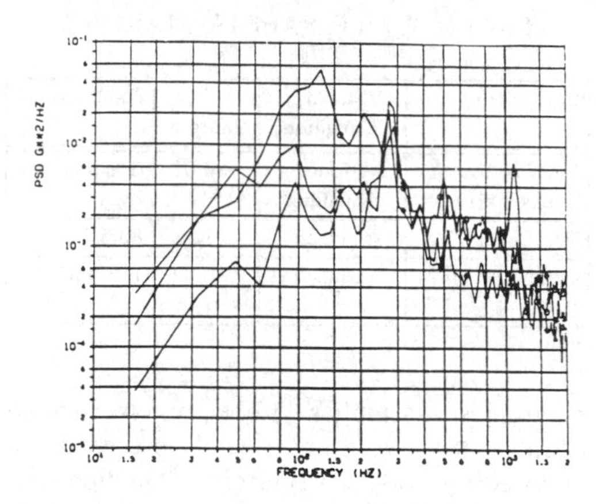

Figure 6: Random Vibration Responses

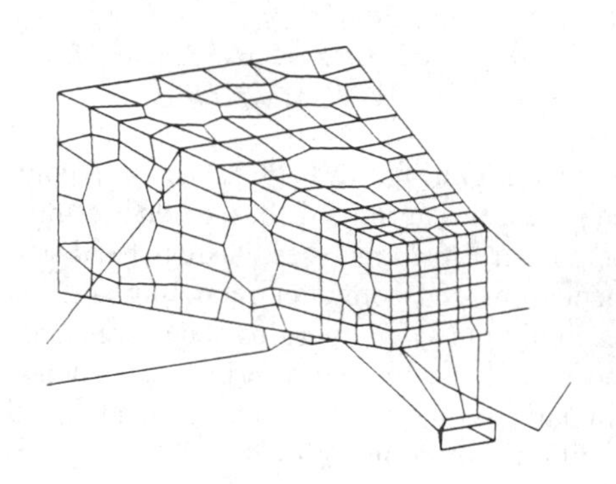

Figure 8: Optical bench FEM

loads analyses. The usage of this FEM also included: track weight, center of gravity, and moments of inertia; determine major load paths for detailed structural analysis; and study changes in optical alignment due to the environmental effects of temperature changes, moisture desorption, and gravity release. The FEM, shown in Figure 7, is constructed from 1721 elements connecting 1145 nodes. The optical bench FEM without the housing and radiator is shown in Figure 8.

To determine dynamic characteristics of the WF/PC, modal analyses were performed. The first four modes of the WF/PC instrument are listed in Table 2. The first two modes of the instrument are bending of the housing. The third and fourth modes describe motion of the optical bench: the third mode is bending of the pickoff mirror arm; the fourth mode is a rigid translation of the optical bench through stretching of the athermalized struts. Low-level sine tests were used to verify these predicted modal frequencies.

WF/PC FRACTURE CONTROL IMPLEMENTATION

Implementation procedures of fracture control for WF/PC are defined in the WF/PC Fracture Control Plan (JPL 1987). Following this plan, all WF/PC hardware components were reviewed and each of these components was classified into one of the following four categories:

1) *Low released mass part*: A component whose failure due to fracture will release

less than 0.25 pounds (113.5 grams) of mass into the Shuttle Cargo bay and will not cause any catastrophic hazard to the Shuttle as a result of subsequent damage to other payloads.

2) *Contained part*: If a component is failed by a fracture, all released fragments not meeting the requirements of a low released mass part will be contained within the payload itself.
3) *Fail-safe part*: A component which can be shown by analysis or test that, after any single fracture, the remaining structure can withstand the redistributed limit loads. In addition, the failure of the part will not result in the release of any fragment that violate the requirements for a low release mass part.
4) *Fracture critical part*: Any part that can not be classified as one of the above three non-fracture-critical parts categories. Table 3 is a partial list of WF/PC fracture-critical parts.

For each of these fracture-critical parts, non-destructive inspection was specified and conducted and a safe-life analysis performed to determine whether the part, containing a pre-existing flaw, could survive a minimum of four lifetimes. The important safe-life analyses features for a Shuttle payload, such as the WF/PC, include:

- The analysis should be based on linear elastic fracture mechanics and quantitatively predict crack growth for specific material, geometry, initial crack size and shape, environment, and loading history.
- It should be assumed that the initial crack is located at the most critical location and orientation. The size and shape of initial cracks is the largest flaw that can remain undetected following the method and level employed to detect the cracks.
- The material properties used to predict crack growth behavior shall be valid for the actual operating environment. If the initial flaw size is determined by non-destructive inspection, the average fracture toughness values should be used. If the initial flaw size is determined by proof testing, the upper bound fracture toughness values should be used.

Table 2: WF/PC Normal Modes

Mode	Frequency (Hz)	Description
1	36.9	Housing + Radiator Pitch
2	40.1	Housing + Radiator Yaw
3	41.6	POM Arm Yaw
4	51.6	Optical Bench Bounce

Table 3: WF/PC Fracture-Critical Parts

Part Name	Static M.S.	# Lives	Type of NDE
Pt A strut support fittings	+4.92	Infinite	Dye Penetrant
Pts B and C support fittings	+11.7	Infinite	Dye Penetrant
Pt B Flexure Beam	+10.2	>100	Dye Penetrant
Optical Bench Strut Aluminum Tube	+14.5	14	Dye Penetrant
Pt A Strut Interface Block Fasteners	+0.62	>100	Proof Test
Purge Tube	+8.1	77	Radiographic

- The loading spectrum defining a lifetime of the component should be composed of all significant load events following the non-destructive inspection or proof testing for crack detection including test, transportation, and launch.
- The effect of crack growth retardation due to intermittent overloads or crack propagation into a hole should not be included in the safe-life analysis.

All safe-life analyses of WF/PC fracture critical

parts were performed employing the NASA/FLAGRO computer program and its material database (JSC 1988).

CONCLUSIONS

Structural safety requirements for Space Shuttle payloads have been discussed, with emphasis placed in three specific areas: (1) structural design and verification; (2) dynamic characterization; and (3) fracture control. An approach employed to meet the safety requirements for the successful structural development of a typical space flight instrument, the WF/PC of HST, has been presented. Implementation details and results have also been summarized.

ACKNOWLEDGEMENT

The work presented in this paper was carried out by the Jet Propulsion Laboratory (JPL), California Institute of Technology, under a contract with the National Aeronautics and Space Administration. The authors wish to thank the members of the JPL Structural and Material Safety Review Board for overseeing and reviewing WF/PC fracture control activities. We also wish to acknowledge the contributions of B. Hoang, J. Zins and P. Rapacz who, at one time or another, have participated in carrying out WF/PC structural analysis and fracture control tasks.

REFERENCES

GSFC GEVS-SE, *General Environmental Verification Specification for STS and ELV Payloads, Subsystems, and Components*, January 1990.

JPL Internal Document D-3993, *Space Telescope Wide-Field/Planetary Camera (WF/PC II) Fracture Control Implementation Plan*, January 1987.

JPL Internal Document D-5882, *Mass Acceleration Curve for Spacecraft Structural Design*, November 1989.

JPL Internal Document D-6820, *Structural Design and Verification Criteria*, December 1989.

JSC-14046, *Payload Verification Require-ments*, Revision A, July 1982.

JSC 22267, *Fatigue Crack Growth Computer Program- NASA/FLAGRO*, Revision A, December 1988.

JSC NSTS-13830, *Implementation Procedure for NSTS Payloads System Safety Requirements*, Revision A, November 1989.

MSFC-HDBK-505, *Structural Strength Program Requirements*, Revision A, January 1981.

NASA NHB 8071.1, *Fracture Control Requirements for Payloads Using the National Space Transportation System (NSTS)*, September 1988.

NASA, NSTS 1700.7, *Safety Policy and Requirements for Payloads Using the National Space Transportation System*, Revision B, January 1989.

Structural Safety & Reliability, Schuëller, Shinozuka & Yao (eds) © 1994 Balkema, Rotterdam, ISBN 90 5410 357 4

Application of H^{∞} control to structural systems

Akira Nishitani & Nariyasu Yamada
Department of Architecture, Waseda University, Tokyo, Japan

ABSTRACT : Considerable and extensive efforts have been recently devoted not only to develop the ideas of active structural control but also to apply control algorithms to existing buildings. Among various control algorithms, optimal control theory has gained much attention in civil engineering field. The optimal control algorithms, however, sometimes have some difficulties regarding both the robust stability and the observers. To provide reappreciation to the classical methodology, the H^{∞} control theory considering the frequency-domain transfer functions is developed. This theory is characterized in terms of:(1)the robust stability;(2)the weighting functions in the frequency domain; (3)no need to assume the statistical properties of excitations. For the H^{∞} control, however, so called trial-and-error procedures are in some cases required, because there are many controllers satisfying the given conditions. The theory has two approaches: one is an analytical approach using the transfer function, and the other is an algebraic approach directly solving the state equation. This paper discusses an application of the latter approach to structural systems.

1. INTRODUCTION

A combination of social demand and modern technological progress are beginning to change the traditional concept of safety. Under such circumstances, considerable and extensive efforts have been recently devoted not only to develop the ideas of active control for structures (Kobori et al. 1991) but also to apply control algorithms to real structures (Ikeda et al. 1991). Among various control algorithms, optimal control theory, which minimizes a performance index in the time domain, has gained much attention in civil engineering field. The optimal control algorithms, however, in some cases have some difficulties such as regarding the robust stability.

The theory of H^{∞} control has been developed to provide reappreciation to classical control theory in the frequency domain(Doyle, Glover, Khargonekar, Francis 1989). The theory is characterized by the following two points: (1)it does not depend on the statistical properties of an external excitation;(2)various types of controls are possible, because this theory can quantify the robust stability. However, in some cases, it sacrifices control efficiency.

This paper discusses an application of the H^{∞} control theory to a structural system. In the H^{∞} control theory, the controlled outputs are represented by the linear combination of the state, control force and external excitation. The solution is to be derived through the Riccati equation. The H^{∞} controls with both the state and output feedbacks (FB) are applied to a single-degree-of-freedom structural system.

2. H^{∞} NORM AND H^{∞} CONTROL THEORY

The H^{∞} control theory employs the H^{∞} norm of the transfer function as a control criterion. In this section, H^{∞} norm is initially defined.

In the case of multiple input and multiple output, transfer function Φ from a input

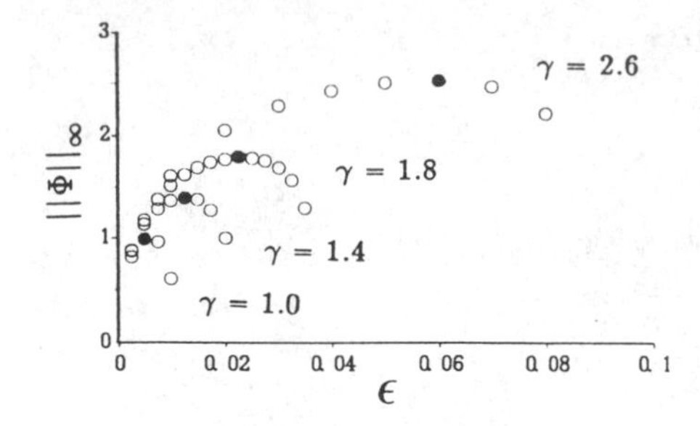

Fig.1 Relationship between $||\Phi||_\infty$ and ε

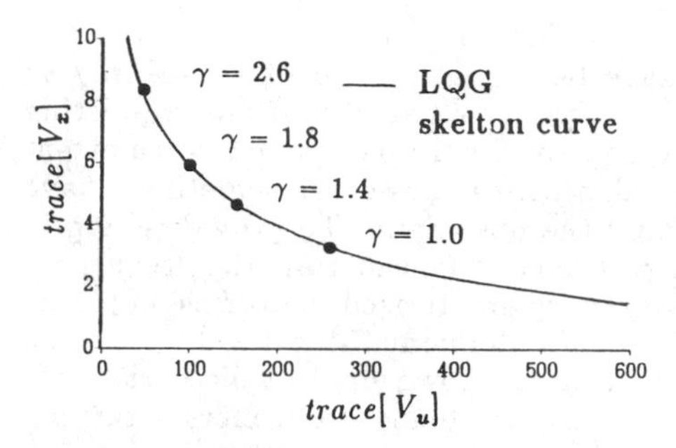

Fig.2 Skelton curve

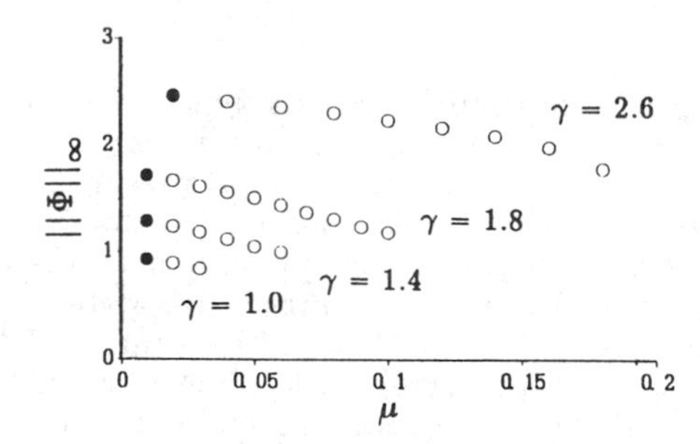

Fig.3 Relationship between $||\Phi||_\infty$ and μ for displacement FB

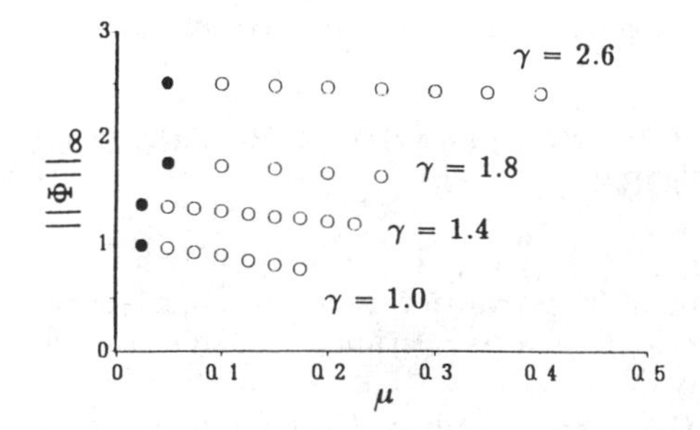

Fig.4 Relationship between $||\Phi||_\infty$ and μ for velocity FB

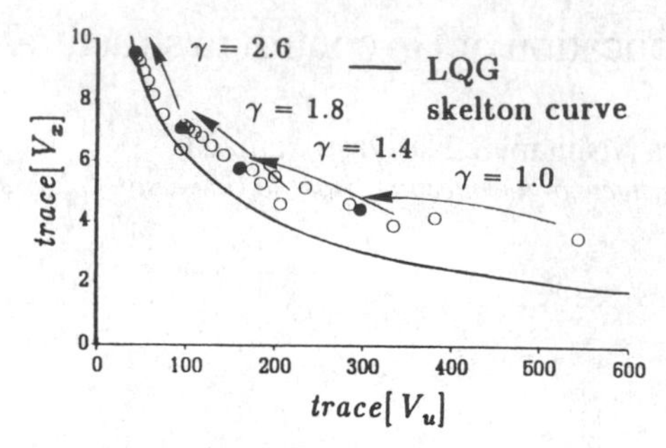

Fig.5 Skelton curve for displacement FB

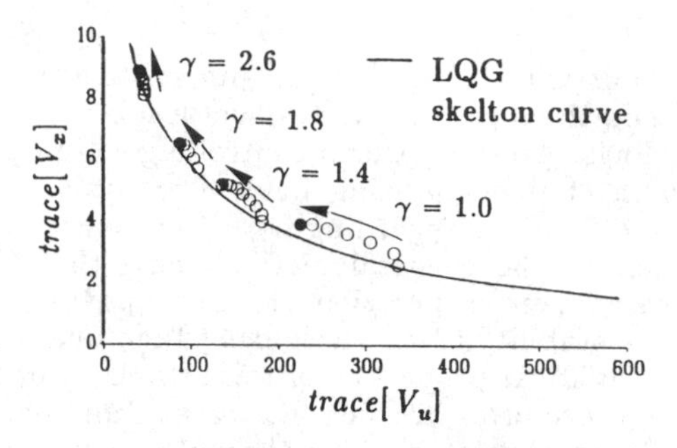

Fig.6 Skelton curve for velocity FB

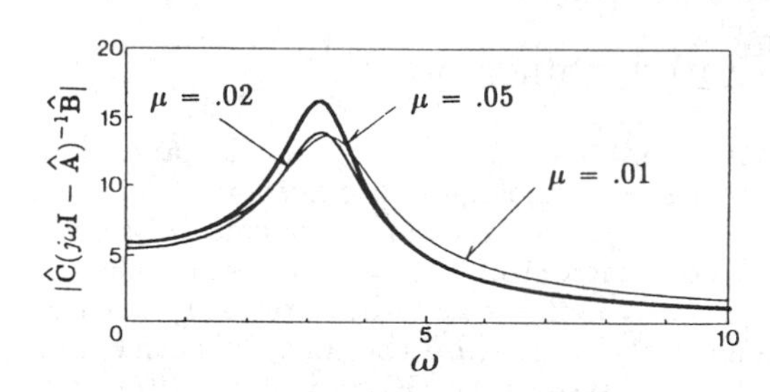

Fig.7 Frequency responses of displacement FB control gain with $\gamma =$ 1.8

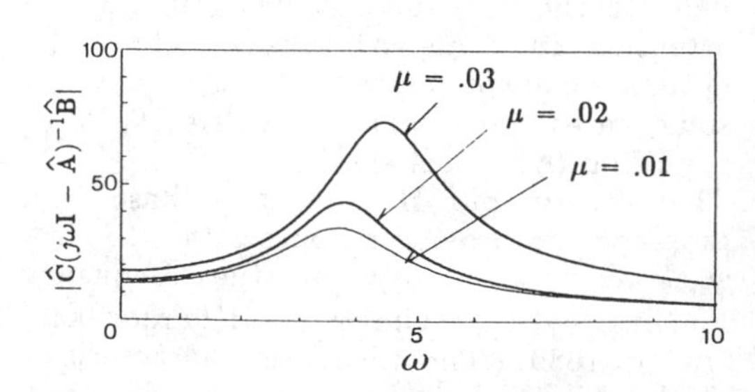

Fig.8 Frequency responses of displacement FB control gain with $\gamma =$ 1.0

vector $\mathbf{w}$ to a output vector $\mathbf{z}$ becomes a matrix. In this case, the H^∞ norm of the transfer matrix is represented by Francis (1987)

$$||\Phi||_\infty = \sup_{||\mathbf{W}(j\omega)||_2} \frac{||\Phi(j\omega)\mathbf{W}(j\omega)||_2}{||\mathbf{W}(j\omega)||_2} \tag{1a}$$

where $\mathbf{W}(j\omega)$ = Fourier transform of $\mathbf{w}(t)$ with $j = (-1)^{1/2}$ and $||\cdot||_2$ denotes the H^2 norm of a vector, which is in the expression, say, for the vector $\mathbf{W}(j\omega)$,

$$||\mathbf{W}||_2 = \sqrt{\frac{1}{2\pi}\int_{-\infty}^{\infty} \mathbf{W}(-j\omega)^T\mathbf{W}(j\omega)\,d\omega} \tag{1b}$$

This can be rewritten:

$$||\Phi||_\infty = \sup_\omega \sqrt{\lambda_{max}[\Phi^T(-j\omega)\Phi(j\omega)]} \tag{2}$$

where $\lambda_{max}[\cdot]$ means the maximum eigen value of matrix $[\cdot]$, superscript T denotes matrix or vector transposition, and bold letters express either matrices or vectors.

The H^∞ control system is described by the following:

$$\dot{\mathbf{x}} = \mathbf{A}_1\mathbf{x} + \mathbf{B}_1\mathbf{w} + \mathbf{B}_2\mathbf{u} \tag{3a}$$

$$\mathbf{z} = \mathbf{C}_1\mathbf{x} + \mathbf{D}_{11}\mathbf{w} + \mathbf{D}_{12}\mathbf{u} \tag{3b}$$

$$\mathbf{y} = \mathbf{C}_2\mathbf{x} + \mathbf{D}_{21}\mathbf{w} \tag{3c}$$

where $\mathbf{x}$:state vector, $\mathbf{u}$:control force vector, $\mathbf{y}$:output vector, $\mathbf{w}$:external disturbance vector, $\mathbf{z}$:controlled output vector.

For the state feedback, the control force $\mathbf{u}$ is regulated by the constant feedback gain $\mathbf{F}$. This gain $\mathbf{F}$ is determined so as to satisfy the following criterion with the H^∞ norm of the transfer matrix Φ from $\mathbf{w}$ to $\mathbf{z}$ (Sampei et al. 1990) :

$$||\Phi||_\infty = ||(\mathbf{C}_1+\mathbf{D}_{12}\mathbf{F})(s\mathbf{I}-\mathbf{A}_1-\mathbf{B}_2\mathbf{F})^{-1}\mathbf{B}_1 + \mathbf{D}_{11}||_\infty < \gamma \tag{4}$$

where $\mathbf{I}$ = identity matrix and γ = positive scalar number. Then,

$$\mathbf{F} = -\{(1/2\varepsilon)\Phi_F{}^T\Phi_F + \Xi_F\}\mathbf{B}_F{}^T\mathbf{P} - \Xi_F\mathbf{F}_F{}^T\mathbf{C}_F \tag{5}$$

where $\mathbf{P}$ is obtained from the Riccati equation (Sampei et al. 1990) involving sufficiently small scalar ε.

For the H^∞ output feedback control, on the other hand, the control force $\mathbf{u}$ is determined from the following system carrying the costate ξ :

$$\dot{\xi} = \hat{\mathbf{A}}\xi + \hat{\mathbf{B}}\mathbf{y} \tag{6a}$$

$$\mathbf{u} = \hat{\mathbf{C}}\xi + \hat{\mathbf{D}}\mathbf{y} \tag{6b}$$

The output feedback control is to determine matrices $\hat{\mathbf{A}}, \hat{\mathbf{B}}, \hat{\mathbf{C}}, \hat{\mathbf{D}}$ in the above system (Eqs.6a-b) so as to satisfy the same condition as Eq.4. To obtain $\hat{\mathbf{A}}, \hat{\mathbf{B}}, \hat{\mathbf{C}}, \hat{\mathbf{D}}$, the following dual system is introduced:

$$\dot{\mathbf{x}} = \mathbf{A}_1{}^T\mathbf{x} + \mathbf{C}_1{}^T\mathbf{w} + \mathbf{C}_2{}^T\mathbf{u} \tag{7a}$$

$$\mathbf{z} = \mathbf{B}_1{}^T\mathbf{x} + \mathbf{D}_{11}{}^T\mathbf{w} + \mathbf{D}_{21}{}^T\mathbf{u} \tag{7b}$$

$$\mathbf{y} = \mathbf{B}_2{}^T\mathbf{x} + \mathbf{D}_{12}{}^T\mathbf{w} \tag{7c}$$

The dual system control force $\mathbf{u}$ is regulated in the form $\mathbf{u} = \mathbf{K}^T\mathbf{x}$ so as to qualify

$$||\Phi_d||_\infty < \gamma \tag{8}$$

where Φ_d = transfer function from $\mathbf{w}$ to $\mathbf{z}$ for the dual system. The control gain $\mathbf{K}^T$ can be given by

$$\mathbf{K}^T = -\{(1/2\mu)\Phi_K{}^T\Phi_K + \Xi_K\}\mathbf{B}_K{}^T\mathbf{Q} - \Xi_K\mathbf{F}_K{}^T\mathbf{C}_K \tag{9}$$

where $\mathbf{Q}$ is also derived from another Riccati equation with sufficiently small μ for the dual system.

Accordingly, $\hat{\mathbf{A}}, \hat{\mathbf{B}}, \hat{\mathbf{C}}, \hat{\mathbf{D}}$ can be derived from $\mathbf{P}, \mathbf{F}, \mathbf{Q}, \mathbf{K}$. When satisfying $(\lambda_{max}[\mathbf{D}_{11}{}^T\mathbf{D}_{11}])^{1/2} < \gamma$ and $\gamma^2\mathbf{Q}^{-1} > \mathbf{P}$ for the strictly proper case (i.e., $\hat{\mathbf{D}} = 0$) the system provides (Sampei et al. 1990) :

$$\begin{aligned}\hat{\mathbf{A}} &= \mathbf{A}_1+\mathbf{B}_2\hat{\mathbf{C}}-\hat{\mathbf{B}}\mathbf{C}_2-(\gamma^2\mathbf{Q}^{-1}-\mathbf{P})^{-1}\mathbf{M}\\ \hat{\mathbf{B}} &= -\gamma^2(\gamma^2\mathbf{Q}^{-1}-\mathbf{P})^{-1}\mathbf{Q}^{-1}\mathbf{K}\\ \hat{\mathbf{C}} &= \mathbf{F}\end{aligned} \tag{10}$$

where

$$\begin{aligned}\mathbf{M} &= \hat{\mathbf{C}}^T\mathbf{B}_2{}^T\mathbf{P}+\hat{\mathbf{C}}^T\mathbf{D}_{12}{}^T\{\mathbf{C}_1+\mathbf{D}_{12}\hat{\mathbf{C}}\}+\varepsilon\mathbf{I}\\ &+\{(\gamma^2\mathbf{Q}^{-1}-\mathbf{P})(\hat{\mathbf{B}}\mathbf{D}_{21}-\mathbf{B}_1)+\hat{\mathbf{C}}^T\mathbf{D}_{12}{}^T\mathbf{D}_{11}\}\\ &\times\{\gamma^2\mathbf{I}-\mathbf{D}_{11}{}^T\mathbf{D}_{11}\}^{-1}\{\mathbf{B}_1{}^T\mathbf{P}+\mathbf{D}_{11}{}^T(\mathbf{C}_1+\mathbf{D}_{12}\hat{\mathbf{C}})\}\end{aligned} \tag{11}$$

Robust stability is one of the beneficial features of the H^∞ control. Brief remarks of the robust stability of H^∞ control are presented.

Mathematical model for a structural system cannot be constructed without involving several kinds of errors such as associated with linearization, model reduction, parame-

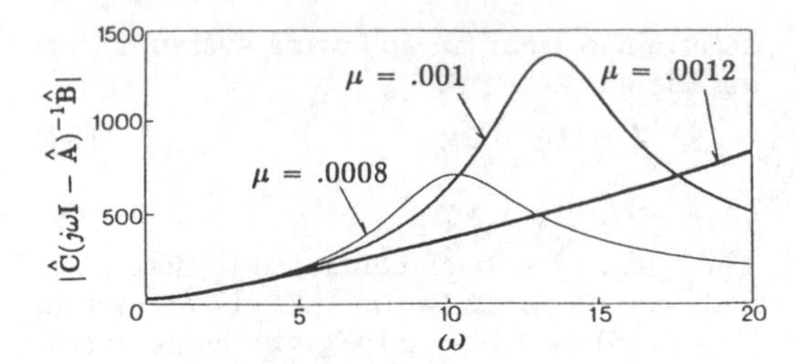

Fig.9 Frequency responses of displacement FB control gain with $\gamma = 0.4$

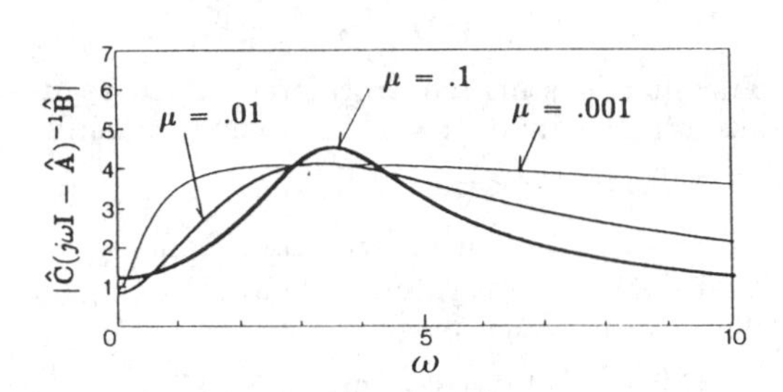

Fig.10 Frequency responses of velocity FB control gain with $\gamma = 1.8$

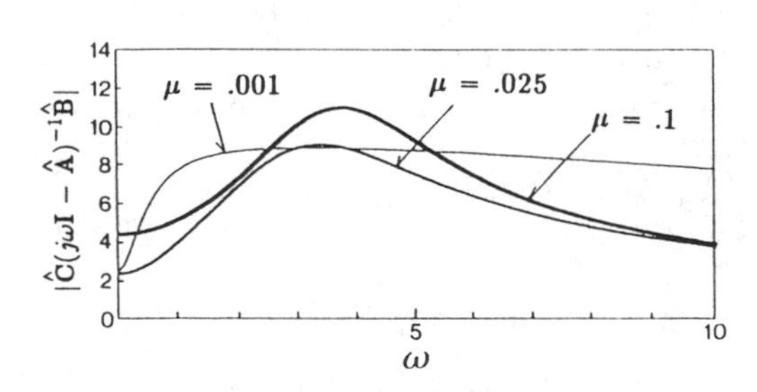

Fig.11 Frequency responses of velocity FB control gain with $\gamma = 1.0$

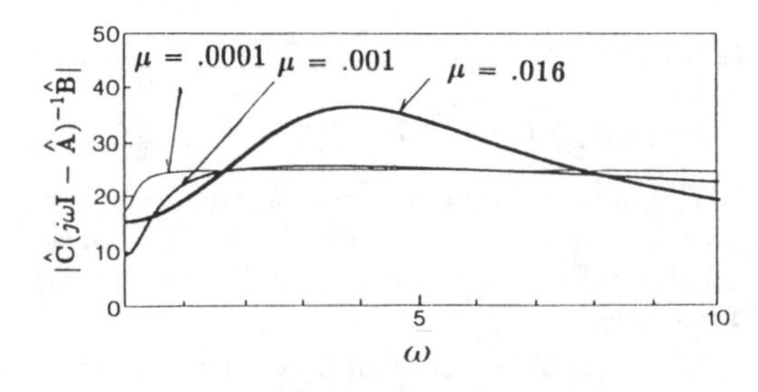

Fig.12 Frequency responses of velocity FB control gain with $\gamma = 0.4$

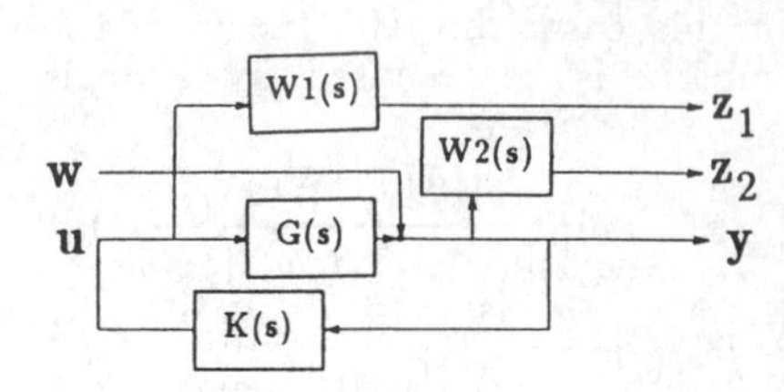

Fig.13 Block diagram

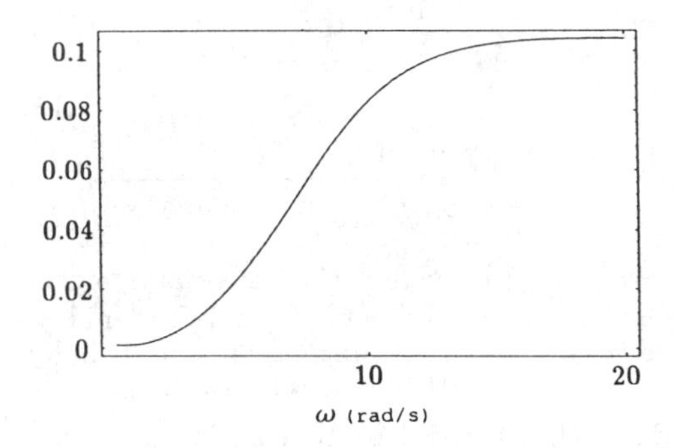

Fig.14 Frequency response of WF

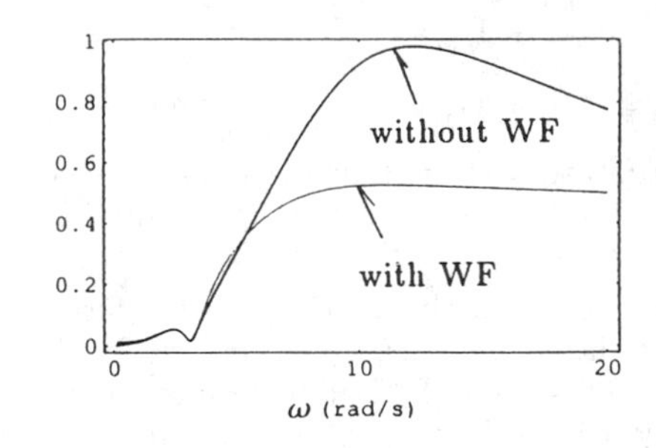

Fig.15 σ plot of $\mathbf{W_1K(I - GK)^{-1}}$

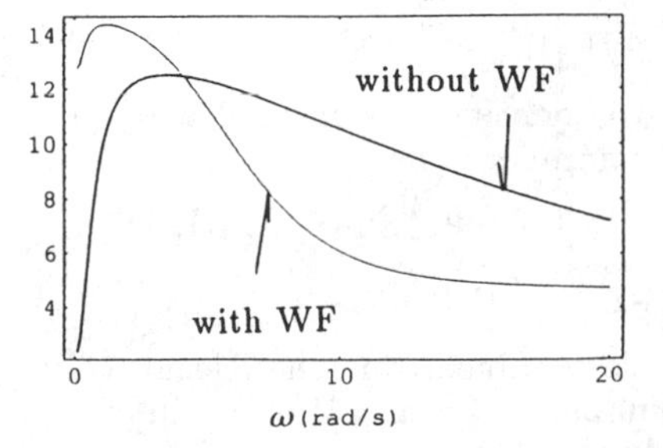

Fig.16 Frequency responses of control gain

ter uncertainty, time delay, measurement noise etc. These errors are likely to exist in the high-frequency regions. On the other hand, those responses which are to be controlled mainly exist in the low-frequency regions. This fact gives a practical significance to take a weighting function in the frequency domain.

3. APPLICATION OF H^∞ CONTROL THEORY

Example 1

The H^∞ control theory is applied to a single-degree-of-freedom (SDOF) structural system.

Consider an earthquake-excited SDOF model with the following notations: m = mass; x = relative displacement of mass; c = damping coefficient; k = stiffness; q_0 = earthquake acceleration as an external disturbance; u = active control force.

Assuming $\mathbf{z} = [x\ \dot{x}]^T$ for simplicity in this example, the state vector expressions like Eq.7 can be obtained. As a result, the coefficient matrices and vectors are:

$$\mathbf{A}_1 = \begin{bmatrix} 0 & 1 \\ -\dfrac{k}{m} & -\dfrac{c}{m} \end{bmatrix}, \mathbf{B}_1 = \begin{bmatrix} 0 \\ -1 \end{bmatrix}, \mathbf{B}_2 = \begin{bmatrix} 0 \\ \dfrac{1}{m} \end{bmatrix}$$

$$\mathbf{C}_1 = \begin{bmatrix} 1 & 0 \\ 0 & 1 \end{bmatrix}, \mathbf{D}_{11} = \begin{bmatrix} 0 \\ 0 \end{bmatrix}, \mathbf{D}_{12} = \begin{bmatrix} 0 \\ 0 \end{bmatrix} \quad (12)$$

and $\mathbf{C}_2 = [1\ 0]$ and $\mathbf{C}_2 = [0\ 1]$ for the displacement and velocity feedbacks, respectively.

State FB control --- The physical data for this example are : $m = 10\ [kg]$, $c = 1.8[N{\cdot}\sec/m]$, $k = 90[N/m]$ and four different values 2.6, 1.8, 1.4, 1.0 are employed for γ. In this case, the transfer function $\mathbf{\Phi}$ from $\mathbf{w}$ to $\mathbf{z}$ becomes a vector.

Figure 1, presenting the relationship between $||\Phi||_\infty$ and ε, indicates that smaller value of γ provides smaller region of ε in which the solution $\mathbf{P} > 0$ of the Riccati equation can be obtained.

In the next stage, the H^∞ control is compared with the optimal control on the basis of the state feedback. In comparing both control systems, earthquake acceleration as an external excitation is set to be a zero-mean white noise, because for a white noise excitation the optimal control can be theoretically performed.

The optimal control system under a white noise excitation $\mathbf{w}_0$ (i.e. $\mathbf{w}_0 = q_0$ for this SDOF system) is defined by:

$$\dot{\mathbf{x}} = \mathbf{A}\mathbf{x} + \mathbf{B}\mathbf{u} + \mathbf{D}\mathbf{w}_0 \quad (13a)$$

$$\mathbf{y} = \mathbf{C}\mathbf{x} \quad (13b)$$

For the optimal control system with the state feedback, the feedback gain $\mathbf{F}$ is regulated so as to minimize the following quadratic performance index:

$$J = E[\mathbf{x}^T(t)\mathbf{Q}_0\mathbf{x}(t) + \mathbf{u}^T(t)\mathbf{R}_0\mathbf{u}(t)] \quad (14)$$

where $\mathbf{Q}_0$ and $\mathbf{R}_0$ are weighting matrices and $E[\cdot]$ denotes expectation of $[\cdot]$. To produce the same circumstances as the H^∞ control, the matrices $\mathbf{A, B, C, D}$ in Eqs.13a-c are set to be identical with the corresponding matrices of the H^∞ control system. Then

$$\mathbf{A} = \mathbf{A}_1\ , \mathbf{B} = \mathbf{B}_2\ , \mathbf{C} = \mathbf{C}_2\ , \mathbf{D} = \mathbf{B}_1 \quad (15)$$

and the spectral density of $\mathbf{w}_0$ is equal to 1. Providing

$$\mathbf{Q}_0 = \begin{bmatrix} 1 & 0 \\ 0 & 1 \end{bmatrix}, \mathbf{R}_0 = r\ (scalar) \quad (16)$$

and having r as a parameter, Fig.2 shows the relationship between $trace[\mathbf{V}_x]$ and $trace[\mathbf{V}_u]$ for the state-feedback H^∞ control system along with the optimal skelton curve. In this example, $trace[\mathbf{V}_x]$ and $trace[\mathbf{V}_u]$ represent:

$$trace\ [\mathbf{V}_x] = E[x^2 + \dot{x}^2]\ , \qquad trace\ [\mathbf{V}_u] = E[u^2] \quad (17)$$

in which $\mathbf{V}_x$ is the covariance matrix of the state vector $\mathbf{x}$, while $\mathbf{V}_u$, in this case, is the variance of the control force u. In passing, it is noted that $\mathbf{V}_u$ for the general case is also a matrix and hence the symbol " $trace[\mathbf{V}_u]$ " is used in Eq.17 as well as Figs.2, 5, and 6.

Output FB --- The following discussion about the output feedback system is limited to the case of strictly proper control system i.e. $\hat{\mathbf{D}} = \mathbf{0}$.

Figure 3 demonstrates the relationship

between $||\Phi||_\infty$ and μ for the displacement feedback, whereas Fig.4 shows the same relationship for the velocity feedback. From these two figures, it is found that smaller values of μ provides $||\Phi||_\infty$ closer to the given value of γ. Judging from the values of μ, the velocity feedback basis has wider range of control force solution than the displacement feedback. In other words, the H^∞ velocity feedback is easier to perform than the displacement feedback.

Substitution of Eq.6 into Eq.3 for the original system with $\hat{\mathbf{D}} = \mathbf{D}_{21} = 0$ yields:

$$\dot{\tilde{\mathbf{x}}} = \begin{bmatrix} \mathbf{A}_1 & \mathbf{B}_2\hat{\mathbf{C}} \\ \hat{\mathbf{B}}\mathbf{C}_2 & \hat{\mathbf{A}} \end{bmatrix} \tilde{\mathbf{x}} + \begin{bmatrix} \mathbf{B}_1 \\ \mathbf{0} \end{bmatrix} \mathbf{w} \qquad (18)$$

$$\textit{with } \tilde{\mathbf{x}} = [\mathbf{x}^T \ \xi^T]^T$$

In the H^∞ control system under a white noise excitation (i.e. $\mathbf{w} = \mathbf{w}_0$) with $\mathbf{D}_{11} = \mathbf{D}_{12} = \mathbf{D}_{21} = \mathbf{0}$, the relationships between $trace[\mathbf{V}_x]$ and $trace[\mathbf{V}_u]$ for the displacement and velocity feedbacks are shown in Figs.5 and 6, respectively. Both figures establish the superiority of the velocity feedback control from the practical view point of control efficiencies.

In the output feedback control, the control force is given in the form $\mathbf{u}(s) = \mathbf{K}(s)\mathbf{y}(s)$ with $\mathbf{K}(s) = \hat{\mathbf{C}}(\mathbf{I} - \hat{\mathbf{A}})^{-1}\hat{\mathbf{B}}$. Corresponding to several values of γ, Figs.7-9 and 10-12 demonstrate the magnitudes of $\mathbf{K}(s)$ for the displacement and velocity feedback control systems, respectively.

Example 2

Assuming modeling errors as additive, the control force $\mathbf{u}(s) = \mathbf{K}(s)\,\mathbf{y}(s)$ is provided so as to satisfy the following criterion:

$$\left\| \begin{matrix} \mathbf{W}_1\mathbf{K}(\mathbf{I} - \mathbf{G}\mathbf{K})^{-1} \\ \mathbf{W}_2(\mathbf{I} - \mathbf{G}\mathbf{K})^{-1} \end{matrix} \right\|_\infty < 1 \qquad (19)$$

where $\mathbf{G}(s)$ represents the transfer function from $\mathbf{u}$ to $\mathbf{y}$, and $\mathbf{W}_1(s)$ and $\mathbf{W}_2(s)$ generally denote the weighting functions (WF) for ensuring the robust stability in the high-frequency regions and for reducing the responses in the low-frequency regions, respectively. Nevertheless, this case does not have to consider $\mathbf{W}_2(s)$ because of treating a SDOF system. Then,

$$\mathbf{W}_1(s) = \mathbf{d}_{w_1} + \mathbf{c}_{w_1}(s\mathbf{I} - \mathbf{A})^{-1}\mathbf{b}_{w_1} \qquad (20a)$$

$$\mathbf{W}_2(s) = \mathbf{I}/\gamma \qquad (20b)$$

$$\mathbf{G}(s) = \mathbf{C}(s\mathbf{I} - \mathbf{A})^{-1}\mathbf{B} \qquad (21)$$

In this control system, whose block diagram is shown in Fig.13, the expression with the transfer matrix for the system is given by:

$$\begin{bmatrix} \mathbf{z}_1 \\ \mathbf{z}_2 \\ \mathbf{y} \end{bmatrix} = \begin{bmatrix} \mathbf{0} & \mathbf{W}_1 \\ \mathbf{I}/\gamma & \mathbf{G}/\gamma \\ \mathbf{I} & \mathbf{G} \end{bmatrix} \begin{bmatrix} \mathbf{w}_f \\ \mathbf{u} \end{bmatrix} \qquad (22)$$

The realization of the above expression leads to:

$$\begin{bmatrix} \mathbf{A}_1 & \mathbf{B}_1 & \mathbf{B}_2 \\ \mathbf{C}_1 & \mathbf{D}_{11} & \mathbf{D}_{12} \\ \mathbf{C}_2 & \mathbf{D}_{21} & \mathbf{D}_{22} \end{bmatrix} = \begin{bmatrix} \mathbf{A}_{w_1} & \mathbf{0} & \mathbf{0} & \mathbf{b}_{w_1} \\ \mathbf{0} & \mathbf{A} & \mathbf{0} & \mathbf{B} \\ \mathbf{c}_{w_1} & \mathbf{0} & \mathbf{0} & \mathbf{d}_{w_1} \\ \mathbf{0} & \mathbf{C}/\gamma & \mathbf{I}/\gamma & \mathbf{0} \\ \mathbf{0} & \mathbf{C} & \mathbf{I} & \mathbf{0} \end{bmatrix} \qquad (23)$$

In this example, the weighting function

$$\mathbf{W}_1(s) = k_{w_1} \frac{s^2 + 2\zeta_2\omega_2 s + \omega_2^2}{s^2 + 2\zeta_1\omega_1 s + \omega_1^2} \qquad (24)$$

is employed with the parameters: $k_{w_1} = 0.1, \omega_1 = 10.0, \zeta_1 = 0.6$, $\omega_2 = 1.0, \zeta_2 = 0.7$. This weighting function thus assumed is shown in Fig.14.

Two kinds of H^∞ velocity feedback control systems are compared: one takes the weighting function and the other does not. Comparison is made between both systems under the condition that the H^∞ norms of the transfer functions from the disturbance $\mathbf{w}$ to the velocity $\dot{x}$ are identical each other. In Fig.15, the robust stabilities are compared between both systems. Figure 16 illustrates schematically the magnitude of the frequency responses of the control systems, demonstrating that the robust stability in those frequency ranges involving the modeling errors improves and that the sufficient conditions for the robust stability be hardly fulfilled without taking the weighting function.

4. CONCLUDING REMARKS

Detailed description of H^∞ control theory has been presented along with the illustration of the H^∞ norm. In this control theory, the H^∞ norm of the transfer function is characteristically employed as a control criterion.

The H^∞ control method has been applied to a single-degree-of-freedom system. The results obtained for both the H^∞ state and output feedback control systems are presented. Compared to the conventional optimal control, a superiority of the H^∞ control is demonstrated in the sense of providing the robust stability. Then, the robust stability has been qualitatively assessed by introducing the weighting function. Either with or without the weighting function employed, the H^∞ control system cuts down the control gain in the high-frequency range where the modeling error is likely to exist. However, this fact does not necessarily provide same significance. For the system without any weighting function, reducing the responses around the resonance frequency eventually leads to the decrease of the control gain in the high frequency range. On the contrary, the system with a weighting function performs the robust control.

The H^∞ control theory has been developed to take the frequency-domain characteristics into account on the basis of the modern control engineering. It is expected that the H^∞ control design can be a powerful and efficient tool to perform the robust stability control.

REFERENCE

Doyle, J.C., Glover, K., Khargonekar, P.P., Francis, B.A. 1989. State-Space Solutions to Standard H^2 and H^∞ Control Problems. *IEEE Transaction on Automatic Control.* Vol.34. No.8: 831-847

Francis, B.A. 1987. A Course in H^∞ Control Theory. Springer-Verlag. New York

Ikeda, Y., et al. 1991. Effectiveness of Realized Seismic-Response-Controlled Structure with Active Mass Driver System. *J. of Struct. Constr. Engng. AIJ.* No.420: 133-141

Kobori, T., Koshika, N., Yamada, K., Ikeda, Y. 1991. Seismic-Response-Controlled Structure with Active Mass Driver System Part1 : Design. *Earthquake Engineering and Structural Dynamics.* Vol.20: 133-149

Sampei, M., Mita, T., Nakamichi, M. 1990. An Algebraic Approach to H^∞ Output Feedback Control Problems. *Systems & Control Letters 14* : 13-24. North-Holland

Yamada, N., Nishitani, A. 1993. Application of H^∞ control to a structural system *J. of Struct. Constr. Engng. AIJ.* No.444: 23-31

Structural Safety & Reliability, Schuëller, Shinozuka & Yao (eds) © 1994 Balkema, Rotterdam, ISBN 90 5410 357 4

Reliability-based measures of stability for actively controlled structures

B.F. Spencer, Jr., M.K. Sain & J.C. Kantor
University of Notre Dame, Ind., USA

ABSTRACT: Uncertainty is inherent in the control of civil engineering structures. To be effective in the presence of uncertainty, a control strategy must have robust performance and stability characteristics. This paper develops quantitative methods for assessing the likelihood that instability will result in a controlled structure due to uncertainty. Numerical examples are given to illustrate the concepts.

1 INTRODUCTION

Remarkable progress has been made in recent years toward the practical implementation of active control for mitigation of excessive structural vibration (see, for example, Soong, 1990; Housner and Masri, 1990; ATC-17, 1993). Although uncertainties are well-known to degrade control performance and to lead to the possibility of system instabilities, an important aspect of the control problem that has received limited attention is the characterization of the effect of uncertainty on the stability of a controlled structure.

The effect of parametric uncertainties on controlled systems is usually defined through certain stability margins (e.g., de Gaston and Safonov, 1988; Sideris and Sánchez Peña, 1989). These approaches generally bound the uncertainty on the model parameters and produce a stability margin that is a *worst-case* measure of the system stability. Stability margins based on such a worst-case analyses do not provide a consistent measure of the relative likelihood of system instability, *i.e.*, one system possessing a larger stability margin than another system does not necessarily imply that the likelihood of system instability is smaller. This limitation results because the stability margin does not explicitly account for the variability or probabilistic structure of the uncertain parameters. Moreover, for many problems in structural control, hard bounds on the uncertain parameters are not meaningful. A probabilistic description provides a more natural and realistic portrayal. In the case of multiple uncertain parameters, a probabilistic description is necessary to avoid unreasonable conservatism.

A systematic approach for determining the probability that instability will result from the uncertainties inherently present in a controlled structure is developed herein. The formulation employs an eigenvalue criterion for characterizing the stability of the system. Efficient first and second order reliability methods (FORM/SORM) are shown to be appropriate for calculating the probability of system instability. A single-degree-of-freedom controlled structure is examined in which the uncertain parameters are mass, damping, stiffness and controller time delay. The methodology is illustrated with several control strategies. The importance of selection of appropriate probability distributions for the parameters is also discussed.

2 PROBLEM FORMULATION

We assume that the uncertainty in the system can be modeled as a random vector of real parameters $\boldsymbol{\Delta}$ with a given mean $\boldsymbol{\mu}_{\boldsymbol{\Delta}}$, covariance $\boldsymbol{\Sigma}_{\boldsymbol{\Delta}}$, and joint probability distribution $F_{\boldsymbol{\Delta}}(\boldsymbol{\delta})$. The state space representation of the equations of motion for an n-degree-of-freedom structure is

$$\dot{\mathbf{z}} = \mathbf{A}(\boldsymbol{\Delta})\,\mathbf{z} + \mathbf{B}(\boldsymbol{\Delta})\,\mathbf{u}, \qquad (1)$$

where the vector $\mathbf{z}$ is a $2n$-dimensional state vector of displacements and velocities, $\mathbf{A}$ is a $2n \times 2n$ matrix composed of the parameters of the structure (masses, stiffnesses, damping values, *etc.*), $\mathbf{u}$ is an r-dimensional vector of control forces, and $\mathbf{B}$ is a $2n \times r$ matrix specifying the points of application of the control forces. The matrices $\mathbf{A}$ and $\mathbf{B}$ are functions of the vector of random parameters $\mathbf{\Delta}$. The structural excitation is neglected in Eq. (1) as it does not influence the (internal) stability of linear systems.

For a broad class of control strategies, the closed-loop state space description can be written as

$$\dot{\tilde{\mathbf{z}}} = \mathbf{A}_{\mathrm{cl}}(\mathbf{\Delta})\,\tilde{\mathbf{z}}, \tag{2}$$

where $\tilde{\mathbf{z}}$ is the state vector (*e.g.*, $\tilde{\mathbf{z}}$ may also contain states of the controller or a disturbance shaping filter), and $\mathbf{A}_{\mathrm{cl}}(\mathbf{\Delta})$ is the closed-loop state matrix. If we assume that the structure state variables are perfectly measurable and that state feedback control is employed, *i.e.*,

$$\mathbf{u} = -\mathbf{K}\mathbf{z}, \tag{3}$$

then $\tilde{\mathbf{z}} = \mathbf{z}$, and the closed-loop state space matrix is given by

$$\mathbf{A}_{\mathrm{cl}}(\mathbf{\Delta}) = \mathbf{A}(\mathbf{\Delta}) - \mathbf{B}(\mathbf{\Delta})\,\mathbf{K}, \tag{4}$$

where $\mathbf{K}$ is the control gain matrix.

The probability that the control of the structure described above is asymptotically stable, p_{st}, depends on the real part of the eigenvalues of the closed-loop state space matrix. The system is asymptotically stable if the real part of all the eigenvalues of the closed-loop state space matrix is negative. The probability of the system being stable p_{st} is then given by

$$\begin{aligned} p_{\mathrm{st}} &= P\{\bigcap_{j=1}^{m} \mathrm{Re}[\lambda_j(\mathbf{\Delta})] < 0\} \\ &= P\{\max_j \mathrm{Re}[\lambda_j(\mathbf{\Delta})] < 0\} \\ &= 1 - P\{\bigcup_{j=1}^{m} \mathrm{Re}[\lambda_j(\mathbf{\Delta})] \geq 0\} \\ &= 1 - P\{\max_j \mathrm{Re}[\lambda_j(\mathbf{\Delta})] \geq 0\} \\ &= 1 - p_{\mathrm{f}}, \end{aligned} \tag{5}$$

where $m = 2n$, λ_j are the roots of $\det[\mathbf{A}_{\mathrm{cl}}(\mathbf{\Delta}) - s\mathbf{I}] = 0$, and p_{f} is the probability that the system is unstable.

In terms of the probability density function $f_{\mathbf{\Delta}}(\boldsymbol{\delta})$, the probability of instability can be written as

$$\begin{aligned} p_{\mathrm{f}} &= \int\!\cdots\!\int_{\bigcup_{j=1}^{m} \mathrm{Re}[\lambda_j(\boldsymbol{\delta})] \geq 0} f_{\mathbf{\Delta}}(\boldsymbol{\delta})\, d\boldsymbol{\delta} \\ &= \int\!\cdots\!\int_{\max_j \mathrm{Re}[\lambda_j(\boldsymbol{\delta})] \geq 0} f_{\mathbf{\Delta}}(\boldsymbol{\delta})\, d\boldsymbol{\delta}\,. \end{aligned} \tag{6}$$

The domain of integration for Eq. (6) is illustrated in Fig. 1 for two uncertain parameters. In cases where the domain consists of disconnected regions, then p_{f} is computed only over the region containing the nominal case.

The multi-dimensional integral in Eq. (6) is generally too complex to evaluate directly for arbitrary probability distributions over arbitrary domains. An upper bound on the probability of the system being unstable when $f_{\mathbf{\Delta}}(\boldsymbol{\delta})$ is jointly Gaussian has been reported in Kantor and Spencer (1992).

The system whose failure probability is described by Eq. (6) can be viewed as a classical reliability problem for a series of components defined as $g_i(\mathbf{\Delta}) = -\mathrm{Re}[\lambda_j]$. If any of the components fail (*i.e.*, $\mathrm{Re}[\lambda_j] \geq 0$, or equivalent-

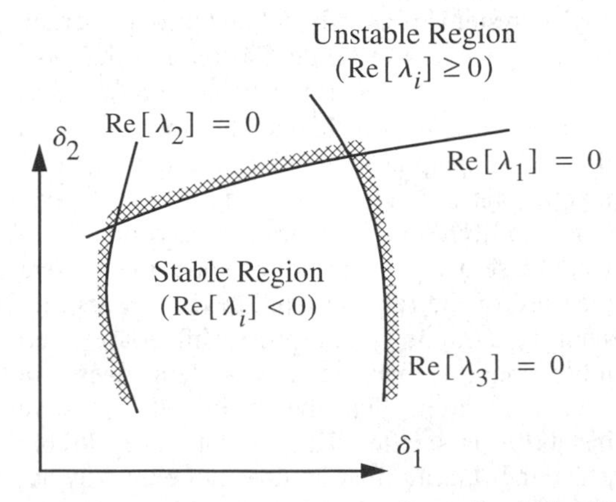

Figure 1. Illustration of the region of stability in terms of eigenvalues for two uncertain parameters.

ly, $g_j(\mathbf{\Delta}) \le 0$), then the entire system is considered failed (*i.e.*, the structure is unstable). Two classes of probability calculations are generally applicable to the problem at hand, Monte Carlo Simulation (MCS) and First- Second-Order Reliability Methods (FORM/SORM). Simple MCS techniques are often computationally intractable for the small probabilities of failure (*i.e.*, probabilities of instability) expected for controlled structures. Alternatively, the FORM/SORM approach, an analytical probability integration scheme, is particularly efficient and accurate for systems with small failure probabilities such as controlled structures (Madsen, *et al.*, 1986; Melchers, 1987).

3 NUMERICAL EXAMPLES

To clarify the above development, consider the single-degree-of-freedom controlled structure reported by Chung, *et al.* (1988). The governing equation of motion is

$$m\ddot{x} + c\dot{x} + kx = -(4k_c \cos\alpha)\,u, \qquad (7)$$

where x is the displacement of the first floor mass with respect to the ground, u is the position of the hydraulic actuator, the walls of the structure have stiffness $k = 7934$ lb/in and damping $c = 9.02$ lb-sec/in, and the floor is of mass $m = 16.69$ lb-sec^2/in. The tendons have stiffness $k_c = 2124$ lb/in and are positioned at an angle $\alpha = 36°$ with the horizontal. Again, we neglect the exogenous structural excitation as it has no effect on the internal stability of a linear system. Control is obtained by positioning the cylinder at the base of the structure, thereby stretching one set of active tendons and releasing the second set of active tendons.

A time delay is assumed to be present in the control loop. For simplicity, we assume that the time delay in the control loop can be modeled as shown in Fig. 2 and has a magnitude $\tau = 20$ msec. A pure time delay has the transfer function $P(s) = e^{-\tau s}$, which has an infinite dimensional state space representation. For this example, a Padé approximant of $P(s)$, denoted $f_p(s)$, provides a useful finite dimensional model of a pure time delay (Sain, *et al.*, 1992).

Using the Padé approximation to model the

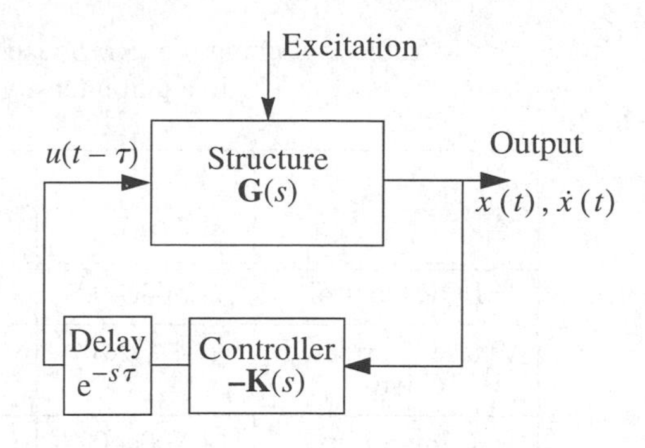

Figure 2. Block diagram representation for the single-degree-of-freedom structure with a delayed controller.

controller time delay for the system in Eq. (7), the characteristic equation is obtained:

$$-\eta\{k_1(s) + sk_2(s)\}f_p(s) + (s^2 + 2\zeta\omega s + \omega^2) = 0, \qquad (8)$$

where $k_1(s)$ and $k_2(s)$ are the components of the transfer function vector for the controller, $\eta = 4k_c\cos\alpha/m$, $\omega^2 = k/m$ and $\zeta = c/(2\sqrt{mk})$. Note that $k_1(s)$ and $k_2(s)$ are constants for perfect state feedback. Herein we employ a third-order Padé approximation,

$$f_3(s) = \frac{1 - 60\tau s + 12\tau^2 s^2 - \tau^3 s^3}{1 + 60\tau s + 12\tau^2 s^2 + \tau^3 s^3}, \qquad (9)$$

which was shown by Sain, *et al.* (1992) to adequately model the time delay in the neighborhood of the crossover frequency for this example structure.

The controller is obtained by minimizing the quadratic performance index given by

$$J = \int_0^\infty (kx^2 + \gamma k_c u^2)\,dt. \qquad (10)$$

The probabilities of instability are examined for three specific control strategies. The first control strategy considered neglects the system time delay and minimizes the performance index given in Eq. (10) to obtain a constant gain, state feedback controller. This control strategy is denoted the Linear Quadratic Regulator (LQR) controller. The second control strategy employs modified

Table 1: Summary of RMS responses for the nominal structure subject to a unit intensity white noise excitation.

	σ_x (in)	$\sigma_{\dot{x}}$ (in/sec)	σ_u (in)	$E[J]$[a] (in-lbs)
LQR Control	1.564e–02	6.124e–01	4.713e–02	6.659
Phase–Corrected Control	1.417e–02	4.931e–01	4.248e–02	5.425
Kalman Filter Control (a)	1.790e–02	4.967e–01	2.716e–02	4.110
Kalman Filter Control (b)	1.417e–02	4.698e–01	3.847e–02	4.736

a. For comparison, all values of the performance index are calculated for $\gamma = 1$.

Table 2: Summary of Reliability Calculations for Normal Variates.

	Reliability Index, β^{SORM}	p_f^{SORM}	Squared Importance Factors (%)			
			τ	m	c	k
LQR Control	3.384	3.571e–04	49	49	0	2
Phase–Corrected Control	3.832	6.354e–05	32	67	0	1
Kalman Filter Control (a)	6.331	1.221e–10	15	84	0	1
Kalman Filter Control (b)	5.111	1.600e–07	22	77	0	1

Table 3: .Summary of Sensitivity Analysis for Normal Variates.

	Sensitivity Factors (FORM)				
	$\partial\beta/\partial\mu_\tau$ $\partial\beta/\partial\sigma_\tau$	$\partial\beta/\partial\mu_m$ $\partial\beta/\partial\sigma_m$	$\partial\beta/\partial\mu_c$ $\partial\beta/\partial\sigma_c$	$\partial\beta/\partial\mu_k$ $\partial\beta/\partial\sigma_k$	$\partial\beta/\partial k_1$ $\partial\beta/\partial k_2$
LQR Control	–3.513e+02 –8.392e+02	4.203e–01 –1.003e+00	1.314e–02 –5.291e–04	–1.513e–04 –6.182e–05	1.926e+00 2.931e+01
Phase–Corrected Control	–2.828e+02 –6.139e+02	4.906e–01 –1.542e–00	7.732e–03 –2.071e–04	–1.237e–04 –4.656e–05	8.753e–01 5.149e+01
Kalman Filter Control (a)	–1.947e+02 –4.809e+02	5.497e–01 –3.199e+00	7.022e–03 –2.825e–04	–1.041e–04 –5.457e–05	
Kalman Filter Control (b)	–2.339e+02 –5.613e+02	5.274e–01 –2.382e+00	6.108e–03 –1.728e–04	–1.018e–04 –4.222e–05	

LQR feedback control gains which are *phase-corrected* to account for the time delay effect (McGreevy, *et al.*, 1988; Chung, *et al.*, 1988; Soong, 1990). Finally, we also consider a control strategy that explicitly accounts for the time delay in the control design. This approach is a loop transfer recovery procedure, is detailed in Sain, *et al.* (1992), and is denoted the Kalman filter controller.

In what follows, we assume the controllers to be known and deterministic. The mass m, stiffness k, damping c, and time delay τ are assumed to have mean values equal to their nominal values given previously and to have a coefficient of variation of 10%. Although not a limitation of the FORM/SORM approach, the variates are assumed to be statistically independent for simplicity.

For the LQR and phase-corrected controllers, we have chosen $\gamma = 1$ for the performance index in Eq. (10). Two realizations of the Kalman filter controller are considered. The first Kalman filter controller minimizes the performance index using $\gamma = 1$. The second Kalman filter control design uses $\gamma = 0.297$ so as to match the RMS displacement response of the phase-corrected controller. For both cases, a loop transfer recovery parameter of $\rho = 10^{-4}$ was chosen (see Sain, *et al.*, 1992). The RMS displacement responses σ_x, $\sigma_{\dot{x}}$, the RMS control action σ_u, and the performance index J for the nominal structure subjected to a unit intensity white noise are given in Table 1. Note that the smallest performance index is obtained by Kalman filter controller (a), although at the expense of a slightly larger RMS displacement response.

The probabilistic analysis software PROBAN (1991) was employed to perform all reliability calculations.

Table 2 provides a summary of the SORM results assuming that all variates are normally distributed. As shown here, the probability of instability is reduced by a factor of 5 when the phase-corrected controller is employed as compared to the uncorrected LQR controller. Note that the nominal phase-corrected controller provides a smaller RMS response while requiring smaller RMS control action.

Results for the two Kalman filter compensators are also listed in Table 2. The first minimizes the same performance index as the LQR and phase-corrected controllers, *i.e.*, for $\gamma = 1$. As shown here, a much lower probability of failure is achieved. To have a more direct comparison of the Kalman filter controller to the phase-corrected controller, we have selected $\gamma = 0.297$ in Eq. (10) so as to match the nominal RMS displacement of the phase corrected case. The results in Table 2 shows that the Kalman filter approach reduces the probability of instability by a factor of nearly 400 over the phase-corrected approach, while achieving the same nominal RMS displacement.

Also provided in Table 2 are the squared *importance factors* for the uncertainties using the various control strategies. The importance factors provide an indication of the relative importance of modeling the uncertainty in the respective parameters. Comparing the importance factors in Table 2 resulting from use of the various control strategies, we see that by correcting for the system time delay, the relative importance of modeling the uncertainty in the time delay is reduced, while the importance factor for the mass increases. This result indicates that reducing the variance in the mass would provide the greatest reduction in the probability of failure.

Somewhat surprisingly, we point out that the importance of the uncertainty in the damping c is negligible. This effect is understood to occur because the structural damping and its associated uncertainty is small in comparison with the additional damping added by the controller. The control also has the effect of modifying the effective structural stiffness and lessening the importance of uncertainty in its nominal value.

Another useful quantity readily obtained in the FORM/SORM approach is the parametric sensitivity factor $\partial\beta/\partial\theta$, where θ is an analysis parameter. The sensitivity factors with respect to the mean and standard deviation of the random variables, as well as the control gains for the LQR and phase-corrected controllers, are tabulated in Table 3. As expected, the results indicate that an increase in any of the standard deviations of the system parameters results in a decrease in β and an associated increase in the probability of instability. In addition, an increase in the mean time delay or the mean stiffness increases the probability of instability, whereas an increase in the mean mass or mean damping constant results

Table 4: Summary of Reliability Calculations for Gumbel Variates.

	Reliability Index, β^{SORM}	p_f^{SORM}	Squared Importance Factors (%)			
			τ	m	c	k
LQR Control	2.817	2.426e–03	93	6	0	1
Phase–Corrected Control	3.396	3.417e–04	90	9	0	1
Kalman Filter Control (a)	5.163	1.216e–07	93	5	0	2
Kalman Filter Control (b)	4.222	1.209e–05	94	5	0	1

Table 5: Summary of Sensitivity Analysis for Gumbel Variates.

	Sensitivity Factors (FORM)				
	$\partial\beta/\partial\mu_\tau$ $\partial\beta/\partial\sigma_\tau$	$\partial\beta/\partial\mu_m$ $\partial\beta/\partial\sigma_m$	$\partial\beta/\partial\mu_c$ $\partial\beta/\partial\sigma_c$	$\partial\beta/\partial\mu_k$ $\partial\beta/\partial\sigma_k$	$\partial\beta/\partial k_1$ $\partial\beta/\partial k_2$
LQR Control	–2.048e+02 –8.178e+02	2.105e–01 –1.522e–01	8.823e–03 –1.607e–03	–1.038e–04 –4.797e–06	1.500e+00 1.211e+01
Phase–Corrected Control	–1.748e+02 –9.226e+02	2.768e–01 –2.578e–01	6.040e–03 –1.081e–03	–1.121e–04 –1.357e–05	8.744e–01 3.607e+01
Kalman Filter Control (a)	–1.197e+02 –1.347e+03	2.155e–01 –2.197e–01	8.591e–03 –1.683e–03	–1.434e–04 –7.486e–05	
Kalman Filter Control (b)	–1.443e+02 –1.138e+03	2.203e–01 –2.041e–01	6.402e–03 –1.185e–03	–1.078e–04 –2.063e–05	

in a decrease in the probability of instability.

To illustrate the effect of choice of the probability distribution for the uncertain system parameters, we now consider the case when the variates follow a Gumbel distribution (see Fig. 3). The means and standard deviations of the system parameters remain as before. The results corresponding to those given previously for the normal case are presented in Tables 4 and 5 for the Gumbel case.

We see here that the choice of probability distribution to model the uncertain parameters has a marked effect on the probability that the system is unstable. In particular, we notice that the probabilities of failure for all control strategies increase significantly when the variates are assumed to follow the Gumbel distribution. In addition, the uncertainty associated with the time delay becomes much more important than when the normal distribution was chosen. The probability density functions for the normal and Gumbel distributions with $\mu = 10$ and $\sigma = 1$ (*i.e.*, a coefficient of variation of 10%) are given in Fig. 3. Here we make two observations: (i) the upper tails of the Gumbel distribution contain much greater probability content than for the normal distribution, and (ii) the lower tails of the gumbel distribution contains less probability content than in the normal distribution. As was discussed previously, larger realizations of the time delay τ are deleterious to system stability, whereas larger values of the mass m are beneficial. Thus, the ad-

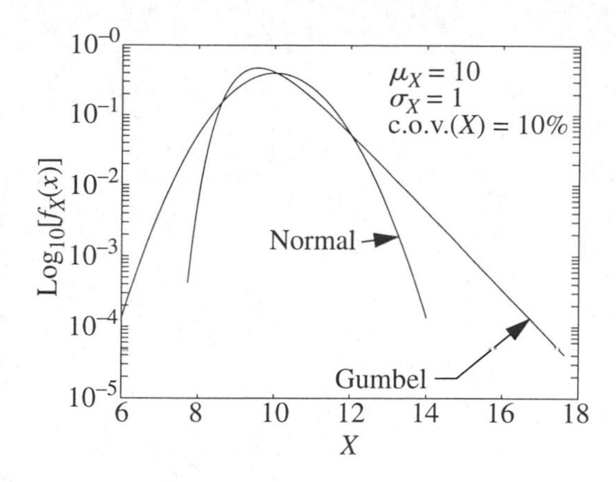

Figure 3. Comparison of the probability density function for normal and Gumbel variates.

ditional probability in the upper tails of the gumbel distribution would tend to increase the importance of uncertainty in the time delay and decrease the importance of uncertainty in the mass.

4 CONCLUSIONS

A systematic and probabilistically consistent approach for determining the likelihood that instability will be produced by the uncertainty inherently present in a controlled structure has been presented. The problem has been formulated based on an eigenvalue criterion. Efficient first and second order reliability methods have been shown to be effective for calculating the probability of instability for the controlled system. Examples employing a single-degree-of-freedom structure have been given and indicate that precise knowledge of system time delay and structural mass are of paramount importance. The choice of probability distributions for modeling the uncertain parameters can greatly influence the probability of instability. In addition, the Padé approximation has been shown to be a useful tool for assessing the effect of time delay on system reliability.

Of the control strategies considered, both the phase-corrected controller and the Kalman filter controller were more robust to parameter uncertainties than the simple LQR controller. However, the probability of instability associated with the Kalman filter controller is significantly smaller than that for the phase-corrected controller, while providing the same nominal RMS displacement response and using smaller RMS control effort.

To develop control strategies which maximize the overall reliability of a controlled structure, consistent, reliability-based measures of performance also need to be developed to assess structural control robustness. Initial efforts in this direction are reported in Spencer, *et al.* (1993a,b).

5 ACKNOWLEDGMENT

This research is partially supported by National Science Foundation Grant No. BCS 90-06781.

6 REFERENCES

ATC 17-1. (1993). *Proceedings on Seismic Isolation, Passive Energy Dissipation, and Active Control*, Applied Technology Council, Redwood City, California.

Chung, L.L., A.M. Reinhorn and T.T. Soong. (1988). "Experiments on Active Control of Seismic Structures," *Journal of Engineering Mechanics, ASCE*, Vol. 114, pp. 241-256.

de Gaston, R.R.E. and Michael G. Safonov. (1988). "Exact Calculation of the Multiloop Stability Margin." *IEEE Transactions on Automatic Control,* Vol. 33, pp. 156-171.

Housner, G.W. and Masri, S.F. (Eds.). (1990). *Proceedings of the U.S. National Workshop on Structural Control Research*, USC Publications No. M9013, University of Southern California, October 25–26.

Kantor, J.C. and B.F. Spencer, Jr. (1992). "On Real Parameter Stability Margins and Their Computation." *International Journal of Control*, Vol. 57, No. 2, pp. 453–462.

Madsen, H.O., S. Krenk and N.C. Lind. (1986). *Methods of Structural Safety*, Prentice-Hall, Inc., Englewood Cliffs, New Jersey.

McGreevy, S., T.T. Soong and A.M. Reinhorn. (1988). "An Experimental Study of Time Delay Compensation in Active Structural Control." *Proceedings of the 6th International Modal Analysis Conference and Exhibits*, Vol. 1, pp. 733-739.

Melchers, R.E. (1987). *Structural Reliability, Analysis and Prediction*, Ellis Horwood Series in Civil Engineering, Halsted Press, England.

PROBAN. (1991). Veritas Sesam Systems, Det norske Veritas, Norway.

Sain, P.M., Spencer Jr.,B.F., Sain, M.K. and Suhardjo, J., 1992, "Structural Control Design in the Presence of Time Delay," *Proceedings of the 9th Engineering Mechanics Conference*, College Station, Texas, May 25–27, pp. 812–815.

Sideris, A. and R.S Sánchez Peña. (1989). "Fast Computation of the Multivariable Stability Margin for Real Interrelated Uncertain Parameters." *IEEE Transactions on Automatic Control,* Vol. 34, pp. 1272-1276.

Soong, T.T. (1990). *Active Structural Control: Theory and Practice*, Longman Scientific and Technical, Essex, England.

Spencer Jr., B.F., M.K. Sain, D.C. Kaspari and J.C. Kantor. (1993a). "Reliability-Based Design of Active Control Strategies," *Proc., ATC-17-1 Seminar on Isolation, Passive Energy Dissipation and Active Control*, San Francisco, California, March 11–12, pp. 761–772.

Spencer, B.F., Jr., M.K. Sain, C.-H. Won, D.C. Kaspari And P.M. Sain (1993b). "Reliability-Based Measures of Structural Control Robustness." *Structural Safety*, to appear.

Structural Safety & Reliability, Schuëller, Shinozuka & Yao (eds) © 1994 Balkema, Rotterdam, ISBN 90 5410 357 4

Elasto-plastic response control by a fuzzy control system of building structures under seismic excitation

T. Yamada, K. Suzuki & H. Kobayashi
Kajima Technical Research Institute, Tokyo, Japan

ABSTRACT: The advantages of fuzzy response control have been reported. To examine structural seismic safety, stable and effective response control must be enabled in the inelastic domain. This study focused on fuzzy response control for inelastic responses and reports the results of a preliminary analytical study and shaking table tests. Although the control effect differed depending on the test cases, favorable results were achieved in the inelastic domain, with a very simple control algorithm. The stability and robustness of the control system were confirmed as well.

1. Introduction

Application of fuzzy set theory, advocated by L. A. Zadeh in 1965, has been increasing in various fields in recent years, especially in the form of fuzzy control. In civil and structural engineering, a number of studies have been carried out on its application to active vibration control[1)-3)]. Fuzzy control has several practical advantages over conventional control methods:

(1) its control algorithm is extremely simple;

(2) it is suitable for a real-time control owing to its rapid response to control commands;

(3) it does not require accurate information on structural and vibrational characteristics of the system controlled;

(4) the analog fuzzy computer enables more speedy operation; and

(5) its control system is more robust both in terms of software and devices.

Furthermore, its control effect in the elastic domain is close to that realized by conventional methods. However, nearly all of the fuzzy and conventional control methods developed so far, aim at response control in the elastic domain, i.e., tremor-like vibrations rather than very large vibrations.

Another application of response control is in the improvement of seismic safety of buildings against strong ground motions. For an active-control-type system, inelastic response control with the sole purpose of reducing the area of the hysteretic loop of the force-displacement relationship can be considered most practical. However, very few studies have been conducted on this kind of system. This is probably because conventional control theory is not suitable for this purpose. The pulse control method[4)] is a conventional method that is known to be relatively simple. However, it is very difficult to apply to control devices for large-scale buildings because of limitations in its manufacture. Other non-fuzzy control theories[5)-7)] have advantages because they can, in a sense, produce an optimal solution. However, there are many problems in applying them:

(1) their control algorithms are more complicated than those of the fuzzy control;

(2) precise structural and vibrational properties of the controlled system are required beforehand usually impossible to obtain, especially the nonlinear stiffness of a structure;

(3) although a probabilistic approach can simplify this problem to some degree, its formulation is very complicated[8)]; and

(4) their strictness in comparison with fuzzy theory engender problems in terms of the system's robustness.

In fuzzy control, there are currently some problems, such as setting the control parameters in advance. However, it remains advantageous as a control system because it can be applied even to an inelastic system, using only the knowledge and empirical rules described as fuzzy control rules.

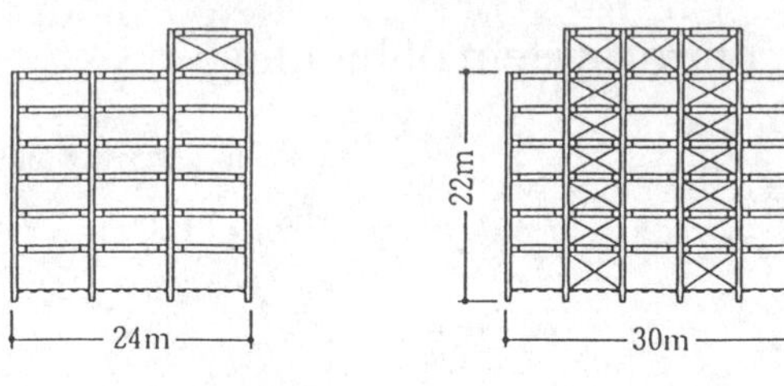

Fig. 1 Model Structure

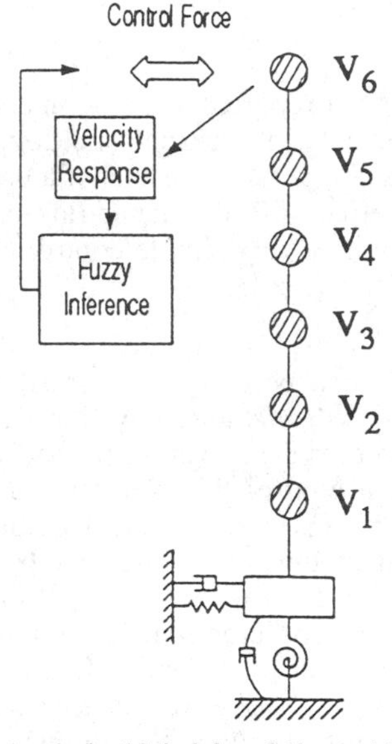

Fig. 2 Analytical Model and Control System at Top Story

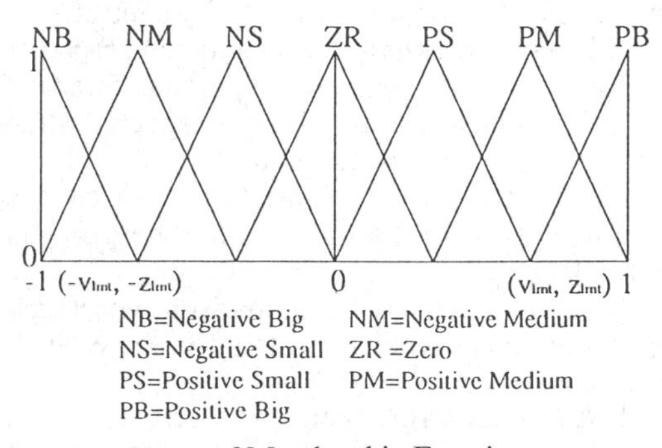

Fig. 3 Shape of Membership Function

Table 1 Rules for Fuzzy Control

(One-dimensional)

Input v6	NB	NM	NS	ZR	PS	PM	PB
Output c	NB	NM	NS	ZR	PS	PM	PB

(Two-dimensional)

Input Variable v \ Input Variable z	NB	NM	NS	ZR	PS	PM	PB
NB	NB		PB		PB		PB
NM	NB		PM		PM		PB
NS	NB		PS		PS		PB
ZR	NB	NM	NS	ZR	PS	PM	PB
PS	NB		NS		NS		PB
PM	NB		NM		NM		PB
PB	NB		NB		NB		PB

Table 2 Analytical Parameters

V_{lmt}(kine)	15, 25, 35, 45, 55
Z_{lmt}(gal)	100, 200, 400, 600, 800
Number of Rules (n)	5, 7, 9, 11
μ_{max}	0.1, 0.24, 0.40, 1.19, 2.44, 9.26
Fuzzy Inference	min-max, algebraic product-max, algebraic product-sum
Defuzzification	gravicenter method, median method

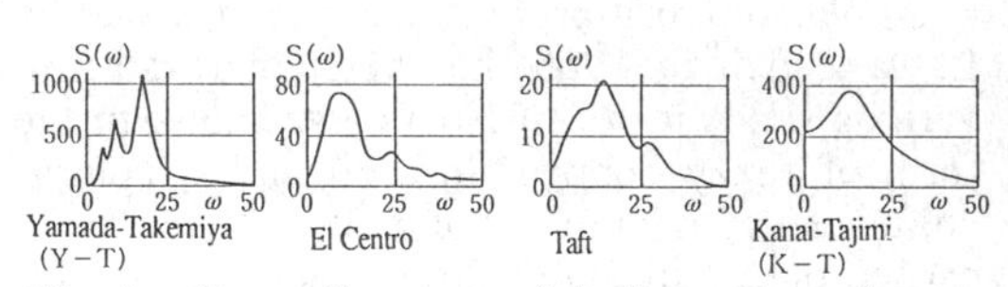

Fig. 4 Power Spectrum of the Input Ground Motions

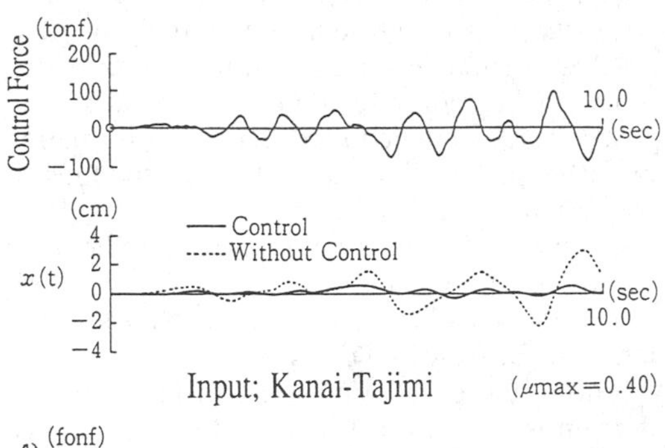

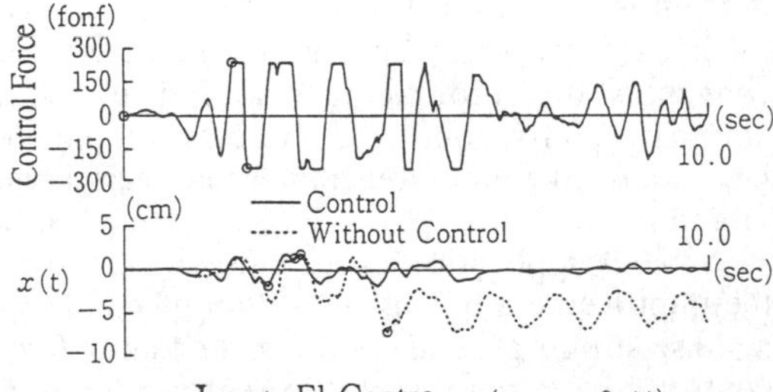

Fig. 5 Comparison of Time History Responses and Control Force

This study takes the above problems into account and aims at applying fuzzy response control to an actual building. It examines the feasibility and applicability of fuzzy control not only to elastic response but also to inelastic response.

2. Analytical Study[9)]

2.1 Structure Model and Control System

The objective building is an existing 6-story steel-frame office building designed according to the current architectural standard. It weighs about 3,100ton, and is outlined in Fig. 1. It was modeled as shear-flexural lumped masses with sway and rocking springs as shown in Fig. 2, and the restoring force characteristic was modeled as bilinear type. The natural period of this morel is 1.08sec. The control system utilizes the response velocity observed either at the top or at individual stories as input information and then a fuzzy inference is done to produce a control force as an output variable at the top of the building or at individual stories. The maximum control force (c_{lmt}) was set to be 240tf.

2.2 Fuzzy Inference and Control Rules

The so called 'If-then' form was adopted for a fuzzy inference. 'Min max' composition and a 'center-of-gravity' method were used as superimposition of rules and for defuzzification, respectively. The membership functions, which were required to construct control rules, were continuous triangular functions, and n corresponding control rules were determined. Taking for an example, for n = 7, membership functions and control rules are shown in Fig. 3 and Table 1.

2.3 Analytical Parameters

To examine the feasibility and robustness of the fuzzy response control system, the number of fuzzy parameters (n), the limit values (v_{lmt} and z_{lmt}) of the variables (v_i, z), the intensity of input seismic motions (indicated as the maximum ductility factor response, μ_{max}, in the bottom story without control), and the spectrum shape of the input motions were selected as analytical parameters. The values used for parameters in the analytical study are shown in Table 2 and Fig. 4.

2.4 Results

The number of fuzzy parameters (n) and the inference type had very little influence on the results. Thus, the following results and discussions are based exclusively on the case of n = 7 using 'min-max' composition for fuzzy inference and a 'center-of-gravity' method for defuzzification.

2.4.1 Response Reduction

Fig. 5 compares the time histories of displacement and control force resulted in each case. Table 3 shows the maximum response values at both the top and the bottom of the specimen with v_{lmt} as a parameter. Table 4 shows the comparison of maximum displacement responses at each story, using one-dimensional rules for severe ground motion (μ_{max}=9.26) with Kanai-Tajimi power spectrum. The followings can be commented according to the results:

a. For the maximum displacement response (x_{max}) and the maximum velocity response (v_{max}), the response reduction was 40 to 85 percent in the whole region from the linear to the nonlinear domain and, with the exception of a few cases, the responses remained linear.

b. A maximum acceleration response (a_{max}) reduction of 60% was recorded. However, it was no more than 30% for other cases. This indicates that the effect of response control depends on the spectrum shape (Table 3).

c. When the displacement response was small, v_{lmt} seemed to greatly affect the response reduction. That is, in terms of x_{max} and v_{max}, the smaller the v_{lmt}, the greater the response reduction at the top story, and the less at the bottom story. Along with the increase in displacement response, the effect of v_{lmt} decreases. This is probably due to the rather high c_{lmt} values primarily taking into account inelastic response control.

d. For the most part, a_{max} was more greatly reduced more by v_{lmt} regardless of the displacement level and the story. This is because if v_{lmt} is augmented, the fuzzy control command becomes less sensitive to the variable v_i. Thus it may contribute to the reduction of acceleration response.

e. The response to relatively small input motions was reduced for the entire duration. However, the response reduction to larger motions was small in the initial stage, but large at the later stage where plastic flow could occur without control. This confirms the great improvement in seismic safety (Fig. 5).

Table 3 Fuzzy Control Effect on Maximum Response

Input Motion [μ_{max}]	v_{lmt} (kine)	xmax (cm) Story 1st	xmax (cm) Story 6th	vmax (kine) Story 1st	vmax (kine) Story 6th	amax (gal) Story 1st	amax (gal) Story 6th
Yamada	15	0.12	0.19	1.4	1.8	24	77
-	25	0.11	0.24	1.2	2.3	3	64
Takemiya	35	0.09	0.28	1	2.5	22	56
(Y-T)	45	0.09	0.3	0.9	2.7	21	49
[0.10]	55	0.08	0.32	0.9	2.9	21	45
	w/o control	0.2	1.3	1.9	8	24	66
ElCentro	15	0.56	1.02	6.2	10.5	129	128
[0.40]	25	0.46	1.26	4.8	13.2	115	114
	35	0.37	1.46	3.8	14.9	103	124
	45	0.36	1.64	4.2	16.4	98	128
	55	0.37	1.88	4.4	17.6	98	132
	w/o control	0.99	6.19	7.6	39	132	290
Taft	15	2.4	10.3	23.8	77.1	576	1111
[1.19]	25	2.2	10.3	20.8	72.5	523	984
	35	2.1	10.2	18.8	73.4	517	950
	45	2.1	11	18.7	73.7	508	923
	55	2	11.2	19.3	74.1	522	909
	w/o control	3.4	17.6	29.4	110.5	671	1112
Kanai	15	3.7	17	41.8	113	1113	1321
-	25	3.6	16.4	43	115	1153	1337
Tajimi	35	3.8	16.5	43.7	117	1164	1352
(K-T)	45	4.4	17.1	43.8	118	1154	1351
[9.26]	55	5.2	18	43.7	119	1163	1361
	w/o control	25	35.6	81.2	159	1416	1476

Table 4 Relationship between Displacement Reduction Ratios

Story	Each-Story Control	Top-Story Control
6	0.75	0.48
5	0.75	0.49
4	0.76	0.57
3	0.79	0.73
2	0.85	0.74
1	0.9	0.76
Average	0.8	0.59

Table 5 Relationship between Control Parameters and Response Reduction Ratios

v_{lmt}	P_{max}	Av. Response Reduction Ratios xmax	Av. Response Reduction Ratios amax
15	0.058	0.63	0.29
25	0.045	0.67	0.37
35	0.038	0.7	0.4
45	0.033	0.69	0.41
55	0.027	0.66	0.4

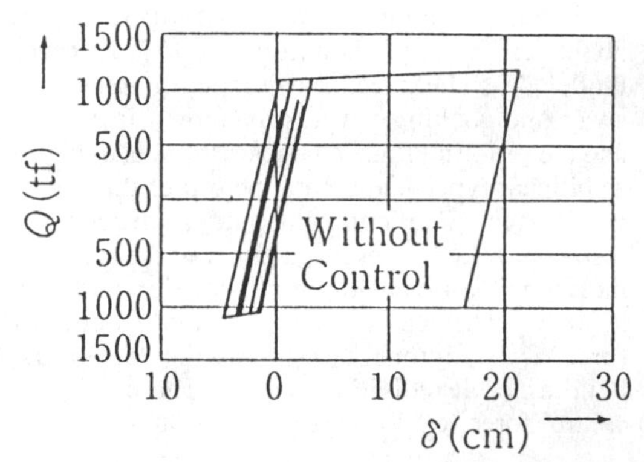

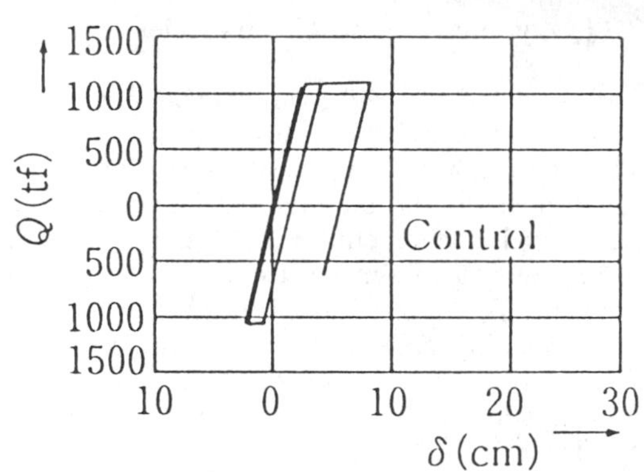

Fig. 6 Comparison of Q-δ Curves in Analysis

Table 6 Maximum Response Ratio of Control with Mis-Rules to without Control

Input Motion (μ max)	Change of Rules	x max 1st Story	x max 6th story	v max 1st Story	v max 6th story	a max 1st Story	a max 6th story	Rule Referring Frequency (%)
K-T (0.24)	NS→PS	0.70	0.79	0.81	0.51	0.78	1.54	23
K-T (9.26)	NM→PM	0.20	0.52	0.59	0.79	0.84	1.00	10
Y-T (0.10)	NS→PS	1.88	2.19	1.86	0.79	2.47	2.81	32

f. When the response control is applied to every story, the reduction of x_{max} is more than double that for top-story control only. In particular, the response reduction at the 1st story, which is assumed to suffer the greatest damage, is significant (Table 4). A similar effect was confirmed both in terms of v_{max} and a_{max}.

g. In every case using two-dimensional rules, the control effect remained almost the same as that using one-dimensional rules. Furthermore, the influence of the variation of z_{lmt} did not influence the response reduction significant more than that of v_{lmt}.

(2) The Relationship between v_{lmt} and Maximum Control Force

Table 5 shows the relationship between v_{lmt} and p_{max}, the maximum control force normalized by c_{lmt}, and the response reduction ratios for x_{max} and a_{max} in terms of the average of those at the top and the bottom story for El Centro with μ_{max}=0.40. As can be seen in Table 5, a more efficient response control can be achieved if appropriate values are set for v_{lmt}. In particular, for x_{max}, approximately the same response reduction can be expected even if p_{max} is slashed in half.

(3) Nonlinear Response Control and Control Stability

To clarify the effect of the inelastic response reduction, the relationships between story shear force (Q) and interstory displacement (d) in the bottom story were compared, as shown in Fig. 6. It can be noted that a stable control was realized even in a fairly large nonlinear domain. Next, Table 6 shows the maximum responses normalized to those without control, with some of the control rules to incorporate mistakes (Mis-rules). This confirms the robustness of the control system. It also shows how frequently the rewritten rules were referred. There is some variation in response levels depending upon the frequency of reference to the Mis-rules and the level of the input motions. The responses for the input motion with Yamada-Takemiya spectrum (μ_{max}=0.10) were not good as those without control, and this can be attributed to the fact that the rule which is especially frequently used was rewritten. Meanwhile, for the input motions with the Kanai-Tajimi spectrum (μ_{max}=9.26), whose intensity make them important in terms of seismic safety, despite the fact that the control became inelastic in almost every story of the system, there is no sign of instability. The response reduction of a_{max} was relatively small because a greater control force was required speculatively to compensate for the higher frequency vibration induced by the Mis-rules. This shows that except when the frequently-used rules for controlling small vibrations are written, the fuzzy response control has sufficient stability and robustness.

These results analytically confirmed the feasibility and applicability of the fuzzy control system to the inelastic response control.

3. Experimental Study

This section examines the feasibility of fuzzy response control based on an experimental approach. Before a practical fuzzy response control system can be realized, its feasibility must be confirmed by an using the actual control system using both software and devices. This is because the characteristics of a control system supposedly affects the response control effects. In addition, the information needed to conduct a simulation study must be obtained.

3.1 Method and System of Fuzzy Response Control

To maintain the relationship between the analytical and experimental studies, the method and system of the fuzzy-based response control system are exactly the same as those that have been used in the analytical study. That is, the 'min-max, center-of-gravity' composition is used for fuzzy inference, and the membership function is a triangular, continuous function with seven fuzzy variables. The control force is applied to the top of the specimen with only the relative velocity response at the top being input information, i.e., top-story control with one-dimensional rules (Fig. 3, Table 1). The actual fuzzy inference operation is executed by an analog fuzzy controller.

3.2 Test Method

3.2.1 Test Specimen

The test specimen is a three-story steel-frame structure as shown in Fig. 7. It is composed of the columns made of steel spring material, other vertical members made of elasto-plastic steel material, and the rigid floor frames; and weighs about 186kg including the control device and the sensors. Two kinds of elasto-plastic members

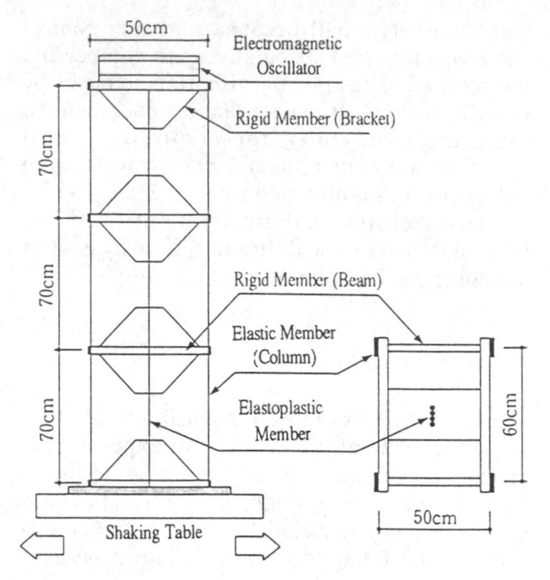

Fig. 7 Test Specimen

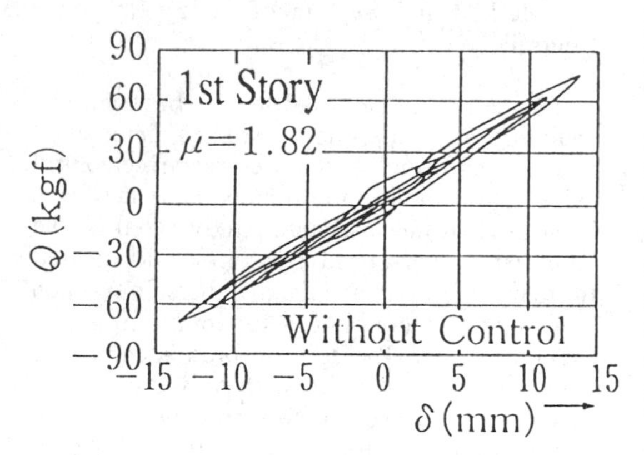

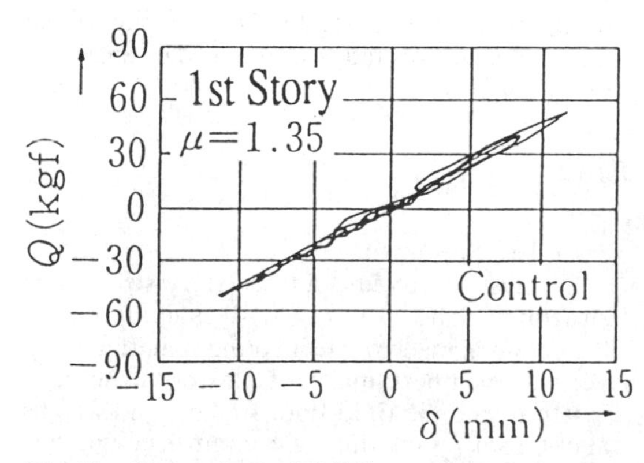

Fig. 8 Comparison of Q-δ Curves in Experiment

Table 7 Influence of Response Delay in Control System

	Specimen	V_{lmt} (kine)	maximum control force (%)	Response Reduction	
				xmax	amax
s-68	Experiment	0.5	1.8	0.71	0.43
	Analysis	1.0	0.9	0.76	0.53
s-49	Experiment	0.1	2.1	0.60	0.45
	Analysis	1.0	1.7	0.60	0.50

Table 8 Influence of Difference of Vibration Properties of Specimen

Story	Response Reduction			
	x_{max}		a_{max}	
	s-68	s-49	s-68	s-49
1	0.67	0.33	0.75	0.30
2	0.67	0.32	0.73	0.33
3	0.59	0.26	0.66	0.17

were used so that the fundamental period of the specimen could be varied. The fundamental period was 0.68sec for one (hereafter it is called s-68) and 0.49 for the other (s-49). The damping coefficient was 0.4 percent.

The control system functioned as follows: The relative response velocity was observed at the top of the specimen and then a fuzzy inference was performed by a fuzzy controller (analog computer) to actuate a control force by an electromagnetic oscillator to the top of the specimen. In performing a fuzzy inference, exactly the same in the analytical study, triangular membership functions, control rules of 'If-then' form, and composition and defuzzification of 'min-max-center-of-gravity' method were used.

3.2.2 Loading

Earthquake motions of a certain maximum displacement was used as input motions and the shaking table was oscillated so that a seismic force could be applied to the specimen fixed on the table. As described above, a fuzzy inference output was used as a control command to produce a control force, i.e., an inertia force, on the specimen by the mass of an electromagnetic oscillator. The weight of the mass was about 10kg, which is about 5% of the total weight of the specimen. In each test case, the displacement, velocity, and acceleration were measured at each story, as well as the control force of the electromagnetic oscillator (hereafter, described as AMD = Active Mass Driver). The test parameters were set on the basis of the results of the analytical study. Two kinds of recorded motions, El Centro (NS) and Taft (EW), were selected for seismic excitation:. The intensity of the input motion was determined in terms of the maximum ductility factor of each story (μ_{max}). The limit values for the input variable (relative velocity of the specimen top) was also an important parameter (v_{lmt}) in this study, and this was varied as 0.5, 1, 5, 10, 30, and 100 percent of the maximum relative velocity response in without-control test cases. The maximum control force was varied according to the total weight of the test specimen. Furthermore, to examine the robustness of the control system, the fuzzy inference rules were partially replaced in some cases (for example, replace NM for PM). Thus, a total of five parameters were selected and test cases were based on various combinations.

3.3 Results and Considerations

3.3.1 Influence of the Control System

Table 7 compares the response reduction of x_{max} and a_{max} (average of those for the first to third stories) in the test cases where v_{lmt} can be considered as nearly optimal, and in the analyses without response delay, for both specimens s-68 and s-49, for Taft as the input motion (μ_{max}=0.49, 0.42). It also shows the maximum control force values normalized according to the total weight of the specimen. According to Table 7, more efficient response control is obtained in the analysis where no response delay was taken into account. In other words, the response delay in the control system influences the control effect.

3.3.2 Dynamic Behavior of the Specimen and the Control Effect

Table 8 shows the effect of differences in the dynamic behavior of the specimens on control effect, taking the examples for Taft input. According to the table, response reduction for s-49 fell by half in comparison with s-68. Since this phenomenon was not apparent in the results of analyses without response delay, this is probably its cause. Above all, as the vibration characteristic of the AMD greatly contributed to the response delay, it is important to optimize them in response to the vibration characteristics of the specimens without increasing the response delay.

3.3.3 Nonlinear Response Control and Control Stability

Fig. 8 shows the stability of the inelastic response control when the input motion was Taft. Although it is in the domain where μ_{max} is pretty small, in spite of the control system response delay described above, it can be seen that the fuzzy response control was achieved. Furthermore, when some of the control rules were replaced, no unstable phenomenon was observed for any input motion. Though several problems were left to be solved, the feasibility of the fuzzy response control was experimentally confirmed.

4. Conclusion

The feasibility and applicability of fuzzy set theory to a seismic response control system which can cover both elastic and inelastic domains was examined both analytically and experimentally.

The following are the concluding remarks of this study:

a. If fuzzy set theory is applied to vibration control, although there is some variation according to input motions and response levels, a favorable control effect can always be obtained using the same rules.

b. The control algorithm is far simpler than the conventional ones and is suitable for real-time control like seismic control.

c. Fuzzy response control has sufficient stability and robustness for any situation.

d. In applying the fuzzy response control, however, optimal control parameters must be applied, taking into account the response delay of the control system.

Control parameters were examined as reported in this paper, but only qualitatively. Future work includes experimental confirmation of the stable response control by fuzzy set theory and quantitative examination of the control parameters.

References

1) Matsumoto, H. et al. 1990. Application of Fuzzy Theory for Vibration Control, Part 1: Vibration Test of 1 Mass Model with Active Mass Damper, Summaries of Technical Papers of Annual Meeting Architectural Institute of Japan (AIJ), Structures I: 843-844.
2) Sakakiyama et al. 1990. On Study of Active Fuzzy Control System for Building Vibration,

Summaries of Technical Papers of Annual Meeting AIJ, Structures I: 841-842.
3) Furuta, H. et al. 1991. Application of Fuzzy Set Theory to Structural Vibration Control, Proc. 7th Fuzzy System Symposium, Nagoya, Japan: 39-42
4) Shimogo et al. 1991. Pulse Control of Elastoplastic Structures under Seismic Excitation, Summaries of Technical Papers of Annual Meeting of AIJ, Structures I: 1027-1028.
5) Kawamoto et al. 1990. Stochastic Active Control of Dynamic Hysteretic Systems, Summaries of Technical Papers of Annual Meeting of AIJ, Structures I: 969-970.
6) Naraoka et al. 1990. Fundamental Study on Active Control for Nonlinear Structures, Summaries of Technical Papers of Annual Meeting of AIJ, Structures I: 827-828.
7) Noda, S. et al. 1990. Instantaneous Optimum Control of Seismic response of Hysteretic Structures, Part 2, Research report, Faculty of Engineering, Univ. of Tottori, Vol. 21: 233-248.
8) Hatada, T. et al. 1989. Generalized Optimal Control Systems for Civil Engineering Structures: Part 2, Summaries of Technical Papers of AIJ, Structures I: 541-542.
9) Yamada, T. et al. 1991: Elasto-Plastic Response Control for Building Vibration using Fuzzy Set Theory, Part 1: Analytical Approach, Summaries of Technical Papers of Annual Meeting of AIJ, Structures I: 1037-1038.

Structural Safety & Reliability, Schuëller, Shinozuka & Yao (eds) © 1994 Balkema, Rotterdam, ISBN 90 5410 357 4

Statistical analysis of hybrid sliding isolation system with new stochastic linearization approach

Ruichong Zhang & Masanobu Shinozuka
Princeton University, N.J., USA

Maria Q. Feng
University of California, Irvine, Calif., USA

ABSTRACT: The statistical properties of the response of a hybrid sliding isolation system that actively controls the structural response due to stochastic earthquake excitation are investigated in this paper. A new stochastic linearization method is proposed to solve such a nonlinear random vibration problem. Numerical computations show that the proposed approach follows closer the results of Monte Carlo simulation than the conventional approach. The statistical properties of the response of the new hybrid system are also compared to the corresponding ones of a passive structural control system, demonstrating that the behavior of the former is superior to that of latter.

1 INTRODUCTION

Active structural control has been attracting the interest of more and more engineering researchers recently. For a comprehensive survey in this area the reader is referred to Soong (1990). Recently, a hybrid sliding isolation system was proposed by Feng et al (1991, 1992) for structures that can be idealized as sliding ones. In this new system, the friction force developed at the sliding interface between the structure and the ground is actively controlled so that the sliding displacement response can be confined within an acceptable range and the transfer of seismic force to the structure is kept to a minimum. Experiments and numerical simulations (Feng et al, 1991 and Shinozuka and Feng, 1993) have shown that the new system is quite efficient compared to the corresponding passive control system. To show that the proposed hybrid system is superior to the conventional passive structural control system, the statistical properties of the response of the system subjected to stochastic external excitations have to be investigated. This investigation is also useful for the structural design.

To solve such a nonlinear random vibration problem, stochastic linearization is perhaps one of the most convenient approaches. Such an approach requires that the mean square of the difference between the original nonlinear and the equivalent linear equation be a minimum. However, this minimization does not imply minimization of the difference between the original structural response and the corresponding one of the equivalent system. By taking into account this aspect, a new stochastic linearization approach is proposed, in which weighted factors are introduced. These weighted factors are determined based on the following criterion. A similar nonlinear system is chosen so that the statistical properties of the response due to a Gaussian white noise excitation can be found by solving the Fokker-Planck equation. The weighted factors are then evaluated by letting the statistical properties of the response, obtained using the proposed approach, be equal to the corresponding ones of the exact solution. Since the statistical properties of the response due to Gaussian white noise excitation are similar to those due to other Gaussian excitations, the proposed approach is expected to be superior to the conventional one for almost all cases.

Using the new approach, the statistical properties of the response of the hybrid system can be obtained. The statistical properties of the new active and the corresponding passive control system are also compared. Finally, a numerical example is provided for illustration.

2. HYBRID SLIDING ISOLATION SYSTEM

A hybrid sliding isolation system is a system in which the structure can slide horizontally and the friction is controlled actively in terms of the structural response. Assuming that the structure is always in sliding phase, while it is subjected to stochastic earthquake ground motion, the governing equation of the structural motion can be expressed as:

$$\ddot{x} + f\,\mathrm{sgn}(\dot{x}) = -\ddot{z} \quad , \quad f = \mu g \tag{1}$$

where $\mathrm{sgn}(\dot{x})$ denotes the sign of $\dot{x}$, f stands for the friction force per unit mass between the structure and the ground, and μ is the kinetic coefficient of friction. $\ddot{z}$ is the (stochastic) ground acceleration that has the following power spectral density proposed by Shinozuka et al. (1993):

$$S_{\ddot{z}}(\omega) = S_{K-T}(\omega)\, m(\omega) \tag{2}$$

where

$$S_{K-T}(\omega) = S_0 \frac{\omega_g^4 + 4\zeta_g^2\omega_g^2\omega^2}{(\omega_g^2-\omega^2)^2 + 4\zeta_g^2\omega_g^2\omega^2} \tag{3}$$

and

$$m(\omega) = \begin{cases} \sin^2(\frac{\omega T_r}{2}) & , \ \omega < \frac{\pi}{T_r} \\ 1 & , \ \omega \geq \frac{\pi}{T_r} \end{cases} \tag{4}$$

In equations (2)-(4), S_{K-T} is the classic Kanai-Tajimi acceleration spectrum with ω_g and ζ_g denoting the dominant frequency and the damping of the ground respectively, S_0 is the magnitude, and T_r is the rise time of the slip function of the seismic source.

The friction is controlled by the pressure in a fluid chamber that is installed in the hybrid system. Experiments (Feng et al, 1991) showed that the friction has the following relation to the pressure p:

$$f = -c_1 p + c_2 \tag{5}$$

where c_1 and c_2 are constants. It is noted that f and p are function of time t. The dynamic characteristics of the pressure control system follow the first-order time-delay model with pressure control signal and are determined based on the structural response (Feng, et al, 1991). For convenience in the analysis, the pressure p is set equal to the pressure control signal u(t), neglecting the time delay effects. There are many control approaches that can generate the pressure control signal in terms of structural response. In this paper, the following two control approaches are considered:

(1) Bang-Bang Control (BBC)

When the structural displacement is in the same direction as its velocity, the pressure control signal is decreased to its minimum, which in terms increases the friction force (cf. equation (5)) so that the structural response is decreased. On the contrary, when the displacement and velocity responses are in opposite directions, the pressure control signal is increased to its maximum. This control algorithm is the so-called Bang-Bang control approach. Mathematically, the pressure control signal is expressed as:

$$u(t) = \frac{u_{min} + u_{max}}{2} + \frac{u_{min} - u_{max}}{2}\,\mathrm{sgn}(x\dot{x}) = \begin{cases} u_{min} \ , & \mathrm{sgn}(x\dot{x}) \geq 0 \\ u_{max} \ , & \mathrm{sgn}(x\dot{x}) < 0 \end{cases} \tag{6}$$

(2) Instantaneous Optimal Control (IOC)

The instantaneous optimal control approach was first proposed by Yang et al. (1986). It requires to minimize the following time dependent objective function J(t) of the control system at every time instant, when subjected to stochastic earthquake ground acceleration:

$$J(t) = \alpha x^2(t) + \beta f^2(t) + \gamma u^2(t) \qquad (7)$$

where α, β and γ represent, respectively, the relative importance of the displacement response, acceleration response and pressure control signal. It is noted that in equation (7), f(t) is associated with the acceleration response, which may be seen clearly in equation (1). The instantaneous optimal pressure control signal can be obtained by solving equations (1), (5) and (7) using the linear acceleration method and a Lagrangian approach (Feng et al, 1991), i.e.

$$u(t) = \frac{c_1 c_2 \beta}{\gamma + \beta c_1^2} - \frac{c_1 \alpha (\Delta t)^2}{6(\gamma + \beta c_1^2)} x \, \mathrm{sgn}(\dot{x}) \qquad (8)$$

where Δt denotes the time increment step, associated with the linear acceleration method. It is interesting to note that the pressure control signal u(t) in equation (8) decreases when $\mathrm{sgn}(\dot{x}x)>0$ while it increases when $\mathrm{sgn}(\dot{x}x)<0$. These properties of the instantaneous optimal control approach are similar to those of the Bang-Bang control approach (cf. equation (6)). It can be shown that the latter is the limit case of the former.

With the aid of equations (5), (6) and (8), the governing equation of motion (Eq. 1) can be rewritten as

$$\ddot{x} + f_d(\dot{x}) + f_r(x) = -\ddot{z} \qquad (9)$$

where the damping and restoring forces are given respectively by:

$$f_d(\dot{x}) = \mathrm{sgn}(\dot{x}) \cdot \begin{cases} -c_1 \dfrac{u_{min}+u_{max}}{2} + c_2 , & \mathrm{BBC} \\ -\dfrac{c_1^2 c_2 \beta}{\gamma + \beta c_1^2} + c_2 , & \mathrm{IOC} \end{cases} \qquad (10)$$

$$f_r(x) = \begin{cases} c_1 \dfrac{u_{max}-u_{min}}{2} \mathrm{sgn}(x) , & \mathrm{BBC} \\ \dfrac{c_1^2 \alpha (\Delta t)^2}{6(\gamma + \beta c_1^2)} x , & \mathrm{IOC} \end{cases} \qquad (11)$$

It is apparent from equation (9) that the active control of the friction actually generates a restoring force which effectively restricts the displacement response. However, this is not the case for the passive control system with constant friction. Another advantage of the hybrid system is that it possesses a so-called failure safe feature, i.e. the hybrid system reduces to the passive control system when the pressure control signal is not actively controlled. Other merits of the hybrid system are mentioned in Feng et al. (1991) and Shinozuka and Feng (1993).

To obtain the statistical response of the hybrid system, a stochastic linearization method is used which solves the following equivalent linearized governing equation, instead of the original one given in equation (9),

$$\ddot{x} + c_e \dot{x} + k_e x = -\ddot{z} \qquad (12)$$

where c_e and k_e are respectively the equivalent damping and stiffness coefficients that are determined based on stochastic linearization criteria described in the following section.

3 NEW STOCHASTIC LINEARIZATION APPROACH

The proposed stochastic linearization criterion requires that:

$$E[(f_d(\dot{x})+f_r(x)-w_c c_e\dot{x}-w_k k_e x)^2] = \text{minimum} \tag{13}$$

where w_c and w_k are weighted factors and E denotes ensemble average. It should be pointed out that when the weighted factors are equal to one, equation (13) degenerates to the conventional criterion. Assume now that the ground acceleration is a stationary Gaussian random process with zero mean. Within the framework of the stochastic linearization, we may assume that x and $\dot{x}$ are also Gaussian zero-mean process. Equation (13) then yields the following values for c_e and k_e:

$$c_e=\frac{E[f_d(\dot{x})\dot{x}]}{w_c\sigma_{\dot{x}}^2},\ k_e=\frac{E[f_r(x)x]}{w_k\sigma_x^2} \tag{14}$$

The weighted factor $w_k(w_c)$ is determined using the following criterion: a similar nonlinear system is chosen so that the displacement (velocity) variance can be found in closed form by solving the Fokker-Planck equation subjected to Gaussian white noise excitation with zero mean. The weighted factor $w_k(w_c)$ is then evaluated by setting the displacement (velocity) variance obtained using the proposed stochastic linearization approach equal to the corresponding value of the exact solution.

The following comments on the new approach are now in order.

(1) The minimization of the mean square of the difference between the original and the equivalent governing equations in the conventional stochastic linearization approach (w_c=w_k=1) does not imply the minimization of the difference between the response statistics of the original and the equivalent system. The weighted factors are therefore introduced so that the statistical response obtained with the new approach can be closer to the actual response when compared to the conventional approach.

(2) The weighted factors are determined using the aforementioned criterion because the exact solution for the response statistics of a nonlinear system is available only for the Gaussian white noise excitation. Furthermore, the basic characteristics of the response statistics of a system subjected to Gaussian white noise excitation are similar to those due to the real excitation with power spectral density defined in equation (2).

Based on the above two comments, it is expected that the proposed approach will be better than the conventional approach.

Specifically, the following similar nonlinear system for determining w_k is selected:

$$\ddot{x} + c_e\dot{x} + f_r(x) = -\ddot{z} \tag{15}$$

The probabilistic density function for the response displacement is given by (Lin, 1967)

$$p_X(x) = A\exp[-\frac{c_e}{\pi S}\int_0^x f_r(u)du] \tag{16}$$

where A is a constant and can be determined based on the condition of $\int_{-\infty}^{\infty} p_X(x)dx = 1$, and S is the constant power spectral density of Gaussian white noise acceleration excitation. The exact solution for the standard deviation of the displacement, when $f_r(x)$ is given by $k_0\text{sgn}(x)$ (k_0 is a constant), is obtained as:

$$\sigma_x = \sqrt{2}\ \frac{\pi S}{c_e k_0} \tag{17}$$

The standard deviation of the displacement obtained using the corresponding equivalent system is

$$\sigma_x = w_k\sqrt{\frac{\pi}{2}}\ \frac{\pi S}{c_e k_0} \tag{18}$$

Equations (17) and (18) yield the value of w_k:

$$w_k = \frac{2}{\sqrt{\pi}} \tag{19}$$

It is interesting to note that the weighted factor w_k is independent of almost all parameters of the similar system of equation (15). Heuristically, let's check a simpler similar system

$$\dot{x} + f_r(x) = -\ddot{z} \tag{20}$$

It is easy to find that the weighted factor w_k is still the same as the one in equation (19). Similarly, in order to determine the weighted factor w_c, the following similar system is used:

$$\ddot{x} + f_d(\dot{x}) = -\ddot{z} \tag{21}$$

The resulting value for w_c will be equal to w_k when $f_d(\dot{x})$ has the form $c_0 \mathrm{sgn}(\dot{x})$ (c_0 is a constant).

The variance of the response displacement and velocity can be computed from the linearized equation (12) as:

$$\begin{aligned}\sigma_x^2 &= \int_{-\infty}^{\infty} \frac{S_{K-T}(\omega)}{(\omega^2-k_e)^2+c_e^2} d\omega \\ &- \int_{-\pi/T_r}^{\pi/T_r} \frac{\cos^2(\omega T_r/2) S_{K-T}(\omega)}{(\omega^2-k_e)^2+c_e^2} d\omega\end{aligned} \tag{22}$$

$$\begin{aligned}\sigma_{\dot{x}}^2 &= \int_{-\infty}^{\infty} \frac{S_{K-T}(\omega)}{\omega^2+c_e^2} d\omega \\ &- \int_{-\pi/T_r}^{\pi/T_r} \frac{\cos^2(\omega T_r/2) S_{K-T}(\omega)}{\omega^2+c_e^2} d\omega\end{aligned} \tag{23}$$

The first integral in equations (22) and (23) can be obtained in closed form using residue theory and the second can always be represented by a hyperbolic-sine or cosine-integral. Substituting equations (22) and (23) into equation (14), it is possible to solve numerically the resulting equation for the equivalent damping and stiffness coefficients c_e and k_e. The root-mean-square (RMS) of the acceleration response can be computed directly from the governing equation (9) as

$$\begin{aligned}a_{RMS} &= \sqrt{E[(\ddot{X}+\ddot{Z})^2]} \\ &= \sqrt{E[f_d^2(\dot{x})]+E[f_r^2(x)]}\end{aligned} \tag{24}$$

where use has been made of orthogonality between the displacement and the velocity in a stationary state (Lin, 1967).

4 MONTE CARLO SIMULATION

To verify the accuracy of the proposed stochastic linearization approach, Monte Carlo simulations are performed. Specifically, the fourth order Runge-Kutta algorithm was applied to solve the governing equation (1) or (9). The earthquake ground acceleration is simulated using the spectral representation method as follows (Shinozuka and Deodatis, 1991):

$$\begin{aligned}\ddot{z}(t) = \sqrt{2} \sum_{k=0}^{M} [2S_{\ddot{z}}(k\Delta\omega)\Delta\omega]^{1/2} \cdot \\ \cdot \cos(k\Delta\omega t+\Phi_k)\end{aligned} \tag{25}$$

where the Φ_ks are independent random phase angles uniformly distributed in the range of $[0,2\pi]$. To provide sufficient accuracy, Δt has to be adequately small. In the example presented in the following section, Δt is fixed at 0.01 sec. A sample of N=500 numerical realizations of the earthquake ground acceleration is created by generating different realizations of Φ_k. For the n-th realization of $\ddot{z}(t,n)$, the structural response $y(t,n)$, $y = x, \dot{x}$ or $\ddot{x}+\ddot{z}$, is determined. Subsequently, the unbiased mean square of the N responses is obtained by:

$$E[Y^2(t)] = \frac{1}{N-1} \sum_{n=1}^{N} y^2(t,n) \tag{26}$$

5 NUMERICAL EXAMPLES

According to the experiments performed in Feng et al. (1991), the parameters of the hybrid system are chosen as: $c_1=2.5\text{x}10^{-6}$ m^3/sec^2kgf, $c_2=1.351$ m/sec^2, $u_{min}=1.0\text{x}10^5$ kgf/m^2, $u_{max}=4.5\text{x}10^5$ kgf/m^2, $c_1c_2\beta/(\gamma+\beta c_1^2) = 1.65\text{x}10^5$ kgf/m^2, $c_1\alpha(\Delta t)^2/[6(\gamma+\beta c_1^2)]=-1.64\text{x}10^7$ kgf/m^3. These parameters corresponds to a range of μ between 2.3% and 11.2%. For the passive control system, three values are chosen for u, corresponding to values for μ equal to 2.3%, 6.1% and 11.2%. The parameters associated with ground acceleration are: $\omega_g=8\pi$ rad/sec, $\zeta_g=0.6$ (rock), $T_r=1$ sec and $S_0=7.253\text{x}10^{-2}$ $m^2/rad/sec^3$ (peak ground acceleration is 0.33g).

The variation of the standard deviation of the displacement with time is shown in Fig. 1. Although the results obtained with the new stochastic linearization approach underestimate the real ones (simulation), they are much closer to the real ones compared to the corresponding results of the

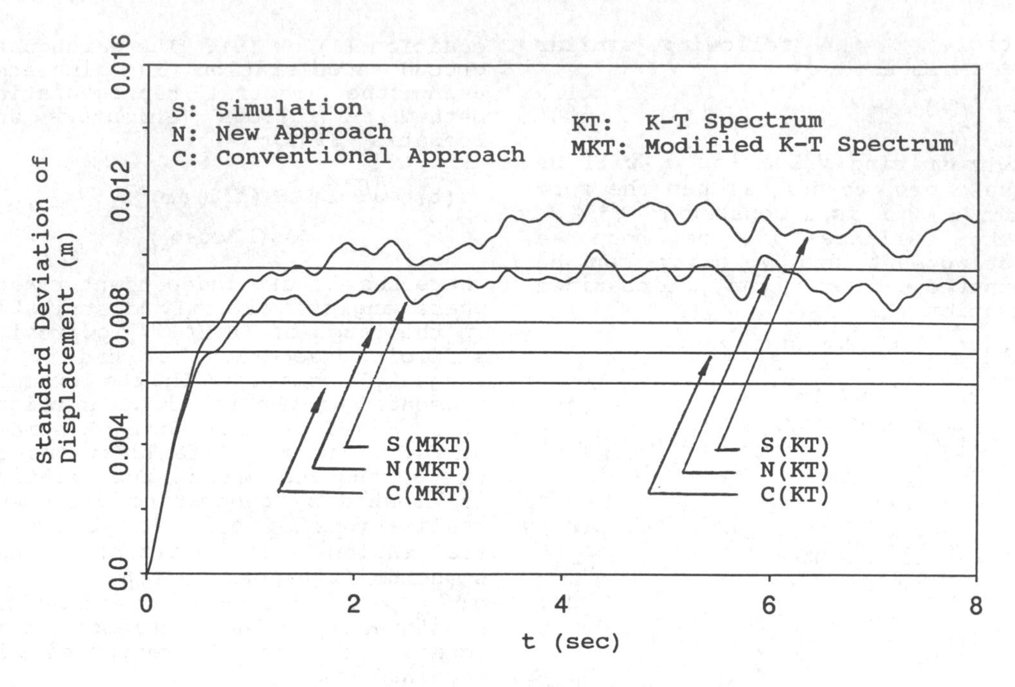

Fig. 1 Standard deviation of displacement vs time (Bang-Bang Control system)

conventional approach. Fig. 1 also displays a comparison of results obtained using the original and the modified Kanai-Tajimi spectrum. These results indicate that there indeed exists a non-negligible difference between the two spectra in the standard deviation of the displacement, although the maximum displacement response may not be affected by the low frequency component of ground acceleration (Seya et al. 1993). A comparison of response statistics between the hybrid and the passive control system is presented in Fig. 2. As explained in the text, the standard deviation of the displacement in the hybrid system is much smaller than the corresponding one in the passive control system. The fact that the standard deviation of the displacement in the passive control system is growing linearly with time was also established by Iwan et al. (1987) and Zhang et al. (1993), among others. The passive control system with a large friction coefficient μ provides a relatively small standard deviation of the displacement, which is comparable to that of hybrid system (cf. Fig. 2). This is more evident for nonstationary ground acceleration (e.g. Shinozuka and Feng, 1993). However, the passive control system with larger μ gives larger RMS for the acceleration response, which is not the case in the hybrid system (see Table 1). Combining the results of Fig. 2 and Table 1, it is obvious that the hybrid system provides both small standard deviation of the displacement and low RMS of the acceleration response, leading to the conclusion that the hybrid system is superior to the passive control system, at least in this respect.

6 CONCLUDING REMARKS

The statistical analysis of the sliding system subjected to stationary earthquake ground motion has shown that the new hybrid system is more effective in resisting

Table 1 Comparison between Active and Passive Systems

	BBC	IOC	P1(μ=11.2%)	P2(μ=6.1%)	P3(μ=2.3%)
a_{RMS}(m/s^2)	0.795	0.957	1.101	0.601	0.226
Simulation	0.755	0.925	1.088	0.598	0.225

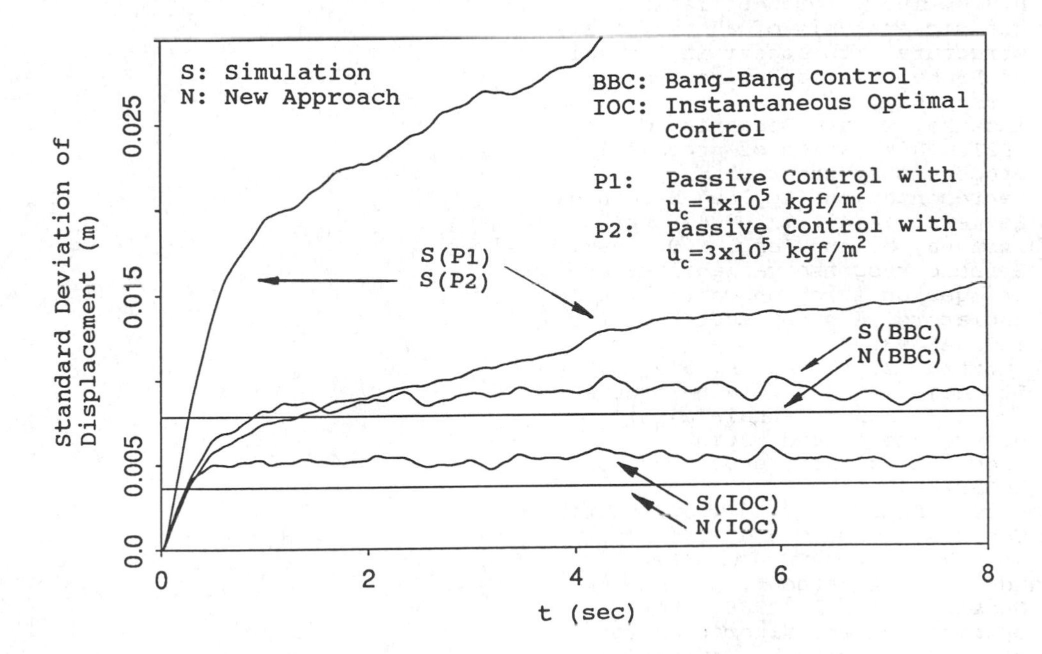

Fig. 2 Comparison of standard displacement deviations in active and passive control systems

earthquake excitation than the passive control system. The new stochastic linearization approach is better in predicting the response statistics than the conventional one. The statistical properties of the sliding structure subjected to nonstationary earthquake ground motion will be examined in other paper.

ACKNOWLEDGEMENT

This work was supported by Contract No. NCEER 91-2031A under the auspices of the National Center for Earthquake Engineering Research.

REFERENCES

Feng, Q., Fujii, S., Shinozuka, M. and Fujita, T. 1991. Hybrid isolation system using friction-controllable sliding bearings, Eigth VPI&SU Symposium on Dynamics and Large Structures, Blacksburge, VA, May 6-8.

Feng, Q., Shinozuka, M., Fujii, S. and Fujita, T. 1992. A hybrid sliding isolation system for bridges, The First US-Japan Workshop on Earthquake Protective Systems for Bridges, Sept. 4-5, SUNY Buffalo, NY.

Iwan, W.D., Moser, M.A. and Paparizos, L.G. 1987. The

stochastic response of strongly nonlinear systems with Coulomb damping elements, Nonlinear Stochastic Dynamic Engineering Systems (eds: E. Ziegler and G.I. Schueller), pp 455-466.

Lin, Y.K. 1967. Probabilistic Theory of Structural Dynamics, McGraw-Hill, Inc.

Seya, H., Talbot, M.E. and Hwang, H.H.M. 1993. Probabilistic seismic analysis of a steel frame structure, (to appear in Journal of Probabilistic Engineering Mechanics).

Shinozuka, M. and Deodatis, G. 1991. Simulation of stochastic processes by spectral representation, Applied Mechanics Reviews, 44(4), 191-204.

Shinozuka, M. and Feng, M.Q. 1993. Seismic response variability of bridges on friction controllable isolators, a paper submitted for ICOSSAR'93.

Shinozuka, M., Zhang, R. and Deodatis, G. 1993. Sine-square modification to Kanai-Tajimi earthquake ground motion spectrum, a paper submitted for ICOSSAR'93.

Soong, T.T. 1990. Active Structural Control: Theory and Practice, John Wiley & Sons, Inc., New York.

Yang, J.N., Akbarpour, A. and Ghaemmaghami, P. 1986. New optimal control algorithms for structural control, Journal of Engineering Mechanics, Vol. 113, No. 9, 1369-1386.

Zhang, R., Elishakoff, I. and Shinozuka, M. 1993. Analysis of nonlinear sliding structures by modified stochastic linearization methods, (to appear in International Journal of Nonlinear Dynamics).

Wind engineering

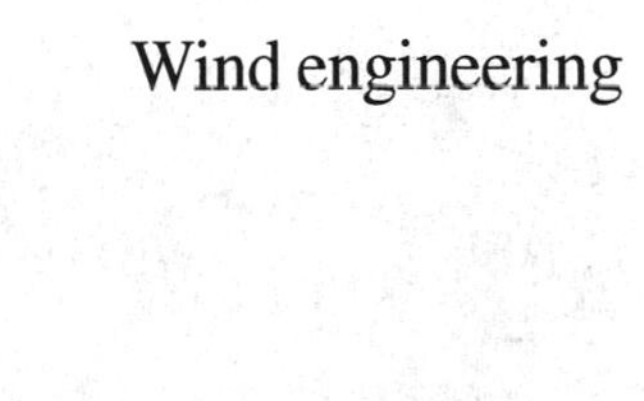

Structural Safety & Reliability, Schuëller, Shinozuka & Yao (eds) © 1994 Balkema, Rotterdam, ISBN 90 5410 357 4

Macro time modeling of wind and snow

R.M.Bennett & K.C.Chou
Department of Civil Engineering, The University of Tennessee, Knoxville, Tenn., USA

R.D.Gilley
HDR Engineering, Inc., Virginia Beach, Va., USA

C.A.Belk
Lockwood Green Technologies, Oak Ridge, Tenn., USA

ABSTRACT: A summary of values of parameters for the load coincidence load combination method for macro variations of wind and snow loads is given. Nine sites across the United States are examined to obtain pertinent macro time variability statistics of wind loads. Daily water equivalent ground snow loads from seventeen sites across the United States are examined to obtain statistics of snow loads. Occurrence dependence between wind and snow loads is also examined. Occurrence dependence is the result of the snow and wind arising from the same storm and from the seasonal variation of the two loads.

1. INTRODUCTION

Significant research has been conducted on modeling loads as stochastic processes and developing methods for combining loads to obtain lifetime maximums. One of the methods of load combination is the load coincidence method (Wen, 1990). The method has been extended to include both within load and between load dependencies (Wen, 1981). The general form of the load coincidence method is that the probability of a load (or combined load) exceeding some value, x, in time t is

$$P(Load\ Exceeds\ x) = 1 - e^{-\nu t} \quad (1)$$

in which ν = the mean upcrossing rate of the combined load above a threshold level x. For two independent loads, ν is obtained as

$$\nu = \lambda_1 G_1(x) + \lambda_2 G_2(x) + \lambda_1 \lambda_2 (\mu_{d1} + \mu_{d2}) G_{12}(x) \quad (2)$$

in which λ_i = the mean rate of occurrence of load i; $G_i(x)$ = the conditional probability of load (or combined load) i exceeding x given the load is on, and μ_{di} = the mean duration of load i.

Hindering the application of advanced probabilistic based load combination methods is the lack of realistic parameters. Pier and Cornell's (1973) statement of twenty years ago that probabilistic structural behavior descriptions are more advanced than the probabilistic description of the loads acting on the structure is still accurate. Some parameters have been suggested for live load (e.g. Corotis and Tsay, 1983). Summarized herein are recent parameters that were developed for wind loads (Belk and Bennett, 1991), snow loads (Gilley and Bennett, 1991), and a discussion of the occurrence dependencies between wind and snow loads. The parameters are developed specifically for the load coincidence method of load combination, but would be applicable to other methods of load modeling (e.g. Larrabee and Cornell, 1981; Grigoriu, 1984).

2. WIND SPEED MODELING

Wind speed data were obtained from continuous gust recording charts for nine cities across the contiguous United States. Two years of records were examined for Austin, Texas; Baltimore, Maryland; Detroit, Michigan; Rochester, New York; Sacramento, California; St. Louis,

Missouri; Tucson, Arizona; and Billings, Montana. Seven years of records were examined for Knoxville, Tennessee. Only "structurally significant" winds, winds above 20 knots, were examined. The 20 knot cutoff is approximately equal to 10% of the minimum design load. Neglecting loads below 10% of the design load is consisten with the development of probabilistic based load factors (Ellingwood et al., 1980), in which a load of 0.1W was ignored. The macro time variations of the wind were fitted with a Poisson square wave pulse process.

From analysis of the data, it was determined that there was significant occurrence dependence. When a wind occurs, it is often followed by several more. The clustered Poisson process model developed by Wen (1990) was adopted. The mean upcrossing rate for a single clustered load process is

$$\nu = \lambda^c \frac{nG(x)}{nG(x)+1} \tag{3}$$

in which λ^c is the mean rate of occurrence of clusters, and n is the mean number of loads per cluster. There was slight positive dependence between individual duration and wind speed. Correlation coefficients ranged from 0.14-0.48. These values are small enough to be neglected in the load combination problem (Wen, 1981).

Average values for the contiguous United States are a mean rate of occurrence of winds of 0.0176 per hour, a mean cluster duration of 5.40 hours, a mean individual duration of 3.76 hours, and a mean number of winds per cluster of 1.36. The intensity can be modeled with an extreme type I distribution with the mode being 0.217 times the 50-year wind, and the shape factor, α, being 11.8 divided by the 50-year wind.

3. SNOW LOAD MODELING

Twenty years of daily water equivalent ground snow data starting with January, 1970 were examined for seventeen sites across the United States - Boise, Idaho; Portland, Maine; Boston, Massachusetts; Detroit, Michigan; Minneapolis, Minnesota; Columbia, Missouri; Great Falls, Montana; Valentine, Nebraska; Pendleton, Oregon; Buffalo, New York; Rochester, New York; Bismarck, North Dakota; Cleveland, Ohio; Knoxville, Tennessee; Stampede Pass, Washington; Green Bay, Wisconsin; and Casper, Wyoming. The data was fitted with a Poisson square wave pulse process. Each pulse enveloped several days of data. Summary statistics are given in Table 1. Like wind loads, there was significant occurrence dependence; hence the adoption of a clustered load process.

Except for Knoxville, Tennessee, there was little dependence between individual duration and intensity. Correlation coefficients ranged from -0.03 to 0.32. The correlation coefficient for Knoxville was 0.44. Typically, snow melts between occurrences in southern U.S. cities. Thus, if a heavy snowfall occurs, it will take longer to melt resulting in a longer duration.

Intensity dependence between subsequent snowfalls was also examined. There was little dependence between clusters, but significant correlation within a cluster. Correlation coefficients between snowfall intensity within a cluster ranged from 0.58 to 0.96.

Six different distributions were examined for snowfall intensity. The Weibull distribution was chosen based on Chi-square tests and the ability to accurately predict 50-year snow loads.

4. WIND AND SNOW LOAD DEPENDENCE

Both intensity and occurrence dependence may exist between snow loads and wind loads. Intensity dependence may be positive when wind causes snow to drift or negative when snow is removed by wind. This dependence would then be a function of roof geometry. Isyumov (1982), O'Rourke et al. (1985), and Templin and Schriever (1982), among others, have examined dependencies of this type.

Occurrence dependence of snow and wind exists in two forms, occurrence of both from the same storm, and seasonal variations in each load. Occurrence dependence is examined in the following.

Table 1. Summary Statistics for Snow Loads

Location	Intensity (mm water)		Mean Time Lag (days)	Mean Cluster Duration (days)	Mean Number in Cluster	Mean Individual Duration (days)
	Mean	Standard Deviation				
Boise, ID	14.5	10.9	156.38	10.48	2.38	4.26
Portland, ME	58.2	55.4	90.94	17.63	4.64	3.79
Boston, MA	31.2	33.3	86.02	6.08	2.32	2.62
Detroit, MI	21.8	21.1	74.19	7.94	2.45	3.21
Minneapolis, MN	43.2	28.7	113.86	28.40	6.30	4.49
Columbia, MO	23.9	19.3	94.07	6.72	2.53	2.61
Great Falls, MT	16.8	14.5	46.48	6.66	2.24	2.91
Valentine, NE	22.9	22.1	84.13	8.92	2.33	3.79
Buffalo, NY	35.6	37.1	55.68	9.31	2.18	4.21
Rochester, NY	28.4	27.7	56.19	9.72	2.16	4.43
Bismarck, ND	31.0	27.4	95.32	19.73	3.21	6.11
Cleveland, OH	20.1	20.6	64.45	7.04	1.66	4.18
Pendleton, OR	14.0	10.4	128.69	6.02	1.52	3.96
Knoxville, TN	10.7	6.4	238.31	2.86	1.17	2.41
Stamp. Pass, WA	571.5	436.6	130.43	77.27	14.53	6.88
Green Bay, WI	40.1	35.3	90.03	19.76	3.55	5.53
Casper, WY	9.4	7.6	37.86	4.68	1.42	3.24

4.1 Wind and Snow Arising from the Same Storm

The intra-load clustering process developed by Wen (1990) is used to model occurrence dependence due to wind and snow arising from the same storm. The model includes a parent generating process which produces the dependent wind and snow occurrences, along with "noise" processes for both the wind and snow, which account for the nondependent occurrences. These processes jointly form a multivariate point process. The parent generating process is characterized by an occurrence rate, ρ, a random delay time, T_i, and a probability that the load is nonzero, p_i. If the delay time is assumed to follow an exponential distribution, the mean upcrossing rate is given by:

$$\nu = \lambda_1^c \frac{n_1 G_1(x)}{n_1 G_1(x)+1} + \lambda_2^c \frac{n_2 G_2(x)}{n_2 G_2(x)+1} + \lambda_1^c \lambda_2^c n_1 n_2 (\mu_{d_1}+\mu_{d_2}) \left[1 + \frac{p_1 p_2 \rho}{\lambda_1^c \lambda_2^c} \frac{1}{a_1+a_2}\right] G_{12}(x) \left[\frac{1}{n_1 n_2 \left(\frac{\mu_{d_1}+\mu_{d_2}}{\mu_{c_1}+\mu_{c_2}}\right) G_{12}(x)+1}\right] \quad (4)$$

in which λ_i^c is the mean rate of occurrence of clustered load i, μ_{ci} is the mean duration of cluster i, p_i is the probability load i is nonzero, and a_i is the mean delay time of load i.

A thorough meteorological analysis would be required to obtain actual values for p_1 and p_2. This would be similar to weather forecasting, that is, determining the probability of wind or snow given a storm occurs. If historical data is examined, then only actual occurrences of dependent loads will be identified. This results in setting p_1 and p_2 equal to one.

Some problems are encountered in using historical data. Wind speed data is recorded at much finer time increments than snow load data; therefore, snow load data that is recorded at time increments finer than one day are necessary. Local Climatological Data Monthly Summaries can be used to obtain water equivalent precipitation data for use in this load model. The data in these summaries are given at one-hour intervals. The water equivalent load data does not separate snow from rain loads; however, the data given at three-hour intervals specifies the type of weather that occurred within the given interval. Thus, hourly snowfall can be inferred from the data given in the monthly summaries. Also, snow depth measurements are made at 7:00 a.m each day to determine whether or not a daily water equivalent snow depth is measured for that particular day (a measurement is made if the ground snow depth is equal to or greater than 2 inches).

Figure 1 gives an example of a snow and wind joint occurrence for Rochester, New York starting at 7:00 a.m. on January 4, 1986. The snowfall data points are actual water equivalent amounts of snowfall that occurred within a given hour. Note that the three wind occurrences on the left side of the graph appear to be clustered with snow occurrences, while the wind occurrences on the right appear not to be clustered with a snow occurrence. Thus, the snow and wind on the left are assumed to be from the same parent process, while the wind occurrences on the right are not.

Figure 2 shows the occurrences for the same time period, but shows the cumulative snow as used in the snow model described in section 3. The snow occurrence is taken to coincide with the parent process with a specific start time of 7:00 a.m. on January 4, 1986. The maximum intensity of the snow occurrence is 36 mm and the duration is 14 days. Three wind occurrences exist within the snow occurrence. However, only the first wind occurrence, which consists of three clustered winds, would have any dependence on the snow process as indicated by Figure 1. This clustered wind occurrence has a delay time of 1.00 days from the starting point of the snow process model. The intensity and duration of this wind occurrence are 67 km/hr and 1.376 days, respectively. The data from this occurrence would then yield one data point describing the parent process.

The time interval for snow loads shown in Figure 2 is too coarse to directly determine between load clustering. Thus, the first step in defining the intra-load clustering model is to examine hourly snowfall and wind occurrences to determine if the snow and wind processes were generated by the same parent process. If the two individual processes appear to be from the same parent process, a model containing the

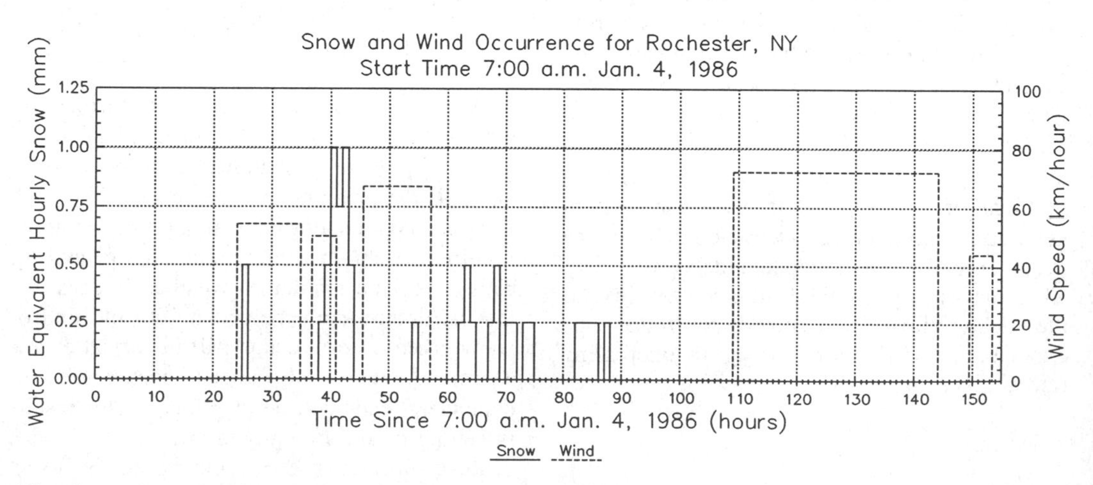

Fig. 1 Snow and Wind Occurrence for Rochester, NY

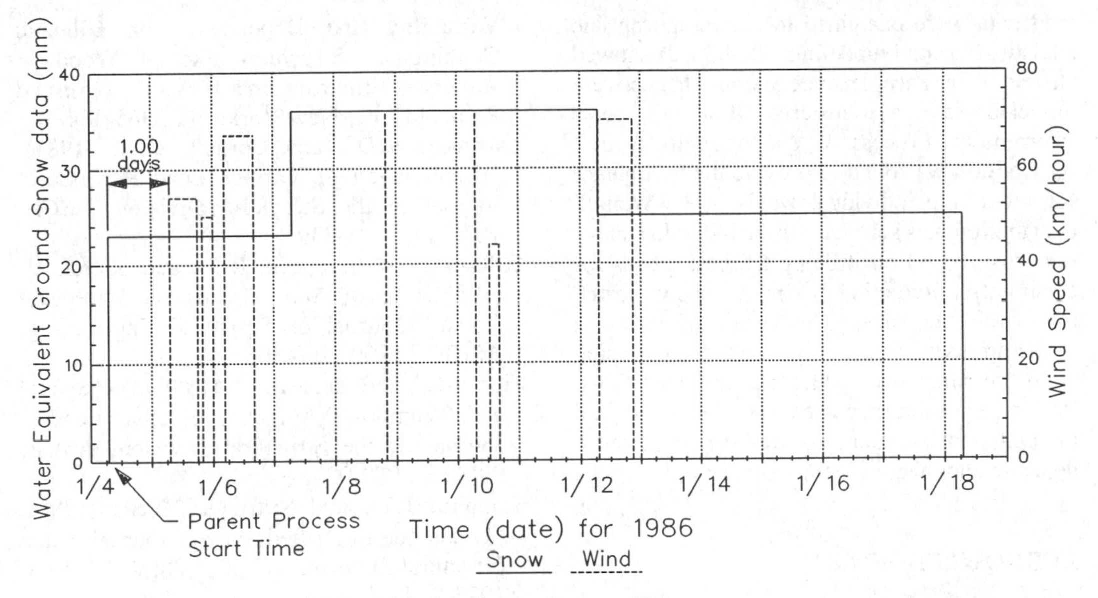

Fig. 2 Example Snow and Wind Model for Rochester, NY.

cumulative snow and the wind occurrence can be developed to determine the statistical parameters necessary to define the process. This two-step process can be extended to cover the length of time necessary to adequately define the parent process.

4.2 Seasonal Variations

Long term seasonal variations is the other form of occurrence dependence between snow and wind loads. The wind speed data indicate that there is a significant drop in wind occurrences during July, August, September, and October for Rochester and Detroit. Figure 3 shows the monthly variation of wind occurrences using two years of wind data collected from continuous gust charts for Rochester and Detroit. Although the two sites differ somewhat in the specific occurrence rates for a given month, the general trend towards increased wind occurrences during the winter months holds true for both cities. Hence, correlation between the two loads will exist.

Long term seasonal variations in the loads can be handled in the load coincidence method by modifying Equation 1 to:

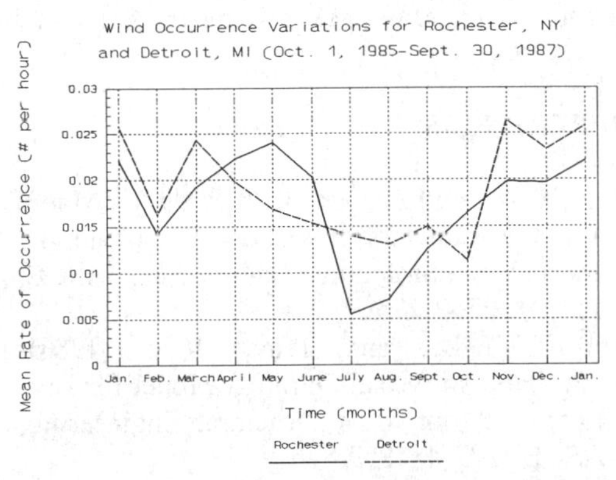

Fig. 3 Wind Occurrence Variation for Rochester, NY and Detroit, MI.

$$P(Load\ Exceeds\ x) = 1 - e^{-N_y \int_0^1 \nu(\tau)\,d\tau} \qquad (5)$$

in which $\nu(\tau)$ is a time varying mean upcrossing rate, the integration is performed over 1 year, and N_y is the number of years of interest.

5. CONCLUSIONS

Wind speed and water equivalent ground snow

load data were examined to obtain appropriate statistics for load modeling. Both loads showed strong occurrence dependence, and hence were modeled as a clustered Poisson pulse intermittent process. An extreme type I distribution wa found to provide the best means for modelling individual winds. The Weibull distribution was selected for the individual snow intensity, based on it being the best means for accurately predicting extreme snow loads. Occurrence dependence between wind and snow exists in two forms, wind and snow arising from the same storm, and seasonal variations. Both types of dependencies were illustrated, and methods for accounting for the occurrence dependencies were given.

ACKNOWLEDGEMENT

This material is based upon work supported by the National Science Foundation under Grant Nos. CES-8706844 and BCS-8921465. This support is gratefully acknowledged.

REFERENCES

Belk, C.A., and Bennett, R.M. (1991). "Macro Wind Parameters for Load Combination." Journal of Structural Engineering, ASCE, 117(9), 2742-2756.

Corotis, R.B., and Tsay, W-Y. (1983). "Probabilistic Load Duration Model for Live Loads." Journal of Structural Engineering, ASCE, 109(4), 859-876.

Ellingwood, B., Galambos, T.V., MacGregor, J.G., and Cornell, C.A. (1980). "Development of a Probability Based Load Criterion for American National Standard A58." U.S. National Bureau of Standards Special Publication 577.

Gilley, R.D., and Bennett, R.M. (1991). "Snow Load Modelling for Structural Load Combination Methods." Civil Engineering Research Series No. 47, The University of Tennessee, Knoxville, Tennessee.

Grigoriu, M. (1984). "Load Combination Analysis by Translation Processes." Journal of Structural Engineering, 110(8), 1725-1734.

Isyumov, N. (1982). "Roof Snow Loads: Their Variability and Dependence on Climatic Conditions." Structural Use of Wood in Adverse Environments, Van Nostrand Reinhold Co., New York, NY, 365-384.

Larrabee, R.D., and Cornell, C.A. (1981). "Combination of Various Load Processes." Journal of the Structural Division, ASCE, 107(ST1), 223-239.

O'Rourke, M.J., Speck, R.S., and Stiefel, U. (1985). "Drift Snow Loads on Multilevel Roofs." Journal of Structural Engineering, ASCE, 111(2), 290-306.

Pier, J-C., and Cornell, C.A. (1973). "Spatial and Temporal Variability of Live Loads." Journal of the Structural Division, ASCE, 99(ST5), 903-922.

Templin, J.T., and Schriever, W.R. (1982). "Loads due to Drifted Snow." Journal of the Structural Division, ASCE, 108(ST8), 1916-1925.

Wen, Y.K. (1981). "Stochastic Dependencies in Load Combination." Structural Safety and Reliability. Elsevier Scientific Publishing Company, Amsterdam, 89-102.

Wen, Y.K. (1990). Structural Load Modeling and Combination for Performance and Safety Evaluation. Elsevier Science Publishers, Amsterdam, Netherlands.

Structural Safety & Reliability, Schuëller, Shinozuka & Yao (eds) © 1994 Balkema, Rotterdam, ISBN 90 5410 357 4

The worldwide increasing windstorm risk: Damage analysis and perspectives for the future

Gerhard Berz
Munich Reinsurance Company, Germany

ABSTRACT: Windstorm catastrophes have, in recent years, produced enormous economic losses and have hit the insurance industry particularly hard. The dramatic increase in windstorm losses is mainly due to the migration of population (and values) into exposed areas, but possibly also to changes in environmental conditions. Climate change will, in the long term, influence all types of windstorms, above all tropical and extratropical cyclones. Stringent countermeasures are imperative, amongst them the reduction of greenhouse gas releases, improved windstorm forecasts, radical land use restrictions, stricter building codes and the prevention of secondary losses. The introduction of substantial deductibles in windstorm insurance will be the most effective way of motivating people to adopt precautionary measures.

1 THE GROWING LOSS BURDEN

In the last six years, many parts of the world have been hit by windstorm catastrophes of an unprecedented severity. Until the mid-1980s, it was widely thought that windstorm losses exceeding US$ 1 billion were only possible, if at all, in the United States, but the gale that hit Western Europe in October 1987 heralded a series of extraordinary windstorm disasters which make US$ 10 billion losses seem almost commonplace nowadays (cf. table 1) and suggest that losses several times that size are well within the realms of possibility.

Take Hurricane Andrew, for instance, which in August 1992 left a trail of destruction fifty kilometres wide across the southern periphery of Miami and southwestern Louisiana, with insured losses that are currently estimated to be at least US$ 16 billion. The amount of damage it caused would have multiplied if its track had been just marginally to the north. Since this kind of loss potential is possible in other coastal regions of the United States, too, and can be generated by other types of windstorm such as tornado and blizzard, it appears to be only a matter of of time before there is a "direct hit", producing a catastrophe event of gigantic dimensions.

Table 1. Some Great Windstorm Disasters since 1987

Year	Event	Economic losses (US-$ m)	Insured losses (US-$ m)	Deaths
1987	Winter Gales Western Europe	3,700	3,100	17
1988	Hurricane 'Gilbert' Jamaica, Mexico	2,000	800	286
1989	Hurricane 'Hugo' Caribbean, USA	9,000	4,500	61
1990	Winter Gales Europe	15,000	10,000	230
1991	Typhoon 'Mireille' Japan	6,000	5,200	51
1992	Hurricane 'Andrew' USA	approx. 30,000	approx. 16,000	44

Losses due to natural disasters have increased dramatically in recent decades (cf. figure 1). The reasons for this have already been discussed in depth in various publications (cf. Munich Re 1990). Table 2 shows that the trend factors established earlier have continued to rise steeply in recent years, especially in the case of windstorm disasters.

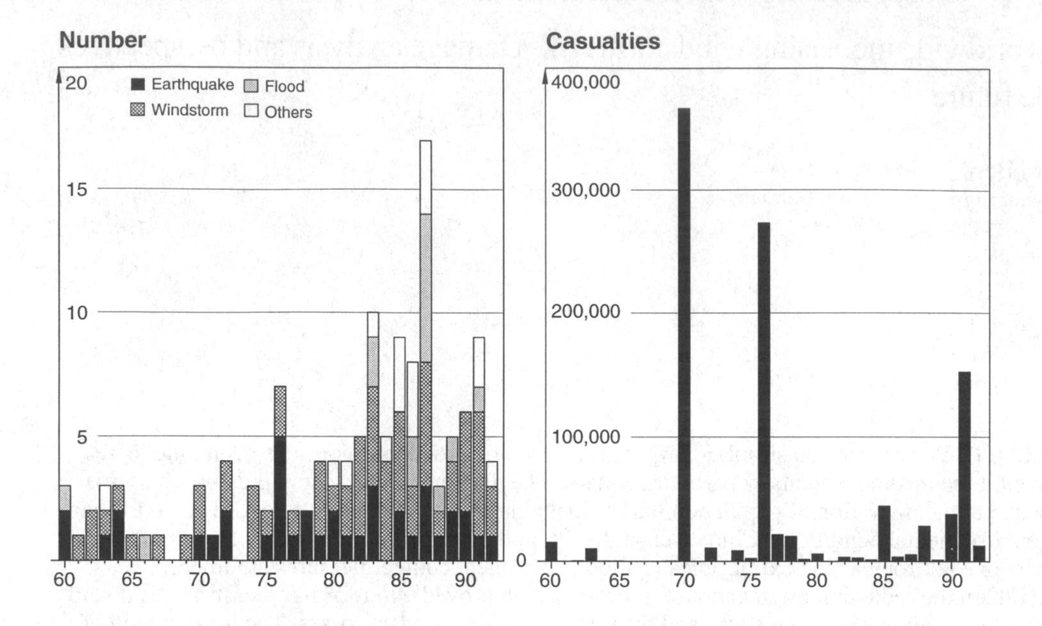

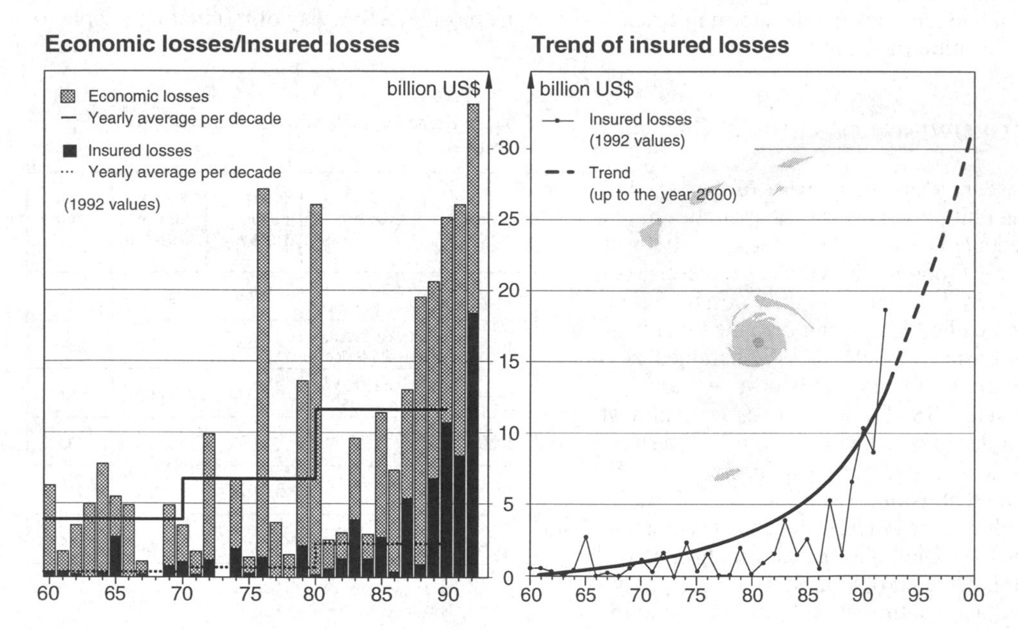

Figure 1. Great Natural Disasters 1960 - 1992

The upper graph on the left shows the increase in the number of great natural disasters in recent decades. There is no recognizable trend in the number of deaths (right). The lower graphs show dramatic increases of economic and even more of insured losses, both already adjusted for inflation.

Table 2. Major Windstorm Catastrophes 1960 - 1992

The number and the extent of major windstorm disasters have both increased dramatically in recent decades. In the period between 1983 and 1992 the number of such disasters and the economic loss they caused rose by a factor of four compared with the 1960s, while the increase in insured losses was almost tenfold.

	(60s) Decade 1960 - 1969	(70s) Decade 1970 - 1979	(80s) Decade 1980 - 1989	**Last ten 1983 - 1992**	**Factor** 80s : 60s	**Factor** Last 10 : 60s
Number	8	14	29	31	3.6	**3.9**
Economic losses *	22.6	33.6	38.0	88.1	1.7	**3.9**
Insured losses *	5.3	8.3	18.9	52.1	3.6	**9.8**

* US-$ billion; extrapolated to 1992 prices

2 THE EFFECTS OF CLIMATE CHANGE

Although the proliferation of values in windstorm-exposed regions can still be regarded as the leading cause of the observed trend, there are growing indications that the steadily emerging phenomenon of global warming is already beginning to have a marked influence on the windstorm risk. Above all, however, we cannot rule out the possibility that what we are currently experiencing is only the initial phase of a really catastrophic development.

Let us look at a number of recent findings and arguments:

- The distinct warming of the atmosphere - with seven of the last ten years being warmer than all other years since worldwide meteorological records began halfway through last century - and the subsequent warming of the oceans raise both sea levels and the water vapour content of the atmosphere (through more evaporation). As a result, there is more energy available for atmospheric processes, which leads to more intensive low-pressure vortices over the tropical and extra-tropical oceans and more vehement thunderstorms (with rainstorms, hailstorms, and tornadoes) over land.

- From the observations made to date, tropical cyclones (hurricanes, typhoons, cyclones) can only develop and gain in strength at ocean surface temperatures exceeding 26°C. If the water temperatures rise, the areas in which cyclones are possible will grow in size (hence threatening in the long term South America, the Mediterranean region, and finally western Europe) and, in particular, the intensity of these cyclones will increase sharply. Given a rise in global temperatures of 3°C, which, with the carbon dioxide content of the atmosphere doubling, is already expected in the second half of next century, pressures as low as 820 hPa may be encountered in the centres of such cyclones (as opposed to the current record of 870 hPa) leading to windspeeds of 360 km/h (as opposed to 300 km/h). Local damage intensities will consequently be 70 % higher since they increase proportional to the third power of the windspeed (cf. figure 2 and table 3).

- The number of strong low-pressure systems over the North Atlantic and Europe has risen by about 40 % since the 1930s, the number of extreme low-pressure systems even more so (cf. figure 3). It is therefore not just a coincidence that the lowest central pressures ever observed in this area (below 920 hPa) were recorded in recent years.

Table 3. Increasing Damage Intensity of Tropical Cyclones

The lowest possible central pressure (P_m) falls with increasing sea surface temperatures (T_s). This causes higher maximum wind speeds (V_m) and much higher damage intensities (D_I).

Ts (°C)	Pm (hPa)	Vm (m/sec)	DI $(Vm/Vm27)^3$
27	911	72	1.00
28	902	75	1.13
29	891	79	1.32
30	879	83	1.53
31	865	88	1.83
32	849	93	2.16
33	829	99	2.60
34	805	106	3.19

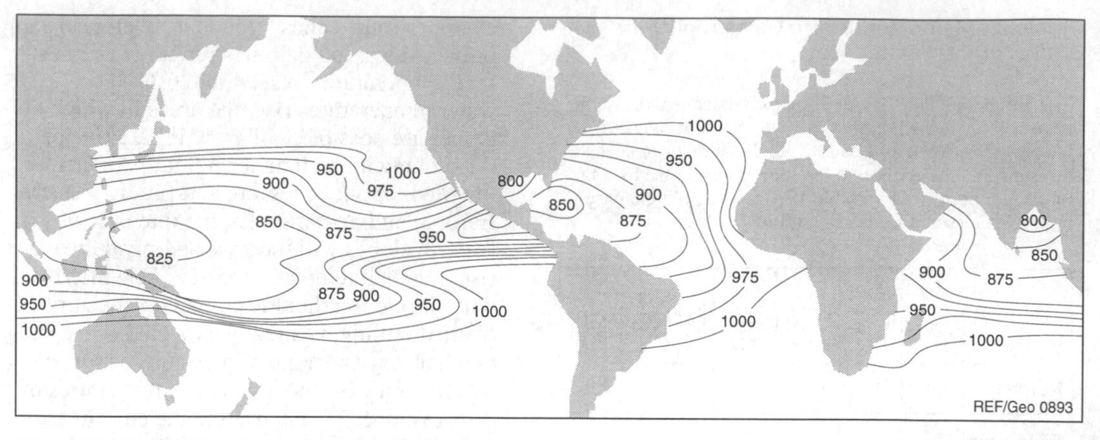

Figure 2. Increasing Intensity of Tropical Cyclones

The carbon dioxide content of the atmosphere in the second half of next century is expected to be twice as high as in pre-industrial times. This will probably cause ocean surface temperatures to rise by several degrees, which in turn will lead to much more intensive cyclones being able to develop over the majority of tropical oceans. The map shows the global distribution of the lowest possible central pressure for a 2 x CO_2 scenario (according to K. Emanuel, 1987). In addition, the areas in which it is possible for cyclones to develop will grow considerably and may then include the South Atlantic, for instance, where none have been recorded to date.

- Also, the cold continental high-pressure system, which in normal winters builds up over the snow surfaces of central and eastern Europe and then acts as a barrier to low-pressure systems that come in from the Atlantic, diverting them towards the north or south before they reach Europe, was much weaker in the last six winters, which were unusually mild with little snow. This meant that its blocking effect was less pronounced so that Atlantic low-pressure systems were able to penetrate much further into central and eastern Europe (cf. figure 4). If winter temperatures in Europe continue to rise, windstorm series like the one in January and February 1990 or most recently in January 1993 may gradually become the norm.

- Due to the combined effect of increasing windstorm force and frequency, rising sea levels, and sinking coastal zones, storm tides are rising dangerously in many regions. In the long term they even threaten the existence of a number of small island states. Expensive coastal protection measures such as those adopted by the countries on the North Sea are too costly for the majority of the countries that are particularly threatened.

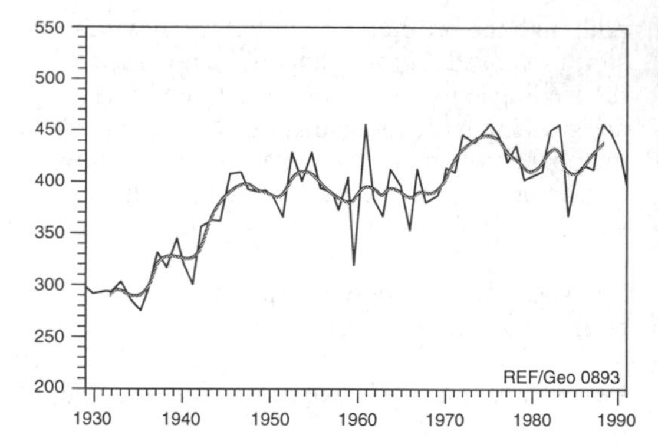

Figure 3. Yearly Number of Strong Extratropical Cyclones below 990 hPa over the Atlantic and Europe (according to H. Schinke, 1992)

3 POSSIBLE COUNTERMEASURES

The mounting windstorm hazard can only be controlled if countermeasures are adopted resolutely throughout the world. Among the measures that are extremely urgent and can be put into practice quite rapidly are the following:

- Everything must be done to curb the further intensification of the man-made greenhouse effect. This includes the immediate implementation of the World Climate Convention agreed on in Rio in 1992, which first of all requires a substantial reduction in the release of greenhouse gases from the industrialized nations mainly responsible for this in the past. Limiting the greenhouse effect can only be achieved in the foreseeable future through measures aimed at reducing energy consumption, especially if Third World countries are to be granted the additional consumption of energy they need for sustainable economic development. At the same time, other energy sources, in particular solar energy, may represent an economical and environmentally sound alternative, especially in less developed regions, and should be promoted more than hitherto by the industrialized countries.

- The meteorological services of forecasting windstorm and severe weather events and of alerting areas that are acutely threatened must be used and improved throughout the world. If forecasts lead to precautions being taken in good time, it is possible, as Hurricane Andrew and the great United States blizzard of March 1993 showed, not only to prevent fatal casualties but also to reduce the extent of damage to property quite significantly.

- In regions with a particularly high exposure, the authorities must impose radical restrictions on land use. In the face of constantly increasing pressure to inhabit these areas, such action, in the medium and long term, is necessary to avoid a situation in which more and more people and industry settle in high-risk areas. In many cases, prohibiting land use directly on the waterfront would be enough to reduce the loss potential dramatically. At least in the more developed regions of the earth it should be possible to prevent large concentrations of people and values in extremely exposed zones.

- In many countries, the building codes governing the windstorm-resistant design of structures must be made more stringent and above all they must be enforced more strictly. As far as normal types of building are concerned, the additional costs of adequate windstorm-resistant design only come to about 1 - 4 % of the total cost of construction. The building regulations must incorporate straightforward, practice-orientated methods of calculation in their consideration of the complicated interplay of pressure and suction forces acting on the exterior parts of buildings, especially on the roofs, facades, windows, and doors. The determining wind force (usually the "fifty-year wind") can no longer be based simply on observations of the past but must also anticipate future trends.

- Damage caused indirectly by windstorm, such as the soaking of buildings whose roofs have been blown off, flying debris, uprooted trees, and cut power lines and telephone cables, are important loss factors and must be included as such in the consideration of precautionary measures.

The International Decade for Natural Disaster Reduction (IDNDR), which the United Nations proclaimed for the 1990s, has made it an aim to gather knowledge on natural hazards from all over the world and convert this into precautionary measures, above all in the Third World. In this way it is hoped that the alarming negative trend in natural disasters can be broken.

4 THE CONSEQUENCES FOR THE INSURANCE INDUSTRY

Precautionary measures, however good they may be, do not provide absolute protection. Insurance cover will therefore continue to be an important means of making provisions of a financial nature. Windstorm cover is one of the most common forms of cover for losses caused by the elements. In Germany, for example, more than 55 % of all residential buildings and 70 % of all households are insured against windstorm damage.

As the insurance density increases, so too does the loss potential of future windstorm catastrophes for insurers and especially for reinsurers, who after the 1990 series of gales, for example, had to carry about two-thirds of the total insured loss of over DM 17 billion. Given the effects of climate change discussed above, much greater loss potentials are to be expected in the future. After the extremely poor results of recent years, these will coincide with shrinking

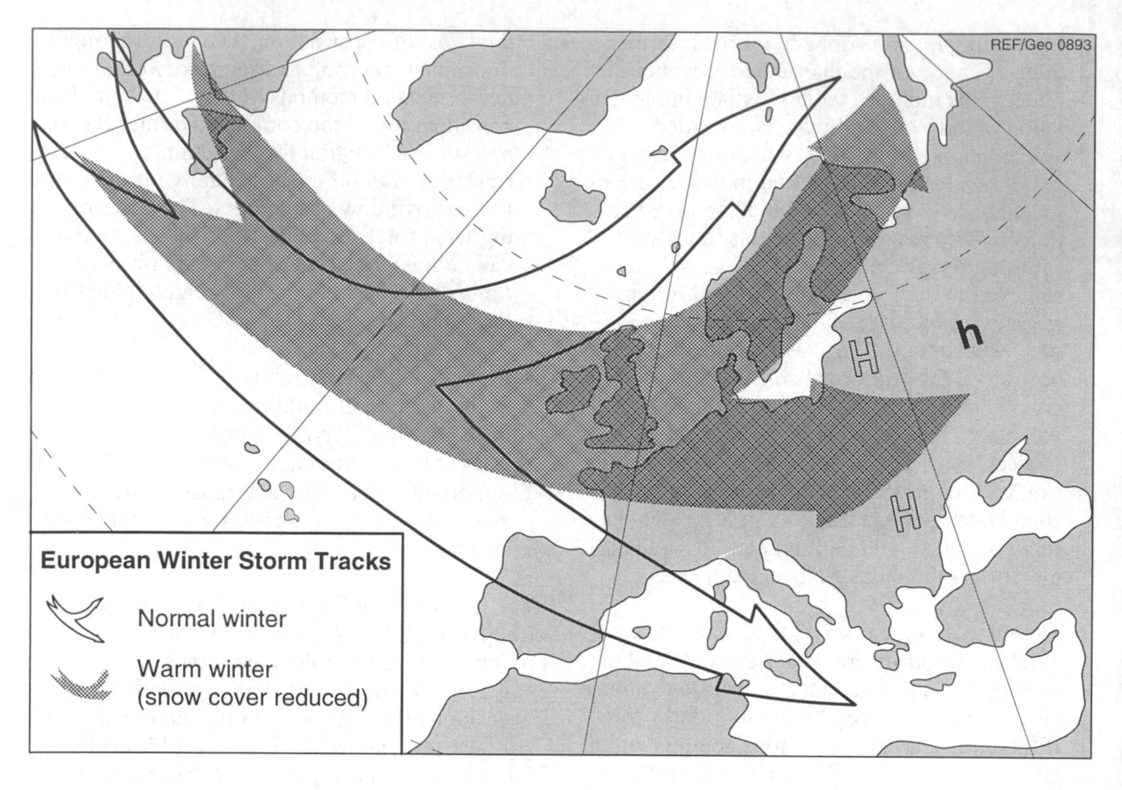

Figure 4. European Winter Storm Tracks

In warm winters with little snow the eastern European cold high-pressure system is weaker and shifts further towards the east. Its effect of acting as a barrier to low-pressure systems coming in from the Atlantic is reduced so that these can penetrate far into western and central Europe. As a result there is a greater hazard of windstorm there even without windstorm activity increasing over the North Atlantic (according to H. Dronia, 1991).

reinsurance capacity throughout the world, which has already led to shortages and rising prices.

The best chance of reducing the insured loss potential and at the same time of motivating insureds to adopt precautionary measures and concentrate on loss prevention is to introduce deductibles of a significant size. Of the millions of claims made in Germany under houseowners' comprehensive insurance policies following the winter gales in 1990, more than 65 % were below DM 1,000. A deductible in the order of 1 % of the insured value, which is relatively small and no problem for the majority of homeowners, would have had a significant effect; above all, it would have reduced the enormous number of minor claims that had to be handled to a manageable and economically reasonable level, involving only those cases in which the policy holders really did suffer. The introduction of an adequate deductible would mean that windstorm premiums, which should really be tending to rise, could be kept at the present level; at the same time, the problems of capacity in the insurance and reinsurance markets would be distinctly reduced.

REFERENCES

Berz, G. 1991. Global Warming and the Insurance Industry. Nature & Resources 27: 19-30.

Dronia, H. 1991. Zum vermehrten Auftreten extremer Tiefdruckgebiete über dem Nordatlantik in den Wintern 1988/89 bis 1990/91. Die Witterung in Übersee 39, 3.

Emanuel, K. 1987. The Dependence of Hurricane Intensity on Climate. Nature 326: 483-485.

Munich Reinsurance Comp. 1988. World Map of Natural Hazards. Wall Map with Explanatory Booklet. Munich.

Munich Reinsurance Comp. 1990. Windstorm - New Loss Dimensions of a Natural Hazard. Munich.

Munich Reinsurance Comp. 1993. Winter Storms in Europe - Analysis of 1990 Losses and Future Loss Potential. Munich.

Schinke, H. 1992. Zum Auftreten von Zyklonen mit Kerndrücken <= 990 hPa im atlantisch-europäischen Raum von 1930 bis 1991. Spezialarbeiten aus der Arbeitsgruppe Klimaforschung des Meteorologischen Instituts der Humboldt-Universität Berlin.

Structural Safety & Reliability, Schuëller, Shinozuka & Yao (eds) © 1994 Balkema, Rotterdam, ISBN 90 5410 357 4

Modeling of artificial stochastic wind loads and interaction forces on oscillating line-like structures

R. Höffer, H.-J. Niemann & J.X. Lou
Building Aerodynamics Laboratory, Ruhr-Universität Bochum, Germany

ABSTRACT: A model for the simulation of stochastic wind forces is discussed. The procedure may serve the purpose of providing time histories of flow forces for the time domain analysis of structures, which tend to perform forced and aeroelastic vibrations. The forces act on vertical or horizontal line-like constructions, which are immersed in a turbulent boundary layer (TBL) flow. The TBL flow can be represented by a numerically generated multi-correlated stochastic field. The buffeting wind loads are obtained from a nonlinear transformation of the wind time histories by means of a consistent extension of the classical quasi-stationary approach. The smoothing effect of turbulence on the aerodynamic force coefficients is taken into account. The spectral band-width of the excitations due to periodic vortex separations is modelled by adapting a stochastic argument to a harmonic function. Constant and fluctuating aerodynamic damping and stiffness forces are calculated simultaneously.

1 INTRODUCTION

For the purpose of the calculation of reliability estimates of oscillating structures, which undergo a broad-band excitation, sophisticated methods of spectral analysis or perturbation theory are available [1]. The applicability of such methods is limited to linear, linearizable or weakly non-linear systems and forced oscillations. Besides the treatment of structures with geometrically [e.g. 2] and physically highly nonlinear restoring forces, e.g in the cases of large displacements and predamaged or degrading components, direct integration is useful in the case of flow-structure interactions. Such approaches require external load processes and a description of flow-induced contributions to the impedance of an aeroelastic system. The flow forces may be calculated from instantaneous velocity components using a quasi-stationary model [3]. The quasi-stationary approach yields correct spectral ordinates of loads up to reduced frequencies $f \cdot b/\bar{u}$ of about 0.2 [8], where f is the frequency in Hz, $\bar{u}$ is the mean wind velocity in the considered height and b is the characteristic aerodynamic width of the body. Beyond this limit, other fluid-mechanical phenomena affect the wind loads, such as the separation induced turbulence. Furthermore, mathematical models exist, which supply time histories of vortex-induced loads.

The required time histories of the wind velocity are obtainable from wind tunnel measurements in a turbulent boundary layer flow, but it often requires much experimental effort to collect long time histories of the wind velocity field at a sufficiently large number of points simultanously. Numerical generation procedures are applied instead, such as the autoregressive method, which allows to simulate important statistical properties of processes of the wind velocity.

Main assumptions for quasistationary wind force models

The turbulent boundary layer flow is given by its longitudinal $\bar{u}+u'$, lateral v' and vertical w' wind velocity components, $\bar{u}$ is the mean wind velocity in a certain height, u', v' and w' are the fluctuating components of velocity with zero mean. It is assumed that the fluctuations of the wind velocity components occur simultaneously in a strip of certain thickness throughout the neigbourhood of the body. This implies that the integral sizes of the turbulent eddies are large compared to the dimension of the cross-section. Merely angles of attack of the wind from flow fluctuations in plane of that considered layer are taken into account for computing the wind force. That means that a two-dimensional flow around the outer surface of the body is assumed.

Since $v', u' << \bar{u}$, the momentary dynamic pressure is approximated by the linearization:

$$q_R(t) = q(t) = \bar{q}\ (1 + \frac{2u'}{\bar{u}}),\quad \bar{q} = \frac{\rho}{2}\bar{u}^2 \quad . \qquad (1)$$

The mean azimuthal angle of attack of the wind due to a reference coordinate system x,y is $\bar{\varphi}$. The actual direction φ' of the turbulent flow deviating from the mean wind can be described approximatively by:

$$\bar{\varphi} + \varphi' = \bar{\varphi} + \arctan \frac{v'}{\bar{u}+u'} \tilde{=} \bar{\varphi} + v'/\bar{u} \qquad (2)$$

The aerodynamic coefficients c_D, c_L, and c_T are constant with respect to time and are determined in smooth flow.

$c_{F,\Theta}(\varphi)$ is often linearized by an expansion Taylor-type series (up to higher order members),

$$c_{F,\Theta}(\varphi) = c_{F,\Theta}(\bar{\varphi}+\varphi') = c_{F,\Theta}(\bar{\varphi}) + \left.\frac{dc_{F,\Theta}}{d\varphi}\right|_{\bar{\varphi}} \varphi' \qquad (3)$$

Time domain calculations are accessible to a direct evaluation of a nonlinear function of the instantaneous angle:

$$c_{F,\Theta}(\varphi) = c_{F,\Theta}(\bar{\varphi} + \varphi') \quad , \qquad (4)$$

F stands for D and L correspondingly.
The fluid forces become:

$$F(t) = b \cdot q_R(t) \cdot c_F(\bar{\varphi}+\varphi'(t)),$$

$$F_\Theta(t) = b^2 \cdot q_R(t) \cdot c_T(\bar{\varphi}+\varphi'(t)) \qquad (5)$$

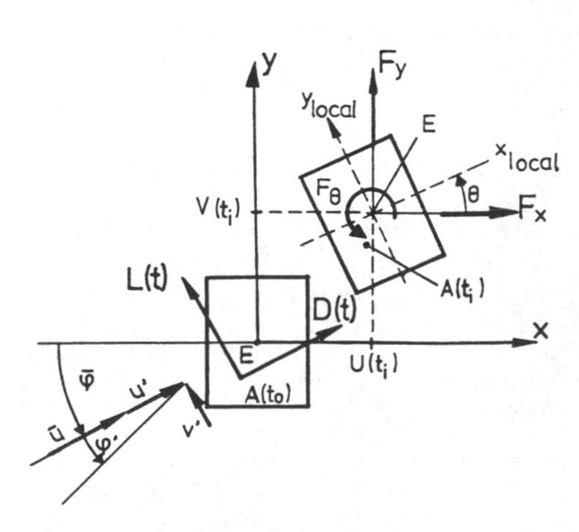

Fig.1: Notations and principal sketch of the cross-sections

The wind forces attack in the aerodynamic center A, which is the point of intersection of the drag- (D) and lift- (L) components [Fig.1]. Generally, its position may be eccentric according to the elastic center of the structure. At the present stage, the spectra and coherences of the strip force components are derived from the turbulence field.

2 NONLINEAR TRANSITION OF FORCE COEFFICIENTS FROM SMOOTH FLOW TO TURBULENT FLOW

Singular points of discontinuity in smooth flow set a limit to the general acceptance of the aerodynamic coefficients' representation as a Taylor-type series, since the differential quotient $dc_{F,\Theta}/d\varphi$ does not exist in any cases. Furthermore, the fluctuations of the angles of attack of the flow typically cover a range of about ±20°, so that $(dc_{F,\Theta}/d\varphi)\cdot\varphi$ would result in significantly erroneous estimates of the coefficients.
8. The consideration of the fluctuation $\Delta(\varphi'(t))$ instead should be a more exact description. $\Delta(\varphi'(t))$ is the variation of a momentary coefficient $c(\bar{\varphi}+\varphi'(t))$ from the coefficient $c(\bar{\varphi})$ at mean flow direction. It is represented by the probability density functions (PDF) $f_\Delta(\Delta)$, which one obtains from a nonlinear transformation of $f_\varphi(\varphi')$ considering the nonlinear interrelation between Δ and φ'.

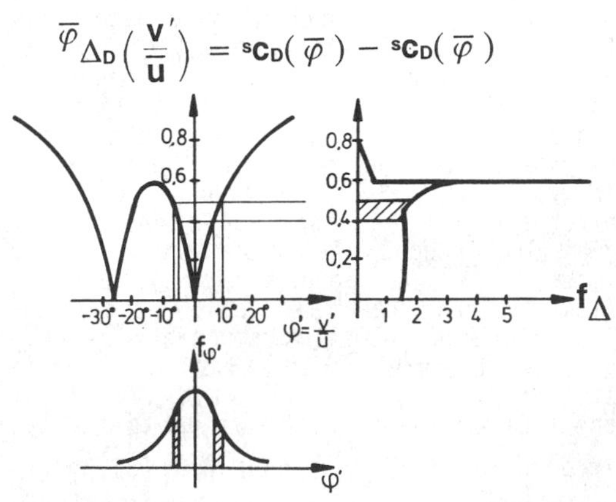

Fig.2: Construction of $f_\Delta(\Delta)$ at $\bar{\varphi}$=13° for the drag strip load on a cylinder of infinite length with quadratic cross-section in turbulent flow

The inverse function $\varphi' = g^{-1}(\Delta)$ may be ambiguous, so that

$$f_\Delta(\Delta) = \sum_n f_\varphi(\varphi'_n) \cdot \left| \frac{d\varphi'}{d\Delta} \right|_{\varphi'_n} \qquad (6)$$

where n is the number of ambiguities at Δ and $\left| \frac{d\varphi'}{d\Delta} \right|$ is the Jacobian, here for the case of a scalar transformation.

A mean aerodynamic strip force is $F = b\bar{q}\,[c(\bar{\varphi}) + \bar{\Delta}]$, where $\bar{\Delta}$ is the difference of a coefficient in smooth flow and its mean value in turbulent flow. $\bar{\Delta}$ may be computed as an time average of $\Delta(\varphi')$ or by evaluating $f_\Delta(\Delta)$.

Experimental verifications show, that even small scales of turbulence shift a mean value [Fig.3]. That can be modeled correctly, even for low- and high-turbulent flows. Following the assumption(1), (5) is

$$\frac{F(t)}{b \cdot \bar{q}} = (1 + \frac{2u'}{\bar{u}})[c_F(\bar{\varphi}) + \Delta(\varphi')],$$

$$\frac{F_\Theta(t)}{b^2 \cdot \bar{q}} = (1 + \frac{2u'}{\bar{u}})[c_T(\bar{\varphi}) + \Delta(\varphi')]. \qquad (7)$$

The forces consist of a gaussian component $(1 + \frac{2u'}{\bar{u}})c(\bar{\varphi})$ with the PDF f_C and the non-gaussian variable $(1 + \frac{2u'}{\bar{u}})\Delta(\varphi')$ with PDF $f_\Delta(\Delta)$. Since the gaussian component is continuous, the PDF of F(t), $f_F(F)$, resp. $F_\Theta(t)$, $f_\Theta(F_\Theta)$, may be calculated as a convolution of PDF of $f_\Delta(\Delta)$ and f_C.

As expected, the statistical moments of any F(t) change compared to linearized evaluations. In the case of drag, the probability distribution remains symmetrically for a mean angle of attack of the flow of 0° [s. Tab. 1] and 45° and nearly of gaussian type for low turbulence intensities. The skewness coefficients increase for ranges of strongly nonlinear interrelations of φ and c_D, eg. at $\bar{\varphi}$=13°, especially in the case of low turbulence. The contribution of cross-wind to the lift forces is often comparable in size to its fluctuating along-wind components. The values of skewness and curtosis mark an increased deviation from a gaussian distribution.

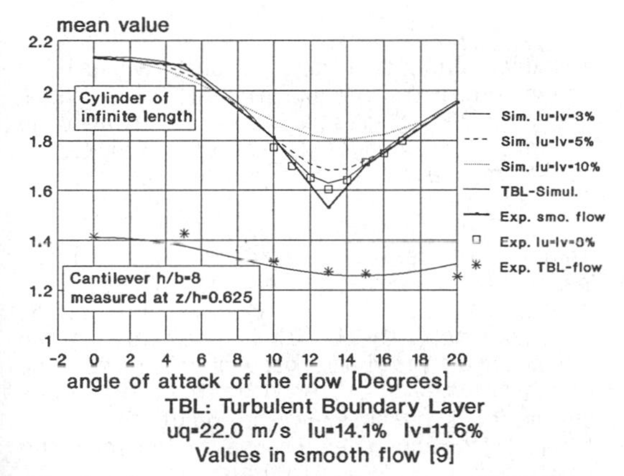

Fig.3: Drag coefficients on a cylinder of infinite length with quadratic cross-section in smooth and turbulent flows

3 VORTEX EXCITATION

In mathematical load models for the time domain the vortex-induced lift is imposed

Tab.1: Coefficients of skewness γ_1 and curtosis γ_2 of the drag strip load on a square cylinder of infinite length

	Wind field characteristics	Wind load	$\gamma_1 = \frac{E[(D-\bar{D})^3]}{\sigma^3}$	$\gamma_2 = \frac{E[(D-\bar{D})^4]}{\sigma^4} - 3$
Simulation 210000 samples	$\bar{u}$ = 22.0 m/s I_u=3.0%, I_v=3.0%	$D(\bar{\varphi}=0°)\ /\ \bar{q}b$	-0.01	0.04
		$D(\bar{\varphi}=13°)\ /\ \bar{q}b$	0.26	0.14
	$\bar{u}$ = 22.0 m/s I_u=14.1%, I_v=11.6%	$D(\bar{\varphi}=0°)\ /\ \bar{q}b$	0.03	-0.07
		$D(\bar{\varphi}=13°)\ /\ \bar{q}b$	0.11	0.04
Experiment 21000 samples		$D(\bar{\varphi}=0°)\ /\ \bar{q}b$	0.03	-0.36
		$D(\bar{\varphi}=13°)\ /\ \bar{q}b$	0.17	-0.04

on cylindrical structures as a transversal force, which runs sinusoidal in time (harmonic model). Originally, the harmonic model serves to consider vortex-induced lateral forces in smooth flow. The corresponding spectral representation is a δ-function at the discrete frequency of vortex-separations. That predominant shedding frequency is calculated from u/b and a dimensionless proportionality constant, the Strouhal number S_r.

Turbulence intensity increases the width of the peak, since in the case of a turbulent flow the quasi-stationary concept would predict that the Strouhal frequency will fluctuate corresponding to the fluctuations of the wind velocities: $f_S(t)=S_r \cdot u(t)/b$. According to u(t) the density function of $f_S(t)$ is of gaussian type. That extension of the quasi-stationary theory leads then to the following model for a regular vortex shedding:

$$\frac{L_S(t)}{\bar{q}\cdot b} = \qquad (8)$$

$$(c_S\ (1+\frac{2u'(t)}{\bar{u}})+\Delta_S(\varphi'))\ \sin(2\pi f_S(t)t)$$

where c_S is the force coefficient of vortex-induced excitation. It depends upon the angle of attack of the flow; c_S is determined from measurements in smooth flow by $\sqrt{2}\cdot\sigma_{c_L}$.

The amplitude of L_S and the frequency of the sinusoidal component are modulated. The momentary frequency of vortex separations is calculated as the constant frequency of a reference sinusoid, which joins closely to the value of the function $\sin(2\pi f_S|_t\ t)$ at time t.

$$\omega_S|_t = \frac{2\pi S_r}{b}\ (\ \bar{u} + u'|_t\ ) = \dot{\Phi}_S|_t \qquad (9)$$

Then, the phase of the frequency-modulated signal is

$$\Phi_S(t) = \bar{\omega}_S\ t + \frac{2\pi S_r}{b} \int u'(t)\ dt \quad . \qquad (10)$$

The argument of the sinusoidal function varies nonlinear with time. In general, the modulating functions are provided as stochastic processes.

$$\Phi_{Sn} = \bar{\omega}_S\ n\ \Delta t + \frac{2\pi S_r}{b}\Delta t \sum_{i=1}^{n} u'_i \ + \Phi_0\ , \qquad (11)$$

where n=1,2,3,...,N, is the number of time steps.

The contributions of periodic vortex shedding to the lateral load on a cylindrical square cantilever have been measured in smooth and turbulent boundary layer flow. As an example, [Fig.4] shows the normalized spectra of the base bending moment. The original set-up produces a spectrum with one vortex-separation-induced peak, which is centered on the Strouhal frequency.

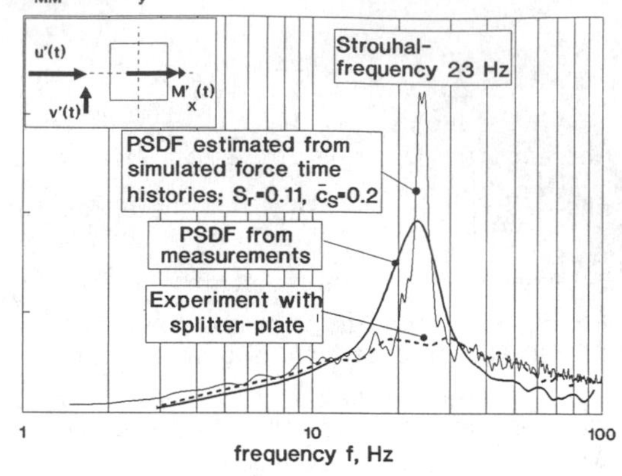

Fig.4: Spectral density functions of the base bending moment of a square cylindrical cantilever in turbulent boundary layer flow

In a special experiment, vortex separations were supressed. A splitter plate was placed along the center-line of the wake near to the bluff-body, separated through a vertical gap of 0,2 b. The height of the splitter plate was two times the model's height and the length had five times b. Around the Strouhal frequency the resulting spectrum follows on the flat course of the turbulence-induced load components. The area between both lines represents a contribution of vortex excitation of 26% to variance.

Outside the Strouhal range the spectral shape of the simulated time histories of the lift force joins closely to the expected shape. Regular eddy shedding dominates around the central Strouhal frequency as desired. The calculated bandwith is in approximate accordance to the estimated values of this experiment. The absolute spectral ordinates and the spatial correlation of vortex-induced forces are slightly overestimated. A reason might be that the vortices separate as a non-stationary process. A coherent vortex-shedding is limited to strips of certain thickness in the case of a high-turbulent flow [4,12] and cannot be derived directly from the wind field.

It is the matter of present investigation

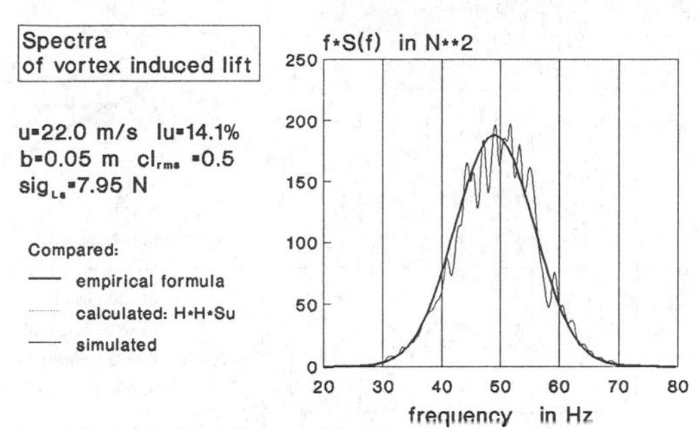

Fig.5: Analytical and simulated spectra of the lift forces due to regular eddy shedding applying an adopted digital filter function

to generate force time histories from a suggested spectral model. Vickery assumes in [4] a gaussian-shaped spectrum of vortex-induced excitations with a local maximum at the Strouhal frequency. Refering on such empirical formulas and with the analytical knowledge of the wind spectrum a digital filter can be designed. A simple model is a symmetrical MA-model. In the time domain it is represented by

$$\mathbf{L_s}(n\cdot\Delta t) = \sum_{k=1}^{P} \mathbf{A}_k \cdot u_{n\Delta t-k} \qquad (12)$$

No difference can be ascertained between the empirical formular and the analytical function which is calculated for the high order of P=80. The spectral ordinates which are estimated from the simulated force time histories are in satisfactory accordance to the empirical formula. That procedure would include e.g. the effect of nonstationarities on the spectrum.

4 MODEL OF AERODYNAMIC IMPEDANCES - AEROELASTIC FORCES

In eq.(13) relative velocities are evaluated from the instantaneous wind velocities and the components of the actual velocities of the structural motion for every strip. The relative wind velocities in mean wind and lateral directions are

$$u_{rel}(t)=\bar{u}+u'(t)-\dot{U}(t)\cos(\bar{\varphi})-\dot{V}(t)\sin(\bar{\varphi}) \quad \text{and}$$

$$v_{rel}(t)=v'(t)+\dot{U}(t)\sin(\bar{\varphi})-\dot{V}(t)\cos(\bar{\varphi}), \qquad (13)$$

where $\dot{U}(t)$, $\dot{V}(t)$ are the translational response velocities of E [s. Fig.1] in the directions of the global x-, y-coordinates. Only translational velocities of the structural response are considered here.

Eq.(14) takes into account the rotation of the body-fixed coordinate system relative to the mean angle of attack of the wind. The fluctuating angle of attack of the relative flow $\varphi'_{rel}=\arctan(v_{rel}/u_{rel})$ and the torsional rotation Θ of the body complete to $\gamma(t)-\varphi'_{rel}\ \Theta(t)$. The angle of attack of the relativ flow about the deflected body, γ, is only insignificantly affected by the contributions of u', $\dot{U}$, and $\dot{V}$ to the denominator and φ'_{rel} remains small, the equation may be linearized to:

$$\gamma(t) = \varphi'(t) + [\dot{U}(t)\sin(\bar{\varphi})-\dot{V}(t)\cos(\bar{\varphi})]/\bar{u} - \Theta(t). \qquad (14)$$

The complete equations for strip forces including external and interactive components flow effects are given by:

$$F_{F,rel}(t)=b\cdot q_{rel}(t)\cdot c_F(\bar{\varphi}+\gamma(t)) \quad \text{and}$$

$$F_{\Theta,rel}(t)=b^2\cdot q_{rel}(t)\cdot c_T(\bar{\varphi}+\gamma(t)), \qquad (15)$$

where $q_{rel}(t)$ can be simplified to

$$q_{rel}(t)=\bar{q}(1+2(u'-\dot{V}\sin(\bar{\varphi})-\dot{U}\cos(\bar{\varphi}))/\bar{u}) \qquad (16)$$

The aeroelastic fluid forces are those terms of the flow forces, which consist of products of $\dot{U}$, $\dot{V}$, and $\Theta(t)$ with wind-related components. They may be looked at as contributions to the damping and stiffness properties of the oscillating system. Since q_{rel} and γ include time-dependent components, the flow-structure-system is subjected to parametric excitations. Lock-in effects are not considered here.

5 SIMULATION OF THE STOCHASTIC PRO-PERTIES OF THE WIND FIELD

The applied numerical method works according to the multidimensional recursive algorithm in [5]. The k processes of the wind field are generated from a k-dimensional AR(p)-model

$$\mathbf{u}(n\cdot\Delta t)=\sum_{p=1}^{P} \mathbf{A}(p)\cdot u(t-p\cdot\Delta t) +\mathbf{Z}(n\cdot\Delta t) \qquad (17)$$

where $\mathbf{u}(n\cdot\Delta t)$ is a vector of k components, $\mathbf{A}(p)$, p=1,2,...,P is a k x k-matrix, $\mathbf{Z}(n\cdot\Delta t)$ is the vector of so called random shocks, Δt is the time increment, and n is

the time step number. **A**(p) is calculated from a assumed matrix of covariances of the process under consideration.
In general, the v.Kármán-power-spectral-density-functions (PSDFs) $S_u(f)$ and $S_v(f)$

$$\frac{f\,S_u(f)}{\tilde{u}^2} = \frac{4\,\hat{f}}{[1+(8.409\,\hat{f})^2]^{5/6}}, \quad \hat{f} = \frac{f\,L_{ux}}{u_{ref}}$$

and

$$\frac{f\,S_v(f)}{\tilde{u}^2} = \frac{4\,\hat{f}(1+755\,\hat{f})^2}{[1+283.12\,\hat{f}^2]^{11/6}}, \quad \hat{f} = \frac{f\,L_{vx}}{u_{ref}} \quad (18)$$

and empirical estimates for the coherence as

$$Coh = e^{-f\frac{\sqrt{c_z^2(z_1-z_2)^2+c_y^2(y_1-y_2)^2}}{u_m}}$$

where $u_m=0.5[\bar{u}(z_1)+\bar{u}(z_2)]$ (19)

may be applied to describe a turbulent boundary layer flow. Herein, z_1,z_2 and y_1,y_2 are ordinates of two locations on the vertical and lateral axis, L_{ux} and L_{vx} are the integral length scales of the u- and v-component in longitudinal direction, and u_{ref} is a reference velocity at reference height. One may obtain the matrix of covariances from the matrix of cross-PSDF by means of inverse fast Fourier transformation. The recursive algorithm adapts **A**(p), P is the order (depth) of the AR-filter.
The natural frequencies of engineering structures often are found beyond in the inertial subrange (Kolmogoroff-range). In such cases, model orders of about 3 to 5 are often employed in structural analysis. Various criteria exist, as the Akaike-FPE-PAIC-, BIC, or AIC-criterion, which unfortunately give contrary predictions [6]. Herein, the applied criterion was a comparative procedure, during that one has to vary the order for adapting the **A**(p), calculate the corresponding PSDF, and compare the various results due to their fitting to the assumed spectra. The point is, of course, the quality of fitting, which is required for the considered mechanism of excitation. If interaction forces are to be dealt with or eigenfrequencies are very low, spectral ordinates down to low frequencies should be representable within the simulated time histories. The fluctuation of the wind velocity components are assumed to be represented by a gaussian joint-probability density function. That means, that gaussian white noise suits as filter input.

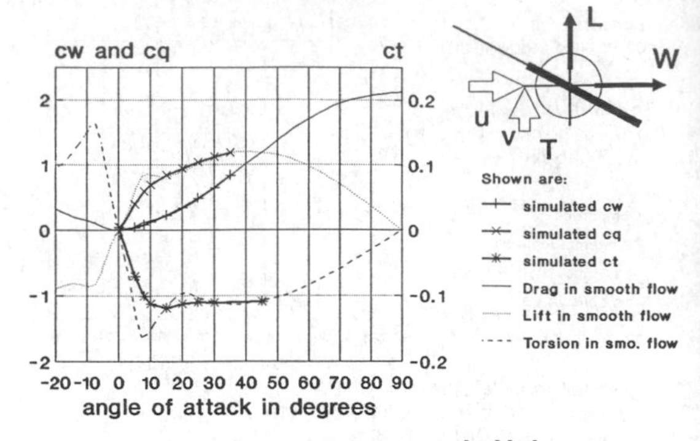

Fig.6: Simulated aerodynamic force coefficients of a flat plate of infinite length in turbulent flow

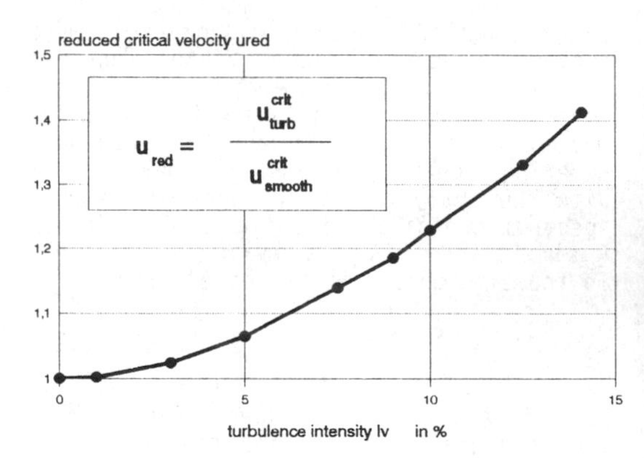

Fig.7: Effect of lateral turbulence on the critical velocity of torsional divergence of a horizontal flat plate of infinite length in turbulent flow

6 APPLICATIONS

Static torsional divergence behaviour

As in smooth flow, aeroelastic instabilities develop from mean flow. Fig.6 shows the prediction of the mean force coefficients for a flat plate of infinite length, subjected to a turbulent boundary layer flow. That allows to estimate boundaries of instability. In the case of a horizontal flat plate, lateral turbulence mainly increases the critical velocity of static torsional divergence [Fig.7].

Tab.2: Statistics of the base moment due to the along wind vibration of the structure

Statistics of the base moments	Statistical parameters			
Types of interaction and load model	$\bar{M}_{yy}$ in MNm	$\tilde{M}_{yy}$ in MNm	$\bar{M}_{xx}$ in MNm	$\tilde{M}_{xx}$ in MNm
const. aerod. damping, lin. load model	4086.0	2456.0	-	-
no add. interaction, nonl. load model	3959.0	2401.0	(-40.0)	1520.0
with interaction, nonlinear load model	3947.0	2059.0	(-54.0)	1177.0

constant aerodynamic damping $\delta_a=\dfrac{\rho\,\bar{u}_H\,b\,c_D}{2\,f_e\,\mu_m}=0{,}022$

Forced oscillations of a high-rise building

As an example the proposed model is applied to a fictitious high-rise building of 300m height and a quadratic cross-section of 50m width. Its structural properties are set to:

$\mu_m=500.0$ t/m, $I_{\theta\theta}=0.225\cdot 10^6$ tm,

$\delta_b=\delta_t=0.126$, $EI_{xx}=EI_{yy}=0.129\cdot 10^{11}$ kNm^2,

$GI=5.0625\cdot 10^9$ kNm^2.

The structure is considered as a spatial beam. Galerkin's method is applied. The interaction forces are considered as out-of-balance forces.

The structure is immersed in a turbulent boundary layer flow with a mean velocity in 10m height of 18.0m/s. For urban terrain a profil exponent 0.25 is assumed, and $\tilde{u}=\tilde{v}=0.25$m/s. A v.Kármán-spectrum and an empirically estimated coherence function, as given in eq.(18), are adjusted by the integral length scales $L_{ux}=100$m and $L_{vx}=120$m, the coefficients of the vertical lateral narrow band coherences are $c_z=c_y=10$. An artificial wind field is generated employing an autoregressive method. The computation of the filter's coefficients is performed with a very high order of 160. The time histories for the levels of the integration points are provided through stochastic interpolation.

The resulting time histories for the base moment are evaluated with regard to its statistical and spectral representation and are compared to results from a linearized approach [3] following the spectral method. The mean values result nearly identically. The r.m.s. moments differ around 10%, which may be found in the approximative consideration of the vertical correlation of the wind loads and the estimation of the r.m.s. drag coefficient for the linearized approach. The action of the complete interaction forces reduces the variance by 14%, whereas the mean response remains unchanged. A comparison $^r\tilde{M}^2_{yy}(\delta+\delta_a)/{}^r\tilde{M}^2_{yy}(\delta) = \delta/(\delta+\delta_a)$ of the resonant components of variance leads to an equivalent aerodynamic coefficient =0.052. Since mainly the fundamental mode contributes to response, that means an increase of nearly 60% of aerodynamic damping. The proposed model allows to consider higher statistical moments. For the discussed example, interaction lead to an higher left-side skewness and a decreased curtosis.

7 CONCLUSIONS

In the study carried out on the modeling of artificial wind loads and interaction forces on line-like structures the following important conclusions can be made:

(1) Calculations in the time domain are often conducted e.g. for the analysis of reliability estimates of structures with highly non-linear restoring forces, e.g. in the case of predamaged or degrading structural components or light-weighted flexible constructions, which tend to perform aeroelastic vibrations. The proposed force models may serve the purpose of providing time histories of flow force.

(2) Autoregressive methods are convenient for the construction of artificial multidimensional load fields. The filter order depends on the considered load mechanism and becomes larger than about 3 to 5 (which conventionally is applied in engineering practice) if interaction forces is to be dealt with or eigenfrequencies are low.

(3) A nonlinear mathematical transition of force coefficients from smooth flow leads to coefficients for the loads in turbulent flow. One obtains the correct mean coefficients from weighted averages over a

certain range of the aerodynamic coefficients functions, which originate from measurements in smooth flow (turbulence-induced smoothing effect on coefficients); the ordinates of f_φ pretend the weights.
(4) Periodic vortex forces on a body in a turbulent boundary layer flow can be considered by use of an adapted harmonic model, but the contributions of periodical vortex-shedding to variance is overestimated. An appropriate digital filter of high order leads to exact estimates for the expected spectral ordinates and the band-width, but requires the knowledge of analytical formulas for the spectra of the vortex induced forces and the wind velocities.
(5) The properties of the proposed force model allow to treat aeroelastic instabilities, at least if the phenomenon is representable by means of quasi-stationary coefficients. Aeroelastic effects, such as fluctuating aerodynamic damping, become predictable.

Acknowledgement

The investigations were conducted as part of a project in frame of the research programme "Dynamics of Structures" at the "Ruhr-Universität Bochum", supported by the "Deutsche Forschungsgemeinschaft". This support is gratefully acknowledged.

REFERENCES

[1] SCHUELLER,G.I.; SHINOZUKA,M. (eds.): Stochastic methods in structural dynamics. Martinus Nijhoff Publ., Dordrecht, The Netherlands, 1987
[2] BORRI,C.; CROCCHINI,F.: Nonlinear dynamic response of large structures to turbulent artificial wind: the new roof of the Olympic Stadium in Rome. pp.A-57 in: Meskouris,K.; Niemann,H.-J.(eds.): Statik und Dynamik im Konstruktiven Ingenieurbau. SFB 151-Berichte Nr.23, Ruhr-Universität Bochum, November 1992
[3] NIEMANN,H.-J.: Dynamic response of cantilevered structures to turbulent wind. Proceedings of the European Conference on Structural Dynamics, EURODYN'90, Bochum 1990, pp.1123. A.A.Balkema, Rotterdam, 1991
[4] VICKERY,B.J.; BASU,R.: Across-wind vibrations of structures of circular cross-section. Journal of Wind Engineering. 1983
[5] IWATANI,Y.: Simulation of multidimensional wind fluctuations having any arbitrary power spectra and cross spectra. Journal of Wind Engineering, Japan Association for Wind Engineering, Vol.11, pp.5, 1982
[6] SCHRADER,P.: Die statistische Stabilität gemessener integralerLängenmaße und anderer Windparameter. Ph.D. Thesis, Ruhr-Universität Bochum, 1993 (in print)
[7] Eurocode EC9, Wind loads - Static action, Working draft 1990 (unpublished)
[8] VICKERY,B.J.; KAO,K.H.: Drag or along-wind response of slender structures. Journal of the Structural Division, ASCE, Vol.98, No.St1, Jan., 1972
[9] SCHEWE,G.: Untersuchung der aerodynamischen Kräfte, die auf stumpfe Profile bei großen Reynolds-Zahlen wirken. DFVLR-Mitteilung 84-19, Wissenschaftliches Berichtswesen der DFVLR Köln, Göttingen, 1984
[10] COOK,N.J.: The designer's guide to wind loading of building structures, Part 2 Static structures. Butterworths, London, 1990
[11] HERLACH,U.: Experimentelle Bestimmung von instationären Strömungslasten an drehschwingenden Profilen allgemeiner, symmetrischer Form. Dissertation, Eidgenössische Technische Hochschule Zürich, Zürich, 1974.
[12] SCHERER,R.J.: Stochastic characteristics of vortex shedding at a 230 m high concrete chimney. EURODYN'93, Trondheim, Norway, 1993 pp.1131. A.A.Balkema, Rotterdam, 1993

Structural Safety & Reliability, Schuëller, Shinozuka & Yao (eds) © 1994 Balkema, Rotterdam, ISBN 90 5410 357 4

Measurement of wind forces on net covered temporary scaffolds

E.J. Hollis
Health and Safety Executive, Sheffield, UK

ABSTRACT: This paper describes how measurements of wind forces on an experimental five-storey temporary scaffold, clad with a range of different net types and sheet, were made. Results, expressed in terms of a Form Factor, are presented. A comparison with wind tunnel modelling results is made.

1 BACKGROUND

In recent years the UK has seen an increase in the use of plastic net and sheeting to clad temporary scaffold structures used in building construction work. This improves the protection of both the public and the workforce from falling debris, and also shields workers from the weather. However, this does increase the wind loading on the scaffold, and, as a number of incidents over the past few years have demonstrated, makes such structures susceptible to damage or collapse under storm conditions.

The Health and Safety Executive (HSE) have responsibility in the UK for helping to ensure the health and safety of people at work in a wide range of industries. Concern over such emerging hazards prompted HSE to undertake a three part programme of research, with collaboration from industry. The work reported here concerns the first phase of the project, the object of which is to provide measurements of those wind forces which act to tear clad scaffold away from a building wall when wind is blowing through the window openings. The only previous reported full scale measurements are those of Gylltoft (1986) for two fully sheeted work place scaffolds.

The two further phases planned will be concerned with the investigation of suction effects that occur about corners, and the lift forces experienced by temporary roof systems often used in conjunction with vertical scaffolding.

2 THE EXPERIMENT

The programme of research set out in Table 1 lists the fifteen configurations selected for investigation. This allowed for the comparison of :

(a) the aerodynamic performance of different net types and sheeting.

(b) the effects of different size window openings (50%, 25% and 12% void area).

(c) the effect of setting the net slack (50% excess material) over taut net.

A suitable location within the 220ha HSE site at Buxton, Derbyshire was selected for unobstructed exposure to the prevailing Westerly winds. Substantial concrete foundations were completed during August 1989. The simulated flat wall of a building is based on three lattice tower units anchored to the foundations and aligned at right angles to the prevailing wind direction. Horizontal steel falsework soldiers link the tower sections on the downwind face, equivalent to the five walkway heights desired, with vertical spacings of 2.4m, as Figure 1. Load cells and special fittings were set at seven equally spaced (2.34m) scaffold attachment points on the soldiers, as shown in Figure 2 & 3, at each walkway height. The mesh of vertical and horizontal scaffold tubes, shown in Figure 2, are connected by 1.0m long horizontal tubes ('putlogs') to the thirty-five load cell attachment points. This type of construction is known as a 'putlog scaffold' in the UK. Wooden boards are laid on the putlogs to form the walkways. Ladders passing through the walkway at one end provided access between

Table 1. Phase 1 configurations investigated (a scaffolded flat wall with window openings)

Config.	Description
A1	net type 1 with 1.5m bottom gap and no wall
A2	net type 1 with no wall
A3	net type 1 with a 50% void wall
A4	net type 1 with a 25% void wall
A5	net type 1 with a 12% void wall
A6	net type 1 set slack with no wall
A7	net type 2 with no wall
A8	net type 2 set slack with no wall
A9	net type 3 with no wall
A10	net type 3 set slack with no wall
A11	sheet with a 12% void wall
A12	sheet with a 12% void wall and a horizontally stiffened scaffold grid
A13	sheet with a 25% void wall
A14	sheet with a 25% void wall and covered ladder access openings
A15*	sheet with a 50% void wall (* this has yet to be investigated)

Fig. 1 The experimental scaffold rig, construction completed (April 1990) [9004-016/10]

different levels.

The soldiers also support vertical timbers at 1.22m spacing to which plywood sheets are fastened in a checker-board pattern to give the desired window void openings, as seen in Figure 4. End walls are also fabricated in a similar way, supported by other horizontal soldiers fastened to the outer faces of the end towers, to prevent air escaping or feeding into the sides of the clad scaffold.

The scaffold, unlike a work place structure, was designed to be as flexible as possible whilst remaining safe. The purpose of this was to attempt to isolate, as far as possible, the localised wind forces transmitted by the putlogs to the load cells. Hence, cross-bracing was omitted. Far more attachment points than necessary were used so that variations in force over the vertical face of the scaffold could be measured. At the sides the horizontal tubes made a sliding contact with the end soldiers in the wind direction but were restrained by fittings from lateral movement.

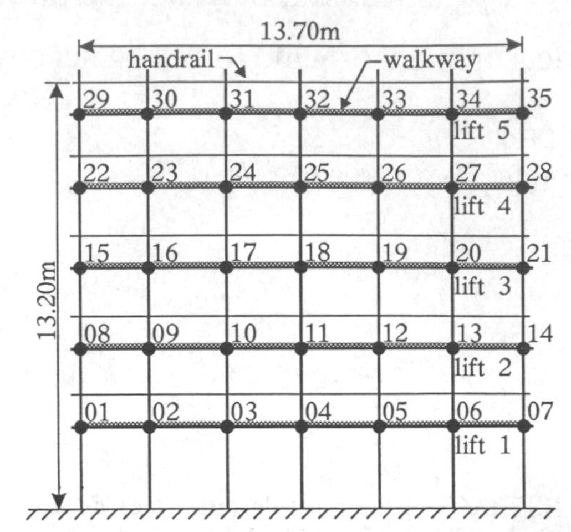

Fig. 2 Schematic diagram of the scaffold, looking upwind, showing the load cell/scaffold attachment point numbering convention [CJA-1]

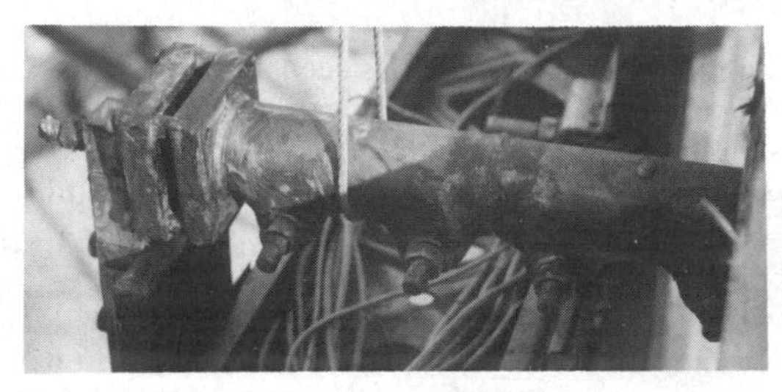

Fig. 3 A load cell attachment point [9003-132/7]

The rig, complete with commissioned instrumentation, was put into operation in June 1990. The two unused towers shown on the right in Figure 1 will be used in phase 2 of the project to simulate the corner of a building. Figure 5 shows the rig operating in storm conditions with a net installed.

The three types of net used in the programme were selected as being representative of the range of commercially available plastic based fabrics.

Fig. 4 The 25% void wall [9012-021/11]

Fig. 5 Storm damage to Configuration A8 (net type 2 set slack with no wall) [9110-097/12]

Net type 1, shown in Figure 6 is of a knitted construction, whilst types 2 & 3 are woven, as Figures 7 & 8. The sheeting is shown in Figure 9.

Fig. 6 Net type 1 [9110-022/4]

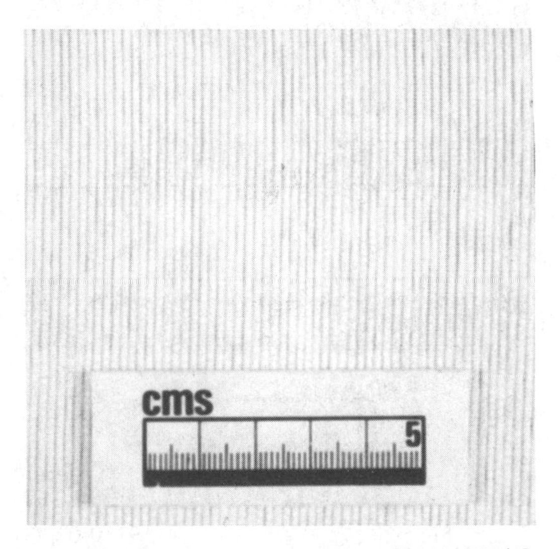

Fig. 7 Net type 2 [9110-022/1]

3 INSTRUMENTATION, DATA SAMPLING AND ANALYSIS

The thirty-five load cells that enable the scaffold wind forces to be measured are a shear beam cantilever type, of either 10kN or 20kN maximum capacity. Unloaded units, one outdoors and the other indoors, provide a means of long term stability assessment and fault monitoring.

Wind speed and direction, measured using a cup anemometer and wind vane, mounted at a height of 10m just upwind and to one side of the rig, are used to trigger automatic sampling. A further eight anemometers are mounted on masts to a height of 20m, some 30m from the rig, to measure the wind speed profile.

Transducer outputs are sampled using a multi-channel digital sampling unit which also provides constant current load cell energisation. This unit incorporates relay controlled switching, which though slower than solid state based units, provides good electrical isolation between chan-

Fig. 8 Net type 3 [9110-022/2]

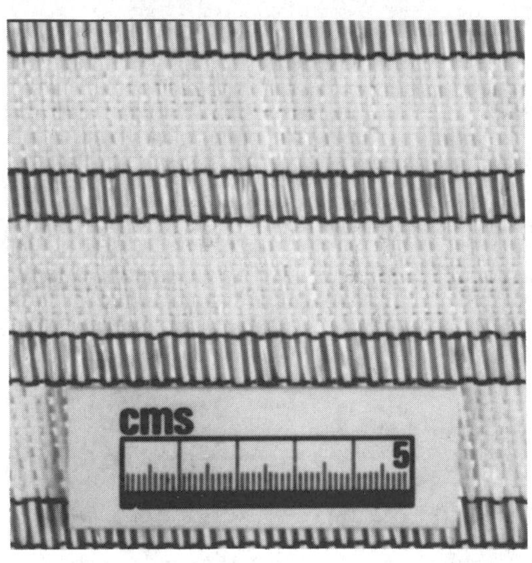

Fig. 9 Sheet [9110-022/3]

nels. Such isolation is an important consideration, as the rig is susceptible to electrical storm damage. This unit is located within the hut shown in Figure 1.

Sampling is controlled by micro-computer (PC) located in a further building, that is linked to the sampling unit by an optical fibre communication line (also for electrical isolation). The PC provides a means of communicating with the sampling unit. Bespoke software allows speed and direction trigger levels to be recalled for setting-up, the management of automatic sampling, and the control of hard disk storage of sampled data. The fastest practical sampling rate for the system described was found to be 0.1Hz. A five minute sampling period has been adopted, which gives a total of 30 records per data file, each record consisting of 48 transducer voltage levels.

As the load cells measure 'attachment' forces in addition to those induced by the wind, the sampling strategy has to include sampling both at high winds, and also during adjacent low wind speed periods to establish a 'datum'. The long term stability of the datum levels, for each load cell, is reviewed during analysis. As an example, for configuration A2 (Table 1), the vertical load cell group 04/11/18/25/32 (Figure 2) indicated an uncertainty in force measurement within +/-0.04kN.

Wind speed profile measurements collected throughout the programme of work have been analysed. These show a distorted wind profile due to the rising slope terrain upwind of the rig. Corrections are made so that the true wind speed at each of the five load cell heights can be found from the 10m height speed measurement.

When it is judged that sufficient data has been collected, for wind speeds in excess of 20m/s and in the design direction (from the West) for a particular configuration, files are categorised according to their average wind direction. The results presented here are only for those data files where the average direction is within +/-7.5deg of the design direction. During the conversion of data to engineering units the true wind forces are found by the subtraction of 'datum' levels (force measurements made at zero or very low wind speeds) from the 'wind' values. Force and wind speed (as measured at the 10m height) values are extracted from each file to compile data specific to each of the thirty-five load cells, so that graphs showing force against (wind speed)^2 can be drawn. The slope of the straight line fitted to each of these plots is found using regression analysis. The slopes are used to calculate Form Factor values. It is therefore important to note that the Form Factor values presented are an average representation of the measured force values.

The same definition of Form Factor as Gylltoft (1986) was adopted:

$$\text{Form Factor} = \frac{\text{wind load}}{\text{speed pressure}} = \frac{(F/A)}{q} \quad (1)$$

where:
F = load cell measured force
A = net/sheet area associated with a load cell

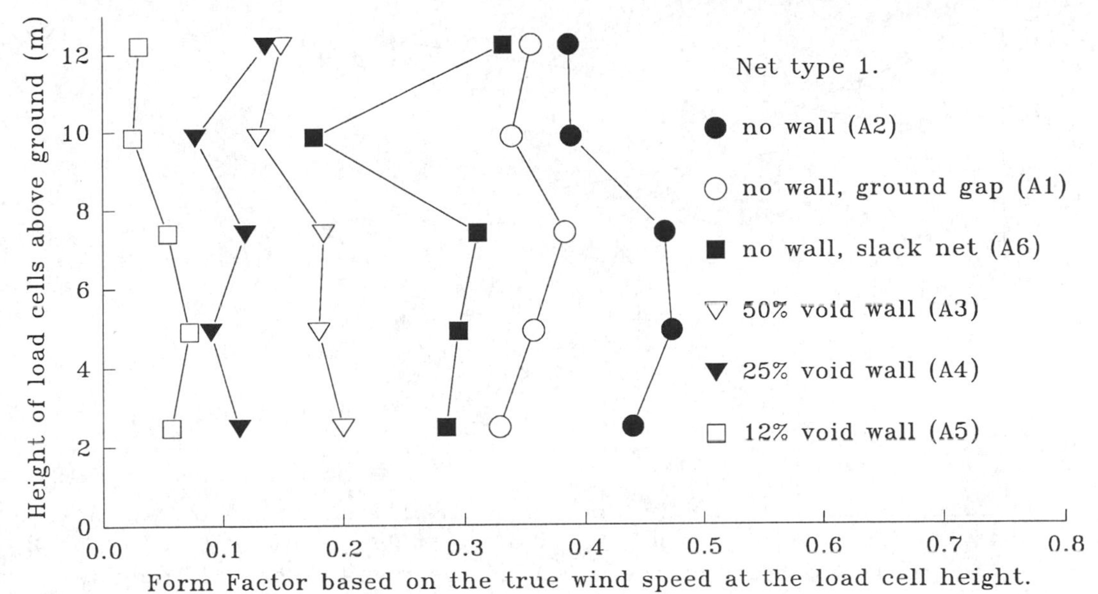

Fig. 10 Comparison of Form Factor profiles for net type 1 [FF12.spg]

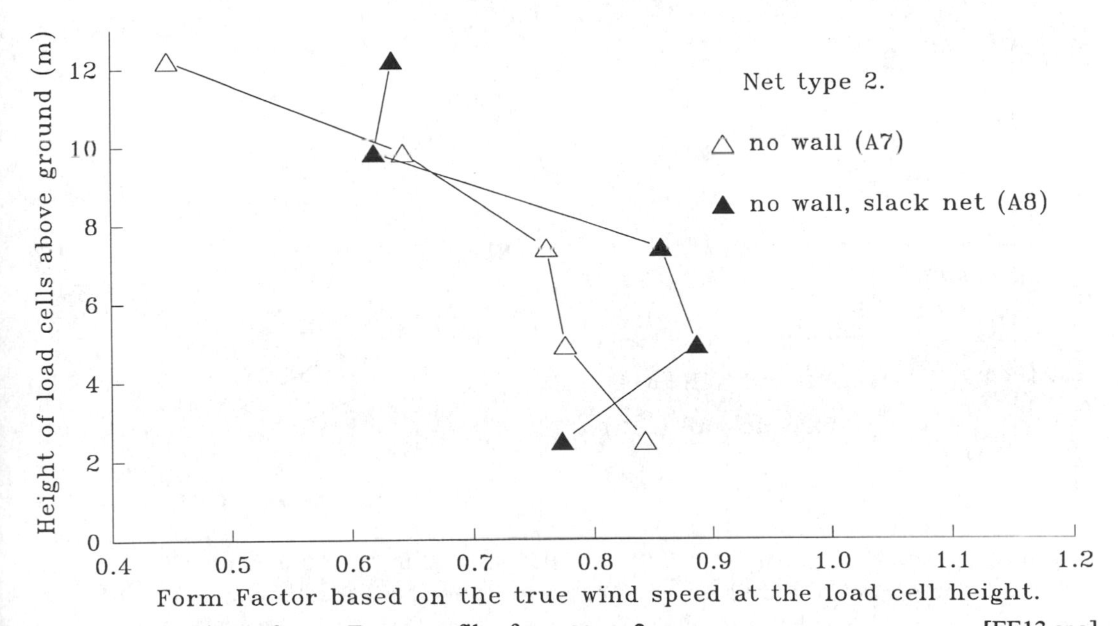

Fig. 11 Comparison of Form Factor profiles for net type 2 [FF13.spg]

U = wind speed
ρ = air density
$q = \frac{1}{2}.\rho.U^2$

For the results presented here the Form Factors are calculated using the wind speed appropriate to the load cell heights.

4 RESULTS

The analysis described above provides a matrix of Form Factor values for the load cell locations shown in Figure 2. To enable a comparison between configurations to be made, an average Form Factor was calculated for the central group

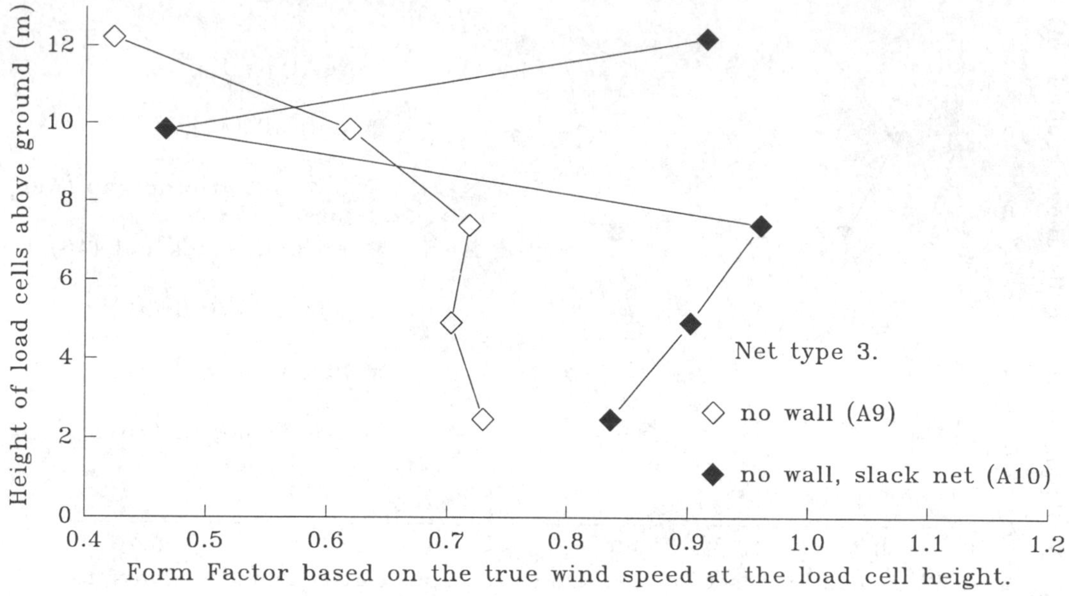

Fig. 12 Comparison of Form Factor profiles for net type 3 [FF14.spg]

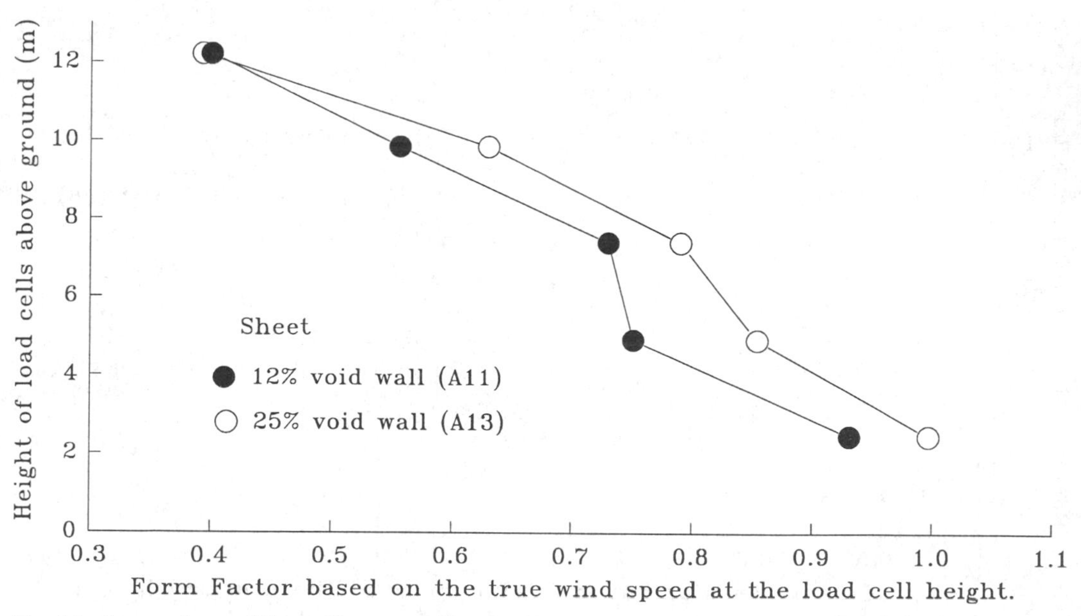

Fig. 13 Comparison of Form Factor profiles for sheet [FF15.spg]

of five load cells, at each height. That is, for the groups (02-06), (09-13), (16-20), (23-27) & (30-34) which correspond to heights above ground level of: 2.47m, 4.91m, 7.40m, 9.85m & 12.21m. The associated horizontal net strip areas are: 28.73m^2, 28.90m^2, 28.90m^2, 28.20m^2, & 25.27m^2, respectively. The variation in these average values with height, and according to cladding type, are shown in the graphs of Figures 10 to 13.

To provide a summary of the Form Factors determined for each load cell location two- and three-dimensional graphics software was used to create contoured surface plots. A selection of the

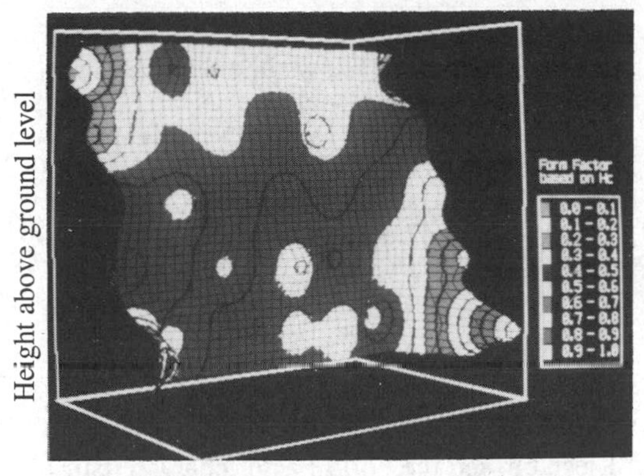

Fig. 14 Form Factor surface plot for Config. A2 (net type 1 with no wall) [9207-118/1]

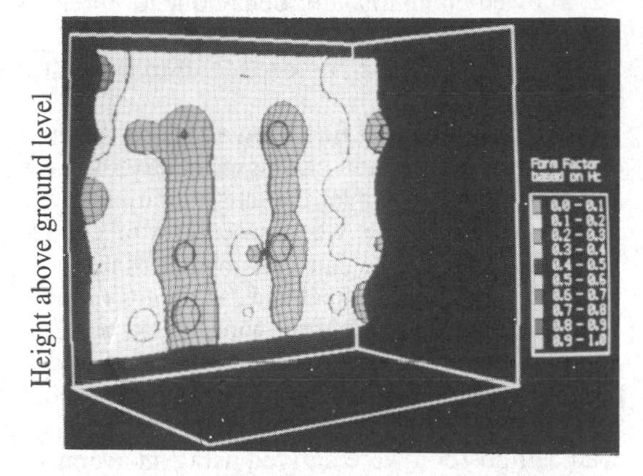

Fig. 15 Form Factor surface plot for Config. A4 (net type 1 with a 25% void wall) [9207-118/3]

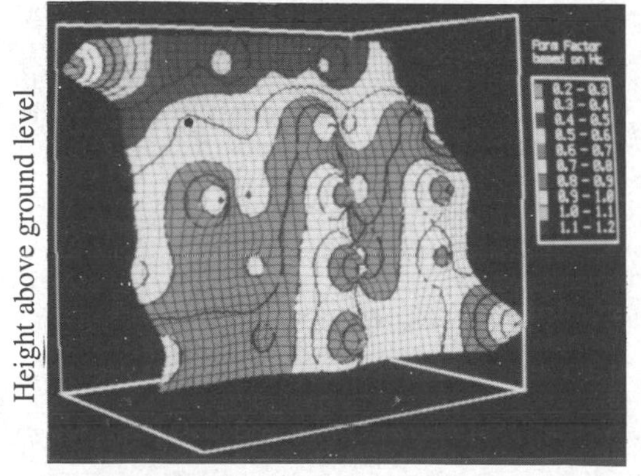

Fig. 16 Form Factor surface plot for Config. A9 (net type 3 with no wall) [9207-118/8]

3D plots is shown in Figures 14 to 17.

Some observations concerning results are:

(1) The graphs in Figure 10, 14 and 15 show, as would be expected, for the same net the Form Factor values are much smaller for a wall with window openings upwind compared to the no wall condition. Also, reduced window areas are accompanied with small Form Factor values.

(2) The use of a 'ground gap' in the netting, as Figure 10 with no wall, provides some reduction in values at lower levels, though of reducing significance with height.

(3) Figures 11 & 12 show that the type 2 & 3 net with no wall, though of different construction, behave in much the same way. The surface plot of Figure 16 is typical of both types.

(4) The 'no wall' profiles of Figures 10 to 12 show that the coarse mesh of net type 1 allows the wind to pass through reasonably unobstructed, whereas types 2 & 3, which are of finer mesh, give rise to significant suction pressure in the wake region, which increases towards the base of the scaffold.

(5) The effect of setting net 50% slack is quite dramatic, as illustrated in Figures 10 to 12. A reduction in loading values at the 9.85m height is observed in all three cases. For the net type 2 & 3 forces became so large during storm conditions that the net material failed, as shown in Figure 5 for configuration A8 (net type 2), or a significant number of fastenings securing the net to the scaffold failed, as occurred with A10 (net type 3). In this latter case additional load was shed onto those areas of scaffold where the net remained attached causing the scaffold clips, which secure the putlogs to the scaffold grid to slip and cause a partial collapse of the scaffold about the top walkway. Previous trials (Rooker(1989)) have shown that forces in excess of 10kN are required to initiate 'clip slip'. Contrary to type 2 & 3, net 1 profile values were found to be less than for the taut condition, attributed to elastic enlargement of the knitted mesh with increased wind load.

(6) The difference between 12% and 25% void wall profiles for sheet, shown in Figure 13, are not large. The difference between them is similar to that of the 'no wall' net type 2 & 3 profiles of Figures 11 & 12.

(7) The subsidiary investigation of Config. A12 was undertaken to see if stiffening of the horizontal tube joints affected the variation in loading across the scaffold face, as observed for earlier configurations, with or without a wall. Comparison of 3D plots for A11 and A12 (sheet with a 12% void wall) have shown this not to be the cause.

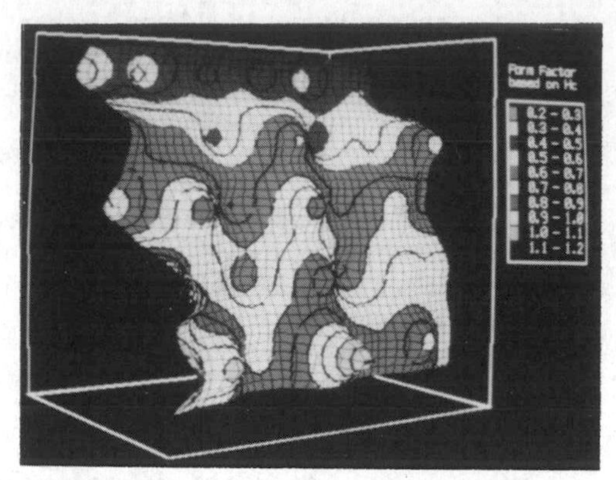

Fig. 17 Form Factor surface plot for Config. A11 (sheet with a 12% void wall) [9207-118/10]

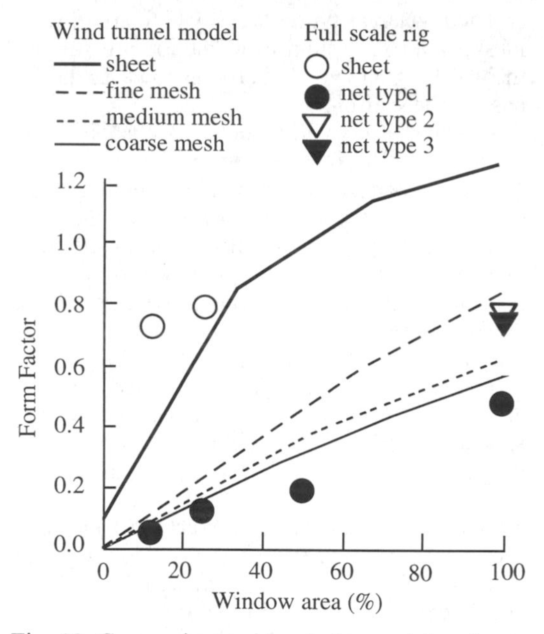

Fig. 18 Comparison with wind tunnel data from Schnabel (1992) [FF18.spg]

(8) Another subsidiary aspect examined with Config. A14 was the effect of covering ladder access openings in the boarded walkways. The 3D plots for A13 and A14 (sheet with a 25% void wall) shows the asymmetry, as seen in Figure 17 for A11, to disappear.

(9) The results of wind tunnel measurements of net clad scaffold forces is reported by Schnabel (1992). Wire meshes were used to simulate a range of net types, mounted as panels about a simple model 0.6m high, and ground plan of 0.6m x 0.2m. The model was unscaled and the boundary layer profile and turbulence were not modelled. This work allows intercomparison of the behaviour of the model meshes, and provides an indication of trends, as shown in Figure 18. Full scale results are also included in Figure 18, derived from the average values from load cell group 16-20, and compare favourably with the model trends observed.

5 CONCLUSIONS

The investigation has:

1. established the Form Factor characteristics over the full window void range for a commonly used knitted net (type 1) and sheet.
2. allowed comparison of behaviour for three net types in the 'no wall' worst case.
3. shown favourable comparison of trend with wind tunnel models.
4. demonstrated that fine weave nets set slack in the horizontal direction can significantly increase the scaffold loading over the taut condition.

Wind loadings for the worst case 'no wall' condition can be such as to cause distortion of the scaffold tubes and the scaffold fastenings (clips) to fail through 'clip slip' following breakage of some of the ties securing the net to the scaffold.

When applying the results of this work it should be remembered that the Form Factors for each load cell position were derived using an averaging procedure. These in turn have been averaged across the face of the scaffold at the five lift heights in Figures 10 to 13.

The information gained from this study will be used by HSE in forming draft clauses for the British Standards Institution Scaffolding Design Code, and subsequently for consideration by the European Committee for Standardization (CEN).

6 REFERENCES

Gylltoft, K 1986. Wind loads on sheeted scaffolds. A field study. Swedish National Testing Institute, Technical Report SP-RAPP 1986:27.

Rooker, K 1989. Unpublished clip slip trials. Health and Safety Executive, RLSD, Buxton.

Schnabel, P 1992. Intermediate report on fluid model experiments to determine wind load on facade scaffolding clad with netting or sheeting. LGA Bayern, München, Report A 18/91.

Structural Safety & Reliability, Schuëller, Shinozuka & Yao (eds) © 1994 Balkema, Rotterdam, ISBN 90 5410 357 4

Time-scale analysis of nonstationary processes utilizing wavelet transforms

Ahsan Kareem, Kurt Gurley & Jeffrey C. Kantor
University of Notre Dame, Ind., USA

ABSTRACT: The Fourier-based approaches in spectral analysis do not preserve time dependence and fail to shed light on the time evolution of spectral characteristics of time localized events. The Fourier transform decomposes a signal into individual frequency components, but does not preserve information as to when the frequencies occurred. The short-term Fourier transform (STFT) provides time and frequency localization to establish a local spectrum of any time instant. The window for the STFT can either be chosen for locating sharp local peaks or for identifying low frequency features, but it is not possible to accommodate both desired features. This drawback can be aleviated if one can have the flexibility to allow the resolution in time and frequency vary in the time-frequency plane to reach a multi-resolution representation of the process. The integral wavelet transform with respect to some basic wavelet provides a flexible time-frequency window which narrows automatically to observe high-frequency contents of a signal and widens to capture low-frequency phenomena. In case of isolating discontinuities in signals, it is essential to have flexibility concerning the resolutions in time and frequency. The wavelets concept can be applied to signal analysis of nonstationary data encountered in engineering mechanics field. The time-scale mapping of a variety of observed processes holds promise to provide evolutionary portraits of the structure of these processes.

1. INTRODUCTION

The analysis of nonstationary processes is frequently encountered in engineering mechanics ranging from the nonstationary signals in atmospheric turbulence to ocean waves in a storm, fluid structure interactions, earthquake induced ground motions and associated structural responses. Similar examples exist in inhomogeneous media, e.g., spatial variation of material properties. Stationary stochastic processes are fully characterized by a unique time independent spectral description. The inability of conventional spectral analysis to describe nonstationary processes requires that specific tools be utilized that go beyond customary Fourier Analysis. This is primarily due to the shortcomings of the Fourier-based approaches which do not preserve time dependence (non-local) and fail to shed light on the time evolution of spectral characteristics of time localized events. The Fourier transform decomposes a signal into individual frequency components, but does not preserve information as to when the frequencies occurred. This is due to the complex emponential nature of the Fourier basis functions that are infinite in extent. If during an evolutionary process the signal is altered at a time instant for a short duration the entire spectrum is affected. An extreme example is the Fourier transform of a Dirac delta function that covers the entire frequency domain. Therefore, for the transient analyses only, the use of the Fourier transform does not provide an adquate tool. This has led to many extensions.

The first of these approaches are based on preserving the classical tools of stationary processes and adapting them to nonstationary cases. This class of approaches may be classified as adaptive methods. An example is reducing the length of an observation to satisfy weak stationarity conditions. This approach has its shortcomings, i.e., *a priori* information regarding the evolution of the process. The other approach concerns evolutive methods aimed at characterizing explicitly the nonstationary attributes of the signal (Flandrin, 1989). Wignor-Ville distribution is an example of this class (Boashash, 1992). Both parametric and non-parametric models are possible in the above categories.

The signal decomposition approaches that can represent a nonstationary signal in terms of a superposition of a number of elementary components that are localized are increasingly becoming very attractive.

2. BACKGROUND

The short-term Fourier transform (STFT) provides time and frequency localization to establish a local spectrum of any time instant (Gabor, 1947). It has

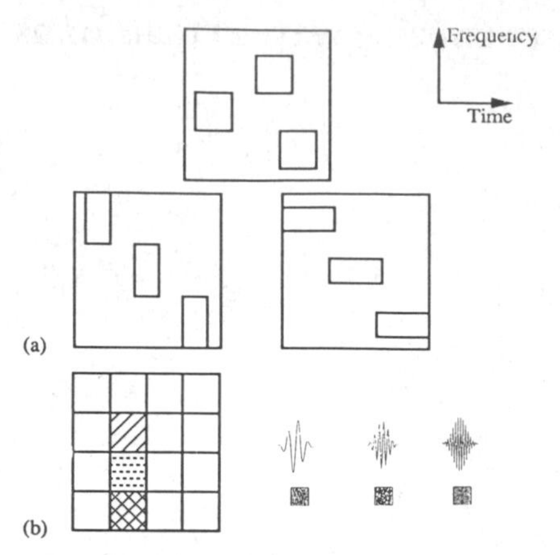

Fig. 1- a: Time-Frequency Representationof STFT.
b: STFT Description of Signal Using a Single Window for all Frequencies.

been extensively used for processing nonstationary signals. The key feature of the STFT is the application of the Fourier transform to a time varying signal, when the signal is viewed through a narrow window, $h(\tau)$, centered at a time or space location,

$$X(t,f) = \int_{-\infty}^{\infty} x(\tau)\, h^*(t-\tau)\, e^{-2j\pi f\tau} d\tau. \qquad (1)$$

The STFT, given by Eq. 1 maps the nonstationary signal into a two-dimensional function in a time-frequency plane. The above process can be viewed in terms of a filter bank. For a given frequency f, Eq. 1 represents filtering of the process at all times with a bandpass filter whose impulse response is the window function modulated to frequency f.

The window for the STFT can either be chosen for locating sharp local peaks or for identifying low frequency features, but it is not possible to accommodate both desired features. This means that the resolution in time and frequency cannot be arbitrarily small since their product has a lower bound. This is referred to as Heisenberg's uncertainty principal or Heisenberg inequality. In Fig. 1, the preceding shortcoming of the STFT is illustrated.

3. WAVELET TRANSFORM

The preceding drawback concerning the resolution limitation of the STFT can be aleviated if one can have the flexibility to allow the resolution in time and frequency vary in the time-frequency plane to reach a multi-resolution representation of the process. This is possible if the analysis is viewed as a filter bank consisting of band-pass filters with constant relative bandwidth (constant-Q filtering). In case of the STFT the analysis filters are regularly spaced over the frequency axis, whereas in the case of wavelet transform these are constant on a logarithmic scale. Therefore, the integral wavelet transform with respect to some basic wavelet, yet to be defined here, provides a flexible time-frequency window which narrows automatically to observe high-frequency contents of a signal and widens to capture low-frequency phenomena. It is very interesting to note that the frequency response of the inner ear to music is naturally distributed into octaves. Similarly, physiology of early mammals' visions indicates that the retinal image is decomposed into spatially-oriented frequency channels with each having a constant bandwidth on a logarithmic scale. In reference to signal processing, it is generally found that naturally occurring transients with larger central frequencies have smaller durations. The wavelet transform approach works best for this type of environment. In Fig. 2, the flexibility of time-frequency resolutions of wavelet transform is demonstrated, i.e., at higher frequencies the wavelet transfer has a higher time resolution, whereas at low frequencies it has a higher frequency resolution.

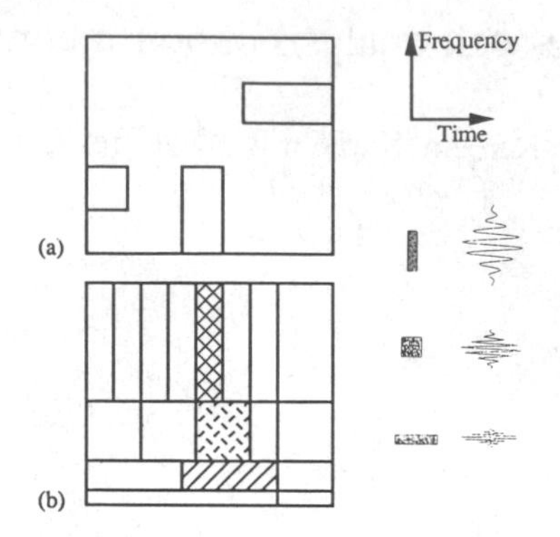

Fig. 2- a: Time-Frequency Resolution of the Wavelet Transform.
b: WT Representation of Functions with Flexible Windows.

A wavelet corresponds to scale a, time or space location b and is given by

$$\psi_{a,b}(t) = \frac{1}{\sqrt{|a|}}\psi\left(\frac{t-b}{a}\right) \qquad (2)$$

where $\psi(t)$ represents a wavelet "prototype" which can be viewed as a band-pass filter. From this function a set of wavelet functions can be constructed by translations and dilations. The wavelets can be viewed as "mathematical microscopes", in which ψ characterizes the optics, while b is the time or position being analyzed and a represents magnification. The parameter a has a dimension of length and is referred to as "scale" in the wavelet literature.

The wavelet function should have the following attributes: admissibility, its average should be zero; similarity, scale decomposition should be obtained by translation & dilation of a prototype or mother function; invertibility, reconstruction from wavelet coefficients and computation of energy and other invariants; regularity, finite spatial support; and cancellations, for some applications we must have vanishing higher order moments (Farge, 1992).

The continuous wavelet transform (CWT) is given by (Goupilland, Grossmann and Morlet, 1984/85),

$$CWT\{x(t);a,b\} = \int x(t)\,\psi^*_{a,b}(t)\,dt,$$

$$= 1/\sqrt{a}\int_{-\infty}^{\infty} x(t)\,\psi^*\left(\frac{t-b}{a}\right)dt$$

$$= \sqrt{2}\int_{-\infty}^{\infty} \hat{x}(w)\,\hat{g}^*(aw)\,e^{ibw}dw. \quad (3)$$

where * denotes the complex conjugate. Typically, CWT is displayed such that the horizontal axis (linear scale) is the shift axis and the vertical axis (logarithmic scale) is the dialation axis using shades of grey or colors. Like the Fourier transform, the wavelet transform is a linear operation as it is an inner product of the signal and the wavelet.

Equation 2 is highly redundant when the parameters (a, b) are continuous (Vetterli & Iterley, 1992). Based on a discrete set of continuous basis functions, the wavelet transform is evaluated on a discrete grid on the time-scale plane. Obviously, one seeks a grid such that the set of basis functions constitutes an orthonormal basis, this implies the absence of redundancy. In the STFT case, oversampling is inevitable and redundant set of points are utilized.

The time-scale parameters (b, a) are sampled on a so-called "dyadic" grid in the time-scale plane. A commonly used definition is (e.g., Chiu, 1991 and Rioul & Vetterli, 1991)

$$C_{j,k} = CWT\{x(t);a=2^j, b=k2^j\} \quad (4)$$

Accordingly, the wavelets are

$$\psi_{j,k}(t) = 2^{-j/2}\psi(2^{-j}t-k) \quad (5)$$

The signal may be reconstructed

$$x(t) = c\sum_j\sum_k C_{j,k}\psi_{j,k}(t). \quad (6)$$

Despite moving to a discrete grid, it turns out that the transform continues to serve as a set of basis functions for square integrable functions.

It is possible to design wavelet functions that are orthonormal, i.e.,

$$\sum_{n=-\infty}^{\infty} \psi^*_{j,k}(t),\psi_{m,n}(t) = \delta_{jm}\delta_{kn} \quad (7)$$

where δ represents the Dirac delta function. The Haar basis is a classical example where

$$\psi(t) = \begin{cases} 1 & 0 \le t < 1/2 \\ -1 & 1/2 \le t < 1 \\ 0 & \text{otherwise} \end{cases} \quad (8)$$

Orthonormality is gauranteed since at a given scale translates are nonoverlapping and as a consequence

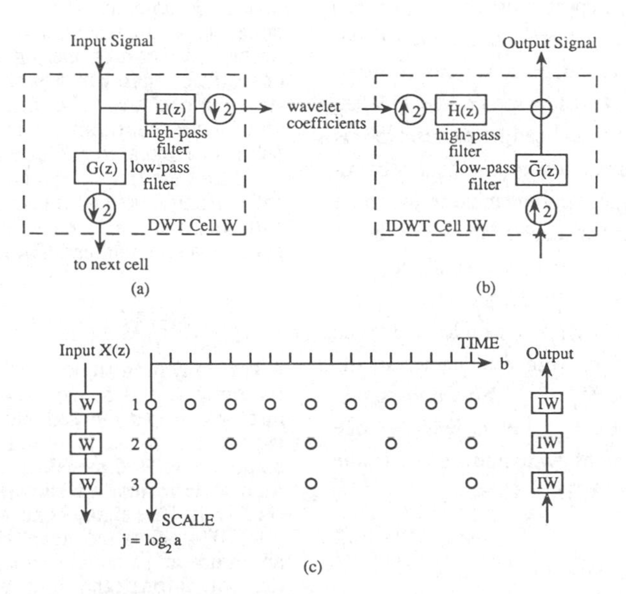

Fig.3 - Schematics of DWT and IDWT.
a&b: Basic Computation Cells of DWT and IDWT.
c: Overall Description of Wavelet Coefficients on a Dyadic Grid in Time-Sacale Plane (after Rioul and Duhamel, 1992).

of scale change by a factor 2 the basis functions are othonormal across the scale (Vetterli & Herley, 1992). Application of the Haar basis is limited in signal processing due to distconinuities. Alternative, continuous sets of basis functions are obtainable from a compactly supported wavelet constructed by Daubechies (1988).

The Discrete Wavelet Transform (DWT) of a sequence $x(n)$ can be realized by dilated or compressed versions of a band pass filter whose relative bandwith are constant. In Fig. 3, an overall schematic diagram of the DWT and IDWT (inverse DWT) is shown utilizing at each cell level a combination of a low pass, $G(n)$, and a high pass filter, $H(n)$, and down-sampling and up-sampling by a factor of 2. The transformed results are in terms of wavelet coefficients that correspond to a dyadic grid in the time-scale plane. The signal may be reconstructed using the IDWT. If $G(n)$ and $H(n)$ represent halfband low pass and halfband high pass filters, then as a result, each iteration in the scheme shown in Fig. 3 halves the width of the low band (increases its frequency resolution by a factor of two), but as a consequence of sub-sampling by two, the resulting time resolution is reduced by a factor of two or halved.

It is important to note that the wavelet coefficients should not be linked to the magnitude of the signal as they represent variations in the signal at a given scale at a given time. For example, if a signal does not exhibit oscillatory patterns at a certain scale and time the value of corresponding wavelet coefficients is equal to zero. For a given locally smooth function the associated wavelet coefficients are small, and for a function containing a singularity the wavelet coefficients increase significantly.

The multiresolution approximation of functions offers an efficacious framework that provides better understanding of wavelet decompositions in terms of a successive approximation procedure (Mallat, 1989). For example, a signal can be represented by an approximation at a resolution plus details corresponding to the orthogonal complement of the space being approximated in the space of the signal being represented. For example, let V_o be the space of all band-limited functions with frequencies in the interval $(-\pi, \pi)$. Similarly, V_{-1} denotes the space of band-limited functions with frequencies in the interval $(-2\pi, 2\pi)$, i.e. $V_o \subset V_{-1}$. Now if the signal $x(t) \in V_o$, then $x(2t) \in V_{-1}$. Let us define a space W_o of bandpass functions with frequencies in the interval $(-2\pi, -\pi) \cup (\pi, 2\pi)$. Then

$$V_{-1} = V_o \oplus W_o \tag{9}$$

where the symbol $\oplus$ means $V_k \perp W_k$ and $V_k + W_k = V_{K+1}$.

Accordingly, W_o is the orthogonal complement of V_{-1}. It can be shown that $\psi(t)$ and its integer translate from an orthonormal basis for W_o. Accordingly, a square integrable signal may be viewed as the successive approximation or weighted sum of wavelets at finer and finer scales (e.g., Vetterli & Herley, 1992).

The wavelet transform is very robust for reconstruction of signals. In case of Fourier based scheme it is difficult to get an exact synthesis of signal due to leakage related problems. Any other error introduced is spread out everywhere in the constructed signal by a Fourier based scheme, whereas in case of wavelets the reconstruction is influenced locally. As the reconstruction process is dependent on the subdomain of the wavelet space also referred to as influence cone the reconstructed signal will be affected locally around the locale of the error or perturbation.

For applications to higher dimensional signals, obviously multi-dimensional wavelets are needed. Both separable wavelets, obtained from products of one-dimensional wavelets and scale functions and two-dimensional non-separable orthonormal scaling functions are being investigated (e.g., Mallat, 1989 and Meneveau, 1991, Kovacevic and Vetterli, 1992).

4. SCALOGRAMS

A very common tool in the processing of signals that provides a distribution of that signal energy in the time-frequency domain is referred to as a spectogram. Similarly, a wavelet spectrogram, or a scalogram is constructed which is the squared modulus of the CWT and represents the signal energy in the time-scale domain. The scalogram provides a two-dimensional representation of the signal and describes variation of the signal energy. The loss of phase information in a scalogram prohibits reconstruction of signal. The phase information provides more accurate information concerning isolated, local bursts in a signal (e.g., Vetterli & Herley, 1992).

Both time-frequency and time-scale descriptions have strong links and transformations from Wigner-Ville distribution to either spectrograms, or scalograms may be obtained (Rioul and Flandrin, 1992).

5. EXAMPLES

A simple demonstration of the application of wavelet transforms is shown below in figs. 4-7. A time history of superimposed sine waves is created (figure 4), the signal is wavelet transformed, and its contour and 3 dimensional representations of the time-scale results are shown in figures 5 and 6 respectively. The signal consists of 2048 points. From 1 to 1024 points the signal is a sine wave with unit amplitude and a radial frequency of .075 rad/s. From half way through the signal to the end a second unit amplitude sine wave is superimposed with a frequency of .01 rad/s. Finally, from 3/4 of the total sig-

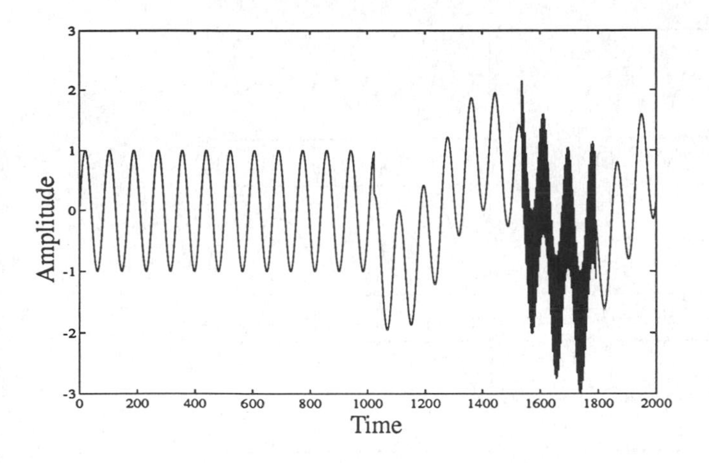

Fig. 4 A Time History Consisting of Several Sine Waves

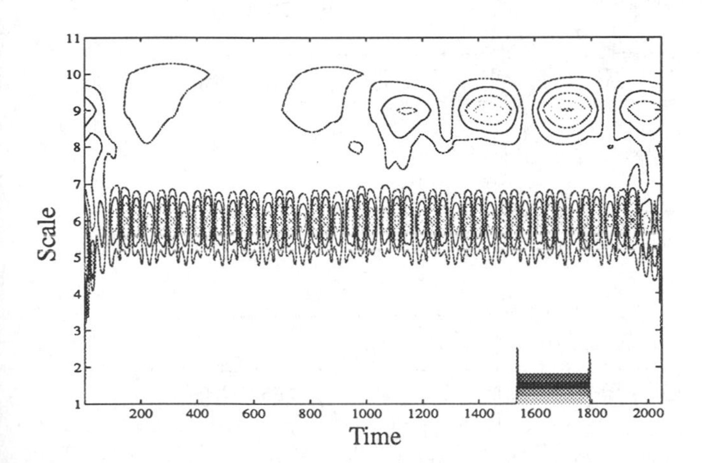

Fig. 5 Time Scale Representation of the Signal in Fig. 4

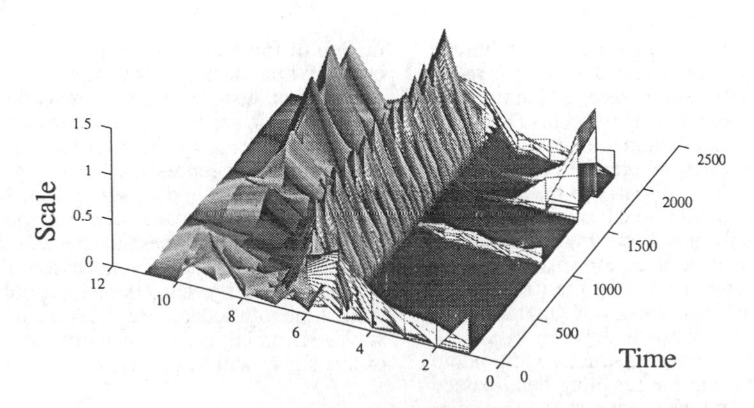

Fig. 6 3-D Representation of Time-Scale Description of the Signal in Fig. 4

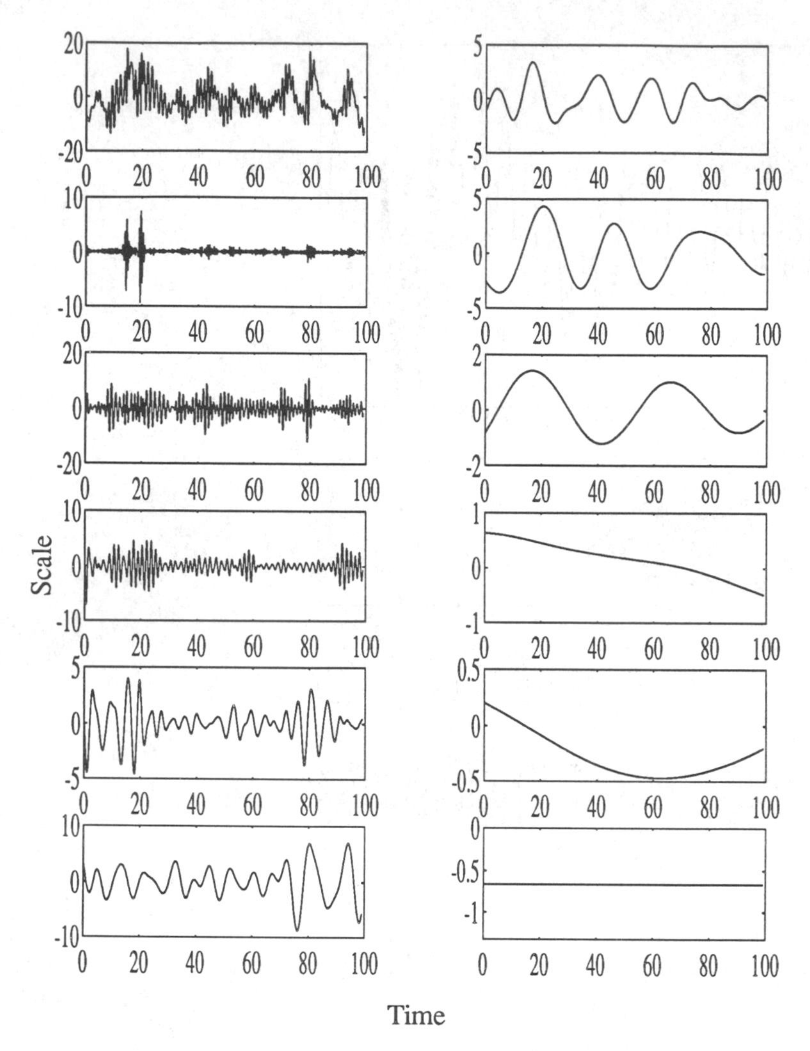

Fig. 7 Multi-Scale Decomposition of Response Time History of a Large Floating Structure

nal length to 7/8 of the signal a third unit amplitude sine wave is added with frequency of $\pi/2$ rad/s. These three frequencies can be seen in figure 5 during the time periods for which they occur. This demonstrates an important feature of the wavelet transform that it identifies the presence of a particular scale for the duration it appears in the signal. As pointed out earlier, conventional spectral analysis fails to provide this feature. Figs. 5 and 6 show the absolute value of the wavelet transform, although by definition the scalogram should be the transformed squared, this is done for purposes of graphical presentation. Figure 7 is a view of the time history of the response of a large floating structure to random environmental loads, and the resulting band-passed time histories using wavelet techniques. The summation of the band-passed histories gives back the original time history. This figure unfolds the response time history into very revealing displays of the time-scale representation. For example, the top block in Fig. 7 shows the response time history. The following block represents the lowest scale or the highest frequency and in subsequent blocks of this figure. the scale increases or the frequency decreases. The last block represents the zero frequency or the mean value. The wavelet based filter bank has helped to identify, e.g., high frequency spikes and their time of occurence. These are attributed to waves slamming the floating structure. More details of this signal will be discussed elsewhere.

6. APPLICATIONS

The wavelets concept discussed in the preceding sections can be applied to signal analysis, data compression, computations and medical images and numerous other immediate applications. The wavelets are making a splash in fields that range from "smart" weapons, detection of stealthy submarines, communcations, human body diagnosis, turbulence and design of more efficient aircrafts (Wavelets, 1992).

The present research concerns the use of wavelets to aid in the analysis of nonstationary data encountered in atmospheric turbulence during thunderstorms, development of shear layer around bluff bodies and the associated evolution of a pressure field beneath these fields and ground motion records. The time-scale mapping of these processes holds promise to provide evolutionary portraits of the structure of these processes.

In case of isolating discontinuities in signals, it is essential to have a flexibility concerning the resolutions in time and frequency. Some basis functions need to be very short, while others must be long to capture the entire range of frequencies. This can be accomplished conveniently by using a short high frequency basis and a long low frequency basis. A wavelet transform offers this flexibility in which the basis functions are derived from a parent or a prototype wavelet by translation and dilation/contraction. When a function is scaled, it either expands or contracts. As the scale gets larger, an increasingly contracted signal is scanned through a constant length filter. This is analogous to geographical maps, as the scale increases one gets a global view, whereas a decrease in scale provides a detailed view of a local region. The local frequency $f = af_o$, where a is scale and f_o is the local frequency associated with the basic wavelet. Therefore CWT are better characterized in terms of scale as opposed to the STFT which is described by frequency. The scale modifies the filter bank impulse responses, i.e., large scales, and small scales correspond to contracted and dilated signals, respectively.

The concept of an "eddy" is known to be a most suitable elementary decomposition of turbulence energy (Tennekes & Lumley, 1972). There is a distinct similarity between the role of eddies in turbulence studies and wavelets in signal decomposition. The wavelets are a very useful diagnostic tool for turbulence as the energy tansfer in turbulence may be local in both space and scale. Some preliminary results using wavelets to analyze turbulence have been reported by Farge (1992). The separation of the flow field dynamics into active and passive components is made possible by the wavelet basis without assuming any hypothetical scale separation which is needed while using the Fourier basis. This assists in distinguishing the low-dimensional dynamically active part of the flow (the coherent structures) from the high-dimensional passive components (the vorticity filaments). The latter can be parametrized, thus significantly reducing the degrees of freedom essential to compute two-dimensional turbulent flows (Farge, 1992). This concept is being examined by the author to study the evolution of a high negative pressure field beneath a separated flow along the sides of a bluff body.

The wavelets will also have an impact on the Large Eddy Simulation techniques in turbulence-structure interaction problems. Presently, the resolvable scales and those modelled are separated in terms of Fourier wave numbers, whereas the very excited regions of wavelet phase space (coherent structures) may be computed explicitly and the background flow field globally parametrized (Farge, 1992 & Liu & Kareem, 1992). The wavelets will also influence the numerical simulation of turbulence through direct solution of a partial differential equation in a wavelet space (e.g., Perrier, 1989 and Meneveau, 1991).

In the analysis of complex motions of coupled buildings under seismic action, the wavelet analysis of signals will provide insight as to how the energy is shifted from one mode to another and at what time instant. The availability of localization and flexiblity in resolutions both in time and frequency will permit identification of the local and global failures and provide a calendar of the sequence of events that leads to such events. This will be a powerful tool for damage diagnostics.

The final paper and presentation will include applications of wavelets to nonstationary signals utilizing software developed in the Matlab environment (Matlab, 1990).

7. CONCLUDING REMARKS

The wavelets offer a new scientific tool to analyze signals associated with evolutionary fields and many tasks in signal processing both in time and space. The wavelets permit a geometrically revealing display of, e.g., one-dimensional signals as a two-dimensional unfolding of space or time and scale as the signal is mapped into a time/space-scale plane. This is accomplished by utilizing dilated or compressed versions of bandpass filters with constant relative bandwidths. The wavelet transform can be compared to a microscope, where the wavelet provides otpics, the enlargement is controlled by scale and position and is fixed by the location parameter. The wavelets are making an impact in a wide range of fields and a high potential exists for them to benefit engineering mechanics field as well. Better understanding of the nonstationary events such as the evolution of turbulent flows, turbulence-structure interactions, ground motions and building responses will become possible through the use of the evolutionary portraits of the structure of sequences of events in time/space-scale plane provided by the wavelets.

The material for this paper has been drawn from a very large number of sources in the fields of engineering, physics and mathematics. It often becomes difficult to assign proper credit to the right paper. Wherever no particular reference is cited, the material is of tutorial nature and has been reported by several authors.

8. ACKNOWLEDGEMENTS

The financial support for this research was provided in part by the NSF Grant BCS-9096274 (BCS-8352223), ONR Grant No. N00014-93-1-0761, and NSF Grant CTS92-08567.

9. REFERENCES

Boashash, B., 1992, Time-Frequency Signal Analysis, J. Wiley

Burrus, C.S. and Gopinath, R.A., 1992, "Wavelet Transforms and Filter Banks," in Wavelets-A Tutorial in Theory and Applications, C.K. Chui (ed.), Academic Press, Inc.

Chui, C.K., 1992, An Introduction to Wavelets, Academic Press.

Daubechies, I., 1988, "Orthonormal Bases of Compactly Supported Wavelets," Commun. Pure Appl. Math., Vol. XLI.

Farge, M., 1992, "Wavelet Transforms and Their Applications to Turbulence," Annual Reviews of Fluid Mechanics, Vol. 24.

Flandrin, P., 1989, "Some Aspects of Non-Stationary Signal Processing with Emphasis on Time-Frequency and Time-Scale Methods," Wavelets, edited by J.M. Combes, A. Grossmann and Ph. Tenamitchian, Springer-Verlag.

Gabor, D., 1946, "Theory of Communcations," J. Inst. Elec. Eng., Vol. 93.

Goupelland, P., Grossman, A. and Morlet, J., 1984/85, "Cycle-Octave and Related Transforms in Seismic Signal Analsis," Geoexploration, Vol. 23.

Kovacevic, J. and Vetterli, M., 1992, "Non-separable Multi-dimensional Perfect Reconstruction Filter Banks and Wavelet Bases for R", IEEE Transactions on Information Theory, Vol. 38, No. 2.

Liandrat, J. and Moret-Bailly, F., 1990, "The Wavelet Transform: Some Applications to Fluid Dynamics and Turbulence," European Journal of Mechanics, B/Fluids, Vol. 9, No. 1.

Liu, Zhendong, and Kareem, A., 1992, "Simulation of Boundary Layer Flow Over a Ridge Many Surforce by LES," Journal of Wind Engineering, JAWE, No. 52, August.

Mallat, S., 1989, "A Theory of Multiresolution Signal Decomposition: The Wavelet Representtion," IEEE Trans. on Pattern Analysis and Machine Intell., Vol. 11, No. 7.

MATLAT, 1990, Pro-Matlab User's Guide, S. Natick, MA.

Meneveau, C., 1991, "Analysis of Turbulence in Orthonormal Wavelet Representation," J. of Fluid Mechanics, Vol. 232.

Perrier, V., 1989, "Toward a Method for Solving Partial Differential Equations Using Wavelet Bases," Wavelets, Edited by J.M. Combes, A Crossmann and Ph. Tenamitchian, Springer Verlag.

Rioul, O. and Vetterli, M., 1992, "Wavelets and Signal Processing," IEEE Sp Magazine.

Rioul, O. and Flandrin, P., 1992, "Time-Scale Energy Distributions: A General Class Extending Wavelet Transforms," IEEE Transactions on Signal Processing, Vol. 10, No. 7.

Rioul, O. and Duhamd, P., 1992, "Fast Algorithms for Discrete and Continuous Wavelet Transforms," IEEE Transactions on Information Theory, Vol. 38, No. 2.

Tennekes, H. and Lumley, J.L., 1972, A First Course in Turbulence, MIT Press, Cambridge, Massachusetts.

Vetterli, M., and Herley, C., 1992, "Wavelets and Filter Banks: Theory and Design," IEEE Transactions on Signal Processing, Vol. 40, No. 9.

Wavelets, 1992, "Wavelets' are causing Ripples Everywhere," Business Week, February, 1992.

Structural Safety & Reliability, Schuëller, Shinozuka & Yao (eds) © 1994 Balkema, Rotterdam, ISBN 90 5410 357 4

Codification of stochastic loads on structures in the Eurocode 1: Wind Action

M. Kasperski
Building Aerodynamics Laboratory, Ruhr-Universität Bochum, Germany

ABSTRACT: The Eurocode 1 - Wind Action, one of the most recent publications of a wind load standard, presents two new features on the example of the pressure coefficients for low-rise buildings. First, a new concept of translating wind tunnel results to extreme pressure distributions has been applied to define the load specifications. This concept is based on the LRC-method which is briefly described. Secondly, the analysis to define the load specifications has been extended to the design effects, i.e. the total effect of load combinations are taken into account. For low-rise buildings, at least the combination of dead load and wind load has to be considered. Beside the commonly known suction on flat roofs, the load specifications of the Eurocode 1 introduces a possible positive wind pressure on flat roofs. For many design situations, this load distribution will become decisive for the design. A comparison to other national codes demonstrate the need of introducing this new load concept.

1 INTRODUCTION

In 1993, the most recent wind code on the world scene will be the Eurocode 1 - Wind Action. It will present the results of a new concept for the description of quasi-static wind loads, introduced with the pressure coefficients for low-rise buildings.

In most codes, the wind loads for non-vibrating structures are described by using a quasi-steady or pseudo-steady assumption. The extreme wind loads are obtained as the product of mean pressure coefficients or - more modern - extreme pressure coefficients, the mean velocity pressure and a gust factor corresponding to an averaging time to be appropriate for the size of the structure. This estimation of the wind effects is conservative if the global or integral wind effects are under consideration, i.e. if one wind load coefficient, e.g. the drag coefficient, is able to describe the total wind action on the structure. If, however, for the design a load distribution is needed, often, parts of the loads will show a favourable i.e. a response-decreasing effect. For low-rise buildings, the latter effect will occur for the bending moment in the downwind corner of the frame (s.a. fig 1.). If the wind is coming from the left, the positive pressure on the upwind wall and the three negative pressures on the two roof areas and the downwind wall respectively will induce effects of the same sign. If, however, the wind is coming from the right, a lower suction on the roof will increase the bending moment in the downwind corner. Then, the quasi-steady approach as described above may lead to an underestimation of the design-decisive reactions.

Further typical examples for structures where the quasi-steady concept may fail are as follows:

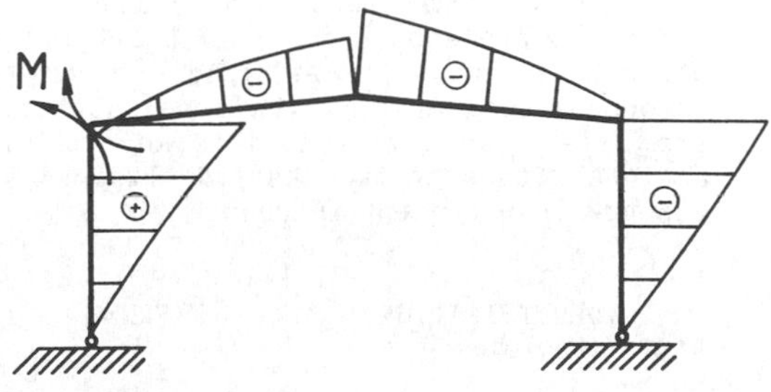

Fig. 1: Influence line for the bending moment in the frame corner (a positive bending moment is defined to induce tension at the bottom)

- low-rise buildings or more general frame structures,
- domed roofs,
- arch-type structures,
- free-standing roofs,
- guyed masts,
- shell structures.

This problem occurs due to neglecting the coincidence of the loads. Often, it is tried to solve this problem by roughly estimating the favourable load effects and introducing a reduction factors which takes into account partial loading effects. If properly done, this will lead to at least a safe but often uneconomic design of the structure.

A more general method - the LRC-method - has recently been published [1] and has been successfully applied for the first time for codification purposes. The design load distributions are identified as extreme load distributions, i.e. instantaneous load distributions causing an extreme effect. They describe the total load as the sum of the mean load and a weighted fluctuating load. The weighting factor is obtained from the correlation of the loads and the influence factors of each load to the response under consideration. In the code, they appear as static loads (identified pressure coefficients times gust response factor times mean velocity pressure), reproducing the extreme response both safely and economically.

In this paper, the LRC-method is shortly presented in chapter 2. The intermediate steps from the identified 'rough' load distributions to the load specifications are described in chapter 3. First, the design-decisive responses have to be defined taking into account at least a load combination of dead load and wind load and/or other live loads such as snow.

The improvement of the reliability of the structure as well as the economy is demonstrated comparing the results obtained from the load specifications of different national Codes to the results obtained with the new load concept in chapter 4.

2 IDENTIFICATION OF EXTREME LOAD DISTRIBUTIONS

The extreme load distribution is defined to be the load pattern which causes a specified extreme response. This response may be a normal force or bending moment as well as a stress. In the time domain, the distribution may be identified using a conditional sampling technique [2]. For a quasi-static linear system, the identification may be based on the LRC (load-response-correlation) method [1]. To obtain the extreme load pattern, the following load data are needed:

- the mean values $\bar{p}_i$
- the standard deviations σ_{p_i}
- the correlations $\varrho_{p_i p_j}$

The extreme load pattern p_e is described to be the sum of the mean load plus the weighted standard deviation:

$$p_{e_i} = \bar{p}_i + g \cdot \varrho_{p_i r} \cdot \sigma_{p_i} \tag{1}$$

with $\varrho_{p_i r}$ - weighting factor

The weighting factor in equation (1) is the correlation of the load p_i and the response r which is obtained from:

$$\varrho_{p_i r} = \frac{\sum_{k=1}^{n} a_k \cdot \sigma_{p_i p_k} \cdot \varrho_{p_i p_k}}{\sigma_{p_i} \cdot \sum_{k=1}^{n} \sum_{l=1}^{n} a_k \cdot a_l \cdot \sigma_{p_k} \sigma_{p_l} \cdot \varrho_{p_k p_l}} \tag{2}$$

with a_i - influence factors of unit load p_i to response r depending on the static system of the structure

n - number of single-load inputs

This analysis leads to a load pattern which is the most frequent load pattern causing the extreme effect. Furthermore, these identified patterns are close approximations to instantaneous load distributions [1].

Generally, the input data describing the loads are obtained from wind tunnel experiments. The number of single-point measurements i.e. the discretization corresponds to the complexity of the investigated flow field, e.g. smaller distances between pressure taps in regions with separating flow. The following analysis with LRC will lead to what may be called a 'rough' pressure distribution, i.e. a detailed pressure distribution allowing the identification of aerodynamic effects. To obtain load patterns which may be used in a code, an intermediate step of smoothing (simplifying) becomes necessary which is orientated more on the structural side to obtain load specifications which are - to a certain degree - as simple as possible.

3 DEFINING LOAD SPECIFICATIONS

As a preparing step to define the wind load distributions for the code, the design situations have to be defined. It is worth mentioning that a restriction to the pure wind effects may lead to a wrongly orientated analysis. For the example of a portal frame in figure 2, dead load and possibly other live loads have to be taken into account, too. The design of the portal frame then depends on the maximum absolute value of the bending moments in the upwind and downwind corner which are due to a combination of at least dead load and wind load.

Depending on the level of the wind velocity and on the weight of the roof construction either the combination of negative bending moment due to dead load plus positive bending moment due to wind load (upwind corner) or negative bending moment due to dead load plus negative bending moment due to wind load (downwind corner) becomes decisive. Hence, two extreme load distributions are needed for codification to allow any combination of light or normal or heavy-weight roof system to any wind climate.

It is worth mentioning that the extreme pressure distributions are depending on the reaction under consideration. For the design, the two wind induced bending moments in the frame's corner and the wind induced support reactions have to be reproduced by the load specifications. As can be seen in fig. 3, the identified pressure distributions for the vertical and

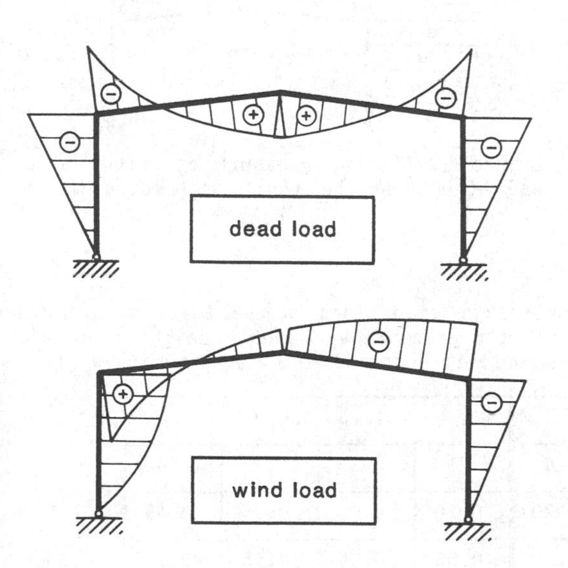

Figure 2: Bending moments due to dead load and wind load

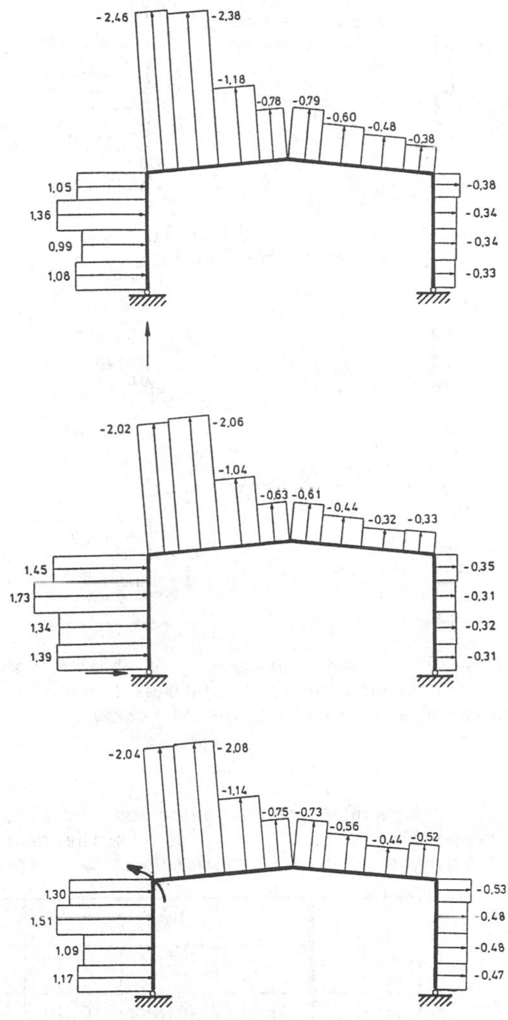

Fig. 3: Identified extreme pressure distributions for the positive bending moment and for the vertical and horizontal support reaction

horizontal support reaction and for the extreme positive bending moment are more or less same.

Strictly speaking, from equation (1) the extreme pressure distributions are depending on the choice of the static system, e.g. a hinged or fixed support. Practically, the influence of the static system tends to be small (s.a. fig. 4). Hence, an enveloping extreme pressure distribution can be found to be equivalent for all static systems.

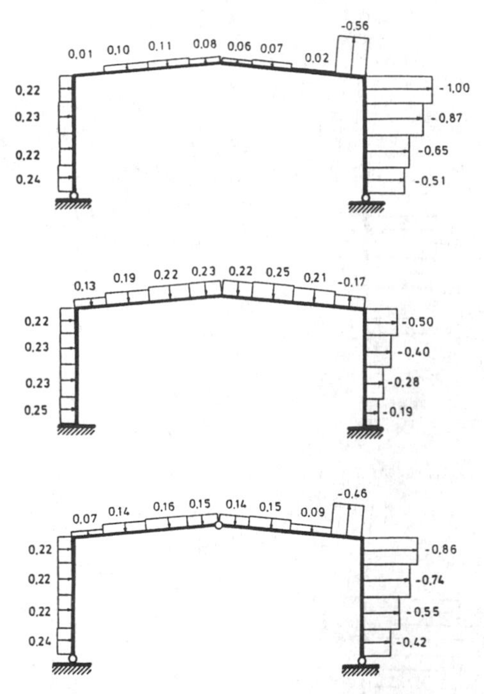

Figure 4: Extreme pressure distributions for different static systems, negative bending moment in the downwind corner

The next step is to check the aerodynamically orientated discretization in relation to the contribution of each single load to the response under consideration. In figure 5, the extreme 'rough' distribution for the positive wind induced bending moment and the single effects (wind load times influence factor for each measuring point) are shown. Obviously, the most spectacular region - the region of high suction induced by the separating flow - will induce negligible contributions to the response.

Hence, from this point of view, a simplification of the 16 measured local loads to four panel loads is reasonable. In table 1, the final load specifications of the Eurocode 1 are summarized. As

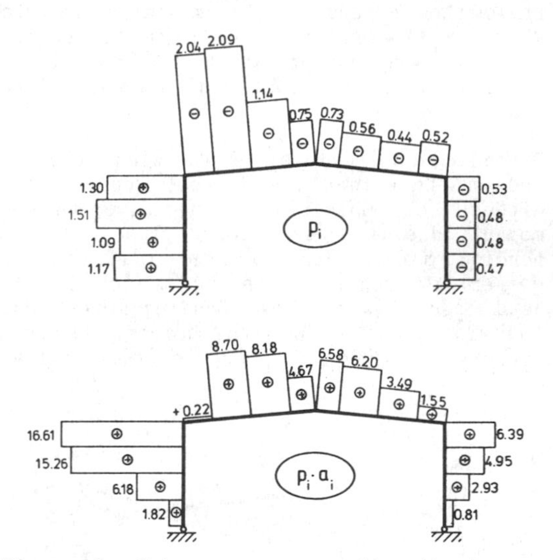

Figure 5: Extreme pressure distribution p_i and effect to the positive bending moment ($p_i \cdot a_i$)

Table 1: Pressure coefficients for low-rise buildings as defined in the Eurocode 1 - the corresponding velocity pressure is the mean velocity pressure at eaves's height times the gust response factor (1: upwind wall; 2: upwind roof; 3: downwind roof; 4: downwind wall)

h/d	load case 1				load case 2			
	1	2	3	4	1	2	3	4
0.4	0.75	-0.95	-0.30	-0.20	0.10	0.05	0.05	-0.45
0.3	0.60	-0.85	-0.40	-0.30	0.20	0.15	0.15	-0.10
0.2	0.40	-0.80	-0.45	-0.30	0.20	0.20	0.20	-0.05

additional parameter influencing the wind load for the design of the frame the aspect ratio height to span (h/d) has been introduced.

4 COMPARISON TO OTHER NATIONAL CODES

The recent draft of the Eurocode 1 has been the first wind load code presenting a positive pressure on flat roofs. It is worth mentioning that the Australian Standard is presenting a similar load case in its amendments of 1993 [8]. To demonstrate the need of this step, several national codes have been investigated:

- Germany: DIN 1055 part 4, August 1986, supplement 1988 [3]
- Austria: ÖNORM B4014, draft Jan. 1992 [4]
- Switzerland: SIA 160, March 1988 [5]
- Canada: National Building Code 1990 [6]
- USA: ASCE-Standard - ANSI/ASCE 7/88 [7]
- Australia: AS 1170.2 amendments 1993 [8]
- ISO 4354 draft 1990 [9]

First, the pure wind effects are compared to the results obtained from wind tunnel experiments. Three different static systems - hinged (a) and fixed (b) support and a 3-pin frame (c) (third hinge at the ridge) - and three ratios for height h to span d (h/d = 0.4, 0.3, 0.2) have been investigated assuming that the rafters are not haunched and columns and rafters are of uniform section size.

In table 2, the results from the wind tunnel experiment are compared to the different codes for the positive bending moment. The best agreement between experiment and code is obtained for the Canadian NBC and the ISO-code, respectively (which are actually same for the load specifications for low-rise buildings). These load specifications are obtained from an extensive analysis of wind induced effects, i.e. this good agreement was somehow expected. The other codes are overestimating this reaction by up to 60% possibly leading to an uneconomic overdesign.

Table 3 gives the results for comparing the negative bending moment. The NBC / ISO-code specifications and the Swiss code are not reproducing a negative bending moment for any of the investigated systems/geometries indicating a shortcoming of safety due to a missing load case. As has been mentioned above, the Australian Standard has recently enclosed an alternative pressure distribution with a partially positive pressure on the roof. The agreement between prediction from these specifications and the wind tunnel experiment is almost sufficient. Only for one case, there is an underestimation of about 50%. The other codes (German, Austrian and U.S.American) are underestimating or neglecting the negative bending moment in most cases.

The influence on the obtained reliability of the designed structure strongly depends on the actual basic load combination and the combination factors, respectively. Assuming the safety concept of the Eurocode 1 [10], the individual safety coefficients γ are as follows:

1.50 for wind load/snow load

Table 2: Positive bending moments applying the pressure coefficients for different national codes (all values in kNm, span d = 27.5 m, distance of frames 5 m, mean wind pressure at eaves height 0.5 kN/m²); Germany (DIN); Austria (ÖNORM); Switzerland (SIA); Canada (NBC) equal to ISO-code; U.S.A (ASCE); Australia (AS); wind tunnel results (WT)

h/d	system	DIN	ÖNORM	SIA	NBC/ISO	ASCE	AS	WT
0.4	hinged	342.2	304.9	272.5	211.2	384.3	291.3	236.3
	fixed	197.2	177.9	178.5	150.6	248.6	170.1	165.3
	3-pin	454.8	397.3	379.3	301.3	532.8	380.9	324.4
0.3	hinged	263.1	218.1	216.2	190.1	308.7	198.0	182.6
	fixed	177.6	146.3	167.2	143.9	231.3	132.7	156.5
	3-pin	359.5	287.2	306.9	253.6	436.5	261.5	261.3
0.2	hinged	206.3	153.7	176.5	150.3	255.1	119.0	155.6
	fixed	161.1	119.2	158.2	137.9	216.1	92.8	147.5
	3-pin	282.2	203.2	247.5	212.7	355.9	153.2	220.2

Table 3: Negative bending moments applying the pressure coefficients for different national codes (all values in kNm, span d = 27.5 m, distance of frames 5 m, mean wind pressure at eaves height 0.5 kN/m²); Germany (DIN); Austria (ÖNORM); Switzerland (SIA); Canada (NBC) equal to ISO-code; U.S.A (ASCE); Australia (AS); wind tunnel results (WT)

h/d	system	DIN	ÖNORM	SIA	NBC/ISO	ASCE	AS	WT
0.4	hinged	-123.9	-161.3	-3.4	33.1	-165.8	-147.8	-91.2
	fixed	29.5	-8.4	10.7	100.8	22.2	-20.9	-38.0
	3-pin	-65.3	-131.2	103.6	124.5	-116.1	-130.1	-105.2
0.3	hinged	-34.1	-76.0	66.6	87.3	-69.0	-90.8	-67.4
	fixed	53.6	13.8	115.9	112.2	52.9	-24.3	-49.1
	3-pin	15.5	-52.6	149.7	157.9	-27.3	-89.8	-92.4
0.2	hinged	29.8	-15.4	109.1	111.1	1.9	-65.5	-59.3
	fixed	67.5	27.7	121.6	115.4	75.2	-41.0	-52.6
	3-pin	68.4	-52.3	174.4	173.5	34.3	-80.9	-84.4

1.35 for dead load

1.00 for favourable dead load

A combination factor Ψ for simultaneous occurrence of an extreme snow load and wind load is reducing the non-dominating life load (in most cases the wind load). Eurocode 1 recommends a combination factor of 0.6 requiring possible modifications for different geographical regions.

Strictly speaking, the dead load of the structure depends on the design to the live loads. Applying an EC-Standard section type HE-A, the actual dead load is obtained iteratively. The weight of the roof construction is varied from light (g = 0.25 kN/m²) to normal (g = 0.5 kN/m²) to heavy roofs (g = 0.75 kN/m²).

The design bending moments obtained applying the load specifications of one of the codes are compared to the results obtained with the original wind tunnel results by normalizing with the wind tunnel results:

$$\text{normalized } M = \frac{M_{Design}\text{ (code)}}{M_{Design}\text{ (wind tunnel)}} \quad (3)$$

A normalized moment greater 1 means a conservative design. Values much greater 1 indicate an uneconomical design. If the normalized moment is less than 1, the design is unsafe.

In figure 6, the results for the Swiss code are shown for two snow climates (no snow s = 0 kN/m² and moderate snow climate s = 0.75 kN/m²) an two different geometries and static systems respectively (h/d = 0.4 and hinged support, h/d = 0.3 and fixed support). The investigated wind climate extends from moderate (mean wind velocity at eaves's height < 25 m/s) to extreme (hurricanes up to 70 m/s).

For lower wind velocities, the real design-decisive effect is due to the combination negative bending due to dead load plus negative bending due to wind in the downwind corner. Since this response is underestimated by the compared code, the normalized bending moment becomes less than 1. For increasing wind velocity, the design-decisive response will occur in the upwind corner. The velocity where this jump of design point occurs, is underestimated by the codes, i.e. the code assumes the upwind corner to be decisive while in reality the downwind corner still shows the greater absolute bending moment. This point is marked by a sudden increase of the normalized bending moment. For higher wind velocities, the estimated design point and the real design point are in the upwind corner, then the normalized bending moment decreases asymptotically to the ratio of the net wind effects.

The degree of under- or overestimating is not only depending on the actual load combination but also on the choice of the static system and the geometry. In the worst case, underestimations up to 35% may occur. For other codes, the general conclusion is the same: the obtained reliability depends on the choice of the static system, the geometry and the actual combination of roof weight and wind and snow climate. Especially, the codes reproducing no negative bending moment should be revised in the near future.

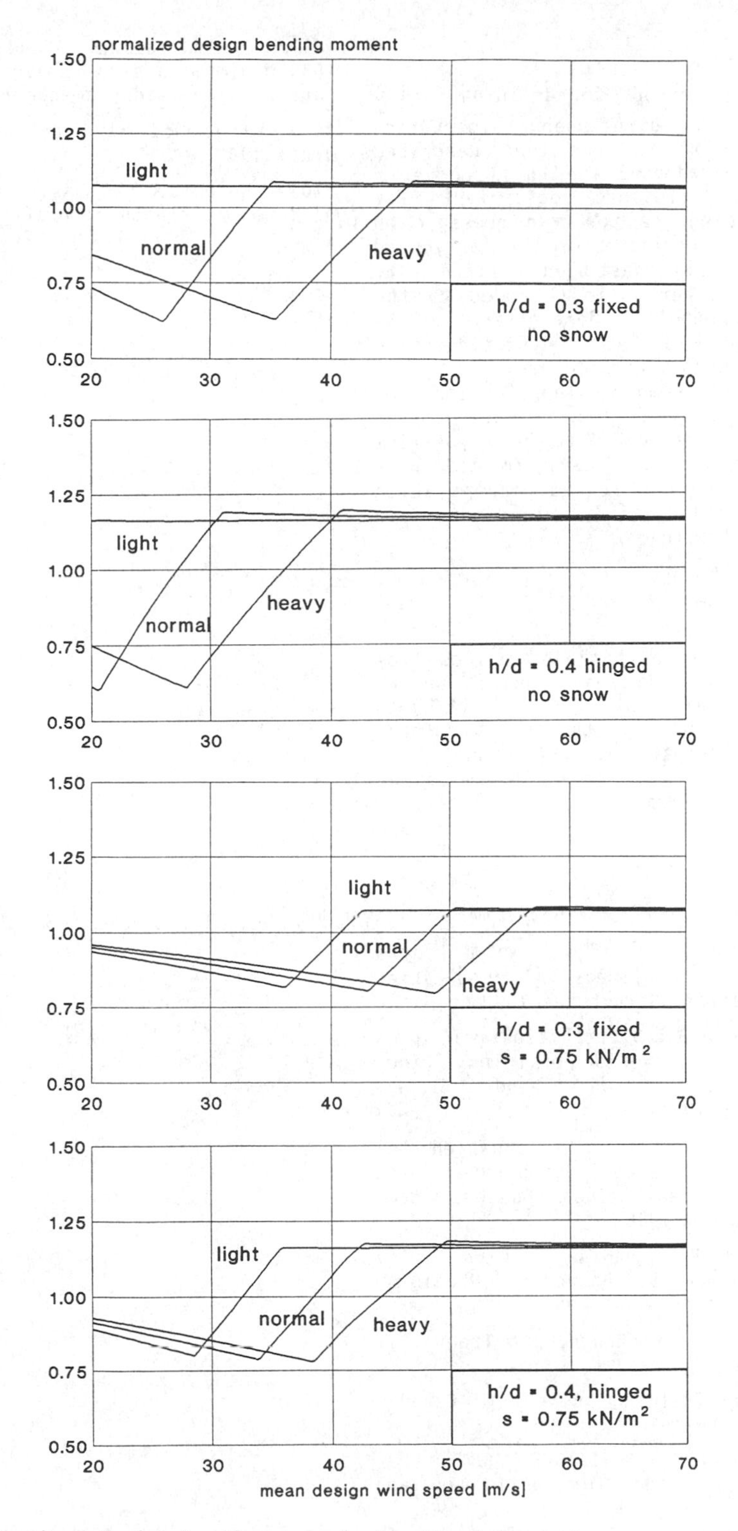

Figure 6: Normalized design bending moments for the Swiss code

5 CONCLUSIONS

A more rational concept for defining design loads due to simultaneous occurring pressure distributions has been described which has been adopted in the Eurocode on the example of pressure coefficients for low-rise buildings. A positive pressure on the roof may be decisive for the design and a second load case has been defined. The comparison of other national codes to the method presented demonstrate the underestimation of the design-decisive response in many practical cases.

In its first complete application for codification of wind load distributions, the LRC-method has proved to be a powerful tool enabling wind engineers to modernize the codes and present a load concept which includes the static and dynamic wind action in a consistent manner.

6 ACKNOWLEDGEMENTS

The author wishes to thank Dr. Holmes from the CSIRO in Australia for several useful discussions. Part of this work was sponsored by the Deutsche Forschungsgemeinschaft in the scope of a special research programme 'Dynamics of Structures' which is gratefully acknowledged.

7 REFERENCES

[1] KASPERSKI, M.: Extreme wind load distributions for linear and non-linear design, Engineering Structures 1992

[2] HOLMES, J.D.: Distribution of peak wind loads on a low-rise building, Proc. 7th Int. Conf. on Wind Engng. Aachen, Germany 1987

[3] DIN 1055 Teil 4 - Lastannahmen bei Bauten; Verkehrslasten; Windlasten bei nicht schwingungsanfälligen Bauten, Aug. 1986 - Ergänzungen 1988

[4] ÖNORM B 4014 Teil 1; Belastungsannahmen in Bauwesen, Statische Windwirkungen, draft Jan. 1992

[5] SIA 160 - Einwirkungen auf Tragwerke, März 1988

[6] National Building Code of Canada, 1990

[7] ASCE Standard - ANSI/ASCE 7/88 Minimum design loads for buildings and other structures

[8] Australian Standard AS 1170.2-1989 Minimum design loads on structures part 2: wind loads; amendments 1993

[9] ISO 4354 Wind Action on Structures draft 1990

[10] Eurocode 1 Basis of Design and Actions on Structure; Draft March 1993

Structural Safety & Reliability, Schuëller, Shinozuka & Yao (eds) © 1994 Balkema, Rotterdam, ISBN 90 5410 357 4

Assessment of recent methods for estimating extreme value distribution tails

J.A. Lechner & N.A. Heckert
National Institute of Standards and Technology, Gaithersburg, Md., USA

E. Simiu
National Institute of Standards and Technology, Gaithersburg, Md., USA & The Johns Hopkins University, Baltimore, Md., USA

ABSTRACT: In the past twenty years a vast new body of extreme value theory has been developed, referred to as 'peaks over threshold modeling.' This theory allows the use in the analysis of all data exceeding a sufficiently high threshold, a feature that may result in improved extreme value estimates. The application of the theory depends upon the performance of methods for estimating the distribution parameters corresponding to any given set of extreme data. We present a comparative assessment of the performance of three such methods. According to our Monte Carlo simulation results the de Haan method method is the best for applications in which the percent estimation errors are comparable to those typical in wind engineering.

1 INTRODUCTION

Classical extreme value theory is based on the analysis of data consisting of the largest value in each of a number of basic comparable sets called epochs (a set consisting, e.g., of a year of record, or of a sample of data of given size; in wind engineering, it has been customary to define epochs by calendar years). For independent, identically-distributed variates with cumulative distribution function F, the distribution of the largest of a set of n values is simply F^n. With proper choice of the constants a_n and b_n, and for reasonable F's, $F^n(a_n+b_nx)$ converges to a limiting distribution, known as the asymptotic distribution. A notable result of the theory is that there exist only three types of asymptotic extreme value distributions, known, in order of decreasing tail length, as the Fréchet (or Fisher-Tippett Type II), Gumbel (Type I), and Weibull (Type III) distributions.

In contrast to classical theory, the theory developed in recent years makes it possible to analyze all data exceeding a specified threshold, regardless of whether they are the largest in the respective sets or not. An asymptotic distribution — the Generalized Pareto Distribution (GPD) — has been developed using the fact that exceedances of a sufficiently high threshold are rare events to which the Poisson distribution applies. The expression for the GPD is

$$G(y) = \text{Prob}[Y \le y] = 1-\{[1+(cy/a)]^{-1/c}\}$$

$$a>0,\ (1+(cy/a))>0 \qquad (1)$$

Equation 1 can be used to represent the conditional cumulative distribution of the excess $Y = X - u$ of the variate X over the threshold u, given $X > u$ for u sufficiently large (Pickands, 1975). $c>0$, $c=0$ and $c<0$ correspond respectively to Fréchet, Gumbel, and Weibull (right tail-limited) domains of attraction. For $c=0$ the expression between braces is understood in a limiting sense as the exponential $\exp(-y/a)$ (Castillo, 1988, p. 215).

The threshold exceedance approach reflected in Eq. 1 can extend the size of the sample being analyzed. Consider, for example, two successive years in which the respective largest wind speeds were 30 m/s and 45 m/s, and assume that in the second year winds with speeds of 31 m/s, 37 m/s, 41 m/s and 44 m/s were also recorded, at dates separated by sufficiently long intervals (i.e., longer than a week, say) to view the data as independent. For the purposes of threshold theory the two years would supply six data points. The classical theory would make use of only two data points. In fact it may be argued that, by choosing a somewhat lower threshold, the number of data points used to estimate the parameters of the GPD could be considerably larger than six in our example. However, whether such an increase in the data sample will improve the quality of the estimates remains to be established. This is one of the issues we address in this paper.

A second important issue is the choice of estimating method. We assess three basic methods which have been proposed for estimating GPD parameters: the Conditional Mean Exceedance method, the Pickands method, and the Dekkers-Einmahl-de Haan method. The second section of the paper briefly describes these methods. The third section describes our Monte Carlo simulations and contains a summary and discussion of the results. The fourth section presents our conclusions.

2 DESCRIPTION OF ESTIMATION METHODS

2.1 Conditional Mean Exceedance Method.

The CME (or mean residual life — MRL — as it it usually termed in biometric or reliability contexts) is the expectation of the amount by which a value exceeds a threshold u, conditional on that threshold being attained. If the exceedance data are fitted by the GPD model and $c < 1$, $u > 0$, and $a+uc > 0$, then the CME plot (i.e., CME vs. u) should follow a line with intercept $a/(1-c)$ and slope $c/(1-c)$ (Davisson and Smith, 1990). The linearity of the CME plot can thus be used as an indicator of the appropriateness of the GPD model, and both c and a can be estimated from the CME plot.

2.2 Pickands Method

Following Pickands' (1975) notation, let $X_{(1)} \geqslant \ldots \geqslant X_{(n)}$ denote the order statistics (ordered sample values) of a sample of size n. For s = 1, 2, .., [n/4] ([] denoting largest integer part of), one computes $F_s(x)$, the empirical estimate of the exceedance CDF

$$F(x;s) = \mathrm{Prob}(X-X_{(4s)}<x \mid X > X_{(4s)}) \qquad (2)$$

and $G_s(x)$, the Generalized Pareto distribution, with a and c estimated by

$$\hat{c} = \frac{\log\{(X_{(s)} - X_{(2s)})/(X_{(2s)} - X_{(4s)})\}}{\log(2)} \qquad (3)$$

$$\hat{a} = \frac{\hat{c}(X_{(2s)} - X_{(4s)})}{2^{\hat{c}} - 1} \qquad (4)$$

One takes for Pickands estimators of c and a those values which minimize (for $1 < s < [n/4]$) the maximum distance between the empirical exceedance CDF and the GPD model. Pickands' method can be shown to be consistent.

Following a critique of an earlier implementation of the Pickands method (Pickands, 1975, Castillo, 1988), an alternative implementation was developed (Lechner, Leigh and Simiu, 1991), which entailed the following steps: (1) choose as threshold u an order statistic of the sample; (2) compute the empirical exceedance CDF for the data above u; (3) nonlinear least-squares fit the GPD model for the parameters c and a; (4) plot the resulting c estimates against u for each order statistic. If the plot of $\hat{c}$ is stable around some horizontal level for most of the order statistic thresholds plotted, then the plot is presumptive evidence for the GPD model being applicable and can be used to yield numerical estimates of c; the distribution is Weibull, Fréchet or Gumbel according as c is negative, positive, or fluctuates around zero.We note that the approach we have just described had been suggested by Bingham (1990).

2.3 Dekkers-Einmahl-de Haan Method

Recent work by Dekkers, Einmahl and de Haan (1989a, 1989b) provides a moment-based estimator which, like Pickands' estimator, is asymptotically unbiased for the true tail parameter and, in addition, is asymptotically normal. We now describe this estimator, using the order-statistic notation introduced above.

Consider an integer-valued function of n, k(n), such that, as $n \to \infty$, $k(n) \to \infty$ and $k(n)/n \to 0$ (e.g., $k(n) = [\sqrt{n}]$). Compute the quantities

$$M_n^{(1)} = \frac{1}{k(n)} \sum_{i=1}^{k(n)} \{\log(X_{(i)} - \log X_{(k(n))}\} \qquad (5)$$

$$M_n^{(2)} = \frac{1}{k(n)} \sum_{i=1}^{k(n)} \{\log(X_{(i)} - \log X_{(k(n))}\}^2 \qquad (6)$$

The estimator of c is

$$\hat{c} = M_n^{(1)} + 1 - \frac{1}{2\{1 - (M_n^{(1)})^2/(M_n^{(2)})\}} \qquad (7)$$

$\hat{a}$ is obtained as the CME value for $X_{(k(n)}$ times 1 − c, where c is given by Equation 7.

3 PERCENTAGE POINT ESTIMATION

For wind engineering purposes the estimates of the wind speeds corresponding to various mean recurrence intervals are of interest. In this section we give expressions that allow the estimation from the GPD of the value of the variate corresponding to any percentage point $1 - 1/(\lambda R)$, where λ is the mean crossing rate of the threshold u per year (i.e., the average number of data points above the threshold u per year), and R is the mean recurrence interval in years. Set

$$\mathrm{Prob}(Y < y) = 1 - 1/(\lambda R) \qquad (8)$$

Thus

$$1 - [1 + cy/a]^{-1/c} = 1 - 1/(\lambda R) \qquad (9)$$

Therefore

$$y = -a[1 - (\lambda R)^c]/c \qquad (10)$$

The value being sought is

$$x_R = y + u \qquad (11)$$

where u is the threshold used in the estimation of c and a.

4 MONTE CARLO SIMULATIONS AND RESULTS

Each of the three estimation methods studied here will produce estimates of c and a for any choice of the threshold; the consistency of these estimates for different thresholds (e.g., for each order statistic) is an important indicator of model appropriateness. However, it is clear that an estimator which performs better on a single determination will perform better in a consistency study. Therefore only the single determination was studied here.

The three estimation methods described earlier were applied to sets of data generated by Monte Carlo simulations from populations with mean $E(X) = 50$ and standard deviation $s(X) = 6.25$, and with (1) Gumbel distribution ($c = 0$) and (2) Weibull distribution with tail length parameter $\gamma = 2$ (i.e., with GPD parameter $c = -0.5$ (Smith, 1989)). The choice $\gamma = 2$ for the Weibull distribution was based on preliminary results obtained for extreme wind speeds recorded in the U.S. (Lechner, Leigh, Simiu, 1991).

The expressions for the Gumbel and Weibull distributions are, respectively,

$$F_G(x) = \exp\{-\exp[-(x-\mu_G)/\sigma_G]\} \qquad (12)$$

$$F_W(x) = \exp\{-[(\mu_W - x)/\sigma_W]^\gamma\}, \quad x < \mu_W \qquad (13)$$

For the Gumbel distribution, the relations between distribution parameters and the expected value E(X) and the standard deviation s(X) are

$$\sigma_G = (6^{1/2}/\pi)s(X) \qquad (14)$$

$$\mu_G = E(X) - 0.57722(6^{1/2}/\pi)s(X) \qquad (15)$$

For the Weibull distribution,

$$E[(X-\mu_W)/\sigma_W] = -\Gamma(1 + 1/\gamma) \qquad (16)$$

$$s[(X-\mu_W)/\sigma_W] = \{\Gamma(1+2/\gamma)-[\Gamma(1+1/\gamma)]^2\}^{1/2} \qquad (17)$$

where Γ is the gamma function (Johnson and Kotz, 1972).

250 samples of size 10000 data were generated from each of the two populations. From these samples, 250 samples of size N = 1000, 250 samples of size N = 250, and 250 samples of size N = 50 were generated by taking from the 10 000 data the largest 1000, 250, and 50, respectively. The same 250 samples of size N were analyzed separately by the three methods. For the sample of size N = 50 we assumed $\lambda = 1.25$/year. Our 10000 data would then correspond to a record length of 40 years. Since λ is proportional to the sample size, we have $\lambda = 6.25$/year and $\lambda = 25$/year for N = 250 and N = 1000, respectively.

The simulation results are summarized in Table 1, where A, C denote the estimated parameters a, c of Eq. 1, E denotes the estimation method (C = CME, P = Pickands, D = Dekkers-Einmahl-De Haan), N denotes sample size, M and SD denote the mean and the standard deviation of the estimators, and R denotes the mean recurrence interval.

A measure of the quality of the performance of the estimators is the mean square error, that is, the sum of the variance and the square of the bias (recall that the bias is the difference between the mean of the estimator and the population mean). It can be seen in Table 2 that the performance of the de Haan and CME estimators was consistently better than that of the Pickands estimator (NIST implementation). For the Gumbel distribution the de Haan estimator was superior to the CME estimator. In this case the r.m.s. estimation errors can be seen from Tables 1 and 2 to be about 5 percent, i.e., comparable to those typical of wind engineering applications.

Table 1. Results of Monte Carlo Simulations for Gumbel and Weibull Distributions

GUMBEL											
		A		C		R = 1		R = 20		R = 200	
N	E	M	SD	M	SD	M	SD	M	SD	M	SD
1000	C	5.26	0.24	-0.06	0.03	73.50	0.61	85.34	2.03	93.17	3.70
	P	5.11	0.29	-0.03	0.06	73.86	0.97	87.53	4.03	97.79	7.92
	D	5.03	0.22	-0.02	0.03	73.94	0.63	88.03	2.21	98.50	4.19
250	C	4.96	0.52	-0.02	0.08	74.00	0.69	88.24	3.11	99.39	8.31
	P	4.93	0.60	-0.01	0.12	74.00	0.75	88.98	4.99	101.95	13.82
	D	4.94	0.44	-0.01	0.06	74.00	0.68	88.30	2.44	98.73	4.93
50	C	5.03	1.06	-0.05	0.17	74.02	0.77	87.94	2.59	98.87	8.94
	P	4.89	1.17	-0.02	0.25	73.99	0.76	88.81	4.65	104.86	21.95
	D	5.02	0.95	-0.05	0.15	74.02	0.75	87.96	2.55	97.29	5.74
Population Values:				0.00		74.10		88.70		99.91	

WEIBULL											
		A		C		R = 1		R = 20		R = 200	
N	E	M	SD	M	SD	M	SD	M	SD	M	SD
1000	C	2.27	0.10	-0.53	0.03	61.09	0.06	61.71	0.08	61.82	0.10
	P	2.27	0.18	-0.52	0.29	61.07	0.10	61.69	0.19	61.80	0.22
	D	2.29	0.12	-0.54	0.05	61.08	0.08	61.69	0.15	61.80	0.18
250	C	1.08	0.09	-0.51	0.06	61.10	0.06	61.76	0.07	61.89	0.10
	P	1.08	0.12	-0.51	0.11	61.09	0.07	61.77	0.17	61.91	0.26
	D	1.09	0.11	-0.52	0.09	61.09	0.06	61.75	0.11	61.88	0.16
50	C	0.49	0.09	-0.52	0.14	61.09	0.07	61.75	0.06	61.88	0.09
	P	0.39	0.24	0.28	1.59	61.07	0.08	61.07	0.08	61.80	0.51
	D	0.49	0.11	-0.54	0.21	61.09	0.07	61.75	0.07	61.88	0.13
Population Values:				-0.50		61.10		61.77		61.90	

Notations: A, C = GPD distribution parameters; N = sample size (sample consists of largest N data out of 10000 data); E = estimation method (C = CME, P = Pickands -- NIST implementation, D = Dekkers, Einmahl, De Haan); M = mean; SD = standard deviation; R = mean recurrence interval, in years.

Having established this, we consider from now on only the de Haan estimates (see Tables 1 and 2). On the average, the estimated values of c are less than the asymptotic values c = 0 for the Gumbel and c = −0.5 for the Weibull distribution. Values closer to the asymptotic values would have been obtained if a higher threshold had been chosen than that implicit in the ratios 50/10000, 250/10000 and 1000/10000 we used. Note that these deviations do not affect significantly the estimates of x_R.

Tables 1 and 2 provide some insight into

Table 2. Estimated Mean Square Errors for Gumbel and Weibull Distributions

		GUMBEL		
		R = 1	R = 20	R = 200
N	E	Mean Square Error		
	C	0.734	15.390	59.140
1000	P	0.987	17.579	67.567
	D	0.420	5.328	19.552
	C	0.481	9.910	69.272
250	P	0.566	24.936	195.269
	D	0.477	6.132	25.663
	C	0.599	7.281	81.001
50	P	0.588	21.654	506.343
	D	0.574	7.073	39.671

		WEIBULL		
		R = 1	R = 20	R = 200
N	E	Mean Square Error		
	C	0.003	0.011	0.015
1000	P	0.011	0.041	0.058
	D	0.006	0.029	0.042
	C	0.004	0.006	0.010
250	P	0.005	0.028	0.068
	D	0.004	0.013	0.027
	C	0.005	0.004	0.009
50	P	0.007	0.493	0.266
	D	0.004	0.006	0.018

Notations: see Table 1.

the effect of lowering the threshold u on the estimates being sought. Consider, for example, the case of the population with a Gumbel distribution for R = 20 years. For N = 1000 and N = 50 the mean square errors are in this case 5.33 and 7.07, respectively, that is, decreasing the threshold resulted in a better estimate of the 20-year value. This is no longer the case for the Weibull distribution. However, this distribution is likely to be of limited interest for wind engineering applications, since it can be easily verified that the percent estimation errors are in this case extremely small, that is, of the order of 0.2 percent.

In general, whether lowering the threshold is advantageous or not depends on the length of the record, the distributional form and parameters of the population, and the mean recurrence interval of the variate of interest. Our results tentatively suggest that in wind engineering applications where records are available for periods of the order of, say, 30 years or more, and extremes with mean recurrence intervals of up to, say, 500 years are sought, the thresholding approach would be most effective for thresholds corresponding to mean crossing rates larger than λ = 1/year. Using actual wind speed data we verified that for a threshold equal to the least maximum yearly extreme during the period of record, the size of the data sample would be about three to six times larger than the size of the sample consisting of the yearly extremes for that period. On the other hand the data for the two samples would be comparable in terms of their position within the distribution tails. Therefore one may expect that in such applications the thresholding approach would result in somewhat better extreme value estimates than the epochal approach, although how significant the improvement would be remains to be determined.

5 CONCLUSIONS

On the basis of our simulation results we tentatively conclude that in cases likely to be of interest in wind engineering applications the de Haan method performs better than the CME method. Both methods are consistently better than the the Pickands method (NIST implementation).

Increasing the number of extreme data by lowering the threshold below a critical value does not improve the quality of the extreme value estimates. The dependence of that critical value on length of record and the mean recurrence interval of the extreme variate being sought needs to be determined by more extensive Monte Carlo experiments with a range of distributions appropriate for the application of interest. Our results suggest that the thresholding approach may be capable of providing better estimates of extremes than the epochal approach, although the extent to which such improvements would be significant also remains to be determined from further Monte Carlo experiments.

6 ACKNOWLEDGMENTS

E. Simiu acknowledges with thanks partial support by the National Science Foundation (Grant MSM-9013116 to Department of Civil Engineering, The Johns Hopkins University), and the helpful interaction on this project with R.B. Corotis of the Johns Hopkins University.

REFERENCES

Bingham, N.H. 1990. Discussion of the Paper by Davisson and Smith. J. Royal Statist. Soc. B 52:431.

Castillo, E. 1988. Extreme Value Theory in Engineering. New York: Academic Press.

Davisson, A.C., and Smith, R.L. 1990. Models of Exceedances Over High Thresholds. J. Royal Statist. Soc. B 52:339-442.

Dekkers, A.L.M. and de Haan, L. 1989a. On the Estimation of the Extreme-value Index and Large Quantile Estimation, Annals of Statistics. 17:1795-1932.

Dekkers, A.L.M., Einmahl, J.H.J., and de Haan, L. 1989b. A Moment Estimator for the Index of an Extreme-value Distribution. Annals of Statistics. 17: 1833-1855.

Lechner, J.A., Leigh, S.D., and Simiu, E. 1992. Recent Approaches to Extreme Value Estimation with Application to Extreme Wind Speeds. Part I: the Pickands Method. J. Wind Eng. Ind. Aerodyn. 41-44:509-519.

Pickands, J. 1975. Statistical Inference Using Order Statistics. Annals of Statistics. 3:119-131.

Smith, R. L. 1989. Extreme Value Theory, in Handbook of Applicable Mathematics. Supplement edited by W. Ledermann, E. Lloyd, S. Vajda, and C. Alexander:437-472, New York: John Wiley and Sons.

Structural Safety & Reliability, Schuëller, Shinozuka & Yao (eds) © 1994 Balkema, Rotterdam, ISBN 90 5410 357 4

Fourier simulation of a non-isotropic wind field model

J. Mann
Risø National Laboratory, Roskilde, Denmark

S. Krenk
University of Aalborg, Denmark

ABSTRACT: Realistic modelling of three dimensional wind fields has become important in calculation of dynamic loads on some spatially extended structures, such as large bridges, towers and wind turbines. For some structures the along wind component of the of the turbulent flow is important while for others the vertical velocity fluctuations give rise to loads. There may even be structures where combinations of velocity fluctuations in different direction are of importance.
Most methods that have been developed to simulate the turbulent wind field are based on one-point (cross-)spectra and two-point cross-spectra. In this paper a method is described which builds on a recently developed model of a *spectral tensor* for atmospheric surface layer turbulence at high wind speeds. Although the tensor does not in principle contain more information than the cross-spectra, it leads to a more natural and direct representation of the three dimensional turbulent flow.
The basis of the model is an application of rapid distortion theory, which implies a linearization of the Navier-Stokes equation, combined with considerations of eddy life times. The physical considerations are quite crude, but the tensor contains essential aspects of the second order structure of atmospheric turbulence. The tensor model has been checked and calibrated with data from different experiments made in connection with estimation of wind loads on a large suspension bridge and on horizontal axis wind turbines.
The wind field can be represented as a generalized Fourier-Stieltjes integral of its spectral components. The necessary factorization (i.e. 'square root') of the spectral tensor can be accomplished in closed form. A numerical simulation algorithm is obtained by recasting the Fourier representation of the wind field in discrete frequency/wavenumber space. The method is considerably faster and simpler than methods based on cross-spectra. A quite similar method has been suggested by Shinozuka and Jan (1972) in general terms, but lack of a realistic spectral tensor of the turbulence in the atmospheric surface layer has prevented its use in wind engineering. The discretization imposes two requirements: If either the width or the height of the domain of the simulated field is not much larger than the length scale of the turbulence care must be taken to represent the energy of the largest scales. Secondly, the space domain must have a large enough margin around the structure of interest in order to avoid effects of the imposed periodicity.

1. DEFINITIONS AND PRELIMINARIES

The atmospheric turbulent velocity field is denoted by $\tilde{\boldsymbol{u}}(\boldsymbol{x})$, where $\boldsymbol{x} = (x_1, x_2, x_3) = (x, y, z)$ is a right-handed coordinate system with the x-axis in the direction of the mean wind field and z as the vertical axis. The fluctuations about the mean wind field, $\boldsymbol{u}(\boldsymbol{x}) = (u_1, u_2, u_3) = (u, v, w) = \tilde{\boldsymbol{u}}(\boldsymbol{x}) - (U(z), 0, 0)$, are assumed to be homogeneous in space, which is often the case in the horizontal directions but is only a crude approximation in the vertical. Since we are interested in shear generated turbulence the mean wind field is allowed to vary as a function of z. Because of homogeneity, the covariance tensor

$$R_{ij}(\boldsymbol{r}) = \langle u_i(\boldsymbol{x}) u_j(\boldsymbol{x} + \boldsymbol{r}) \rangle \tag{1}$$

is only a function of the separation vector $\boldsymbol{r}$ ($\langle\ \rangle$ denotes ensemble averaging).

We shall use *Taylor's frozen turbulence hypothesis* (see e.g. Panofsky and Dutton, 1984) to interpret time series as 'space series' and to change between frequency and wavenumber. Since the mean wind speed is not constant in space, the wind speed U in the Taylor relation $\tilde{\boldsymbol{u}}(x, y, z) = \tilde{\boldsymbol{u}}(-Ut, y, z)$ must chosen as a vertical average of $U(z)$.

We only aim at simulating turbulence which has the right second order statistics, such as variances, cross-spectra etc. The velocity field is otherwise assumed to be Gaussian, which is not a bad approximation for strong winds. All second order statistics can be derived from the covariance tensor, (1), or its Fourier transform, the spectral tensor:

$$\Phi_{ij}(\boldsymbol{k}) = \frac{1}{(2\pi)^3} \int R_{ij}(\boldsymbol{r}) \exp(-\mathrm{i}\boldsymbol{k} \cdot \boldsymbol{r}) \mathrm{d}\boldsymbol{r}, \tag{2}$$

where $\int \mathrm{d}\boldsymbol{r} \equiv \int_{-\infty}^{\infty} \int_{-\infty}^{\infty} \int_{-\infty}^{\infty} \mathrm{d}r_1 \mathrm{d}r_2 \mathrm{d}r_3$. The spectral tensor is the basis of the Fourier simulation and we shall briefly describe the simple isotropic tensor model in this section and a more advanced non-isotropic model in section 2.

The stochastic velocity field can be represented in terms of a generalized stochastic Fourier-Stieltjes integral:

$$\boldsymbol{u}(\boldsymbol{x}) = \int \mathrm{e}^{\mathrm{i}\boldsymbol{k}\cdot\boldsymbol{x}} \mathrm{d}\boldsymbol{Z}(\boldsymbol{k}) \tag{3}$$

where the integration is over all wave-number space (see Batchelor, 1953). The orthogonal process $\boldsymbol{Z}$ is connected to the spectral tensor by

$$\langle \mathrm{d}Z_i^*(\boldsymbol{k})\mathrm{d}Z_j(\boldsymbol{k})\rangle = \Phi_{ij}(\boldsymbol{k})\mathrm{d}k_1\mathrm{d}k_2\mathrm{d}k_3 \quad (4)$$

which is valid for infinitely small $\mathrm{d}k_i$ and where * denotes complex conjugation.

Is it very difficult to measure the spectral tensor directly. Instead cross-spectra defined as

$$\chi_{ij}(k_1, y, z) = \frac{1}{2\pi}\int_{-\infty}^{\infty} R_{ij}(x, y, z)\mathrm{e}^{-\mathrm{i}k_1x}\mathrm{d}x \quad (5)$$

are often measured and are often used in practical applications, such as estimation of loads on structures. The connection between the components of the spectral tensor and the cross-spectra is

$$\chi_{ij}(k_1, y, z) = \int_{-\infty}^{\infty}\int_{-\infty}^{\infty} \Phi_{ij}(\boldsymbol{k})\mathrm{e}^{\mathrm{i}(k_2y+k_3z)}\mathrm{d}k_2\mathrm{d}k_3. \quad (6)$$

$F_i(k_1) = \chi_{ii}(k_1, 0, 0)$ (no summation) is the one-point spectrum.

1.1 *The isotropic tensor model*

The spectral tensor of incompressible isotropic turbulence is

$$\Phi_{ij}(\boldsymbol{k}) = \frac{E(k)}{4\pi k^4}\left(\delta_{ij}k^2 - k_ik_j\right) \quad (7)$$

(Batchelor, 1953) where the so-called energy spectrum, $E(k)$, can be chosen to be

$$E(k) = \alpha\varepsilon^{2/3}L^{5/3}\frac{L^4k^4}{(1+L^2k^2)^{17/6}} \quad (8)$$

as suggested by von Kármán (1948) (L is a length scale, α the von Kármán constant and ε the rate of viscous dissipation of specific turbulent kinetic energy). The variance of the wind velocity fluctuations whose magnitude of the wave vector is in the range $(k, k+\mathrm{d}k)$ is $2E(k)\mathrm{d}k$.

Using (6) with $y = 0$ and $z = 0$ we get the one-point u-spectrum

$$F_1(k_1) = \frac{9}{55}\alpha\varepsilon^{2/3}L^{5/3}\frac{1}{(1+L^2k_1^2)^{5/6}} \quad , \quad (9)$$

the v- and w-spectra

$$F_i(k_1) = \frac{3}{110}\alpha\varepsilon^{2/3}L^{5/3}\frac{3+8L^2k_1^2}{(1+L^2k_1^2)^{11/6}} \quad (10)$$

(for $i = 2, 3$) and all the one-point cross-spectra are zero.

The advantage of the isotropic turbulence model is that it describes the spectra and cross-spectra well for high frequencies or small separations compared to the length scale of the turbulence. The disadvantages are that the variances of the velocity components are equal, which is not supported by data from the atmospheric surface layer. In fact $\sigma_w^2/\sigma_u^2 \approx 0.25$ and $\sigma_v^2/\sigma_u^2 \approx 0.5 - 0.7$ depending on the averaging time. Furthermore, isotropy also implies that the cross-spectrum, χ_{13}, must be zero which is certainly not the case in shear generated turbulence at the lower frequencies.

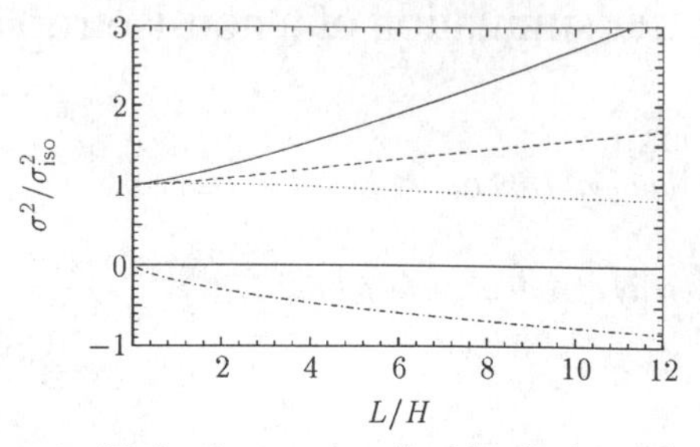

Figure 1: (Co-)variances normalized by the isotropic variance, σ^2_{iso}, as a function the anisotropy parameter L/H. The solid line is $\sigma^2_{11}/\sigma^2_{\text{iso}}$, the dashed $\sigma^2_{22}/\sigma^2_{\text{iso}}$, the dotted $\sigma^2_{33}/\sigma^2_{\text{iso}}$ and the dash-dotted is $\langle u_1u_3\rangle/\sigma^2_{\text{iso}}$. The shear flow experiments described in the text has values of L/H ranging from 3.1 to 8.8. Isotropic turbulence has $L/H = 0$.

2. THE 'SHEARED' SPECTRAL TENSOR

Only an outline of the derivation of the sheared spectral tensor will be given since more detail can be found in Townsend (1976) or more specifically in Mann (1993).

To model the spectral velocity tensor in a shear flow we linearize the Navier-Stokes equation to estimate the effect of the shear on the turbulence. If we assume the shear to be linear such that $\partial U/\partial z$ is constant we obtain a simple linear differential equation for the time evolution or the 'stretching' of the spectral tensor. If the statistics of the initial condition is the isotropic von Kármán tensor (7) with energy spectrum (8), then the tensor $\Phi_{ij}(\boldsymbol{k}, t)$ will become more and more 'anisotropic' with time. The linearization is unrealistic, however, and at some point the stretched 'eddies' will break up. To close the problem an equilibrium is postulated where eddies of size $\propto |\boldsymbol{k}|^{-1}$ are stretched by the shear over a time proportional to their life time, θ. At least for relatively high frequencies (the inertial subrange), $\theta \propto k^{-2/3}$ (Landau and Lifshitz, 1987), and we introduce a parameter, H, with the dimension of a length such that the non-dimensional life time, β, can be written as $\beta \equiv \frac{\partial U}{\partial z}\theta = (kH)^{-2/3}$. For a more realistic modelling of the eddy life time outside the inertial subrange see Mann (1993).

To the present purpose it is most convenient to present the results in terms of the stochastic process $\mathrm{d}\boldsymbol{Z}(\boldsymbol{k})$. For a derivation see Mann (1993). We write $\boldsymbol{k}_0$ for (k_1, k_2, k_{30}) with $k_{30} = k_3 + \beta k_1$. If $\mathrm{d}Z_i^{\text{iso}}$ has the statistics of the isotropic von Kármán tensor, (7), then the sheared tensor may be found from (4) and the following equations

$$\begin{pmatrix} \mathrm{d}Z_1(\boldsymbol{k}) \\ \mathrm{d}Z_2(\boldsymbol{k}) \\ \mathrm{d}Z_3(\boldsymbol{k}) \end{pmatrix} = \begin{bmatrix} 1 & 0 & C_1 - k_2C_2/k_1 \\ 0 & 1 & k_2C_1/k_1 + C_2 \\ 0 & 0 & k_0^2/k^2 \end{bmatrix} \begin{pmatrix} \mathrm{d}Z_1^{\text{iso}}(\boldsymbol{k}_0) \\ \mathrm{d}Z_2^{\text{iso}}(\boldsymbol{k}_0) \\ \mathrm{d}Z_3^{\text{iso}}(\boldsymbol{k}_0) \end{pmatrix} \quad (11)$$

where

$$C_1 = \frac{\beta k_1^2(k_0^2 - 2k_{30}^2 + \beta k_1k_{30})}{k^2(k_1^2 + k_2^2)} \quad (12)$$

and

$$C_2 = \frac{k_2k_0^2}{(k_1^2+k_2^2)^{\frac{3}{2}}}\arctan\left[\frac{\beta k_1(k_1^2+k_2^2)^{\frac{1}{2}}}{k_0^2 - k_{30}k_1\beta}\right]. \quad (13)$$

Compared to the isotropic model (Eqs. (7) and (8)) we have an extra parameter, L/H (L is the length scale from

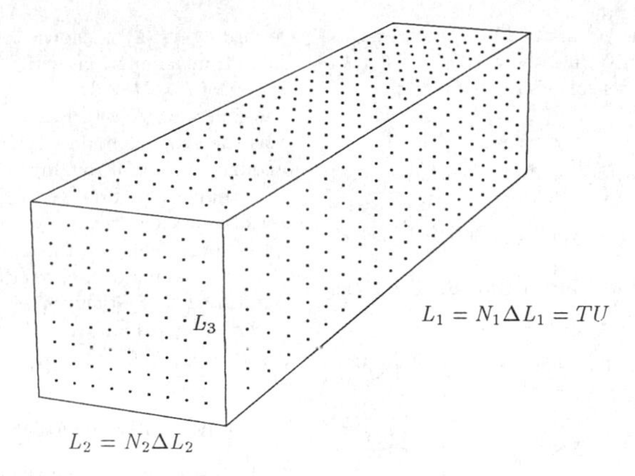

Figure 2: The box B consists of $N_1 \times N_2 \times N_3$ points and has side lengths L_i, $i = 1, 2, 3$, so the separation between the points in the i-direction is $\Delta L_i = L_i/N_i$. U is the mean wind speed and T is the simulation time.

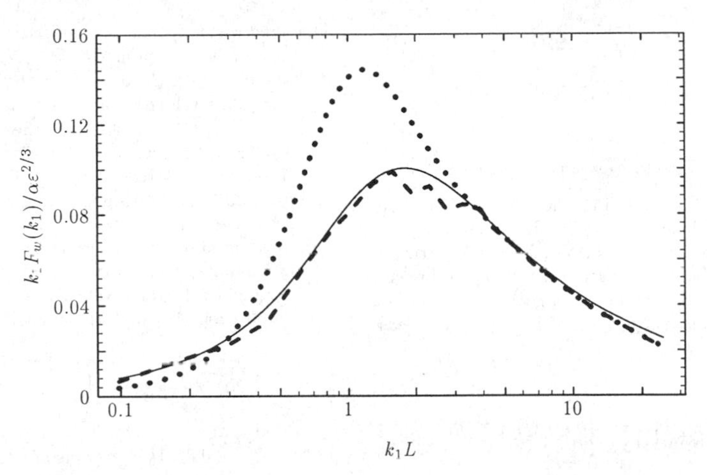

Figure 3: Illustration of the discretization problem by an isotropic w-spectrum. See the text for a discussion.

(8)), which determines the anisotropy of the tensor. Integrating the spectral tensor over the entire wavevector space gives the (co-)variances as a function of L/H (see fig. 1). It is seen that when anisotropy is introduced, $\sigma_u^2 > \sigma_v^2 > \sigma_w^2$ and $\langle uw \rangle < 0$ as confirmed by observations. Examples of one-point spectra and cross-spectra derived from the model may be seen in Mann (1993).

Three experimental tests of the model has been made. Two are atmospheric, one over water and one over flat terrain giving the parameters $L/z = 0.73$, $L/H = 8.8$ and $L/z = 0.87$, $L/H = 5.3$, respectively. The third is based on data from a boundary layer wind tunnel giving $L/z = 0.60$, $L/H = 3.1$, implying that the turbulence is closer to being isotropic compared to the atmospheric turbulence.

3. Fourier simulation

Having discussed the spectral tensor, Φ, both for the isotropic and the non-isotropic case we shall now describe how to simulate a velocity field, $\boldsymbol{u}(\boldsymbol{x})$, from it.

We would like to approximate the integral Eq. (3) by a discrete Fourier series:

$$u_i(\boldsymbol{x}) = \sum_{\boldsymbol{k}} e^{i\boldsymbol{k}\cdot\boldsymbol{x}} C_{ij}(\boldsymbol{k}) n_j(\boldsymbol{k}) \tag{14}$$

where the i'th component of $\boldsymbol{x}$ is $x_i = n\Delta L_i$ with $n = 1, ..., N_i$. The symbol $\sum_{\boldsymbol{k}}$ denotes the sum over all wavevectors $\boldsymbol{k}$ with components $k_i = m2\pi/L_i$, with the integer $m = -N_i/2, ..., N_i/2$, $n_j(\boldsymbol{k})$ are independent gaussian stochastic complex variables with unit variance and $C_{ij}(\boldsymbol{k})$ are coefficients to be determined. See figure 2. The great advantage of (14) is that, once the coefficients are known, it can be evaluated very fast by the fast Fourier transform (FFT). Solving (14) we get

$$C_{ij}(\boldsymbol{k}) n_j(\boldsymbol{k}) = \frac{1}{V(B)} \int_B u_i(\boldsymbol{x}) e^{-i\boldsymbol{k}\cdot\boldsymbol{x}} d\boldsymbol{x}, \tag{15}$$

where $V(B) = L_1 L_2 L_3$ is the volume of B and $\int_B \mathrm{d}\boldsymbol{x}$ means integration over the box B. To find the coefficients $C_{ij}(\boldsymbol{k})$ we calculate the covariance tensor of (15) obtaining

$$\begin{aligned} &C^*_{ik}(\boldsymbol{k})C_{jk}(\boldsymbol{k}) \\ &= \frac{1}{V^2(B)} \int_B \int_B \langle u_i(\boldsymbol{x}) u_j(\boldsymbol{x}')\rangle \, \mathrm{e}^{\mathrm{i}\boldsymbol{k}\cdot\boldsymbol{x}} \mathrm{e}^{-\mathrm{i}\boldsymbol{k}\cdot\boldsymbol{x}'} \mathrm{d}\boldsymbol{x}\mathrm{d}\boldsymbol{x}' \quad (16) \\ &= \frac{1}{V^2(B)} \int\int R_{ij}(\boldsymbol{x}-\boldsymbol{x}') 1_B(\boldsymbol{x}) 1_B(\boldsymbol{x}') \mathrm{e}^{\mathrm{i}\boldsymbol{k}\cdot(\boldsymbol{x}-\boldsymbol{x}')} \mathrm{d}\boldsymbol{x}\mathrm{d}\boldsymbol{x}' \end{aligned}$$

where 1_B denotes the indicator function of B. Using the convolution theorem we get

$$C^*_{ik}(\boldsymbol{k})C_{jk}(\boldsymbol{k}) = \int \Phi_{ij}(\boldsymbol{k}') \prod_{l=1}^{3} \operatorname{sinc}^2\left(\frac{(k_l - k'_l)L_l}{2}\right) \mathrm{d}\boldsymbol{k}' \quad (17)$$

where $\operatorname{sinc} x \equiv (\sin x)/x$. For $L_l \gg L$, assuming $\mathrm{d}k_l = 2\pi/L_l$, we have

$$C^*_{ik}(\boldsymbol{k})C_{jk}(\boldsymbol{k}) = \frac{(2\pi)^3}{V(B)} \Phi_{ij}(\boldsymbol{k}) = \langle \mathrm{d}Z^*_i(\boldsymbol{k}) \mathrm{d}Z_j(\boldsymbol{k})\rangle . \quad (18)$$

The relation $C_{ij}(\boldsymbol{k}) n_j(\boldsymbol{k}) = \mathrm{d}Z_i(\boldsymbol{k})$ then implies

$$C_{ij}(\boldsymbol{k}) = \frac{(2\pi)^{3/2}}{V(B)^{1/2}} A_{ij}(\boldsymbol{k}) \quad (19)$$

with $A^*_{ik} A_{jk} = \Phi_{ij}$, which is expected comparing (3) to (14).

3.1 *Problems with discretization and periodicity*

Two problems occur by simulating a field by the Fourier series (14) with the coefficients (19). The first is that for many applications the dimensions of the simulated box need not to be much larger than the length scale of the turbulence model, L. Thereby (18) may not be a good approximation to (17). However, almost always $L_1 \gg L$, so we can at least reduce (17) to

$$\begin{aligned} &C^*_{ik}(\boldsymbol{k})C_{jk}(\boldsymbol{k}) \quad (20) \\ &= \frac{2\pi}{L_1} \int \Phi_{ij}(k_1, k'_2, k'_3) \prod_{l=2}^{3} \operatorname{sinc}^2\left(\frac{(k_l - k'_l)L_l}{2}\right) \mathrm{d}\boldsymbol{k}'_\perp \end{aligned}$$

This integration, which has to be done numerically is here limited to wavevectors, $\boldsymbol{k}$, obeying $k = |\boldsymbol{k}| < 2.5/L$. Outside this volume we consider (18) to be a good approximation to (17) regardless of the dimensions of the box. This discretization problem is illustrated by Figure 3. The thin line is the target spectrum (10), the dotted line is the average spectrum obtained by using (19) with dimensions $64L \times 2L \times 2L$, that is $L_l \gg L$ is not true for $l = 2, 3$. The dashed line is an average of ten $512 \times 32 \times 32$ points simulations using (20) as described above.

The second problem is that the simulated velocity field (14) is periodic in all three directions. Shinozuka and Jan (1972) suggested to perturb the wavevectors in (14) to avoid this problem. This would however corrupt the efficiency of the FFT. Our solution to the problem is to use a larger spatial window. points. In Figure 4 the coherence of vertical velocity fluctuations for a vertical separation

$$\mathrm{coh}_{ww}(k_1, z) \equiv \frac{|\chi_{33}(k_1, z)|^2}{\chi_{33}(k_1, 0)^2} \quad (21)$$

calculated from the sheared velocity tensor with $L/H = 8$ according to (6) is shown together with coherences calculated from simulations with $2048 \times 32 \times 32$ points and dimensions $256L \times 3L \times 3L$. Since the simulated field is periodic the coherence goes to 1 as $z \to L_3 = 3L$. In a response analysis the space domain (L_2 and L_3) should be chosen large enough to contain roughly *twice* the structure of interest in each dimension. However, if $L_l \gg L$ or if the structure is insensitive to low frequency fluctuations the structure might cover more than half the simulated field in each direction.

An example of a simulated non-isotropic velocity field with $L/H = 8$ is shown in Figure 5. It is seen that the shear elongates and tilts the fluctuations.

3.2 *Implementation and speed*

The implementation of the model includes three steps:

1. Evaluate the coefficients $C_{ij}(\boldsymbol{k})$, either by (19) or if necessary by (20).
2. Simulate the Gaussian variable $n_j(\boldsymbol{k})$ and multiply.
3. Calculate $u_i(\boldsymbol{x})$ from (14) by FFT.

The time consumption in the first step is proportional to the total number of points $N = N_1 N_2 N_3$. It is well known that the required time to perform the FFT is $O(N \log_2 N)$.

If only one velocity component, i, is needed then $A(\boldsymbol{k})$ is simply $\sqrt{\Phi_{ii}(\boldsymbol{k})}$. In this case the first two steps in the $2048 \times 32 \times 32$ simulation used in Figure 4 took $3\frac{1}{4}$ minutes on a Intel 486 25 MHz computer. The FFT used used $2\frac{1}{4}$ minutes and I/O 45 seconds giving a total execution time of $6\frac{1}{4}$ minutes. If the integral (20) is not used only half a minute is saved on the execution time while the spectra are poorly simulated as illustrated by Figure 3.

To simulate all three velocity components $\mathrm{d}Z^{\mathrm{iso}}_i$ is first calculated from (19) with the explicit factorization

$$A(\boldsymbol{k}) = \frac{E^{1/2}(k)}{(4\pi)^{1/2} k^2} \begin{pmatrix} 0 & k_3 & -k_2 \\ -k_3 & 0 & k_1 \\ k_2 & -k_1 & 0 \end{pmatrix} \quad (22)$$

of Φ^{iso}. Then $\mathrm{d}Z^{\mathrm{iso}}_i(\boldsymbol{k}_0)$ is transformed into $\mathrm{d}Z_i(\boldsymbol{k})$ by (11) and (14) is used to get $\boldsymbol{u}(\boldsymbol{x})$.

4. Conclusion

In this paper we simulate a stochastic field from a spectral tensor. This has not been done before for a realistic spectral velocity tensor for atmospheric surface layer turbulence, apparently because no such tensor model has been available. This paper is in principle a special case of the general paper by Shinozuka and Jan (1972) There are some differences, however: Shinozuka and Jan use (19) which in our specific study is shown *not* to be a good approximation to the exact (17). Furthermore, we use evenly spaced wavevectors while Shinozuka and Jan suggest to wobble the wavevectors in order to avoid the imposed periodic boundary conditions.

The two most time consuming steps in our algorithm are the evaluation of the spectral tensor using time proportional to the number of points in the simulations $N = N_1 N_2 N_3$, and the FFT using $O(N \log_2 N)$.

Methods bases on cross-spectra often has the decomposition of N_1 cross-spectral matrices as the most time consuming step. The fastest algorithm to do this known to the

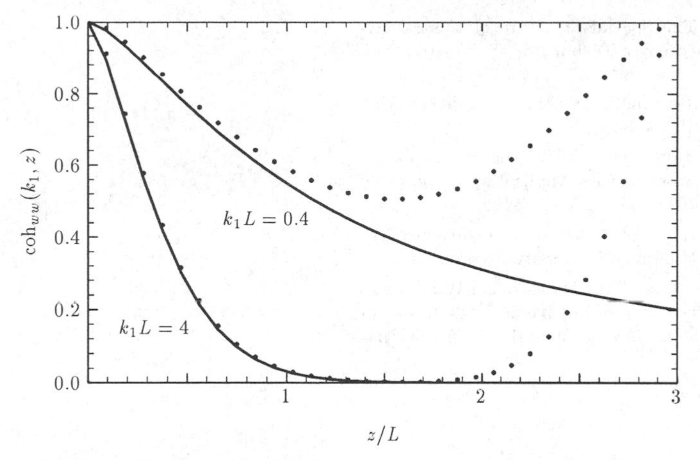

Figure 4: Illustration of periodicity. Simulated (dots) and model w-coherences (curves) as functions of vertical coordinate z. The vertical dimension of the box is $L_3 = 3L$.

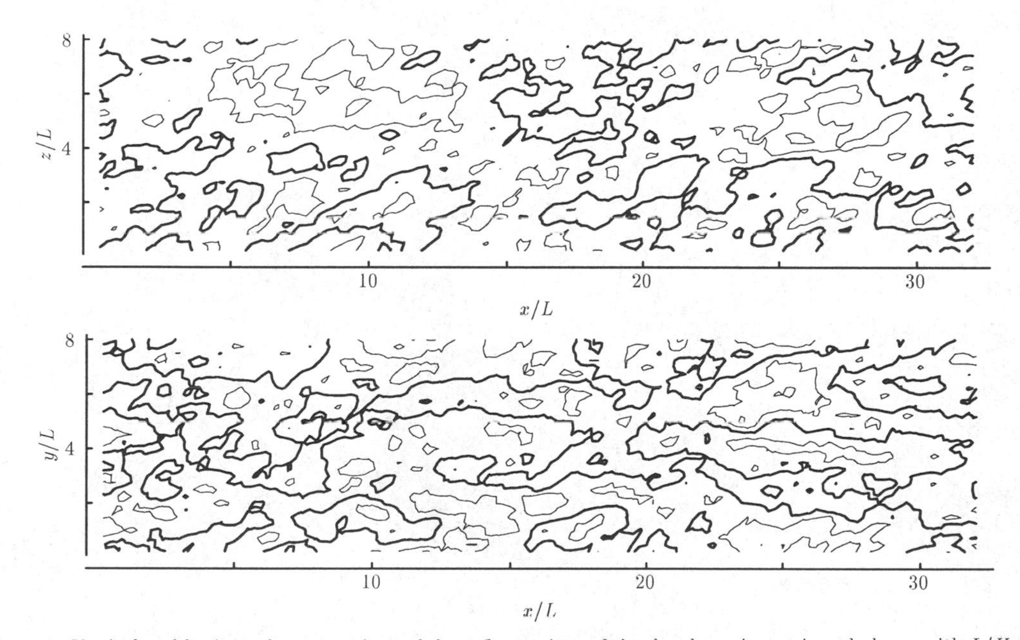

Figure 5: Vertical and horizontal cross sections of the u-fluctuations of simulated non-isotropic turbulence with $L/H = 8$. The broad contours have $u_1 = 0$ and the thin contours have $u_1 = \pm\sigma_u$ or $\pm 2\sigma_u$.

authors is proportional to $N_1 N_2^2 N_3^2$, Winkelaar (1991), i.e considerably slower than our simulation method. Winkelaar's study uses coherences and not complex cross-spectra in which case important phase information is lost. Simulations based on Winkelaar (1991) would thus never show tilted velocity fluctuations as in the upper contour plot of Figure 5.

REFERENCES

Batchelor, G.K.: 1953, *The theory of homogeneous turbulence*, Cambridge.

Kármán, Th. von: 1948, Progress in the Statistical Theory of Turbulence, *Proc. National Akad. Sci.* **34**, 530-539.

Landau, L.D. and Lifshitz, E.M.: 1987, *Fluid Mechanics*, Pergamon Press.

Mann, J.: 1993. 'The Spatial Structure of Neutral Atmospheric Surface-Layer Turbulence', Submitted to *J. Fluid Mech.*

Panofsky, H.A. and Dutton, J.A.: 1984, *Atmospheric Turbulence*, John Wiley & Sons.

Shinozuka, M. and Jan, C.-M.: 1972, Digital Simulation of Random Processes and its Applications, *Journal of Sound and Vibration*, **25** (1), p. 111-128.

Townsend, A.A.: 1976, *The Structure of Turbulent Shear Flow*, 2nd ed., Cambridge University Press.

Winkelaar, D.: 1991, Fast Three Dimensional Wind Simulation and the Prediction of Stochastic Blade Loads, In *Proceedings from the 10'th ASME Wind Energy Symposium, Houston.*

Structural Safety & Reliability, Schuëller, Shinozuka & Yao (eds) © 1994 Balkema, Rotterdam, ISBN 90 5410 357 4

Reliability-based design of transmission line structures under extreme wind loads

Yoshisada Murotsu, Hiroo Okada & Shaowen Shao
University of Osaka Prefecture, Sakai, Japan

ABSTRACT: This paper deals with a methodology for probabilistic design of transmission line structures under stochastic extreme wind loading conditions. A stochastic model for the lifetime maximum wind load is derived. Reliability analysis of a realistic transmission tower is presented, by generating dominant structural failure modes which have high probabilities of occurrence. Design modifications are made to improve the reliability of the tower structure.

1 Introduction

Transmission line structures are very sensitive to meteorological loads such as wind, snow, ice and temperature loads, etc. In recent years, transmission lines in Japan suffered several extensive damages caused by violent typhoons in south area and by severe snow conditions in north area. Transmission tower structural designers have recognized that the current design criteria need modification. A newly designed transmission structure should have a reasonable reliability level over its lifetime. This leads to the requirement for a methodology of probabilistic design[1~4]. This paper deals with such a methodology, taking account of transmission line structures under stochastic extreme wind loading conditions. The basic steps to assessment of reliability for transmission tower structures involve (1) modeling the random properties of the extreme loadings acting on the structure and estimating the structural member strengths, and (2) performing a structural reliability analysis to compute the failure probabilities or reliabilities of structural members and of the overall structural system. In the following sections, the methods are first proposed for dealing with these problems. Then, a case study is presented on a realistic steel transmission tower which was damaged during a typhoon in 1991. The dominant failure modes which have high probabilities of occurrence in this tower structure are found. Attempts are made to improve the reliability of the tower.

2 Structural Reliability Analysis

A transmission tower can be modelled as a frame structure. For this type of structures, Murotsu, et al. have proposed a method to automatically generate probabilistically dominant failure modes [5,6]. Based on the method, a computer program package called STRELAS (STructural RELiability Analysis System) has recently been developed for the reliability assessment of general spatial framework structures[7,8]. In this study, STRELAS is used to carry out reliability analysis of a transmission tower. A brief explanation of the method is given in the following[5 ~ 8]:

- It is assumed that the framework structure consists of uniform and homogeneous elements, and only concentrated loads and moments are applied to it. In such a structure, critical sections where plastic hinges may form are the joints of elements and the places where the concentrated loads act. These points are taken as nodes while the parts between the nodes are taken as elements. Then, each element end is considered as a potential plastic hinge. A set of such element ends constitutes a failure mode.

- The combined load effects of axial force, shear, bending moment and twisting moment on yielding of an element end are approximately expressed by a linear relation. Therefore, the plasticity condition of element end i becomes

$$Z_i = R_i - \boldsymbol{C}_i^T \boldsymbol{X}_t = 0 \qquad (1)$$

where

Z_i: safety margin,
$Z_i > 0$: safety; $Z_i \leq 0$: failure
R_i: reference strength of element end i
$\boldsymbol{X}_t$: nodal force vector (axial force, shear, bending moment and twisting moment)
$\boldsymbol{C}_i$: load-strength correlation vector (linear relation is assumed in the load effects)

- All elements are assumed to have perfectly elasto-plastic behavior. After element ends $r_1, r_2, ..., r_{p-1}$ have failed, the safety margin of surviving element end r_p becomes

$$Z_i^{(p)} = R_i + \sum_{k=1}^{p-1} \left| a_{ir_k}^{(p)} \right| R_{r_k} - \left| \sum_{j=1}^{6l} b_{ij}^{(p)} L_j \right| \qquad (2)$$

where R_{r_k} are the residual strengths of the failed element ends, L_j the external loads, l the number of nodes, and $a_{ir_k}^{(p)}$, $b_{ij}^{(p)}$ the influence coefficients of R_{r_k}, L_j, respectively.

- The plastic collapse of the structure is determined by investigating the property of the reduced total structural stiffness matrix. For example, after element ends $r_1, r_2, ..., r_{p_q}$ have failed, the reduced total stiffness matrix $\left[\boldsymbol{K}^{(p_q)} \right]$ is checked by the following condition:

$$\left| \left[\boldsymbol{K}^{(p_q)} \right] \right| / \left| \left[\boldsymbol{K}^{(0)} \right] \right| \leq \varepsilon \qquad (3)$$

where $\left[\boldsymbol{K}^{(0)} \right]$ denotes the stiffness matrix in elastic condition, and ε is a specified constant. If Eq. (3) is satisfied, element end failure sequence $r_1 \to r_2 \to \cdots \to r_{p_q}$ forms a complete failure path reaching a structural collapse. It constitutes a structural failure mode.

- There are numerous failure modes in a structure with high redundancy. A so called branch-and-bound technique[5 $\sim$ 8] is adopted to generate probabilistically dominant failure paths. The probability of occurrence for a failure path $r_1 \to r_2 \to \cdots \to r_p$ is expressed as follows:

$$P_{fp(q)}^{(p)} = P \left[\bigcap_{i=1}^{p} \left(Z_{r_{i(q)}}^{(i)} \leq 0 \right) \right] \qquad (4)$$

Eq. (4) is approximately evaluated step by step until the structural failure modes with high probabilities of occurrence are found.

3 Model of Wind Load

A model for lifetime maximum wind load is considered in the present paper. According to JEC-127-1979[9], the wind pressure on conductors and a transmission tower is evaluated by

$$L_w = C \times \frac{1}{2} \rho v^2 \times (0.5 + 40/S) \times \left(\frac{H}{10} \right)^{0.25} \times A \qquad (5)$$

where C, ρ, v, S, H, $(H/10)^{0.25}$, and A are drag coefficient, air density, wind speed, span length, height, height coefficient, and area, respectively. In Eq. (5), the wind speed v is the most important random variable, although many other factors also have some uncertainties. In this study, wind load L_w is expressed as a function of n-year lifetime maximum wind speed V_n:

$$L_w = k V_n^2 \qquad (k : constant) \qquad (6)$$

Consequently, the random properties of L_w is determined by those of V_n. Weather observatory station data are available to establish the probability distribution of annual maximum wind speed V_1. The present study uses a Gumbel distribution as shown in the following:

$$F_{V_1}(v_1) = \exp\left(-\exp\left(-A\ (v_1 - B) \right) \right) \qquad (7)$$

$$A = \frac{1}{0.780 \sigma_{V_1}}, \qquad B = \mu_{V_1} - 0.450 \sigma_{V_1}$$

where μ_{V_1}, σ_{V_1} denote the mean and the standard deviation of V_1 which are obtained from statistical data[10].

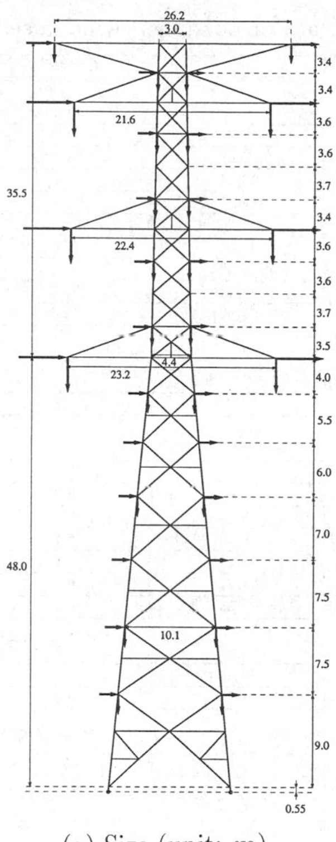

(a) Size (unit: m)

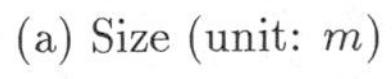

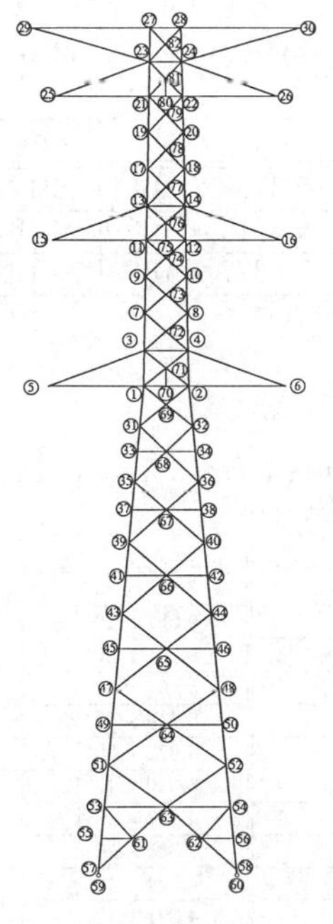

(b) Node No.

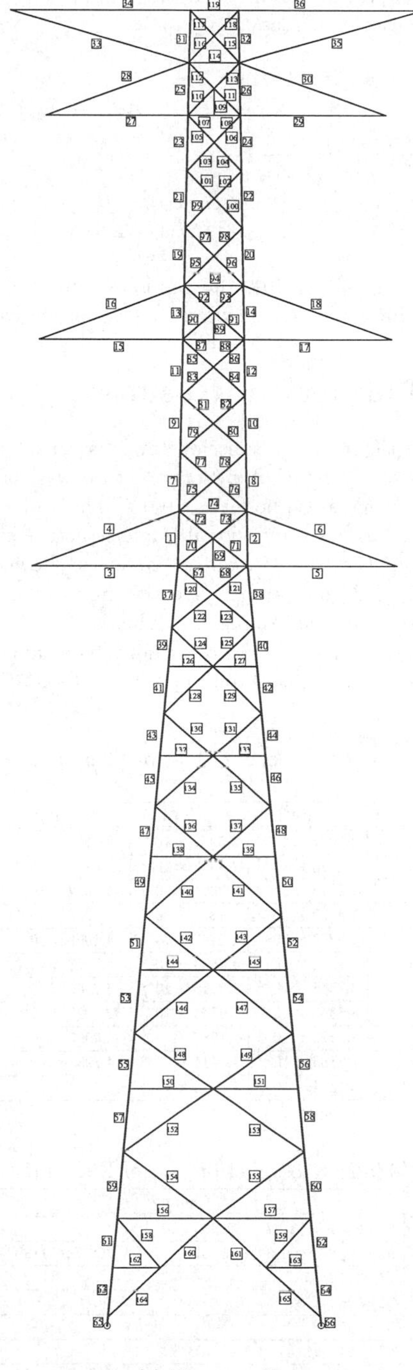

(c) Element No.

Figure 1: Model of Tower Structure

Then, the n-year lifetime maximum wind speed distribution becomes

$$F_{V_n}(v_n) = \{F_{V_1}(v_n)\}^n = \exp(-n\exp(-A(v_n - B))) \quad (8)$$

$$f_{V_n}(v_n) = \frac{d}{dv_n}F_{V_n}(v_n) = An\exp[-A(v_n - B) - n\exp(-A(v_n - B))] \quad (9)$$

From Eqs. (6) ~ (9), the random properties of n-year lifetime extreme wind load L_w are derived.

4 Numerical Results

A realistic steel transmission tower is considered in this study which type of structure was damaged during a typhoon in 1991. The tower is modelled as a 2-dimensional frame structure consisting of 165 elements and 82 nodes as shown in Fig. 1. Tables 1 and 2 give the cross-sectional data of elements. Young' modulus is $E = 2.1 \times 10^8\ kN/m^2$. Yield stress σ_y is taken as a random variable whose mean and coefficient of variation

Table 1: Cross-sectional Data of Structural Elements

P.no	A_p (m^2)	I_z (m^4)	AZ_{yp} (m^3)
1	4.6027×10^{-3}	2.5232×10^{-5}	3.0676×10^{-4}
2	3.4815×10^{-3}	1.4862×10^{-5}	2.0476×10^{-4}
3	2.2718×10^{-3}	7.3394×10^{-6}	1.1624×10^{-4}
4	1.4987×10^{-3}	3.4826×10^{-6}	6.5036×10^{-5}
5	4.9273×10^{-3}	4.2107×10^{-5}	4.1005×10^{-4}
6	5.7265×10^{-3}	4.8573×10^{-5}	4.7477×10^{-4}
7	6.7546×10^{-3}	8.2019×10^{-5}	6.7006×10^{-4}
8	8.6356×10^{-4}	7.9761×10^{-7}	2.3623×10^{-5}
9	9.8922×10^{-4}	1.1985×10^{-6}	3.0995×10^{-5}
10	1.2183×10^{-3}	1.8715×10^{-6}	4.2983×10^{-5}
11	6.4654×10^{-4}	4.3723×10^{-7}	1.5134×10^{-5}

Table 2: Cross-section No. of Elements

P.no	elem.no
1	1 ~ 8, 15 ~ 18, 27 ~ 30, 33 ~ 36
2	9 ~ 12, 67 ~ 69, 74, 87 ~ 89, 94, 107 ~ 109, 114, 119
3	13, 14, 19, 20
4	21 ~ 26, 31, 32
5	37 ~ 46,
6	47 ~ 54
7	55 ~ 66
8	70 ~ 73, 75 ~ 86, 90 ~ 93, 95 ~ 106, 110 ~ 114, 115 ~ 118, 120 ~ 133, 138, 139, 144, 145, 150, 151, 158, 159
9	134 ~ 137, 140 ~ 143, 148, 149, 156, 157, 160, 161, 164, 165
10	146, 147, 152 ~ 155
11	162, 163

Table 3: Data of Loadings (wind speed: 60m/s)

node	load (kN)
W_t	
3, 4	18.9
9, 10	5.9
13, 14	15.3
19, 20	3.3
23, 24	23.7
31, 32	10.6
39, 40	16.4
47, 48	22.5
51, 52	18.0
V_c	
5, 6, 15, 16, 25, 26	15.2
29, 30	1.2
H_t	
3, 4	8.1
9, 10	6.0
13, 14	7.1
19, 20	5.1
23, 24	7.1
31, 32	6.9
35, 36	6.0
39, 40	6.9
43, 44	7.3
47, 48	7.8
51, 52	14.1
H_c	
5, 6	40.2
15, 16	42.9
25, 26	45.2
29, 30	5.7

Table 4: Random Properties of Maximum Wind Speed

years	μ_{v_n} (m/s)	σ_{v_n} (m/s)
1	5.3757×10^1	6.1298×10^0
10	6.4713×10^1	6.1073×10^0
50	7.2377×10^1	6.1020×10^0

Table 5: Failure Probabilities of Element Ends for a 50-year Lifetime

elem.end.no	P_{fe}	CSF
$46S$	6.205×10^{-1}	0.9414
$44E$	6.081×10^{-1}	0.9475
$54S$	5.164×10^{-1}	0.9921
$42S$	5.164×10^{-1}	0.9921
$46E$	5.144×10^{-1}	0.9930
$44S$	5.072×10^{-1}	0.9965
$40E$	5.040×10^{-1}	0.9981
$52E$	4.987×10^{-1}	1.001
$54E$	4.365×10^{-1}	1.031
$50S$	4.304×10^{-1}	1.034
$52S$	4.291×10^{-1}	1.035
$48E$	4.135×10^{-1}	1.043
$42E$	4.021×10^{-1}	1.048

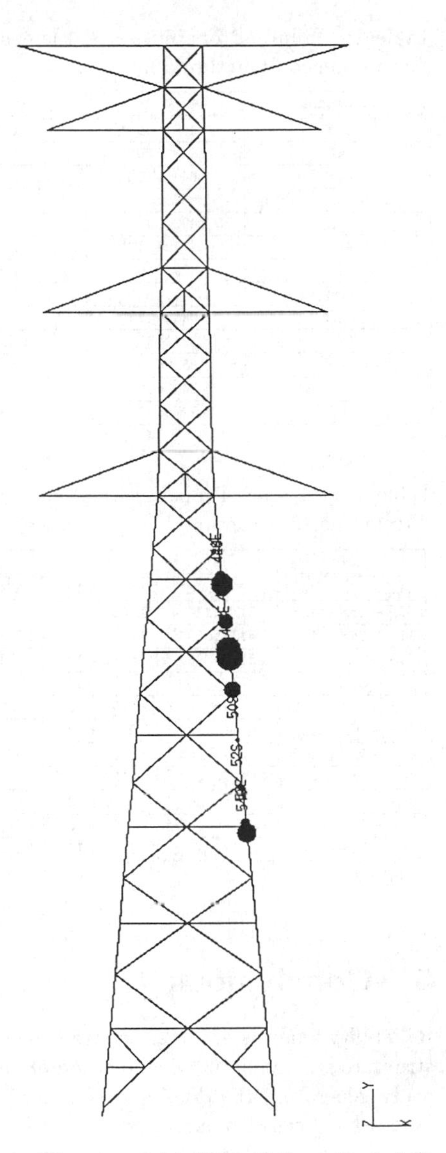

Figure 2: Elements with High Failure Probabilities

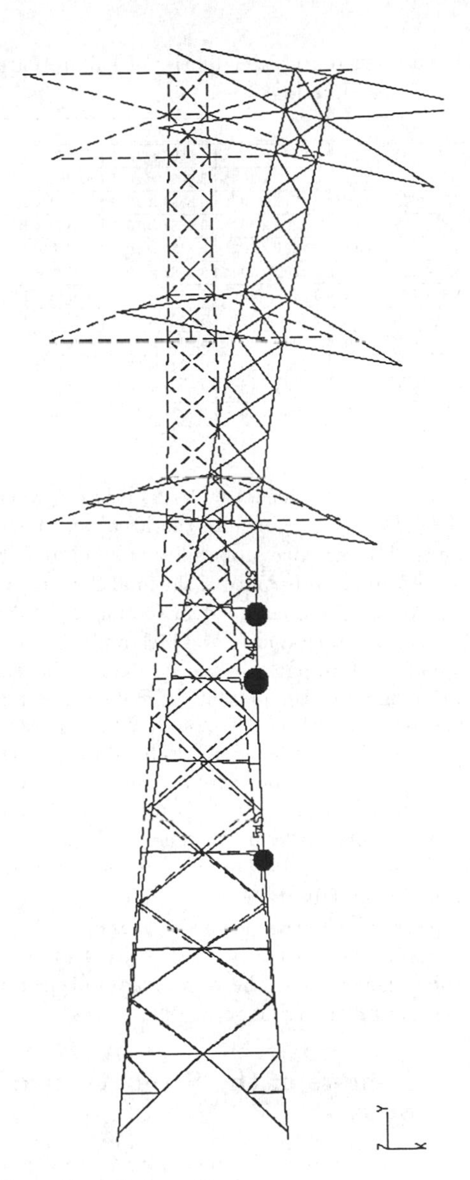

Figure 3: A Failure Mode

are 4.9×10^5 kN/m^2 and 0.1, respectively. The loads acting on the structure involve wind pressures H_c, H_t, and gravities V_c, W_t on the conductors and the tower, respectively. Table 3 shows the loading data corresponding to a $60m/s$ wind speed. Table 4 lists 1, 10, 50-year maximum wind speeds. The 50-year maximum wind speed is used in the calculation.

4.1 Results of the Realistic Tower

The element ends with high failure probabilities are listed in Table 5. S and E denote the two ends of an element, and CSF denotes the central safety factor. It is noted that the failure probabilities of those element ends are very high, which obviously could not bear a 50-year lifetime. The positions of those element ends in the structure are shown in Fig. 2. All of them are in the main leg of the tower. Table 6 shows the probabilistically domi-

Table 6: Ocurrence Probabilities of Failure Paths for a 50-year Lifetime

path	$P_{fP(L)}$	$P_{fP(U)}$	P_{fm}
$44E-54S-42S$	3.2079×10^{-1}	3.9560×10^{-1}	5.1383×10^{-1}
$46S-54S-42S$	3.2592×10^{-1}	3.9552×10^{-1}	5.1368×10^{-1}
$44E-54S-40E$	3.1719×10^{-1}	3.8890×10^{-1}	5.0132×10^{-1}
$46S-54S-40E$	3.2229×10^{-1}	3.8884×10^{-1}	5.0119×10^{-1}
$54S-46E-40E$	2.7301×10^{-1}	3.8589×10^{-1}	5.0378×10^{-1}
$44E-42S-52E$	3.1702×10^{-1}	3.8582×10^{-1}	4.9792×10^{-1}
$46S-42S-52E$	3.2183×10^{-1}	3.8574×10^{-1}	4.9779×10^{-1}
$54S-46E-44S$	2.6876×10^{-1}	3.8409×10^{-1}	5.0684×10^{-1}
$42S-46E-52E$	2.7314×10^{-1}	3.8381×10^{-1}	4.9844×10^{-1}
$46E-44S-52E$	2.6724×10^{-1}	3.8044×10^{-1}	4.9841×10^{-1}
$46E-40E-52E$	2.6669×10^{-1}	3.8000×10^{-1}	4.9836×10^{-1}
$44E-40E-52E$	3.1057×10^{-1}	3.7934×10^{-1}	4.9785×10^{-1}

nant failure paths generated by STRELAS, where $P_{fP(L)}$, $P_{fP(U)}$, P_{fm} denote the lower and upper bounds of occurrence probability of a failure path and the failure probability of the final element end in the path. As mentioned in section 2, STRELAS uses a branch-and-bound technique to search dominant failure paths in the structure. The calculation time for the present example is about 10 hours by a Sun SPARCstation 2. The parameter ε in Eq. (3) is set to 0.05. Only bending moment and axial force are considered in plasticity condition in which buckling is also included. One of the most probable failure modes is schematically illustrated in Fig. 3. The element ends forming the dominant failure modes are all in the leg of the tower, which are the same as those in Table 5. This result is consistent with the actual situation of the damage where the tower was broken from its leg during the typhoon in 1991.

4.2 Results of the Strengthened Tower

In order to improve the reliability of the tower structure, the pipe thicknesses of elements are increased up to 2 times for the leg and 1.5 times for the brace, with the outside diameters being kept the same as before. The results from STRELAS are listed in Tables 7 and 8. The failure probabilities of the element ends and the failure modes decrease to a very low level. There is a little change in the orders of the element ends and failure modes, compared with those in Tables 5 and 6, while the most dangerous sections are still in the middle part of the leg as shown in Fig. 2. Increasing the pipe thickness, especially in the leg, seems to be effective for improving the structural reliability of the transmission tower.

Table 7: Failure Probabilities of Element Ends (Strengthened Structure)

elem.end.no	P_{fe}	CSF
$46S$	4.499×10^{-4}	1.807
$44E$	3.918×10^{-4}	1.820
$54S$	2.044×10^{-4}	1.875
$52E$	1.687×10^{-4}	1.893
$46E$	1.492×10^{-4}	1.907
$42S$	1.471×10^{-4}	1.902
$44S$	1.419×10^{-4}	1.905
$40E$	1.281×10^{-4}	1.915
$54E$	8.985×10^{-5}	1.952
$52S$	8.394×10^{-5}	1.949
$50S$	7.952×10^{-5}	1.957
$48E$	6.540×10^{-5}	1.974
$66E$	5.227×10^{-5}	1.997

Table 8: Ocurrence Probabilities of Failure Paths (Strengthened Structure)

path	$P_{fP(L)}$	$P_{fP(U)}$	P_{fm}
$44E-54S-42S$	8.6402×10^{-12}	2.9755×10^{-6}	1.5741×10^{-4}
$46S-54S-42S$	4.5475×10^{-12}	2.9537×10^{-6}	1.5543×10^{-4}
$44E-54S-40E$	8.6402×10^{-12}	2.6556×10^{-6}	1.3688×10^{-4}
$46S-54S-40E$	4.5475×10^{-12}	2.6362×10^{-6}	1.3515×10^{-4}
$44E-52E-42S$	1.2733×10^{-11}	2.5481×10^{-6}	1.5726×10^{-4}
$46S-52E-42S$	4.5475×10^{-12}	2.5297×10^{-6}	1.5529×10^{-4}
$54S-46E-42S$	2.0464×10^{-12}	2.2841×10^{-6}	1.7405×10^{-4}
$44E-52E-40E$	1.2733×10^{-11}	2.2752×10^{-6}	1.3674×10^{-4}
$46S-52E-40E$	4.5475×10^{-12}	2.2587×10^{-6}	1.3501×10^{-4}
$54S-46E-40E$	2.0464×10^{-12}	2.0360×10^{-6}	1.5187×10^{-4}
$44E-54S-50S$	8.6402×10^{-12}	1.7450×10^{-6}	1.2729×10^{-4}
$46S-54S-50S$	4.5475×10^{-12}	1.7414×10^{-6}	1.2732×10^{-4}

5 Conclusions

Reliability analysis of a realistic transmission line structure is carried out. The method of automatically generating the dominant structural failure modes has been demonstrated to be effective for the reliability analysis of complex structural systems. A probabilistic model of the lifetime maximum wind load is derived based on statistical weather data. The results for the existing structure obtained from computer program STRELAS show that the tower is very weak in its main legs, and the reliability improvement by increasing the pipe thickness is effective.

6 Acknowledgement

This research has been supported in part by Research Aid of Inoue Foundation for Science which is gratefully acknowledged.

References

[1] Mozer, J. D., Orde, C. I., and Hribar, J. A., Reliability of Existing Transmission Line Towers, Proc. ASCE, 893, 1980, pp. 28-43.

[2] Ghannoum, E., Probabilistic Design of Transmission Lines design criteria corresponding to a target reliability, IEEE Transactions on Power Apparatus and Systems, Vol. Pas-102, pp. 3057-3079, 1983.

[3] Mozer, J. D., Perrot, A. H. and Digioa, A. M. Jr., Probabilistic Design of Transimission Line Structures, J. Structural Engineering, ASCE, Vol. 110, No. 10, pp. 2513-2528, 1984.

[4] Murotsu, Y., Okada, H., Matsuzaki, S., and S. Nakamura, On the Probabilistic Collapse Analysis of Transmission Line Structures, Proc. of the First International Symposium on Probabilistic Methods Applied to Electric Power Systems, Toronto, Canada, 1986, pp.53-64.

[5] Murotsu, Y., Okada., H., Yonezawa, M., and Kishi, M., Identification of Stochasticcally Dominant Failure Modes in Frame Structure, 4th Inter. Conf. on Application of Statistics and Probability in Soil and Structural Engineering, Universita di Firenze, Italy, 1983, Pitagora Editrice, pp. 1325-1338.

[6] Thoft-Christensen, P. and Murotsu, Y., Application of Structural Systems Reliability Theory, Springer-Verlag, 1986.

[7] Murotsu, Y., Okada, H., Oishi, T., Niho, O., Nishimura, K., and Kaminaga, H., Reliability Assessment Techniques Applied to the Design of Marine Structures, Proc. of 10th OMAE, ASME, 1991, Vol. 2, pp. 229-235.

[8] Murotsu, Y., Okada, Matsuda, A., T., Niho, O., Kobayashi, M., and Kaminaga, H., Application of the Structural Reliability Analysis System (STRELAS) to a Semisubmersible Platform, Proc. of 11th OMAE, ASME, 1992, Vol. 2, Safety and Reliability, pp. 209-217.

[9] The Japanese Electrotechnical Committee, Design Standards on Structures for Transmission, JEC-127-1979, (in Japanese), Denkishoin, 1979.

[10] The Japan Meterological Agency, Annual Maximum Wind Speed (in Japanese), Technical Data Series, No. 34, 1971.

Structural Safety & Reliability, Schuëller, Shinozuka & Yao (eds) © 1994 Balkema, Rotterdam, ISBN 90 5410 357 4

Statistical uncertainty of sample integral scales

P. Schrader
Building Aerodynamics Laboratory, Ruhr-Universität Bochum, Germany

ABSTRACT: The paper treats the statistical uncertainty of samples of the longitudinal integral scale in the atmospheric boundary layer; turbulent wind speed fluctuations are modelled as a stationary stochastic process. The effect of longitudinal integral scale errors on the buffeting response of lightly damped structures is illustrated. Using the Millionshchikov hypothesis or other appropriate assumptions, the coefficient of variation and the asymtotic probability distribution of sample integral scales can be estimated by AR simulation. The coefficient of variation of sample integral scales is estimated for a number of special cases in the atmospheric boundary layer.

MAIN NOTATIONS

f_1 first natural frequency of structure
f frequency
$r_{uu}(\tau)$ sample autocorrelation of u component of turbulent wind velocity; $r_{uu}(0)=1$
$g_{uu}(f)$ one-sided sample spectral density of u component of turbulent wind velocity
$G_{uu}(f)$ one-sided spectral density of u component of turbulent wind velocity
T_{rec} record length of wind speed record (unit: time)
T_{ux} longitudinal integral time scale of u component of turbulent wind velocity
$t_{ux,o}$ a sample of T_{ux} obtained by integration over the sample autocorrelation from $\tau=0$ up to its first zero crossing
$t_{ux,e}$ a sample of T_{ux} calculated by setting $t_{ux,e} = \tau$ for $r_{uu}(\tau)=e^{-1}$
$t_{ux,f}$ a sample of T_{ux} obtained by fitting a Kaimal spectral density to a sample spectral density (Schrader (1993))
V mean wind speed at height z
V_{10} mean wind speed at 10 meters' height
z height above ground
α mean wind speed profile exponent
τ time lag (for autocorrelation)
ζ modal damping as percentage of critical damping
σ_u standard deviation of turbulent wind speed

1. INTRODUCTION

Integral scales of the longitudinal component of wind velocity are important for predicting the response of wind sensitive structures to wind loading and for scaling when carrying out wind tunnel tests (Simiu (1981a),(1981b), Niemann (1990)). Measurements of these scales are subject to a very large scatter in the atmospheric boundary layer (Schrader (1990),(1993)).

2. INFLUENCE OF INTEGRAL SCALE ERRORS ON STRUCTURAL RESPONSE

2.1 *Basic assumptions and preliminary remarks*

The structural analysis results (Schrader (1993)) given in this paper are based on a special version (Soize (1975),(1977)) of the classical spectral method (buffeting response in along-wind direction, no aeroelastic effects, linear elastic structure, very small damping, quasi-stationary theory for pressure coefficients, height-independent spectral density, first natural frequency of the structure in the -5/3-range of the spectral density, etc....).

In this paper, when varying the longitudinal integral scale in the structural analysis, the lateral integral scales (and lateral correlations of turbulent wind speed) remain unchanged; consequently, the quasistatic (and static) structural responses are not affected here by varying the longitudinal integral scale. It is not implied that no physical relation exists between the longitudinal integral scale and the lateral ones. This independent variation of the longitudinal integral scale is warranted by the fact that purely statistical errors of measured integral scales are studied.

On the other hand, of course, increasing/decreasing the longitudinal integral scale implies decreasing/ increasing the resonant responses of structures.

The influence of integral scale variations on structural analysis has also been dealt with in Kanda (1983) (three high-rise buildings, integral scale changes of up to 50%) and in Schuëller (1983) (a high-rise building and a chimney, statistical variation of integral scale exhibiting a coefficient of variation of 28%). In this paper, changes of integral scales by a factor of up to three are taken into account.

2.2 *Analysis of three structures*

Results for three different structures (Schrader (1993)) are presented in this paper. Some important data, on which the analysis is based, are shown in table 1.

Stochastic means of relevant hourly maximum responses (Soize (1977)) are given for three structures in figures 1,2 and 3. I.e., trippling the integral scale produces roughly a 15% to 30% decrease in the relevant responses; reducing it by a factor of three produces roughly a 20% to 40% increase in the relevant responses.

3. ESTIMATING THE STATISTICAL UNCERTAINTY OF SAMPLE INTEGRAL SCALES

3.1 *Theoretical remarks on the capabilities of AR models and periodogram simulation*

In this section, it is supposed that a gaussian AR model is available which posesses the same spectral density and the same mean as some given stationary stochastic wind process (Schrader (1993)).

It is often assumed that wind speed records stem from a gaussian stationary stochastic process; in the atmospheric boundary layer, this model is mostly used for detrended wind records and sometimes for undetrended wind records. Under this assumption, the probability distribution of the sample autocorrelation or sample spectral density would be the same for our AR model and the (possibly detrended) wind process.

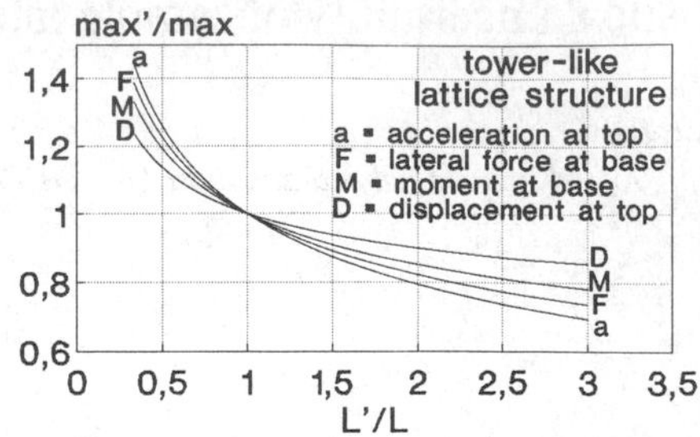

fig.1 Change of maximum responses when changing the integral scale

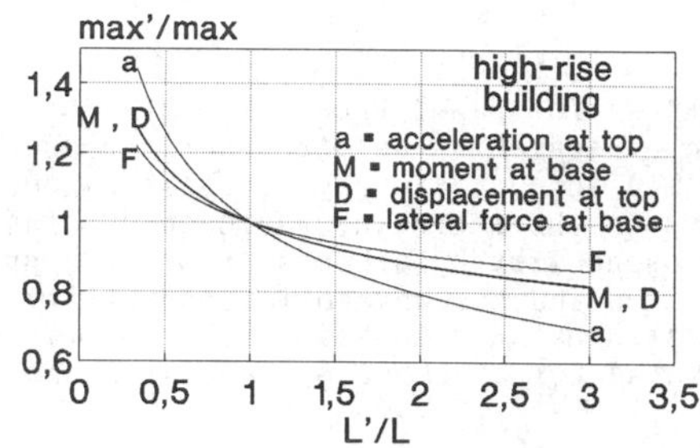

fig.2 Change of maximum responses when changing the integral scale

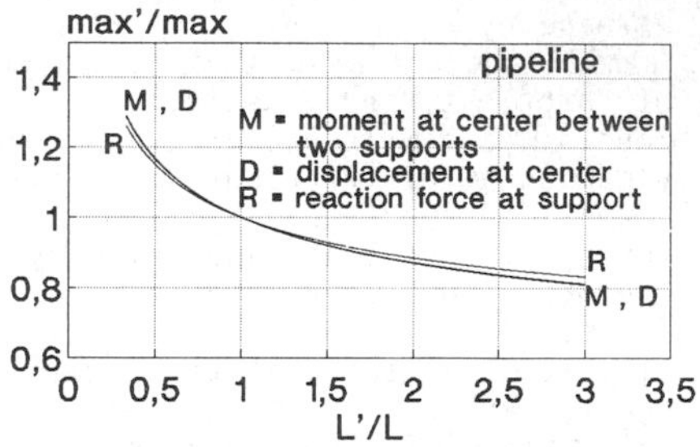

fig.3 Change of maximum responses when changing the integral scale

Table 1. Some data on the structures and the wind environment - from Soize (1975),(1977)

	tower-like lattice structure	high-rise building	pipeline
f_1	1.347Hz	0.209Hz	0.327Hz
ζ	1.5%	3.0%	1.0%
height	30m	180m	
length			68m
V_{10}	20.0m/s	20.0m/s	32.68m/s
σ_u	6.0m/s	10.0m/s	5.37m/s
α	0.28	0.40	0.15

For a non gaussian wind process, the validity of the (frequently applied (Monin (1975)) Millionshchikov hypothesis would warrant agreement (Schrader (1993)) between our AR model and the wind process as far as the asymptotic probability distributions of $r_{uu}(\tau)$ (for not too large a τ) and $g_{uu}(f)$ are concerned; at least, this is applicable to the type of sample autocorrelation and sample spectral density on which this paper is based (Schrader (1993)).

Some experimental confirmation of the theoretical probabilty distribution of the periodogram of turbulent wind in the atmospheric boundary layer can be found in Prenninger (1988). This probability distribution is largely insensitive to deviations of the wind process from a gaussian model (Brillinger (1981)).

Therefore, there is a sound theoretical basis for estimating the coefficient of variation of $t_{ux,o}$, $t_{ux,e}$ and $t_{ux,f}$ (and, presumably with less precision, their probability distribution) for a given stationary stochastic wind process by means of a suitable AR model (Schrader (1993)).

The usual nonstationarities mainly affect the low frequency ordinates of $g_{uu}(f)$; hence, usually, nonstationarities are likely to affect $t_{ux,o}$ and $t_{ux,e}$ very strongly while their influence on $t_{ux,f}$ is likely to be much weaker (Schrader (1993)).

3.2 *On autoregressive simulation and periodogram simulation*

The AR simulation technique used in this paper (see Schrader (1993)) is largely based on Iwatani (1982),(1988),(1990). The AR models are driven by gaussian white noise.

The estimates $t_{ux,o}$ and $t_{ux,e}$ are both under the very strong influence of the low frequency part of $g_{uu}(f)$. Consequently, in order to estimate coefficients of variation of $t_{ux,o}$ and $t_{ux,e}$, it is essential for the AR model to accurately reflect the low frequency spectral ordinates of the turbulent wind speed process to be simulated. However, customary criteria for determining AR order are usually inadequate for achieving this, and much larger AR orders are required instead (Schrader (1993)).

In this paper, starting with some spectral density given in the literature (Panofsky (1984)) for a continuous wind process, the aliased spectral density of the sampled wind process is calculated by means of a suitable formula; the corresponding autocovariance is then computed by the fast Fourier transform technique, and the Yule-Walker equations are solved successively for a number of AR orders; for each order, the AR spectral density is computed; the order is judged to be satisfactory if agreement between the AR spectral density and the spectral density of the sampled wind process is satisfactory. This method is described more extensively in Schrader (1993).

For a turbulent wind record $u(k\Delta t)$ $(k=0,...,K-1)$ with $K=2^n$, applying the fast Fourier transform technique, the periodogram $p_{uu}(f)$ is easily obtained in the Fourier frequencies $f_i=i/K\Delta t$ $(i=1,..., (K/2)-1)$. Even for a non gaussian stationary wind process, this extremely widespread technique implies that, asymptotically, $p_{uu}(f_i)/(G_{uu}(f_i)/4)$ follows a chi-square distribution with 2 degrees of freedom (Brillinger (1981)); furthermore, asymptotically, $p_{uu}(f_{i1})$ and $p_{uu}(f_{i2})$ are uncorrelated for distinct Fourier frequencies. Obviously, employing a standard pseudo random number generator, it is easy to generate an artificial periodogram for a fictitious wind speed record that stems from a stationary stochastic wind process whose spectral density is known. This technique was used successfully in Schrader (1993) to estimate the scatter of $t_{ux,f}$.

3.3 *Examples of estimated scatter of sample integral scales in the atmospheric boundary layer*

The atmospheric integral time scale and wind spectral density were calculated according to Panofsky (1993) (for neutral stratification, homogeneous terrain). The spectral density is of the Kaimal type (Schrader (1993)):

$$G_{uu}(f) = 4\sigma_u^2 \cdot T_{ux} / [1+6(f \cdot T_{ux})]^{5/3} \quad (1)$$

The formula used for T_{ux} is :

$$T_{ux} = 2 / .1818V \quad (2)$$

In this paper, it does not matter very much whether formula (2) is exact, and the only purpose of the T_{ux} values obtained from it is to illustrate the scatter of sample integral scales stemming from one and the same stationary stochastic process.

The data on which the following examples (Schrader (1993)) are based are shown in table 2. The roughness length z_0 given in table 2 corresponds to a terrain with hedges, many trees and few buildings. The

Table 2. Data on wind environment and wind speed recording (for simulating the scatter of sample integral scales); variants A, B, D

	z_0 [m]	z [m]	V [m/s]	T_{ux} [s]	σ_u^2 [m^2s^{-2}]	Δt [s]	T_{rec} [s]	AR order
A	.30	10.00	10.00	5.50	7.44	.200	1228.8	128
B	.30	10.00	10.00	5.50	7.44	.200	4915.2	128
D	.30	10.00	10.00	5.50	7.44	.100	1228.8	256

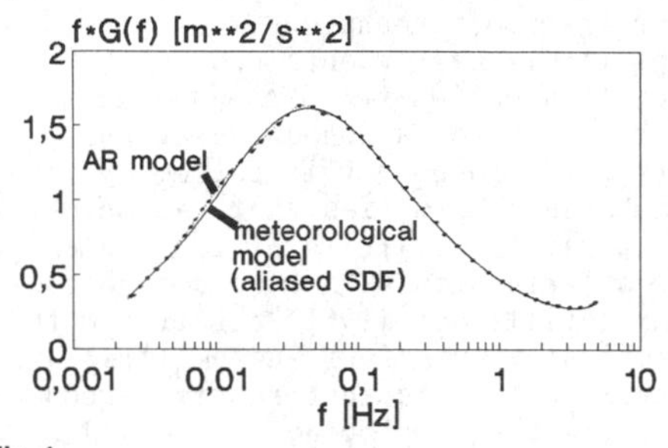

fig.4

wind spectral density and the AR spectral density used for the computer simulation are shown in figure 4.

Three simulation variants were computed. As for variant A, the total record length of one wind speed record is approximately 20 minutes (a realistic value). As for variant B, its total record length, T_{rec}, was quadrupled with respect to variant A (an unrealistically large value). As for variant D, its sampling frequency was doubled with respect to variant A. For each variant and each investigated calculation method of T_{ux}, approximately 1650 sample integral scales were calculated.

The basic results of the computer simulation can be seen in table 3. Quadrupling the wind record length leads to halfening the C.O.V. of t_{ux}. Indeed, basically, the C.O.V. is proportional to $(T_{rec}/T_{ux})^{-1/2}$ (Schrader (1993),(1990)). Doubling the sampling frequency has almost no effect (if any) on the C.O.V. of t_{ux}.

The means of $t_{ux,o}$ and $t_{ux,e}$ are much smaller than T_{ux}, and depend on the block length (here: 204.8s) (Schrader (1993)).

For a realistic wind record length of about 20 minutes, the C.O.V. of $t_{ux,o}$ is very large. However, for larger T_{ux}, this problem will necessarily become much worse because, basically, quadrupling T_{ux} implies doubling the C.O.V. of t_{ux} (Schrader (1993),(1990)). E.g., in the situation described in table 2, according to formula (2), T_{ux} could be about four times as large at z=60m, and the C.O.V. of $t_{ux,f}$ would become about 25% for a 20 minutes' record; the C.O.V. of $t_{ux,o}$ would reach about 40%.

Table 3. Simulation results : scatter of sample integral scales and mean wind speed

		$t_{ux,f}$	$t_{ux,o}$	$t_{ux,e}$	V
A	mean	5.42s	3.73s	3.19s	9.99m/s
	C.O.V.	.130	.211	.154	.0246
B	mean	5.41s	3.62s	3.15s	9.99m/s
	C.O.V.	.062	.107	.079	.0124
D	mean	5.45s	3.65s	3.02s	9.99m/s
	C.O.V.	.123	.222	.157	.0243

The histograms of $t_{ux,f}$ and $t_{ux,o}$ were lognormal or nearly gaussian, and those of $t_{ux,e}$ were roughly lognormal or roughly gaussian (see figures 5 to 7).

In case of variant A and $t_{ux,f}$, besides AR simulation, "periodogram simulation" was tried too. The results could hardly be distinguished from those obtained by AR simulation.

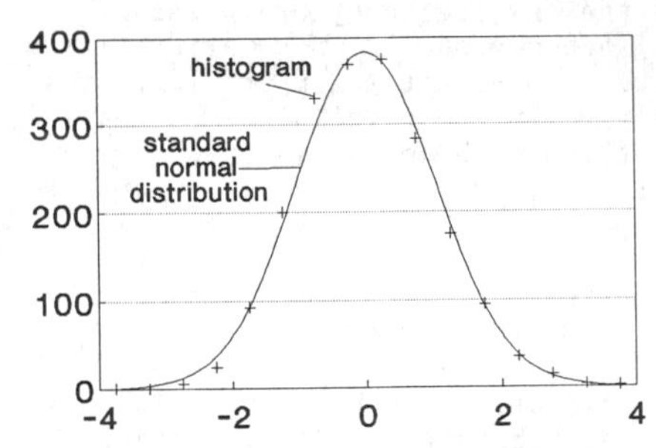

fig. 5 Normalized histogram of log of sample integral scale $t_{ux,o}$

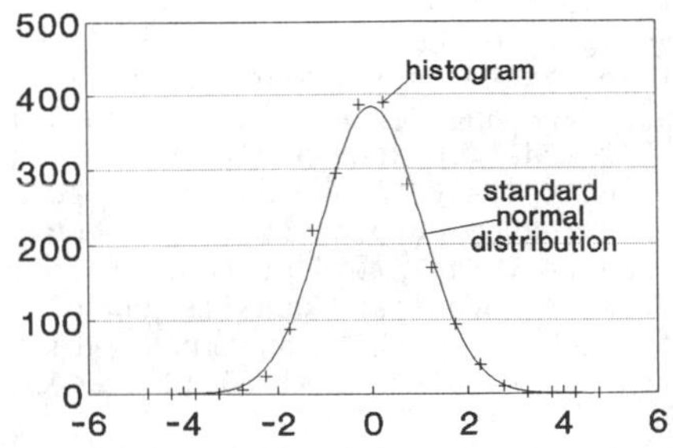

fig. 6 Normalized histogram of log of sample integral scale $t_{ux,f}$

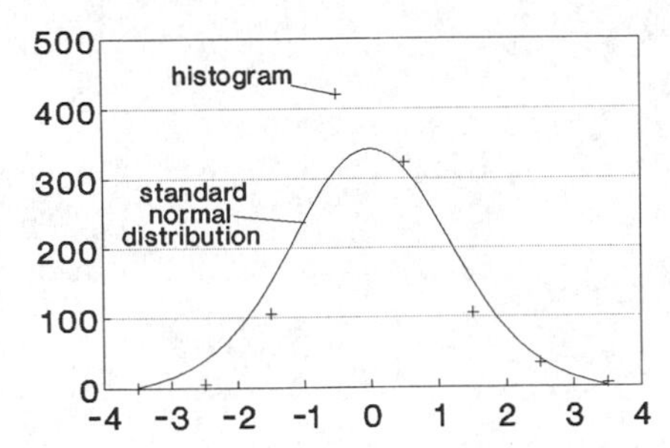

fig. 7 Normalized histogram of log of sample integral scale $t_{ux,e}$

4. MAIN CONCLUSION

In the atmospheric surface layer, for a 20 minutes' stationary wind record, the coefficient of variation of the sample integral scale is often larger than 20% and sometimes much larger (40% or even more).

ACKNOWLEDGEMENTS

The assistance of Prof. H.-J. Niemann (head of the Building Aerodynamics Laboratory at Ruhr-Universität Bochum) and his staff are gratefully acknowledged. The support of the SFB 151 is gratefully acknowledged too.

REFERENCES

Brillinger, D.R. 1981. *Time Series - Data Analysis and Theory.* Holden-Day, Inc..

Iwatani, Y. 1982. Simulation of multidimensional wind fluctuations having any arbitrary power spectra and cross spectra (in Japanese). *Journal of Wind Engineering, Japan Association for Wind Engineering,* vol. 11: 5-15.

Iwatani, Y. 1988. Simulation of multidimensional wind fluctuations associated with given power spectra and cross spectra and its accuracy (in Japanese). *Journal of Wind Engineering, Japan Association for Wind Engineering,* vol.36: 11-26.

Iwatani, Y.: On the order of autoregressive model for spectral estimation (in Japanese). *Journal of Wind Engineering, Japan Association for Wind Engineering,* vol. 45: 57-64.

Kanda, J. 1983. Reliability of gust response prediction considering height dependent turbulence parameters. *Journal of Wind Engineering and Industrial Aerodynamics,* vol. 14: 455-466.

Monin, A.S., Yaglom, A.M. 1975. *Statistical Fluid Mechanics : Mechanics of turbulence,* vol.2. The MIT-Press.

Niemann, H.-J. 1990a. Dynamic response of cantilevered structures to wind turbulence. Proc. 1st EURODYN, vol.2: 1123-1131. Bochum (Germany): Ruhr-Universität Bochum.

Niemann, H.-J. (ed.) 1990b. Beiträge zur Bauwerksaerodynamik. *SFB 151-Berichte Nr. 17,* Bochum (Germany): Ruhr-Universität Bochum.

Panofsky, H.A., Dutton, J.A. 1984. *ATMOSPHERIC TURBULENCE - Models and methods for engineering applications.* John Wiley & Sons, Inc..

Prenninger, P.H.W., Schuëller, G.I., Lin, Y.K. 1988. Experimental verification of the probability distribution of sampled wind spectra. *Journal of Wind Engineering and Industrial Aerodynamics,* vol. 31: 1-7.

Schrader, P. 1990. On the Statistical Accuracy of Integral Time Scale Measurements. In *Niemann (1990b)*: 73-121.

Schrader, P. 1993. *Die statistische Stabilität gemessener integraler Längenmaße und anderer Windparameter.* Ph.D. thesis. Bochum (Germany): Institut für Konstruktiven Ingenieurbau, Ruhr-Universität Bochum.

Schuëller, G.I., Hirtz, H., Booz, G. 1983. The effect of uncertainties in wind load estimation on reliability assessments. *Journal of Wind Engineering and Industrial Aerodynamics,* vol. 14: 15-26.

Simiu, E. 1981a. Modern developments in wind engineering: part 1. *Engineering Structures,* vol.3: 233-241.

Simiu, E. 1981b. Modern developments in wind engineering: part 2. *Engineering Structures,* vol.3: 242-248.

Structural Safety & Reliability, Schuëller, Shinozuka & Yao (eds) © 1994 Balkema, Rotterdam, ISBN 90 5410 357 4

Reliability-based local design wind pressures for simple rectangularly-shaped buildings

J.Wacker
Wacker & Partner, Büro für angewandte Strömungsmechanik, Birkenfeld, Germany

E.J. Plate
Wind Engineering Division, Institute for Hydrology and Water Resources Planning, University of Karlsruhe, Germany

ABSTRACT: Standard assessment methods for local peak wind pressures are compared and additionally a reliability-based assessment method is provided. The latter combines second moment reliability principles with wind pressure data from wind tunnel experiments and with the simplified Cook-Mayne approach to calculate local peak wind pressures on buildings with a consistent safety level. In an example, local peak wind pressures on front and roof of simple rectangularly-shaped buildings were calculated. The aerodynamic coefficients were determined in wind tunnel experiments. Differences are quantified. Local peak wind pressures in separated flow regions may be underestimated significantly if the pressures are assessed by means of the quasi-steady approach. The more sophisticated Cook-Mayne approach and the second moment reliability approach (with wind direction reduction factor included) yield higher peak pressures. The peak factor approach was calibrated against the second moment reliability approach and appropriate mean peak factors for the design of small building surface elements are estimated. Finally, local wind loads for rectangular buildings specified in the draft of the new Eurocode EC Wind loads are compared with reliabilty-based wind loads calculated from wind tunnel experiments. The corresponding safety level of local wind loads specified in the draft of the new EC Wind loads was estimated in terms of safety indices. For some configurations in separated flow regions the safety index takes values lower than 3, resulting in a design load below typical values for ultimate limit states design of usual structures.

1 INTRODUCTION

Many modern buildings are covered with light-weight facade elements, glass panels and roof pavers, which are directly exposed to wind action. Usually, cladding of large buildings is quite expensive, and the potential losses due to wind damage are high. Hence, the designer aims at a safe and economic construction and design of building cladding. Nevertheless, in many countries wind damage to buildings mainly consists of failure of individual elements of the building surface. A rational design of small cladding elements, glass panels and roof pavers against wind require the realistic assessment of the local peak wind pressures acting on the building surface and, possibly, the reliable estimation of the fatigue life due to repetitive loads. In this paper only local peak wind pressures are considered.

In recent years, different approaches for the assessment of local design wind loads were developed, which take into account the random character of wind load. Nevertheless, there exist some uncertainties involved in wind load prediction. In this paper commonly used approaches for the assessment of local peak wind pressures - the quasi-steady approach, the peak factor approach and the simplified Cook-Mayne approach - are considered and compared with an approach which is based on second moment reliability principles. As an example, local peak wind pressures on front and roof of simple rectangularly-shaped buildings were calculated. The aerodynamic coefficients were determined in wind tunnel experiments. Differences are quantified. Then, the peak factor method was calibrated against the second moment reliability approach and appropriate mean peak factors for local peak wind pressures are estimated. Finally, local wind loads for rectangular buildings specified in the draft of the new EC Wind loads (1992) are compared with reliabilty-based wind loads calculated from wind tunnel experiments. The corresponding safety level of local wind loads specified in this code was estimated in terms of safety indices.

2 STANDARD ASSESSMENT METHODS FOR LOCAL PEAK WIND PRESSURES

2.1 Quasi-steady approach

For the quasi-steady approach the local design wind pressure with a mean recurrence interval of 50 years, $p_{50y}(x,y,z)$, is given by the equation

$$p_{50y} = q_{50y} \cdot c_e(H) \cdot G(z) \cdot c_p \quad , \qquad (1)$$

where q_{50y} is the mean (hourly) dynamic pressure with a mean recurrence interval of 50 years, $c_e(H)$ is the exposure and height factor at building height H, $G(z)$ is the appropriate gust response factor at height z, and c_p is the mean pressure coefficient:

$$c_p = \frac{\overline{p} - p_{ref}}{q(H)} \quad , \tag{2}$$

where $\overline{p}$ is the time-averaged local mean wind pressure, p_{ref} is the reference static pressure of the undisturbed approach flow and $q(H)$ is the mean dynamic pressure at building height H. According to the linear quasi-steady theory, the gust factor $G(z)$ is proportional to the longitudinal turbulence intensity $T_u(z)$ and may be estimated (see also EC Wind loads, 1992) by

$$G(z) = 1 + 2 \cdot k \cdot T_u(z) \quad \text{, EC Wind loads: } k = 3.5 \ . \tag{3}$$

It is obvious, that building-generated contributions to the peak wind loads are not included in this quasi-steady approach.

2.2 Simplified Cook-Mayne approach

In contrast to the quasi-steady approach the method proposed by Cook and Mayne (1980) combines the statistics of the dynamic pressure $q(H)$ of extreme hourly wind speeds with the statistics of peak pressure coefficients:

$$c_{ppeak} = \frac{p_{peak} - p_{ref}}{q(H)} \quad , \tag{4}$$

where p_{peak} is the local peak pressure, which also includes load fluctuations due to building-generated flow disturbances. Both wind climate and peak pressure coefficients are treated as random variables. Currently the Cook-Mayne method is one of the most powerful standard approaches to calculate local wind loads to be expected during the lifetime of the structure with a given probability of occurence. For convenient practical application of the method, Cook (1982) simplified the original method by making some assumptions. Applying the simplified Cook-Mayne method, local peak wind loads with a mean return period R of 50 years, p_{50y}, may be assessed by

$$p_{50y} = q_{50y} \cdot c_e(H) \cdot \left(U_{c_{ppeak}} + \frac{1,4}{a_{c_{ppeak}}} \right) \quad , \tag{5}$$

where $U_{c_{ppeak}}$ and $1/a_{c_{ppeak}}$ are the mode and dispersion of the Fisher- Tippett Type I (FT1) extreme value distribution $F(c_{ppeak})$ for the largest peak pressure coefficients:

$$F(c_{ppeak}) = e^{-e^{\left(-y_{c_{ppeak}}\right)}} \quad , \tag{6}$$

where the reduced variate is given by

$$y_{c_{ppeak}} = a_{c_{ppeak}} \left(c_{ppeak} - U_{c_{ppeak}} \right) \quad . \tag{7}$$

The peak pressure coefficient, which for design should be averaged over the area of the element under consideration depends on the size of the element. Usually, for cladding design this effect is considered by low-pass filtering the local pressure fluctuations. Lawson (1980) considered aerodynamic admittance functions and the coherency of pressure fluctuations and suggested that the required smallest load duration, which corresponds to the averaging time t_{av}, may be estimated by

$$t_{av} = \frac{4,5 \cdot l}{u} \quad , \tag{8}$$

where l is the characteristic length of the element under consideration and u is the typical design mean wind velocity. From Equation 8 it follows that for typical design mean wind speeds and typical cladding elements t_{av} takes a value of about 0.5 s (rather than 1 s, as is specified in several codes of practice).

2.3 Peak factor approach

For the peak factor approach the difference of the local peak pressure coefficient from the local mean pressure coefficient is assessed in terms of the rms pressure coefficient:

$$p_{50y} = q_{50y} \cdot c_e(H) \cdot \left(c_p \pm k \cdot c_{prms} \right) \quad , \tag{9}$$

where k is the peak factor, which depends on the importance of the structure under consideration and the size of the element or on t_{av}; c_{prms} is the rms pressure coefficient:

$$c_{prms} = \frac{p_{rms} - p_{ref}}{q(H)} \tag{10}$$

Despite the major advantages of the Cook-Mayne method and - with some restrictions - the peak factor approach, in most wind loading standards the quasi-steady approach is used to calculate design wind loads. Two reasons are its simplicity and the availability of many more mean pressure coefficients than rms and peak loading coefficients.

3 LOCAL PEAK WIND PRESSURES BASED ON SECOND MOMENT RELIABILITY PRINCIPLES

In the previous chapter local design wind pressures were calculated by multiplication of a reference dynamic pressure with an appropriate exposure factor, a gust factor and a mean pressure coefficient. The latter two factors may be replaced by a peak pressure coefficient. For instance, the local peak wind pressures specified in the draft of the new EC Wind loads (1992) are given by

$$p_s - q_s \cdot c_{e,s} \cdot G_s \cdot c_{p,s} \tag{11}$$

where q_s, $c_{e,s}$, G_s and $c_{p,s}$ are assumed implicitly to be deterministic parameters and to be known exactly. In reality, their magnitudes vary randomly and are full of uncertainties, i.e. parameters like wind velocity pressure, exposure factor, pressure coefficient among others may be regarded as random variables (see Figure 1).

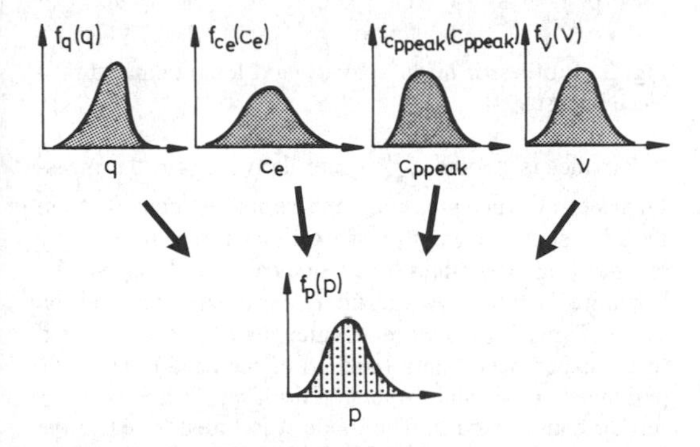

Fig. 1 Statistical distributions of the local peak wind pressure and the parameters relevant to the assessment of local peak wind pressures

To account for the uncertainties and the variability of the parameters and consequently of the local wind pressures, principles of second moment analysis (see Hasofer and Lind, 1974, Ravindra and Galambos, 1978, Schueller, 1981, Davenport et al., 1985, Holmes, 1985, and Plate, 1993, among others) may be applied. Then it is possible to calculate wind loads with a consistent safety level which may be expressed in terms of safety indices. Second moment analysis (design level II) is based on the assumption that both load p and resistance r are Gaussian or lognormal distributed random variables and consequently are described completely by the corresponding mean values μ_p and μ_r or $\mu_{\ln p}$ and $\mu_{\ln r}$ and standard deviations σ_p and σ_r or $\sigma_{\ln p}$ and $\sigma_{\ln r}$.

The structure under consideration fails if r - p < 0 or ln(r/p) < 0. The limit state corresponds to m = r - p = 0 or m = ln(r/p) = 0. The so-called safety margin m is also Gaussian distributed (see Figure 2). If load p and resistance r are affected by multiplicative effects, then p and r tend to be lognormal distributed and the safety index which is a measure for safety may be defined as:

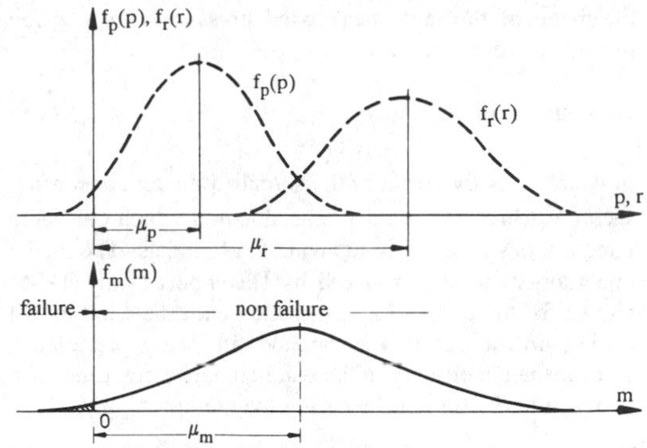

Fig. 2 Probability densities of load p, resistance r and safety margin m

$$h = \frac{\mu_{\ln\left(\frac{r}{p}\right)}}{\sigma_{\ln\left(\frac{r}{p}\right)}} \tag{12}$$

The corresponding probability of failure is

$$F_f = 1 - \Phi\left(\frac{1}{V_m}\right) = 1 - \Phi(h) = \Phi(-h) \quad , \tag{13}$$

where Φ denotes the Gaussian distribution. Typical relations between h and F_f are:

$$\begin{aligned} h = 2: F_f = 2,27 \cdot 10^{-2} \\ h = 3: F_f = 1,35 \cdot 10^{-4} \end{aligned} \tag{14}$$

With $\sigma^2_{\ln r} \approx V_r^2$ and $\sigma^2_{\ln p} \approx V_p^2$ (for V < 0,3) (see Ravindra and Galambos, 1978) and linearization of the problem for $0,25 < V_p / V_r < 4$ (see Lind, 1971) we obtain

$$h = \frac{\ln\left(\frac{\mu_r}{\mu_p}\right)}{\alpha(V_r + V_p)} \quad , \tag{15}$$

where α is a numerical constant with α = 0,75 if only wind loads are considered. Equation 15 allows the separation of load and resistance components:

$$\mu_r \cdot e^{(-0,75 \cdot h \cdot V_r)} = \mu_p \cdot e^{(0,75 \cdot h \cdot V_p)} \quad , \tag{16}$$

where $\tau_0 = e^{(0,75 \cdot h \cdot V_p)}$ is sometimes denoted as mean load factor. One major advantage of this separation is that load factors can be determined independently of the resistance uncertainties. In Equation 16 μ_p corresponds to the mean of the local peak wind pressure with a given mean recurrence interval:

$$\mu_p = \overline{q} \cdot \overline{c}_e \cdot \overline{c}_{ppeak} \cdot \overline{\nu} \quad , \qquad (17)$$

in which $\overline{\nu}$ is the mean of the so-called model uncertainty factor ν which is a random variable and which considers uncertainties inherent in the wind load models. The model uncertainty factor introduced by Davenport et al. (1985) allows for the uncertainty associated with the wind tunnel testing procedures in representing full scale phenomena and for the uncertainty in the wind climate estimation.

The introduction of the partial load factor τ_p yields:

$$\tau_p = \left(\frac{\mu_p}{p_{design}} \right) \cdot \tau_0 \quad . \qquad (18)$$

Thus, the design wind load may be calculated by:

$$p_{design} = \left(\frac{\mu_p}{\tau_p} \right) \cdot e^{(0,75 \cdot h \cdot V_p)} \quad . \qquad (19)$$

If the factors involved in Equation 17 are statistically independent, the coefficient of variation of the wind pressure may be calculated by

$$1 + V_p^2 = \left(1 + V_q^2\right)\left(1 + V_{c_e}^2\right)\left(1 + V_{c_{ppeak}}^2\right)\left(1 + V_\nu^2\right) \quad . \qquad (20)$$

Combining Equation 19 with Equations 17 and 20 we obtain

$$p_{design} = \frac{1}{\tau_p} \cdot q_s \cdot c_{e,s} \cdot \left(\frac{\overline{q}}{q_s} \right) \cdot \left(\frac{\overline{c}_e}{c_{e,s}} \right) \cdot \overline{\nu} \cdot \overline{c}_{ppeak} \cdot e^{\left(0,75 \cdot h \cdot \sqrt{\left(1+V_q^2\right)\left(1+V_{c_e}^2\right)\left(1+V_{c_{ppeak}}^2\right)\left(1+V_\nu^2\right) - 1}\right)} \quad , \qquad (21)$$

in which p_{design} may be for instance the local peak wind pressure with a mean recurrence interval of say 50 years, or the specified design wind load in a wind load standard.

4 EXAMPLE: ASSESSMENT OF LOCAL PEAK WIND PRESSURES BY MEANS OF DIFFERENT APPROACHES

The afore-mentioned methods will be applied and compared with each other in an example for application, in which local peak wind pressures on front and roof of simple rectangularly-shaped buildings are estimated.

4.1 Aerodynamic Coefficients: Basic Values

At Karlsruhe University an extensive research program on mean and fluctuating forces and pressures on different cubical forms has been started. The sharp-edged building models covered a wide range of side and aspect ratios and were exposed to different wind directions and wind velocity and turbulence intensity profiles which correspond to three different roughness pattern (see Figure 3). The longitudinal integral length scales $L_{ux}(z=15cm)$ covered a range from 25cm to 30cm. Modeling laws as given in detail by Plate (1982) were obeyed in this study.

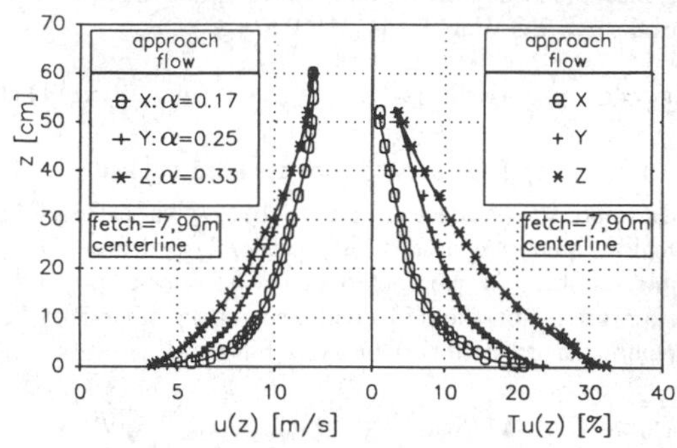

Fig. 3 Profiles of mean velocity and longitudinal turbulence intensity

Parameters c_p and c_{prms} were derived from the measurements of wind pressure time series. Moreover, from the time series of wind pressure fluctuations statistically independent peak values (one value per record, t_{av} = 0,5 s in nature, moving average filter) were extracted and the Fisher-Tippet Type I extreme value distribution was fitted to the experimental data (method of moments). The FT1 parameters mode and dispersion are used as the basis for further considerations. The mode was transformed to the 1 hour standard observation period. The dispersion is unchanged. Peak loading coefficients c_{ppeak} corresponding to a design load with a mean return period R of 50 years were calculated by applying the simplified Cook-Mayne method.

It is beyond the scope of this paper to discuss all results in detail. The results of local peak wind pressures are described in Wacker et al. (1991) and Wacker and Plate (1992). All results of the experiments and the experimental equipment and procedure will be summarized in Wacker (1993). In Table 1 experimental results used in this study are summarized, including the peak pressure coefficients specified in the draft of the EC Wind loads (1992). The aerodynamic pressures are non-dimensionalized by the mean velocity in model height.

4.2 Estimation of other required parameters

In addition to the aerodynamic coefficients and their coefficients of variation magnitudes of $\overline{q}_{50} / q_{50,s}$, $\overline{c}_e / c_{e,s}$, $\overline{\nu}$, V_q, V_{c_e}, and V_ν have to be estimated. Fürniß (1991) showed that for Northern and Central Europe typical values for $\overline{q}_{50} / q_{50,s}$ and V_q are 1,07 and 0,15. For design purposes the wind is usually assumed to blow

Table 1 Results of the wind tunnel experiments and wind pressures specified in EC Wind loads (1992) required for the example of application

roof, center line, leading edge region, wind direction ß = 0°													
		BL X				BLY				BL Z			
L/B/H model [cm]	x/L [-]	cp wt [-]	cprms wt [-]	cppea CaM w [-]	EC cp*G(z [-]	cp wt [-]	cprms wt [-]	cppea CaM w [-]	EC cp*G(z [-]	cp wt [-]	cprms wt [-]	cppea CaM w [-]	EC cp*G(z [-]
15/15/10	0.0133	0.83	0.15	1.7	2.5	0.86	0.23	2.3	2.7	1.04	0.39	3.4	3.1
	0.133	0.91	0.17	2.0	1.5	0.96	0.26	2.5	1.7	1.03	0.42	3.6	1.9
	0.266	0.96	0.19	2.0	1.5					0.81	0.37	2.8	1.9
	0.4	0.84	0.23	1.9	1.5	0.68	0.29	2.1	1.7	0.52	0.31	2.3	1.9
15/15/30	0.0133	0.76	0.12	1.3	2.3	0.84	0.15	1.6	2.4	0.85	0.16	1.8	2.6
	0.133	0.78	0.12	1.4	1.4	0.89	0.16	1.7	1.5	0.93	0.17	2.1	1.6
	0.266	0.80	0.13	1.5	1.4	0.93	0.16	1.8	1.5	0.96	0.21	2.1	1.6
	0.4	0.85	0.15	1.7	1.4	0.93	0.18	1.9	1.5	0.94	0.22	2.0	1.6
	0.4666									0.88	0.23	1.9	1.6

front, stagnation point, wind direction ß = 0°													
		BL X				BLY				BL Z			
L/B/H model [cm]	z/H [-]	cp wt [-]	cprms wt [-]	cppea CaM w [-]	EC cp*G(z [-]	cp wt [-]	cprms wt [-]	cppea CaM w [-]	EC cp*G(z [-]	cp wt [-]	cprms wt [-]	cppea CaM w [-]	EC cp*G(z [-]
15/15/30	0.8	0.91	0.12	1.3	1.6	0.84	0.18	1.5	1.7	0.82	0.23	1.6	1.9
5/15/10	0.8	0.83	0.21	1.7	1.9	0.80	0.29	2.0	2.1	0.82	0.40	2.7	2.5

from the most critical wind direction. To obtain more realistic and economical wind loads the statistics of wind direction must be included in the design process. Davenport (1977) has suggested a wind direction reduction factor of 0,8. On this basis the values for $\overline{q}_{50} / q_{50,\varepsilon}$ and V_q should be adjusted to 0,86 and 0,185, respectively. Values for $\overline{c}_e / c_{e,s}$ and V_{c_e} are assumed to be 1,0 and 0,06 based on the three different roughness pattern considered. According to Davenport et al. (1985) $\overline{v}$ and V_v may be assumed to be 1,05 and 0,14, respectively. These tentative values are partly based on the results provided by the full scale Aylesbury experiment and its wind tunnel counterpart, and include also uncertainties in the wind climate estimation.

The partial load factor τ_p is chosen to be 1,5. This is in agreement with several wind load standards. The required safety index h is assumed to be 3. Often it is recommended that h = 3 should be retained for the routine design situations (Ravindra and Galambos, 1978).

4.3 Calculation of local peak wind pressures

Local peak wind pressures were calculated by means of the Equations 1, 5 and 21 and the magnitudes of the parameters given above. The results are described in terms of c_{ppeak}:

- quasi-steady approach:

$$c_{ppeak,qsa} = c_{p_{wt}} \cdot \left(1 + 2 \cdot 3{,}5 \cdot T_u(z)_{wt}\right) \quad , \qquad (22)$$

- simplified Cook-Mayne approach:

$$c_{ppeak,CaM} = \left(U_{c_{ppeak_{wt}}} + \frac{1{,}4}{a_{c_{ppeak_{wt}}}} \right) \qquad (23)$$

- second moment reliability approach:

$$c_{ppeak,SMR} = \frac{1}{\tau_p} \cdot \left(\frac{\overline{q}}{q_s}\right) \cdot \left(\frac{\overline{c}_e}{c_{e,s}}\right) \cdot \overline{v} \cdot \left(c_{ppeak,CaM} + \frac{0{,}5772}{a_{c_{ppeak_{wt}}}} \right) \cdot e^{\left(0{,}75 \cdot h \cdot \sqrt{(1+V_q^2)(1+V_{c_e}^2)(1+V_{c_{ppeak}}^2)(1+V_v^2)-1}\right)} \qquad (24)$$

- EC Wind loads:

$$c_{ppeak,EC} = c_{p_{EC}} \cdot \left(1 + 2 \cdot 3{,}5 \cdot T_u(z)_{EC}\right) \qquad (25)$$

The peak factor approach (Equation 9) was calibrated by

$$k = \frac{c_{ppeak,SMR} - c_{p_{wt}}}{c_{prms_{wt}}} \qquad (26)$$

The safety level of the peak wind loads specified in the EC Wind loads was calculated in terms of the safety index h by replacing p_{design} in Equation 21 by the wind pressure p_s from Equation 11.

4.4 Results

In Figure 4 local peak pressure coefficients for the stagnation point on the front wall are plotted for different model configurations and approach flows. It becomes clear that the local peak pressure coefficients increase from BL X (BL = boundary layer) to BL Z (see Figure 3) and with decreasing building height. Reasons are the increasing turbulence intensity from BL X to BL Z and with decreasing building height. Wacker and Plate (1992) confirmed that local peak wind pressures on the front wall may be predicted well by quasi-steady theory. Nevertheless, the magnitudes of c_{ppeak} are systematically lower than the values calculated by means of the simplified Cook-Mayne approach or the second moment reliability approach. This is due to the fact that $k = 3{,}5$ was used in the quasi-steady approach; $k = 3{,}5$ seems to be more appropriate when peak pressures averaged over a period of 1 sec rather than 0,5 sec are considered. However, for the design of small cladding elements, glass panels and roof pavers peak pressures averaged over a period of 0,5 sec are more useful. Therefore the peak factor used in the quasi-steady approach should be raised when used in combination with wind tunnel experiments. The application of Equation 26 to 18 different building - approach flow - configurations (see Wacker, 1993) showed that the peak factor k for the stagnation point on the front face of rectangular buildings varies between 3,8 and 5,4. The corresponding mean peak factor was calculated to be 4,2. Due to the consideration of parameter variability and model uncertainty the peak pressure coefficients calculated by means of the second moment reliability approach are slightly higher than those proposed by the Cook-Mayne approach, although in the second moment reliability approach the wind direction reduction factor was included.

In most cases the peak pressure coefficients specified in the draft of EC Wind loads (1992) are higher than the

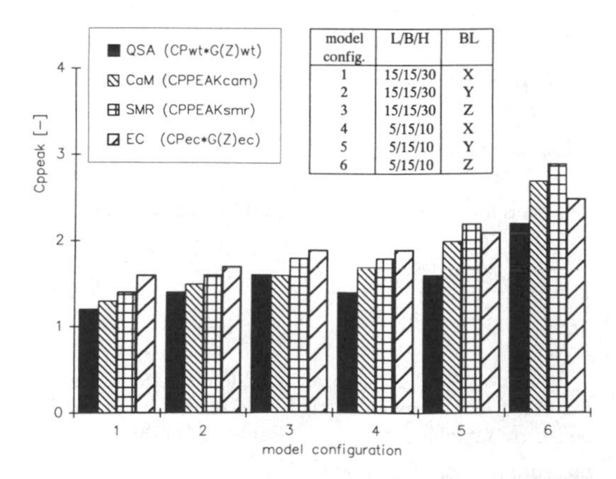

model config.	L/B/H	BL
1	15/15/30	X
2	15/15/30	Y
3	15/15/30	Z
4	5/15/10	X
5	5/15/10	Y
6	5/15/10	Z

Fig. 4 Local peak pressure coefficients at the stagnation point of the front wall of different building models and different approach flows; calculated by different approaches

values calculated with the second moment reliability approach, i.e. $h > 3$. An exception may be very low buildings in rough flow. The mean safety index was calculated to $h = 3{,}2$. In the EC a peak factor $k = 3{,}5$ is used to calculate the gust factor. From a physical point of view this value is too low (see above). This again suggests that in the EC the required safety level is provided by other specified parameters than the peak factor: higher turbulence intensities than in nature or conservative mean pressure coefficients.

In Figures 5 and 6 local peak pressure coefficients along the roof centerline (wind direction $\beta = 0°$) are plotted for different model configurations and approach flows. While a significant peak pressure recovery with increasing x/L occurs for the low building exposed to rough flow (see Figure 5, BL Z), this is not the case in all other cases shown in Figures 5 and 6 ($x/L < 0{,}5$). The results suggest that the pressure recovery of local peak pressures is - compared to the pressure recovery of the corresponding mean pressures - shifted further downstream. In particular for the lower of the two buildings under consideration and for flow over rough terrain (BL Z) there are major discrepancies between the results of the quasi-steady approach and those proposed by the simplified Cook-Mayne approach and the second moment reliability approach. Again the second moment reliability approach yields higher peak pressures than the Cook-Mayne approach. Reasons were given above. The results in Figures 5 and 6 indicate that local peak wind pressures in separated flow regions may be underestimated significantly when predicted by means of the quasi-steady approach. There are two main reasons: first, the used peak factor ($k = 3{,}5$ in agreement with EC Wind loads, 1992) is not an appropriate value, and second, building-generated contributions to the peak wind pressures in separated flow regions are neglected in the quasi-steady approach. This is confirmed by the calibration of the peak factor approach against the second moment reliability approach for 15 different building - approach flow - configurations (totally about 50 measuring points on the leading edge and mid region ($x/L < 0{,}5$) of the roof). The calculation resulted in a mean peak factor k of about 6,7. Concerning the magnitude of k there was no definite difference between leading edge and mid region ($x/L < 0{,}5$) of the roof.

The peak pressures in the mid region of the roof ($x/L < 0{,}5$) estimated with the quasi-steady approach are in good agreement with the values specified in the EC wind loading standard, whilst in the leading edge region EC values are significantly higher than the quasi-steady values calculated from wind tunnel experiments. Local peak wind pressures, which were calculated from wind tunnel measurements by means of the Cook-Mayne approach or the second moment reliability approach, show higher values than those proposed by the draft of the EC Wind loads (1992). This holds true in particular in the mid region ($x/L < 0{,}5$) of the roof, whilst in the leading edge region of the roof the calculated values often are lower than the specified EC values. This is due to the higher magnitudes of mean pressure coefficients specified in the code for this edge region. An exception may be very low buil-

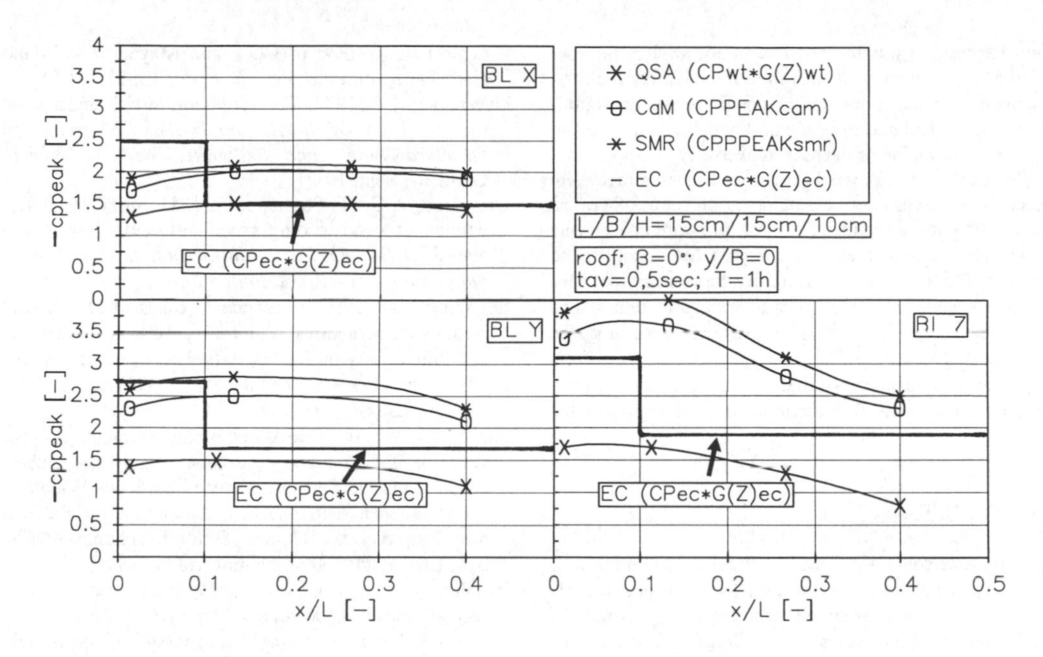

Fig. 5 Local peak pressure coefficients for the leading and mid region of the roof, calculated by different approaches; L/B/H = 15cm/15cm/10cm, different approach flows

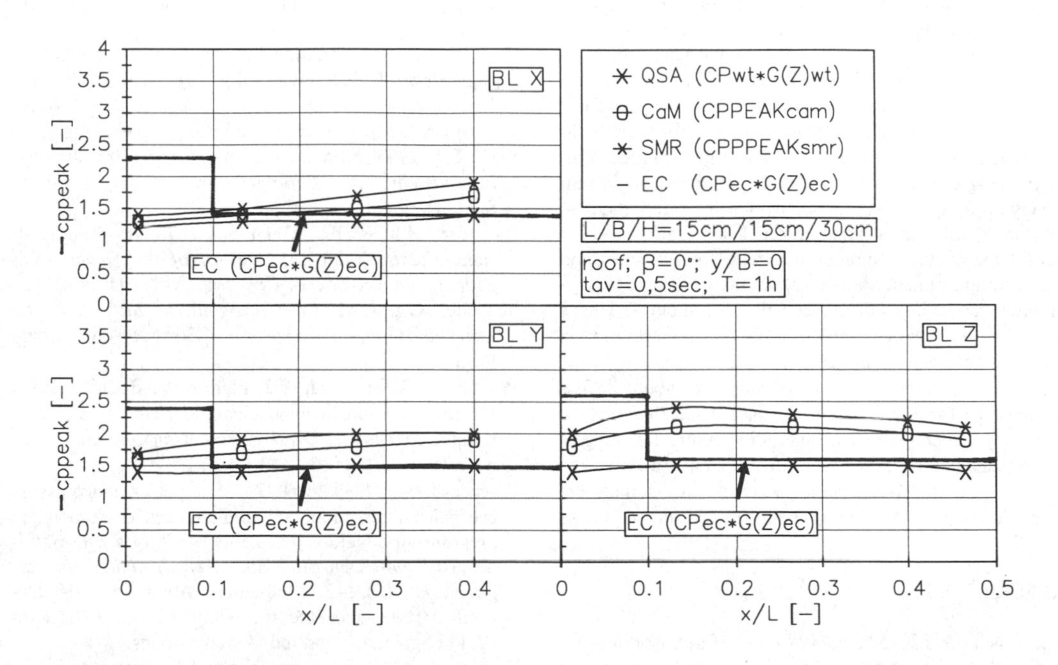

Fig. 6 Local peak pressure coefficients for the leading and mid region of the roof, calculated by different approaches; L/B/H = 15cm/15cm/30cm, different approach flows

dings exposed to rough flow. The safety level of the specified EC peak wind pressures in terms of safety indices - averaged over many measuring positions and configurations - was calculated to be about 4 and 1,5 in the leading edge and mid region of the roof, repectively.

The local roof peak wind pressures discussed here were selected from pressure measurements on the roof centerline for 0° wind direction. The peak pressure coefficients in the leading corner and edge region for oblique wind direction (45°), not shown here, are more negative. Moreover, for very low buildings with very large dimensions no wrong conclusions should be drawn concerning the peak wind pressure coefficients on the roof: if the roof is long enough, it is supposed that a strong recovery of peak pressure coefficients will occur in the mid region of the roof.

5 CONCLUSIONS

Some general conclusions may be drawn. Second moment reliability principles are a powerful tool to handle the variability of the parameters relevant for the assessment of local peak wind pressures on buildings. Application of simplified second moment reliability approaches and combination with wind pressure data from properly conducted wind tunnel experiments and with the simplified Cook-Mayne method provide local peak wind pressures on buildings with a consistent safety level. Accepting that reliability-based local peak wind pressures are most reliable, the results of the calculations in the example show that in separated flow regions local peak wind pressures may be underestimated significantly when applying the quasi-steady approach. Consequently, the comparison of the reliability-based local peak wind pressures with the corresponding local design wind loads given in the draft of the new Eurocode "Wind Loads", in which the quasi-steady approach is used, indicates that the safety index at least in the mid region of the roof is lower than 3 (h = 3 is considered to be a appropriate value for ultimate limit states design of usual buildings). In the leading edge region the mean safety index takes values of about 4. This is due to the higher mean pressure coefficients specified in the code for this region. At the stagnation point on the front wall the safety index takes values of about 3. The calibration of the peak factor approach against the method based on second moment reliability principles suggest mean peak factors of k = 4,2 for the front wall and k = 6,7 for separated flow regions when small cladding elements, glass panels and roof pavers are considered.

REFERENCES

Cook, N.J. & J.R. Mayne 1980. A refined working approach to the assessment of wind loads for equivalent static design. *Journal of Wind Engineering and Industrial Aerodynamics* 6: 125-1374.

Cook, N.J. 1982. Calibration of the quasi-static and peak-factor approaches to the assessment of wind loads against the method of Cook and Mayne. *J. of Wind Engineering and Ind. Aerodynamics* 10: 315-341.

Davenport, A.G. 1977. The prediction of risk under wind loading. *Proc. of the Second Intern. Conference on Structural Safety and Reliability*, held at Munich, Germany, Sept. 19-21.

Davenport, A.G., T. Stathopoulos & D. Surry 1985. Reliability of wind loading specifications for low buildings. *ICOSSAR'85, Proc. 4th International Conference on Structural Safety and Reliability.*

EC Wind loads 1992. Eurocode 1, Basis of Design and Actions on Structures, Volume 1, Actions on Structures, Part 2.7, Wind loads, Static and dynamic actions. *Draft of Project Team PT5*, February 1992, CEN/TC250/SC1/92/N57.

Fürniß, J. 1991. Extreme Winddrücke auf quaderförmige Gebäude und exemplarische Abschätzung des Sicherheitsniveaus der entsprechenden Eurocode-Windlastspezifikationen. Diplomarbeit, angefertigt im Rahmen des Teilprojektes B7 des Sonderforschungsbereichs 210, Universität Karlsruhe, unveröffentlicht.

Hasofer, A.M. & N.C. Lind 1974. Exact and invariant second-moment code format. *Journal of the Engineering Mechanics Division, Proceedings of the American Society of Civil Engineers*, Vol. 100, No. EM1.

Holmes, J.D. 1985. Wind loads and limit states design. *Civil Engineering Transactions*, The Institution of Engineers, Australia.

Lawson, T.V. 1980. *Wind effects on buildings, volume 1: design applications*. London, Applied Science Publishers.

Plate, E.J. 1982. Wind tunnel modeling of wind effects in engineering. In E.J. Plate (ed.), *Engineering Meteorology*, chapter 13, pp. 573-639, Amsterdam-Oxford-New York: Elsevier Scientific Publishing Company.

Plate, E.J. 1993. *Statistik und angewandte Wahrscheinlichkeitslehre für Bauingenieure*. Verlag Ernst & Sohn, Berlin.

Ravindra, M.K. & T.V. Galambos 1978. Load and resistance factor design for steel. *J. of the Struct. Div., Proc. of the Am. Soc. of Civ. Eng.*, Vol. 104, No. ST9.

Schueller, G.I. 1981. *Einführung in die Sicherheit und Zuverlässigkeit von Tragwerken*. Ernst & Sohn Verlag, Berlin.

Wacker, J., R. Friedrich, E.J. Plate & U. Bergdolt 1991. Fluctuating wind load on cladding elements and roof pavers. *Journal of Wind Engineering and Industrial Aerodynamics* 38: 405-418.

Wacker, J. & E.J. Plate 1992. Local peak wind pressure coefficients for cuboidal buildings and corresponding pressure gust factors. *Preprints of Second International Colloquium on Bluff Body Aerodynamics and Applications*, BBAA2, Melbourne, Australia, 7-10 December 1992, accepted for publication in Journal of Wind Engineering and Ind. Aerodynamics.

Wacker, J. 1993. Prognose lokaler Windlasten auf quaderförmige Gebäude durch Synthese von Windkanaluntersuchungen und probabilistischen Ansätzen. Sonderforschungsbereich 210, Universität Karlsruhe, zur Veröffentlichung.

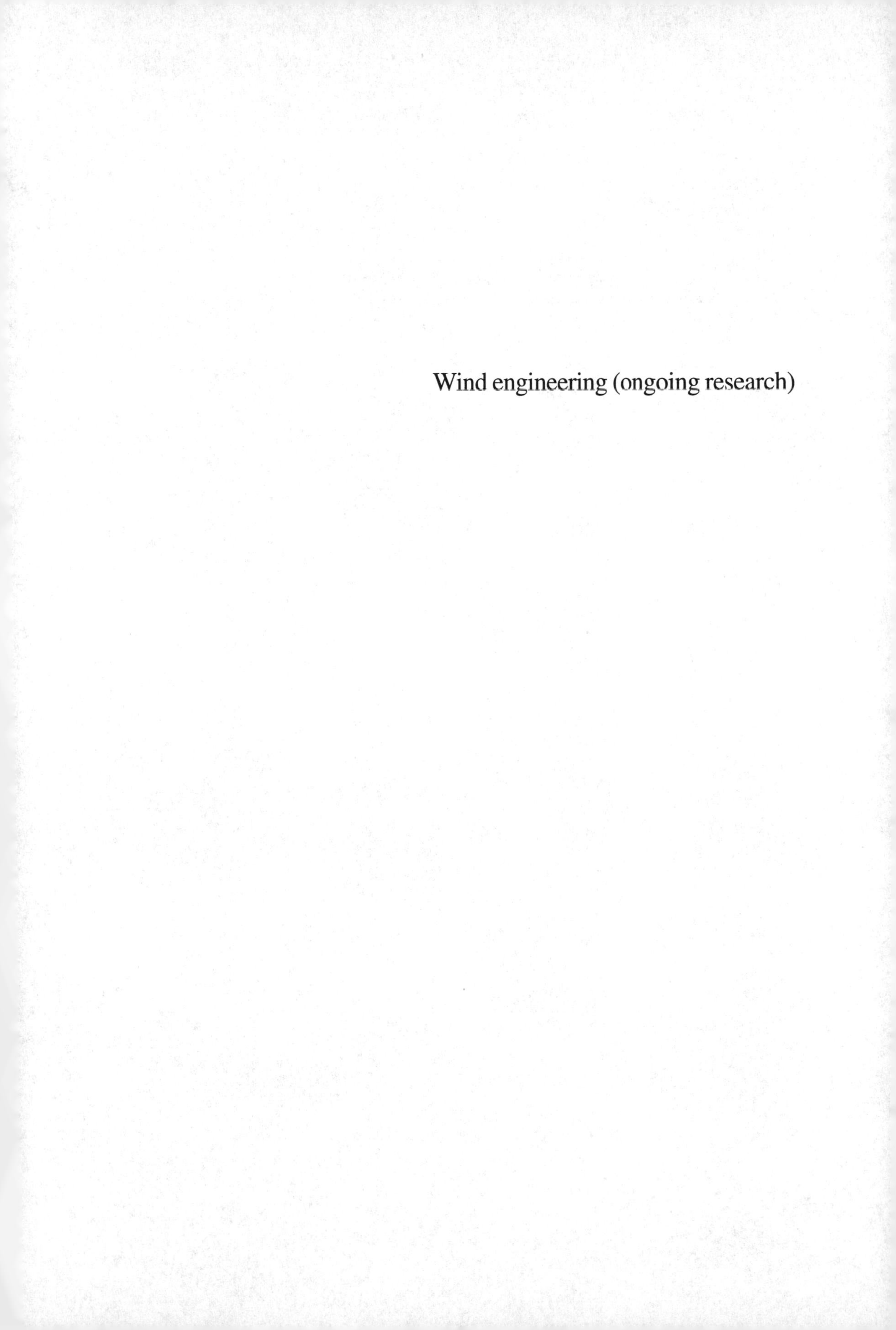

Wind engineering (ongoing research)

Structural Safety & Reliability, Schuëller, Shinozuka & Yao (eds) © 1994 Balkema, Rotterdam, ISBN 90 5410 357 4

Optimal damping of multi-story buildings under wind excitation

D. Bestle & W. Schiehlen
Institute B of Mechanics, University of Stuttgart, Germany

ABSTRACT: Wind excited vibrations of high-rise buildings can be reduced by installing additional damping devices like tuned mass dampers. The design of tuned mass dampers on different levels of the building is performed via a computer–aided modeling and design approach. A multibody system model is used for describing the dynamic behavior of the structure, and the problem of optimizing the parameters of the damping devices is formulated as a nonlinear programming problem. An application to a building of the University of Stuttgart shows that optimal designs with minimal accelerations of the building can easily be obtained.

1 INTRODUCTION

Wind excited vibrations of high-rise buildings impair the structural safety as well as the well–being of the residents. For reducing the vibrations to a tolerable size additional damping devices can be installed, see e.g. Hirsch (1983). As a matter of fact, tuned mass dampers are widely used to suppress vibrations of civil engineering structures. In most cases, the design of such tuned mass dampers is based on a single degree–of–freedom model of the structure on the basis of the first natural frequency, e.g. Fujino and Abe (1992).

For designing tuned mass dampers on different levels and taking into account the multi–frequency response of the building, the multibody system approach can be used. The dynamic behavior of the structure is then described by a parametrized model where the stiffness and damping coefficients of the tuned mass dampers can be chosen as design variables for optimizing the dynamic behavior of the building with respect to wind excitation.

2 MODELING AND IDENTIFICATION

The method of designing multiple tuned mass dampers is demonstrated for a building of the University of Stuttgart with fourteen stories. The eigenfrequencies of the first two bending modes have been measured as $f_1 = 0.76\ Hz$ and $f_1 = 2.56\ Hz$ (Luz and Wallaschek 1992).

A model being closely related to the structure of the building consists of fourteen rigid floors connected by elastic columns which result in forces depending on relative displacements of adjacent floors only (Luz 1991). Numerical studies, however, have shown that additional beams have to be included due to rather stiff cores like a staircase running through the whole height of the building (Obermüller 1992).

For optimization purposes the model can be reduced to a multibody system with four degrees of freedom by summarizing four and three stories to single bodies, respectively, Fig. 1. The inertia is then represented by four rigid bodies while the stiffness is modeled by absolute and relative springs. The absolute springs resulting in a moment proportional to the absolute rotation of a single body are related to the elastic columns in each story, the relative springs are related to stiff cores.

The masses M_i and moments of inertia I_i with respect to the center of gravity of each body can be found from mass distribution and geometrical data whereas the stiffness coefficients c_{abs} and c_{rel} may be identified via an eigenvalue analysis. It turns out that for this special combination of

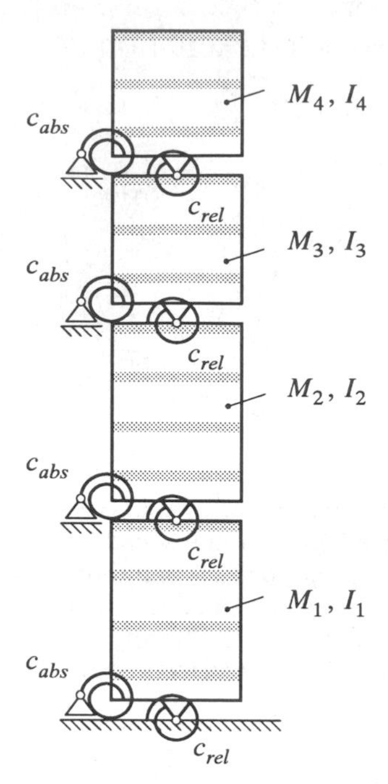

Fig. 1: Multibody system model of a 14–story building

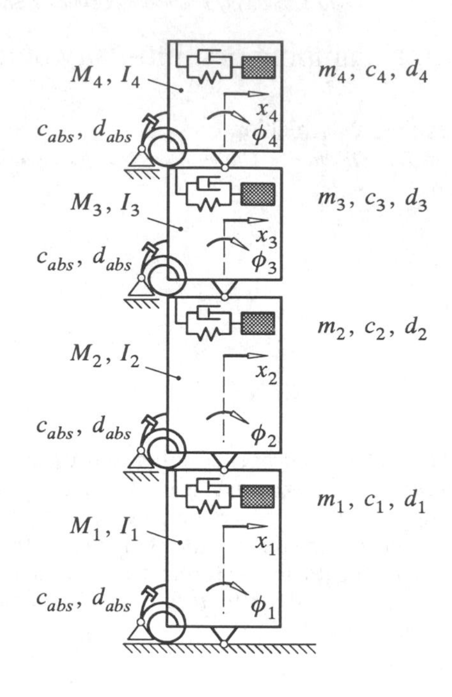

Fig. 2: Multibody system model with mass dampers

stories to rigid bodies the relative springs can be neglected (Reichert 1992). For considering damping effects absolute dampers in parallel to the springs are added.

Tuned mass dampers are most effective if they are mounted to the top of each body. The resulting model with eight degrees of freedom is shown in Fig. 2. The kinematics of the multibody system is described by absolute angles ϕ_i and relative displacements x_i of the tuned mass dampers which are summarized in a vector $\boldsymbol{y} \in \mathbb{R}^8$ of generalized coordinates. Due to small displacements the equations of motion can be linearized with respect to the equilibrium position (Schiehlen 1986):

$$\boldsymbol{M}(\boldsymbol{p})\,\ddot{\boldsymbol{y}} + \boldsymbol{D}(\boldsymbol{p})\,\dot{\boldsymbol{y}} + \boldsymbol{K}(\boldsymbol{p})\,\boldsymbol{y} = \boldsymbol{h}(t) \tag{1}$$

where the mass matrix $\boldsymbol{M}$, damping matrix $\boldsymbol{D}$, and stiffness matrix $\boldsymbol{K}$ depend on the design parameters $\boldsymbol{p}$.

Generation of equations of motion for such multibody systems can be automated by computer formalisms (Schiehlen 1990). In particular, a symbolic formalism like NEWEUL (Kreuzer and Leister 1991) offers great advantages in the subsequent optimization step.

3 OPTIMIZATION APPROACH

The comfort of the residents of a multi–story building may be evaluated by the horizontal accelerations of the individual stories. For the reduced model an integral type criterion can be formulated as

$$\psi(\boldsymbol{p}) = \int_0^T \sum_{i=1}^{4} w_i\, \ddot{y}_i^2 \, dt \tag{2}$$

where $\ddot{y}_i$ are the horizontal accelerations at the top of each body, w_i are weighting factors, and T is a finite time of interest. Via the equations of motion kinematic quantities like the accelerations are completely determined by the parameters of the model. Therefore, the objective function can be considered as a function of the design variables only.

The task of finding optimal values for the design variables may be formulated as a nonlinear programming problem which has to be solved in an iterative process starting with a given design. In each iteration step at least one function evaluation has to be performed which is a time–consuming numerical simulation of the dynamic behavior of the structure due to wind excitation. To reduce the number of iteration steps sequential quadratic programming (SQP) methods can be used, e.g. Fletcher (1987). The remaining problem with these algorithms is the requirement of gradients of the objective function with respect to the design variables.

In principal, there are two different approaches for computing gradients: numerical differentiation via finite differences and analytical sensitivity analysis via the adjoint variable approach (Haug 1987). Numerical studies have shown the adjoint variable method to be superior to numerical differentiation with respect to reliability, accuracy and efficiency (Bestle and Eberhard 1992). Therefore, the latter approach is used for optimizing the tuned mass dampers. Based on symbolical equations of motion the sensitivity analysis can be performed on the computer automatically.

4 NUMERICAL RESULTS

In the following, the tuned mass dampers are optimized with respect to the performance function (2) for typical wind loadings. If a single gust, Fig. 3, is applied to the building without additional damping devices it will respond with low–damped vibrations almost in its first eigenmode, Fig. 4. Installing tuned mass dampers and optimizing them with respect to the accelerations at the top of the building, i.e. $w_1 = w_2 = w_3 = 0$, $w_4 = 1$, helps to damp out long term vibrations drastically, Fig. 4.

If we choose all mass dampers to be identical, i.e.

$$m_1 = m_2 = m_3 = m_4\,,\quad c_1 = c_2 = c_3 = c_4\,,\quad d_1 = d_2 = d_3 = d_4 \qquad (3)$$

we have to minimize criterion (2) with respect to three design variables. It turns out that the mass of the additional damping devices has to be bounded for obtaining reasonable results. For upper bounds of 1%, 5% and 10% for the ratio of the total weight of the tuned mass dampers and the total weight of the building we find $\psi_{1\%}/\psi_0 = 0.33$, $\psi_{5\%}/\psi_0 = 0.17$, and $\psi_{10\%}/\psi_0 = 0.12$, respectively. In all cases, like in the 5%–case in Fig. 4, long term vibrations can be damped out whereas the first acceleration peak cannot be influenced very much by mass dampers.

If only the mass of the additional damping devices is chosen to be identical and the stiffness and damping coefficients are optimized individually, the additional increase of the performance is not significant: $\psi_{5\%}/\psi_0 = 0.15$, $\psi_{10\%}/\psi_0 = 0.11$.

In both cases, the comfort of the building is improved by increasing the weight of the mass dampers. This dependence changes for a more realistic broad band excitation, Fig. 5, where a local minimum of the performance function (2) with $w_i = 1$, $i = 1(1)4$, exists for the mass ratio

$$\frac{\sum m_i}{\sum M_i} \approx 2.25\% \qquad (4)$$

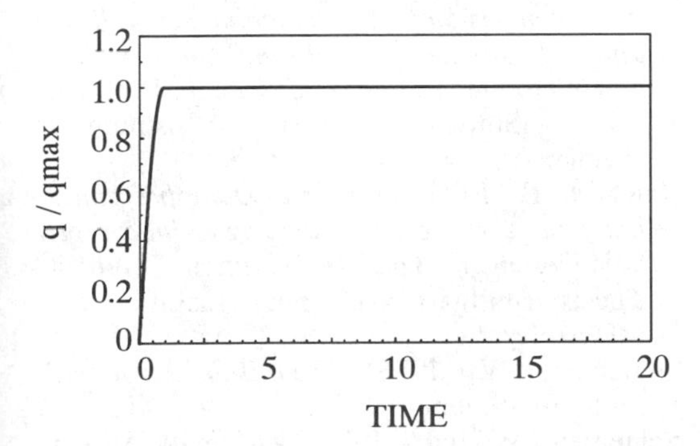

Fig. 3: Specific wind loading for a single gust

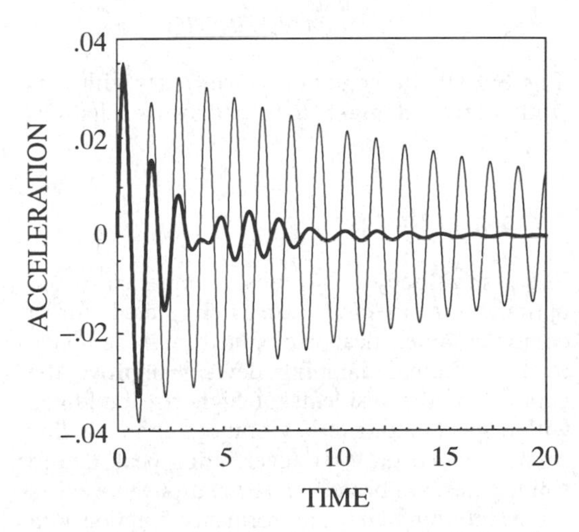

Fig. 4: Acceleration of the top of the building due to a single gust with (—) and without (—) tuned mass dampers

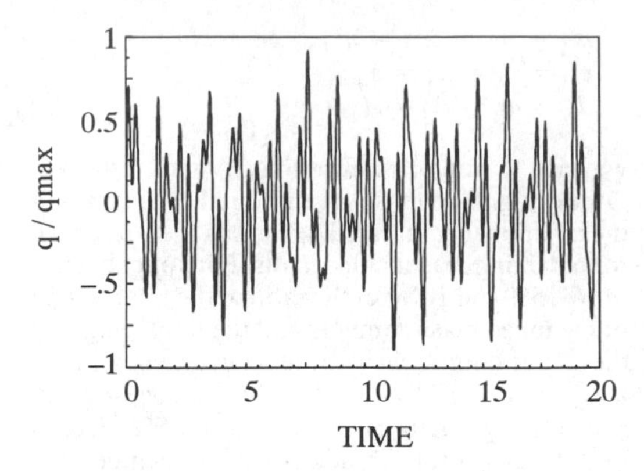

Fig. 5: Broad band excitation

as shown in Fig. 6. If the mass ratio is limited to less then about 10% optimization will automatically provide this optimal value.

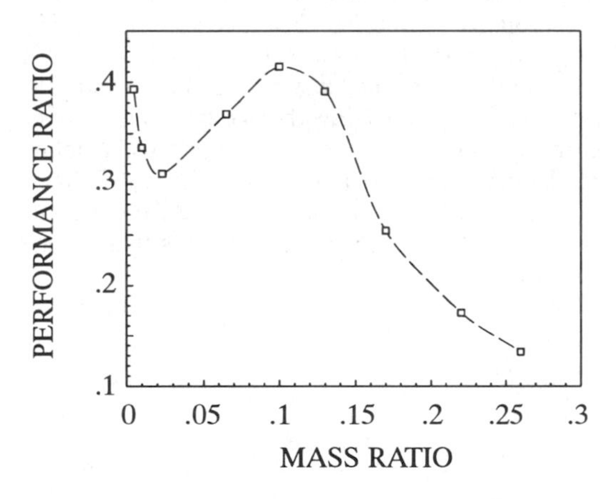

Fig. 6: Ratio of the performance values with and without mass dampers for broad band excitation

5 CONCLUSIONS

The multibody systems approach and optimization methods are well suited for a computer–aided design of tuned mass dampers. Such additional damping devices improve the comfort of the residents of high–rise buildings with respect to long term vibrations although they cannot reduce the first acceleration peak due to single gusts. For broad band excitation there exists a local minimum of the performance function with respect to the mass ratio of the total weight of the tuned mass dampers and the total weight of the building. Further numerical studies have to show the influence of excitation frequencies on this minimum.

REFERENCES

Bestle D. & P. Eberhard 1992. Analyzing and Optimizing Multibody Systems. *Mech. Struct. & Mach.* 20: 67–92.

Fletcher, R. 1987. *Practical Methods of Optimization*. Chichester: Wiley.

Fujino, Y. & M. Abe 1992. Dynamic Characterization of Multiple Tuned Mass Dampers. *Proc. of the 1st Int. Conf. on Motion and Vibration Control*: 176–181. Tokyo: The Japanese Soc. of Mech. Eng.

Haug, E.J. 1987. Design Sensitivity Analysis of Dynamic Systems. In C.A. Mota–Soares (ed.), *Computer–Aided Optimal Design*, p. 705–755. Berlin: Springer.

Hirsch, G. 1983. Herabminderung winderregter Schwingungen durch Dämpfungsmaßnahmen. In R. Frimberger & G.I. Schueller (eds.), *Beiträge zur Anwendung der Aeroelastik im Bauwesen*, p. 127–146.

Kreuzer, E. & G. Leister 1991: *Programmsystem NEWEUL'90*. Anleitung AN–24. Stuttgart: University, Institute B of Mechanics.

Luz, E. 1991. Ermittlung der Erdbebenbeanspruchung von Hochhäusern mittels einfacher mechanischer Modelle. In: *Publikation Nr. 5 der Deutschen Gesellschaft für Erdbeben–ingenieurwesen und Baudynamik*, p. 205–220.

Luz, E. & J. Wallaschek 1992. Experimental Modal Analysis Using Ambient Vibration. *Int. J. of Automatical and Experimental Modal Analysis*: 29–39.

Obermüller, J. 1992. *Berechnung der Erdbeben–ersatzlasten nach der Antwortspektrenmethode an einem räumlichen Hochhausmodell bei unterschiedlichen Einfallswinkeln der Erd–bebenbeschleunigung im Grundriß*. Diploma–Thesis. Stuttgart: University, Institute of Mechanics.

Reichert, E. 1992. *Optimale Dämpfungsmaß–nahmen zur Herabminderung winderregter Schwingungen von Hochbauten*. Student–Thesis. Stuttgart: University, Institute B of Mechanics.

Schiehlen, W. 1986. *Technische Dynamik*. Stuttgart: Teubner.

Schiehlen, W. (ed.) 1990. *Multibody Systems Handbook*. Berlin: Springer.

Structural Safety & Reliability, Schuëller, Shinozuka & Yao (eds) © 1994 Balkema, Rotterdam, ISBN 90 5410 357 4

Combination of wind and thermal loads on cooling towers

A. Flaga
Institute of Structure Mechanics, Cracow Technological University, Poland

ABSTRACT: The basic loads in cooling tower design are dead load g, wind load w and thermal load t Results of static calculation of a cooling tower at an independent treatment of thermal and wind loads permi to formulate the statement that the most unfarvourable combination of loads g, w and t is, as a rule, th combination of loads in case of a cooling tower operating in winter. For the winter period events maximu wind velocity V_{max} in winter and minimum air temperature T_{min} in winter are not simultaneous events no even totally independent also. Therefore, for the sake of structural safety the following two sets of rando events should be analysed: 1. Minimum air temperature T_{min} from winter and corresponding with it win velocity V, defined by the joint probability density function $p_1(t_{min}, v)$; 2. Maximum wind velocity V_{max} fro winter and corresponding with it temperature T, defined by the joint probability density function $p_2(v_{max}, t)$ To establish possible values of coefficients of coincidence for wind and thermal loads in a winter seaso adequate probabilistic considerations were carried out. Moreover, some examples of static calculation result of a cooling tower H = 150 m for different combinations of dead load, wind load and thermal load were als presented and analysed.

1 INTRODUCTION

The basic loads in cooling tower design - apart from dead (constant) load g - are variable loads from wind w and temperature t. In static-strength analysis of cooling towers generally three combinations of these loads should be analysed; these are: 1-g and w, 2-g and t, and 3-g, w and t. Whereas in cases 1 and 2 the loads g, w, and t should be taken with their extreme values, so in case 3 for wind and thermal loads so called coefficients of load coincidence (conjunction) should be applied, since these loads do not, as a rule, occur simultaneously with their extreme values. Thermal loads on cooling towers usually refer to four operation states: 1 - cooling tower in operation, summer season; 2 - cooling tower in operation, winter season; 3 - cooling tower in standstill, summer season; 4 - cooling tower in standstill, winter season. A separate problem, however, not analysed in this paper, is the possibility of a planned or caused by break-down switching a cooling tower off or into operation both in summer and winter.

As a matter of fact, the thermal fields in a cooling tower are unsteady fields, whereas, wind load is a dynamic load. For the sake of simplification of considerations equivalent substitutional loads were adopted in both cases and treated as static loads. The magnitude and distribution of these loads on both sides of the cooling tower shell were determined basing mainly on papers: Flaga (1991), Dulińska and Flaga (1992, 1992), Tarczyński (1989) and Stoffregen (1984). Moreover, the main characteristics of dead, wind and thernal loads were assumed according to the current Polish Standards: PN-82/B-02000, PN-82/B-02001, PN-77/B-02011 and PN-86/B-02015.

For four kinds of loads treated separately i.e. dead load g, wind load w and thermal load for a cooling tower working in winter (t_w) and cooling tower operating in summer (t_s), detailed static calculations of internal force distributions in the cooling tower shell were carried out for a cooling tower H= 150 m and geometrical parameters as in Figure 1. Two sample calulation results with reference to it are shown in Figures 2 and 3.

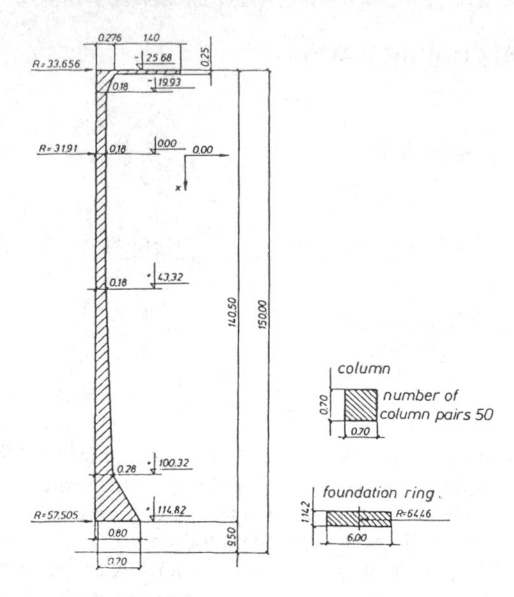

Fig.1. Cooling tower geometry assumed in calculations.

2 PROBLEMS OF WIND AND THERMAL LOAD COMBINATIONS IN COOLING TOWERS

The problems of combination of wind and thermal loads are subsequently given on an example of a cooling tower operating in winter. It results from observations of the Polish climate that maximum wind velocites are observed in the autumn and spring season, whereas, minimum air temperatures occur in winter. Moreover, in winter the periods of heavy wind are separated from periods of very low temperatures with a time lapse of about ten or more days. So, for the winter period events (V_{max}, T_{min}) i.e. the maximum wind velocity V_{max} in winter and minimum air temperature. T_{min} in winter are not simultaneous events not even totally independent also. Therefore, for the sake of structural safety the following two sets of random events (variables) should be analysed: 1. Minimum air temperature T_{min} from winter and corresponding with it wind velocity V (i.e.(T_{min}, V)), defined by the joint probability density function $p_1(t_{min}v)$; 2. Maximum wind velocity V_{max} from winter and corresponding with it temperature T (i.e.(V_{max},T)), defined by the joint probability density function $p_2(v_{max},t)$. Probability density distributions of wind velocity and air temperature in these two systems are presented

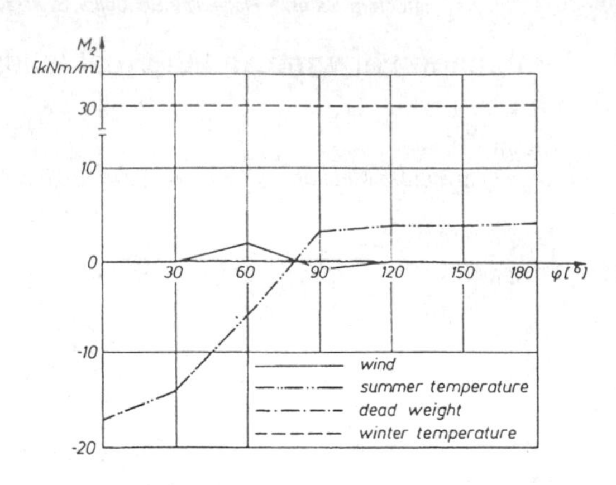

Fig.2. Comparison of circumferential bending moments M_2 in the lower zone of the cooling tower shell (100.32 m).

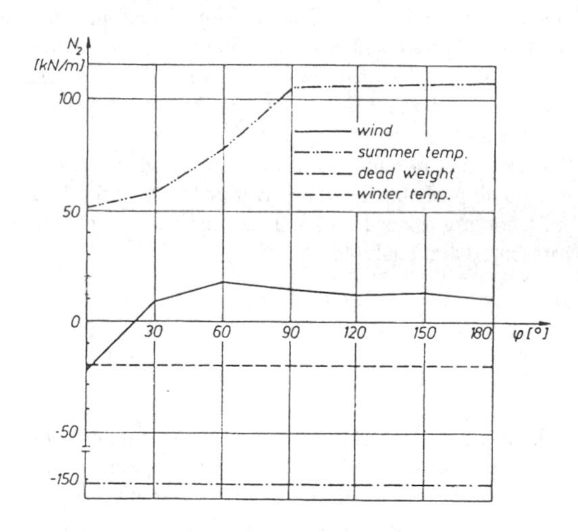

Fig.3. Comparison of circumferential forces N_2 in the lower zone of the cooling tower shell (100.32 m).

schematically in Figure 4. For these distribution functions at an a priori assumed small probability of exceeding $p \ll 1$, so that $p = P(V_{max}>v, T<t)$ or $p=P(T_{min} <t, V>v)$ we will obtain various sets of pairs of the magnitudes (v,t) and corresponding with them pairs of coefficients of coincidence (conjunction) and subsequently combinations of loads w and t. Probability p of exceeding the load combination (V_{max},T) or (T_{min}, V) concerns every arbitrary year (winter). Hence, the probability of not exceeding these combinations within the course of N years will be $W = (1-p)^N$ and the probabily of a reverse event $P = 1-W = 1-(1-p)^N$. Adopting the

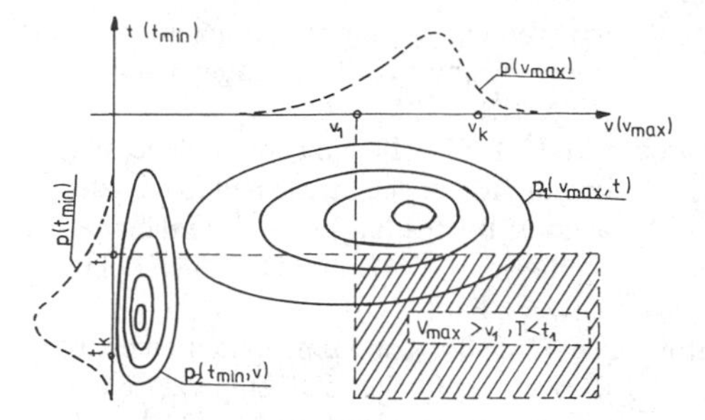

Fig.4. Probability density distributions of wind velocity and air temperature in winter.

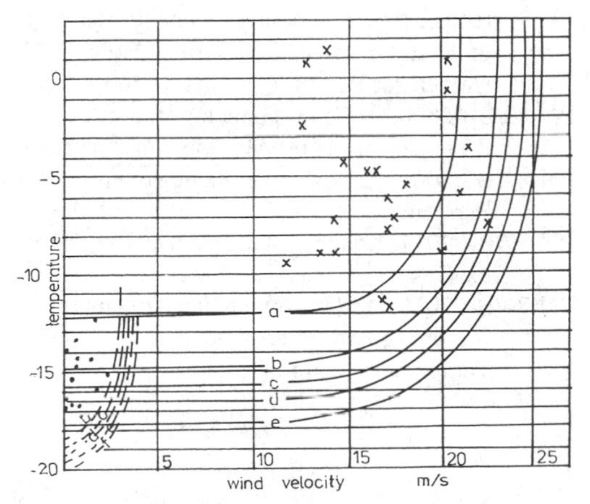

Fig.5. Measurement results of different probability levels p of exceeding the load combination (V_{max},T) or (T_{min},V) after Huang (1984).

serviceability period of a cooling tower L = N = 50 years and P = 0.05 (W=0.95) - as it is usually assumed for variable design loads - then will be p = 0.001. For characteristic loads p = 0.02 can be adopted (i.e. return period T_p= 1/p =50 years) and then P = 0.63. Typical measurement results in this aspect, elaborated statistically (approximating the real distribution by normal distribution) are shown after Huang (1984) in Figure 5. Curves a,b,c,d,e,f correspond respectively to the following levels of probability p: 0.05; 0,01; 0.005; 0.0025; 0.001; 0.0005. Adopting a respective level p (i.e. curve in Figure 5) an infinite set of pairs (v,t) will be obtained. On this basis respective pairs of coefficients of coincidence (ψ_w,ψ_t) for combination of loads w and t can be determined and so e.g.
- adopting: v = v_k= 20 m/s (v_k - characteristic wind velocity); t_k = -24°C (t_k - characteristic air temperature); p = 0,001 - from Figure 5 ψ_{wd}=v/v_k=1,0 and ψ_{td}=t/t_k ≅-15/-24= 0.625 will be obtained,
-adopting: v=v_k=20m/s; t_k=-24° C; p=0.02 - from Figure 5 ψ_{wk}=v/v_k=1.0 and $\psi_{t,k}$≅ −9/−24=0,375 will be obtained.

According to the current Polish Standards: PN-82/B-02000; PN-82/B-02001, PN-77/B-02011 and PN-86/B-02015 with respect to the pairs: dead and wind loads and dead and thermal loads the following combinations should be taken into account, namely:
- in the serviceability limit state combinations of characteristic loads

$$q_{k1} = g_k + w_k; \quad q_{k2} = g_k + t_k, \tag{1}$$

- in the ultimate limit state combinations of design loads

$$q_{d1} = q_d + w_d = \gamma_g g_k + \gamma_w w_k, \tag{2}$$
$$q_{d2} = q_d + t_d = \gamma_g g_k + \gamma_t t_k, \tag{3}$$

where load coefficients γ_g, γ_w and γ_t are as follows: γ_g=0.9 or 1.1; γ_w= 1.3; γ_t= 1.1.

Two samples of static calculation results for four different load combinations i.e. :
- combinations of design loads (p=0.001)
1. q_{d1}= 1.1 g_k+ 1.3 w_k; 2. q_{d2}=1.1 g_k +1.1 t_k;
3. q_{d3}= 1.1 g_k+1.0 w_k+ 0.625 t_k,
- combination of characteristic loads (p=0.02)

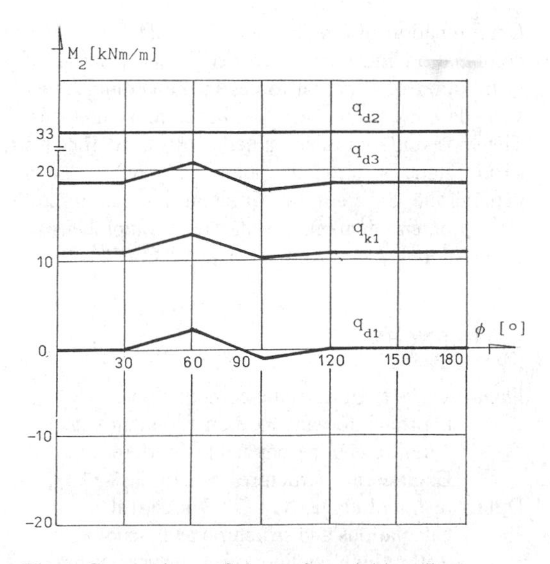

Fig.6. Comparison of circumferential bending moments M_2 in the lower zone of the cooling tower shell (100.32 m) for four different load combinations (note: t_k- for winter period).

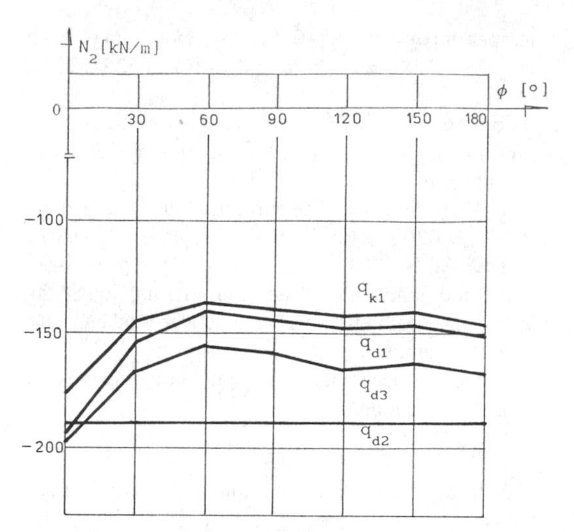

Fig.7. Comparison of circumferential forces N_2 in the lower zone of the cooling tower shell (100.32 m) for four different load combinations (note: t_k - for winter period).

4. $q_{k1} = 1.0\ g_k + 1.0\ w_k + 0.375\ t_k$
are presented in Figures 6 and 7.

3 FINAL REMARK

Determination of coefficients of coincidence at load combination must be preceeded by an analysis of distribution of internal forces in the cooling tower shell deriving from each of the component loads. The most unfavourable values of these coefficients (from static - strength point of view) will, in general, be different for different internal forces. This problem demand a separate , comprehensive elaboration.

REFERENCES

Flaga, A. 1991. Quasi-static computational approach to wind load on cooling towers in conditions of aerodynamic interference. Engineering Structures, vol.13: 317-328.

Dulińska, J. and Flaga, A. 1992. Numerical calculations and structural analysis of a 150 m high cooling tower for various types of wind load. Archives of Civil Engineering, XXXVIII, 3: 205-222.

Dulińska, J. and Flaga, A. 1992. Static analysis of the influence of thermal load on hyperboloidal reinforced concrete cooling tower. Archives of Civil Engineering, XXXVIII, 4: 301-322.

Tarczyński, L. 1989. The temperature load of cooling tower shell subjected to climatic actions. Proceedings of 3rd International Symposium on Natural Draught Cooling Towers, Paris: 187-196.

Stoffregen, U. 1984. Long time maesurements of the effects of water and temperature on cooling towers. Proceedings of 2nd International Symposium, Ruhr-Universitat Bochum, Germany: 487-500.

Huang, W. 1984. An evaluation method of load combination of windstorm and low temperature on cooling towers. Proceeding of 2nd International Symposium, Ruhr - Universitat Bochum, Germany: 466-473.

PN-82/B-02000: Action on building structures. Principles of the establishment of the values.

PN-82/B-02001: Actions on building structures. Permanent actions.

PN-77/B-02011: Loads in static calculations. Wind loads.

PN-86/B-02015: Actions on building structures. Variable environmental actions. Temperature action.

Structural Safety & Reliability, Schuëller, Shinozuka & Yao (eds) © 1994 Balkema, Rotterdam, ISBN 90 5410 357 4

Wind-induced dynamic response and the safety index

J.D. Holmes & Lam Pham
CSIRO Division of Building, Construction and Engineering, Australia

ABSTRACT: This paper considers the safety of structures subjected to dynamic response to wind loads. Resonant dynamic response results in wind load effects which vary with wind speed to a higher power than two. It is shown that provided a nominal design wind speed of high return period is adopted, the safety index is insensitive to the exponent of wind load variation.

1 INTRODUCTION

Previous studies (Pham *et al.* 1983; Holmes 1985; Holmes *et al.* 1985) have described approaches to the safety of structures for which wind is the dominant loading. Safety indices were calculated (Leicester 1985) to determine appropriate load factors for wind load effects. Separate studies for wind loads produced by tropical cyclones (i.e. hurricanes or typhoons) and by non-cyclonic windstorms were carried out (Pham *et al.* 1983). Approximately uniform safety was obtained when the wind load acts in combination with dead load of a range of magnitudes. Wind loading parameters (e.g. topographic and shelter effects, and pressure coefficients) obtained either from a direct application of a wind loading code or standard, or from special wind tunnel tests, were also considered separately (Holmes 1985). These studies showed that a single load factor for wind loads could be applied, provided that a nominal wind speed for ultimate limit states design of sufficiently low probability of exceedence (i.e. high return period) was chosen. In the case of the new Australian loading standards (Standards Australia 1989a; 1989b), a wind load factor of 1.0 and wind speeds with a nominal probability of exceedence of 5% in a 50-year lifetime have been adopted.

In the previous papers, it has been assumed that wind load effects are proportional to the square of the wind speed. This assumption is valid for the majority of structures for which the wind action is quasistatic. However, for structures such as towers and tall buildings, resonant dynamic response may be significant or even dominant. This results in wind load effects which vary with wind speed to a higher power than two. In the case of along-wind response, the fluctuating part of the response typically increases with wind speed to a power of 2.5 to 3. This is caused by an increase in the correlation of wind gusts producing excitation at the natural frequency of the structure, with increasing wind speed. For cross-wind response, which is often dominant for tall enclosed structures, the wind load effects may vary according to wind speed raised to a power of 3 to 4.

2 PROBABILISTIC MODELS

To enable a safety index to be computed, it is necessary to have probabilistic models of the wind loads and of the structural resistance (in the present paper, other loads have been disregarded). For the wind loads, the following model of wind loading has been adopted:

$$W = B\,V^{n} \qquad (1)$$

In equation (1), V is a reference wind speed which, in the case of the current Australian Standards, is a maximum gust, in a 50-year reference period, at a height of 10 m above ground level, in open country terrain.

B is a factor which incorporates all other wind loading parameters *except* the reference wind speed, i.e. velocity multipliers for terrain, topography, height, and for the conversion from a gust wind speed to a mean wind speed; it also includes turbulence intensity and structural size effects and, when resonant dynamic response is important, the spectral density of wind velocity fluctuations, and the structural and aerodynamic damping.

Equation (1), with a fitted value of n, is a sufficiently accurate model for the present purposes. n will have a value exceeding 2, and may exceed 3 in some cross-wind dynamic response situations. For this paper, n is treated as deterministic.

In the present paper, the following were assumed for the distribution type, mean/nominal value, and coefficient of variation, of structural resistance, R, the factor B, and the wind speed, V.

Table 1. Resistance and wind load models

	Distribution type	Mean/ nominal	Coeff. of variation
Resistance R	Lognormal	1.18	0.10
Wind load factor, B	Lognormal	0.75	0.4
Wind speed, V*	Extreme type I	1.05	0.11
Wind speed, V†	Extreme type I	0.85	0.11

* 'Working stress' nominal wind speed, i.e. 50-year return period (2% exceedence in 1 year).
† 'Ultimate limit state' nominal wind speed, i.e. 5% exceedence probability in a 50-year reference period.

The distributions and values of mean/nominal and coefficient of variation in Table 1 are similar to those used in the previous work (Pham *et al.* 1983; Holmes 1985; Holmes *et al.* 1985). The 'ultimate limit state' nominal wind speed is that used in current Australian Standards (Standards Australia 1989a; 1989b); the 'working stress' nominal wind speed is that used in earlier Australian Standards, and is equivalent to the nominal design wind speeds used in many other national wind loading codes.

3 SAFETY INDEX

Safety indices were calculated for the following limit state condition:

$$R = W \quad (2)$$

and for the following design criterion:

$$0.9R_N > \gamma_W W_N \quad (3)$$

where the value of the wind load factor, γ_W is taken as 1.5 when a 'working stress' wind load is used, and 1.0 when an 'ultimate limit state' wind load or wind speed is used. The subscript N denotes the nominal value.

Thus, dead and live loads were ignored in the present work. This is a reasonable assumption for load effects in structures which are dynamically sensitive to wind loads.

Safety indices, β, were computed using the 'advanced' method described by Leicester (1985), using the following definition of β:

$$p_f = \Phi(-\beta) \quad (4)$$

where p_f is the theoretical probability of failure, and $\Phi(\ )$ is the cumulative distribution function of a unit normal variate.

However, very close approximations to equation (4) can be obtained by using the following closed form:

$$\beta = \ln\{(\bar{R}/\bar{W})\sqrt{[(C_W^2+1)/(C_R^2+1)]}\} \Big/ \sqrt{\{\ln[(C_W^2+1)/(C_R^2+1)]\}} \quad (5)$$

where the overbar denotes the mean value, and C denotes coefficient of variation.

Equation (5) is exact for the case of R and W both being lognormal variables. Although the distribution of V in equation (1) was taken as Type I extreme for the present work, equation (5) also gives quite accurate approximations for that case. However, it should be noted that the commonly used alternative approximation:

$$\beta = \ln(\bar{R}/\bar{W})/\sqrt{(C_W^2 + C_R^2)} \quad (6)$$

does not give accurate results in the present case, due to the high value of C_W.

4 DISCUSSION

The computed safety indices as a function of the exponent, n, for the two alternative nominal wind speeds and wind load factors, are shown in Figure 1. Figure 1 clearly shows that when the 'ultimate limit state' nominal wind speed is used, the safety index is insensitive to the wind speed exponent, n. However, the level of safety reduces with increasing n when the traditional 'working

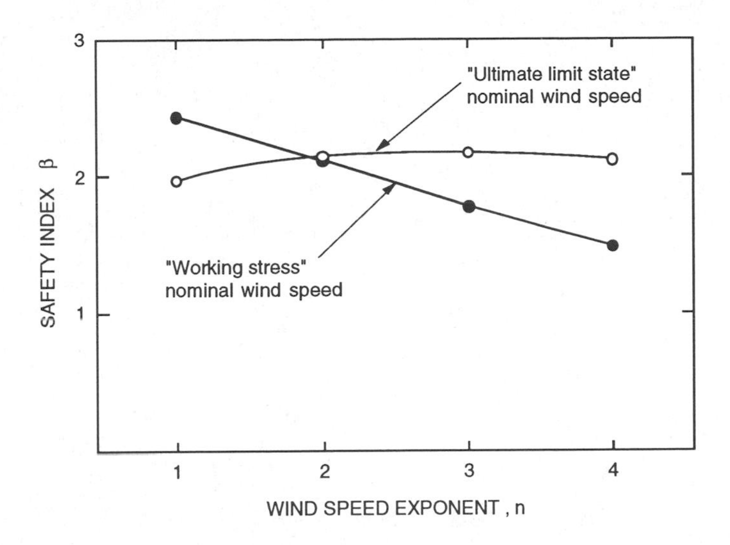

Figure 1. Safety index versus exponent, n.

stress', i.e. 50-year return period, wind speed is adopted. This reduction in safety could be avoided by using extra multipliers for dynamic response to wind, in a similar way to 'cyclone factors' or 'hurricane factors', and for the same reason – to counteract the greater rate of increase of wind load, or wind speed, with risk of non-exceedence (or return period). However such factors are not part of current practice, and a simpler approach is the use of an 'ultimate limit state wind speed' of high return period as the nominal wind, as used in Australia.

5 CONCLUSIONS

The paper has considered the safety of structures subjected to dynamic response to wind loads. It has been shown that provided a high return period nominal design wind speed is adopted, the safety index is insensitive to the exponent of wind load variation with wind speed. However, the use of the traditional 50-year return period wind speed can lead to large variations in safety unless special factors are applied to design wind loads acting on structures with significant dynamic response.

REFERENCES

Holmes, J.D. 1985. Wind loads and limit states design. *Civil Engineering Transactions, Institution of Engineers, Australia* CE27: 21–26.

Holmes, J.D., Pham, L. & Leicester, R.H. 1985. Wind load estimation and the safety index. *Proc. 4th International Conference on Structural Safety and Reliability, Kobe, Japan, 27–29 May 1985* Vol. 3: 131–139.

Leicester, R.H. 1985. Computation of a safety index. *Civil Engineering Transactions, Institution of Engineers, Australia* CE27: 55–61.

Pham, L., Holmes, J.D. & Leicester, R.H. 1983. Safety indices for wind loading in Australia. *Journal of Wind Engineering and Industrial Aerodynamics* 14: 3–14.

Standards Australia 1989a. *SAA loading code: Part 1 – Dead and live loads and load combinations: AS 1170.1–1989.* Standards Australia, North Sydney.

Standards Australia 1989b. *SAA loading code: Part 2 – Wind loads: AS 1170.2–1989.* Standards Australia, North Sydney.

Structural Safety & Reliability, Schuëller, Shinozuka & Yao (eds) © 1994 Balkema, Rotterdam, ISBN 90 5410 357 4

Random vibration of a galloping oscillator in wind

R.N. Iyengar
Indian Institute of Science, Bangalore, India

K. Popp
Institute of Mechanics, University of Hannover, Germany

ABSTRACT: A method is proposed to analyse the random response of a galloping oscillator in turbulent wind. The key step is a transformation of the well known homogeneous equation of motion to an inhomogeneous one, where the lateral turbulent velocity fluctuation occurs as external excitation which drives the system. A comparison of numerical and experimental results is shown.

1 INTRODUCTION

Studies of the response of prismatic structures to turbulent wind have important engineering applications. A type of self excited oscillation, called galloping has been extensively studied when the flow is uniform. Parkinson and Smith (1964) and Novak (1969) developed the quasi-steady theory for galloping, where the system is modelled as a nonlinear self excited oscillator. The nonlinearity arises due to the velocity dependent aerodynamic forces. The lateral force coefficient is measured on a static body in a wind tunnel such that the angle of incidence of the flow is the same as the relative wind direction during the oscillation of the body. When the flow is turbulent, it has been demonstrated that the quasi-steady assumption is still valid, provided the lateral force characteristic is measured in a wind tunnel with the corresponding level of turbulence. However, it is of considerable practical interest to know whether the smooth flow lateral force coefficients can still be used for arriving at useful solutions under the presence of turbulence. Previously Spanos and Chen (1981) have suggested that by replacing in the fluid dynamic force expressions, the uniform velocity by turbulent velocity terms, one can get plausible engineering approximations. Lindner (1992, 1993) has addressed this question and found encouraging results for some particular cross sections. In the present study, we consider the random vibration of the galloping oscillator in turbulent flow, following the above assumptions.

2 EQUATIONS OF MOTION

For a single degree-of-freedom system and galloping with velocity $\dot{y}$ under the effect of a uniform wind with velocity U, the quasi-steady theory postulates that the aerodynamic force is

$$F(t) = 0.5\, C_f\, \rho h L U^2 , \tag{1}$$

where the lateral force coefficient

$$C_f = \sum_{j=1,3...}^{7} a_j\, (\dot{y}/U)^j \tag{2}$$

is the one measured on a static body in the wind tunnel. Now under the effect of turbulence, we take the longitudinal flow velocity to be (U+u) and the cross velocity to be v, where u, v are zero mean stationary uncorrelated gaussian random processes. The galloping oscillator under the effect of turbulence can be modelled as

$$\ddot{y} + 2\xi\omega\dot{y} + \omega^2 y - (2n/h)(1+\epsilon w)^2\, S = 0,$$
$$S = \sum_j (a_j/U^{j-2})\, [(\dot{y}+v)^j/(1+\epsilon w)^j] . \tag{3}$$

Here, n is a mass parameter and h is the width of the section exposed to the flow. The longitudinal random component u(t) of the velocity is normalized as $u = \sigma_u w$, so that $\epsilon = \sigma_u/U$ is a small parameter called turbulence intensity. Thus, in eq. (3) the relation $u/U = \epsilon w$ has been used. When u = v = 0, the system is known to have a limit cycle oscillation. Now, under the effect of stochastic terms (u, v) the limit cycle will be perturbed. Since eq. (3) is not in a form which can be analytically handled a transformation is introduced. Let

$$(\dot{x}/U) = (\dot{y}+v)/(U+u) . \quad (4)$$

This gives

$$y = x - \int v dt + \epsilon \int w \dot{x} dt \quad (5)$$

and eq. (3) becomes

$$\ddot{x}(1+\epsilon w)+\dot{x}[2\xi\omega(1+\epsilon w)+\epsilon\dot{w}]+\omega^2(x+\epsilon\int w\dot{x}dt)$$

$$-(2nU^2/h)(1+\epsilon w)^2 \sum_j a_j (\dot{x}/U)^j = f(t) , \quad (6)$$

$$f(t) = (2\xi\omega v + \dot{v} + \omega^2 \int v dt) .$$

This equation clearly shows that the lateral turbulent velocity fluctuation v(t) drives the system as an external agency f(t), whereas the component u(t) arises as a parametric term which is of secondary importance in finding the amplitudes. It is remarkable that beside v(t) there appears also the derivated and integrated process on the right hand side of eq. (6).

3 ANALYTICAL CONSIDERATIONS

In the deterministic case, i. e. when $u = v = 0$, eq. (6) has a periodic solution of the type $x = R \cos(\omega t-\theta)$ as shown by Novak. Hence, it is reasonable to assume the solution in the stochastic case in the form (Manohar and Iyengar 1991)

$$x(t) = R \cos(\omega t-\theta) + z(t), \quad (7)$$

where R is the known deterministic limit cycle amplitude and θ is an arbitrary phase angle in $(0, 2\pi)$. In the steady state z(t) is taken as a gaussian stationary zero mean random process. Since θ is uniformly distributed in $(0, 2\pi)$, x(t) itself is strongly nongaussian. We find the variance and other properties of the unknown process z(t) using the method of equivalent linearization. The product terms like $\dot{x}w$, $\ddot{x}w$, $\dot{x}\dot{w}$ are linearized first, e.g.

$$\dot{x}w = A\dot{x} + Bw . \quad (8)$$

From error minimization we get

$$A<\dot{x}^2> + B<w\dot{x}> = <\dot{x}^2 w> ,$$

$$A<\dot{x}w> + B<w^2> = <\dot{x}w^2> . \quad (9)$$

If we take the joint probability density function p(z, w) to be gaussian the third order moments on the right hand side of eq. (9) will vanish indicating that in this order of approximation the above product terms can be neglected. However, we have still the term $[(1+\epsilon w)^2 \dot{x}^j]$ to be linearized. Now, since $(1+\epsilon w)^2$ is always positive, it appears reasonable to select the linear form as

$$(1+\epsilon w)^2\dot{x}^j = C_j\dot{x} . \quad (10)$$

Error minimization leads to

$$C_j = <(1+\epsilon w)^2 \dot{x}^{j+1}>/\sigma_{\dot{x}}^2 . \quad (11)$$

Now, with the help of eq. (7) and the assumed gaussian properties of z und w, all coefficients C_j can be found. For example

$$C_1 = 1+\epsilon^2(1+2R_2^2/\sigma_{\dot{x}}^2) ,$$

$$\sigma_{\dot{x}}^2 = (R^2\omega^2/2)+\sigma_2^2 , \quad \sigma_2^2 = <\dot{z}^2> ,$$

$$R_2 = <\dot{z}w> ,$$

$$C_3 = 3(\sigma_{\dot{x}}^2 - \frac{R^4\omega^4}{8\sigma_{\dot{x}}^2}) + 3\epsilon^2[\sigma_{\dot{x}}^2 - \frac{R^4\omega^4}{8\sigma_{\dot{x}}^2} + 2R_2^2 (1 + \frac{R^2\omega^2}{2\sigma_{\dot{x}}^2})] . \quad (12)$$

Similar expressions can be derived for C_5 and C_7. The cross correlation R_2 between $\dot{z}$ and w and the variance σ_2^2 of $\dot{z}$ are the two unknowns. Since the equivalent linearized equation for z(t),

$$\ddot{z} + 2\xi\omega\dot{z} + \omega^2 z - (2nU^2/h)\dot{z} \sum_j (a_j C_j/U^j) = f(t) , \quad (13)$$

can be solved in closed form, for the steady state moments, one can derive two nonlinear equations for the unknowns R_2 and σ_2. The solution of these simultaneously would provide an approximate random vibration result. As already pointed out w is a parametric excitation term and hence the correlation between $\dot{z}$ und w will be very small. Thus, if we further take $R_2 = <\dot{z}w> = 0$, considerable simplifications are possible. The response variance can be expressed as

$$\sigma_1^2 = <z^2> = \int_0^\infty S_{ff}(\Omega)|H(\omega,\Omega,\sigma_2^2)|^2 d\Omega, \quad (14)$$

$$\sigma_2^2 = <\dot{z}^2> = \int_0^\infty \Omega^2 S_{ff}(\Omega)|H(\omega,\Omega,\sigma_2^2)|^2 d\Omega, \quad (15)$$

where $S_{ff}(\Omega)$ denotes the single sided spectral density function of the excitation process f and H is the transfer function. Here, σ_2^2 is the unknown which can be found by solving eq. (15). As an alternative, the covariance analysis can be used as will be shown later. Further, σ_1^2 and σ_x^2 can be found. Finally, the variance of the response will be given by

$$\sigma_y^2 = \sigma_x^2 + \sigma_{v1}^2 - 2 <zv_1> . \tag{16}$$

Here, σ_{v1}^2 is the variance of $v_1 = \int v dt$, and $<zv_1>$ is a cross correlation term, both of which can be found using standard procedures of linear random vibration theory.

4 TURBULENCE MODELLING

Lindner (1993) has shown in wind tunnel experiments that the assumption of a homogeneous and isotropic turbulence with $\epsilon = \sigma_u/U = \sigma_v/U$ is justified, and the correlation functions R_{uu}, R_{vv} of the velocity fluctuations u, v can be approximated very accurately by

$$R_{uu}(\tau) = \sigma_u^2 \exp[-U|\tau|/L_x] , \tag{17}$$

$$R_{vv}(\tau) = \sigma_v^2 \exp[-U|\tau|/L_y] , \tag{18}$$

where L_x, $L_y = L_x/2$ denote the scale length of turbulence in longitudinal and lateral direction, respectively, and ϵ is known as turbulence intensity. The random process v with R_{vv} according to eq. (18) can be modelled by a first order shape filter

$$\dot{v}(t) = -\delta v(t) + \zeta(t) \ , \ \zeta(t) \sim (0,q) \ , \qquad \delta = U/L_y = 2U/L_x . \tag{19}$$

Here, $\zeta(t)$ is a zero mean stationary white noise process with intensity $q = 4\epsilon^2U^3/L_x$. The random process $v_1 = \int v dt$ can be gained from v(t) by a low pass filter,

$$\dot{v}_1(t) = -\frac{\delta}{N} v_1(t) + v(t) \ , \quad N >> 1 . \tag{20}$$

5 COVARIANCE ANALYSIS OF THE RANDOM VIBRATION

For an approximate covariance analysis of the galloping oscillator under turbulent wind the equivalent linearized eq. (13), where f(t) follows from (6), is combined with the filter eqs. (19) and (20) resulting in the forth order state equation

$$\underline{\dot{z}}(t) = \underline{A}(\sigma_2^2)\, \underline{z}(t) + \underline{b}\zeta(t) , \tag{21}$$

where

$$\underline{z} = [z \ \ \dot{z} \ \ v_1 \ \ v]^T$$
$$\underline{b} = [0 \ \ 1 \ \ 0 \ \ 1]^T$$
$$\underline{A} = \begin{bmatrix} 0 & 1 & 0 & 0 \\ -\omega^2 & -D & \omega^2 & (2\xi\omega-\delta) \\ 0 & 0 & -\delta/N & 1 \\ 0 & 0 & 0 & -\delta \end{bmatrix} \tag{22}$$

$$D = D(\sigma_2^2) = 2\xi\omega - (2nU^2/h) \sum_j (a_j C_j^*/U^j),$$
$j = 1, 3, 5, 7.$

Here, the coefficients C_j^* are taken as $C_j^* = C_j(\epsilon^2{=}o)$ for simplicity and read using the abbreviations $r = R^2\omega^2/2$, $s = \sigma_x^2 = r + \sigma_2^2$, cp. eqs. (11), (12) :

$$\begin{aligned} C_1^* &= 1 , \\ C_3^* &= 3s - 3r^2/(2s), \\ C_5^* &= 15s^2 - 45\, r^2/2 + 10\, r^3/s, \\ C_7^* &= 105s^3 - 315\, r^2 s + 280\, r^3 - 525\, r^4/(8s). \end{aligned} \tag{23}$$

These coefficients depend on the unknown quantity $\sigma_2^2 \equiv P_{22}$, which is the 2,2-element of the covariance matrix $\underline{P} = < \underline{z}\, \underline{z}^T >$. Thus, the covariance analysis of system (21) requires the iterative solution of the Ljapunov equation

$$\begin{aligned} &\underline{A}(P_{22}^{(i)})\underline{P}^{(i+1)} + \underline{P}^{(i+1)}\underline{A}^T(P_{22}^{(i)}) + \underline{Q} = \underline{0} , \\ &\underline{Q} = q\underline{b}\underline{b}^T,\ i = 0(1)n,\ \underline{P}^{(0)} = \underline{0}, \\ &\underline{P} := \underline{P}^{(n)} \text{ if } (P_{22}^{(n)} - P_{22}^{(n-1)})/P_{22}^{(n)} \le 10^{-6} . \end{aligned} \tag{24}$$

With respect to applications the quantity

$$y(t) = R\cos(\omega t - \theta) + \Delta(t) \tag{25}$$

is of interest. Here, $\Delta(t)$ is the random fluctuation of the limit cycle due to turbulence and reads using eqs. (5) and (7),

$$\begin{aligned} \Delta(t) &= y(t) - R\cos(\omega t - \theta) \\ &= x(t) - R\cos(\omega t - \theta) - v_1(t) \\ \Delta(t) &= z(t) - v_1(t). \end{aligned} \tag{26}$$

In our experiments $\ddot{\Delta}(t) \equiv a(t)$ has been measured and σ_a^2 has been determined. On the other hand, $\ddot{\Delta}(t)$ can be calculated regarding eqs. (13) and (19),

$$\begin{aligned} \ddot{\Delta}(t) &= \ddot{z}(t) - \ddot{v}_1(t) \\ &= -\omega^2 z(t) - D\dot{z}(t) + \omega^2 v_1(t) + 2\xi\omega v(t) \\ &= \underline{c}^T \underline{z}(t) , \\ \underline{c}^T &= [-\omega^2 \ \ -D \ \ \omega^2 \ \ 2\xi\omega] . \end{aligned} \tag{27}$$

Thus, σ_a^2 can easily be calculated using the covariance matrix $\underline{P}$, cf. eq. (24),

$$\sigma_a^2 = <\ddot{\Delta}^2> = <\underline{c}^T \underline{x}\, \underline{x}^T \underline{c}> = \underline{c}^T \underline{P}\, \underline{c} . \qquad (28)$$

6 EXPERIMENTAL AND NUMERICAL RESULTS

Experiments have been performed on a galloping oscillator consisting of a square prism (length $L = 0.92$ m, width $h = 0.1$ m, mass $m = 8.97$ kg, circular frequency $\omega = 30.78$ rad/s, damping ratio $\xi = 0.003$, mass parameter $n = 3.09 \cdot 10^{-4}$) under turbulent cross flow in a wind tunnel. The lateral force measured in smooth flow can be approximated by a polynomial of order seven, where the corresponding coefficients read

$$a_1 = 3.5 \; ; \; a_3 = -130.3 \; ; \; a_5 = 2858.2 \; ; \; a_7 = -18157. \qquad (29)$$

Table 1: Experimental and numerical results

	ϵ [%]	L_x [m]	U [m/s]	σ_a^2 [m^2/s^4] exp.	σ_a^2 [m^2/s^4] num.	σ_2^2 [m^2/s^2] num.
I	2.3	0.0491	10.5	1.24	0.34	0.06
II	4.1	0.098	10.6	1.24	0.79	0.19
	6.9	0.0729	9.4	1.45	0.37	0.42
	9.8	0.064	8.7	0.63	0.13	0.73
III	6.5	0.132	11.4	1.50	2.35	0.55
	8.5	0.122	11.5	1.61	2.24	0.96
	11.5	0.105	10.6	1.63	0.02	1.49
	15.5	0.094	12.6	2.55	13.40	3.80
	18.0	0.090	15.0	3.36	135.23	7.22

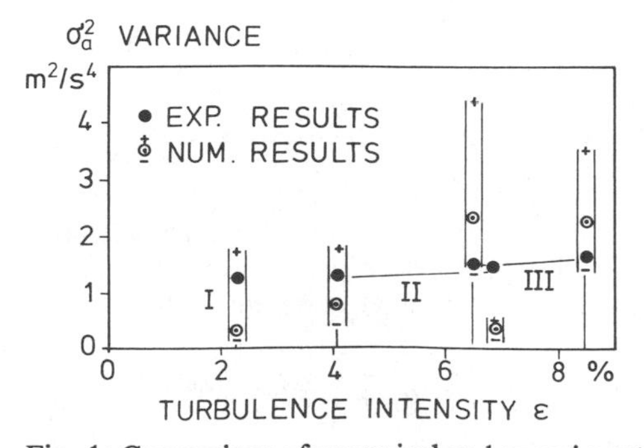

Fig. 1: Comparison of numerical and experimental results

The turbulence has been generated by grids made of bars with different width (I, II, III). The experimental set-up and the measurement devices are found in Lindner (1993). At certain wind speeds U galloping vibration occur, where U has been adjusted in such a way, that the average vibration amplitude was $R = 0.03$ m = const for different turbulence intensities ϵ. Also numerical calculations have been carried out using the same system parameters and data ϵ, L_x, U, R as in the experiments. The corresponding results are listed in Table 1 and shown in Fig. 1. The results are sensitive against changes in U, thus, in Fig. 1 numerical limits have been plotted for U (1 ± 0.06). It can be seen that most of the experimental results are within the calculated margins for turbulence intensities $\epsilon \lesssim 10$ % .

CONCLUSIONS

It has been shown, that the wellknown non-linear homogenous equation of motion of a galloping oscillator in turbulent wind can be transformed into an inhomogenous one, where the lateral turbulent velocity fluctuation drives the system. On this basis an analysis has been carried out. A comparision with some experimental results shows an overall good agreement for turbulence intensities smaller than about 10 %. However, further research is necessary to find out the limits of the theory shown.

REFERENCES

Lindner, H. 1992. Simulation of the turbulence influence on galloping vibrations. J. Wind Engrg. and Industr. Aerodyn.: 41-44

Lindner, H. 1993. Untersuchungen zum Turbulenzeinfluß auf die Galloping-Schwingungen rechteckiger prismatischer Körper. Dr.-Ing. Diss., Inst. of Mechanics, University of Hannover

Manohar, C. S. and Iyengar, R. N. 1991. Narrowband random excitation of a limit cycle system. Arch. Appl. Mech., Vol. 61, 2

Novak, M. 1969. Aeroelastic galloping of prismatic bodies. ASCE, J. Eng. Mech. Divn., Vol. 95, EMI: 115 - 142

Parkinson, G. V. and Smith, J. D. 1964. The square prism as a nonlinear oscillator. Quart. J. App. Math., Vol. 17: 225 - 259

Spanos, P. and Chen, T. W. 1981. Random response to flow-induced forces. ASCE, J. Eng. Mech. Divn., Vol. 107, EM6

Structural Safety & Reliability, Schuëller, Shinozuka & Yao (eds) © 1994 Balkema, Rotterdam, ISBN 90 5410 357 4

The fully correlated gust: A new concept for separating a high-frequent gust loading process and a low-frequent, spatially correlated gust load

M. Kasperski & H.-J. Niemann
Building Aerodynamics Laboratory, Ruhr-Universität Bochum, Germany

ABSTRACT: A new method is presented to split the gust process into a low-frequent component - the fully correlated gust which is defined as a static load in the sense of a stochastic trend - and a high-frequent dynamic component. The new averaging time is considerably shorter as the classical averaging time of 10 minutes. The fully correlated gust is defined as the instantaneous velocity profile corresponding to an averaging time T showing nearly simultaneous peak values over a certain range of heights. The instantaneous velocity distributions are identified by use of the VCI-method (Velocity Correlation Identification). As input, a theoretical description of the turbulence is used. On the example of flat open country, new averaging times of 20 to 60 seconds are found. Based on full-scale experiments, the identification method is verified by a comparison to a time domain identification.

1 INTRODUCTION

In modern wind load theory, the load process due to gusts is defined as the high-frequency component of wind speed fluctuations with periods less than 600 or 3600 seconds, respectively. This concept has been derived from meteorological observations, dividing the wind into the mean component related to macro-scale fluctuations and the wind turbulence which is handled as a stochastic process. With this concept, a more precise calculation of the responses of a linear system becomes possible in principle, although the amount is large due to the great dimensions of the load process. For non-linear systems, this concept is hardly practicable. Especially for non-linear calculations, usually, rough simplifications are introduced to describe the complex wind load, leading to doubts in both the reliability and the ability of interpretation of the so obtained results.

On the other hand, most structures experience the gust load quasi-statically and linearly. So from this point of view, there is no need for a full dynamic and non-linear calculation within the 10-minute-average. However, the classical concept does not allow to separate that part of the wind load which is really transmitted resonantly or non-linearly.

A new method has been developed to split the gust process in a fundamental different way into a low-frequent component which is defined as a static load in the sense of a stochastic trend and a high frequent dynamic component. The new averaging time is considerably shorter as the classical averaging time of 10 minutes. Among others, the great advantage of this approach lies in a practicable calculation of the responses of a non-linear system in the time domain.

2 IDENTIFICATION METHOD

The fully correlated gust is defined as the instantaneous velocity profile corresponding to an averaging time T showing nearly simultaneous peak values over a certain range of heights. The instantaneous velocity distributions are identified by use of the VCI-method (Velocity Correlation Identification), which is similar to the LRC-method [1]. As input, the mean values, the standard deviations and the correlations are needed for averaging times in the range of 10 minutes to 1 second. They are obtained by low-pass filtering of the cross-spectral densities of the turbulent flow which are adopted from a theoretical turbulence model [2].

For flat, open terrain, the input parameters are as follows:

mean velocity profile

$$\bar{u}(z) = (z/10)^{0.16}$$

turbulence intensity

$$I_u(z) = 0.187 \cdot (z/10)^{0.16}$$

spectral density function

$$\frac{S_u(f) \cdot f}{\sigma_u^2} = \frac{4 \cdot f'}{[1 + (8.409 \cdot f')^2]^{5/6}}$$

where f' - normalized frequency

$= f \cdot {}^xL_u(z)/\bar{u}(z)$

${}^xL_u$ - integral length of u-component in x-direction

coherence in vertical direction

$$R(\Delta z, f) = \frac{2}{\Gamma(5/6)} \left[\left(\frac{\mu}{2} \right)^{5/6} \cdot K_{5/6}(\mu) - \left(\frac{\mu}{2} \right)^{11/6} \cdot K_{11/6}(\mu) \right]$$

$$\mu = 0.747 \frac{\Delta z}{L(f)} \left[1 + 70.8 \left(\frac{f \cdot L(f)}{u_m} \right)^2 \right]^{1/2}$$

$$L(f) = \begin{cases} 2 \cdot {}^zL_u(z) & \text{for } f \le 0.008 \text{ Hz} \\ 2 \cdot {}^zL_u(z) \cdot 0.04/f^{2/3} & \text{for } f > 0.008 \text{ Hz} \end{cases}$$

$$u_m = [\bar{u}(z_1) \cdot \bar{u}(z_2)]^{1/2}$$

where K_n are the Besselfunctions of second type and order n:

$$K_n = \frac{I_{-n}(x) - I_n(x)}{\sin(n\pi)}$$

$$I_n(x) = (\tfrac{1}{2}x)^n \sum_{k=0}^{\infty} \frac{(\tfrac{1}{4}x^2)^k}{k!\,\Gamma(n+k+1)}$$

The statistical parameters of the remaining dynamic gust are obtained by a corresponding high-pass filtering. They are the new data base for the numerical generation of the load time series.

The extreme velocity distributions are obtained as:

$$u(z, z_{ref}, T) = \bar{u}(z) + g(z, T) \cdot \sigma_u(z, T) \cdot r(z, z_{ref}, T)$$

where $\bar{u}(z)$ - mean value of velocity at height z above ground

z_{ref} - reference height showing the extreme velocity

T - averaging time

g - peak factor

σ_u - standard deviation of the velocity

r - correlation coefficient

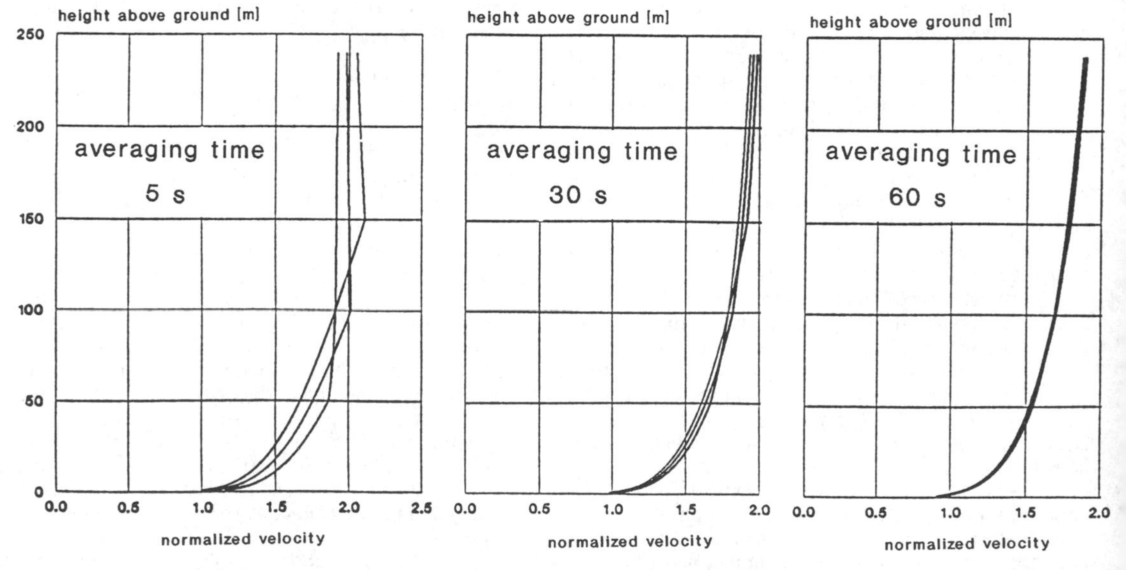

Fig. 1: Identified instantaneous extreme velocity distributions for flat open country

The so estimated velocity profiles are the most frequent velocity distributions for an extreme velocity occurring at reference height z_{ref}. Furthermore, they are very close to instantaneous extreme velocity profiles, if the probability density of the process is more or less gaussian. This can be shown by a comparison to velocity profiles obtained from a time domain identification using results from a full-scale experiment (s.a. chapter 3).

The basic gust is defined to be nearly independent from the reference height. Using in the turbulence model the parameters for flat, open country as described above, the basic gust distribution is identified to correspond to an averaging time of 20 seconds to 60 seconds relating to a range of heights from 50 to 150 m (fig. 1).

For the 60s gust, the correlation/coherence is considerably decreased to about 50% of the original level . Furthermore, the amplitude of the remaining dynamic gust is reduced to about 50%. For the base-moment of a 150m high cantilevered beam, the ratio of the static to dynamic response now becomes 4.9 compared to 1.4 with the classical separation using the 10min mean.

The remaining high-frequent gust now may be described as follows:

spectral density function

$$S \cdot f/\sigma_u^2 = 0.074 \cdot f'^{-2/3}$$

coherence:

$$R(\Delta z) = 1 - \mu \cdot \Delta z^{2/3}$$

using the reduced integral length which is 50% of the original one.

So, the accuracy of the resonant and/or non-linear part of the response is improved due to an improved description of the loads.

The newly defined averaging time has to be checked especially in respect to the following two restrictions:

- the remaining gust should have statistically weakly stationary characteristics

- the basic gust should not induce considerable dynamic effects.

Otherwise, the initial conditions have to be identified or the averaging time has to be increased, respectively.

3 VERIFICATION OF THE IDENTIFICATION MODEL

The full-scale data are obtained from velocity measurements at the mast Gartow II [3]. During a strong storm at the beginning of 1990, the velocities are sampled for 30 minutes with a frequency of 10 Hz for 21 channels (17 velocities, 4 wind directions). The mean values of the velocities and the turbulence intensity of the resulting wind velocity are shown in figure 2.

The time domain identification of the extreme velocity distributions uses time series obtained as true moving averages over different gust duration (1 - 300 seconds). For each reference height, the extreme velocity distribution is sampled. To compare the theoretical model, the statistical parameters (standard deviation and correlation) are calculated for each averaging time.

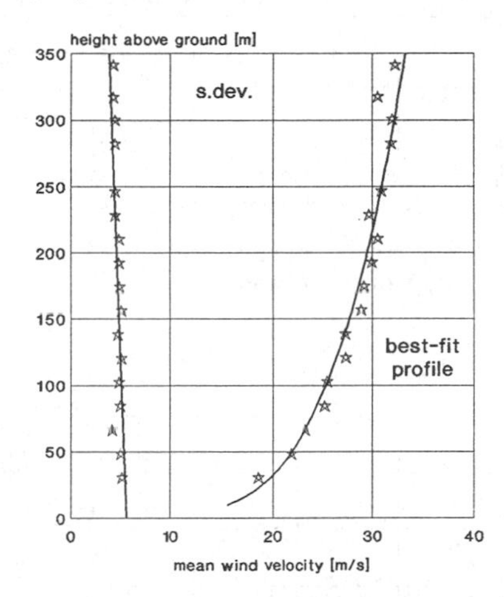

Fig. 2: Mean wind velocity, turbulence intensity

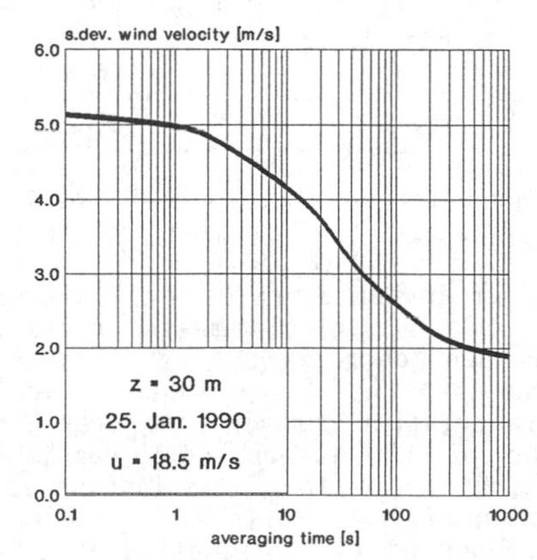

Fig. 3: Standard deviation for different averaging times

Figure 3 shows the standard deviation of the velocity fluctuations for the lateral component depending on the averaging time. Even for 5 minutes averaging time, there is a considerable fluctuation of the mean values which are due to a low-frequent trend. This trend will influence the probability density to deviate considerably from the supposed gaussian model. Hence, as a preparing step, a trend removal becomes neccessary.

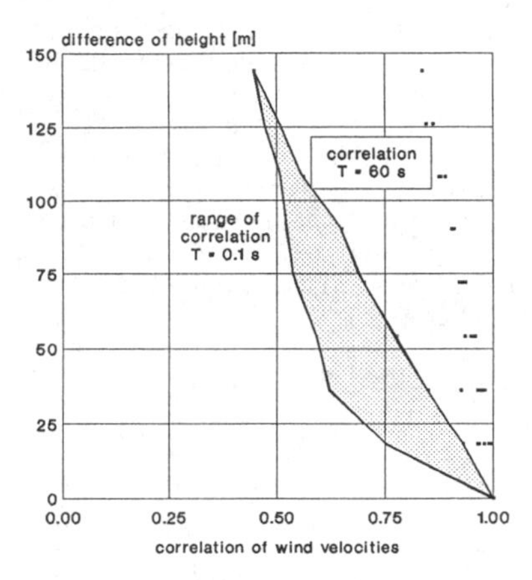

Fig. 4: Correlation structure for different averaging times

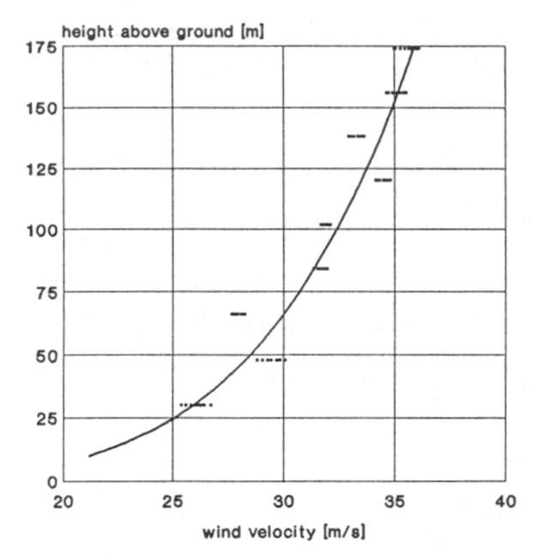

Fig. 5: Instantaneous velocity profiles compared to a best-fit Hellmann-profile

As trend, a moving average of 120 seconds is used. This averaging time is identified by a sudden decrease of the standard deviations of the cross wind velocity fluctuations. After removing the trend, the comparison of the theoretically estimated distributions to instantaneous distributions is satisfactory. In figure 4, the dependancy of the correlation structure on the averaging time is shown. For the new averaging time, the correlation depends only on the relative distance. In figure 5, the best-fit profile of the fully-correlated gust is compared to instantaneous extreme velocities for heights up to 175 m. The scatter of the instantaneous velocities (9 samples of extreme distribution) tends to be small. However, the Hellmann-profile is not the best approach to the instantaneous velocity distribution in this case.

4 CONCLUDING REMARKS

On the example of flat open country, a new averaging time of 20 to 60 seconds is proposed for separating the static and the dynamic wind action. A verification of the identification method based on full-scale measurements on the 344 m high guyed mast Gartow II shows a sufficient agreement between the velocity-correlation-identification and sampled extreme velocity distributions.

For a non-linear calculation, time series of the remaining dynamic gust have to be generated. Based on the shorter averaging time, this can be done more efficiently and accurately since the statistical parameters needed as input are described more precisely.

5 ACKNOWLEDGEMENTS

The authors wish to thank Professor Dr. Peil and Dr. Nölle from the University in Karlsruhe for offering their full-scale measurements to check the reliability of the identification model. Part of this work was sponsered by the Deutsche Forschungsgemeinschaft in the scope of a special research programme 'Dynamics of Structures' which is greatfully acknowledged.

6 REFERENCES

[1] KASPERSKI, M.: Extreme wind load distributions for linear and non-linear design, Engineering Structures 1992

[2] ESDU - Engineering Science Data Unit: Characteristics of atmospheric turbulence near the ground, ITEM 85020, Oct.1985, ITEM 86010, Oct. 1986

[3] PEIL, U; NÖLLE, H.: Wind velocity measurements on the mast Gartow II, University Karlsruhe, January 1990

Structural Safety & Reliability, Schuëller, Shinozuka & Yao (eds) © 1994 Balkema, Rotterdam, ISBN 90 5410 357 4

On fatigue of guyed masts due to wind load

U. Peil
Institute for Steel Structures, Technical University of Braunschweig, Germany

H. Nölle
Institute for Steel Structures, University of Karlsruhe, Germany

ABSTRACT: Lifetime predictions of guyed masts are rather complicated due the complexity of the stochastic wind process, the nonlinearities of the dynamic system behaviour and due to uncertenties of the prediction of fatigue of lokal details. The way of calculation is briefly demonstrated.

1 INTRODUCTION

Several tall guyed masts have collapsed in the past few years all over the world. In Germany inspections on a large number of guyed masts were carried out to detect damage. These checks showed damage on a large number of masts due to fatigue, especially at the connection of ropes with the mast shaft (Scheer 1985). In England, fatigue cracks were found in corner leg joints of guyed masts (Lambert 1987). The fatigue is caused by natural wind. In general there are a number of wind induced excitation mechanisms, like drag- and lift instabilities, vortex excitations and buffeting exitations due to the gusty wind. In this paper the influence of wind buffeting on the fatigue of masts is investigated.

Prediction of fatigue damage or life time must take into account the influence of

- longitudinal and lateral gusts at all wind speeds
- system response
- fatigue of the local detail

It is briefly demonstrated how to calculate the fatigue state and some important problems are discussed.

2 WIND LOAD

The wind load is a stochastic process. Nearly all investigation into wind effects on buildings look at the behaviour of the wind under extreme stormy conditions which are rare events. Fatigue problems usually arise due to the large number of lower wind speed conditions. The prediction of lower wind speed processes is complicated due to thermal convection effects of the atmosphere. The influence of non neutral atmosheric conditions on the wind profile is large. Fig. 1 shows a comparision of measured hourly mean wind profiles and profiles calculated by a logarithmic law. The continuous line represents the usual logarithmic law for the site and a measured wind speed at 10m height. The cross marked points represents measurements. The circle marked points represents a logarithmic law with an additional stability term which describes thermal effects (Troen 1990):

$$v(z) = \frac{v^*}{\kappa} \cdot \left[\ln\left(\frac{z}{z_0}\right) - \psi\left(\frac{z}{L}\right) \right]$$

ψ is an empirical stabilty function which describes the influence of thermal effects.

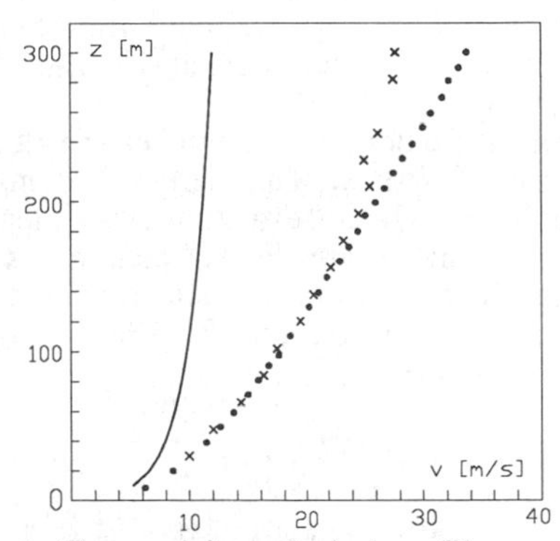

Fig.1 Comparision of wind profiles

It is obvious that the usual logarithmic law underestimates dramatically the real wind speed distribution over height while the extended logarithmic law describes the wind profile adequately. It should be noted that the usual logarithmic law or the exponential Hellmann law are restricted to a neutral atmosphere which only occurs in extreme gale conditions with very high wind speeds. The turbulence characteristic of lower wind speed depends on the thermal effects as well. Fig. 2 shows the rms values over height of the measured wind velocities. The distribution of the rms over height is not a constant function as often assumed.

A theoretical calculation of thermal effected wind profiles is possible (Troen 1990). The stability term of the extended logarithmic law depends on the Obukhov-Length L, which is a function of the vertical and horizontal wind speed gradient and temperature gradient of the atmosphere. Normally these input functions and their probability of occurence over time are unknown.

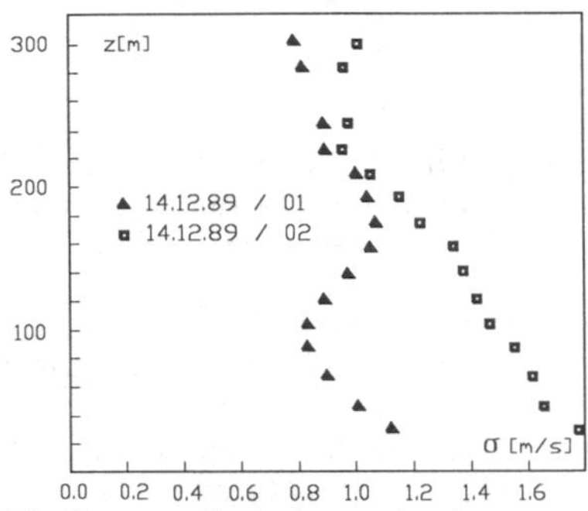

Fig 2. rms of wind speed

In a research project we started to measure the profiles of wind speed, the rms and their ocurrence over time up to the height of 341m. The profiles are classified using neural networks. After some time we hope to be able to present wind speed profiles, rms-distributions and probabilities of ocurrence. The measurements are performed at a 344m mast. The wind speed, temperatures and directions and the mast response are measured at different levels in small distances (fig. 3), (Peil 1991, 1992).

Our measurements of the wind process and mast response also show that there is a remarkable response in the lateral direction of the mean wind even in symmetric situations. Calculations, which do not take account of this effect, underestimate the variance of the response and the damage state of the structure.

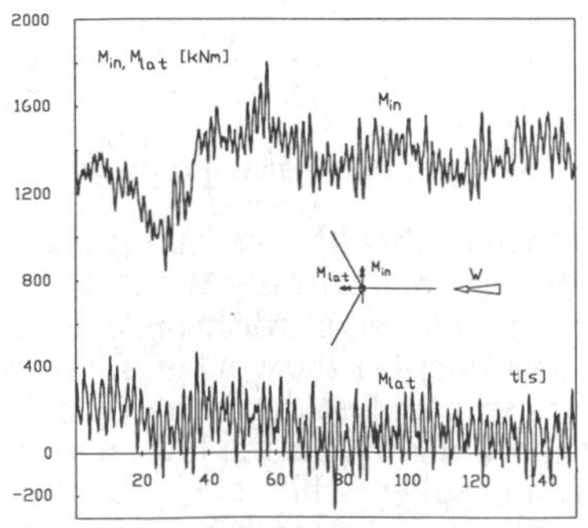

Fig 3. Inline and across bending moments

The lateral response may have different causes:
-lateral turbulence of the wind
-galloping excitations
-energy transfer between close eigenfrequencies
-chaotic motion of the guys.

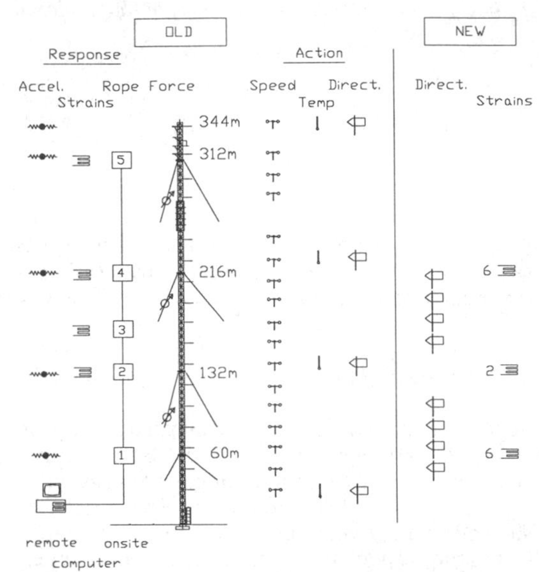

Fig.4 Extended Measurement Equipment

The influence of galloping excitation, energy transfer and chaotic motion is studied theoretically. The lateral component of the turbulent wind speed due to rotating eddies is measured with an extended measurement equipment. The lateral component of the wind speed is measured by directionmeters of short wavelenght (3m) which are installed clusterwise in distances of 18m. Additionally the windforces which act directly at the guys are measured by calibrated strain gauges implemented at the rope connection plate (fig.4). The influence of local oro- and topography is taken into consideration following the method of the European-Wind-Atlas (Troen 1990), which contaminates the global wind process with the local surface effects, using a calculation of a threedimensional potential flow over the local oro- and topography.

3 DYNAMIC RESPONSE

To determine the response in terms of the local stresses the aerodynamical and mechanical transfer functions must be known. Guyed masts show a number of difficulties in estimating the

mechanical transfer function. This is due to
-Nonlinearity of the guys behaviour
-Nonlinearity due to 2nd order theory
-Nonlinear dynamic motion of the guys
-Damping behaviour of guys and mast shaft

Due to the nonlinearities nonlinear time-step calculations are necessary. If the nonlinearities are not too high, a linearized calculation is sufficient, if the amplitudes are small (Peil 1991,1992). The linearisation must take into account:

-linearisation around the operating point
-dynamics of the ropes
-equivalent damping.

A comparision of measured and calculated mast responses shows that the agreement between measured and calculated eigenfrequencies is good if the dynamic behaviour of the ropes is taken into account in a linearized manner. Measured and calculated resonance amplitudes do not fit well however. This is due to the nonlinear behaviour of the guys with large resonance amplitudes. The nonlinearities of the guy behaviour can be taken into account in a linearized calculation, using a frequency dependent higher damping decrement for the guys (Peil 1992, 1992).

Within a small frequency band near the resonance peak the system shows chaotic motions (Petersen 1992, Peil 1992). Although the symmetric system is symmetrically excited chaotic unsymmetric motions take place as shown in fig. 5. Fig. 6 shows a Poincaré-map of the motion.

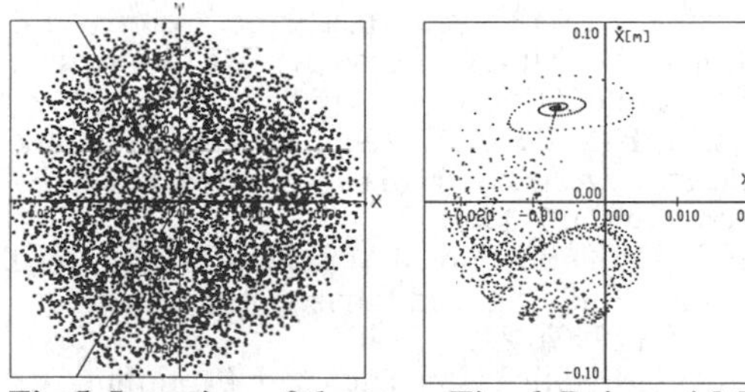

Fig.5 Location of the top   Fig. 6 Poincaré-Map

Because of the complexity of the problem full scale measurements are necessary to verify the theoretical calculations. System identifications are performed to get precise information about the dynamic behaviour of guyed masts. For dynamic excitation either the natural wind is used or the mast is excitated by cutting a pretensioned auxiliary rope. To take account of the nonlinear effects different rope forces were choosen. The method performing dynamic exitations using unbalanced masses is complicated and costly because big masses are necessary to induce low frequency vibrations.

If the wind is used as a dynamic force a sufficient number of anemometers has to be installed to be able to describe the microstructure of the gusty wind. This procedure is used for the system identification of our measurement mast (fig.4) (Peil 1992).

Usually measurement equipment as described above is not available. In this case the second procedure, cutting a pretensioned auxiliary rope, is used. The frequency content of an excitation in this manner is characterised by high amplitudes at the lower frequency domain, where guyed masts have the most important eigenfrequencies. Fig. 7 shows a comparison of measured and calculated amplitude spectra of a 200m mast in Heidelstein/Rhön (Peil 1993).

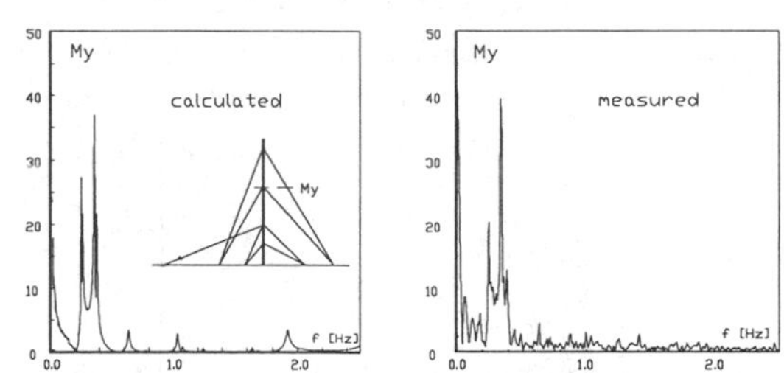

Fig. 7 Measured and calculated amplitude spektra (auxiliary rope)

4 DAMAGE DETERMINATION

With the (linearized) identified system an accurate calculation of the system response can be performed. Using the random vibration theory with the linearized mechanical transfer function of the identified guyed mast it is possible to calculate the rms-response of the structure. For the determination of the damage state an incremental cumulative damage rule is used, with the collective of the interesting response cycles (double amplitudes) as input. The collective of the interesting response of a stochastic Gaussian process is known only for the case of a narrow band process. In this case the amplitudes are Rayleigh distributed, the magnitude and numbers of response cycles are known. For non narrow band processes, like the response processes of guyed masts with a lot of eigenfrequencies, the distribution of the response and the number of cycles are not known. To get information about the distribution of the response, measured strains-samples are classified using the rainflow method. Fig. 8 shows a typical strain time signal and the fig. 9 the distribution of response cycles.

To take account of local stress concentrations in a damage calculation the so called hot-spot-concept is used. The local stress peaks are determined using the Finite-Element-Method (ANSYS). A fine mesh should be chosen near the critical points to get the real stress peaks (Peil 1993).

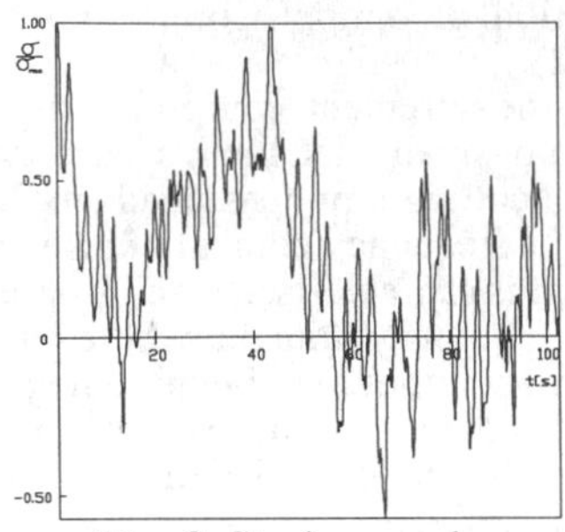

Fig. 8 Strain sample

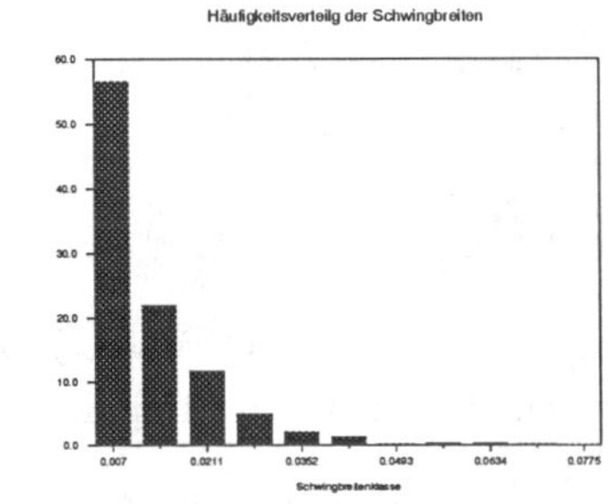

Fig. 9 Distribution of stress cycles

The damage calculations of the connections of the rope to the mast shaft is much more complicated, because the nonlinearity of the ropes is not neglectible. To dermine the dynamic rope support reactions of ropes hit directly by the wind, nonlinear time step calculations with correlated wind time histories are performed. Fig. 10 shows the procedure, fig. 11 the hours of occurence in a year of different transverse guy forces, depending on the angle the wind attacks from.

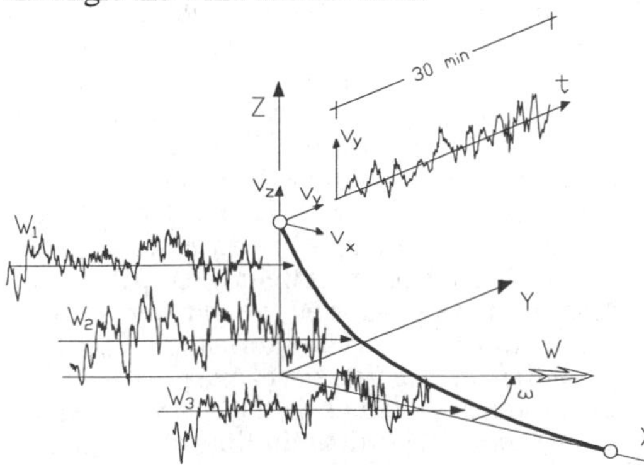

Fig. 10 Calculation of guy transverse forces

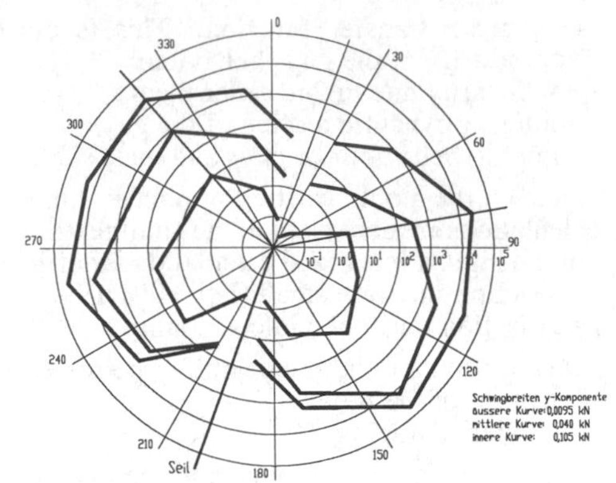

Fig. 11 Probabilty of transverse rope forces

5 ACKNOWLEDGEMENTS

The financial support of the "Volkswagen-Foundation", the "Deutsche Forschungsgemeinschaft" and the German Telekom is gratefully acknowledged.

REFERENCES

Lambert,M., M.H.Ogle, B.W.Smith: Die Untersuchung windinduzierter Ermüdung in hohen abgespannten Stahlmasten. Proc. IASS-Working Group No. 4, London Ontario, 1987.

Peil,U., H.Nölle: Measurement of Wind Load and Response of a Guyed Mast. Proc. Europ. Conf. Struct. Dynamics, Eurodyn '90, Rotterdam, Brookfield, (1991).

Peil,U., H.Nölle, Z.H.Wang: Dynamisches Verhalten abgespannter Maste. VDI-Berichte Nr. 978, (1992), 1-19.

Peil,U., H.Nölle: Guyed Masts under Wind load. Journ. Wind Eng. and Industr. Aerodynamics ', 41-44, (1992), 2129-2140.

Peil,U., H.Nölle: Gutachten zur Ermittlung der Lebensdauer des Funkmastes Heidelstein / Rhön. unveröffentlich, 1993.

Scheer,J., Peil,U.: Diverse Gutachten zu Fragen der Standsicherheit abgespannter Maste. unpublished, 1985.

Petersen,C.: Chaotische Taumelschwingungen abgespannter Maste. Stahlbau 61 (1992), 179-185.

Troen,J., E.L.Petersen: Europäischer Windatlas, Risø National Laboratorium, Roskilde, Denmark, 1990, including the program WASP.

Structural Safety & Reliability, Schuëller, Shinozuka & Yao (eds) © 1994 Balkema, Rotterdam, ISBN 90 5410 357 4

Reliability analysis of the towers in the transmission line Rio Grande-S.J.do Norte crossing in southern Brazil

J.D.Riera & J.L.D.Ribeiro
CPGEC/DECIV, Universidade Federal do Rio Grande do Sul, Porto Alegre, Brazil

V.R.da Silva, W.O.Simão & R.C.Ramos de Menezes
Companhia Estadual de Energia Elétrica, Porto Alegre, Brazil

ABSTRACT: An experimental program to measure the actual structural response of a 132m high reinforced concrete tower of a transmission line in southern Brazil is presented. The program also aims to confront observed data with original design assumptions in order to provide information on model uncertainties.

1 INTRODUCTION

A complete reliability assessment of one of the towers of the Rio Grande - São José do Norte crossing is presently under development. This evaluation is taken place simultaneously with the implementation of an experimental program aimed at the verification of the structural response to wind action. The crossing comprises a 120m high concrete tower with a 12m high steel truss on top, which supports the cables at one side of the 1050m long main span. Figure 1 shows schematically the crossing. The tower at the other end of the main span, as well as the anchor towers, are conventional steel trusses and will not be dealt with in this work.

2 EXPERIMENTAL PROGRAM

The steel rebars at elevation of 10m above ground level were already instrumented with resistance strain gages to measure the total bending moments in the longitudinal and transverse directions. The wind force per unit tower length at elevation 76m in both directions will be measured by means of a pneumatic averaging technique and recorded in two separate channels. The effectiveness of the averaging technique and its frequency response have already been studied in laboratory experiments. In this project, by distributing the taps (Figure 2) so that the product of the pressure at each point times its area of influence is proportional to the projection in the desired direction of the resultant force in the area, the total force in half the section will be measured. Cable force transferred to the tower will also be measured and recorded. Figure 3 shows schematically the setup of the equipment, which includes a wind velocity registering station at the top of the steel truss that allows the continuous recording of mean, rms and peak values of both the wind modulus and direction at 5 min intervals. The arrows show the points of data acquisition. Observed values will be confronted with wind tunnel data as well as with original design assumptions in order to provide additional information on model uncertainties. The probability distribution of the tower resistance to wind action will be referred to the mean wind velocity at elevation 10m above ground level, and determined by equations of motion of the conical shell that conforms the tower. This solution, presently being subjected to verification and calibration as described in section 3, is based on a very detailed discrete representation of the structure, which includes all important rebars, and considers fracture of concrete and reinforced concrete as possible failure modes. The determination of the tower strength will be performed by simulation, in which the variability of all material properties, as well as foundation properties can be included. The response of a series of sample towers to a slowly increasing turbulent wind will be then evaluated up to the collapse of the towers, furnishing simulated strength samples, that should allow the estimation of its probability distribution. In view of

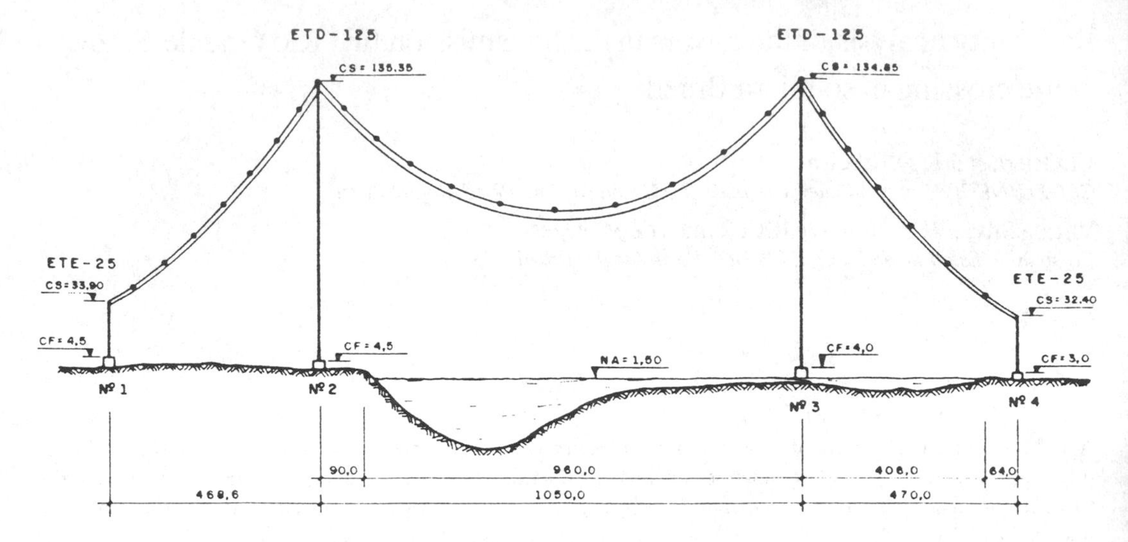

Figure 1: Scheme of the crossing

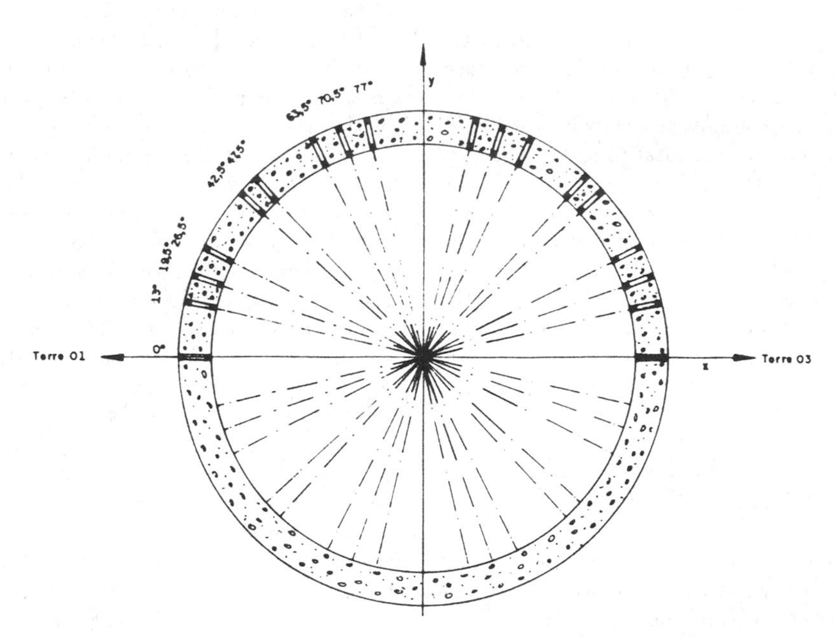

Figure 2: Pressure taps distribution in a cross section (tubes in only half the section shown).

the high computational cost of each simulation, it is expected that at most 10 sample points will be available. The theorectically determined response of the sample towers should also allow the determination of the evolution of damage indices, such as crack development, with the reference wind velocity. Such information may be vary valuable in the future to assess the performance of the structure under strong winds. Finally, the annual probability of failure can be obtained by the convolution of the strength distribution with the annual extreme wind density function at the site, modeled as type I. After the experimental stage is completed, the reliability analysis will be redone on

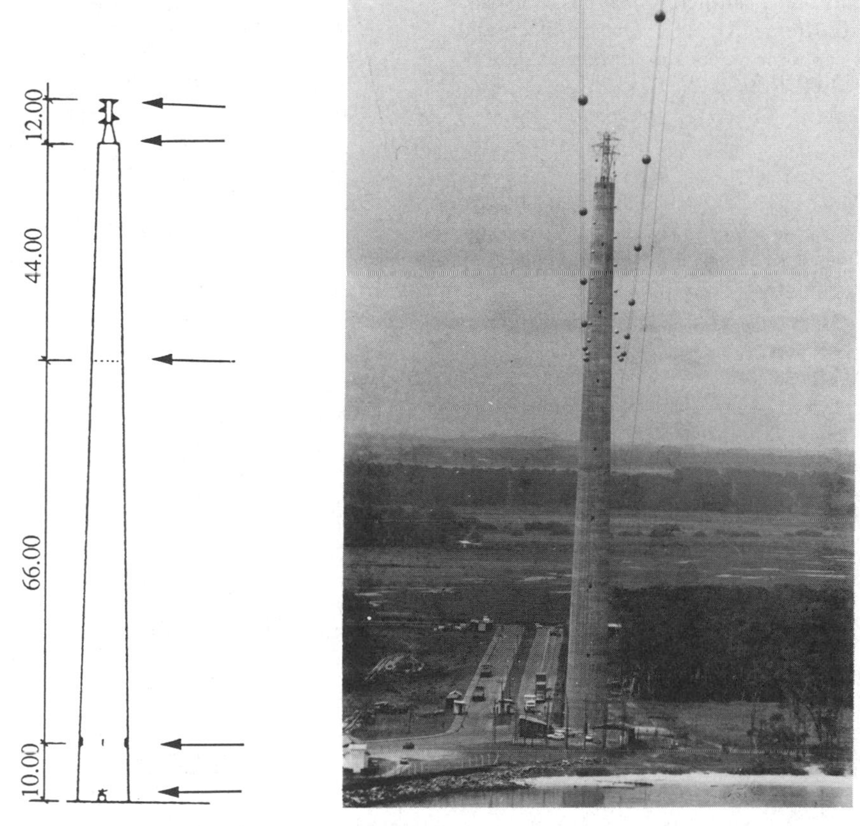

Figure 3: (a) Setup of equipment and (b) View of the concrete structure.

the basis of parameters measured at the site, such as actual drag coefficient, damping, etc. In addition to valuable information on these factors, the project should provide useful evidence on the influence of model uncertainties on reliability assessments (Ramos de Menezes (1992)).

3 MECHANICAL MODEL

An accurate determination of the concrete tower behavior up to complete failure is of paramount importance in the reliability analysis of the system. In addition to the variability of cross-sectional dimensions and material properties along the length of the structure, items such as geometric nonlinearity (P-Δ effect), physical nonlinearities, including fracture, as well as dynamic effects, must be taken into account. An approach that allows the consideration of all these factors is presently being tested against experimental evidence. The scheme, based on a discrete representation of an elastic continuum, has been successfully applied to the prediction of fracture of plain concrete (Rocha and Riera (1991)) and to the evaluation of the impulsive response of beam and plate structures (Rocha et al. (1991)). It has also shown a satisfactory correlation with measured impact response of plastic shells (Iturrioz and Riera (1993)). Presently the correlation of fracture properties of concrete, and of the yield strength of steel along and between rebars is under study.

REFERENCES

Ramos de Menezes, R.C. 1992. *Failure-data-based reliability assessment considering mechanical model uncertainties*. Thesis, Inst. Engrg.

Mech., Univ. of Innsbruck, Innsbruck, Austria.

Rocha, M.M.; Riera, J.D. 1991. On size effects and rupture of nohomogeneous materials. In J.G.M. VanMier, J.G. Rots & A. Bakker (eds.) *Fracture Process in Concrete, Rock and Ceramics.* Chapman & Hall / E. & F.N. Spon, London: 451-460.

Rocha, M.M.; Riera, J.D.; Krutzik, N.J. 1991. Extension of model that aptly describes fracture of plain concrete to the impact analysis of reinforced concrete. *Transactions 11th International Conference on Structural Mechanics in Reactor Technology (SMiRT 11)* Tokyo, Japan, Vol.J. 03/04: 51-56.

Iturrioz, I.; Riera, J.D. 1993. Determinacion de la respuesta de una cáscara elastoplástica a cargas impulsivas mediante un modelo de elementos discretos (D.E.M.). To be published, *Anales, XXVI Jornadas Sudamericanas de Ingeniaria Estructural*, Montevideo, Uruguay, 1993.

Structural Safety & Reliability, Schuëller, Shinozuka & Yao (eds) © 1994 Balkema, Rotterdam, ISBN 90 5410 357 4

Reliable prediction of wind induced oscillations of a steel suspension bridge by identification of wind load parameters in wind tunnel section model test

F.Thiele & B.Wienand
Universität Gesamthochschule Kassel, Fachgebiet Stahlbau, Germany

ABSTRACT: The aerodynamic stability of a river Rhine suspension bridge altered for enlarged traffic area has been investigated in wind tunnel section model tests at the University of Kassel. Parameter identification techniques to measure unstationary aerodynamic forces on the wind exposed model have been used in the tests. These techniques have established favorable chances for reliable prediction of aerodynamic stability of the bridge structure during reconstruction time and in the final condition by means of analytical calculation with respect to meteorological data.

1 AERODYNAMIC SAFETY

The critical wind velocity is a decisive factor in the consideration of bridge safety. Excited oscillations may take place by exceeding the critical velocity. The oscillation amplitudes depend on the magnitude and duration of the excitation.

This connection is shown in Figure 1. It can be applied to any wind exposed object. Meteorological windstatistical data composed by Caspar (1970) and Wyatt (1991) in the environment of the examinated object show the limited duration of time of constant wind velocity:

$$T_W(v_2) < T_W(v_1)\ ,\ \text{for}\ v_2 > v_1.$$

Wind tunnel tests on a model of the object supply a correlation between the identified logarithmic decrement ϑ of oscillation damping and the wind velocity v.

Calculations can be made to find out the time T_A necessary to bring up oscillations from an amlitude A to a more dangerous amplitude X * A, if the damping decrement ϑ is negative:
$T_A = -\ln X/(f\vartheta)$. The Figure shows that the amplitude magnifying factor X doesn't exceed a limited amount X_{Max}, which is an important factor in the safety of the object.

2 ORIGINAL BRIDGE

The method for evaluating results of wind tunnel tests reported here had been tested on the Rhine river suspension bridge Köln-Rodenkirchen. The traffic area of the forty year old bridge will be enlarged. The proposed design consists of an additional bridge deck being supported by the existing cable structure plus one additional main cable (Figure 2). In order to reduce the stress in the central cable, a hinge along the central axis of the bridge was provided to connect the two bridge decks and to allow suitable prestressing of the cables during construction time.

3 LOAD PARAMETER IDENTIFICATION

Aeroelastic investigations had been carried out at Kassel University by Thiele (1992) using wind tunnel section model tests to protect the bridge from aerodynamic instability during and after construction. Investigations concentrated on unstationary bridge oscillations caused by wind. These oscillations depend on stationary aerodynamic forces correlated to the amplitudes of bridge oscillations. These aerodynamic forces can be understood as mass-, damping- and stiffness parameters of the oscillating structure and thus can be measured on the wind exposed oscillating model by changing normal system parameter identification methods to special "load parameter identification"

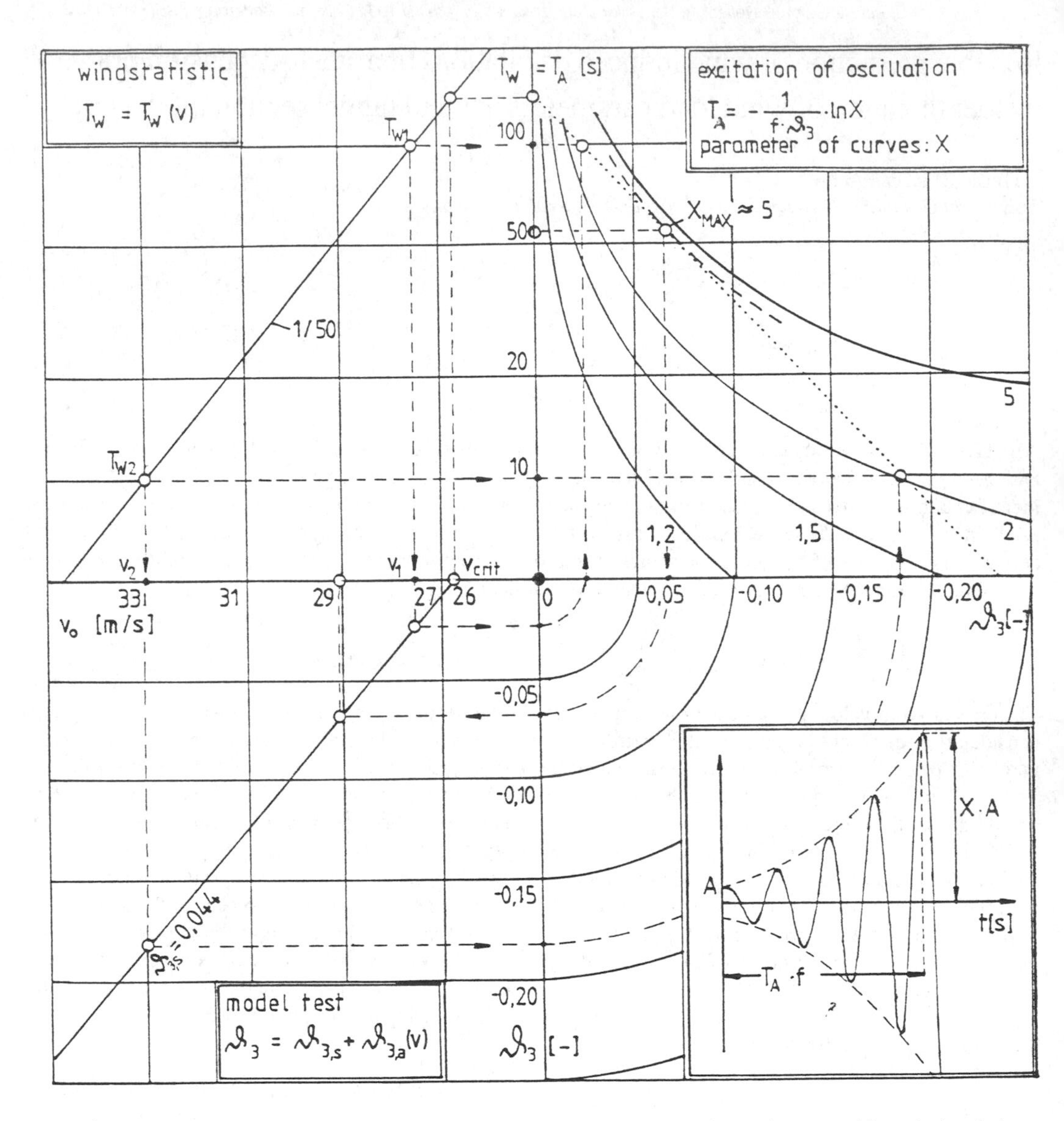

Fig. 1 Investigation of the excitation of oscillation by the results of model tests and by windstatistic

techniques. Load parameters had been identified as forces and moments per length unit of longitudinal bridge axis and then had been available for analytical dynamic calculations on the model and on the full scale bridge.

4 TESTING PROCEDURE

In addition to translational and rotational movements the model had been given the facility to vibrate in the hinge-activating third degree of freedom. In correspondence three accelerometers had been attached to the model in direction of the physical degrees of freedom. After a constant wind velocity was adjusted in the wind tunnel, the model was given a defined deflection by using a lightweight thread. After cutting this thread, free model oscillations occured and accelerometer signals were recorded during a defined time interval. The signals had been evaluated by software tools first developed for structural

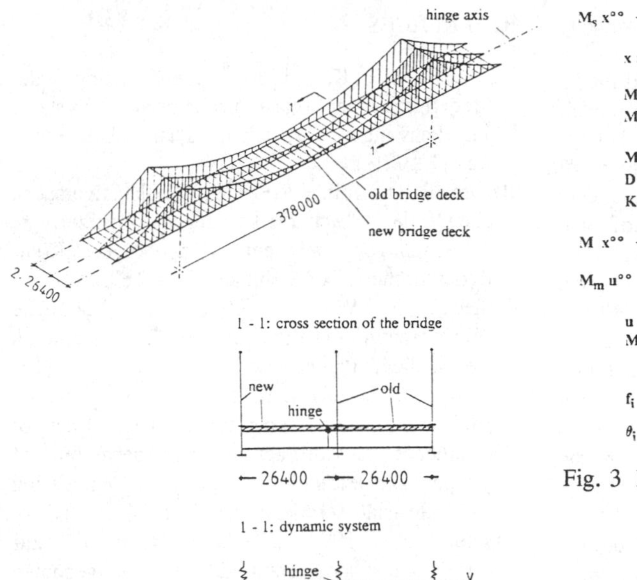

Fig. 2 System sketch of the new suspension bridge

$$M_s x^{\circ\circ} + D_s x^{\circ} + K_s x = M_a x^{\circ\circ} + D_a x^{\circ} + K_a x \qquad (1)$$

x :	vector of displacements (physical coordinates)
M_s, D_s, K_s :	system parameter matrices
M_a, D_a, K_a :	wind load parameter matrices

$$M = M_s - M_a$$
$$D = D_s - D_a \quad \} \text{ parameter matrices with wind} \qquad (2)$$
$$K = K_s - K_a$$

$$M x^{\circ\circ} + D x^{\circ} + K x = 0 \qquad (3)$$

$$M_m u^{\circ\circ} + D_m u^{\circ} + K_m u = 0 \qquad (4)$$

u :	vector of modal coordinates
M_m, D_m, K_m :	modal parameter matrices (diagonal)
$f_i = \sqrt{\frac{k_i}{m_i}}$	eigenvalue frequency
$\vartheta_i = d_i / (2 m_i f_i)$	logarithmic damping decrement

Fig. 3 Equation of motion and parameters

dynamics by Badenhausen (1986).

The tests had been carried out under smooth flow conditions. Investigations by Wardlaw (1983) have shown, that the turbulent flow has a stabilizing effect on systems with cross sections simular to the one, that is discussed here.

5 ANALYTICAL FORMULATION

The equation of motion in Figure 3 represents the sum of all forces acting either upon the oscillating test model or upon the bridge. The wind induced forces are assumed to be linear functions of the deflections x and their derivatives. By transposing the wind load matrices M_a, D_a and K_a to the corresponding matrices M_s, D_s and K_s the homogeneous equation then defines a complex eigenvalue-problem with wind velocity v_{crit} and frequency f_{crit} as eigenvalue twin-set.

The matrices can either be evaluated in terms of modal coordinates or in terms of physical coordinates. Logarithmic decrements of modal damping $\vartheta_i = d_i/(2m_i f_i)$ as functions of wind velocity may be used to define critical velocity v_{crit} such as $\vartheta_i(v_{crit}) = 0$.

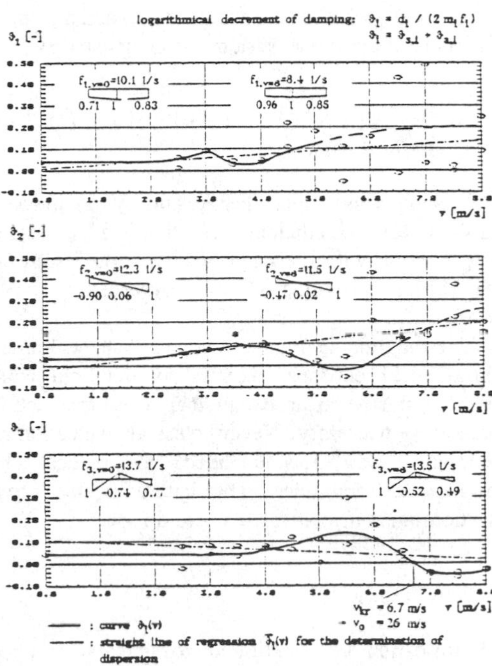

Fig. 4 Identified decrements of damping

6 RESULTS OF IDENTIFICATION

The test results reported here show only small variation of mode shapes with increasing wind velocity. The load parameter coefficients of mass and stiffness connected matrices M_a and K_a turned out to be very near constant. Damping connected matrix D_a however varied if wind velocity changed.

The diagrams in Figure 4 show identified modal damping parameters for the hinge open model. Logarithmic decrements ϑ (v) = ϑ_s + ϑ_a (v) are understood as the sum of system damping ϑ_s identified at v = 0 independent of wind velocity plus aerodynamic damping ϑ_a as a function of wind velocity.

A decrease of damping decrement ϑ_3 connected to the third mode is found, when wind velocity increases. For wind velocity v_{crit}=6.7 m/s the damping decrement ϑ_3 equals zero, thus defining critical wind velocity and remains negative when wind velocity increases. For wind velocity v=8 m/s the measurement had to be stopped before the end of the identification-time intervall due to large oscillation amplitudes of the model.

In additional tests a suitable increase of critical velocity v_{crit} had been achieved by adding small amounts of additional system damping capacity.

7 APPLICATION OF THE RESULTS TO THE ORIGINAL BRIDGE

Following wind tunnel test results, wind induced unstationary oscillations of the bridge under construction with the hinge not yet closed were to be expected if wind velocity exceeded v_{crit}=95 km/h or 26 m/s.

The diagram in Figure 1 shows that the limited duration of exposure reduces the required wind speed criterion to an extent that no further action should be necessary. Nevertheless augmentation of damping was recommended to reduce the oscillation amplitudes. The nature of the cross section made it relatively easy to do so.

8 CONCLUSIONS

Identification of wind load parameters in wind tunnel model tests proved to be a reliable method of experimentation with regard to prediction of wind induced oscillations of bridges. In the bridge investigation project reported here the load identification method delivered a new basis for discussion about aerodynamic stability in context with wind statistical data not possible so far (Barbré (1958)).

REFERENCES

Badenhausen, K. 1986. Identifikation der Modellparameter elastomechanischer Systeme aus Schwingungsversuchen. Diss., Universtiät Kassel.

Barbré, R. and Ibing, R. 1958. Windkanalversuche über die Sicherheit gegen winderregte Schwingungen bei der Hängebrücke Köln-Rodenkirchen. Der Stahlbau 27: p. 169.

Caspar, W. 1970. Maximale Windgeschwindigkeiten in der Bundesrepublik Deutschland. Bautechnik 47.

Wardlaw, R. L., Tanaka, H. and Utsunomiya, H. 1983. Wind tunnel experiments on the effects of turbulence on the aerodynamic behaviour of bridge road decks. Journal of Wind Engineering and Industrial Aerodynamics, Vol. 14, p. 247.

Thiele, F. 1990. Teilmodellversuche zur aerodynamischen Stabilität der umgebauten Rheinbrücke Köln-Rodenkirchen. Versuchsbericht 8/1990, Fachgebiet Stahlbau, Universität Kassel.

Wyatt, T. A. 1991. Appreciation of aerodynamic stability of Köln-Rodenkirchen Bridge; Stellungnahme.

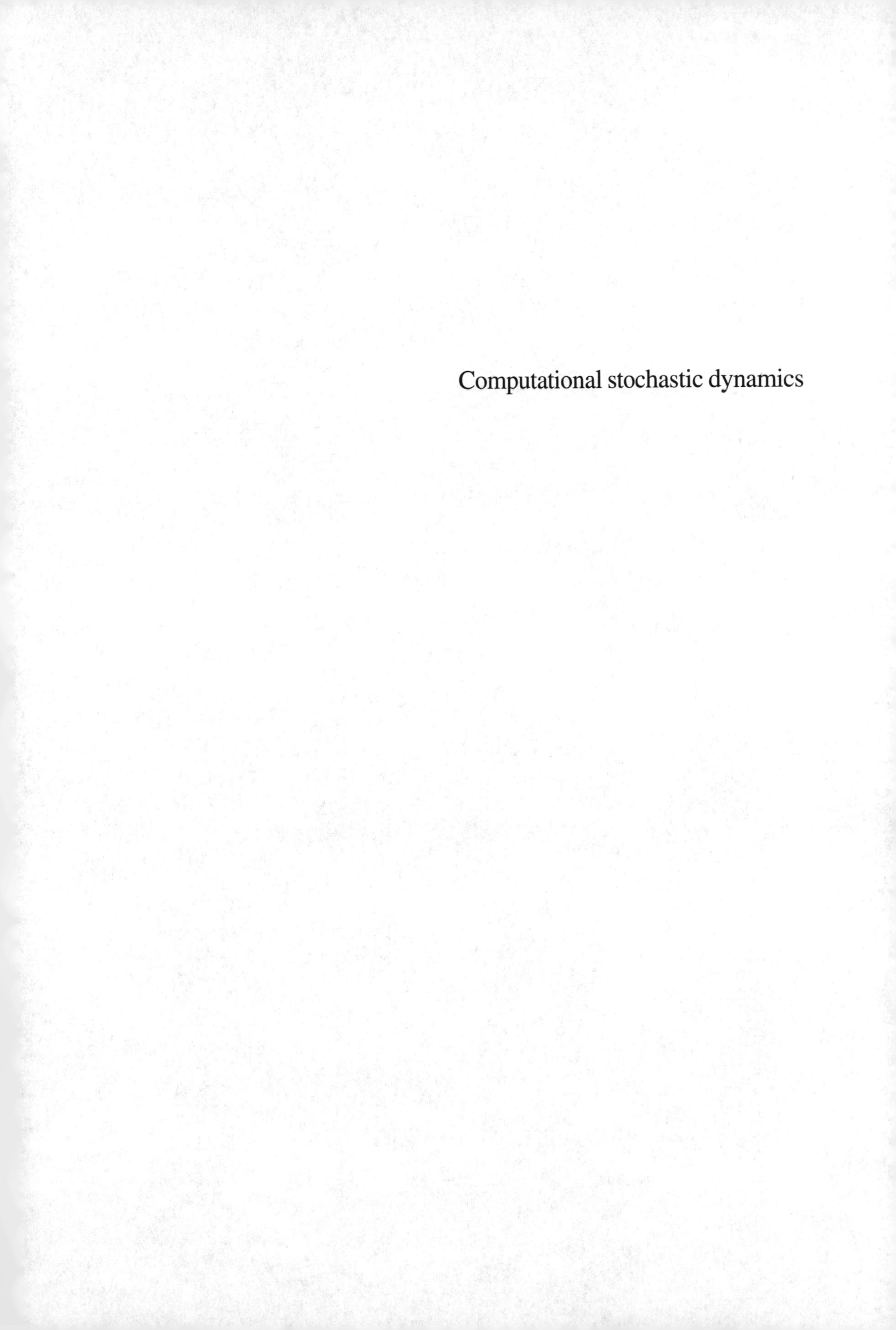

Computational stochastic dynamics

Structural Safety & Reliability, Schuëller, Shinozuka & Yao (eds) © 1994 Balkema, Rotterdam, ISBN 90 5410 357 4

COSSAN – Computational stochastic structural analysis perspectives of software developments

C.G. Bucher, H.J. Pradlwarter & G.I. Schuëller
Institute of Engineering Mechanics, University of Innsbruck, Austria

ABSTRACT: The paper presents a concept for software designed specifically for reliability based structural analysis. While incorporating traditional FE analysis, the COSSAN-software concentrates on the treatment of the randomness of loads and structural parameters as extension of the usual design variables. A completely interactive program structure enables the transfer of data between different computational tasks. In this context, the utilization of third-party FE codes is possible and has been successfully tested. In its final form, the software will rely on specifically designed FE software currently under development.
Several examples show the realization of the software and application to engineering problems.

1 INTRODUCTION

In engineering analysis, generally there are various stages in using computer resources. Generally, the computer analysis starts out with the gathering of information on important quantities like e.g. structural data or material properties. Then the scope of both required and possible solutions is decided on. Based on previous experience and new requirements as given by the available information, an existing computer code is chosen or the decision is made to a write new code from scratch. In both cases, a significant amout of time is spent for testing and debugging of both code and data. Also, structural data etc. may have to be preprocessed for computer use. After the algorithms are coded and tested, updates follow as necessary.
Now the production stage can begin. The questions as posed by the design process give rise to a number of computer tasks to be performed. Computationally extensive calculations can provide a variety of data on the mechanical behavior of the structure. The next stage assembles these data systematically during postprocessing. This includes final verification checks regarding input data and algorithm performance. The last step through the analysis cycle is the interpretation stage, that is the attempt to answer the questions asked in the very beginning of the design.The production stage has traditionally been the domain of high-speed computers, it is the area where they have been utilized for several decades. Typically, in this stage little interaction with the user is required, batch mode of operation prevails. In both testing and preprocessing stages, which more recently became computer-aided, the demand for user interaction is far more pronounced. This is of significance during algorithm testing, i.e. the

process of identifying the most suitable (efficient) algorithm for the particular problem. It requires the possibility of closely monitoring the progress during the steps of the analysis. Commercially available packages frequently do not support this level of interactivity.
The above mentioned statements appear to be of paramount importance in context with software being developed for the purposes of structural reliability analysis. This is related to the fact that the engineering community as whole does not have sufficient experience with the application of reliability methods.
Consequently, a COSSAN software must provide means to aquire experience with and confidence in the tools of stochastic methods.
It should be mentioned that on a world-wide basis considerable efforts are put into such software developments [1 - 3].

2 ANALYSIS TOOLS

The tools required for stochastic structural analysis basically must serve three purposes [4]. Those are
(1) Structural Analysis - Calculation of displacements or stresses under various loading conditions.
(2) Failure Criteria - Decision on safety in terms of serviceability or failure (collapse).
(3) Probabilistic Assessment - Computation of probabilites associated with safety or failure taking into account the avilable information.
Utilizing the most developed tools from those three categories ensures that the COSSAN software can be utilized advantageously. In reviewing the available methods, it appears that for item (1) the **Finite Element Methods** provides the necessary modeling flexibility. For item (2) the **Response Surface Methods** [5, 6] give the desired computational efficiency, and finally, for item (3) the **Advanced Monte Carlo Simulation Methods** [7 - 11] are reasonably accurate with respect to probability calculations.
Equivalent linarization is utilized as supplementary tool to provide an overall estimate of the dynamic response of nonlinear MDOF-systems [9, 12].
However, such a collection of tools does not in itself constitute a workable software package. In order to make the interrelations sufficiently clear, the software must provide logically clear and easily accessible connections. This has been achieved by developing a language interpreter (**S**tructural **Lang**uage - SLANG) which incorporates the above mentioned tools as parts of its syntactical elements.

3 STRUCTURAL LANGUAGE (SLANG)

As mentioned above, the design and implementation of a command interpreter for structural analysis and reliability methods allows the transparent stochastic analysis of structural systems. SLANG incorporates the construction of commands sequences including conditional branches and loops similar to BASIC. Structural and reliability analysis can be performed on a step-by-step basis, or - if desired - in terms of predefined macros which can be called during execution. The data objects which the commands operate on, are accessible to the user at any time during execution of the analysis, so that error checking and debugging is greatly simplified.
The syntactical rules are very rigid, so that a formal association of the commands with their meaning becomes feasible. Again this greatly facilitates debugging. Finally, SLANG allows issuing of operating system commands (e.g. UNIX commands) to the shell enviroment. This

can be utilized for "remote control" of third-party software, e.g. commercial FE packages. Listing 1 shows a fragment of a SLANG-file which builds element stiffness matrices within a loop, builds the table of global degrees-of-freedom, assembles the element matrices into the global stiffness matrix, solves for the displacement under static loading and plots the deformed structure onto a window.

```
* build element stiffness matrices/
object modify, set, loop_counter 0,/
# label stiffness
  object modify,  add,  loop_counter
      1, /
  element build,  stiffness,
      loop_counter, k_mat /
  control if,     integer less,
    loop_counter number_of_elements
    stiffness, /
* build global matrices & vectors/
  global build, restraints,, /
  global build, stiffness full,,
  glob_k /
  global build, load, 1 1 1.,
     glob_l /
*  solve for displacements/
  linalg solve,, glob_k glob_l,
     glob_w /
  global tonode, displacements,
     glob_w, /
* plot deformed geometry/
  structure view, nodes deformed,
     2  -15   20   10., /
* pass control to shell/
  control exit,,,    /
```

Listing 1: SLANG sample file

The user interface is designed to support

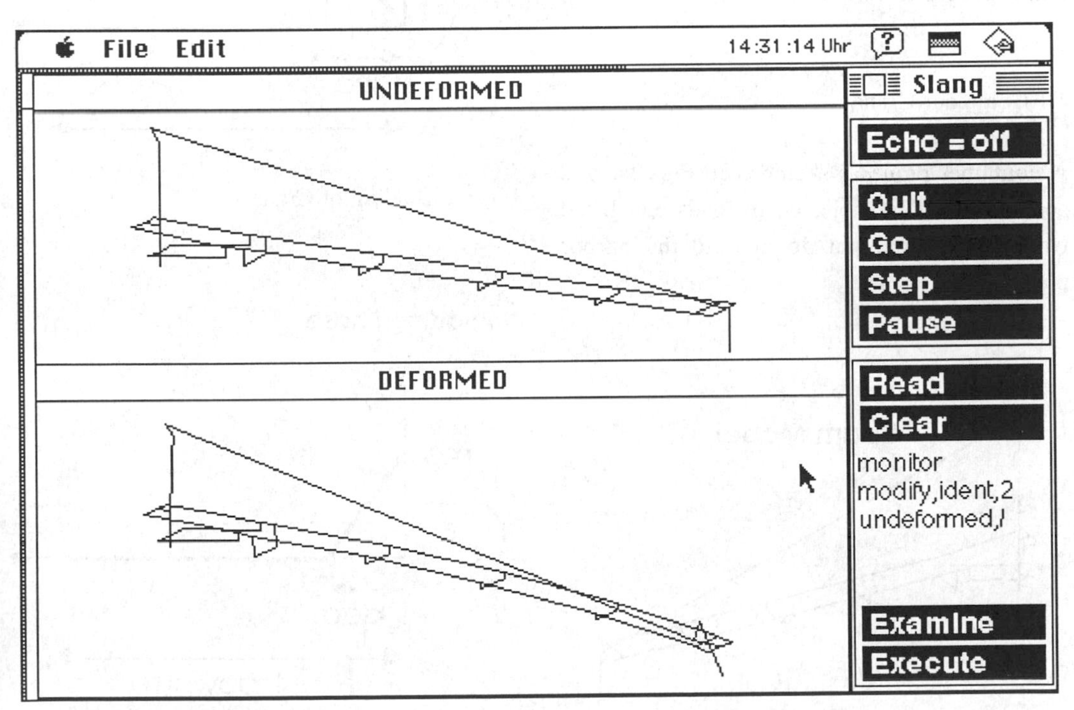

Fig. 1: SLANG user interface

multiple simultaneous windows which can be opened, addressed, drawn into, and closed from SLANG. In addition to a command sequence read from a file, SLANG accepts commands interactively, so that predefined flow of the program can be changed interactively. A sample showing the realization on a desktop computer is shown in Fig.1.

4 NUMERICAL EXAMPLES

The following two examples are chosen mainly to show the capabilites of the software package to deal with structural problems involving random changes in the system configuration (geometry, degrees of freedom) and the randomness of both loading and system data. Both examples also show how to arrive at probabilistic measures of the response, either in terms of first and second moments, or in terms of failure probabilities.

4.1 Container Crane

A container crane as sketched in Figs.2a, b is considered. It is subjected to loads e.g. lifted from a ship and put down onto the harbor platform.

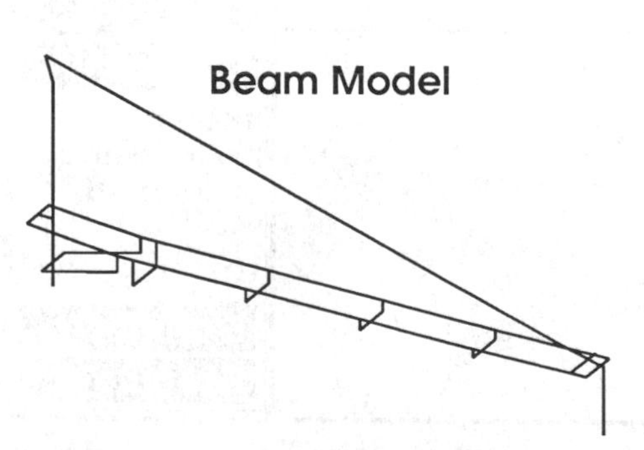

Fig. 2a: Container Crane - Beam Model

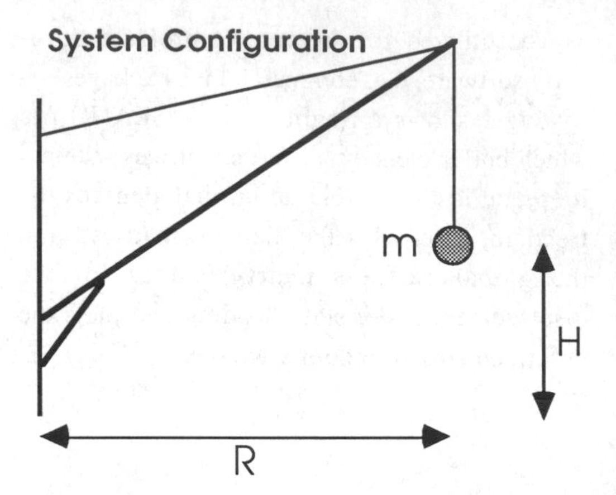

Fig. 2b: Container Crane - Configuration

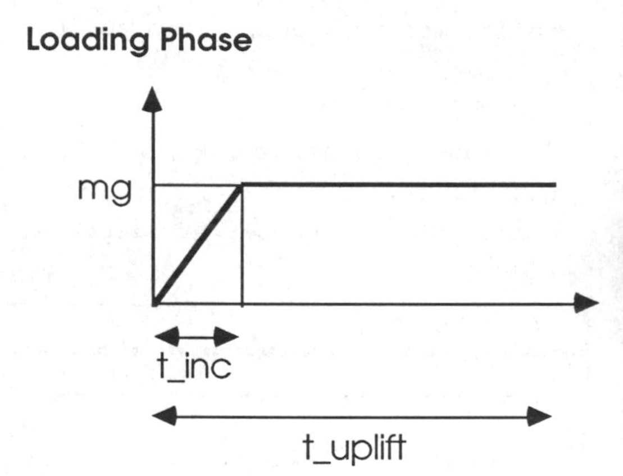

Fig.3a: Loading of Crane

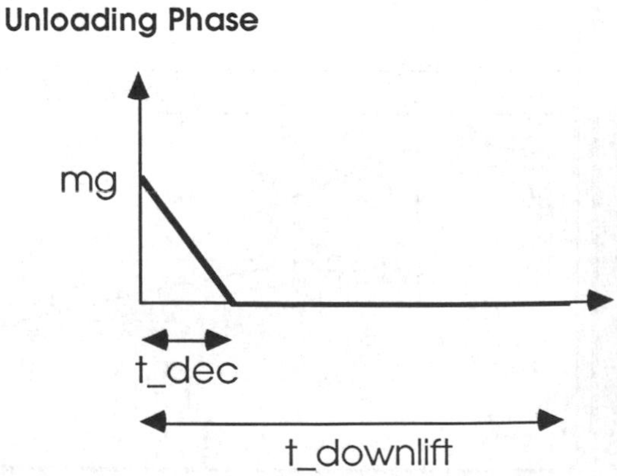

Unloading Phase

mg

t_dec

t_downlift

Fig.3b: Unloading of Crane

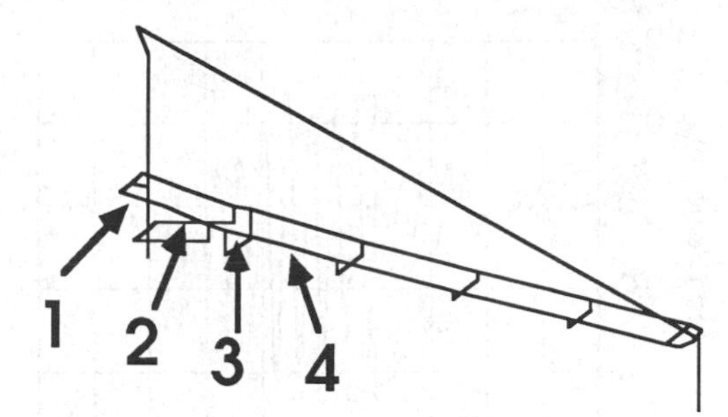

Fig.4: Cross sections analyzed

These loads obviously are random in nature. Also, the system configuration (due to its kinematics) as defined by the outreach during loading and unloading is randomly distributed. It should be noted that due to the increase of the hook load in the loading phase (see Fig. 3), there is some dynamic amplification which causes stress peaks in the crane structure.

Four cross sections as indicated in Fig.4 have been investigated with respect to the maximum stress (in terms of the v.Mises stress).

11 random variables describing the variability of system and load were considered in the analysis. The meaning of these random variables can be seen from comparing the notations given in Tab.1 and in Figs.2a, b, 3a, b. The variables fac_x, fac_y, which are not defined in these figures, are horizontal load components (out of the vertical) as fraction of the vertical load. Their types of distribution, mean values and standard deviations are given in Tab.1.

The probability of failure in terms of the probability of exceeding the yield stress of 300 N/mm^2 was calculated using Importance Sampling based on 256 samples. The values of p_f are listed in Tab.2 for the four cross sections (cf. Fig.4).

Tab.1: Statistical Parameters of Random Variables

Variable	Type	Mean	Std. Dev.
m [kg]	lognormal	15000	6000
R_up [m]	uniform	15.5	7.5
R_down[m]	uniform	15.5	7.5
t_inc [s]	lognormal	0.3	0.2
t_dec [s]	lognormal	0.5	0.3
t_uplift [s]	lognormal	5.0	3.0
t_downlift [s]	lognormal	5.0	3.0
fac_x [-]	normal	0.0	0.1
fac_y [-]	normal	0.0	0.1
H_up [m]	uniform	0.0	1.66
H_down [m]	uniform	0.0	1.66

Tab.2: Resulting failure probabilities

Section	p_f
1	$1.2 \cdot 10^{-10}$
2	$1.7 \cdot 10^{-5}$
3	$5.6 \cdot 10^{-7}$
4	$5.6 \cdot 10^{-7}$

4.2 Frame Structures Under Earthquake

Two frames, closely adjacent to each other and subjected to an earthquake type ground acceleration are considered (see Fig. 5). They can impact each other during the earthquake, so that nonlinear effects occur in the otherwise linear structures.

The structural elements are divided into 11 groups (see Table 3), whose elastic moduli are considered to be independent normally distributed random variables, with a coefficient of variation of 10% and fully correlated within each group.

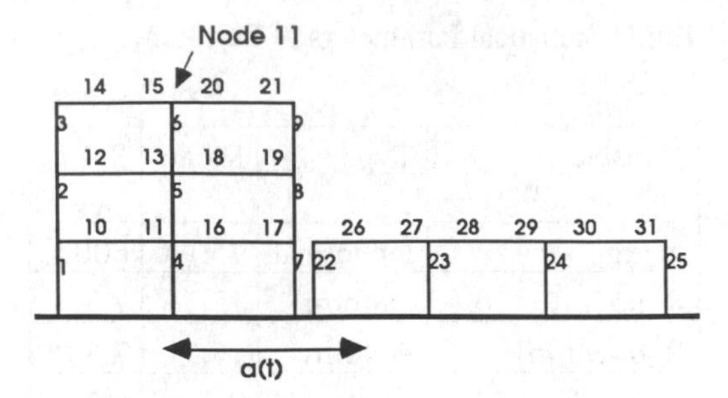

Fig.5: Two adjacent Frames

Table 3: Groups of Elements

Group	Elements	Group	Elements
1	1, 2, 3	2	4, 5, 6
3	7, 8, 9	4	10, 11, 16, 17
5	12, 13, 18, 19	6	14, 15, 20, 21
7	22	8	23
9	24	10	24
11	26 - 31		

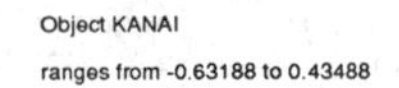

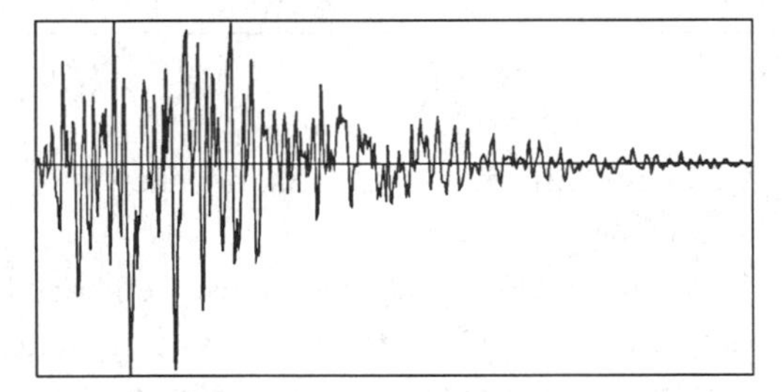

Fig. 6: Ground acceleration

In order to show the effect of randomness of the system properties, the ground acceleration a(t) is assumed to be deterministic (see Fig.6). Again, Monte Carlo Simulation is performed to evaluate the statistical properties of the response of the structures. The following two figures show the mean value and the standard deviation of the horizontal displacement response at node 11.

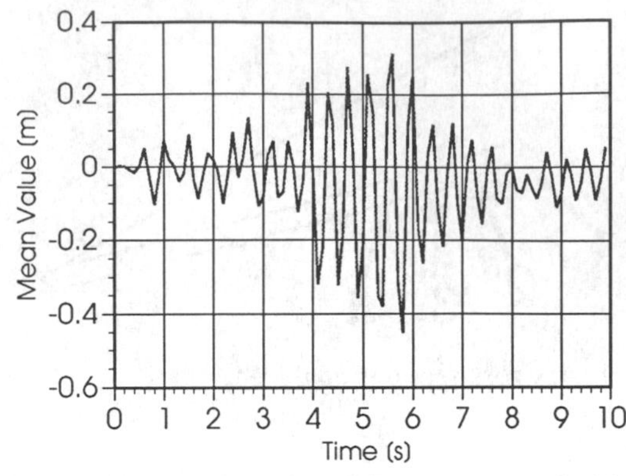

Fig. 7: Mean Value of Displacement at node 11

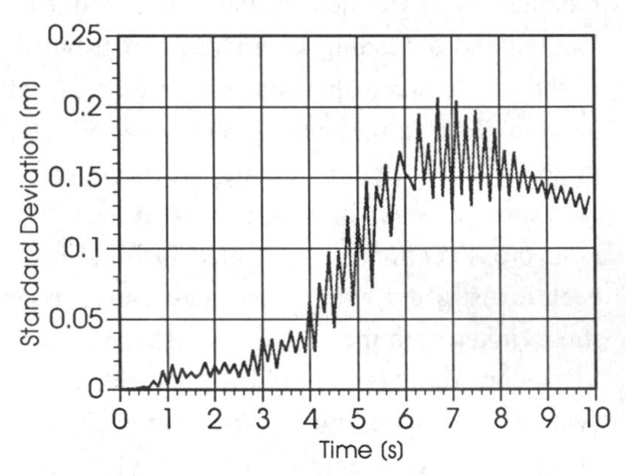

Fig. 8: Standard Deviation of Displacement at node 11

It is seen that the variability of the elastic properties has a quite dramatic effect on the response. This is mainly due to the relatively large sensitivity of the impact times to the variation of the natural frequencies of the two structures.

5 CONCLUSIONS

The realization of COSSAN as based on the structural language SLANG provides an interactive tool for stochastic structural analysis. Both state of the art FE-codes and advanced reliability methods are easily accessible to the user. However, if desired, the full utilization of a user-preferred third-party FE system is also possible. It should be mentioned that various implementations from desktop computers to vector machines show that the code is easily portable, and extendable. Finally, the fact that interfaces to current FE-Pre/Postprocessing and Desktop-Publishing software are implemented, makes the software especially attractive for professional use.

ACKNOWLEDGEMENTS

This research has been partially supported by the *Austrian Industrial Research Promotion Fund* under contract no. 6/636 which is gratefully acknowledged by the authors. The authors also acknowledge the assistance of *Markus Gasser* and *Wolfgang Wall* in context with the numerical computations.

REFERENCES

[1] Madsen, H.O.; "PROBAN: Theoretical Manual for External Release", Technical Report No 88-2005, AS Veritas, Oslo, Norway, January 1988.

[2] Liu, P.-L., DerKiureghian, A.: "Finite-Element Reliability Methods for Geometrically Nonlinear Stochastic Structures", Report No. UCB/SEMM-89/05, Dept. of Civil Engineering, University of California, Berkeley, USA, January 1989.

[3] Millwater, H.R., Wu, Y.-T., Dias, J.B., McClung, R.C., Raveendra, S.T., Thacker, B.H.: "The NESSUS Software System for Probabilistic Structural Analysis", A.H-S. Ang et al. (eds.): Strcutural Safety and Reliability, ASCE, New York, N.Y., 1990, Vol III, pp 2283 - 2290.

[4] Bucher, C.G., Pradlwarter, H.J., Schuëller, G.I.: "COSSAN - Ein Beitrag zur Software-Entwicklung für die Zuverlässigkeitsbewertung von Strukturen", VDI-Bericht Nr. 771, 1989, pp. 271-281.

[5] Bucher, C.G., Bourgund, U.: "A fast and efficient response surface approach for structural reliability problems", *J. Structural Safety*, Vol. 7, Nr. 1, 1990, pp. 57-66.

[6] Bucher, C.G., Chen, Y-M., Schuëller, G.I.: "Time variant reliability analysis utilizing response surface approach", in: Reliability and Optimization of Structural Systems '88, P. Thoft-Christensen (Ed.), Lecture Notes in Egr., 48, Springer Verlag, Berlin, pp. 1-14, 1989.

[7] Schuëller, G.I., Stix, R.: "A Critical Appraisal of Methods to Determine Failure Probabilities", *J. Structural Safety*, 4(1987), pp. 293-309.

[8] Bucher, C.G.: "Adaptive Sampling - An Iterative Fast Monte Carlo Procedure", *J. Structural Safety*, Vol. 5, No. 2, June 1988, pp. 119-126.

[9] Schuëller, G.I., Pradlwarter, H.J., Bucher, C.G.: "Efficient Computational Procedures for Reliability Estimates of MDOF Systems", *J. Nonlinear Mechanics*, Vol. 26, No. 6, 1991, pp. 961-974.

[10] Pradlwarter, H.J.: "A Selective MC Simulation Technique for Nonlinear Structural Reliability", Proc. ASCE Specialty Conference, Denver, Colorado, July 8-10, 1992, ASCE, New York, N.Y., USA, 1992, pp 69-72.

[11] Pradlwarter, H.J., Schuëller, G.I., Melnik-Melnikov, P.G.: "Reliability of MDOF-Systems", *J. Prob. Eng. Mech.*, to appear.

[12] Pradlwarter, H.J., Schuëller, G.I.: "Equivalent Linearization - A Suitable Tool to Analyze MDOF-Systems",*J. Prob. Eng. Mech.*, 8, 1992, pp 115 - 126.

Structural Safety & Reliability, Schuëller, Shinozuka & Yao (eds) © 1994 Balkema, Rotterdam, ISBN 90 5410 357 4

A new algorithm for simulating stationary Gaussian processes based on sampling theorem

Mircea Grigoriu & Stavroula Balopoulou
Cornell University, Ithaca, N.Y., USA

ABSTRACT: An algorithm is proposed for generating realizations of a stationary Gaussian process with band limited spectral density. The algorithm is based on global and local parametric models depending on a finite number of values of the process at equally spaced instances, referred to as nodal points. The standard spacing between nodal points is equal to a half of the inverse of the largest frequency of the process. The proposed simulation algorithm has the attractive features of the spectral representation and the ARMA models. The definition of the global and local models does not require any calculations and the sample generations based on these models can be performed on-line so that the storage requirement is minimum. Several examples are presented to illustrate the proposed algorithm. The examples involve stationary Gaussian and lognormal processes.

1 INTRODUCTION

A new algorithm is developed for generating realizations of a stationary Gaussian process $X(t)$ with band limited spectral density. The algorithm is based on global and local parametric models depending on a finite number of random variables. The parametric models are defined according to the sampling theorem and provide approximate representations of $X(t)$ to any desired accuracy.

Several examples are presented to illustrate and evaluate the proposed algorithm. The examples involve a narrow band stationary Gaussian process and a non-Gaussian lognormal process. Results show that the algorithm is accurate for Gaussian processes and can be extended to also generate realizations of non-Gaussian processes.

2 GLOBAL MODEL

Let $X(t)$, $t \in R$, be a real-valued zero-mean stationary process $X(t)$ with covariance function $c(\tau) = EX(t+\tau)X(t)$ and spectral density $s(f) = \int_R e^{-i2\pi f\tau}c(\tau)$ vanishing outside a bounded interval $(-\bar{f}, \bar{f})$, $0 < \bar{f} < \infty$. According to the sampling theorem (Brigham 1974, Wong & Hajek 1985),

$$c(\tau) = \sum_{k=-\infty}^{\infty} c(\tau_0+kT)\, \alpha_k(\tau-\tau_0;\, T) \tag{1}$$

for any real τ_0 because $c(\tau)$ is a deterministic function whose Fourier transform coincides with the spectral density $s(f)$ that is concentrated on $(-\bar{f}, \bar{f})$, $0 < \bar{f} < \infty$, in which

$$\alpha_k(u;\, T) = \frac{\sin[\pi(u-kT)/T]}{\pi(u-kT)/T} \tag{2}$$

and $T = 1/(2\bar{f})$.

Define the global model of $X(t)$ as a family of parametric stochastic processes

$$X_N(t) = \sum_{k=-N}^{N} X_k\, \alpha_k(t;\, T) \tag{3}$$

in which $N = 1,, 2, \ldots$ and $X_k = X(kT)$ are random variables that are fully defined by the finite dimensional distributions of $X(t)$. The instances kT, $k = -N, \ldots, N$ are called the nodal points. The parametric process $X_N(t)$ has the same first two moments as $X(t)$ asymptotically as $N \to \infty$ so that it approaches this process in the mean square sense with increasing order (Wong & Hajek 1985). Indeed, the mean of $X_N(t)$ is zero for any value of N because $EX_k = EX(kT) = 0$, $k = -N, \ldots, N$.

The cross-covariance of two parametric representations of order N and M is

$$c_{N,M}(t+\tau,t) = EX_N(t+\tau)X_M(t) \quad (4)$$
$$= \sum_{k=-N}^{N} \alpha_k(t+\tau;\ T) \sum_{l=-M}^{M} c((k-l)T)\ \alpha_l(t;\ T)$$

Denote by

$$h_k(t;\ M) = \sum_{l=-M}^{M} c((k-l)T)\ \alpha_l(t;\ T) \quad (5)$$
$$= \sum_{l=-M}^{M} c(-kT+lT)\ \alpha_l((t-kT)+kT;\ T)$$

the second sum after index l in Eq. 4 and let M approach infinity. From Eq. 1, $h_k(t;\ M)$ converges to $c(t-kT)$ as $M \to \infty$. Using again Eq. 1 it can be shown that the remaining sum in Eq. 4 after index k, approaches $c(\tau)$ as $N \to \infty$.

The covariance function in Eq. 4 shows that the family of processes $X_N(t)$ with $N < \infty$ is not stationary in the wide sense. This is caused by the dependence of $X_N(t)$ on a finite number of values of $X(t)$ in the bounded range $(-NT, NT)$ when $N < \infty$. However, $X_N(t)$ approaches stationarity in the wide sense as $N \to \infty$. The result also shows that $X_N(t)$ provides a satisfactory approximation of $X(t)$ for large values of N. The approximation does not hold when t is close to the boundary of the interval $(-NT, NT)$ or outside it. In fact, $X_N(t)$ for $N < \infty$ approaches zero as t increases indefinitely because the functions $\alpha_k(t;\ T)$ in the representation of the process vanish as $t \to \infty$. This problem can be eliminated by letting N increase indefinitely. However, the resultant model $X_N(t)$ would become impractical for simulation. An alternative local representation of $X(t)$ is defined in the next section for efficient simulation.

Suppose now that the process $X(t)$ is Gaussian. Then, $X_N(t)$ in Eq. 3 is a version of $X(t)$ asymptotically as $N \to \infty$. From the previous properties, $X_N(t)$ is equal to $X(t)$ in the second-moment sense asymptotically as $N \to \infty$. From Eq. 3 and the hypothesis that $X(t)$ is Gaussian, the parametric representations $X_N(t)$ are Gaussian processes for any value of N as linear combinations of the Gaussian variables $\{X_k\}$. Therefore, all finite dimensional distributions of $X(t)$ and $X_N(t)$ coincide asymptotically as $N \to \infty$ because they only depend on the mean and covariance functions of these processes.

There is no simple extension of this statement to the case of non-Gaussian processes. When $X(t)$ is not Gaussian the finite dimensional distributions of any order of $X(t)$ and $X_N(t)$ coincide provided that (i) the instances $\{t_i\}$ at which these distributions are calculated coincide with nodal points and (ii) N is sufficiently large such that $t_i \in (-NT, NT)$ for all indices i. The coincidence of these distributions follows from Eq. 3 showing that $X_N(t_i) = X(t_i)$ when t_i is a multiple of T and belongs to $(-NT, NT)$. The distribution of $X_N(t)$ at instances different from nodal points can be calculated but its determination is complex.

Suppose that the zero-mean, stationary, band-limited process $X(t)$ is narrow band with power concentrated in the frequency range $(-\bar{f}-f_0, -f_0+\bar{f}) \cup (-\bar{f}+f_0, f_0+\bar{f})$, in which $0 < \bar{f} \ll f_0 < \infty$. The representation in Eq. 3 is still valid but becomes impractical because the required sampling rate $1/(2f_0)$ can be very dense when f_0 is large. An alternative parametric representation of this process can be based on the observation that $X(t)$ is (Cramer & Leadbetter 1967, Davenport & Root 1958)

$$X(t) = V(t)\cos(2\pi f_0 t + \psi(t)) \quad (6)$$

and

$$X(t) = X_c(t)\cos(2\pi f_0 t) + X_s(t)\sin(2\pi f_0 t) \quad (7)$$

in which

$$V(t) = \left(X_c(t)^2 + X_s(t)^2\right)^{1/2}$$
$$\psi(t) = tan^{-1}\left[-\frac{X_s(t)}{X_c(t)}\right] \quad (8)$$

Processes $X_c(t)$ and $X_s(t)$ are linear transformation of $X(t)$ such that their mean is zero, are stationary, and have the covariance functions

$$\tilde{c}(\tau) = EX_c(t+\tau)X_c(t) = EX_s(t+\tau)X_s(t)$$
$$= 4\pi \int_{-\bar{f}}^{\bar{f}} d\nu\ s(\nu+f_0)\cos(2\pi\nu\tau) \quad (9)$$

and

$$\hat{c}(\tau) = EX_c(t+\tau)X_s(t) = -EX_c(t)X_s(t+\tau)$$
$$= 4\pi \int_{-\bar{f}}^{\bar{f}} d\nu\ s(\nu+f_0)\sin(2\pi\nu\tau) \quad (10)$$

These processes are uncorrelated when the spectrum $s(f)$ is symmetric about the central frequency f_0. Consider the parametric family of stochastic processes

$$X_N(t) = X_{N,c}(t)\cos(2\pi f_0 t) + X_{N,s}(t)\sin(2\pi f_0 t) \quad (11)$$

in which

$$X_{N,c}(t) = \sum_{k=-N}^{N} X_{c,k}\, \alpha_k(t;\, T)$$
$$X_{N,s}(t) = \sum_{k=-N}^{N} X_{s,k}\, \alpha_k(t;\, T) \qquad (12)$$

$X_{c,k} = X_c(kT)$, $X_{s,k} = X_s(kT)$, and parameters $\{\alpha_k,\ T\}$ are defined in Eq. 2. Processes $X_{N,c}(t)$ and $X_{N,s}(t)$ are Gaussian as linear combinations of discrete values of $X_c(t)$ and $X_s(t)$ and are fully defined by the probability of these processes. The sampling rate, $T = 1/(2\bar{f})$, used to define processes $X_{N,c}(t)$ and $X_{N,s}(t)$ is much lower than the required sampling rate, $1/2(f_0 + \bar{f})$, corresponding to a direct use of Eq. 3 for process $X(t)$. This parametric representation has properties similar to the model in Eq. 3.

3 LOCAL MODEL

An alternative representation of $X(t)$ is

$$Y_n(t) = \sum_{k=n_t-n}^{n_t+n+1} X_k\, \alpha_k(t;\, T), \qquad n_t T \le t \le (n_t+1)T \qquad (13)$$

in which $n_t = [t/T]$ = the largest integer smaller that t/T and n is a positive integer. The representation has a local character because it involves $2(n+1)$ nodal values of $X(t)$ centered about the active cell $[n_t T, (n_t+1)T]$, i.e. the cell containing current time t. The choice of n defines the size of the window or vicinity about the active cell. This local representation has similar asymptotic properties as $X_N(t)$ in Eq. 3.

There is a notable difference between the representations in Eqs. 3 and 13. Although they both involve values of $X(t)$ equally spaced at $T = 1/(2\bar{f})$, these values are centered about zero for $X_N(t)$ and about t for $Y_n(t)$. As t increases N must take large values to assure that t is included in $(-NT, NT)$ and $X_N(t)$ provides an accurate approximation of $X(t)$. On the other hand, the representation $Y_n(t)$ involves $2(n+1)$ values of $X(t)$ at any time t. This feature is particularly attractive in simulation because the generation of samples of $Y_n(t)$ depends on a relatively small number of random variables that can be generated sequentially as time t increases. In contrast, simulation based on $X_N(t)$ requires to generate samples of all variables $\{X_k\}$, $k = -N, \ldots, N$ and store their values prior to the determination of a realization of $X_N(t)$. The local representation in Eq. 13 can be extended without difficulties to the case in which $X(t)$ is a narrow-band stationary Gaussian process with power centered at the frequencies $\pm f_0$. A global approximation of the process is $X_N(t)$ in Eqs. 11 and 12. A local approximation of the process can be provided by the model

$$Y_n(t) = Y_{n,c}(t)\cos(2\pi f_0 t) + Y_{n,s}(t)\sin(2\pi f_0 t), \qquad n_t T \le t \le (n_t+1)T \qquad (14)$$

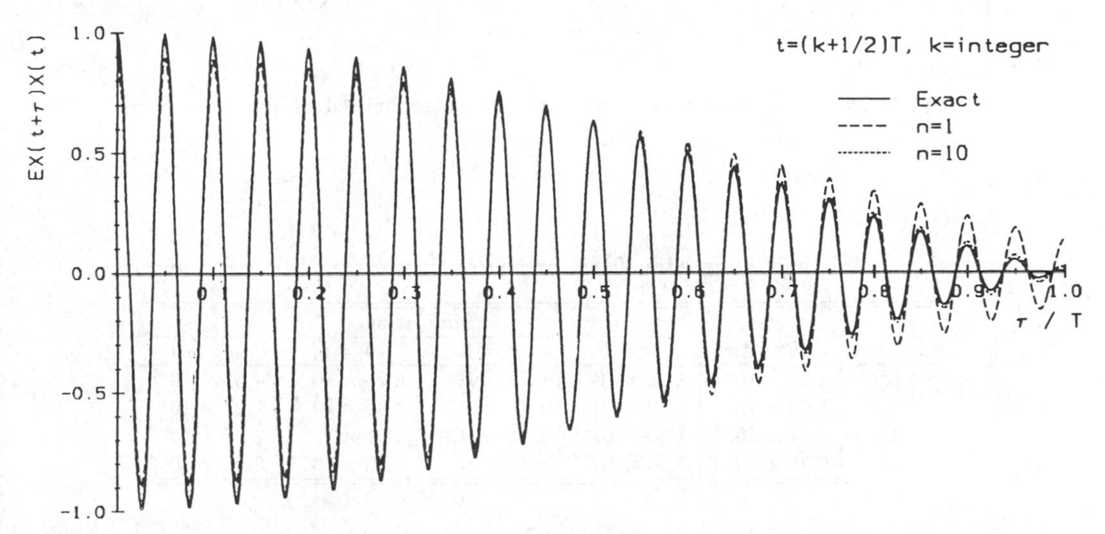

Figure 1. Exact and approximate covariances of a narrow band Gaussian process.

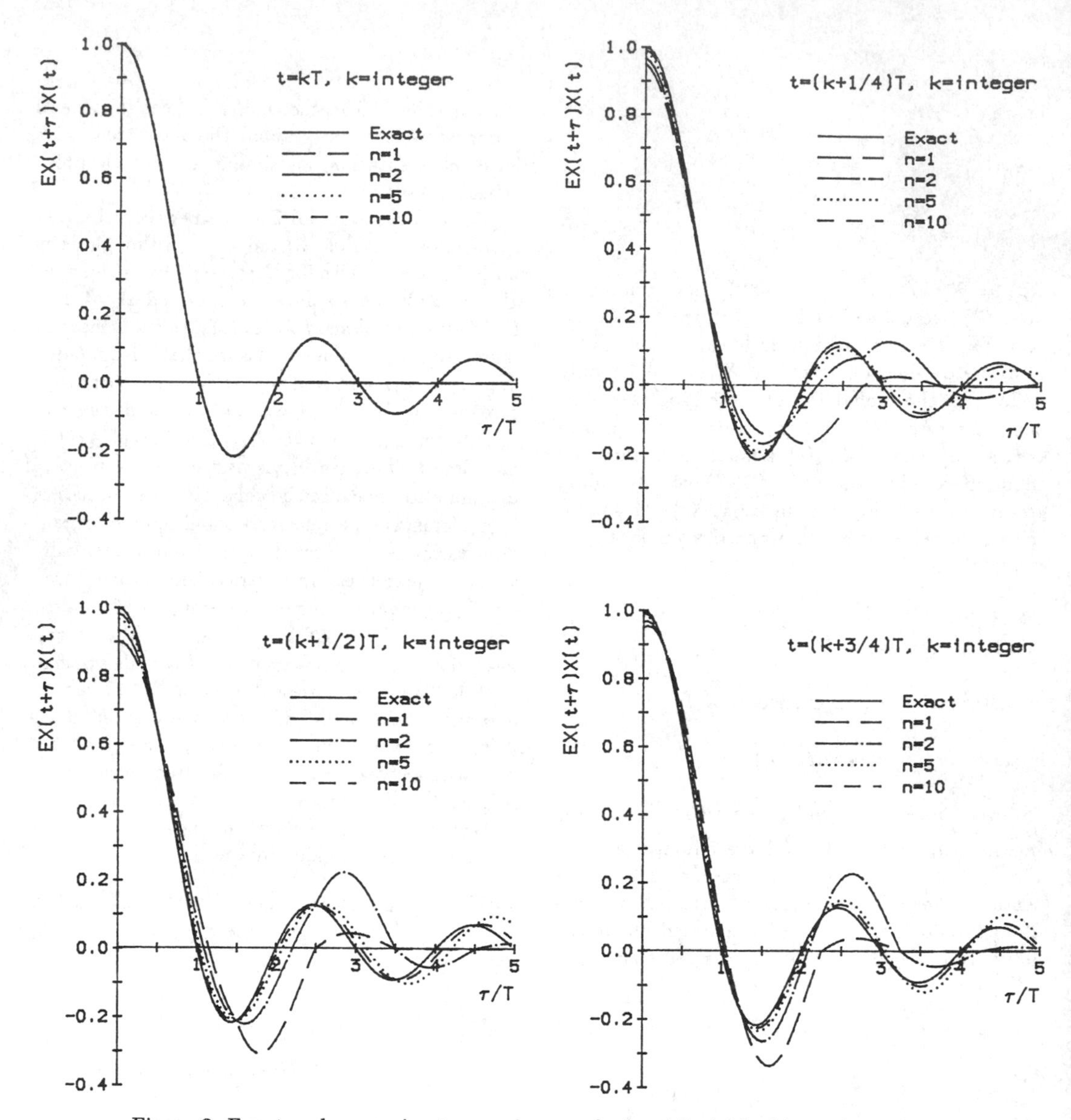

Figure 2. Exact and approximate covariances of a band limited white noise process.

Table 1. Mean upcrossing rates of the envelope of a narrow band Gaussian process.

Level v	Exact	Simulation			
		$n=1$	$n=2$	$n=5$	$n=10$
1.0	4.39×10^{-1}	4.13×10^{-1}	4.21×10^{-1}	4.28×10^{-1}	4.31×10^{-1}
2.0	1.96×10^{-1}	1.72×10^{-1}	1.79×10^{-1}	1.86×10^{-1}	1.92×10^{-1}
3.0	2.71×10^{-2}	1.80×10^{-2}	1.93×10^{-2}	2.21×10^{-2}	2.27×10^{-2}
4.0	9.71×10^{-4}	4.60×10^{-4}	6.80×10^{-4}	8.20×10^{-4}	6.60×10^{-4}

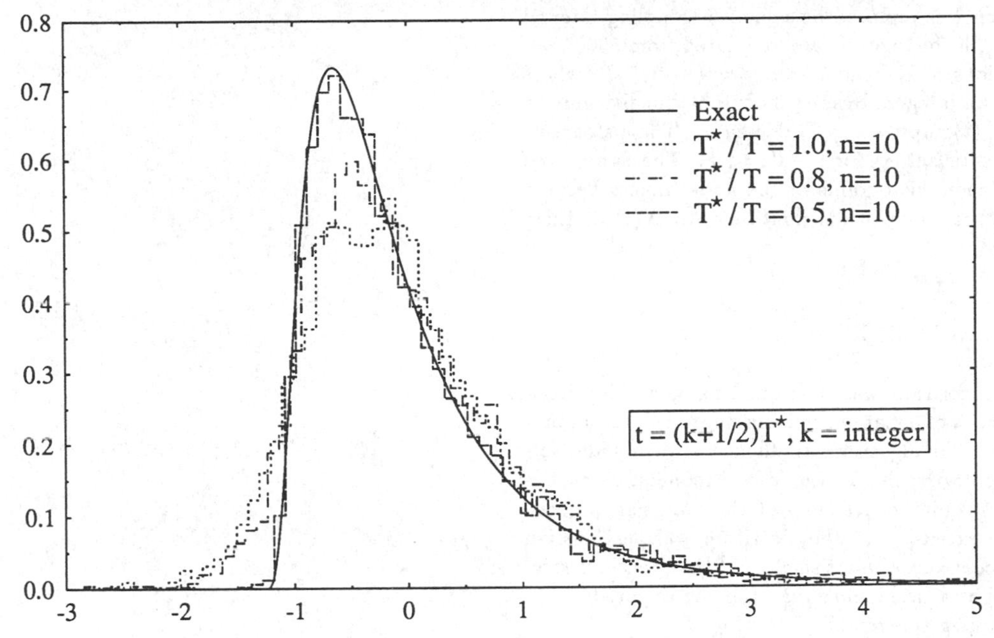

Figure 3. Marginal density and histograms of the local model in Eq. 13 for a lognormal translation process.

in which

$$
\begin{aligned}
Y_{n,c}(t) &= \sum_{k=n_t-n}^{n_t+n+1} X_{c,k}\ \alpha_k(t;\ T) \\
Y_{n,s}(t) &= \sum_{k=n_t-n}^{n_t+n+1} X_{s,k}\ \alpha_k(t;\ T)
\end{aligned}
\tag{15}
$$

with n_t and n as in Eq. 13. The local representation $Y_n(t)$ of $X(t)$ has the same asymptotic properties as $X_N(t)$ in Eqs. 12 and 13.

4 EXAMPLES

Suppose that $X(t)$ is a zero mean stationary Gaussian process with a narrow band spectrum $s(f) = s_0 = 0.5$, for $|f \pm f_0| < \bar{f}$, $0 < \bar{f} \ll f_0$, $\bar{f} = 0.5$, $f_0 = 20$, and zero otherwise. Figure 1 shows the exact and approximate covariances of $X(t)$. The approximate covariances correspond to the local model $Y_n(t)$ in Eqs. 14–15 for several values of n. The approximations improve with n. Differences between the exact and approximate covariances can only be observed for relatively large values of the time lag. The peaks of narrow band processes tend to cluster in time such that maxima of $X(t)$ can be estimated more accurately from upcrossings of the envelope $V(t)$ of the process defined in Eq. 8. Table 1 gives exact and approximate mean v-upcrossing rates of $V(t)$. The approximations are estimates of the mean v-upcrossing rate of $V(t)$ and are based on 5000 realizations of the local model in Eqs. 14–15 for several values of n. They are satisfactory for values of n larger than 2.

As a second example consider a lognormal translation process

$$X(t) = a + \exp(\sigma Z(t)) \tag{16}$$

in which $a = -\sqrt{(1+\sqrt{5})/2}$, $\sigma^2 = \log\left((1+\sqrt{5})/2\right)$, and $Z(t)$ is a zero mean unit variance stationary Gaussian process with covariance function

$$\rho(\tau) = \frac{\sin\left(2\pi \bar{f} \tau\right)}{2\pi \bar{f} \tau} \tag{17}$$

The process $X(t)$ has mean zero and variance one. Figure 2 shows the exact and approximate covariance functions of the band limited white noise $Z(t)$ for $\bar{f} = 0.5$. The approximations are based on the local model in Eq. 13. They improve with increasing n and depend on the value of t.

Figure 3 shows the marginal density of $X(t)$ and histograms of the local model $Y_n(t)$ in Eq. 13 in which T is replaced by a smaller sampling interval T^*. The histograms are calculated from 5000 realizations of $Y_n(t)$ at instant $t = (k+1/2)T^*$ where k is an integer. Results show that the distribution of $Y_n(t)$ improves as T^* decreases. The histograms are satisfactory for $T^*/T = 0.5$. The figure does not show histograms at the nodal points because $Y_n(t)$ is equal in distribution with $X(t)$ at these points.

5 CONCLUSIONS

An algorithm was developed for generating realizations of a stationary band limited Gaussian process. The algorithm is efficient and provides satisfactory results for the second-moment characteristics and the extremes of the Gaussian process. The extension of the algorithm to non-Gaussian processes can be satisfactory if the process is sampled at a rate higher than the one required by the sampling theorem.

REFERENCES

Brigham, E. O. 1974. *The Fast Fourier Transform.* Englewood Cliffs, New Jersey: Prentice-Hall, Inc.

Cramer, H. and Leadbetter, M. R. 1967. *Stationary and Related Processes.* New York: John Wiley & Sons.

Davenport, W. B. and Root, W. L. 1958. *An Introduction to the Theory of Random Signals and Noise.* New York: McGraw-Hill Book Co., Inc.

Grigoriu, M. and Balopoulou, S. 1992. *A Simulation Method for Stationary Gaussian Random Functions Based on the Sampling Theorem.* Report No. NCEER-92-0015, National Center for Earthquake Engineering Research, State University of New York at Buffalo

Wong, E. and Hajek, B. 1985. *Stochastic Processes in Engineering Systems.* New York: Springer Verlag.

Structural Safety & Reliability, Schuëller, Shinozuka & Yao (eds) © 1994 Balkema, Rotterdam, ISBN 90 5410 357 4

Computational stochastic mechanics: Recent and future developments

Masanobu Shinozuka
Princeton University, N.J., USA

ABSTRACT: This paper deals with three subjects in computational stochastic mechanics that appear to be particularly promising from the research and applications point of view: the concept of variability response function, the buckling of imperfection-sensitive structures and some computational aspects of stochastic dynamics.

1 INTRODUCTION

Three areas of research and applications in computational stochastic mechanics that appear to be particularly promising are examined in this paper: the concept of variability response function, the buckling of imperfection-sensitive structures and some computational aspects of stochastic dynamics.

2 CONCEPT OF VARIABILITY RESPONSE FUNCTION

2.1 Introduction

The challenge of realistically accounting for uncertainties that are inherent in all structural systems has attracted a lot of attention in recent years. Considerable material and/or geometric random inhomogeneities are either introduced during the manufacturing process or can simply be found as an inherent characteristic of the material itself. The latter case is especially true for concrete and soil. A rigorous description of these random properties can only be achieved using stochastic fields. Because of the system stochasticity outlined above, the response of the system is going to be stochastic as well. The analysis of the response variability of stochastic structural systems, in general, consists of evaluating the first and second moment of their response.

As analytical solutions are restricted to simple linearly elastic structures, numerical techniques have to be used to deal with more complex stochastic structural systems. In most cases, the only available numerical approach is a class of methods known as "Stochastic Finite Element Methods" (SFEM). Four papers containing literature reviews of SFEM are mentioned here: Vanmarcke et al. (1986), Benaroya and Rehak (1988), Brenner (1991) and Der Kiureghian et al. (1991).

One of the most important concepts in understanding the underlying mechanisms of the response variability of such systems is the variability response function (Shinozuka, 1987a; Deodatis and Shinozuka, 1989; Deodatis, 1990).

2.2 Definition

For the purpose of demonstrating the concept, one-dimensional stochastic structural systems (e.g. trusses and frames) are considered with elastic modulus varying over the area of a finite element (e) as:

$$E^{(e)}(x) = E_0^{(e)}[1 + f^{(e)}(x)] \qquad (1)$$

For two-dimensional systems (e.g. plane stress/plane strain problems), the corresponding expression is:

$$E^{(e)}(x,y) = E_0^{(e)}[1 + f^{(e)}(x,y)] \quad (2)$$

while for three-dimensional problems:

$$E^{(e)}(x,y,z) = E_0^{(e)}[1 + f^{(e)}(x,y,z)] \quad (3)$$

where $E_0^{(e)}$ = mean value of elastic modulus and $f^{(e)}(x)$, $f^{(e)}(x,y)$ and $f^{(e)}(x,y,z)$ = 1D, 2D and 3D zero-mean, homogeneous stochastic fields, respectively, describing the (stochastic) fluctuation of the elastic modulus around its mean value. In order to prevent the occurrence of non-positive values of the elastic modulus, bounds are imposed on $f^{(e)}(x)$, $f^{(e)}(x,y)$ and $f^{(e)}(x,y,z)$ (e.g. Wall and Deodatis, 1993). Using the principle of stationary potential energy, the stochastic element stiffness matrix can be expressed as (Deodatis, 1990; Deodatis, 1991; Deodatis and Shinozuka, 1991 and Wall and Deodatis, 1993):

$$\mathbf{K}^{(e)} = \mathbf{K}_0^{(e)} + \sum_{k=1}^{M} X_k^{(e)} \Delta \mathbf{K}_k^{(e)} \quad (4)$$

where all matrices in the right-hand-side of Eq. 4 are deterministic and the weighted integrals $X_k^{(e)}$ are random variables defined as:

$$X_k^{(e)} = \int_{L^{(e)}} x^i f^{(e)}(x) dL^{(e)} \quad (5)$$

$$X_k^{(e)} = \int_{A^{(e)}} x^i y^j f^{(e)}(x,y) dA^{(e)} \quad (6)$$

$$X_k^{(e)} = \int_{V^{(e)}} x^i y^j z^k f^{(e)}(x,y,z) dV^{(e)} \quad (7)$$

Equations 5, 6 and 7 correspond to the 1D, 2D and 3D cases, respectively, with $L^{(e)}$, $A^{(e)}$ and $V^{(e)}$ denoting the length, area and volume of the (e)-th finite element, respectively. Due to the zero-mean property of the weighted integrals $X_k^{(e)}$, $\mathbf{K}_0^{(e)}$ represents the mean value of the stochastic element stiffness matrix. Using standard finite element analysis methodology, the equations of equilibrium can be written as:

$$\mathbf{KU} = \mathbf{P} \quad (8)$$

where $\mathbf{K}$ is the stochastic global stiffness matrix assembled from the stochastic element stiffness matrices, $\mathbf{U}$ is the stochastic global displacement vector and $\mathbf{P}$ is the deterministic global force vector.

To analyze the response variability, the displacement vector $\mathbf{U}$ is approximated by a first-order Taylor expansion around the mean values of the weighted integrals:

$$\mathbf{U} \approx \mathbf{U}_0 + \sum_{e=1}^{N} \sum_{k=1}^{M} \left(X_k^{(e)} - \bar{X}_k^{(e)} \right) \left[\frac{\partial \mathbf{U}}{\partial X_k^{(e)}} \right]_{\mathcal{E}} \quad (9)$$

where both subscripts $_0$ and $_{\mathcal{E}}$ denote evaluation at the mean values of the weighted integrals, N is the total number of finite elements and $\bar{X}_k^{(e)}$ denotes the mean value of $X_k^{(e)}$.

It can be shown (Shinozuka, 1987a; Deodatis and Shinozuka, 1989; Deodatis, 1990; Deodatis, 1991; Deodatis and Shinozuka, 1991 and Wall and Deodatis, 1993) that the variance vector of $\mathbf{U}$ can be expressed in the following form:

$$\text{Var}\{\mathbf{U}\} = \int_{-\infty}^{\infty} S_{ff}(\kappa_x)\, \mathbf{VRF}(\kappa_x) d\kappa_x \quad (10)$$

$$\text{Var}\{\mathbf{U}\} = \int_{-\infty}^{\infty} \int_{-\infty}^{\infty} S_{ff}(\kappa_x, \kappa_y)\, \mathbf{VRF}(\kappa_x, \kappa_y) d\kappa_x\, d\kappa_y \quad (11)$$

$$\text{Var}\{\mathbf{U}\} = \int_{-\infty}^{\infty} \int_{-\infty}^{\infty} \int_{-\infty}^{\infty} S_{ff}(\kappa_x, \kappa_y, \kappa_z)\, \mathbf{VRF}(\kappa_x, \kappa_y, \kappa_z) d\kappa_x\, d\kappa_y\, d\kappa_z \quad (12)$$

Note again that Eqs. 10, 11 and 12 correspond to the 1D, 2D and 3D cases, respectively. The above three equations are derived under the assumption that the elastic modulus of all finite elements is characterized by the same stochastic field, having power spectral density function $S_{ff}(\kappa_x)$, $S_{ff}(\kappa_x, \kappa_y)$ and $S_{ff}(\kappa_x, \kappa_y, \kappa_z)$, respectively, with κ_x, κ_y and κ_z denoting the wave numbers.

$\mathbf{VRF}(\kappa_x)$, $\mathbf{VRF}(\kappa_x, \kappa_y)$ and $\mathbf{VRF}(\kappa_x, \kappa_y, \kappa_z)$ in Eqs. 10, 11 and 12, respectively, are vectors whose components are the first-order approximations of the variability response function (Deodatis and Shinozuka 1989) at the corresponding degree of freedom of the structure. In the following, all three quantities will be simply called variability response function. Closed-form analytic expressions have been established for the variability response function for several 1D and 2D problems (Shinozuka, 1987a; Deodatis and Shinozuka, 1989; Deodatis, 1990; Deodatis, 1991; Deodatis and Shinozuka, 1991 and Wall and Deodatis, 1993).

At this point it should be mentioned that the variability response function is non negative and symmetric with respect to the origin of the coordinate system.

2.3 Importance of variability response function

Equations 10, 11 and 12 indicate that the variance vector of $\mathbf{U}$ will be a function of the particular form of the power spectral density function (note the similarity of the variability response function to the frequency response function in random vibration analysis). Unfortunately, beyond a reasonable estimation of its mean value (which is assumed to be zero without loss of generality) and coefficient of variation, very little information is usually available about the probabilistic characteristics of the stochastic field describing the random material property. It is this unavailability of detailed information that makes it engineering-wise very significant to establish "spectral-distribution-free" bounds of the response variability of the system. The paramount importance of the variability response function lies in the fact that it can be used to establish such bounds for cases where the power spectral density (or autocorrelation) function of the stochastic field is not known. The calculation of these "spectral-distribution-free" bounds is achieved following a procedure similar to the one described in Shinozuka (1987a) and Deodatis and Shinozuka (1989).

The upper bound for the coefficient of variation of the i-th nodal displacement U_i can be expressed as:

$$\mathrm{COV}[U_i] \leq \sigma_{ff} \sqrt{\frac{\text{maximum}\{VRF_i\}}{|\mathcal{E}\{U_i\}|^2}} \tag{13}$$

where σ_{ff} is the coefficient of variation of the elastic modulus, VRF_i is the variability response function corresponding to the i-th degree of freedom U_i and $\mathcal{E}$ denotes the expectation. Since closed-form analytic expressions are available for the variability response function, it is straightforward to estimate the maximum of VRF_i as indicated in Eq. 13.

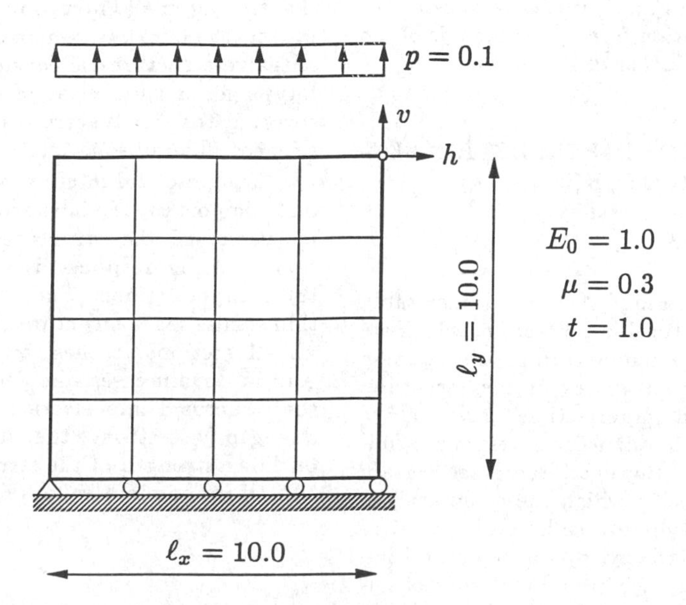

Fig. 1 Square Plate Subjected to In-Plane Loading

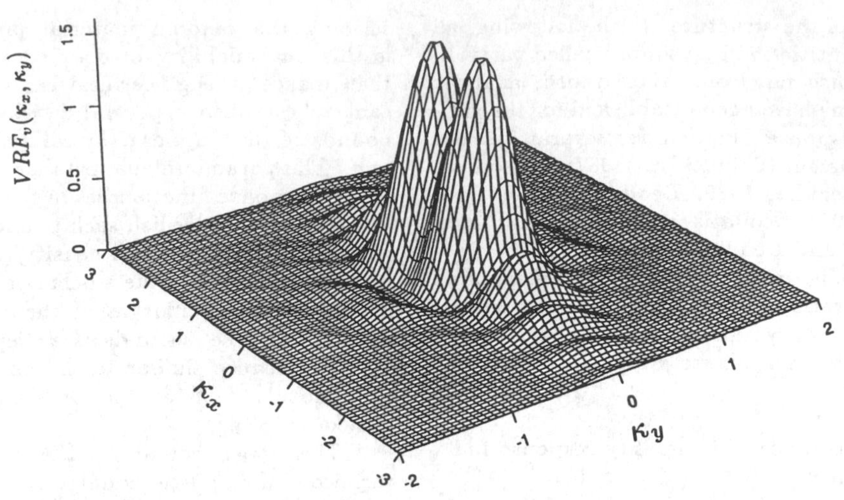

Fig. 2 Variability Response Function of Vertical Displacement v as a Function of κ_x and κ_y

2.4 Numerical Example

A square plate subjected to in-plane loading is considered in this numerical example. The structure is shown in Fig. 1 along with numerical values of several parameters involved in the problem. The variability response function VRF_v of the vertical displacement of the upper right node v (see Fig. 1) is plotted as a function of wave numbers κ_x and κ_y in Fig. 2. Based on Fig. 2, it is easy to establish spectral-distribution-free upper bounds of the coefficient of variation of v.

3 BUCKLING OF IMPERFECTION-SENSITIVE STRUCTURES

3.1 Introduction

Practical engineering structures inevitably carry small structural imperfections which are inherent in the manufacturing, the erection and the loading processes of the structure. In general, such imperfections have only a mild effect on the structural response (Shinozuka, 1987b). However, there are certain exceptional cases, in which the structural response is altered dramatically, both quantitatively and qualitatively, by the structural imperfections. One such prominent case of imperfection sensitivity is the bifurcation buckling of structures (Bazant and Cedolin, 1991).

The study of the effect of structural imperfections on the bifurcation buckling of structures has both great theoretical and practical interest. The theoretical interest stems from the sensitive dependence of the response (buckling strength of the structure) on the initial data (loads and material and geometric properties), so that the response exhibits certain characteristics of chaotic systems (Thompson and Hunt, 1984). In fact, the buckling response of imperfection-sensitive structures is considered as the prototype for a large class of similar problems, covering the full spectrum of the engineering sciences (Thompson, 1982).

The practical interest stems from the primary importance in advanced engineering applications of the structures whose bifurcation buckling response is sensitive to structural imperfections. Such structures include thin shells, space structures, beams with thin-walled sections, arches, trusses and frames. And in certain cases, the detrimental effect of the structural imperfections on the buckling strength is so strong, that it can decrease the buckling strength of the structure even below 50% of its nominal value (Yamaki, 1984).

3.2 Current state-of-the-art

Up to date, practically all the investigations on the effect of structural imperfections on the buckling response of structures have been based on Koiter's theory (Bazant and Cedolin, 1991; Thompson and Hunt, 1984; Thompson, 1982 and Yamaki, 1984). This theory employs an asymptotic expansion of the structural response around the bifurcation buckling load of the corresponding perfect structure, i.e. the idealized structure with no structural imperfections. It results in simple formulas for the pre-buckling and the post-buckling response of the structure and for the effect of one-mode shape imperfections on the buckling strength. Koiter's theory has been very successful in its qualitative predictions about the various types of buckling responses of imperfection-sensitive structures.

However, there have been some reservations with respect to certain qualitative and quantitative predictions by this theory, especially about the effect of structural imperfections on the buckling response. From the analytical point of view, these reservations stem from the asymptotic nature of the theory, which means that it is accurate enough only in a small region around the bifurcation buckling load of the corresponding perfect structure (Palassopoulos, 1991). From experimental point of view, the main prediction by this theory, about the dominance of the effect of a certain one-mode shape imperfection on the buckling strength, has not been verified. On the contrary, the experimental results indicate that the buckling strength does depend on a multitude of imperfection components, in addition to this one-mode shape imperfection (Arbocz et al., 1987).

The present day situation, however, is best exemplified by the current design practice for structures which are subjected to bifurcation buckling (Kollar and Dulacska, 1984). Koiter's theory is used only qualitatively for a gross assessment of the imperfection sensitivity of the structure. Then, very large (of the order of 10) empirical safety factors are employed, mainly in order to cover our present-day limitations that exist in analyzing the effect of the structural imperfections on the buckling strength.

3.3 Recent theoretical developments

An alternative approach has been recently proposed by Palassopoulos (1991, 1992a,b) to solve such problems. This approach is based on an alternative conceptualization of the structural response, employing a universal imperfection magnitude parameter as control variable for the problem. Then, using perturbation theory for the analysis of the structural response, the buckling strength and mode of the structure are obtained from the solution of a generalized eigenvalue problem.

Research is currently conducted along the lines of this alternative approach to introduce a stochastic description of these imperfections. The objectives of this research effort are:

a. Systematic search and identification of all components of stochastic structural imperfections which affect significantly the buckling response.
b. Identification and understanding of qualitatively new phenomena, associated with the interaction among these random imperfection components and their effect on the buckling response.
c. Development of appropriate methods for the simulation of the significant components of the structural imperfections as stochastic fields.
d. Development of reliable methods for the accurate evaluation of the effect of imperfections on the buckling response, incorporating the recent theoretical developments into the general framework of the Stochastic Finite Element Method.
e. Establishment of correlation between the imperfection and the manufacturing, erection and loading processes of the structure.

4 COMPUTATIONAL ASPECTS OF STOCHASTIC DYNAMICS

In recent times there has been a surge of interest in the so-called "colored noise" (non-white noise) studies in stochastic mechanics (Moss and McClintock, 1989). This interest is the outcome of the intent to depict physical mechanisms of a variety of engineering systems. However, when non-white noise is invoked, the process becomes non-Markovian and theoretical solutions are in general difficult to obtain, even in approximation. For such physical systems, noise simulated from the power spectral density using FFT on the basis of spectral representation of stochastic processes (Shinozuka and Deodatis, 1991) can be applied with substantial merit. The advantage of the method lies in the ease of generation of non-white and Gaussian noise along with substantial control on noise parameters like duration and time discretization (Shinozuka and Deodatis, 1991). More importantly, the generated noise can be tailored to the need of particular system parameters.

As a first example, results are presented on the application of the spectral representation (SR) method to simulate the mean first passage time (MFPT) T_F for an overdamped particle in bistable potential (Billah and Shinozuka, 1990). The dynamic equation is:

$$\dot{x}(t) = \frac{dU}{dx} + f(t) = -x + x^3 + f(t), \quad (14)$$

where $U(x) = x^4/4 - x^2/2$ (Fig. 3), with an unstable position at $x = 0$ and two stable ones at $x = -1, 1$; where $f(t)$ is a zero-centered, stationary Gaussian noise with autocorrelation function:

$$< f(t)f(t') >= (\frac{D}{\tau}) \exp\,(-|t - t'|/\tau), \quad (15)$$

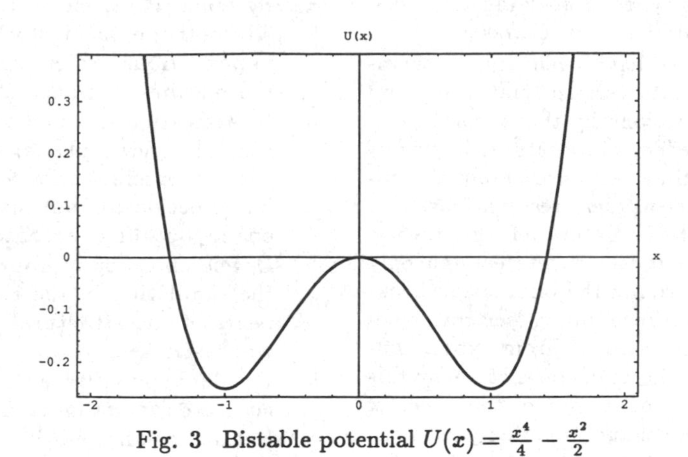

Fig. 3 Bistable potential $U(x) = \frac{x^4}{4} - \frac{x^2}{2}$

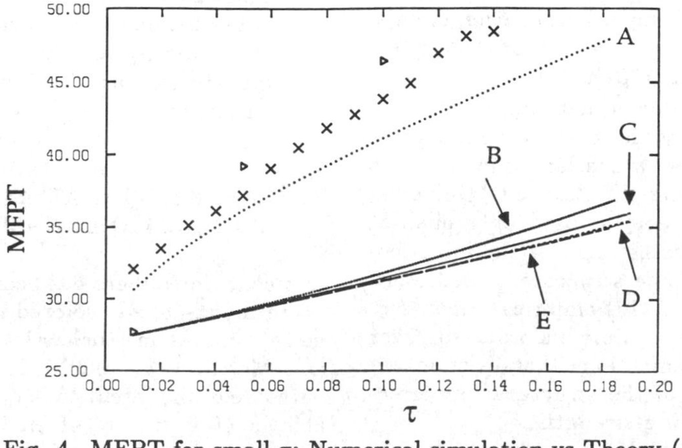

Fig. 4 MFPT for small τ; Numerical simulation vs Theory (A to D). Numerical – triangle: SDE method; cross: PSD method

where D is the noise intensity and τ the correlation time. Using the SR method, noise can be directly generated from $S(\omega)$ corresponding to the exponentially correlated noise.

On the other hand noise can also be generated by the following equation:

$$f'(t) = -\frac{f(t)}{\tau} + \frac{\xi(t)}{\tau}, \qquad (16)$$

where $< \xi(t)\xi(t') >= 2D\delta(t - t')$. This method is often termed as the Stochastic Differential Equation (SDE) method. It is important to note that in this method Eq. 14 and Eq. 16 together represent the system under noise.

On the other hand, for the SR method, Eq. 14 along with noise characteristics given by $S(\omega)$ represent the physical phenomenon; the method is effectively one-dimensional. This characteristic of the SR method is substantially advantageous compared to the SDE method. This can be seen as follows: For the SDE method, the system has to be solved in the enlarged space of variables x and f. Therefore, even when the noise $\xi(t) = 0$, the system stays two-dimensional. This aspect has generated the controversy (Moss and McClintock, 1989) in the literature on the interpretation of the system when $f(t) = 0$.

It is to be noted that in the theoretical solution to the problem, both Eqs. 14 and 16 are used which means that the uncoupled state of the system of $x(t)$ and $f(t)$ has to be carefully interpreted. Different theoretical analysis has been carried out to date using different types of the "extended Fokker-Planck" equation but none could be verified by the SDE system, since $f(t)$ was also generated by a similar two-dimensional system.

Using the SR method, the validity of the different theories for MFPT under colored noise can be ascertained quite easily. One such example is shown in Fig. 4 in which A to D represents the plot of MFPT for five different theories. The "triangle" represents the numerical SDE approach which is inconclusive (it jumps up and down the figure). The "cross" represents the result using the SR method; the results are consistent in that it follows a particular curve. Also, the theory represented by Curve A (Doering et al., 1987) shows substantial matching with the SR method especially for very small τ.

Another important application of the SR method is for systems where stability is of concern. A particular case of this is the stability study of suspension bridges (Shinozuka and Billah, 1993). The peculiar nature of this problem is that the control parameter (the parameter by which turbulence enters the system), namely, the upstream flow velocity is nonlinear which results in nonlinear noise. For this nonlinear noise problem it is agreed upon today that white noise cannot be used as it brings in physical and mathematical inconsistency.

Experimental observations on section model tests indicate a duality – stability as well as instability – in the s.d.o.f. torsional instability under turbulence and a lack of satisfactory interpretation of this duality has existed in the literature for certain time. Due to the lack of theoretical results numerical simulation studies were carried out using the SR method which for the first time gave a physical interpretation of the duality: The stabilization and destabilization is due to the correlation of the turbulence. Also, the study showed that the effect of the nonlinear noise is substantially important for stability.

Further, a recent numerical study (Billah and Shinozuka, 1991) using the SR method has verified a complicated theoretical study which deals with the effect of colored noise on a Duffing oscillator with a negative stiffness (a physical representation of a buckled beam or plate when only one mode of vibration is considered).

ACKNOWLEDGMENTS

This work was supported by Contract No. NCEER 92-3302A under the auspices of the National Center for Earthquake Engineering Research. The author also acknowledges the assistance by K.Y.R Billah, G. Deodatis and G. Palassopoulos in preparation of this paper.

REFERENCES

Arbocz, J. et al. (1987). "Buckling and Post-Buckling," Springer, New York, N.Y.

Bazant, Z.P. and Cedolin, L. (1991). "Stability of Structures," Oxford University Press, New York N.Y.

Benaroya, H. and Rehak, M. (1988). "Finite element methods in probabilistic structural analysis: A selective review," *Applied Mechanics Reviews*, 41(5), 201–213.

Billah, K.Y.R. and Shinozuka, M. (1990). "Numerical method for colored-noise generation and its application to a bistable system," *Physical Review A*, Vol. 42, pp. 7492.

Billah, K.Y.R. and Shinozuka, M. (1991). "Stabilization of nonlinear system by multiplicative noise," *Physical Review A*, Vol. 44, p. R4779.

Brenner, C.E. (1991). "Stochastic finite element methods," *Internal Working Report No. 35-91*, Institute of Engineering Mechanics, University of Innsbruck, Austria.

Deodatis, G. (1990). "Bounds on response variability of stochastic finite element systems," *Journal of Engineering Mechanics*, ASCE, 116(3), 565–585.

Deodatis, G. (1991). "Weighted integral method. I: Stochastic stiffness matrix," *Journal of Engineering Mechanics*, ASCE, 117(8), 1851–1864.

Deodatis, G. and Shinozuka, M. (1989). "Bounds on response variability of stochastic systems," *Journal of Engineering Mechanics*, ASCE, 115(11), 2543–2563.

Deodatis, G. and Shinozuka, M. (1991). "Weighted integral method. II: Response variability and reliability," *Journal of Engineering Mechanics*, ASCE, 117(8), 1865–1877.

Der Kiureghian, A., Li, C.C. and Zhang, Y. (1991). "Recent developments in stochastic finite elements," *Lecture Notes in Engineering IFIP 76, Proceedings of Fourth IFIG WG 7.5 Conference*, (Eds. Rackwitz and Thoft-Christensen), Germany, Springer-Verlag.

Doering, C.R., Hagan, P.S. and Levermore, C.D. (1987). "Bistability driven by weakly colored Gaussian noise: the Fokker-Planck boundary layer and mean first-passage times," *Physical Review Letter*, Vol. 59, p. 2129.

Kollar, L. and Dulacska, E. (1984). "Buckling of Shells for Engineers," Wiley, New York, N.Y.

Moss, F. and McClintock, P.V.E. (Editors) (1989). "Noise in nonlinear dynamical systems," Vol. 1-3, Cambridge University Press, Cambridge.

Palassopoulos, G.V. (1991). "Reliability-Based Design of Imperfection Sensitive Structures," Journal of Engineering Mechanics, ASCE, Vol. 117, No. 6, pp. 1220-1240.

Palassopoulos, G.V. (1992a). "Response Variability of Structures Subjected to Bifurcation Buckling," Journal of Engineering Mechanics, ASCE, Vol. 118, No. 6, pp. 1164-1183.

Palassopoulos, G.V. (1992b). "A New Approach to the Buckling of Imperfection-Sensitive Structures," Journal of Engineering Mechanics, ASCE, accepted for publication, Manuscript No. 003667-EM.

Shinozuka, M. (1987a). "Structural Response Variability," Journal of Engineering Mechanics, ASCE, Vol. 113, No. 6, pp. 825-842.

Shinozuka, M. (1987b). "Stochastic Mechanics," Vols. I, II, III, Department of Civil Engineering and Engineering Mechanics, Columbia University, New York, N.Y.

Shinozuka, M. and Billah, K.Y.R. (1993). "Stability of Long-span bridges in turbulent flow," Proc. of 7th US National Conference on Wind Engineering, Vol. 2, p. 663.

Shinozuka, M. and Deodatis, G. (1991). "Simulation of stochastic processes by spectral representation," *Applied Mechanics Reviews*, Vol. 44, No. 4, pp. 191–204.

Thompson, J.M.T. (1982). "Instabilities and Catastrophes in Science and Engineering," Wiley, New York, N.Y.

Thompson, J.M.T. and Hunt, G.W. (1984). "Elastic Instability Phenomena," Wiley, New York, N.Y.

Vanmarcke, E., Shinozuka, M., Nakagiri, S., Schuëller, G.I. and Grigoriu, M. (1986). "Random fields and stochastic finite elements," *Structural Safety*, 3(3+4), 143–166.

Wall, F.J. and Deodatis, G. (1993). "Variability response functions and upper bounds of response of 2D stochastic systems," *submitted for publication to the ASCE Journal of Engineering Mechanics.*

Yamaki, N. (1984). "Elastic Stability of Circular Cylindrical Shells," North-Holland, New York, N.Y.

Structural Safety & Reliability, Schuëller, Shinozuka & Yao (eds) © 1994 Balkema, Rotterdam, ISBN 90 5410 357 4

Indirect sampling method for stochastic mechanics problems

P.D. Spanos & B.A. Zeldin
Rice University, Houston, Tex., USA

ABSTRACT: A new numerical method for problems of stochastic mechanics and other areas involving a small number of random parameters is presented. In essence, it is a stratified sampling method but more efficient. Alternatively, this new method can be viewed as a Galerkin type approximation in the sample space. The developed solution exhibits certain optimality features due to the orthogonality of the Galerkin projection. Therefore, the proposed method is superior in terms of the sampling error and of the requisite computational time over the straightforward Monte Carlo method and the stratified sampling method. The usefulness of the proposed method is examined by analyzing the behavior of a beam with random rigidity; the determination of the natural frequencies and of the seismic response of the beam is discussed.

1 INTRODUCTION

The Monte-Carlo simulation method has been widely used in the field of stochastic mechanics and others fields, primarily because of its versatility. Often it is the only option available to solve complex problems. However, indiscriminate use of the method can not be advocated due to its considerable computational cost. In fact, several variance reduction techniques have been developed in this regard. They involve importance sampling, stratified sampling, and others [1].

A numerical method for problems of stochastic mechanics and other areas representing the solution by a small number of random parameters is presented. In essence, it is a stratified sampling method, but more efficient. Alternatively, this new method can be viewed as a Galerkin approximation in the sample space. Several examples are considered involving the use of the Loeve-Karhunen expansion for stochastic fields approximation [2,3]. The examples deal with the evaluation of natural frequencies and seismic response of beams with random rigidity.

2 FORMULATION

2.1 Solution representation

Consider a problem with random parameters governed by the equation

$$L(\underline{\xi})\, u = f(\underline{\xi}), \qquad (1)$$

where $\underline{\xi} = (\xi_1, \xi_2, \ldots \xi_M)$ is a random vector, and $L(\underline{\xi})$ is a mathematical operator describing the performance of the system. Further, $f(\underline{\xi})$ describes the load, and $\xi_1, \xi_2, \ldots \xi_M$ are statistically independent random variables.

Solving equation (1) is equivalent to finding a function $u = u(\underline{\xi})$ such that for every realization $\underline{\xi}^* = (\xi^*_1, \xi^*_2, \ldots \xi^*_M)$ of the random vector $\underline{\xi}$ there exists a deterministic function $u(\underline{\xi}^*)$ which satisfies equation (1). Consider the space R^M of $\xi_1 \times \xi_2 \times \ldots \xi_M$ as the sample space Ω. This space can be divided into N subdomains or strata $\{\Omega_i, i = 1, N\}$ having the shape of M-dimensional disjoint rectangles with prescribed probability mass as shown in Figure 1 for M=2. Next, introduce the set of functions or spline basis $\{\varphi_i, i = 1, \ldots N\}$ such that

$$\varphi_i(\underline{\xi}) = \begin{cases} 1, & \text{if } \ \underline{\xi} \in \Omega_i \\ 0, & \text{otherwise} \end{cases}. \qquad (2)$$

Clearly, $\varphi_i(\underline{\xi})$ and $\varphi_j(\underline{\xi})$ have disjoint supports. That is,

$$\int_{\Omega} q(\underline{\xi})\, \varphi_i(\underline{\xi})\, \varphi_j(\underline{\xi})\, p_{\underline{\xi}}(\underline{\xi})\, d\underline{\xi} = 0 \qquad \text{if} \quad i \neq j, \qquad (3)$$

where $p_{\underline{\xi}}(\underline{\xi})$ is the probability density function of $\underline{\xi}$, $q(\underline{\xi})$ is an arbitrary random variable. Thus, the set $\{\varphi_i, i = 1, \ldots N\}$ is an orthogonal basis for the class of random variables which are constant for every Ω_i. Any random variable can be approximated adequately by the use of these basis functions, provided the partition of Ω is fine. The solution of equation (1) can be represented as a linear combination of the functions $\{\varphi_i, i = 1, \ldots N\}$. That is,

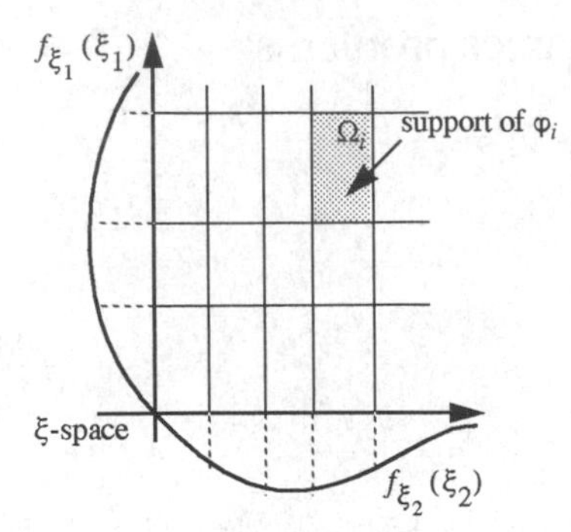

Figure 1. Stratified sample space

$$u(\xi) = \sum_{i=1}^{N} c_i \varphi_i(\xi), \quad (4)$$

where the coefficients c_i are to be determined.

Then, the solution given by equation (4) can be construed as a projection of the exact solution into the space spanned by $\{\varphi_i, i = 1, \ldots N\}$. Expressing the solution in the form of the equation (4), the induced error in the equation (1) can be made orthogonal to the space spanned by $\{\varphi_i, i = 1, \ldots N\}$. That is, using the operator of mathematical expectation, < >, it can be found

$$\langle L(\xi) \sum_{i=1}^{N} c_i \varphi_i((\xi) \varphi_j(\xi)) \rangle = \langle f((\xi) \varphi_j(\xi)) \rangle$$

$$j = 1, \ldots, N, \quad (5)$$

which, because of equation (3), leads to

$$\langle L(\xi) c_i \varphi_i^2(\xi) \rangle = \langle f(\xi) \varphi_i(\xi) \rangle. \quad (6)$$

This sequence of deterministic equations can be solved to find the coefficients c_i. Upon deriving c_i, the statistical properties of the solution can be estimated by relying on equation (4). Specifically,

$$\langle u(\xi) \rangle = \sum_{i=1}^{N} c_i p_i, \text{ and } \langle u^2(\xi) \rangle = \sum_{i=1}^{N} c_i^2 p_i, \quad (7)$$

where

$$p_i = \langle \varphi_i(\xi) \rangle = \langle \varphi_i^2(\xi) \rangle. \quad (8)$$

Similarly, the distribution function of the solution can be found using the equation

$$P_u(v) = Pr(u < v) = \sum_{i=1}^{N} \chi_v(c_i) p_i, \quad (9)$$

where

$$\chi_v(x) = \begin{cases} 1 & \text{if } x \le v \\ 0 & \text{if } x > v \end{cases}. \quad (10)$$

2.2 Solution interpretation

The proposed method may by viewed as a Galerkin-type procedure for random media. Several authors have explored the idea of using projection procedures in conjuction with random variables. In references [2,3,4,5,6] this procedure has been applied for stochastic mechanics problems with randomness in the spatial domain. Due to the correlation between the solution and the random parameters describing the properties of the structures, this class of problems is especially difficult to solve. In this regard, the stochastic field has been discretized by the use of the Loeve-Karhunen expansion in references [2,3,4] or of the midpoint method in reference [5]. In this manner the problem is first characterized by a finite set of random variables. Then, the solution can be derived by a Galerkin projection into finite dimensional spaces spanned by orthogonal chaos polynomials as in references [2,3,4], or just linear functions as in reference [5]. However, these bases can yield a large order system of equations which must be solved to determine the solution.

Another possible basis for the representation shown in equation (4) is given by equation (2); see also reference [6]. The concept of using spline type approximation has been discussed widely in the area of computational mechanics in connection with the finite element method. From this perspective, the system of functions $\{\varphi_i\}_{i=1}^{N}$ represents the simplest spline of piecewise-constant functions. Then, the proposed method involves approximation of the random variables in the finite dimensional subspace of splines defined by some partition of the sample space. Additional advantages of this representation relate to equation (3) since each term c_i in the expansion (4) can be found independently. Therefore, every term c_i in equation (4) can be readily determined.

From another perspective, the use of piece-wise constant functions makes this method a generalized sampling procedure. Indeed, examining equation (6) one can deduce that this equation is equivalent to the following

$$\int_{\Omega_i} [L(\xi) c_i - f(\xi)] p_\xi(\xi) d\xi = 0. \quad (11)$$

If the Lebesgue integral involved in equation (11) can be interpreted in the Riemann sense and all pertinent quantities are adequately smooth, the mean-value theorem states that there exists some $\xi^* \in \Omega_i$ such that

$$L(\xi^*) c_i = f(\xi^*). \quad (12)$$

This equation shows that c_i represents just a solution of equation (1) for the realization ξ_* of the random vector ξ. Then, the sequence of equations (12) can be interpreted as a sequence of samplings. In fact, this "indirect sampling" is optimal in the sense that every element of this sequence represents a certain region of the sample space and can be interpreted as the only outcome with a given probability. Moreover, as $\xi_* \in \Omega_i$, the proposed method can be viewed as analog to stratified sampling [1]. But unlike the stratified sampling method the point inside every stratum is computed to make some error of the approximation of the given numerical problem (1) orthogonal to the chosen space and minimal for a given stratification.

To show this properly regression analysis can be applied [7]. Any random variable $\Theta(\xi)$ which is a function of ξ on the sample space can be estimated by $\hat{\Theta}$ using the set of random variables $\{\varphi_i, i = 1, \ldots N\}$ defined by the equation (2), where $\hat{\Theta}$ is an arbitrary function of $\{\varphi_i, i = 1, \ldots N\}$ rather than ξ. This estimate provides a minimal variance for the difference $\Theta - \hat{\Theta}$. That is,

$$\langle(\Theta - \hat{\Theta})^2\rangle \text{ is minimal.} \tag{13}$$

It can be shown, that as $\{\varphi_i, i = 1, \ldots N\}$ are indicator functions of disjoint sets, the estimate $\hat{\Theta}$ can be found using linear regression analysis and the solution can be expressed as

$$\hat{\Theta} = \sum_{i=0}^{\infty} \theta_i \varphi_i(\xi), \tag{14}$$

where $\theta_i = \langle\Theta|\Omega_i\rangle = \langle\Theta\varphi_i\rangle$, and $<|>$ denotes conditional expectation.

Let $u(\xi)$ be the exact solution of the equation (1), and let $\tilde{u}$ be an approximation of this solution. Define the error of such an approximation by

$$\varepsilon = L(\xi)\tilde{u} - L(\xi)u = L(\xi)\tilde{u} - f(\xi). \tag{15}$$

Then, the estimate $\hat{\varepsilon}$ of ε can be derived using equation (14). If the approximation $\tilde{u}$ is taken from a system of equations (5), then $\hat{\varepsilon} = 0$. Thus, the proposed method ensures that the error defined by the equation (15) has a zero mean square estimate from the indicator functions of chosen stratification.

Related perspective can be generated using some algebra concepts [7]. The vector ξ defines a sigma-algebra $\boldsymbol{G}$ in the sample space, and the stratification shown in Figure 1 defines a more coarse pure atomic σ-algebra $\boldsymbol{G}_1 \subset \boldsymbol{G}$ with indicator functions $\{\varphi_i, i = 1, \ldots N\}$. Then, the above regression analysis applied to the error ε leads to

$$\langle\varepsilon\rangle = 0 \ , \hat{\varepsilon} = \langle\varepsilon|\boldsymbol{G}_1\rangle = 0. \tag{16}$$

In other words, this method yields the minimal error defined by equation (15) with respect to the coarse σ-algebra of given stratification.

3 EXAMPLES

3.1 Preliminary remarks

The proposed method is applied for the analysis of the dynamic behavior of a beam of unit length. The beam problem can be described by the equation

$$(EI(x)u''(x,t))'' = q(x,t), \tag{17}$$

where u is the beam deflection, and q denotes the distributed force acting on the beam which in general is taken as a stochastic process. The symbol *EI(x)* denotes the beam bending rigidity which is assumed to be a normal homogeneous stochastic process with mean equal to 1 and autocorrelation function

$$R_{EI}(x_1, x_2) = \sigma^2 exp\left(-\frac{(x_2 - x_1)^2}{c}\right) \tag{18}$$

where σ and c are constants. Thus, randomness is manifested in this problem through the operator and the load.

In implementing the proposed method, first the approximation of the stochastic field *EI(x)* through a finite set of random variables is derived. For this purpose, the Loeve-Karhunen expansion is deemed especially effective. It is an optimal, in the mean square sense, representation of the field over the set of random variables. Subsequent application of the finite difference scheme [8] or of any alternative discretization scheme leads to the system of linear algebraic equations

$$(A_0 + \xi_1 A_1 + \ldots \xi_M A_M)\underline{u} = \underline{f}, \tag{19}$$

where $A_1, A_2, \ldots A_M$ are matrices the dimension of which depends on the number of nodes used for the discretization, $\underline{u}$ is a vector representing the solution at the nodal points, and $\underline{f}$ corresponds to the force.

Next, specific numerical examples of application of the proposed method are presented; the numerical values $\sigma = 0.3$ and $c = 0.5$ are used.

3.2 Beam eigenvalue problem

The eigenvalue problem of a clamped-clamped beam is considered first. Then, the force in equation (17) takes the form $q(x) = \lambda u(x)$. This kind of problem is quite difficult either for an analytical or for a numerical treatment. Only a few articles are available on this topic. A description of pertinent analytical methods was presented by Boyce [9]. In the papers of Goodwin and Boyce[10], Hasselman and Hart [11] some numerical examples of solution of stochastic eigenvalue problems can be found.

Note, that these algorithms can only be applied in the case of small randomness and can be computationally costly.

In implementing the proposed method, the random domain is divided into a set of rectangles $\{\underline{\xi} \mid \xi_i^j \le \xi_i \le \bar{\xi}_i^j,\quad i=1,\dots M;\quad j=1,\dots N\}$ of prescribed equal probability mass. Then, the basis $\{\varphi_i, i = 1, \dots N\}$ can be constructed to conform with equation (2). Next, u and λ can be expressed in the form

$$\underline{u}(\underline{\xi}) = \sum_{i=1}^{N} \underline{v}_i \varphi_i(\underline{\xi}), \quad \lambda(\underline{\xi}) = \sum_{i=1}^{N} c_i \varphi_i(\underline{\xi}). \quad (20)$$

Substituting equations (20) into equation (19), multiplying it by $\varphi_i(\underline{\xi})$, and taking the mathematical expectation of the result yields

$$(A_0 + \xi_{1,i} A_1 + \dots \xi_{M,i} A_M)\underline{v}_i = c_i \underline{v}_i,$$

$$i = 1, \dots N, \quad (21)$$

where

$$\xi_{k,i} = \frac{1}{p_i}\int_{\Omega} \xi_k \varphi_i(\underline{\xi}) p_{\underline{\xi}}(\underline{\xi})\, d\underline{\xi} = \frac{1}{p_i}\int_{\Omega_i} \xi_k p_{\underline{\xi}}(\underline{\xi})\, d\underline{\xi},$$

and $$p_i = \int_{\Omega_i} p_{\underline{\xi}}(\underline{\xi})\, d\underline{\xi}. \quad (22)$$

Finally, statistical analysis can be performed in conjunction with equation (7) to estimate analytically the moments of the first two eigenvalues and eigenvectors.

The theoretical values of the first two eigenvalues for the deterministic case, when the beam rigidity is set equal to the mean rigidity of the problem under consideration, are $\lambda_1^{det} = 22.37$ and $\lambda_2^{det} = 61.67$. The corresponding deterministic finite difference approximation with 41 node points gives $\lambda_1^{findif} = 22.32$ and $\lambda_2^{findif} = 61.33$. It was found that despite the considerable variability in the rigidity of the beam, the variability in the eigenvectors is negligible. However, the variability in the eigenvalues is essential and it is of the order of 10%; see Figure 2. The influence of different number M of used random variables ξ_i has been studied. It was found that for this problem the contribution of the terms beyond ξ_2 in equation (19) is negligible. The results in terms of convergence of this method for different order of partition of axes ξ_1 and ξ_2, that is for different values of number N or indirect samplings, are plotted in Figure 2(a) for the standard deviation of λ_1, and in Figure 2(b) for the standard deviation of λ_2. It is seen that the proposed method yields quite good approximations even when the value of number N is quite small. However, the Monte-Carlo method, that is when the parameters were sampled arbitrarily, yields reliable results only if the number of the used simulations is large.

3.3 Beam response to deterministic load

The second problem involves continuous systems with random parameters exposed to deterministic excitation. Specifically, the dynamic response of a cantilever beam to earthquake-type base excitation is considered. In this case the force term in equation (17) can be expressed as

$$q(x,t) = a_g(t) - \ddot{u}(x,t) - \alpha \dot{u}(x,t), \quad (23)$$

where α is a coefficient of damping, u represents the displacement of the beam relative to the base, and $a_g(t)$ is taken as the time history of the ground acceleration produced by the North-South component of El Centro earthquake recorded on station No 117 and reported in the reference [12]. It is shown on Figure 3 for comparisons with the beam response. Further, it is assumed that the beam has unit mass per length. The discretization of the beam by a finite difference scheme built upon 20 nodes in the spatial domain in conjuction with the Loeve-Karhunen expansion of the bending rigidity is used. Then, the solution is taken in the form of equation (20) where in this case $\underline{v}_i = \underline{v}_i(t)$ are deterministic vector-functions. Substituting this expression into the resulting equation, multiplying it by $\varphi_i(t)$, and averaging, an uncoupled system of deterministic

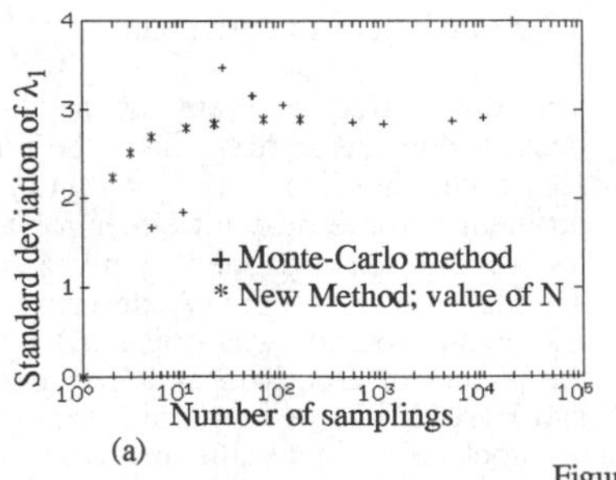

(a)

Standard deviation of λ_2
+ Monte-Carlo method
* New Method; value of N
Number of samplings

(b)

Figure 2.

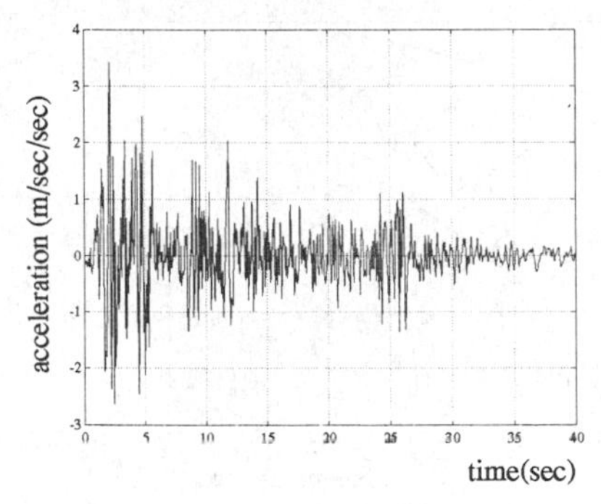

Figure 3. Earthquake-type excitation

ordinary differential equations is derived. Each equation of this system is solved numerically using the central difference scheme. Finally, the mean value and the standard deviation of the free end displacement are determined by relying on equation (7).

The time history of the free end displacement of the cantilever beam having the mean characteristic for the stiffness and damping is plotted in Figure 4(a). Further results of the calculations are shown in Figure 4(b,c) for different value of number N of indirect samplings. In Figure 4, N_1, N_2, and N_α denote the number of strata in the domain of ξ_1, ξ_2, and α, respectively. The case with $\alpha = 0.4$ which corresponds to damping of approximately 6% of critical for the first mode of the system with deterministic rigidity equal to the mean of the corresponding stochastic problem is considered. Also the case where α is a random variable statistically independent from ξ and uniformly distributed between 0.1 and 0.7 is examined. The computations show that 3 strata in the domain of α can adequately represent the dependence of the solution on the damping variability. Also, the calculations reveal that only the first two components of the vector ξ influence significantly the beam response. It can be seen that the dynamic response of the beam to the deterministic excitation is strongly affected by the rigidity variability.

3.4 Beam response to stochastic load

The third problem is described again by equations (17) and (23) but it involves stochastic base excitation. Specifically, $a_g(t)$ is taken as a stationary random process. The proposed indirect sampling method can be readily applied for treating this problem. The solution is expressed in the form of equation (20). In this case $\underline{v}_i = \underline{v}_i(t)$ is not a deterministic vector-function, but a stochastic vector-process. That is, $\underline{v}_i(t)$ is the response of a deterministic system to random excitation. A number of techniques exist for the solution of this problem. In particular, a spectral approach can be applied provided that $a_g(t)$ is a second order stochastic process. In the latter case the second-order

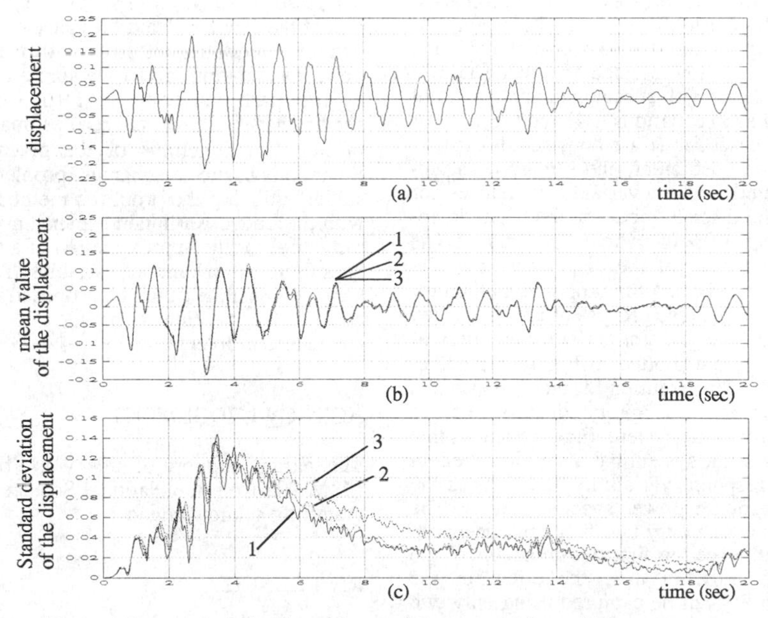

Figure 4. Base excited vibration of the top of the beam: (a) deterministic system with mean characteristics; (b) mean value; (c) standard deviation, stochastic system

1) —— N = 63 (N_1=9, N_2=7) 2) - - - - N = 7 (N_1=7)
3) - - - -N = 45 considering random damping (N_1=5, N_2=3, N_α=3)

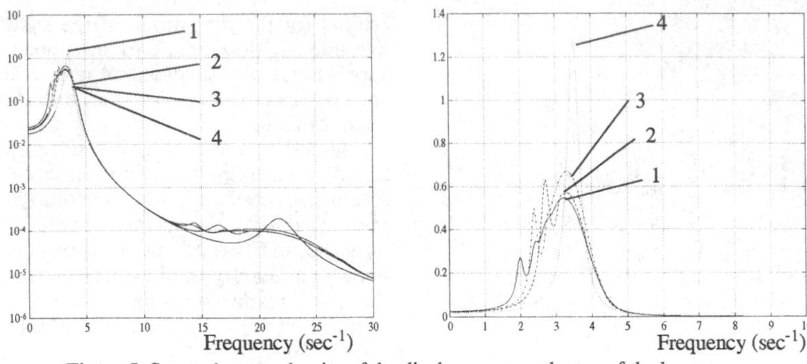

Figure 5. Spectral power density of the displacement on the top of the beam.

1) —— N = 63 (N_1=9, N_2=7) 3) - - - -N = 42 considering random damping (N_1=7,N_2=3, N_α=2)
2) - - - -N = 7 (N_1=7) 4) .... deterministic system with mean characteristics

characteristics of the solution can be determined from the formulae

$$E[\underline{u}] = \sum_{i=1}^{N} E[\underline{v}_i(t)]\,p_i \quad \text{, and} \tag{24}$$

$$R_{\underline{u}}(t_1, t_2) = \sum_{i=1}^{N} R_{\underline{v}_i}(t_1, t_2)\,p_i. \tag{25}$$

Then, this approach can be viewed as semi-analytical method analogous to the directional sampling. At first, the vector ξ is simulated, and then, the solution for the given simulation is calculated using known analytical techniques with subsequent application of some averaging as in equations (24) and (25).

Again, two cases for α are considered. First, α is a deterministic coefficient, and second α is uniformly distributed random variable; it induces for each mode of the discrete model of the beam damping 6%, and 2% to 10% of critical, respectively. The power spectral density of the displacement of the free end is calculated using the proposed method for $a_g(t)$ being a white noise process of unit two sided spectral density. The data for different numbers of indirect samplings are plotted in Figure 5 together with the corresponding solution for the response of a deterministic system with rigidity equal to the mean rigidity of the stochastic system. Figure 5 shows that the randomness of the system has a significant effect on the system response variability and reduces the peaks of the response power spectral density. The calculations show that only two first components of the vector ξ influence the first two moments of the solution significantly. Further, the effect of the damping variability can be captured using only two strata in its domain.

4 CONCLUDING REMARKS

A Galerkin-type numerical method for stochastic mechanics problems has been presented. Specifically, it has been proposed to use a Galerkin projection into the space of simple random variables. This space can be spanned by the piece-wise constant spline functions with a chosen partition of the sample space. Further, it has been shown that the proposed method can be construed a generalized sampling; it is proposed to call it indirect sampling. Indeed, it has been shown that this method is closed to the stratified sampling method and it is optimal in the sense of equation (16). That is, the approximation of the problem from the space of simple random variables produces an error with mean and conditional expectation, given the sigma-algebra induced by this partition, equal to zero. It has also been shown that this error has zero estimate from the set of indicator functions of the given stratification. Some stochastic mechanics problems have been studied utilizing the proposed method in conjuction with the Loeve-Karhunen expansion which is a versatile tool for the approximation of a stochastic field by a finite set of random variables. These examples have demonstrated that the proposed method can be applied for treating a broad class of stochastic mechanics problems.

ACKNOWLEDGEMENT

The partial support of this work from the grant MSM-902 from the National Science Foundation is gratefully acknowledged.

REFERENCES

1. Mckay M.D., Beckman R.J. and Conover W.J. "A Comparison of Three Methods for Selecting Values of Input Variables in the Analysis of Output

from a Computer Code", Technometrics, vol. 21, No. 2, May 1979.

2. Ghanem, R.G., Spanos, P.D. and Akin, E. "Orthogonal Expansion for beam Variability", Proceedings of the Conference on Probabilistic Method in Civil Engineering, ASCE, Blacksburg, Va., May 25-27, pp.156-159, 1988.
3. Ghanem,R.G., Spanos, P.D. "Stochastic Finite Elements: A Spectral Approach", Spring-Verlag, New-York, 1991.
4. Jensen, H, Iwan, W.D. "Response of Systems with Uncertain Parameters to Stochastic Excitation". Journal of Engineering Mechanics, Vol. 118, No 5, pp.1012-1025, May, 1992.
5. Grigoriu, M. "Statistically Equivalent Solutions of Stochastic Mechanics Problems". Journal of Engineering Mechanics, Vol. 117, No. 8, pp. 1906-1919, August, 1991.
6. Zeldin B.A., Spanos P.D. "Pseudo-Simulation Method for Stochastic Problem". Proceedings of the 6th ASCE Conference on Probabilistic Mechanics and Structural and Geotechnical Reliability, July 8-11, pp. 37-40, 1992.
7. Pfeiffer P.E. "Probability for Applications", Spring-Verlag, New York, 1990.
8. Spanos, P.D., Zeldin, B.A. "Mixed Finite Difference Method". Proceedings of the Ninth Engineering Mechanics Conference of ASCE, College Station, pp. 804-807, May 24-27, 1992.
9. Boyce, W.E. "Random Eigenvalue Problem". In "Probabilistic Methods in Applied Mathematics"(A.T.Bharucha-Reid, ed.), Vol. I, pp.1-73. Academic Press, New-York, 1969.
10. Goodwin, B.E. and Boyce, W.E. "The Vibration of a Random Elastic String: The Method of Integral Equations". Quarterly of Applied Mathematics, 22, pp. 261-266, 1964.
11. Hasselman, T.K. and Hart, G.C. "Modal Analysis of Random Structural Systems". Journal of the Engineering Mechanics Division, Vol. 98, No EM3, pp. 561-579, 1979.
12. "Strong Motion Earthquake Accelerogram Index Volume", Earthquake Engineering Research Laboratory, California Institute of Technology, Pasadena, California, EERL 76-02, August, 1976.

Structural Safety & Reliability, Schuëller, Shinozuka & Yao (eds) © *1994 Balkema, Rotterdam, ISBN 90 5410 357 4*

Probability of damage excursion in a continuum with random field properties

Y. Zhang & A. Der Kiureghian
Department of Civil Engineering, University of California, Berkeley, Calif., USA

ABSTRACT: A probabilistic method for predicting the excursion of accumulated local damage above a specified threshold in a continuum with random field properties is introduced. The active-set gradient projection method is introduced, which finds the local design points on the limit-state surface in a decreasing order of their importance. A first-order series system reliability analysis yields an approximation to the probability of fracture initiation within the random field domain of the continuum. An example illustrates the methodology.

1. INTRODUCTION

Methods to predict local fracture in 2- or 3-dimensional solids have been of interest for a long time. The occurrence of such fracture in many cases is tantamount to the failure of a structure. One is then interested in assuring that the likelihood of fracture at any location within the structure is acceptably small.

It is generally believed that fracture is initiated by accumulating damage in a material. An initial damage state in the form of pre-existing micro-fractures or cavities in the microscopic structure of the material grows under the action of loads, and a fracture is initiated when the accumulated damage at a location reaches a critical threshold. The initial damage distribution in a structure usually is unknown because of the complex and inherently random micro-structure of the material. Furthermore, significant uncertainties exist in the process of damage accumulation due to randomness in material properties or loads. It is clear, therefore, that a stochastic approach to predicting the accumulation damage and occurrence of fracture in a structure is essential.

In recent years, several mathematical models describing damage mechanics in continua have been developed (Lemaitre and Chaboche 1990). These models are easily implementable in finite element codes capable of analyzing complex inelastic structures with arbitrary loading and boundary/initial conditions (Simo and Ju 1987). The advent of stochastic finite element methods (Der Kiureghian and Ke 1988), as well as efficient techniques for computing the gradient of inelastic response (Zhang and Der Kiureghian 1991), now make it possible to investigate the the process of damage accumulation and fracture within a probabilistic framework, fully accounting for all uncertainties involved.

Randomness in the damage process arises from several sources: The initial damage at each location, material properties, loading, and boundary or initial conditions. In the most general case, these uncertainties are random fields in space and time. The statistical characteristics of these fields (e.g., distributions, moment functions) can be estimated from data generated by testing and measurement. The ultimate difficulty lies in the probabilistic analysis for predicting the evolution of damage and initiation of fracture in the structure.

This paper describes a first-order, finite-element based reliability method for estimating the probability that the accumulated damage in an inelastic body, during the course of application of loads, will exceed a specified threshold anywhere within the random field domain of the material. Use is made of a recent method for discretization of random fields (Li and Der Kiureghian 1992), and a first-order reliability method for computing the first-excursion probability of uncertain dynamical systems (Zhang and Der Kiureghian 1992). An example application to a two-dimensional continuum with a random field initial damage state demonstrates the methodology.

2. PROBABILISTIC FORMULATION OF MATERIAL DAMAGE

The structural reliability problem is defined by the probability integral

$$P = \int_{g(\mathbf{x})\leq 0} p(\mathbf{x})d\mathbf{x} \tag{1}$$

where $\mathbf{x}$ is a vector of random variables with probability density function $p(\mathbf{x})$, and $g(\mathbf{x})$ denotes a limit-state function that defines the event of interest (e.g., failure of the structure) as $g(\mathbf{x}) \leq 0$. Among the approximation methods for solution of this integral, the first-order reliability method (FORM) has gained wide popularity because of its computational efficiency, as well as its accuracy for most structural reliability problems of interest. In this method, an approximation to the integral is obtained by linearizing the integration boundary $g(\mathbf{x}) = 0$, denoted limit-state surface, at one or more points known as design points $\mathbf{x}_i^*$, $i = 1, 2, \ldots$ The approximation is carried out in a standard space obtained by transforming the random variables into independent normal variates. Any random fields involved must be discretized so as to be represented in terms of the random variables $\mathbf{x}$.

Without loss of generality, it is assumed that $\mathbf{x}$ here denotes the transformed normal variates. In the standard normal space, $\mathbf{x}_i^*$ are points with minimum distance to the origin. These points are found by solving a nonlinear programming problem:

$$\min\{ f(\mathbf{x}) \mid g(\mathbf{x}) = 0 \} \tag{2}$$

where $f(\mathbf{x}) = \frac{1}{2}||\mathbf{x}||^2$ is the objective function. At each design point, the normalized negative gradient row vector $\alpha_i = -\nabla g(\mathbf{x}_i^*)/||\nabla g(\mathbf{x}_i^*)||$ and the reliability index $\beta_i = \alpha_i \mathbf{x}_i^*$ are computed. If there is a single significant design point, the first-order approximation of the probability is given by $P \approx \Phi(-\beta_1)$, where $\Phi(.)$ is the standard normal cumulative probability. In the case of multiple design points, provided certain convexity requirements are satisfied, first-order approximate bounds to P that are given in terms of α_i and β_i can be used (Ditlevsen 1979).

As is clear from the above discussion, solution of the design points is the main task in FORM analysis. Two numerical procedures are necessary for this purpose: a constrained optimization algorithm for solving Eq. 2, and a procedure for evaluating the limit-state function $g(\mathbf{x})$ and its gradient $\nabla g(\mathbf{x})$. For the latter, an extended finite element formulation that includes computation of the gradients by direct differentiation is employed in this study (Zhang and Der Kiureghian 1991). For the former, a new algorithm is introduced in the subsequent sections.

For FORM analysis of the problem of damage accumulation and initiation of fracture in a structure, the limit-state function describing this mode of failure must be formulated. Account must be made of the time and space variabilities. Assume $\mathbf{x}$ represents all uncertain variables in the structure, including those describing random fields. In the context of the finite element method, which discretizes the structure domain into small elements, a measure of the local damage at time t may be defined as the average damage accumulated over the element, here expressed as $s_n(\mathbf{x}, t)$. The subscript n, denoting the element number, is treated as a location variable. Let $r_n(\mathbf{x}, t)$ denote the corresponding damage threshold for initiation of fracture, which for the purpose of generality is also taken to be a function of $\mathbf{x}$, t and n. Define the point-in-time-and-space safety margin as

$$w_n(\mathbf{x}, t) = r_n(\mathbf{x}, t) - s_n(\mathbf{x}, t) \tag{3}$$

This function takes on a negative value whenever a fracture occurs in element n at time t. To describe the event of fracture within the structure during an interval $t \in [0, T]$, the limit-state function is defined as

$$g(\mathbf{x}) = \min_{n \in \mathcal{N},\ t \in [0,T]} w_n(\mathbf{x}, t) \tag{4}$$

where $\mathcal{N}$ is an index set containing all the element numbers. Clearly, $g(\mathbf{x}) \leq 0$ if a fracture initiates anywhere within the structure during the specified interval. The minimization over time involved in the definition of the limit-state function can be handled by a method recently developed by the authors for the general case where both $s_n(\mathbf{x}, t)$ and $r_n(\mathbf{x}, t)$ are allowed to fluctuate with time. In the present case, $s_n(\mathbf{x}, t)$ is a non-decreasing function of time, provided the material does not possess a "self-healing" property. If $r_n(\mathbf{x}, t)$ is assumed to be a non-increasing function of time (i.e., no possibility of the material becoming stronger with time), then it is clear that the minimum of $w_n(\mathbf{x}, t)$ over time will occur at $t = T$. Hence, the minimization over time is accomplished by replacing t by T in the expression for w_n. For the sake of simplicity of notation, in the following analysis the reference to time is omitted. It is noted that the algorithm described is not restricted to this special, but practically relevant, case. A more general formulation involving a safety margin that fluctuates in time can be solved by complementing the method developed here with that developed in the earlier study (Zhang and Der Kiureghian 1992).

The form of $g(\mathbf{x})$ implies that, for any given $\mathbf{x}$, there exists at least one $\hat{n} \in \mathcal{N}$ such that

$$g(\mathbf{x}) = w_{\hat{n}}(\mathbf{x}) \tag{5}$$

$\hat{n}$ is called the active set index. The active set index is a function of $\mathbf{x}$ and is written as $\hat{n} = \hat{n}(\mathbf{x})$. At the given $\mathbf{x}$, the gradient of the limit-state function is given by

$$\nabla g(\mathbf{x}) = \nabla w_{\hat{n}}(\mathbf{x}) \tag{6}$$

However, it should be noted that the function $g(\mathbf{x})$ is not continuously differentiable at all $\mathbf{x}$. The continuity is broken at values of $\mathbf{x}$ where the active set index changes. This presents a major difficulty in solving the optimization problem in Eq. 2 by conventional algorithms. In the following section, a special constrained optimization algorithm is developed to solve this problem for the global design point.

3. THE ACTIVE SET GRADIENT PROJECTION ALGORITHM

In a typical gradient projection method (Luenberger 1986), a sequence

$$\{ \mathbf{x}_k \mid g(\mathbf{x}_k) = 0,\ f(\mathbf{x}_k) \leq f(\mathbf{x}_{k-1}) \}_{k=1}^{\infty}$$

is generated, and the objective function is minimized as $k \to \infty$. Here, the active set scheme is activated during the enforcement of $g(\mathbf{x}_k) = 0$. At a point $\mathbf{x}_k$, the active set index $\hat{n}_k = \hat{n}(\mathbf{x}_k)$ is identified according to Eq. 5, and the constraint $w_{\hat{n}_k}(\mathbf{x}) = 0$ is activated at $\mathbf{x}_k$ and its neighborhood. To find an improved point $\mathbf{x}_{k+1}$, a search direction $\mathbf{d}_k$ is obtained by projecting the negative gradient of the objective function, $-\nabla f(\mathbf{x}_k)$, onto the tangent plane of the active constraint surface. $\mathbf{d}_k$ is written as

$$\mathbf{d}_k = -\nabla f(\mathbf{x}_k) + \frac{(\nabla f(\mathbf{x}_k), \nabla w_{\hat{n}_k}(\mathbf{x}_k))}{||\nabla w_{\hat{n}_k}(\mathbf{x}_k)||^2} \nabla w_{\hat{n}_k}(\mathbf{x}_k) \tag{7}$$

where $(.,.)$ denotes the vector inner product and $||.||$ is the Euclidean vector norm. Along the direction $\mathbf{d}_k$, an initial trial point

$$\mathbf{y} = \mathbf{x}_k + \gamma \mathbf{d}_k \quad \text{and} \quad \gamma \in (0, 1] \tag{8}$$

is obtained. The selection of γ will be explained shortly.

As $w_{\hat{n}_k}(\mathbf{x})$ is a nonlinear function, the point $\mathbf{y}$ may not be on the constraint surface. One attempts to return to a point on the constraint surface which is closest to $\mathbf{y}$. The iterative solution of a nonlinear equation is necessary for this purpose. Furthermore, the active set may have changed at the new point. Thus, identifying the new active set, a move is made towards the new constraint surface. This process is repeated until a point $\mathbf{x}_{k+1}$ is obtained satisfying $g(\mathbf{x}_{k+1}) = 0$.

The point $\mathbf{x}_{k+1}$ may not be acceptable, as the condition $f(\mathbf{x}_{k+1}) \leq f(\mathbf{x}_k)$ is not automatically assured. If satisfied, one proceeds to find the next point of the sequence; otherwise, the above procedure is repeated by starting from a trial point $\mathbf{y}$ which is closer to $\mathbf{x}_k$. Hence, if necessary, γ in Eq. 8 is reduced recursively until a reduced objective function is obtained, yielding the desired point $\mathbf{x}_{k+1}$. The convergence criterion used to terminate the sequence is

$$\frac{\|\mathbf{x}_{k+1} - \mathbf{x}_k\|}{\|\mathbf{x}_k\|} \leq \epsilon_1 \quad \text{and} \quad \frac{|\,f(\mathbf{x}_{k+1}) - f(\mathbf{x}_k)\,|}{|\,f(\mathbf{x}_k)\,|} \leq \epsilon_2 \qquad (9)$$

where ϵ_1 and ϵ_2 are relative precisions for $\mathbf{x}$ and $f(\mathbf{x})$ respectively.

As noted earlier, the move from the trial point $\mathbf{y}$ to the active constraint surface requires the solution of a nonlinear equation. This solution is necessary at each step of the active set gradient projection algorithm. Hence, a robust nonlinear equation solver is critical for successful implementation of the algorithm. The problem is to solve $w_{\hat{n}}(\mathbf{x}^*) = 0$ for $\mathbf{x}^*$ in the neighborhood of a trial point $\mathbf{x}^0$. An alternative and more effective approach is to solve the unconstrained optimization problem

$$\min v(\mathbf{x}), \quad \text{where } v(\mathbf{x}) = \frac{1}{2} w_{\hat{n}}^2(\mathbf{x}) \qquad (10)$$

The steepest descent method is employed to solve this problem. At each iteration point $\mathbf{x}^i$, the steepest descent direction is calculated from

$$\mathbf{h}^i = -\nabla v(\mathbf{x}^i) = -\nabla w_n(\mathbf{x}^i) w_n(\mathbf{x}^i) \qquad (11)$$

Along this direction, a line search is performed to find the local minimum, $\min_\lambda v(\mathbf{x}^i + \lambda \mathbf{h}^i)$. An exact line search, which can be costly, is avoided by using an approximate line search aimed at reducing the objective function by a certain amount in a finite number of calculations. The Armijo rule (Luenberger 1986) is found to be particularly effective for this purpose.

Experience shows that FORM analysis by linearization of the limit-state surface at a single point, the global design point, may not give sufficiently accurate results for problems involving spatial or temporal variabilities. In the following, the above algorithm is extended to find the local solutions of the problem in Eq. 2. The solution points, denoted local design points, are found in the order of increasing distance from the origin, which coincides with the decreasing order of their importance.

4. FINDING LOCAL DESIGN POINTS

When there is an element in the structure that experiences a damage level substantially higher than all other elements outside of its immediate neighborhood, the limit-state function defined by Eq. 4 tends to be reasonably smooth due to the minimization process in its definition, and a FORM analysis with a single design point as outlined above provides a reasonably accurate estimate of the damage-excursion probability. However, when damage is likely to exceed specified thresholds at several different locations within the structure, the limit-state surface tends to be strongly nonlinear and possess multiple design points, each point being associated with the excursion event in a given element. In such cases, the probability estimate based on a single (global) design point may not be accurate. An improved approximation is obtained by linearizing the limit-state surface at each local design point. Since the event of damage excursion within the structure is equivalent to the union of the excursion events of individual elements, a series system reliability that treats each linearized surface as a failure mode may be performed to compute a first-order approximation to the probability of interest. Simple bounds for series system reliability, given in terms of the α_i and β_i values of individual design points, are used for this purpose (Ditlevsen 1979).

The optimization algorithm introduced in the previous section is now extended to find the local design points. The main difficulty is that the sequence generated by the algorithm tends to converge to the global design point, no matter where the sequence is initiated. One possible approach is to construct barriers around previously solved design points so that the sequence is forced to converge to a new design point. If constraint equations are established in the random variable space, the numerical procedure will be further complicated. By the nature of the problem, it is found that an implicit barrier can be conveniently constructed by reducing the index set $\mathcal{N}$ in the definition of $g(\mathbf{x})$. This is explained below.

Let $\mathbf{x}_1^*$ denote the first (global) design point. This point defines the most likely damage excursion event within the structure. The active set index corresponding to this design point is $\hat{n}_1 = \hat{n}(\mathbf{x}_1^*)$ and element $\hat{n}_1$ is the most likely element to fracture. If another design point $\mathbf{x}_2^*$ is to be significant, its active set index $\hat{n}_2 = \hat{n}(\mathbf{x}_2^*)$ must be outside the immediate neighborhood of $\hat{n}_1$. Therefore, if the neighborhood of element $\hat{n}_1$ is excluded from consideration, $\hat{n}_2$ will be the most likely element to fracture in the remaining part of set $\mathcal{N}$.

Suppose the design point $\mathbf{x}_1^*$ is found and the corresponding most likely fracture element is $\hat{n}_1$. Let the subset $\mathcal{N}_1^*$, $\mathcal{N}_1^* \subset \mathcal{N}$, define the immediate neighborhood of $\hat{n}_1$. The limit-state function is redefined as

$$g(\mathbf{x}) = \min_{\substack{n \in \mathcal{N} \\ n \notin \mathcal{N}_1^*}} w_n(\mathbf{x}) \qquad (14)$$

Using this definition, a local solution $\mathbf{x}_2^*$ is obtained by solving Eq. 2. Next, a region $\mathcal{N}_2^*$ around the second most likely fracture element $\hat{n}_2$ is considered, and the problem is solved again using a limit-state function defined by excluding from $\mathcal{N}$ the subsets $\mathcal{N}_1^*$ and $\mathcal{N}_2^*$. The result is $\mathbf{x}_3^*$. This process is continued until all significant local design points are obtained.

The significance of each design point $\mathbf{x}_k^*$ is determined from its contribution to the total probability of damage excursion. This contribution is directly related to the reliability index $\beta_k = \alpha_k \mathbf{x}_k^*$ of the design point. Hence, an appropriate rule for terminating the analysis is when β_k is sufficiently greater than β_1.

The size of each subset $\mathcal{N}_k^*$ should be selected with caution. If it is too small, the new design point will be too close to the preceding point, in which case the two fracture events will be strongly correlated and the new design point will not significantly add to the system probability, even though β_k is not much greater than β_{k-1}. On the other hand, if $\mathcal{N}_k^*$ is too large, a potentially significant design point might be missed, in which case the fracture probability would be underestimated. For practical implementation, we find that an examination of the function $w_n(\mathbf{x}_k^*)$ at each design point provides helpful insight for selecting a proper size for $\mathcal{N}_k^*$.

5. DAMAGE MODEL

For the analysis in this paper, we consider a model for ductile plastic damage in elasto-plastic materials caused by accumulation of large plastic strains (Lemaitre and Chaboche 1990). A local damage variable D is introduced and the stress-strain relations are written in the form

$$\frac{\sigma_{ij}}{1-D} = C_{ijkl}(\varepsilon_{kl} - \varepsilon_{kl}^p) \tag{15}$$

where C_{ijkl} are the elastic coefficients and ε_{ij} and ε_{ij}^p are the total and plastic strains, respectively. The variable D conceptually represents the density of the micro-fractures and voids in the material. We assume D_0 denotes its initial value, and D_c denotes the threshold at which a macro-fracture initiates. Plastic strains are derived from the flow rule:

$$\dot{\varepsilon}_{ij}^p = \dot{\lambda}\frac{\partial f}{\partial \sigma_{ij}} \tag{16}$$

where $\dot{\lambda}$ is a rate factor and f is the yield function, which for J_2 plasticity with a linear kinematic hardening rule, can be expressed as

$$f = \frac{1}{(1-D)^2}\left(s_{ij} - \frac{2}{3}H_{kin}\varepsilon_{ij}^p\right)\left(s_{ij} - \frac{2}{3}H_{kin}\varepsilon_{ij}^p\right) - \frac{2}{3}\sigma_y^2 \tag{17}$$

where H_{kin} is the coefficient of kinematic hardening, σ_y is the yield stress, and s_{ij} are the deviatoric stresses

$$s_{ij} = \sigma_{ij} - \delta_{ij}\sigma_{kk}/3 \tag{18}$$

The rate equation of of D can be expressed as

$$\dot{D} = -\frac{\partial \varphi^*}{\partial Y} \tag{19}$$

where φ^* is a damage dissipation potential and Y is a variable associated with D and defined by

$$-Y = \frac{1}{2E(1-D)^2}[(1+\nu)s_{ij}s_{ij} + (1-2\nu)(\sigma_{kk})^2/3] \tag{20}$$

where E and ν are the Young's modulus and Poisson's ratio, respectively. For the ductile plasticity damage model,

$$\varphi^* = \frac{S}{2}\left(\frac{-Y}{S}\right)^2 \frac{\dot{\lambda}}{1-D} \tag{21}$$

where S is a material and temperature-dependent parameter and influences the damage rate.

This model has been implemented in a finite element code and used to analyze the example structure described in the following section.

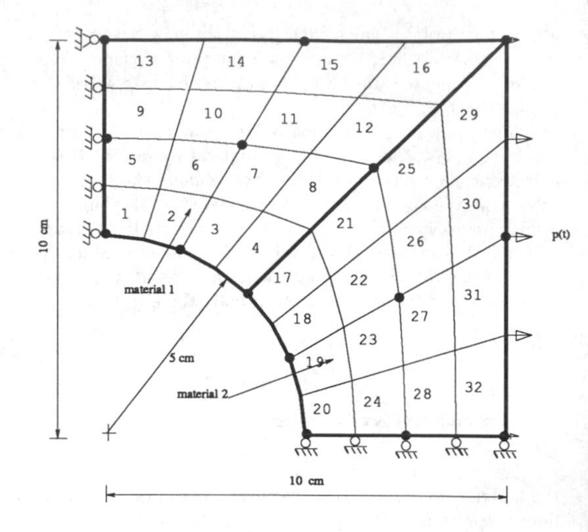

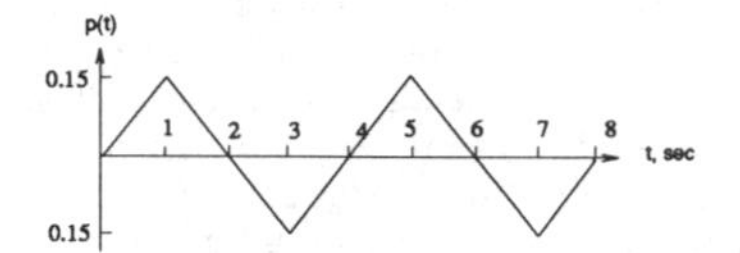

Figure 1. Example plate with applied load $p(t)$

6. NUMERICAL EXAMPLE

The plate in Fig. 1 under the cyclic loading shown is considered. The plate has a thickness of 1 cm and is assumed to be under plain stress conditions. The domain of the plate is discretized into 32 quadrilateral four-node elements for finite-element analysis.

The plate is made of two materials with different properties: Material 1 for elements 1-16 has the properties $E = 70.0\ MPa$, $\nu = 0.2$, $H_{kin} = 2.24\ MPa$, $S = 0.05$, $D_c = 0.27$, and a random yield stress σ_y having a normal distribution with mean 0.243 MPa and standard deviation 0.036 MPa, and material 2 for elements 11-32, which has the same E, ν, H_{kin} and $S = 0.02$, $D_c = 0.19$, and mean 0.243 MPa and standard deviation 0.049 MPa of σ_y. Note that material 2 has a lower damage threshold and a smaller S in comparison to material 1, indicating that damage will accumulate faster in material 2 under identical load conditions. The initial damage, D_o, is assumed to be a homogeneous Gaussian random field within each material domain with mean 0.05, standard deviation 0.02, and autocorrelation coefficient function $\rho = exp(-l/5.0)$, where l is the distance between any two points within each material. Using the OLE method of Li and Der Kiureghian (1992), the random field is represented by the nodal random variables at the 15 nodes indicated by solid dots in Fig. 1.

For reliability analysis, the limit-state function defined in Eqs. 3 and 4 with $r_n = D_c$ and $s_n = D$ within each element is considered. The first design point is found with $\hat{n}_1 = 20$ and $\beta_1 = 2.224$. Excluding $\hat{n}_1$ from $\mathcal{N}$, the second design point is found with $\hat{n}_2 = 1$ and $\beta_2 = 2.306$. Also excluding $\hat{n}_2$ from $\mathcal{N}$, the third design point is found with $\hat{n}_3 = 32$ and $\beta_3 = 2.988$. Since $\beta_3 >> \beta_1$, subsequent design points are not significant and the procedure is terminated. Based on the three design points, a series system analysis is performed, resulting in $P = 0.0195$ and the generalized reliability index $\beta_g = \Phi^{-1}(1-P) = 2.063$. FORM analysis can also produce sensitivities of the probability with respect to each random variable. The analysis indicates that the damage-excursion probability for the plate is most sensitive to the yield stresses and to the initial damage variables around elements 1 and 20, as expected. Contours of accumulated damage for the mean and design points of the random variables are shown in Fig. 2, where the region of damage excursion in each case is shaded. Different modes of fracture are evident in the damage contours for the three design points.

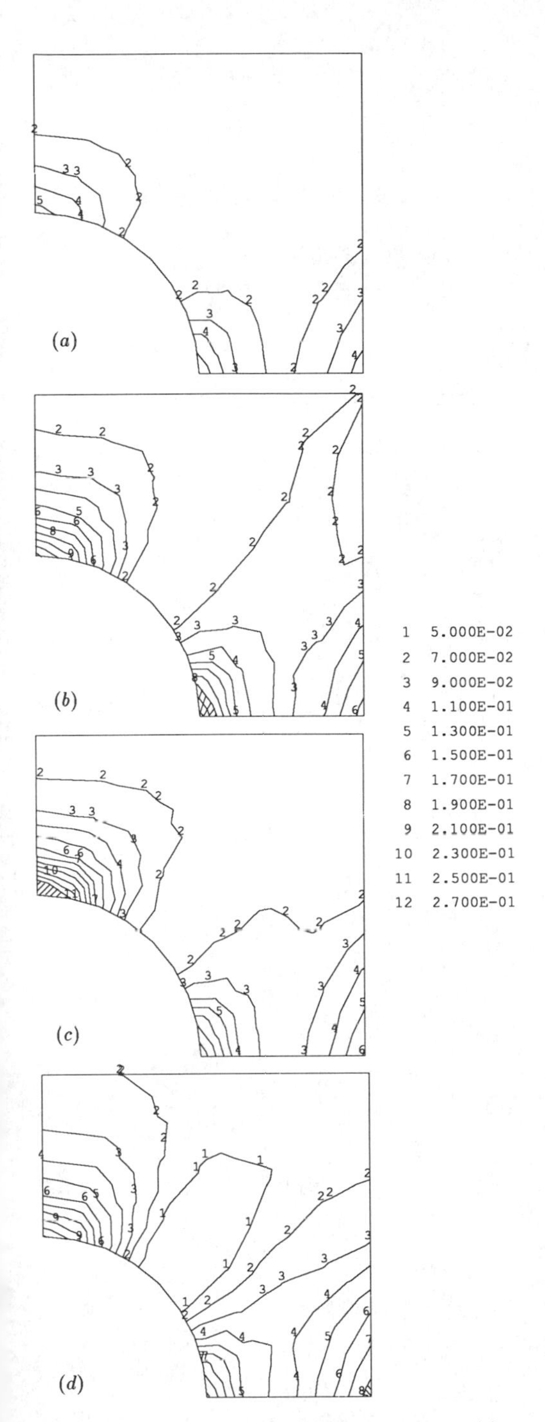

Figure 2. Contours of accumulated damage: (a) mean point, (b) design point $\mathbf{x}_1^*$, (c) design point $\mathbf{x}_2^*$, (d) design point $\mathbf{x}_3^*$

7. SUMMARY AND CONLUSIONS

A first-order reliability analysis for a continuum with a random field damage mechanics model is developed. The method identifies the local design points, which are associated with likely modes of localized damage excursion above specified thresholds. A series system reliability analysis yields the probability of damage excursion within the continuum, accounting for the random field properties of the material. The approach takes advantage of several recent developments in stochastic finite element analysis.

8. ACKNOWLEDGMENT

Support from the U.S. National Science Foundation under Grant No. MSM-8922077, with Dr. Ken Chong as Program Director, is gratefully acknowledged.

REFERENCES

[1] A. Der Kiureghian and B. J. Ke, "The Stochastic Finite Element Method in Structural Reliability," *Probabilistic Engineering Mechanics*, 3(2), 83–91, 1988.

[2] O. Ditlevsen, "Narrow Reliability Bounds for Structural Systems," *J. Struct. Mech.*, 7(4), 453–472, 1979.

[3] J. Lemaitre and J. L. Chaboche, *Mechanics of Solid Materials*, Cambridge University Press, 1990.

[4] C. C. Li and A. Der Kiureghian, "An Optimal Discretization of Random Fields," UCB/SEMM Report No. 92/04, Department of Civil Engineering, University of California at Berkeley, 1992.

[5] J. Lubliner, J. Oliver, S. Oller and E. Onate, "A Plastic-Damage Model for Concrete," *Int. J. Solids Structures*, 25(3), 299–326, 1989.

[6] J. C. Simo and J. W. Ju, "Strain and Stress Based Continuum Damage Model: Part I: Formulation," *Int. J. Solids Structures*, 25(3), 8–15, 1987.

[7] Y. Zhang and A. Der Kiureghian, "Dynamic Response Sensitivity of Inelastic Structures," UCB/SEMM Report No. 91/06, Department of Civil Engineering, University of California at Berkeley, 1991.

[8] Y. Zhang and A. Der Kiureghian, "First-Excursion Probability of Uncertain Structures," *Proceedings*, ASCE Specialty Conf. on Prob. Mechanics and Struct. and Geotech. Reliab., Denver, CO, 531–534, 1992.

Damage assessment and limit states

Structural Safety & Reliability, Schuëller, Shinozuka & Yao (eds) © 1994 Balkema, Rotterdam, ISBN 90 5410 357 4

Stochastic evaluation of system states for predamaged structural components

G. Bertrand & R. Haak
Deutsche Aerospace Airbus

Abstract

Safety relevant mechanical components and structures in aircraft are designed according to damage tolerant principles. Conventional design practice does not take into account any functional dependencies between the state parameters (stochastic variables) of the basic pre–damaged conditon and the residual safety margin.

It is sometimes not possible to decide which of two potential failure modes (eg. fracture or buckling of slender composite structures) has the higher probability of occurrence.

The above problem can be solved by a probabilistic representation of the system states of a structural component taking into account the stochastic nature of the damage process and the parameters driving the damage mechanism.

The analytical formulation of the evolutionary collapse mechanism is given by an incremental representation of the mechanical function with reference to an updated state. The probability of transition into the critical failure mode is then determined by the superpositioning of a stochastic pertubation vector on a balanced reference system state.

Integration of the failure integral in the region of the limit function determines the probability of transition from the partially failed into the collapsed state.

Values of the critical pertubation vector reveal information about the degradation of reliability of successive damaged states and their residual safety.

1. Deterministic mechanical presentation of system states

Following analytic presentation of shell equation with incompatibility terms and deterministic structural parameter is an extension of /1/. The formulation used is an adequate representation of the highly physical nonlinear damage process. Introducing incremental unknowns physical and statistical nonlinearity is eliminated. The governing incremental equilibrium– and compatibility equations are extended to deal with strain defects. this is done using an analogy between stress and strain tensor.

Substituting deterministic unknowns by stochastic variables the remaining capacity of successive damaged system state can be investigated in a probabilistic manner. Allowable stochastic disturbancies which are superimposed on a reference state which is in equilibrium in the mean are defined from explicitly solvable limit state functions build up from Green's influence functions.

Linearization of highly nonlinear equations describing system states with predeformation and defects is best performed in an incremental decomposition of the momentum state ($\overset{o}{-}$) into deformations v_m of the reference state and increments $\overline{v}_m$ to the neighbouring state (–).

Metric decomposition:

$$\overset{o}{\overline{a}}_i = \vec{a}_i + \overline{\vec{a}}_i \qquad 1.1$$

Tensor decomposition:

$$\overset{o}{\overline{C}}{}^{\alpha m} = \dot{C}^{\alpha m} + \overline{C}^{\alpha m} \qquad 1.2$$

Metric of the neighbouring state (–) is obtained by a transformation rule.

$$\vec{a}_{\overline{i}} = (\delta_i^m + L_i^m) \, \vec{a}_m \qquad 1.3$$

$$L_i^m = v^m \mid \alpha \qquad 1.4$$

Solutions to the stated problem are simplified using a dual representation of the governing vector equation for equilibrium and compatibility (see /1/):

$$C^\alpha \mid \alpha = \chi \qquad 1.5$$

$$F^\alpha \mid \alpha + \vec{a}_\alpha x C^\alpha = K \qquad 1.6$$

Substituting eq. 1.1–1.4 into 1.5–6. together with definitions for stress and strain tensor $n^{\alpha m}$, $k^{\alpha m}$ (see App.) and eliminating the reference state which is in equilibrium, result in extended equilibrium and compatibility equations.

1.1 Analogy between stress and strain tensor with perturbation terms

$$\vec{a}_\beta : (\chi^{\alpha\beta} \mid \alpha - \chi^{\alpha 3}\, b^\beta_2\,) = \pi^\beta \qquad 1.1.1\text{–}2$$

$$\vec{a}_3 : (\chi^{\alpha 3} \mid \alpha + b_{\alpha\beta} * \chi^{\alpha\beta}) = \pi^3 \qquad 1.1.3$$

and with

$$\alpha^*_{\alpha\beta} = \varepsilon^{\alpha\varrho} * \varepsilon^{\beta\lambda} * \alpha_{\varrho\lambda}$$

$$\chi^{\alpha 3} = d^{*\alpha\beta} \mid \alpha - \tau^\beta$$

result in strain equations for strain tensor $\alpha^{\alpha\beta}$ and curvaturetensor $k^{\alpha\beta}$ together with perturbation terms I^m:

$$\vec{a}_\beta : (\chi^{\alpha\beta} \mid \alpha - b^\beta_\alpha * \alpha^{*\alpha\beta} \mid \gamma) = \pi^\beta - b^\beta_\alpha * \tau^\beta = I^\beta \qquad 1.1.4\text{–}5$$

$$\vec{a}_3 : (\alpha^{*\alpha\beta} \mid \alpha\beta + b_{\alpha\beta} * \chi^{\alpha\beta}) = (\pi^3 + \tau^\beta \mid_\beta) = I^3 \qquad 1.1.6$$

Simultaneous set of stress tensor equations reads:

$$\vec{a}_\beta : (n^{\alpha\beta} \mid \alpha - n^{\alpha 3} * b^\beta_\alpha) = -p^\beta \qquad 1.1.7\text{–}8$$

$$\vec{a}_3 : (n^{\alpha 3} \mid \alpha + b_{\alpha\beta} * n^{\alpha\beta}) = -p^3 \qquad 1.1.7\text{–}9$$

substituting $n^{\alpha 3}$:

$$n^{\alpha 3} = m^{\alpha\beta} \mid \alpha - M^\beta$$

results for eq. 1.1.9 in:

$$\vec{a}_3 : (m^{\alpha\beta}|\alpha\beta + b_{\alpha\beta} * n^{\alpha\beta}) = M^\beta|_\alpha - p^3 \qquad 1.1.10$$

Analogy between $\alpha^{\alpha\beta} \sim m^{\alpha\beta}$ and $\chi^{\alpha m} \sim n^{\alpha m}$ together with external vectors $I^\beta \sim p^\beta$ and $I^3 \sim p^3$ is obvious.

Load singularity correspond with strain incompatibility terms which will be useful when constructing strain influence functions from known relations for singular loads.

1.2 Incremental equilibrium– and extended compatibility equations.

Following the process of decomposition and canceling the known reference state which is in equilibrium linearized incremental equations are obtained. Introduction of stress function $\psi = \psi^m * \bar{a}_m$ and deformation function $MD = v^m * \bar{a}_m$ and substitutions into stress– and strain tensor gives:

$$n^{\alpha\beta} = -\varepsilon^{\alpha\mu} * \varepsilon^{\beta\lambda} * \Psi^3|_{\lambda\mu} - \zeta^\alpha * \vec{a}^\beta \qquad 1.2.1$$

$$\bar{\chi}^{\alpha\beta} = -\varepsilon^{\alpha\mu} * \varepsilon^{\beta\lambda} * \bar{\upsilon}^3|_{\lambda\mu} + II^\alpha * \vec{a}^\beta \qquad 1.2.2$$

Together with eq. 1.1.6 and eq. 1.1.10 we obtain a dual set for equilibrium and compatibility:

$$-\varepsilon^{\alpha\mu} * \varepsilon^{\beta\lambda} * b_{\alpha\beta} * \bar{\psi}^3_{i\lambda\mu} + (n^{\alpha\beta};\alpha * L^3_\beta) +$$
$$2n^{\alpha\beta} * L^3_{\beta};\ \alpha + B * a^{\alpha\beta} * a^{\lambda\mu} * \bar{\upsilon}^3_{i\alpha\beta\lambda\mu} = \bar{p}^3 \qquad 1.2.3$$

$$-\varepsilon^{\alpha\mu} * \varepsilon^{\beta\lambda} * b_{\alpha\beta} * \bar{\upsilon}^3_{i\lambda\mu} + \frac{1}{D} a^{\alpha\beta} * a^{\lambda\mu} * \bar{\Psi}^3_{i\alpha\beta\lambda\mu} +$$
$$\varepsilon^{\alpha\mu} * \varepsilon^{\lambda\beta} * \chi_{\lambda my};\ \alpha * L^3_\beta +$$
$$2 * \varepsilon^{\alpha\mu} * \varepsilon^{\lambda\beta} * \chi_{\lambda\mu} * L^3_{\beta i\alpha} = \bar{\pi}^3 \qquad 1.2.4$$

which leads with following simplifications to:

Equilibrium equation:

$$B_n\ (L^{-1} * \bar{v}^3 \mid {}^{\alpha\beta})_{,\alpha\beta} -$$
$$\delta^{\alpha\mu}_{\beta\lambda} * (\dot{b}^\beta_\alpha * \bar{\psi}^3 \mid {}^\lambda_\mu - \dot{\Psi}^3 \mid {}^\lambda_\mu * \bar{v}^3 \mid {}^\beta_\alpha) = \bar{p}^3 \qquad 1.2.5$$

Compatibility equation:

$$(L * \bar{\psi}^3 \mid {}^{\varrho\pi}),\ \varrho\pi +$$
$$D_n\ \delta^{\alpha\mu}_{\beta\lambda}\ (\dot{b}^\beta_\alpha * \bar{v}^3 \mid {}^\lambda_\mu - v^3 \mid {}^\lambda_\mu * \bar{v}_3 \mid {}^\beta_\alpha) = \bar{\pi}^3 \qquad 1.2.6$$

1.3 Integralequations for deformation v^m and stress function Ψ^m

1.3.1 Integrodifferentialequation for v^m

The system of equation 1.2.5–6. consists of bending and membrane actions which are separated using Green's influence functions for plate bending $g_{(3)3}(\xi, o)$ and membrane shell actions $g_{(3)\alpha}(\xi, o)$:

$$G_{(3)3}(\xi, \theta) = g_{(3)3} + \tilde{g}_{(3)3} \qquad 1.3.1$$

$$G_{(3)\alpha}(\theta, \xi) = -\tilde{g}_{(3)\alpha}\ (0, \xi)\ ;\alpha = 1.2 \qquad 1.3.2$$

Total deformation v^m (0) of the shell is obtained from resulting integral equation based on eq. 1.2.5–6:

$$v^m(\theta) = \int_F p^{(3)}(\xi) * G_{(3)m}(0,\xi)dF \quad 1.3.3$$

$m = 1, 2, 3$

The kernel $k_{(3)j}$ of the membrane influence function $g_{(3)3}$ results from:

$$\bar{g}_{(3)3} = \int_F K^n_{(3)} \ (\eta,\xi) * G_{(3)n}(\eta,\theta) \ dF \quad 1.3.4$$

for ex. :

$$K^3_{(3)} = \delta^{\alpha\mu}_{\beta\lambda} * \overline{\psi} \mid ^{\lambda}_{\mu} * \dot{b}^{\beta}_{\alpha} \ g_{(3)3} \triangleq \frac{E * h}{1 * \nu^2} (\dot{\chi}_1^2 + \dot{\chi}_2^2) * g_{(3)3} \quad 1.3.4a$$

where, χ_α – mean curvature

1.3.2
Integrodifferential equation for stress function Ψ^3 with residual deformation

According to the analogy between v^3 and ψ^3 in eq. 1.2.5–6 a similar integral equation for ψ^3 with singular deformation terms (π^3 – defects) can be constructed:

$$\psi^3(\theta) = \int_F \pi^3(\xi)H_{(3)} \ (\theta,\xi) \ dF$$

where $H_{(3)}$ (Θ, ξ) plays the role of an influence function for singular deformations (defects) derived from corresponding functions for singular loads.

In addition the influence function $H_{(3)}\alpha$ for membrane stress resultants $n^\alpha = \int \overline{\psi}^3 \mid \alpha\beta d\beta$ are derived from:

$$H_{(3)\alpha}(\xi,\theta) = -\int K^3_{(\alpha)}(\eta,\xi) * G_{(3)3}(\eta,\theta)ds$$

$$K^3_{(\alpha)} = (D * \dot{b}_{\alpha\alpha}) * \overline{g}_{(3)3\mid\beta} * \dot{b}_{\alpha\beta}$$

$$n^\alpha(\xi) = D * \int (\dot{b}_{\alpha\beta} * v_{3\mid\alpha\beta}) * H_{(3)\alpha} * dF$$

1.3.3
Equations for stability

Internal shell reactions $\overline{q}$ are defined by eq. 1.2.5):

$$\overline{q} = S^{\alpha\mu}_{\beta\lambda} \ (\dot{b}^{\beta}_{\alpha} * \overline{\psi}^3 \mid^{\lambda}_{\mu} - \dot{\psi}^3 \mid^{\lambda}_{\mu} * \overline{v}^3 \mid^{\beta}_{\alpha}) \quad 1.3.3.1$$

Solutions of eq. 1.2.5–6 have the form

$$\overline{v}^3(x,y) = \Sigma^n_{k=1} \ f_k * \phi_k(x,y) \quad 1.3.3.2$$

where ϕ_k are buckling eigenmodes of the shell. If bilinear forms are used to represent influence functions $G_{(3)}$ m (x, y,ξ, η):

$$G_{(3)m} = \Sigma^n_{m,k=1} \frac{\phi_M(x,y) * \phi_K(\xi,\eta)}{m * \omega^2_K} \quad 1.3.3.3$$

The integralequation for $v^3(x,y)$ reads:

$$\overline{v}^3(x,y) - m * \omega^2_k \int G_{(3)m}(x,y,\xi,\eta) * \overline{q}(\xi,\eta)dF_\xi = 0 \quad 1.3.3.4$$

Substituting eq. 1.3.3.2–3 into eq. 1.3.3.4 and comparing coefficients of ϕ_k,the equation for stability is given by

$$[f_i - \Sigma^n_{k=1}C_{ik} * f_k] = 0 \quad 1.3.3.5$$

Coefficients C_{ik} of eq. 1.3.3.5 are given by

$$C_{ik} = \frac{1}{m\omega^2_i} \int \phi_i \ (\delta^{\alpha\mu}_{\beta\mu} \ (-\dot{\psi}^3 \mid^{\lambda}_{\mu} * \phi_k \mid^{\beta}_{\alpha} + \dot{b}^{\beta}_{\alpha} \ (b^{\lambda}_{\mu} * \phi_n * D))dF \quad 1.3.3.6$$

Characteristic buckling parameter (α) is obtained from the determinant of the coefficients F_K:

$$\mid E - \alpha \ \Sigma^n_{k=1} \ C_{ik} \mid = 0 \quad 1.3.3.7$$

2.
Stochastic system states

2.1
Introduction of stochastic perturbations

Identifying the reference state (o) as equilibrated in the mean with deformations v^m and stresses $\psi^3/\alpha\beta$ the neighbouring state (–) results from stochastic perturbation vector $p^3(\xi)$ with density $f(p^3)$ and internal stochastic variation of system parameters. The incremental decomposition of shell equations (1.2.5–6.) with physical and statistical linearization allows to compute the stochastic system states characterized by stochastic structural parameter.

Green's influence function is updated step by step accounting for degradation of structural parameter which define membrane and bending resistance.

Disturbances of the initial conditions of the reference state in a stochastic way lead to reliability estimates of the stability of the reference state. Application of a stochastic external incremental load define a residual safety margin in a probabilistic manner with respect to a known in the mean equilibrated state.

Introduction of stochastic state variable x_i the system state function (1.3.3) reads for the time t=τ :

$$L(\bar{x},t) : v^m(\bar{x},t) - \int \bar{p}^{(3)}(t)G_{(3)m}(\bar{x},t)d\tau = 0 \quad 2.1.1$$

Occurrence probability P of the limit state (t=τ) defined by eq. 2.1 is obtained from the Jointdensity–Integral I $(f_i(x))$ with density functions f_i:

$$P(X_i < x_i) = \int_{L(x,t) \le 0} \pi(f_i(\tilde{x}_i)dx_i \quad 2.1.2$$

First and second order reliability method can be applied to solve eq. 2.1.2.

Expressing the system state function eq. 2.1.1 in terms of safety margins the collapse probability of the reference state – which may be a state with defects – is obtained with respect to perturbations of the state variables x_i. Reliability of predamaged structures is thus given through probabilistic interpretation of safety factors.

2.2 Probability density function of incremental structural response

Stochastic loading term $p^3(\xi)$ in eq. 1.2.5 can be developed by a series of trigonometric functions with stochastic parameter p_α:

$$p^3(\xi,\eta) = \Sigma^n_{\alpha=1} p_\alpha * \sin\frac{\alpha\pi\xi}{a} * \sin\frac{\alpha\pi\eta}{b} \quad 2.2.1$$

Solving function $v^3(\xi,\eta)$ satisfying boundaries are expressed correspondingly with stochastic parameter v_α

$$v^3(\xi,\eta) = \Sigma^n_{\alpha=1} v_\alpha * \sin\frac{\alpha\pi\xi}{a} * \sin\frac{\alpha\pi\eta}{b} \quad 2.2.2$$

Substituting eq. 2.2.1–2 into eq. 1.2.5–6 and similar expressions for $\psi^{3}(\xi,\eta)$ algebraic equations are obtained (after application of the Galerkin method) of the form:

$$\tilde{p}_\alpha = P(\tilde{v}_\alpha, \tilde{g}_i) \quad 2.2.3$$

Conditional probability density f_v^3 $(v_3 \mid g_i)$ for structural response v^3 with the assumption that structural parameter $\bar{g}_i$ have occurred reads:

$$f_{v_\alpha}(v_\alpha \mid \bar{g}_i) = f_{p_\alpha}(\tilde{p}_\alpha \mid \bar{g}_i) \left| \frac{\delta p_\alpha}{\delta v^3_\alpha} \right| \quad 2.2.4$$

Absolute density function $f_{v\alpha}$ (v_α) accounting for stochastic structural properties represented by f_{gi} (g_i) follows by integration from eq. 2.2.4:

$$f_{v_\alpha}(v_\alpha) = \int_{\infty}^{\infty} f_{v_\alpha}(\hat{v}_\alpha \mid \bar{g}_i) * f_{g_i}(\tilde{g}_i)dg_i \quad 2.2.5$$

where eq. 2.2.4 has to be substituted for the first term under the integral. Probability distribution for v_α follows from integration of eq. 2.2.5:

$$F(v_\alpha) = \int f_{v\alpha}(\tilde{v}_\alpha)\, dv_\alpha \quad 2.2.6$$

2.3 Reliability of system state (n) against bifurcation

The system stability of the reference state (x) is evaluated by the probability (P) of returning into this position after application of a perturbation vector x_i so that

$$x_i = (\dot{X}_i + \dot{x}_i) < \varepsilon$$

$$P\ (\text{sub}\ \|x_i\| < \varepsilon) \quad 2.3.1$$

With known relation between perturbation x_i and stochastic system parameter g_i and their density function f (x, g_i) an expression for transition probability P_r is evaluated on the basis of the stability criterion (eq. 1.3.3.5) which is written in the form $\Psi(\dot{x}_i, \bar{g}_i) = 0$:

$$P_T = \int_{\Psi=0} \int f(\tilde{x}_i, \tilde{g}_i)d\tilde{x}_i d\tilde{g}_i : (\|x_i\| < \varepsilon) \quad 2.3.2$$

Eq. 2.2.10 defines the probability of a stabile system state (n) to move into collapsed state under the condition $(\|x_i\|<\varepsilon)$ for the perturbation vector.

3. Examples

3.1 Transition probability for two–mode failure of a cylindrical rudder shell–component

Tensor quantities for the quadratic (a,a) cylindrical shell surface are

$$\theta^1 = \chi, \theta^2 = y$$

$$\dot{b}_{11} = \dot{v}^3_{11}; \dot{b}_{22} = \frac{1}{R} = \dot{b}^2_2 = \frac{1}{R} + \dot{v}^3_{,22}$$

$$\nabla_{\dot{b}}(...) = \dot{b}_{11}\nabla() + \dot{b}_{22} * \nabla()$$

$$\nabla_{\dot{\psi}}(...) = \dot{\psi}^3_{,22} * \nabla() + \dot{\psi}^3_{,22} * \nabla() \quad 3.1.1\text{–}2$$

with these definitions the incremental relations for ^{3} and ^{3} are obtained.

Corresponding integro–differential equations read with Green's influence functions G_1, G_2:

$$B * \nabla\nabla_v^{-3} - (\nabla_{\dot{\psi}}\bar{\nabla}^3_v + \nabla_{\dot{b}}\bar{\nabla}\psi^3) = \bar{q}$$

$$\nabla\nabla\bar{\psi}^3 + D(\nabla_{\dot{b}} * \nabla\bar{v}^3) = 0 \quad 3.1.2\text{–}3$$

$$\bar{v}^3 - \int_F G_1(\bar{q} + \nabla_\psi\nabla\bar{v}^3 + \nabla_b * \nabla\bar{\psi}^3)dF = 0$$

$$\bar{\psi}^3 - D * \int_F G_2 * \nabla_b\nabla\bar{v}^3 dF = 0 \quad 3.1.24\text{–}5$$

Solution for $\bar{v}^3$, ψ^3 which satisfy the condition

are used to represent the kernel $G_{1/2}$:

$$\bar{v}_i^3 \ (x,y) \ = \ \sum_i v_2 * \sin\frac{\pi x}{\alpha} \ * \ \cos\frac{\pi y}{\alpha} \qquad 3.1.7\text{–}8$$

$$\bar{\psi}_i^3(x,y) = \sum \psi_i \sin\frac{\pi * x}{a} * \sin\frac{\pi y}{a} + \frac{x^2}{2} + \frac{y_2}{2}$$

$$G_{iK}(x,y,\xi,\eta) = \frac{1}{m} \overset{m}{v_i} \ (x,y) \ * \ \overset{m}{v_n}(\xi,\eta) \quad w^2$$

$$G_{in}(x,y,\xi,\eta) = \frac{1}{n} \overset{n}{\psi_i}(x,y) \ * \ \overset{n}{\psi_n}(\xi,\eta) \quad w^2 \qquad 3.1.9\text{–}10$$

Substitution of eg. 3.1.7–10 into eg. 3.1.4–5 results after application of Green's theorem in solutions for perturbation parameter v_i and ψ_i:

$$[v_i \ - \ (Q_{iK} \ - \ q_{iK}) \ * \ v_K] \ = \ 0 \qquad 3.1.11\text{–}12$$

$$[\psi i \ - \ \phi_{iK} \ * \ \psi_u] \ = \ 0$$

where

$$\int \overset{n}{v_r} \cdot \overset{m}{v_r} \ dF \ = \ \delta_{nm} \qquad 3.2.6$$

$$\bar{q} \ = \ \sum qn \ \cdot \ \sin\frac{\Pi x}{a}$$

$$Q_{iK} \ = \ \int_F v_i \ (\nabla_\psi * \nabla v_K \ + \ \nabla_b * \nabla\psi_K) \ dF$$

$$q_{iK} \ = \ \int v_i \ * \ q_K dF \qquad 3.1.13\text{–}16$$

$$\phi_{iK} \ = \ \int_F \psi_i \ * \ \nabla_b \nabla v_K dF$$

Stability condition of the reference state (°) results from the determinant of the coefficients of perturbation vector v_i, ψ_i:

$$|\delta_{iK} \ - \ (Q_{in} \ \lambda * \bar{q}_{in})| \ = \ 0 \qquad 3.1.17\text{–}18$$

$$|\delta_{iK} \ - \ \phi_{iK}| \ = \ 0$$

Introducing a stochastic perturbation vector v_i, ψ_i quantil values for the safety factor v of the reference state (o) are calculated from

$$P(\bar{v} \ < \ \bar{v} \ = \ \int \pi \ (fi(\bar{v}_i, \bar{\psi}_i) dv_i \qquad 3.1.19$$

$$G(v) \leq 0$$

with G(v) from eq. 3.1.11 – 12

Transition probability for instability of the reference state (o) with respect to a perturbation quantity is obtained from

$$P(\bar{\lambda} \ < \ \varepsilon) \ = \ \int_S f \ (\bar{\lambda}) d\lambda$$

$$S \leqq 0$$

where $S(\lambda)$=o results from eg. 3.1.17–18.

Results of eg. 3.1.19 and 3.1.20 give information about the relative influence of stochastic singular loads (represented by q_{in}) on failure probability with respect to bifurcation and strength. Results are documented in the appendix.

3.1.1 Results

A series of fin box skin panels and wing shell components (see Fig. 1) were tested. Characteristic remaining load carrying capacity after buckling of thin shell components is shown with scatter of tests results for buckling and strength failure. Determination prediction of panel buckling for an orthotropic composite (see Ref. /2/) fits well the test values in the mean. Prediction of strength failure (ca. 40% about critical buckling load for shell thickness t = 1.6 mm) succeeds with eq. 3.1.4-5 which takes into account prestress and predeformation of the buckled state. Derived safety factors must relate to this partially "damaged" state (see Fig. 2 + 3). Results of the reliability analysis on the basis of eq 3.1.19-20 is shown in Fig. 4. It confirms the well known remaining structural load carrying capacity of slender structures and their high reliability after introduction of defects (see also Fig. 5). For non-slender structures the mechanical behaviour is quite different. Spacing between buckling mode and strength-failure does not occur for a tested wing shell component (t = 4.25 mm). Conventional design analysis based on safety - factors give no general answer which mode would be critical - as necessary information for the design base - and it cannot establish a reliability-related quantity for two-mode failure. Reliability investigation (see Fig. 5) give answer on the transition probability (eq 3.1.20) of a predamaged reference state into an instabil collaps state. Parametric sensitivity analysis reveal information on important design variables.

Reference

(1): Bertrand, G.

Analytische Untersuchung ohne Verbund vorgespannter Schalen.
Habilitation, Berg. Univ. Wuppertal (1985)

(2): Spanink A.W.

"Technical Handbook Strength Data"
TH 3.372, Fokker Amsterdam (1987)

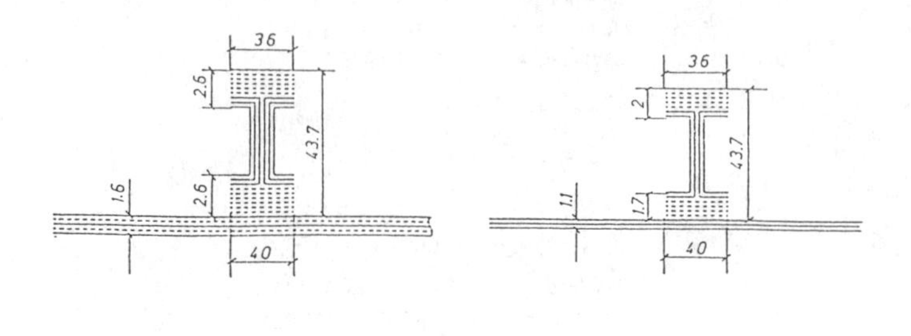

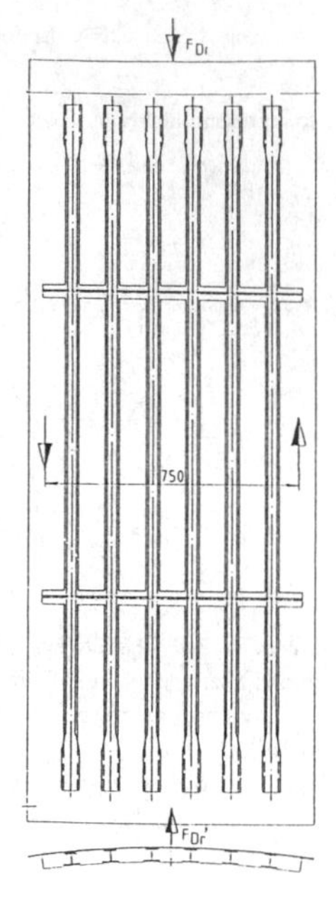

Fig.1 :Fin box side shell construction

Panel and stringer from composite

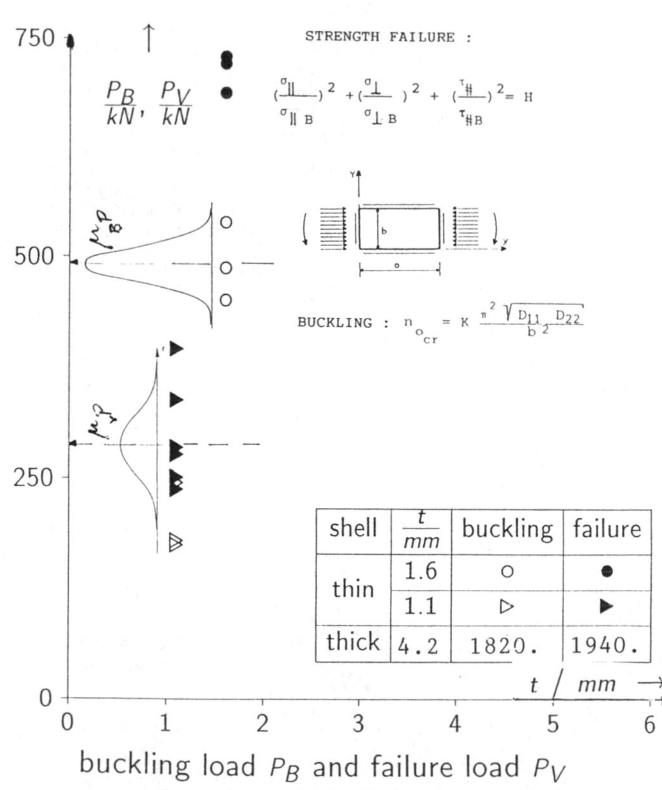

shell	t / mm	buckling	failure
thin	1.6	○	●
	1.1	▷	▶
thick	4.2	1820.	1940.

Fig.2 :Scatter of test results for Fin Box skin panel.

Deterministic prediction of buckling and failure

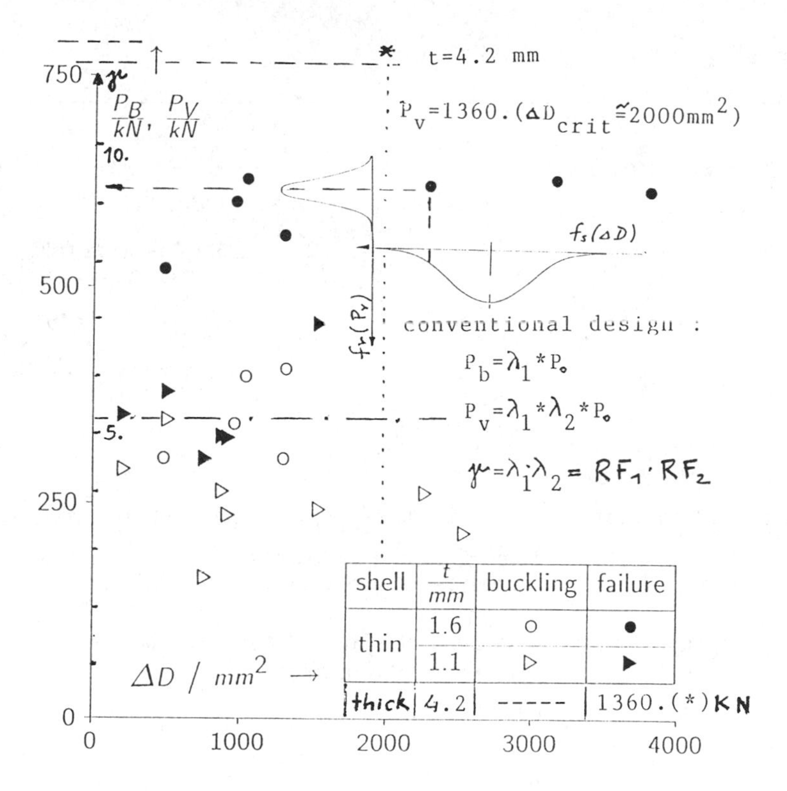

Fig.3: buckling load P_B and failure load P_V as function of area of delamination ΔD

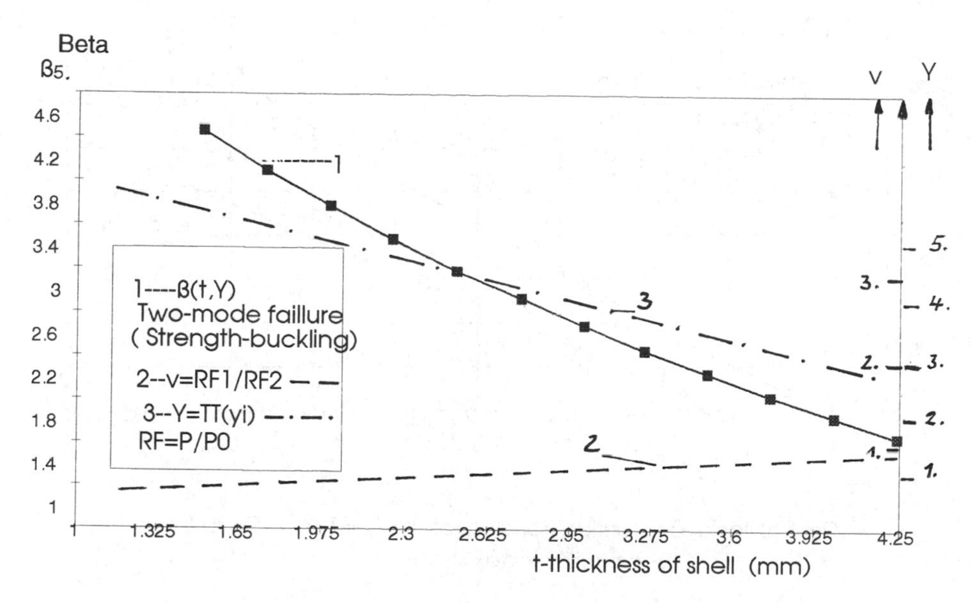

Fig.4 : Reliability analysis of slender fin box(β-Reliability index)
RF-Reserve factor (RF1-Buckling;RF2-Strength-failure)
v=RF1/RF2 ; Y-Product of partial reserve-factor (TT(yi)

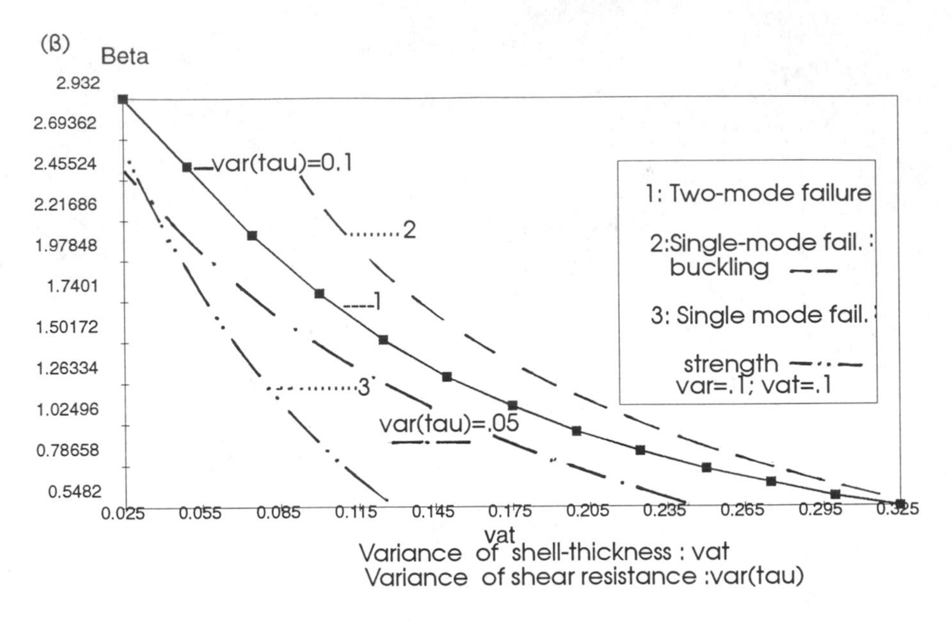

Fig.5:Parameter study on shell with defect D=10%
β – Reliability Index

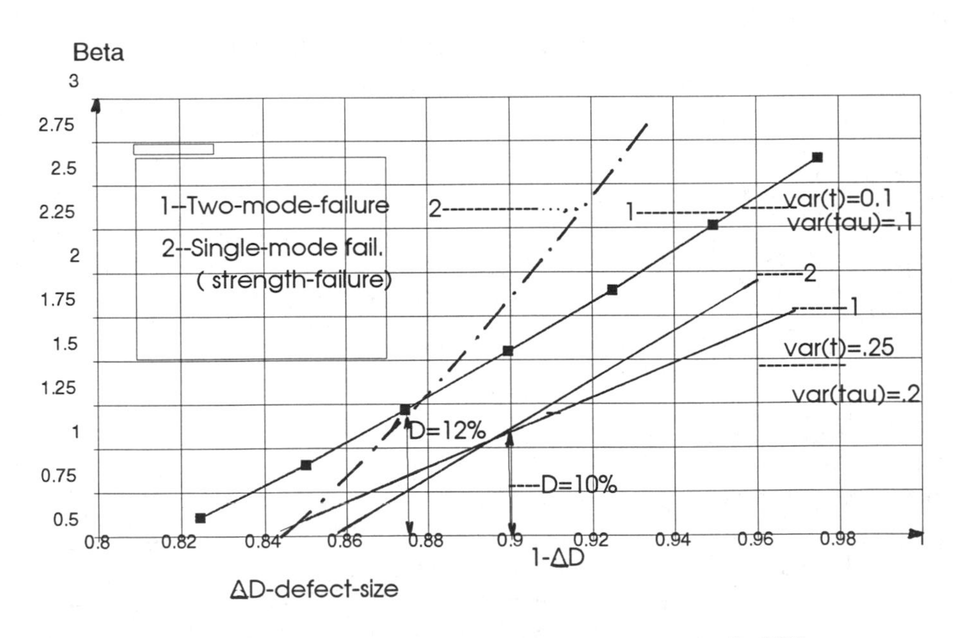

Fig.6 : Mode -Bifurcation for defect size D=12% ; D=10%
ß-Reliability-index

Structural Safety & Reliability, Schuëller, Shinozuka & Yao (eds) © 1994 Balkema, Rotterdam, ISBN 90 5410 357 4

Reliability-based condition assessment of concrete structures

Bruce R. Ellingwood & Yasuhiro Mori
Johns Hopkins University, Baltimore, Md., USA

ABSTRACT: Concrete structures may be affected by aging which may cause changes in strength and stiffness beyond the baseline conditions assumed for design. These changes should be considered as part of the process of evaluating a structure for possible continued future service. Methods are being developed to evaluate time-dependent reliability of concrete structures subjected to future extreme events. These methods enable the impact on safety and serviceability of uncertainties in operating conditions, structural strength, and strength degradation due to aggressive environmental conditions to be assessed quantitatively. The role of periodic inspection and maintenance in enabling a target reliability to be met over a period of continued service also is considered.

1 INTRODUCTION

Concrete structures in service may be affected by aging, involving changes in strength and stiffness beyond the baseline conditions that are assumed in structural design. Some of these aging effects are benign; others may cause component or system strengths to degrade over time, particularly when the concrete is exposed to an aggressive environment. Such aging effects may accelerate the risk of structural failure. Deterioration of infrastructure from effects of aging is a widespread problem in many countries, affecting buildings, bridges and other civil structures. Many deteriorated concrete structures must be evaluated for possible repair and continued service, since their replacement would be economically infeasible. Decisions to continue service and/or to perform inspection and maintenance should be supported by quantitative evidence that the strength of concrete structural systems is sufficient to withstand future extreme events within the proposed service period with an acceptable level of reliability.

Research is being conducted as part of the U.S. Nuclear Regulatory Commission's Structural Aging Program (Naus, 1986) to develop methods to facilitate quantitative assessments of current and future reliability and performance of concrete structures in nuclear power plants (NPPs) in support of decision making with regard to license renewal issues. This methodology takes into account the stochastic nature of past and future loads due to operating conditions and the environment, randomness in strength and in degradation resulting from environmental stressors. While the current research is focussing on concrete containments and Category I structures in NPPs, the time-dependent reliability analysis and other aspects of the methodology are equally applicable to other buildings and facilities in hostile environments, given the necessary statistical data base.

2 COMPONENT AGING AND STRUCTURAL RESISTANCE

The initial strength of concrete structural members and components can be described statistically by data that have been gathered in research over the past decade to develop improved bases for structural design of common reinforced concrete buildings and concrete structures in NPPs (Ellingwood and Hwang, 1985). However, environmental stressors may attack the integrity of the concrete and/or steel reinforcement in concert with or independent of operating, environmental or accidental loads (Siemes, et al., 1985), causing the strength to deteriorate over time. The most likely sources of deterioration of concrete structures are corrosion of reinforcement, followed by sulfate attack, freeze-thaw cycling, and reactive aggregate reactions within the concrete. The strength of a structure or component may degrade in time according to,

$$R(t) = R_0 \cdot G(t) \tag{1}$$

in which R_0 = component capacity in the undegraded (original) state, and $G(t)$ = time-dependent degradation function defining the fraction of initial strength remaining at time t. The degradation mechanisms are uncertain, experimental data are lacking, and thus the function $G(t)$ should be treated as stochastic. Most significant strength degradation mechanisms have been identified qualitatively (Clifton and Knab, 1989).

3 STOCHASTIC LOAD MODELS

Events giving rise to significant structural loads occur randomly in time and are random in intensity. When viewed on a timescale of 40 years or more, the duration of design-basis events generally is very short, and thus such events occupy only a small fraction of the total life of a component. With these assumptions, a structural load that varies in time can be modeled as a sequence of randomly occurring pulses with random intensity, S_j, and duration, t, as illustrated in Fig. 1. The event duration is assumed to be sufficiently short that any change to the state of the structure occurs only during the application of the load (i.e., cumulative damage during a load event does not occur). A simple stochastic load model is obtained by assuming that the occurrence of the load events is described by a Poisson process with mean occurrence rate, λ, and that the intensities, S_j, are identically distributed and statistically independent random pulses of constant amplitude, described by the distribution function, $F_S(s)$. Many of the operating, environmental and accidental loads that act on NPP structures can be modeled by such processes (Hwang, et al, 1987).

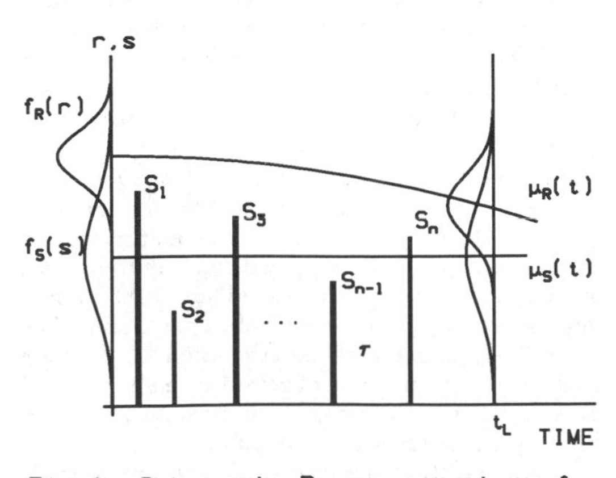

Fig.1: Schematic Representation of Load Process and Degradation of Resistance

4 TIME-DEPENDENT RELIABILITY ANALYSIS

4.1 *Component*

Structural loads and strength vary in time, making structural reliability time-dependent (Kameda and Koike, 1975). To illustrate the basic concepts of time-dependent reliability analysis, consider a structural component subjected to the sequence of discrete stochastic load events illustrated in Fig. 1. The strength of the component deteriorates with time due to environmental factors according to Eqn.(1). We assume that $G(t)$ is independent of the load history; with this assumption, the subsequent formulation can address deterioration due to corrosion, sulfate attack and similar environmental effects. Assume for the present that $G(t)$ is deterministic and equal to mean $E[G(t)] = g(t)$. This assumption will be examined later. If n events occur within time interval $(0, t_L)$ at deterministic times t_j, $j = 1, \ldots, n$, the reliability function is represented as follows:

$$L(t_L) = P[R_0 \cdot g(t_1) > S_1 \cap \ldots \cap R_0 \cdot g(t_n) > S_n] \tag{2}$$

In general, the number of loads that occur in $(0, t_L)$ and the times at which they occur is random. Taking the randomness in load events into account, the time-dependent reliability function becomes (Ellingwood and Mori, 1990),

$$L(t_L) = \int_0^\infty \exp\left[-\lambda t_L \cdot \left[1 - \frac{1}{t_L}\int_0^{t_L} F_S\{r \cdot g(t)\}dt\right]\right] \times f_{R_0}(r)dr \tag{3}$$

in which λ is the mean occurrence rate of load events and $f_R(r)$ is the probability density function (pdf) of R_0, expressed in units that are dimensionally consistent with S. The limit state probability, or probability of failure, is given by,

$$F(t_L) = 1 - L(t_L) \tag{4}$$

When several time-dependent load processes act on a component, the probability distribution of the structural action due to the combined loads can be obtained approximately as a function of the statistical descriptors of the individual loads

(Wen, 1977; Hwang, et al, 1987). For example, if a component is subjected to two time-dependent load processes with intensity S_1 and S_2, respectively, failure may occur due to one of the following three mutually exclusive non-zero load combinations: only S_1 is non-zero (S_1); only S_2 is non-zero (S_2); or both S_1 and S_2 are non-zero (S_1+S_2). The failure probability of the component then is,

$$P_f = P_f^{(S_1)} + P_f^{(S_2)} + P_f^{(S_1+S_2)} \tag{5}$$

in which $P_f^{(q)}$ is the limit state probability under the load combination q. This probability can be computed approximately from Eqn.(3) with $F_S(s) = F_q(s)$ and $\lambda = \lambda_q$, in which F_q is the cdf of the intensity of the load combination q and λ_q is the incidence rate of the load combination. The incidence rate of combination S_1+S_2 is evaluated approximately as,

$$\lambda_{S_1+S_2} \approx \lambda_{S_1}\lambda_{S_2}(\tau_{S_1} + \tau_{S_2}) \tag{6}$$

in which λ_{S_i} and τ_{S_i} are the mean occurrence rate and the mean duration of load process S_i.

4.2 *Structural system*

A structural system composed of beams, columns, slabs, and walls can be modeled by a combination of two fundamental subsystems: series systems and parallel systems (Melchers, 1987). The reliability of a structure modeled as a series system of components generally provides a conservative estimate of the system reliability.

Assume that n discrete load events occur at deterministic times t_j, $j = 1,\ldots,n$. Load S_j acting on the structure induces in member i a structural action c_iS_j (e.g., moment, axial force, shear, etc.). The S_j's are statistically independent and identically distributed random variables. The change in strength of each component in time is described by a degradation function, $g_i(t_j)$. If the initial strength of the structural components, $\underline{R}_0 = \{R_1,\ldots,R_m\}$, is deterministic and equal to $\underline{r} = \{r_1,\ldots,r_m\}$, the time-dependent reliability function for a series system, $L_S(t_L)$, is represented as,

$$L_S(t_L) = P\begin{pmatrix} r_1g_{11} > c_1S_1 & \cap\cdots\cap & r_1g_{1n} > c_1S_n \\ \cap\ r_2g_{21} > c_2S_1 & \cap\cdots\cap & r_2g_{2n} > c_2S_n \\ \vdots\qquad\vdots & \vdots\ \ddots\ \vdots & \vdots \\ \cap\ r_mg_{m1} > c_mS_1 & \cap\cdots\cap & r_mg_{mn} > c_mS_n \end{pmatrix}$$

$$= \prod_{j=1}^{n} F_S\left(\min_{i=1}^{m} \frac{r_i \cdot g_{ij}}{c_i}\right) \tag{7}$$

in which $g_{ij} = g_i(t_j)$

The conditioning on the number of load events, $N(t_L) = n$, the times at which they occur, $\underline{T} = \underline{t}$, and the initial strengths, $\underline{R}_0 = \underline{r}$ must be removed to take into account the randomness in occurrence of load events and in strength. The time-dependent reliability function for a system becomes (Ellingwood and Mori, 1990),

$$L_S(t_L) = \underbrace{\int_0^{\infty}\cdots\int_0^{\infty}}_{m\text{-fold}} L_S(t_L|\underline{R}_0 = \underline{r}) \cdot f_{\underline{R}}(\underline{r})d\underline{r} \tag{8}$$

in which

$$L_S(t_L|\underline{R}_0 = \underline{r}) = \exp\left[-\lambda t_L \cdot \left\{1 - \frac{1}{t_L}\int_0^{t_L} F_S\left(\min_{i=1}^{m} \frac{r_i \cdot g_i(t)}{c_i}\right) dt\right\}\right] \tag{9}$$

Eqn.(8) cannot be integrated in closed form because of the (m+1)-fold integration. However, performing the m-fold integration in Eqn.(8) by Monte Carlo simulation while evaluating Eqn.(9) numerically in closed form is relatively efficient. An importance sampling technique (e.g., Rubinstein, 1981; Verma, et al, 1989; Melchers, 1990; Mori, 1992) is used in the following section to determine the time-dependent system reliability.

4.3 *Illustration of time-dependent reliability*

Component reliability

The effect of degradation in component strength on the component reliability is illustrated in the following paragraphs using several simple models of time-dependent strength. These degradation models, mean values of which are summarized in Table 1, were selected to examine the sensitivity of the time-dependent reliability to the type of degradation. Those factors that appear to have a significant impact on time-dependent reliability should receive special attention in any data acquisition program supporting condition assessment and service life prediction of reinforced concrete structures. The linear and parabolic functions model, in an approximate manner, degradation in strength due to corrosion of reinforcement and sulfate attack, respectively. The square root

Table 1: Degradation model

Shape of the degradation function	Degradation rate $g(40)$	Corresponding degradation mechanism
Linear: $g(t) = 1 - at$	0.9	Corrosion
Parabolic: $g(t) = 1 - at^2$	0.9	Sulfate attack
Square root: $g(t) = 1 - a\sqrt{t}$	0.9	Diffusion controlled degradation

Table 2: Load process parameters

	Mean*	C.o.v.	Pdf	$\lambda(\text{yr}^{-1})$	τ**
Dead load	$1.00D_n$	0.07	Normal	–	40 years
Live load	$0.40L_n$	0.50	Type I	0.5	3 months
Earthquake load	$0.08E_{ss}$	0.85	Type II	0.11	30 sec.

* D_n and L_n are nominal loads and E_{ss} is safe shut-down earthquake load specified for design.

** τ is the mean duration of a load event.

function represents a degradation process that is diffusion-controlled and where the rate decreases in time. The degradation rate parameter is given with reference to the residual strength at 40 years.

The combination of time-varying live load, L, and dead load, D, is considered first. The current design requirement for concrete structures in flexure is (ACI 318; 1989),

$$0.9R_n = 1.4D_n + 1.7L_n \tag{10}$$

in which R_n = nominal strength and D_n and L_n = nominal dead and live loads. The probabilistic models of load intensity are summarized in Table 2. It is assumed that the initial strength has a lognormal distribution with $\mu_R = 1.15R_n$ and coefficient of variation (c.o.v.) $V_R = 0.15$. These statistics are selected to be representative of existing strength data for reinforced concrete components subjected to flexure or compression (Ellingwood and Hwang, 1985). It is assumed that $D_n = L_n$. The effect of variability in dead load was found to be negligible compared with the effect of other factors, and thus the dead load is treated as deterministic in the following illustrations. In each analysis, the service period, t_L, is assumed to be 60 years.

The effect of variability in degradation function, $G(t)$, on the component failure probability is presented in Fig. 2 using the linear degradation model,

$$G(t) = 1 - At \tag{11}$$

where A is a random variable normally distributed

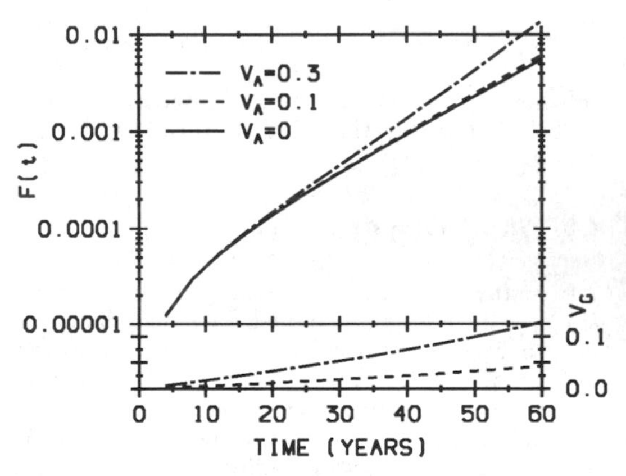

Fig.2: Dependence of Failure Probability on $V_{G(t)}$

with mean value, $\mu_A = 0.005$, corresponding to $g(40) = 0.8$, and with coefficient of variation, V_A = 0.1 or 0.3. The corresponding c.o.v.'s in the degradation functions are also presented in the figures. The failure probability is evaluated by Monte Carlo simulation. The variability in $G(t)$ has a second-order effect on $F(t)$ compared with other time-dependent factors such as degradation models and statistical parameters of load processes if $V_{G(60)}$ is less than about 0.15. Accordingly, the variability in degradation function is not considered in subsequent illustrations.

The effect of the shape of the degradation function on the limit state probability, $F(t)$, is presented in Fig. 3 for $g(t) = 0.9$. The failure probability assuming that no degradation occurs is also shown as a point of reference. The shape

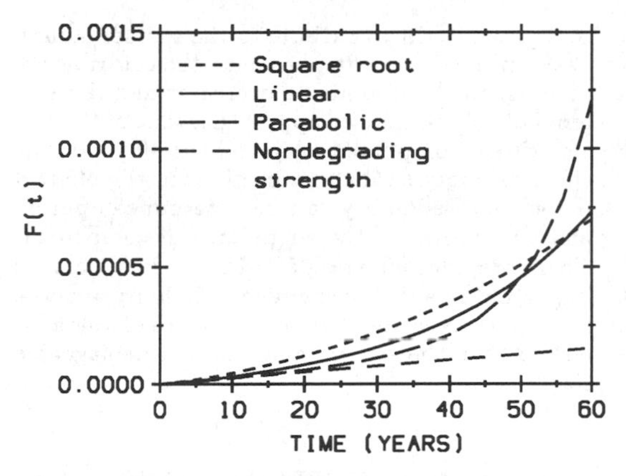

Fig.3: Dependence of single component failure probability on degradation model: D+L, g(40)=0.9

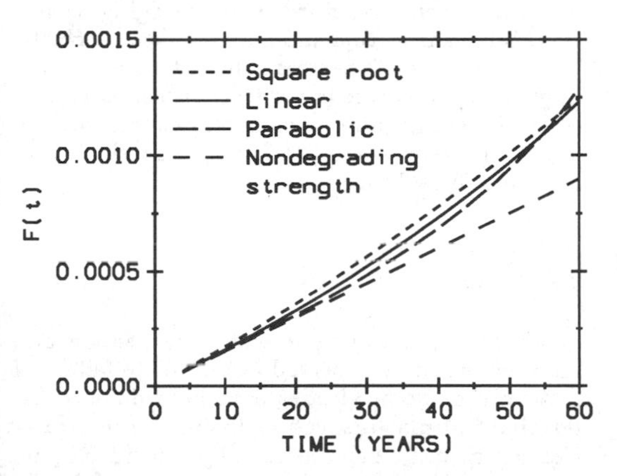

Fig.4: Dependence of single component failure probability on degradation model: D+L+E, g(40)=0.9

of the degradation functions has a significant impact on reliability for longer service lives. Moreover, for lower residual strength at time t, the failure probability is sensitive to small changes in the strength; this effect accumulates with time.

The time-dependent reliability of components has been shown to be sensitive to the mean rate of occurrence of load events (Ellingwood and Mori, 1990; Mori, 1992). The combination of dead, live, and earthquake loads ($D + L + E$) is used to illustrate this point further. The mean value of initial strength is determined from the following current design requirement for concrete structures in NPPs (ACI 349, 1976),

$$0.9R_n = D_n + L_n + E_{ss} \tag{12}$$

in which E_{ss} = structural effect of the safe shutdown earthquake. It is assumed that $E_{ss} = 2.0D_n$. The mean occurrence rates for the combination of $L + E$ is $\lambda_{L+E} = 0.014$ from Eqn.(6). The shape of the degradation function affects the limit state probability, shown in Fig. 4, in a manner similar to that under $D + L$ in Fig. 3. However, the failure probability under $D + L + E$ is less sensitive to the shape of the degradation function, and with $g(40) \geq 0.9$, the effect of the shape of the degradation function is negligible up to 50 years. This insensitivity is due to the small mean occurrence rate and large variability in load E.

System reliability

The effect of degradation in component strength on the system reliability is illustrated for the load combination $D + L$ using the same load and resistance parameters as in the illustrations of component reliability and the linear degradation model. Without loss of generality, the c_i's in Eqn.(9) are assumed to be unity for all components. In each analysis, a five-component system is considered. The degradation rate equals 0.8 at 40 years for all components unless otherwise specified.

The strengths of the components in a system are correlated because of common construction materials, fabrication and construction practices. The sensitivity of the system reliability to the stochastic dependence among component strengths is investigated by assuming that the strengths of the components are identically distributed and equally correlated pair-wise with

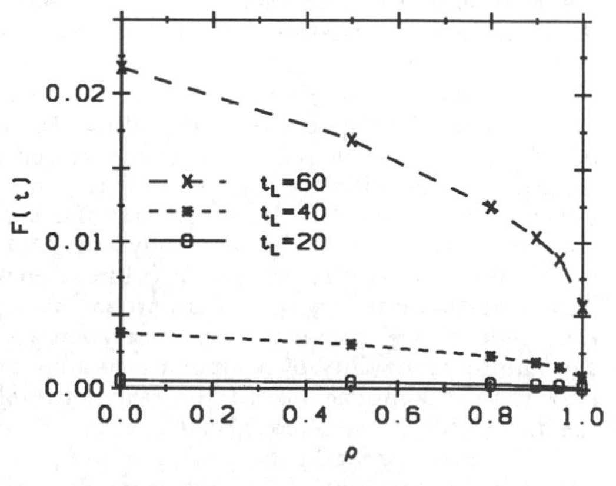

Fig.5: Dependence of System Failure Probability on ρ

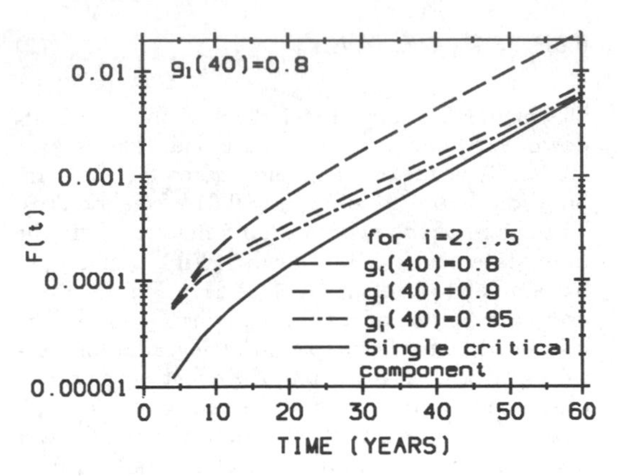

Fig.6: Dependence of System Failure Probability on Degradation Rate

correlation coefficient, ρ. The failure probability increases as the correlation coefficient decreases, as shown in Fig. 5. However, the effect of ρ is relatively small, particularly when $\rho < 0.8$, and decreases as the coefficient of variation of load intensity increases. The reduction in the sensitivity of the reliability function to ρ is due to the dependence among component failure events that arises from common random loads with relatively large variabilities, and is consistent with the results of a previous study (Thoft-Christensen and Sorensen, 1982). In light of the relative insensitivity to ρ, it seems appropriate to model systems such as those considered herein as being composed of components with independent strengths, particularly when the coefficient of variation of load intensity is large relative to that of resistance.

In general, a system consists of components with different characteristics of strength and because of this, some components affect the system failure probability more than others. The effect of component degradation rates on system reliability are shown in Fig. 6. The degradation rates are set so that $g(40) = 0.8$ for one component (critical component) while $g(40) = 0.9$ or 0.95 for the other components (noncritical components). The initial strengths of all the components are assumed to be identically and lognormally distributed and statistically independent of one another. These system failure probabilities are bounded by the failure probability of the single critical component in the system and the failure probability of a system consisting of only critical components. At the early stage of the service life when all components remain close to their initial strength, the system failure probability is close to that of the system in which all components degrade equally and the noncritical components still contribute to the system failure. Later in the service life, however, the contribution of noncritical components decreases and the system failure probability approaches that of the single critical component. After about 40 years, the effect of noncritical components is negligible and the system reliability can be determined approximately as that of the single critical component. Therefore, identification of critical components is important in a time-dependent reliability analysis of a complex system because it enables the analyst to keep the system analysis to a manageable size.

5 IN-SERVICE INSPECTION AND MAINTENANCE STRATEGIES

Periodic in-service inspection followed by suitable maintenance may restore a degraded reinforced concrete structure to near-original condition. Since inspection and maintenance are costly, there are tradeoffs between the extent and accuracy of inspection, required level of reliability, and cost. To design an optimum inspection/maintenance program, the following optimization problem must be solved:

$$\begin{aligned} &\text{Minimize} && C_T \\ &\text{Subject to} && F(t) < P_f' \end{aligned} \tag{13}$$

in which C_T is the total cost of inspection and maintenance plus expected losses due to failure of a structure. Some studies have been done to determine optimal inspection/maintenance strategies for metallic structures subjected to fatigue, assuming that a component is replaced if the intensity of detected damage exceeds a critical value (e.g., Thoft-Christensen and Sorensen, 1987). However, a component may not be replaced following inspection; instead, only those damages that are detected might be repaired. In this case, the strength of the component may not be fully restored to its original level. The effect of damage overlooked at an inspection, i.e., the extent of inspection, also should be considered in designing the optimal strategies.

The time-dependent reliability analysis described in this paper provides a tool for designing optimum inspection/maintenance strategies. To illustrate this with a simple example, two strategies are considered: (1) infrequent inspection and maintenance carried out at 20 and 40 years, with strength restored to its original level following maintenance, and (2) frequent inspection and maintenance performed at 10, 20, 30, 40 and 50 years, with strength restored to only

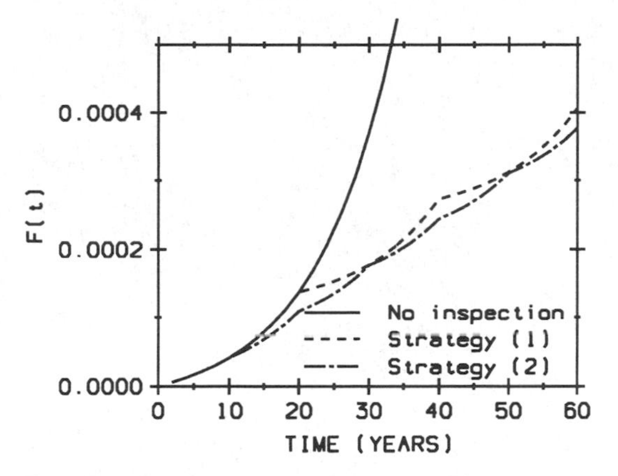

Fig.7: Failure Probability with Repair

97 % of its original level following maintenance. Strength degradation is assumed to occur linearly, with $g(40) = 0.8$. The component strengths are assumed to degrade at the same rate following maintenance. The failure probabilities with these strategies are illustrated in Fig. 7 for the combination of dead and live loads with a mean live load occurrence rate of 0.5/year. At the time of inspection and maintenance, the failure probability changes its slope; this change is more distinct when the component is inspected more thoroughly. If $F(40) \leq 0.00025$, strategy (2) would be acceptable in terms of risk, whereas strategy (1) would be unacceptable. Alternate acceptable strategies should be prioritized on the basis of cost, as suggested by Eqn.(13). Methods to deal with randomness in inspection and effectiveness of repair have been developed but are considerably more complex (Mori, 1992).

6 CONCLUSIONS

The reliability of components and systems is sensitive to the choice of models of strength degradation. Dependence in component strengths within a system has a lesser effect on system reliability unless the component strengths are highly correlated. If the critical components within a system can be identified, condition assessment can focus on those and the scope of the time-dependent reliability analysis can be reduced. Appropriate degradation models and load process statistics must be identified to utilize the above methodology for realistic condition assessment of structures. Time-dependent reliability analysis can provide a tool to determine strategies for inspection and maintenance that are necessary to maintain structural reliability at an acceptable level during a period of extended service. However, additional data on relative costs of inspection, repair and maintenance are required to optimize these inspection/maintenance policies with respect to life cycle costs.

ACKNOWLEDGEMENT

Support of this research through Grant 19X-SD684V from Oak Ridge National Laboratories is gratefully acknowledged.

REFERENCE

ACI 1989. "Building code requirements for reinforced concrete (ACI Standard 318-89)." American Concrete Institute, Detroit, MI.

Clifton, J. R. and L. I. Knab 1989. *Service life of concrete.* National Bureau of Standards. NUREG/CR-5466.

Ellingwood, B. and H. Hwang 1985. "Probabilistic descriptions of resistance of safety-related structures in nuclear plants." *Nuclear Engineering and Design* 88(2): 169 - 178.

Ellingwood, B. and Y. Mori 1990. "Probabilistic methods for condition assessment and life prediction of concrete structures in nuclear power plants." *Proc. of Water Reactor Safety Information Meeting.* Rockville, MD.

Hwang, H., B. Ellingwood, M. Shinozuka, and M. Reich 1987. "Probability-Based Design Criteria for Nuclear Plant Structures." *J. of Str. Engr. ASCE* 113(5): 925 - 942.

Kameda, H. and T. Koike 1975. "Reliability analysis of deteriorating structures." *Reliability approach in structural engineering.* Maruzen Co., Tokyo, Japan. pp. 61 - 76.

Melchers, R.E. 1987. *Structural reliability; analysis and prediction.* Ellis Horwood Ltd., West Sussex, England.

Melchers, R.E. 1990. "Search-based importance sampling." *Structural Safety* 9: 117 - 128.

Mori, Y. 1992. *Reliability-based condition assessment and life prediction of concrete structures.* PhD thesis, Johns Hopkins University, Baltimore, MD.

Naus, D. J. 1986. *Concrete component aging and its significance relative to life extension of nuclear power plant.* Oak Ridge National Laboratory, NUREG/CR-4652.

Rubinstein, R.Y. 1981. *Simulation and the Monte Carlo method.* John Wiley, New York.

Siemes, A. J. M., A. C. W. M. Vrouwenvelder, and A. van den Beukel 1985. "Durability of buildings: a reliability analysis." *Heron* 30(3): 1 - 48.

Thoft-Christensen, P. and J. D. Sorensen 1982. "Reliability of structural systems with correlation elements." *Applied Mathematical Modeling* 6: 171 - 178.

Thoft-Christensen, P. and J. D. Sorensen 1987. "Optimal strategy for inspection and repair of structural systems." *Civil Engr. Syst.* 4: 94 - 100.

Verma, D., Fu, G. and F. Moses 1989. "Efficient structural system reliability assessment by Monte Carlo methods." *Proc. ICOSSAR'89:* 895 - 901.

Wen, Y. 1977. "Statistical combination of extreme loads," *J. of Str. Div. ASCE* 103(6): 1079 - 1093.

Structural Safety & Reliability, Schuëller, Shinozuka & Yao (eds) © 1994 Balkema, Rotterdam, ISBN 90 5410 357 4

Automated damage assessment system using neural network

Hitoshi Furuta & Naruhito Shiraishi
Kyoto University, Japan

Hiroki Ohtani
Kinki Nippon Railway Co., Japan

ABSTRACT: In this paper, an attempt is made to develop an automatic damage assessment system by introducing the technique of neural network. In order to reduce the maintenance load, information regarding crack width and configuration will be input and stored as visual information, which is input to a computer through an image scanner. Using the ability of pattern recognition of neural network, it is possible to evaluate the damage state of existing structures automatically. Several examples are presented to demonstrate the applicability of the system developed here.

1 INTRODUCTION

To establish a rational maintenance program, it is inevitable to evaluate the damage state of existing structures (Shiraishi, Furuta & Sugimoto 1985). In bridge structures, RC (Reinforced Concrete) decks have been suffering from damage, because they resist the applied loads directly (Furuta & Shiraishi 1989, Shiraishi, Furuta, Umano & Kawakami 1991, Furuta, Shiraishi, Umano & Kawakami 1991). In the damage assessment of RC bridge decks, crack width and configuration play important roles, since they provide us of useful information in estimating the remaining life. However, the investigation on cracks is not easy, because it should be performed by experienced engineers through chalking and measuring. The chalking and measuring require a lot of working time and load.

In this paper, an attempt is made to develop an automatic damage assessment system by introducing the technique of neural network (Rumelhart, McClelland & PDP Research Group 1986). The neural network is useful for pattern recognition and voice recognition. Using the ability of pattern recognition, it is possible to make the automatic evaluation of the damage state, which can save the working time and load necessary in the inspection and analysis. In order to reduce the maintenance load, information regarding crack width and configuration will be input and stored as visual information, which is transferred to a computer through an image scanner. The visual information of crack pattern is transformed into a matrix pattern and interpreted by using a damage criterion obtained through the neural computing. The present damage assessment system has such a characteristic as that it can be used at any time and place, because it is built on a micro computer. Several examples are presented to demonstrate the applicability of the system developed here.

2 AUTOMATED DAMAGE ASSESSMENT SYSTEM BASED ON NEURAL NETWORK

As well known, the neural network computation has such advantages that it has the ability of machine learning from the past records and data and it is suitable for the pattern recognition of visual data like pictures and photographs. Utilizing the above advantages of neural network, it is possible to develop an evaluation system of damage state of RC bridge deck, so as to reduce the maintenance load. The present system consists of three subsystems; 1)system for making learning data, 2)learning system and 3) damage assessment system. Fig. 1 shows the architecture of the system.

2.1 System for making learning data

The necessary learning data are made in the following. At first, sample photographs of crack states of RC decks are collected. They are evaluated by an experienced engineer according to some manual of maintenance. Here, the manual (The Kinki Branch of Ministry of Construction, Japan 1985) presented by the Kinki Branch of Ministry of Construction, Japan is used to classify the damage state into five ranks; rank 0(no damage), rank I(slight damage), rank II(moderate damage), rank III(severe damage), and rank IV(collapse). As seen in Fig. 1, this subsystem consists of a micro computer(NEC PC-9801 VX) and an image scanner(OMRON HS-10RII), which is controlled by a program written in Quick-BASIC. Using the program, the visual information given by the photographs

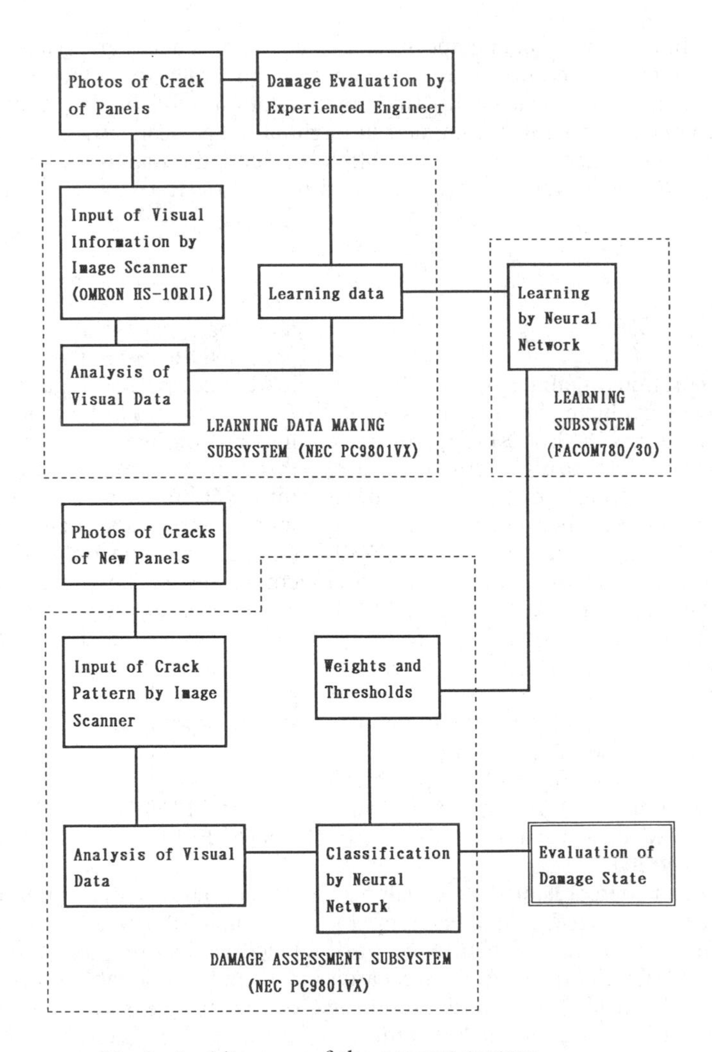

Fig.1. Architecture of the present system

is transformed into a pattern image of 640 X 400 dots. Each dot has such binary information as black and white. Because of the limitation of memory capacity of micro computer, it is impossible to treat the pattern information as the dot image. It is, therefore, attempted here to express the dot image in terms of matrix pattern. Namely, the dot image is divided into N X M matrix meshes (see Fig. 2). Each element of the matrix is calculated as the ratio of the area of cracks involved in the mesh to the all area (see Fig. 3).

2.2 Learning system

In the neural computation, the learning is executed by solving a problem in which the mean error between the target values and estimated values should be minimized by controlling adequate explanatory parameters. In general, the learning process in a neural network requires a lot of computation time. For instance, the number of explanatory parameters to be adjusted becomes 31,115 for a case with 1000 nodes for the input layer, 30 nodes for the first intermediate layer, 30 nodes for the second intermediate layer and 5 nodes for the output layer. Therefore, the learning procedure is implemented on the main frame computer(FACOM M-780/30) installed at the Information Processing Center of Kyoto University. It is noted that the learning process is executed using th back propagation algorithm. The computer program is made paying attention to the following items:

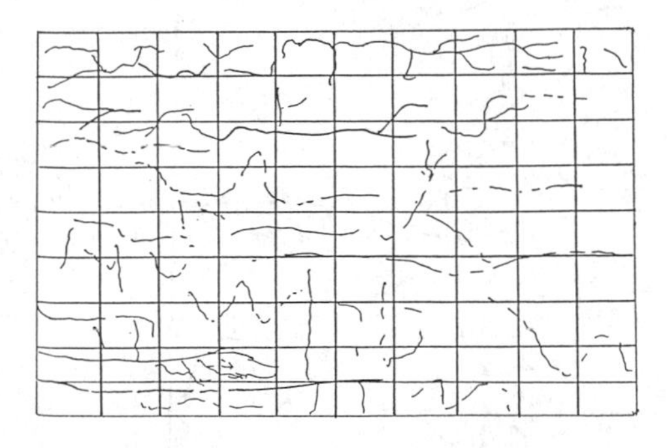

Fig.2. N X M matrix form

1.The increments of explanatory variables are

$$W = 0.1S \tag{1}$$

W: increment of weight

$$h = 0.2S \tag{2}$$

h:increment of threshold value
S: mean error

These variables are determined empirically through some numerical computations and the past experience. The learning data are transferred from a micro computer to the main frame computer.

2.The learning process is terminated when the mean error S becomes smaller than a prescribed small value S0. Although the value of S0 is better as it is closer to zero, the value of S0 is given as 0.1.

2.3 Damage assessment system

For a bridge deck, the damage assessment is performed using a micro computer and an image scanner. The photographs of the deck in consideration is read by the scanner and stored in the micro computer. Using the weights and threshold values obtained through the back propagation algorithm, the damage pattern is judged as one of the five damage categories.

2.4 Characteristics of the present system

The present system has such advantages:

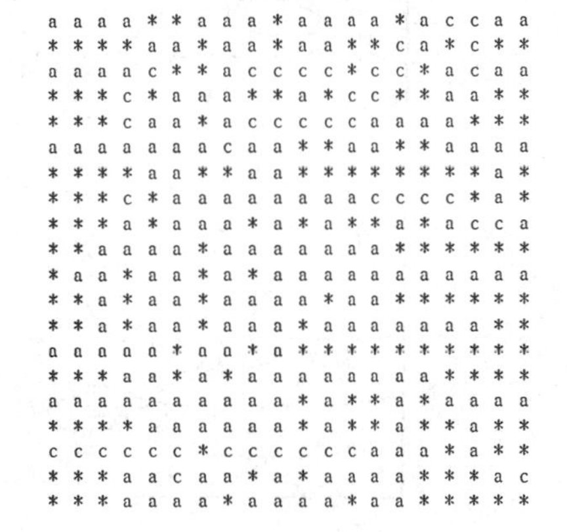

Fig.3. Symbolic expression

1.It is possible to make a remarkable reduction of the input work, because the crack width and configuration can be measured and analyzed as the visual information with the aid of scanning process.

2.It is possible to acquire the necessary knowledge through the learning ability of neural network based on the past data.

3.It is possible to learn a great deal of data within rather shorter time by using the main frame computer.

4.It is possible to judge the damage state at site because one can use a micro computer.

5.It is possible to evaluate the damage state of panels with water or gas pipes with the aid of the learning ability of neural network.

On the other hand, the system has the following problems to be overcome:

1.It is necessary to improve the computer program of the back propagation process so as to reach the optimum solution for any case.

2.It is desirable to introduce three such dimensional information as crack depth into the damage assessment.

3.It is necessary to increase the number of learning data to be available in the damage assessment.

4.It is necessary to find some useful means to obtain photographs even for panels with obstacles. It is considered that a video camera with flexible extension rods or a miniature helicopter is promising.

3 ILLUSTRATIVE EXAMPLES

Consider 15 panels of an existing bridge. Panel means the area surrounded by girders and cross beams. In general, the damage assessment of RC bridge deck is made panel by panel. Fig. 4 shows the crack pattern of panel 1. Among 15 panels, 12 panels are used as learning data. each dot pattern of panel cracks is expressed by a 40 X 40 matrix. For each panel, the evaluation result should be provided by an experienced engineer, which is given in terms of such five ranks as rank 0, I, II, III and IV, according to the manual given by the Kinki Branch of Ministry of Construction, Japan.

Using the above data, the learning process is performed through the neural computation. The structure of neural network used is as follows:

1.the number of layer is 4.

2.the number of unit involved in the input layer is 1600.

3.the number of unit involved in the first intermediate layer is 30.

4.the number of unit involved in the second intermediate layer is 30.

5.the number of unit involved in the output layer is 5.

Here, the numbers of input and output layers

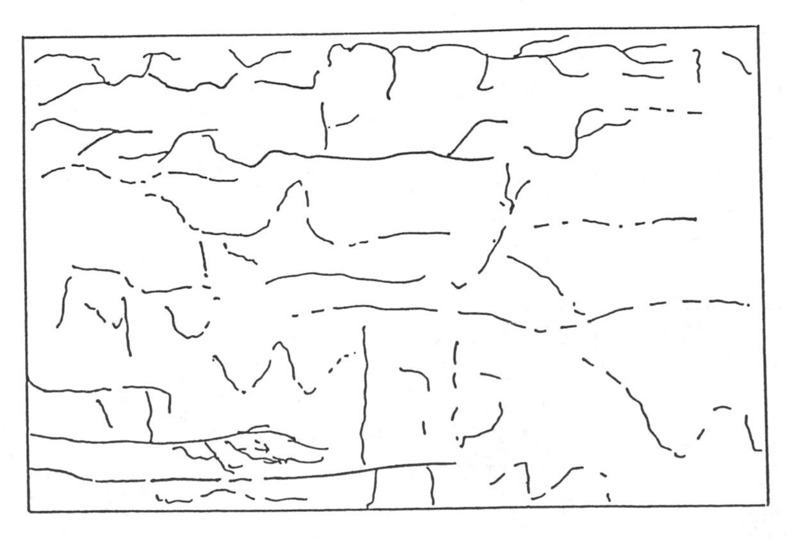

Fig.4. Crack configuration of panel 1

are determined by the conditions that the crack pattern is expressed by a 40 X 40 matrix and the damage state is classified into the five categories. For the learning process, 12 data is used. Table 1 presents the numerical results obtained by the neural computation. The closer to unity the value is, the more possible the rank is as the damage state. For panel 3, the result is given as (0, I, II, III, IV) = (0.0, 0.047, 0.075, 0.961, 0.0). This implies that the damage rank III should be chosen because the values for rank III is the closest to unity. The results for panel 1 to panel 12 show well correspondence to the results by the experienced engineer. On the contrary, the results of panel 13 to 15 do not present such clear classification as those of panel 1 to panel 12. Nevertheless, it is possible to judge the damage state properly based on the results provided by the present system. Although the largest values of 0.413, 0.466 and 0.322 for panel 13 to panel 15 are not close to unity, the obtained ranks are the same as the expert's judgment.

Next, consider the damage assessment of panels with attachments. Fig. 5 shows the crack pattern with a gas pipe and Fig. 6 shows the crack pattern with two straight lines which denote concrete joints. For panel 2 through panel 15 used in the previous example, the images of gas pipe and concrete joints are added, respectively. As a result, we can obtain 45 data as a total. Table 2 presents the numerical results which show a good correspondence to the judgment by an experienced engineer. From this fact it is concluded that the present system can evaluate the damage state of RC bridge decks with such attachments as gas and water pipes.

4 CONCLUSIONS

In this paper, an attempt was made to develop an automatic evaluation system of the damage state of RC bridge decks. Through the numerical simulation, the following conclusions were derived:

1. Using the present system, it is possible to evaluate the damage state of RC bridge decks with shorter working time and less working load. Any engineer can reach a meaningful classification of the damage state when he can

Table 1. Numerical results by neural network

Panel No.	Rank 0	Rank I	Rank II	Rank III	Rank IV	Expert's Opinion
Panel 1	0.000	0.249	0.743	0.051	0.000	II
Panel 2	0.000	0.073	0.218	0.904	0.000	III
Panel 3	0.000	0.047	0.075	0.961	0.000	III
Panel 4	0.000	0.072	0.062	0.958	0.000	III
Panel 5	0.000	0.073	0.628	0.133	0.000	II
Panel 6	0.000	0.287	0.330	0.851	0.000	III
Panel 7	0.000	0.077	0.316	0.855	0.000	III
Panel 8	0.000	0.050	0.080	0.952	0.000	III
Panel 9	0.000	0.095	0.660	0.070	0.000	II
Panel 10	0.000	0.780	0.021	0.137	0.000	I
Panel 11	0.000	0.135	0.041	0.854	0.000	III
Panel 12	0.000	0.023	0.001	0.987	0.000	III
Panel 13	0.022	0.124	0.145	0.413	0.012	III
Panel 14	0.001	0.233	0.331	0.466	0.133	III
Panel 15	0.120	0.132	0.322	0.233	0.111	II

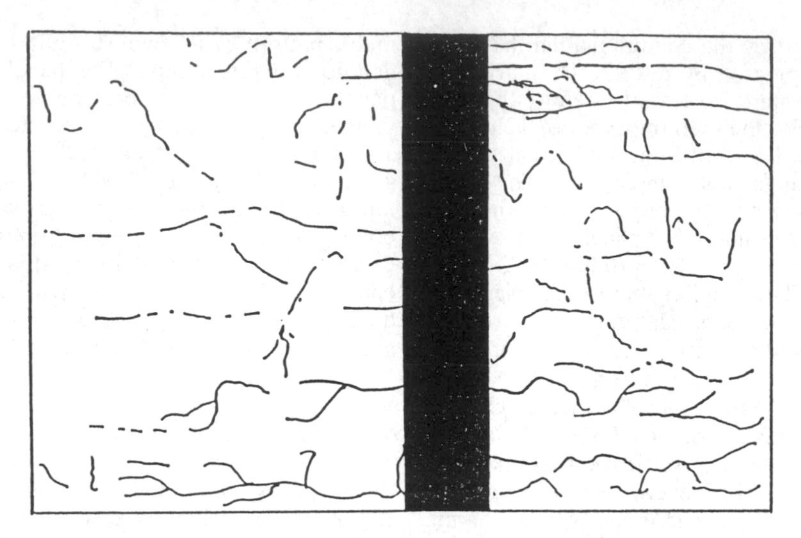

Fig.5. Crack configuration with gas pipe

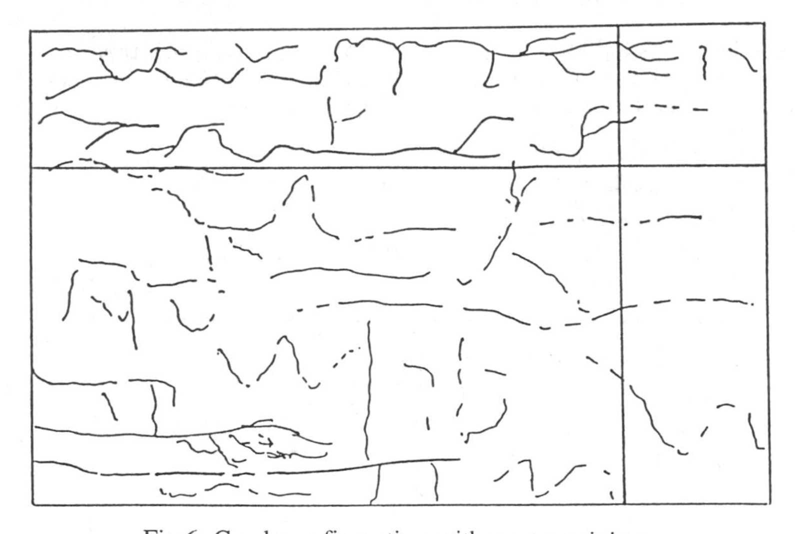

Fig.6. Crack configuration with concrete joints

obtain the photographs of the crack state of RC decks.

2.In the neural computation, it is possible to utilize the past records and the intuition and engineering judgment of experienced engineers as learning data. Using the learning ability of the neural network, it can avoid the knowledge acquisition which is a bottle neck of knowledge-based damage assessment systems.

3.In the present system, machine learning is performed using a main frame computer so as to reduce the computation time. Once the learning is completed, the damage assessment for new RC bridge decks can be made by using a micro computer. This implies that a maintenance engineer can use the present system at site with ease.

4.Using the learning ability of neural network, it is possible to evaluate the damage state of RC bridge decks with concrete joints or gas and water pipes.

Table 2. Numerical results of panel with attachments

Panel No.	Rank 0	Rank I	Rank II	Rank III	Rank IV	Expert's Opinion
Panel 1	0.000	0.211	0.811	0.131	0.000	II
Panel 2	0.000	0.133	0.218	0.903	0.000	III
Panel 3	0.000	0.127	0.011	0.984	0.000	III
Panel 4	0.000	0.111	0.012	0.928	0.000	III
Panel 5	0.000	0.081	0.821	0.121	0.000	II
Panel 6	0.000	0.215	0.127	0.991	0.000	III
Panel 7	0.000	0.111	0.221	0.899	0.000	III
Panel 8	0.000	0.251	0.011	0.972	0.000	III
Panel 9	0.000	0.076	0.986	0.010	0.000	II
Panel 10	0.000	0.980	0.111	0.116	0.000	I
Panel 11	0.000	0.131	0.040	0.940	0.000	III
Panel 12	0.000	0.013	0.100	0.922	0.000	III
Panel 13	0.000	0.332	0.343	0.901	0.000	III
Panel 14	0.000	0.356	0.222	0.802	0.000	III
Panel 15	0.100	0.109	0.876	0.211	0.000	II

REFERENCES

Furuta, H.& N. Shiraishi 1989. An expert system for damage assessment of bridge structures Uuing fuzzy production rules, *Proc. of IABSE Colloquium on Expert Systems in Civil Engineering*, .197-206.

Furuta, H., N. Shiraishi, M. Umano & K. Kawakami 1991. Knowledge-based expert system for damage assessment based on fuzzy reasoning, *Computers and Structures*, .40, 137-142.

Rumelhart, D., J. McClelland and PDP Research Group 1986. *Parallel distributed processing*, MIT Press.

Shiraishi, N., H. Furuta, & M. Sugimoto, 1985. Integrity assessment of bridge structures based on extended multi-criteria analysis, *Proc. of ICOSSAR*, 1: 505-509.

Shiraishi, N., H. Furuta, M. Umano & K. Kawakami 1991. An expert system for damage assessment of a reinforced concrete bridge deck, *Fuzzy Sets and Systems*, 44:.449-457.

The Kinki Branch of Ministry of Construction, Japan 1985. *Manual for Maintenance* (in Japanese).

Structural Safety & Reliability, Schuëller, Shinozuka & Yao (eds) © 1994 Balkema, Rotterdam, ISBN 90 5410 357 4

Seismic lifetime prediction of existing buildings using Markov chain model

T. Hamamoto
Department of Architecture, Musashi Institute of Technology, Tokyo, Japan
T. Takahashi
Department of Architecture, Tokyo Polytechnic College, Japan
T. Morita
Taisei Corporation, Tokyo, Japan

ABSTRACT: This paper presents a Markov chain model to predict the lifetime of existing buildings in seismically active regions. Variability in earthquakes is described as that which fluctuates with short-term and long-term. The damage state prediction (DSP) model is introduced to quantify multiple levels of damage state in terms of the stiffness degradation and strength deterioration of structural system. The DSP model is based on the Iwan's distributed element model and capable of predicting permanent system damage in terms of a prescribed maximum ductility factor of each element. Every earthquake intensity, a short-term damage transition probability (SDTP) matrix is constructed by a series of nonlinear response analyses using a DSP model. Then a long-term damage transition probability (LDTP) matrix is constructed by multiplying the element of the SDTP matrix by the corresponding earthquake occurrence rate and integrating the result with respect to all possible earthquake intensities. The LDTP matrix is used to predict the future damage state of structural system in the Markov chain model. For illustration, the damage evolution of a reinforced concrete frame building is presented for different seismic activities and soil conditions.

1 INTRODUCTION

To predict the lifetime of existing buildings located in the seismic regions, it is useful to construct a mathematical model which can deal with the evolution of structural damage in time and the time to reach a prescribed level of damage. However, there are many uncertainties and randomnesses in the modeling of earthquake excitation, dynamic characteristics of structural system, nonlinear response analysis and also in the description of damage state. Under these uncertain circumstances, a probabilistic approach can provide the theoretical basis for dealing explicitly with uncertainties and synthesize different sources of uncertainty systematically.

Subjected to strong ground motion, structures are often likely to undergo response in the inelastic range and the stiffness and strength of structural system occasionally degrade or deteriorate particularly at significant level of damage. A number of models have been proposed for investigating nonlinear response, ranging from a simple bilinear model to more sophisticated hysteretic models. In the stochastic response analysis, one of the most frequently used hysteretic models is a differential equation model which is mathematically motivated and introduced by Wen(1976) and developed for damage analysis by Baber and Wen(1980). Another popular hysteretic model is a distributed element model which is physically motivated and proposed by Iwan(1966) and developed for damage analysis by Iwan(1973) and Iwan and Cifuentes(1986).

To quantify the seismic damage of existing structures, it is necessary to select a pertinent measure indicating structural damage state as a nonlinear response process. For reinforced concrete structures, Banon and Veneziano(1982) and Park and Ang(1985) have proposed their damage measures in terms of ductility ratio and hysteretic energy index. Uncertainty in the description of structural damage has been expressed by a damage probability matrix which describes the probability of different damage states at a specific earthquake intensity(Whitman et al.,1975) or a fragility curve which describes the damage probability corresponding to a specific damage state at various earthquake intensities(Hwang and Jaw,1990). However, both descriptions are not suitable to deal with the evolution of damage state in time. As an alternative, a damage transition probability matrix which is used in the Markov chain model (Bogdanoff and Kozin,1985) seems to be a good selection to predict the future damage state of structural system.

This paper introduces a damage state prediction (DSP) model which is based on the Iwan's distributed element model and capable of reproducing the system deterioration in terms of the yielding and failure of each element. Multiple levels of damage state are described by the stiffness and strength deterioration of the system. For a prescribed earthquake intensity, a short-term damage transition probability (SDTP) matrix is constructed by a series of nonlinear response analyses using the DSP model. To construct a long-term damage transition probability (LDTP) matrix, the element of the SDTP matrix is multiplied by the occurrence rate associated with the earthquake intensity and the result is integrated with respect to all possible earthquake intensities. The LDTP matrix is used to predict the future damage state in the Markov chain model. To illustrate the applicability of this method, the damage evolution of a reinforced concrete frame building is presented for different seismic activities and soil conditions.

2 DAMAGE STATE PREDICTION MODEL

Structural damage is strongly related to the nonlinear and history-dependent behavior of structural system. To evaluate the evolution of structural damage, it is necessary to appropriately characterize and idealize the phenomenon of hysteresis. In this study, a distributed element (DE) model is adopted to represent the hysteretic behavior of reinforced concrete buildings to earthquake excitation. The original DE model (Iwan,1966) was only composed of a series of spring-slider element which produces bilinear behavior as indicated in Fig.1(a). The DE model has been improved to reproduce the deteriorating behavior of structural system by adding a series of spring-slider-slider element which produces slip behavior as indicated in Fig.1(b) (Iwan,1973) or by cutting the spring-slider element one by one if the element reaches a specified level of plastic deformation (Iwan and Cifuentes,1986).

The damage state prediction (DSP) model to be used in this study consists of a system composed of multiple pairs of a spring-slider element and a spring-slider-slider element, a linear spring element and a viscous damping element as shown in Fig.2. The yielding of a pair of elements causes the instant stiffness reduction of the system. The breaking of a pair of elements causes the permanent stiffness

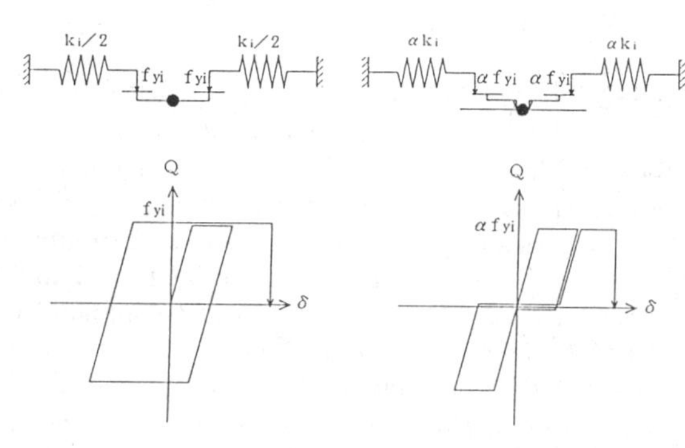

Fig.1 Two elements in DSP model

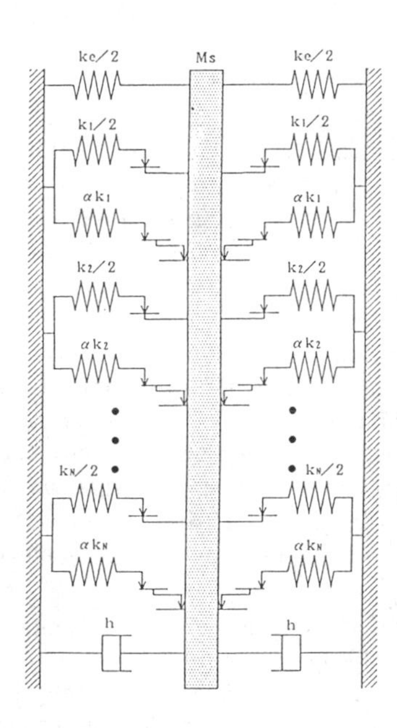

Fig.2 System of DSP model

loss and strength deterioration of the system. In this way, the system behavior may be completely controlled at the element level. This is very convenient to quantify the system deterioration.

The spring-slider element in the i-th pair has a yielding force equal to $k_i X_{yi}$, where k_i is the spring constant and X_{yi} is the yielding displacement. If the relative displacement becomes larger than βX_{yi}, the i-th pair of elements breaks and the contribution to the total restoring force becomes zero. The spring-slider-slider element in the i-th pair is set to have a spring constant, αk_i (α is the stiffness ratio). The DSP model accounts for the progressive loss of stiffness and strength of structure with large amplitude oscillation. Each pair of elements is arranged so that $X_{y1} < X_{y2} < \cdots < X_{yN}$. Since each pair of elements breaks when the displacement exceeds the value βX_{yi}, they will break in ascending order. Each pair of element is completely defined in terms of two parameters, k_i and X_{yi}. Since there are N pairs of elements, there will be 2N parameters. It is assumed that the yielding force is constant for all pairs of elements. This reduces the number of parameters from 2N to N+1. The initial stiffness, K_0, of structural system can be related to the parameters of the DSP model by the relationship

$$K_0 = k_e + \sum_{i=1}^{N}(1+\alpha)k_i \; , \tag{1}$$

in which k_e is the stiffness of a linear spring. The ultimate strength, F_u, of the DSP model can be expressed as

$$F_u = k_e \cdot X_{yN} + \sum_{i=1}^{N}(1+\alpha)f_{yi} \; , \tag{2}$$

in which f_{yi} is the yielding force of the i-th pair. Thus, both the initial stiffness and ultimate strength of a DSP model may be expressed by the contribution of N pairs of elements and a linear spring element.

3 DESCRIPTION OF EARTHQUAKES

The future nature of the earthquake ground motion expected at a particular site is unpredictable in the deterministic sense and realistically represented by a probabilistic point of view. According to the time scale of fluctuation, the variability of earthquake force may be described as that which fluctuates with short-term and long-term. The long-term variation may be approximately described by a pulse process characterized by random occurrence. A seismic hazard model is used to calculate the occurrence rates of all relevant earthquake intensities over a specified time period. The short-term variation, on the other hand, may be described by a continuous random process. A ground motion model is constructed as a quasi-nonstationary random process which is characterized by a stationary power spectral density function and a deterministic envelope function. Sample ground motions are generated with all possible earthquake intensities and used as the inputs to structural system.

The Kanai-Tajimi power spectrum is used with the density function of ground acceleration given by

$$G(\omega) = G_0 \cdot \frac{1 + 4h_g^2(\omega/\omega_g)^2}{[1-(\omega/\omega_g)^2]^2 + 4h_g^2(\omega/\omega_g)^2} \qquad (0 < \omega < \infty) \; , \tag{3}$$

in which ω_g is the predominant circular frequency, h_g is the damping coefficient and G_0 is the spectral intensity which is a measure of earthquake intensity. To simulate the transient character of real earthquakes, the Jennings-type envelope function is assumed as

$$g(t) = \begin{cases} (t/t_1)^2 & : 0 \le t \le t_1 \\ 1 & : t_1 \le t \le t_2 \\ \exp[-c(t-t_2)] & : t \ge t_2 \; , \end{cases} \tag{4}$$

where $t_2 - t_1$ is the duration of stationary part and c is a decay coefficient. The method used for artificial ground motion generation is the superposition of sinusoids having random phase angles and amplitudes devised from a stationary power spectral density function of the motion. The final simulation motion, $a(t)$, then becomes

$$a(t) = g(t) \cdot \sum_{i=1}^{M} \sqrt{2G(\omega_i)\Delta\omega}\cos(\omega_i t + \phi_i) \; , \tag{5}$$

where ω_i is the i-th circular frequency, ϕ_i is the i-th random phase angle, uniformly distributed in the range of 0 to 2π, and M is a large integer.

A seismic hazard curve is described through a plot of annual exceedance probability versus peak ground acceleration, A_p. It is represented as

$$P[A_p > a] = 1 - \exp[-(v/a)^k] \; , \tag{6}$$

which has a form of a type II extreme value distribution. v and k are the size and shape parameters of the distribution, respectively. The integral over frequency of $G(\omega)$ equals the variance of ground acceleration. The variance, σ_A^2, of Eq.(3) may be obtained as

$$\sigma_A^2 = \frac{\pi G_0 \omega_g (1 + 4h_g^2)}{4h_g} . \quad (7)$$

The expected peak ground acceleration may be expressed as the standard deviation of ground acceleration, σ_A, multiplied by a peak factor, Z, assumed here to be 3.0:

$$A_p = Z \cdot \sigma_A . \quad (8)$$

Thus, the relationship between a spectral intensity, G_0, and a peak ground acceleration, A_p, may be obtained as

$$G_0 = \frac{4h_g (A_p/Z)^2}{\pi \omega_g (1 + 4h_g^2)} . \quad (9)$$

4 SHORT-TERM DAMAGE PREDICTION

The range of damage state is divided into N damage levels corresponding to the dimension of a two-dimensional Markov matrix which is referred to as a short-term damage transition probability (SDTP) matrix, as shown in Fig.3. Each column and each row in the matrix contains transition probabilities, e.g., element P_{ij} is the probability that the post-earthquake damage level D_{post} is level j, given that the pre-earthquake damage level D_{pre} is level i, or

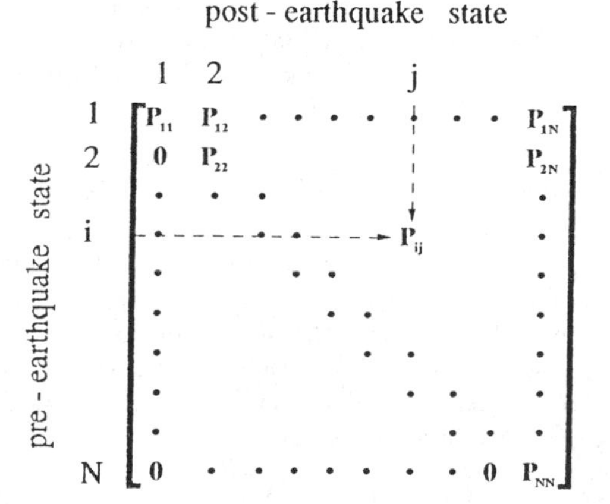

Fig.3 SDTP matrix

$$P_{ij} = P(D_{post} = D_j \mid D_{pre} = D_i) . \quad (10)$$

The elements in the matrix may be determined numerically. Every earthquake intensity, sample time histories are generated and used as the inputs to structural system. Using a DSP model, a series of nonlinear dynamic analyses are carried out to find the post-earthquake damage state for each of the time series. To construct a SDTP matrix, the pre-earthquake damage state is varied from level 1 to N. Depending on earthquake intensity, the stiffness and strength of structural system deteriorate during an earthquake. The damage level gradually evolves in time from a pre-earthquake damage state to a post-earthquake damage state. Since structural damage is an irreversible process, the resulting SDTP matrix becomes a triangular matrix.

5 LONG-TERM DAMAGE PREDICTION

The SDTP matrix is conditional on a specific level of earthquake intensity. To construct a LDTP matrix, which is independent of earthquake intensity, the element P_{ij} in the SDTP matrix is multiplied by the occurrence rate associated with the earthquake intensity level and then the result is integrated with respect to all possible earthquake intensities. Assuming that the occurrence of earthquake follows the Poisson process, the t-year prediction of damage state is given by

$$\{P(t)\} = \sum_{K=0}^{\infty} \{P(0)\}[\Phi]^K \frac{(\nu t)^K}{K!} \exp(-\nu t) , \quad (11)$$

in which $\{P(0)\}$ is an initial damage state vector, $\{P(t)\}$ is a t-year future damage state vector, $[\Phi]$ is a LDTP matrix whose element Φ_{ij} gives the probability of being in damage level j having started in damage level i, ν is the mean occurrence rate of earthquake and K is the number of occurrence.

6 NUMERICAL EXAMPLES

To illustrate the proposed method, the seismic damage evolution of a twelve-story reinforced concrete frame building is investigated. The building is modeled as a single degree of freedom system. The initial natural period of the structure is 1.70sec (3.69rad/sec). Two different locations (*site*A and *site*B) and nine levels of earthquake intensity are assumed as shown in Table 1. The parameters in the Type II extreme value distribution of peak ground acceleration are assumed as v=60.8cm/sec^2 and k=3.3 at *site*A and v=12.3 cm/sec^2 and k=1.9 at *site*B. The mean occurrence rate is assumed as ν=1.0. The parameters in Kanai-Tajimi spectrum are assumed as follows: ω_g=15.7rad/sec and h_g=0.6 for firm soil condition, ω_g=7.85rad/sec and h_g=0.8 for soft soil condition. The parameters in the envelope function are fixed to be t_1=1.5sec, $t_2 - t_1$=7sec, c=0.18 and the total duration is 20 sec. 50 sample ground motions are generated for each earthquake intensity. The DSP model parameters are assumed as follows: the stiffness of a linear element; k_e=2000kg/cm, the yielding displacement for each pair of elements; X_{y1}=3cm, X_{y2}=6cm, X_{y3}=9cm, X_{y4}=12cm, X_{y5}=15cm, X_{y6}=18cm, X_{y7}=21cm, X_{y8}=24cm, X_{y9}=27cm, X_{y10}=30cm, X_{y11}=33cm, X_{y12}=36cm, X_{y13}=39cm, X_{y14}=42cm, X_{y15}=45cm, X_{y16}=48cm, X_{y17}=51cm, X_{y18}=54cm, X_{y19}=57cm, X_{y20}=60cm, the yielding force for each pair of elements; f_y=4000kg(constant), the stiffness ratio; α=1.0, the maximum ductility factor; β=2.0, the total mass of structural system; M_s=490t, and the damping ratio; h=0.05. The number of damage levels is set to be 21.

Fig.4 shows hysteretic behavior of the DSP model, which is compared with a published experimental result (Wakabayashi,1975). It can be seen that a DSP model is capable of reasonably reproducing the hysteretic and deteriorating behavior of structural system. Fig.5 shows the hysteretic behavior of an example structure resting on firm soil for earthquake intensities 5 and 7 at both sites. It is observed that the damage increases with earthquake intensity. Fig.6 shows the evolution of stiffness degradation and strength deterioration of structural system during 50 simulated earthquakes of level 7. In the figure, k_a and k_b denote the structural stiffness after and before earthquake, and f_a and f_b represent the structural strength after and before earthquake. The variability in the post-earthquake damage state is due to the randomness of earthquake ground motion. Fig.7 shows the evolution of damage state probability in time for firm and soft soil at both sites. Damage levels

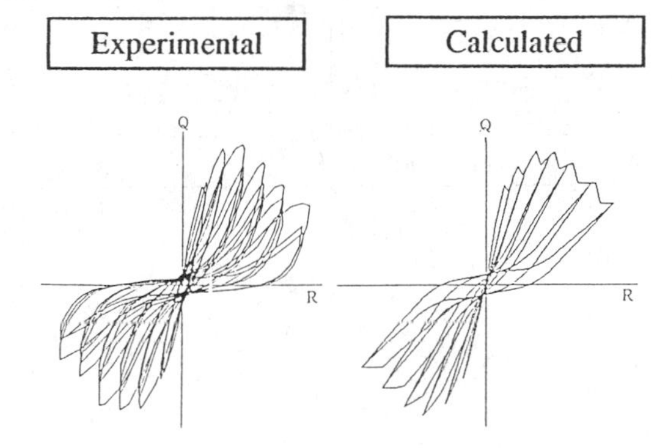

Fig.4 Hysteretic behavior of DSP model

Table 1 Earthquake intensities at sites A and B

int. level	Site A				Site B				occurce rate
	range of A_p (cm/sec^2)	median of A_p (cm/sec^2)	G_0 (cm^2/s^3)		range of A_p (cm/sec^2)	median of A_p (cm/sec^2)	G_0 (cm^2/s^3)		
			firm	soft			firm	soft	
1	~ 96.1	62.4	8.6	15.8	~ 27.5	12.9	0.4	0.7	0.800
2	~120.8	105.9	24.8	45.4	~ 41.3	32.7	2.4	4.3	0.100
3	~150.7	132.5	38.9	71.1	~ 60.7	48.5	5.2	9.5	0.050
4	~200.1	168.3	62.8	114.7	~100.1	73.8	12.1	22.1	0.030
5	~247.7	218.7	106.0	193.7	~145.6	117.0	30.3	55.4	0.010
6	~306.3	270.6	162.3	296.5	~211.4	170.0	64.0	117.0	0.005
7	~405.5	341.7	258.7	472.9	~346.0	256.1	145.3	265.6	0.003
8	~501.1	442.7	434.3	793.7	~501.9	403.8	361.3	660.3	0.001
9	501.1~	619.4	850.1	1553.7	501.9~	728.1	1174.7	2146.9	0.001

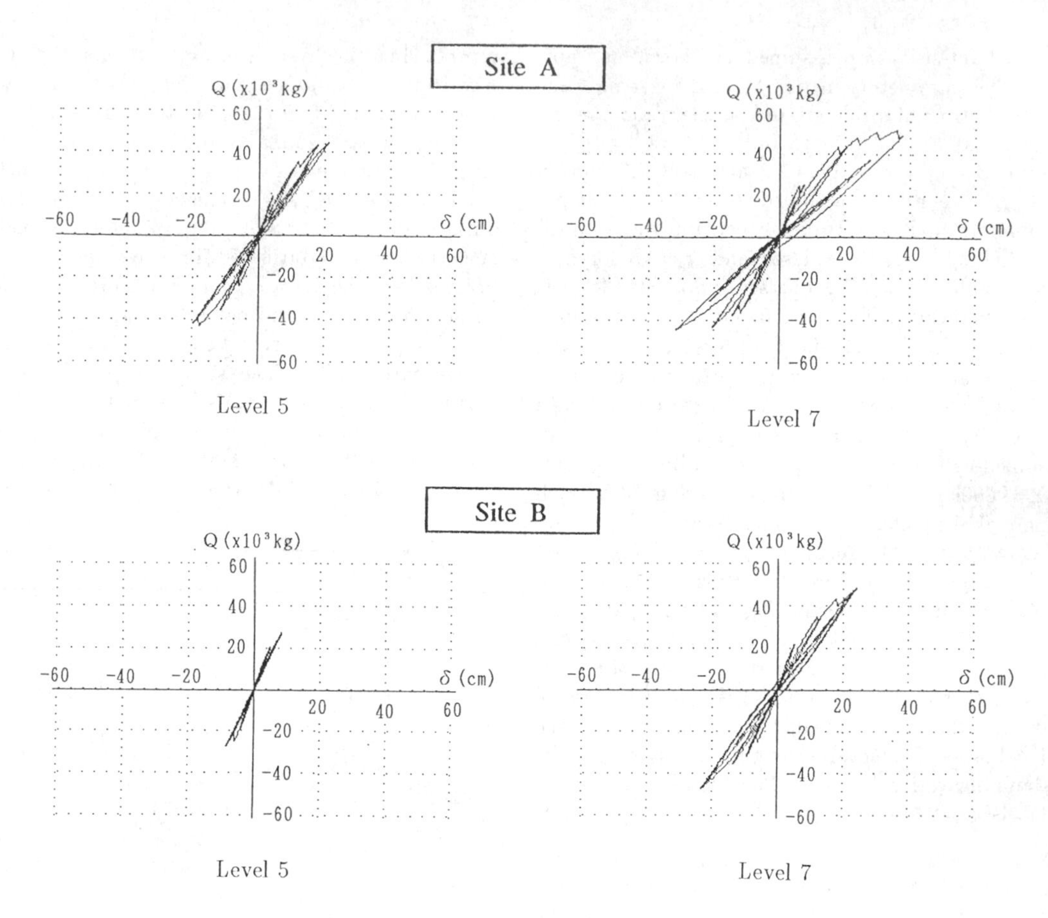

Fig.5 Hysteretic behaviors on firm soil

10 and 20 correspond to 50 percent and total loss of system strength, respectively. The variability of damage on soft soil is larger than that on firm soil at *site*A, while the variability of damage on firm soil is larger than that on soft soil at *site*B. The location of the peak corresponds to the most probable damage state after t-years. The steep peak represents a small variability in the possible damage state, while the gentle peak represents a large variability in the possible damage state. It can be seen that the variability in damage state increases with time. The shift of the peak corresponds to the evolution of damage in time. Structural damage at *site*A evolves more rapidly than that at *site*B. The amount of damage on soft soil is larger than that on firm soil at both sites.

7 CONCLUSIONS

A Markov chain model has been developed for predicting the future damage state of existing structures in the seismic regions. The proposed model is capable of dealing with multiple levels of damage state and evolution of the damage in time in terms of stiffness degradation and strength deterioration. Based on the numerical results of an example structure, it is concluded that the Markov chain model is a useful and powerful tool to predict the lifetime of existing buildings for different seismic activities and soil conditions. Furthermore, it is proved that a damage state prediction model can reproduce the nonlinear degrading dynamic behavior of structural system during earthquakes reasonably and relate the stiffness degradation to strength deterioration of structural system uniquely.

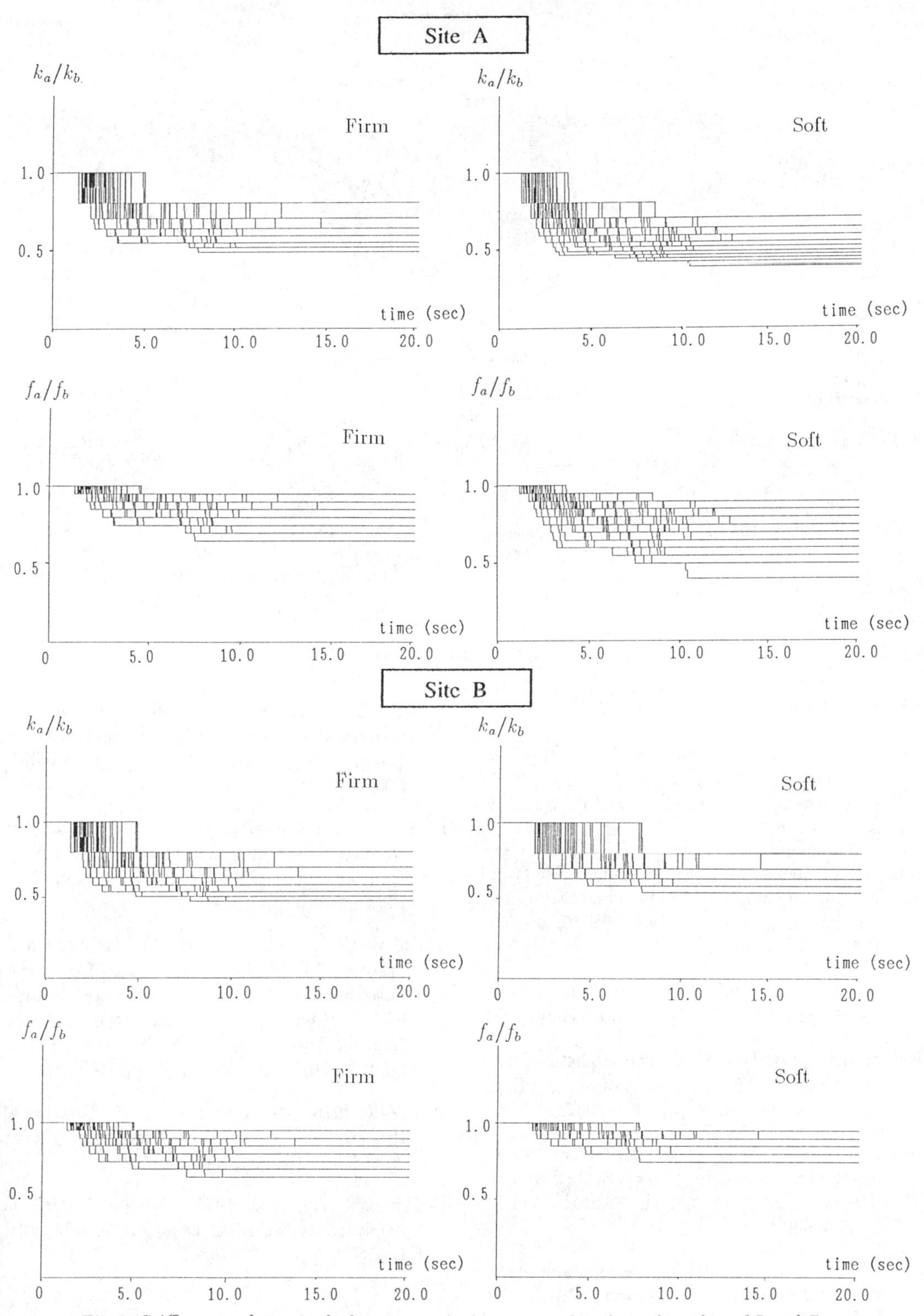

Fig.6 Stiffness and strength deterioration during simulated earthquakes of Level 7

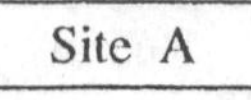

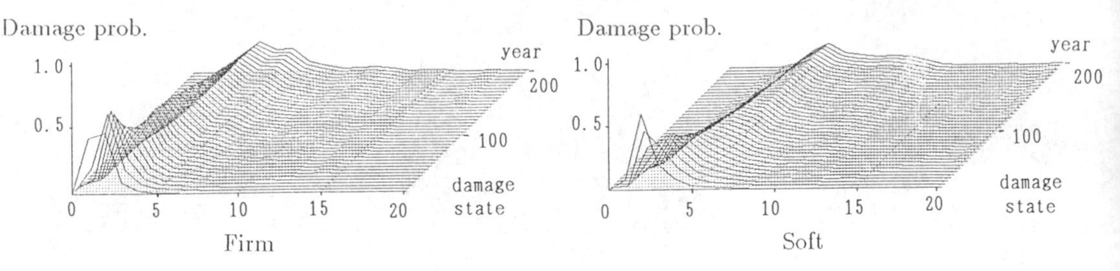

Site B

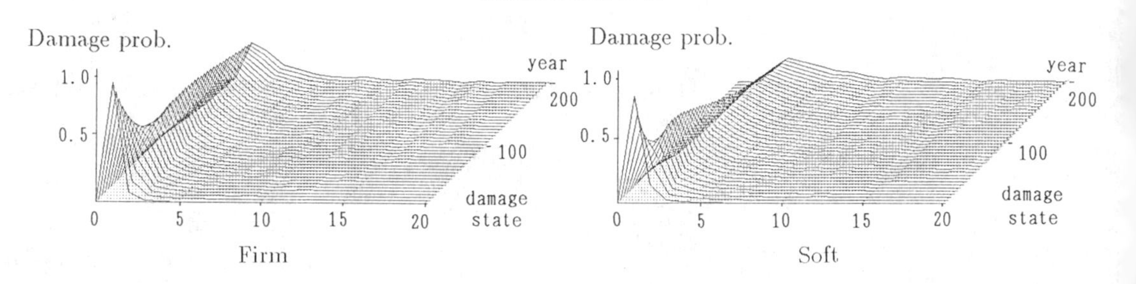

Fig.7 Evolution of damage state in time

REFERENCES

Baber,T.T., Wen,Y-K. 1980. Random Vibration of Hysteretic, Degrading Systems, J. of Eng. Mech. Div., ASCE, Vol.107, No.EM6, pp.1069–1087.

Banon,H., Veneziano,D. 1982. Seismic Safety of Reinforced Concrete Members and Structures, Earthq. Eng. Struct. Dyn., Vol.10, pp.179–193.

Bogdanoff,J.L., Kozin,F. 1985. Probabilistic Models of Cumulative Damage, John Wiley.

Hwang,H.H., Jaw,J-W. 1990. Probabilistic Damage Analysis of Structures, J. of Struct. Eng., ASCE, Vol.116, No.7, pp.1992–2007.

Iwan,W.D. 1966. A Distributed-Element Model for Hysteresis and Its Steady-State Dynamic Response, J. Appl. Mech., ASME, Vol.33, pp.893–900.

Iwan,W.D. 1973. A Model for the Dynamic Analysis of Deterioratig Structures, Proc. 5th WCEE, Rome, pp.1782–1791.

Iwan,W.D., Cifuentes, A.O. 1986. A Model for System Identification of Degrading Structures, Earthq. Eng. Struct. Dyn., Vol.14, pp.877–890.

Park,Y.J., Ang,A.H-S. 1985. Mechanistic Seismic Damage Model for Reinforced Concrete, J. Struct. Eng., ASCE, Vol.111, No.4, pp.722–739.

Wakabayashi, M., et al. 1975. An Experimental Study on Elastic-Plastic Behavior of ⊐-Shape Subassembly Reinforced Concrete Frames with Emphasis on Shear Failure of Column, Disaster Prevention Research Institute Annual, Kyoto Univ., Vol. 18B, pp. 99–121.

Wen,Y-K. 1976. Method for Random Vibration of Hysteretic Systems, J. of Eng. Mech. Div., ASCE, Vol.102, No.EM2, pp.249–263.

Whitman, R. V., et al. 1975. Seismic Decision Analysis, J. of Struct. Div., ASCE, Vol. 101, No. ST5, pp. 1067–1084.

Structural Safety & Reliability, Schuëller, Shinozuka & Yao (eds) © 1994 Balkema, Rotterdam, ISBN 90 5410 357 4

Methods to assess time-dependent serviceability of concrete structures

Chun Q. Li & Robert E. Melchers
Department of Civil Engineering and Surveying, The University of Newcastle, N.S.W., Australia

ABSTRACT: Time–dependent structural response due to creep and shrinkage effects is of considerable importance. This structural response is also highly uncertain and as such ought to be amenable to analysis using probabilistic methods. In the present paper, two methods, namely a simulation approach and an analytical approach, will be used to predict the time–dependent serviceability of a simple reinforced concrete structure. The advantage of the simulation approach is that all the random variability can be included in the structural response at any point in time, and in analytical approach, the stochastic nature of the problem can be explored by using first passage probability and the Rice formula.

KEYWORDS: Time–Dependent, Serviceability, Creep and Shrinkage, Analytical, Concrete.

1. INTRODUCTION

This paper deals with the time–dependent structural serviceability problem of reinforced concrete structures when they are subject to stochastic applied load processes and when they may change their material properties with loading or with time or both. A typical example of this problem type is the deflection of a reinforced concrete beam under a stochastic load process, when the effect of concrete creep must be considered. The prediction of this time–variant structural response in reinforced concrete structures is of practical importance. Considerable research has explored the time–variant behaviour of reinforced concrete structures[1]. However, most of this work was performed in a deterministic framework[2].

Creep and shrinkage in reinforced concrete structures are phenomena about which uncertainty exists in prediction due to incomplete understanding and hence modelling, and in relation to uncertainties in the data to be used for prediction models. Some previous attention has been given to creep and shrinkage modelling under uncertainty[1], but the estimation of time–variant structural response under stochastic applied load process has not received much attention. In this paper the serviceability of a simple structure (a reinforced concrete beam) and the probability of serviceability violation under two types of stationary stochastic loadings — a Gaussian process load and a Poisson process load will be explored.

2. MATHEMATICAL MODELS

Models for Creep and Shrinkage. Creep usually is described by the creep coefficient $\phi(t, t')$[1]. At time t, this is the ratio of the creep strain to elastic strain since the first loading occurred at time t'. Shrinkage at time t is described by the shrinkage strain $\varepsilon_{sh}(t, t_0)$ measured to the time t_0 when curing ended. Many models for creep and shrinkage have been given in the literature[2]. Those selected for use herein are due to Bazant and Panula[1]

$$\phi(t, t') = \varphi_1(t'^{-m} + \alpha)(t - t')^n \quad (1a)$$

$$\varepsilon_{sh}(t, t_0) = \varepsilon_u \cdot k_h \cdot S\,(t, t_0) \quad (1b)$$

where φ_1, m, n and α are parameters depending on various properties of the concrete material (see below), ε_u is the ultimate shrinkage strain and k_h and $S(t, t_0)$ are parameters accounting for relative humidity and the shape of the structural element respectively.

The parameters used to describe $\phi(t, t')$ and $\varepsilon_{sh}(t, t_0)$ in equations (1) can be expressed in terms of four basic random variables as well as times t and t'[1]

$$\phi(t, t') = f(\gamma, f'_c, s/c, w/c, t, t') \quad (2a)$$

where; f'_c = concrete compressive strength, w/c =

water–cement ratio, s/c = sand–cement ratio, γ = parameter depending on cement type. Also, $\varepsilon_{sh}(t, t_0)$ can be expressed as

$$\varepsilon_{sh}(t, t_0) = \varepsilon_u(1 - h^3)[1 - \frac{0.36(k_s D)^2}{t - t_0}]^{-1/2} \qquad (2b)$$

where; ε_u = ultimate shrinkage strain, h = mean relative humidity of environment, k_s = shape factor, D = effective thickness.

The statistics of $\phi(t, t')$ and $\varepsilon_{sh}(t, t_0)$, at any time t and for any given t′ and t_0, may be obtained from above models by simulation, i.e., by sampling from the (presumed) known probability distribution functions for each of the basic random variables and by using equations (2). In particular, the mean and standard deviation for both $\phi(t, t')$ and $\varepsilon_{sh}(t, t_0)$ may be calculated. This information will be used as indicated below.

Gaussian Loading Process. A Gaussian stationary loading process is described by its mean μ_q (which is constant) and the correlation function $R_q(\tau)$, which is a function only of the time interval $\tau = t_1 - t_2$. If the two arbitrary time points t_1 and t_2 are sufficiently far apart, it is reasonable to allow correlation to be ignored. With this interpretation, it is relatively easy to simulate the load process by sampling from the load amplitude distribution (with Gaussian density) a given number of times, each representing a time point in a given time period.

Poisson Loading Process. A Poisson loading process is a pulse process $\mathbf{Q}(t)$, in which the load occurrence time, duration and intensity are all treated as random variables. A pulse process can be described by its mean occurrence rate υ, a random variable duration d with mean duration μ_d and an intensity random variable $\mathbf{Q}$ with a density function $f_Q(\mathbf{q})$[1]. The generation of a realisation of a Poisson loading process requires the determination of the arrival time of the kth pulse (i.e., an event), the real acting period of the pulse and the intensity (level) of loading for each pulse. If τ_k denotes the time between the $(k-1)^{st}$ and the kth pulses, it can be shown that the τ_k, k=1, 2, ..., m, are independent, identically distributed exponential random variables. Therefore, the arrival time (waiting time) of the kth pulse can be expressed as

$$t_k = \sum_{i=1}^{k} \tau_i, \qquad k \geq 1 \qquad (3)$$

where t_k denotes the kth arrival time. It is easy to show that t_k has a gamma distribution with parameters k and υ. Since the duration and intensity are assumed independent of arrival time, they may be assumed to be independent. Also, for simplicity in the present study, they are assumed to be normal random variables.

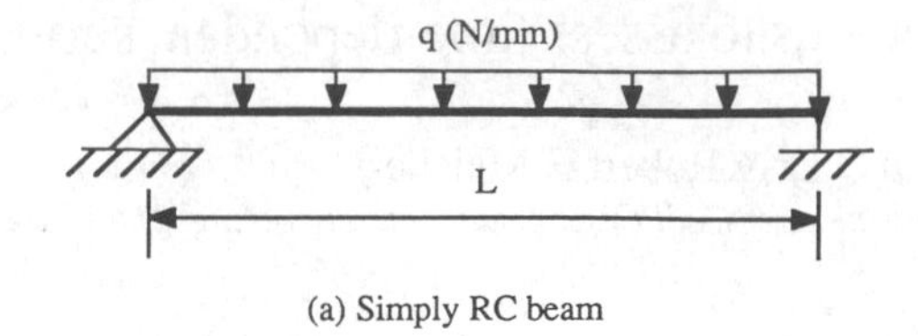

(a) Simply RC beam

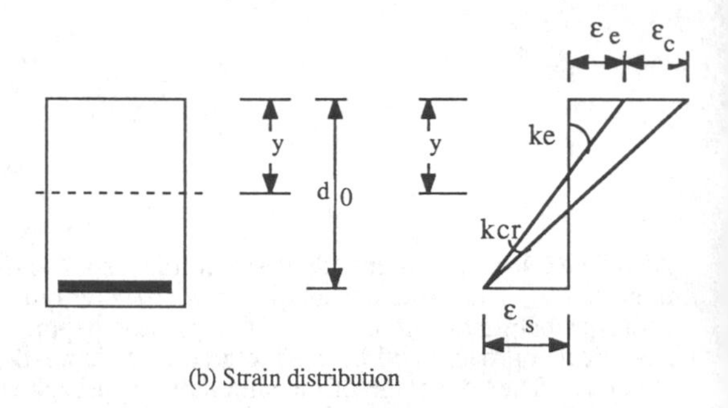

(b) Strain distribution

Figure 1. Reinforced Concrete Beam and Its Cross–Section

3. SIMULATION APPROACH

The deformation of a simple reinforced concrete beam (see Figure 1) due to creep and shrinkage caused by loading applied to the structure may be expressed in terms of strain ε as[2]

$$\varepsilon(t) = \varepsilon_e(t) + \varepsilon_{cr}(t) + \varepsilon_{sh}(t) \qquad (4)$$

where $\varepsilon_e(t)$, $\varepsilon_{cr}(t)$ and $\varepsilon_{sh}(t)$ are elastic, creep and shrinkage strains respectively, all treated as random variables and as functions of time. For sufficiently low levels of load, $\varepsilon_e(t)$ and $\varepsilon_{cr}(t)$ may be replaced by $\sigma(t)/E$ and $\sigma(t)\phi(t, t')/E$ respectively (where $\sigma(t)$ is the stress caused by the applied loading and E is the elastic modulus of concrete). In evaluating equation (4) the main difficulty is to calculate the second term, i.e., the creep coefficient $\phi(t, t')$, since it is related to the applied loads.

Gaussian Loading. When the stress varies continuously with time, such as is the case for a continuous Gaussian loading process, $\phi(\,)$ changes due to the changing load as well as changing as a function of time. This situation can be modelled using the principle of superposition[2]. Accordingly, the creep strain may be written as

$$\varepsilon_{cr}(t) = \frac{\sigma(t')}{E}\phi(t, t') + \frac{1}{E}\int_{t'}^{t} \frac{\partial\sigma(\tau)}{\partial\tau}\,\phi(t, \tau)\,d\tau \qquad (5)$$

where the first term refers to the effect of the first application of the load and the second term to all load changes thereafter. The closed form integration of this Stieltjes integral is known to be difficult. It is usual to employ a numerical method, such as the "step–by–step" method, for its solution[1,2]. This is based on replacing the second part of expression (5) by

$$\varepsilon_{cr2}(t) = \frac{1}{E}\sum_{j=1}^{k}\frac{1}{2}[\phi(t, t_j) + \phi(t, t_{j-1})][\sigma_j - \sigma_{j-1}] \qquad (6)$$

in which the creep strain at time t depends on all previous stress changes. Because of the complexity of equation (6), the most direct way to obtain the statistics of $\varepsilon_{cr}(t)$ and hence of $\varepsilon(t)$ is by simulation. Such an approach has been adopted herein.

Poisson Loading. Discrete Poisson loading processes are already discretised (even though the time steps are not fixed *a priori*) and this is a potential advantage. For the situation considered herein, for each pulse the loading and hence stress is assumed to remain constant with time, so that equation (4) can be used immediately for the creep strain calculation. Also note that after each discrete pulse there could be an unloading effect due to the next pulse being of lower value. For creep, the residual creep strain at time t due to the kth pulse alone is then given by

$$\varepsilon'_{crk}(t_L) = \frac{\sigma_k}{E}[\phi(t_L, t_k) - \phi(t_L, t_k+d_k)] \qquad (7)$$

where; σ_k = stress caused by k^{th} pulse, t_k = arrival time of kth pulse, and d_k = duration of kth pulse. As implicit in equation (5), the total creep effect up to time t is given by the sum of the residual creep strains for all (m) pulses which have acted since t′ and prior to time t

$$\varepsilon'_{cr} = \sum_{k=1}^{m} \varepsilon'_{crk} \qquad (8)$$

When the three strain terms in equation (4) are known, the deflection of the beam can be obtained, under some assumptions[1], as

$$\Delta(t) = \frac{5L^2}{48y}\varepsilon_e + \frac{5L^2}{48d_0}\varepsilon_{cr} + \frac{L^2}{8d_0}\varepsilon_{sh} \qquad (9)$$

For given sample set of these strains, the beam deflection can be calculated and compared to the deformation limit δ_L to see whether the serviceability failure condition has been attained. This procedure is then repeated a sufficient number of times to allow the probability of serviceability failure to be determined. Details of the complete simulation procedure are given in Li and Melchers[1].

4. ANALYTICAL APPROACH

In this approach the models for creep and shrinkage effects have to be expressed analytically. To achieve this, $\phi(t, t')$ can be described, for a given first loading time t′, as

$$\phi(t, t') = \phi(t) = f_\phi(t) \cdot \xi_\phi \qquad (10)$$

where ξ_ϕ is a random variable used to describe the uncertainty associated with creep such that $E[\xi_\phi] = 1$, and $f_\phi(t)$ is a pure time function to describe the time–dependent behaviour of the creep effect. Then for a given t′, the mean and correlation functions of $\phi(t, t')$ are

$$E[\phi(t, t')] = f_\phi(t)E[\xi_\phi] = f_\phi(t) \qquad (11)$$

$$R_\phi(t_1, t_2) = (\lambda_\phi^2 + 1) \cdot f_\phi(t_1)f_\phi(t_2) \qquad (12)$$

Using simulation results as a guide, it is convenient to assume that $f_\phi(t) = m \cdot t^n$, where m, n are parameters. In the same way the shrinkage strain for a given t_0 is expressed as

$$\varepsilon_{sh}(t, t_0) = \varepsilon_{sh}(t) = f_s(t) \cdot \xi_s \qquad (13)$$

with mean function and correlation function

$$E[\varepsilon_{sh}(t, t_0)] = f_s(t)E[\xi_s] = f_s(t) \qquad (14)$$

$$R_s(t_1, t_2) = (\lambda_s^2 + 1) \cdot f_s(t_1)f_s(t_2) \qquad (15)$$

As noted, the serviceability criterion of interest herein is the deformation of a beam, taking into account the effects of creep and shrinkage. For sufficiently low level of the instantaneous loading Q(t), superposition is valid sothat the time–dependent deformation can be expressed, for given t′ and t_0, by[3]

$$X(t) = a \cdot q(t) + b \cdot [q(t')\phi(t, t') + \int_{t'}^{t} \frac{\partial q(\tau)}{\partial\tau}\phi(t, \tau)d\tau] + c \cdot \varepsilon_{sh}(t, t_0) \qquad (16)$$

where a, b and c are parameters determined from a structural analysis. It is evident that the structural response is non–stationary process since $\phi(t, t')$ is a function of time (and of the loading). However, shrinkage is independent of applied load. It follows that in general the structural response X(t) will be neither Gaussian nor stationary. However, to render the problem tractable, it will be assumed herein that the structural response X(t) in equation (16) can be modelled approximately as a continuous Gaussian process. This allows the statistics of the parameters of X(t) for a given t′ to be obtained as follows

$$\mu_X(t) = a \cdot \mu_q + b \cdot \mu_q \cdot f_\phi(t) + c \cdot f_s(t)$$

$$C_X(t_1, t_2) = a^2[R_q(\tau) - \mu_q^2]$$

$$+ b^2[R_q(0)R_\phi(t_1, t_2) - \mu_q^2 f_\phi(t_1)f_\phi(t_2)]$$

$$+ c^2[R_s(t_1, t_2) - f_s(t_1)f_s(t_2)]$$

$$\mu_{\dot{X}} = b \cdot \mu_q f'_\phi(t) + c \cdot f'_s(t) \tag{17}$$

$$C_{\dot{X}}(t_1, t_1) = - a^2 \cdot R''_q(\tau)$$

$$+ b^2[R_q(0)R''_\phi(t_1, t_2) - \mu_q^2 f'_\phi(t_1)f'_\phi(t_2)]$$

$$+ c^2[R''_s(t_1, t_2) - f'_s(t_1)f'_s(t_2)]$$

With the statistics of the structural response X(t) known, the probability that X(t) upcrosses the deformation limit δ_L can be bounded by[3]

$$p_f(t) \le p_f(0) + [1 - \exp(-\int_0^t \nu d\tau)] \tag{18}$$

where ν is the upcrossing rate. Note that $\dot{\delta}_L = 0$, thus the solution for equation (18) is

$$\nu(t) = \frac{\sigma_{\dot{X}|X}(t)}{\sigma_X(t)}\phi[\frac{\delta_L - \mu_X(t)}{\sigma_X(t)}]\{\phi[-\frac{\mu_{\dot{X}|X}(t)}{\sigma_{\dot{X}|X}(t)}]$$

$$+ \frac{\mu_{\dot{X}|X}(t)}{\sigma_{\dot{X}|X}(t)}\Phi[\frac{\mu_{\dot{X}|X}(t)}{\sigma_{\dot{X}|X}(t)}]\} \tag{19}$$

Details of the derivation for all above expressions are given in Li and Melchers[3].

5. WORKED EXAMPLE

Consider the serviceability problem of the reinforced concrete beam shown in Figure 1. The values of basic random variables are shown in Table 1 and other parameters, used in the problem including those for the applied loading, are shown in Table 2. The allowable deflection limit δ_L was set at 35 mm.

Simulation Approach. For a stationary Gaussian load process, one realisation may be obtained by sampling n times from the probability density function for the load, representing n different points in the time interval [0, t]. With this realisation the creep strain at all times t can be obtained using equations (6) and the deflection computed through equation (9). This procedure was repeated 400 times to obtain the probability of serviceability failure at all time t. For the Poisson loading process, the general procedure was to sample for arrival time, duration and intensity to obtain the kth pulse of one realisation of the process. Otherwise the procedure for obtaining deflection $\Delta(t)$ and the probability of serviceability failure paralleled that for the Gaussian load process. Typical result of the simulation is shown in Figure 2. The results are found not to be sensitive to greater sample size.

Analytical Approach. The expression for the deflection of the beam is of the form given in equation (16) with

$$a = \frac{5L^4}{384EI},\ b = \frac{5L^4 y}{384EI \cdot d_0} \text{ and } c = \frac{L^2}{8d_0}$$

Table 1. Values of Basic Random Variables

VARIABLE	MEAN	STANDARD DEVIATION
ε_u	0.00075	0.00024
h	0.65	0.13
k_s	1.15	0.0575
D	100	5
w/c	0.56	0.056
f'_c	25	2.5
s/c	3.08	0.308
γ	1	0.05

Table 2. Parameters of Cross–Section and Loading

PARAMETER	UNIT	MEAN	STANDARD DEVIATION
Q	N/mm	15	1.5
ν	per year	5	–
d	year	0.2	0.02
E_c	N/mm^2	2500	956.3
I_e	mm^4	3.9×10^9	–
y	mm	221.1	–
L	mm	8000	–

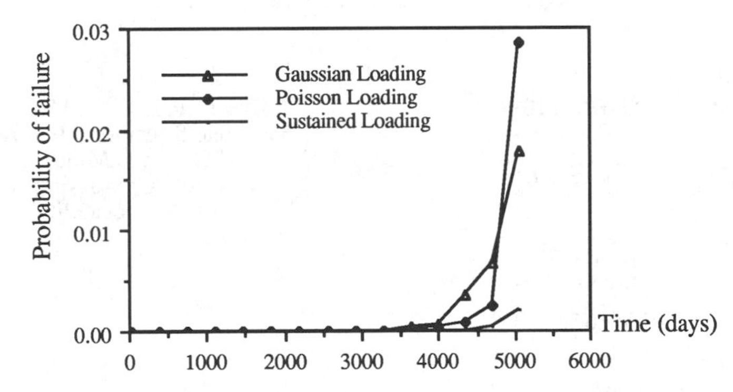

Figure 2. Failure Probability As A Function of Time (Simulation Results)

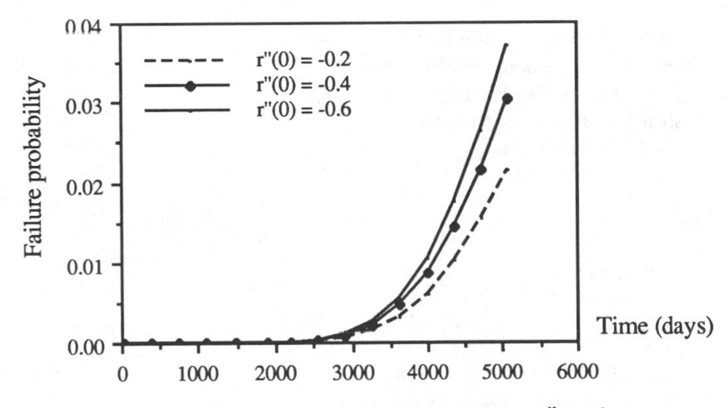

Figure 3. Failure Probability For Different r″( ) of Gaussian Load ($\lambda_\phi = 0.1$, $\lambda_s = 0.1$)

The stationary Gaussian load process has a mean μ_q, C.O.V. λ_q and (auto) correlation function $r(\tau)$ defined such that $r(\infty) = 0$, $r'(0) = 0$ and $r''(0)$ exist. Hence $r(0) = \sigma_q^2 + \mu_q^2 = \mu_q^2(\lambda_q^2 + 1)$. The creep coefficient is assumed to have the form $f_\phi(t) = 0.82 \cdot t^{\,0.29}$, obtained from curve fitting to limited experimental data for a time of first loading $t' = 28$ days, and with $\lambda\phi = 0.1$. The shrinkage strain is assumed to have a form $f_s(t) = 3.675 \times 10^{-6} \cdot t^{\,0.553}$, also obtained from curve fitting with $t_0 = 7$ (days), and with $\lambda\sigma = 0.05$. It then follows readily from equation (17) that the parameters required for the solution of equation (19) can be obtained as follows

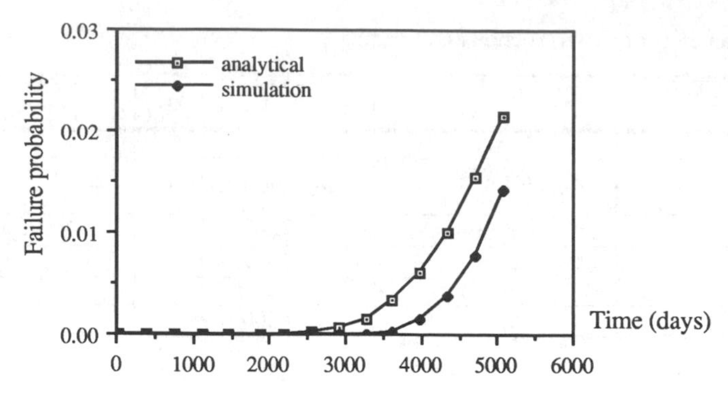

Figure 4. Comparison of Two Approaches

$$\mu_x = a\mu_q + 0.82b\mu_q t^{0.29} + c3.675 \times 10^{-6} t^{0.553}$$

$$\sigma_x^2 = a^2\mu_q^2\lambda_q^2 + b^2\mu_q^2 f_\phi^2(t)(\lambda_q^2\lambda_\phi^2 + \lambda_q^2 + \lambda_\phi^2)$$

$$+ c^2 f_s^2(t)\lambda_s^2$$

$$\mu_{\dot{x}} = 0.24b \cdot \mu_q t^{-0.71} + 2.03 \times 10^{-6} c \cdot t^{-0.447}$$

$$\sigma_{\dot{x}}^2 = -a^2 r''(0) + b^2\mu_q^2 f_\phi'^2(t)(\lambda_q^2\lambda_\phi^2 + \lambda_q^2 + \lambda_\phi^2)$$

$$+ c^2 f_s'^2(t)\lambda_s^2$$

The failure probability $p_f(t)$ was computed using equation (18) with $p_f(0) = 0$. Typical results are shown in Figure 3. Finally the results obtained from both analytical and simulation approaches under Gaussian load are compared in Figure 4.

6. CONCLUSION

Two approaches have been presented in the paper for the time–dependent structural serviceability problem, involving creep and shrinkage deflection of a reinforced concrete beam. The advantage of the simulation approach is that all the random variability can be included in the structural response at any point in time, and in analytical approach, the stochastic nature of the problem can explored by using first passage probability and hence the Rice formula.

7. REFERENCES

1. Li, C.Q. and Melchers, R.E., (1992), "Reliability Analysis of Creep and Shrinkage Effects", J. Struct. Engg., ASCE, 118, (9), 2323 – 2337.
2. Gilbert, R.I., (1988), Time Effects in Concrete Structures, Elsevier.
3. Li, C.Q. and Melchers, R.E., (1993), "Gaussian Upcrossing Rate Solution for Structural Serviceability", Struct. Safety, (To Appear).

Structural Safety & Reliability, Schuëller, Shinozuka & Yao (eds) © 1994 Balkema, Rotterdam, ISBN 90 5410 357 4

One and two-dimensional global damage indicators for R/C structures under seismic excitation

S. R. K. Nielsen
Department of Building Technology and Structural Engineering, University of Aalborg, Denmark
H. U. Köylüoğlu & A. Ş. Çakmak
Department of Civil Engineering and Operations Research, Princteon University, N.J., USA

ABSTRACT: The maximum softening concept is based on the variation of the vibrational periods of a structure during a seismic event. Maximum softening damage indicators, which measure the maximum relative stiffness reduction caused by stiffness and strength deterioration of the actual structure, are calculated for an equivalent linear structure with slowly varying stiffness characteristics. In the paper, a one dimensional scalar valued and a two-dimensional vector-valued maximum softening damage indicator are defined. The one-dimensional damage indicator, defined considering the variation in the first period of the structure, analogous to a SDOF system, is a genuine global damage index representing the average damage throughout the whole structure. The two-dimensional damage indicator is defined considering the variations in the first and second periods, and thus is analogous to an equivalent linear two degree-of-freedom system. The components of the damage vector can be interpreted as damage indicators for the lower half and the upper half of a structure, hence representing a simple local description of the damage state of the structure. Since statements on the post earthquake reliability should be obtained solely from knowledge of the latest recorded values of the damage indicators defined above, these are required to possess a Markov property. This problem has been investigated based on numerical Monte-Carlo simulations, and it is observed that the Markov assumption of the one-dimensional damage indicator is justified for the mean value of the transition probability density function (tpdf), whereas some deviations are observed for the variance and higher order statistical moments. For the two-dimensional damage indicator the Markov assumption seems justified for both the mean values and the covariances of the tpdf. From these, it is concluded that the Markov properties of the mean and covariances of the two-dimensional indicators are superior to the Markov property of the first and second order moments of the one-dimensional damage indicator.

1. INTRODUCTION

Local damage in reinforced concrete structures can be attributed to micro-cracking, bond deterioration at the steel-concrete interfaces and yielding of the reinforcement bars. To the extent that framed reinforced concrete structures can be modelled by conventional non-linear beam theories, local damage at a cross-section of the structure can adequately be measured by the degradation of bending stiffness and moment capacity of the cross-section. In any case, the overall effect of local damage is deterioration of stiffness and strength of the structure. A global damage indicator can then be defined as a scalar or vector-valued function of such continuously distributed local damages which characterize the overall damage state and serviceability of the structure.

Global damage indicators are response quantities characterizing the damage state of the structure after earthquake excitations, and as such can be used in decision-making during the design phase or in case of post-earthquake reliability and repair problems. In serving these purposes, a global damage indicator should ideally fulfil at least the following requirements:

1. The damage indicator should be observable by measurements.
2. The damage indicator should be a non-decreasing function of time unless the structure is repaired or strengthened.

3. Dependent on the definition of the failure event, a well-defined deterministic failure surface in the space of the component damage indicators, separating safe states from unsafe states of the structure, should exist.

4. It should be possible to derive post-earthquake reliability estimates for a partly damaged structure solely from the latest recorded value of the damage indicator.

The fourth requirement implies that two initially identical structures, for which the same present value of the scalar or vectorial global damage indicator is recorded, should be considered to have the same structural reliability, when exposed to the same future earthquake process, independent of preceding different loading histories. Mathematically, this means that the values of the global damage indicator after each earthquake are assumed to form a Markov chain. If the local damages, as measured by the stiffness and strength deterioration of all beams and columns, are point-wise the same throughout the two structures, the probability distributions of local damages certainly will be identical after the next earthquake, irrespective of the different preceding loading histories. This establishes the Markov property for the finite or infinite dimensional vector process made up of all local damage measures. A sufficient condition for this case is obtained, if the local damages have the same spatial distribution in all loading histories, so that the damage state can be described by a single scaling parameter. Then the local damage is identical in the two structures if the global damage indicators are the same. In this case the sequence of scalar global damage indicator values forms a Markov chain. In general, the Markov property does not apply very well to a scalar global damage indicator process. However, if the components of the vector damage indicator process can be interpreted as a measure of average damage in various parts of the structure, and hence represent a crude description of local damages, a better fit to the Markov property may be expected.

The maximum softening concept is based on the variations of the vibrational periods of a structure during a seismic event. The variation in the vibrational periods is partly caused by the degradation of the incremental stiffness in the elastic range, and partly by the averaged effect of the loss of incremental stiffness at excursions into the plastic range at various positions in the structure. The maximum softening damage indicators are defined as the maximum relative stiffness reduction caused by stiffness and strength deterioration. The numerical values for the damage indicators are calculated, dependent on the variation in the vibrational periods, for an equivalent linear system with slowly varying stiffness characteristics.

The one-dimensional maximum softening concept based on an equivalent linear SDOF system was introduced by DiPasquale and Cakmak (1987) as a global damage indicator for reinforced concrete structures. This index has been calibrated based on an analysis of data from shake table experiments with reinforced concrete frames performed by Sozen and his associates at the University of Illinois at Urbana Champaign in the 1970's (Healey and Sozen (1978), Cecen (1979)). It was demonstrated that the maximum softening values at failure showed relatively small variability independently of the limit state definition, which ranged from slightly damaged structures to total collapse. Hence the one-dimensional maximum softening damage indicator fulfils the third of the indicated requirements within acceptable limits. A comparison of the one-dimensional maximum softening with other global damage indicators showed the versatility of the maximum softening in separating failed from unfailed structures. (Rodriquez-Gomez (1990)). Nielsen and Cakmak (1991) investigated the Markov property of the sequence of one-dimensional maximum softening values based on Monte-Carlo simulations using the SARCF-II program (Rodriguez-Gomez et al. (1990)), and found that the Markov property was fulfilled for the mean value of the tpdf, but that deviations were observed for higher order conditional moments. Because of the inability of the applied structural analysis program to handle severely damaged structures, the considered samples might have been biased to some extent. For this reason this analysis will also be re-iterated in the present study with a robust structural analysis program SARCOF developed by Mørk (1992).

In the present study, a generalization of the one-dimensional maximum concept to two dimensions has been suggested, where the components of the damage indicator can be interpreted as the maximum softening of, respectively, the lower and the upper half of the structure. Hence, the two-dimensional maximum softening provides a simple description of the distribution of damages in the structure. This generalization can be carried out to multi-dimensions.

The Markov property of the one-dimensional and the two-dimensional maximum softening damage indicators is tested numerically by means of Monte Carlo simulation. Structural analysis is performed with the computer program SARCOF, which has been specially developed for stochastic analysis of reinforced concrete frames under seismic excitation based on Monte-Carlo simulations. The theoretical background and the capabilities of this program are given in Mørk (1992). The program considers moment-curvature relation as an extended version of the model developed by Roufaiel-Meyer (1987), taking into account the transition from uncracked to cracked sections. Strength deterioration is assumed to be related to the crushing of concrete in the compression zone, and is functionally correlated to the hysteretic energy accumulated in the cross-section subsequent to the first exceedance of a critical curvature. The finite length of plastic zones is taken into account, considering the plasticity at the end sections and at 3 internal cross-sections. The incremental bending stiffness field is next obtained by linear interpolation between the value at these sections. Finally, a system reduction scheme, based on a truncated expansion of external nodal point degrees-of-freedom in the linear eigenmodes of the initial undamaged structure, has been implemented. However, full description is maintained for the internal degrees-of-freedom controlling the hysteresis at the control sections.

2. MAXIMUM SOFTENING DAMAGE INDICATORS

Figure 1a shows the time-variation of the first and second eigenperiod of the structure during the ith earthquake, initiated at the time t_i and of the length l_i. Both time-series have been normalized with respect to the corresponding eigenperiods $T_1(0)$ and $T_2(0)$ of the initial referential state, where the structure is exposed to gravity loads alone. Partial cracking of the cross-sections due to gravity loads has been considered at the evaluation of $T_1(0)$ and $T_2(0)$. The increase of the eigenperiods from the value $T_j(t_i)$, $j = 1,2$ at the start of the ith earthquake to the final value $T_j(t_i + l_i)$, $j = 1,2$ at the end of the excitation can be attributed to the gradual deterioration of incremental bending stiffness at elastic branches of the moment-curvature relations. The local peaks of the eigenperiod time-series are caused by the complete loss of incremental stiffness during momentary excursions into the plastic range. In figure 1a, the moving time-averages $\hat{T}_j(t_i)$, $j = 1,2$ of the eigenperiod time-series are also shown. As suggested by Rodriguez-Gomez (1990), these time-averages may be identified as the eigenperiods of the equivalent linear systems with slowly varying parameters as shown in figure 1b and figure 1c. Due to the plastic deformations, $\hat{T}_j(t_i)$, $j = 1,2$ attains a maximum value during the ith earthquake at the instants of time $t_{j,i}$, $j = 1,2$. In general $t_{1,i} \neq t_{2,i}$ as is also the case in figure 1a.

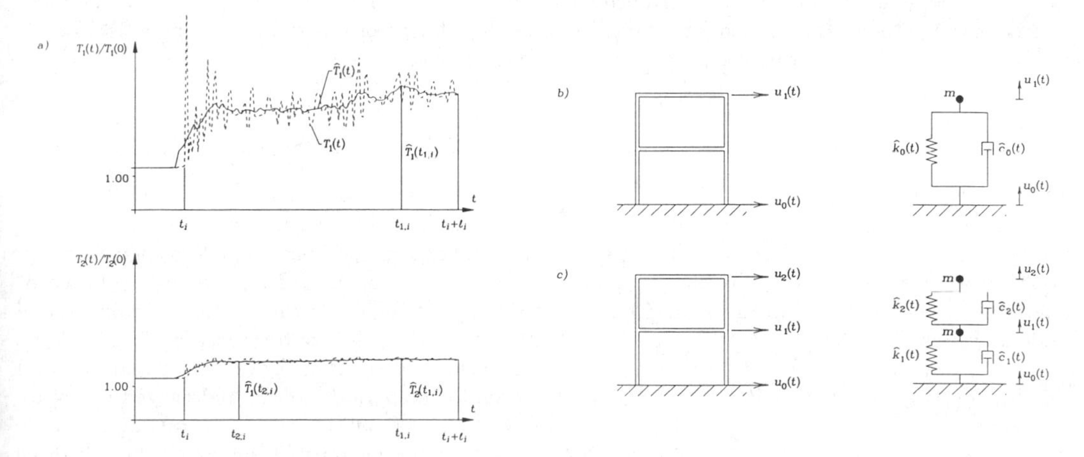

Figure 1: a) Time variation of first and second eigenperiods. b) Equivalent linear SDOF system. c) Equivalent linear two-degrees-of-freedom system.

The one-dimensional maximum softening damage indicator is then defined as follows (DiPasquale and Cakmak (1987, 1990))

$$D_{0,i} = 1 - \left(\frac{\hat{k}_0(t_{1,i})}{k_0(0)}\right)^{\frac{1}{2}} = 1 - \frac{T_1(0)}{\hat{T}_1(t_{1,i})} \tag{1}$$

where $\hat{k}_0(t)$ and $k_0(0)$ signify the spring constant of the equivalent linear SDOF system in figure 1b at the time t, and the corresponding spring constant at $t = 0$ for the initial undamaged structure.

In figure 1c, a generalization of this concept is shown, where the structure is replaced by an equivalent linear two-degrees-of-freedom system. The time-dependent spring constants of the equivalent system represent, respectively, the equivalent stiffness of the lower half and the upper half of the structure. The masses of the equivalent linear system are assumed to be of equal magnitude m. The following damage indicators can then be introduced, indicating respectively the maximum softening after the ith earthquake of the lower half and the upper half of the structure

$$D_{j,i} = 1 - \left(\frac{\hat{k}_j(t_{1,i})}{k_j(0)}\right)^{\frac{1}{2}}, \quad j = 1,2 \tag{2}$$

where $\hat{k}_1(t_{1,i})$ and $\hat{k}_2(t_{1,i})$ signify the spring constants of the equivalent linear system at the instant of time $t_{1,i}$, where the one-dimensional maximum softening $D_{0,i}$ in the ith earthquake occurs. Because the initial values $k_0(0)$, $k_1(0)$ and $k_2(0)$ of the equivalent spring constants are needed in the definitions, both the one-dimensional and the two-dimensional maximum softening indicator require that the structure has been instrumented from its construction. The following quantities are introduced to define spring constants

$$\hat{\omega}_j = \frac{2\pi}{\hat{T}_j(t_{1,i})}, \quad j = 1,2 \tag{3}$$

Notice, that $\hat{\omega}_2$ has been defined from $\hat{T}_2(t_{1,i})$ rather than from $\hat{T}_2(t_{2,i})$. $\hat{k}_1(t_{1,i})$ and $\hat{k}_2(t_{1,i})$ can then be identified from the following expressions, derived from the solutions to the characteristic equation of the equivalent linear two-degrees-of-freedom system

$$\frac{\hat{k}_1(t_{1,i})}{m} = \frac{1}{2}\left(\hat{\omega}_1^2 + \hat{\omega}_2^2 - \sqrt{\hat{\omega}_1^4 - 6\hat{\omega}_1^2\hat{\omega}_2^2 + \hat{\omega}_2^4}\right) \tag{4}$$

$$\frac{\hat{k}_2(t_{1,i})}{m} = \frac{1}{4}\left(\hat{\omega}_1^2 + \hat{\omega}_2^2 + \sqrt{\hat{\omega}_1^4 - 6\hat{\omega}_1^2\hat{\omega}_2^2 + \hat{\omega}_2^4}\right) \tag{5}$$

(4) and (5) can only be used, if $\hat{\omega}_2 \geq \sqrt{3+\sqrt{8}}\,\hat{\omega}_1$. If $\hat{\omega}_2 < \sqrt{3+\sqrt{8}}\,\hat{\omega}_1$, the following expressions are used instead

$$\frac{\hat{k}_1(t_{1,i})}{m} = \frac{1}{2}\left(\hat{\omega}_1^2 + \hat{\omega}_2^2\right) \tag{6}$$

$$\frac{\hat{k}_2(t_{1,i})}{m} = \frac{1}{4}\left(\hat{\omega}_1^2 + \hat{\omega}_2^2\right) \tag{7}$$

3. TEST OF MARKOV PROPERTY OF DAMAGE INDICATORS

The maximum softening damage indicators fulfil the first and second indicated requirements as well as the third requirement, with sufficient reliability. In what follows, the fourth requirement, the Markov property of the damage indicator sequences will be tested.

The Markov property of the sequences of maximum softening values $\{D_{0,i}, i = 1,2,\dots\}$ and $\{\mathbf{D}_i, i = 1,2,\dots\}$, $\mathbf{D}_i^T = [D_{1,i}, D_{2,i}]$ can equivalently be tested from the following conditional joint moment relations

$$\forall k,i \;:\; E[D_{0,i}^k \mid D_{0,i-1} = d_{0,i-1},\dots,D_{0,1} = d_{0,1}] = E[D_{0,i}^k \mid D_{0,i-1} = d_{0,i-1}] \tag{12}$$

$$\forall k,l,i \;:\; E[D_{1,i}^k D_{2,i}^l \mid \mathbf{D}_{i-1} = \mathbf{d}_{i-1},\dots,\mathbf{D}_1 = \mathbf{d}_1] = E[D_{1,i}^k D_{2,i}^l \mid \mathbf{D}_{i-1} = \mathbf{d}_{i-1}] \tag{13}$$

The order k and $k+l$ for which (12) and (13) are valid will be tested empirically by means of Monte Carlo simulation. If the damage indicator sequences possess the Markov property, they must satisfy the criteria listed in (12) and (13) for arbitrary k and $k+l$.

4. NUMERICAL RESULTS

The considered structure for numerical study is the planar 3 storey 2 bay reinforced concrete frame shown in Fig. 2, which has been designed according to the UBC specifications for earthquake zone 4. The geometrical and structural details of the frame are indicated in Fig. 2. The linear first and second eigenperiods of the undamaged structure are $T_1(0)$ =0.695 s and $T_2(0)$ =0.221 s, corresponding to the initial angular eigenfrequencies $\omega_1(0)$ =9.04 s^{-1} and $\omega_2(0)$ =28.4 s^{-1}.

This structure is excited by a horizontal acceleration process at the ground surface, which is determined as the response process of an intensity modulated Gaussian white noise, filtered through a Kanai-Tajimi filter with parameters ω_0 and ζ_0 (Tajimi (1960)). The bedrock acceleration $\{\ddot{r}_b(t),\ t \in [0, \infty)\}$ is modelled as a time modulated unit Gaussian white noise process, i.e.

$$\ddot{r}_b(t) = \beta(t)W(t) \tag{14}$$

$\{W(t),\ t \in [0, \infty)\}$ is a unit white noise process, which is a zero mean Gaussian process with the auto-covariance function

$$E[W(t_i)W(t_j)] = \delta(t_i - t_j) \tag{15}$$

where $\delta(\cdot)$ is the Dirac delta-function. Realizations of a unit Gaussian white noise process are generated by the method of Ruiz and Penzien (1969). $\beta(t)$ is a deterministic intensity function, defined as (Jennings et al. (1968))

$$\beta(t) = \beta_0 \begin{cases} \frac{t^2}{t_1^2} & , 0 \le t \le t_1 \\ 1 & , t_1 < t < t_0 + t_1 \\ \exp(-c(t - t_0 - t_1)) & , t_0 + t_1 \le t \end{cases} \tag{16}$$

where β_0 is a given amplitude.

Application of the Kanai-Tajimi filter implies that the displacement of the earth surface $r_0(t)$ relative to the bedrock displacement $r_b(t)$ is governed by the differential equation

$$\ddot{r}_0 + 2\zeta_0\omega_0\dot{r}_0 + \omega_0^2 r_0 = -\ddot{r}_b(t) \tag{17}$$

3 modes are maintained in the expansion of external degrees-of-freedom at system reduction. The modal damping ratios are taken as $\zeta_i = 0.05$, $i = 1, 2, 3$. The eigenperiods $T_j(t)$ are calculated with an interval of $0.24\ T_1(0)$, using a Rayleigh-Ritz method with the first 5 linear eigenmodes of the initial undamaged structure as shape functions in combination with Jacobi iteration. Moving time-averages $\hat{T}_j(t)$ of the instantaneous eigenperiods are calculated using an averaging interval of $2.4\ T_1(0)$.

In the earthquake model only the intensity parameter β_0 of the earthquake process and the angular eigenfrequency ω_0 of the Kanai-Tajimi filter is varied. Below $\beta_{0,i}$ signifies the intensity of the ith earthquake in the sequence. The values ω_0=8.8s^{-1} and ω_0=27.8 s^{-1} are considered for ω_0, which implies resonance in respectively the 1st and the 2nd mode, as the angular eigenfrequencies of the structure are diminishing from their initial values due to the stiffness deterioration of the structure. In all cases a damping ratio of $\zeta_0 = 0.3$ of the Kanai-Tajimi filter is applied. The other parameters used in earthquake generation are $t_1 = 3$ s, $t_0 = 20$ s and $c = 0.2$ s^{-1}. $\beta_0 = 0.6$m/s$^{\frac{3}{2}}$, ω_0=8.8 s^{-1} corresponds approximately to a peak surface acceleration of 0.5 g.

Conditional mean values $E[D_{0,3} \mid d_{0,2}, d_{0,1}]$ and conditional variational coefficients $V[D_{0,3} \mid d_{0,2}, d_{0,1}]$ in the 3rd earthquake are estimated for the one-dimensional damage indicator for different observations of $D_{0,1} = d_{0,1}$ in the first earthquake, but for same or close observations of $D_{0,2} = d_{0,2}$ in the second earthquake. Similarly, conditional mean values $E[D_{j,3} \mid \mathbf{d}_2, \mathbf{d}_1], j = 1, 2$, condi-

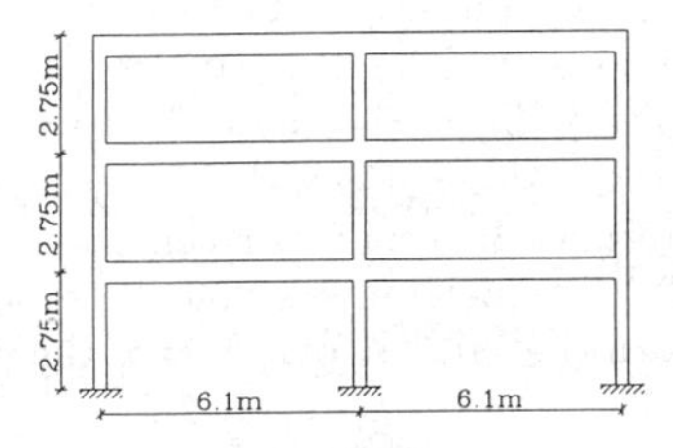

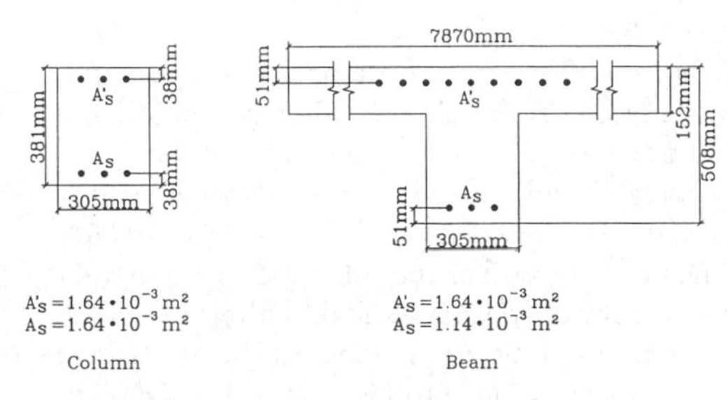

Figure 2: Details of 3 storey 2 bay reinforced concrete frame designed according to UBC zone 4. (Rodriquez-Gomez (1990)).

Table 1: Damage history, conditional mean value and conditional variational coefficient of one-dimensional maximum softening after 3rd earthquake. $\beta_{0,1}$=0.2 m/s$^{\frac{3}{2}}$, $\beta_{0,2}$=0.2 m/s$^{\frac{3}{2}}$, $\beta_{0,3}$=0.4 m/s$^{\frac{3}{2}}$, ω_0 =8.8 s^{-1}.

1st earthquake $d_{0,1}$	2nd earthquake $d_{0,2}$	3rd earthquake $E[D_{0,3}]$	3rd earthquake $V[D_{0,3}]$
0.000	0.445	0.690	0.138
0.201	0.444	0.634	0.124
0.201	0.444	0.654	0.118
0.201	0.444	0.708	0.121
0.235	0.444	0.652	0.131
0.235	0.444	0.623	0.117
0.256	0.444	0.646	0.123
0.257	0.445	0.665	0.153
0.308	0.444	0.666	0.131
0.309	0.444	0.670	0.127
0.309	0.444	0.662	0.112
0.309	0.445	0.664	0.111
0.309	0.445	0.631	0.130
0.310	0.445	0.697	0.132
0.445	0.445	0.690	0.138

Table 2: Damage history, conditional mean values, conditional variational coefficients and conditional correlation coefficient of two-dimensional maximum softening after 3rd earthquake. $\beta_{0,1}$=0.2 m/s$^{\frac{3}{2}}$, $\beta_{0,2}$=0.2 m/s$^{\frac{3}{2}}$, $\beta_{0,3}$=0.4 m/s$^{\frac{3}{2}}$, ω_0 =8.8 s^{-1}.

1st earthquake		2nd earthquake		3rd earthquake		3rd earthquake		3rd earthquake
$d_{1,1}$	$d_{2,1}$	$d_{1,2}$	$d_{2,2}$	$E[D_{1,3}]$	$E[D_{2,3}]$	$V[D_{1,3}]$	$V[D_{2,3}]$	$\rho[D_{1,3}, D_{2,3}]$
0.000	0.000	0.470	0.226	0.710	0.406	0.128	0.228	0.874
0.215	0.130	0.470	0.224	0.656	0.366	0.115	0.182	0.894
0.214	0.129	0.472	0.222	0.677	0.383	0.110	0.195	0.844
0.214	0.129	0.470	0.229	0.727	0.429	0.112	0.235	0.849
0.252	0.140	0.467	0.224	0.669	0.383	0.122	0.208	0.880
0.251	0.144	0.470	0.219	0.645	0.366	0.109	0.159	0.951
0.274	0.153	0.469	0.234	0.667	0.382	0.114	0.224	0.892
0.272	0.168	0.470	0.234	0.686	0.391	0.142	0.245	0.931
0.328	0.187	0.470	0.219	0.687	0.388	0.121	0.205	0.932
0.331	0.164	0.470	0.225	0.682	0.396	0.104	0.159	0.898
0.331	0.164	0.473	0.226	0.685	0.381	0.103	0.171	0.893
0.331	0.164	0.472	0.220	0.688	0.392	0.118	0.196	0.883
0.331	0.164	0.469	0.227	0.657	0.376	0.120	0.202	0.899
0.332	0.168	0.472	0.230	0.721	0.419	0.123	0.217	0.861
0.470	0.226	0.470	0.226	0.710	0.406	0.128	0.228	0.874

Table 3: Damage history, conditional mean value and conditional variational coefficient of one-dimensional maximum softening after 3rd earthquake. $\beta_{0,1}$=0.3 m/s$^{\frac{3}{2}}$, $\beta_{0,2}$=0.3 m/s$^{\frac{3}{2}}$, $\beta_{0,3}$=0.5 m/s$^{\frac{3}{2}}$, ω_0 =27.8 s^{-1}.

1st earthquake $d_{0,1}$	2nd earthquake $d_{0,2}$	3rd earthquake $E[D_{0,3}]$	3rd earthquake $V[D_{0,3}]$
0.000	0.442	0.636	0.142
0.000	0.442	0.643	0.140
0.221	0.444	0.603	0.132
0.228	0.442	0.598	0.112
0.300	0.444	0.587	0.107
0.300	0.444	0.581	0.105
0.307	0.442	0.633	0.144
0.307	0.442	0.637	0.125
0.307	0.443	0.613	0.145
0.332	0.443	0.584	0.132
0.332	0.444	0.590	0.121
0.332	0.443	0.603	0.141
0.333	0.444	0.611	0.137
0.442	0.442	0.636	0.142
0.442	0.442	0.643	0.140

Table 4: Damage history, conditional mean values, conditional variational coefficients and conditional correlation coefficient of two-dimensional maximum softening after 3rd earthquake. $\beta_{0,1}$=0.3 m/s$^{\frac{3}{2}}$, $\beta_{0,2}$=0.3 m/s$^{\frac{3}{2}}$, $\beta_{0,3}$=0.5 m/s$^{\frac{3}{2}}$, ω_0 =27.8 s^{-1}.

1st earthquake		2nd earthquake		3rd earthquake		3rd earthquake		3rd earthquake
$d_{1,1}$	$d_{2,1}$	$d_{1,2}$	$d_{2,2}$	$E[D_{1,3}]$	$E[D_{2,3}]$	$V[D_{1,3}]$	$V[D_{2,3}]$	$\rho[D_{1,3}, D_{2,3}]$
0.000	0.000	0.467	0.230	0.657	0.380	0.132	0.232	0.876
0.000	0.000	0.466	0.258	0.663	0.402	0.132	0.194	0.836
0.235	0.141	0.469	0.229	0.625	0.365	0.123	0.194	0.856
0.240	0.162	0.468	0.224	0.620	0.357	0.105	0.183	0.890
0.321	0.165	0.470	0.222	0.609	0.357	0.100	0.146	0.875
0.321	0.165	0.470	0.221	0.603	0.356	0.099	0.156	0.842
0.327	0.175	0.468	0.216	0.654	0.378	0.135	0.200	0.891
0.327	0.179	0.468	0.228	0.658	0.380	0.117	0.196	0.913
0.327	0.179	0.469	0.234	0.634	0.386	0.136	0.180	0.892
0.352	0.200	0.468	0.231	0.606	0.363	0.123	0.179	0.896
0.352	0.200	0.469	0.239	0.612	0.363	0.113	0.173	0.883
0.352	0.200	0.469	0.232	0.625	0.369	0.132	0.181	0.898
0.355	0.187	0.470	0.232	0.632	0.381	0.146	0.214	0.900
0.467	0.230	0.467	0.230	0.657	0.380	0.132	0.232	0.876
0.466	0.258	0.466	0.258	0.663	0.402	0.132	0.194	0.836

tional variational coefficients $V[D_{j,3}|\mathbf{d}_2, \mathbf{d}_1], j = 1, 2$ and conditional correlation coefficients $\rho[D_{1,3}, D_{2,3} \mid \mathbf{d}_2, \mathbf{d}_1]$ in the 3rd earthquake are estimated for the two-dimensional damage indicator for different observations of $\mathbf{D}_1 = \mathbf{d}_1$ in the first earthquake and same or close observations of $\mathbf{D}_2 = \mathbf{d}_2$ in the second earthquake. In all cases conditional moment estimates are based on 100 independent simulations. A remarkable observation is, that $\rho[D_{0,i}, D_{1,i} \mid \mathbf{d}_{i-1}, \dots, \mathbf{d}_1] = 1.000$ in all analysed cases, independently of the damage history. $D_{0,3}$ and $D_{1,3}$ can consequently be assumed to be proportional.

Tables 1 and 2 show the obtained conditional moments in the 3rd earthquake for the one-dimensional and two-dimensional damage indicator, respectively. In all cases $d_{0,2} \approx 0.444$, whereas $d_{0,1}$ is varied. The intensities of the earthquakes follow from the legend of the tables. The angular eigenfrequency of the Kanai-Tajimi filter is ω_0 =8.8 s^{-1}, corresponding to resonance in the first mode.

No attempt is made to keep $\mathbf{d}_2^T = [d_{1,2}, d_{2,2}]$ in the second column of table 2 constant. These values show relatively high variability, which is due to the weaker correlation between $D_{1,2}$ and $D_{2,2}$ ($\rho[D_{1,2}, D_{2,2}] \approx 0.94$). Mean value and variational coefficient of the 15 indicated data in table 1 for $E[D_{0,3}]$ are $E[E[D_{0,3}]] = 0.6635$ and $V[E[D_{0,3}]] = 0.0362$. If the results of table 1 at which $d_{2,2}$ values in table 2 differ significantly from the others are filtered out (these are the rows 4, 6, 7, 8, 13, 14), the corresponding results become $E[E[D_{0,3}]] = 0.6447$ and $V[E[D_{0,3}]] = 0.0253$. The 30 per cent decrease in the variational coefficient of $E[D_{0,3}]$ implies that conditioning on $D_{0,2}$ and $D_{2,2}$ will improve the Markov property of the mean of $D_{0,3}$, compared to conditioning solely on $D_{0,2}$. Surely, the same compaction is observed for the variational coefficient of $E[D_{1,3}]$, since $D_{0,3}$ and $D_{1,3}$ are perfectly correlated. When $V[D_{0,3}]$ is subjected to the same filtering, there is 9 per cent compaction in $V[V[D_{0,3}]]$, which implies that there is also improvement in the Markov property of the variance of $D_{0,3}$.

Similar comparative analysis for the data in table 1 at which $d_{0,2} = 0.444$ (rows 2, 3, 4, 5, 6, 7, 9, 10, 11 of table 1) subjected to filtering out the rows 4, 6 and 7 at which the $D_{2,2}$ value differs significantly from the others, shows 50 per cent compaction in $V[E[D_{0,3}]]$ and 13 per cent increase in $V[V[D_{0,3}]]$. This implies that the Markov property of the mean of $D_{0,3}$ has been improved based on such conditioning, whereas the Markov property of the variance has debased. However, the latter observation is quite unique in the material, and is believed to be attributed to statistical scatter due to the very small sample size considered.

When the data in table 1 at which $d_{0,2} = 0.445$ (rows 1, 8, 12, 13, 14, 15 of table 1) is analysed similarly (rows 8 and 13 are filtered out), the corresponding results show 45 per cent compaction in $V[E[D_{0,3}]]$ and 9 per cent compaction in $V[V[D_{0,3}]]$, which implies that conditioning on $D_{0,2}$ and $D_{2,2}$ will improve the Markov property of both the mean and the variance of $D_{0,3}$ compared to conditioning solely on $D_{0,2}$.

Table 3 and table 4 show corresponding results, when the angular eigenfrequency of the Kanai-Tajimi filter is $\omega_0 = 27.8\ \mathrm{s}^{-1}$, corresponding to resonance in the second mode. The intensities of the earthquakes have been increased in order to obtain comparable damage levels.

$E[D_{0,3}]$ and $V[D_{0,3}]$ in table 3 display the same tendency as the corresponding results in table 1. In table 4 major differences compared to table 2 are that the $d_{2,2}$ values have become more variable, $E[D_{2,3}]$ has increased slightly in proportion to $E[D_{1,3}]$, and the correlation coefficient of $D_{1,3}$ and $D_{2,3}$ has decreased to $\rho[D_{1,3}, D_{2,3}] \approx 0.88$. All these effects are due to the increased excitation of the second mode, which will induce relatively higher damage in the upper part of the building compared to the first mode.

Similar analysis of the data listed in table 3 and table 4 more clearly shows that conditioning on $D_{0,2}$ and $D_{2,2}$ will improve the Markov property of both the mean and variance of $D_{0,3}$, compared to conditioning solely on $D_{0,2}$, [9].

For the cases studied, conditioning on $D_{1,2}$ and $D_{2,2}$ reduces the variability of the observed mean values and of the observed variational coefficients of $D_{0,3}$ compared to conditioning solely on $D_{0,2}$. Further, this observation has been supported by a number of similar investigations not reported in this paper. Due to the proportionality of $D_{0,i}$ and $D_{1,i}$ it can then be concluded that the Markov properties of the mean and covariances of $[D_{1,i}, D_{2,i}]$ are superior to the Markov property of the mean and variances of the one-dimensional damage indicator $D_{0,i}$.

5. CONCLUSIONS

In this study, a one dimensional scalar valued and a two-dimensional vector valued maximum softening damage indicator are defined. The one-dimensional damage indicator, defined considering the variation in the first period of the structure, analogous to an SDOF system, is a genuine global damage index representing the average damage throughout the whole structure. The two-dimensional damage indicators are defined considering the variations in the first and second periods of the structure, and are thus analogous to an equivalent linear two degrees-of-freedom system. The components of the two dimensional damage vector can be interpreted as damage indicators for the lower half and the upper half of the structure, hence representing a simple local description of the damage state of the structure.

Since statements on the post earthquake reliability should be obtained solely from knowledge of the latest recorded values of the damage indicators above defined, these damage indicators are required to possess Markov property. This problem has been investigated based on numerical Monte-Carlo simulations. A sample plane frame which has been designed according to the UBC specifications for earthquake zone 4 is subjected to realizations of different earthquakes and the value of the damage indicators is obtained using the computer program SARCOF. For the studied cases, it is observed that the Markov assumption of the one-dimensional damage indicator is justified for the mean value of the transition probability density function (tpdf), whereas some deviations are observed for the variance and higher order statistical moments. Tests checking whether conditioning on a two-dimensional damage vector improves the Markov property of the one-dimensional damage indicator demonstrated that the Markov properties of the mean and covariances of the two-dimensional indicators are superior to the Markov property of the first and second order moments of the one-dimensional damage indicator.

Finally, following the results of this study, it is expected that the Markov property of n dimensional damage indicators should have better Markov property compared to the $n-1$ dimensional ones.

6. REFERENCES

Cecen, H. 1979. Response of ten storey, reinforcedconcrete model frames to simulatedearthquakes. Ph.D.-thesis. University of Illinois at Urbana Champaign.

DiPasquale, E. and Cakmak, A.S. 1987. Detection and Assessment of Structural Damage. Technical Report NCEER-87-0015, National Center for Earthquake Engineering Research, State University of New York at Buffalo.

DiPasquale, E. and Cakmak, A.S. 1989. On the derivation between local and global damage indices. Technical Report NCEER-89-0034, National Center for Earthquake Engineering Research, State University of New York at Buffalo.

DiPasquale, E. and Cakmak, A.S. 1990. Detection of Seismic Structural Damage Using Parameter-Based Global Damage Indices. Probabilistic Engineering Mechanics, Vol. 5, 60-65.

Healey, T.J. and Sozen, M.A. (1978): Experimental study of the dynamic response of a ten storey reinforced concrete frame with a tall first storey. Report No. UILU-ENG-78-2012, SRS 450, University of Illinois at Urbana-Champaign.

Jennings, P.C., Housner, G.W. and Tsai, N.C. 1968. Simulated Earthquake Motions. Report of the Earthquake Engineering Research Laboratoria, California Institute of Technology.

Mørk, K.J. 1992. Stochastic Analysis of Reinforced Concrete Frames under Seismic Excitation. Soil Dynamics and Earthquake Engineering, Vol. 11, No. 3, 145-161.

Nielsen, S.R.K. and Cakmak, A.S. 1991. Evaluation of Maximum Softening as a Damage Indicator for Reinforced Concrete Under Seismic Excitation. Proc. 1st Int. Conf. on Computational Stochastic Mechanics, ed. Spanos, P.D. and Brebbia, C.A., 169-184.

Nielsen, S.R.K., Köylüoğlu, H.U. and Cakmak, A.S. 1992. One and Two-Dimensional Maximum Softening Damage Indicators for Reinforced Concrete Structures under Seismic Excitation. Structural Reliability Theory, Paper No. 92, Dept. of Building Technology and Structural Engineering, University of Aalborg, Denmark.

Rodriguez-Gomez, S. 1990. Evaluation of Seismic Damage Indices for Reinforced Concrete Structures. M.S. in Engn. Dissertation. Dept. of Civil Engineering and Operations Research, Princeton University, Princeton, N.J.

Rodriguez-Gomez, S., Chung,Y.S. and Meyer, C. 1990. SARCF-II User's Guide. Seismic Analysis of Reinforced Concrete Frames. Technical Report NCEER-90-0027, National Center for Earthquake Engineering Research, State University of New York at Buffalo.

Roufaiel, M.S.L. and Meyer, C. 1987. Analytical Modeling of Hysteretic Behaviour of R/C Frames. J. Struct. Engng., ASCE, Vol. 113, 429-444.

Ruiz, P. and Penzien, J. 1969. Probabilistic Study of Behavior of Structures during Earthquakes. Report No. EERC 69-3, University of California, Berkeley, Cal.

Tajimi, H. 1960. Semi-Empirical Formula for the Seismic Characteristics of the Ground. Proceedings of the 2nd World Conference on Earthquake Engineering, Vol. II, 781-798, Tokyo and Kyoto.

Uniform Building Code 1988. Earthquake Regulations.

Structural Safety & Reliability, Schuëller, Shinozuka & Yao (eds) © 1994 Balkema, Rotterdam, ISBN 90 5410 357 4

'VaP' a tool for practicing engineers

M. Petschacher
Institute of Structural Engineering, ETH-Zürich, Switzerland

ABSTRACT: Solving reliability problems involves handling a truly dynamic task. It can be seen as a repetition of definition, analysis of problems, assessment of results, and feedback. One method of solution to this task is by the means of the so-called Variables Processor, briefly described as VaP. It is an interactive computer program for working with stochastic quantities in widely differing branches of technology. In VaP all definitions such as limit state functions or basic variables are made during run-time. Some well known methods of analysis are available to calculate statistical moments or failure probabilities. An illustrative example of cracking widths in reinforced concrete demonstrates the interacting method of working with VaP. It is shown how different types of questions can be formulated in an easy manner.

1 THE NEED OF SIMPLE TOOLS?

Developments in reliability theory helped bring about a breakthrough in thinking in terms of probabilities and furthered its acceptance. An important milestone of this kind was the development of FORM or SORM. Using this method the calculation of the probability integral of complicated problems in n-dimensional space have been transformed to a sequence of simple mathematical operations. Thus the calculation of the probability of failure p_f has become a straight forward task.

Today reliability theory is often applied for complex structures, especially for example in areas of high damage potential as in the case of offshore structures. This probabilistic modelling results, in contrast to the traditional deterministic approach, in a greater understanding of the actual problems. The design of such complex structures has, to be sure, to be left in the hands of specialists, because the use of sophisticated software requires very specialized knowledge. With these special cases the need to apply probabilistic considerations is indeed emphasized, but they do not contribute to a general acceptance of the theory in practice. In the same way as the finite element method could only attain its position as a general calculation method through the availability of suitable software, this support must also be provided in the field of reliability theory. The practicing engineer is quite often faced with simple problems and needs support visual aids in the calculation process. Only the provision of suitable software will lead to the introduction of reliability analysis into the everyday work of the practicing engineer.

The reliability theory provides a basis for conceptual decisions. The working process becomes a dynamic task, which to a certain extent involves a repetitive process. New information has to be taken into consideration continuously, so that definitions of limit states, extensions to the stochastic model and calculations follow one another in succession. Starting with these considerations and the desire to promote reliability theory in practice a computer program was developed as an interactive tool for the engineer.

2 THE VARIABLES PROCESSOR - VAP

In [Petschacher, 1993] the concept of a so-called Variables Processor, shortly named as *VaP*, is presented. The name describes the basic functionality of the program in which stochastic values are processed. The solution presented is seen as a feasible approach and the form a suitable compu-

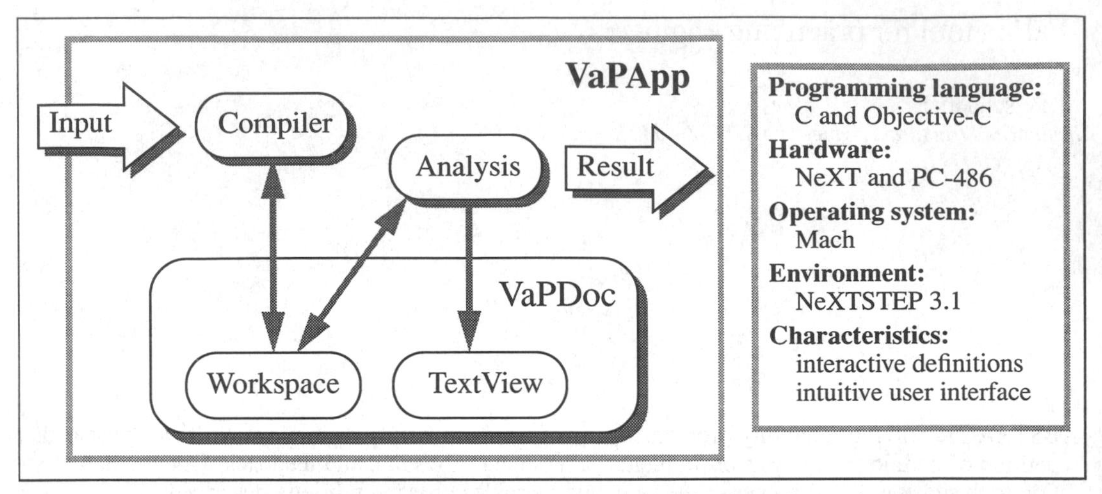

Fig.1 The concept of the Variables Processor and some details about the environment.

ter program should take for engineering practice. The concept of the program supports the typical working process, as given by the treatment of a reliability problem. The design of the user interface to the computer was paid special attention. Its structure is logical and provides dialogue windows that are intuitive and increase the working efficiency. In the usual mathematical notation the user inputs a limit state function G(**X**), without the inconvenience of having to write the function in a programming language or as a batch file. For the basic variables **X** concrete descriptions are used, to make the limit state function easier to be read.

One characteristic of the software is that all definitions of limit state functions and basic variables are made during run-time of *VaP* and do not have to be known at the time of compilation. The components of *VaP* are shown schematically in figure 1. The tasks of the areas *Compiler*, *Workspace* and *Analysis* correspond to the functions of acquiring, managing and processing data.

In the area *Compiler* all input from the user is processed. A scanner reads a limit state function G(**X**), which is written as an algebraic expression and reduces the string of characters to symbols. Then a parser processes the symbols from the scanner according to the grammar of a special meta-language. By this means the correct spelling and consistency of the expression are checked. Errors are immediately reported to the user. The end product of the parser is a syntax tree, which represents G(**X**) as a binary tree. In the realisation of *VaP* there was one important boundary condition, i.e. to keep to a minimum the time expenditure for the calculation of the limit state function. The solution with the syntax tree as a form of storage uncouples the given expression from the parser. This is in contrast to solutions in the form of an interpreter, in which for every evaluation of the expression it is syntactically reanalysed. Another factor is the number of calculations of G(**X**). This depends on the selected method of calculation and the number of basic variables. Therefore, the internal data structure of the syntax tree was realised in a form that allowed a recursive pass through the binary tree. All names used in the limit state function, the so-called identifiers, represent in general a random variable. In one of the steps following the definition of G(**X**) a probability density function with the corresponding parameters is assigned to each random variable. For this purpose a series of distribution types is available, which can be used selectively, as required. Likewise, individual variables can be declared to be deterministic for the purpose of parametric studies. By choosing a new type a change in the distribution function results in a simple manner, whereby the mean value and standard deviation remain unchanged. Of course, parameters of random variables may be changed easily while working on the screen.

VaP offers the possibility of defining an arbitrary number of limit state functions and also of basic variables. All defined elements are stored in a *Workspace*, see figure 1. The *Workspace* is part of a document, designated by *VaPDoc*, and serves to manage all identifiers put in. The external ap-

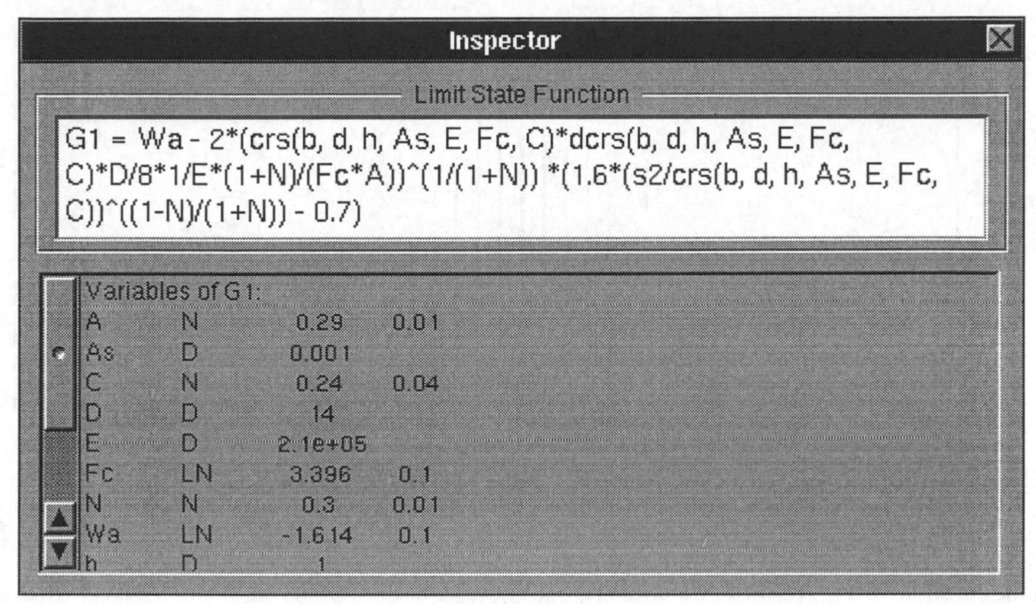

Fig.2 Input of the limit state function G_1 by writing in a natural style.

pearance of the document is a window on the screen, which contains a text editor called *TextView*. Therein all results are presented in chronological sequence, together with the input data. Commentaries can also be added to the results. Since a document forms a complete unit several problems can be handled in parallel.

In the area *Analysis* several methods of reliability analysis have been implemented, as each method has its own advantages and disadvantages. Before calculating the limit state function a test is carried out to check that all basic variables have already been defined. The question to be clarified by the calculation can concern the statistical moments or the probability of failure. In addition, either an approximation method like FORM, a simulation method like crude Monte Carlo Method, or a method of numerical integration can be chosen. In the case of time-consuming analyses, as e.g. when using the Monte Carlo Method, the user has control over the calculation and can stop it at any time.

VaP was programmed in the object-oriented language Objective-C, which has derived many features of Smalltalk-80. In its present form the program runs on a NeXT workstation under the operating system MACH and the graphical user interface NeXTSTEP. An older version of the program [Petschacher, 1990] exists for the Macintosh. *VaP* fits modern requirements to functionalism and to graphical user interface, which enables intuitive working.

3 AN ILLUSTRATIVE EXAMPLE

The problem of cracking in reinforced concrete is used to demonstrate the use of *VaP*. In the example a slab of thickness 300 mm is selected. It has a concrete quality of B30/20 according to the Swiss Standard [SIA 162, 1993] and reinforcement Ø14/125 (As = 1231 mm^2/m). The subject of the following discussion is the calculation of the mean cracking width.

An important influence factor in cracking theory is the tensile strength of concrete, f_{ct}. The calculation of this quantity is based on a formula containing an empirically obtained factor C and the concrete strength f_c:

$$f_{ct} = C f_c^{2/3} \qquad (1)$$

Based on equation (1), information on the tensile strength in bending can be derived, see also [CEB, 1990]. A well-known empirically determined formula is:

$$f_{cb} = f_{ct}\,(0.6 + 0.4h^{-0.25}) \qquad (2)$$

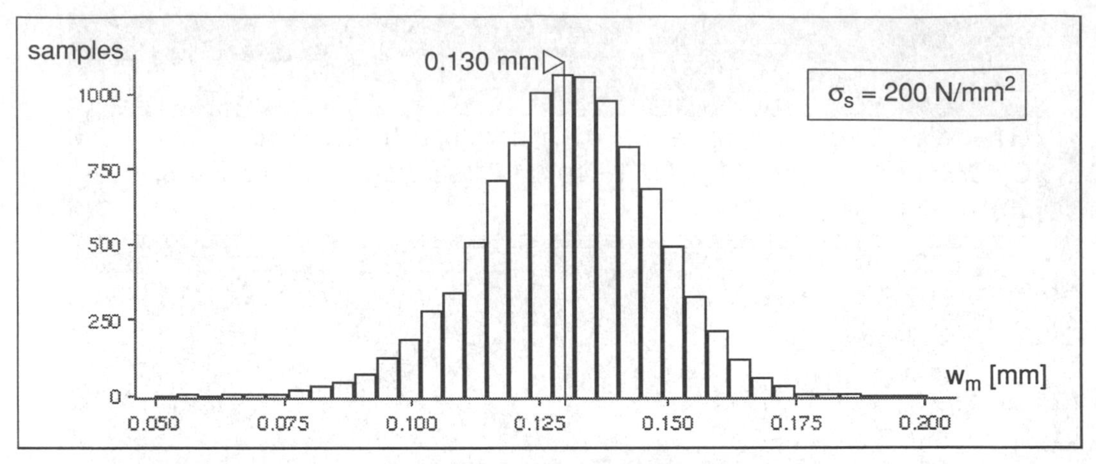

Fig.3 Density of the mean cracking width W_m, given a particular stress level $\Sigma_s = \sigma_s$.

Another important factor of the cracking theory is the constitutive law $\tau(\delta)$, with which the relative displacement δ between the reinforcing bars and the surrounding concrete is described. A possible description is taken from [Noakowski, 1978]:

$$\tau(\delta) = f_c A \delta^N \qquad (3)$$

The chosen relation in equation (3) contains the material constants A and N. They are obtained from tests and take into account the position of the bond, the concrete cover and the surface roughness of the reinforcing bars.

The theory developed by [Krips, 1984] of progressive cracking formation allows a description of the creation of cracks at a stress level above the critical cracking stress. Because of the way in which the theory is formulated it is possible to treat any arbitrary loading, independent of the steel reinforcing content. An already simplified formula from [Krips, 1984] for the mean cracking width w_m is:

$$w_m = w_1 [1.6\alpha - 0.7] \qquad (4)$$

$$\text{with} \quad w_1 = 2\left[\sigma_{sr}\Delta\sigma_{sr}\frac{d_s}{8}\frac{1}{E_s}\frac{1+N}{f_c A}\right]^{\frac{1}{1+N}}$$

$$\text{with} \quad \alpha = \left(\frac{\sigma_s}{\sigma_{sr}}\right)^{\frac{1-N}{1+N}}$$

In equation (4) σ_{sr} is the stress in the reinforcing steel at the creation of the first crack, while $\Delta\sigma_{sr}$ is the stress jump in the reinforcing steel due to the transition from states I to II at the creation of the first crack. For both these quantities an analytical expression is available for the case of pure bending. Further, d_s is the diameter of the bar, A and N are the bonding parameters in the constitutive law and f_c is the compressive strength of the concrete. The steel stress σ_s describes a loading above the critical cracking stress.

The cracking width at critical stress level σ_{sr} is denoted by w_1. In equation (4) the mean cracking width w_m depends on a linear term with respect to α, which is a measure for the stress level. In a similar way the theoretical upper and lower cracking widths can be described.

A limit state is reached when an unacceptable cracking width w_a is exceeded. It should be noted, however, that the values given in the Standards for w_a are not generally valid. Their derivation is not clear and takes no account of the random character associated with this limit value. The admissible cracking width, beyond which a crack is deemed unacceptable, is thus defined as a random variable W_a. The limit state function is described as follows:

$$G_1(\sigma_s) = W_a - W_m(\sigma_s) \leq 0 \qquad (5)$$

The input of the limit state function is evident from figure 2. The identifier for equation (5) is taken to "G1" and all basic variables will be for-

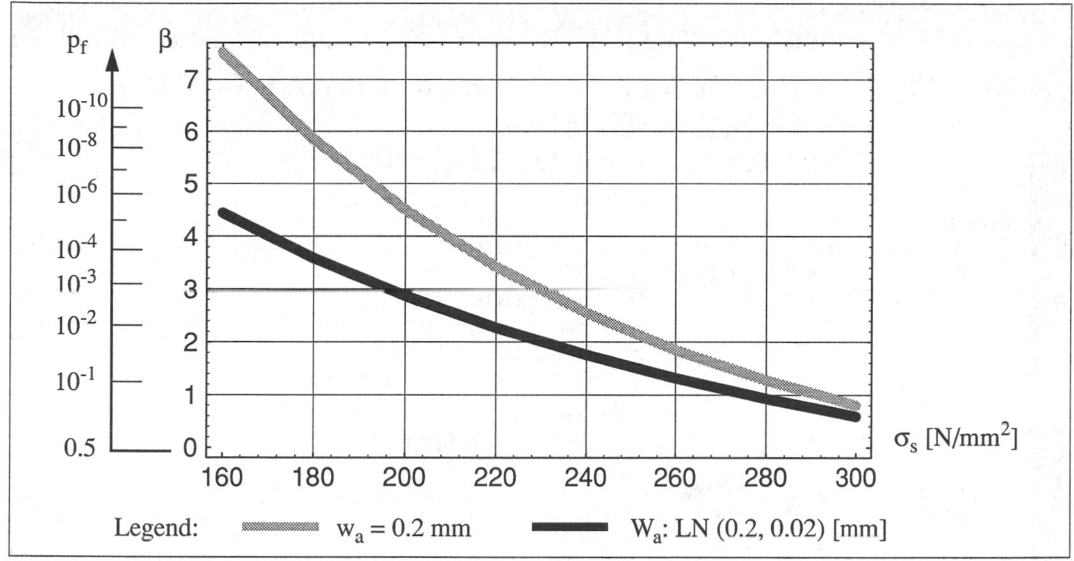

Fig.4 Variation of β for different stress levels, limit state function G_1.

mulated in upper case letters. The expression is written in the upper sub-window and will be parsed by pressing the RETURN-key. The right hand side " ≤ 0" of the equation can be left out as it is implicitly assumed. If there are no syntactical errors then all relevant identifiers are shown in the lower sub-window, although without its parameters.

Table 1 Summary of basic variables used.

X	f_X	μ_X	σ_X	Unit
F_c i)	LN	30	3.0	[N/mm²]
C	N	0.24	0.03	[-]
A	N	0.29	0.01	[-]
N	N	0.30	0.01	[-]
W_a ii)	LN	0.2	0.02	[mm]
Σ_s iii)	T1L	250	30	[N/mm²]

i) F_c is taken as recommended by [JCSS, 1992] with μ = nom + 10 and with $\sigma = 3$ in N/mm².

ii) W_a describes a transition zone from desired to undesired state, associated with different human sensitivities to cracking from a visual point of view.

iii) Between stress level and load action a linear dependency is assumed, for the stress level is small compared with the ultimate strength.

The following action is to introduce the parameters of each variable. The stochastic variables A, N, C and f_c in equation (4) are assumed to be independent. A summary of the parameters for the basic variables is given in Table 1. A special window guides the user through the problem asking to choose the type and giving the necessary values. After successful definition the probability density function of the basic variable will be shown. In addition, a variable can be set to deterministic, which reduces the dimension of the problem.

The first step in analysis might be to carry out a Monte Carlo simulation for the random variable W_m only, by which the resulting histogram gives an indication of the probability density of W_m. The result of such a simulation given a stress level $\sigma_s = 200\text{N/mm}^2$ is shown in figure 3. As additional information, the expected value and the standard deviation are given. With the given relations, these quantities become $\mu = 0.130\text{mm}$ and $\sigma = 0.016\text{mm}$.

Next an analysis with FORM for equation (5), $p_f(\sigma_s) = P(G_1(\sigma_s) \leq 0)$, is carried out with different realisations of $\Sigma_s = \sigma_s$. The variation of $\beta(\sigma_s)$ for two cases are shown in figure 4. The grey curve represents the formulation as analysing equation (5). It shows a weighted average with reference to the subjective acceptable cracking width W_a, a random variable defined in Table 1. The upper black curve is obtained if the output.rtf — ~/ICOSSAR93

G1 = Wa - 2*(crs(b, d, h, As, E, Fc, C)*dcrs(b, d, h, As, E, Fc, C) * D/8*1/E*(1+N)/(Fc*A))^(1/(1+N)) * (1.6*(s2/crs(b, d, h, As, E, Fc, C))^((1-N)/(1+N)) - 0.7)

Variables of G1:

A	N	0.29	0.01	As	D	0.001	
C	N	0.24	0.04	D	D	14	
E	D	2.1e+05		Fc	LN	3.396	0.1
N	N	0.3	0.01	Wa	LN	-1.614	0.1
b	D	1		d	D	0.3	
h	D	0.26		s2	D	200	*(given stress level)*

Results with FORM for G1: β = 2.872 -> p_f = 0.002040

Name	Alpha	x-Design
A	-0.1964	0.284
C	0.6027	0.309
Fc	-0.2005	28.180
N	0.1641	0.305
Wa	-0.7287	0.162

G2 = Wa - 2*(crs(b, d, h, As, E, Fc, C)*dcrs(b, d, h, As, E, Fc, C) * D/8*1/E*(1+N)/(Fc*A))^(1/(1+N)) * (1.6*(S2/crs(b, d, h, As, E, Fc, C))^((1-N)/(1+N)) - 0.7) | S2

Results with FORM for G2 conditioned by {S2}: β = 1.370 -> p_f = 0.08540

Name	Alpha	x-Design	
A	-0.1433	0.288	
C	0.5857	0.272	
Fc	-0.1159	29.382	
N	0.0939	0.301	
Wa	-0.5373	0.185	
Parameter			
S2	0.5704	269.393	*(sensitivity of stress level)*
G2	-0.8214	-1.125	*(sensitivity of inner part of G2)*

G3 = Wa - E[2*(crs(b, d, h, As, E, Fc, C)*dcrs(b, d, h, As, E, Fc, C)* D/8*1/E*(1+N)/(Fc*A))^(1/(1+N)) * (1.6*(S2/crs(b, d, h, As, E, Fc, C))^((1-N)/(1+N)) - 0.7) | S2]

Results with FORM for G3: β = 1.545 -> p_f = 0.06117

Name	Alpha	x-Design
A	-0.1807	0.287
C	0.6862	0.282
Fc	-0.1554	29.143
N	0.1260	0.302
Wa	-0.6756	0.179

Fig.5 FORM results for limit state functions G_1, G_2, and G_3.

variable is defined as a deterministic value, $W_a = w_a$.

The expected value of the probability of failure $E[p_f]$ is obtained by integrating the probabilities $p_f(\sigma_s)$ over the probability density function of the random variable Σ_s. This integration is carried out by defining a new limit state function G_2:

$$G_2 = W_a - W_m(\sigma_s) \leq 0 | \Sigma_s \tag{6}$$

The input is such that if the expression for G_1 is taken, see figure 2, only small changes are necessary. The identifier for the limit state function "G1" is changed to "G2" and "s2" will be written in upper case "S2", to indicate a stochastic quantity. The expression " | S2" is appended to the text, which is to be read as - "... conditioned by S2". After successful input of this expression the new identifier "S2", standing for random variable Σ_s, has to be defined, with parameters given in Table 1.

As written in equation (6), an averaging of the limit state function with respect to Σ_s is represented. This means that at each stress level σ_s a specific value of W_a is assumed as solution. This is not correct, since a subjective feeling about the acceptable cracking width is independent of the stress level. Special consideration is necessary by the formation of the expectation operator, therefore equation (6) will be rewritten. The formulation given in equation (7) forms the expected value over Σ_s within the limit state function:

$$G_3 = W_a - E[W_m(\sigma_s) | \Sigma_s] \leq 0 \tag{7}$$

Also equation (7) is input in an analogous manner to the previous examples, in that the part of the mean cracking width W_m in the expression, see also figure 2, is enclosed by the characters "E[" and "]".

The result of a FORM analysis for equation (6) and equation (7), identified as "G2" and "G3", are presented in figure 5. The differences between both of these approaches can be seen. Calculating equation (6) a nested FORM analysis is performed, where in case of equation (7) the inner expectation operator will be calculated by the Gauss-Hermite formula, [Zhou et al., 1988]. *VaP* offers such possibilities to verify rules and inequalities for expectation operators.

4 CONCLUSIONS

The application of reliability theory to simple problems must be made possible in order to convince the practicing engineer of the importance of a probability approach. The way to achieving this goal can only be through suitable computer means, which do not present the user with additional hurdles. *VaP*, an interactive computer program for processing stochastic quantities, offers such a solution. It has a simple and intuitive user interface. All definitions are made during run-time and are input in the usual manner.

By developing such interactive tools the application of the reliability theory can be promoted and a better acceptance by engineers will be reached.

5 ACKNOWLEDGMENT

The author thanks Professor J. Schneider for reviewing the text and his fruitful comments on the content of this paper.

6 REFERENCES

CEB-FIP 1990. Model Code: final draft. *Comité Euro-International du Béton*. Lausanne.

JCSS 1992. Modelling of Random Variables in Euro-Codes: first draft. *Joint Committee on Structural Safety*: Zurich.

Krips, M. 1984. *Rißbreitenbeschränkung im Stahlbeton und Spannbeton*. Dissertation: Technischen Hochschule Darmstadt.

Noakowski, P. 1978. Die Bewehrung von Stahlbetonbauteilen bei Zwangsbeanspruchung infolge Temperatur. *Deutscher Ausschuß für Stahlbeton*: Heft 296.

Petschacher, M. 1990. *MacVaP - Computer Program for Processing Stochastic Variables*. Institute of Structural Engineering: unpublished. ETH-Zurich.

Petschacher, M. 1993. *Zuverlässigkeit technischer Systeme - Computerunterstützte Verarbeitung von stochastischen Größen mit dem Programm VaP*. Institut für Baustatik und Konstruktion: Bericht Nr. 199. ETH-Zürich.

SIA 162, 1993. *Betonbauten*. Norm: Schweizer Ingenieur- und Architekten-Verein, Zürich.

Zhou, J. & Nowak, A.S. 1988. Integration Formulas to Evaluate Functions of Random Variables. *Structural Safety* 5: 267 - 284.

Structural Safety & Reliability, Schuëller, Shinozuka & Yao (eds) © 1994 Balkema, Rotterdam, ISBN 90 5410 357 4

Uncertainty modelling of remaining life of pavements due to misestimation of pavement parameters in NDT

K.M.Vennalaganti, S.Nazarian & C.Ferregut
Department of Civil Engineering, The University of Texas at El Paso, Tex., USA

ABSTRACT: The recent mechanistic-empirical approaches for predicting the remaining life of flexible pavements are mainly based upon the predicted strains at the interfaces of different layers. To determine these strains, nondestructive testing (NDT) techniques are utilized. Unfortunately, uncertainties in determining the strains may result in significant errors in the predicted remaining life. A number of major factors that contribute to these inaccuracies include the imprecise knowledge of thickness and Poisson's ratio of each pavement layer and the inaccuracies in measuring the loads and the deflections using the NDT device.

A methodology which allows to account for the variability in the assumed pavement parameters and measured load and deflections, and to quantify their influences on the calculation of the remaining life of the pavement is presented herein. The proposed methodology has been used to analyze four pavement sections, representing a wide range of highways from secondary to interstate. Results of the probabilistic analysis show that the variability in the pavement parameters increases the probability of failure of the pavement.

1. INTRODUCTION

The state of the art in the prediction of the remaining life (number of 18-kip loads that can be further applied to the pavement) of an existing flexible pavement involves nondestructive testing (NDT). NDT is carried out to determine the stiffness of existing pavements. To obtain accurate pavement layer moduli, the falling weight deflectometer (FWD) device is used throughout the world.

A FWD device impacts a load on the flexible pavement and thus creates deflections on the surface of the pavement. These deflections are measured by sensors or geophones of the device. The pavement layer moduli are backcalculated by numerical methods using the measured deflections. This method involves an iteration process from which the moduli are found by comparing the measured deflections of the FWD with the computed deflections. The computed deflections are determined using a linear-elastic analysis of the pavement. The deflections are computed on the surface pavement for known (assumed) layer thicknesses, Poisson's ratios and initial, estimated layer moduli. If the measured deflections are different from the computed deflections then the moduli of the pavement layer are changed and the process is repeated. This process is continued until an acceptable tolerance level is achieved.

Using the backcalculated moduli and the pavement parameters linear elastic-theory is employed to compute the critical strains, which are then used to determine the remaining life for two failure criteria: fatigue and rutting.

Unfortunately, during the entire process of backcalculation and estimation of the remaining life, the layer thicknesses and the Poisson's ratios of the existing pavement have to be assumed. Generally, these assumed values correspond to those specified in the design of the pavement. However, these design values rarely correspond to the actual values of the existing pavements since variability is associated with these parameters. The variability in the layer thickness and Poisson's ratio is due to the inconsistency in the construction, improper compaction and material inconsistency. In addition to these variabilities, uncertainties in the measurements of the load and deflections during NDT also add to misestimating the backcalculated moduli and the remaining life. From a pavement management point of view, understanding and quantifying how these uncertainties influence the estimation of the remaining life is very important, since the design of a new pavement system or an overlay depends on the accurate estimation of the remaining life.

Several researchers have attempted to quantify these uncertainties. Hudson et al. (1986), and Bensten et al. (1989) show that the deflections and loads are known within an accuracy of 2 to 5 percent. Rodriguez-Gomez et al. (1991) show that the backcalculated moduli may be underestimated between 40% and 60% of the time.

In this study, an uncertainty analysis was conducted to asses the influence of the variability in the thicknesses, Poissson's ratios and the load and deflections of the FWD on the remaining life of the pavement.

2. METHODOLOGY

The methodology used to study the influence of variability of the pavement parameters is based on Monte Carlo simulation. The backcalculated moduli, the critical strains, and the remaining life of the pavement were systematically simulated. The models used and the procedure adopted are further discussed in detail.

2.1 Pavement Systems

Four three layer flexible pavement systems representing a wide range of highways were studied. The representative high-traffic volume pavement system, pavement 5-12, had asphalt concrete (AC) layer thickness of 5 in. and a base layer of 12 in. Similarly, pavement 5-6, pavement 3-12 and pavement 3-6, had 5 in. and 3 in. thick AC layers, respectively. The thickness of the base layer for these pavements were either 6 in. or 12 in. The thickness of the subgrade layer was considered as 240 in. as recommended by Bush (1985). The four pavement sections are illustrated in Figure 1. Poisson's ratios and moduli were assumed to be similar for all pavement systems. The moduli of AC, base and subgrade layers were assumed to be 400 ksi, 30 ksi and 10 ksi, respectively. The Poisson's ratios were assumed to be 0.35, 0.40, and 0.45 for the three layers, respectively. These values are typical of flexible pavements and are commonly used by pavement engineers.

2.2 Determination of Deflections and Backcalculation

Theoretical surface deflection basins were developed using the linear elastic-theory program, BISAR (Dejong et al., 1973). The surface deflections were assumed to correspond to the field measurements that

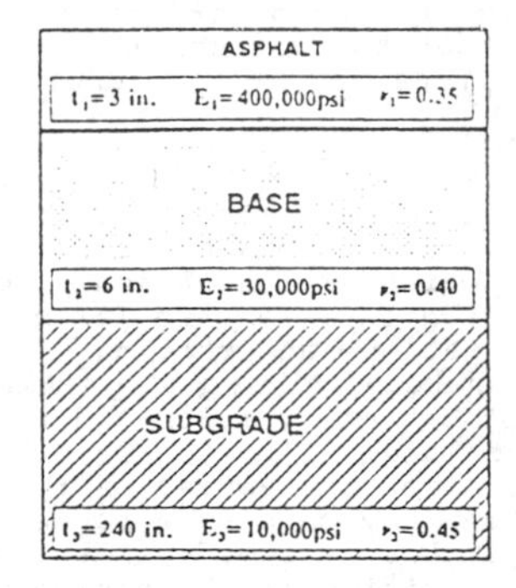

PAVEMENT 3-6

ASPHALT
t_1=3 in. E_1=400,000psi ν_1=0.35
BASE
t_2=12 in. E_2=30,000psi ν_2=0.40
SUBGRADE
t_3=240 in. E_3=10,000psi ν_3=0.45

PAVEMENT 3-12

ASPHALT
t_1=5 in. E_1=400,000psi ν_1=0.35
BASE
t_2=6 in. E_2=30,000psi ν_2=0.40
SUBGRADE
t_3=240 in. E_3=10,000psi ν_3=0.45

PAVEMENT 5-6

ASPHALT
t_1=5 in. E_1=400,000psi ν_1=0.35
BASE
t_2=12 in. E_2=30,000psi ν_2=0.40
SUBGRADE
t_3=240 in. E_3=10,000psi ν_3=0.45

PAVEMENT 5-12

Figure 1. Pavement Systems used in this study

would be measured by a Falling Weight Deflectometer. These avoided device-related and site-related errors. A standard FWD load of 9000 lb. was used to develop these deflections.

As mentioned earlier the backcalculation of moduli is an iteration process and can be written as a function of layer thicknesses, t_i, Poisson's ratios, ν_i, load, l, surface deflections, D_l, and $\hat{E}_i$ the initial, minimum and maximum acceptable moduli within which the backcalculated moduli is defined. To evaluate this function, program BISDEF (Bush, 1985) was used. This program is widely used. The function for the backcalculated moduli can be formulated as;

$$f(E_i) = f\{\hat{E}_i,\ t_i,\ \nu_i,\ l,\ D_k\};\ i = 1,3;\quad k = 1,7 \ldots\ldots\ldots\ldots\ldots (1)$$

2.3 Computation of Critical Strains and Remaining Life

The critical strains at the interfaces of the pavement layers are determined by program BISAR (Dejong et al., 1973). The tensile strain at the bottom of the AC layer, ε_t (in./in. X 10E6 or micro-strains), and the compressive strain at the top of the subgrade, ε_{vs} (in./in. X 10E6 or micro-strains), were computed with a dual wheel load of 4500 lb. on each wheel. The number of extra 18-kip loads (remaining life) that are required to cause the failure of the pavement due to fatigue and rutting were determined using the following mechanistic-empirical relationships. For fatigue Finn et al, 1987:

$$Log\ N_F = 15.947 - 3.291 \log\left(\frac{\varepsilon_t}{10^{-6}}\right) - 0.854 \log\left(\frac{E_R}{10^3}\right) \ldots\ldots (2)$$

where E_R (psi) is asphalt layer modulus and N_F, number of 18-kip loads to fatigue failure, and for rutting Shook et al, 1982:

$$N_R = 1.077\ X\ 10^{18} \left(\frac{10^{-6}}{\varepsilon_{vs}}\right)^{4.4843} \quad . \quad (3)$$

where N_R is number of 18-kip loads to rutting failure. Equation (2) and (3) are accepted by pavement engineers of the TxDOT and several other pavement organizations.

2.4 Uncertainty Modelling Methodology

Monte Carlo simulation techniques were used to estimate the uncertainties associated with the remaining life. The methodology used is illustrated in the following steps:

1) The pavement parameters, t_i, (except for t_3), ν_i, l, D_l, were assumed as random variables. Since no data were available, the variability and the statistical distribution were assumed for each of the variables using the authors' experience. All the pavement parameters were assumed to be statistically independent and normally distributed. The coefficient of variation (COV) of 0.2, 0.1,0.05, and 0.02 was assumed for the thickness, Poisson's ratio, load and deflections, respectively. Using the above variabilities and the mean values of the pavement parameters shown in Figure 1, 10,000 sets of pavement parameters were numerically generated for each pavement system. The theoretical deflection basins and a 9000 lb. load were used as the mean values to generate the normal distributions of deflections and load, respectively.

2) For each generated set of pavement parameters, the layer moduli were backcalculated. Using the backcalculated moduli and the pavement parameters, the critical strains and the remaining life was computed.

3) Finally, probabilistic distributions were fitted to E_i, ε_t, ε_{vs}, N_F and N_R to describe the variability of the simulated data. In addition, the probability of failure of the pavement was used to study the influence of variability of pavement parameters on the predicted remaining life of the flexible pavement systems.

2.5 Traffic Data

Statistical traffic data for four types of highways: Interstate Highway (IH), US roads (US), State Highway (SH) and Farm to Market road (FM) were obtained from the TxDOT. The data comprises the number of ESALs (18-kip loads), n, that are applied in one year. The mean values of n of IH, US, SH, and FM highways are 711, 126, 55, and 10, thousand ESALs respectively. The coefficients of variation for the number of applied ESALs applied in one year in each of these highways are 0.69, 1.42, 1.79, and 0.60, respectively. The best fit probability distribution to the traffic data for each of these four pavements was found to be a lognormal distribution.

3. RESULTS

Several probabilistic distributions were fitted to each of the simulated variables (E_i, ε_t, ε_{vs}, N_F and N_R). The best fit distributions of all the simulated variables were found to be similar for all the pavement systems considered in this study. The parameters and the distribution for each simulated variable are summarized in Table 2. The distribution of N_F was found to be an extreme Type II distribution for large values and N_R was found to be lognormal. In the case of N_R, thinner pavements exhibit more variability, this conclusion can be made observing the parameters of N_R. Physically, it is obvious that the thinner the pavements get, the more sensitive they are to compressive strain (consolidation of subgrade soil). Furthermore, the pavements with same base layer thickness exhibit similar

Table 2 - Parameters of Distributions of Simulated Variables

Simulated Variables / Distribution Type		Pavement Systems			
		5-12	5-6	3-12	3-6
E_1, psi	λ	12.94	12.94	13.01	13.02
Lognormal	ζ	0.522	0.563	0.575	0.594
E_2, psi	λ	10.28	10.10	10.27	10.15
Lognormal	ζ	0.373	0.666	0.271	0.489
E_3, psi	μ	10118	10182	10134	10143
Normal	σ	632.1	733.3	636.1	689.6
ε_t, micro-strain	α	0.050	0.033	0.027	0.019
Type I S	ν	220.9	247.7	301.0	349.5
ε_{vs}, micro-strain	μ	431.8	579.0	598.6	873.4
Normal	σ	58.40	62.70	77.00	95.50
N_F	α	4.040	2.814	2.963	2.699
Type II L	ν	1082.0	706.2	344.1	225.9
N_R	λ	7.440	6.009	5.968	4.229
Lognormal	ζ	0.616	0.536	0.607	0.538

variability in N_F. This can be concluded by observing the parameters of the distribution of remaining life due to fatigue, N_F.

To study how the variability in the pavement parameters influence the performance of the pavement, a scalar that measures the performance of the pavement under random loads is needed. The probability of failure was chosen as such a quantity. A failure event of the pavement was established such that when the remaining life N_c (where c = F for fatigue or R for rutting) is less than the number of traffic loads, n, failure occurs. Accordingly, the probability of failure, P_f, is $P(N_c - n < 0)$. The probability of failure of each of the pavement was computed under the basic assumption that the remaining life of the pavement estimated on the day of the NDT remains the same through out the life of the pavement (even though the remaining life changes with time). Conversely, the distribution of the traffic was considered to change with time. The mean value of the traffic distribution was assumed to change every year by a certain percentage growth in the traffic. For example, for certain percentage growth of traffic, G, the mean value of the traffic distribution of the kth year is

$$\overline{n}_{k^{th}\,year} = \left[\frac{(1+G)^k - 1}{G}\right] x\, \overline{n}_{1st} \quad . \qquad (4)$$

where $\bar{n}_{1st}$ is the first year's mean value of the traffic data. A value of 5% for G was used in this study. Further, a COV of the traffic distribution was assumed to remain constant as of the first year's distribution. This assumption was made since no data were available to predict the changes in the COV for the future years.

Probabilities of failure were computed using the classical method proposed by Rosenblueth and Esteva (Ang and Tang, 1984) to handle lognormal distributions of N_R. Rackwitz and Fiessler (1978) method was used to compute the probabilities of failure for the non-normal distributions of N_F. A span of 20 years was chosen and the probability of failure for each year and for each pavement system considered were computed.

A comparative study between the probabilities of failure computed with the remaining life as non-random variable and considering it a random variable was conducted. To accomplish this Figures 2 and 3 were developed for both pavement failure criteria, rutting and fatigue,respectively. In the Figures, the curves labeled without variability represent the probabilities of failure computed with N_c as non-random variable.

In the case of rutting (Figures 2 a-d), significant difference is noticed between the curves labeled with and without variability. For example, in Figure 2a, the difference in the predicted life of the pavement between the curves labeled with and without variability for US traffic for $P_f = 0.2$ is 4 years. A difference of about 9 years is noticed in the case of the SH traffic at the same

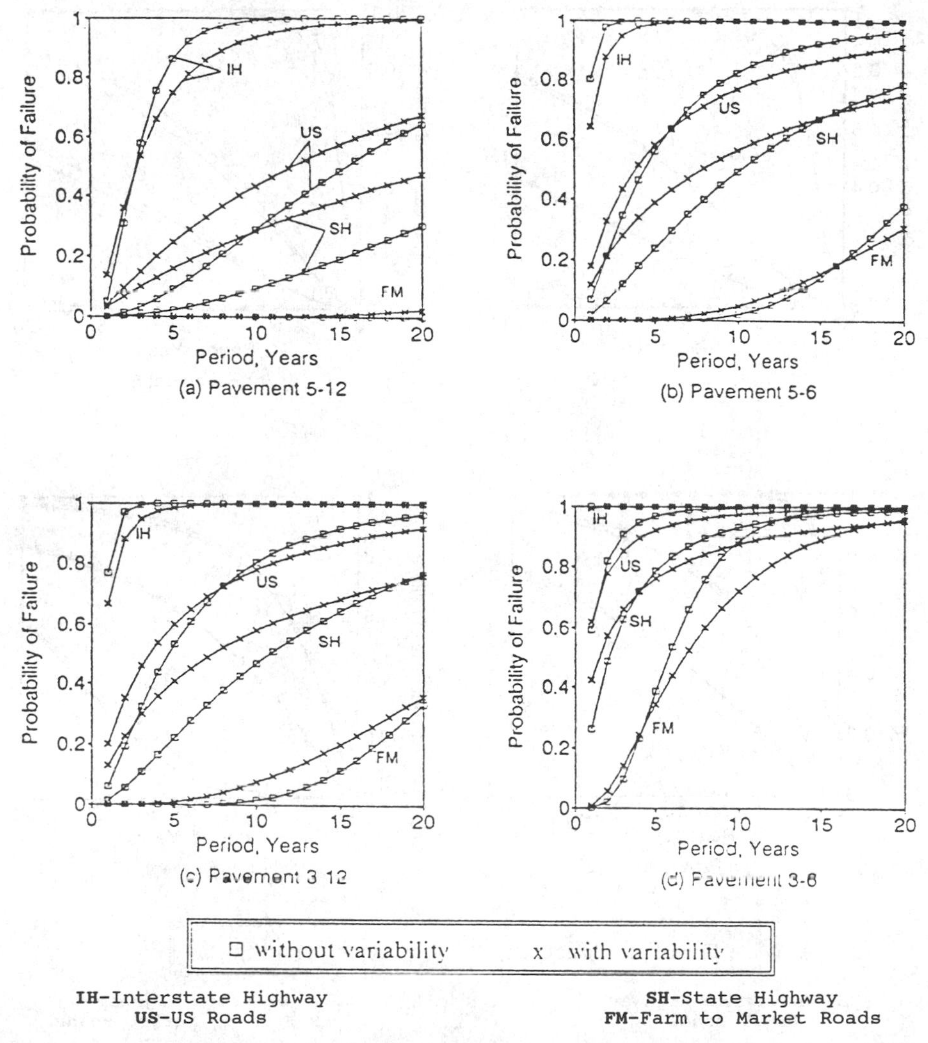

Figure 2. Probabilities of Failure of Pavement Systems (rutting)

level of probability. This shows that accounting for the uncertainties in the pavement parameters significantly decreases the estimated remaining life of the pavement. However, in the case of IH traffic the curves cross each other showing an inverse effect of the variability. This happens when the mean value of the pavement exceeds the mean value of the remaining life, an undesirable situation since the failure of the pavement is practically confirmed according to the definition of probability of failure of the pavement. Intuitively, when the mean value of n exceeds the mean value of N_c, failure of the pavement may be considered and beyond this point the results are of no practical value.

In the case of fatigue (Figures 3 a-d), the difference between the probabilities of failure computed with and without variability in the pavement parameters is negligible. This indicates that the fatigue mode of failure of the pavement is mostly influenced by the distribution of the traffic loads.

4. CONCLUSIONS

This paper illustrates a methodology to predict the remaining life of a pavement accounting for the variability in the pavement parameters. This methodology is a step towards achieving more realistic values for the remaining life of a pavement. The authors hope that this paper will encourage researchers to account for

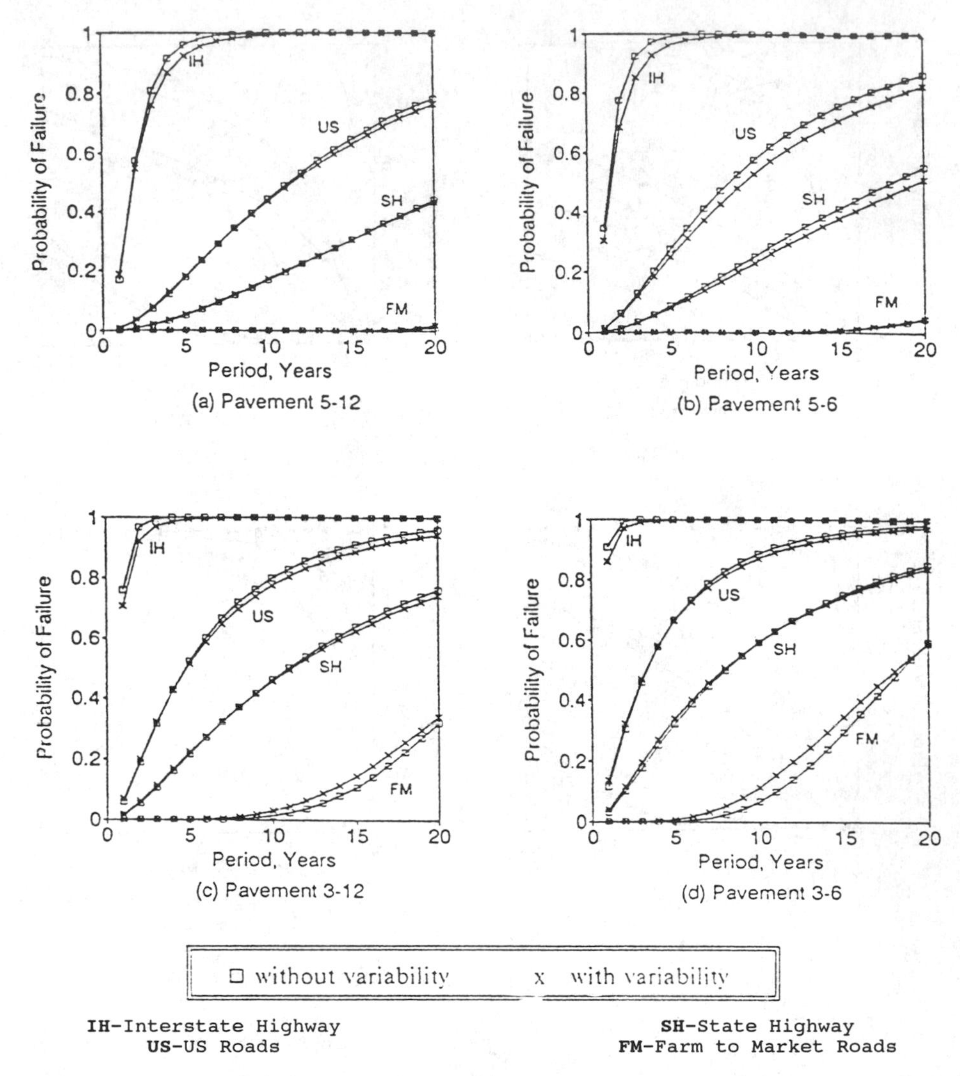

Figure 3. Probabilities of Failure of Pavement Systems (Fatigue)

the variability of the pavement parameters, since significant differences are noticed when all the uncertainties are taken into account. It was observed that the best-fit distribution of the predicted remaining life due to fatigue cracking criteria was an extreme type II distribution, since this distribution has long been assumed normal by pavement engineers during pavement rehabilitation or maintenance.

REFERENCES

Ang, A.H-S., and Tang, W.H., 1984. Probability Concepts in Engineering Planning and Design, Vol. II - Decision, Risk and Reliability , John Wiley and Sons, New York.

Bensten, R.A., Nazarian, S., and Harisson, J.A., 1989. "Reliability Testing of Seven Nondestructive Pavement Testing Devices, " STP 1026, American Society for Testing and Materials, Philadelphia, pp 278-290.

Bush III, A.J. and Alexander, D.R., 1985. "Computer Program BISDEF", U.S.Army Corps of Engineers WES, Vicksburg, MS, November.

Dejong, D.L., Peutz, M.G.F. and Korswagen, A.R., 1973. "Computer Program BISAR, "External Report, Koninklijkel/ Shell-laboratorium, Amsterdam, Netherlands.

Finn, F., Saraf, C., Kulkarini, R., Nair, K., Smith, W. and Abdullah, A. 1977. "The Use of Distress Prediction Subsystems for the Design of Pavement Structures", Proceedings, Fourth

International Conference on the Structural Design of Asphalt Pavements, pp 3-37.
Hudson, W.R., Elkins, G.E., Uddin, W., and Reilley, K.T., 1986. "Pavement Condition Monitoring Methods and Equipment Phase I - Part I Evaluation of Deflection Measuring Equipment, " Report No FH67/1, ARE Inc., Austin Tx., pp. 113.
Rackwitz, R., Fiessler, B., 1978. "Structural Reliability Under Combined Random Load Sequence", Computer and Structures. Pergamom Press, Vol.9. pp. 484-494.
Rodriguez-Gomez, J., Ferregut, C., Nazarian, S., 1991. "Impact of Variability in Pavement Parameters on Backcalculated Moduli", Road and Airport Pavement Response Monitoring Systems, Special Publications, American Society of Civil Engineers, Edited by Vincent C. Janoo and Robert A. Eaton, Mar., pp. 261-275.
Shook, J.F., Finn, F.N., Witczak, M.W., and Monismith, C.L., 1982. "Thickness Design of Asphalt Pavements - The Asphalt Institute Method", Proceedings, Fifth International Conference on the Structural Design of Asphalt Pavements.

Structural Safety & Reliability, Schuëller, Shinozuka & Yao (eds) © *1994 Balkema, Rotterdam, ISBN 90 5410 357 4*

A study on the feedback technique of damage information

Norio Yamamoto
Research Institute of Nippon Kaiji Kyokai, Tokyo, Japan

ABSTRACT: In order to assess the reliability of large scale structures such as ship structure, it is indispensable to utilize data which is obtained in-service because there are many uncertain factors concerned with the reliability assessment at the design stage. When to utilize the data for the assessment of structural reliability, information on damage is considered to be the highest potential data source. But from the practical view point, it is difficult to utilize the damage data for the analysis because the amount of usable data is limited in general. In this paper, discussion is made on how to feedback the damage information to the process of decision to cope with the damage by applying Bayesian method.

1 INTRODUCTION

In the ship structural design, there are various uncertainties due to the complexity of the structures, the severity of the environment and the randomness of the wave induced loads, etc.

The developments of the conventional design procedure have been generally achieved by the relative comparison of the damaged ships with not damaged ships.(Pattofatto, 1991) But the evaluations of load and strength according to these conventional design procedures are sometimes insufficient because of the lack of quantitative evaluation on the damage phenomenon.

On the other hand, more accurate evaluation of load and strength have been able to be performed in accordance with the recent development of analytical techniques and fabrication skills. (Machida, 1992) According to such a technical background, design by positive comparison based on the consideration of probabilistic characters of load and strength have been aimed.

When the reliability assessment is aimed, accumulation of the data on load and strength is necessary to reveal probabilistic character of these factors. And it is also necessary to feedback damage data to the process of reliability assessment. But the damage phenomena occurred in the hull structures are made up many events which include uncertainty factors. And the amount of usable data is limited in general. Consequently, how to draw an effective information from the limited amount of data is dominant problem on the investigation of the feedback technique.

In this study, discussions were made on how to feedback the damage information to the decision to cope with the damages by applying the method based on Bayesian procedure to the fatigue cracked damages occurred in the ship structures.

2 ANALYTICAL METHOD BASED ON BAYESIAN MODEL

2.1 Basic procedure of Bayesian method

In the practical application of Bayesian procedure, how to compose prior distribution is important subject. For the effective composition of prior distribution, Akaike(1980,1982) proposed the following procedure.

1. Compose the distribution which explain the objective event by using many unknown parameters as data distribution.
2. Introduce the probabilistic distribution as the restriction, which the unknown parameters should fulfill. Then compose the prior distribution with an uncertain parameter which represents the strength of the restriction.
3. Determine the parameter which governs the

form of the prior distribution of unknown parameters to minimize the ABIC(Akaike's Bayesian Information Criterion), while the other unknown parameters are evaluated by the likelihood method.

2.2 Application to the fatigue cracked damages occurred in ship structures

Ship structures are the welded structures which are made up of plates and frames. And longitudinal-transverse stiffener connections of these structures are the fatigue sensitive points. Although the structural type of these points are regarded as identical structures, working stresses at each point are different from each other because of the difference of load acting on each member or the difference of the environmental condition surrounding each member.

Figure 1 shows the fatigue design procedure generally applied to the ship structural design. Although unknown factors are existed in each step of evaluation, working stresses at each point due to wave induced loads are assumed as the most dominant factor in the fatigue design in this study.

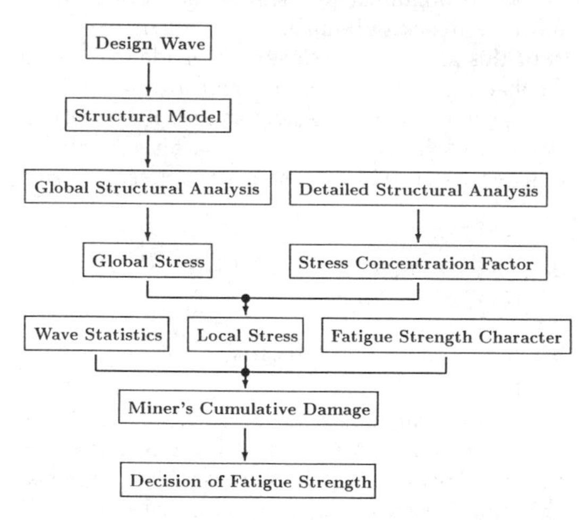

Fig.1 Schematic flow of fatigue design

To deduct the undue difficulty, simplified models which can explain the event of fatigue damages were introduced. And then the most suitable model which expresses the actual event was chosen selectively through the Bayesian system.

1. Define the location of each point in the part of a ship $i(i = 1, 2, ..., N_L)$ in longitudinal direction and $j(j = 1, 2, ..., N_V)$ in vertical direction.

2. Define the local stress $s_{ij}^{(m)}$ working at each point i, j of ship m $(m = 1, 2, ..., M)$ has a value of the average nominal stress multiplied by a stress concentration coefficient $\kappa_{ij}^{(m)}$. The coefficients depend on scantling and details of local structures.

3. Assume that the long term distributions of wave induced stress at each point of each ship is governed by Eqn.(1). In this study, scale parameter β_{ij} is assumed as unknown parameter and define a set of unknown parameters as shown in Eqn.(2).

$$F_S(s_{ij}^{(m)}|\beta_{ij}) = 1 - exp\left(-\frac{s_{ij}^{(m)}}{\kappa_{ij}^{(m)}\beta_{ij}}\right) \qquad (1)$$

$$\mathbf{B} = (\mathbf{B}_1, \mathbf{B}_2, ..., \mathbf{B}_i, ..., \mathbf{B}_{N_L})^t \qquad (2)$$

where $\quad \mathbf{B}_i = (\beta_{i1}, \beta_{i2}, ..., \beta_{ij}, ..., \beta_{iN_V})$

4. The relation between stress range(S) and average fatigue life($\bar{T}$) of the material used for the members is assumed to given by Eqn.(3) and assumed that fatigue life distribution is governed by Eqn.(4).(Yoneya, 1993)

$$S = C\bar{T}^{-1/n} \qquad (3)$$

where $\quad n = 3.84$
$\quad C = 4697.59$

$$F_T(t) = 1 - exp\left\{-\left(\Gamma(1 + 1/\gamma)\frac{t}{\bar{T}}\right)^\gamma\right\} \qquad (4)$$

where $\quad \gamma = 1.65$

5. An equivalent constant stress range, which gives equivalent Miner's damage under the variable stress condition, is evaluated as shown in Eqn.(5). And assume that the fatigue life distribution under the variable stress condition is defined by introducing an equivalent constant stress range($S_{eq\ ij}^{(m)}$) to Eqn.(4). Then fatigue life distribution at each point of a ship is obtained as shown in Eqn.(6).

$$S_{eq\ ij}^{(m)} = \kappa_{ij}^{(m)}\beta_{ij}\Gamma(1+n)^{1/n} \qquad (5)$$

$$F_T(t_{ij}^{(m)}) = 1 - exp\left\{-\left(\frac{t_{ij}^{(m)}\beta_{ij}^n}{\xi_{ij}^{(m)}}\right)^\gamma\right\} \qquad (6)$$

where $\quad \xi_{ij}^{(m)} = \left(\frac{C}{\kappa_{ij}^{(m)}}\right)^n \frac{1}{\Gamma(1+1/\gamma)\Gamma(1+n)}$

6. When the information about the fatigue damages of each point i, j of each ship m are obtained

from the results of inspection conducted at time $T^{(m)}$ from the beginning of its service, the likelihood function of the uncertain parameter β_{ij} will be formulated as following.

$$L(\beta_{ij}) = \prod_{m \in M_{n\ ij}} Pr[not\ cracked|m,i,j] \times \prod_{m \in M_{c\ ij}} Pr[cracked|m,i,j] \quad (7)$$

where $Pr[cracked|m,i,j] = f_T(T^{(m)})dT^{(m)}$

$$Pr[not\ cracked|m,i,j] = 1 - F_T(T^{(m)})$$

In Eqn.(7), $M_{c\ ij}$ and $M_{n\ ij}$ denote the set of ships that the damage at point i,j were reported and not reported respectively. And the maximum likelihood estimator of β_{ij} can be evaluated as following.

$$\bar{\beta}_{ij} = \left\{ M_{c\ ij} \ / \ \sum_{m=1}^{M} (T^{(m)} \ / \ \xi_{ij}^{(m)})^{\gamma} \right\}^{1/n\gamma} \quad (8)$$

7. The error of estimation defined by subtracting the estimated value of the unknown parameter from the true value are assumed to follow a normal distribution whose mean is 0 and variance is σ^2, regardless of i,j.

$$p(\bar{\mathbf{B}}|\mathbf{B},\sigma^2) = \left(\frac{1}{2\pi\sigma^2}\right)^{\frac{N_L N_V}{2}} exp\left\{-\left(\frac{||\mathbf{B}-\bar{\mathbf{B}}||^2}{2\sigma^2}\right)\right\} \quad (9)$$

8. The value of stress at each point in the considered area are closely corresponding to the wave induced loads distributing smoothly along ship hull. Therefore, it is reasonable to expect that the changes of the values between adjacent points are gradual. The gradualness can be expressed as following Eqn.(10) to (12) and it is assumed that the square sum of these values is expected to be small enough.(Akaike 1986)

- on the border of considered area

$$\Delta^2\beta_{iN_V} = \beta_{i-1N_V} - 2\beta_{iN_V} + \beta_{i+1N_V} \quad (10)$$

$$\Delta^2\beta_{N_Lj} = \beta_{N_Lj-1} - 2\beta_{N_Lj} + \beta_{N_Lj+1} \quad (11)$$

- inside of considered area

$$\Delta\beta_{ij} = \beta_{i-1j} + \beta_{i+1j} - 4\beta_{ij} + \beta_{ij-1} + \beta_{ij+1} \quad (12)$$

9. Each quantity defined in Eqn.(10) to (12) is assumed to follow a normal distribution with 0 mean and a certain standard deviation. Then the following probability distribution as the restriction, which the unknown parameters **B** should fulfill, is obtained.

$$q(\mathbf{B}|\epsilon) = \left(\frac{1}{2\pi}\right)^{\frac{N_L N_V - 4}{2}} \epsilon^{N_L N_V - 4} \phi exp\left\{-\left(\frac{||\epsilon \mathbf{D}\mathbf{B}||^2}{2}\right)\right\} \quad (13)$$

In this equation, ϵ represents the strength of the restriction and the matrix $\mathbf{D}$ is defined as followings.

$$\mathbf{D} = \begin{bmatrix} \mathbf{D}_1 & & & & \\ \mathbf{I} & \mathbf{D}_2 & \mathbf{I} & & \\ & \cdot & \cdot & \cdot & \\ & & \cdot & \cdot & \cdot \\ & & \mathbf{I} & \mathbf{D}_2 & \mathbf{I} \\ & & & & \mathbf{D}_1 \end{bmatrix} \quad (14)$$

$$\mathbf{D}_1 = \begin{bmatrix} 1 & -2 & 1 & & \\ & \cdot & \cdot & \cdot & \\ & & \cdot & \cdot & \cdot \\ & & 1 & -2 & 1 \end{bmatrix}$$

$$\mathbf{D}_2 = \begin{bmatrix} -2 & & & & \\ 1 & -4 & 1 & & \\ & \cdot & \cdot & \cdot & \\ & & \cdot & \cdot & \cdot \\ & & 1 & -4 & 1 \\ & & & & -2 \end{bmatrix}$$

10. Introduce a parameter ω to combine Eqn.(9) and Eqn.(13). The posterior distribution of unknown parameters **B** can be obtained by the Bayes' theorem.

$$\omega = \epsilon\sigma \quad (15)$$

$$h(\mathbf{B}|\bar{\mathbf{B}},\omega) = \frac{p(\bar{\mathbf{B}}|\mathbf{B},\sigma^2)q(\mathbf{B}|\omega)}{\int p(\bar{\mathbf{B}}|\mathbf{B},\sigma^2)q(\mathbf{B}|\omega)\ d\mathbf{B}} \quad (16)$$

11. The parameter ω, which governs the form of the prior distribution of unknown parameters **B**, is determined to minimize the ABIC which is defined by Eqn.(17). While another parameter σ is evaluated by the likelihood method.

$$ABIC(\omega,\sigma^2) = -2ln\int p(\bar{\mathbf{B}}|\mathbf{B},\sigma^2)q(\mathbf{B}|\omega)\ d\mathbf{B} \quad (17)$$

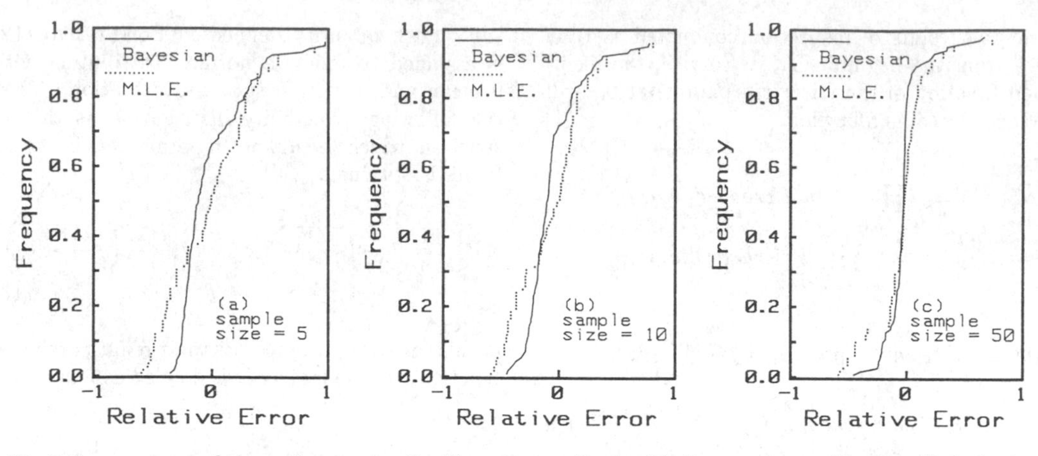

Fig.2 Comparison of the estimates by the Bayesian method with the maximum likelihood method

Following these procedure, it will be possible to select the most suitable model which explain the event of actual occurrence of damages. In term, with the results obtained using so selected model, various investigations can be made.

2.3 Verification of applied method

To verify the Bayesian method described above, comparison between the estimates by the Bayesian method and the estimates by the maximum likelihood method were performed by applying both methods to the simulated damage data.

Simulated damage data were generated by following procedures.

1. Set the point to be evaluated as 12 points in the vertical direction and 9 points in the longitudinal direction.

2. Assume the smooth distribution of stress over the area.

3. Set the sample size as 5, 10 and 50.

4. Generate the fatigue failed number for each point according to the Eqns.(3) and (4) by using Monte Carlo method.

Results are shown in Figs.2-(a), (b) and (c). In these figures, a horizontal axis shows the value which is obtained by dividing the error of estimation by the true value. And a vertical axis shows the cumulative frequency of these values.

From the results of example calculations, more accurate estimations were made by applying the Bayesian method to the data of 5 samples as compared with the estimations by applying the maximum likelihood method to the data of 10 samples. Although the case of enough amount of data such as 50 samples, the estimations with highly accuracy can be made by applying the Bayesian method.

When the smooth distribution of stress is expected, the Bayesian method, which is composed of smoothness prior distribution, will be helpful to perform an effective investigation

3 FEEDBACK OF DAMAGE INFORMATION

3.1 Example of damages

The damages adopted in the examples of analyses were the fatigue cracked damages on side longitudinals(hereinafter, it is called as SL) at the intersections with transverse bulkhead(hereinafter, it is called as T.Bhd) and transverse ring which had reported from the survey(Yoneya, 1993) of very large crude oil carriers(hereinafter, it is called as VLCC). Figure 3 shows the schematic structural type of these intersections and typical example of fatigue cracked damage.

To cope with the damage, how to repair the damages is the primary subject. But some difficult problems are involved in this subject such as:

• How to decide the degree of reinforcement (Is it all right to restore to the original condition? or is it necessary to improve the strength?)

• How to decide the area which need to be reinforced (Is it enough to repair where the damages were occurred? or is it necessary to reinforce the part where the damage has not occurred yet?) Moreover, from the practical view point, rapid treatment for these problems is required.

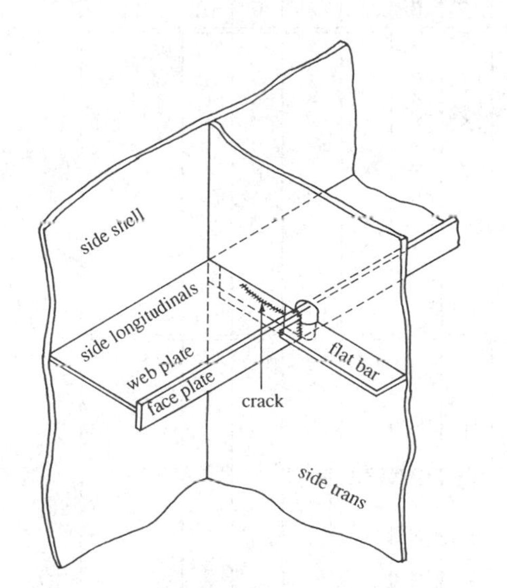

Fig.3 Structural type & typical cracked damage

3.2 Damages on 3 VLCCs

When rapid treatment to cope with the damages is required, the amount of available damage information is usually limited. Assuming such a situation, damages on 3 VLCCs, which have same structural arrangement of the members but different scantling, were adopted in this example. In this study, damages occurred in port side of No.3 tank were chosen. Longitudinal extent to be considered is the intersections from forward T.Bhd to afterward T.Bhd. And vertical extent is the intersections from SL nearby the load water line(hereinafter, it is called as LWL) to the 12-th one below LWL. Locations where fatigue cracked damages were reported on 3 VLCCs are summarized in Table 1.

Figure 4 shows the result of estimations. This figure shows the ratios of the estimated stress range, which is calculated as the mode value of the posterior distribution of the unknown parameters, to the maximum value of these estimates. It is said that this result show the state of stress distribution on the SLs in this hold which is chosen selectively to explain the state of actual fatigue damages.

Consequently, based on these results, it will be possible to make quantitative decisions to cope with the damages such as necessary degree of reinforcement or necessary area to be reinforced. Table 2 shows the results of required degree of stress relaxations for ship A-1. These ratios are determined to prevent the initiation of fatigue crack in service through 20 years with the reliability of 50%.

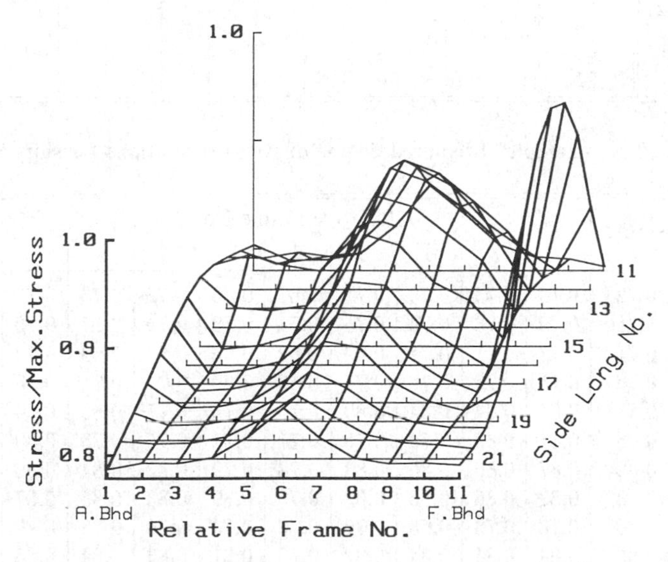

Fig.4 Example of estimated stress on the presence of finding damages

Table 1 Cracks detected in No.3 tank of ships A-1, A-2 and A-3

	SL No.	A.Bhd 1	Relative Frame No. 2	3	4	5	6	7	8	9	10	F.Bhd 11
	11						×					
Ship A-1	12						×	×				
	13			×			×	×	×	×		×
age of ship	14		×	×		×	×	×	×	×		×
at	15	×					×	×	×	×		×
inspection	16	×					×	×		×		×
is	17						×	×				×
3.75 years	18						×	×				
	19						×	×				
	20											
	21											
	22											
	11											
Ship A-2	12											
	13											
age of ship	14											
at	15											
inspection	16						(NO DAMAGE)					
is	17											
1.5 years	18											
	19											
	20											
	21											
	22											
	11											
Ship A-3	12						×					
	13						×					×
age of ship	14						×					×
at	15	×					×					×
inspection	16						×					×
is	17											×
2.5 years	18											
	19											
	20											
	21											
	22											

Table 2 An example of required degree of stress relaxations for ship A-1

SL No.	A.Bhd 1	Relative Frame No. 2	3	4	5	6	7	8	9	10	F.Bhd 11
11	0.72	0.73	0.73	0.73	0.71	0.67	0.70	0.72	0.73	0.73	0.74
12	0.78	0.79	0.78	0.78	0.75	0.71	0.72	0.75	0.77	0.79	0.74
13	0.71	0.71	0.71	0.71	0.69	0.64	0.65	0.67	0.69	0.73	0.63
14	0.76	0.75	0.76	0.76	0.73	0.68	0.69	0.71	0.74	0.78	0.64
15	0.81	0.84	0.84	0.84	0.80	0.75	0.75	0.77	0.80	0.84	0.68
16	0.76	0.81	0.80	0.79	0.76	0.71	0.72	0.75	0.75	0.80	0.65
17	0.82	0.87	0.86	0.85	0.83	0.78	0.79	0.82	0.84	0.86	0.73
18	0.83	0.87	0.86	0.86	0.83	0.79	0.80	0.83	0.85	0.87	0.80
19	0.76	0.78	0.78	0.78	0.76	0.73	0.73	0.76	0.78	0.79	0.77
20	0.83	0.84	0.84	0.84	0.82	0.79	0.81	0.83	0.84	0.85	0.85
21	0.78	0.78	0.78	0.78	0.76	0.74	0.76	0.77	0.78	0.78	0.79
22	0.91	0.90	0.90	0.90	0.88	0.85	0.88	0.90	0.90	0.90	0.91

Table 3 Cracks reported from 22 VLCCs

ship	age (year)	relative SL No. 1	2	3	4	5	6	7	8	9	10
A	3.75			×	×	×	×	×			
B	4.75			×	×	×	×	×	×		
C	16.83					×	×				
D	18.0										
E	4.75	×	×	×	×	×	×	×	×		
F	5.00										
G	3.92		×								
H	3.00			×			×				
I	3.92		×	×	×		×				
J	5.00										
K	3.00										
L	17.00										
M	10.00										
N	16.00										
O	16.75		×	×	×	×	×		×	×	
P	18.58				×	×	×	×			
Q	15.00										
R	18.42			×	×	×	×				
S	2.33			×		×	×	×	×		
T	2.33		×	×	×	×	×		×		
U	1.83				×	×					
V	2.00										

3.3 Damages on 22 VLCCs

In this example, adopted data were the damages on SLs at the intersection with forward T.Bhd of 22 VLCCs which were chosen randomly from the fleet of VLCCs including first generation(VLCC built before 1983 is calls as First Generation VLCC) and second generation(VLCC built after 1983 is calls as Second Generation VLCC). Vertical extent to be considered was from SL nearby the LWL to the 10-th SL below LWL. Table 3 shows the locations where fatigue cracked damages were reported.

In this analysis, wave induced pressure, which was considered to be the most relevant to the stress worked on SLs, was regarded as the unknown parameters to be evaluated. Because tank arrangement and structural arrangement of the members of these ships are different from each other. In this case, distribution of fatigue life given by Eqn.(6) is modified as shown in Eqn.(18).

$$F_T(t_j^{(m)}) = 1 - \left\{ - \left(\frac{t_j^{(m)} \beta_j^{(m)n}}{\xi_j^{(m)}} \right)^{\gamma} \right\} \qquad (18)$$

where $\beta_j^{(m)} = \dfrac{\Delta H_j l_T^{(m)2} l_L^{(m)}}{12 Z_j^{(m)}}$

ΔH_j ; wave induced pressure worked on j-th SL
$Z_j^{(m)}$; section modulus of j-th SL of ship m
l_T ; space between T.Rings
l_L ; space between SLs

Table 4 shows the estimated maximum stress ranges during 20 years and the required stress relaxation ratios to prevent the initiation of crack in service through 20 years with the reliability of 50%. These values were calculated as the average value for First Generation VLCCs and Second Generation VLCCs respectively. For example, this table also show the ratios for ship A-1 together with the ratios shown in Table 2.

Table 4 Summary of estimated max. stress range during 20 years & stress relaxation ratio

SL No.	1st Generation		2nd Generation		relaxation ratio (A-1)	
	max. stress (MPa)	relaxation ratio	max.stress (MPa)	relaxation ratio	results from 3 VLCCs	results from 22 VLCCs
1	214.2	1.54	415.6	0.78	0.74	0.71
2	242.5	1.35	449.1	0.72	0.74	0.74
3	276.2	1.18	457.4	0.71	0.63	0.68
4	290.8	1.12	475.6	0.68	0.64	0.64
5	302.5	1.07	472.8	0.69	0.68	0.65
6	288.1	1.13	464.9	0.70	0.65	0.63
7	261.6	1.24	441.8	0.74	0.73	0.69
8	256.9	1.26	410.8	0.79	0.80	0.73
9	226.2	1.75	344.9	0.94	0.77	0.91
10	238.0	1.66	308.5	1.05	0.85	1.00

From these results, there are no significant defferences between the results deduced from the data of 3 VLCCs and 22 VLCCs. From this fact, it is considered to be possible to perform sufficient estimation even though the amount of available damage information is limited.

4 CONCLUSIONS

In this study, investigations were made on how to feedback the damage information to the quantitative decisions to cope with the damages. The method based on Bayesian procedure was introduced and was applied to the analysis of fatigue cracked damages which had occurred on the SLs of VLCCs.

As the results, following remarks were obtained.

(1) It is possible to draw an effective results from the data only concerning with the existence of damage.

(2) The effectiveness of analysis based on the Bayesian method is verified although the amount of usable data is limited.

(3) By feeding back the damage information to the process of the analysis and the evaluation, it will be possible to evaluate the uncertainties which have not been evaluated enough in the design stage and also it will be possible to make quantitative decision.

ACKNOWLEDGEMENTS

The author wishes to express sincere appreciation to Prof. H.Itagaki of Yokohama National University for his valuable suggestions and encouragement during this study.

REFERENCES

Akaike, H., "Likelihood and the Bayesian Procedures",Bayesian Statistics, 1980.

Akaike, H., "On the Fallacy of the Likelihood Principle",Statistics & Probability Letters, 1, 1982.

Akaike, H., "The Selection of Smoothness Priors for Lag Estimation", Bayesian Inference and Decision Techniques, 1986.

Machida,S., et al., "Recent Japanese Research Activities on Structural Reliability of Ships and Offshore Structures", Int. Conf. on Behavior of Offshore Structures, 1992.

Pattofatto,G., "The Evolution of Inspection and Repair Procedures for Ship Structures", Marin Structural Inspection, Maintenance and Monitoring Symposium, 1991.

Yoneya,T., et al., "Hull Cracking of Very Large Ship Structures", Integrity of Offshore Structures, 1993

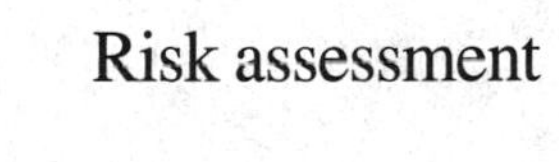
Risk assessment

Structural Safety & Reliability, Schuëller, Shinozuka & Yao (eds) © 1994 Balkema, Rotterdam, ISBN 90 5410 357 4

Risk comparison: One way of risk evaluation

H. P. Berg
Bundesamt für Strahlenschutz, Salzgitter, Germany

D. Gründler & F. Lange
Gesellschaft für Reaktorsicherheit, Köln, Germany

Abstract: Risk evaluations are an important tool in order to assess the safety of nuclear power plants or other types of nuclear installations. One helpful way can be the comparison of the expected risk of the facility to be evaluated with the conventional risk in the respective area from the same type of event. The general procedure of such a radiological - conventional risk comparison based on cumulative complementary frequency distributions of fatalities is explained by the example of a plane crash.

1 INTRODUCTION

Risk is not a physical variable which can readily be defined, compared and evaluated against standards. The acceptance or rejection of an activity which includes some risk requires a decision based on assessments made under uncertainty. This is difficult enough when it comes to personal choices in everyday life, which are often made in an intuitive manner. When it comes to policy decision concerning industrial safety, public health and environmental protection, risk assessment and decision-making can be problematic and, with a view to public debate and accountability, need to proceed in a methodical manner.

On the other hand, risk estimations are important in order to evaluate the safety of industrial facilities, in particular of nuclear installations, and to support decisions of the licensing authorities. Taking risk evaluations as an additional tool into account, it may be helpful to contrast the expected risk of the respective facility to be investigated with a comparable conventional risk. The conventional - and in general socially accepted - risk can then be used as a measure.

If it can be demonstrated that the risk of the investigated facility lies below - and, maybe, far below - the conventional risk, it can be assumed that sufficient precautions are taken with respect to the public in the environment of the planned facility. Otherwise, additional measures (for example constructional ones or changes of the industrial process or of the operational procedures) are necessary in order to ensure a sufficient design of the plant, and thus, the protection of the public.

The general procedure of such a radiological - conventional risk comparison will be discussed by comparing as an example the radiological risk from a plane crash onto a nuclear installation to the population in the vicinity of this facility with the risk of this population from plane crashs independent from the existence of the nuclear facility.

2 GENERAL PROCEDURE

As a basis for the comparison addressed above, on the one hand the conventional risk must be determined, i.e. the risk of secondary deaths from plane crash which exists in the region of the planned facility independent from the industrial installation considered. This risk has to be investigated for a collective of people living around the envisaged site. The radius of this area must be chosen in such a way that a sufficient large area with respect to the population distribution and the characteristics of the buildings is taking into account in a representative manner.

In a first step of a site analysis, the use of the area and the kind of buildings of the region to be investigated have to be determined, because possible fatalities due to a plane crash depend on the characteristics of the buildings. Therefore it is reasonable to subdivide the relevant region into seven building classes:

1. no buildings (field, wood, meadow, fallow land, water areas),
2. suburb built-up area with one-storied and two-storied buildings (single houses for one family),
3. area of a town with larger buildings,
4. village centre (one-to-two-storied densely built-up area),
5. town centre consisting of large apartment

houses, office buildings and companies,

6. industrial plants as well as

7. special local buildings and areas with an expected large number of persons (hospitals, schools, swimming pools, railway stations).

In order to get a realistic picture about the occurrence of a certain building class it must be taken into account that in practice a mixture of different building classes exists. For example, in the second class some buildings may have been constructed which must be assigned to the fifth class because of the number of storeys. Moreover, in some cases industrial plants are part of the town centre.

For each of these seven building classes, damages by plane crash and the corresponding frequency of occurrence must be investigated. The frequency of occurrence is determined by the occurrence of a plane crash in the relevant area, the fraction of the respective building class area of the total area, the percentage of the actual built-up area of the building class, the time period of public residence in the building class as well as the probability of an occurrence of a damage due to a plane crash onto the built-up area of the respective building class.

As a mean value for the Federal Republic of Germany, the frequency of occurrence of plane crashs with sufficient damage potential can be estimated as 10^{-10} per m^2 and year (GRS 1980). If no local accumulations of the air traffic in the respective region are recognised, the area-weighted frequency of occurrence of a plane crash in the environment of the plant site can be assumed as constant.

The risk from a plane crash can then be described as a cumulative complementary frequency distribution (ccfd) of expected fatalities in the population around the site.

Possible consequences of a plane crash onto the plant are determined by the existing danger potential of the facility. It should be pointed out that this danger potential may - depending e.g. on the radioactive inventory - be relatively constant in time as in case of a nuclear power plant or variable as in case of a storage building for radioactive waste or transportation of radioactive waste. In the latter case, an appropriate probabilistic modelling must be chosen.

In order to define a measure of the radiological consequences, in a first step the potential radiation exposures of people in the surroundings are calculated. These exposures result from the release of radioactive matter out of the planned nuclear facility. From this, in order to determine potential late fatalities collective doses are computed taking into account the population density distribution.

Expected late fatalities are assigned to the calculated collective doses by the use of dose-risk coefficients relating dose to expected late facilities.

Finally, the ccfd's of conventional and radiological fatalities are compared.

The procedure can be applied to accident analyses of different kinds of nuclear facilities, in particular analyses of rare events, transport accidents etc.. Moreover, the application of the explained risk comparison is not limited to nuclear aspects, this method can also be used to evaluate the risk potential of all kinds of hazards.

3 EXAMPLE OF A RISK COMPARISON

In order to show the procedure described above in practice, examples like the buffer hall of a repository for radioactive waste or the hall of an interim storage facility may be considered. If a plane crash onto such a facility is assumed, the possible damage spectrum ranges from plane crashs without any activity release up to such events where many waste packages are destroyed with corresponding activity release; this wide spectrum of different possible hits can be assigned to a limited number of damage classes like:

- indirect hit of waste packages which are temporarely outside of buildings by wreckage; a fire is assumed but without affecting waste packages and consequently without release,
- direct hit of waste packages outside of buildings; besides mechanical impact on the waste packages, an additional thermal impact from fire is assumed,
- direct hit of waste packages inside the facility by the plane taking into account additional mechanical impacts by the drop of equipment and the roof of the building as well as the burning of the kerosene; in this case, the largest damages have to be expected.

The assignment of events to the damage classes results from the configuration of the plane crash.

The release of radioactive material out of the investigated plant is then determined by the damage class and the release behaviour of the respective plant. In order to assess the spectrum of different releases depending on which part of the plant and of the radionuclide inventory are affected by plane crash, a large number of source terms must be generated by applying probabilistic safety analysis techniques.

A source term represents the released activities of individual radionuclides for the simulated plane crash configuration. The radionuclide-specific activities are determined by the activity content involved in the accident and the fraction assumed to be released into the atmosphere.

For the purpose of subsequent analysis of possible radiological consequences and their expected frequencies of occurrence the large number of source terms have to be reduced in a representative way to a limited number of source terms, so-called release categories. This can be achieved by ranking the original large number of source terms according to radiological importance. As an approximate but sufficient measure of the radiological importance of a source term, a so-called hazard index can be associated with each source term (Lange et al 1992).

This radiological hazard index of a source term is calculated by summation of the activity of the various radionuclides multiplied by nuclide-specific weigthing factors. The weigthing factors can be determined for each radionuclide by calculating the total effective dose resulting from unit release for standardized conditions, taking into account the exposure pathways inhalation, groundshine, ingestion and cloudshine. The weighting factors used are intended to be an adequate measure of the relative radiological significance of the individual radionuclides released in an accident.

To facilitate the analysis of environmental consequences, the large number of source terms must first be appropriately grouped into a limited number of source term groups. In a next step for each source term group a representative source term is determined designated as release category. The procedure to determine representative source term groups is described taking reference to Fig. 1 (Gründler et al 1991).

The source terms are then arranged in ascending order according to the radiological hazard index and in a next step the ccfd of the radiological hazard index as shown in Fig. 1 is constructed.

In a next step source term groups are formed by combining source terms with approximately equal hazard indices. This is done on the basis of the cumulative probability in a way that the range of radiological hazard indices of source terms having high hazard indices is not too large. This procedure is intended to assure representativeness particularly for the source terms resulting in higher radiological consequences. In the example illustrated in Fig. 1 the spectrum of source terms is grouped into 12 source term groups.

Finally, for each source term group a release category is derived. By taking an average of the activities of individual radionuclides of the source terms within a group, the radionuclide composition and activities of a release category are calculated. The probability of each release category conditional that a plane crash onto the plant resulted in a release is also recorded.

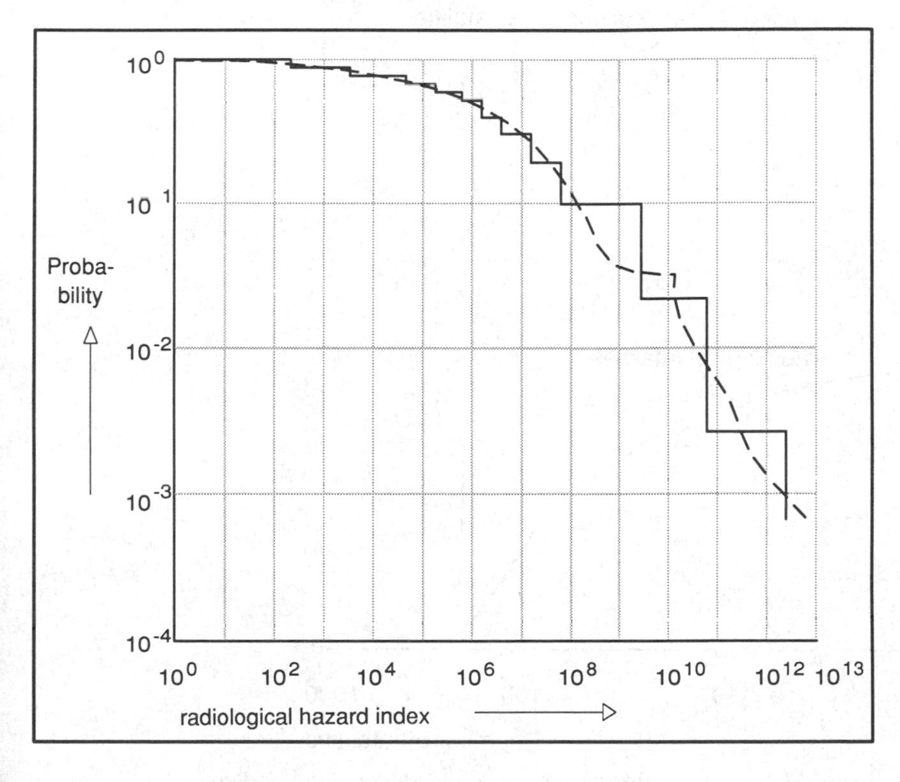

Fig. 1: Cumulative complementary frequency distribution of the radiological hazard index of source terms.

For each of the 12 release categories radiological consequences expressed as collective effective dose are calculated by using the computer code UFOMOD (Ehrhardt 1987). This accident consequence code takes into account the population distribution around the nuclear facility and the probabilities of different atmospheric dispersion conditions derived from local weather statistics. The results for the 12 release categories are then superposed to a joint cumulative complementary frequency distribution of collective effective dose or of associated late fatalities by taking into account the conditional probability of each of the 12 release categories. The final ccfd relating the expected frequency per year of radiological late fatalities is then immediately determined by the expected frequency per year of plane crashes onto the plant.

An example of such a comparison of radiological and conventional risks from plane crash is shown in Fig. 2. The expected frequency per year of fatalities from plane crash in the population around the nuclear facility is expressed as ccfd of conventional fatalities. This risk to the population exists irrespective of the planned nuclear facility. The radiological risk associated with possible plane crash onto the nuclear facility is also expressed as ccfd but of expected late fatalities from potential radiation exposures from activity releases. When comparing the conventional risk from plane crash with the radiological risk it has to be pointed out that in one case the end point of the ccfd are immediate secondary deaths from plane crash, in the other case expected late fatalities from radiation exposures assuming no countermeasures after the event.

The validity of such an approach depends on the parameters and boundary conditions for the calculation of the different risks. In particular, the radiological risk has to be determined in a conservative manner. This can be ensured by

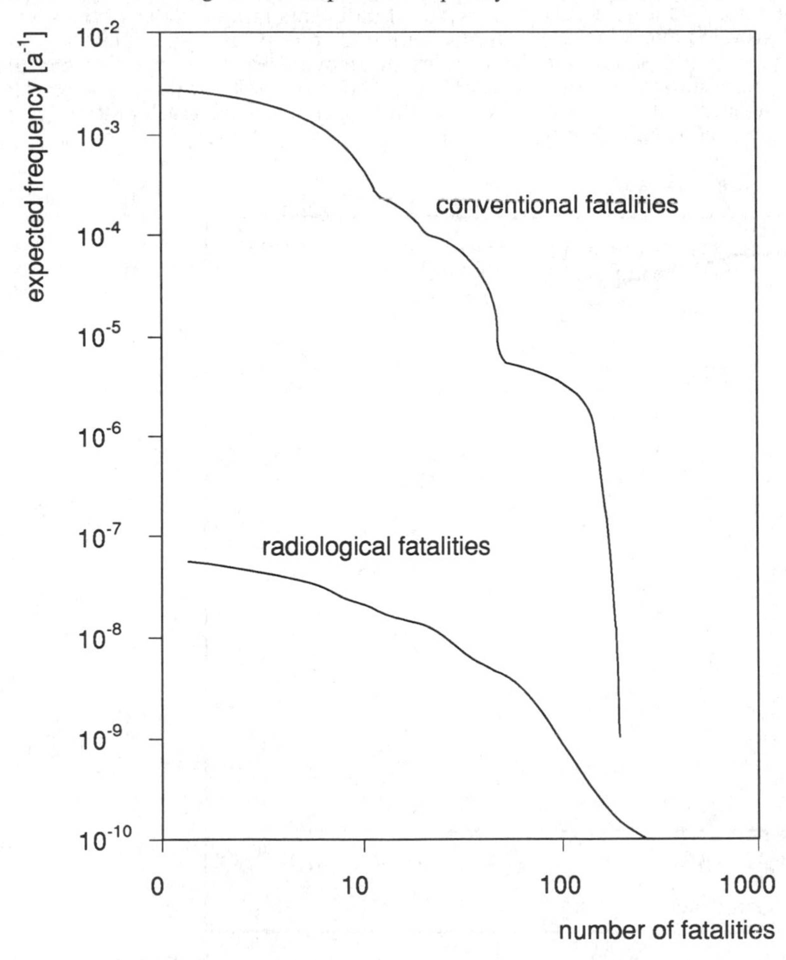

Fig. 2: Example of a comparison of the probability of conventional and radiological fatalities in case of a plane crash.

upperbound assumptions about activity release behaviour and by the procedure of calculating distributions of collective doses. One important parameter is the size of the area taken into account in the risk considerations.

On the one hand, the area must be large enough to cover the representative building classes in the neighbourhood of the planned industrial facility but not too large in order to avoid an increase of the conventional risk due to the enlargement of the investigated area.

On the other hand, the determination of the radiological risk is based on the population living in the considered area. Again, a conservative assumption of the size of the area is necessary.

This may lead in praxi to the fact that the size of the area which must be taken into account is different for both parts of the radiological conventional risk considerations.

From a comparison of the frequency distributions displayed in Fig. 2 it is evident that in this special case the conventional risk is many orders of magnitude higher than the radiological risk.

Due to the large distance between the two curves, the result of the comparison supports the safety evaluation of the planned facility and may be helpful in discussions with the public.

REFERENCES

Gesellschaft für Reaktorsicherheit (GRS) 1980. Deutsche Risikostudie - Kernkraftwerke, Fachband 4 - Einwirkungen von außen. Der Bundesminister für Forschung und Technologie (ed.), Köln, Verlag TÜV Rheinland.

Lange, F., D. Gründler & G. Schwarz 1992. Methods and Results of a Probabilistic Risk Assessment for Radioactive Waste Transports. In Proceedings of the 10th International Symposium on the Packaging and Transportations of Radioactive Materials, Yokohama.

Gründler, D., G. Philip & W. Wurtinger 1991. Probabilistic Source Term Determination. In R. Baker, J. Tuohy, L.C. Oyen, K.J. Lee, W.Z. Oh, (eds.), Proceedings of the 1991 Joint International Waste Management Conference, Seoul 1991, Korean Nuclear Society, Vol. 1, pp. 293-296.

Erhardt, J. 1987. The program system UFOMOD to assess the consequences of nuclear accident. KfK report - 4330, Kernforschungszentrum Karlsruhe.

Structural Safety & Reliability, Schuëller, Shinozuka & Yao (eds) © 1994 Balkema, Rotterdam, ISBN 90 5410 357 4

Acceptance criteria for high consequence risks: A critical appraisal

K.van Breugel
Delft University of Technology, Netherlands

ABSTRACT: There is a tendency that authorities use the risk concept as a basis for legislation. This tendency is evaluated in this contribution. In this evaluation some aspects of the traditional risk concept are discussed. The interpretation and valuation of both theoretical probabilities and risks are considered to be two major reasons for the development of a type of legislation which, to the authors' opinion, will result in a legitimation of hazards rather than an elimination of them. Aware of the fact that absolute safety is impossible, it is recommended not to tune legislation to what seems to be technically and economically achievable, but to a norm which reflects what ought to be or ought to happen. Early examples of "normative thinking" are mentioned and it is indicated how normative thinking can be made operational for a practical problem, viz. storage of hazardous waste.

1 INTRODUCTION

Our world seems to be a world that believes in figures. Figures would help us to valuate what we are doing and form an objective basis for decision making. Particularly the objectivity of figures seems to be one of the major arguments for the development of the Quantitative Risk Analysis (QRA) and, implicitly, for adoption of the risk concept. In some countries an increasing trend can be perceived that authorities apply the risk concept as an instrument to control industrial and environmental risks.

The impact of QRA and the risk concept can hardly be overestimated. Whether this impact is to be judged positive in all respect, and whether the risk-concept in its traditional form is always the most adequate tool to guarantee safety, is judged differently. In several articles the traditional risk concept has been criticized (Bomhard 1992, Reid 1989, Eibl et al. 1992). The critics refer to either the weaknesses, accuracy and reliability of the QRA or the way in which the results of QRA are used in decision making. Some of these critics will be briefly discussed and evaluated in the following. In this evaluation emphasis will be on the applicability of the risk concept in cases of Low Probability/High Consequence Risks and environmental problems.

2 THE RISK CONCEPT

In its traditional form the risk R is defined as the product of the probability of failure P{F} and the associated consequences C. The consequences, expressed in the number of fatalities or in monetary losses, form the basis for the calculation of the fatality rate per year and the potential losses per year, respectively. By checking whether the risk R does not exceed a certain predefined value an activity is considered acceptable or not. Tedious problems associated with the application of the risk concept refer to:

1. The reliability of predicted event probabilities and the interpretation of these figures;
2. The completeness and consistency of the consequence analysis and;
3. The problem of criteria setting.

2.1 *Event probability*

The calculation of the event probability, i.e. the probability of failure, requires a description of the materials properties and actions. The properties of most of the building materials are well known. This does not necessarily mean that it is always easy to describe these properties mathematically as the many discussions on "tail-problems" demonstrate (Melchers 1985, Sexsmith 1969). It should further be born in mind that for artificial building materials like steel and concrete, the material properties also reflect the characteristics of the manufacturing process. Thanks to the increasing simulation potential in computational materials science, we are able today to predict the standard deviation of a composite material like concrete as a function of variations in the major raw materials from which concrete is made. With the help of adequate simulation programs it can be made plausible that extremely low strength values are a function of the manufacturing process like poor workmanship and factors of that kind rather than of natural variations in the raw materials (Fig. 1, lower part). The frequency, with which the former factors may give rise to extremely low strength values, is hard to predict. Consequently the accuracy and eloquence of the lower tail of the strength distribution function is questionable.

Even more complicated than the description of the strength are the descriptions of the actions, particularly of accidental loads. Although quite an amount of statistical data is available about hazardous loads like industrial fires, impact, explosions and earthquakes, specification of extreme loads for a particular case is still more an art than a science. In case of non-physical actions, i.e. chemical reactions and nuclear radiation, specification of the actions is not less complicated (van Breugel 1992).

The foregoing implies, that caution must be exercised with respect to the reliability of the theoretical "frequency" with which low probability events are predicted to occur. Stating that those low probability figures should be considered as a degree of belief rather than a frequency does not solve this problem adequately. In section 2.3 this topic will be discussed in more detail.

2.2 *Consequences*

As far as the consequences are concerned the focus of attention in the traditional risk concept is on either the number of fatalities or financial losses. Other consequence aspects like political, social and cultural aspects, environmental impact and also genetic effects have hardly been paid attention to. A fundamental problem is how to value those aspects objectively, if this were possible at all. For attempts to account for these aspects reference is made to (Ditlevsen 1991). Most of these attempts are modifications of the traditional risk concept in the sense that the number of fatalities and financial losses are increased by inserting arbitrarily chosen coefficient with which the actual fatality numbers and losses are weighted. It is noticed that these modifications do not alter the basic structure of the risk concept!

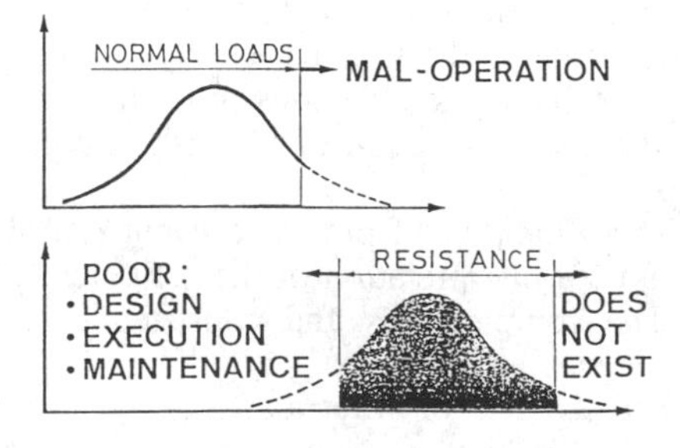

Fig. 1 Schematic representation of input for probability assessment.

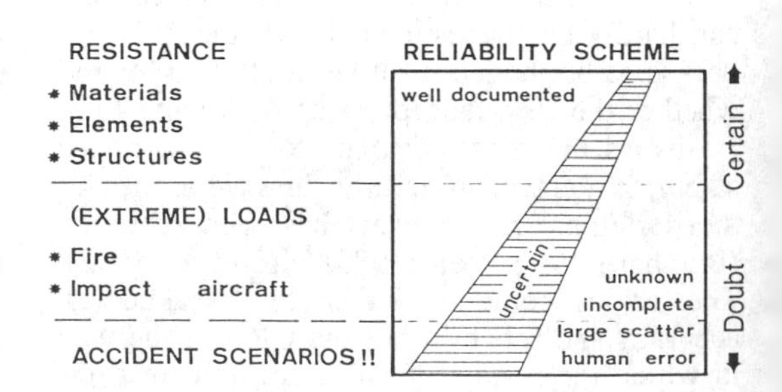

Fig. 2 Diagrammatical representation of degree of uncertainty in different input catagories of a riks analysis (van Breugel 1992).

An important aspect of consequence prediction refers to the completeness and consistency of the hazard scenarios which are taken into account. Compared to the inaccuracies and uncertainties in the material properties and actions, much less is known about the uncertainties in hazard scenarios except that they are greater (see Fig. 2). The scenarios considered in a risk analysis have been estimated to represent about 70% of the actual number of them (Goossens 1987). Subjectivity and the human factor have a major influence on the completeness and consistency of hazard scenarios and of consequence analysis (Koster 1987).

2.3 *Acceptance Criteria*

Among the problems associated with the application of the risk concept, that of fixing an acceptance criteria is certainly the most fundamental one. A risk criterion, for example an individual risk of 10^{-6} pp/py, is a value set by a qualified authority and is considered to be an acceptable risk people may be exposed to (Suurland 1987, Vrijling 1989). Justification of a risk-based criterion is based on a comparison of this criterion with other risks, for example the risk to which a person is exposed to in travelling.

By comparing a calculated risk with a defined criterion this risk is implicitly considered to be an absolute risk, i.e. a risk with a frequentist meaning. If not, a comparison of theoretical risks with a predefined criterion that - and this must be emphasized - does have a statistical basis would become meaningless! Here we come across a major problem, viz. the discrepancy between the interpretation of theoretical risks by politicians and criteria setting authorities on the one hand and scientists on the other hand. This interpretation problem rises the question whether the risk concept is really capable to keep technical activities and developments on the right tracks. Application of the risk concept presupposes that all consequences can, finally, be expressed in figures and that on the basis of a comparison of figures well considered decisions concerning the acceptability of risk bearing activities are possible.

3 ACCEPTANCE CRITERIA: A CRITICAL APPRAISAL

3.1 *Rational or confusing*

One of the major advantages of the risk concept would be the rationalization of the decision process. There is room, however, for the statement that risk figures confuse the decision process rather than rationalize them. Let us, for example, consider the topical question whether a nuclear power plant shall be designed for an aircraft impact or not. The probability of aircraft impact is about 10^{-7} per year (van Breugel 1992). Based on this figure the individual risk would be, for an arbitrarily chosen location, 10^{-9} pp/py. Compared with a single-valued risk criterion of, say, 10^{-8} pp/py, there seems to be no reason to specify an aircraft impact since this loading can be ignored without coming in conflict with the specified risk criterion. The facts that in case of an aircraft impact an area of 3,000 square kilometres may be rendered unfit for agriculture for many decades, some people say hundreds of years, and that hardly predictable genetic damage may occur with consequences for several generations remain outside the decision process as long as we stick to the single-valued risk criterion. In this and similar cases the risk criterion is confusing and clouds the decision process rather than rationalizing it. The forgoing brings us to the statement that the risk concept acts here as a vehicle for introduction of risk-bearing activities, which should have been judged unacceptable for may be a number of reasons which are not, and cannot be, explicitly and adequately accounted for in a risk concept that operates with a single-valued risk criterion.

3.2 *Objectivity*

Objectivity in decision making would be another advantage of the risk concept. The supposition of objectivity is based on the idea that figures, i.e. fatality rates or other risk parameters, are mutually comparable. A classic objection against this procedure is that voluntary and non-voluntary risks are not comparable by just comparing hard risk figures. This problem can easily be solved by defining the acceptance level for non-voluntary risks one to several orders higher than for voluntary risks (Vrijling

Table 1. Individual risks according to different authors (Melchers 1985, Lees 1980, author).

Activity	Individual risk pp/py
Voluntary risk:	
car driving	20.10^{-5}
air travel	2.10^{-5}
smoking	500.10^{-5}
alpine climbing	4.10^{-5}
contraceptive pill	2.10^{-5}
Non-voluntary risk:	
meteorite	6.10^{-11}
lightning	1.10^{-7}
structural failures	1.10^{-7}
construction work	2.10^{-4}
abortion	1.10^{-1}

1989). If we subsequently consider the non-voluntary individual risks listed in table 1 and if it were logical and rational to strive to a specific and acceptable level for all non-voluntary risks, the primary task of society and criteria setting authorities would be to reduce the extreme high risk figures for abortion. Probably this is the most penetrating example showing the inadequacy of figures to give guidance for the decision process in an unambigeous way. Figures, however hard they may be, are ignored as soon as they bring people in conflict with their conception of the world and may become the victim of their own concept. In conclusion, figures do not tell the whole story. They are only useful within the boundaries of a well-defined model (Ditlevsen 1991, Häfele 1990), but are, therefore, not necessarily applicable for decision making in the real world.

3.3 *Multi-aspect approach*

In the foregoing there has been pointed to some fundamental shortcomings of the risk concept, particularly when it operates with a single-valued acceptance criterion. This situation can be substantially improved by moving to a multi-aspect approach. Each consequence aspect (see section 2.2) is than judged by appropriate criteria which refer to the aspect in view. While doing so the multidisciplinary character of the majority of the man-made hazards is emphasized, which offers the possibility to evaluate risk-bearing activities in a more balanced way. It is noticed, however, that adoption of a multi-criteria procedure does not imply a fundamental change in the basic structure of the risk concept. It remains a concept in which risk bearing activities are judged by comparing the theoretical risks with arbitrarily fixed criteria.

3.4 *Ethics*

Defining acceptance criteria, i.e. risk criteria, is considered to be a matter of ethics. It is noticed, however, that ethics are involved already in the decision to adopt the risk concept for judging risk bearing activities! This decision implies, that it is considered sufficient to compare the theoretically obtained risk values with criteria which are (estimated to be) accepted by society. These criteria reflect, in essense, what is considered to be technically and economically achievable. The adoption of criteria of this type means an implicit legitimation of that what has been achieved in past decennia, including its inherent risks. It disregards, however, a fundamental reflection on the nature and the roots of our technique-shaped society. "Normative thinking" is replaced by mathematics. Although a mathematical aspect cannot be ignored in normative thinking, normative thinking is more!

4 NORMATIVE THINKING

It is not the privilege of a modern society to envisage the reality of man-made risks and the fact that the environment needs protection. The old Babylonian building code, issued by King Hammurabi in 2200 B.C., contains the well known clause: "If a builder builds a house for a man and not make its construction firm and the house which he has built collapses causing the death of the owner of the house - that builder shall be put to death". In the Jewish book Deuteronomium we read, that an owner was obliged to provide the flat roof of his house with a parapet. If not, and somebody would fall from the roof and died, his death was considered bloodshed. Further, a strong protest was heard against damaging of the environment by unchecked cutting of trees in Lebanon for building purposes. Obviously the environment has its own rights! In times of war - even then! - is was forbidden to destroy fertile land of the enemy so as to make it unfit for growing crops. Finally we mention to the old Chinese saying: "Don't start a fire that you cannot stop". We may not start a process, of which the consequences are unpredictable and uncontrollable.

The keytone in the quoted regulations and sayings is, that the increase in welfare and prosperity, and even making war!, was not considered illegal, but might not be realized at the cost of an other persons life or the environment. In these clauses the notion of an "absolute norm" is found. This notion is still heard in the more contemporary clause in the Dutch constitution: "The authority's care is focused on the habitability of the country and the protection and improvement of the environment". This type of legislation is identified here as "legislation with a high profile". It reflects a norm. That what "ought to be or ought to happen". This high profile of legislation is given up as soon as acceptance criteria are tailored to what is technically of economically achievable. With setting a non-zero risk criterion we have not defined a norm, but only the margin between the norm and that what is considered to be just acceptable and achievable. The non-zero risk criterion defines the price a society accepts to pay for the preservation of the present level of prosperity without a critical reflection on the legitimacy of this level and the way along which this level has been reached!

4.1 *Unrealistic*

It would be unrealistic to speak about absolute norms and zero-risk criteria in a world in which it can easily be demonstrated that, with imperfect people and imperfect technique as input, absolute norms and zero-risk criteria can never be met. There is a sympathetic, but at the same time a confusing, or even haughty, element in this reasoning. The sympathy concerns to the acknowledgement of imperfectness of men and technique and, consequently, the inability to guarantee absolute safety. The confusing element in this reasoning is that people, who don't want to take responsibility for absolute safety, do take responsibility for criteria which actually reflect the price, in terms of fatality rates and environmental impact, that is considered acceptable for preservation of the present level of prosperity. From which authority one has got the mandate for fixing this price is left an open question, if it is considered a question at all. Whether it is really rational to refuse to take responsibility for an absolute norm, but to accept to take responsibility for a certain number of fatalities and environmental damage, is another open question. If it are statistical facts, reflecting the imperfectness of men and technique, which are considered to provide us with a rational basis for acceptance criteria in the form of individual risks, then we might wonder whether we will ever arrive at a fundamental discussion about the direction in which our industrialized society shall move in the future!

4.2 *A shift of the points in legislation*

The question arises whether it would really make sense to identify a non-zero risk criterion as the margin between what is considered the norm and what is technically and economically achievable. At first sight it might even seem to be a play upon words without any benefit for the practice.

With respect to these remarks it is noticed, that with setting a non-zero risk criterion we are defining a criterion that can be satisfied. Satisfying criteria implies psychological satisfaction. With a non-zero risk criterion we provide ourself with a certificate of good behaviour. But, meanwhile, the high profile of legislation encountered in the classic "not at the cost of another persons life or the environment" is replaced by "if the price is not too high". The keytone of legislation is a complete different one. This is not a play of words, but a shift of the points! We make for a society, in which man is removed from the centre of it to make place for technique and the industrial complex. Technique is no longer for the benefit of people, but, reversely, people live for the maintenance of a technique-stamped world. The crucial point here is not that the existence and the adoption of technique and risk bearing activities would be immoral because it appears to be impossible to avoid accidents and fatalities. The crucial point is, that the focus of attention in risk-based legislation is, in essence, no longer the protection of people and the environment, but the preservation of a certain level of prosperity in a technique-stamped society, which is considered to be justifiable provided that the price is not too high. With legislation of this type the thread, emanating from risk bearing activities, is not controlled but, in essence, sanctioned. Hazards are given a legal basis. In legislation, however, one shall find the norm, that what ought to happen and ought to be. In that way the appeal-function of the norm remains. This appeal-function of the norm is more than a subjective, emotional and non-rational voice, but may turn out to be a more rational basis for decision making than hard figures and is, in that way, certainly of practical importance.

4.3 *Two Worlds*

In the discussion about the role of figures, rules, norms and criteria, Staudinger (1979) has introduced the "concept of two worlds". In these two worlds the meaning of the word "norm" is different. One meaning is "aim" and "direction". The other meaning is that of "rule", according to which people have to act.

The norm as aim and direction refers to welfare and prosperity of mankind. Not only the parts, but the total is considered. In that sense the norm is holistic. The norm expresses what ought to happen and ought to be. It defines the difference between good and evil. This norm concern the world in which ethic decisions are made. This world is identified as the "primary world".

Although it is quite obvious that the contents of such a norm depend on factors like religious beliefs, local traditions, education and political and economic constellation, this norm is evidently different from the norm as rule. The norm as rule refers to the means and methods utilized to achieve certain aims. It concerns regulations for methods, procedures, acts, etc., which are in force within a certain discipline. This norm is "method-bound" and "method-oriented". It exhibits the same constraints which are characteristic for the discipline to which it belongs. It concerns parts of the total. Not men, but acts of men are considered. This norm is technique-oriented. It refers to the world, which is the model of the primary world: the "secondary world".

What happens today is that norms which belong to the secondary world are used for decision making in the primary world. Also the calculated individual risk is a method-bound and method-oriented operator which, just like the risk concept, belongs to the secondary world. With respect to the application of the risk concept in the primary world Häfele (1990) states: "Der bisher in Rede stehende Risikobegriff (the individual risk; vB) ist für die Diskussion technischer Systeme im aussertechnischen Bereich nicht oder allenfalls bedingt geeignet". What Häfele explains here is, that a one-dimensional risk criterion has not the quality so as to make it an adequate parameter to communicate risks interdisciplinary. If we nevertheless do that, the primary world is going to exhibit all the (negative) features of the secondary world. It becomes impersonal. The model of the primary world is taken to be the world which it represents. This means a shift of the points. Not the nature of developments and facts, but merely the functional aspects of them are paid attention to and become major factors in decision making.

5 ALTERNATIVES

5.1 *Conceptual Thinking*

In those cases where the traditional risk concept is inadequate for judging risk-bearing activities alternatives should be given. Without giving up the positive features of the risk concept a first step in the direction of an alternative should be that acceptance criteria are not merely tailored to the figures that can be produced by a risk analysis, but to all the relevant consequence aspects irrespective of the probability of occurrence of a catastrophic event. Emphasis on consequence aspects requires a multi-disciplinary judgement by the parties involved in a particular activity. There is no doubt that in many cases consequence control instead of risk control will result in higher initial costs. If a society cannot afford the costs for adequate protective measures it shall refrain from the technology or activity in view (Bomhard 1992).

With the concept of consequence control the focus of attention is on conceptual thinking. It exhibits a strong deterministic component with emphasis on passive safety and "system technology" rather than on active safety and materials technology (Bomhard 1992, van Breugel 1992).

5.2 *Storage of hazardous waste*

For an example of conceptual thinking and system technology the attention is drawn to the waste problem. Montague (1984) states that the majority of traditionally designed dumps for permanent storage of (hazardous) wastes is leaking. Either due to mechanical or chemical action the lining systems, generally synthetic liners, are damaged and leak. Liners which are not leaking yet are expected to leak in the near or far future, as with elapse of time thermodynamic instability will cause degradation of liner systems. A solution merely based on materials technology must be considered inadequate (Bomhard 1992a, van Breugel 1992). Whether leakage will result in pollution of the environment will depend on the local absorption capacity of the biosphere. Irrespective of the probability of leakage it shall be investigated whether the consequences of possible uncontrolled leakage are acceptable or not. If not, storage concepts shall be designed which do not prodominantly rely on materials technology, but on "system technology" (Bomhard 1992). A system, built according to the system technology concept, is characterized by its:

1. Inspectability;
2. Controllability;
3. Repairability;
4. Renewability.

A storage systems that meets these criteria is the double-walled storage system shown in Fig. 3. The primary container is inspectable and controllable. If due to any unforeseeable reason leakage should occur, the primary container is repairable and, in the ultimate case, renewable. An example of a renewal operation is shown schematically in the bottom part of Fig. 3.

Adoption of a double walled storage system will, of course, result in an increase of the investment costs. These costs, in a normative assessment of the waste problem, shall not in the first place be compared with the cost of cleaning of the soil in case of leakage, although this might already result in the decision in favour of higher initial investments, but must be considered justifiable because of the fundamental norm that activities may not be performed at the cost of the environment.

6 CLOSURE

In a modern industrialized society the presence of man-made hazards is unavoidable. To control these hazards authorities are inclined to adopt the risk concept, even a single-valued risk criterion, as a basis for legislation. An attempt was made in this paper to demonstrate the inadequacy of risk-based legislation to protect people and environment. Particularly in case of Low Probability/High Consequence risks it is considered evident, that the consequences can not be described adequately in

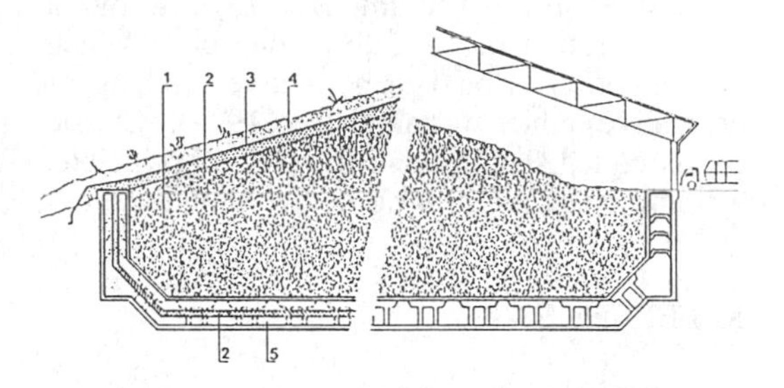

legend
1 Waste
2 Mineral sealing
3 HDPE liner
4 Vegetation layer with drainage
5 Filling of inspection space with environmetally neutral material (optional)

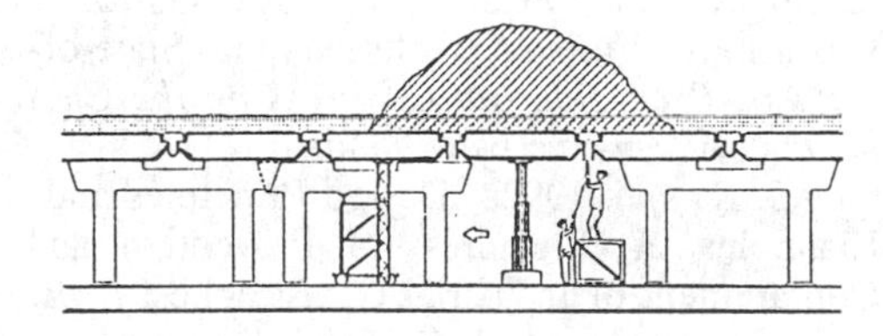

Local stabilization of waste above elements to be replaced

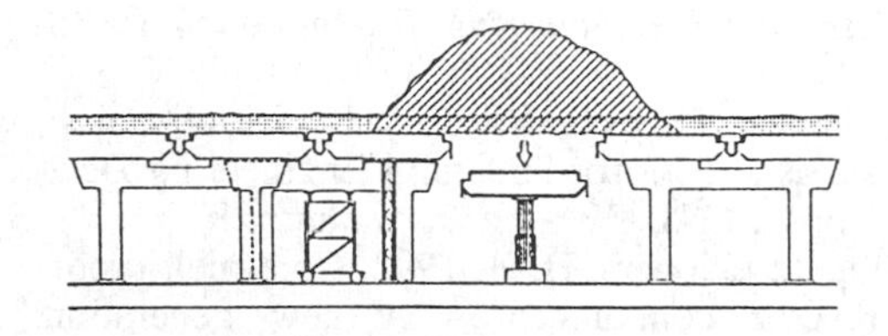

Removal/replacement of damaged slab element.

Fig. 3 Double-walled storage system for hazardous waste. Inspectable, controllable, repairable and renewable (after Rudat 1992).

quantitative terms (see also Chakraborty 1991). In those cases - in essense in all cases, also in those where the consequences are relatively small - the consequences require normative contemplation. Normative thinking, that is narrowed to a comparison of risk figures, is inappropriate as it ignores the fact that reality is more than what can be measured and counted and would keep us away from a fundamental and real normative reflection on modern developments in technique and technology.

There seems to be ample room for the statement that for judging of complex risk bearing activities we do not require more sophisticated procedures for generating risk figures, but a reconsideration of the safety concept in which due attention is paid to normative thinking or normative ethics (Chakraborty 1992). Classic forms of legislation seem to offer, in principle, a sound basis for such normative thinking.

REFERENCES

Bomhard, H. 1991. Prevention and containment of large-scale technology-related accidents - Shell and spatial structures as protective and safety measures. *IASS-Bulletin*. Vol. 32, no. 3: 138-148.

Bomhard, H. 1992. Concrete and Environment - An introduction. *Proc, FIP-Symposium:* Vol. 1. 51-69. Budapest.

Breugel, K. van. 1992a. Storage Systems for Hazardous Waste - Evolution and State-of-the-Art. *Proc. 9th. Int. Congress on the Chemistry of Cements:* New Delhi.

Breugel, K. van. 1992b. Design Principles and Examples of Structures for Prevention and Containment of Industrial Catastrophes, *Proc. FIP- Symposium:* Vol. 1, 71-90. Budapest.

Chakraborty, S. 1991. Ethical and Social Aspects in Comprehensive Risk Assessment. *Structural Engineering International* 1: 38-41.

Ditlevsen, O. 1991. Bayesian Decision Analysis as a Tool for Structural Engineering Decisions. *JCSS, Working Document:* 28.

Eibl, J., Cüppers, H.-H. 1992. Core-melt-proof reactor containment - A new generation. *Proc, FIP-Symp.* Vol. 1: 129-147: Budapest.

Goossens, L.H.J. et al. 1987. Accident Sequence Precursor Methodology Using Generic Operating Experience to Evaluate Risky Plants. *IFHP-IULA Symp. Prevention and Containment of Large-Scale Industrial Accidents:* 55-62. Rotterdam.

Häfele, W. 1990. *Energiesysteme im Uebergang.* Landsberg/Lech: Poller AG & Co., Buchverlag.

Koster L. 1987. Approach to Safety by Industry. *IFHP-IULA Symp. Prevention and Containment of Large-Scale Industrial Accidents:* 121-126. Rotterdam.

Lees, F.P. 1980. *Loss Prevention in the Process Industry*. Vol. I,II. London: Butterworths.

Melchers, R.E. 1985 *Structural Reliability: Analysis and Prediction.* John Wiley & Sons.

Montague, P. 1984. The limitation of Landfilling. *Beyond Dumping*, London, 3-18.

Reid, S.G. 1989. Risk Acceptance Criteria for Performance-Oriented Design Codes. *ICOSSAR'89*. San Francisco. 1911-1918.

Rudat, D. 1992. Design Principles and Examples of Structures for Treatment and Containment of Hazardous Waste. *FIP-Symposium, Budapest*, Vol. 1, 91-128. Budapest.

Sexsmith, R.G., et al. 1969. Limitations in Application of Probabilistic Concepts. *ACI-Journal*, 823-828.

Staudinger, H., 1979. *Chance und Risiko der Gegenwart - Eine kritische Analyse der wissenschaftlich-technischen Welt.* Paderborn: Inst. für Bildung und Wissen.

Suurland, J.A. 1987. Risk criteria and their use in environmental regulations. *IFHP-IULA Symp. Prevention and Containment of Large-Scale Industrial Accidents:* 83-88. Rotterdam.

Vrijling, J.K. 1989. Some considerations on the acceptable probability of failure. *ICOSSAR'89*: Vol. III, 1919-1925. San Francisco.

Structural Safety & Reliability, Schuëller, Shinozuka & Yao (eds) © *1994 Balkema, Rotterdam, ISBN 90 5410 357 4*

How to build and exploit overstressed tests to improve the knowledge of reliability and reduce development

L. Delange & T. Le Fèvre
Rams Group, Société Européenne de Propulsion S.A., Forêt de Vernon, France

ABSTRACT : Growing competition within the civilian launch market has created the need to develop a more reliable product in less time and at a lower cost.: for example, the Ariane 4 reliability objective is 0,90 while Ariane 5 objective is 0,98, the launcher being 10 % less expensive.
To reach this goal on the Ariane 5, an innovative approach has been used in the design and exploitation of the test program for the Vulcain engine.
The objective has been to detect and correct failure modes at the earliest possible point through extensive use of overstress testing.
This has resulted in :
a) a significant reduction in the amount of development hardware and number of test required,
b) the capability to calculate demonstrated reliability as a function of the test conditions,
c) a hierarchization of observed failure modes permitting to optimize the redesign effort,

1. INTRODUCTION

The Ariane 5 launcher is designed to a reliability target an order of magnitude higher than that of the Ariane 4. This requires a reliability, of 0,9946 for the Vulcain engine being developed by SEP. The demonstration of this reliability value using the binomial law would require in the order of 5 flight–equivalent missions to be demonstrated on each of 170 development engines.
In today's economic environment, the reliability goal must be demonstrated using a reduced number of engines and tests. This can be accomplished by employing advanced strategies in designing the test program, and in quantifying the reliability. These are based on an extensive use of overstress testing, the precise exploitation of failure modes found, and demonstrated reliability growth.

2. DESIGN OF OVERSTRESS TESTS

A program of overstress testing is used to demonstrate a high rate of reliability growth during the development program. While these tests increase the risk of hardware failure, the early detection and correction of inherent failure modes are the key elements in reaching the reliability objective in the most efficient manner.

The basic approach consists of 3 parts :

a) Failure modes not time–dependent, e.g. fragile rupture, start and shutdown failures.

   Margins are demonstrated by testing at conditions well in excess of the normal regime : at 5 standard deviations for example.
   Tests are first made varying single parameter at a time, to determine their relative influences. Then, interactions are investigated using tests in which multiple parameters are varied, using design–of–experiment techniques.
   Analysis of test results is performed using the stress–resistance method applied to failure detection thresholds and to the demonstrated

margins. The result is the estimated failure probability, based on the number of tests and the overstress coefficient. This probability is used for the basic decision to accept or reject the design for flight

b) Time dependent failure modes in the normal flight domain

Testing is performed for extended duration at operating levels within the normal domain with the objective of inducing successive failures. Upon the occurrence of a failure (or incipient failure), the component is replaced and testing continues until the appearance of successive failure modes.
Analysis of the failures provides the basis for defining Survival Laws for successive modes. These laws permit assessment of the acceptability of the design for flight, or, whether a modification of the design, or life-cycle duration is required.

c) Time dependent failure modes outside the normal flight domain

Long-duration tests are performed at overstress conditions in order to identify modes which can occur when the flight engine is operating in a low probability regime. Different complete engines are tested at different average operating conditions (e.g. thrust chamber pressure and mixture ratio) and beyond limit conditions.
A regression analysis is performed on the observed life duration values, from which one can generate a model of the engine survival law as a function of the flight conditions.

The three techniques described above are complementary methodologies for determining the limits of the hardware, see fig.1

The effectiveness of the overall approach relies heavily on the quality of the FMECA

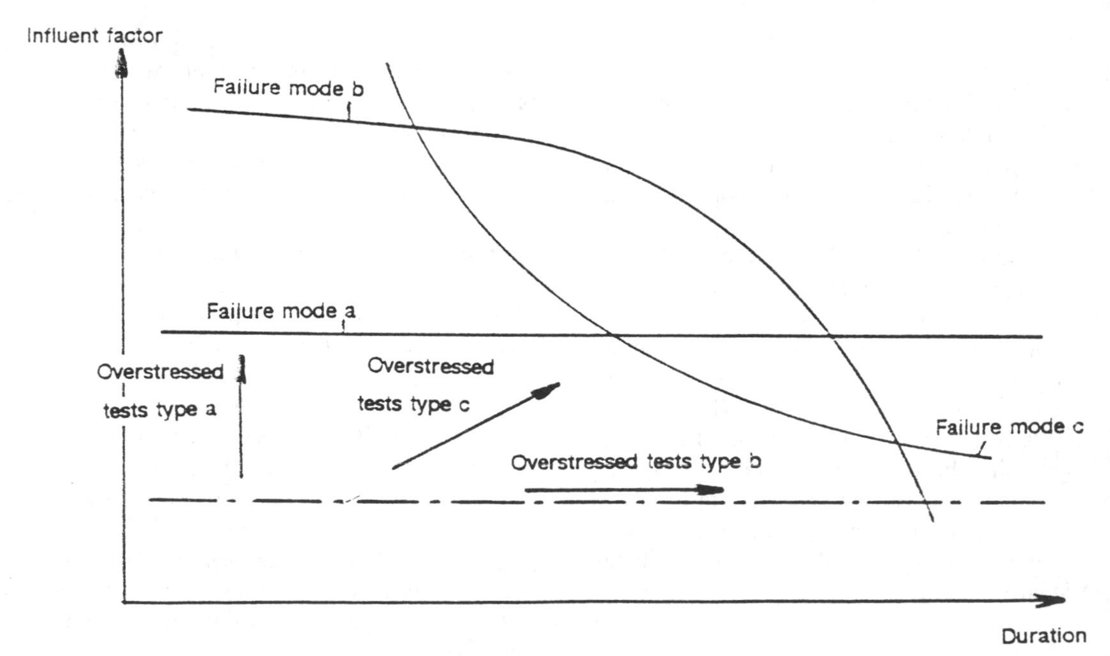

Legend : stress in nominal conditions
: resistance for a hardware failure mode

Fig. 1 : three methodologies for overstressed tests

and the risk analyses, which identify critical failure modes and influence factors. The preventive analyses are essential in organizing the tests into a hierarchy, taking into account past technical experience.
All this information can be processed in a reliability calculation.

3. RELIABILITY ESTIMATION

The calculation of reliability is based on 3 contributions :

a) No failures during test (R_{NO})

The binomial law is applied assuming no failures will occur during the test program. The probability value calculated, R_{NO}, is a function of the number, and accumulated duration, of tests ; it permits definition of the minimum number of tests at the beginning of the development phase.

b) Failures during test (R_O)

The voluntary overstressed test stragegy will lead to some incipient or destructive failures. The careful exploitation of these failures will permit to understand the failure mode mechanism.

For each failure mode observed, a reliability model can be derived, describing the life and failure probability as a function of certain influent parameters (operating conditions, dimensional characteristics, external factors).
This concept is illustrated by the example given in section 5 : a rocket engine thrust chamber life duration is given as a function of mixture ratio, chamber pressure and test duration.
For each failure mode, i, occurring during test, a probability of occurrence, Pi, is calculated as a function of the influent parameters.
The global reliability, Ro, is then computed for all observed modes by :

$$R_O = (1-P_1) \times (1-P_2) \times \ldots \times (1-P_N)$$

$$R_O = \prod_i (1-P_i) \qquad (1)$$

c) Known hardware limitations (R_L)

Past experience indicates design limits and risks for which relevant margins have to be implemented in the design.
Typically these margins are justified in the design files.
It is then easy to evaluate the risk of outpassing the limitation during nominal functionning given the strength and stress values and dispersions.
This gives, for each limitation, a probability of tresspassing the limitation, P_J, which is assumed to be the probability of failure of the hardware given the stress conditions.
The global reliability, R_L, is then computed for all known limitations.

$$R_L = \prod_J (1-P_J) \qquad (2)$$

d) Total Reliability Estimation

The total reliability, R, of the item can then be estimated as :

$$R_{NO} \times R_O \leq R \leq R_O \times R_L \qquad (3)$$

Where $R_O \times R_L$ is the most favourable estimation based on all what is known, assuming no failure mode has been ignored.
Where $R_{NO} \times R_O$ is the unfavourable estimation based on what has been observed, assuming that some "random" failures have been ignored in the analysis.

The total reliability is, then, a function of :
- the number of tests and accumulated duration
- the operating point chosen
- the design margins implemented

If the calculated reliability does not satisfy the specified requirement, it is necessary to modify either the design, in order to eliminate failure modes, or the operating point, in order to reduce the failure rate of predominant modes.
The validity of the reliability calculation depends on the physical understanding of the failure modes and on the degree to which they

are represented by the model.
Equally, it is necessary to test the statistical validity of the hypothesis, and to assure that the influence factors truly have the assumed effect on the physical phenomenon.
Similar to some Bayesian approaches, "a priori" information can be integrated into the calculation of occured failure modes probability (R_O) or into the calculation of limitation tresspassing probability (R_L). Thus past experience is taken into account in the calculation.

4. ANALYSIS OF OVERSTRESS TESTS SURVIVAL LAWS

The use of the Weibull laws in connection with failure incidents is well recognized by statisticians. During development of the HM7 engine for the third stage of the Ariane 4, it has been demonstrated that the incidence of cracks in the combustion chamber follows the law : (see fig.2) :

$$R = \mathrm{Exp}\,(-N/N_C)^{\beta} \qquad (4)$$

With : N = Number of cycles
Nc = Mean number of cycles before first crack
β = 2, 1 shape parameter

Since the Vulcain combustion chamber is similar in concept to the HM7, it is likely to follow a similar law. This assumption can be refined by taking into account influent parameters such as combustion pressure (PC), mixture ratio (RM), and test duration (T).

The added precision from this refinement is equivalent to that from increasing the "β" parameter. By better describing the process,

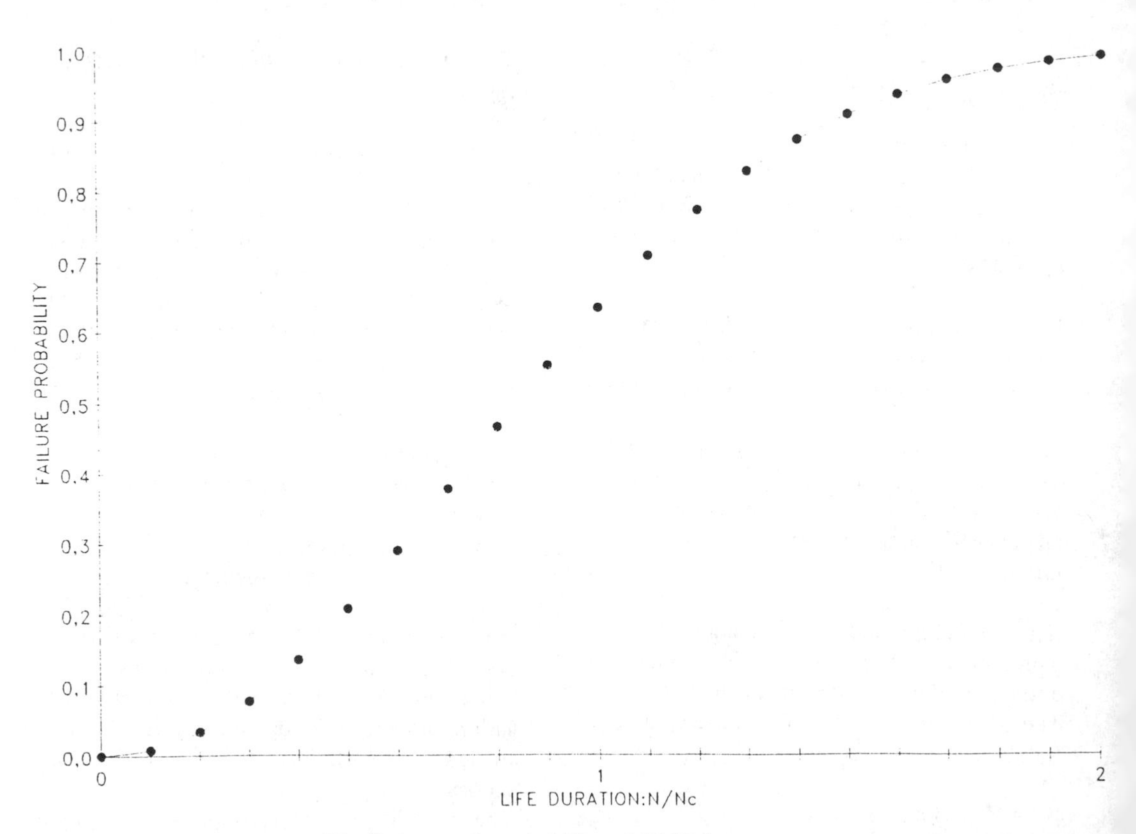

Fig. 2 :1st crack probability of HM7 thrust chamber at reference working conditions.

the estimated reliability is improved, and confidence in the success of the ultimate mission is increased.

Thus we exploited data coming from a thrust chamber tested for long duration in overstressed conditions.

Fig 3. displays the results of a regression analysis performed on these data. Although not optimized by the test method described above, the reliationship between cycle life and the pertinent influent factors is shown. The regression analysis gives the number of cycles before failure N, as a function of the quoted influent factors :

$$N = a + b\,(PC) + c\,(RM)^n + d\,(T)^m \qquad (5)$$

with : a,b,c,n,m, = constants
PC = Thrust Chamber Pressure
RM = Mixture Ratio
T = Test duration

This law explains 85 % of the variation observed in life duration. The residual scatter permits evaluation of the failure probability for a flight type mission (N = 5 cycles)

All these calculations for all observed failures and known limitations can be computed with equation (1), (2) and (3), thus giving a complete reliability figure as a function of working conditions.

Fig.4 shows the whole engine reliability estimation as a function of pressure of combustion (PC) and mixture ratio (RM), all other influent parameters being in reference conditions.

Past development programs can give precise knowledge of the behavior of a hardware item with respect to a given failure mode, and the parametric survival law. This can provide a

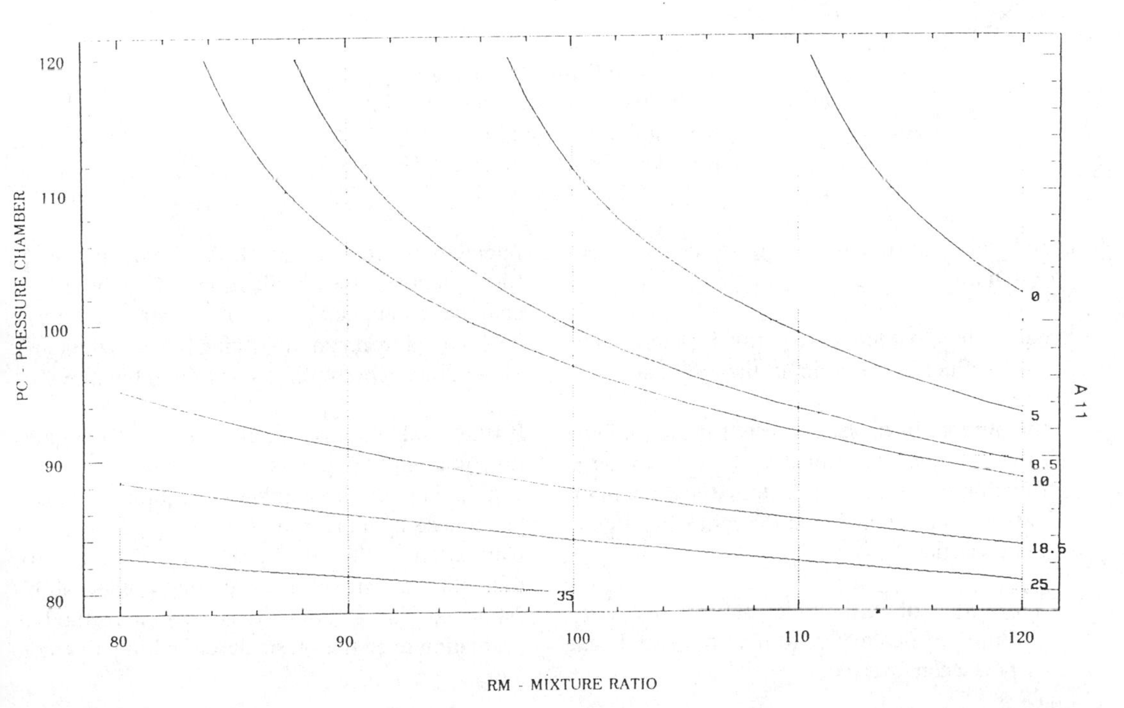

Legend : PC = Chamber Pressure in %
RM = Chamber Mixture ratio in %

Fig.3. Thrust Chamber life duration in cycles as a function of pressure of combustion (PC) and Mixture Ratio (RM).

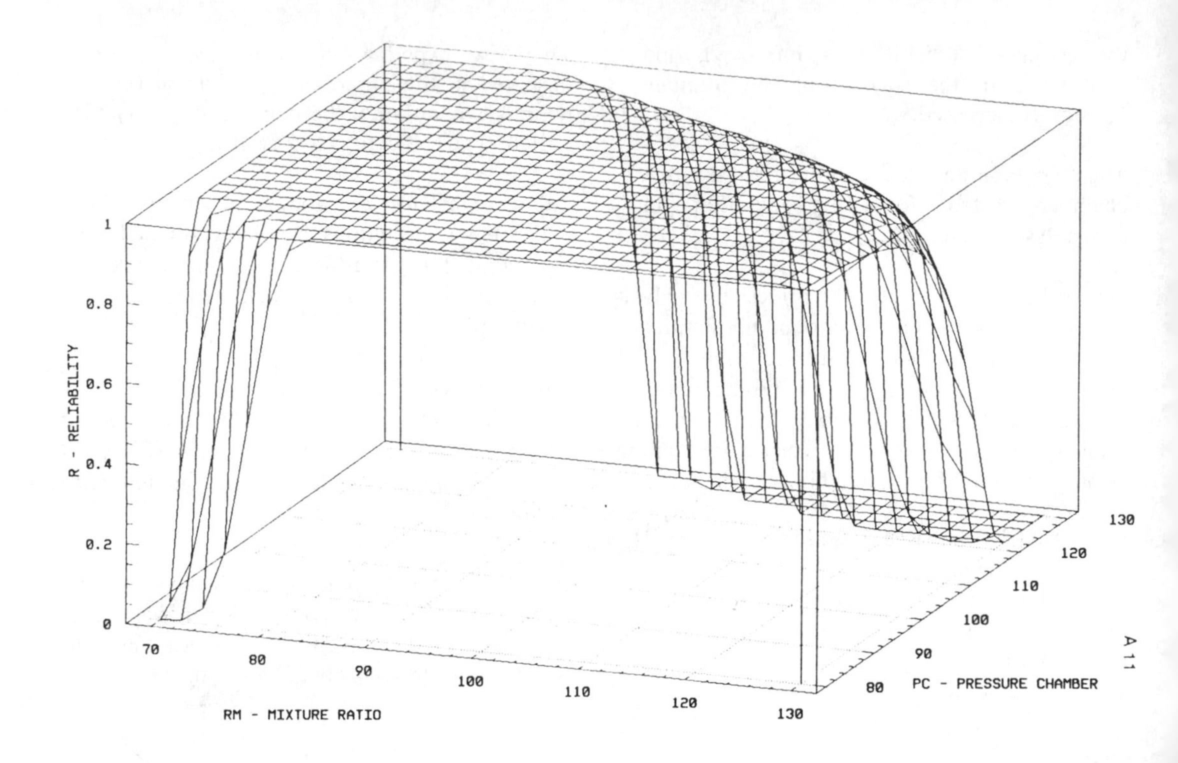

Fig. 4 : whole engine reliability estimation
Legend : R = Whole engine estimated reliability
PC = Chamber Pressure in %
RM = Chamber Mixture ratio in %

useful tool for the design of a development program :

a) parameters to control are defined. Determining their influence is the aim of the test plan.

b) the number of overstress tests for each failure mode can be determined with certainty as a function of the influent factors, the domain of variation considered, and the reliability figure to be verified.

c) the order of magnitude of the necessary variation of parameters to obtain a significant response can be fixed.

This enable design of a minimum experiment to meet a development requirement. By this approach, the use of fractional plans is possible and risks linked to unknown or biased interactions are minimized. As shown in fig. 5, for example, the evaluation of combustion chamber endurance by use of overstress tests can be obtained by a plan involving 3 parameters in 2 modes, thus representing, 4 tests combinations.

If the evaluation is performed on 4 engines functionning at points compatible with the engine domain, the number of required tests can be reduced by a factor of 2, in comparison with a plan based on the referenced point. Additionally reliability at different functionning points can be estimated. This example illustrated the method of evaluating overstress tests described in paragraph 3 c).

Conversely, if it appears that complementary objectives must be included during the life-evaluation testing, "soft" (non-overstress) operating points can be defined.

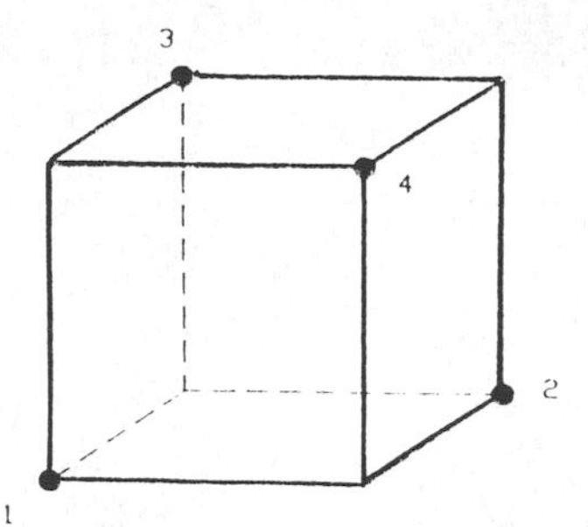

≠OF TESTS	X1	X2	X3
1	–	–	–
2	+	+	–
3	–	+	+
4	+	–	+

Fig.5 : Choice of a fractional plan 2^{3-1} for the design of experiment

5. CONCLUSION

The initial reliability of a product is a function of its conceptual design ; reliability growth occurs during development ; reliability is maintened thereafter through manufacturing and maintenance.

This paper has presented a suggested approach to reliability growth during development following the critical analysis of the initial design. This method affords significant savings in the cost of tests and development hardware. A method of calculating reliability figures is given.

This approach is being implemented in the development of the Vulcain engine which will be used on the future Ariane 5 launcher. For this project, the reliability objective appeared in the early stage of development as the major element in defining the number of tests required. The diversity of the possible test conditions available and the need to know the reliability precisely without excessive numbers of test articles and tests has motivated us to use an approach based on overstress testing integrated into an organized test plan.
In order to maintain the reliability achieved during development, it is important to improve manufacturing processes by process control and reduction of human errors. Today, these activities become our highest priority.

Structural Safety & Reliability, Schuëller, Shinozuka & Yao (eds) © 1994 Balkema, Rotterdam, ISBN 90 5410 357 4

Probabilistic analysis of labour accidents in workplaces

S. Hanayasu
Research Institute for Industrial Safety, Ministry of Labour, Tokyo, Japan

ABSTRACT: To perform probabilistic safety performance analyses of occupational accidents in workplaces, a statistical analysis of the frequency distributions of labour accidents was conducted. From accident investigations, it was found that Poisson and negative binomial distributions agreed with actual accident distributions. The frequency distribution of the number of occurrences of accidents can be represented as a mixture of the probability distribution function of accident rate to the probability distribution function of the number of accidents, which reaches the negative binomial distribution. The distribution of accident frequency rate, parameter of the accident frequency distribution, can be determined by making use of accident data. If sufficient data are not available, the Bayesian statistical inference scheme can be employed as an alternative method to estimate the parameter of the distributions.

1 INTRODUCTION

Every year many occupational accidents take place in the construction industry in Japan. According to the recent governmental statistics on occupational accidents in construction work, the number of new comers who received workmen's accident compensation insurance in the year of 1990 was as many as to 113,364. Among them, the number of serious accidents which kept workers away from work duties for 4 days or more were 60,900 including 1,075 fatal accidents (JISHA Report 1992).

The number of occurrences of occupational accidents as well as various accident indicators such as the accident frequency rate or the accident severity rate in the construction industry are decreasing in a long-term tendency. However, due to the current economic activities in the construction industry, the number of occurrences of fatal accidents are, in turn, on the increase.

Labour accidents constitute not only occupational injuries and illness for workers but also property damage, product performance downgrades and other associated losses with a large amount of direct and indirect financial losses. Therefore, the role in implementing safety programs for preventing accidents and controlling other losses is becoming imperative.

Recognizing these serious accident situations, various social levels from government agencies to individual establishments have been striving to prevent occupational accidents from taking place by making use of various safety programs on a compulsory and voluntary basis.

In carrying out the safety programs in an effective way, it is necessary in advance to obtain basic knowledge of the accident situations in workplaces. For this purpose, various statistical accident investigations are conducted annually by the Japanese government. According to them, we can easily find, for example, the annual change in the number of occupational injuries as well as the trends of the accident frequency rate and the accident severity rate in many industrial sectors. Also agencies causing such accidents and the types of accidents are reported as basic labour accident statistics (JISHA Report 1992).

However, statistical investigations from the stochastic viewpoint, such as the frequency distribution of accident occurrences or the time intervals between occupational accidents are not included in the governmental statistical survey programs.

With this thinking, in this paper, accident investigations and statistical analyses have been carried out with particular emphasis on the probabilistic analysis of the frequency distributions of the occurrence of occupational accidents for the purpose of providing a more appropriate probabilistic safety performance evaluation method for workplaces including many construction sites, together with a more detailed understanding of the probabilistic nature of accident occurrence situations.

2 STATISTICAL ANALYSIS OF ACCIDENT FREQUENCY DISTRIBUTIONS

There are several accident indicators to assess the safety performance of workplaces such as 1) injuries with loss of workdays, 2) injuries without loss of workdays, 3) property damage due to accidents, 4) workdays lost due to accidents, 5) total accident cost, etc. (Laufer et al. 1986). Among them, the number of occupational accidents and the accident (occupational injury) frequency rate has been widely used as a basic measurement of safety performance in many undertakings for a long time.

The accident frequency rate is defined as the number of occurrences of occupational accidents for a certain unit of manpower or employee hour exposure.

In the governmental survey programs in Japan, there is not any investigation on the stochastic or probabilistic analysis of occupational accidents. However, depending on the statistical literature to date, we can say that if some events are taking place at random and the expectation of the events per unit time is constant, then the frequency distribution of the occurrences of events in a fixed interval of time has the Poisson distribution and time intervals between successive events becomes the exponential distribution (Ang et al. 1975, Maguire et al. 1952). Poisson distribution and negative exponential distribution are expressed in the following equations.

$$P(x \mid \lambda) = \frac{(\lambda t)^x}{x!} \exp(-\lambda t) \qquad x=0,1,2,\ldots \qquad (1)$$

$$E(X) = \lambda t, \quad V(X) = \lambda t$$

where x : number of events observed
t : time duration considered
λ : number of events per unit time
$E(X)$: expectation
$V(X)$: variance

$$f(t \mid \lambda) = \lambda \cdot \exp(-\lambda t) \qquad (2)$$

$$E(T) = 1/\lambda, \quad V(T) = 1/\lambda^2$$

where t : time period to events
λ : number of events per unit time
$E(T)$: expectation
$V(T)$: variance

Parameter λ in each probability function is defined as the number of events per unit time, which, in this case, corresponds to the accident frequency rate. Hence, if the accident frequency rate in a workplace were given, then the probability having a specific number of occurrences of accidents in a given time t as well as the one that an accident will take place within a certain period of time t can be calculated by these equations.

In other words, we can conduct a probabilistic safety performance analysis for a workplace whether there is any significant tendency for changing accident situations in a given time period, by analyzing the probability having observed specific number of occurrences of accidents or the time intervals between accidents.

Namely, if the number of occurrence of accidents observed is very large and the probability having such a number of accidents is significantly small, then we can conclude that the accident situation has changed to worse. Similarly, if the time intervals between accidents has a significantly long time period, then we can conclude that the accident frequency rate has changed into a smaller rate than the initial rate, and consequently, the accident situation has been improved (Hanayasu 1983).

To find the actual accident situations, several investigations on the frequency distribution of the number of accidents in various industrial sectors were conducted.

Figure 1 shows an example of the results of the investigations on the distribution of the number of occurrences of fatal accidents within one day in 1982 classified by building construction sectors. This figure shows that each

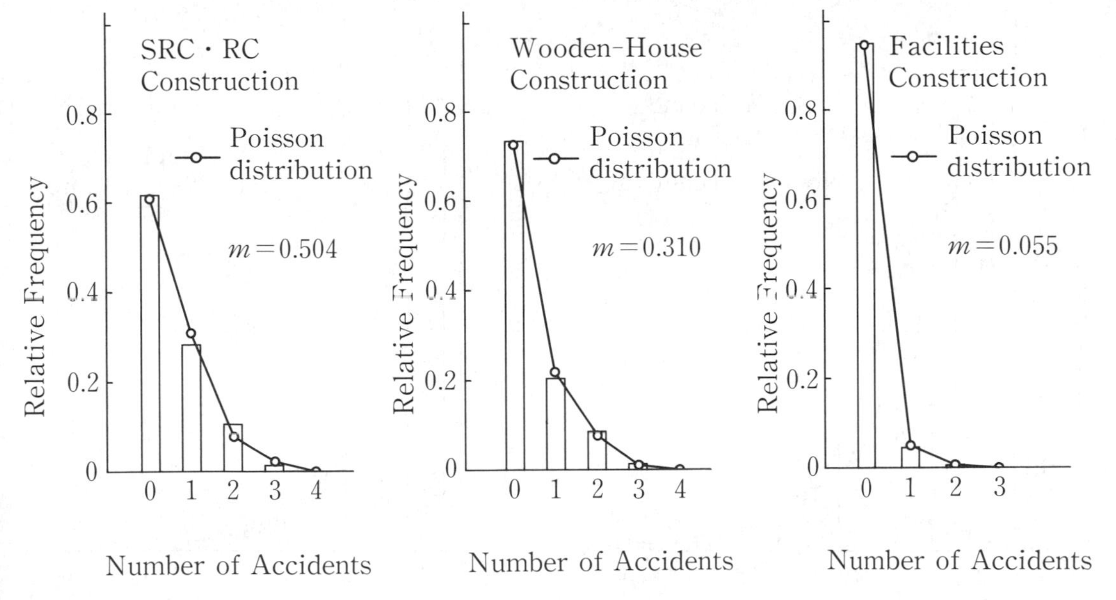

Fig. 1. Frequency distributions of fatal accidents in building construction sectors in 1982.

building construction sector had its frequency distribution according to the Poisson distribution. In addition to this figure, many examples can be shown in which their accident frequencies depend on the Poisson distribution.

However, there are some instances in which their accident frequencies disagree with the Poisson distribution. Figure 2 represents an example whose distributions differ form the Poisson distribution. This figure gives the frequency distributions of accidents within one month, which were taken place during the New Sanyo and the New Joetsu bullet train tunnel construction work.

The New Sanyo bullet train trunk line form Okayama to Hakata (444 km length) was constructed from the year of 1970 to 1974 covering 2157 working months associated with 1812 injuries including 55 deaths for tunnel construction. The New Joetsu bullet train trunk line from Omiya to Niigata (304 km length) was also constructed during the period from 1971 to 1982 totaling 3002 working months with 1416 casualties including 84 fatalities for tunnel construction work.

In order to clarify the difference between Fig. 1 and Fig. 2, it was assumed, in this paper, that the parameter of the probability function of the number of accidents, which is identical to the accident frequency rate in this case, is not constant but varies in accordance with another probability function.

Here, for the convenience of mathematical calculation, the probability distribution function of the parameter or the accident frequency rate was assumed to be a gamma distribution as given below.

$$h(\lambda) = \frac{(c\lambda)^{k-1}c}{\Gamma(k)} \exp(-c\lambda) \qquad (3)$$

Then, the frequency distribution of the number of occurrences of accidents considering the variation in the frequency accident rate can be obtained by compounding the accident rate distribution with the Poisson distribution, which finally reaches the following equation.

$$P(x) = \frac{\Gamma(x+k)}{\Gamma(k)} \frac{1}{x!} \left(\frac{c}{c+t}\right)^k \left(\frac{t}{c+t}\right)^x \qquad x=0,1,2,\ldots \qquad (4)$$

$$E(X) = \frac{kt}{c}, \quad V(X) = \frac{kt}{c}\left(\frac{c+t}{c}\right)$$

The above equation is referred as the nega-

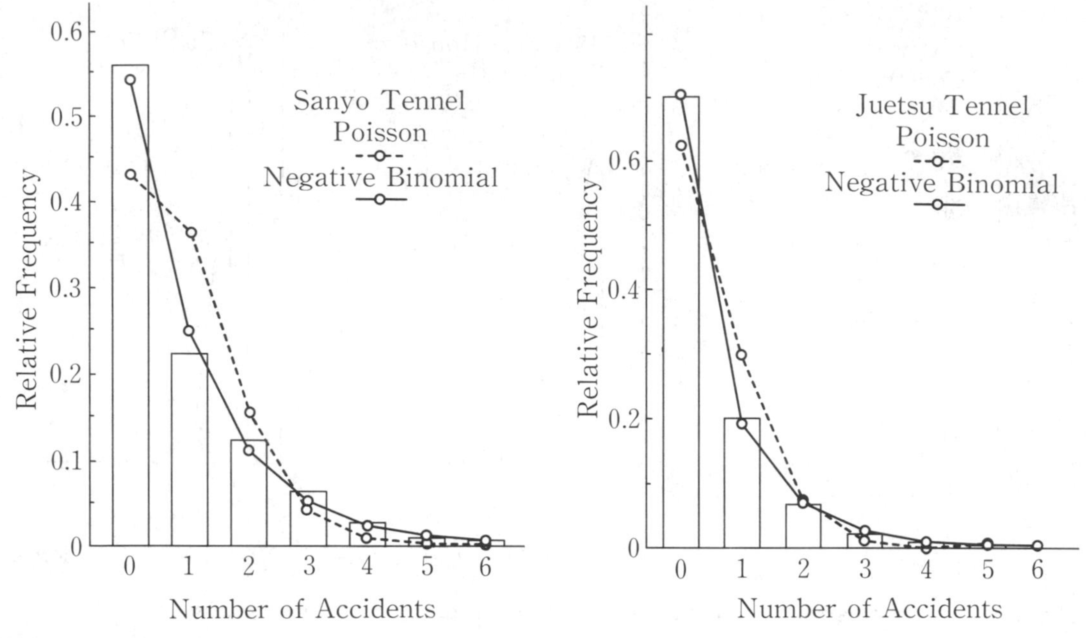

Fig. 2. Frequency distributions of the number of accidents in Sanyo and Joetsu bullet train tunnel construction work.

tive binomial distribution. In a similar way, the time interval between accidents taking the fluctuation of the accident frequency rate into consideration, can be obtained by mixing the distribution of accident frequency rate into exponential distribution and integrating the entire distribution domain, which ultimately yields the composite exponential distribution, or sometimes is referred as the Pareto distribution.

$$f(t) = \frac{k}{c}\left(\frac{c+t}{c}\right)^{K+1} \tag{5}$$

$$E(T) = \frac{c}{k-1}, \quad V(T) = \frac{c^2 k}{(k-2)(k-1)^2}$$

For the purpose of calculating the probability functions given above, the observed sample average(mean) M and the sample variance S were employed to estimate the parameter of these probability density functions, provided that sufficient accident data are available. The estimated parameters c and k from available accident data are given as follows.

$$c = \frac{Mt}{S\text{-}M}, \quad k = \frac{M^2}{S\text{-}M} \tag{6}$$

Using these parameters, the negative binomial distributions in Fig. 2 were analyzed. From this figure, we can see that the actual accident frequency distributions have a good agreement with the negative binomial distributions rather than the Poisson distribution. Hence, it is appropriate to assume that the parameter itself varies and depends on the probability distribution.

Therefore, when we conduct a safety performance evaluation of accidents in terms of probability, both Poisson and negative binomial distributions should better be calculated.

3 STATISTICAL ANALYSIS OF ACCIDENTS BY BAYESIAN INFERENCE

As shown in the preceding section, when sufficient accident data are available, estimation of parameters can be conducted by making use of these data. However, since the occurrence of accidents is rare, it is difficult to obtain much accident data enabling the estimating of reliable parameters, especially for an individual workplace.

In such a case, the Bayesian statistical inference scheme can be employed as an alternative method to estimate the parameter of the probability distribution functions.

Suppose the situation in an operating workplace where N_1 labour accidents have taken place within time period T_1, then using Bayes' theorem, the distribution of parameter λ, which is called the posterior distribution in the area of Bayesian statistics, can be deduced as in the following equation. This distribution is also referred as the gamma distribution whose parameters N_1 and T_1 have already been given (Benjamin et al. 1970).

$$h(\lambda) = \frac{(\lambda T_1)^{N_1} T_1}{N_1!} \exp(-\lambda T_1) \qquad (7)$$

Furthermore, if other N_2 accidents have taken place within the succeeding time period T_2 after T_1, then the renewed posterior distribution can be derived by making use of Bayes' theorem assuming that the former posterior distribution given by Eq. (7) to be an a-priori distribution.

Similarly, whenever new accident information are in hand, the renewal distribution of the accident frequency rate can be acquired by similar repetition. Namely, when accident data denoted by (N_1, T_1), (N_2, T_2), ... (N_n, T_n) were obtained, the renewal up-to-date accident frequency rate distribution that based on the Bayes' theorem can be derived as in the following gamma probability distribution function.

$$h(\lambda) = \frac{(\lambda T)^N T}{N!} \exp(-\lambda T) \qquad (8)$$

$$E(\lambda) = \frac{N+1}{T}, \quad V(\lambda) = \frac{N+1}{T^2}$$

$$\text{where: } N = \sum_{i=1}^{n} N_i, \quad T = \sum_{i=1}^{n} T_i$$

Hence, the gamma distribution has a useful advantage not only for the mathematical deduction convenience but also for giving a more detailed understanding on the accident occurrence situation from the probabilistic viewpoint.

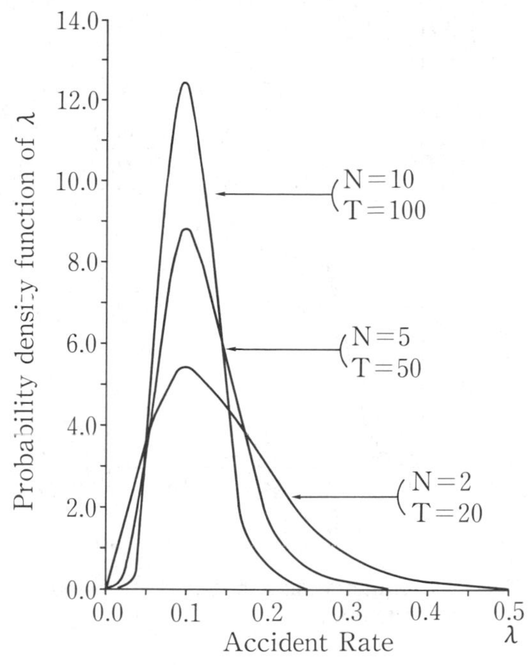

Fig. 3. Posterior distribution of accident frequency rate by Bayes' theorem.

Finally, by the mixing the posterior distribution with the Poisson distribution we can obtain the frequency distribution of the number of occurrences of accidents within a certain period of time t. This mathematical deduction also reaches the negative binomial distribution whose parameters N and T are already known.

$$P(x) = \frac{(x+N)!}{x!N!} \left(\frac{T}{T+t}\right)^{N+1} \left(\frac{t}{T+t}\right)^x \qquad x = 0, 1, 2, \ldots \qquad (9)$$

$$E(X) = \frac{(N+1)t}{T}, \quad V(X) = \frac{(N+1)(T+t)t}{T^2}$$

Fig. 3 shows an example of the analysis of the probability distributions of the accident frequency rate by Bayes' theorem. In this example, the posterior distributions were analyzed for three different prior accident situations (N, T)=(10,100), (5,50), (2,20), where the average accident rate $\lambda = N/T = 0.1$ has been kept constant. From this figure, we

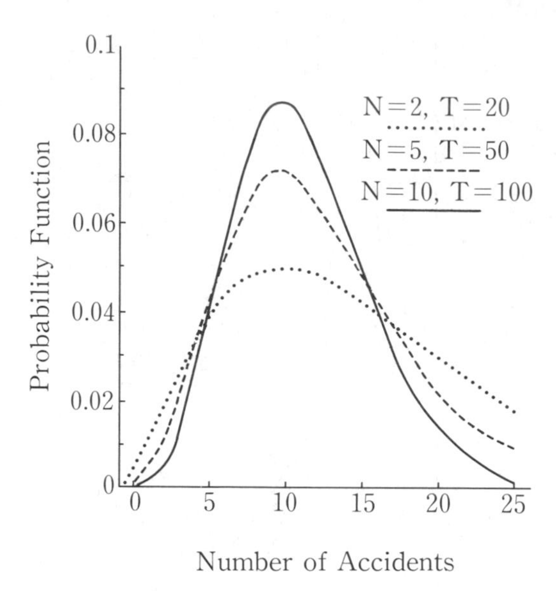

Fig. 4. Frequency distributions of accidents by Bayes' theorem.

can see that the more detailed accidents information acquired, the more precise the accident frequency rate can be estimated, viz. the variance of the posterior distributions becomes smaller in accordance with the length of the time period observed.

As another example of the analysis on the frequency distributions of accidents considering the variance of accident frequency rate by making use of the Bayesian inference, Fig. 4 presents the frequency distributions of the number of occurrences of accidents within 100 unit periods. Prior accident information for the analysis of the posterior accident frequency rate distributions were the same one that were analyzed in Fig. 3. From Fig. 4, we can see that even though the average accident rate is identical for each accident case, the feature of the accident frequency distribution is different according to the length of the observed time, similar to Fig. 3. Namely, according to the increment of the accident information, the variance of the renewal accident occurrences frequency distribution is getting smaller.

Here, keep the average accident rate constant and let $T \rightarrow \infty$ so that $N \rightarrow \infty$, then the limiting from of the negative binomial distribution approaches the Poisson distribution expressed as follows.

$$P(x) = \frac{(tN/T)^x}{x!} \exp(-tN/T) \qquad x = 0, 1, 2, \ldots \qquad (10)$$

Thus, it should be kept in mind that the probabilistic evaluation of accidents using the Poisson distributions is an approximate method compared with the negative binomial distribution.

As shown in this study, the Bayesian statistical inference scheme has played an important role in conducting the probabilistic safety performance analyses of occupational accidents in workplaces. This statistical methodology can be used as extensive method applicable to the various accident situations in workplaces from the scarce accident data available stage to the sufficient accident data available stage.

4 CONCLUDING REMARK

The main points obtained in this study can be summarized as follows:

(1) From the analysis of frequency distribution of the number of occurrences of accidents within a fixed time interval, it was found that the frequency distributions of many accidents agree with the negative binomial distribution as well as the Poisson distributions

(2) Depending on the results of frequency distributions of accidents, it was recognized that many accidents take place at random with a rough approximation. However, the accident frequency rate itself depends on the probability distribution whose probability function can be expressed as a gamma distribution. Hence, the frequency distribution of the number of occurrences of accidents as well as the frequency distribution of time intervals between accidents are represented as a mixture of the probability density functions. This mixing of the probability function of accident rate to the distribution function of the number of occurrences of accidents ultimately reaches the negative binomial distribution. Also, the mixture of the frequency distribution of accident rate to the frequency distribution of time intervals between accidents finally reaches the composite exponential or Pareto distribution.

(3) If sufficient accidents data are available, parameters of the probability distribution function of the negative binomial distribution or the composite exponential distribution, can be estimated by making use of these accident data.
(4) In case sufficient data are not available, the Bayesian statistical inference can be employed as an alternative method to estimate the parameters of the distribution functions. By making use of Bayes' theorem, a posterior distribution of the accident rate can be obtained as a gamma distribution whose parameters are the number of accidents and observed time. Then, the frequency distribution of the number of occurrences of accident within a given period of time, can be obtained by mixing the probability distribution function of accident rate to the Pōisson distribution, which also yields a negative binomial distribution. Similarly, by mixing the accident rate distribution with the exponential distribution, the frequency distribution of time periods between accidents, which considers the variance of accident rate, can be represented as a composite exponential distribution.
(5) Utilizing Bayes' theorem, the renewed posterior distribution of accident rate can be estimated every time when new accident data are acquired. Then by making use of this up-to date accident rate distribution, the renewal of the probability distribution functions of number of occurrences of accidents as well as the frequency of the time periods between accidents, can be achieved.
(6) Limiting from of the negative binomial distribution becomes to the Poisson distribution. Hence, analysis of the Poisson distribution is an approximate method compared with the negative binomial distribution.

REFERENCES

Ang, A. H-S. and Tang, W. H. 1975. Probability Concepts in Engineering Planning and Design, Vol. 1 : 114-124 : John Wiley & Sons.

Benjamin, J. R. and Cornell, C. A. 1970. Probability Statistics and Decision for Civil Engineers : 632-638 : McCraw-Hill.

Hanayasu, S. 1983. Stochastic Analysis of Accidents and of Safety Problems. Proceedings of IABSE Report 44 : 17-29.

Laufer, A. and Ledbetter, W. B. 1986. Assessment of Safety Performance Measures at Construction Sites : Journal of Construction Engineering, 12/4, ASCE : 530-542.

Maguire, B. A., Pearson, E. S. and Wynn, A.H. 1952. The Time Intervals between Industrial Accidents : Biometrika 39 :168-180.

Safety and Health Data Book in Japan. (1991 edition) 1992. Japan Industrial Safety and Health Association : 1-9.

Structural Safety & Reliability, Schuëller, Shinozuka & Yao (eds) © 1994 Balkema, Rotterdam, ISBN 90 5410 357 4

Discussion on safety and reliability of JR Maglev

Y. Hatano & H. Takeuchi
Central Japan Railway Co., Japan
S. Sone
The University of Tokyo, Japan

ABSTRACT : JR Maglev is a completely new super high-speed railway system. Upon successful completion of the first stage of development, a larger scale test track is now being constructed for development and confirmation for a real application to an important intercity railway. This paper discusses a method and reports some results of acquiring required safety and reliability level of a new and big system from experience of conventional and proven system.

1 INTRODUCTION

1.1 *Brief history of the development of JR maglev*

30 years has passed since Japanese National Railways (JNR) started research for linear-motor propelled magnetic levitation (maglev) system. In 1970 basic facilities for superconductive maglev was completed in Railway Technical Research Institute (RTRI) of the then JNR. The first levitated run was achieved in 1972 by a linear synchronous motored (LSM) experimental vehicle, and a 7 km long test track was completed in 1977 in Miyazaki, southern part of Japan. At this track a test vehicle ML500 attained the world speed record, 517 km/h, of guided vehicle of any kind in 1979. In order to allow passengers' space, guideway was converted from inverted T shape to U shape in 1980, since then manned or coupled runs were made by 3 car MLU001 and MLU002 test vehicles with various track conditions.

When JNR was split into 7 operating companies, 1 institute (new RTRI) and some supporting organisations in 1987, the test and development facilities were taken over by RTRI and keen discussion on the future development of superconductive maglev was made nationwide.

All major ministries of Japanese government responsible to transportation, i.e. Ministry of Transport, Ministry of Construction and National Land Agency were involved and the following basic national strategy was adopted: Super-high speed maglev is strongly worth developing for Japanese society primarily for inter city traffic of 300-800km distance, with possible application to a shorter distance shuttle service between city centre and airport apart far from each other. The second stage test track of about 40km long, having double track section with around 4 percent grade and tunnels should be constructed and basic application tests which was impossible at Miyazaki, should be carried out before decision of the system to be adopted for the new high-speed railway, Chuo (Central) Shinkansen, is made. Test site was decided to Yamanashi Prefecture out of 18 candidates, because of geographical reason and the fact that it is located on a most suitable direct line connecting Osaka with Tokyo.

1.2 *Present status of the test track*

Construction work started in April 1992. Research and development of the application tests are now shared by, and closely coordinated with, RTRI, Central Japan Railway Company (CJRC), operator of the new Central Shinkansen, and Japan Railway Construction Public Corporation, JRCC, with items concerning system's safety and reliability responsible mainly to CJRC.

2 TECHNOLOGY OF JR MAGLEV

2.1 *System features*

The vehicle of the JR Maglev system is loaded with super conductive magnets (SCM) and controlled by a linear synchronous motor with the primary side on the ground. It is designed to travel at the maximum speed of 550km/h without a driver on board. The main functions of the individual subsystems are as follows.

2.2 *Traffic control system and train operation control system*

The traffic control system is in charge of the management of the operating schedule and route control. The train control system generates the location-speed curve according to the timetable made by the corresponding local traffic control system.

2.3 *Drive control system*

The drive control system gives ground coils the VVVF current which is synchronised with the train location in order to drive the vehicle according to the location-speed curve made by the train operation control system.

2.4 *Ground coils and guideway*

On the sidewall of the "U"-shaped guideway, the ground coils for propulsion, levitation and guidance are installed and the current which runs in these coils makes up the moving magnetic field for travelling vehicles.

2.5 *Vehicle*

Vehicles equipped with SCM are driven by the moving magnetic field mentioned above. Vehicles levitate with no contact to the guideway at high speed and travel with wheels at low speed. So these wheels are lifted or lowered according to the velocity.

2.6 *Train locating system and safety control system*

The radio waves of certain frequencies which are transmitted by antennae on board are detected by the inductive wire with transposition laid along the guideway. The position of trains is identified with cm order accuracy. The velocity of the train is also calculated from the change of the trains' location.

The safety control system has a fail-safe structure and it is designed so that its output signal is fixed to the safety state when troubles occur in its hardware, software or datum. This subsystem is mainly in charge of an interval control and a route control and monitors the trains' location, velocity and the conditions of the routes with the interlocking for safety. Its function corresponds to that of the ATC (Automatic Train Control) adopted by the high speed railway system such as the Tokaido-Shinkansen in Japan.

3 SAFETY

3.1 *Safety brake control*

The safety control system continuously monitors train location, velocity and the conditions of switches throughout the main track and orders the safety brake system which consists of the aero-dynamic brake and the wheel disc brake. This brake system works independently from the regenerative brake system which is the JR Maglev's fundamental one.

3.2 *Operational control modes*

Considering the features of the JR Maglev and the

operational rules for conventional railway systems, we examined the possible hazards which could occur to the JR Maglev system and established 5 operational control modes as follows.

Mode 1: Vehicle stops immediately without designating the stopping point

Even though the possibility of failures is reduced by increasing the individual equipments' reliability, we cannot eliminate it completely. Even for an unexpected failure, we should prepare the effective strategies. Also in the case that the guideway is broken down or obstacles including suicides invade the guideway, the mode 1 should be selected. The other cases for adoption of this mode are as follows.

a) Regenerative brake failure caused by trouble with the converters.
b) Damage to the rolling stock or the failure of many super conductive magnets.
c) Failure of the safety control system.

Mode 2: Vehicle travels toward the designated point and stops there

Refuges for passengers and facilities for electric power supply to the rolling stock are constructed at the designated point. The cases for adoption of this mode are as follows.

a) Fire on the vehicle or along the guideway. (Stop outside of tunnels)
b) Forward hindrance caused by trouble from the preceding train, switches or equipments laid along the guideway.
c) Demand for repair of equipments on board.

Mode 3: Travel at the restricted speed / levitation

The vehicle continues to travel levitating at the restricted speed in the following cases.

a) Unusual vibration which occurs at high speed.
b) Slight impediments caused by natural condition.

Mode 4: Travel at the restricted speed / with wheels

The vehicle continues to travel with wheels at the restricted speed in the following cases.

a) Damage to the wheels used for suspension or guidance.
b) Inability to levitate caused by damage of ground facilities along the guideway.

Mode 5: Travel by inertia without braking

The vehicle passes by the area by inertia without braking in case of failure of the ground coils or the feeding system.

3.3 *Defensive structures for the vehicle*

In order to prevent obstacles or suicides from entering the track, defensive structures equalled to those of the Shinkansen are constructed. In addition a shock-absorber is installed at the head of the vehicle considering its maximum speed of 550 km/h.

3.4 *Reduction of magnetic influence to passengers*

In order to reduce the magnetic influence to passengers on board, the super conductive magnets are installed at the bogie frames. Steel shelters against the magnetic fields are also provided for the seats.

On the platform, gates and steel shelters protect the passengers from the magnetic fields made by SCM installed on the vehicle.

4 RELIABILITY

4.1 *Definition of operational service level*

The actual transportation system is designed or operated not to stop completely because of the local hindrances caused by the natural disturbance or the troubles of facilities even though its operational service level may go down.

Preparing for the discussion on MTBF and MTTR of the JR Maglev, we now define the 5 grades of operational service levels as the criteria for decision whether the system is available or not. (Table 1)

We regard of the level A, B or C as available. The kinds of failures which lower the operational service level to D or E are classified as shown in Table 2. Generally the area 2, 3 and 4 satisfy the

Table 1. Operational service levels

Level	Definitions	State
Level A	The transportation system works completely.	Available
Level B	A part of the system is in trouble and some trains are delayed. But no train gives up its operation.	Available
Level C	Travelling velocity is restricted or some trains give up their operation. But the trains can travel throughout the main track.	Available
Level D	There is a place where trains cannot pass by. The transportation performance has gone down considerably.	Unavailable
Level E	The system has gone down at many places, where the trains cannot pass by.	Unavailable

Table 2. Classification of the facilities' failures

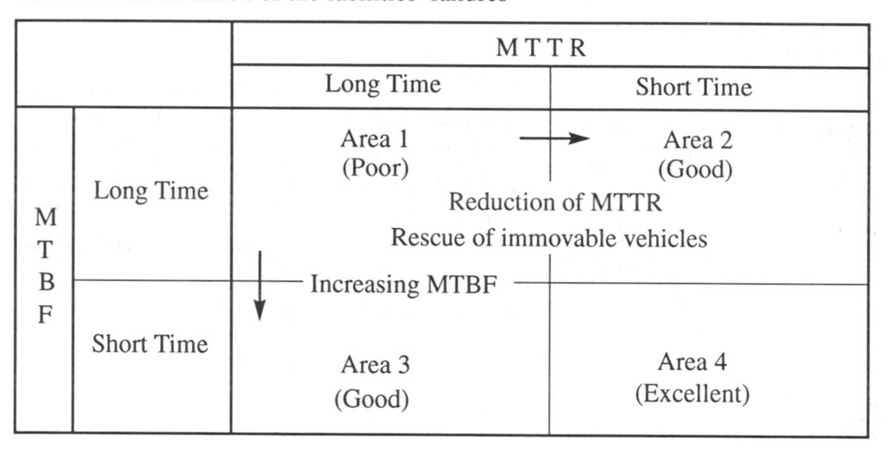

		MTTR	
		Long Time	Short Time
MTBF	Long Time	Area 1 (Poor)	Area 2 (Good)
MTBF	Short Time	Area 3 (Good)	Area 4 (Excellent)

requirements for reliability. The facilities which belong to the area 1 should be improved by reducing MTTR or increasing MTBF so that they could belong to the area 2 or 3. The basic concept is stated with some examples in 4.2 and 4.3.

4.2 *Increasing MTBF*

4.2.1 *Reliability Block Diagram*

In order to clear up the weak facilities from the viewpoint of system reliability, we have made a reliability block diagram which shows how each facility contributes to the whole system reliability. The reliability block diagram varies according to the operational service level. Fig. 1 shows the reliability block diagram which corresponds to the operational service level C which is the target to be realised. In this diagram facilities which can have redundant structures or backup facilities are expressed as follows.

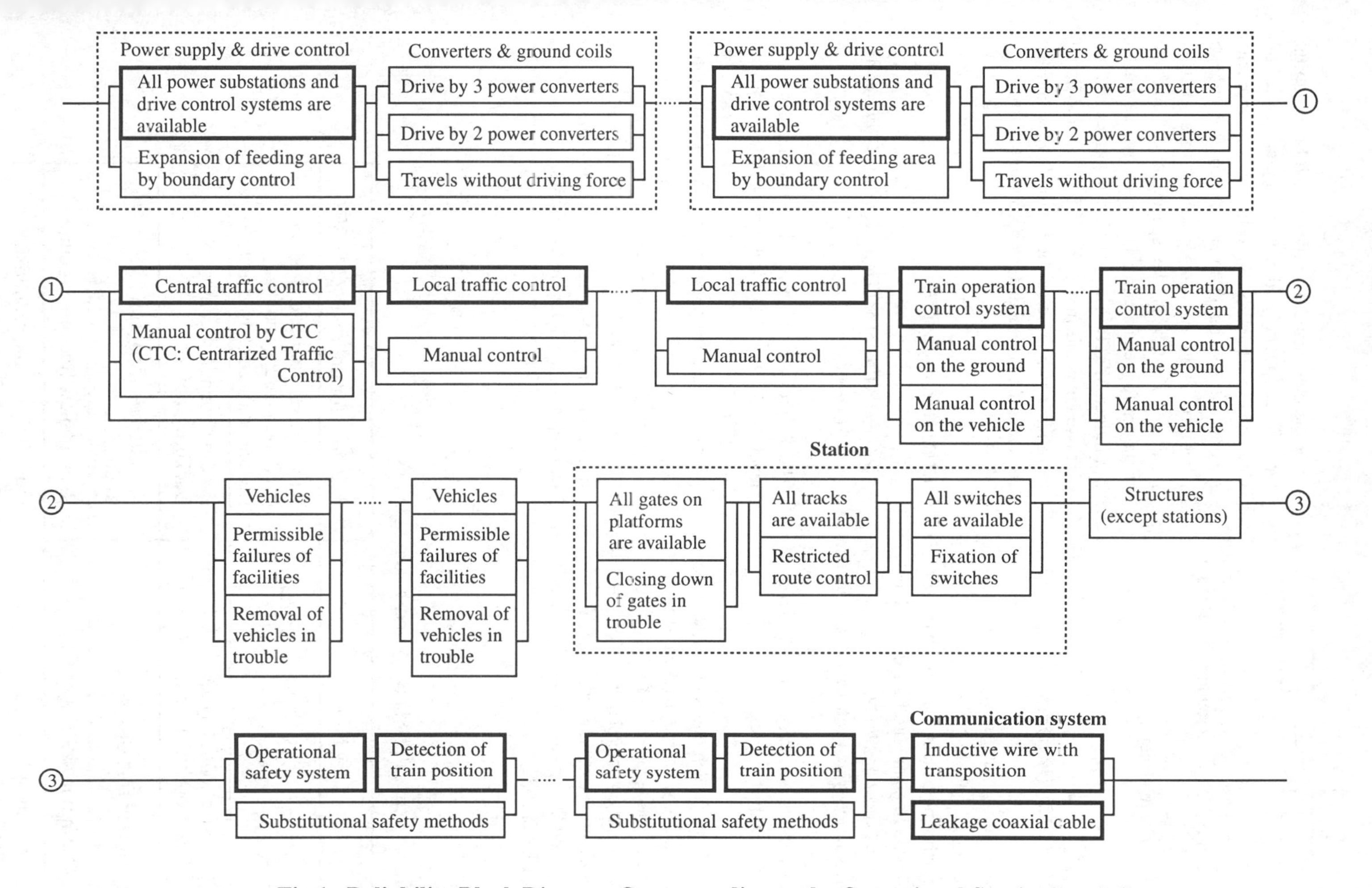

Fig 1. Reliability Block Diagram Corresponding to the Operational Service Level C

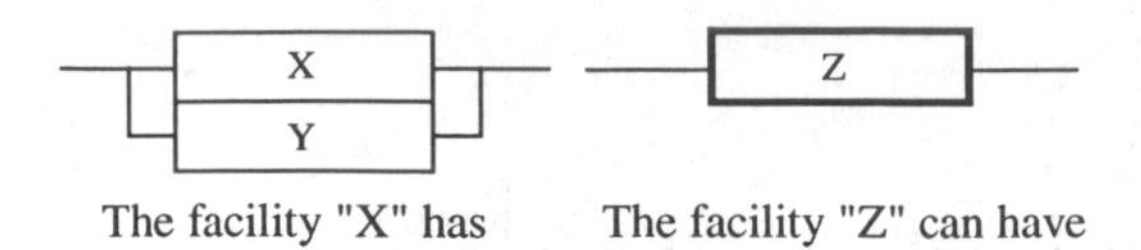

The facility "X" has a backup facility "Y"

The facility "Z" can have a redundant structure

Fig.2 Backup or redundancy expression

4.2.2 *MTBF of JR Maglev system*

In order to improve MTBF of the JR Maglev system, we have clarified the facility whose reliability is the bottleneck on the basis of the reliability block diagram. Individual facilities are divided into the following two types.

Type α : Facilities which can have the redundant or backup structures.

Type β : Facilities which cannot have the redundant or backup structures.

For the facilities which belong to the type α, adequate redundancy or the backup structure is realised within the economic limitation by the construction of the commercial line, on the basis of their contributions to the whole system reliability and the test results obtained from the new Yamanashi Test Track.

For the other hand, facilities which belong to the type β are possibly the weak points from the viewpoint of reliability. So technical developments are concentrated on them in order to give enough reliability. It is also important for this kind of facilities to establish the specifications and the strategies for repair and passengers' rescue which can reduce MTTR. The concrete solutions will be discussed in detail in 4.3. Table 3 shows the list of facilities which are classified into 2 types and the corresponding methods for improving MTBF.

4.3 *Reduction of MTTR*

4.3.1 *Definition of MTTR for estimated failures of equipments*

The failures which lower the operational service level to A, B or C are not serious from the viewpoint of MTTR because repair can be postponed until the night repair time. On the other hand, failures which lower the operational service to D or E need immediate repair. In this case MTTR of the whole system may be defined as the time spent for repairing or removing them. Four examples for the latter are as follows.

Table 3. Improvement of MTBF of individual facilities

Type	Facilities	Methods for improving MTBF
α	1. Power supply & drive control 2. Converters & ground coils	Three-power converter system, Feeding boundary control, Redundancy, Backup structure
	3. Traffic control system 4. Train operation control system	Feeding boundary control, Backup structure
	5. Safety control system	Feeding boundary control, Redundancy
	6. Communication system	Redundancy, Backup structure
β	1. Vehicle system 2. Station 3. Structures (except stations)	Independence and redundancy of components, Enough safety factor

(1) Hindrance of the main track by the vehicle in trouble

Because the operational service revel becomes rank D in this case, the vehicle blocking the main track have to be repaired or removed immediately. MTTR in this case is defined as the time spent for repair of removal.

(2) Hindrance of the main track by ground coils

When the ground coils fall off from the sidewall of guideway, vehicles cannot pass by the place and the operational service revel becomes rank D. In this case MTTR is defined as the time spent for repairing or removing these broken coils.

(3) Hindrance of the main track by switches

When a switch stops halfway, trains cannot pass by it. So MTTR in this case is defined as the time spent for throwing and fixing it completely by hand-operation or repairing it.

(4) Simultaneous failure of neighbouring power conversion substations.

When over two neighbouring power conversion substations fall in trouble at the same t ime, the operational service level becomes rank D because the uncontrollable area cannot be eliminated even by the feeding-boundary changing control. In this case MTTR is defined as the time spent for repairing them.

4.3.2 *Reduction of MTTR*

We will discuss the methods for reducing MTTR in the cases of troubles mentioned in 4.3.1.

(1) Hindrance of the main track by the vehicle in trouble

Equipments on board should be constructed with the redundant structure and reliable components so that vehicles can reach the closest station by themselves, even in the case of trouble.

Example :

When suspension wheels cannot be lowered because of some trouble, the vehicle lands with the emergency landing device and its braking force decreases. So the wheel disc brake system has been designed to ensure the braking force with the other available wheels. By this countermeasure, the vehicle can travel toward the closest station at relatively high speed and quick removal of the vehicle in trouble is realised.

We will also establish the strategy for the rescue of the immovable train with another train in commercial use.

(2) Hindrance of the main track by ground coils

The specifications and construction methods of ground coils should be determined so that their repair or exchange can be completed in short time.

(3) Hindrance of the main track by switches

Because switches of the JR Maglev are larger and more complicated than conventional ones, their structures have been determined so that they can be thrown by hand-operation even in the case of trouble. They will be made fault-tolerant with the redundant driving and control system.

(4) Simultaneous failures of neighbouring power conversion substations

The vehicle is driven by the three power conversion system. High reliability of the total system is achieved by cutting off the broken devices used in the main circuits by the converter and driving the vehicle by the other two converters.

The troubles of equipments that lower the operational service level to D or E and the methods for reducing their MTTR are shown in Table 4.

5 THE CHECKPOINTS ON SAFETY AND RELIABILITY AT THE YAMANASHI TEST TRACK

In order to decide the specifications of equipments and confirming the convenience of installation and maintenance, components will be tested under various operational conditions. Checkpoints concerned with safety and reliability are as follows.

(1) Functions of the operational safety system and the braking system

(2) Methods for monitoring and detection of troubles

(3) The operational control modes which cope with the various failures of the facilities, natural harn and obstruction.

(4) Durability

Table 4. The serious troubles that lower the operational service level to D or E, and the methods for reducing MTTR

Troubles of facilities	Methods for reducing MTTR
1. Hindrance of the main track by the vehicle in trouble	1. The fault-tolerant structure which keeps the vehicle from complete standstill. Rescue by another train.
2. Hindrance of the main track by ground coils' falling off	2. Establishment of the methods for quick repair of the ground coils. Specifications of the coils which is convenient for immediate repair.
3. Hindrance of the main track by failures of switches	3. Independent and redundant structure of the driving and control units. Mechanisms for hand-operation.
4. Simultaneous failures of the neighbouring power conversion substations	4. Driving with two power converters by cutting off the broken one. Preparing the spare units.

(5) Means to give the refuge to the passengers and operate the rescue train

6 FURTHER PROGRAMS

On the basis of test results of the fundamental performance, safety and reliability at the Yamanashi test track, the system will be optimised for the construction of the commercial line in the future.

REFERENCES

Kaminishi, K., A. Seki & H. Tsuruga 1993. The development of the Superconducting Maglev System. *13th International Conference on Magnetically Levitated Systems and Linear Drives* : 19-21. May, Argonne, Ill. USA.

Kubota, K., T. Tanaka, Y. Osada, S.Sasaki & Y. Yokota 1993. Train Control Systems for Superconductive Magnetic Levitation System. *ibid.*

Miyama, S., K. Matsuda & M. Minemoto 1993. Status Quo of Development of Super-conducting Maglev System in Japan. *ibid.*

Okada, K., A. Takano & M. Yamazaki 1993. Guideway and Infrastructure in JR Maglev. *ibid.*

Takao, K. & N. Shirakuni 1993. Vehicles for Superconducting Maglev System on the Yamanashi Test Line (Except Linear Synchronous Motors). *ibid.*

Structural Safety & Reliability, Schuëller, Shinozuka & Yao (eds) © 1994 Balkema, Rotterdam, ISBN 90 5410 357 4

Budgeting for risk reduction

Oswald Klingmüller
Gesellschaft für Schwingungsuntersuchungen und Dynamische Prüfmethoden mbH, Mannheim, Germany

ABSTRACT ; Risk analysis as a convolution of hazard and consequences has been extensively used in connection with costly investments with high risk. As an extension to the determination of a risk it is useful to identify the sources of risk that can lie either on the hazard side or on the consequences' side. In the classical economical environment only a limited budget is available for a production or a public investment. If a limited budget has to be distributed among activities for risk reduction, the optimal strategy has to observe the stochastic nature of the decision problem. A solution is proposed and discussed for application in three problems : transport over a bridge, waste deposit base barrier, earthquake hazard to a lifeline.

KEYWORDS

Risk analysis; decision theory; stochastic programming; heavy weight transport; waste deposit; lifeline.

1 INTRODUCTION

Risk analysis as a convolution of hazard and consequences has been extensively used in connection with costly investments with high risk. As an extension to the determination of a risk it is useful to identify the sources of risk that can lie either on the hazard side or on the consequences' side. In the classical economical environment only a limited budget is available for a production or a public investment. Furthermore only a limited budget will be available to reduce a risk that is found not to be acceptable. The possible actions to reduce a risk are of very different nature:

1. On the hazard side it can be the reduction of an existing failure probability, where first the confidence into the failure probability as depend of stochastic parameters on the resistance side or on the load effect side has to be established.

2. On the consequences' side it can be a preparation for rapid evacuation of a possibly endangered neighbouring population or the removal of valuable goods from the respective area.

It has been shown (Klingmüller 1985, Klingmüller 1986), how the formulation of a budgeting problem for optimal risk reduction can lead to a consequent collection and structurisation of data. In refs. (Gossow and Klingmüller 1989) and (Klingmüller and Bourgund 1992) it has been extensively elaborated, how the concept can be applied to actual problems of civil engineering.

Because of the many assumptions and the fuzziness of the data applied, a risk analysis should not be finished by the definition of a risk in terms of expected losses, but the most benefit can be gained if additional evaluations are carried out to answer the following questions :

1. Are there components or elements in the analysed system that have more influence

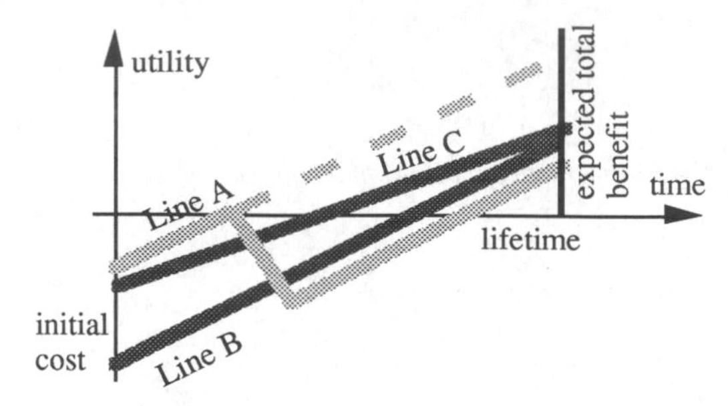

Figure 1 : Utility, cost and overall expected benefit

on the overall risk than others and where is this influence arising of?

This evaluation is called sensitivity analysis.

2. What changes in components or elements serve to reduce the risk ?

Of course, the answer to these questions is closely connected to the budgeting problem as defined above. This evaluation is called risk reduction strategy.

Whether or not the possibility of a reduction of risk by additional spending is reasonable and how large a eventually available budget will come out of the analysis of the expected overall benefit of an investment.

Figure 1 gives a schematic overview over the cost time relationship. From an economic point of view every installation starts with a considerable loss in form of the initial cost. During the time of use this loss has to be compensated by the accumulated net utility, i.e. utility minus maintenance and service cost. In general after the lifetime the accumulated utility will even exceed the initial cost because interests and initial costs for a follower installation have to be covered.

Line A is representing the low cost situation where in case of no failure (interrupted line) the overall benefit after the lifetime will be maximum. In case of failure the line is deflected because of additional cost of repair and delay in utility. Line B is representing the situation where higher initial cost leads to a greater reliability. As the risk will be smaller because of the higher availability thanks to additional initial costs the total expected benefit might be equal. Line C is representing the situation where enhanced inspection and maintenance activities will reduce the utility but at the same time the reliability is increased and overall expected benefit is equal to lines A and B.

The overall benefit after the total lifetime must be assumed to be a stochastic variable because of the risk associated the probability of failure and respective consequences.

A budget available for risk reduction can only be defined if this expected overall benefit is considered.

The risk reduction strategy has to take care of the fact, that the data on which ground the decision has to be taken are uncertain and given in terms of probabilities or fractiles of stochastic quantities, and additionally some activities cannot be described by a simple analytic function, but steps or edges must be described.

For the understanding of the character, first the solution of the decision problem by stochastic optimisation is demonstrated, and second for application problems suitable risk reduction functions in relation are discussed.

2 THE DECISION PROBLEM

It is assumed that a risk reduction is linearly dependent on the amount that is spend on a certain activity.

$$R_R = \Sigma (a_i \cdot C_i) \qquad (1)$$

where a_i are the stochastic coefficients of effectiveness,
C_i is the budget associated with activity "i".

To have the maximum possible risk reduction, the available total budget C_0 has to be distributed among the actions "i", as formulated by the stochastic optimisation problem :

$$\text{maximise} \quad R_R = \Sigma (a_i \cdot C_i) \qquad (2)$$

subject to

$$\Sigma\, C_i = C_0,$$
$$0 \le C_i \le C_0 \quad \forall\ i.$$

As the solution of problem (2) is dependent upon the realisation of the stochastic cost coefficients, there is no unique solution. As optimal it can be accepted to maximise the expected risk reduction. For normal-distributed stochastic cost coefficients this problem can be described by an equivalent deterministic quadratic optimisation problem by means of a utility function (Faber 1972).

To investigate the character of such a problem an example for the equivalent deterministic optimisation problem is given in two variables :

$$\text{maximise} \quad \bar{a}_1 \cdot C_1 + \bar{a}_2 \cdot C_2 - \frac{b}{2}\left((\sigma_1 \cdot C_1)^2 + (\sigma_2 \cdot C_2)^2 \right) \qquad (3)$$

subject to

$$C_1 + C_2 = C_0,$$
$$0 \leq C_1 \leq C_0,$$
$$0 \leq C_2 \leq C_0.$$

$\bar{a}_1, \bar{a}_2$ are the mean values of the effectiveness coefficients a_1 and a_2,
σ_1 and σ_2 the respective standard deviations,
"b" is the "risk attitude" coefficient, that is introduced by the utility function. A high value is chosen for a high confidence towards the description of the stochastic effectiveness coefficients.

In figure 2 an example for mean values of the coefficients of effectiveness : $\bar{a}_1 = 0.2$; $\bar{a}_2 = 1.0$ is given.

By inspection of figure 2, it can be recognised that, if the stochastic character of the effectiveness parameters is neglected, a simple linear programming problem has to be solved where the solution is given by an intersection of the linear boundaries. There is a sure optimum by dedicating the total budget towards the action "1".

For the given standard deviations, the stochastic cost coefficients are changing the objective function to quadratic, but still for low risk attitude values, i.e. risk aversion, the deterministic solution stays optimal. For a higher risk attitude value, the optimum shifts towards a partitioning of the budget to the two different actions.

In financial matters, the shown problem can be seen at as associated with the well known investment strategy, i.e. to give one third of the money to a fixed interest investment, another third to buy long term state loans, and the last third to let make risky money at the stock exchange.

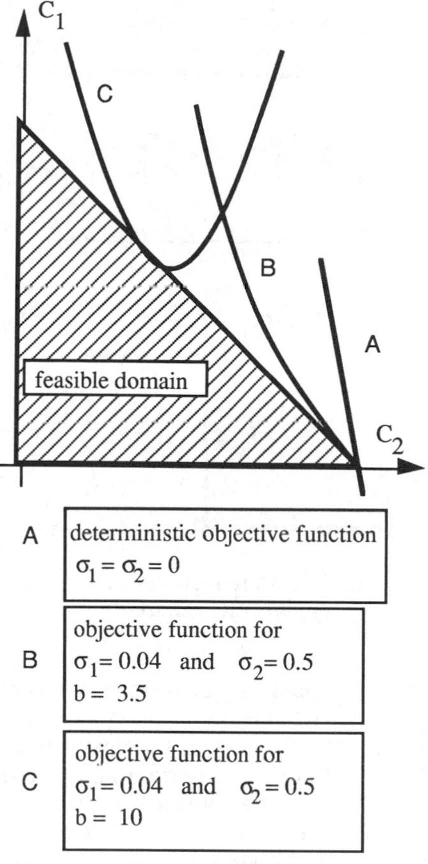

Figure 2 : Stochastic optimisation in 2 variables

For general engineering problems, the situation is not as simple as given by the above example. Especially there is almost no situation with a linear benefit function. The common situation is, that the risk reduction can be either described by an exponential function or by a step-wise function (fig. 3).

The exponential function (fig. 3, R_1) is appropriate if there is a high effectiveness for the initial spending, but for higher sums only a limited increase will be given, as may be the case in material testing, where the testing of samples and the installation of a sound quality assessment programme is of utmost importance. But when it comes to test all material ever used for a construction, the expenses will only lead to a stabilisation of the fourth digit in the fractile values. A step function (fig. 3, R_2) is used, if a certain action is associated with a fixed sum, e.g. to add an

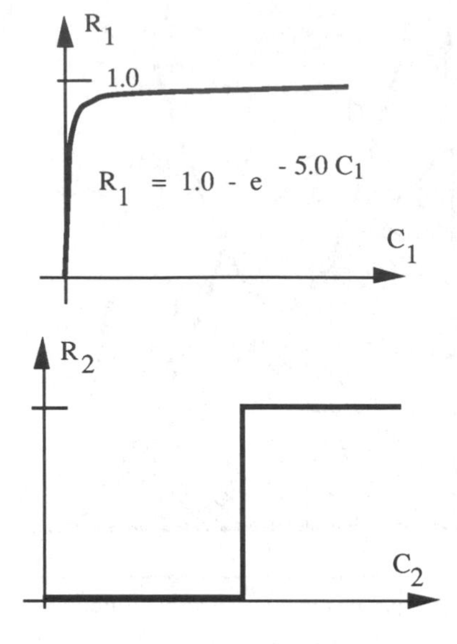

Figure 3 : Risk reduction functions in engineering reliability

additional support to a bridge will cost a certain amount. To buy a half support by spending half the money is a useless action.

For this kind of general risk reduction functions problem (2) will not be linear even for fixed effectiveness coefficient. The general form of (2) will then read as

$$\text{maximise} \quad R_R = \Sigma\,(a_i \cdot R_i\,(C_i)\,) \qquad (4)$$

subject to

$$\Sigma\, C_i = C_0,$$
$$0 \le C_i \le C_0 \quad \forall\, i.$$

In the following examples the effectiveness coefficients are assumed as given probability values and not as stochastic variables, so that problem (4) is solved as a deterministic problem, where the expected risk reduction is maximised.

3 APPLICATION TO HEAVY WEIGHT TRANSPORT OVER A BRIDGE

In a fictitious scenario a two span bridge is to be crossed by a heavy weight transport. The risk analysis showed a risk for persons to be

$$R = (10 + 2) \cdot 1{,}27 \cdot 10^{-6} \;.$$
$$= 15{,}24 \cdot 10^{-6}$$

where

the 2 is referring to the driver and his companion in the truck,

the 10 is referring to possibly endangered pedestrians,

$1{,}27 \cdot 10^{-6}$ is a probability of failure as evaluated by structural analysis.

The material losses will not be discussed herein (cf. Klingmüller and Bourgund 1972). To reduce the risk, several activities are in competition.

1. A prohibitive system with respect to parties not concerned will cost 20 monetary units, if properly executed. For less expenditure the prohibitive system will be not as effective. And e.g. for a very low expenditure (publication in the local newspaper) the effect can be neglected. An expenditure in excess of 20 monetary units will not increase the effectiveness accordingly. At best, there will be no harm to the unconcerned parties. Thus the risk is reduced to $1{,}397 \cdot 10^{-5}$ and the risk reduction is $12{,}7 \cdot 10^{-6}$. The associated risk reduction function is given in fig. 3.
   As people may be attracted by additional activities a probability of 90 % to be fully effective will be assigned to this action.
2. A careful inspection of the bridge can be executed to have a more precise idea of the actual carrying capacity of the bridge. This activity will increase the confidence level of the failure probability, and thus the risk will be reduced by $13{,}76 \cdot 10^{-6}$ to $1{,}48 \cdot 10^{-5}$ for the spending of 30 monetary units. From zero spending onwards, a certain proportionality can be assumed between expenditure and effectiveness. Thus for this activity, the risk reduction function is assumed exponential (cf. fig. 4).
3. Increasing the strength of the mid support by injecting grouting into the soil would cost 30 monetary units. The increase in the carrying capacity would reduce the failure probability, and thus the risk will be reduced by $13{,}74 \cdot 10^{-6}$ to a new value of $1{,}46 \cdot 10^{-6}$. As there is only an "either-or"

in this decision, a step function will be assumed (cf. fig. 3). As soil improvements are always very difficult engineering tasks, a probability of 90 % to be fully effective will be assigned to this action.

4. The installation of temporary supports can be obtained at the cost of 20.000,- monetary units. With respect to total failure the strengthening only concerns one of several failure modes and therefor with respect to harm to unconcerned parties the action is assumed to reduce the risk by $4{,}9 \cdot 10^{-6}$ to a new value of $10{,}34 \cdot 10^{-6}$. As there is only an "either-or" in this decision a step function will be assumed (cf. fig. 3). As temporarily installed supports are mostly executed by used material a probability of 98 % to be fully effective will be assigned to this action.

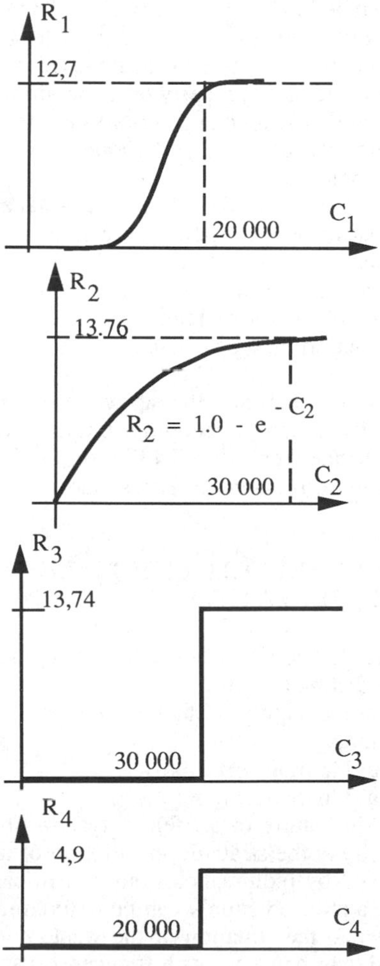

Figure 4 : Risk reduction functions for heavy weight transport

Only four of a large number of different activities are included in this scenario, but it can be recognised, that totally different activities have to be compared, if an optimum decision is wanted. The comparison of the possible activities also shows the character of the data, that are needed for the risk reduction budgeting decision. In table 1 the data are summarised.

Table 1: Risk reduction and survival costs

action	cost	eff.	risk red.	surv. cost
(1)	20	0.9	12.7	0.174
(2)	30	1.0	13.76	0.218
(3)	30	0.9	13.74	0.243
(4)	20	0.98	4.9	0.41

The quantity "risk reduction over effectiveness times cost" is known as the survival costs and is a first indicator, where a budget is to be spend most effectively.

With respect to the more fuzzy situation of a gradual effectiveness and different shapes of risk reduction curves (see fig. 3) an optimisation problem as given by

maximise

$$R(c1,c2,c3,c4) = 0.9 \cdot RR1(c1) + RR2(c2) + 0.9 \cdot RR3(c3) + 0.98 \cdot RR4(c4)$$

subject to

$$c1 + c2 + c3 + c4 \leq c0$$
$$c1 \geq 0.0, \quad c2 \geq 0.0,$$
$$c3 \geq 0.0, \quad c4 \geq 0.0.$$

must be solved. The solution for increasing available budget is given in table 2.

By the budget allocation as given in table 2, determined by the optimisation procedure, it can be seen, that the actions indicated by the survival costs are chosen. The quantification however of the budget allocation has to be determined by the optimisation procedure.

Table 2 : Budget allocation for optimal risk reduction

	available budget				
action	10	20	30	40	50
	budget allocation				
1	-	-	20	20	20
2	10	20	10	20	10
3	-	-	-	-	-
4	-	-	-	-	20

4 APPLICATION TO BASE BARRIER OF WASTE DEPOSIT

For the risk minimisation of waste deposits, a very complex situation of influences to the risk and actions for risk reduction has to be taken care of. In contrast to structures where mainly the ultimate limit state is guiding the design with respect to the consequences the states from complete tightness to total loss of serviceability and harm to third parties should be fuzzified in the sense that e.g. minor leakages will lead to minor consequnences, This gradual decrease of serviceability is shown in figure 4

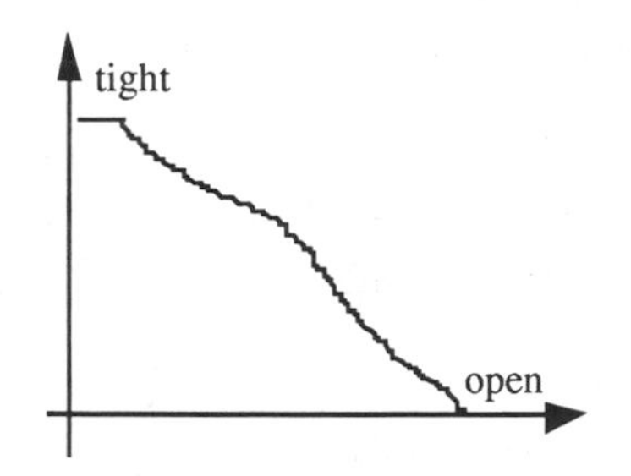

Figure 4 : Serviceability of base barrier

In addition it must be observed that the severity of consequences depend on the duration of a leakage, i.e. of the total amount of contaminating substances released and the spatial distribution in ground water and soil.

Possible actions for risk reduction are :

1. Hazard or probability of a leakage
   - increase in the thickness of a mineral clay barrier as the barrier is always built in layers of finite thickness, there a multiple step function may be chosen,
   - quality control for the clay material used with more care with soil investigations at the possible resource site monotonous risk reduction function can be appropriate
   - quality control for installation of the clay material monotonous risk reduction function can be appropriate
   - addition of one or even two layers of plastic foils to the clay barrier a step function is appropriate
   - control system for the tightness of the barrier and inspection for the installation a step function is appropriate and
   - for inspection the effectiveness will increase with the expenditure up to a certain limit.
2. Consequences
   - control of wastes to be deposed, so that the chemical property of a contaminating liquid is known and actions can be taken the effectiveness will increase with the expenditure
   - replacing of drinking water intakes, so that there is no pollution a step function may be chosen
   - additional drinking water resources as stand by redundancy a multiple step function may be chosen.

Although the actions above are chosen arbitrarily, it can be recognised, that there are more chances for risk reduction on the hazard side than on the consequences side.

5 APPLICATION TO EARTHQUAKE HAZARD TO A LIFELINE

With respect to lifelines the following hazards can be distinguished :

- hazard to supply or the resource being empty
- hazard to transport (lines)
- hazard to distribution.

A main feature of a lifeline system of high reliability is the strategic installation of active and stand-by redundancies, so that in case of failure a reduced supply can be maintained. In contrary to the situation of the waste deposit, the risk can be very much reduced by actions with reference to consequences. These may be either actions, that guarantee the most effective distribution of a reduced supply and a layout of the lifeline system, that helps to detect local failures and supports immediate refurbishment.

In a case study the energy supply of a population of approximately 5 million people by a gas pipeline is investigated. This pipeline is leading over a distance of appr. 2000 km from a resource to the intermediate storage and distribution nodal point. A portion of less than 100 km of the pipeline is passing an area of very high seismicity.

In the scenario for risk analysis a time dependence of consequences is assumed that takes account of the fact that after failure of the pipeline the supply can be maintained for a short duration by emergency systems. After that time losses increase severely as illustrated by figure 5.

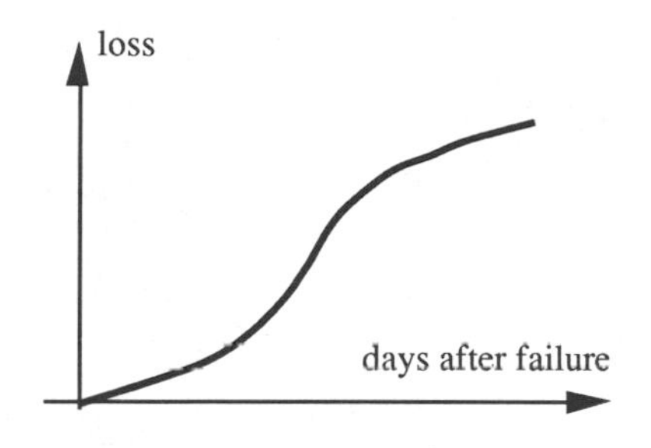

Figure 5 : Time dependent losses

With such a time dependence of losses the possibilities of quick leak detection and repair become most important.

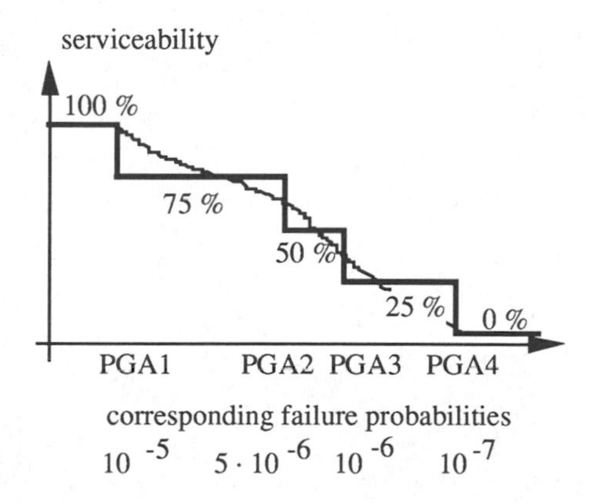

Figure 6 : Degrees of loss of serviceability

For the evaluation of consequences the gradual decrease of serviceability is discretized into four steps where the degree of loss of serviceability is accepted to be directly dependent upon the earthquake peak ground acceleration (PGA, see figure 6).

In a simplified scenario four activities for risk reduction are included :

1. improved inspectability for fast leak detection
2. enhanced inspection and maintenance for lower failure probability
3. increase in wall thickness of pipe for lower failure probability
4. deviation of seismically hazardous area

In a first evaluation it has been determined that for the second activity - improved inspection and maintenance - survival cost (cost to risk reduction ratio) will be lowest. On the other hand as this activity is to be carried out over the total service life of the pipeline the overall cost in absolute monetary units will be high and might not be available. The optimal risk reduction has to be found with respect to the expected overall benefit of the pipeline that defines the available budget and by the solution of the suitable stochastic optimisation problem which leads to the budget distribution.

In a parametric study it will be investigated which amount of losses are implicitly considered in the applicable codes for buried lifelines and how an optimal design concept might be derived.

First results showed that the losses that are to be prevented by adequate design with corresponding failure probabilities in the range of 10^{-6} are highly in excess of the installation costs, the exact value depending on the degree of development of the nations.

Indirect consequences such as individual and/or social instability (unemployment after earthquake damage) are rated much higher and will in general justify high safety levels.

6 CONCLUSIONS

The formulation of the risk reduction problem can be seen as an important tool as to what are the important questions and which data are needed for the answer. In many situations actions to be taken for a reduction of a risk are of totally different nature but can be combined if monetary units are assigned to them. Thus, if a risk reduction problem is formulated, the solution for different budgets will provide an optimum strategy for the distribution of the budget, but on the other hand can be utilised to reveal a lack of knowledge with respect to important data.

7. REFERENCES

FABER, M.M. 1972. Stochastische Programmierung, Physica Verlag Würzburg.

GOSSOW, V., and KLINGMÜLLER, O. 1989 Deponiebautechnik und Risikoanalyse, Baumarkt 10.

KLINGMÜLLER, O. 1982. Influence of structural safety on overall risk analysis of LNG storage facilities, 4th ICOSSAR '85, ed. by Konishi/Ang/Shinozuka, IASSAR.

KLINGMÜLLER, O. 1986. Collection and usage of reliability data for risk analysis of LNG storage tanks, 5th EuReData Conf., Proc. ed. by H.J.Wingender, Springer Verlag, Berlin Heidelberg.

KLINGMÜLLER, O., and BOURGUND, U. 1992. Sicherheit und Risiko im Konstruktiven Ingenieurbau, Vieweg Verlag Wiesbaden.

ACKNOWLEDGEMENT

Part of the work has been supported by the European Community under the EPOCH programme project "Seismic behaviour and vulnerability of buried lifelines".

Structural Safety & Reliability, Schuëller, Shinozuka & Yao (eds) © 1994 Balkema, Rotterdam, ISBN 90 5410 357 4

An integrated approach in probabilistic modelling of hazardous technological systems with emphasis on human factor

K.T. Kosmowski & K. Duzinkiewicz
Technical University of Gdansk, Poland

ABSTRACT: This paper is devoted to probabilistic modelling of hazardous systems. Some more important issues are described which have been encounter during the conceptual design stage of a PSA/HRA (Probabilistic Safety Analysis/Human Reliability Analysis) software system applying the expert system technology. Taking into account two basic objectives of its development, namely, the training and decision making, at different levels of details, with regard to the reliability and safety of technological systems, as well as current limited research and financial resources to be involved in the project, it was decided to follow a step by step procedure to cover gradually in a balanced manner the important issues of the PSA of level 1 (PSA1). It means that the analysis concerns mainly the internal potential plant's damages, i.e. the software system is not designed at present to enable quantitative assessments of releases of toxic or radioactive materials to the environment. The emphasis is placed on the human factor reliability aspect being lately recognized as one of the most important contributor to the risk associated with operation of hazardous technological systems.

1 INTRODUCTION

It has been lately recognized that so called human factor is one of the most significant contributor to the risk associated with operation of hazardous systems (Dougherty 1988). On the other hand the state of the art of human reliability modelling and available at present techniques are not satisfactory (Dougherty 1990). As significance of Probabilistic Safety Analyses (PSAs) performed to support the safety related decision making is becoming more and more accepted, there is lately a considerable interest to computerize relevant assessments and, in particular, to apply the expert system technology (IAEA 1990). One of the most difficult problem in designing the relevant software system is associated with modelling the human factor reliability and integrating events of human induced errors into the logic structure of equipment oriented probabilistic models.

The purpose of this paper is to present an integrated approach in probabilistic modelling of hazardous technological systems which has been work out during the conceptual design of an expert system giving support to PSAs of level 1 which is being developed at the Technical University of Gdansk (Kosmowski, et al. 1991). The software system is designed gradually taking into account three basic scopes of PSA/HRA to cover important issues of safety related training and decision making. These scopes differ to some respects, as regards e.g. details of data and knowledge bases, details of inferring and evaluations supervised by experts/users, representing and treating of uncertainties.

The emphasis is placed on computer aided modelling the human factor reliability. It is one of the most difficult software design task. A classification scheme of human actions/errors is proposed to be useful for selecting appropriate modelling technique for the case analyzed. Errors of commission due to mistakes are difficult to model (Dougherty 1988). In the paper an approach is outlined aimed at reducing subjectivity of probabilistic assessments of the confusion matrix components thanks to previous evaluation of similarity measures based on vectors of important variables of the plant response for initiating events considered.

2 CLASSIFICATION OF HUMAN ACTIONS AND ERRORS

Human actions/errors can be divided to be related to the phases of an accident into three categories (Dougherty 1988, IAEA 1992):

(A) Actions/errors in planned activities, i.e., so-called pre-initiator events that cause equipment (systems) to be unavailable when required post initiator.

(B) Errors in planned activities that lead directly, either by themselves or in combination with equipment failures, to initiating events/faults (e.g. unplanned power plant shutdown), i.e. human induced initiators.

(C) Actions/errors in event-driven (off-normal) activities, i.e. post-initiator events; these can be either safety actions or errors that aggravate the fault sequence. Interactions of this category can be further subdivided into three different types for incorporation into the PSA, namely: (C1) procedural safety actions, (C2) actions/errors aggravating the accident progression and (C3) improvising recovery/repair actions.

Interactions (C2) are a special set of commission errors that occur post-fault and can significantly aggravate the accident progression. They are the most difficult to identify and model. Recovery actions (C3) are included usually only in accident sequences that dominate risk profiles. They may include the recovery of previously unavailable equipment or the use of non-standard procedures to ameliorate the accident conditions.

Described above behavior types seem to involve different error mechanisms, which may mean radically different reliability characteristics. Human errors are often classified to be one of two kinds (Reason 1990):

I. slip - (1) an error in implementing a plan, decision or intention (the plan is correct, its executing is not), or (2) an unintended action; a type of slip is lapse, an error in recall, e.g. of a step in a task;

II. mistake - an error in establishing a course of actions, e.g. an error in diagnosis, planning or decision making.

Errors are often classified as errors of commission or omission. Error of commission is often understood as incorrect performance of a system-required task or action, or the performance of some extraneous task or action that is not required by the system and which has the potential for contributing to some system-defined failure. Error of omission is a failure to perform a task or action (Dougherty 1988). In Fig. 1 a classification tree of human event-driven errors (category C) is proposed which enables to select appropriate technique for modelling the human reliability. Another tree was proposed for pre-initiating (latent) actions/errors (Kosmowski 1992).

As it was mentioned three levels of effort to carry out the PSA/HRA have been distinguished: I, II and III which correspond to the PSA/HRA methodological issues (methods applied, details of modelling and the contribution of experts required) and relevant scopes of the computer aided analyses. Assumed features of the software system supporting PSA1 and HRA using the expert system technology and conventional computer programs are presented in Table 1. The software system development has been scheduled to enable gradual and balanced realization of research and designing works with regard to resources available.

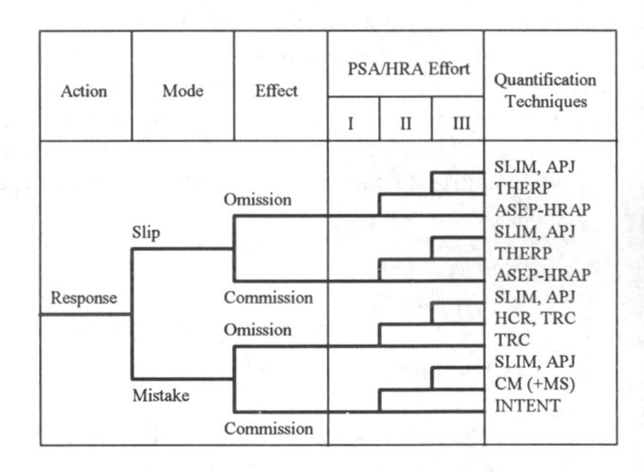

Abbreviations of the Human Reliability Analysis (HRA) techniques:

APJ - Absolute Probability Judgment

ASEP-HRAP - Accident Sequence Evaluation Procedure, Human Reliability Analysis Procedure (NUREG/CR-4772, 1987)

CM (+MS) - Confusion Matrix (with Modelling Support)

HCR - Human Cognitive Reliability

INTENT - A method for estimating HEP for decision-based errors (Gertman 1992)

SLIM - Success Likelihood Index Method

THERP - Technique for Human Error Rate Predictions

TRC - Time Reliability Correlation

Remarks on applications some of these techniques from the PSA perspective can be found in (Cacciabue 1988, Humphreys 1988); an outline of CM (+MS) method in this paper.

Fig.1. Classification of human event-driven errors and some related quantifying techniques for different HRA/PSA effort

3 CONFUSION MATRIX WITH MODELLING SUPPORT (CM-MS)

Mistakes, especially errors of commission due to misdiagnosis are considered as most difficult to model and quantify. The assessed probabilities of misdiagnosis errors of accident situations in a short period after initiating events are usually placed as elements of so called confusion matrix. They represent the probability of confusing a transient j with another transient k, possibly leading to erroneous actions (Hannaman and Spurgin 1984, Wakefield 1988). The probabilities p_{jk} of such confusion depend on the similarity of symptoms such as alarms, enunciators, values of various process variables or directions and rates of their changes (IAEA 1992). The approach described herein contains a method, easy for computer implementation, to calculate two measures of similarity of symptoms which can be then used to support the expert judgments concerning mentioned probabilities.

Symptoms of abnormal situations are recognized in operation practice taking into account important variables of the process. Correct selection of observed variables is crucial for successful diagnosis. During the training process the operators acquire knowledge or can get personal impression on the importance of these variables. The training related importance of a particular process variable can be represented by weight coefficient (Wenda 1975). Another factor influencing the confusion probability is the degree of training and simulator practice given to the operator with respect to each transient. If training on transient j is more frequent than on transient k, it is likely that transient k, when it actually occurs, is believed by the operator to be transient j. The layout of a control room, especially the arrangement of indicators and a manner of presentation of information, the available time for diagnosis of the plant and the stress level of operators should be also considered by experts in the evaluation process.

On the other hand it is possible to identify a previous misdiagnosis if new clear symptoms appear which alert the operator. The possibility to recover depends strongly on time available to correct errors and contradiction of goals, to be reach by implementing of operational procedures (previous and new), for initiating events considered. It is known that for some initiating events relevant sets of procedures to be executed can consist of the same ones (Wakefield 1988). In such situations of non-contradictory goals the decreasing of the probability p_{jk} can be proposed.

The information about transients of important process variables obtained from simulation is used for evaluation of the possibility of misdiagnosis during the observation time T_d. The list of initiating events is fixed. The spectrum of plant transients is considered for the first minutes of an accident progression. Ideally, a more extensive list of accident situations would be compared to each other to assess the possibility of misdiagnosis. It is assumed as follows:

- For all m listed accident situations the uniform set of n process variables x_i for diagnosis purposes is defined. This set does not contain such process variables which can not be observed in the control room by operator and it should not be too numerous. With each variable is associated a coefficient $\alpha_i, i = 1,\dots,n, \alpha_i \in [0,1]$, that characterizes the importance of the variable x_i in the diagnosis process. These coefficients are related to the operators training.
- For all m listed accident situations and all n process variables, defined for diagnosis purposes, the transients are known (similarly as those represented in Fig.2). Transients can be approximated to facilitate further evaluation of similarity measures.
- Transient of each continuous variable x_i is characterized by two quantities: values of variable x_i and value of its derivative $\dot{x}_i$.
- Values of minimum $x_{i,min}, \dot{x}_{i,min}$ and values of maximum $x_{i,max}, \dot{x}_{i,max}$ are evaluated, e.g.

$$x_{i,min} = \min_j \left\{ x_{i,min}^{(1)}, \dots, x_{i,min}^{(j)}, \dots, x_{i,min}^{(m)} \right\}$$

The following describes the steps that are performed in the proposed approach to calculate the distance or similarity measures for pair (j,k) of accident situations:

(i) In the time interval T_d, r points of observation at the moments t_p, $p = 1,\dots,r$, are selected.

(ii) For each point t_p the observation vector $D_p^{(j)}$ is created, and

$$D_p^{(j)} = \left(x_{1,p}^{(j)}, \dot{x}_{1,p}^{(j)}, \dots, x_{i,p}^{(j)}, \dot{x}_{i,p}^{(j)}, \dots, x_{n,p}^{(j)}, \dot{x}_{n,p}^{(j)} \right)^T$$

$$j = 1,\dots,m; \quad p = 1,\dots,r$$

(iii) Normalization of the observation vector is performed according to formulas:

$$x_{i,p}^{*(j)} = \frac{x_{i,p}^{(j)} - x_{i,min}}{x_{i,max} - x_{i,min}}; \quad \dot{x}_{i,p}^{*(j)} = \frac{\dot{x}_{i,p}^{(j)} - \dot{x}_{i,min}}{\dot{x}_{i,max} - \dot{x}_{i,min}}$$

After this normalization:

$$0 \le x_{i,p}^{*(j)} \le 1; \quad 0 \le \dot{x}_{i,p}^{*(j)} \le 1$$

(iv) The distance measure for each pair (j,k) after selected observation time t_p is calculated:

$$\rho_{t_p}(j,k)=\frac{1}{2np}\left\{\sum_{s=1}^{p}\sum_{i=1}^{n}\alpha_i\left(\left(x_{i,s}^{*(j)}-x_{i,s}^{*(k)}\right)^2+\left(\dot{x}_{i,s}^{*(j)}-\dot{x}_{i,s}^{*(k)}\right)^2\right)\right\}^{1/2}$$

$$p=1,\ldots,r$$

The distance measure satisfies:

$$0\le\rho_{t_p}(j,k)\le 1$$

(v) The similarity measure for each pair (j,k) after the observation time t_p is calculated:

$$s_{t_p}(j,k)=1-\rho_{t_p}(j,k);\quad p=1,\ldots,r$$

The similarity measure also satisfies:

$$0\le s_{t_p}(j,k)\le 1$$

The calculated distance or similarity measures form the basis for creating distance or similarity tables for all pairs (j,k) of accident situations. Depending on the obtained value of distance or similarity measure, the probability of confusion is then judged by experts, e.g. using SLIM or APJ techniques. Linguistic statements concerning confusion based on similarity measure can be also proposed, e.g. high, medium, low or insignificant which can be then a basis for evaluation of probability (Wakefield 1988).

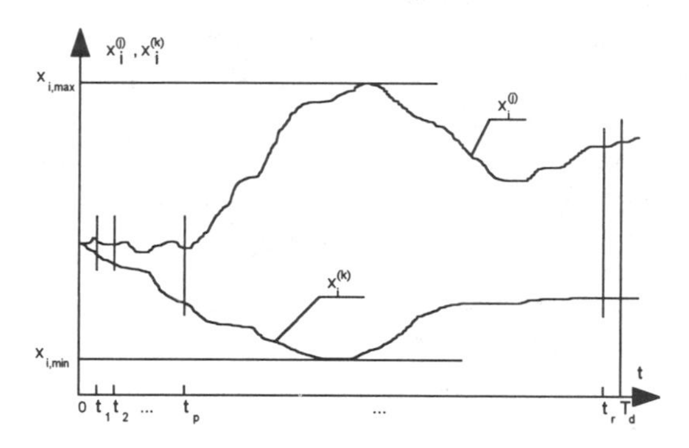

Fig.2 Example of transients for variable x_i in two accidental situations j and k

4 INTEGRATION OF HRA/PSA MODELS AND PROBABILISTIC UNCERTAINTY

4.1 Integration of probabilistic models

In the HRA/PSA screening process (Hannaman 1984, Dougherty 1988) only more probable human failure events are taken into account for further consideration. Reliability models of human factor and probabilistic models of the plant's equipment are integrated at different levels of hierarchy of the plant structure: components, subsystems and systems. Common practice in performing PSA is to construct first the equipment failure oriented event and fault trees for initiating events of interest. The next step is to modify these fault and event trees with regard to events representing human errors. During the quantitative evaluation only more significant human errors are included.

The methodology of the probabilistic modelling of complex systems is based usually upon the concept of logical models with regard to a hierarchy of events. The high hierarchy events are split into subevents for which basic probabilistic models can be established with reference to statistical data (often only generic reliability data are available) or assessments given by experts. The basic models and intermediate probabilistic results are then combined with regard to logical modelling framework to obtain final probabilistic results.

In our expert system there are two knowledge bases: Components Knowledge Base and Systems Knowledge Base. Their declarative part is created based respectively upon topological and functional-logical information. Then in reasoning process using backward/forward chaining the structures of fault/event trees are created. In the next step these trees are modified with regard to events representing human errors of different psychological mechanisms. These events are placed into the structure of the logic model for relevant initiating event at appropriate level of hierarchy. The fault/event trees are then quantified. The reliability data associated with failures of equipment and human errors are taken from relevant project specific data bases: Reliability Data Base (RDB) and Human Reliability Data Base (HRDB). These reliability data bases are created respectively by correction or fusion of components general reliability data or from human reliability models. The process of acquiring of plant specific reliability data and selecting the appropriate reliability models is menu-driven and supervised by the user.

In quantifying of fault and event trees the relevant reliability models, equipment and human factor oriented, are combined to obtain final results (probabilities of top events and accident sequences), optionally with associated uncertainty measures. Thus, more or less "hard" and "soft" probabilistic data and intermediate results are combined at different levels of the model hierarchy to obtain the final probabilistic results. In this process the possible recovery events can be taken into account. There is an option to introduced them during the minimal cut sets review. The relevant procedure is menu driven.

4.2 Representing uncertainties in HRA/PSA

Current PSA/HRA methodologies have been developed adapting the Bayesian subjective probability framework (Apostolakis 1989, Wu 1990) which requires precise defining of events. On the other hand there are encountered cases of events in HRA/PSA practice which can not be straightforwardly quantitatively assessed, due to insufficient knowledge (e.g. concerning the progression of physical processes during some accident conditions or dominant failure phenomena) or imprecision of propositions (using often by experts in evaluations some linguistic statements). For dealing with cases of approximated evaluations other theoretical frameworks can be considered, e.g. the theory of possibility based on the fuzzy set theory or Shafer's theory of evidence (Zadeh 1978, Dubois & Prade 1986, 1988). Some researchers are skeptical as regards applying these new theories for representing and combining information under uncertainties in PSA (Wu 1990). We share this opinion when combining of information from non-equivalent or contradictory sources, including experts, is of interest. In such cases much more attractive is the Bayesian probability framework (Wu 1990).

On the other hand there are known some drawbacks of the Bayesian framework which can lead, in more complex cases, to violating its basic principle (Lee 1987). Therefore, we propose to apply alternatively another framework for representing uncertainties within PSA, based on the possibility theory, in which values of probability will be represented as the fuzzy numbers. Such framework seems to be justified especially in cases when the computerized PSA/HRA studies can not be supported, in some important issues analyzed, by high quality opinions obtained from several domain experts (Table 1, scope I and II).

Our opinion is that it is justified to apply this new framework for initial or preliminary probabilistic assessments. The advantage is that the calculations including uncertainty measures represented by fuzzy numbers are very simple. More important findings from such preliminary analysis can be then analyzed in details with the contribution of several experts. When more precise data and high quality expertise given by several domain experts are available, we think that the Bayesian probability framework might be applied (Table 1, scope III). As it was mentioned this framework is attractive when conflicting expert opinions and data are to be aggregated.

There are expressed often opinions that the fuzzy set theory is especially suited to model man-machine systems. Fuzzy probability measures represented by fuzzy numbers were already used to model human reliability (Terano 1983). The fuzzy probabilities represented as fuzzy numbers can be used for quantitative evaluation of fault and event trees as well as the Human Reliability Analysis Event Trees (HRAETs) and Operation Action Trees (OATs). It has been already suggested that results of probabilistic modelling of complex technological system are rather fuzzy (Volta 1986). Some references on application of this theory, e.g. for human problem solving and advanced modelling of operator behavior, are quoted and relevant methods shortly described in a report (Kosmowski 1992).

The probabilistic data used in conventional PSA, especially concerning human factor reliability, are often subjective and represented by quasi-random variables assuming the log-normal distribution. Such variable Q is usually defined by the median value $q_{50\%}$ and so called error factor EF: $Q=(q_{50\%}, EF)$. The error factor is related to the percentiles of the log-normal distribution (e.g. 10% and 90%) as follows:

$$EF = q_{90\%}/q_{50\%} = q_{50\%}/q_{10\%}.$$

Taking into account the subjectivity of assessments and the fact that often some attributes (shaping factors) are not included systematically, we think that it is more justified to describe such situation using e.g. a triangular fuzzy number: $\tilde{Q}=(q_l, q_m, q_n)$ (Terano, et al. 1983), which might be related to the parameters of described above distribution as follows: $q_m = q_{50\%}$, $q_l = q_m/EF$ and $q_n = q_m EF$.

In general, the conversion of the a quasi-random variable Q_i, characterized by the pdf $\delta_i(q)$, into a fuzzy variable $\tilde{Q}_i$, having the membership function $\mu_i(q)$, can be obtained dividing the pdf $\delta_i(q)$ by $\max_q \delta_i(q)$ (Dubois & Prade 1986):

$$\mu_i(q)=\delta_i(q)/\max_q \delta_i(q).$$

This membership function can be then approximated by a triangular or trapezoid fuzzy number (Tanaka 1983). Other assignments can be proposed, e.g. based on the mean value and standard deviation of a quasi-random variable (Terano 1983). In some cases the possibilistic measures can be obtained directly by combining multi-expert judgments and then represented by a fuzzy number (Dubois 1986).

Fuzzy quantification of a fault tree (or an event tree) is quite simple. Having for each i-th error/failure event the fuzzy numbers $\tilde{Q}_i$, the fuzzy probability of a defined event $\tilde{Q}$ can be calculated using the fuzzy numbers calculus (Tanaka 1983). For instance, for a fault tree having m minimal cut sets M_j, the fuzzy probability of the top event can be calculated according to the formula

$$\tilde{Q} = 1 - \prod_{j=1}^{m} (1 - \prod_{i \in M_j} \tilde{Q}_i).$$

Multiplying fuzzy numbers in this formula and their subtraction from 1, is performed in a similar way as described in (Tanaka 1983). The calculation cost of the fuzzy probability $\tilde{Q}$ (with uncertainty range) is much lower than in the case of quasi-random variable (Monte Carlo simulation usually required), although the uncertainty range will be wider than in the case of most random variables. It can be shown that multiplying two triangular fuzzy numbers: $\tilde{Q}_1 = (q_{l1}, q_{m1}, q_{n1})$, $\tilde{Q}_2 = (q_{l2}, q_{m2}, q_{n2})$ gives the result $\tilde{Q} = (q_{l1}q_{l2}, q_{m1}q_{m2}, q_{n1}q_{n2})$, thus the error factor represented the uncertainty range is a product of the uncertainty ranges of these two fuzzy numbers. In the case of two variables with log-normal distribution the effect of uncertainty propagation in a aggregated variable is lower.

Although some authors are skeptical as regards the application of new non-Bayesian theories in PSA practice, they do not exclude this in the future, when this new theories will become more mature (Wu 1990). On the other hand there have been already proposals to compare some PSA results with thresholds limits of defined quantitative safety criteria represented by the fuzzy membership functions (Apostolakis 1989, Wu 1989). The possibilistic approach seems to be attractive for representing uncertainties in expert system (Zadeh 1983). In the presence of linguistic variables their membership functions must be carefully quantitatively defined.

The research effort is needed to propose methods for dealing with uncertainties in HRA/PSA systematically, especially in quantifying expert opinions (Mosleh 1988).

5 FEATURES OF THE SOFTWARE SYSTEM

Assumed features of the software systems supporting PSA1 and HRA using the expert system technology are described in Table 1. The prototype of the system of scope I has been developed to be applied at the university in a computer laboratory for training purposes (the subject: Reliability and Diagnosis of Technological Systems). The software system of the scope II, at present under development, is foreseen for training and safety related decision making of moderate complexity.

Table 1. Assumed features of the computer aided PSA1/HRA of different scopes using the expert system technology

Features of analyses and software	Scope I	Scope II	Scope III
Exhaustiveness of analyses	Rather low	Moderate	High
Cost of assessments	Relatively low	Moderate	High
Minimum users (Us)/ experts (Es) involved	Single U/E	Single E/U	Several Es/Us
Menu-driven help and explanation	Extensive	Moderate	Moderate
Heuristics	Simple	Detailed	Consensus
Simulation codes	Indirect access	Simplified models	Direct access
Object oriented modelling	Some	Limited	Advanced
Dependent failure analyses	Simplified	Limited	Exhausti-ve
HRA models or techniques used	Simplified (Fig.1/I)	More detailed (Fig.1/II)	Multiple experts (Fig.1/III)
Uncertainty treating frameworks	Fuzzy or Bayesian probability	Fuzzy or Bayesian probability	Bayesian or fuzzy probability
Topological and functional-logical data	Simplified	Detailed	Detailed
CAD graphical representation	Simplified	Detailed	Detailed
Semi-automatic event/fault trees construction	Some	Most	Most
Using for HRA/PSA training	Initial training	Detailed training	Research/ advanced training
Using for safety related decision making	Initial decisions	Prelimi-nary decisions	Particular issues and final decisions
Hardware required	PC-386	PC-486	Work-station

The design of the software system of the third level is foreseen in the future for dealing with particular issues of HRA/PSA and final decision making. Development of its knowledge base, object oriented models and external data bases requires considerable research and engineering effort with contribution of experienced PSA/HRAs experts. It is obvious that expert contribution in supervising of the modelling process and a high quality expertise will be required. It was assumed that for some more important and difficult issues the opinions of several expert will be available (e.g. using such HRA quantification techniques as SLIM or APJ - Fig1., effort III). The whole analysis process and PSA/HRA results will be documented to enable scrutinizing or auditing..

Till now our main effort was aimed at developing: the CAD graphical support, topological and functional-logical data bases (the approach is general and useful also for the scope III), reliability data bases, HRA methods (computerized THERP technique have been developed) and probabilistic modelling of the plant using the expert system technology. The heuristics associated with fault/event trees construction is represented in rules appropriately selected in inferring process. Several sets of rules have been distinguished and meta-rules proposed that control the inferring process.

Our previous designing and testing effort have been directed to integrate data bases, models and software packages including AutoCAD with a C based PC shell (KAPPA PC from Intellicorp, Inc.) to be run under MS Windows 3.1. It included also the interface module to the Reliability Data Base (RDB) of IAEA enabling correction of data and module of a computerized version of the THERP technique. The edition of P&I diagrams of front-line and supporting systems (topological data) is performed using a computer program TSIDE written in AutoLISP to be run within the AutoCAD system, version 12, generating the ASCII files that create the Topological Data Base. A special program has been also developed to represent graphically functional-logical information for each initiating event and to create the Functional-logical Data Base.

Our research plans for the near future include object oriented deterministic modelling to be applied to support the PSA. The purpose is to base the logical modelling and human reliability modelling, on object oriented hierarchical models of the system analyzed. The modelling concept is similar to that described in (Dobrzeniecki 1989). Relevant models should posses a deep knowledge properties to be verified with regard to results of simulations obtained from reference deterministic simulation codes.

6 CONCLUDING REMARKS

The development of the software system for computer aided PSA/HRA using the expert system technology require considerable research and designing effort. In the paper an integrated approach is proposed to carry out research and design works with regard to resources planned to be involved in the project to cover gradually in a balanced manner the important issues of the PSA1. Three scopes of the PSA/HRA analyses and related software have been distinguished. A tree was proposed to facilitate the selection of human reliability techniques for a psychological error mechanism of interest and the scope of the analysis.

To realize this concept an integrated software environment has been designed including a CAD system, data bases, a shell for building expert systems and conventional modules. The system consists of general and plant specific data and knowledge. It helps to acquire new data and knowledge from domain experts. Some tasks within PSA can be partly automated, e.g. the construction of fault and event trees. These trees are then modified with regard to human failure events to be included at different levels of hierarchy of the model logic structure. The process of human reliability modelling is menu-driven with a significant contribution of expert(s). A method is proposed to support probabilistic evaluation of mistake (an error of commission) in the event driven situation. The process of PSA/HRA is documented to enable easy scrutinizing and auditing of results. The system can be used for training purposes and safety related decision making.

Additional research effort is required related mainly to PSA/HRA of the scope III. It should include such topics as:

- an advanced framework for representing and treating of imprecision and uncertainties at different levels of the model structure,
- combining the quantitative information from different quality sources including experts,
- effective probabilistic evaluation of accident scenarios under uncertainties with regard to the equipment oriented logic models and with the inclusion of human induced failures events, including intention failures, as well as recovery events.

ACKNOWLEDGMENTS

The authors would like to thank the IAEA, Vienna for supporting the research works and Dr. M. Dusic, Division of Nuclear Safety for valuable discussions and written materials concerning Human Reliability

Analysis (HRA) performed within Probabilistic Safety Analyses (PSA) as well as the Committee of Scientific Research in Warsaw for financial support of the research and designing works aimed at developing a PSA/HRA expert system.

REFERENCES

Apostolakis, G.E. 1989. Uncertainty in probabilistic safety assessment. Nuclear Engineering and Design, Vol. 115, pp. 173-179.

Apostolakis, G., P. Kafka 1992. Advances in probabilistic safety assessment. Nuclear Engineering and Design, Vol. 134, pp.141-148.

Cacciabue, P.C. 1988. Evaluation of human factors and man-machine problems in the safety of nuclear power plants. Nuclear Engineering and Design, Vol. 109, pp. 417-431.

Dobrzeniecki, A.B., L.M. Lidsky1989. Modelling an analysis of complex physical systems using model-based reasoning with constraint satisfaction. A paper presented at the IAEA Control Systems Society, Knoxville, TN, May 1989.

Dougherty, E.M., J.R. Fragola 1988. Human Reliability Analysis: A Systems Engineering Approach with Nuclear Power Plant Applications. A Wiley-Interscience Publication, John Wiley & Sons Inc., New York.

Dougherty, E.M. 1990. Human reliability analysis - where should thou turn ? (Guest Editorial). Reliability Engineering and System Safety, Vol. 29.

Dubois D., H. Prade 1986. Fuzzy sets and statistical data. European Journal of Operational Research. Vol. 25, pp. 345-356.

Dubois, D., H. Prade 1988. Possibility Theory: An Approach to Computerized Processing of Uncertainty. Plenum Press, New York.

Gertman, D.I. et al. 1992. INTENT: a method for estimating human error probabilities for decisionbased errors. Reliability Engineering and System Safety, Vol. 35, pp. 127-136.

Hannaman, G.W., A.J. Spurgin 1984. Systematic Human Action Reliability Procedures (SHARP). EPRI NP-3583, Research Project 2170-3.

Humphreys, P. (ed.) 1988. Human Reliability Assessor Guide. Safety and Reliability Directorate, UK, RTS 88/95Q.

IAEA 1990 (International Atomic Energy Agency). Use of Expert Systems in Nuclear Safety. IAEA-TECDOC-542, Vienna.

IAEA 1992 (International Atomic Energy Agency). Procedure for Conducting Human Reliability Analysis in Probabilistic Safety Assessment (Draft of a Report).

Kosmowski, K.T., Z. Beker, W. Chotkowski, K. Duzinkiewicz 1991. Reliability Evaluation and Probabilistic Safety Analysis (level 1) Expert System: REPSA1ES. Report of the Research Contract No. 6070/RB, Institute for Electrical Power and Control Engineering, Technical University of Gdansk, September 1991.

Kosmowski, K.T. 1992. Assessment of human factor reliability modelling techniques for application within an PSA expert system. An Internal Report, Technical University of Gdansk, Department of Electrical Engineering, Division of Control Engineering, Gdansk.

Lee, N., Y.L. Grize, K. Dehnad 1987. Quantitative models for reasoning under uncertainty in knowledge-based expert systems. International Journal of Intelligent Systems, Vol. II, pp. 15-38.

Mosleh, A., V.M. Bier, G. Apostolakis 1988. A Critique of current practice for the use of expert opinions in probabilistic risk assessment. Reliability Engineering and System Safety, Vol. 20, pp.63-85.

Reason, J. 1990. Human Error, Cambridge University Press.

Tanaka H. et al. 1983. Fault-tree analysis by fuzzy probability. IEEE Transactions on Reliability, Vol. R-32, No.5, pp. 453-457.

Terano T., Y. Murayama, N. Akiyama 1983. Human reliability and safety evaluation of man-machine systems. Automatica, Vol.19, No.6, pp.719-722.

Volta, G., H. Otway 1986. The logic of probabilistic risk assessment versus decision levels. Nuclear Engineering and Design, Vol. 93, pp.329-334.

Wakefield, D.J. 1988. Application of the human cognitive reliability model and confusion matrix approach in a Probabilistic Risk Assessment. Reliability Engineering and System Safety, Vol. 22, pp. 295-312.

Wenda, W.F. 1975. Engineering Psychology and Synthesis of Systems of Information Presentation (in Russian). Machinostroyenie, Moscow.

Wu, J.S., G.E. Apostolakis, D. Okrent 1989. Probabilistic Risk Assessment and intelligent decision support systems. Nuclear Engineering and Design, Vol. 113, pp. 269-282.

Wu, J.S., G.E. Apostolakis, D. Okrent 1990. Uncertainties in system analysis: probabilistic versus nonprobabilistic theories. Reliability Engineering and System Safety, Vol. 30, pp. 163-181.

Zadeh, L.A. 1978. Fuzzy sets as a basis for theory of possibility. Fuzzy Sets and Systems (1), pp. 2-28.

Zadeh L.A. 1983. The role of fuzzy logic in the management of uncertainty in expert systems. Fuzzy Sets and Systems, Vol. 11, pp. 199-227.

Structural Safety & Reliability, Schuëller, Shinozuka & Yao (eds) © 1994 Balkema, Rotterdam, ISBN 90 5410 357 4

Target reliability levels from social indicators

Niels C. Lind
Instituto de Matemáticas y Física Fundamental, Consejo Superior de Investigaciones Científicas, Madrid, Spain

ABSTRACT: Social indicators that include life expectancy and gross domestic product can be used to to select acceptable reliability levels and other criteria of "enough safety". The safety level is acceptable according to a social indicator if its increment from expected life time lost and cost are not negative. It is argued that risk of loss of life in the future should be discounted at the same rate as finances. An example, selection of safety level of a dam in an earthquake zone, shows how the optimal target safety level is calculated from the Life Product Index, and illustrates the importance of discounting.

1. INTRODUCTION

Now that we (think we) can calculate the reliability of a structure, we must admit that this great skill is not matched by much knowledge of what reliability value to aim for. Socio-economic optimization (Level IV design) hasn't been accomplished as yet. It has at best only been approached by calculating the level at which safety costs about the same per "life saved" as in other contexts where health and safety is being managed. Whether that cost is too high (or lives lost too cheaply, conversely) is hard to say.

Professionals who manage health and safety don't openly render account of their key decisions. The arena is left open and attractive for pressure by special-interest advocates, e.g. people who think that the economy is being choked by costly regulation, or others who call *any* risk, however small, an "unacceptable" violation of their rights.

Social indicators may give a fresh insight in this problem by facilitating quantitative open accounting, laying bare inconsistencies and inefficiencies, and supporting unification of public health and safety policy. The promised payoff is greater health and safety for all at less cost.

Social indicators are statistics that reflect some aspect of the quality of life in a society or group of individuals. In the public management of risk in a society the overall objective should be to serve the common good in a consistent and defensible manner. Tolerable levels of risk or, equivalently, acceptable levels of safety should be determined in relation to costs of all kinds, supported by explicit accounting using a broadly acceptable scale. This paper shows how quantitative safety criteria can be derived from compound social indicators that aim to reflect broadly accepted goals that may carry labels such as "national development", high expectancy of "quality-adjusted life", the "common good" or the "public interest".

Basic social indicators are statistical time series, eg. life expectancy, gross domestic product, population, and adult literacy rate. Various compound social indicators are actively being developed and validated (See e.g. the Journal *Social Indicators Research*).

Compound social indicators are generally

compiled to reflect the common good. Compound social indicators are often aimed to portray a nation's performance in serving its people. The Human Development Index (HDI) of the United Nations Development Program (UNDP 1990) is an example. Its composition is briefly described in the Appendix.

Any undertaking (project, program or regulation, adoption of a new therapy, etc.) that affects the public by changing health or risk and expenditure will have an expected impact on a a compound social indicator. Suppose that this indicator is accepted as a valid indicator of some aspect of the common good. If the net impact of the undertaking is negative, then the proponent of the undertaking ought to explain why, nevertheless, the public interest is served by the undertaking. Conversely, a positive net impact on the valid social indicator would lend support to acceptance.

A general criterion of the *acceptability* of an undertaking or the *tolerability* of a risk has been derived by this reasoning (Lind *et al.* 1991, Lind 1992b). The criterion compares net economic benefit or cost with net risk or gain in the expectancy of quality-adjusted life. The criterion was specialized for four measures, of quite different origin: The HDI, the Life Product Index (LPI), the Life Time Efficiency (LTE), and the GDP limit. For a broad set of 26 implemented health and safety programs documented in the U.S. Federal budget these criteria are remarkably consistent; equivalent judgment was passed by the HDI, the LPI and the two other criteria on 25 out of the 26 programmes (Lind *et al.* 1991).

2. DERIVATION OF A SAFETY CRITERION

A *compound social indicator f* is a function of other social indicators *a, b, ...*:

[1] $f = f(a,b,\ldots,e,\ldots).$

Suppose that f is differentiable and denote increments by d and partial derivatives by subscripts. This gives for infinitesimal increments

[2] $df = f_a\,da + f_b\,db + \ldots + f_e\,de + \ldots.$

In particular, if there is no change to the variables other than *b* and *e*, then df vanishes if

[3] $db/de = -\,f_b/f_e.$

Two particular social indicators considered here are the HDI and the LPI, described in the Appendix. For both, [3] may be specialized into the form

[4] $db/de = b/E,$

in which *E* is a compound statistical time series in the same units as *e*, ie. years.

E in [4] is specific to each indicator. For the Life Product Index LPI, *b* denotes the gross domestic product per person per year and *e* denotes the life expectancy at birth. Further, *w* denotes the proportion of total life time spent purely in economic activity, i.e. net of any "work satisfaction",

[5] $f_{(LPI)} = b^w e.$

Then for the LPI *E* equals *we*. For Canada in the 1990s this gives (see Appendix)

[6] $E = we = (0.1)(80\text{ years}) = 8.0\text{ years}.$

For the HDI ca. 1990 the value of E equalled about 8.3 years. In view of the differences between the rationales of these two indicators, the agreement is remarkable. Some confidence in the value of E in [5] has been gained also in the context of health and safety regulation.

3. STRUCTURAL SAFETY CRITERION

[4] is specialized into a safety criterion for a structure as follows. Denote the probability of failure by P and the number of fatalities expected in case of failure by Z. The infinitesimal change in the expected number of fatalities is then *Z dP*. If the average life expectancy of the population exposed is e_{ave}, then the total expected number of life years lost

is $e_{ave}ZdP$, and so

[7] $de = - e_{ave} Z dP/N$

where N is the population. Let the total cost of the structure be c, and consider an infinitesimal change dc. This change does not contribute materially to the economy, because the structure will render the same service etc. as before. The increased cost reduces the GDP per person by

[8] $db = - dc/N$.

[6] and [8] transform [4] into

[9] $db/de = (-dc/N)/(- e_{ave} Z dP/N) = b/E$,

giving the criterion of optimality in the form

[10] $A dc + Z dp = 0$,

specifically

[11] $(w/b)(e/e_{AVE}) dc + Z dP = 0$.

The values we = 8 years, b = US$ 20,000 /year and e_{ave} = 40 years are inserted in [11] to give $A = 10^{-5}$ person-yr $\$^{-1}$. This value would be appropriate for developed countries in the 1990s.

4. DISCOUNTING OF RISKS

Discounting of future quantities is crucial in the evaluation of any proposal. While discounting is an economical necessity when dealing with monetary quantities, it is not widely accepted for non-monetary entities such as human life or health. For example, the U.S. Environmental Protection Agency in evaluating regulations to ban asbestos, simply added up the number of lives that would be "saved" regardless of *when* they would be saved, while in other regulations (controlling the risk of chemical plant accidents) it dealt with immediate benefits.

The principal arguments against discounting future quantities of risk are (a) that it "monetarizes" human life because risk concerns possible loss of life, and (b) that it serves to justify a disregard of future generations and therefore is selfish or unfair. The first argument suggests that money is lower than risk to life in some imagined hierarchy, so that no amount of money can compensate for any amount of risk to life. Since many people risk their lives in economic activity (few occupations are as safe as staying home if you can avoid the stairs), this is not borne out by observed human behaviour. The second argument, similarly, does not hold. It would require those who hold it true to explain why discounting of money is not also selfish and unfair. Still, discounting of future risk is an issue that deserves serious scrutiny before adoption.

Future generations have no bargaining power, and one can only speculate about their preferences. How can one then justify decisions that are binding upon them? The moral problem concerning future generations is not different than the general moral problem concerning persons now living. The solutions, from (i) moral philosophy, (ii) economics, and (iii) asking members of the public, are unanimous: *Risk should be discounted at the same rate as money.*

(i) The moral philosophy solution to this problem is well established. It is the symmetry requirement of the "golden rule" ("Do unto others as you wish to be done unto" or the like). The golden rule is a fundamental ethical truth, prominent in all major religions and ethical systems. In its most developed modern form it is expressed in the *categorical imperative* of Kant (1786), from which all specific moral duties can be derived: "Act only on that maxim which you can will to be a universal law".

As presented by Paté-Cornell (1984), discounting follows from opportunity cost as a fact of life and may be based on the following ethical principles of *equal health and safety opportunity*: "(1) A [human] life at all times is equally valuable to society at any other time, and should be equally protected under the same basic laws and principles that prevail when the decision is made. (2) We should make provision so that each human being can receive in the future an equal amount of lifesaving technology at the time when it is needed, and it should be at least equal to that available to current generations. (3) The capital set aside for future safety will be used by future generations according to their preferences, that is, they may choose to actually spend it for their health safety, or to consume it, or to invest it further."

Among the implications of this are that current generations have the right to use resources and create risks for the future, but only to the extent that they would do the same if they were to live with the consequences. Also, according to Paté-Cornell (1984), if economic efficiency is the accepted criterion then it is essential that compensation actually occur in the future in one form or another, for example in the form of accumulated wealth or increased longevity. In particular, if we accept that today different persons may incur and share risks and benefits, then this must be acceptable for the future as well.

In this reasoning there is no imposition of our values on the future, no presumption that posterity may not have different values. Indeed, their values will likely be different from ours. "The Kantian aspect ... is to treat future generations in the same way as we want to be treated today" (Paté-Cornell 1984). Many issues are irrelevant, such as the qualitative difference between catastrophic and conventional risk, reversibility or irreversibility of risks, or time horizon.

The main result that follows from these principles and from the requirement of self-consistency independent of when the analysis is performed, is: *All costs and benefits, including future risks to life or health, economic or cultural values, heritage, and the environment should be discounted at the same rate.*

(ii) Similar results have been obtained by the requirement of consistent economics (eg. Keeler and Cretin 1983) on the basis of economic dominance among different life-saving policies. Bordley (1990) recently applied economic reasoning and obtained the same result for life risk reduction policies, showing that discounted longevity as a measure of benefits can be deduced from a utility-maximization model.

(iii) Discounting is a matter of values, and some people are only persuaded if the requirements of ethical theory and of rational accounting are also in agreement with public perception. Recently Cropper *et al.* (1992) have investigated how members of the public feel about discounting of human lives saved. The central question they asked was what fraction of a life saved today (*X*) counts the same as a life saved at some time in the future (*Y*). They surveyed some 1,600 households in Maryland and the District of Columbia and 1000 households nationwide in the USA. They reported the mean values given in row (4) of Table 1. Their results may be surprising but appear consistent between the three polls. They can be summarized as simple interest at 20 %pa for the first 50 years and 2.81 %pa compounded annually for the next 50 years.

Table 1. The *Lives Ratio* function, i.e. the Number of Lives 'Saved' in the Future Equivalent to One Life 'Saved' Today (Cropper and Portney 1992).

(1) Time Horizon, yrs	5	10	25	50	100
(2) Poll	USA	USA	MD	DC	MD
(3) Sample Size	475	480	462	528	442
(4) Lives Ratio Y/X	2	3	6	11	44
(5) Discount Rate, %	17	11	7	5	4

Although these pioneering results should be regarded as preliminary, and perhaps specific to the USA in the late 20th century, it it difficult to argue against or ignore this *vox populi* because of the large sample size and the consistency. There is also excellent agreement with people's discount rates for money, revealed in the national survey and inferred independently in other studies from purchasing behaviour about energy-saving appliances and military re-enlistment bonuses.

Moreover, people with high discount rates for money have also high rates for saving lives (Cropper *et al.* 1992). So, public opinion appears self-consistent: *discount future lives like money*. It follows that life risk should be discounted similarly.

How is a social index affected? That depends, of course, on the composition of the index. The HDI and the LPI are not changed by the need to discount risk. Although they incorporate life expectancy, it must be remembered that life expectancy is merely a weighted sum of *present* mortalities, which are not subject to discounting.

Yet, discounting has a profound influence on the quantities that enter a risk management decision, as illustrated in the following example.

5. EXAMPLE

A dam is to be constructed in an earthquake zone. Note: the data in this section are fictitious (though meant to be realistic), intended only to illustrate the procedure. The maximum ground acceleration, S, in a 5-year period is given (using suitable units) by the exponential distribution

$$[12] \quad F_S(s) = 1 - \exp(-s);$$

the net resistance of the dam, R, has a Gaussian density function

$$[13] \quad f_R(r) = (2\pi)^{-1/2}(av)^{-1}\exp[-(r/a-1/v)^2/2)$$

in which the coefficient of variation, v, is given as 0.25, while a is the mean resistance and the central safety factor; a is the design parameter to be determined.

Failure in any 5-year period has the probability

$$[14] \quad P = \int_0^\infty [1 - F_S(x)]\, f_R(x)\, dx$$

$$= N(1/v-av)\exp[-a(1-av^2)],$$

in which N(.) is the standardized normal distribution function. P is assumed independent of failure in any other 5-year period.

If failure occurs, the expected loss is 2000 lives and US$1875 million. The dam has a projected service life of 125years. Maintenance at 1 % of initial cost is scheduled every 20 years respectively.

The analysis was carried out in tabular form for several combinations of the rates of discounting and design safety factors. One particular instance of the analysis is shown in Table 2. The initial cost of that dam was $M 1500. The summed costs in the last line are carried into line 3 of Table 3.

Table 3 gives the results: The change of sign in columns J and L show that the optimal value of the safety factor changes from about 11 to 8.5 if risk is discounted like money. The required probability of failure is increased about threefold.

6. CONCLUSIONS

Some compound social indicators are functions of the level of economic activity and the mortality of the population, among other variables. Any such social indicator implies a relative value placed upon wealth in relation to safety and life expectancy in good health. Correspondingly, there is an implied value of the optimal safety level for a structure, reflecting the *marginal* cost of risk reduction.

Two recently proposed social indicators are considered in this paper and explained briefly in the appendix, namely the Human Development Index and the Life Product Index. These two indices are in agreement that the economic equivalent of a health-related quality-adjusted life year in technologically advanced countries in the last decade of the 20th century is roughly equal to US$ 100,000.

Future risk should be reduced to a present value by discounting at the same rate as finances.

7. ACKNOWLEDGMENTS

Work reported in this paper was carried out partly with the financial support of the Natural Sciences and Engineering Research Council of Canada and the Consejo Superior de Investigaciones Científicas, Spain.

Table 2. Example - Sample Calculations (see text).

COV(R) = 0.25;
Central Safety Factor a = 6;
5-year conditional Probability of Failure = 0.0075870;
5-year conditional Probability of Survival = 0.9924130;
Compound interest rates: 3.8 %pa financial risk, 3.8 %pa for life risk.

Time Period, Years:	Probability of: Survival	Failure in 5-year Per.	Discount Factor for: $	Risk	Cost or Loss: Not discounted $M	Life	Discounted $M	Life
- 0	1.0000000	0.0000000	1	1	1500.0	20.0	1500.0	20.0
0 - 5	0.9924130	0.0075294	0.83	0.83	14.1	15.1	11.7	12.5
5 - 10	0.9848836	0.0074723	0.69	0.69	14.0	14.9	9.6	10.3
10 - 15	0.9774113	0.0074156	0.57	0.57	13.9	14.8	7.9	8.5
15 - 20	0.9699956	0.0073594	0.47	0.47	13.8	14.7	6.5	7.0
20 - 25	0.9626363	0.0073035	0.39	0.39	163.7	14.6	64.4	5.7
25 - 30	0.9553328	0.0072481	0.33	0.33	13.6	14.5	4.4	4.7
30 - 35	0.9480846	0.0071931	0.27	0.27	13.5	14.4	3.7	3.9
35 - 40	0.9408915	0.0071385	0.22	0.22	13.4	14.3	3.0	3.2
40 - 45	0.9337530	0.0070844	0.19	0.19	163.3	14.2	30.5	2.6
45 - 50	0.9266686	0.0070306	0.15	0.15	13.2	14.1	2.0	2.2
50 - 55	0.9196380	0.0069773	0.13	0.13	13.1	14.0	1.7	1.8
55 - 60	0.9126607	0.0069244	0.11	0.11	13.0	13.8	1.4	1.5
60 - 65	0.9057363	0.0068718	0.09	0.09	162.9	13.7	14.4	1.2
65 - 70	0.8988645	0.0068197	0.07	0.07	12.8	13.6	0.9	1.0
70 - 75	0.8920448	0.0067679	0.06	0.06	12.7	13.5	0.8	0.8
75 - 80	0.8852769	0.0067166	0.05	0.05	12.6	13.4	0.6	0.7
80 - 85	0.8785603	0.0066656	0.04	0.04	162.5	13.3	6.8	0.6
85 - 90	0.8718946	0.0066151	0.03	0.03	12.4	13.2	0.4	0.5
90 - 95	0.8652796	0.0065649	0.03	0.03	12.3	13.1	0.4	0.4
95 -100	0.8587147	0.0065151	0.02	0.02	12.2	13.0	0.3	0.3
100 -105	0.8521996	0.0064656	0.02	0.02	162.1	12.9	3.2	0.3
105 -110	0.8457340	0.0064166	0.02	0.02	12.0	12.8	0.2	0.2
110 -115	0.8393174	0.0063679	0.01	0.01	11.9	12.7	0.2	0.2
115 120	0.8329495	0.0063196	0.01	0.01	11.8	12.6	0.1	0.1
120 125	0.8266299	0.0062716	0.01	0.01	161.8	12.5	1.5	0.1
0 -125		0.1720547			2722.6	364.1	1676.9	90.3

Table 3. Costs and Losses vs. Central Safety Factor

Compound interest rate: 3.8 %pa financial, i %pa for risk.

Central Safety Factor a	Prob.of Failure in 5yrs P	Initial Cost c,$M	Costs and losses: No Disc. $M	LL	With Disc. $M	LL	Increments: (H) 3.8% dc	(I) 3.8% ZdP	(J) 3.8% D	(K) 0% ZdP	(L) 0% D
2	0.153319	1300	3434	1687	1985	744					
4	0.030157	1400	3148	1058	1720	268	-265	-476	-24340	-629	-30460
6	0.007587	1500	2723	364	1677	90	-43	-178	-7980	-694	-28620
8	0.002422	1600	2677	137	1742	43	65	-47	-580	-227	-7780
10	0.000964	1700	2771	68	1836	29	94	-14	1320	-69	-880
12	0.000465	1790	2890	43	1928	24	92	-5	1640	-25	+840
14	0.000263	1880	3023	33	2022	24	94	0	1880	-10	+1480
16	0.000167	1960	3146	28	2107	22	85	-2	1620	-5	+1500
18	0.000117	2040	3271	26	2192	21	85	-1	1660	-2	+1620
20	0.000088	2120	3398	24	2278	21	86	0	1720	-2	+1640

8. APPENDIX

8.1 *Life Product Index*

The Life Product Index (LPI) is derived from three basic aggregated indicators that are widely available and accurate (Lind 1992b). The LPI is intended as a model of "quality-adjusted life expectancy". It is a function $b^w e$ of (1) the real gross domestic product, GDP, per person per year, denoted by b, and (2) the life expectancy at birth, denoted by e (see [5]). The parameter w is a reflection of the value placed on a reduction of mortality in terms of economic expenditure. This value is practically a constant, evident from the time budget of a nation; it equals the proportion of total life time spent purely in economic activity, i.e. net of any "work satisfaction" etc.

The Life Product Index is an approximate measure of benefit to the public. The real gross domestic product per person, b, is a measure of the average share of the production of wealth available to persons to spend on whatever they find will add most to the enjoyment of life. The life expectancy of those persons at birth, e, is then an appropriate factor on b to account for the *duration* of that enjoyment. Conversely, the enjoyment of life has two dimensions: duration and intensity. If the duration is measured by e, then b can serve as weighting factor to express the *intensity* of enjoyment that an average person can expect from a life spent in that society.

Work consumes roughly 1/7 of the average person's life nowadays in North America. For a developing society an estimate gave w closer to 20%; w does not vary much between nations. When normalized with respect to a reference nation the life product is insensitive to the value of w (Lind 1992b).

Time spent at work produces - together with invested capital (that can be measured by time spent in the past) - the average person's share of the GDP, but it also produces some "work satisfaction", which is difficult to define and measure. However, most work in this world is hard, dull, dangerous or uncomfortable, and little work would get done if it were not for economic benefit. The gross pay represents a large and quantifiable proportion, estimated at roughly 3/4 of the total benefit a person receives from work. The remainder may be called the "work satisfaction". Thus $w = (3/4)(1/7) = 0.1$ approx. For less developed countries w would likely be greater; the value $w = 1/6$ has been suggested for broad international comparisons (Lind *et al.* 1991).

Equation [5] gives the *Life Product Criterion*

$$\text{[A1]} \quad D \equiv de/e + w\, db/b > 0$$

suitable for comparison of a pair of options that differ by de and db. D is stationary at optimum, giving [6].

The LPI criterion [A1] rests on a relationship between four seemingly unrelated entities: the public good, the time budget of the public, tolerable risk, and optimality of an undertaking. "Optimality" is thus linked to the goal of serving the "public good".

8.2 *Human Development Index*

The Human Development Index (HDI) was produced by the United Nations Development Programme to rank the nations of the world (UNDP 1990). The HDI reflects the view that development enlarges people's choices. The HDI is composed of three basic indicators of longevity, knowledge, and purchasing power. Life expectancy at birth, e, quantifies longevity. Literacy is quantified in terms of estimated adult literacy, a. Command over resources needed for a decent living is expressed in Purchasing Power Parity of Real GDP per capita, adjusted to account for national differences in exchange rates, tariffs and tradeable goods. It is also truncated approximately at the level of Mexico. The common logarithm of this quantity is b.

The three basic indicators a, b and e are compounded into the HDI in a simple way. First, the range of each for all 130 nations (the set under study) is calculated, and the relative position of the country in each range is scored. The HDI of a country is the average of its three scores.

The validity of the HDI has been questioned

by many. Much of the criticism falls on the adult literacy component. However, adult literacy is rarely affected by safety and health policy options. The HDI is then analogous to the LPI in analysis. For further details about the HDI see UNDP (1990), Lind *et al.* (1991) and Lind (1992a).

9. REFERENCES

Bordley, R. F., "Measuring Risk Reduction Benefits with Discounted Longevity," *Operations Research*, 38, 5, September-October 1990, pp. 815-819.

Cropper, M. L. and Portney, P. R., "Discounting Human Lives," *Resources*, 108, pp. 1-4, Summer 1992.

Kant, I., *Grundlegung der Metaphysik der Sitten*, 2nd. ed., Riga, 1786, see Edwards, P. *The Encyclopedia of Philosophy*, 3, p. 95, Macmillan Publishing Co., New York, NY, 1967.

Keeler, E. B., and Cretin, S., "Discounting of Life Saving and other Nonmonetary Effects," *Management Science*, Vol. 29, No. 3, pp. 300-306, March 1983.

Lind, N. C. (1992a) "Some Thoughts on the Human Development Index," *Social Indicators Research*, Kluwer Academic Publishers, 27, pp. 89-101.

Lind, N. C. (1992b) "A Compound Index of National Development," *Social Indicators Research*, Kluwer Academic Publishers, 28, pp. 325-342.

Lind, N. C., Nathwani, J. S. and Siddall, E., *Managing Risks in the Public Interest*, Institute for Risk Research, University of Waterloo, Waterloo, ON, 1991.

Lind, N. C., Nathwani, J. S. and Siddall, E., "Management of risk in the public interest," *Canadian Journal of Civil Engineering*, 18, 1991, pp. 446-453.

Paté-Cornell, M. E., "Discounting in Risk Analysis: Capital vs. Human Safety," in *Proceedings of the Symposium on Structural Technology and Risk*, Grigoriu, M. (ed.),July17-20, 1983, University of Waterloo Press, Waterloo, ON, 1984.

UNDP (United Nations Development Programme), *Human Development Report 1990*, Oxford University Press, 1990.

United States. Office of Management and Budget, *The Federal Budget of the United States for Fiscal Year 1992*, Washington, DC, 1991, part 2, IX.C. Reforming Regulation and Managing Risk-Reduction Sensibly, pp. 367-376.

Structural Safety & Reliability, Schuëller, Shinozuka & Yao (eds) © 1994 Balkema, Rotterdam, ISBN 90 5410 357 4

Quantitative risk assessment of large structural systems

J.B.Mander
Department of Civil Engineering, State University of New York at Buffalo, N.Y., USA

D.G.Elms
Department of Civil Engineering, University of Canterbury, Christchurch, New Zealand

ABSTRACT: A methodology and the underlying principles for the use of quantitative risk assessment techniques are presented in this paper. Multiple fault and/or event trees may be used, and the process by which a believable result is obtained from sparse and sometimes low-quality data is set forth: it requires an iterative approach bearing in mind a set of guiding criteria, which are described. Case studies are presented concerning the general risk exposure levels faced by (i) locomotive engineers, (ii) motor vehicle users of bridge structures that are faced with collapse by catastrophic earthquake motions, and (iii) the personal risk to occupants in buildings that require seismic retrofitting.

1 INTRODUCTION

Failure of a large engineering or structural system may lead to: loss of life or limb; loss or severe damage to the constructed facility itself; material and commercial loss; and down-time within the system leading to revenue loss and hence profit. Each one of these losses could be the focus of a risk analysis. For structural systems, engineers generally focus on the economical minimization of the probability of structural failure, often without due regard to the consequences of such a failure. Importance may be placed on: the number of people who may use the structure; ownership, whether public or private; and if the system is an essential or non-essential facility after a catastrophe. However, the measures of importance used are somewhat arbitrary and are based on collective experience and engineering judgment, rather than rigorous analysis. The problem with this conventional first order-second moment reliability approach is when there is little or no historical experience in a given field, it is difficult to obtain a reliable and calibrated model for future design applications.

An alternative approach is to focus on the risk exposure to the human user of the constructed facility. In this paper the risk to life and limb is examined using a number of existing quantitative risk assessment techniques. A methodology is developed which uses multiple fault and/or event trees and its utility demonstrated through a series of case studies.

2 GENERAL RISK

Risk is a complex notion, a binary idea with two distinct components: the probability of an event, and its consequences. A Quantitative Risk Assessment (QRA), therefore, becomes the product of the two and needs two data sets: (i) probabilities and (ii) consequences. For environmental risk situations a major problem often lies in deciding upon a common and acceptable measure of the consequences of an event.

Traditional first-order-second-moment reliability analysis implicitly focuses on failure probabilities and results in a comparative reliability index (β), and is only loosely concerned with post-failure consequences. These are usually tied to forms of economic or financial loss as the primary measure of the consequences. In this paper, economic loss will be neglected, and the loss of life taken as the primary measure of the consequences of accidents resulting from failure due to environment effects.

The Fatal Accident Rate (FAR) is adopted herein as the basic measure of risk. FAR can be heuristically thought of as: the number of fatalities per 1000 working lives over 40 years, each person working 2500 hours per year. FAR can be defined more formally as:

$$FAR = 10^8 \, P[F] \, / \, T_h \qquad (1)$$

in which $P[F]$ = probability of a fatality, and T_h = the exposure time in person hours.

Table 1 presents some typical FAR values found in the literature (see Kletz (1978), Lees (1980), Elms and Mander (1990)). It should be noted that these values show the transient nature of risk. Clearly ones risk exposure depends on what one is doing at a specific time. For example, a coal miner might commence the day at home, travel to work by car, and then work as a miner resulting in FAR values of 1, 57, and 50, respectively. FAR values for occupational risk appear to fall in the range of 5 to 70.

According to statistics in the United States, the death rate from accidents and adverse effects is 37.2 per 100,000 population of which 18.9 is from motor vehicle accidents. Assuming a uniform risk exposure, these accident rates translate into FAR values of 4.2 and 2.1, respectively.

Table 1. Typical FAR values.

Activity or risk exposure	FAR
Construction	67
Travelling by car	57
Coal mining	50
Railway shunters (UK)	45
Metal manufacturing	45
Locomotive engineers (NZR)	13
Agriculture	10
Chemical processing	5
Sleeping	1

3 COMPONENTS OF RISK

General risk for a specific activity or occupation is given by a composite FAR. This can be disaggregated into components of risk which may arise from different accident types or failure modes. Fig. 1 shows a general approach to modelling the components of risk. The process is cyclical and proceeds as follows.

Firstly specific failure modes are identified and corroborated by a synthesis of historical accident records, if any. Secondly, these components of risk can be separately modelled with the use of fault-trees and/or event-trees. Fault-trees disaggregate risk into lower levels of detail. Event-tree models explain how an initiating event may lead to failure resulting in a probabilistically quantifiable outcome. The tree models should be checked against four criteria which are necessary conditions for well-formed models. The trees should be checked for a consistent level of *completeness* throughout. They should also be checked for *correctness* with respect to QRA objectives and data availability. Thirdly, the components of risk determined from the tree models are then aggregated to provide an assessed level of general risk. At this stage the overall model is examined for global *balance* and the *appropriateness* of this result is then examined with respect to the observed FAR. If necessary the fault-trees are refined to improve the outcome.

The data in such a model falls into one of two categories: *hard* or *soft*. Hard data is derived from well documented historical information from which probabilities and their variances may be determined. Soft data results from incomplete, sparse or loose information and may also constitute anecdotal information which can only lead to statistical inferences or probabilistic estimates. The art of developing a complete, consistent and correct model is to appropriately balance the required input data with the accuracy

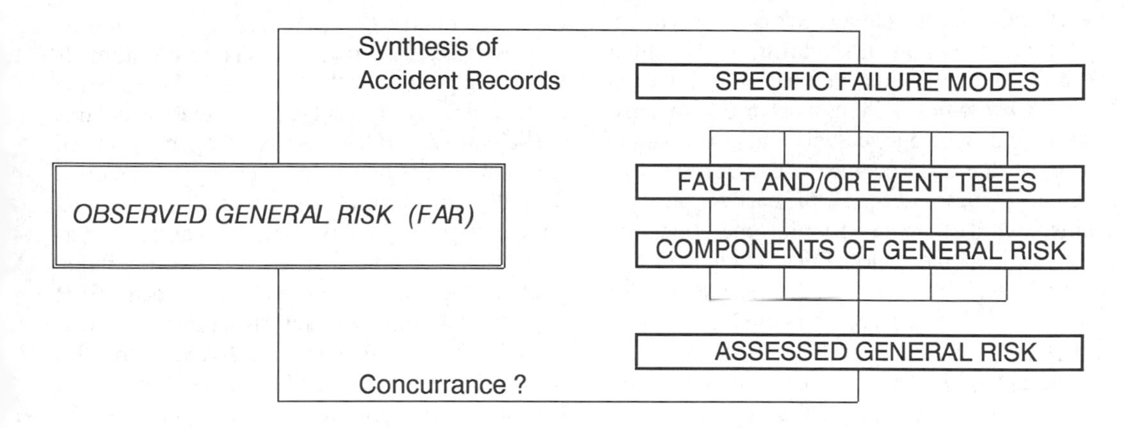

Fig. 1 A quantative risk assessment (QRA) methodology

required as output. This is done through *pegging* (or fixing) portions of a fault or event tree within the total data structure while *hard* and *soft* data is shifted between defined limits. This sensitivity approach ensures that the results match the computed level of risk to the general risk based on the Fatal Accident Rate (FAR).

4 APPLICATIONS

4.1 *Locomotive Engineer Hazards*

A study to assess the risk exposure to New Zealand Railways locomotive engineers (train drivers) was undertaken by Elms and Mander (1990). The principal use of the study was to verify that driving trains is a relatively safe occupation, and that by introducing single manning of trains would not lead to unsafe or risky operating practices. Results of a QRA for the 164 km section of the electrified North Island Main Trunk Railway between Palmerston North and Taihape is shown in Fig. 2. There are four basic accident types: (I) *Train-Train Collision* which may result from either head-on or tail on collision at a crossing loop, a tail-on collision on a single line half-block, collisions resulting from special train control orders given for pilot running or other signal-out conditions, or collision with equipment for on-track maintenance; (II) *Derailment* which may be caused by either mechanical rolling stock faults, track faults, or both; (III) *Tip-Over Accidents* which may occur at tight radius main line

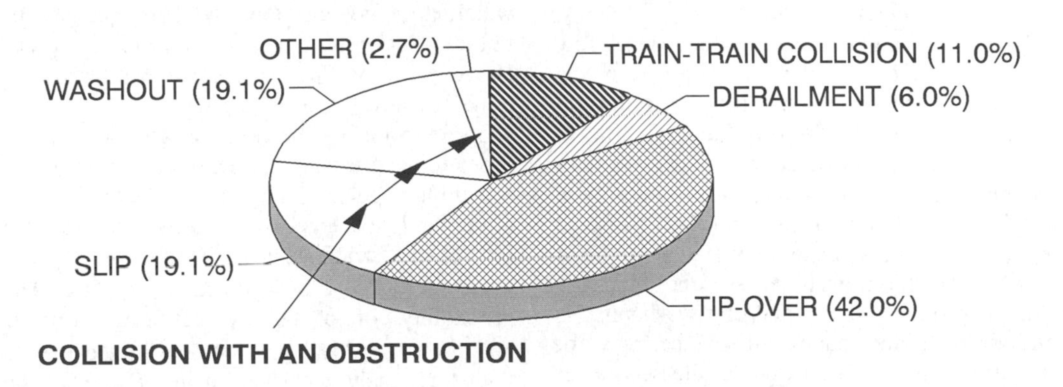

Fig. 2 Distribution of fatalities for NZR locomotive engineers

"speed-trap curves" due to excessive speed, or at turn-outs which may arise from either speeding or incorrect switching; (IV) *Collision With Obstruction* which may occur due to slips in cuttings, or bridge, culvert and embankment washouts due to flooding. The former two accident types depend primarily on the sophistication of the train running operations and the condition of the mechanical equipment and the track, whereas the latter two accident types depend on the track geometry and the terrain over which the railway passes. From Fig. 2 it will be noted that it is the latter two accident types that dominate the make up of the overall FAR. There are a number of reasons for this, which may not be so for other railways throughout the world. Firstly, the New Zealand Railway system has a narrow 1067 mm track gauge with many tight radius curves over mountainous terrain traversing many flood-prone rivers. Secondly the terrain is geologically young and relatively unstable with washouts occurring frequently.

Fig. 3 shows the fault and event trees that were developed to determine the probability of a *collision with an obstruction* per km of track between Palmerston North and Taihape. This component of risk resulted in a FAR of 6.7 or 41 percent of the total FAR for this track section. Fig. 2 also shows that this general *collision with an obstruction* accident category can be further disaggregated into: (a) washouts of bridges and culverts, (b) slips in cuttings, and (c) other mishappenings which include level (grade) crossing accidents, livestock on the track, and mischief.

Results of the study raise the question: How might the safety of the railway system be improved? As the probability of train-train collision and fatality is very low, significant improvement in vigilance and early-warning systems may be costly and result in only a marginal reduction in the general FAR. Instead the best risk mitigation strategy appears to be to ensure that the probability of line blockage either through slips and/or washouts is minimized. In fact much civil engineering work has already been undertaken on the section of track considered by eliminating all but three curves, and by widening cuttings and improving bridge and culvert waterways.

This raises a second and related question: Is the return period for the design flood used in the structural design of bridge and culvert structures sufficient? A 1,000 year return period is used for new NZR bridge and culvert structures. This design return period is based on avoiding economic (rolling stock, structure, and revenue) loss, rather than the risk exposure to operating personnel. This translates into a local FAR exposure of 0.5 for new structures, which is considerably less than an assessed value of 3.0 for the existing system. Thus over time, as the bridge and culvert inventory is upgraded, the risk exposure to its principal operators, the locomotive engineers, will reduce.

4.2 *Risk Exposure to Motor Vehicle Users in Earthquakes*

Earthquake resistant bridge structures are designed in accordance with code loadings which require that the structure be able to: (i) withstand minor earthquakes without being damaged; (ii) withstand moderate earthquakes with minor damage (yielding); (iii) withstand major earthquakes but without collapse. The first two design limit states are based on serviceability and strength, respectively. Conventional first order second moment reliability based limit states design procedures can be adopted for these levels of ground shaking, though with difficulty. The third limit state is based on structural deformability and cannot be handled by the conventional wisdom which does not consider the consequences of failure, such as the loss of life or limb. However, using Eq. 1 a target return period for a design level risk exposure (FAR) can be adopted to derive inelastic design spectra. In order to ensure that the incremental risk to the travelling public does not exceed the aforementioned level of uniform risk exposure for vehicle accidents, it is recommended that a value of $FAR = 2$ be adopted in Eq. (1). The probability of a fatality resulting from a collapse, $P[F \mid EQ]$, may be determined from a fault-tree analysis. Thus, using Eq. (1), the design return period in years can be found from:

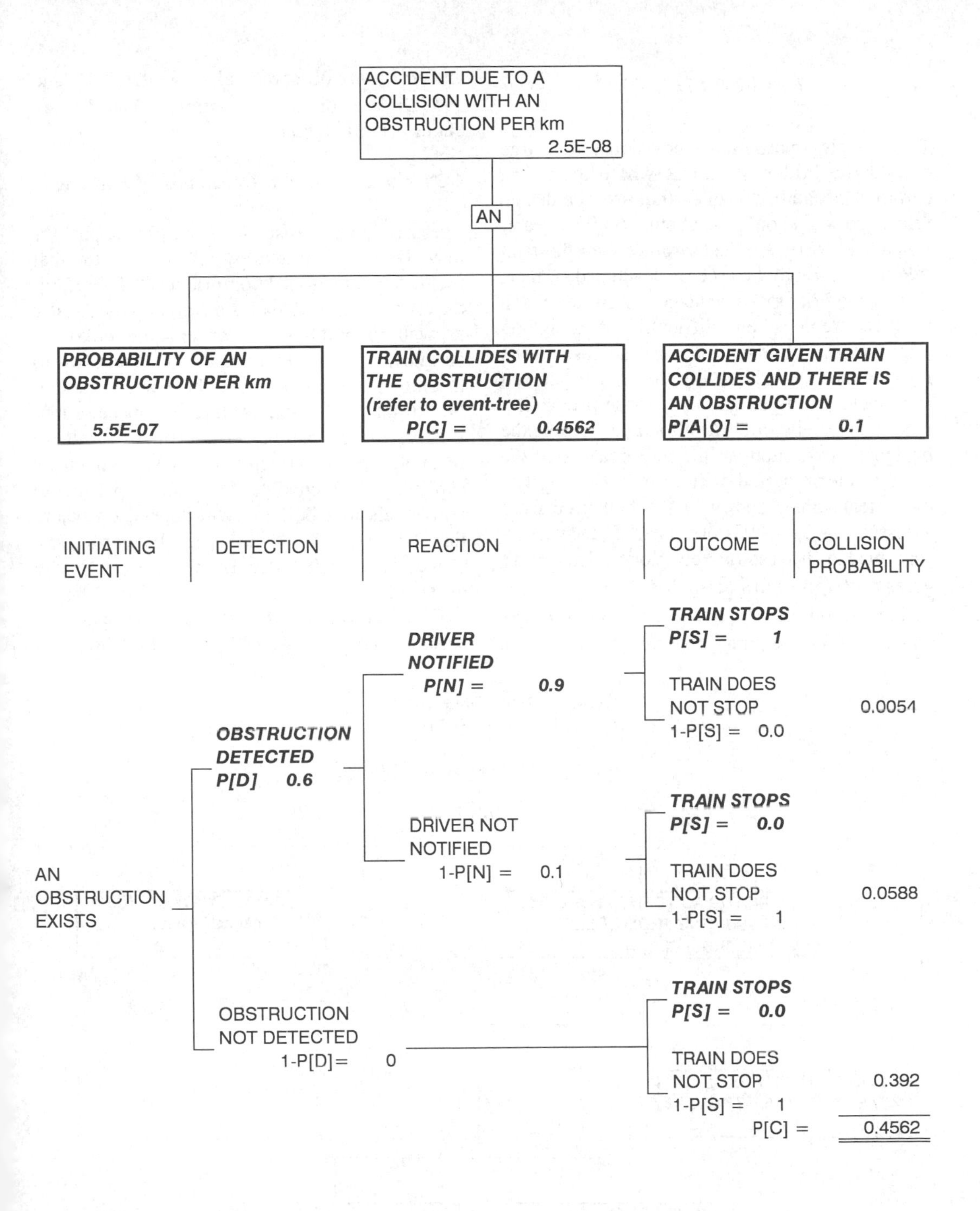

Fig. 3 Fault and event trees for a locomotive collision with an obstruction

$$T_y = 6000\ P[F\,|\,EQ] \qquad (2)$$

This simple relationship shows that for large non-ductile bridge structures where there is a certainty of fatality in an earthquake, the design return period should be about 6,000 years. For more lightly trafficked routes then the fault tree in Fig. 4 may be used or developed further. One of the principal advantages of this approach is it overcomes the difficulty of assigning importance to seismic design forces. The importance of a bridge with the present method is reflected in its usage. Thus if the probability of vehicles being on, beneath or approaching the bridge is low, such as in rural areas, then the design return period will also be low. The numerical values shown in the fault tree were assessed for an I-290 interstate highway-over-road bridge in western New York. This result gives a design return period of $T_y = 420$ *years*. In Fig. 4 note that the boxes with bold outlines and italic script require user defined input. $P[A]$ and $P[B]$ are observed values while $P[C]$ and $P[F\,|\,C]$ are conservative estimates based on an engineering judgment.

4.3 *Risk Exposure to Occupants of Buildings*

The earthquake resistant design philosophy for new building structures is similar to that previously mentioned for bridges. Therefore, the aforementioned analysis could apply equally as well to buildings. For existing buildings designed only for gravity loads that are in low to medium seismic risk zones there are a number of differences that require further consideration. Firstly, the probability of a collapse will be sensitive to the structures' intrinsic ductility capability. Secondly, the risk exposure to individuals in a building will depend on where they are located at the time the earthquake strikes and their state of readiness to take evasive action.

Fig. 5 shows a risk assessment for a ductile seismic retrofit designed for a medium rise

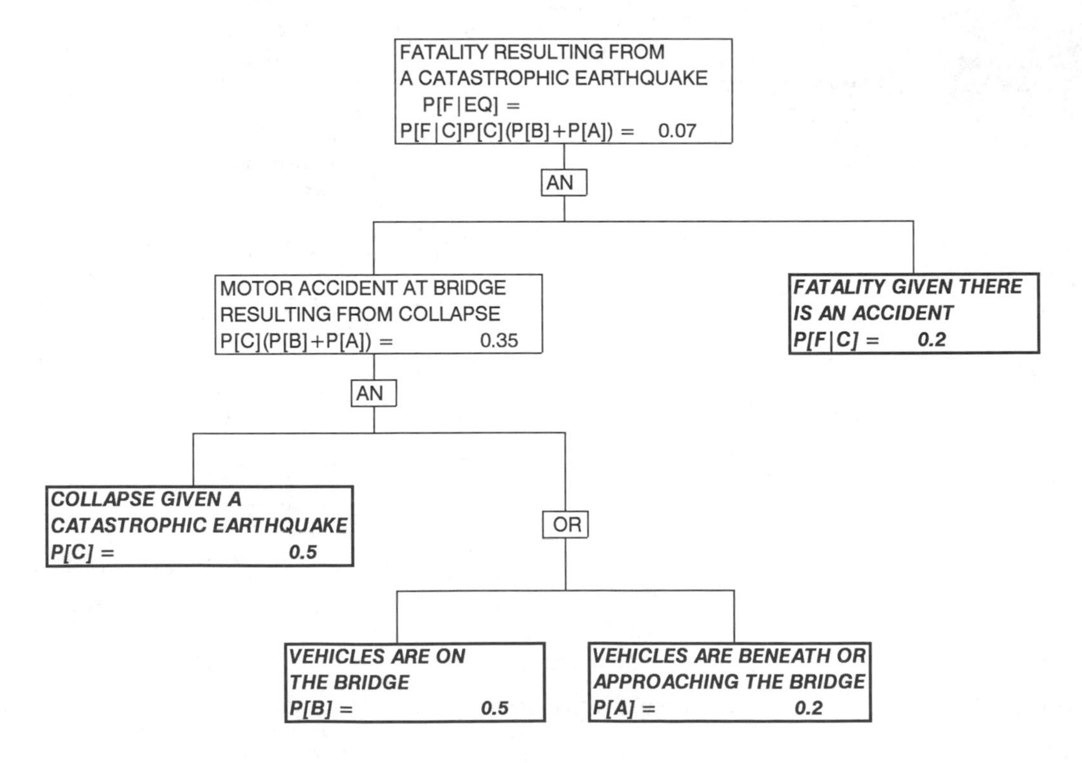

Fig. 4 Fatal accident probability for a bridge collpase due to a catastrophic earthquake

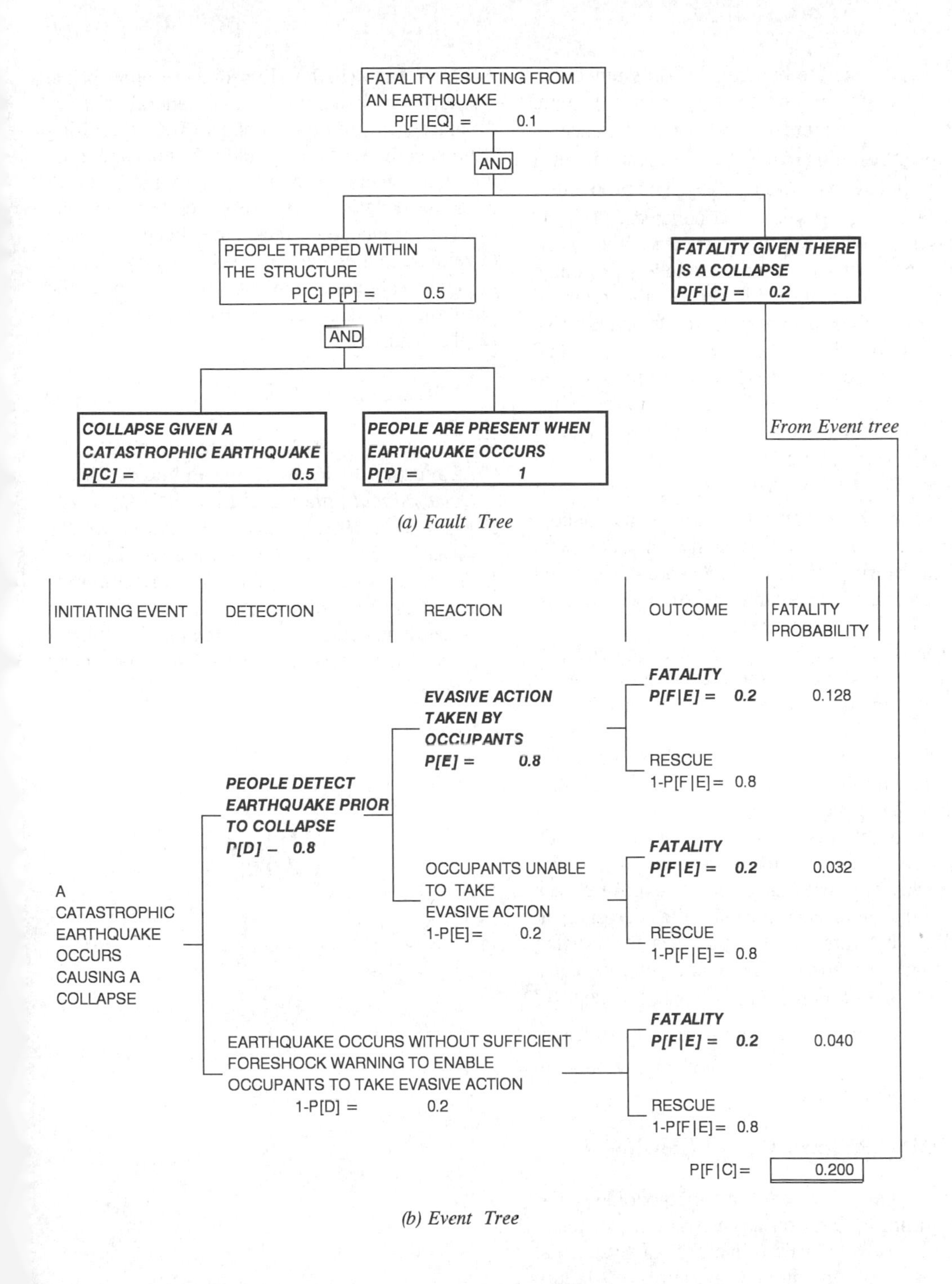

Fig. 5 Fatal accident probability for a building collapse due to a catastrophic earthquake

reinforced concrete building. It is assumed that the structure is constantly occupied with $P[P] = 1$, and there is a 50 percent chance of collapse, resulting from $P-\Delta$ effects, given a catastrophic event, $P[C] = 0.5$. The probability of a fatality given there is a collapse, $P[F|C]$, is assessed by using the event tree, Fig. 5(b). From the fault and event tree models presented in Fig. 5, together with Eq. 2, the required return period to sustain an equitable level of risk exposure to individual occupants should be 600 years. For retrofit designs of limited or no ductility, higher fatal probabilities will result due to the propensity of sudden collapse without warning. For example $P[D] = 0.2$, $P[E] = 0.2$ and $P[F|E] = 0.2$, $P[C] = 0.9$ giving $P[F|C] = 0.8$ and hence $P[F|EQ] = 0.72$ for a non-ductile structure. The present study indicates that design return periods of 2000 and 4,000 years should be used for building retrofits with limited or no ductility, respectively.

From this example, it can be concluded that existing non-ductile structures require a great deal of strength if they are to ensure that the risk exposure to their occupants is kept within target bounds. Conversely, the strength requirements can be minimized based on serviceability requirements (say a 150 year return period), and a seismic retrofit provided to permit inelastic response in order to resist strong ground motions. The present uniform exposure risk modelling approach permits the designer to explore different retrofit strategies for existing building structures. This could imply return periods and ductility factors of $T_y = 4000\,yrs$, $\mu = 1.5$, $T_y = 2000\,yrs$, $\mu = 4$ and $T_y = 600\,yrs$, $\mu = 8$ for non, limited and highly ductile retrofits, respectively.

5 SUMMARY AND CONCLUSIONS

This paper advances a methodology for performing a QRA on large structural systems. Unlike conventional reliability analysis that focuses only on implicit measures of failure probabilities through a reliability index (β), the present QRA approach directly considers probabilities and consequences. The Fatal Accident Rate (FAR) is used as an appropriate measure of general risk. General risk is disaggregated into components of risk resulting from specific types of accident or mishappening that could lead to a fatality. Fault and/or event trees are developed to assess these components of risk. Once the trees have been *correctly* developed to a *consistent* level of *completeness, pegging* techniques can then be applied to the *hard* and *soft* data sets to give the *appropriate* results required.

REFERENCES

Kletz, T.A. 1978. Hazard Analysis, Its Application to Risks to the Public at Large, *Occupational Safety and Health*, UK.

Lees, F.P. 1980. *Loss Prevention in the Process Industries*. UK: Butterworth.

Elms, D.G. & J.B. Mander 1990. Locomotive Engineer Hazards - A Quantitative Risk Assessment Study, *Proc. Annual Conference of the Institution of Professional Engineers New Zealand*.

Structural Safety & Reliability, Schuëller, Shinozuka & Yao (eds) © 1994 Balkema, Rotterdam, ISBN 90 5410 357 4

The conceptual flow of PSA on railway train crashing

H. Shibata
Yokohama National University, Japan

ABSTRACT: This paper deals with the conceptual flow of PSA on high-speed railway train crashing. Recently, even commercial trains, their top speed in service is arround 300 km/h. In high seismisity countries like Japan, the seismic event may cause its derailing, and its probability is not so low in the engineering sense. The author tries to develope the conceptual flow for PSA under seismic events as well as other causes.

1 INTRODUCTION

This paper deals with the safety of the railway train. Recently, the speeds of main railway systems in the world such as Shinkansen, TGV, ICE and so on have a tendency to go up to 300 km/h, and on the other hand, body of their coaches to be lighter. The systems have been planed without the assumption of crash-type accidents, because they provide the well designed safety system.

Also in Japan, up to the summer of 1992, the Shinkansen system was completely isolated from ordinary traffic devices like road crossings, and there had been never crash-type accident since 1964 from the bigining. However new section open last summer has several railway crossings in its section.

According to the statistics of 1991 (Uchida, 1992), there were 48 serious accidents in all railways in Japan. Two numbers are significant, one is events of derailing, its number is 38, and another one is responsible events to operating systems such as dispatchers, operators, drivers and so on, the number of them is 27. Even though there was no accident induced by natural hazards like earthquakes, storms, avalanches in 1991, there are many records such as Kwanto earthquake-1923, Tokachi-oki earthquake-1968 and so on.

These were three serious accidents after the second world war. Hachiko-line accident in 1946, Mikawashima accident in 1962 and Tsurumi accident in 1963 (JNR, 1968). The number of casualties of Hachiko-line was approximately 1000. All three were started from derailing of the train, and the first two were induced by human errors, however, on the last one no deterministic cause of derailing was found. And also Mikawashima and Tsurumi accidents were triple accidents.

According to observation of previous major accidents and recent stochastic, three major points should be considers as followo:

a) derailing, cause and mechanism,
b) natural hazard as its cause and
c) human error as its cause.

In this paper, the author discusses the conceptual flow on all types of railway accidents, however, more emphasizing on derailing events. And also, he tries to refer to the difference between the stochastic approach as PSA and the deterministic safety logics which have been using in the railway engineering.

2 DETERMINISTIC SAFETY SYSTEM AND RAILWAY ENGINEERING

Basically the safety of trains has been ensured by a chain of deterministic safety logics, like railway signal, air brake, automatic train control and so on. Their systems consist of fundamentally fail safe logics. The engineers working for these areas have been asking the absolute safety always, and if any accidental event may occur by vilating their safety logic,then they try to add a new element to improve their logic towards the absolute safety. Even though, there are many railway accidents every year in the world against such situations, the reasons have been discussed because of poor systems, poor maintenance, poor educations and so on. And they believe that such accidents could be eliminated from the annual record, if those are carefully made.

The author can find such a example in a recent accident of a suspended cable car. The system of its automatic operating system by counting its number of revolutions of its wheel, and it converted to the distance which cargo runs from its starting point, and it will reach to a certain value at the point near to the station, the control system cut the driving power once and it goes slowly and the cargo automatically stop at the pre-determined point at the station A. In the case of miscounting and the cargo would not stop at the scheduled point near to the station A, the operator should push the emergency button to stop and adjusting stopping position manually. The operator failed to push the emergency button, and the cargos were crashed into the stoppers. There were no over-run limit switch. The designer believed the counting system almost absolutely exact. And he assumed that even the counter would fail to count exact distance of the cargo, it would be back-upped by the operator. A simple over-run limit switch is a very common device for such a system, but he substitutes only the operator for the counting device in the system. He couldn't recognize the necessity of a limit switch indicating the absolute position of the cargo. The second point of this accident is the requirement for the operator. If the system is completely manual, the operator must control its velocity, when the cargo approaching to the station. Therefore he operates carefully its system always. But in this system, if the cargo would not stop in the position his role is only for watching, and he would have to push the emergency button, if necessary. "To equip the limit switch fits the principle above-mentioned, because the better training of operators are effective for the safety always, but not absolute." This statement is that of the absolute safety logics and not the sense of the probabilistic evaluation. The author will discuss on this subject in the followings.

What is "reliable" at the statement above? The author has been working for the reliability and safety since 1970, and in his sense "reliable" has been used in the sense of probabilistic approach as mostly used. However, for the safety engineer above-mentioned or public, it is more subjective sense. A "reliable" system means the system which is well organized to ensure their function by the adequately designed back-up system and by consisting of well-qualified elements. So, therefore, there is no probabilistic sense in their minds in general. If the would meet an event of malfunction of the system, they must improve the system by adding some new system as mentioned.

For non-engineering people especially, the way of thinking like PSA is not easy to be understood. And one of the serious examples is the criminal law of Japan on the response for the death or injure caused by an accident. The responsible person for the accident may be guilty, of the cause related to his action. And this relation is examined deterministically in the court.

On the other hand, the probabilistic approach is always related to the work to foresee a future event based on the experienced data as shown in Fig. 1. From the raw data, which we observed, we try to establish the model, and estimate future behavior. Such type of analyses can be made both probabilistically and deterministically. The latter approach gives the mean value of probabilistic approach in general, or one sampled result. Recently "the safety goal" is discussed in the field of nuclear accidents, and this value is treted as a deterministic value as Design Basis Accident. The way of evaluating such a value is probabilistic, but the final value "safety goal" is treated as deterministic one. Monte Carlo simulation is one of the methods to evaluate the probabilistic feature of the modes. An actual event, accident, is one sample from the family of events which we can foresee through such a simulation, and DBA might be an uferbound event. According to Japanese oriminal law, if some one would

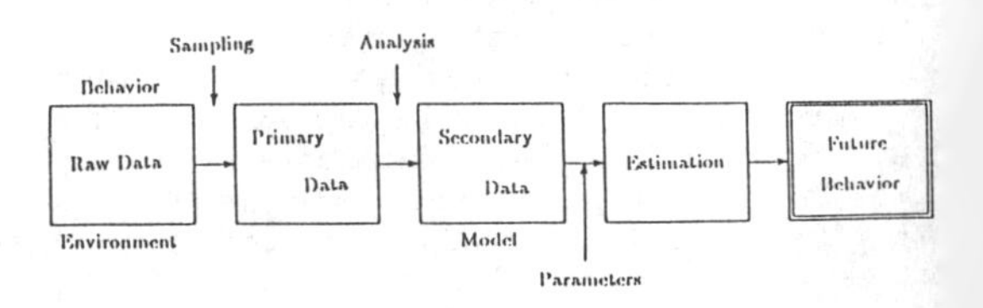

Fig.1 Flow of Data Processing for Predicting an Event from Real Events

be responsible for the event, accident which was happened, he should know the system completely and should be able to foresee its behavior under any circumstance, and the jadge doesn't like to discuss his case in probabilistic way. Back to the safety engineering problem, the engineer must foresee the behavior of the system and design the system with well organized back-up system in Japan. As the author mentioned before, the word "reliable" means only "well organized and complete" in this context.

3 RAILWAY ACCIDENTS AND THEIR CAUSES

Most of railway accident, started from three major causes, Hachiko-line Acc. in 1946 Mikawashima Acc. in 1962 and Tsurumi Acc. in 1963, and the categorization of their causes is rather complicated. There are multiple accidents, which were started from the first accident and the disaster was enlarged by the following second and third accidents as shoun in Fig. 2. Such a type of accident is not rare like Tsurumi event 1963 in Tokyo, or the recent colliding accident in the Northern Germany in the fall of 1992.

So, three major causes are an external cause including natural event, a system malfunction cause and an operational cause. Also the first accident may be a cause for the second accident. As a mechanism to induce an accident by one of these causes, a blockage of the way is the first, and a mis-alignment of rail induced by such a cause is the next one. An event caused by a malfunction of a locomotive or a coach may be considered; there are two or three types, that is, loosing their normal running state, losing the control of their speed and failing to stop at the required point. Malfunction of a railway signal system or a train control system is often observed in the field. This caused by an operational error and a system failure as well as a mis-design of the system. It is difficult to show these in one figure, but the author dare to try it as Fig. 2.

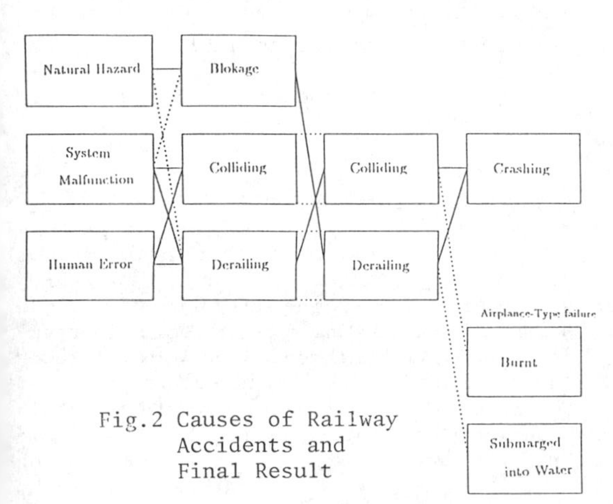

Fig.2 Causes of Railway Accidents and Final Result

JNR made the complete study on Tsurumi accident-1963, and they reported it in 1968, 5 years later the accident. In the report, they said that the cause might be associated with several causes, and only one of them could cause no initial event. Also said that accident itself was unfortunate, probabilistic event, however, for the safety, all possible counter measures had to be done, and 25 counter measures were mentioned and they had been carried out by 1967. And also Karikachi test line, approximately 10 km length, was built in 1966, and for several years the surveys on field tests of derailing were done on the deterministic base. They said that the direct cause of the first derailing had to be a probabilistic one, but they considered that the counter measures had to be done completely and absolutely in engineering sense.

4 SCENARIOS OF RAILWAY ACCIDENTS

We can find some accident news in news papers very often. In Japan, the situation has been much improved since 1963, Tsurumi Accident. However, in the early summer of 1991, there was one serious accident. Scenarios of these two accidents are quite different. And another accident occurred in the fall of 1992 was almost the same as the latter one. The Tsurumi accident consists of three events, the first event was derailed of the freight train and then the passenger train hit them as referred in Section 1. And the second train was derailed and hit the coaches of the other passing passenger train. Three coaches of both trains were seriously damaged. Approximately 350 persons were killed like an air crash. On the other hand, recent two ones, in 1991 and 1992 in Japan were head colliding, and they were not so serious like the previous one in the view point of victims. However, latter two accidents have the same mode as their scenarios. Both accidents occurred in the local railways operated by small companies whose economical situations are not well balanced. And their systems were poor and caused by human errors. This is one of

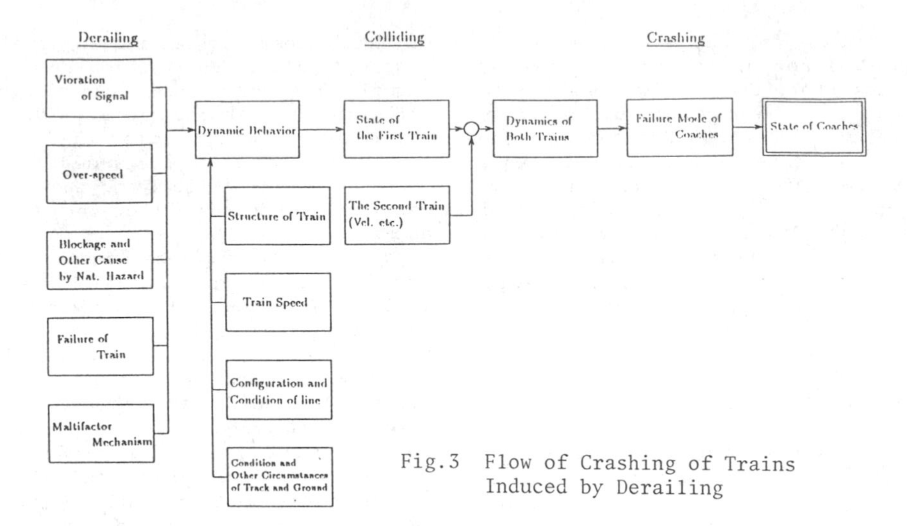

Fig.3 Flow of Crashing of Trains Induced by Derailing

new trends in Japan. Last ten years, most of local lines of the previous Japan National Railway separated from the main body and operated by local sectors, and they employ a simplified signal system and the conditions for their employees are not qualified well including their professional educations.

Thus the mode of the serious accident has been gradually changing according to the social environment. However, the most serious accident like an air crash accident, in which more than one hundred persons would be killed, might be a double or triple accident. The reason is the following: The second or third accident will occur without any cause of the second and third trains, and it is no way for directly protect by a system. Then in some cases the relative speed becomes very high, even the first train is almost stopping. One of the reasons for this, such an accident has a high probability of occurrence at the main trunk lines more than two double track portion. So we should expect trains run with high speed, large number of coaches, large number of passengers, mixing lighter body coaches and heavier, new shell-type structural design of coaches, frequent services, that is, high density of trains, mixed traffic condition of a passenger train and a freight train is a cause of serious accidents. Therefore we can introduce one scenario as shown those of three accidents. In Europe and other regions, projects of high speed trains like TGV have been gradually developed, and scenario of such a type of an accident have to be examined, even the probability of occurrence is different region by region.

The probability of occurrence of this type of an accident will be discussed later, but the scenario should be examined in relation to the structural dynamics model. This will be done according to Fig.3. To save the space, the author explains the figure briefly. To establish the scenario, two train colliding will be discussed caused by a derailing of one train.

a) a cause of derailing of the first train by any cause,
b) the relative position and situation of two trains,
c) their relative velocity,
d) the mode of colliding,
e) the dynamics of coaches of both trains,
f) the final states and positions of coaches,
g) failure mode of an individual coach and
h) state of passengers in an individual coach.

5 RAILWAY ACCIDENT AND EARTHQUAKE

Tokaido-shinkansen run along the southern coast of the Honshu island, and some part is passing through the high seismicity zone where a seismic induced accident will be expected during a future Tokai earthquake predicted by the seismologists officially and other local high intensety earthquakes. In this zone, the national warning system has been established. And the system will be closed, if the warning will be made. However, the author assumes that an earthquake whose intensity would exceed MMI IX, then to try to estimate how the event would be developed. Kameda, Kyoto University has been working this subject in the view point of a civil engineer since 1979 (Kameda, 1979). He mentional the probability of running into a failed section for the high-speed train was very high under an event which we should expect.

The author evaluated the situation simpler than Kameda in the follouing: Between Tokyo to Nagoya, 40 trains run both direction at rash hours. The most of trains don't stop in-between. Their average speed is approximately 170 km/h, and the maximum speed is 210 km/h except a few trains. Distance between two cities is 366 km. The density of trains are 0.1 train/km on both ways, and their length is 368 m. Each train can stop within 120 sec by cutting electric power supply which activate its emergency brake automatically. The run curve is like in Fig.4. The first 60 second it decreases only 50 km/h from top speed. Most of tracks between Tokyo and Nagoya are running on soil surface directly, embankment, or cutting, and partially ordinary bridge or in tunnel. This is different from the other newly constructed lines.

It is well know that the track on soil easily buckle and form snake shape under a condition of an earthquake exceed MMI VII. It is said that if the amplitude of this snake shape exceeds 10 cm, the train will

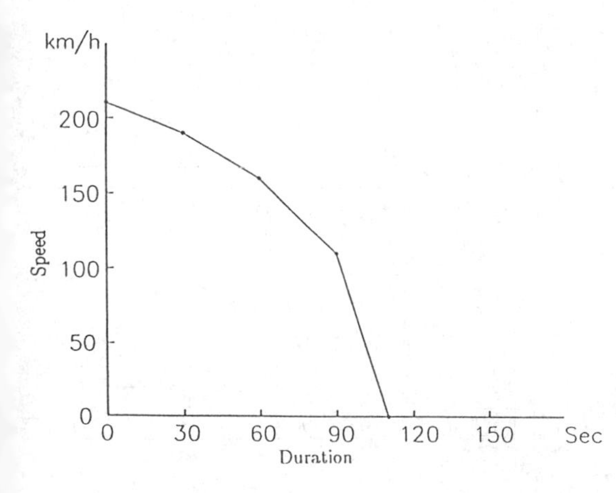

Fig.4 Run Curve of Tokaido-shinkansen with Emergency Break, observed by the Author

be derailed easily, and the records of the previous earthquakes since 1923 shows that the amplitude might exceed 30 cm. The behavior of a train was analyzed including the stability caused by horizontal seismic acceleration was analyzed in Japan (Kunieda, 1971) and Russia (Kostarev, 1991). It is almost definite a train would be derailed under the condition above-mentioned.

By a future Tokai earthquake which was predicted that the possibility of occurrence is more than 40% in next ten years by Rikitake, the section from Atsugi to Mikawa, approximately 200 km length is covered by the warned area out of 366 km above, whose intensity may reach to MMI VII. At the assumed event 29 trains (at 9.00 am by the current time table) would be in this sections. Most of trains run over the average speed, then we assume that 21, 80% of trains run over 200 km/h. As analyzed by Kameda the probability of derailing in such circumstances, we should expect more than 50% of trains in this section would be derailed. That means, 10 trains would be derailed in total. Kameda tried to consider the train failures at 50% of 100 points of track failures in this section, that is, 0.25 point per one kirometer length.

The probability of hitting together to a derailed train is not so high. If we assume that two trains, which would be in the distance 200 km, will be passing by in 200 km/h each and within 60 second, and have a possibility of colliding each other, the probability is 0.03. This value is not so high, but as PSA not so low. This value comes from that two trains within 5931 m distance shall collide in next 60 sec in this section, 200 km. Then each of 10 trains have such a probability on one direction, then approximately 30% of one train will be expected to have a chance to collide under a circumstance we assumed.

In each one train approximately 1600 passengers occupies their seat in rush hour, and some of them will be killed as an airplane crash. The rate may be lower than airplane crash, because of lower speed and no fuel fire in general. The exact rate of deaths could be estimated by the analysis which will be discussed in Section 6. According to the recent records, it may reach to 10% of total number of passengers, if the coach would be rugged completely.

This means that the expected deaths reach to 480 per each a seismic event. Of course, this number is a quite rough estimation, however, it should be mentioned that this figure is not so low compare to the statistics of the average annual number of deaths in Japan. All types of accidents brought 386 deaths in 1991, in the Japan Railway Companies which formed the previous Japan National Railway, JNR. Before Tsurumi Accident, the average annual number by their responsible accident was 30∿35 and almost none from 1964 to 1990, but it was 42 in 1991.

6 SIMULATION ON CRASH DYNAMICS

Here two types of crashing will be treated. One is blocking the run of the train and derailing. Blocking the run of train consists of more than several coaches is for the fundamental simulation. Derailing may cause various modes of train behaviors, and blocking is included as one of most serious causes. The modes are listed as Table 1. These modes are related to their motions after derailing. If it is blocked completely, the train becomes shape like zig-zag horizontally, or vertically sometimes, by the axial force. In this mode, each coach may be broken by its buckling, telescopic colliding adjacent two coaches or hit them side-by-side. Whether or not it will be blocked completely depends on the way of derailing. If it derails and run down completely out side of its track, it is almost sure to be blocked, or in the similar situation as shown in Fig.5. The author may refer to the simulation of crashing of coach body, and this may be done a super-confuter, but it is not practical for the evaluation in this paper. Really what we need to calculate is energy which is possible to be absorbed by its body deformation and is supplied by the dynamic motion of the train. We may assume that the number of casualty could be a function of the absorbed energy of each coach, if their designs are the same. Also high deacceleration may cause another type of casualty, however, the rate of deacceleration is reverse to energy absorbing capacity of coach structure very often. Therefore, the structure of a coach should be carefully examined in this viewpoint. In Japan, aluminum alloy has been used since late 1960's for structural material, and such structures are less capacity than steel structure.

A model for simulating a dynamics of a train is as shown in Fig.5. A train is assumed to be a chain of rigid bars on flat plate with the friction coefficient μ, and run straight. Two types of events can be introduced. One is a sudden

Table 1. Mode and Cause of Derailing

A. Blockage
1) Derailed Train
2) Over-run at the Stopping-point
3) Land-slide
4) Aavalanche
5) Structure Failure along Track
6) Railway-crossing trouble

B. Track Misalignment
1) Subsidence of Foundation
2) Heat Effect in Summer Season
3) Break of Rail
4) Bridge-connection Misalignment
5) Snake-shape Misalignment Induced by an Earthquake

C. Malfunction of Coach and Locomotive
1) Shaft Failure
2) Tire Failure
3) Decoupling of Coaches
4) Failure of Connection of Track and Body

D. Operational Error
1) Over-run
2) Over-speed
3) Head Colliding
4) Mis-selection of Route (Switch)

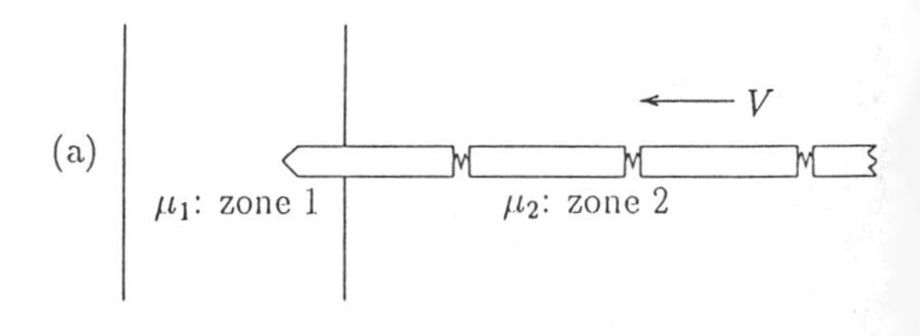

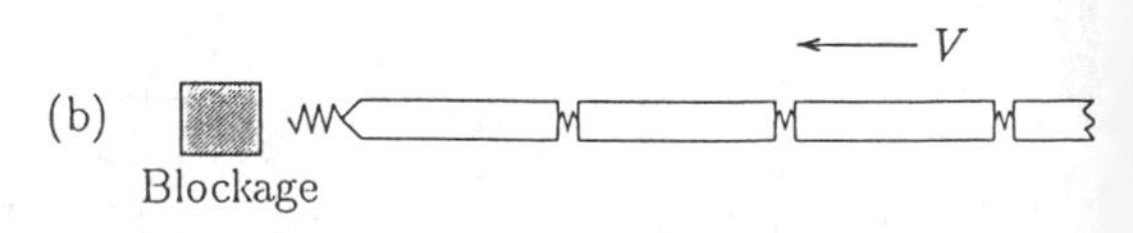

Fig.5 Models for Behavior of Train afeter Derailing

increase of the friction coefficient μ, for example, 0.01 to 0.1, as Fig.5(a). And the other model is a sudden blockage of the head of the train, as Fig.5(b). In this case the train is assumed to have a non-linear spring with damping element at its head. This is coming from the deformation of its body mainly. The nature of this spring, which is not only a single spring, but also more complicated, decides the behavior of the train after blockage, as well as the strength of bodies and the mechanism of couplers, that is, the three modes mentioned above.

7 UNCERTAINTY AND IMPLEMENTATION

Currently there are too many uncertainties for implementing the whole process as shown in Fig.6. The major uncertainties are following three. The first one depends on the details of track and its surroundings, and the second one is to predict the random behaviors of each a physical phenomenon like crashing. If we have the detailed information on a particular line, for example, the ratio of tunnel length to the total distance, we can gradually solve the problem on the train dynamics except quite local conditions. Those local conditions may be molded and the distribution of these models can be numerized. The third one is the relation to other trains.

Random behavior of each model can be studied like buckling of a coach body. If it is symmetric, the probability of forming its deformation to one side to other side is almost equal, but unless otherwise it has a tendency according to its shape, and the value of its probability may be analyzable. An event inducing a series of events may be evaluated such as a destructive earthquake in a certain section. Also the probabilistic value of the probability to induce crashing of a train to other derailed train can be calculated by a simple mathematics and some feasible assumptions, but we must recognized that there would be some uncertainties still.

These three types of uncertainties are indicated in Fig.6. The most unsolved uncertainties lie in the first one, for example, if there is a pole or obstacle along the track, and a derailed train hit it as Fig.5(b). The point of hitting of its head is deviated from the center of the head, the train may initiate to rotate its first coach. Its behavior may be simulated by setting some parameters, but this is one of samples which might occur

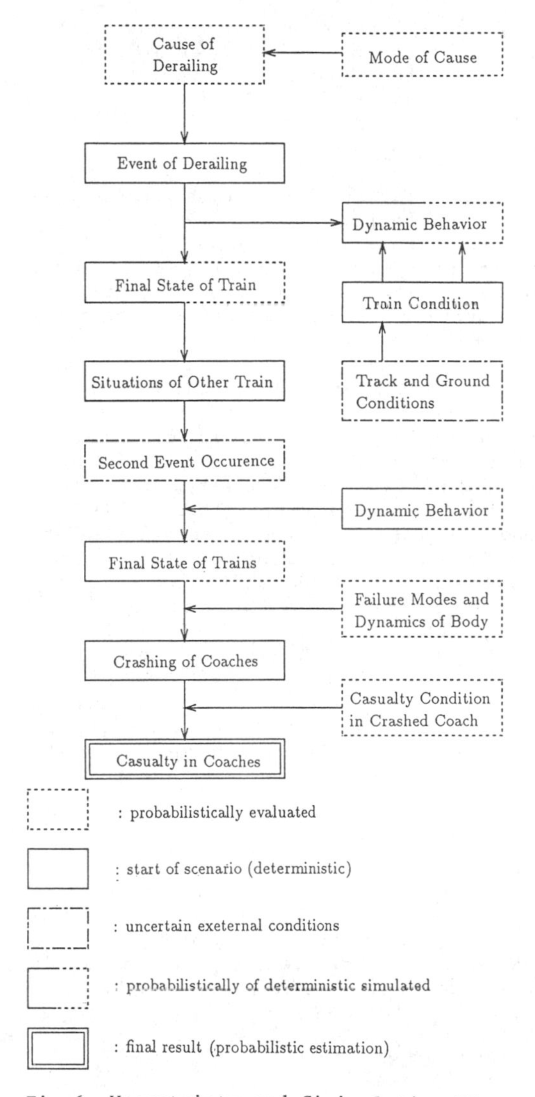

Fig.6 Uncertainty and Siminulation Flow of Derailing of Trains

at a particular pole along 100-200 km length section of the line. It is difficult to obtain such data from previous events because of less frequent events. At this moment, the uncertainty of this category might be decided by a subjective way. The second one, the train dynamics itself and the pattern of crashing, is easier to obtain its value by an idealized model and its dynamics. Only elaborate efforts to solve one situation by other situation are necessary. Third one, the probability of occurrence of an initial event, is a simple stochastic problem, if we treat it mathematically, but there might be some uncertainties, if we deal with an engineering way.

8 CONCLUDING REMARKS

During writing this paper, there was an accident, similar to the subject which the author discussed, at Northein in North Germany (Nov.16,1992). In a some developing country, the top cause of accidental deaths is the railway accident, and the order of some tens thousand deaths per year is reported. It clearly shows the necessity of such a study, even its implement is very difficult. By considering the process of railway accidents in this way, we can find more points to improve its safety, even if we can't obtain the value itself. Railway engineer has gone to more absolute safety logic way, according to the author observation. He doesn't intend to say "no" on the absolute safety logic approach. We must considered such approaches more and it is dangerous to believe the stochastic way only, but the effort to evaluate it stochastic way is significant to improve the safety of railways from ultra-high speed one to local one.

9 ACKNOWLEDGEMENT

Professor Kunieda, Meisei University, gave the information on derailing-type accidents which he had been working for in JNR. The author greatly appreciates his guidance in this field since 1960's.

10 REFERENCE

Survey Committe on Tsurumi Accident, JNR (1968), Technical Report on Tsurumi Accident (April,1968) 49pp.+16 Appendicies, Total 1250 pp (in Japanese).

Kameda, Hiroyuki and others(1979), Risk Assessment of Running Vehicles against Randomly Occuring Structural Failures during Earthquakes, *Proc. of 7 WCEE*, Vol.1, (Jan. 1979) p.363

Kunieda, Masaharu and others(1971), Theoretical Analysis of Over-Turnning of Railway Coaches by Earthquake Ground Motions, Vol.47, No.414 (Feb. 1971) p.164

Kostarev, V.V.(1991), CKTI-Vibroseism, St. Petersburg, Russia, *Private letter*

Uchida, Yoshiaki(1992), Accidents caused in Japanese Railways in 1991, *Denkisha-no-Kagaku(J. of Electric Rolling Stock)*, Vol.45, No.11 (Nov. 1992) p.37 (in Japanese)

Structural Safety & Reliability, Schuëller, Shinozuka & Yao (eds) © 1994 Balkema, Rotterdam, ISBN 90 5410 357 4

Hazard engineering and learning from failures

J.R. Stone & D.I. Blockley
Department of Civil Engineering, University of Bristol, UK

ABSTRACT: Hazard engineering is concerned with the identification (through hazard audit) and treatment (through hazard management) of exceptional circumstances where the hazards need to be controlled using specialist skills.
A 'systems' view is taken in which failure is seen as the culmination of a series of pressures (both social and technical), and as a process which may be studied at different levels of detail. The hierarchical description allows lessons to be learned though the detection of similar patterns in outwardly different failures.

1 DEFINITIONS

Many of the terms commonly used in discussion of engineering safety are subject to varying meanings and interpretations. The following definitions are used in this paper:-

- Safety: freedom from unacceptable risk / personal harm (Fido & Wood, 1989)
- Risk: the combined effect of the chances of occurrence of some undesirable event and its consequences in a given context.
- Hazard: a set of preconditions in the operation of a product or system with the potential for initiating an accident sequence (Turner, 1992).

2 INTRODUCTION

The moral and legal requirements that the engineering professions should learn from failures implies the need for a constant search for similarities between past failures and current projects. Failures should be seen as opportunities for learning rather than occasions for attributing blame (Turner, 1992). One question, therefore, is "at what level of definition should these studies be implemented?" At one level of definition, it is simple (though unhelpful) to state that a specific failure can never recur: Chernobyl, for example, happened at a specific point in time and space which will not be repeated. It is clear that the more we increase the detail of investigation of failures, the more individual and distinct they appear. Looking in the other 'direction', however, towards greater generality of description, more similarities appear. At the highest, most general level of description, all failures of engineered artefacts are the same—they all failed in some way to reach a specified fitness for purpose.

In the UK, the Engineering Council has recently published a code of professional conduct entitled 'Engineers and Risk Issues' (Engineering Council, 1992), covering the duties and responsibilities of professional engineers in these areas, and with the aim of encouraging a greater awareness, understanding and effective management of risk issues. The Code argues that risk management should be an integral part of all aspects of engineering activity, that it should be conducted systematically and should be auditable. It goes on to say that engineers should discuss failures so that lessons can be learned.

Society expects engineers to recognise their failures and to attempt to learn from them. Learning is traditionally accomplished through the process of inquiry—formal or informal—leading to amendments to statutory requirements, Codes of Practice and methods of working. Professional bodies also maintain a continuing audit of safety performance and may publish reports highlighting areas of growing concern (e.g. Standing Committee on Structural Safety, 1992).

A new discipline of Hazard Engineering described here represents an attempt to formalise this learning process.

3 THE DEVELOPMENT OF FAILURES

Turner's model of system failures (Turner, 1978) is a useful tool when considering the development of failures. It is based on the observation that most system failures are not caused by a single factor and that conditions for failure do not develop instantaneously. Rather, multiple causal factors may accumulate, unnoticed and not fully understood over a considerable period of time. This time is called the 'incubation period'.

Within the incubation period a number of types of conditions can be found, in retrospect. Firstly, events may be unnoticed or misunderstood because of wrong assumptions about their significance: those dealing with them may have an unduly rigid outlook, brushing aside complaints and warnings, or they may be misled or distracted by nearby events. Secondly, dangerous preconditions may be unnoticed because of the difficulties of handling information in complex situations: poor communications, ambiguous orders and the difficulty of detecting important signals in a mass of surrounding noise may be all-important here. Thirdly, their may be uncertainty about how to deal with formal violations of safety regulations that are thought to be outdated or discredited because of technical advance. Fourthly, when things do start to go wrong, the outcomes are typically worse because people tend to minimise danger as it emerges, or to believe that failure will not happen.

The incubation period, in which interlinking sets of such events build up, is brought to a conclusion either by taking preventative action to remove one or more of the dangerous preconditions that have been noticed, or by a trigger event after which harmful energy is released. The previously hidden factors are then brought to light in a dramatic and destructive way, which provides an opportunity for a review and a reason for a reassessment of the reasons for failure. There can then be an adjustment of the precautions to attempt to avoid a recurrence of similar incidents in the future.

4 THE BALLOON MODEL

Imagine the development of an accident (failure, disaster) as analogous to the inflation of a balloon (Blockley, 1992). The start of the process is when air is first blown in to the balloon, when the first preconditions are established and the incubation period is initiated. Consider the pressure of air as analogous to the 'proneness to failure' of the project. Events accumulate to increase the predisposition towards failure. The size of the balloon can be reduced by lowering the pressure and letting air out, and this parallels the effect of management decisions which remove some predisposing events and thus reduce the proneness to failure. If the pressure of such events builds until the balloon is stretched then only a small trigger event, such as a pin or lighted match is needed to release the energy pent up in the system. The trigger is often confused with the cause of the accident. The trigger is not the most important factor, the overstretched balloon represents an accident waiting to happen: it is a hazard or set of preconditions to failure. In accident prevention, it is thus important to recognise the development of pressures in the balloon, i.e. to audit the hazard. The symptoms which characterise the incubation of an accident need to be identified and checked.

5 HAZARD ENGINEERING

There is therefore a need to define a discipline that is particularly concerned with the development and practice of safety and hazard management. The words safety, hazard and risk have many different interpretations and are used in many different contexts. There is a need for a discipline that would unify all the disparate ideas and techniques in various hazard and safety audits and risk and reliability assessments across the various sectors of the engineering industry.

Consider, for example, the range of different and complementary approaches to risk control employed in the process (petrochemical), structural engineering and nuclear power industries. Kletz (1992) describes the steps taken by many companies in the process industries, including the use of well developed safety audits in conjunction with specific hazard-spotting techniques such as hazard and operability studies (HAZOP) and hazard analysis (HAZAN). Many of the analytical techniques employed in structural engineering, both for reliability analysis and for the development of design codes of practice, can be grouped under the general heading of reliability theory, and categorised as level 1, 2 or 3 according to the detailed treatment of notional probabilities of failure (Melchers, 1987, Galambos, 1992). The pursuit of safety in the nuclear power industry has lead to many advances in the study of complex systems, including the development of event tree and fault tree analysis (Garrick, 1992).

In reliability theory the probability of failure is a measure which is a restricted case of the more general notion of proneness to failure. The probability of failure is a measure of the distance between the design point and the nearest limit state boundary of an multi-dimensional hyper-volume, for a closed world model containing only idealised technical models of behaviour. It is suggested that the evidence which may be used to obtain a more general estimate of the 'distance' between the design point and the nearest constraint may be collected by an appropriate 'hazard audit', in which calculated failure probabilities would constitute part (but not all) of the data. For large unusual or risky projects the audits may be detailed and regular, whereas for small projects the audits may be small and fairly general. The important point is to establish types of audits which are appropriate for the particular project. The audits would necessarily involve a consideration of both technical and human factors and in particular would examine their interaction.

The objective of a hazard audit is therefore very clear: it is to identify a set of conditions which reflect the state of the project at a given time. If things are departing from plan, the hazard audit offers an opportunity to consider the state of development of the incubation period by providing an estimate of the distance between the design state point and the nearest constraint boundary of the hyper-volume. The axes of the hyper-volume may be both nominal and ordinal to represent quantitative and qualitative data. Hazard management is then the set of actions which will maintain an acceptable distance between the design point and the constraint boundary. Hazard engineering is therefore equivalent to navigating the passage of the design state point through the constraint hyper-volume as it changes with time, and is illustrated in Figure 1.

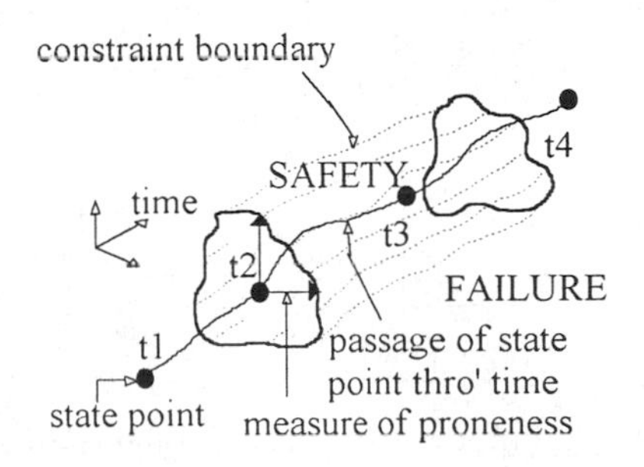

Figure 1. Project state point within hyper-volume

The state point is seen at times t_1, t_2, t_3, t_4 as it progresses through an irregular and changing space with possibly ill-defined constraint boundaries. Evidence for proneness to failure is accumulated by measuring the distance to the different boundaries.

Hazard engineering should be concerned with the identification (by hazard audits and other means) and the treatment (through hazard management) of exceptional circumstances where hazards exist which need to be controlled by the use of specialist skills.

Failures are often the result of the unintended consequences of human action and therefore almost impossible to predict. However they may be possible to control if the symptoms are identified early enough in the incubation period by a hazard audit. Hazard engineering is therefore defined as being concerned with ensuring the appropriate resolution of conflicting requirements placed upon the design, construction and use of an artefact by a range of uncertainties about its likely future operation. These uncertainties are to be identified by a hazard audit.

The pressures on an engineering project accumulate over time, insidiously increasing the level of proneness to failure. Figure 2 illustrates that gradual accumulation of pressure leading to eventual failure, and contrasts it with the way in which the processes of hazard engineering (audit and management) can control the situation.

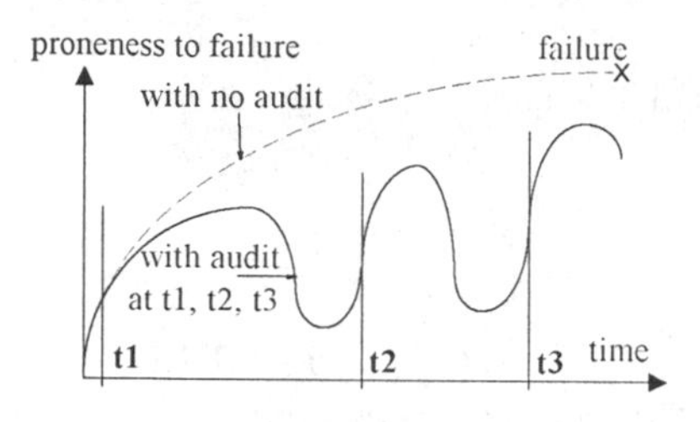

Figure 2. Influence of Hazard Engineering on proneness to failure

If no measures are taken to reduce pressure, then failure could occur as indicated by the dashed line. The project is audited to assess proneness to failure ("balloon pressure") at discrete intervals (t_1, t_2, t_3), and appropriate hazard management actions taken.

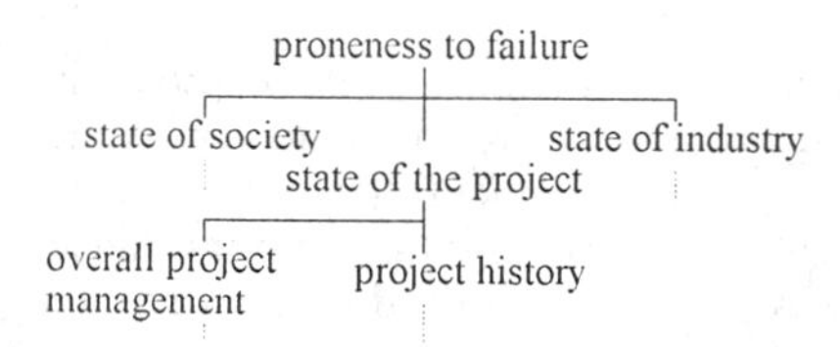

Figure 3. Top levels of the 'proneness to failure' hierarchy
(from Dester, 1992).

A measure of proneness against time might therefore be expected to show firstly an increasing proneness, then a decline as 'depressurising' is undertaken, followed in turn by further increase until the next audit. This cycle may be seen to repeat throughout the duration of a project.

A Hazard Audit has been described (Dester, 1992) as "the formal, systematic gathering of evidence to indicate how prone a project is to failure", and constitutes the investigation of a project in terms of a hierarchy of concepts. Each of these concepts has the potential to provide evidence of proneness to failure. Figure 3 shows the concepts at top levels of the hierarchy. The dependability or otherwise of concept in respect of a particular project is established through the auditor answering a series of questions relating to both technical and management factors.

6 REFLECTIVE PRACTICE

Existing approaches to safety auditing have been seen (Dester, 1992) as being restricted largely to checklists designed to establish the current project state. There is often little or no attempt to ensure that the audit conclusions are implemented. This can be therefore be seen as an open-ended system with no feedback. One objective of hazard auditing is to ensure that recommendations are implemented and maintained. The audit system must be a closed loop with regular feedback and monitoring. We can compare this requirement with the reflective practice model of problem solving (Blockley, 1992).

This model is shown in Figure 4, and consists of the stages [world - sense - think (mind) - act -

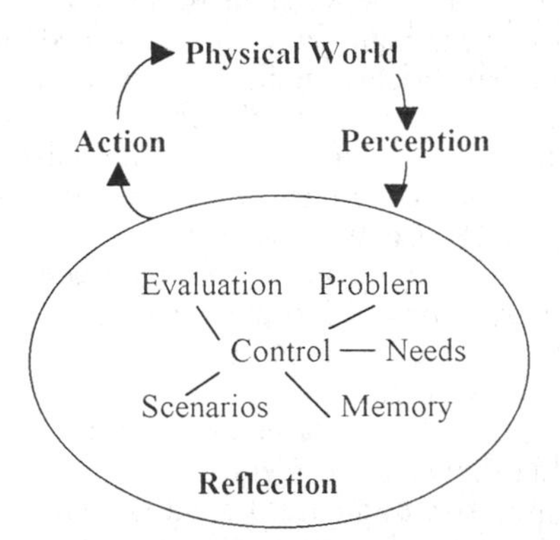

Figure 4. The reflective practice problem solving loop

world] linked in a continuous loop. In overview, the problem solving loop is simply this: we perceive the world (the senses) as sets of patterns in the brain, we interpret our perceptions (think / reflect), we act upon the world (behave), and finally we perceive a new state of the world. Hazard auditing can then be though of as the process of sensing the current state of a closely defined 'world' (i.e. a project), and hazard management as the related process of acting upon that world to ensure continuing safety.

7 DEVELOPMENT OF KNOWLEDGE BASE FOR USE IN AUDIT

The process of learning from failures is concerned largely with the search for regular patterns, since the identification of commonly occurring patterns in previous failures allows their use as diagnostic tools in the study of future projects.
A considerable amount of information on failures is held in various databases, such as that developed during the AEPIC Project (Loss, 1987). In order to learn from these failures it is necessary to be able to search for patterns at varying levels of detail. For example, although two failures may exhibit no obvious similarities in their 'low level' technical details, a study of 'higher level' concepts such as social context, 'safety culture' (Pidgeon, Turner, Blockley and Toft, 1991) and management organisation may reveal common patterns in their incubation. In order to facilitate this form of analysis, the failures database needs to include more structured information than simply the basic details of each case. We are currently developing a knowledge based system (KBS) incorporating not only case-specific data as a series of records, but also a collection of structured representations of the failure histories in the form of Event Sequence Diagrams (ESDs), together with their accompanying textual narratives. The ESD representations facilitate the machine learning of new concepts for inclusion in the hazard audit vocabulary, through the identification of common patterns as previously described (Stone and Blockley, 1990). The textual narratives will enable users querying the KBS for advice on similarities between representations of current projects and those of the failures database to receive advice 'grounded' in the original data. Current work is therefore concerned with the interaction between the learning algorithms and failure records, and the way in which new concepts learned from detailed studies of previous failures can be incorporated into the hazard audit vocabulary. It is believed that a KBS in this form will constitute a powerful tool for use in hazard auditing.

8 CONCLUSIONS

Failures of engineered artefacts frequently occur through the gradual accumulation of small unnoticed 'pressures'. The study of a number of failures has indicated that certain common patterns of contributory factors may be found. We propose that hazard engineering should be a new discipline which envisages the use of hazard audits to measure these pressures, using a structured, hierarchical vocabulary based partly upon identified common factors, and hazard management to take appropriate action to reduce the pressures.

REFERENCES

Blockley, D.I. (1992). *Engineering safety*. London: McGraw-Hill.

Blockley, D.I. (1992). Structural failure and reflective practice. In M. Drdácký, *Proc. Second Int.Conf. on Lessons from Structural Failures*, Prague.

Comerford, J.B. and Stone, J.R. (1992). AI in risk control. In D.I. Blockley (Ed.), (1992). *Engineering safety*. London: McGraw-Hill.

Dester, W.S. (1992). *The development of a structure for the design of hazard audits*. PhD Thesis, University of Bristol.

Engineering Council (1992). *Engineers and risk issues* (Code of Professional Practice). London: The Engineering Council.

Fido, A.T. and Wood, D.O. (1989). *Safety management systems*. London: Further Education Unit.

Galambos, T.V. (1992). Design codes. In D.I. Blockley (Ed.), (1992). *Engineering safety*. London: McGraw-Hill.

Garrick, B.J. (1992). Risk management in the nuclear power industry. In D.I. Blockley (Ed.), (1992). *Engineering safety*. London: McGraw-Hill.

Kletz, T.A. (1992). Process industry safety. In D.I. Blockley (Ed.), (1992). *Engineering safety*. London: McGraw-Hill.

Loss, J. (1987). AEPIC Project : Update. *J.Performance of Constructed Facilities*

A.S.C.E., 1(1), Feb , 11-29.
Melchers, R.E. (1987). *Structural reliability - Analysis and prediction.* Chichester: Ellis Horwood.
Pidgeon, N.F., Turner, B.A., Blockley, D.I. and Toft, B. (1991). Corporate safety culture: improving the management contribution to system reliability. *Proc. European Reliability 1991 Conf.* (in press).
Standing Committee on Structural Safety. (1992). *Ninth report.* London: Institutions of Civil and Structural Engineers.
Stone, J.R. and Blockley, D.I. (1990). Structural reliability through learning from case histories. In A.H-S. Ang, M. Shinozuka and G.I. Schuëller (Eds.). *Structural safety and reliability: Proc. ICOSSAR '89, Fifth International Conference on Structural Safety and Reliability*, San Francisco, 7-11 August, 1989, Vol 3, 1755-1761.
Turner, B.A. (1978). *Man-made disasters.* London: Wykeham.
Turner, B.A. (1992). The sociology of safety. In D.I. Blockley (Ed.), (1992). *Engineering safety.* London: McGraw-Hill.

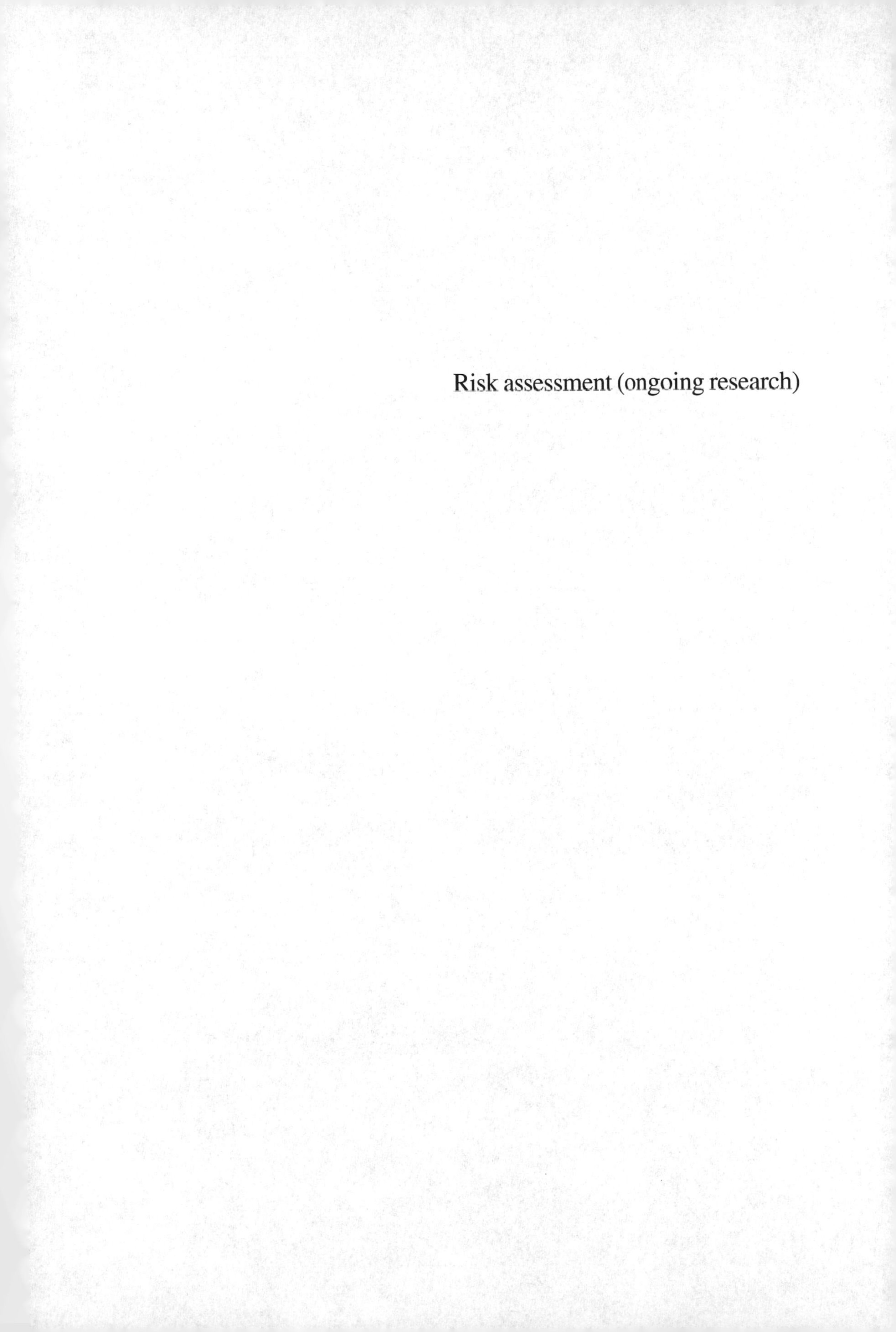

Risk assessment (ongoing research)

Structural Safety & Reliability, Schuëller, Shinozuka & Yao (eds) © 1994 Balkema, Rotterdam, ISBN 90 5410 357 4

Accidental explosions near buildings, methodology of the valuation of the effects

P. Bailly, J. Brossard, C. Desrosier & J. Renard
Orléans University, France

ABSTRACT : Explosions are one of the risks induced by the industrial activity. When an accident is considered, the knowledge of the involved energy and the reactive process can allow a deterministic analysis of the consequences. The modelling of the shock wave is presented, based on experimental data. The loading of a structure, which is the interaction between the wave and that one, is also modelled using recent results. Examples of computation of the transient response of structures are presented and discussed.

1 THE TYPE OF THE CONSIDERED AGGRESSION

Explosion is one of the risks induced by industrial activity. The use and the stocking of chemical products which may react violently is frequent, so it is necessary to take in account the explosion hazard for the safety of people and installations. In many cases the explosion occurs after an unconfined gas cloud is formed (Lannoy 1983). When an accident is considered, some parameters, as the reactive process, the involved energy, the position of the gas cloud, are known and the other ones are unknown, as the shape of gas cloud, the primer...Nevertheless it can be chosen, among all the possible sketches, some typical ones to work out a deterministic analysis of the consequences of the accident. The engineering problem, which is the calculation of the building to resist the effects of an explosion or the evaluation of the consequences of a failure, involves physical and mechanical topics. In a reactive gas cloud, the chemical reaction begins near the primer, and propagates in all the volume. The propagation of the reaction creates a wave, the characteristics of which define the two types of explosions : detonation and deflagration (table 1).In fact, only the detonation is considered, it is assumed that it is the worse case, so the potential chemical energy and the position of the centre of the cloud is necessary to determine the external effects of the explosion and to perform a mechanical study.

Table 1. characteristics of explosions.

	detonation	deflagration
front wave velocity (m/s)	1500-2000	0.01-100
displacement of the product of the combustion / wave	same direction	opposite direction
pressure behind the front	high 5 MPa	low <<p. atm
volume expansion	no	yes
external blast wave	supersonic shock wave	sonic wave

2 THE EXTERNAL EFFECTS OF AN EXPLOSION

The external effects of an explosion are a heat flux and a pressure wave. We counted only the pressure wave, the temporal evolution of this one in a point of the space has a typical aspect in the case of a detonation, as shown in Fig. 1. The characteristic parameters, pressures, duration and impulses, are notified in the same picture. It has been proved experimentally that these parameters could be correlated through a similarity law using the reduced distance λ which is the ratio of the distance from the centre of the explosion and the cubic root of the chemical energy of the explosion (Brossard et al. 1985). It has been collected information about explosions from 0.001 to 10,000 m^3 of gas. Results show that the pressure evolution is different from this coming from an explosion of T.N.T. for an equivalent mass of this one (Baker et al. 1983). A recent study of the effect of asymmetric ignition on the vapour spatial blast shows that the distance to consider for the determination of the reduced distance λ is the distance from the centre of the cloud and not the distance from the ignition point (Desrosier and al. 1991). In the case of an ignition in the periphery of the cloud, the pressure wave has not a perfect symmetry in the nearly field. So a modelling of the pressure evolution has been worked out, only the chemical energy and the distance from the centre of the cloud are necessary to get the static pressure in any point of the space. In the case of the deflagration, there is not a typical aspect of the evolution of the pressure in a point and similarity laws could not be found, so the modelling is now impossible.

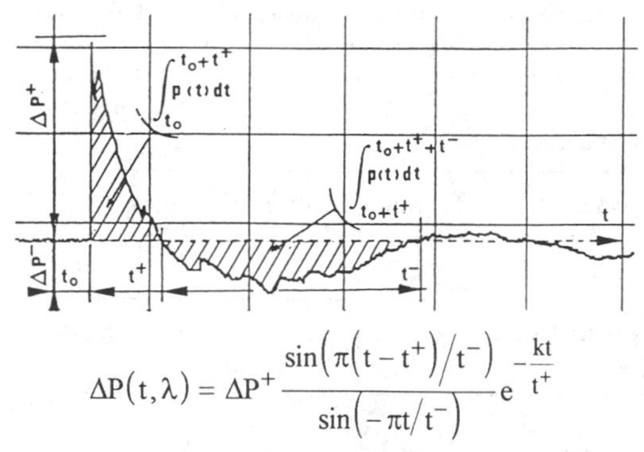

$$\Delta P(t,\lambda) = \Delta P^+ \frac{\sin\left(\pi\left(t-t^+\right)/t^-\right)}{\sin\left(-\pi t/t^-\right)} e^{-\frac{kt}{t^+}}$$

Fig. 1 Typical pressure record and modelling of the signal .

The six parameters of the blast wave generated by the detonating vapour cloud are correlated with the reduced distance λ. Then, for purposes of practical calculus it is interesting to model the pressure signal of the incident wave by means of the sine-exponential function (Desrosier et al. 1991). This modelling takes into account the negative phase which is not always considered and the mechanical effects of which are quite important (Bailly 1988).

3 VALUATION OF THE LOADING ON A BUILDING

The main aerodynamic phenomena observed in an interaction of a blast wave with a building are the reflection and the diffraction. A theoretical modelling of the pressure field near the building is difficult because the usual assumptions of the acoustic waves cannot be made. Some numerical computations has been worked out and the agreement with experiment is good for sonic waves and low pressures. Our way is to get the pressure on the faces of a building using only experimental data. In the case of the building has an elementary geometry as a parallelepiped (or a cylinder) experimental data has been collected and will be presented in this paper. Fig. 2 shows the interaction of a shock front with an obstacle (Ps is the static pressure, Pr the reflected one, Pa the stag. press., Pt the drag. press). In the case of a complex geometry or of a set of buildings, like in a plant, a special experimentation, on a reduced model, is defined and conducted in our laboratory. The scaling laws allow to get the real pressure.

3.1 Reflection

The simple approach of the reflection is to consider the incident signal, and to multiply the over pressure by the reflection ratio (Kinney and Graham 1985). This is not convenient if the signal has a negative phase. It has also been observed that the duration of the over pressure is altered by the reflection. The reflection of the shock wave is different from the reflection of a sonic wave or an ideal shock (a discontinuity of pressure). In peculiar the reflection ratio ($\Delta Pr/\Delta Ps$) has been found experimentally different from the theoretical one. A modelling, based on the experimental data, has been worked out with characteristics defined in the function (Brossard et al. 1988). In fact this modelling is available for the normal or the oblique reflection if this one is a regular reflection and the Mach effect does not occurs (the Mach reflection depends on the angle of incidence and on the value of the over pressure).

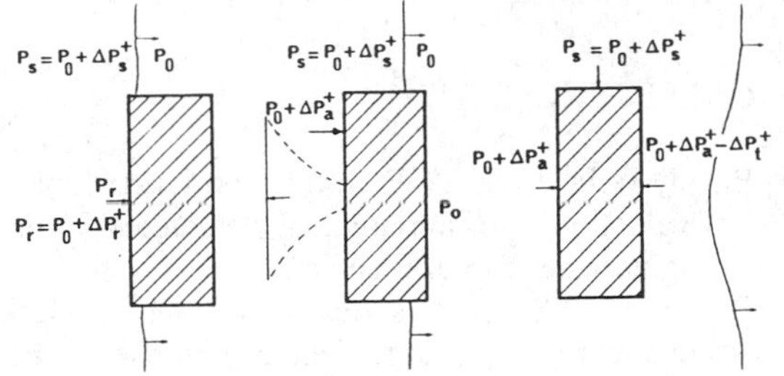

Fig. 2 Interaction of a shock front with an obstacle.

3.2 Diffraction

The interaction of the wave with the building creates diffraction on the edges. A rarefaction wave is propagated from an edge and the reflection pressure on a wall is altered (Fig. 3). In a first approach, when this wave is coming in a point, the pressure decreases from the reflected value to the stagnation one (Kinney and Graham, 1985]. The interaction of the rarefaction wave with the reflected one is an intricate problem, so it is assumed that the altered signal of pressure could be modelled by the same type of function as the incident one (a) with appropriate characteristics. These characteristics are the results of the correlation of experimental data using the reduced distance and the distance from the edge. An example of modelling and measured pressure in shown in Fig. 4. This modelling is able to give the pressure history in any point of a building.

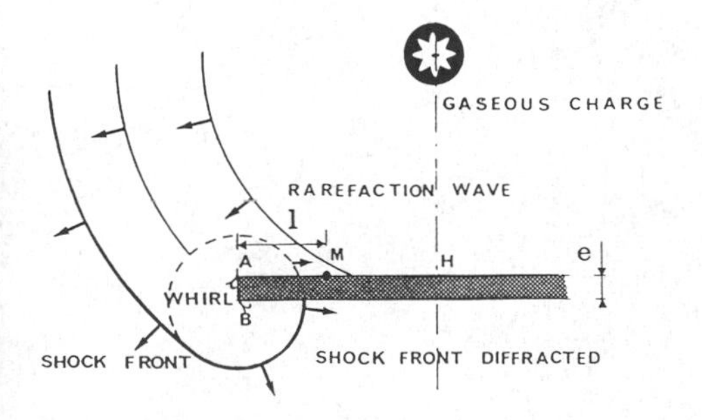

Fig 3 : Diffraction of a shock wave .

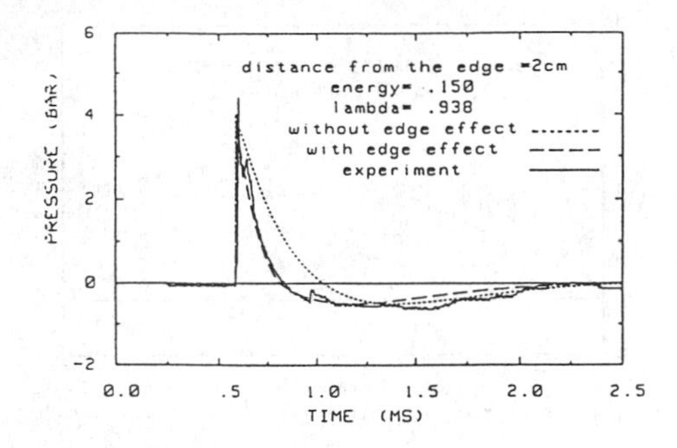

Fig. 4 : experimental and modelled pressures near the edge of a wall.

4 ESTIMATION OF THE MECHANICAL EFFECTS ON BUILDINGS

Two ways are possible to investigate the mechanical response of a structure : the computational way which is now usual for an engineer (the difficulty is to introduce the precise loading defined behind which gives the accuracy of the result) ; and the experimental way which is possible according to the scaling laws of the explosion and these of the mechanic to use a reduced model (the model is not always perfect and similar to the real building). In fact no way is perfect and we always carry out the computation and the experiment. The comparison of the results avoids important mistakes. Fig. 5 shows examples of computations and the accuracy of the modelling of the loading on a simple structure.
The difficulty of the experimentation is to respect the mechanical scaling laws for the structure. But the loading is independent of the mechanical response of structure. So a very interesting result is to get the experimental pressure on the faces of a model. It is easy to find the real pressure and to introduce it in the computation. Several results of the mechanical response could be compared : the computation with the modelled loading, the computation with the experimental pressure and the experimental response of a model. An example of this methodology is given by (Bailly et al. 1989).

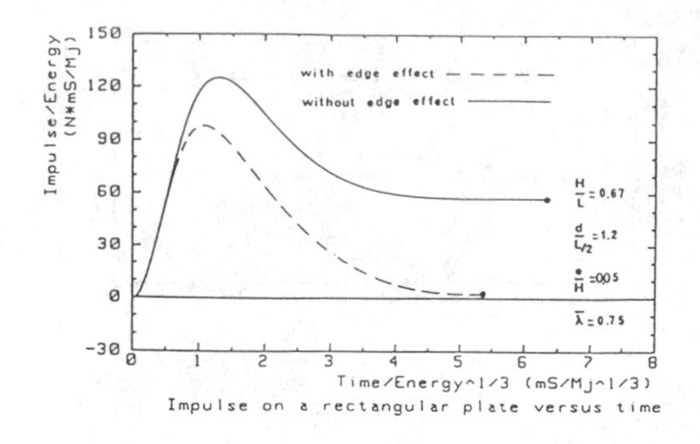

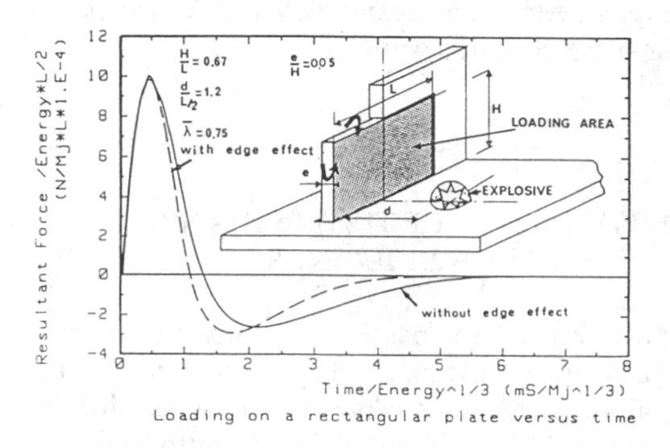

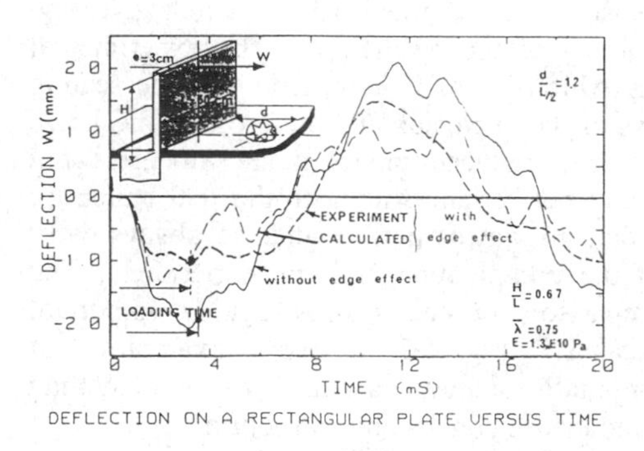

Fig 5 : Impulse and resultant force on a wall and mechanical response

5 CONCLUSION

The recent conclusion of the industrial studies worked out on our laboratory is that the classical modelling of the loading is not always able to get a good mechanical response allowed by the preciseness of the computation. The modelling of the diffraction and the experimental contribution are necessary to carry out a real investigation of the effects of an accidental explosion. In particular two important aspects must be promptly investigated : the pressure loads on various geometry and the effects of the spatial non uniformity of the explosible gas cloud on the blast wave properties.

6 REFERENCES

Baker, W.E. et al. 1983. Explosion Hazard and Evaluation, Fundamental studies in engineering, 5, Ed. Elsevier.

Bailly, P. 1988. Effets des explosions sur les constructions : chargement et réponse de la structure. Thèse, University of Orléans.

Bailly, P. et al. 1989. Prediction of Ariane IV explosion effects on mechanical structures on the launch pad. Loss prevention and safety promotion in the process industries. Oslo.

Brossard, J. et al. 1985 Air blast unconfined gaseous detonation. Dynamics of shock waves, vol. 94 of Progress in Astronautics and Aeronautics, pp. 556-566.

Brossard, J., Bailly, P., Desrosier, C. and Renard, J. 1988. Overpressures imposed by a blast wave. Dynamics of explosion, vol. 114 of Progress in Astronautics and Aeronautics, pp389-400.

Desrosier, C., Reboux, A., Brossard, J. 1991. Effect of asymmetric ignition of the vapour cloud spatial blast. Dynamics of detonations Astronautics and Aeronautics, pp. 21-37.

Kinney and Graham 1985. Explosive shocks in air. Mac Millan Ed.

Lannoy, A. 1983. Analyse des explosions air-hydrocarbure en milieu libre : étude détèrministe et probabiliste du scénario d'accident, prévision des effets des surpressions. Thèse, University of Poitiers.

Structural Safety & Reliability, Schuëller, Shinozuka & Yao (eds) © 1994 Balkema, Rotterdam, ISBN 90 5410 357 4

Integrating failure modes, effects, and criticality analysis with fault tree analysis

Jürgen Deckers & Hendrik Schäbe
Agentur für Sicherheit von Aerospace Produkten GmbH

ABSTRACT: In this paper a method for integrating FMECA and Fault Tree Analysis is described. A software tool FTREE is presented and the algorithm it is working on explained. The approach leads to higher quality and efficience of both analyses.

1 INTRODUCTION

Failure Modes, Effects, and Criticality Analysis (FMECA) and Fault Tree Analysis (FTA) can be combined. ASAP has developed software for both analysis methods. We give a short description of our method.

2 FMECA

The FMECA can be considered as a systematical approach to collect all possible failures on component level together with the distribution function of their interarrival times. Usually an exponential distribution is assumed and the failure rate is given. For each failure mode of a component consequences are assessed and possible redundancies must be indicated. Moreover a severity has to be indicated according to one of the following classes:

1 loss of life,
2 serious injury of persons, serious damage of technical objects,
3 minor injury of persons, minor damage to technical objects,
4 negligible.

In order to assess the severity classes correctly, an information about the probability of system failure caused by a certain failure mode is required.

The information collected within the FMECA can be used as basis events within a fault tree analysis.

3 GENERATING A FAULT TREE BASED ON AN FMECA

For FMECA a software tool FMODE designed and programmed by ASAP is used for detailed information, see Deckers & Schäbe (1992). For FTA a tool FTREE was developed. This program comprises the following options

- Editing a fault tree,
- Generating a fault tree report,
- Displaying and drawing the tree,
- Append another fault tree to an existing one,
- Update of Fault Tree.

This program has the following <u>advantages</u>:

- the time to failure can have a distrubution belonging to one of 15 two parametric families and is not restricted to the exponential distribution
- the fault tree can consists of maximal 64000 basic events or gates
- each gate can have up to 30 entries
- fault tree construction is arbitrary as well as bottom-up, top-down or by mixed technique,
- computer constructed the fault tree automatically, idenifies unused or undefined events and annunciates the operator about errors.

<u>Editing</u> of a Fault Tree is supported as follows. Before creating a new fault tree the computer builds up an initial version of the tree by

- transferring from an FMECA all lines as basic events,
- building up a simple tree by OR gates coinciding with consequences used in the FMECA. If the consequences are distiguished into "local effects", "superior effects" and "end effects" this construction can be applied as well. All events leading to one and the same consequence of next higher level are collected into one OR gate.

The user can correct and improve the Fault Tree by introducing new events, deleting existing, changing gates from OR to AND, adding or deleting events from gates.

Generation of a Fault Tree Report is accompanied with identification of all irrelevant events, i.e. events not leading to the TOP event by any logical combination. A complete printout of all events is made. Irrelevant events are printed out first. Unnecessary gates are deleted, i.e. gates with only one entry.

Reduction of the fault tree is possible, i.e. an OR gate of events having exponential failure time distributions can be transferred into a basic event with exponential failure time distribution, provided the events involved in the gate are not used within another gate.

A fault tree is constructed consisting of basic events and gates, leading to the TOP event. The fault tree is plotted.

It is also possible to plot subtrees of the original Fault Tree. An example of a fault tree is given in figure 1.In the next step the Fault Tree is converted into a reliability block diagram.

Events connected by an OR gate will be transferred into elements (boxes on the reliability block diagram) connected in series and events leading to an AND gate yield elements connected in parallel.

An element in the reliability block diagram can be a single failure mode of a component rather than the component itself. Figure 2 presents an example of a reliability block diagram.

The reliability structure expressed by means of the reliability block diagram is now analysed. Reliability Analysis of the system is performed within

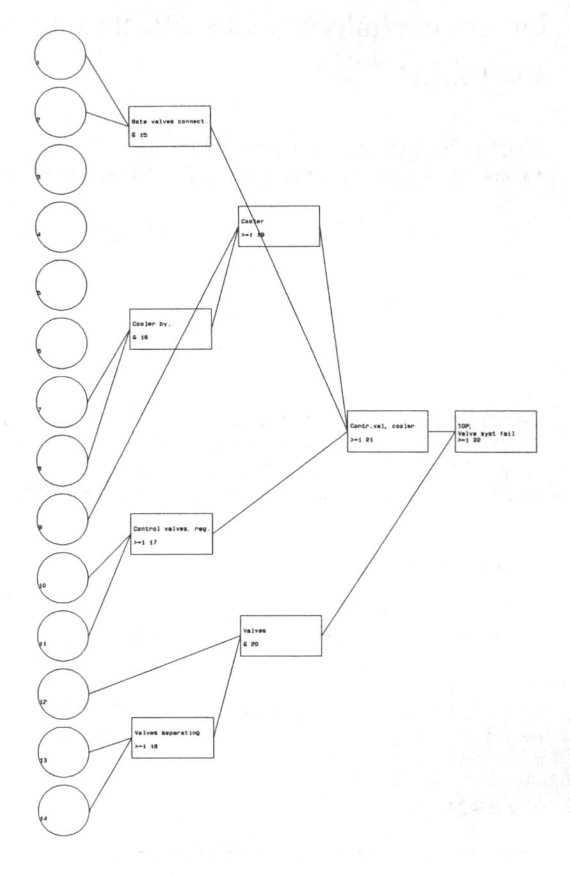

Figure 1 Fault Tree

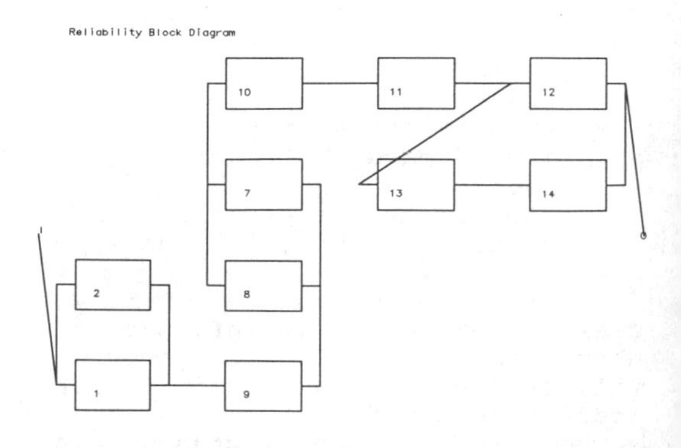

Figure 2 Reliability Block Diagram

three steps:
- Finding path sets
- Finding Cut sets
- Computing Barlow ProschanImportances and System failure probability.

Finding of Path Set is carried out by a simple algorithm.

The computer traces the paths through the network starting at the input. At branching points the computer starts with the first element, depicted in the reliability block diagram as the lowest connected with the node. The next path is found going one step back along the actual path. The element after the node is excluded from the considerations for the next path and a new connection is traced to the output.

If no connection is found, an additional step back is performed to the next node into direction to the input. The first element after _this_ node belonging to the actual path is excluded and the path is traced again. This procedure is repeated until all paths are found. This algorithm is fast enough and, depending on the problem, may be faster than the algorithm presented by Aggarwal et al. (1973)used in [2]. It has to be noted that always only one elements is excluded from path tracing.

From the path set, the cut set can be evaluated easily. Cut sets of size one are obtained taking all elements occuring in all paths. Cut sets of size two are all pairs of elements occuring in all paths etc. By this procedure all cuts are found beginning with the cut sets of smallest cardinality.

In the next step, the failure probability of the system is computed. Instead of evaluating the reliability terms of the system, as e.g. by Aggarwal et al. (1973), Heidtmann (1989), Locks & Wilson (1992), Chen, Xu & Tong (1991), an inclusion - exclusion method is used, see Fong & Buzacott (1987). Two formulae are used to evaluate the system failure probability using path sets and cut sets, respectively. For path sets we have the series expansion:

$$P_{fail}=1 - \Sigma \underset{P_i}{\pi} (1-p_i) + \Sigma\Sigma \underset{P_i u P_j}{\pi} (1-p_i) - \Sigma\Sigma\Sigma \underset{P_i u P_j u P_k}{\pi} (1-p_i)+\ldots, \quad (1)$$

where Σ is a sum over all path sets, P_i, P_j, P_k denotes paths, and "u" denotes the union of sets. By P_{fail} and p_i failure probabilities of system and i-th element are denoted. For evaluation of failure probability based on cut sets an analogous expression can be used.

$$P_{fail}= \Sigma \underset{C_i}{\pi} p_i + \Sigma'\Sigma' \underset{C_i u C_j}{\pi} p_i - \Sigma'\Sigma'\Sigma' \underset{C_i u C_j u C_k}{\pi} p_i+\ldots \quad ,(2)$$

Here C_i denotes the i-th cut set and Σ' denotes a sum over all cut sets. The sums (1) and (2) are evaluated until the error, given by the last term in the sum is sufficiently small ($<10^{-7}$) or a certain limit of computing time is exceeded. In the latter case the operator is annunciated that an approximation is used. Pure parallel or series structures are evaluated exactly by an extra routine. If the failure probability is evaluated on an interval $[x_u,x_o]$ then $p_i=F_i(x)$, and $x_u \le x \le x_o$, where $F_i(.=)$ denotes the distribution of the time to failure of the i-th element in the system. Then (1) and (2) can be further simplified. All elements fulfilling

$$1-F_i(x_o)>1-\epsilon \quad (3)$$

are deleted from the path sets, all path sets fulfilling

$$\underset{P_i}{\pi} (1-F_i(x_u))<\epsilon \quad (4)$$

are deleted. Then the number of terms in (1) decreases. The following conditions are used to simplify (2):

$$\pi_{P_i} F_i(x_o)<\epsilon$$

to delete a cut set and

$$F_i(x_u)>1-\epsilon$$

to delete an element from all cut sets.

Formulae (1) and (2) are used depending on whether the number of paths is smaller than the number of cuts, or for almost euqal numbers of paths and cuts for $P_{fail}>0.5$ and $P_{fail}<0.5$, respectively.

By this procedure the failure probability for a system can be evaluated on an interval $[x_u,x_o]$. Figure 3 shows the failure probability of the system.

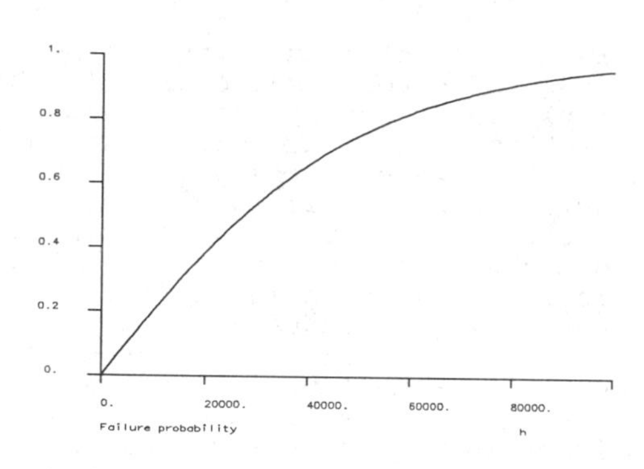

Figure 3 Failure Probability of System

Additionally, the Barlow-Proschan importances, see Barlow & Proschan (1975) and Aven (1987) can be computed. They allow a ranking of the system's components of the with respect to their influence on system failure probability. The results of the FTA can be used to perform a backfitting for the FMECA. Statements such as severity of failure, possible consequences, and compensating provisions can be corrected using results of the Fault Tree Analysis.

4. REFERENCES

Aggarwal, K.K., J.S.Gupta & K.B.Misra 1973. A new Method for System Reliability Evaluation. *Microelectronics & Reliability* 12: 435-440.

Aven, T. 1987. On the Construction of Certain Measures of Importance of System Components. *Microelectronics & Reliability*. 26: 279-281.

Barlow, R. & F.Proschan 1975. *Statistical Theory of Reliability*. New York: Holt, Rinehart and Winston Inc.

Chen S., D. Xu & S. Tong 1991. A New Algorithm for Minimal Disjoint Sum of Products. *Microelectronics & Reliability*. 31: 817-822.

Deckers, J. & H. Schäbe 1992. FMECA rechnergestützt erstellen. *QZ* 37: 366-369

Fong, C.C. & J.A. Buzacott 1987. Improved Bounds for System-Failure Probability. *IEEE Trans. Reliability* R-36: 454-458.

Heidtmann, K. 1989. Smaller Sums of Disjoint Products by Subproduct Inversion. *IEEE Trans. Reliability* 38 : 305-311.

Locks, M.O. & J.M.Wilson 1992. Note on Disjoint Product Algorithms. *IEEE Trans. Reliability*. 41: 81-84.

Structural Safety & Reliability, Schuëller, Shinozuka & Yao (eds) © 1994 Balkema, Rotterdam, ISBN 90 5410 357 4

Reliability-based design of crane structures

M.Gasser, M.Macke & Y.Schorling
Institute of Engineering Mechanics, University of Innsbruck, Austria

J.Hartl
Palfinger AG, Bergheim/Salzburg-Kasern, Austria

ABSTRACT: Safety standards of modern structural and mechanical systems require more and more the development of reliability oriented concepts of analysis and design processes. A case study of a hydraulic wood loading crane based on this concept is presented. Field measurements of the operating cranes were performed, providing a realistic data base of the kinematics of the crane and the hook loads for the structural and consequently the reliability analysis. Ultimate load and fatigue failure due to the upcrossing of stress levels as well as crack growth in welding seams are considered. Some preliminary results are presented.

1 INTRODUCTION

Particulary recently, the quantification of safety and risk of failure of mechanical components, systems and structures gained considerable attention and importance. This is on one hand due to increasingly competitive markets which require reliability products and on the other hand due to the newly introduced legislation on product liability. In this context more information on the quality of the respective product is needed. This can be provided by an improvement of the structural analysis and design process. Therefore it is necessary to take into account random characteristics of the system, which forms the basis to assess the reliability of structures and mechanical components.

In the case study as presented in this paper a reliability-based design process is applied to hydraulic wood loading cranes. The mechanical model of the structure considered is established by means of the finite elements. The loading history is based on full scale measurements taken at various crane structures during operation. So a realistic load model could be derived. The reliability analysis encompasses both, ultimate load and fatigue failure. The associated probabilities of failure serve as an appropriate operative quantity for the design engineer to allow him to interactively redesign - and ultimately optimize - the structure.

2 METHOD OF ANALYSIS

2.1 *General remarks*

The design procedure of crane structures or components is generally carried out in several stages. Starting with a general layout i.e. predesign of the product, an increasingly refined design is obtained. Full scale measurements of loads during operation form the basis for a realistic analysis. These load histories replace the synthetic loads that are utilized so far for design and testing and which only remotely have anything to do with the actual performance of cranes. This, along with a detailed investigation of workmanship, i.e. quality in terms of characterizing initial crack distributions, allows a realistic quantification of the reliability of the structural types under investigation.

2.2 *Load analysis*

Although quite generally applicable, the design process as presented here is developed for hydraulic wood loading cranes mounted on trucks (Figure 1). During operation the operator guides the motion of the crane to follow an apparantly arbitrary path by activating a lever system controlling hydraulic cylinders for a lift motion, and thereby adjusting the angles of the jibs as well as the extension of the telescope jib. The loading of the crane depends on one hand on the type of operation of the crane (e.g. moving long timber or short timber) and on the other hand on the skills of the operator. In context with the design process - so far - the latter was taken into account only in an extremly simplified way. Therefore full scale measurements over a period of 30 days, for different types and locations of operation were performed. The measured data consist of parameters such as cylinder pressures, angular adjustments and extensions of the telescope jib. The load process as derived from these data is regarded as a sample of the actual loading characteristics of the crane structure and hence provides the data base for the subsequent analysis.

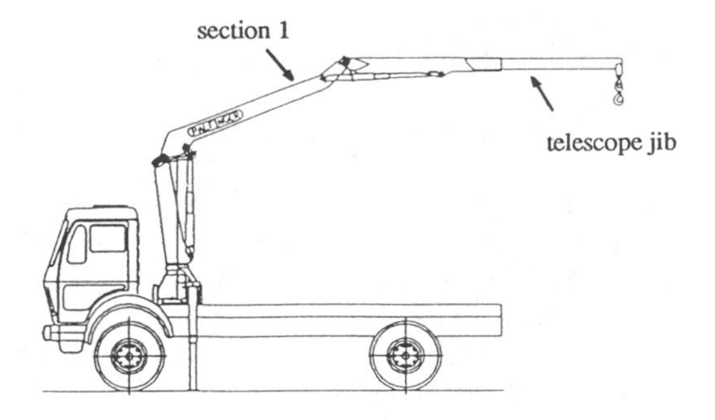

Fig.1 Schematic sketch of a wood loading crane

2.3 *Structural modeling and analysis*

The mechanical modeling for the subsequent reliability analysis reflects the respective failure modes to be analyzed. This implies that for ultimate load failure a simplified beam model (Figure 2) suffices, while for investigating fatigue failure detailed finite element modeling (Figure 3) is required, which even includes the modeling of welding seams. In addition special FE models of welding seam types of cracked configurations are also used. Based on these models the subsequent structural analysis is performed.

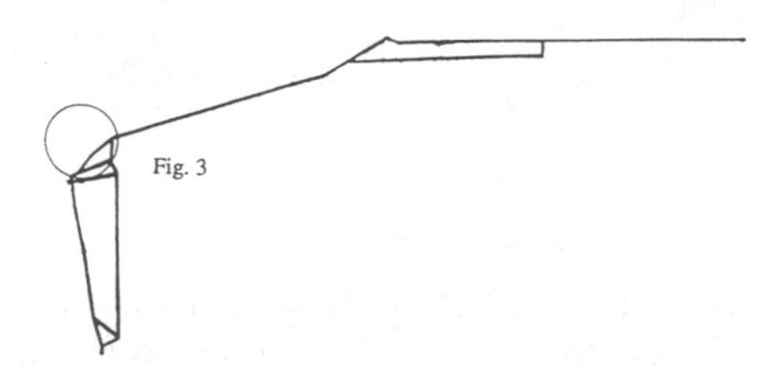

Fig.2 Simplified FE model

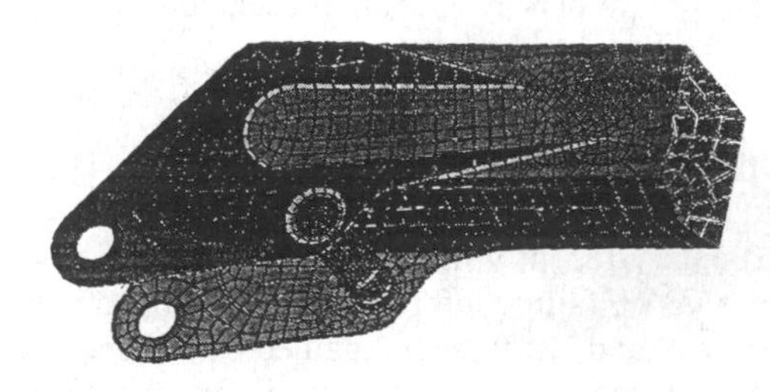

Fig.3 Detailed FE model

Utilizing the full scale measurements, i.e. transforming them into the appropriate kinematics as well as hook loads, a dynamic simulation of the sequences of crane operations can be performed. For the following structural analysis - utilizing the beam model - the forces of inertia are taken into account. By this procedure information on the sectional forces of all crane components at each time step is available, i.e. the entire loading process of each component.

In a next step these loading processes are used in context with detailed FE models consisting of shell elements (up to 5000 d.o.f.). For linear (deterministic) systems the displacement vector has to be computed only once, i.e. for unit loads. (Thus the actual loading processes acting on element level are derived for each time step from a simple multiplication of the sectional forces with a stress tensor). Similar considerations are valid for the calculation of the stress intensity factors (SIFs), governing fatigue crack growth. The computations for cracked welding seam type models, consisting of solid elements, are carried out only once for unit loads and the effective SIFs in time are then derived by multiplication of the SIF tensor utilizing factors obtained from the element loading process.

2.4 *Failure analysis*

An important step within the reliability analysis is the identification of the dominant failure modes. Analogously to the conventional design process (global predesign succeeded by a detailed analysis) the structural failure modes of consideration are ultimate load failure and fatigue failure.

The analysis of ultimate failure is carried out by using the results of the dynamic simulation as described above. This is followed by a statistical analysis of the calculated sectional forces. The dynamic response characteristica at selected critical points of the structure are then obtained by employing techniques of estimating the expected rates of upcrossings of a given threshold, i.e. the number of excursions of the loading process from the safe domain (Grigoriu (1984), Nataf (1962), Lin (1976)).

Fatigue failure is analysed by using the derived loading process at (the detailed FE) element level. By this the force distribution as well as the particular welding procedure may be taken into account. Consequently the crossing of a critical stress threshold and fatigue crack growth is considered. The exceedance rates of relevant material stress levels, i.e. the upcrossings of certain stress levels up to the yield limit, are estimated analogously to the procedure mentioned above (Grigoriu (1984), Nataf (1962), Lin (1976)). In addition the first and second statistical moments of the stress processes are calculated.

The analysis is complemented by taking into account fatigue crack growth (FCG) within the welded regions. FCG is determined by the loading process, the geometry factor (both combined in the SIF) and the (initial) crack size distribution (Kanninen and Popelar (1985)). To relate actual loading processes to laboratory fatigue experiments of constant load amplitude, equivalent load cycles are derived by using a cycle-counting method. By means of the initial crack size distribution, describing the damaged state of the structure, FCG curves for arbitrary cycles of the lifetime of the structure can be determined by applying the Markov chain technique. Finally the time-variant probability of failure due to unstable crack growth is calculated.

3 PRELIMINARY RESULTS

During the full scale tests strain gauges have been attached at different locations of the crane structure. By comparing the measured and computed stress components, relative differences less than 10 % are found. Hence, the proposed structural model provides a good representation of the actual structural behavior. In figures 4 and 5 the respective stress components at section 1 on the top of the main jib (Figure 1) are illustrated for 1000 timesteps of 0.2 seconds.

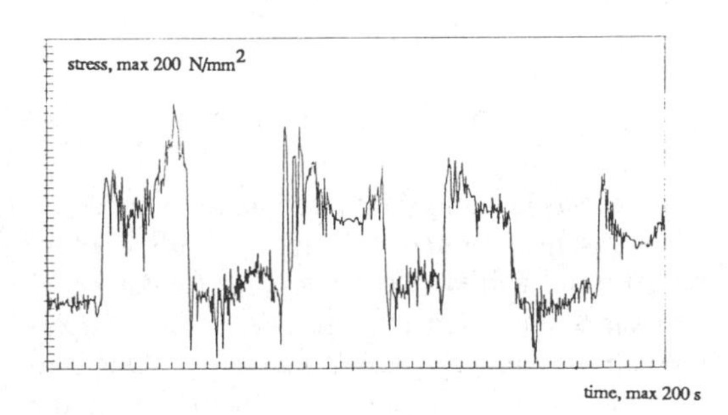

Fig.4 Measured stress components

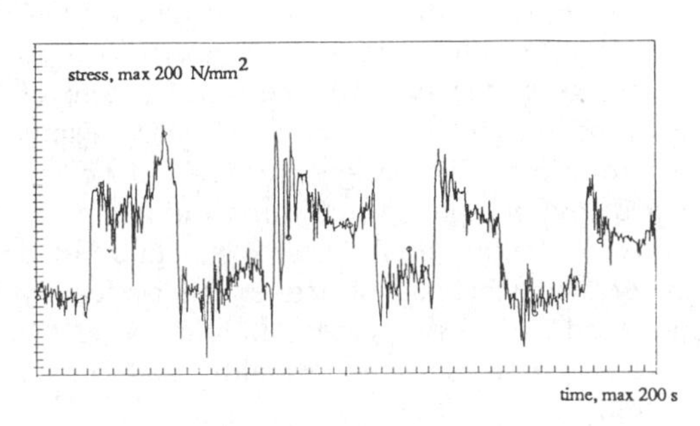

Fig.5 Calculated stress components

For the detailed FE model, exeedance rates of the yield stress were calculated. The results are - due to the amount of data to be processed - visualized by using a post processor (SDRC (1991)). This allows a quick survey of the complex structure with regard to its reliability. In Figure 6 a particular area i.e. locations at the bottom of the detailed model of the main jib can be seen, where the exceedance rates take on relatively high values as compared to the remaining structure. It is interesting to note that this location is identical to the position, that caused in the past repeatedly failure problems during operation.

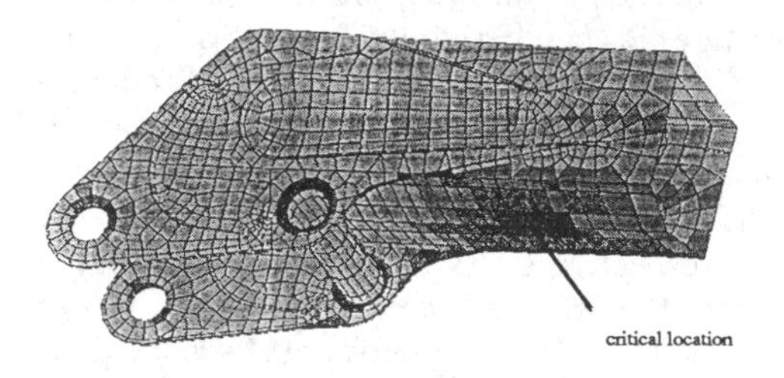

Fig.6 Exceedance rates of yield stress

4 CONCLUSIONS AND OUTLOOK

The proposed analysis procedure is considered to represent an important step towards a more realistic reliability assessment of (crane type) structures and components. Due to the implementation within the design process - taking into account realistic loading and kinematics of the crane, supported by full scale measurements - the design engineer can interactively improve and consequently optimize the structure without extensively testing different prototypes. This reduces both, the time for the development of new products and also costs. Preliminary results are promising. The design process will be extended by computing the time-variant probability of failure of the crane structure. In a further step the crane structure will be locally optimized considering fatigue failure due to the upcrossing of stress thresholds and crack growth.

ACKNOWLEDGEMENT

This research is partially supported by the Austrian Industrial Promotion Fund (FFF) under Contact No. 6/636, which is gratefully acknowledged by the authors. This project is supervised by *M. Glück* of Palfinger AG, Salzburg, and *G.I. Schuëller*, Institute of Engineering Mechanics, University of Innsbruck.

REFERENCES

Grigoriu, M. 1984. Crossings of non-Gaussian translation processes, Journal of Structural Engineering, Vol. 110, No. 4, p.610-620.

Kanninen, M.F., Popelar, C.H. 1985. Advanced fracture mechanics, New York: Oxford University Press, Oxford: Clarendon Press.

Lin, Y.K. 1976. Probabilistic theory of structural dynamics, New York: Krieger, Reprint.

Nataf, M.A. 1962. Détermination des distributions de probabilités dont les marges sont donnés, Compte Rendus de l'Academie des Science, Paris, 225, p.42-43.

SDRC - Structural Dynamics Research Corporation, 1991. I-DEAS finite element modeling user's guide, Milford: SDRC.

Structural Safety & Reliability, Schuëller, Shinozuka & Yao (eds) © 1994 Balkema, Rotterdam, ISBN 90 5410 357 4

Safety and reliability problems in chemical plant recognized during preoperational phase and operating experience backfiting

M. Kožuh, B. Mavko & Dj. Vojnović
'Jožef Stefan' Institute, Ljubljana, Slovenia

ABSTRACT: For Resin Synthesis Chemical Plant HAZOP analysis was performed in order to identify the potential threats to plant safety and reliability. A number of issues were identified on the hardware as well as in human interactions with the system. After the six months of operation some of the predictions became through and the discussion is given in the paper.

INTRODUCTION

For the Resin Synthesis Chemical Plant we have performed HAZOP analysis to identify the potential threats to safety of the plant and quality of the product. Interest for such an analysis was shown from the part of the plant because the old plant was burn in a fire and during licencing process regulatory body demanded some evidence that safety was taken into acount for new one. The team of the plant experts was formed and was lead by our team from the institute. With sistematic work we have identified around hundred potentialy dangerous topics which can develop in consequences regarding safety or quality of the product.

We can divide the topics into three cathegories according to the cause of the threat. The first one was the hardware problems which were more or less connected to the fact that heating media is thermooil, which can cause fire threat and which was the reason for the fire in old plant. The second threat was the group of man interface problems due to improper behaviour of the operators controling the process and workers during various operations of the process. The third one were problems because of lack of knowledge and experience with the new introduction of process computer which will control certain parts of the batch process and which will be interfaced by the operators after completion of each phase.

PERFORMING HAZOP ANALYSIS

For the chemical factory Resin synthesis HAZOP analysis was performed. The old one was burned down and for the new one the owner wanted to demonstrate that safety issues were addressed properly. A team was formed which consisted from the experts from factory and as the leaders of the process people from IJS were appointed. Start of the process was initiated with the short course on the methods used in safety analyses for the people from the plant to give them basic knowledge on the subject. The next step was getting familiar with the process so that later on there were no problems in understanding the details of the production. Since the batch process was in question, we focused ourselves on different phases or unit operations. We studied all the substance that goes into the process and that comes out as well as the energy flows on the border of the system. Reffering to all relavant parameters we sistematicaly studied all possible deviations in the process. We made some benefit from the fact that there was twenty years of experience from the old resin synthesis, but the differences had to be taken into account.

About hundred issues were identified as potential chalenge for safety or as a threat to product quality. Not all the scenarios had the same consequences. Only some of them could led to consequences which would have the impact to the environment. There was great care given to all the possibilities of fire since it has the major contribution to the environment risk.

We identified some scenarios where human interaction was not covered enough with the procedures and where there was credit given to the fact that personnel is trained enough to deal with the problems. The suggestions were made to write appropriate procedures to cover the anticipated problems identified in the HAZOP analysis. Most of the problems connected to the process automatic control we could not solve because at the time we did not have the detailed description of how the computer is controling the process during transient conditions.
The identified key human interactions were not analyzed in normal HRA manner but were pointed out to show the importance of these actions.

Hazop analysis as beeing suggested by prof. T.Kletz is donne by the questions which are refering to the plant process parameters. As such it is very formal and it can be done in very straight forward manner. There are allready some attempts to automize the process by use of expert system. Regarding the process and hardware there is not much to add except that the leader of the Hazop team should have wast experince in engineering and in troubleshooting.

There is much greater problem how to handle the human interactions within the process. There is some evidence that this has been identified, but as far we can see none of the solutions has been addopted for general use.

PLANT OPERATION BACKFITING

From the first operating experience we would like to propose some we hope general solutions which can be used in the future.

Most of the problems are conected to human errors and are following:

- identification where human interaction can cause serious problems,

- what can be donne with such actions,

    o introduce it in the procedures and how to control them,

    o use some kind of automatic action to avoid the human failures,

    o prepare special training to cover these problems,

    o or to live with it hopping nothing would happen,

- how to present the process to operators to give them appropriate information problem of components in the process on which the operator does not have influence.

- serious problem is the communication problem between the operators and the workers on site.

The working personnel should be trained and equiped with the operating and emergency procedures. Since there are different professions involved with the process they are trained separately. On the site they most of the time work separately and their communication is limited to their initiative, they are not tought to communicate during the process. Next serious problem is schematical presentation of the process (man-mashine interface) in the control room where only the active components are presented on the control board diagram and where only those components have status lights (open, closed). In such a case there are two options for the operators either they forget that there are some components which are omitted or to add the component in the procedures to check the component status before the start of the process. In such a case there should be checklist according to which the status of the non active components should be determined and verified or all the non active equipment should added to the plant schematic and be equiped with status indication even though operator have to open or close it manualy on site.

One other problem is, where to put the borders of the system which has connections to some support systems which seems to be out of the scope. From our experience we can claim that all the components which are operated from the control room dependless from their physical location should be within the borders of the system analyzed.

When there is an overlap in working responsabilities between operators and other workers this overlap should be avoided by making clear cut between their responsabilities.

When we did our HAZOP study we were mostly depending on the knowledge of the participating exeperts and on the design descriptions of the process. At that time there was no detailed job descriptions and responsabilities of the crew responsible for operation of the process. During the HAZOP process all the participants learned a great deal about the operation which hopefully would end in an apprehensive set of operating procedures.

The analysis was used for licensing purpose so after the licencing audit we finished with our work for the plant.

After six months of operation, people from the plant returned and proposed to continue with our cooperation on the analysis of the process. It tourned out that some of the issues addressed in our analysis really happened and that there were some more things which due to our limited time were not recognized. Based on operating experience, we can now reflect what was good and where more emphasis should be given in the future.

The first thing to mention is that the operating procedures shoud be written in such a way that operator and other personnel have enough material for mitigating the abnormal conditions. Where there is interaction between different people their relations should be specified. Based on the past events we can claim that communication is one of the key problems, because it cannot be detected properly during the HAZOP process. We have found that human actions can be assessed during HAZOP by looking closely the procedures. We think that this aproach is correct, but cannot be used in the design phase to create the relevant procedures. The HAZOP process can be though used to identify the steps in human actions which can be in the later process transformed into the operating procedures with all the possible outcomes. After the completed HAZOP the plant walk through is beneficial since the real plant can have some differences from the design.

Normaly differences are minor but can drasticaly afect the safety. For example during the HAZOP all the experts claimed that the air lines beneath the thermooil lines are from such a material that they are safe, even if the oil would spill over the air lines, which would give operator the ability to operate with the necesary valves. During plant walk through we found out that these lines are made of the plastic which is not heat resistant.

The other thing that happened during the operation was that one of the manual valves on the feeding line of the vesel stayed closed, when new charging of the vessel started. The result was the water hammer efect which broke the line in the expansion joint region. This accident was the result of lack of communication between the workers on site and operators together with the ommission of this valve on the control board schematics.

CONCLUSIONS

HAZOP as organized way to review process systems is very good in predicting hardware problems. With the knowledgeble team of experts a lot of valuable informations can be put together for use in writing plant operating and emergency procedures and for operator training. The weak point of the HAZOP is how to deal with human actions and posible human errors. Identifying possible errors is not enough. We should qulitatively try to distinguish between errors of commission and omision and what we found very important the problems with communication. Sometimes it is very difficult with the human errors, where there is some automatic control performed by process computer. With this problems we can cope if we emphasize them in the initial phase and then watch them during operation. We think that HAZOP is first of all important for plant safety and not so important to the authority. Since the plants are more or less reluctant to do such an analysis, there is somethimes good that authority encourage them to start the process.

REFERENCES

M.Kožuh, J.Sušnik, D.Vojnovič: Hazard and Operability Study for the Resin Synthesis of Color Medvode, "Jožef Stefan" Institute, Ljubljana, April 1991

Kletz, T.A.: HAZOP & HAZAN. Notes on the Identification and Assessment of Hazards. The Institution of Chemical Engineers,1986

The 1988 European Summer School on Major Hazards. Christ's College Cambridge, July 1988

King R.: Safety in the Process Industries, Butterworth-Heinemann 1990

Guidelines for Chemical Process Quantitative Risk Analysis, Center for Chemical Process Safety of the American Institute of Chemical Engineers 1989

Arendt J.S. et al: Evaluating Process Safety in the Chemical Industry, A Manager's Guide to Quantitative Risk Assessment, Chemical Manufacturers Association, December 1989

Kandel A. Avni E.:Engineering Risk and Hazard Assessment, CRC Press Vol.1 1988 Risk Management News, Vol.1, No.2, October 1991, JBF Associates, Inc.

Structural Safety & Reliability, Schuëller, Shinozuka & Yao (eds) © 1994 Balkema, Rotterdam, ISBN 90 5410 357 4

Automatic fault tree construction using a new component behavioral model

T.S. Liu & J.D. Wang
National Chiao Tung University, Taiwan

ABSTRACT: This study presents a new method for fault tree construction based on a new definition of events, an extended decision table, and two pruning procedures. The behavioral models of components are accounted for by extended decision tables. To demonstrate the methodology, a computer program is developed and implemented for a digital tape transport.

1 INTRODUCTION

The construction of a fault tree is the most time-consuming task in fault tree analysis. In fault trees, an event denotes a state that occurs to a system component. Thus, the component model and the event definition are crucial for the efficiency and accuracy of the fault tree construction. Efforts have been directed toward automatic fault free construction in computer implementation. (Fussell,1973) and (Taylor,1982) adopted transfer functions as models for component failures. The Lapp-Powers algorithm (Lapp,1977) based on a directed graph model of cause and effect for individual component was employed to facilitate construction of fault trees. (Salem et al,1977) developed a method for constructing fault trees of general complex systems. Components were modeled by decision tables. (Camarda et al,1978) and (Gough et al,1990) converted reliability block diagrams into fault trees. (De Vries,1990) presented a quantitative methodology for the automated generation of fault trees for electrical/electronic circuits.

The objective of this study is to present a new event definition, a virtual transfer component, and a extended decision table to effectively model the behavior of components. Events are defined in terms of both qualitative descriptions and quantitative event types. Furthermore, the k-out-of-n and standby redundant sub-systems are modeled by virtual transfer components. Two pruning procedures dealing with production of transfer symbols and successive logic gates with the same type are proposed for the generated tree during fault and after the tracing process. Finally, a digital tape transport is employed to demonstrate the proposed method.

2 EVENT DEFINITION

In this study system states are represented by events consisting of the event description and the event type. Event descriptions are qualitatively verbal descriptions of events, such as "failed to open", "rupture", "stuck", etc. In contrast, the event type is represented by the relation between system variables and quantitative intervals of the system variables. The system variables may be any physical quantities, such as torque, pressure, position, etc., or variations of any physical quantities, such as variations of pressure or temperature. Moreover, for a variable with a continuous domain, the entire domain is subdivided into a finite number of nonoverlapping subintervals, and the subintervals are in turn represented by the lower and upper bounds of the variable value. The system variable may be "S" that denotes smaller than, "A" amid, or "L" larger than a prescirbed interval.

3 COMPONENT BEHAVIORAL MODEL

Decision tables, combining states of inputs and internal operational and failed states, are used to describe each possible output state of components. In contrast to (Salem et al,1977), this work presents three steps in developing an extended decision table of a component behavioral type:

- Enumerate all combinations of input states and internal modes for a component and determine the corresponding output states using failure mode and effect analysis, each row represents a combination.
- Split each row into rows with identical inputs and internal modes, but each row contains only one respective output.
- Utilize a decision table reduction method to produce a compact table which can save computer time and memory in the process of constructing the fault trees.

These steps require that if m rows contain identical elements in all columns except for one and only one of the input or internal mode columns, and if m equals to the number of all the possible events of the exceptional column, then these rows combine into a single row by eliminating all but one in which a "don't care" event is substituted for the element of the exceptional column.

4 CONSTRUCTION OF FAULT TREES

The construction of fault trees begins with the TOP event and traces backward, according to the system structural configuration and decision tables in a database, to the primary and undeveloped events of components. Editing effort is required to prune the generated fault trees.

To model a system in a computer program, nodes are defined as any points at which the outputs of one or more components are connected to the input of the preceeding component, and vice versa. System states are the specified conditions of a system, defined either at nodes or as internal modes of components. When a specific event is being traced, the event must be checked if it is compatible with system states already defined. Moreover, the present system states are to be updated according to the event if the event is beneath an AND gate. Otherwise,the system states that are defined in developing an event beneath an OR gate must be reset as undefined. Another important system states are boundary conditions, which remain to exist as the fault tree is constructed and may be defined at any node of the system, or as any existing component internal modes. For components in the presence of initial conditions, component conditions may vary with forthcoming events.

The construction process begins with the undeveloped events in the TOP event. For each event to be developed, an appropriate decision table is used to determine the required internal events and input events. Each input event is in turn traced back to the outputs of succeeding components. In addition to the procedures proposed by (Salem et al,1977), transfer symbols are proposed to simplify the generated fault trees and save constructing time during the constructing process. When any event being developed is compatible with system states and there is another identical event which occurred in the same component and had already been developed completely in other branches of the fault tree, the previously developed event will be denoted a numberred transfer symbol. The input to the event being developed is thus replaced by the numbered transfer symbol.

After the tracing process is completed, in order to minimize the number of branches in the generated fault tree, it is proposed that if two successive logic gates have the same type,

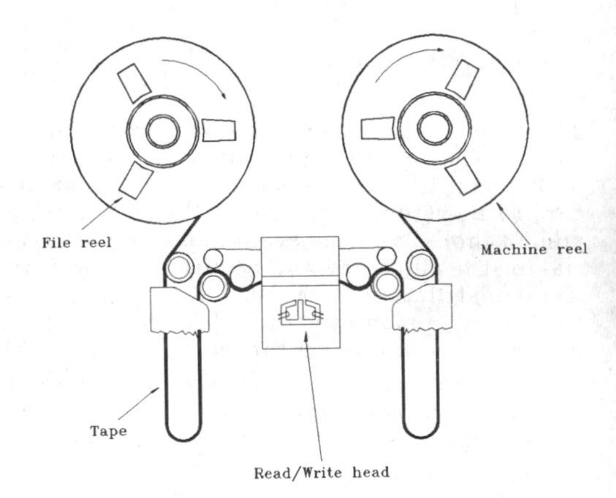

Fig.1 Tape transport system

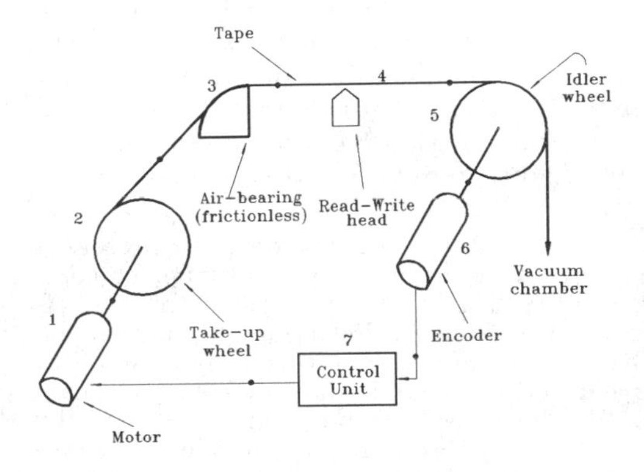

Fig.2 Schematic diagram of capston control system

logic gate can the input events of the lower be moved up to become additional input events of the upper logic gate; moreover, the lower logic gate and its output event are deleted. However, if a transfer symbol is assigned to the output event of the lower logic gate, this combination task may not be executed for the sake of minimizing the total number of events in the generated fault tree.

5 EXAMPLE

As shown in Fig.1, a tape transport is designed with a small capstan to pull the tape past a read/write head with a take-up wheel turned by a DC motor. This structure includes capstan, vacuum, and reel. The capstan motion is isolated from the reels by vacuum columns that provide constant tension to the tape at the head. A schematic of the capstan control is shown in Fig.2.

It is assumed that the vacuum chamber and control unit are faultless and the purpose of the capstan system is to control the velocity of the tape at the read/write head about 5 m/s measured from the encoder. The control unit provides a negative feedback of the velocity measured from the encoder to the DC motor.

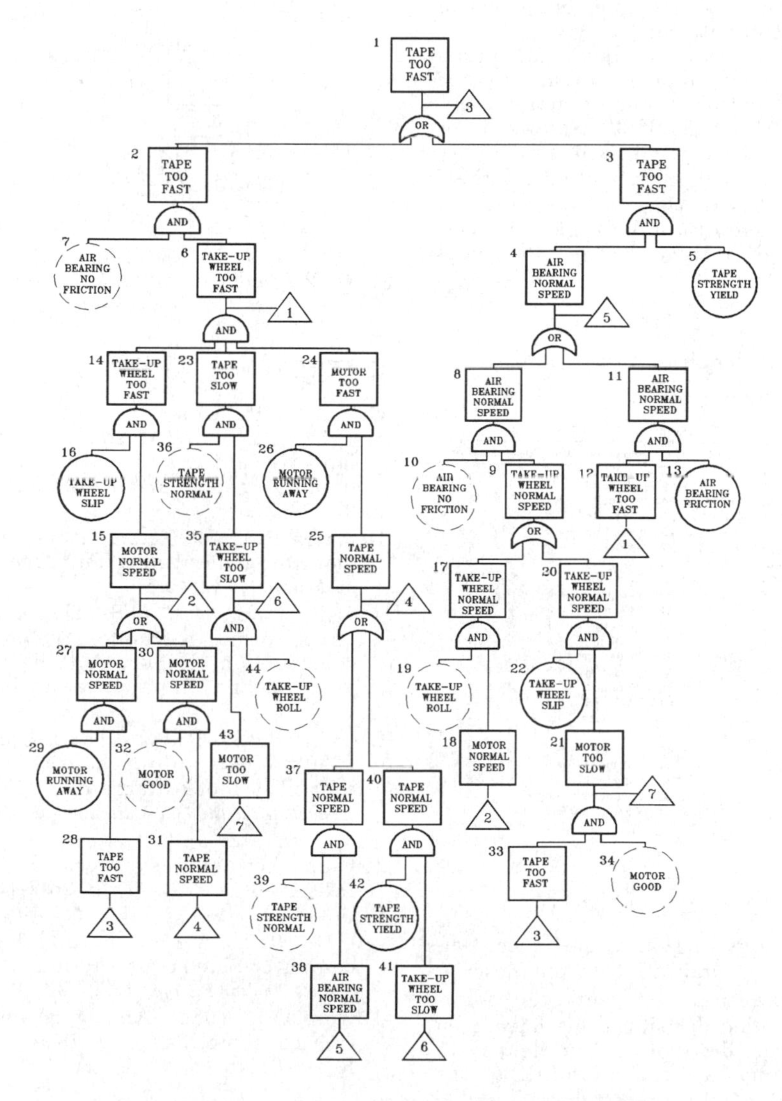

Fig.3. Fault tree of capston control system

The take-up wheel which truned by DC motor drives the tape by means of the viscous friction (nominal value at 0.01 N·m· s).

To analyze and quantify the possible fault of the capstan system, it is assumed that the components of the capstan system have associative internal modes and corresponding event types as shown in Table 1. The TOP event of the fault tree is defined as "tape too fast", the generated fault tree using the proposed method is shown in Fig.3.

By inspecting the transfer symbol, one can find that there are some cause and effect loops in this fault tree. The reasion is that the capstan system is an closed feedback system and the computer program dealing with any component will return to the same component in some steps of computation.

The pruning procedure reduces the orginal fault tree of more than 90 events to one comprising 44 events. Furthermore, event numbers 7,10,19,32,34,36,39, and 44 hardly fail and may be treated as of probability one (component good). By eliminating these events and reconstructing the fault tree, a more succinct tree with only 5 events is generated as depicted in Fig.4. Hence, it becomes easier to trace the failure cause.

Table 1. Event definition in the fault tree

Event No.	Event Type				Event No.	Event Type			
	Variable	Min.	Max.	Relation		Variable	Min.	Max.	Relation
1	Velocity	4.5	5	L	23	Velocity	4.5	5	S
2	Velocity	4.5	5	L	24	Velocity	4.5	5	L
3	Velocity	4.5	5	L	25	Velocity	4.5	5	A
4	Velocity	4.5	5	A	26	Situation	2	2	A
5	Strength	0	0	A	27	Velocity	4.5	5	A
6	Velocity	4.5	5	L	28	Velocity	4.5	5	A
7	Bearing	0	0.005	A	29	Situation	2	2	A
8	Velocity	4.5	5	A	30	Velocity	4.5	5	A
9	Velocity	4.5	5	A	31	Velocity	4.5	5	A
10	Bearing	0	0.005	A	32	Situation	1	1	A
11	Velocity	4.5	5	A	33	Velocity	4.5	5	L
12	Velocity	4.5	5	L	34	Situation	1	1	A
13	Bearing	0	0.005	L	35	Velocity	4.5	5	S
14	Velocity	4.5	5	L	36	Strength	1	1	A
15	Velocity	4.5	5	A	37	Velocity	4.5	5	A
16	Friction	0.009	0.01	S	38	Velocity	4.5	5	A
17	Velocity	4.5	5	A	39	Strength	1	1	A
18	Velocity	4.5	5	A	40	Velocity	4.5	5	A
19	Friction	0.009	0.01	A	41	Velocity	4.5	5	S
20	Velocity	4.5	5	A	42	Strength	0	0	A
21	Velocity	4.5	5	S	43	Velocity	4.5	5	S
22	Friction	0.009	0.01	S	44	Friction	0.009	0.01	A

6 CONCLUSION

For fault tree construction, a new event definition in both a qualitative (event description) and a quantitative (event type) forms, and an extended decision table have been proposed. Event descriptions are defined to provide with physical understanding to the events, whereas event types are defined to distinguish the signal magnitude and facilitate the back-tracking process. Two procedures dealing with production of transfer symbols and successive logic gates with the same type are proposed to prune the generated fault tree. The proposed pruning procedures can not only save constructing time but also refine the generated fault tree, and hence save the computer memory. The example shows that the pruning process reduces the number of events from 90 to 44 and it takes only 0.22 second for constructing the fault tree using the 80486 IBM/PC.

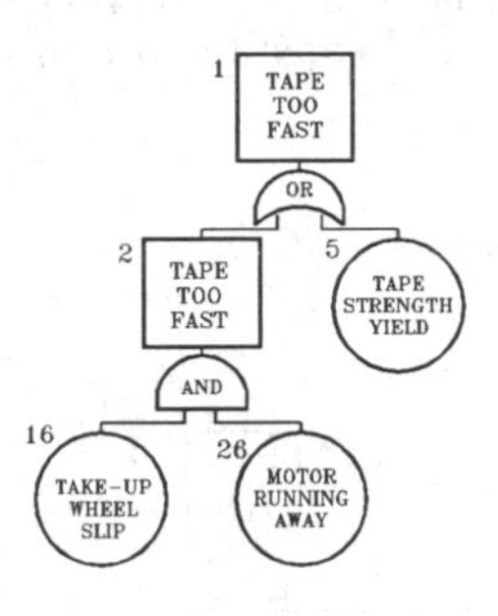

Fig.4 P···ned tree of capston control system

REFERENCES

Camarda, P., Corsi, F. and Trentadue, A. 1978. An efficient simple algorithm for fault tree automatic synthesis from the reliability graph, *IEEE Trans. on Reliability.* R-27:215-21.

De Vries, R. C. 1990. An automated methodology for generating a fault tree, *IEEE Trans. on Reliability.* 39:76-86.

Fussell, J. B. 1973. Synthetic tree model: A formal methodology for fault tree construction, Report ANCR-1098, Aerojet Nuclear Company, Idaho Falls, Idaho USA.

Gough, W. S., Riley, J. and Koren, J. M. 1990. A new approach to the analysis of reliability block diagrams, p.456-64. *Proc. Annual Reliability and Maintainability Symposium.*

Lapp, S. A. and Powers, G. J. 1977. Computer-aided synthesis of fault-trees, *IEEE Trans. on Reliability.* R-26:2-13.

Salem, S. A., Apostolakis, G. E. and Okrent D. 1977. A new methodology for the computer-aided construction of fault trees, *Annals Nuclear Energy.* 4:417-33.

Taylor, J. R. 1982. An algorithm for fault-tree construction, *IEEE Trans. on Reliability.* R-31:137-46.

Structural Safety & Reliability, Schuëller, Shinozuka & Yao (eds) © 1994 Balkema, Rotterdam, ISBN 90 5410 357 4

Integral safety plans as new strategy in Switzerland

Miroslav Matousek
Dr Matousek Engineering, Schwerzenbach, Switzerland
Hermann Egli
Swiss National Accident Insurance Organization, Lucerne, Switzerland

ABSTRACT: The different safety problems need to be considered integrally and appropriatc instruments are to be developed. The integral safety plan is such an instrument and presents a new strategy to reduce failures, accidents and damage. The integral safety plan gives information mainly regarding which hazards are associated with building or technical facilities during all phases - from construction to demolition and disposal - and which safety measures are to be established against these. The importance of integral safety plans led in Switzerland to the project "Integral Safety Plans for Civil Engineering Structures and Plants". This project is supported by the Swiss National Accident Insurance Organization (SUVA). The schedule of activities is divided in four phases and the results of the first phase are described.

1 THE NEED FOR INTEGRAL SAFETY PLANS

The idea of safety plans was briefly presented on ICOSSAR'89 in San Francisco (Matousek 1989). In the meantime new laws, standards and directives have been issued, e.g. the new Swiss technical codes (SIA-Norm 1989), the ISO standards, the Swiss law of "Environmental Protection", the "Swiss Ordinance on the Prevention of Major Accidents" (Handbuch 1991), the European directives "Safety and Health Protection on Building Site" (EG-Einzelrichtlinie 1992) and "Safety in Use". The various legal requirements as well as the complexity of the safety problems represent a new task and hence a new challenge for the responsible persons.

The elaboration of the integral safety plan already in the planning stage has to assure, that all four safety areas - technical safety, security, accident prevention and health protection, environmental safety - are systematically considered in all following phases of building, structure and technical facilities: construction, use, monitoring and maintenance, reconstruction and restoration, demolition, recycling and disposal of the waste.

The integral safety plan should be useful as a management tool for the responsible persons. It should enable them to identify the critical situations and to apply the necessary safety measures in an optimal manner. The established safety measures are to be planned systematically, with a call for tenders and also realized. The procedure to assure safety has to be transparent.

Already in the last 10 years some safety plans have been elaborated (Matousek 1989). Depending on the problem, the safety plans deal with only one hazard (e.g. impact of lorries, fire in tunnels) or an extensive hazard investigation for complex technical facilities (e.g. chemical facilities, water power station). The safety plans have proved useful and the experiences gained are now under consideration. The idea of safety plans is therefore not new (Matousek 1983). In some areas (e.g. structural system, sewerage system, piping tunnel) the safety plans belong already to the standard of technology and science. What is new is the integral consideration of the safety problem.

2 ELEMENTS OF THE INTEGRAL SAFETY PLANS

2.1 Safety goals

Safety goals are first to be seen as the fulfilment of the safety requirements established in laws, standards, recommendations etc. Furthermore, safety goals should be established on the basis of risk acceptance and acceptable safety costs.

2.2 Building or technical facilities as a system

For each phase, the building and the technical facilities are defined as a system and its components need to be described in detail. Generally, the following phases have to be considered: construction, use, monitoring and maintenance, reconstruction and restoration, demolition, recycling and disposal. The components of the system are building-elements (e.g. load-bearing elements, finishings, utilities), components of building site (e.g. auxiliary structures, site installations), procedures and works, people involved, natural and technical environment.

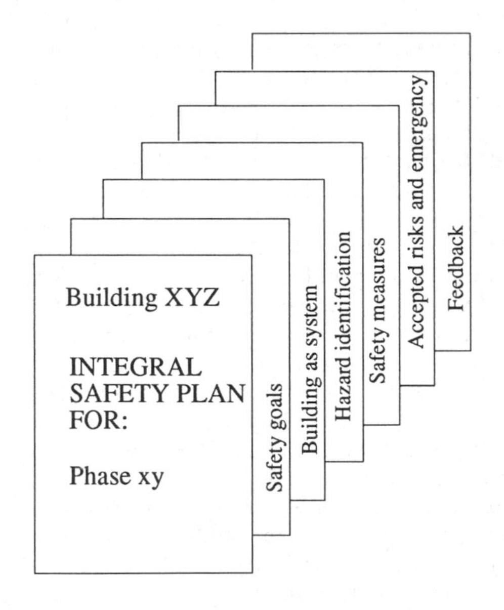

Elements of the integral safety plan

2.3 Hazard identification and risk evaluation

Hazards have to be identified for each phase of the building and the technical facilities. In each phase one or more hazards may exist simultaneously. In order to define a logical structure for these hazardous situations they are described by so-called hazard scenarios. Each hazard scenario consists of a main hazard and an accompanying circumstance. The main hazard is always assumed to be extreme in its effect, form, and magnitude. The accompanying circumstance characterizes any hazard and factor which occurs at the same moment and hence accompany the main hazard. Generally, the hazard scenarios are to be identified in the four safety areas. Any hazard scenario found has to be evaluated as a risk. The risk evaluation relates to the effect of damage to people and the environment as well as to an estimation of the probability.

2.4 Safety measures

Identified risks evaluated in terms of safety goals and safety measures for their reduction are to be investigated. This concerns building, technical, organizational, management and personal precautions. The possible safety measures are to be evaluated in respect to the reduction of risk as well as to the costs and then optimal precautions should be chosen.

2.5 Accepted risks and the emergency

Not all risks can be eliminated. Some risks must be consciously accepted and taken into account. The accepted risks have to be summarized and the risk bearer (client, tenderer, other person etc.) has to be identified. Then it should be established how the risks are to be monitored with the help of the risk indicators, which procedures have to be implemented in dangerous situations and if failure occurs how the damage is to be repaired. Appropriate emergency measures are to be elaborated and an emergency service has to be organised and put on the alert.

2.6 Feedback

The feedback assures that changes of the hazard-environment, the building, the building procedures etc. have to be identified and considered in good time. The safety plan should be adapted and the safety measures have to be improved or new safety measures are to be applied. Accidents and incidents have to be investigated and the safety measures are to be kept up to date with the newest technologies.

3 SUVA-PROJECT "INTEGRAL SAFETY PLANS"

The European directive "Safety and Health Protection on Building Site" (EG-Einzelrichtlinie 1992) as well as the importance of integral safety plans led to the project "Integral Safety Plans for Civil Engineering Structures and Plants". This project is supported by the Swiss National Accident Insurance Organization (SUVA). From the SUVA point of view, the application of the integral safety plans and the realization of established safety measures is a new strategy for further reduction of accidents in civil engineering. The SUVA-project not only corresponds with the the European directive "Safety and Health Protection on Building Site", but it also covers all safety areas as well as all phases of building and technical facilities.

For the project a working group of specialists has been formed. The schedule of activities is divided in four phases.

The scope of the 1st phase - accomplished in June 1992 - was to describe the basic concept and to develop rudimentary tools for the application of integral safety plans. Based on these a working seminar was organized with specialists from government agencies, regional administration, professional organizations, the technical university, engineering and consulting firms, contractors and others. In this seminar, the SUVA-project and the basic principles were presented and the safety problem explained by the presented papers. The content of the integral safety plans was discussed, especially in relation to their elaboration and realization.

In the 2nd phase the basic concept is to be developed. The elaboration of the integral safety plans will be investigated and the content established for new buildings as well as for existing buildings. Guidelines, checklists and examples will be prepared for the practical elaboration of the integral safety plans.

In the 3rd phase typical buildings shall be chosen as pilot-projects and for these integral safety plans are to be elaborated. The working group supports this elaboration, giving professional input. Results will be analysed to ensure feedback from the experiences gained.

In the 4th phase adaptation and completion of the procedure for the elaboration of the integral safety plan will be established and realized. Subsequently, seminars will be organized for all those involved in the building process, e.g. clients, architects, engineers, contractors etc., to inform them about the integral safety plans and their application.

After the accomplishment of phase 1 in June 1992 the second phase was begun. For the realization of the three following phases a time schedule of three years was foreseen.

4 ACTUAL RESULTS AND CONCLUSIONS

The results of the first phase were analysed and documented (SUVA-Projekt 1992). The results show very clearly, that integral safety plans for civil engineering are very important. They help those involved to see and to carry their responsibility regarding safety. Therefore, integral safety is becoming a very interesting question and is receiving wide acceptance. Safety plans are also gaining importance in other countries. Safety seminars over the last few years confirm this. The European directive "Safety and Health Protection on Building Site" is a very important document, which will widely influence the building industry in the European Community. The SUVA-project considers not only this directive in good time and in practice, but contributes largely to the real improvement of safety in the civil engineering.

5 REFERENCES

"EG-Einzelrichtlinie 92/57/EWG des Rates vom 24.6.1992 über die Sicherheit und den Gesundheitsschutz auf Baustellen, gemäss Artikel 16 der EG-Richtlinie 89/391/EWG

Handbuch I zur Störfallverordnung, StFV, BUWAL, Eidg. Drucksachen- und Materialzentrale, 3000 Bern, Juni 1991

Matousek M., Schneider J.: Gewährleistung der Sicherheit von Bauwerken - Ein alle Bereiche des Bauprozesses erfassendes Konzept, Institut für Baustatik und Konstruktion, ETH Zürich, Bericht Nr. 140, Birkhäuser-Verlag Basel und Stuttgart, 1983

Matousek M.: Safety Plans for Buildings, Structures and Technical Facilities, 5th International Conference on Structural Safety and Reliability, San Francisco, USA, August 1989

SIA-Norm 160 "Einwirkungen auf Tragwerke" und die Empfehlung SIA 169 "Erhaltung von Ingenieurbauwerken", Referate und Beispiel der Studientagungen vom 30. und 31.8.1989 bzw. 19.9.1989 ETH-Zürich, SIA-Dokumentation D 041, Schweizerischer Ingenieur- und Architekten-Verein, Postfach, 8039 Zürich

SUVA-Projekt: Grundlagen, Rigi-Arbeitsseminar - Referate, Auswertung des Arbeitsseminars, Schlussbericht Phase 1, SUVA-Bulletin Nr. 50/1992, Schweizerische Unfallversicherungsanstalt SUVA - Abteilung Bau, Luzern, 1992

Structural Safety & Reliability, Schuëller, Shinozuka & Yao (eds) © 1994 Balkema, Rotterdam, ISBN 90 5410 357 4

Importance of semi-rigid connections parameters in the reliability of metal structures

A. Mébarki
Laboratoire de Mécanique et Technologie (LMT), ENS Cachan/CNRS/Université Paris 6, Cachan, France

A. Colson
ENSAIS, Strasbourg, France

ABSTRACT: The authors consider metal frames with semi-rigid connections. The random variables are the geometrical dimensions, the mechanical characteristics of the structural members and the connections parameters. The applied loads may also be considered as random variables.

The results afforded are the histograms and the statistical parameters of the ultimate load multiplier λ which ccoresponds to the occurrence of the failure by either buckling or plastic mechanism events. Monte Carlo simulations are ran. The collected histograms show the multi-modal failure of the structures considered herein. They demonstrate also the sensitivity of the reliability to the characteristics of both the connections and the constitutive members.

1- INTRODUCTION

To describe semi-rigid connections behaviour, sophisticated models may be used. But, regarding the global structural reliability involving semi-rigid connections, some of the model parameters may not require complex formulations if the reliability is not very sensitive to them. To obtain homogeneous reliability for the various structural members and the connections, probabilistic studies are done. The present paper studies the sensitivity of metal frames reliability to the random variables involved, focusing particularly on the connections parameters, [1]. The structures under study herein are simple frames involving semi-rigid connections. The basic random variables are the yielding stress f_y of the steel and the ultimate carrying capacity of the connections, M_u. The connections stiffness, K, and the shape factor, a, are considered as having deterministic values.

The structural reliability analysis is performed, by running Monte Carlo simulations. It is expressed through the histograms obtained for the ultimate values, λ_{ult}, of the loads multiplier which correspond to the occurrence of the structural failure by either plastic mechanisms or buckling.

2- STRUCTURE UNDER STUDY

The selected structure is a metal frame, see Fig 1, under the effect of a uniform vertical load p=50 kN/ml and an horizontal load F=1.3 kN. Its supports are two hinges located at A and D supports. The nodes B and C are semi-rigid connections. The structure is analysed while the loads are increasing proportionally by discrete increments ($\Delta p=\Delta\lambda.p$, $\Delta F=\Delta\lambda.F$) where λ is the loads multiplier.

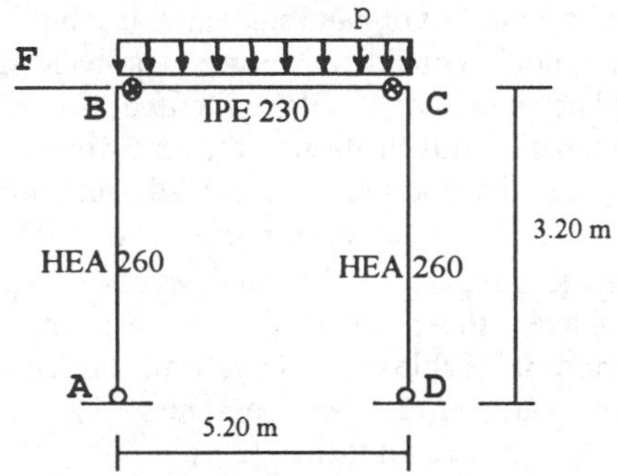

Figure 1- Definition of the structure

This latter increases until its ultimate value, λ_{ult}, for which the structure fails by either buckling or plastic mechanism. The elasto-plastic analysis of the structure, with second order effects, is done by running PEP-program, [2]. The sensitivity analysis of the structural reliability is based on λ_{ult} values. It must be noticed that this is a first step before considering the probabilities of failure.

The steel behaviour for the structural members is assumed to be elasto-plastic with. a yielding stress denoted f_y. We have adopted as M-Φ curve (bending moment M - rotation Φ) describing the semi-rigid connections behaviour, the following definition, [2]:

$$\Phi = \frac{M}{K} \cdot \frac{1}{1-\left(M/_{Mu}\right)^a} \quad (1)$$

where: Φ=rotation value, M= bending moment, K= initial stiffness, Mu= ultimate carrying capacity of the connection and a= shape parameter governing the non linear part of M-Φ curve.

3- PROBABILISTIC DESCRIPTION OF THE STRUCTURAL BEHAVIOUR

According to the current geometrical tolerances, we have neglected their randomness since several studies of similar structures have shown that the dimensions may be regarded as deterministic parameters, [3]. We have then selected two basic random variables: the yielding strength, f_y, of the steel, and the connections carrying capacity, M_u. We have supposed also that the stiffness K and the shape parameter a are both deterministic. The laods p and F are random variables but we have not considered any probabilistic distribution, herein. We focused on the preliminary informations drawn from the statistical distribution of the loads multiplier values, in order to detect the most influent parameters regarding the final values of λ_{ult}. We have, thus, selected the distributions reported in Table 1. For all the structures studied hereafter, we assume that the parameter a is equal to 1.5, [2,4].

Monte Carlo simulations are used. At the k-th simulation among a total number N_{sim}, λ_{ult} value and the mode of failure (buckling or plastic mechanism) are stored in order to draw the histograms.

Table 1- Mechanical parameters distributions

Parameter	Distrib.	μ (Mpa)	σ (Mpa)
f_y	Gaussian	280	27
M_u	Gaussian	μ_{Mu}	σ_{Mu}

with μ, σ= mean and standard deviation values

4- RESULTS

For all the cases we have considered, the number of performed simulations is N_{sim}=100 since the mean value and the standard deviation of λ_{ult} do not vary significantly from 100 simulations to 2000.

4.1- Effect of the carrying capacity M_u

The steel strength f_y is first considered as having a deterministic value. It is assumed to be equal to its mean value, i.e. 280 MPa giving then M_p=258 kN.m for the columns (HEA 260) and M_p=176 kN.m for the beam (IPE 230); M_p being the plastic bending moment of the cross-sections. Four mean values have been successively considered for M_u as illustrated in Table 2. The collected results ($\mu_{\lambda ult}$, $\sigma_{\lambda ult}$= mean and standard deviation values of the ultimate loads multiplier λ_{ult}) are given in Table 3 where Mp=176 kN.m for the beam.

Table 2- Values considered for M_u

Parameter (kN.m)	Case 1	2	3	4
μ_{Mu}	100	140	180	220

with σ_{Mu}=0.1 μ_{Mu}; K=30 000 kN.m/rd; a=1.5; f_y=280 MPa.

Table 3- Collected results for λ_{ult}

Case		Mode of failure M[1].	B[2].	Global failure (M.+B.)
1	$\mu_{\lambda ult}$	**1.25**	**1.373**	**1.372**
	$\sigma_{\lambda ult}$	**0**	**0.049**	**0.05**
2	$\mu_{\lambda ult}$	**-**	**1.579**	**1.579**
	$\sigma_{\lambda ult}$	**0**	**0.062**	**0.062**
3	$\mu_{\lambda ult}$	**1.718**	**1.648**	**1.702**
	$\sigma_{\lambda ult}$	**0.011**	**0.025**	**0.034**
4	$\mu_{\lambda ult}$	**1.720**	**1.654**	**1.719**
	$\sigma_{\lambda ult}$	**0.007**	**0.0002**	**0.004**

where Case 1: (μ_{Mu}=100 kN.m)<Mp; Case2: (μ_{Mu}=140 kN.m)<Mp; Case 3: (μ_{Mu}=180 kN.m)=Mp; Case 4: (μ_{Mu}=220 kN.m)>Mp; M[1].= Mechanism; B[2].= Buckling.

These results show that when the mean value μ_{Mu} increases, the resistance of the structure (expressed through $\mu_{\lambda ult}$) also increases, reducing then the risk of failure. Furthermore, the standard deviation $\sigma_{\lambda ult}$ decreases when the mean value μ_{Mu} increases. When μ_{Mu} < Mp (of the IPE-230 beam), the failure occurs almost by buckling as the first plastic hinge appears in the semi-rigid connections and the collapse follows then by instability of the structure, but when μ_{Mu} > Mp (of the IPE-230 beam), the first plastic hinge appears at the mid-span of the beam and the collapse occurs when the structure is transformated into a plastic mechanism.

4.2- Effect of the semi-rigid connections stiffness K

We assume again that f_y=280 MPa while M_u is a random variable and we consider successive values for the semi-rigid connections stiffness, i.e. K=5000 kN.m/rd, K=30,000 kN.m/rd, K=40,000 kN.m/rd, K=60,000 kN.m/rd, and K=90,000 kN.m/rd. The carrying capacity M_u is considered as having a mean value equal to 100 kN.m and a standard deviation value equal to 10 kN.m. The results obtained are given in Fig.2 through 6 as indicated in Table 4.

Table 4- Values considered for K (kN.m/rd).

K:	5000	30,000	40,000	60,000	90,000
Fig.	2	3	4	5	6

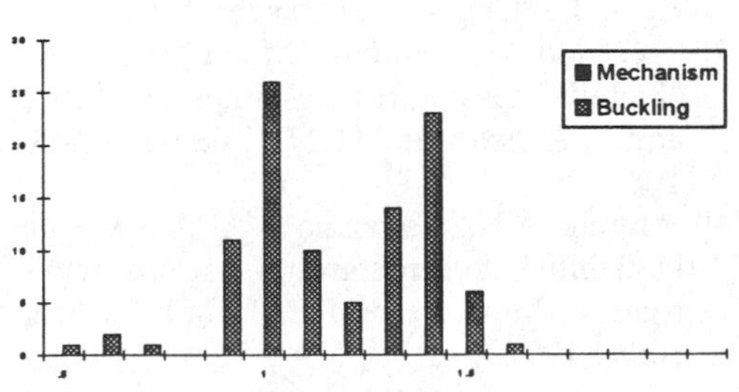

Figure 2- Histograms of λ_{ult} values.

From the results reported for these various cases, where the semi-rigid connections stiffness K is varying, it can be observed that the failure mode changes when this stiffness varies. Actually, when K<60,000 kN.m/rd, the failure occurs essentially by buckling but when K>60,000 kN.m/rd, the transformation of the structure into plastic mechanism becomes the predominant mode of collapse. When the stiffness increases, the mean value of the critical loading factor varies reaching a maximum value, $\mu_{\lambda ult}$=1.547, for K=40,000 kN.m/rd.

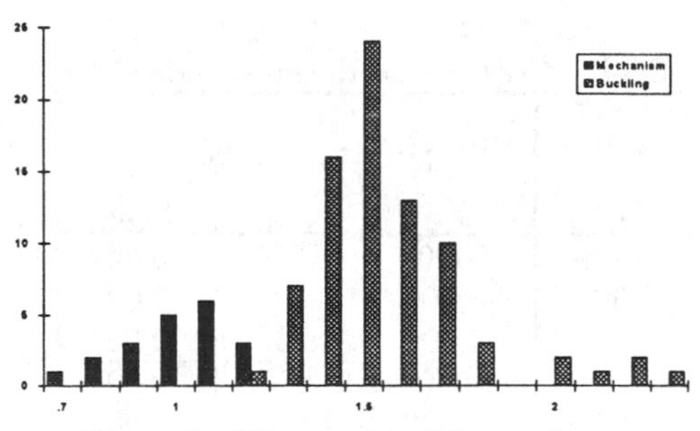
Figure 3- Histograms of λ_{ult} values.

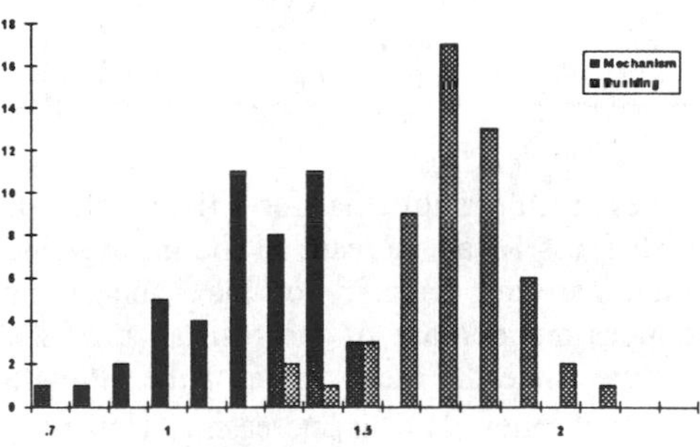
Figure 4- Histograms of λ_{ult} values.

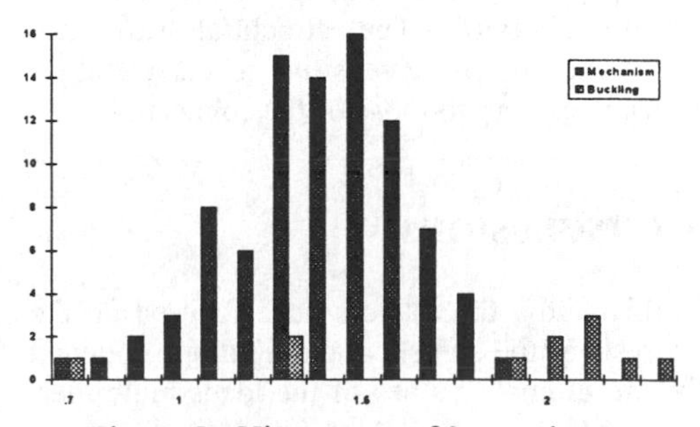
Figure 5- Histograms of λ_{ult} values.

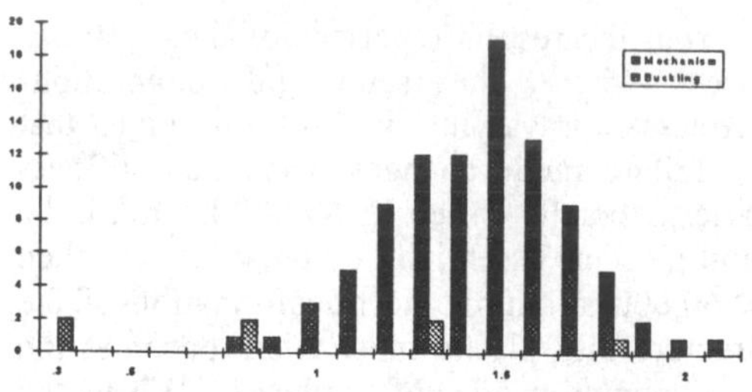

Figure 6- Histograms of λ_{ult} values.

We have also considered the influence of the stiffness K when the mean value of the carrying capacity of the connections is changing. The results are given in Table 5.

Table 5- Effects of the mean value μ_{Mu}

μ_{Mu} (kN.m)	Stiffness K	λ_{ult}	
		$\mu\lambda_{ult}$	$\sigma\lambda_{ult}$
100	5000	1.444	0.04
	30,000	1.537	0.04
	90,000	1.536	0.06
180	5000	1.74	0.09
	30,000	1.931	0.08
	90,000	1.934	0.09
220	5000	1.60	0.06
	30,000	2.027	0.04
	90,000	2.045	0.03

These results show that for a fixed value of the stiffness K, an increase of the mean value of the carrying capacity of the connections produces an increase of the resistance of the structure since the mean value of the ultimate loads multiplier, λ_{ult}, is increasing. However, this latter is not really sensitive to an increase of μ_{Mu} when the stiffness is greater than 30,000 kN.m/rd. The structural resistance might then be more sensitive to the members resistances than to those of the connections.

5- CONCLUSIONS

In this study, the authors have focused on the analysis of the statistical distributions obtained for the ultimate values of the loads multiplier, λ_{ult}, which expresses the resistance of the steel frame against either buckling or plastic mechanism events. The random variables are the carrying capacity, M_u, of the semi-rigid connections and the yielding stress, f_y, of the constitutive material of the structural members. This study is a first step in the reliability analysis of steel frames with semi-rigid connections. It concerns for the moment only results about the sensitivity, of the statistical distributions of λ_{ult} values, regarding the structural members and connections parameters. A more complete study is required to assess and analyse he probabilities of failure.

Monte carlo simulations are ran and histograms of λ_{ult} values are established. The collected results show that an increase of the mean value of M_u reduces the risk of failure. When the connections are less resistant than the beam (mean value of Mu < Mp which is the plastic resistance of the beam cross-section), the structure fails almost by buckling; but when the connections are more resistant than the beams then the collapse by plastic mechanism is predominant. The failure mode is also sensitive to the connections stiffness, K. For large values of this stiffness, the structure fails by plastic mechanism event while it collapses by buckling for small values of the stiffness.

REFERENCES

[1]- Colson A., Bjorhovde R.; "Intérêt économique des assemblages semi-rigides", Construction Métallique, CTICM (Paris), n°2, 1992.

[2]- Galéa Y, Bureau A; "Programme PEP Micro. Notice d'utilisation", CTICM, St-Rémy Lès Chevreuse (France), 1990.

[3]- Mébarki A., Pinglot M., Lorrain M.; "Fiabilité des poutres isostatiques en béton armé", Annales de l'ITBTP (Paris), n°443, 1986.

[4]- Amrouche N., Lattanzio W.; "Analyse de la fiabilité de structures à liaisons semi-rigides", Mémoire de DEA, LMT Cachan, Juillet 1991.

Structural Safety & Reliability, Schuëller, Shinozuka & Yao (eds) © 1994 Balkema, Rotterdam, ISBN 90 5410 357 4

Association of serviceability and strength limit states in the evaluation of the reliability in nonlinear behaviour

Jean-Pierre Muzeau
Laboratory of Civil Engineering, Blaise Pascal University, Clermont-Ferrand, France
Maurice Lemaire
LaRAMA, Blaise Pascal University, Clermont-Ferrand, France
Pierre Besse
DTO, Bureau Veritas, Paris, France

ABSTRACT : This paper presents the SRQ method which allows to evaluate the reliability index when the limit state function depends on mechanical implicit variables as it can be found when buckling can occur for example. Based on a quadratic function and on the use of the least squares method, the SRQ method gives an approximation of the limit state surface. As an application, the authors emphasize the interest of using the SRQ method to combine two kinds of limit states : an ultimate limit state and a serviceability limit state in view to obtain an efficient design of compression members. An example shows the interest of the proposed procedure to study the reliability offered by the AISC-LRFD Specification in the case of a pin-ended column.

1 INTRODUCTION

To evaluate the reliability of compression members with large values of slenderness, it is obviously necessary to take into account the ultimate limit state of strength but it can also be beneficial to consider the effect of displacements due to buckling by the calculation of a serviceability limit state. The association of these two limit states will then give a reliable design for columns.

Nevertheless, when a structure buckles, the effect of geometrical nonlinearity does not allow the explicit calculation of the value of the bending moment due to the axial load. The use of a nonlinear mechanical model is a way to obtain the exact relationship describing the behaviour of the structure but this relationship is implicit. It is then necessary to establish a method of calculating the structural reliability because the limit state function and its derivatives are not available under an explicit analytical form.

A response surface method called the SRQ method (Quadratic Response Surface in French) associated with a software of evaluation of reliability (as FORM/SORM for example) has been developed to solve the problem. It consists in approaching locally, near a critical point, the limit state function $G(X_i)$ using a well-known mathematical function.

The first part of this paper presents the SRQ method. The second one gives an example of the association of the two limit states in the case of a pin-ended compression member designed in application to the AISC-LRFD Specification (AISC 1986).

2 THE SRQ METHOD

The quadratic response surface method or SRQ method (El-Tawil, Lemaire & Muzeau 1991) is based on the technique of making explicit the implicit limit state function from its numerical values obtained with a very efficient mechanical model. It allows the use of the classical algorithm of calculation of the reliability index β which runs when the limit state function is explicit or, in other words, when all its terms can be directly calculable.

To limit computation time, a quadratic function $Q(X_i)$ is adopted as a basis of approximation.

In a space of n random variables X_i, the k approximation possesses the following form :

$$G(X_l) \approx Q^{(k)}(X_l) = c + \sum_{i=1}^{n} a_i X_i + \cdots$$

$$\cdots + \sum_{i=1}^{n} b_{ii} X_i^2 + \sum_{i=1}^{n-1} \sum_{j=i+1}^{n} b_{ij} X_i X_j \qquad (1)$$

where : c, a_i and b_{ij} are L constant coefficients.

The function $Q^{(k)}(X_l)$ is taken as an interpolation function of the hypersurface of the limit state in a space of n random variables. It is defined by a minimum number of points :

$$R_{min} = L = \frac{(n+1)(n+2)}{2}$$

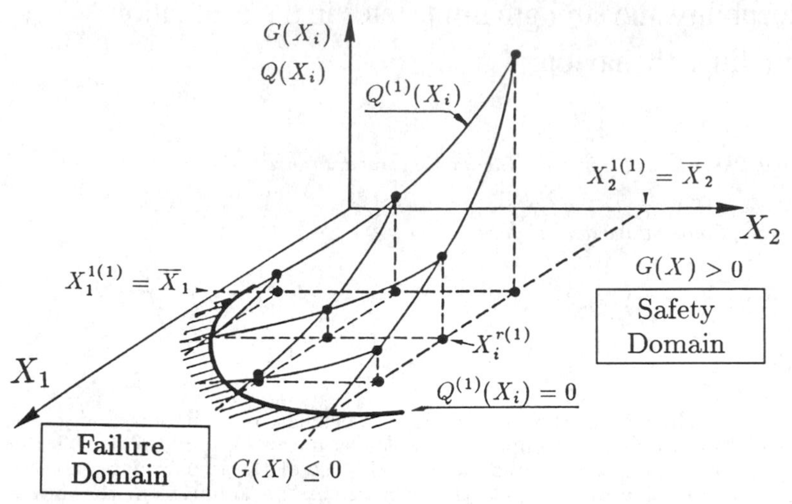

Figure 1. Example of interpolation in the case of two variables (n=2, L=6 and R = 7)

A larger number of points, R, associated with the least squares method, stabilizes the solution.

The calculation of the coefficients consists in the minimization of :

$$\sum_{r=1}^{R} \left| Q^{(k)}(X_l^r) - G(X_l^r) \right|^2 \tag{2}$$

where $G(X_l^r)$ is a realization of the limit state function and $Q^{(k)}(X_l^r)$ its approximation.

If $\{C\}$ is the vector of the unknown coefficients, it is solution of the linear system :

$$[P]\ \{C\} = \{H\} \tag{3}$$

in which $[P]$ is a matrix function of the variables X_l^r, and $\{H\}$ is a vector function of the realizations $G(X_l^r)$ of the limit state.

Let $\langle C \rangle$ be the transposed vector of the unknown coefficients :

$$\langle C \rangle = \langle c,\ a_i\ (i=1,n), b_{ii}\ (i=1,n), \cdots$$
$$\cdots,\ b_{ij}\ (i=1,n-1, j=i+1,n) \rangle$$

and $\langle X^r \rangle$ the transposed vector of combinations of degree $d \leq 2$ of the realizations of random variables :

$$\langle X^r \rangle = \langle 1,\ x_i\ (i=1,n),\ {x_i}^2\ (i=1,n), \cdots$$
$$\cdots, x_i x_j\ (i=1,n-1, j=i+1,n) \rangle$$

it comes :

$$Q(X_l^r) = Q^r = \langle X^r \rangle\ \{C\} \tag{4}$$

G^r is the r realization and it is necessary to minimize, relatively to $\{C\}$ (equation (2)) :

$$\sum_{r=1}^{R} (Q^r - G^r)^2$$

or :

$$\sum_{r=1}^{R} (Q^r - G^r) \left\langle \frac{\partial Q^r}{\partial \{C\}} \right\rangle d\{C\} = 0$$

$$\sum_{r=1}^{R} \Big(\{X^r\}\ \langle X^r \rangle \Big)\ \{C\} = \sum_{r=1}^{R} \Big(G^r \{X^r\} \Big) \tag{5}$$

$$[P] = \sum_{r=1}^{R} \Big(\{X^r\}\ \langle X^r \rangle \Big) \tag{6}$$

$$\{H\} = \sum_{r=1}^{R} \Big(G^r \{X^r\} \Big) \tag{7}$$

The solution of equation (3) gives the coefficients of the function $Q(X_i)$. The approximate cartesian equation of the limit state function is obtained by the intersection of the hypersurface $Q(X_i)$ with the hyperplane of the variables X_i, or, in other words, $Q(X_i) = 0$ is the approximate explicit function which is sought (figure 1).

Then, the explicit form of Q allows the use of a classical algorithm for the calculation of β (for example FORM/SORM) which gives $x_i^{*(k)}$ (an approximation of the most probable failure point)

Table 1. Random variables

Random variables	Distribution	Mean value	COV	Tolerance
Cross-sectional dimensions :				
Δh, Δb	Uniform	$0\,mm$		± 0.2 mm
Δt	Uniform	$0\,mm$		± 0.1 mm
Out-of-straightness u_0	Gaussian	$0\,mm$	$0.1\%l$	$\pm 0.3\%l$
Eccentricity E_{cc}	Gaussian	$0\,mm$	$0.175\%l$	$\pm 0.35\%l$
Yield stress F_y	Log-normal	263.2 MPa	6%	
Young modulus E	Log-normal	21 10^4 MPa	6%	
Axial load F	Log-normal	$\overline{F}$	10%	

(Muzeau, Lemaire & El-Tawil 1992).

To improve the approximation in the neighbourhood of the point $x_i^{*(k)}$, iterations are necessary. The iteration $k+1$ starts with a new set of points $x_i^{l(k+1)}$ slightly closer to the failure surface calculated during the previous iteration k but staying in the safety domain because $g(x_l) > 0$ may be not defined.

3 EXAMPLE OF APPLICATION

For the combination of two limit states (ultimate limit state and serviceability limit state), it is possible to obtain curves describing the evolution of the reliability of a simple compressed member to take into account the effect of nonlinear displacements (very important for large values of slenderness). The SRQ method allows the values of the reliability index β to be obtained when a member may buckle and the results can be compared.

This example concerns the calculation for a pin-ended IPE column axially loaded (Figure 2). It is carried out using Fe 360 steel grade (yield stress $F_y = 235$ MPa) with effective length equal to l. Table 1 shows the random variables taken into account.

If $\overline{\lambda}$ is the reduced slenderness parameter and A_g the gross area of the member, the LRFD Specifications (AISC 1986) give the maximum factored compressive load P_n such as :

$$P_n = A_g F_{cr} \tag{8}$$

where F_{cr} takes buckling into account. It is obtained with the following relationships :

$$F_{cr} = 0.658^{\lambda_c^2} F_y \quad \text{if } \lambda_c \leq 1.5 \tag{9}$$

$$F_{cr} = \left(0.877/\lambda_c^2\right) F_y \text{ if } \lambda_c > 1.5 \tag{10}$$

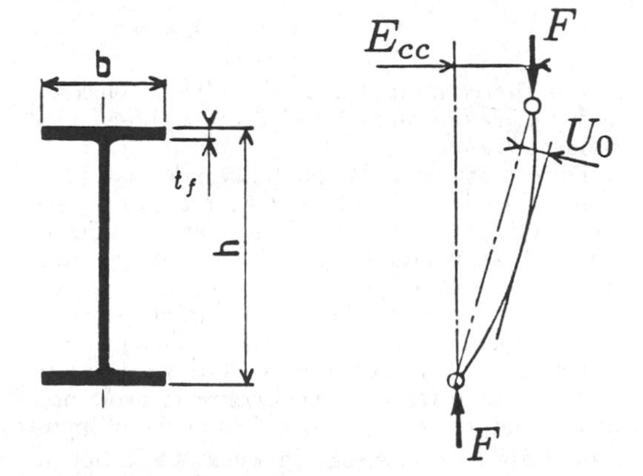

Figure 2. Pin-ended compression member

Then, the mean value $\overline{F}$ is obtained by :

$$\overline{F} = P_n/\gamma \tag{11}$$

if γ is the considered safety factor chosen here equal to 1.5 ($\overline{F}$ is an unfactored axial load).

With IPE shapes, the ultimate limit state function has the form :

$$U_i = 1 - \frac{N}{\widetilde{N}_p} \pm \frac{M}{\alpha \widetilde{M}_p} = 0 \tag{12}$$

if N and M are respectively the axial load and the bending moment and if N_p and M_p are the plastic resistances ($\alpha = 1.22$ with this kind of shape).

The serviceability limit state function is :

$$S_i = f_{300} \pm f = 0 \tag{13}$$

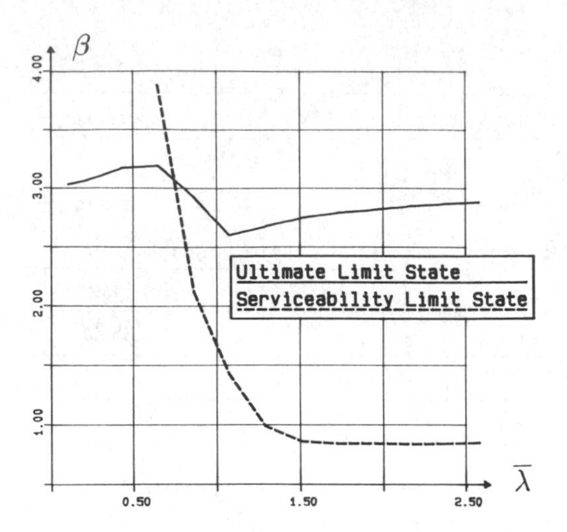

Figure 3. AISC-LRFD : Curves $\beta = f(\overline{\lambda})$ obtained with two kinds of limit states

if f is the computed deflection due to buckling and f_{300} is the allowable deflection chosen here equal to $l/300$.

The evolutions of the reliability index as a function of the slenderness are shown in Figure 3. The values of β depends of the limit state considered. To reach an ultimate limit state is always much more severe than to reach a serviceability limit state and the probabilities of failure attached to each of them are always different. This example shows that it is very important to be cautious with the large values of slenderness to avoid problem of large displacements. If β of the ultimate limit state increases slightly when $\overline{\lambda} > 1$, but, in this range, the reliability attached to the serviceability limit state decreases very significantly. So, the choice of values of β linked to different kinds of limit states may lead to define some limits of acceptability of compressed columns from the point of view of strength but also for displacement.

4 CONCLUSION

The Quadratic Response Surface (SRQ) method offers an efficient solution to evaluate the reliability index when the limit state function depends on implicit variables or when it is implicit by itself. It allows the evaluation of the reliability index with a sufficient precision without requiring simplifying hypotheses.

Its use in a study of Standard Codes where instabilities lead to nonlinear behaviours allows the analysis of a serviceability limit state in association with the classical ultimate limit state to give a very reliable design. It can be linked to very sophisticated mechanical models, an essential condition to a rational use of reliability methods.

ACKNOWLEDGEMENTS

This work was carried out in the frame of the Task 1.1 of the EPSOM programme financially supported by the CLAROM. The authors gratefully acknowledge it.

REFERENCES

AISC 1986. *Load and resistance factor design.* American Institute of Steel Construction : 1st edition.

El-Tawil, K., M. Lemaire & J.P. Muzeau 1991. Reliability method to solve mechanical problems with implicit limit states. *Proc. 4th IFIP WG 7.5 :* 181-190. Munich : Springer-Verlag.

Muzeau, J.P., M. Lemaire & K. El-Tawil 1992. Méthode fiabiliste des surfaces de réponse quadratiques (SRQ) et évaluation des règlements. *Construction Métallique* 3 : 41-52.

Structural Safety & Reliability, Schuëller, Shinozuka & Yao (eds) © 1994 Balkema, Rotterdam, ISBN 90 5410 357 4

Decision making in IRM using reliability

M. Vasudevan & G. M. Zintilis
Billington Osborne-Moss Engineering Limited, Maidenhead, UK

ABSTRACT: The wide range of data used in IRM activities are subject to uncertainties due to human and measurement errors. This results in inaccurate inspection data as well as analytical model uncertainty. As a consequence, there is uncertainty not only in the prediction of the condition of the defect but also its consequence on the integrity of the structure as a whole. These uncertainties lend themselves well to a probabilistic description of the problem and its possible outcomes.

An integrated approach towards reliability-based decision making and strategy planning of IRM activities is presented. The decisions are carried out within a knowledge-based expert system using inspection results and methods, environmental conditions as well as structural and reliability analysis results.

1 INTRODUCTION

IRM activities are carried out as a means to monitor and maintain the integrity of offshore structures. These activities are important not only as a means to safeguard the lives of personnel aboard offshore installations but also as a way of protecting the considerable financial and technical investments represented by the installations themselves.

Currently, it is required that the entire structure should be inspected over a five year period. This requirement is a lower-bound of what is needed to effectively monitor the structural integrity of an offshore structure. This is because, based on the CA's recommendation, an average of only 20% of the structure needs to be examined annually. However in a five year period, special circumstances could occur, as for example those following a supply boat collision or a hurricane, which would require new inspections or detailed monitoring of a damaged or suspect part of the structure. This has recently been the case in several fields such as those in the Gulf of Mexico and Thailand.

Current practice is to delegate IRM activities to a multi-disciplinary team comprising inspection personnel (divers/ROV operators), structural engineers, metallurgists as well as maintenance and repair specialists. In theory this team has all the knowledge and expertise required to monitor and carry out remedial measures to maintain an acceptable level of structural integrity. However, in practice the situation can be less than ideal. Managing activities with such diverse technical backgrounds implies that the team's efficiency depends on a base of knowledge and an information exchange capable of efficiently delivering the required inputs to each discipline in an appropriate and familiar form. What is required is a set of options to remedy the potential loss of structural integrity due to the presence of a defect.

The conventional approach used in the design and assessment of offshore structures is deterministic. Despite widespread acceptance of the probabilistic nature of engineering, the lack of comparisons between the results of probabilistic approaches and of the more familiar deterministic methods, have suppressed the use of reliability methods in design offices. This reluctance has been further fuelled by the fact that the probability distributions of rare events and their associated parameters used in reliability methods have, to an extent, not generally been based on actual data specific to each structure in question.

2 EXPERT SYSTEMS : A TOOL FOR IRM

An expert system is a computerised representation of the knowledge of specialists and practitioners ("experts") within a technical area, intended to perform a task or to deal with a situation. The system is capable of storing not only a variety of data but also the rules which relate the data to the solution or objective of the task being performed. In this way, a centralised, accessible and updateable source of knowledge can be created as the core for the above-mentioned information exchange system. By customising the user access, the system can then cater to the types and quantity of information required by each of the IRM team members. Therefore the flexible data transfer concept mentioned above is realised.

Billington Osborne-Moss Engineering Limited (BOMEL) are engaged in the development of such an expert system, for the Assessment and Management of Defects on Offshore Structures (AMaDeOS) with the support of the Commission of the European Communities and AMOCO (UK) Exploration Company. Figure 1 illustrates the system architecture and hierarchy that is embodied within the AMaDeOS system.

The expert system is built around three knowledge bases namely (1) the Structural Knowledge Base, (2) the Inspection/Maintenance Knowledge Base and (3) the Diagnostic Knowledge Base. These knowledge bases correspond to the data and rules which each of the three main disciplines utilise during the execution of their tasks.

3 OPERATIONAL CASE STUDY

Consider the case of an offshore platform subjected to hurricane loading of a magnitude which exceeded that considered during the design. Subsequent inspection revealed numerous cracks which had developed at various locations throughout the structure.

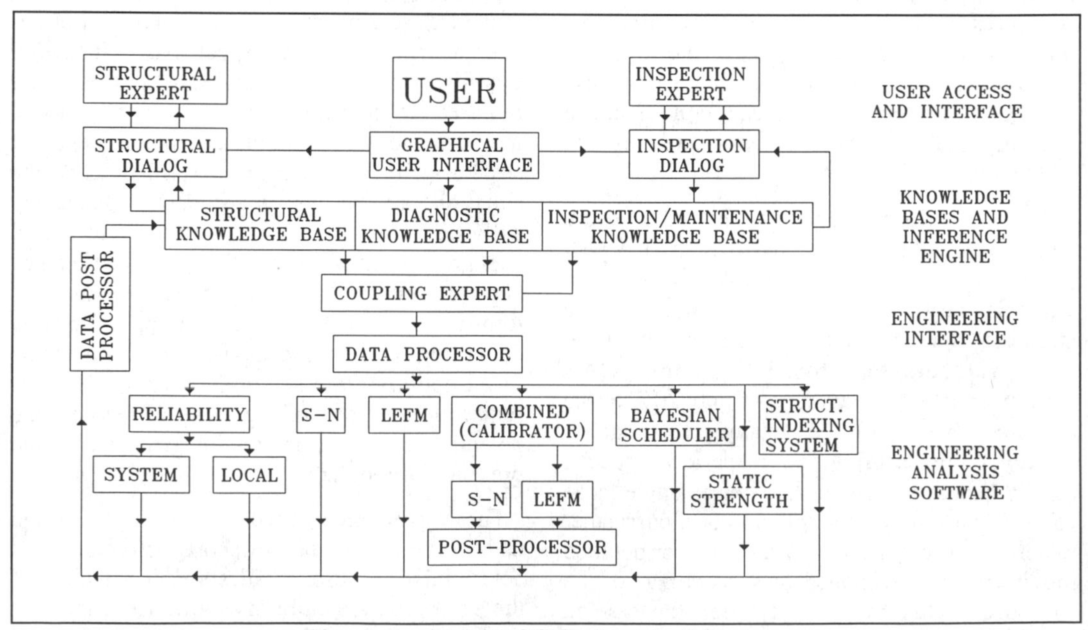

Figure 1 : AMaDeOS System Architecture

The following is a description of the way in which AMaDeOS handles fracture assessment. The procedure is illustrated by the flow chart in Figure 2. A corresponding assessment for fatigue is also necessary so that the above questions can be dealt with in full.

3.1 Data collation and defect categorisation

In order to formulate a solution various items of data are required such as the structural model and environmental conditions assumed in the original design, measured (or estimated) environmental data during the hurricane, material parameters and past inspection reports. From this, all data pertaining to locations of, and in close proximity to, the damaged members are extracted.

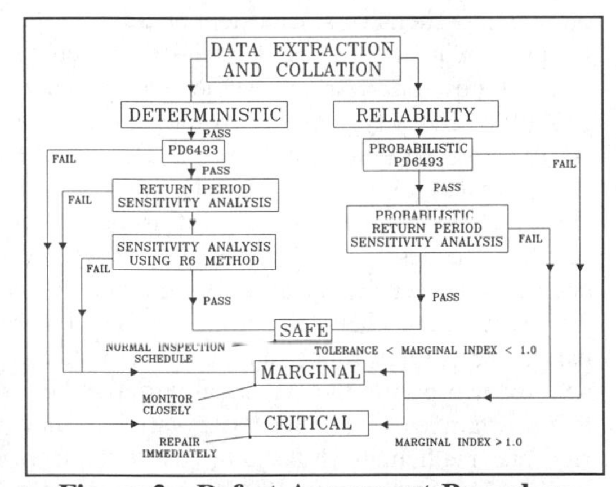

Figure 2 : Defect Assessment Procedure

The defects are then categorised according to their type, location and importance. Their individual importance is assessed on the basis of the structural function of the member on which the defect occurred. In this example, global jacket analyses were performed whereby joint failure was assumed to occur so as to estimate the effect on the rest of the structure using the member removal technique.

On the basis of the above categorisation, the available repair options in the inspection database are matched with the defects. This produces a reference repair list which will act as a solution source for the following assessments.

3.2 Deterministic defect assessment

Firstly, the defects are categorised according to their immediate criticality. The most critical defects are those which may cause member severance or plastic collapse. These assessments should be carried out according to specific in-house or other general procedures (PD6493 1991) with loads appropriate for a damaged structure. The magnitude of these loads could be specified by the regulatory authority or in accordance with in-house specifications (eg 1 year return period). The defects that are identified as being critical in this first pass analysis should be repaired immediately.

Those defects not identified as being critical according to the above procedure are then subjected to a return period sensitivity analysis. With this, a relationship is developed between the return period and the stress at the defect location. Each defect is then subjected to the failure assessment procedure described in the paragraph above under wave loading of increasing return periods up to, for example, the remaining life of the structure. If failure is found to occur, the defect can then be categorised as marginal. The criticality of these marginal defects can be assessed for example using the conventional check:

(Return Period first causing failure / Factor of Safety) > Remaining Life of structure

or by calculating a more rigorous Marginal Index such as

$I_{marginal}$ = (1 - (Return Period first causing failure)/(Remaining Life of structure)) x Factor of Safety

where the Factor of Safety is assigned a value agreed between the Operator and the Regulatory Authority.

The Marginal Index shown above enables the prioritising of marginal defects.

Under the deterministic analysis carried out thus far, the joints still categorised as safe may in fact be marginal based on their sensitivity to changes in material toughness or crack geometry. As an example, let the observed crack depth be a_{obs} and the accuracy of the measurement be $acc_{measured}$. The crack depth is then assumed to range between a_{obs} - $acc_{measured}$ and a_{obs} + $acc_{measured}$. Using the Level 2 Failure Assessment of PD6493, each of the cracks is subjected to a sensitivity analysis similar to that recommended in the CEGB 'R6' method. As each crack has a range of geometries, it will also have a range of load factors (ie. the F^Ls as defined in 'R6'). Based on the number of discrete checks carried out for each crack geometry, a crack is categorised as marginal if any of the results fall below a cut-off load factor limit F^L_{min} which may be a user specified value (eg 1.2) or a default value of 1.0. These marginal cracks should be monitored more closely during future inspections.

3.3 Reliability defect assessment

The reliability method of analysing a similar situation mirrors the above deterministic procedure more rigorously. The defects encountered are assigned default Target Failure Probabilities based on their importance, location, and accessibility (eg Failure Probability of 10^{-5} for a joint for which access is restricted). The user can however change these default values on the basis of experience or in order to meet regulatory or in-house specifications.

All basic variables pertaining to the PD6493 assessment procedure (eg loading, crack geometry, material properties etc) are assigned probability distributions and the relevant statistical parameters (ie mean and standard deviation). Again, where this is justifiable or in order to perform sensitivity analyses, the default values for PDFs suggested may be changed by the user. A Reliability Level 3 (Monte-Carlo) simulation is carried out to determine the actual reliability of the damaged structure. A defect is categorised for immediate repair if the calculated reliability index is below the target value. The loading used for this step is a probabilistic version of the load appropriate for a damaged structure, as used in the deterministic procedure above (eg 1 year or the regulatory requirement). It should be pointed out that this first step only tags those defects which are susceptible to imminent failure thus requiring repair.

In order to identify marginal defects, the above reliability analysis is repeated for waves of various return periods up to, say, the remaining life of the structure. As before and in a way which mirrors the deterministic approach, a defect is categorised as being marginal if any of the return periods up to the remaining life of the structure results in a reliability less than the target value. The return period which causes this is then a measure of the criticality of the marginal defect. All remaining defects are then categorised as being safe. A fatigue assessment is then performed on all the defects in order to further ascertain the need for repair or additional inspection.

4 CONCLUSIONS

It is clear that the reliability method involves a more comprehensive stochastic modelling of the system whilst the deterministic method concentrates on a more subjective pass/fail procedure and a type of partial factor approach (similar to Reliability Level 1). The attractiveness of the deterministic approach is that it is simpler and its format is based on familiar codes of practice. However, deterministic sensitivity analyses are limited in the way they handle variations in basic parameters. Only reliability based methods are able to incorporate the statistical variations due to uncertainties in all the basic variables in a rigorous mathematical fashion. As data from measurements of the actual performance of each structure are collected and stored in the database of the expert system, the criteria and the results of reliability-based assessments will become more dependable. In the meantime the user should employ both methods until a satisfactory level of calibration is achieved.

REFERENCES

Guidance on methods for assessing the acceptability of flaws in fusion welded structure, PD6493, BSI Standards 1991.

Assessment of the integrity of structures containing defects, R/H/R6 - Revision 3, CEGB 1986.

Structural Safety & Reliability, Schuëller, Shinozuka & Yao (eds) © 1994 Balkema, Rotterdam, ISBN 90 5410 357 4

Safety concept for constructions serving environmental protection purposes

J. D. Wörner & D. Kiefer
Technical University of Darmstadt, Germany

ABSTRACT: With the help of a safety concept for constructions serving environmental protection purposes the possible danger can be minimized. With the help of a risk analysis the overall risk will be assessed. Further, the comparison of the actual risk with the tolerable risk should lead to the appropriate safety for this kind of constructions.

1 INTRODUCTION

In today's industrial age the production of certain chemicals is indispensable, at the same time one must make sure that these chemicals will not attack and cause harm to our environment. For instance, it is necessary with the help of construction measures to hold back substances that might pollute the ground water. A concept will be developed in order to minimize the danger of pollution by stipulating safety precautions. This can be achieved by lying down individual steps, beginning with the planning and the execution of the plans up to the operation of the system.

The following concept serves for the dimensioning of constructions like chemical plants, filling station, waste deposits or waste utilization plants.

The primary aim of the development of a safety concept is not to protect the concrete against environmental influences or to ensure the durability of the concrete. It is much more the intention to protect soil and ground water against polluting matter. This protection may follow from the application of concrete structures as protective barriers. With the help of resistances, such as primary and secondary barriers, the ground water can be protected against dangerous substances.

Unfortunately, the tightness of the tank (primary barrier) can not be solely relied upon, and therefore a tight catch basin (secondary barrier) is demanded. Normally the primary barrier is expected to be tight. However, in case of damage of the primary barrier, the secondary barrier must be able to prevent the leakage of polluting matter to continue into the subsoil (ground water) for the time until countermeasures are implemented.

The entire system may be assessed with the help of a risk analysis. This risk-calculation may be done in two different ways. The risk determined with the help of the probabilistic method is the product of the failure occurrence probability multiplied by the degree of damage. In this risk-calculation, however, the biggest accident thinkable remains possible. Therefore, in a concept for concrete structures serving environmental protection purposes a deterministic method must also be taken into account parallel to the probabilistic method. This means that the degree of damage must be kept within certain limits. This limitation of damage follows from the application of the secondary barrier.

2 DESIGNS OF TIGHTNESS FOR CONCRETE STRUCTURES

Three different designs can be applied in proving the tightness of the secondary barrier (concrete structure):

- Design of uncracked sections
- Calculation of the compressive zone in flexural members
- Crack width calculation

3 RISK-ANALYSIS

3.1 *Targets and limits of a risk-evaluation*

In order to limit the failure of the catch basin there is no point in applying the probability of failure, which was worked out in proving the tightness on the basis of statistical safety theory. In setting up the safety criteria it is more reasonable to operate with the overall risk of the system.

Risk is related to the likelihood of damage caused by an accident. This element of damage must be related to a numeric value. In other words, damage must be measurable in standardized units.

The relation between the degree of damage S of accident and the likelihood of damage W_s leads to risk value R:

$$\text{Risk } R = \sum S \times W_S \quad (1)$$

If all possible degrees of damage of an accident and the respective probabilities are known, then the risk can be deduced from this equation (Hauptmann 1987). In order to determine the risk, it is necessary to draw up a sequence of events analysis for the system. The sequence of events analysis describes how the system reacts in case of damage of the primary barrier. The construction of the analysis of sequences of events should consider the likelihood of failure of all system components.

The size of the safety factor in proving the tightness of the concrete structure depends on the estimation of risk. The risk for all three designs of tightness should be within the same order. This may result in different levels of safety for each of the three designs of tightness.

3.2 *Identification of risk*

In order to carry out the risk calculation of a given system serving environmental protection purposes, an global identification of risk has to be completed in advance. The identification of risk serves for the function of drawing up all possible dangers in their entirety (Franck 1989). In this connection the following questions should be answered:

1. Where are the releasing risk factors situated?
2. What events might occur?
3. Who and what is exposed to damage?

A structure serving environmental protection purposes is made up of at least two protection barriers.

The risk factors are put together of those components, which are expected to cause failure of the primary or secondary barrier. The following risk factors apply to the primary barrier:

- failure probability which results from the design of the primary barrier
- reasons for failure probability, depending on construction material and its age.
- occurrence probability of leakage during filling.
- the distribution of spilled amount of dangerous substances for the respective cause of defect.

On the basis of obligatory notification to the appropriate authorities in case of accident and the central recording of all registrated accidents in the Federal Republic of Germany the Federal Department for Statistics prepares and publishes accident statistics (Statistisches Bundesamt 1992). The data does not represent the actual accidents in proportion to all possible accidents. The missing relevant data concerning the total population of tanks prevents the development of this proportion. Thus, the quantification of the occurrence probabilities for different causes of accident is not possible.

The following risk factors account for failure of the concrete catch basin:

- occurrence probability and distribution of the decisive dimensioning loads.
- distribution of the characteristic values of concrete (concrete tensile strength etc.)
- distribution of medium's penetration depth and depth of damage in the uncracked and cracked concrete
- distribution of the critical crack width w_{crit}, at which, depending on the medium, the thickness of the structure is just penetrated in the time of admission
- occurrence probability of the estimated time of admission
- occurrence probability of tightness failure in construction details such as joints and connections
- inaccuracies in the model for the proof of tightness.

3.3 *Conception of the risk-analysis*

Figure 1 shows a structural diagram concerning a tank storage-system with an uncoated concrete catch basin, the latter divided into its different construction elements. For simplicity the primary barrier is assumed to experience only a single failure, and therefore it is not divided into different risk components.

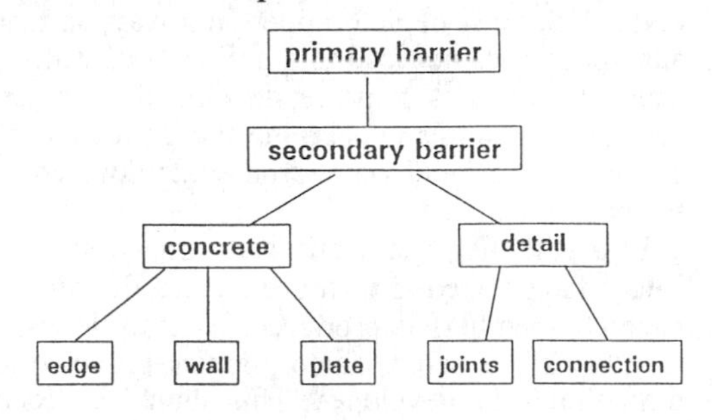

Figure 1 Structural diagram for the entire system

In case the tightness of the concrete structure is dimensioned according to the design of uncracked sections, figure 2 helpfully demonstrates all possible outcomes of a failure of this design. If the tightness of a concrete structure is dimensioned according to the crack width calculation, then it is likely, that the appearing stresses may be smaller than the strength. Each event is related to a probability of success or failure. Each event in figure 2 is numbered for better readability.

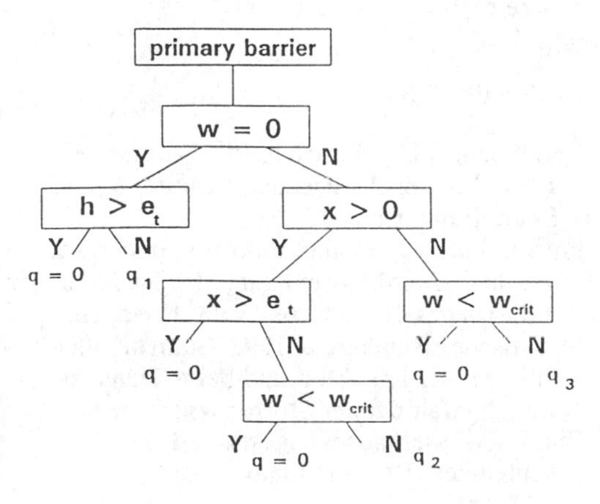

Figure 2 Sequence of events diagram for concrete structure

It is proposed that the damage can be derived by multiplying the probable volume with a factor representing the toxicity. With reference to equation (1), the computing of risk at each branch in the sequence of events diagram for a concrete structure may follow from equation (2):

$$\text{Risk} = p_{f1} \sum_{i=1}^{n} p_{2,i} * \int_{0}^{\infty} f_{2,V} V dV * A \qquad (2)$$

p_{f1}: failure probability of the primary barrier
$p_{2,1}$: occurrence probability of each event in the sequence of event diagram for the secondary barrier
n: number of events in the sequence of event diagram for the secondary barrier
$f_{2,V}$: probability distribution function of the volume which leaks through the secondary barrier
V: actual volume of substance which leaks through the secondary barrier
A: toxicity parameter for the medium

The overall risk for the entire system follows from the addition of all individual risks. The toxicity parameter depends on the following points:

- degree of danger of the substance to water
- amount of substance
- environment, location of the installation
- infrastructure
- operation- and failure surveillance
- precautionary measures such as quality control, maintenance, and control of the concrete constructions.

3.4 *Tolerable risk*

The calculated overall risk for an system serving environmental protection purposes must be compared with the tolerable risk. The tolerable risk must adhere to the principle, that a pollution of the water is not to be feared (Bundes Wasserhaushaltsgesetz 1986). In other words, human beings and environment must be protected against damaging effects as far as possible.

In absence of other definitions the German § 4 Draft of the model code VAwS (Jan. 1992) is used to define the toxicity parameter and the tolerable risk. In this model code compounds are classified in toxicity class 0-3, the so called WGK (degree of danger of substance to water). According to regulation concerning surface storage a graduation of the toxicity parameter A depending of toxicity class may be proposed (table 1).

Table 1. Determination of toxicity parameter A and Calculation of the tolerable risk.

Toxicity class	0	1	2	3
Toxicity parameter A	1	10	100	1000
Tolerable stored volume [m³]	100	10	1	
Tolerable risk [m³/a]	10^{-4}	10^{-4}	10^{-4}	

The tolerable storage volume depends on the toxicity class of the compound. The different quantities are listed in table 1. The failure probability of the primary barrier is assumed to be 10^{-6}/a. The tolerable risk is derived at by the multiplication of the assumed failure probability of the primary barrier with the tolerable quantity and with the toxicity parameter A. The tolerable risk emerges from the depending toxicity class.

The regulation concerning surface storage does not specify an tolerable storage volume of compounds in toxicity class 3.

3.5 *Interpretation of risk-evaluation*

The estimated safety factor γ_r for the crack width calculation is derived by the following procedure: The critical crack width for a medium is the crack width at which a penetration volume of zero is produced within a given time of admission. With the help of the risk calculation the crack width w_{cal} is determined, at which the overall risk is smaller than the tolerable risk. The safety factor γ_r is obtained by the quotient of critical crack width w_{crit} to the calculated crack width w_{cal}.

The estimation of the safety factors for the two other tightness designs follows in the same way.

4 CONCLUSION

The concept for constructions serving environmental protection purposes should be based on a risk-calculation. A sole estimation of the failure probability according to the design applied in proving the tightness of concrete is granted exception to. The application of this approach does not make a quantification of the exiting amount of hazardous substance possible. The risk is developed by the multiplication of failure probability with the possible amount and with a toxicity factor. The toxicity factor is related to the danger of the medium, which acts upon the concrete in case of failure. The sum of all the individual risks makes up the overall risk, which must be smaller than the tolerable risk. The tolerable risk must be determined in a way, so that adherence does not lead to a pollution of waters. The tolerable risk may be developed from the various regulations concerning the protection of soil and the ground water from hazardous compounds.

With the help of limit state functions the safety factor needed in the tightness design for concrete members is produced. Next to the estimation of the safety factor, a risk-minimizing draft should be developed. This should be done using comparative techniques and the utilization of all known relevant constructions in order to achieve the smallest risk possible.

The limitation of risk is advanced by the application of a barrier system. A secondary barrier (an uncoated concrete basin e.g.) is needed around the primary barrier (the storage tank) in order to limit the damage caused by failure of the primary barrier.

A complete calculation of the above presented concept for constructions serving environmental protection purposes is not possible yet. For this more accurate data is needed covering the probability distribution function for failure of the primary barrier, and the critical crack width for a medium and its penetration depth in uncracked concrete.

REFERENCES

Hauptmanns, U., Herttrich, M., Werner, W. 1987. Technische Risiken, Ermittlung und Beurteilung. Berlin.

Franck, E. 1989. Risikobewertung in der Technik. In Gerhard Hosemann (ed.), Risiko in der Industriegesellschaft, p. 43-93. Erlangen.

Statistisches Bundesamt 1992. Statistik über Unfälle bei der Lagerung und beim Transport wassergefährdender Stoffe. Wiesbaden.

Bundes Wasserhaushaltsgesetz § 19, 23. September 1986, Germany.

Geotechnical engineering

Structural Safety & Reliability, Schuëller, Shinozuka & Yao (eds) © 1994 Balkema, Rotterdam, ISBN 90 5410 357 4

Reliability assessment in dynamic analyses of earthquake-resistant design of concrete gravity dams

Houqun Chen, Shunzai Hou & Aihu Liang
Institute of Water Conservancy and Hydroelectric Power Research

ABSTRACT: Based on the statistics analyses of the data of site–specific seismic hazard assessment of more than twenty important hydraulic projects designed or under construction in China, the probabilistic level of fortification against earthquake and the probabilistic model of seismic actions are proposed. Using the obtained results, a systematic seismic reliability analysis of a series of concrete gravity dams designed under existed Code were carried out. Some basic considerations and algorithm for analysis are discussed. Finally some conclusions and recommendations are provided for the new 《Earthquake Resistant Design Code for Hydraulic Structures of China》, which is no being revised.

1 INTRODUCTION

The analytical techniques and corresponding criteria presently used to evaluate the seismic performance of concrete dams are deterministic. Since characterizing the occurrence of earthquakes is most meaningful in terms of probabilities of occurrence, use of probability–based algorithm and criteria to evaluate the seismic performance of concrete dams is an appropriate approach. It is expected that reliability–based aseismic design of concrete dams will become more common in engineering practice. In China, reliability analyses have been involved in the aseismic design of building according to the 《Aseismic Design Regulation of Building of China (GBJ11–89)》 in 1989. Recently, the 《Unified Standard for Structure Reliability Design of Hydrualic Engineering of China》 is under publication. Correspondingly, dynamic reliability analysis of dams is to be considered in the new 《Earthquake Resistant Design Code for Hydraulic Structures of China》 which is now being revised.

As is well known from the engineering practice, both the level of fortification against earthquake and approach of aseismic design for such important structures as large dams are distinguished from those of ordinary buildings.

In this paper two important considerations will be addressed:

* First the probabilistc level of fortification against earthquake and probabilistic model of seismic actions for major hydraulic projects and
* Sencond, the seismic reliability analysis and reliability–based design of concrete gravity dams.

2 PROBABILISTIC LEVEL OF FORTIFICATION AGAINST EARTHQUAKE

The probabilistic level of fortification against earthquake and probabilistic model of seismic actions are important problems to be solved for dynamic reliability analysis of major hydraulic structures.

The probabilistic analysis of seismic actions is based on the seismic hazard assessment at the site of an engineering project in terms of a ground motion parameter versus annual probability of exceedance.

In China, the seismic hazard evaluation has been used in Er–Tan large hydroelectric power project for the first time in the early eighties. After then, site–specific seismic hazrd evaluations have been carried out almost for every important hydraulic or hydroelectric project in China. In each case not only the "basic intensity" of a site was defined as the maixmum intensity expected at the dam site for a certain period, but also the curve in terms of annual probability of exceedance to peak ground acceleration or to intensity rating in 12–grade scale additionally for some of projects.

According to the 《Unified Standard for Structure Reliability Design of Hydraulic Engineering of China》, the design reference periods for 1 class retaining hydraulic structures and for others are 100 and 50 years respectively, while a design reference period of 50 years is stipulated uniformly for all kinds of buildings.

The relationship of the annual probability of exceedance (P_1) and that for design reference period T_o (P_T) can be established as

$$P_{T_o} = 1-(1-P_1)^{T_o} \quad (1)$$

In Fig. 1 the curves of peak ground acceleration versus probability of exceedance for T_o = 100 of 23 designed major hydraulic projects located in areas of different seismicity in China are illustrated. The curves have been corrected to the uncertainties of attenuation functions. In 1992 a new 《Seismic Zoning Map of China》 was issued, which gives the "basic intensity" of the whole territory of China with a probability of exceedance of 0.1 for 50 years (Bao 1985). The corresponding recurrence interval is 475 years. However, it is not recommended to be used for important hydraulic projects.

Based on the defined basic internsities of the specific sites of hydraulic projects and results of site-specific seismic hazard evaluation, a statistic calibration has been carried out. It is discovered, that the "basic intensity" at sites of hydraulic projects has a probability of exceedance of 0.1 for 100 years. Its return period is 950 years, which is much longer than in 《Seismic Zoning Map of China》.

In accordance with the practice and the requirement of the "Earthquake Resistant Design Code for Hydraulic Structures of China", the design seismic intensity for 1 class retaining hydraulic structures should be in 1 grade higher than its basic intensity. It means that the design peak ground acceleration should be doubled. The probabilistic level of fortification against earthquake for 1 class retaining hydraulic structures can be derived through the same calibration as used for "basic intensity". Finally, a preliminary proposed probabilistic level of fortification is recommended with P_{100} = 0.02, the corresponding recurrence interval is 4950 years. The recommended levle is smaller than a level of $P_1 = 10^{-4}$, which is commonly used for important hydraulic projects in practice of China. Actually, the level of $P_1 = 10^{-4}$ is directly borrowed from the probabilistic level of the "safe shotdown earthquake" for nuclear power plants without sufficient basis for hydraulic projects. It is also worthy to be noted that the proposed recurrence interval is much longer than (1600 – 2475) years provided for intensity of "seldomly occurred earthquake" in 《Aseismic Design Regulation of Buildign of China (GBJ11–89)》.

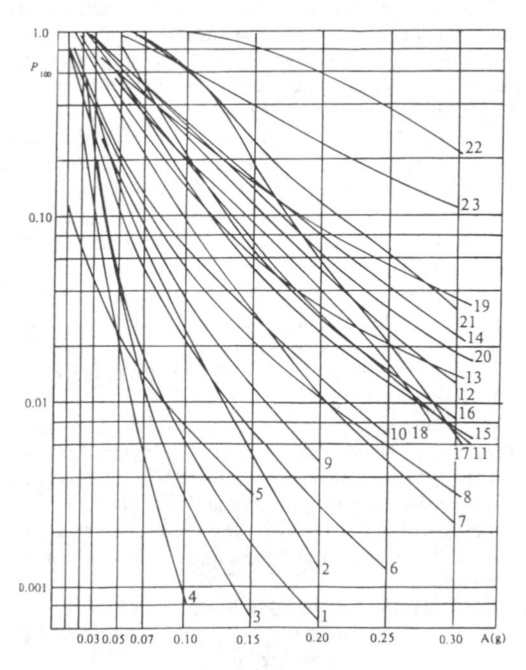

Fig.–1 The curves of probability of exceedance to peak ground acceleration for 100 years of important hydraulic projects

3 PROBABILISTIC MODEL OF PEAK GROUND ACCELERATION

It is obviously that the acceleration has more uncertainities than any other factor involved in seismic actions. Based on the results shown in Fig. 1, the probabilistic model and their statistical parameters in 100 years for peak ground acceleration at dam site are studied by different methods. By using the K–S goodness–of–fit test at the significance level of 5%, it is shown that the Type II asymptotic distribution of extremes with a shape parmeter K is an acceptable model for the peak ground acceleration of dam site, that is,

$$F_{II} = e^{-(A/A_\varepsilon)^{-k}} \quad (2)$$

in which A_ε is the location parameter, usually it takes the mode acceleration with a probability of exceedance of $1-e^{-1} = 0.632$.

The value of K for each project can be esitmated by using least square method. A mean value of 2.3 is obtained, and the correspnding coefficient of variation $V_A = 1.38$.

As the peak acceleration with a probability of exceedance of $P_{100} = 0.1$ is considered as characteristic value A_K, then for engineering use a value of $A_k / A_\varepsilon = 3$ and a mean value of $m_A = 0.6\,A_k$ may be taken in design practice of dams. It should be noted that for 1 class retaining hydraulic structures the value of A_k should be doubled.

It is assumed that both horizontal and vertical components of seismic actions have the same statistic characteristics.

4 SOME BASIC CONSIDERATIONS FOR SEISMIC RELIABILITY ANALYSIS OF CONCRETE GRAVITY DAMS

The following basic considerations are taken in the

seismic reliability analysis of concrete gravity dams:

1) As the reliability analysis has to be consistent with the basic design code of concrete gravity dams, in seismic reiability analysis, only the safety of tensile and compressive strength of concrete at different dam sections and the safety of stability against sliding along the foundation surface are checked using 2–dimensional method of structural mechanics.

However, it is recognized, that tensile stresses are hardly avoidable for gravity dams during strong earthquake and some reasonable tensile capacity of dam concrete must be expected.

2) According to the 《Unified Standard for Structure Reliability Design of Hydraulic Engineering of China》, seismic actions are taken as accidental actions with a partial sefety coefficient $\gamma_f = 1.0$ and a coefficient of accidental design state $\psi = 0.90$. The ultimate limit state may be determined by

$$\gamma_o \psi S(F_d, \alpha_k) \leqslant \frac{1}{\gamma_d} R(F_d, \alpha_k)$$

in which

S(*), R(*) — functions of action effect and structural resistance respectively;

F_d, f_d — design values of actions and resistances respectively;

α_k — characteristic value of geometric parameters;

γ_d — structural coefficient related to reliability index of structure;

γ_o — coefficient of importance.

3) To keep the continuity of design code, only the actual reliability indexes (β) of different gravity dams designed under existed code are calibrated and corresponding structural coefficients γ_d are suggested for reliability–based design of concrete gravity dam.

4) The seismic effects on dams are determined by the response spectrum analysis including the dynamic interactions of dam–reservoir–foundation. The maximum dynamic responses are obtained by the supperposition method taking the first five vibration modes into consideration. Both acceleration and coefficient of dynamic amplification D obtained from code are taken as random variable. Based on the existed statistic data, it is assumed that the D has a Type I asymptotic distribution of extremes with a simplified average coefficient of variation V = 0.3 in a period range of 0.2 – 2.0 second.

5) At present, there are no sufficient statistical data of dynamic properties for dam concrete and foundation rock. Therefore, the probabilistic model and parameters of thier static properties are adopted in seismic reliability analyses except that the dynamic tensile and compressive strength are assumed to be increased in 30% above their static values in the light of preceding researchers.

5 ALGORITHM OF ANALYSIS AND CALIBRATED RESULTS

The limit–state equation is solved by geometric method. Expand the performance function G(Y) at the failure point of ith iteration $\{Y\}_i = \{Y_{1i}, Y_{2i}, \dots Y_{ni}\}^T$ in a Taylor series truncated at the first–order term

$$G(Y) = G(Y_1, \dots Y_n) \approx G(Y_{1i}, \dots Y_{ni}) + \frac{\partial G}{\partial Y_1}(Y_1 - Y_{ni}) + \dots + \frac{\partial G}{\partial Y_n}(Y_n - Y_{ni}) \quad (3)$$

Let

$$G(Y_i) = G(Y_{li}, \dots Y_{ni}) \quad (4)$$

$$\nabla G(Y_i) = \left\{ \frac{\partial G}{\partial Y_1}, \frac{\partial G}{\partial Y_2}, \dots \frac{\partial G}{\partial Y_n} \right\}_{Y = Y_i} \quad (5)$$

and thus

$$G(Y) = G(Y_i) + \nabla G(Y_i)\{Y\} - \nabla G(Y_i)\{Y\}_i \quad (6)$$

Table 1 Basic Characteristics of Gravity Dams (upstream slope 0.0)

No.	Height (M)	Rock class	Depth of water (M)	Down–stream slope	Concrete class	Basic freq. (Hz)
1	190	I	182.5	0.78	C20	1.47
2	190	II	182.5	0.88	C15	1.45
3	170	I	163.0	0.74	C15	1.59
4	170	II	163.0	0.85	C15	1.61
5	150	I	143.5	0.74	C15	1.81
6	150	II	143.5	0.80	C15	1.77
7	130	I	124.0	0.74	C15	2.08
8	130	II	124.0	0.75	C15	1.97
9	110	I	104.5	0.74	C10	2.46
10	110	II	104.5	0.74	C10	2.31
11	90	I	85.0	0.74	C10	3.00
12	90	II	85.0	0.74	C10	2.82
13	90	III	85.0	0.78	C10	2.36
14	70	I	65.5	0.73	C10	3.80
15	70	II	65.5	0.73	C10	3.58
16	70	III	65.5	0.73	C10	2.94
17	50	I	46.0	0.73	C10	5.24
18	50	II	46.0	0.73	C10	4.95
19	50	III	46.0	0.73	C10	4.10
20	30	I	26.5	0.70	C10	7.94
21	30	II	26.5	0.70	C10	7.56
22	30	III	26.5	0.70	C10	6.42

Table 2 Statistic Parameters of Random Variables

Random variable	Probab. model	Mean value			Coef. of variation
α	Normal	0.185			0.30
A	Frechet	$0.60A_K$			1.38
D	Gumbel	$1.052D_K$			0.30
		Class	Tensile	Compr.	
[σ] (MPa)	Log. normal	C10	0.92	11.47	0.284
		C15	1.35	16.85	0.267
		C20	1.76	22.00	0.251
		C25	2.14	26.72	0.235
		C30	2.52	31.55	0.220
f	Normal	I	1.35		0.20
		II	1.20		0.21
		III	1.00		0.22
C (MPa)	Log. normal	I	1.40		0.36
		II	1.20		0.36
		III	0.90		0.40

Note: α = Uplift coef., f = Friction coef.of rock, C = Cohesion of rock, D_K = Standard value of D, [σ] = Dynamic strength of concrete.

Table 3 Calibrated Results (with seismic intensity of 9)

No	Stability against sliding			Tensile strength			Compressive strength		
	K	β	γ_d	K	β	γ_d	K	β	γ_d
1	1.27	1.61	0.54	1.05	1.13	0.80	3.91	3.06	2.99
2	1.29	1.62	0.57	1.07	1.15	0.82	3.22	2.79	2.46
3	1.18	1.51	0.48	1.01	1.13	0.77	3.05	2.79	2.33
4	1.22	1.55	0.52	1.03	1.21	0.79	3.39	2.82	2.59
5	1.14	1.45	0.43	1.02	1.17	0.78	3.32	2.83	2.53
6	1.15	1.47	0.47	1.00	1.07	0.77	3.55	2.88	2.71
7	1.17	1.47	0.43	1.01	1.04	0.77	3.65	2.90	2.79
8	1.07	1.38	0.40	1.00	1.12	0.77	3.73	2.92	2.85
9	1.25	1.53	0.45	1.01	1.13	0.78	2.70	2.50	2.06
10	1.11	1.41	0.40	1.03	1.20	0.79	2.74	2.56	2.09
11	1.39	1.63	0.54	1.01	0.98	0.85	3.03	2.49	2.54
12	1.22	1.51	0.48	1.01	1.00	0.85	3.08	2.52	2.59
13	0.98	1.21	0.37	1.01	1.13	0.85	3.39	2.70	2.84
14	1.59	1.75	0.61	1.03	0.94	0.86	3.43	2.59	2.89
15	1.37	1.61	0.52	1.00	0.83	0.82	3.49	2.62	2.93
16	1.09	1.32	0.41	1.03	1.02	0.86	3.75	2.74	3.15
17	2.00	1.98	0.76	1.00	0.95	0.84	4.09	2.68	3.44
18	1.66	1.81	0.63	1.01	0.91	0.85	4.04	2.66	3.39
19	1.29	1.50	0.47	1.04	0.96	0.87	4.36	2.78	3.67
20	3.43	2.48	1.43	1.01	0.96	0.94	5.65	2.79	5.27
21	2.85	2.34	1.19	1.01	0.93	0.95	5.57	2.78	5.20
22	1.98	1.97	0.80	1.01	1.03	0.94	5.48	2.77	5.12

Then, the limit–state surface is

$$G(Y_i) + \nabla G(Y_i)\{Y\} - \nabla G(Y_i)\{Y\} = 0 \quad (7)$$

The distance from origin (O) to the point (P) on tangent plane is

$$|OP| = \frac{G(Y_i)\{Y\}_i - G(Y_i)}{\|\nabla G(Y_i)\|} = \{\alpha\}_i^T\{Y\}_i + \frac{G(Y_i)}{\|\nabla G(Y_i)\|} \quad (8)$$

where

$$\{\alpha\}_i = -\frac{\nabla G(Y_i)}{\|\nabla G(Y_i)\|} \quad (9)$$

The coordinates of the point P are

$$\{Y\}_{i+1} = |OP|\{\alpha\}_i \quad (10)$$

Repeating the above step with $\{Y\}_{i+1}$ until convergence, the final failure point $\{Y\}^*$ can be obtained, and reliability index is

$$\beta = \{Y\}^{*T}\{\alpha\}^* \quad (11)$$

A series of gravity dams with heights from 30 m to 190 m under diffirent kinds of rock conditions, designed according to the existed code, were selected for calibration analysis. Their basic design characteristics are shown in Table 1.

The probabilistic models and their statistical parameters of all random variables in static and dynamic actions and resistances are listed in Table 2.

For each dam the conbined stresses at upstream and downstream sides of 20 horizontal sections as well as total stability of the whole dam were calculated. For calibration purpose both deterministic safety factors (K) and reliability indexes (β) with its related structural coefficient (γ_d) at every calculated point for each dam were obtained. Part of results calculated for seismic actions with design intensity of 9 are shown in Table 3.

6 CONCLUSION AND RECOMMENDS

As the safety factors of stability against sliding and of tensile and compressive stresses are 1.0, 1.0 and 2.5 respectively, it is obviously from the results obtained, that both the stability against sliding and the compressive stress can satisfy the requirements even during earthquake with an intensity of 9. However, the safety factors of tensile stresses at upper part of dam are insufficient. Those parts need to be strengthened by increasing class of concrete during strong earthquake. Such conclusion is fully consistent with the damage during past events.

It is worthy to be noted that the calibrated relability indexes of gravity dams are significantly larger than those for building under the seismic actions with the same intensity.

For aseismic reliability design purpose the structural coefficients (γ_d) of 0.40, 0.85, and 1.90 are proposed for stability against sliding, tensile and compressive stresses respectively. The corresponding reliability indexes are 1.30, 1.05 and 2.33 respectively.

REFERENCES

Bao Aibin, Lee Zhongxi, Gao Xiaowang, Zhou Xiyuan, 1985, Probabilistic calibration of basic intensity for parts of China, ACTA seismological Sinica, Vol. 7, NO. 1.

Aseismic design regulation of building of China (GBJ11–89), 1989, Architectural industry press.

Structural Safety & Reliability, Schuëller, Shinozuka & Yao (eds) © 1994 Balkema, Rotterdam, ISBN 90 5410 357 4

Measuring uncertainty correction by pairing

Ove Ditlevsen
Department of Structural Engineering, Technical University of Denmark, Lyngby, Denmark

ABSTRACT: Under certain circumstances a pairing of measurements from two independent and possibly inaccurate measuring methods reveals information about the population of measuring objects as well as about the two measuring error populations. This pairing method in combination with stochastic interpolation principles is the basis for a method of getting information about the uncertainty related to measurements of spatial variations of soil properties and of separating the uncertainty population and the property population from each other. The method is developed for the following particular application. Elasto–plastic continuum mechanics predicts that there is almost proportionality between the undrained shear strength c_v as measured by the vane test and the cone tip resistance q_c, both imagined to be measured at the same point of an ideal saturated clay. Taking this as a "law", observed deviations of the measured pairs (c_v,q_c) from being situated on the same straight line through the origin in the (c_v,q_c) coordinate system must be attributed to measuring uncertainty. In fact, large deviations are observed in practice because in the pair (c_v,q_c) assigned to a given point the value of q_c, say, must be obtained by interpolation between values of q_c measured at other points of the soil body. Acceptance of the proportionality law makes it possible by the pairing method to estimate the proportionality constant as well as the measuring uncertainty of both c_v and q_c in terms of probability distributions. This leads to the transformation of the random field of "measured" q_c values into the random field of "true" undrained shear strength values. The developed technique has been successfully applied to an extensive set of filtered CPT (cone penetration test) cone tip resistance measurements made in the Storebælt clay till in Denmark. This investigation has served the preparation of the anchor block design for the suspension bridge presently under construction.

Introduction

The problem treated in this paper is a special case of the following general measuring uncertainty evaluation problem: A sample is drawn from some unknown population Ω (object population) and each element of the sample is characterized by a measured value. The measurement procedure is assumed to be less than perfect. On each measured value it introduces an error drawn from some unknown population M_1. Without knowing anything about population M_1 it is clearly not possible on the basis of the obtained sample of values to infer anything about the properties of population Ω. However, the situation is different if each element of the sample from Ω also is characterized by a measured value obtained by use of another independent measuring method with error population M_2. It is shown in the next section that if both measuring methods besides being independent are such that the two mean errors for a given object are independent of the error–free value that should be assigned to the object then it is possible to estimate the variances of each of the three value populations corresponding to Ω, M_1, M_2 on the basis of the sample of pairs of measured values. In order to obtain estimates of the mean values of the three value populations it is necessary to assume that at least one of the measuring methods deliver unbiased measurements, that is, that the mean error is zero.

The principle of using two inaccurate measuring methods on the same sample of objects from Ω to obtain estimates of population parameters that characterize Ω may be relevant for developing measuring procedures by which large samples of pairs of observations can be obtained at low costs. This may turn out to be particularly relevant in connection with data collection for reliability evaluation of existing structures.

Under certain conditions the principles of the method are even applicable in cases of destructive testing. This is the situation in this paper. Assume that some material property varies in space as a random field. By the process of measuring the property at a point the material is changed irreversibly or even destructed within a certain neighbourhood of the point. Therefore the property cannot be remeasured by some physical measuring device applied at the same point. However, if the first measurement method is applied for one set of points and the second measurement method is applied for another set of points these two sets of measurements can in a certain sense be paired by use of a stochastic interpolation procedure (kriging) based on the random field model of the property variation. At each point of measurement by the first method a measured value is paired with a probability distribution derived by stochastic interpolation between the measured values obtained by the second method. By suitable assumptions about the joint probabilistic structure of the measurement error fields and the "true" property field it is then possible to pull out the "true" field from the measured field. These assumptions are generalizations of the simple independence type assumptions made in the elementary pairing method described in the next section.

Pairing method

Let a random quantity Z be defined on the object population Ω and let it be measured by two independent measuring methods giving the measures X_1 and X_2 respectively. Then the measuring errors are Y_1 and Y_2 and $Z = X_1 + Y_1 = X_2 + Y_2$. The conditional covariance between Y_1, Y_2 given Z is zero, that is, $Cov[Y_1,Y_2|Z] = 0$, which implies that $Cov[X_1,X_2|Z] = Cov[Z-Y_1,Z-Y_2|Z] = 0$.

Under the assumption that the conditional means $E[Y_1|Z]$ and $E[Y_2|Z]$ do not depend on Z, it then follows from the total representation theorem (Ditlevsen (1981), p. 56: $Cov[X_1,X_2] = E[Cov[X_1,X_2|Z]] + Cov[E[X_1|Z],E[X_2|Z]] = E[0] + Cov[Z,Z]$) that

$$Cov[X_1,X_2] = Var[Z] \tag{1}$$

Thus the variance of the "true" quantity Z can be estimated by estimating the covariance between the results of the two measuring methods. Since $Cov[X_1,Y_1] = E[Cov[X_1,Y_1|Z]] = E[Cov[X_1,Z-X_1|Z] = E[-Var[X_1|Z]] = -Var[X_1] + Var[E[X_1|Z]] = -Var[X_1] + Var[E[Z-Y_1|Z]] = - Var[X_1] + Var[Z]$ it follows from $Var[Z] = Var[X_1] + Var[Y_1] + 2Cov[X_1,Y_1]$ that $Var[Z] = Var[X_1] - Var[Y_1]$. Thus the variance of the measuring error Y_1 by use of (1) is obtained as

$$Var[Y_1] = Var[X_1] - Cov[X_1,X_2] \tag{2}$$

By symmetry, $Var[Y_2]$ is obtained by interchanging X_1 and X_2.

The following generalizes this idea of pairing the results of two different inaccurate measuring methods and never the less obtain accurate information about the variability of the actual random object. The idea is specifically extended to a situation where the objects are the values of a random field realization. Herein the object population is represented by the "true" field of undrained shear strengths in a saturated clay. Two different inaccurate measuring methods are used at two different sets of points of which the one, S_1, may be more dense than the other, S_2. The pairing is made by stochastic interpolation to the points of S_2 between the measured values at the points of S_1. Rather than being a number, at least one of the elements in the pair becomes a probability distribution obtained by the interpolation. Also there will be stochastic dependence between the different pairs. Herein the two measuring methods are the CPT (cone penetration test) method and the vane test method respectively.

Measurement error modeling

Let $\mathbf{Z} = \mathbf{X} + \mathbf{Y}$ be a vector of logarithms of cone tip resistances obtained by imagined perfect CPT measurements. The vector $\mathbf{X}$ corresponds to the imperfectly measured CPT values while $\mathbf{Y}$ contains the measuring errors. The CPT values refer to the points at which the vane tests are made. The CPT values are therefore not obtained directly but are results of suitable interpolations between physically measured CPT values at other points. Thus the interpolation results are considered as imperfect measurements that can be paired with the physical measurements obtained by the vane test. The logarithms of the vane test measurements are contained in the vector $\mathbf{X}_v$.

Modeling assumption 1a: $E[Y|Z]$ *is independent of* Z.

Since $Cov[Z,Y'|Z] = \mathbf{0}$ (where prime ' means "transpose") it follows by use of this assumption in the total representation theorem (Ditlevsen (1981), p. 90) that $Cov[X+Y,Y'] = Cov[E[Z|Z], E[Y|Z]'] = \mathbf{0}$ implying that

$$Cov[X,Y'] = -Cov[Y,Y'] \tag{3}$$

Substituting this into $Cov[Z,Z'] = Cov[X+Y,(X+Y)'] = Cov[X,X'] + Cov[Y,X'] + Cov[X,Y'] + Cov[Y,Y']$ gives

$$Cov[Z,Z'] = Cov[X,X'] - Cov[Y,Y'] \tag{4}$$

Modeling assumption 2: $Cov[Y,Y']$ *is proportional to* $Cov[X,X']$.

For the homogeneous case of common standard deviation σ of all elements of X we have

$$Cov[X,X'] = \sigma^2 P_X \tag{5}$$

where P_X is the correlation matrix of X. It then follows from (4) and the modeling assumption 2 that

$$Cov[Z,Z'] = \delta^2 P_X, \quad \delta^2 \leq \sigma^2 \tag{6}$$

Remark Let $Z_1,...,Z_n$ and $X_1,...,X_n$ be the elements of Z and X respectively. For the interpretation of the modeling assumption 2 it is interesting to note that it is equivalent to the assumption that the linear regression of Z_i on X depends solely on X_i and that the corresponding regression coefficient is independent of i. In fact, since $\hat{E}[Z|X] = E[Z] + Cov[Z,X']\, Cov[X,X']^{-1} (X-E[X])$ this assumption implies that $Cov[Z,X']\, Cov[X,X']^{-1} = aI$ for some constant a, where I is the unit matrix. Substituting $Z = X + Y$ and using (3) then give the equation $I - Cov[Y,Y']\, Cov[X,X']^{-1} = aI$ from which it follows that $Cov[Y,Y'] = (1-a)\, Cov[X,X']$. Since the diagonals on both sides must be non-negative it follows that $a \leq 1$. By use of (4) it follows similarly that $a \geq 0$. Setting $a = (\delta/\sigma)^2$ we obtain (6). □

Elasto–plastic continuum mechanics predicts that there is almost proportionality between the undrained shear strength c_v as measured by the vane test and the cone tip resistance q_c, both imagined to be measured at the same point of an ideal saturated clay. Taking this as a "law", there is a constant c such that for all practical purposes we have

$$Z = X_v + Y_v + ce \tag{7}$$

where $e' = [1 \dots 1]$ and where Y_v contains the measuring errors by the vane tests.

Modeling assumption 1b: $E[Y_v|Z]$ *is independent of* Z.

As in the deduction of (4) it follows from (7) that the relations obtained under the modeling assumption 1a hold with X and Y replaced by X_v and Y_v respectively. In particular we have $Cov[Z,Y_v'] = \mathbf{0}$.

Modeling assumption 3: The measuring errors in different vane tests are mutually independent and the logarithms of the measuring errors have common standard deviation γ.

This assumption says that $Cov[Y_v,Y_v'] = \gamma^2 I$ where I is the unit matrix.

Modeling assumption 4: The triple (Z,Y,Y_v) *is jointly Gaussian.*

This assumption implies that $X_v = Z - Y_v - ce$ is Gaussian with mean and covariance matrix

$$E[X_v] = E[Z] - E[Y_v] - ce \tag{8}$$

$$Cov[X_v,X_v'] = Cov[Z,Z'] + Cov[Y_v,Y_v'] \tag{9}$$

Information from measurements related to X (CPT data)

Let ζ be a Gaussian vector which together with X, Y, X_v, Y_v is jointly Gaussian. It is imagined that ζ is a vector of measurements that contains information about X given through the conditional means and covariances $E[X|\zeta]$ and $Cov[X,X'|\zeta]$. Since the two measuring methods are independent, the information contained in ζ and obtained solely by the CPT method carries no information about the outcome of the measuring error vector Y_v related to the vane test. Therefore all covariances between elements of Y_v and elements of ζ are zero. From the joint Gaussianity and $Cov[Z,Y_v'] = 0$ (modeling assumption 1b) it then follows by use of the linear regression theory apparatus that $Cov[Z,Y_v'|\zeta] = \mathbf{0}$. Moreover it follows that $Cov[Y_v,Y_v'|\zeta] =$

$Cov[Y_v, Y_v']$. Using the information contained in ζ, the mean vector $E[Z]$ in (8) and the covariance matrix $Cov[Z,Z']$ in (9) should therefore simply be replaced by the corresponding conditional quantities $E[Z|\zeta]$ and $Cov[Z,Z'|\zeta]$ respectively. In order to determine these conditional means and covariances we first determine $E[Z|X]$ and $Cov[Z,Z'|X]$ by the theory of linear regression of Z on X. Noting that the homogeneity of X implies that $E[X] = E[Z] = \mu e$ (setting $E[Y] = 0$) where μ is the common mean, we have from the previous remark that $E[Z|X] = \lambda^2 X + \mu(1-\lambda^2) e$, $\lambda = \delta/\sigma$, while the residual covariance matrix is $Cov[Z,Z'|X] = Cov[Z,Z'] - Cov[Z,X'] Cov[X,X']^{-1} Cov[X,Z'] = Cov[Z,Z'] - \lambda^2 Cov[Z,Z'] = \delta^2(1-\lambda^2) P_X$. The total representation theorem next gives the results

$$E[Z|\zeta] = \lambda^2 E[X|\zeta] + \mu(1-\lambda^2) e \tag{10}$$

$$Cov[Z,Z'|\zeta] = \lambda^2 (1-\lambda^2) Cov[X,X'] + \lambda^4 Cov[X,X'|\zeta] \tag{11}$$

Thus (8) and (9) are replaced by

$$E[X_v] = E[Z|\zeta] - ce \tag{12}$$

$$Cov[X_v, X_v'] = Cov[Z,Z'|\zeta] + \gamma^2 I \tag{13}$$

setting $E[Y_v] = 0$. (If $E[Y]$ and $E[Y_v]$ are not zero, their contributions may be included in the constant c).

Information from measurements of X_v (vane test data) leading to the likelihood function of the three unknown parameters c, γ, and λ

Let x_v be the observation of X_v and write

$$\xi(c,\lambda) = x_v - E[X_v] \tag{14}$$

$$R(\gamma,\lambda) = Cov[X_v, X_v']^{-1} \tag{15}$$

Then the Gaussian density of X_v computed at x_v is

$$f_{X_v}(x_v) \propto \sqrt{\det R(\gamma,\lambda)} \cdot \exp\left[-\frac{1}{2}\xi(c,\lambda)' R(\gamma,\lambda)\, \xi(c,\lambda)\right], \quad c \in \mathbb{R},\ \gamma \in \mathbb{R}_+,\ \lambda \in [0,1] \tag{16}$$

($\propto$ means "proportional to"). The right side of (16) defines the likelihood function $L(c,\gamma,\lambda;x_v)$ of c, γ, and λ. Let (C,Γ,Λ) be the set of Bayesian random variables corresponding to (c,γ,λ), and adopt the non–informative prior for which $(C, \log\Gamma, \log(\Gamma^2+\sigma^2\Lambda^2))$ has a diffuse prior over $\mathbb{R} \times \{\log\gamma, \log(\gamma^2+\sigma^2\lambda^2) \,|\, \gamma \in \mathbb{R}_+, 0\le\lambda\le1\}$. Then the prior density of (C,Γ,Λ) is proportional to $\lambda/[\gamma(\gamma^2+\sigma^2\lambda^2)]$ and we get the posterior density $f_{C,\Gamma,\Lambda}(c,\gamma,\lambda|x_v)$ by multiplying this prior with the likelihood function.

Before the likelihood function (16) can be used to infer about the parameters c, γ, and λ there is a further step to be taken because the set of vane test undrained shear strength observations in practice often will be imperfect in the sense that some of the test results are reported not by their values but by the information that the values are larger than the measuring capacity of the applied vane (censored data). This is expressed by saying that each of the random variables representing the vane test measurements is *clipped* (or censored) at a given value x_{0i}, i = 1, ...,n. Thus the sample is given as $x_1 = x_{01}, x_2 = x_{02}, ..., x_r = x_{0r}, x_{r+1} < x_{0r+1}, ..., x_n < x_{0n}$, where the sample has been ordered such that the first r vane test shear strengths are larger than the respective measuring capacities while the remaining n − r tests are "well–behaved". For this clipped sampling case the likelihood function is obtained by integrating the joint density of X_v in (16) with respect to x_i from x_{0i} to ∞ for i = 1, ..., r. Thus the likelihood function becomes

$$L(c,\gamma,\lambda;x_v) \propto \sqrt{\det R(\gamma,\lambda)} \cdot \int_{x_{01}}^{\infty} \cdots \int_{x_{0r}}^{\infty} \exp\left[-\frac{1}{2}\xi(c,\lambda;x)' R(\gamma,\lambda)\, \xi(c,\lambda;x)\right] dx_1 \cdot \ldots \cdot dx_r \tag{17}$$

The numerical studies of the posterior density of (C,Γ,Λ) obtained from (17) after multiplication by the prior density must in practice be based on Monte Carlo integration except if r = 1 or 2.

Specific application to the clay till at Anchor Block West of the Great Belt in Denmark

The mean vectors $E[X] = \mu e$ and $E[X|\zeta]$ as well as the covariance matrices $Cov[X,X']$ and $Cov[X,X'|\zeta]$ have been obtained from the random field modeling of the logarithm of the CPT cone tip resistance q_c for the Anchor Block West area of Storebælt. The details of this modeling is reported in Ditlevsen and Gluver (1991). Interpreting the random field modeling as a pragmatic interpolation tool, Ditlevsen (1991a), the choice of model is made such that it becomes practicable to compute values of the likelihood function (17) a large number of times. Four parameters define the log q_c field. These are μ (mean of field), σ (standard deviation of field) (both used explicitly herein), ρ (vertical correlation parameter), and κ (horizontal correlation parameter). A joint Bayesian distribution of these four parameters has been determined on the basis of 60 CPT cone tip resistance profiles obtained in a rectangular mesh of 6 times 10 points in the horizontal plane with a distance of 20 m between the points in both directions of the mesh. The original cone tip logarithmic resistance profiles have been filtered and averaged in a way as described in Ditlevsen (1991b). The resulting "interpreted" logarithmic CPT profiles are over a depth of 15 m each made up of 150 values that are modeled as an outcome of a homogeneous Gaussian Markov sequence of mean μ, standard deviation σ, and correlation coefficient ρ between successive random variables. In the horizontal planes the correlation structure is modeled such that the correlation coefficient between two interpreted log q_c random variables is $\kappa^{(\mathrm{dist})^2}$ where "dist" is the distance between the two points at which the log q_c values are considered, and dist is measured in the unit of the mesh point distance 20 m.

The statistical evaluation of the parameters c,λ, and γ has been made conditional on given values of μ,σ,ρ, and κ. Due to a relatively small statistical uncertainty of these last parameters their values are first put to their maximum posterior density estimates $\mu = 0.946 + \log[\mathrm{MPa}]$, $\sigma = 0.650 + \log[\mathrm{MPa}]$, $\rho = 0.985$, $\kappa = 0.0156$, Ditlevsen and Gluver (1991).

Fig. 1 shows three sets of contour curves in full line for the posterior joint density of (Λ,Γ) given that C = 2.100, 2.255, 2.400 respectively. The intermediate value is close to the point at which the posterior density of (C,Λ,Γ) is maximal, while the two other values are in the lower and upper tail respectively of the distribution of C, see Fig. 2. The contour curves for the density as obtained by the product of the marginal distributions of Λ and Γ given C = 2.255 are shown by dotted lines. It is seen that (Λ,Γ) is only weakly dependent on C and that there is only a weak dependence between Λ and Γ. In each diagram the marginal densities of Λ and Γ given C = 2.255 are shown along the abscissa axis and ordinate axis respectively.

Fig. 2 shows four conditional density functions for C given (Λ,Γ) = (0.70,0.40), (0.90,0.30), (0.88,0.29) and (0.80,0.25) respectively. It is seen that these distributions are almost identical. This indicates that C in practice can be modeled

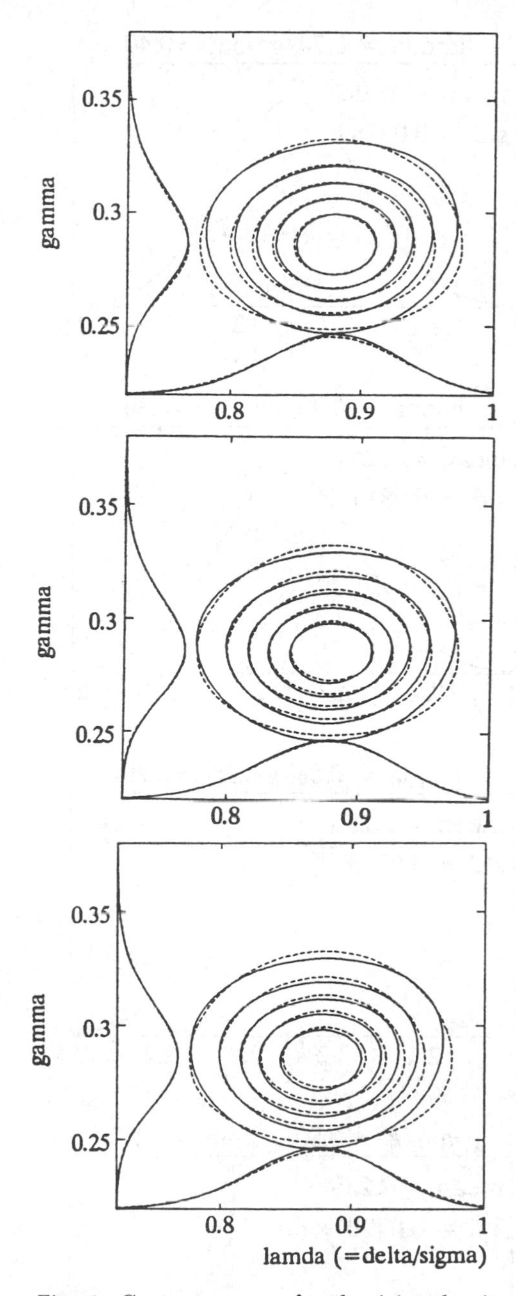

Fig. 1. Contour curves for the joint density of (Λ,Γ) *given that* C = 2.100, 2.255, 2.400 *from top down (full line). The marginal densities of* Λ *and* Γ *are shown along the abscissa axes and the ordinate axes respectively (full lines). The approximating marginal density of* Λ *is shown in the diagrams with dotted line. For* Γ *the approximating density is a truncated Gaussian density. The dotted contour curves correspond to the product of the two marginal densities corresponding to* C = 2.255. *The unit of* exp(Γ) *is* MPa.

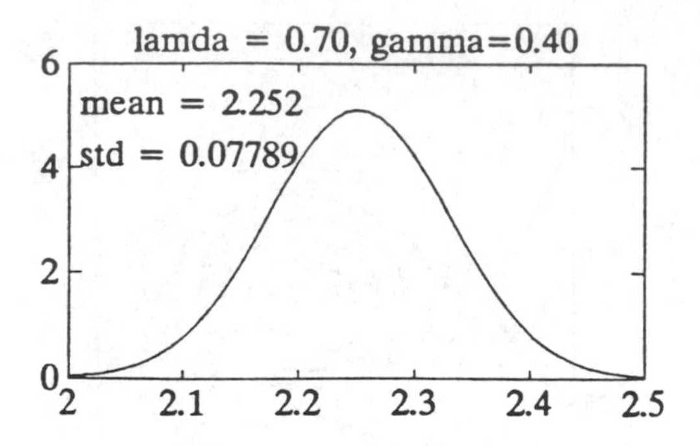

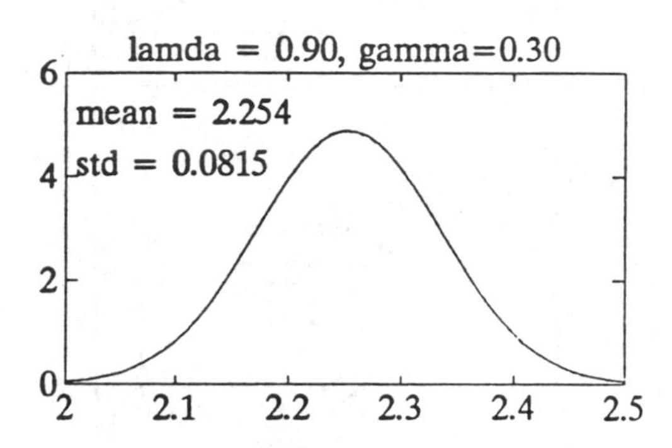

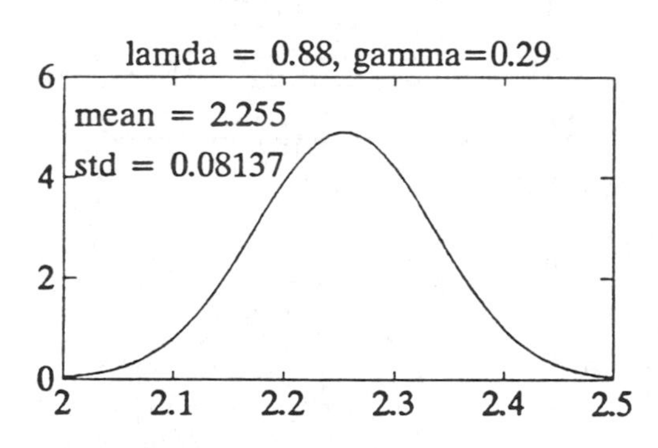

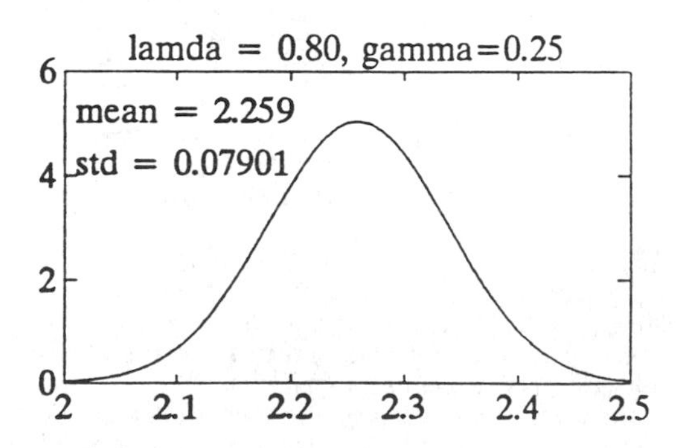

Fig. 2. Conditional density functions for C *given* (Λ,Γ). *These density functions show that* C *is almost independent of* (Λ,Γ). *The unit of* $\exp(\gamma)$ *is* MPa.

to be independent of (Λ,Γ). In conclusion it can be stated that C, Λ, and Γ are practically mutually independent for the given point estimate values of μ,σ,ρ,κ.

The analysis of the posterior density of the Bayesian field parameters M,Σ,P,K (corresponding to μ,σ,ρ,κ respectively) reported in Ditlevsen and Gluver (1991) shows that P can be put equal to a function $\psi_3(\Sigma)$ of Σ. Therefore ρ is eliminated from the numerical studies of the posterior density of (C,Λ,Γ) by substituting $\rho = \psi_3(\sigma)$. Numerical studies with variations of μ,σ,κ in the vicinity of their maximum posterior density estimates show that C depends on μ while Λ depends on σ and κ. Other dependencies are negligible.

With the present data it can be assessed that the parameter C can be modeled to have a Gaussian density with parameters $E[C|\mu] = 1.827 + 0.453\mu$ (unit of $\exp(\mu)$ is MPa), $D[C|\mu] = 0.081$. In particular, $E[C|\mu = 0.946] = 2.255$. With respect to Λ the numerical studies show that the random variable $1/(\Lambda^2+8.05)$ has a probability density that can be approximated by a truncated Gaussian density with mean $a = 10^{-2}[11.34+0.4(\sigma-0.647) + 3(\kappa-0.017)]$ (unit of $\exp(\sigma)$ is MPa), standard deviation $b = 10^{-3}[1.19 + 0.9(\sigma-0.647) - 5(\kappa-0.017)]$, and lower truncation point at $1/(1+8.05) = 0.1105$. Thus Λ^2 can be approximately represented as $[1/(a+bV)]-8.05$ where V is a truncated standard normal random variable with lower truncation point at $(0.1105-a)/b$ ($= -2.439$ for $\sigma = 0.647$, $\kappa = 0.017$) and V is independent of all other random variables in the problem. Fig. 1 (middle) shows the density function for Λ given $C = 2.255$, $\Sigma = 0.647$, $K = 0.017$ together with this approximation.

The results of this numerical study of the likelihood function form the basis for an elaborate reliability analysis of the western anchor block of the Great Belt suspension bridge taking due account of both the measuring uncertainty and the statistical uncertainty as evaluated by the method presented herein, Ditlevsen and Gluver (1991), AS Storebæltsforbindelsen (1991).

Acknowledgement This work has been financially supported by the Danish Technical Research Council and by AS Storebæltsforbindelsen.

References

AS Storebæltsforbindelsen (1991). "Anchor blocks foundation failure. In–situ cast type. Reliability study". Document No. 20000–X11–2–001. "Anchor blocks foundation failure. Caisson study. Reliability study". Document No. 20000–X11–002. Joint Venture: COWIconsult, B. Højlund Rasmussen, Rambøll & Hannemann in cooperation with O. Ditlevsen and B. Hansen.

Ditlevsen, O. (1981). "*Uncertainty Modeling*". McGraw–Hill Inc, New York.

Ditlevsen, O. and Gluver, H. (1991). "Parameter estimation and statistical uncertainty in random field representations of soil strength". Proc. CERRA–ICASP6: Sixth international conference on applications of statistics and probability in civil engineering (L. Esteva, S.E. Ruiz, eds.), Institute of Engineering, UNAM, Mexico City, 691–704.

Ditlevsen, O. (1991a). "Random field interpolation between point by point measured properties". *Computational Stochastic Mechanics* (P.D. Spanos, C.A. Brebbia, eds.), Computational Mechanics Publications, Southampton, Boston, Elsevier Applied Science, London, New York, 801–812.

Ditlevsen, O. (1991b). "Partitioning of Storebælt clay till cone tip resistance records in population components". Report to AS Storebæltsforbindelsen, Copenhagen, Denmark, 1991.

Structural Safety & Reliability, Schuëller, Shinozuka & Yao (eds) © 1994 Balkema, Rotterdam, ISBN 90 5410 357 4

Quality assurance of geomembrane liners

R.B.Gilbert
University of Texas, Austin, Tex., USA

W.H.Tang
University of Illinois, Urbana, Ill., USA

ABSTRACT: A probabilistic methodology is developed to evaluate the quality of installed geomembrane liners. Statistics are derived for the fraction of incompetent seams, and the size and frequency of defects in the liner after implementation of a standard construction quality assurance (CQA) program including destructive seam strength and non-destructive seam continuity testing. CQA results from case history projects are presented to gain insight into these parameters for typical installations. Finally, a reliability based approach is proposed for designing CQA programs to achieve desired levels of performance reliability in the final product.

1 INTRODUCTION

The environmental impact of a waste containment facility depends on the ability to minimize leakage from the facility. Lining systems containing soil and geosynthetic components are placed between the waste and the environment to minimize leakage. The state of the art in lining system design includes the use of geomembrane liners (e.g. high density polyethylene or polyvinyl chloride) either alone or overlying compacted clay. In the United States, geomembranes are required by regulations for both solid and hazardous waste landfills. The total rate of leakage through a geomembrane liner is related to the size and frequency of holes in the geomembrane, however very little information is currently available to predict the size and frequency of holes.

Holes in the geomembrane are primarily due to inadequate seaming but may also be a result of punctures and tears. During installation, a construction quality assurance (CQA) program is implemented to minimize the frequency and size of geomembrane holes. A typical CQA program will consist of destructive seam testing to evaluate the quality of seams, non-destructive seam testing to evaluate the continuity of seams, and visual inspection for defects. All inadequate seams and defects that are detected during installation are repaired. A typical as-built geomembrane with destructive test locations, reconstructed seam lengths and repair locations is shown on Figure 1.

It is of interest to provide answers to the following questions: "What is the reliability of the final product and how do uncertainties affect the reliability?" and "How can we design a CQA program to control the reliability?" In this paper, probabilistic procedures are developed to evaluate the statistics for 1) the fraction of incompetent seam length and 2) the frequency and size of defects. Data from case study geomembrane installations are analyzed and presented. A reliability based approach is proposed for designing CQA programs to achieve specified levels of performance reliability.

2 DESTRUCTIVE SEAM TESTING

During installation, samples are removed at a specified interval from seams and subjected to strength testing. At each location of a failing destructive test (i.e. inadequate seam strength), the length of incompetent seam is determined through additional testing on either side of the initial test location. Once the length of seam containing the failing test sample is bound by passing test locations, the incompetent seam length is reconstructed (see Figure 1). Documented destructive testing results from 26 geomembrane installations are presented in Table 1. These results include the total length of seam (column 2), the number of destructive

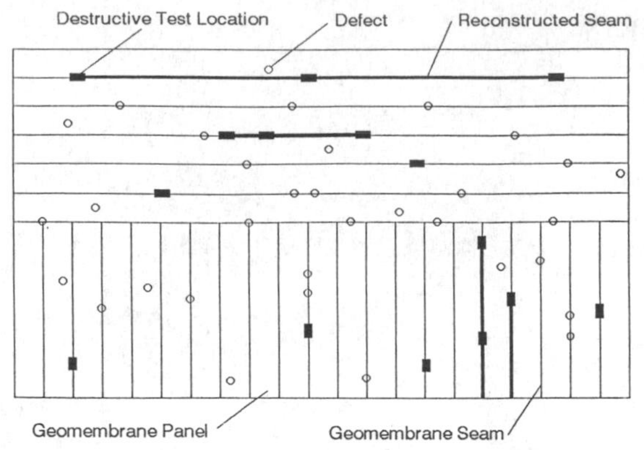

Figure 1 Schematic of As-Built Geomembrane Liner

Table 1 Seam Destructive Testing Results for Case Study Geomembranes

							Fraction of Incompetent Seam					
							Prior to Repair				After Repair	
Cell	Seam Length (m)	Tests	Average Test Interval (m)	Failing Tests	Total Repaired Length (m)	Average Repaired Length (m)	Mean	Random c.o.v.	Stat. c.o.v.	Total c.o.v.	Mean	c.o.v.
(1)	(2)	(3)	(4)	(5)	(6)	(7)	(8)	(9)	(10)	(11)	(12)	(13)
1-P	5306	50	106	7	253	36	0.160	0.225	0.321	0.392	0.112	0.558
1-S	5306	51	104	13	253	19	0.275	0.112	0.225	0.252	0.227	0.305
2-P	5708	59	97	17	776	46	0.305	0.152	0.195	0.247	0.169	0.446
2-S	5708	57	100	5	160	32	0.105	0.262	0.383	0.464	0.077	0.632
3-P	7303	68	107	7	156	22	0.118	0.185	0.330	0.378	0.096	0.462
3-S	7303	66	111	2	58	29	0.045	0.320	0.560	0.645	0.038	0.781
4-P	13162	149	88	46	1548	34	0.315	0.086	0.120	0.148	0.198	0.235
4-S	13162	144	91	30	1101	37	0.215	0.123	0.159	0.201	0.132	0.329
5-P	2221	26	85	1	29	29	0.077	0.365	0.667	0.760	0.064	0.915
5-S	2221	28	79	2	44	22	0.107	0.303	0.536	0.616	0.087	0.756
6-P	2221	24	93	2	22	11	0.125	0.192	0.529	0.563	0.115	0.610
6-S	2221	26	85	6	120	20	0.269	0.170	0.317	0.360	0.215	0.450
7/8-P	3934	33	119	2	36	18	0.091	0.229	0.542	0.589	0.082	0.655
7/8-S	3934	37	106	6	54	9	0.189	0.114	0.336	0.355	0.175	0.382
9/10-P	3249	29	112	0			0.034	0.000	0.966	0.966	0.034	0.966
9/10-S	3249	32	102	2	23	11	0.094	0.197	0.541	0.576	0.087	0.623
11/12-P	5129	46	111	6	377	63	0.152	0.307	0.344	0.461	0.079	0.892
11/12-S	5129	53	97	18	997	55	0.358	0.151	0.182	0.236	0.164	0.517
13-P	5917	88	67	6	120	20	0.080	0.246	0.361	0.436	0.059	0.587
13-S	5917	80	74	2	20	10	0.038	0.232	0.563	0.609	0.034	0.669
14-P	6185	62	100	3	21	7	0.065	0.151	0.480	0.503	0.061	0.531
14-S	6185	71	87	3	31	10	0.056	0.195	0.482	0.520	0.051	0.570
15-P	6170	56	110	3	104	35	0.071	0.313	0.478	0.571	0.055	0.747
15-S	6170	49	126	0			0.020	0.000	0.980	0.980	0.020	0.980
16-P	5678	48	118	1	10	10	0.042	0.191	0.685	0.711	0.040	0.742
16-S	5678	48	118	6	52	9	0.146	0.112	0.346	0.363	0.137	0.387

tests (column 3), the number of failing tests (column 5), and the length of reconstructed seam (column 6). In addition, the average testing interval (column 4) and the average length of reconstructed seam per failing test (column 7) are calculated and compiled in Table 1.

A random two-state occurrence process described in Tang, et al (1989) is used to model the occurrence of competent and incompetent seam lengths in the geomembrane seam. Suppose state 1 represents competent seam and state 2 represents incompetent seam. The duration of each occurrence of a given state is assumed to be exponentially distributed with mean values $1/\mu_1$ and $1/\mu_2$ for states 1 and 2, respectively. The distributions of measured state 2 durations were analyzed for the case study liners, and the exponential distribution hypothesis was not rejected at the 10 percent significance level. Since μ_1 and μ_2 are not known, the destructive test results will be used to evaluate these parameters. The joint distribution of μ_1 and μ_2 can be obtained using a Bayesian approach [see Gilbert and Tang (1993)] based on 1) the number of failing test results relative to the number of passing test results, and 2) the total length of state 2 segments encountered.

After completing the destructive testing program, we are interested in evaluating the fraction of incompetent seams, p, in the installation. The statistics of the average fraction of state 2 over a length of seam L, p(L), are derived in Tang, et al (1989) for known values of μ_1 and μ_2. Since μ_1 and μ_2 here are represented by a distribution of possible values, the statistics of p(L) will also be represented by distributions. The expected value of the variance in p(L) accounts for the random spatial variability of p along the length of the seam, and it is inversely related to L due to averaging effects. In addition, the variance in the mean of p(L) accounts for the statistical uncertainty in evaluating p(L) based on a limited number of tests, and it is inversely related to the number of destructive tests. The calculated statistics of p(L) are summarized in columns 8 through 11 of Table 1 for the case study liners. Generally, the magnitude of the random variability (column 9) is about one-half the magnitude of the statistical uncertainty (column 10) for these cases. If an individual seam segment on the order of 100 m were considered instead of the entire seam length in a given project, the magnitude of the random variability would be comparable with that of the statistical uncertainty.

Once incompetent seam segments are detected through destructive test results and their lengths are determined, the seam is repaired. Ultimately, we are interested in the fraction of incompetent seam lengths in the installation after repair. The statistics of this fraction, denoted by p*, can be estimated from those for p and the length of reconstructed seam. The mean fraction is essentially reduced by the fraction of seam length that is reconstructed, while the total variability in the fraction remains constant. The resulting statistics for the case study liners are presented in columns 12 and 13 of Table 1. On average, the CQA destructive testing programs resulted in a reduction in the mean fraction of incompetent seam from 0.183 before repair to 0.138 after repair, or roughly a 25 percent reduction. The effect that the testing interval τ has on this reduction can be evaluated from the data presented in Table 1. For example, τ averages 111 m for Cell 3-S yielding a reduction in the mean fraction of 16 percent, while it averages 88 m for Cell 4-P yielding a 37 percent reduction. These data confirm that the ability of the destructive testing program to minimize the remaining fraction of incompetent seam lengths after repair depends on the testing interval; therefore, we will be able to control the effectiveness of the CQA program by specifying τ.

3 NON-DESTRUCTIVE TESTING AND INSPECTION

Geomembrane seams are non-destructively tested for continuity over their entire length. Non-destructive testing methods typically consist of a vacuum device to detect leaks in extrusion welds and an air pressure method to detect leaks in fusion welds; although, other methods such as ultrasound and electrical resistivity are also available but less widely used. In addition, geomembrane seams are visually inspected for defects such as holes and tears. All detected defects are repaired with geomembrane patches or with welding extrudate. The location, size and type of repairs are documented (e.g. see Figure 1) in most cases.

The probability of detecting a defect during non-destructive testing and visual inspection is related to two considerations: 1) the ability to detect a defect as a function of the size of the defect, and 2) the distribution of defect sizes in the installation. The distribution of defect sizes detected during installation represents a biased sample of actual defect sizes because we are more likely to detect large defects than smaller ones. The relationship between the probability of detection and the size of the defect is given by the detectability function. The distribution of actual defect sizes can be inferred from the distribution of detected defect sizes using a Bayesian approach to filter it through the detectability function. An approach developed by Tang (1973) for analyzing defects in steel welds is extended to evaluate the statistics of defect size and frequency.

Very little information is available concerning the detectability function for non-destructive testing and inspection methods used during geomembrane installation. Conceptually, there will be some defect size for which the defect will be detected nearly 100 percent of the time. This size will be denoted d_{100}. Further, as the defect size approaches zero, the detection probability will approach zero. Between these two limits, the shape of the detectability function could conceivably take on any form. A simple power function is proposed to describe the shape of the detectability function, as shown on Figure 2. The parameter d_{50} represents the defect size for which there is an equally likely chance of detection or non-detection, while the shape factor m is a constant controlling the shape of the curve. The magnitude of d_{50} will be related to the type and level of non-destructive testing and visual inspection performed. For example, Giroud and Bonaparte (1989) contend that the size of defects remaining after a typical CQA program (i.e. vacuum box or air pressure testing) is on the order of 1 to 5 mm based on interviews with CQA personnel. In that case, d_{50} is likely to be within that range of values.

The distribution of detected defect sizes is not typically reported but can be estimated based on the distribution of documented repair sizes where this information is available. Repair size distributions have been compiled for the case study liners and can generally be approximated by lognormal distributions with coefficient of variation (c.o.v.) values ranging from 0.8 to 0.85. Defect sizes will actually be smaller than repair sizes. In most projects, the minimum allowable repair size must extend at least 0.15 m beyond the defect in all directions. Based on practical considerations, the maximum possible defect size beneath a repair with a diameter of 0.30 m is estimated to be 25 mm; therefore, the probability that a detected defect is less than or equal to 25 mm in diameter, $P(d_1 \leq 25mm)$, can be estimated by the proportion of repair sizes less than or equal to 0.30 m. If we further assume that the distribution of detected defect size is also lognormal with a c.o.v. value of about 0.8, we can obtain the distribution of detected defect sizes, $f_1(d)$, using $P(d_1 \leq 25mm)$. Finally, the initial un-biased distribution of defect sizes, $f_0(d)$, may then be evaluated by filtering $f_1(d)$ through the detectability function, and the probability of detection is obtained by integrating $f_0(d)$ over the detectability function.

Seam non-destructive testing data for the case study geomembranes are compiled in Table 2. The total seam length is presented in column 2, while the number of detected defects are presented in column 3. The frequency of detected defects and $P(d_1 \leq 25mm)$ are compiled in columns 4 and 5, respectively. Using an assumed d_{50} value of 10 mm and m equal to 0.5 (Figure 2), the detection probabilities for the case study liners have been calculated and are presented in column 6 of Table 2.

The updated distribution of remaining defect sizes, $f(d^*)$, is obtained from $f_0(d)$ using a Bayesian approach and assuming that all detected defects are repaired. The calculated mean value for defect size after repair, $E(d^*)$, is presented in column 7 of Table 2 for the case study liners. Very little information concerning the actual size of defects in the final product after CQA is available to compare with these estimated mean values. Giroud and Bonaparte (1989) estimate that the average defect size after CQA ranges from 2 to 10 mm. Laine (1991) analyzed several defects that were detected in a completed installation using electrical resistivity subsequent to a standard CQA program. These defects were a variety of shapes and sizes, but their approximate diameters were on the order of 1 mm. Based on this limited information, the estimated expected defect sizes given in Table 2 are slightly conservative, but reasonable.

The uncertainty in d* has also been calculated for the case study liners. First, there is uncertainty in d* due to spatial variability. The total rate of leakage through the geomembrane will be related to the spatial average defect size. Hence, the inherent variability in the spatial average, not that for a single defect, is considered. This spatial average variability is estimated as the variability for a single defect divided by the total number of defects remaining after repair. The expected number of defects

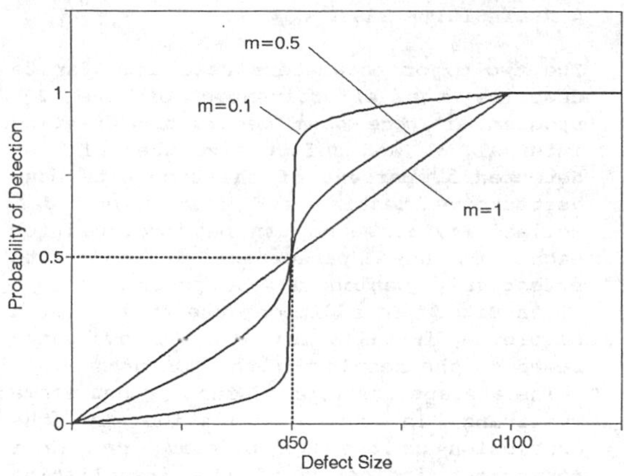

Figure 2 Non-Destructive Testing Detectability Function

Table 2 Seam Non-Destructive Testing Results for Case Study Geomembranes

Cell	Seam Length (m)	Detect. Defects	Detect. Freq. (1/m)	Detect. Size Fraction <25 mm	Detect. Prob.	Defect Size d*			Defect Frequency y*		
						Mean (mm)	Spatial c.o.v.	Total c.o.v.	Mean (1/m)	Spatial c.o.v.	Total c.o.v.
(1)	(2)	(3)	(4)	(5)	(6)	(7)	(8)	(9)	(10)	(11)	(12)
2-P	5708	316	5.54E-02	NA	0.912	8.62	0.0711	0.256	5.36E-03	0.189	0.865
2-S	5708	341	5.97E-02	NA	0.912	8.62	0.0684	0.255	5.78E-03	0.182	0.864
3-P	7303	372	5.09E-02	0.155	0.959	9.26	0.0949	0.263	2.18E-03	0.256	0.882
3-S	7303	337	4.61E-02	0.104	0.979	9.74	0.137	0.281	9.93E-04	0.375	0.924
4-P	13162	1057	8.03E-02	0.143	0.965	9.37	0.0609	0.253	2.92E-03	0.164	0.860
4-S	13162	919	6.98E-02	0.102	0.979	9.75	0.0828	0.259	1.50E-03	0.228	0.874
5-P	2221	162	7.30E-02	0.307	0.076	0.00	0.083	0.260	1.04E-02	0.222	0.873
5-S	2221	185	8.33E-02	0.382	0.821	7.92	0.0644	0.254	1.83E-02	0.173	0.862
6-P	2221	87	3.92E-02	0.364	0.835	8.01	0.0976	0.264	7.83E-03	0.262	0.884
6-S	2221	159	7.16E-02	0.375	0.826	7.95	0.0704	0.256	1.52E-02	0.190	0.865
7/8-P	3934	208	5.29E-02	0.258	0.907	8.57	0.0851	0.260	5.45E-03	0.227	0.874
7/8-S	3934	272	6.91E-02	0.227	0.925	8.76	0.0829	0.259	5.63E-03	0.221	0.873
9/10-P	3249	228	7.02E-02	0.192	0.943	8.99	0.103	0.267	4.26E-03	0.277	0.888
9/10-S	3249	206	6.34E-02	0.251	0.911	8.61	0.0874	0.261	6.22E-03	0.233	0.876
11/12-P	5129	208	4.06E-02	NA	0.912	8.62	0.0875	0.261	3.93E-03	0.233	0.876
11/12-S	5129	242	4.72E-02	NA	0.912	8.62	0.0812	0.259	4.57E-03	0.216	0.871
13-P	5917	370	6.25E-02	0.394	0.812	7.86	0.0443	0.250	1.45E-02	0.120	0.853
13-S	5917	291	4.92E-02	0.366	0.833	8.00	0.0533	0.252	9.89E-03	0.143	0.856
14-P	6185	285	4.61E-02	0.129	0.970	9.48	0.126	0.276	1.43E-03	0.341	0.911
14-S	6185	248	4.01E-02	0.237	0.919	8.70	0.083	0.260	3.55E-03	0.223	0.873
15-P	6170	183	2.97E-02	0.168	0.953	9.16	0.126	0.276	1.47E-03	0.340	0.910
15-S	6170	168	2.72E-02	0.243	0.916	8.67	0.0996	0.265	2.51E-03	0.265	0.885
16-P	5678	154	2.71E-02	0.181	0.948	9.07	0.131	0.279	1.50E-03	0.352	0.915
16-S	5678	172	3.03E-02	0.428	0.783	7.69	0.0599	0.253	8.44E-03	0.163	0.860

after repair is compiled in column 10 of Table 2, and will be discussed subsequently. The resulting spatial average c.o.v. values are presented in column 8 of Table 2. Second, there will be systematic uncertainty in the expected value of d* due to uncertainty in the form of the detectability function (i.e. in d_{50} and m). Assuming that a c.o.v. value of 0.25 accounts for the uncertainty in d_{50} and that m is uniformly distributed between 0 and 1, this systematic uncertainty has been evaluated using a first order Taylor series approximation to be about 0.24. The expected value of d* is much more sensitive to d_{50} than to m. This systematic uncertainty is combined with the spatial variability to obtain the total c.o.v. in d*, and these total c.o.v. values are compiled in column 9 of Table 2.

Consider next the frequency of defects in the installation. The total number of defects in a given site is an unknown value N, which can be inferred using a Bayesian procedure (Gilbert, 1993) based on the number of defects detected (column 3 in Table 2) and the corresponding detection probability (column 6). All defects that are detected will be repaired, and the distribution of the number of defects remaining after repair, f(N*), will be the distribution of N adjusted by the number of detected defects. The resulting frequency of defects after repair, y*, is given by N*/L where L is the total length of seam (column 2).

The statistics of y* are summarized in columns 10 through 12 of Table 2 for the case study liners. The range of calculated mean values compares favorably with typical frequencies reported in the literature. Giroud and Bonaparte (1989) report a typical seam defect frequency of $3.3x10^{-3}$ defects per m after CQA based on forensic analyses for a limited number of case histories. In addition, Darilek, et al (1989) estimate the seam defect frequency to range from $4.5x10^{-3}$ to $1.4x10^{-2}$ defects per m based on electrical resistivity surveys.

The spatial average c.o.v. values (column 11) for the defect frequency are greater than those for the defect size. In addition, the systematic uncertainty in y* due to uncertainty in d_{50} and m is about 0.84, or more than 3 times that for d*. Therefore, E(y*) is more sensitive than E(d*) to the form of the detectability function. Finally, the total c.o.v. values for y* are compiled in column 12 of Table 2 for the case study liners.

4 RELIABILITY BASED CQA

The two major parameters used thus far to describe the effectiveness of the CQA program are the seam destructive testing interval, τ, and defect size that will be detected 50 percent of the time with non-destructive testing and inspection, d_{50}. Reliability measures can be developed for each of these parameters in order to effectively design a CQA program for a given site. In addition, the quality of a completed installation can be evaluated based on the results of the CQA program.

The average fraction of incompetent seams remaining in an installation at the conclusion of the CQA program, p*, is a convenient indicator of the installation quality. A reliability measure of interest for p* is the probability that p* exceeds a target level, say 10 percent. In Figure 3, this probability is plotted as a function of τ for three different values of the fraction of failures observed during destructive testing. A lognormal distribution was assumed for p* to calculate the exceedance probability. The curves shown can be used to determine the destructive testing interval based on a target level of reliability. For example, if a target failure probability of 25 percent is selected, then the required destructive testing intervals would be 150 m for an observed failure fraction of 0.1, 65 m for 0.2, and 50 m for 0.3. In addition, the quality or reliability of a completed installation at some intermediate stage of installation can also be evaluated. For example, if an average testing interval of 100 m were used and the frequency of failures was running at approximately 0.25, then the probability that p* exceeds 10 percent would be equal to approximately 0.6. If a target failure probability of 0.25 were specified, then the testing frequency should be increased accordingly for the remainder of the installation to achieve this level.

In addition to the fraction of incompetent seam, both the size and frequency of defects in the final product also serve as useful indicators of quality. Plots similar to Figure 3 could be developed for the size and frequency of defects as a function of d_{50}. Unlike the destructive testing interval, however, d_{50} is not as easily quantified. More research is required to develop the non-destructive testing detectability functions in detail for different non-destructive testing methods.

Ultimately, the CQA program should be designed relative to an allowable leakage

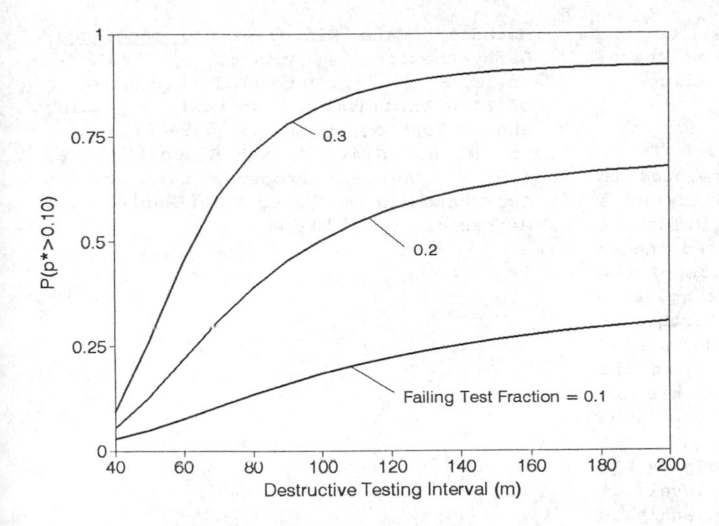

Figure 3 Incompetent Seam Fraction Reliability Versus Destructive Testing Interval

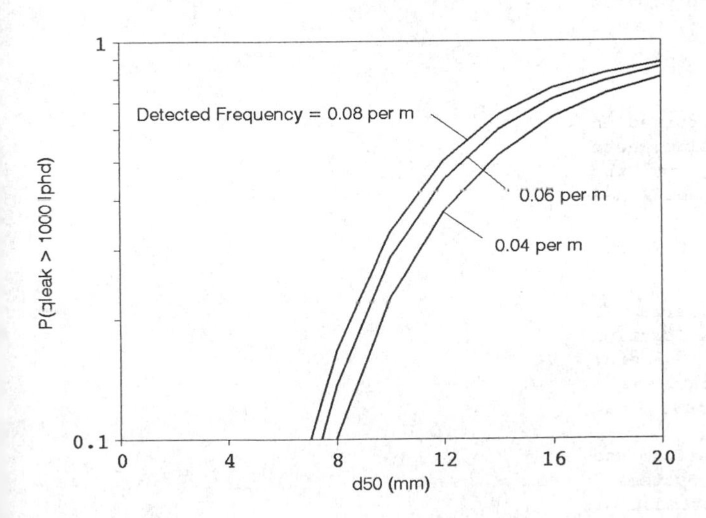

Figure 4 Liner Leakage Reliability Versus Non-Destructive Testing Resolution

rate for the completed liner. The probability that the leakage rate exceeds a target value can be plotted as a function of d_{50} in order to design a CQA program to achieve a specified level of reliability. As an example, the failure probability for a target leakage rate of 1000 lphd (liters per hectare per day) is plotted as a function of d_{50} on Figure 4 for a given head of fluid on the geomembrane liner. Curves are presented for three different detected defect frequencies. If a target failure probability of 0.5 is desired, the CQA program specifying a non-destructive testing method with a d_{50} value of 12 mm will be acceptable if the detected defect frequency is 0.08 per m or less. This plot also shows how the quality of the initial installation before non-destructive testing will affect the reliability. For a given d_{50} of 10 mm, the failure probability increases from 0.25 to 0.4 if the frequency of detected defects increases from 0.04 per m to 0.08 per m (Figure 4). This detected frequency will be related to the conditions under which the geomembrane is installed (i.e. temperature, wind, etc.), to the method used to create geomembrane seams, and to the capability of the installer.

5 CONCLUSIONS

A probabilistic approach is developed to evaluate the quality or reliability of a geomembrane liner installation after a standard CQA program. The required inputs to obtain the pertinent statistics of the remaining fraction of incompetent seam length, defect size and defect frequency consist of data gathered and documented during a routine installation. With the proposed procedure, it is possible to evaluate quality of a completed installation based on site specific CQA results, and alternatively to design a CQA program to achieve a target level of performance reliability. Finally, analysis of field data from a number of sites show wide variation in the quality of the completed installation. A reliability based procedure can be expected to provide a more consistent CQA program.

ACKNOWLEDGEMENTS

The research reported here is supported in part by the Illinois Office of Solid Waste Research (Project # OSWR-09-002) and the National Science Foundation (Project # MSS 92-04433).

REFERENCES

Darilek, G. T., Laine, D. L. and Parra, J. O. (1989), "The Electrical Leak Location Method for Geomembrane Liners: Development and Applications," Proceedings, Geosynthetics '89, San Diego, pp. 456-466.

Gilbert, R. B. (1993), "Performance Reliability of Landfill Lining Systems," Ph.D. Thesis submitted to Department of Civil Engineering, University of Illinois at Urbana-Champaign.

Gilbert, R. B. and Tang, W. H. (1993), "Reliability Based Quality Assurance for Waste Containment Liners," Research Report, Department of Civil Engineering, University of Illinois at Urbana-Champaign.

Giroud, J. P. and Bonaparte, R. (1989), "Leakage Rates Through Liners Constructed with Geomembranes - Part I. Geomembranes and Part II. Composite Liners," Geotextiles and Geomembranes, 8: 1 and 2, pp. 27-67 and 78-111.

Laine, D. L. (1991), "Analysis of Pinhole Seam Leaks Located in Geomembrane Liners Using the Electrical Leak Location Method: Case Histories," Proceedings, Geosynthetics '91, Atlanta, pp. 233-253.

Tang, W. H. (1973), "Probabilistic Updating of Flaw Information," Journal of Testing and Evaluation, 1: 6, pp. 459-467.

Tang, W. H., Sidi, I. and Gilbert, R. B. (1989), "Average Property in a Random Two-State Medium," Journal of Engineering Mechanics, pp. 131-144.

Structural Safety & Reliability, Schuëller, Shinozuka & Yao (eds) © *1994 Balkema, Rotterdam, ISBN 90 5410 357 4*

Safety control of arch dams by using possibility regression models

Ichiro Kobayashi & Ryoji Miike
Kumamoto University, Japan

ABSTRACT: The safety control of arch dams can be carried out using three control chart methods. The possibility regression models are applied on the basis of the fuzzy set theory. The first two methods, which follow the theory of interval estimation in multivariate regression models, are a kind of residual analysis. In the third method the difference between the estimates of fuzzy regression analysis for the preliminary data and for the moving controlled data is examined. Several numerical examples are illustrated showing how these control chart methods could be applied to the maintenance and safety control of arch dams.

1 INTRODUCTION

Because dam failure causes catastrophic loss of life and damage to property, many papers have been presented on the evaluation of dam safety (Giampoli 1980).

The safety of arch dams can be estimated by using the control chart method which has been presented by the authors (Miike and Kobayashi 1985, 1991). This uses the multivariate regression model (i.e. the probability model); its control limits can be decided by assuming that random errors are distributed normally in the regression model. Even if random errors are not distributed normally, any response variables still tend to the normal distribution due to the central limit theorem, provided the sample size is sufficiently large. But if the sample size is not so large, we cannot completely accept the assumption about the normality of random errors (Miike and Kobayashi 1989). Furthermore, those who monitor the behavior of arch dams find the evaluation of data through statistical analysis, using the theorem about the normal distribution, studentized residuals and other statistical problems very difficult to understand.

To overcome these difficulties by using probability models, this paper presents three control chart methods using fuzzy multivariate regression analysis (i.e. the possibility model) for the safety control of arch dams. In possibility regression models, it

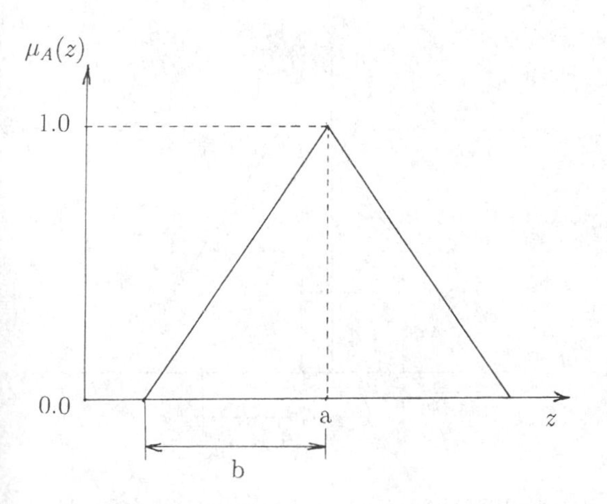

Fig.1 Symmetrical fuzzy number of $A = (a,\ b)$

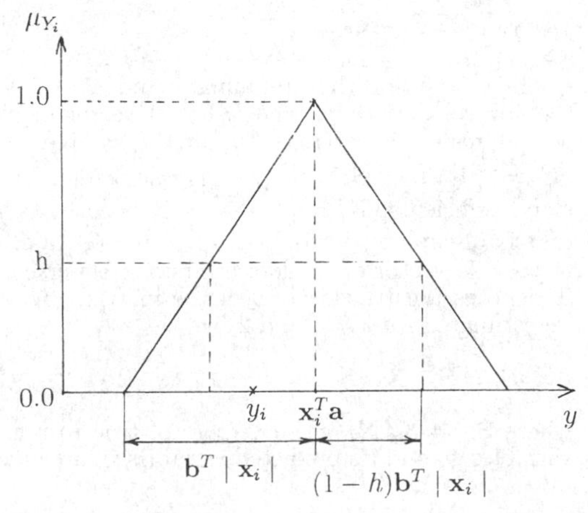

Fig.2 Membership function of Y_i

is assumed that the gap between the data and the estimated value in the regression model is due to an ambiguity in the system. The results of numerical examples show that the proposed control chart methods are not only simple to handle, but also very practical for the maintenance and safety control of arch dams.

2 THEORIES

The aim of this paper is to apply the control method to the safety control and maintenance of arch dams, whose behavior can be expressed by the regression model. The authors have proposed three control chart methods to find the variation of the regression model. The first group of two methods is based on the theory of interval estimation (Control charts I and II) and the third method on the theory of the test of the difference between multivariate regression models (Control chart III).

Control chart methods are based on the assumption that the parameters are usual in the control period, if the control parameters are inside both control limits in the control chart and, in addition, the parameters were inside the control limits in the preliminary period, too. The parameters outside the control limits are called "unusual" ones.

By using the non-dimensional forms of parameter in this paper, control charts II and III are obtained with a pair of control limits parallel to the time axis of the abscissa as well as the ordinary quality control chart method.

2.1 Probability regression model

Suppose two sets of multivariate regression models in matrix notation (Cramer 1951, Draper and Smith 1967), for example, the one being for the preliminary period before the earthquake and the other for the control period after the earthquake, as follows:

$$\mathbf{y}_\ell = \mathbf{X}_\ell \boldsymbol{\beta}_\ell + \mathbf{e}_\ell \tag{1}$$

where $\ell = 1$ for the preliminary and $\ell = 2$ for the control period; $\mathbf{y}_\ell$ = a $(n_\ell \times 1)$ vector of observed response variables; n_ℓ = the sample size; $\boldsymbol{\beta}_\ell^T = [\beta_0\ \overline{\boldsymbol{\beta}}_\ell^T]$; $\overline{\boldsymbol{\beta}}_\ell$ = a $(p \times 1)$ vector of regression coefficients; $\mathbf{X}_\ell = [\ \mathbf{I}_{n\ell}\ \overline{\mathbf{X}}_\ell\ ]$; $\overline{\mathbf{X}}_\ell$ = a $(n_\ell \times p)$ matrix of explained variables; $\mathbf{I}_{n\ell}^T = [1\ 1 \cdots 1]$; $\mathbf{e}_\ell$ = a $(n_\ell \times 1)$ vector of random errors; the superscript T denotes the transpositon of the matrix. $\boldsymbol{\beta}_\ell$ can be estimated for $\ell = 1$ and 2 by

$$\widehat{\boldsymbol{\beta}}_\ell = \mathbf{S}_\ell^{-1} \mathbf{X}_\ell^T \mathbf{y}_\ell \tag{2}$$

where $\mathbf{S}_\ell = \mathbf{X}_\ell^T \mathbf{X}_\ell$. The vectors of true response variables $\boldsymbol{\eta}_\ell$ and its predicted variables $\widehat{\boldsymbol{\eta}}_\ell$ are given by

$$\boldsymbol{\eta}_\ell = \mathbf{X}_\ell \boldsymbol{\beta}_\ell, \quad \widehat{\boldsymbol{\eta}}_\ell = \mathbf{X}_\ell \widehat{\boldsymbol{\beta}}_\ell \tag{3}$$

Suppose that the vectors of true response and predicted variables for the control period can be expressed, using $\boldsymbol{\beta}_1$ and $\widehat{\boldsymbol{\beta}}_1$ for the preliminary period as follows:

$$\boldsymbol{\eta}_2' = \mathbf{X}_2 \boldsymbol{\beta}_1, \quad \widehat{\boldsymbol{\eta}}_2' = \mathbf{X}_2 \widehat{\boldsymbol{\beta}}_1 \tag{4}$$

The residual vector $\mathbf{v}_2'$ and the residual sum of squares RSS_ℓ are

$$\mathbf{v}_2' = \mathbf{y}_2 - \widehat{\boldsymbol{\eta}}_2' \tag{5}$$

and

$$RSS_\ell = \mathbf{y}_\ell^T \mathbf{y}_\ell - \widehat{\boldsymbol{\beta}}_\ell^T \mathbf{X}_\ell^T \mathbf{y}_\ell \tag{6}$$

The mean square of random error is

$$\widehat{\sigma}_\ell^2 = \frac{RSS_\ell}{n_\ell - p'}, \quad p' = p + 1 \tag{7}$$

If the vector of random errors $\mathbf{e}_\ell$ is distributed normally about mean $\mathbf{0}$ with the variance-covariance matrix of $\sigma_\ell^2 \mathbf{I}$ such that $\mathbf{e}_\ell \sim N(0, \sigma_\ell^2 \mathbf{I})$, then both $\widehat{\boldsymbol{\eta}}_2'$ and $\widehat{\boldsymbol{\eta}}_2$ are also distributed normally such that

$$\widehat{\eta}_{2,j} = \mathbf{x}_{2,j}^T \widehat{\boldsymbol{\beta}}_2 \sim N(\eta_{2,j}, D_{2,j}\sigma_2^2),$$

$$\widehat{\eta}_{2,j}' = \mathbf{x}_{2,j}^T \widehat{\boldsymbol{\beta}}_1 \sim N(\eta_{2,j}', D_{2,j}'\sigma_1^2) \tag{8}$$

where $D_{2,j} = \mathbf{x}_{2,j}^T \mathbf{S}_2^{-1} \mathbf{x}_{2,j}$ and $D_{2,j}' = \mathbf{x}_{2,j}^T \mathbf{S}_1^{-1} \mathbf{x}_{2,j}$; $\mathbf{x}_{2,j}^T$ is the jth row vector of $\mathbf{X}_2$; $\widehat{\eta}_{2,j}$ and $\widehat{\eta}_{2,j}'$ being the element of $\widehat{\boldsymbol{\eta}}_2$ and $\widehat{\boldsymbol{\eta}}_2'$.

Under the hypothesis that $\boldsymbol{\beta}_1 = \boldsymbol{\beta}_2$, $\eta_{2,j}' = \eta_{2,j}$ and $\sigma_1^2 = \sigma_2^2$ we have

$$-\tau \leq T \equiv \frac{\widehat{\eta}_{2,j}' - \widehat{\eta}_{2,j}}{\sqrt{(D_{2,j}' + D_{2,j})\widehat{\sigma}_1^2}} \leq \tau \tag{9}$$

where $\tau = t(n_1 - p'; 0.05)$ is the value of t-

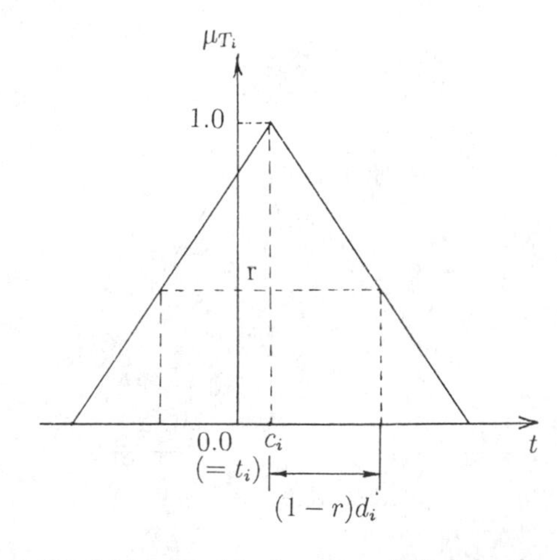

Fig.3 Membership function of T_i defined "about zero"

Table 1 Results of possibility regression analysis on significant independent variables

	K	A_1	A_2	A_3	B_1	B_2	B_3	C
1	(-4.595, 0.490)		(-1.765, 0.470)					
2	(-4.680, 0.871)	(1.529, 0.076)	(-1.807, 0.186)					
3	(-4.251, 1.129)	(1.915, 0.096)	(-1.173, 0.059)	(-7.582, 0.379)				
4	(-4.835, 0.920)	(2.006, 0.100)	(-1.237, 0.062)	(-7.730, 0.387)	(-2.287, 0.114)		(0.001, 0.000)	
5	(-4.374, 0.934)	(2.204, 0.110)	(-1.250, 0.062)		(-7.812, 0.391)	(-1.594, 0.080)	(0.129, 0.006)	(0.001, 0.000)
6	(-4.698, 0.923)	(2.105, 0.105)	(-1.234, 0.062)		(-7.963, 0.398)	(-2.216, 0.111)		(0.001, 0.000)
7	(-4.448, 0.932)	(2.309, 0.115)	(-1.223, 0.061)		(-8.500, 0.425)	(-2.123, 0.106)		(0.001, 0.000)
8	(-5.012, 0.937)	(2.049, 0.102)	(-1.212, 0.061)		(-8.241, 0.412)	(-2.661, 0.133)		(0.001, 0.000)
9	(-8.123, 0.406)	(1.709, 0.085)	(-1.948, 0.097)			(-3.713, 0.186)		(0.002, 0.000)
10	(-6.349, 1.705)	(1.327, 0.066)	(-1.585, 0.079)			(-5.337, 0.267)		(0.001, 0.000)
11	(-2.032, 0.385)	(1.172, 0.513)	(-1.383, 0.069)		(-2.610, 0.743)	(-2.987, 0.149)		(0.001, 0.000)
12	(-7.152, 0.358)	(1.711, 0.086)	(-1.621, 0.325)			(-3.774, 0.189)		(0.002, 0.001)
13	(-2.191, 1.022)	(1.543, 0.600)	(-1.765, 0.088)			(-2.562, 0.128)		(0.001, 0.000)
14	(-3.719, 0.231)	(0.980, 0.912)	(-1.380, 0.069)		(-2.475, 0.124)	(-3.544, 0.177)		(0.001, 0.000)

distribution with $n_1 - p'$ degrees of freedom in a confidence level of 95 %. When using T in Eq.(9) as the parameter in the control chart, $\pm t(n_1 - p'; 0.05)$ gives the upper and lower control lines in the chart. This is the theory of the test of the difference between two multivariate regression models.

The confidence interval can be also given according to statistical analysis by

$$-\tau \leq V'_{2j} \equiv \frac{v'_{2j}}{\sqrt{(1 + D'_{2,j})\hat{\sigma}_1^2}} \leq \tau \tag{10}$$

$\pm\tau$ in Eq.(10) gives the upper and lower control limits in the control chart for parameter V'_{2j} .

The safety of dams can be estimated by these control equations (9) and (10).

2.2 Possibility regression model

2.2.1 Interval estimation

Assume that the fuzzy estimated value of observed response variable y_i can be expressed by the regression model as follows:

$$Y_i = A_0 + A_1 x_{i1} + \ldots + A_p x_{ip} \tag{11}$$

where $A_j = (a_j, b_j)$ is a symmetrical fuzzy number.

Fig.1 shows the possibility distribution of A, in which a is the center and b is the width.

Using the extension principle of the fuzzy set theory, Y_i can be expressed by a symmetrical fuzzy number such that

$$Y_i = (\mathbf{x}_i^T \mathbf{a},\ \mathbf{b}^T \mid \mathbf{x}_i \mid) \tag{12}$$

where $\mathbf{a}^T = [a_0\ a_1 \ldots a_p]$, $\mathbf{b}^T = [b_0\ b_1 \ldots b_p]$ and $\mathbf{x}_i^T = [1\ x_{i1} \ldots x_{ip}]$.

Fig.2 shows the possibility distribution of Y_i, in which the center of distribution is $\mathbf{x}_i^T \mathbf{a}$ and its width $\mathbf{b}^T \mid \mathbf{x}_i \mid$. The fuzzy coefficients a_i and b_i are determined as follows (Tanaka 1987):

1. The given data y_i must be included in the range such that

$$\mu_{Y_i} \geq h \tag{13}$$

in which μ_{Y_i} = the membership function of Y_i as shown in Fig.2.

2. The sum of any width of Y_i is minimized such that

$$J(\mathbf{b}) = \sum_i \mathbf{b}^T \mid \mathbf{x}_i \mid \rightarrow min. \tag{14}$$

Therefore, the linear possibility regression problem can be interpreted as the following LP problem:

$$\sum \mathbf{b}^T \mid \mathbf{x}_i \mid \rightarrow min. \tag{15}$$

$$\mathbf{x}_i^T \mathbf{a} - (1-h)\mathbf{b}^T \mid \mathbf{x}_i \mid \leq y_i \tag{16}$$

$$\mathbf{x}_i^T \mathbf{a} + (1-h)\mathbf{b}^T \mid \mathbf{x}_i \mid \geq y_i \tag{17}$$

$$\epsilon \mid a_i \mid \leq b_i \tag{18}$$

where ϵ = a positive constant, which must not be zero. From Eqs.(16) and (17) we obtain the non-dimensional form of the control equation as follows:

$$-(1-h) \leq U_i \equiv \frac{y_i - \mathbf{x}_i^T \mathbf{a}}{\mathbf{b}^T \mid \mathbf{x}_i \mid} \leq (1-h) \tag{19}$$

Control chart I can be made according to Eqs.(16) and (17). Control chart II corresponds to Eq.(19), where the upper and lower limits $\pm(1-h)$ in Eq.(19) correspond to $\pm t(n_1 - p'; 0.05)$ in Eq.(10).

2.2.2 Test of difference

Divide the observed data into two sets, the preliminary ($\ell = 1$) and the controlled ($\ell = 2$) data. For these two blocks of data, we postulate the following possibility regression models

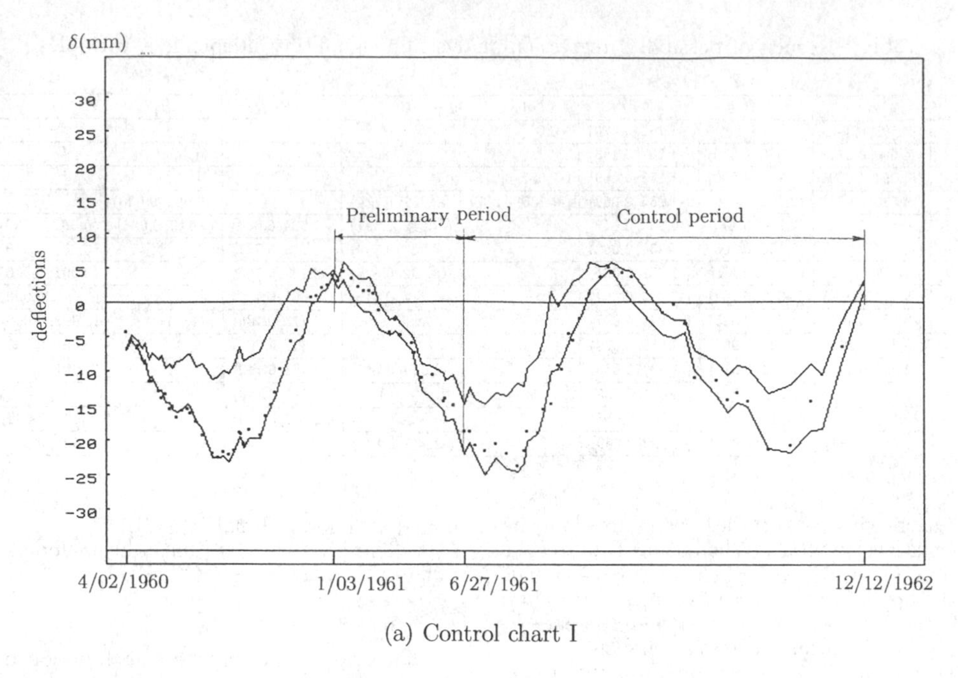

(a) Control chart I

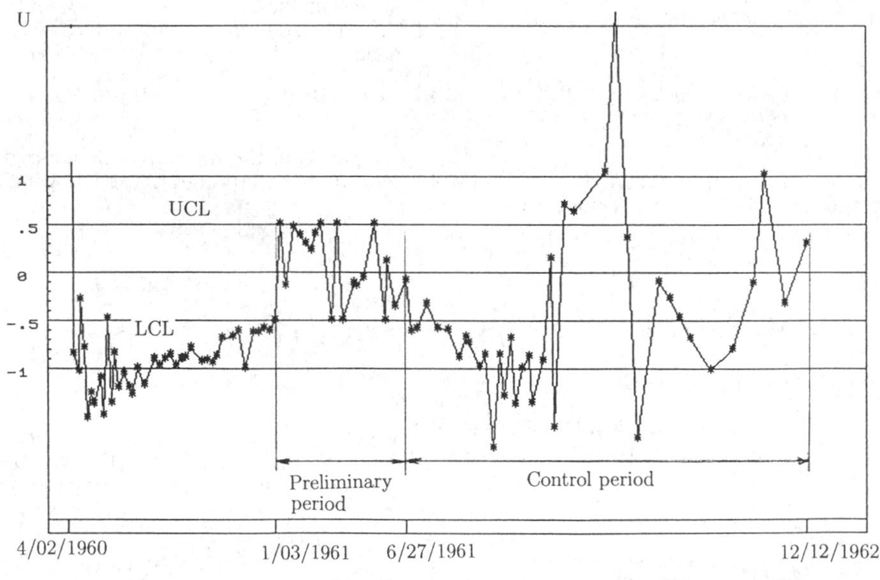

(b) Control chart II

Fig.4 Control charts I and II by interval fuzzy estimation

$$Y_{\ell i} = \mathbf{x}_{2i}^T \mathbf{a}_\ell, \quad (i = 1, 2, \cdots, n) \tag{20}$$

where $\mathbf{x}_{2i}$ = the ith vector of explained variables for the controlled data and n = the data size.

If there is no difference between models for the preliminary data and for the controlled data, $\mathbf{a}_1$ should be equal to $\mathbf{a}_2$, thus $Y_{1i} = Y_{2i}$. If Y_{1i} is nearly equal to Y_{2i}, we have

$$T_i = Y_{1i} - Y_{2i} = (c_i, d_i) \tag{21}$$

where $c_i = \mathbf{x}_{2i}^T \mathbf{a}_1 - \mathbf{x}_{2i}^T \mathbf{a}_2$, $d_i = \mathbf{b}_1^T \mid \mathbf{x}_{2i} \mid + \mathbf{b}_2^T \mid \mathbf{x}_{2i} \mid$ and T_i must be a fuzzy number which means "about zero".

Fig.3 shows the membership function of T_i, in which t_i is the crisp number of 0. From Fig.3 we obtain the upper and lower control limits for the

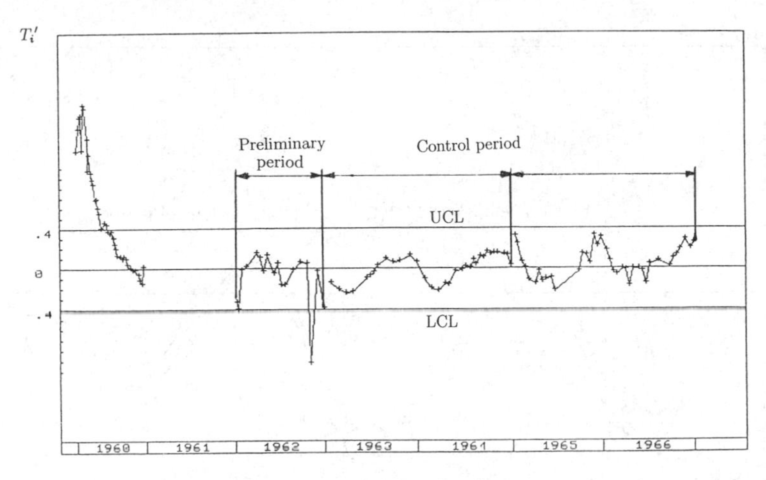

Fig.5 Control chart III by test of difference between regression models

test of the difference by using the fitness parameter r as follows:

$$c_i - (1-r)d_i \leq t_i \leq c_i + (1-r)d_i \tag{22}$$

Transforming Eq.(22) into non-dimensional form we obtain

$$-(1-r) \leq T_i' \equiv \frac{t_i - c_i}{d_i} \leq (1-r) \tag{23}$$

Thus, we obtain control chart III with this pair of upper and lower limits for the test of the difference between two multivariate possibility regression models.

3 APPLICATIONS

The safety control of arch dams has been carried out by applying the statistical method to process and interpret the measured data like the deflection of dam δ, mean temperatures t's, temperature gradients α's at several sections along the cantilever element of the dam, the water level of reservoir h, and so on .

Assume that deflection δ can be expressed by the regression model as follows:

$$\delta_j = K + \sum_i A_i t_{ij} + \sum_i B_i \alpha_{ij} + C(h_j - h_0)^2 + e_j$$

$$(j = 1, 2, \cdots, n) \tag{24}$$

where subscripts i and j denote the ith section at which the temperatures are measured and the jth set of observations, respectively; K= a constant; A_i, B_i and C = regression coefficients; $h_j - h_0$ = the water depth of the reservoir; e_j= a random error which is not considered in the possibility regression analysis.

In order to compare the results, the observed data is referred to Miike and Kobayashi (1985) in which probability regression analysis was carried out. The height of the A dam is 75m and reservoir filling started in April 1960. The irreversible trend of deflections in dams is remarkable in the initial period after the beginning of reservoir filling.

3.1 Control chart methods for possibility regression models

The results of fuzzy regression analysis for 14 cases using significant independent variables are shown in Table 1. The data size for the preliminary period varies from 12 to 40. Since deflections in the middle-scale arch dams such as the A dam are supposed to be affected chiefly not by the water depth but by temperature, the fuzzy coefficients of C are very small.

Fig.4 shows control charts I and II for Case 12 ($h = 0.5$). In Fig.4(a), the dots show the observed deflections and two solid lines are the upper and lower control limits. In Fig.4(b) the asterisks * are non-dimensional observed deflections U_i which are defined by Eq.(19). UCL and LCL denote the upper and lower control limit, respectively. These methods are based on the evaluation that dams are safe in the control period when the controlled data belong to the same population as the preliminary data and, in addition, the dam was safe in the preliminary period. But we cannot decide immediately that the dams are becoming dangerous without investigating of the mechanism of the fracture of structures, merely because the observed deflections were plotted outside the control limits.

The control chart III for Case 14 ($r = 0.6$) is shown in Fig.5. The data size in the preliminary data is 30 and the moving controlled data are taken from each different two-year-periods. The effect of the difference between the preliminary and moving control period does not appear in Fig.5.

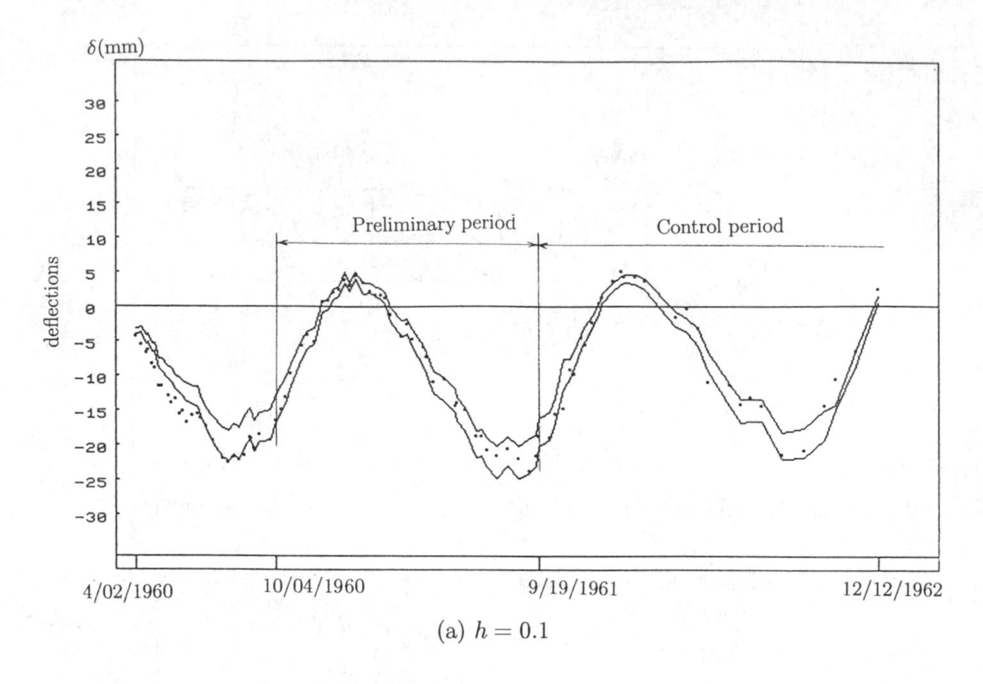

(a) $h = 0.1$

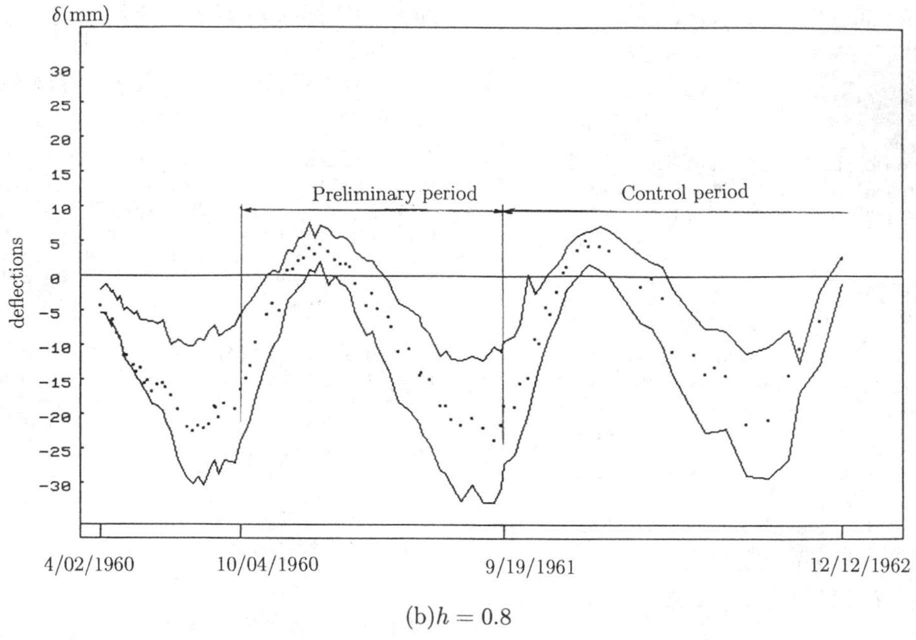

(b)$h = 0.8$

Fig.6 Relation between h and width of control limits for Control chart I

3.2 Relation between fitness parameter and width of interval

Fig.6 shows the relation between the fitness parameter h in Eq.(11) and the width of the control limits for Case 11. As the value of h draws near to zero, safety control must be performed very rigorously. In order to carry out the safety control of dams, we have to use language, which always contains ambiguity and multiplicity of meaning. The degree of linguistic expression like "very rigorous", "normal" and so on can be qualified over a specific range by using the membership function of the fuzzy set theory (Zadeh 1975, Dubois and Prade 1980).

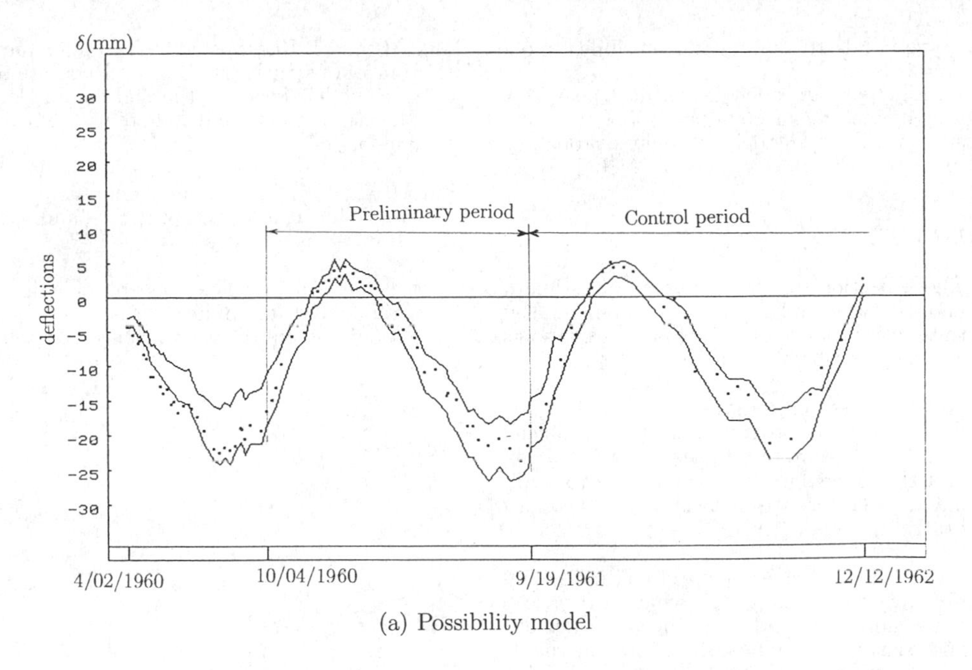

(a) Possibility model

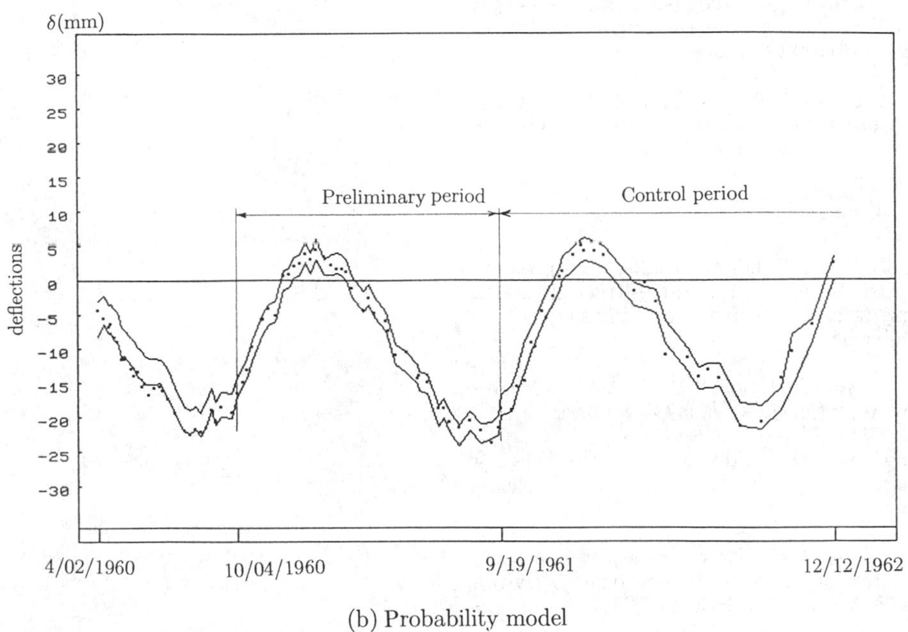

(b) Probability model

Fig.7 Comparison of possibility model and probability model

We have to consider not only statistics and mechanics but socio-psychological factors, therefore possibility models are more effective and applicable to the safety control of arch dams than probability models.

3.3 Comparison between possibility methods and probability methods

Fig.7 compares the results of the possibility model for $h = 0.5$ and the probability model for control chart I. In this numerical example the safety control of the probability method with a confidence level of

95% corresponds to that of the possibility model with $h = 0.2$ to 0.4.

From Fig.7 we can conclude that it is possible to perform the maintenance and control of arch dams by using possibility models as well as probability models.

4 CONCLUSION

1. The deflections of arch dams can be estimated and safety control can be carried out by using the control chart methods of possibility models as well as probability models.

2. Because possibility regression methods are reduced to linear programming (LP) problems, existing LP subroutine packages are available and the algorithms are very simple to implement. Therfore we can carry out the safety control in a short time.

3. Since, in the safety control of arch dams, we have to use language which always contains ambiguity and multiplicity of meaning, and consider not only statistics and mechanics but also socio-psychological factors, possibility models are a more effective and applicable tool than probability models.

The authors express their deep gratitude to Enterprise Bureau, Miyazaki prefecture that has given them the precious observed data.

REFERENCES

Cramer, H. 1951. Random variables and probability distributions, Cambridge Tracts in Mathematics and Mathematical Physics, No.36: Cambridge.

Draper, N. R. and H. Smith 1967. Applied Regression Analysis: John Wiley & Sons.

Dubois, D. and H. Prade. 1980. Fuzzy Sets and systems, Theory and Applications, Academic Press.

Giampoli, D.A. 1980. New perspectives on safety of dams, Journal of Professional Activities, ASCE, Vol.106, No.E13: 265-273.

Miike, R. and I. Kobayashi 1985. Safety control of dams by multivariate regression model, Proc. of the 4th Int. Conf. on Structural Safety and Reliability, Vol.III, Kobe, Japan: III 383-III 392.

Miike, R. and I. Kobayashi 1989. Safety control of arch dams by regression model, Design of Hydraulic Structures 89, (Ed.Albertson,M. and Kia,R.), Balkema: 91-96.

Miike, R and I. Kobayashi 1991. Maintenance and control of structures using multivariate regression model, Proc. of the 2nd Japan Conf. on Structural Safety and Reliability : 341-348 (In Japanese).

Tanaka, H. 1987. Fuzzy data analysis and possibilistic linear models, Fuzzy Sets and Systems, 24: 363-375.

Zadeh, L.A. 1975. The concept of a linguistic variable and its application to approximation reasoning-I, Information Sciences, 8: 199-249.

Structural Safety & Reliability, Schuëller, Shinozuka & Yao (eds) © 1994 Balkema, Rotterdam, ISBN 90 5410 357 4

Probabilistic foundation stability analysis: Mobilized friction angle vs available shear strength approach

Farrokh Nadim & Suzanne Lacasse
Norwegian Geotechnical Institute, Oslo, Norway
Tom R.Guttormsen
Saga Petroleum, Sandvika, Norway

ABSTRACT: Probabilistic bearing capacity analyses for foundations of two offshore gravity base structures were performed. The analyses were performed based on both the mobilizid friction angle and the available shear strength approaches. The results showed that, depending on the soil type, the computed probability of failure may vary drastically for the two formulations. This poses a dilema for the design engineer: for a given site condition, one approach may predict an adequate design, while the other may predict an unsafe foundation.

1 INTRODUCTION

Offshore gravity platforms exert large static and cyclic loads on the foundation soil. An essential part of the design of the platforms is to ensure that the foundation soil has sufficient capacity to carry these loads. Foundations of gravity base structures and jack-up platforms offshore Norway are designed using the concept of limit state analysis for bearing capacity. The bearing capacity analyses may be done by two approaches: 1) mobilized friction angle, or 2) available shear strength. The two approaches define the safety factor (or material coefficient) differently and the numerical values obtained for a given foundation may differ significantly.

This issue is not adequately addressed in the Norwegian codes (e.g. NBR, 1989) or in other international codes used in practice (Høeg, 1986). Tjelta et al. (1988) illustrated this problem for the Gullfaks C gravity base structure installed in the North Sea in May 1989. For this platform the material coefficient based on the mobilized friction angle approach was calculated to be $\gamma_{me} = 1.60$, whereas the material coefficient obtained with the available shear strength method was $\gamma_{mu} = 1.26$. The minimum material coefficient at the time was 1.3 (NPD, 1985). The results revealed that the actual safety level cannot be uniquely determined with only partial load and material coefficients: one analysis method implies that the foundation has an ample margin of safety, while the other method predicts that the foundation safety margin is not adequate.

To gain better insight into the problem, a probabilistic formulation of the two method was developed because the probability of foundation non-performance is independent of the method of analysis. In principle, one should obtain a consistent measure of foundation safety margin with the probabilistic formualtion using either analysis procedure. Probabilistic bearing capacity calculations were performed for two gravity base structures in the the North Sea, one located at a soft clay site and the other one located at a dense sand site.

2 DEFINITION OF MATERIAL COEFFICIENT

The design codes for fixed offshore structures on the Norwegian Continental Shelf use partial safety factors (e.g. NPD, 1992). The Norwegian Standard NS 3481 (NBR, 1989) provides the rules for soil investigation and geotechnical design of marine structures offshore Norway:

> "The material coefficient for soil may be expressed either as the ratio between the undrained shear strength and the shear stress mobilized for equilibrium, or as the ratio between the characteristic friction coefficient (i.e. tangent of the characteristic friction angle) and the friction coefficient being mobilized for equilibrium (i.e. the tangent of the mobilized friction angle)."

In the available shear strength method, the material coefficient, γ_{mu}, is defined as the ratio between the undrained shear strength and the shear stress mobilized for equilibrium. The effect of cyclic loading on the soil shear strength must be taken into account. This effect is usually a reduction in the reference static strength. The reduction in static strength will depend on the average shear stress prior to cyclic loading, the magnitude of the cyclic shear stresses, the number of cycles and the overconsolidation ratio (OCR) prior to cyclic loading (Andersen and Høeg, 1991). Figure 1 shows the definition of γ_{mu} for a dilative soil (e.g., dense sand, heavily overconsolidated clay) and for a contractive soil (e.g., loose sand, normally consolidated clay).

The mobilized friction angle method uses the Mohr-Coulomb failure criterion to define the material coefficient. The Mohr-Coulomb failure criterion states that the shear strength, τ_f, is proportional to the effective normal stress acting on the failure plane:

$$\tau_f = c + \sigma_c' \cdot \tan\phi' \tag{1a}$$

or

$$\tau_f = (a + \sigma_c') \cdot \tan\phi' \tag{1b}$$

where ϕ' is the characteristic friction angle, σ_c' is the effective normal stress on the failure plane, c is the cohesion, and $a = c/\tan\phi'$ is the attraction.

The material coefficient for the mobilized friction angle method, γ_{me}, is expressed as the ratio between the tangent of the characteristic friction angle, $\tan\phi'$, and the friction coefficient mobilized for equilibrium, $\tan\rho$:

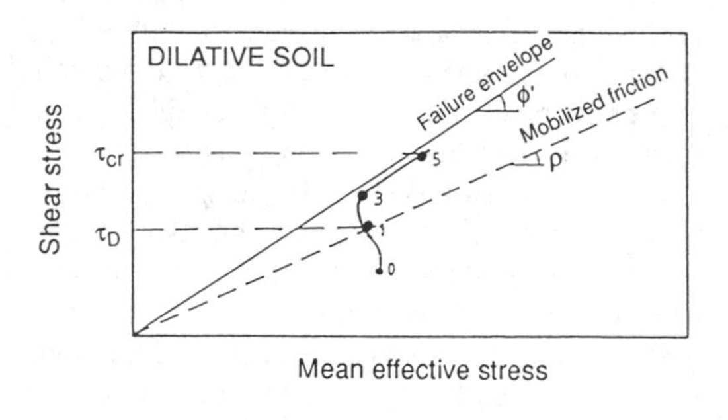

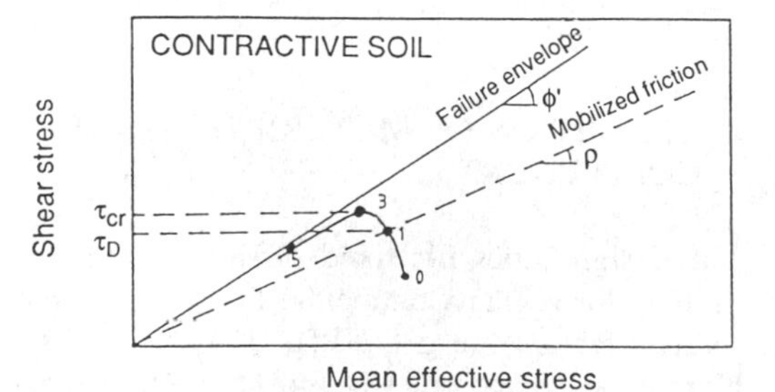

Fig. 1 Determination of material coefficient from different approaches to stability analyses.

$$\gamma_{me} = \tan\phi' / \tan\rho = 1/f \qquad (2)$$

where f is the degree of soil strength mobilization.

Figure 1 shows the definition of γ_{me} for a dilative soil and a contractive soil. It is obvious that, if one follows the code specifications, the computed safety margin depends on soil type. For a dilative soil, the material coefficient, γ_{mu}, for an "available shear strength" analysis may be larger than the material coefficient, γ_{me}, for a "mobilized friction angle" analysis. For a contractive soil, the situation may be the opposite.

3 PROBABILISTIC BEARING CAPACITY ANALYSIS

The first software for doing probabilistic bearing capacity analyses with the available shear strength method was based on NGI's limiting equilibrium method (Lauritzsen and Schjetne, 1976) and the first- and second-order reliability methods (NGI, 1988; 1990). The probabilistic bearing capacity analysis procedure comparing the results of available shear strength and mobilized friction angle analyses was based on the General Procedure of Slices (GPS). The GPS method was originally proposed by Janbu (1954) and further extended by Svanø (1981). This approach was selected to avoid an additional model error to account for the use of different calculation methods.

Figure 2 shows the GPS idealization of the bearing capacity problem for a general case. The loading, the soil profile, and the slip surface may be irregular. The slip surface that gives the maximum degree of soil mobilization is termed the critical slip surface. When the degree of mobilization on the critical surface is equal to unity, the corresponding surface is called the failure surface. The soil mass located between the ground surface and the assumed slip surface is divided into a number of slices by vertical lines. The cross-section of such a slice is indicated by the shaded area on Fig. 2. The same slice is shown in an enlarged scale in Fig. 3 together with the forces acting on the slice. The solution procedure is based on assuming a line of thrust for the inter-slice forces, and the same degree of shear mobilization, f, along the slip surface for all slices. These two assumptions provide enough equations to solve for the degree of mobilization, although an iterative solution procedure is necessary.

In its original version (Janbu, 1954) the degree of mobilization was based on the mobilized friction angle concept (Eqs. 1 and 2). Svanø (1981) used Janbu's pore pressure equation (Janbu, 1976) to modify the GPS for undrained conditions. By predicting the pore pressure change one could predict the effective stress path to failure, and hence the undrained (available) shear strength. Thus, for the available shear strength analysis with the GPS method, the degree of mobilization is simply defined as the shear stress required for equilibrium along the slip surface divided by the available shear strength along the projected effective stress path to failure.

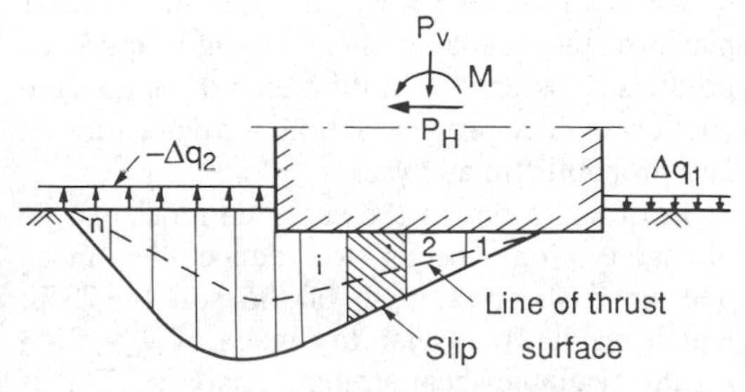

Fig. 2 General procedure of slices for stability analyses (Svanø, 1980)

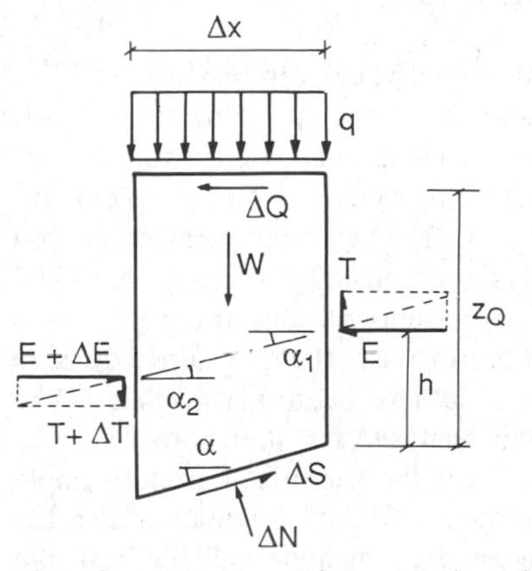

Fig. 3 Forces acting on the shaded slice in Fig. 2

To perform the probabilistic calculations, the first-order reliability method (FORM) was used to estimate the probability of non-performance, P_f, the reliability index, β, and the sensitivity factors for the uncertain parameters. Details of this method are available elsewhere (e.g., Gollwitzer et al., 1988), and will not be repeated here. The FORM approximation was chosen for its simplicity and quickness. Earlier studies at NGI (1988, 1990) indicated that FORM gives essentially the same results as the more advanced second-order and improved second-order reliability methods (SORM).

The RELGPS program was developed to perform probabilistic bearing capacity analysis with the General Procedure of Slices. The program provides access to the deterministic program ESAU (Svanø, 1980) and to the subroutine FORM5 (Gollwitzer et al., 1988) where the first-order reliability estimates and design point values are calculated. The limit state function g(x) is defined as:

$$g(X) = 1 - f \tag{3}$$

where f is the degree of mobilization and X is a vector of random variables, such as soil shear strength, storm loads, etc. Unsatisfactory performance (i.e., failure) occurs when the limit state function, g(X), becomes negative.

4 EXAMPLE CALCULATIONS

4.1 *Contractive soil*

Two example calculations were performed. The first case studied a gravity platform at a soft clay site. The foundation geometry, soil profile, and the RELGPS model of the problem are shown in Fig. 4. The key input parameters used in the analysis are listed in Table 1. The uncertainty associated with the soil properties is typical for a North Sea soft clay where extensive site investigation has been performed. The results of the analysis are shown in Table 2. For this platform the mobilized friction angle approach predicts a "safer" foundation than the available shear strength approach for both the deterministic and probabilistic analyses.

The "true" probability of failure of the foundation is, however, independent of the method of analysis. The probability of failure inferred from the reliability index is a "nominal" probability of failure.

In order to obtain the same nominal margin of safety for this platform, one needs to increase the mean of the undrained shear strength of the soil by 20%, or increase significantly the excess pore pressure. The former results in a reliability index of $\beta = 4.21$ for the available shear strength analysis. A bias of 20% on the mean undrained shear strength can be plausible since sampling disturbance, stress relief, and specimen handling on board and in the laboratory tend to reduce the undrained shear strength measured in the laboratory. Increasing the predicted excess pore pressure is even more plausible because of the shape of the effective stress path for contractive soils (rapid increase in excess pore pressure close to failure, after about 2% shear strain in the case of Fig. 1). It can therefore be concluded that, for this particular soil type, the probabilistic formulation may resolve the differences between the two analysis procedures.

4.2 *Dilative soil*

The second case studied was a gravity platform at a dense sand and overconsolidated clay site. The foundation geometry, soil profile, and the RELGPS model of the problem are shown in Fig. 5. The key input parameters used in the analysis are listed in Table 3. Again, the uncertainty associated with the soil properties is typical for a North Sea site with extensive site investigation on these types of soil. The results of the analysis are shown in Table 4. For this platform the available shear strength approach predicts a "safer" foundation than the mobilized friction angle approach for both the deterministic and probabilistic analyses.

In order to obtain the same nominal margin of safety, one needs to reduce the mean undrained shear strength of the soil by 23%, which results in a reliability index of $\beta = 2.43$ for the available shear strength analysis. This is not surprising, because in effect with the large

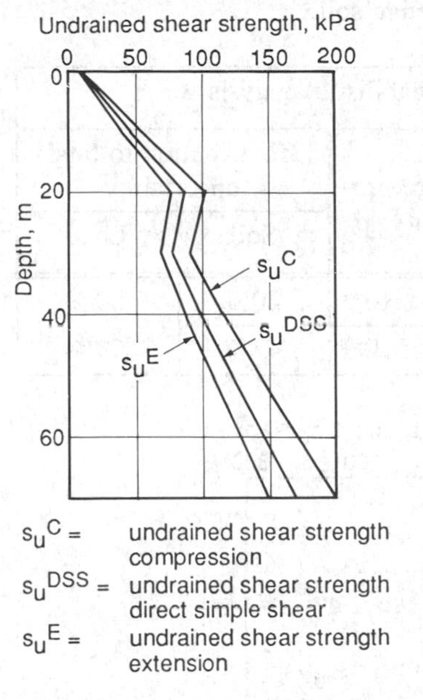

s_u^C = undrained shear strength compression
s_u^{DSS} = undrained shear strength direct simple shear
s_u^E = undrained shear strength extension

a) Soil shear strength profile

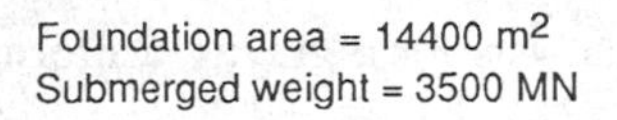

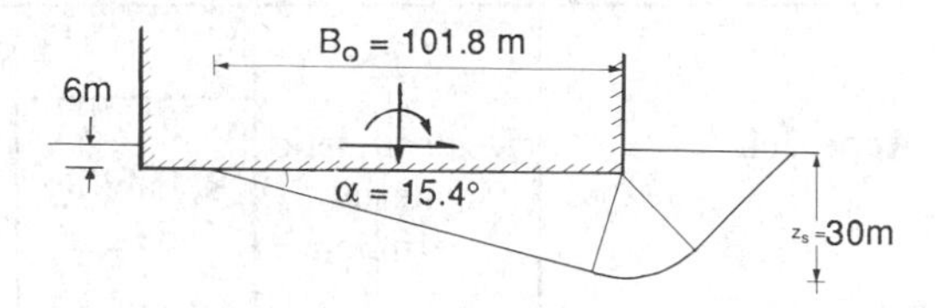

b) Critical slip surface computed by CAP (Lauritzsen and Schjetne, 1976)

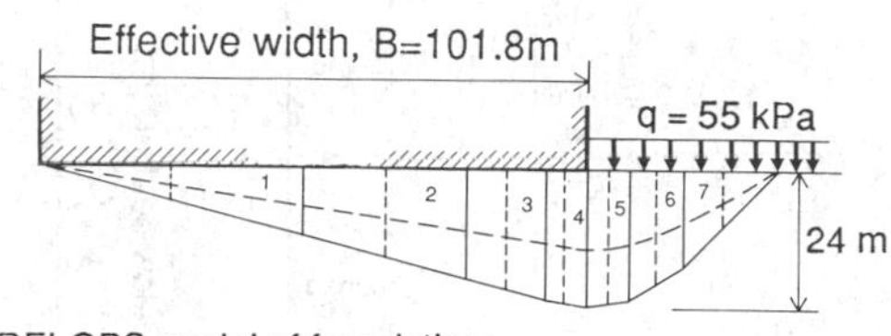

c) RELGPS model of foundation

Fig 4 Foundation model and soil characteristics for the gravity platform on contractive soil.

Table 1 Key parameters for platform on contractive soil.

	Parameter	Mean, μ	Standard deviation, σ	Distribution function
Load	Submerged weight, P_v (MN)	3500	175	Normal
	Horizontal force, P_H[1] (MN)	250	25	Gumbel
	Overturning moment, M[1] (MNm)	30000	3000	Gumbel
	Model error	1.0	0.1	Normal
Soil	Undrained shear strength	See Fig. 4	0.1 μ	Log-normal
	Friction coefficient, tanϕ'	0.532	0.027	Log-normal
	Attraction, a (kPa)	6	0.6	Log-normal
	Effective unit weight, γ' (kN/m^3)	10	0.5	Normal
	Janbu's (1976) pore pressure parameter, D	-0.3 → 0.5	0.44	Normal
	Model error	1.0	0.1	Normal

[1] P_H and M are perfectly correlated

Table 2 Results of the analysis for a gravity platform on contractive soil.

Approach	Deterministic material coefficient	Probabilistic analysis			
		Reliability index, β	Failure probability, p_f	Contribution to total uncertainty	
				Soil	Load
Mobilized friction angle	1.90	4.15	1.65×10^{-5}	20%	80%
Available shear strength	1.35	2.81	2.47×10^{-3}	47%	53%

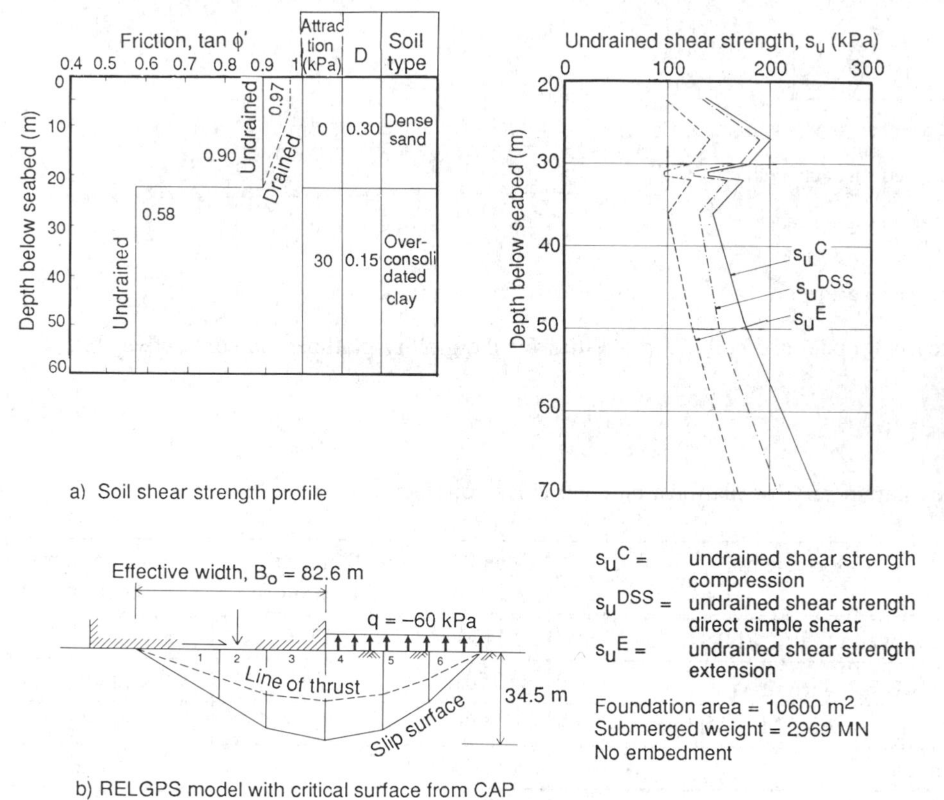

Fig. 5 Foundation model and soil characteristics for the gravity platform on dilative soil.

reduction factor put on the undrained shear strength, both methods are defining the same reference soil strength; thus the effects of the soil dilative behaviour on its shear strength are neglected. One may question the definition itself of material coefficient in mobilized frictionangle analysis for dilative soils. It can be seen from Fig. 1 that in terms of the mobilized friction angle, it is not possible to account for the large increase in the shear strength caused by the dilative tendency because the stress path to failure falls on the failure envelope, and often at very small shear strains.

In other words, the material coefficient defined in terms of the mobilized friction is not a good measure of the performance for dilative soils. Hence, a probabilistic approach cannot resolve the difference in the deterministic material coefficients obtained by different analysis procedures as the definition of the

Table 3 Key parameters for platform on dilative soil.

	Parameter	Mean, μ	Standard deviation, σ	Distribution function
Load	Submerged weight, P_v (MN)	2960	148	Normal
	Horizontal force, P_H[1)] (MN)	1089	108	Gumbel
	Overturning moment, M[1)] (MNm)	21412	2141	Gumbel
	Model error	1.0	0.10	Normal
Soil	Undrained shear strength	See Fig. 5	0.1 μ in sand 0.1 μ in clay	Log-normal
	Friction coefficient in sand, tanϕ'	0.90	0.09	Log-normal
	Friction coefficient in clay, tanϕ'	0.58	0.03	Log-normal
	Attraction in clay, a (kPa)	30	10	Log-normal
	Pore pressure parameter, D, in clay	0.15	0.05	Normal
	Pore pressure parameter, D, in sand	0.30	0.10	Normal
	Model error	1.0	0.10	Normal

[1)] P_H and M are perfectly correlated

Table 4 Results of the analysis for a gravity platform on dilative soil.

Approach	Deterministic material coefficient	Probabilistic analysis			
		Reliability index, β	Failure probability, p_f	Contribution to total uncertainty	
				Soil	Load
Mobilized friction angle	1.35	2.48	6.65 x 10^{-3}	79%	21%
Available shear strength	1.54	4.54	2.32 x 10^{-6}	75%	25%

performance function in the mobilized friction angle approach appears to be inadequate (too conservative).

5 DISCUSSION

Based on the example calculations, it may be concluded that for a gravity platform on contractive soil, the probabilistic formulation seems capable of providing a consistent measure of safety based on either approach. For a gravity platform on dilative soil, however, the probabilistic formulation cannot resolve the differences between the two approaches, as the definition of the margin of safety based on the mobilized friction angle method is inadequate for this soil type.

In the case studies, the critical failure surface was determined before hand using a deterministic approach based on the available shear strength method. The mobilized friction angle analysis may result in a different critical surface than the available shear strength analysis. Because of the limitations in the deterministic software used and the complexity of the problem, no search was performed to find the critical surface based on the mobilized friction angle method. This may explain part of the difference between the results of the two

methods. Ideally, the search for the critical failure surface should be part of the probabilistic analysis. The relevant critical surface is that with the largest failure probability, which is not necessarily the same as the one obtained by the deterministic approach.
In the "calibration" runs, the results of the mobilized friction angle analyses were chosen as "bench mark" values, and the available shear strength parameters were varied until the same reliability index was obtained. This choice was arbitrary. Similar results could be obtained by choosing the results of the available shear strength analyses as bench mark values, and varying the parameters related to the mobilized friction angle approach. In order to identify the appropriate model(s) for different soil types, more attention should be paid to the "model error" associated with the two analysis methods. The statistics of this "model error" variable may be found by comparing the predicted results with carefully controlled experimental results (laboratory or model tests, or measurements on prototypes). The appropriate use of the model error variables could harmonize the results of the reliability analyses based on the two methods.

ACKNOWLEDGEMENTS

The work described in this paper was supported by Statoil, Elf Petroleum Norge A/S, A/S Norske Shell, Mobil Research and Development Corporation, U.K. Health and Safety Executive, the Norwegian Petroleum Directorate, and NGI. Their support and permission to publish this paper are gratefully acknowledged. The opinions expressed in the paper are those of the authors and do not necessarily reflect the views or policy of the supporting organisations.

REFERENCES

Andersen, K.H. and K. Høeg 1991. Deformations of soils and displacements of structures subjected to combined static and cyclic loading. *Proc. 10th European Conf. on Soil Mech. and Found. Eng.*

Gollwitzer, S. T. Abdo and R. Rackwitz 1988. *FORM manual*. Reliability Consulting Programs GmbH, Munich, Germany.

Høeg, K. 1986. *Geotechnical issues in offshore engineering*. Marine geotechnology and nearshore/offshore structures. ASTM, Spec. Tech. Publ., 923.

Janbu, N. 1954. Application of composite slip surface for stability analysis. *Proc. European Conf. on Stability of Earth Slopes*, Stockholm.

Janbu, N. 1976. *Soil properties, their determination and use in stability analysis of clays*. NIF-Course, Gol, Norway.

Lauritzsen, R. and K. Schjetne 1976. Stability calculations for offshore gravity structures. *Proc. OTC'76*, Houston, Texas.

Norges Byggstandardiseringsråd (NBR) 1989. NS 3481 - *Soil investigations and geotechnical design for marine structures*.

Norwegian Geotechnical Institute 1988. *Probabilistic bearing capacity analysis of gravity platform*. Report 51411-8.

Norwegian Geotechnical Institute 1990. *Probabilistic analysis of bearing capacity using cyclic shear strength*. Report 514160-10.

Norwegian Petroleum Directorate 1985. *Regulations for structural design of load bearing structures intended for exploitation of petroleum resources*.

Norwegian Petroleum Directorate 1992. *Regulations concerning loadbearing structures in the petroleum activities*.

Svanø, G. 1980. *Effective stress analysis, undrained. Program documentation for "ESAU"*. Internal report, Norwegian Institute of Technology.

Svanø, G. 1981. *Undrained effective stress analyses*. Doctoral thesis. Norwegian Institute of Technology.

Tjelta, T.I., A.Å. Skotheim and G. Svanø 1988. Foundation design for deepwater gravity base structure with long skirts on soft soils. *Proc. BOSS'88* Trondheim, Norway.

Structural Safety & Reliability, Schuëller, Shinozuka & Yao (eds) © 1994 Balkema, Rotterdam, ISBN 90 5410 357 4

Probabilistic evaluation of lateral liquefaction spread under earthquake loading

J.A. Pires & A.H-S. Ang
Department of Civil Engineering, University of California, Irvine, Calif., USA

I. Katayama
TEPSCO, Advanced Engineering Operation Center, Chiyoda-ku, Tokyo, Japan

ABSTRACT: Monte Carlo simulation is used to calculate the probability of liquefaction in a soil layer spreading over a finite area or a fraction of a finite area. Horizontally layered soil deposits in which the ground motion is propagated upwards from the bedrock through the soil layers are considered. The vertical propagation of the ground motion through the soil layers is analyzed using random vibration analysis. The statistical spatial correlation of the undrained shear strength of the sand against liquefaction is taken into consideration, and the ground motions over the area of interest are assumed to be perfectly correlated. The method is used to compute the probability of liquefaction spreading over a specified area under a given earthquake loading and is illustrated with a specific problem in two dimensions. Soil deposit fragility curves against liquefaction are obtained, which give the probability of liquefaction spreading over a given area or a fraction of a given area for a specified peak ground acceleration and strong motion duration at the bedrock interface.

1 INTRODUCTION

In general, the problem of liquefaction spreading spatially over a finite area is three-dimensional and requires determination of the spatial variation of the relevant soil properties as well as of the spatial variation of earthquake ground motions, the latter involving three-dimensional wave propagation analysis. In this study, the problem is simplified in two dimensions based on certain idealizations. In general, the spread of liquefaction, i.e., the lateral extent of liquefied soil, over an area will depend on the following: the profile of the soil deposit, the geology of the area, the soil properties, namely, the undrained shear strength of the sand, and the ground motion in the particular zone. Evaluation of the contiguous soil volumes that are likely to liquefy under an earthquake loading is useful for the safety assessment of buried lifelines, e.g., conduits for high-voltage electric power transmission lines, gas and water lifelines.

A Monte Carlo simulation method to compute the probability of liquefaction in a soil layer spreading over a finite area is presented. Horizontally layered soil deposits in which the ground motion is propagated upwards from the bedrock through the soil layers are considered. The vertical propagation of the ground motion through the soil layers is analyzed using random vibration analysis. To account for the variability of soil properties with depth, the soil deposit is divided into layers with assumed constant soil properties. The horizontal spatial correlation of the relevant soil properties is taken into consideration to compute the probability of liquefaction spreading over a specified area under a given earthquake loading. The principal characteristics and assumptions of the method are: (i) ground motions are specified at the bedrock and are spatially correlated over the area of interest for a given earthquake; (ii) soil layers are horizontal; (iii) one-dimensional site amplification analysis is used to account for the effect of local site conditions on the ground accelerations and stresses in the soil deposit; (iv) uses in-situ soil properties and past data on the occurrence and non-occurrence of liquefaction to characterize the soil resistance against liquefaction; (v) the soil properties, e.g., SPT-blowcount are spatially correlated in the horizontal direction and are represented by a two-dimensional homogeneous random field through a specified correlation function and probability distribution function; and (vi) the upward propagation of ground motion

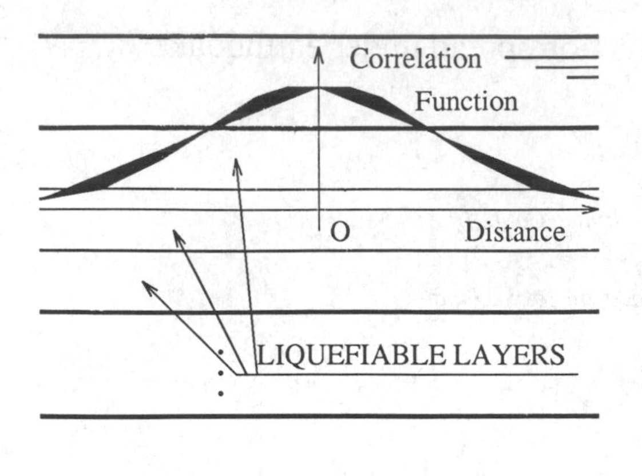

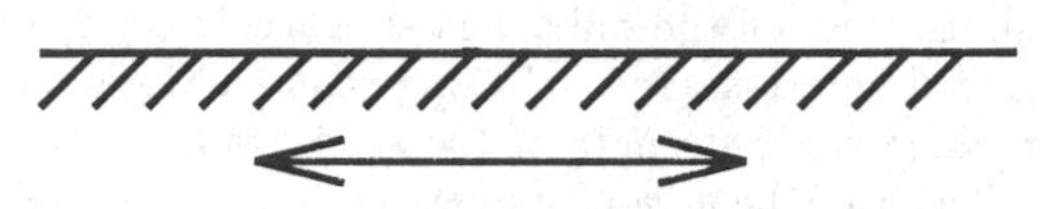

Fig. 1 Horizontally layered soil deposit

through the soil layers is not significantly affected by the random spatial variability of the undrained shear strength against liquefaction.

Mathematically, the probabilitiy that liquefaction in the layer will spread over a given area is equivalent to the probability that the soil resistance against liquefaction within the area of interest is everywhere less than the seismic load effect that induces liquefaction. This problem, is similar to a first passage problem. However, the random field of interest is a non-zero mean random field and the threshold level, although constant over the entire area of interest, is a random variable and can take values that are less than the mean of the random field. Analytical solutions for this problem are not available and, accordingly, the Monte Carlo simulation appears to be a viable alternative.

The probabilities of the onset of liquefaction for any given point at a given depth in the soil deposit are obtained using the method proposed by Pires, Ang and Katayama (1989). The principal features of this method are: (i) ground accelerations are represented by stochastic processes and methods of structural stochastic dynamics are used to compute the statistics of the ground motion acceleration and shear stresses throughout the deposit; (ii) nonlinear and hysteretic soil behavior is considered and a transmitting boundary is included in the one-dimensional lumped-mass model used in conjunction with the random vibration analysis for site amplification studies; (iii) in-situ soil properties and past historical data on field evidence of liquefaction are used to define the in-situ soil resistance against liquefaction.

2 METHODOLOGY

Consider an horizontally layered soil deposit susceptible to liquefaction such as the one shown in Fig. 1. The probability that liquefaction will occur at a specific point in the critical layer (say point O in Fig. 2), independently of liquefaction occurring anywhere else in the layer, is given by(Pires, Ang and Katayama, 1991 and 1993)

$$P[E_0] = \int_0^\infty F_{W_u}(q) f_{W(a,t)}(q) dq \qquad (1)$$

where F_{W_u} is the probability distribution function of the soil resistance against liquefaction at that location, W_u, and $f_{W(a,t)}$ is the probability density function of the earthquake load effect at point O, $W(a,t)$, for a ground excitation with expected maximum ground accelration, $A = a$, and strong motion duration $T_E = t$. The quantity W_u measures the soil ressitance against liquefaction as a function of its hysteretic energy absorption capacity under undrained loading conditions and, accordingly, the load effect, $W(a,t)$, is related to the hysteretic energy absorption demands of the earthquake loading as well as to the maximum earthquake induced shear stresses at the point of interest. To simplify the notation, introduce the random variables $S = W_u$ and $Q = W(a,t)$.

Let the soil resistance in the layer be described by a two-dimensional homogeneous lognormal random process with mean $\bar{S}$, variance σ_S^2 and autocorrelation function $R_{SS}(r_{ij})$, where r_{ij} is the horizontal distance between any two points in the layer. To compute the probability that liquefaction will spread over a rectangular area with dimensions $B \times L$, the area is divided into elementary squares with side D numbered i=1,2,...,N, as shown in Fig. 2. The soil resistances against liquefaction at the center of the elementary squares are random variables denoted by S_i, $i = 1, ..., N$. The random variables S_i are identically distributed but are not statistically independent. The coefficient of correlation for a pair of random variables (S_i, S_j) is

$$\rho_{SS}(r_{ij}) = \frac{R_{SS}(r_{ij})}{\sigma_{S_i}\sigma_{S_j}} \qquad (2)$$

where $\sigma_{S_i} = \sigma_{S_j} = \sigma_S$.

The probability that liquefaction will extend over the entire area $B \times L$ is the probability that all elements within that area will liquefy. Therefore,

$$P_L(B,L) = P[(S_1 < Q) \dots \cap (S_N < Q)] \quad (3)$$

which can be writen as

$$P_L(B,L) = \int_0^{\infty} F_{S_1 \dots S_N}(q, \dots, q) f_Q(q) dq \quad (4)$$

where

$$F_{S_1 \dots S_N}(q, \dots, q) = \int_0^q \dots \int_0^q f_{S_1 \dots S_N}(s_1, \dots, s_N) dS_1 \dots dS_N \quad (5)$$

Monte Carlo simulation is used to evaluate the integral in Eq. 5 for each value of q. To perform the Monte Carlo simulation it is convenient to transform the random process $S(x,y)$ as well as the random variables S_i and Q as follows. First, transform $S(x,y)$ to

$$C(x,y) = \ln[S(x,y)] \quad (6)$$

where $C(x,y)$ is a normal random process with mean and variance

$$\bar{C} = \ln \bar{S} - 0.5\sigma_C^2; \; \sigma_C^2 = \ln(1+\delta_S^2) \quad (7)$$

respectively, where $\delta_S = \sigma_S/\bar{S}$. The autocorrelation function of $C(x,y)$ is

$$R_{CC}(r) = \ln[R_{SS}(r)/\bar{S}^2 + 1] \quad (8)$$

Then, define the random process $C'(x,y)$ as follows

$$C'(x,y) = [C(x,y) - \bar{C}]/\sigma_C$$

which is a normal random process with zero mean, unit variance and autocorrelation function

$$R_{C'C'}(r) = \frac{\ln[\delta_S^2 \rho_{SS}(r) + 1]}{\ln[1+\delta_S^2]} \quad (9)$$

where $\rho_{SS}(r) = R_{SS}(r)/\sigma_S^2$.

Furthermore, the random variables Q and S_i are transformed to

$$L' = \frac{\ln Q - \overline{\ln Q}}{\sigma_{\ln Q}} \quad (10)$$

$$C_i' = \frac{\ln S_i - \overline{\ln S_i}}{\sigma_{\ln S_i}} \quad (11)$$

which are normal random variables with zero mean and unit variance. Then, the probability of liquefaction at a specific point in the layer is given by

$$P[E_i] = P[C_i' \le \frac{\sigma_L}{\sigma_C} L' - \frac{\bar{C} - \bar{L}}{\sigma_C}] \quad (12)$$

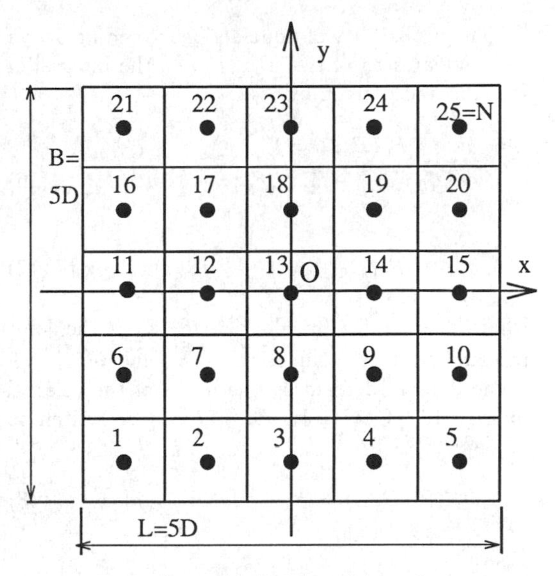

Fig. 2 Discretization of random field into correlated random variables

where

$$\begin{aligned} \bar{L} &= \ln \bar{Q} - 0.5\sigma_L^2 \\ \sigma_L^2 &= \ln(1+\delta_Q^2) \\ \bar{C}_i &= \ln \bar{S}_i - 0.5\sigma_{C_i}^2 \\ \sigma_{C_i}^2 &= \ln(1+\delta_{S_i}^2) \end{aligned}$$

It is convenient to define the normal random variable Z with mean

$$\bar{Z} = -\frac{\bar{C}_i - \bar{L}}{\sigma_{C_i}} = -\frac{\ln(\bar{S}_i/\bar{Q}) - 0.5\ln[(1+\delta_{S_i}^2)(1+\delta_Q^2)]}{\sqrt{\ln(1+\delta_{S_i}^2)}} \quad (13)$$

and standard deviation

$$\sigma_Z = \frac{\sigma_L}{\sigma_C} = \sqrt{\frac{\ln(1+\delta_Q^2)}{\ln(1+\delta_{S_i}^2)}} \quad (14)$$

such that $P[E_i] = P[C_i' \le Z]$. Then, Eq. 4 for the probability of liquefaction extending over an area of size $B \times L$ is replaced by

$$P_L(B,L) = \int_{-\infty}^{\infty} F_{C_1' \dots C_N'}(z, \dots, z) f_Z(z) dz \quad (15)$$

The coefficient of correlation for a pair of random variables C_i' and C_j' is given by Eq. 9 with the distance r equal to the distance between the points where the random variables C_i' and C_j' are sampled, i.e., r_{ij}.

The probability of liquefaction extending over a rectangular area of size $B \times L$, i.e., the integral of Eq. 15, is also given by

$$P_L(B,L) = \int_{-\infty}^{\infty} \int_{-\infty}^{z} \cdots \int_{-\infty}^{z} f_{C'_1,\ldots,C'_N,Z}(c'_1,\ldots,c'_N,z) dc'_1 \ldots dc'_N dz \quad (16)$$

or,

$$P_L(B,L) = \int_{g(\vec{x})\leq 0} f_{\vec{X}}(\vec{x}) d\vec{x} \quad (17)$$

where $\vec{X} = [C'_1\ C'_2 \ldots C'_N\ Z]^T$, $f_{\vec{X}}(\vec{x})$ is the joint probability density function of $\vec{X}$, and $g(\vec{x}) \leq 0$ is the domain defined by the limits of the integral in Eq. 16. Obviously, Eq. 17 can be writen as follows:

$$P_L(B,L) = \int_{\text{all } \vec{x}} I[g(\vec{x})] f_{\vec{X}}(\vec{x}) d\vec{x} \quad (18)$$

where

$$I[g(\vec{x})] = \begin{cases} 1 & \text{if } g(\vec{x}) \leq 0 \\ 0 & \text{otherwise.} \end{cases}$$

Once the correlation matrix for the standard normal random variables $C'_1, \ldots, C'_N$ is obtained, the integral in Eq. 18 is computed, in approximation, by

$$\hat{P}_L(B,L) = \frac{1}{N_S} \sum_{j=1}^{N_S} I[g(\vec{x}_j)] \quad (19)$$

where N_S is the number of random vectors $\vec{x}_j$ sampled from $f_{\vec{X}}(\vec{x})$.

3 EXAMPLE APPLICATION

The method described in the previous section is used to compute liquefaction fragilities for the soil deposit shown in Fig. 3. The frequency content of the earthquake ground motion is represented by the Clough-Penzien power spectrum (Clough and Penzien, 1975) and the strong motion duration is assumed to be 8.0 seconds. Each liquefaction fragility curve shows the probability that liquefaction in the layer will spread over a specified area as a function of the expected maximum ground acceleration of the ground motion specified at the base of the soil deposit.

3.1 Correlation function of soil properties

The correlation function of the soil resistance against liquefaction is assumed to be of the form

Depth (m)	Name of Layer	N SPT *	Shear Wave Velocity (m/sec)	Percent Fines
0.5 – 6.0	Upper Yurakucho (SAND)	10	175	21
11.0	Lower Yurakucho (CLAYED SAND)	3	119	44
*UNCORRECTED		>50	>450	--

Fig. 3 Soil deposit analyzed

$$\rho_{S,S}(r) = exp[-(r/b)^2] \quad (20)$$

where b is a positive parameter and r is the horizontal distance between any two points in the layer. This form of the autocorrelation function has been suggested for some soil properties, e.g., relative density, grain-size distribution and shear strength (Yamazaki and Shinozuka, 1989; Vanmarcke, 1977a). Other possible forms for the correlation function have also been used (Vanmarcke, 1977a; Fardis and Veneziano, 1981); e.g.,

$$\rho_{S,S}(r) = exp(-|r|/a) \quad (21)$$

with $a > 0.0$. This second form has the disadvantage that its second derivative at x=0.0 does not exist which implies that the random field representing the soil property is not differentiable. This problem is not present with the form of Eq. 17 which is adopted in this study. Here, b is treated as a parameter and the results are presented as a function of the dimensionless parameters B/b and L/b.

The actual values of the correlation function parameters, b or a, should be evaluated for a particular project as the adverse consequences of liquefaction will depend on the size of the lateral extent of liquefaction relative to the diameter of the buried conduit and the manhole dimensions. Various methods have been proposed to estimate the parameters b or a for the correlation function of soil properties (e.g., Fardis and Veneziano, 1981; Vanmarcke, 1977a). Vanmarcke(1977a) introduced the correlation length, δ_u, for a random field which can be related to the parameters b and a as follows

$$\delta_u = 2a, \quad \delta_u = \sqrt{\pi}\, b \tag{22}$$

This correlation length defines a distance such that the average number of uncorrelated observations of the soil property within it is 1.0.

The following horizontal correlation lengths have been reported for some soil properties in selected soil deposits: (a) for the relative density in a sand deposit (Fardis and Veneziano, 1981), δ_u = 67.0 m, which implies b = 37.8 m, with similar results obtained for the SPT-blowcount and the grain-size distribution; and, (b) for the shear strength of a varved clay for an enbankment stability study (Vanmarcke, 1977b), δ_u = 46.0 m, which implies b = 26.0 m. Although the results appear to indicate horizontal correlation lengths for soil properties of about 50.0 m to 60.0 m, it is recommended that the correlation lengths should be evaluated on a case by case basis.

A promising sampling technique for the estimation and construction of confidence intervals for the parameters of the whole joint distribution of a n-dimensional homogeneous random field, e.g., the power spectral density function or the autocorrelation function of the 2-dimensional random field used to represent the undrained shear strength of the sand against liquefaction, is the bootstrap sampling technique (Chrysikopoulos, 1992), in particular, the nonparametric sampling technique for homogeneous random fields proposed by Politis and Romano (1992).

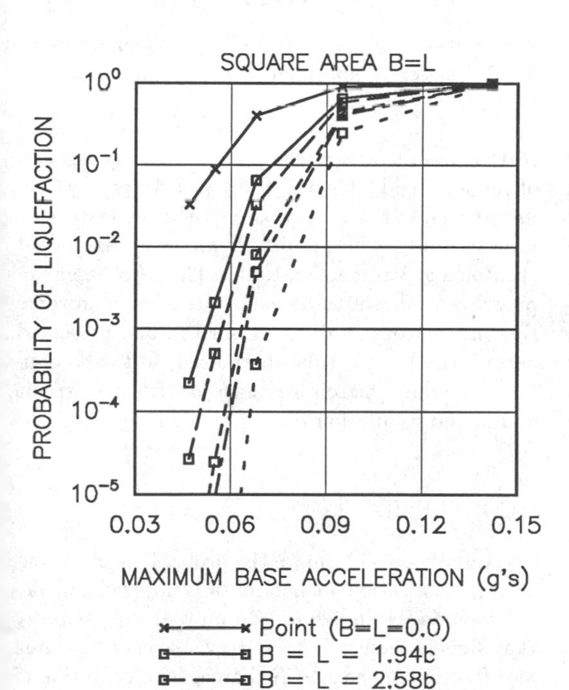

Fig. 4 Liquefaction fragilities for the soil deposit analyzed

3.2 Liquefaction fragilities

The liquefaction fragilities for square areas with sizes ranging from B = 1.94b to B = 5.81b are shown in Fig. 4. The probability of liquefaction spreading over a given area decreases very rapidly as the area increases. It should be noted that the size, D, of each elementary square in which the area $B \times L$ is divided is chosen as small as possible provided that the covariance matrix of the random variables C'_k, $k = 1, \ldots, N$ is positive-definite (Liu and Der Kiureghian, 1991). At this point, it should be noted that the computational efficiency and accuracy of the method may be improved if more efficient methods for discretizing the random field are used, e.g., optimal discretization methods of random fields used in conjunction with SFEM (Li and Der Kiureghian, 1992).

The probabilities of liquefaction spreading over a given area conditional on the occurrence of liquefaction at a specific point within the area are shown in Fig. 5 for various values of the maximum acceleration at the base of the soil deposit. It is noted that the probabilities that liquefaction in the layer will spread over a given area ($B \times L$), $B = L$, increase as the probability of liquefaction at a specific point within the area increases, i.e., as the severity of the loading increases. The results clearly illustrate the intuitive concept that greater loads are more likely to result in liquefaction spreading over greater areas.

Using the proposed method it is also possible to compute the probabilities that at least a specified fraction of an area $B \times L$ will liquefy (Hoshiya, 1991). When computing the probability that at least a specified fraction of an area will liquefy, the segments of area that liquefy can be anywhere within the area of interest and are not required to be contiguous to each other. Such probabilities for the soil deposit and earthquake loading considered in the previous example, and for an area with $B = L = 3.87b$ are shown in Fig. 6. Also shown

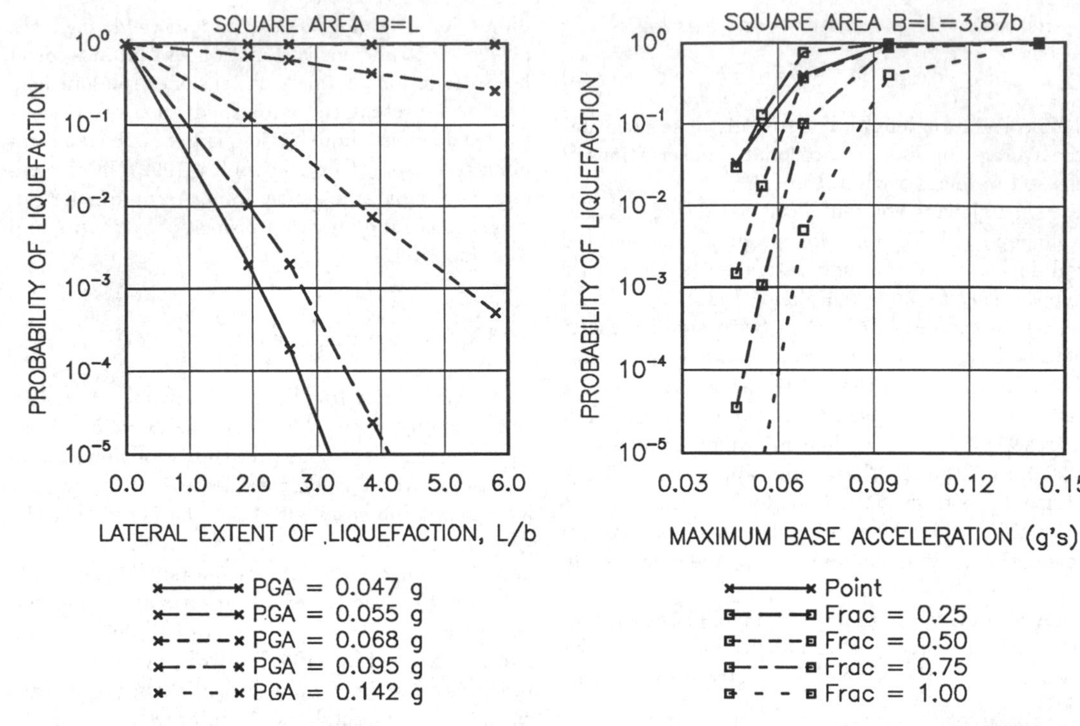

Fig. 5 Conditional probabilities of liquefaction spread

Fig. 6 Fragilitiy curves for soil deposit analyzed – fraction of specified area that liquefies

in Fig. 6 is the fragility curve corresponding to the probability of liquefaction at a specific point within the area of interest.

Note that, for the earthquake loading intensities considered, the probabilities that at least 25 percent of the specified area will liquefy are greater than the probability of liquefaction at a specific point within the area. These results are expected because the probability of liquefaction anywhere in the area of interest is greater than the probability of liquefaction at a specific point within that area. By comparing the results shown in Figs. 4 and 6, it is also noticed that, for a given earthquake loading, the probability that at least a specified percentage of a square area will liquefy is greater by approximately an order of magnitude than the probability that a square area with size equal to the fraction of the original area considered will liquefy in its entirety. As an example, consider the probabilities that at least 75 percent the area of interest will liquefy (see Fig. 6) and the probabilities that a square area with a side length equal to 3.39b (see Fig. 4) will entirely liquefy.

The possibility of extending the method for microzonation purposes is being investigated. In this regard, a simulation technique for the generation of random fields denoted the *Local Averaging Subdivision* (LAS) has been used to investigate spatial variability effects in liquefaction risk assessment (Fenton and Vanmarcke, 1991). The LAS-based approach was shown to be well-suited for microzonation studies but it differs from the one presented here in the liquefaction index used, in the computation of the dynamic response of the soil deposit and in the simulation technique used.

4 CONCLUSIONS

A method to calculate the probability of the onset of liquefaction in horizontally layered soil deposits is is extended to compute the probability that liquefaction will spread over a specified area or a fraction of the specified area. The definition of liquefaction fragilities is generalized to be the probability that liquefaction in the layer will spread over a specified area or a fraction of that area as a function of the expected maximum ground acceleration of the earthquake loading specified at the base of the soil deposit. Those probabilities are useful for the seismic safety assessment of buried lifelines such as conduits for high-voltage electric power transmission lines.

It is observed, that the probability that liquefaction will spread over a given area decreases substantially as the contiguous lateral extent of liquefaction increases. Also, the probability of liquefaction in the layer spreading over a given area increases as the probability of liquefaction for a specific point within the area of interest increases, i.e., as the severity of the load increases. Finally, it is noted that the probability that liquefaction will spread over a small fraction of a specified area can exceed the probability of liquefaction at a specific point within that area, and that the probability of liquefaction anywhere within an area can be significantly greater than the probability of liquefaction at a specific point within that area.

REFERENCES

Chrysikopoulos, C. 1992. *Personal Communication.*

Clough R. and J. Penzien 1975. *Dynamics of Structures*, McGraw-Hill, New York.

Fardis, N. M. and D. Veneziano 1981. Estimation of SPT-N and relative density. *J. Geotech. Engrg.*, ASCE, 107(10) 1345-1359.

Fenton, G. A. and E.H. Vanmarcke 1991 Spatial variation in Liquefaction risk assessment. *Geotechnical Engineering Congress, Vol. I*, eds. F.G. McLean, D.A. Campbell and D.W. Harris(ASCE, New York) 594-607.

Hoshiya, M. 1991. Probability of liquefaction spread. in: *Stochastic Structural Dynamics 2: New Practical Applications*, eds. I. Elishakoff and Y.K. Lin (Springer-Verlag, Heidelberg) 45-54.

Li, C-C. and A. Der Kiureghian, A. 1992. Optimal discretization of random fields for SFEM. in: *Probabilistic Mechanics and Structural and Geotechnical Reliability*, ASCE, ed. Y. K. Lin, 29-32.

Liu, P. L. and A. Der Kiureghian 1991. Finite element reliability of geometrically nonlinear uncertain structures. *J. Engrg. Mech.*, ASCE, 117(8) 1805-1825.

Pires, J. A., Ang, A. H-S. and I. Katayama 1989. Probabilistic analysis of liquefaction. in: *Structural Dynamics and Soil Structure Interaction*, eds. A.S. Cakmak and I. Herrera(Computational Mechanics Publications, England, 1989), pp.155-168.

Pires, J. A., Ang, A. H-S. and I. Katayama 1993. Probabilistic evaluation of lateral extent of liquefaction under earthquake loading. *submitted, Nuclear Engrg. Design.*

Politis, D. N. and J. P. Romano 1992. Nonparametric resampling technique for homogeneous strong mixing random fields. Research Report, Department of Statistics, Purdue University, West Lafayette, Indiana.

Vanmarcke, E. H. 1977a. Probabilistic modeling of soil profiles. *J. Geotech. Engrg. Div.*, ASCE, 103(11) 1227-1246.

Vanmarcke, E. H. 1977b. Reliability of earth slopes. *J. Geotech. Engrg.*, ASCE, 103(11) 1247-1265.

Yamazaki, F. and M. Shinozuka 1989. Statistical preconditioning in simulation. in: *Structural Safety and Reliability, Vol. II*, eds. A. H-S. Ang, M. Shinozuka, G. I. Schueller(ASCE, New York, USA) pp. 1129-1136.

Structural Safety & Reliability, Schuëller, Shinozuka & Yao (eds) © 1994 Balkema, Rotterdam, ISBN 90 5410 357 4

Reliability-based limit state design of pile foundations of highway bridges

Wataru Shiraki & Shigeyuki Matsuho
Tottori University, Japan

ABSTRACT: A lot of studies on reliability analysis of superstructures of highway bridges have been performed, and the introduction of reliability-based limit state design is discussed in many countries. For substructures and foundations, however, not so much attention has been focused on them. In this study, the necessity and effectiveness of reliability-based design of foundations of highway bridges are discussed. In analysis four typical types of steel pier and pile foundation systems supporting three-span continuous box girder bridge are selected out of the existing actual highway bridge structures in Japan. The both design results by the reliability-based limit state design and the conventional allowable stress design method are compared.

1. INTRODUCTION

In many countries, nowadays, the Limit State Design Method (LSDM) based on the reliability theory is introduced into the structural design standards instead of the conventional Allowable Stress Design Method (ASDM). Especially, the Ontario Highway Bridge Design Code (OHBDC) of Canada (Ministry of Transportation and Communications 1983) is well known as the first design code adopted the reliability-based LSDM.

In Japan, however, designs of structures except concrete structures have still being made according to the ASDM. In order to introduce the reliability-based LSDM into the Japanese Steel Highway Bridge Specification(JSHBS) (Japanese Society of Highway 1990), the authors are investigating the reliability-based design method of highway bridges since 1980. In ICOSSAR'89, we presented a study on probabilistic evaluation of load factors for steel rigid-frame piers on urban expressways (Shiraki, Matsuho and Takaoka 1989).

In this study, our attention is focused on the pile foundation systems supporting highway bridges. The effectiveness of reliability-based design of pile foundations is discussed comparing the design results by the OHBDC and the JSHBS.

2. LIMIT STATE DESIGN METHOD OF PILE FOUNDATION

2.1 *Current code of based on allowable stress design method in Japan*

In the JSHBS, the design formulas for pile foundations are expressed as follows;

$$\sum_{j} P_j \leq R'(=R/\alpha) \qquad (1)$$

where R' is allowable bearing capacity, R is ultimate bearing capacity, α is safety factor, and P_j is external force acting on the pile top due to the load j. P_j is evaluated by the theory of elasticity. The current JSHBS has various shortcomings such that the safety margins of piles designed by Eq.(1) are constant regardless of structural size and the safety factor α is determined mainly by experiences and so on.

2.2 *Limit state design method in Canada*

In the OHBD code (Ministry of Transportation and Communications 1983), ultimate limit states are defined theoretically or experimentally and design calculations of pile foundations are performed using load and resistance factor design formulas. One of them is given as

$$f \cdot R \geq \gamma_D \cdot D + \eta \cdot (\gamma_L \cdot L + \gamma_E \cdot E) \quad (2)$$

where f, R and η are the resistance factor, bearing capacity of pile and importance factor (=1.0, usually) of the structural system, respectively. And D, L and E are the nominal values of dead, live and earthquake loads. Moreover, γ_D, γ_L and γ_E are the load factors corresponding to D, L and E. In the OHBDC, the resistance and load factors are provided as follows;
For load factors;

γ_{DC}=1.25 (for cast-in-place concrete),

γ_{DB}=1.20 (for banking),

γ_L =1.40 and γ_E=1.30.

For resistance factors;

f_ϕ=0.80 (for frictional force),

f_C =0.50 (for cohesion), f_S = 0.90 (for steel).

3. DESIGN OF PILE FOUNDATION BY LIMIT STATE DESIGN METHOD

3.1 *Structures and ground under consideration*

The design is made for four typical types of steel rigid frame pier structures and pile foundation systems supporting a three-span continuous box girder bridge of span length L= 40 (m) and effective road width 18.2 (m) (see Figs.1 and 2). These four types of structural models were provided based on the existing bridge

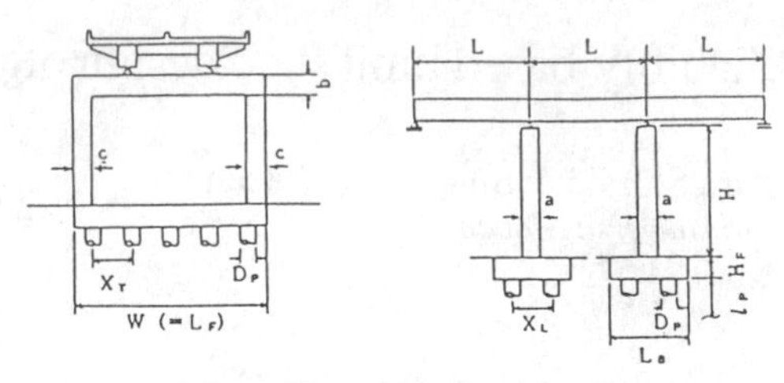

Fig.1 Rigid-Frame Pier and Pile Foundation System Supporting Three-Span Continuous Box Girder Bridge

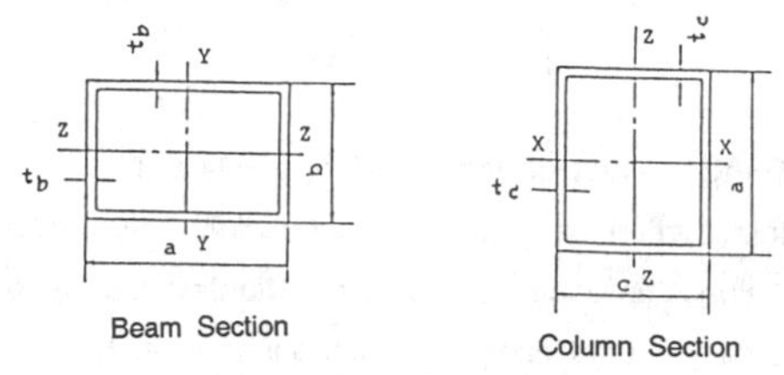

Fig.2 Cross Section of Rigid-Frame Pier

structures constructed on the Hanshin (Osaka-Kobe area in Japan) Expressway Network. Table 1 shows the configuration of each model determined by considering the combination of two basic parameters such as the total height of pier H=10, 20 (m), and the total width of pier W=20, 30 (m). Superstructures of these structural models are all the same. The pile foundation is assumed to be continuous footing foundation type whose earth covering is 1.0 (m). Fig.3 shows boring log at the site.

3.2 *Definition of limit state*

Ultimate limit state of pile foundation systems is defined according to the OHBDC as follows. Design is made considering combination of dead and earthquake loads (D+E). In this case, pile foundation is subjected to

Table 1 Four Models for Pier Structure

Model No.	L(m)	H(m)	W(m)	h(m)	l (m)	a(m)	b(m)	c(m)	t_b(mm)	t_c(mm)	Weight(t)
1	40.0	10.0	20.0	9.17	18.5	2.00	1.67	1.5	22.5	28.8	62.5
2	40.0	10.0	30.0	8.75	28.0	2.00	2.50	2.0	29.0	39.3	118.7
3	40.0	20.0	20.0	19.17	18.0	2.00	1.67	2.0	20.9	23.0	91.3
4	40.0	20.0	30.0	18.75	27.5	2.00	2.50	2.5	24.0	26.6	138.7

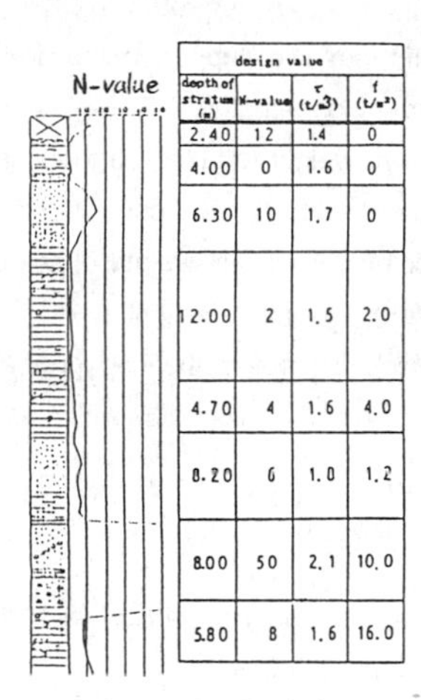

design value			
depth of stratum (m)	N-value	τ (t/m3)	f (t/m²)
2.40	12	1.4	0
4.00	0	1.6	0
6.30	10	1.7	0
12.00	2	1.5	2.0
4.70	4	1.6	4.0
8.20	6	1.0	1.2
8.00	50	2.1	10.0
5.80	8	1.6	16.0

Fig. 3 Boring Log Foundation

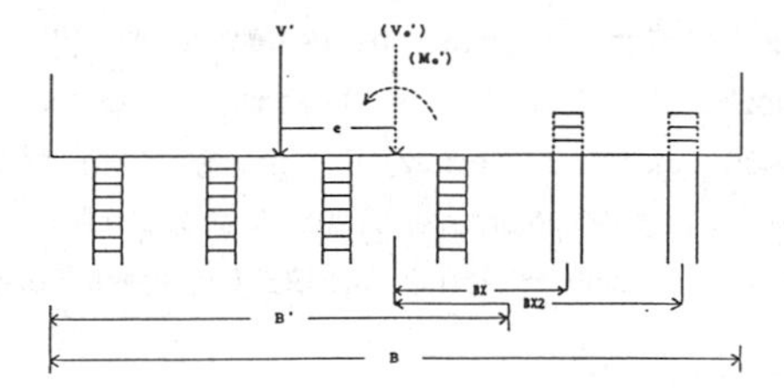

B : Width of Footing
B' : Width of Footing Subjected to Compressive Force
n : Number of Piers Subjected to Compressive Force
n' : Number of Piers Subjected to Pull-Out Force
V_0' : Design Value for Vertical Force (= V')
M_0' : Design Value for Moment
e : Eccentric Distance
BX : Distance from Centre of Footing to Centre of Pier Subjected to Pull-Out Force

Fig. 4 Ultimate Limit State Based on the OHBDC

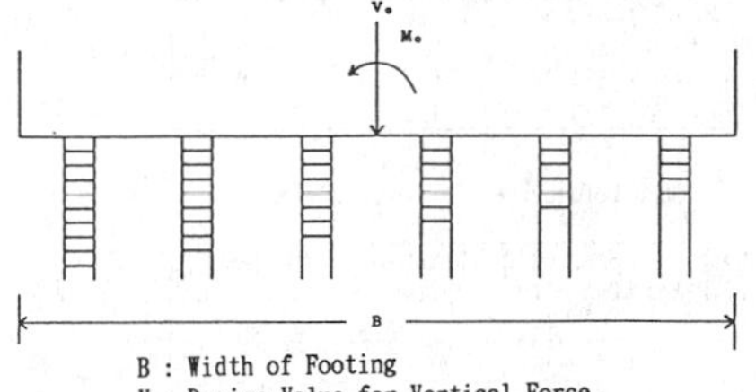

B : Width of Footing
V_0: Design Value for Vertical Force
M_0: Design Value for Moment

Fig. 5 Ultimate State Based on the JSHBS

eccentric load V' which can be considered as vertical force V_0' and moment M_0' as shown in Fig.4. In the OHBDC, total width of footing foundation is divided into two zones, assuming that compressive force acts on one side and that tensile force acts on the other side. Moreover, resistant forces piles in compressive zone are assumed to be the same. Resistant forces of piles in tensile zone are also assumed to be the same. In Fig.4, the compressive zone is expressed by effective width B', and is derived from equilibrium of vertical force and moment as follows;

$$V = n \cdot Q_f - n' \cdot Q'_f$$
$$V(B/2-e) = n \cdot Q_f \cdot B'/2 - n' \cdot Q'_f \cdot B' \qquad (3)$$
$$B' = n \cdot B/(n+n')$$

where n is the number of the piles in compressive zone, Q_f is the design value of their ultimate vertical load, n' is the number of the piles in tensile zone, Q_f' is the design value of pull-out resistance, and V is vertical component at footing edge of the side where compressive force due to eccentric load V' is induced.

The limit state in the current JSHBS corresponding to the limit state of the OHBDC is shown in Fig.5. The design of pile foundation for this limit state is performed using Eq.(1).

3.3 *Design of pile foundation*

The design of steel rigid frame pier structures is carried out under the above-mentioned conditions so that vertical force acting on top of pile do not exceed bearing capacity of pile. The bearing capacity of pile is estimated by formula or load test. Vertical force acting on the top of each pile can be considered to consist of two components as mentioned in section 3.2. These two vertical components is expressed as follows;

(1) The component due to vertical force V_0' is expressed by

$$P_{NV} = V_0'/n + (n'/n) \cdot P_{NV}' \qquad (4)$$

where P_{NV}' is pull-out force acting on each pile due to vertical component.

(2) The component due to moment M_0' is expressed by

$$P_{NM} = M_0'/(BX+BX2) - P_{NM}' \quad (5)$$

where P_{NM}' is pull-out force acting on each pile due to moment component. In this study, design calculations are performed for both cases considering and neglecting the influence of pull-out force. Moreover, pile foundation is designed based on the same assumptions as in our early study (Shiraki, Matsuho and Takaoka 1987) in order to compare design results with each other.

These assumptions are as follows; 1) depth of embedment into bearing layer is constant (2m), 2)interval of pile centres is 2.5 times of pile diameter, 3)each pile is fixed to the footing in the vertical direction, 4) negative frictional force along the circumference of pile is not considered, 5) design of footing is not carried out, but its thickness is determined according to footing width in the transverse direction of the bridge, 6) wall thickness of steel circular pile is 14(mm), and 7) design variable is diameter of the pile.

Parameter values using in the design are as follows; reaction force due to dead load of superstructure R_D=1105(tonf) (per two supports); reaction force due to live load of superstructure R_L=1105(tonf) (per two supports); own weight of pier 63(tonf) (for model 1), 119(tonf) (for model 2), 92(tonf) (for model 3), and 139(tonf) (for model 4); unit weight of soil on the footing γ_S=1.4(tonf/m^3); unit weight of the footing γ_F=2.5(tonf/m^3). Moreover, bearing ground is assumed to be sand layer (N-value>50) which is deeper than 37.6(m).

3.4 *Design results and comparison with the design results by the JSHBS*

According to the OHBDC, the design of pile foundation are carried out. The results are shown in Fig.6, Tables 2 and 3. Table 4 shows the design results (Shiraki, Matsuho and Takaoka 1987) according to the old JSHBS (Japanese Society of Highway 1980) before revision at 1990. Moreover, Table 5 shows the design results according to the new JSHBS (Japanese Society

Table 2 Design results of pile foundation by the OHBDC (not considering pull-out)

Model No.	Footing Size $L_F \times L_B \times H_F$ (m×m×m)	Size of Pile: Diameter D (mm)	Thickness t_p (mm)	Length l_p(mm)	Interval x_T (m)	x_L (m)	Number of Piles	Bearing Capacity (t)	Natural Period (sec): Longitudinal Direction	Transverse Direction
1	20× 9.07×5.0	907	14.0	33.6	2.50	2.27	8×4=32	598.1	0.5	0.5
2	30×10.92×7.5	874	14.0	31.1	2.50	2.18	12×5=60	613.5	0.5	0.5
3	20×11.29×5.0	904	14.0	33.6	2.50	2.26	8×5=40	593.1	0.7	1.0
4	30×12.46×7.5	831	14.0	31.1	2.50	2.08	12×6=72	542.1	1.0	1.0

Table 3 Design results of pile foundation by the OHBDC (considering pull-out)

Model No.	Footing Size $L_F \times L_B \times H_F$ (m×m×m)	Size of Pile: Diameter D (mm)	Thickness t_p (mm)	Length l_p(mm)	Interval x_T (m)	x_L (m)	Number of Piles	Bearing Capacity (t)	Natural Period (sec): Longitudinal Direction	Transverse Direction
1	20× 8.59×5.0	859	14.0	33.6	2.50	2.15	8×4=32	523.6	0.5	0.5
2	30×10.30×7.5	824	14.0	31.1	2.50	2.06	12×5=60	531.6	0.5	0.5
3	20×10.82×5.0	866	14.0	33.6	2.50	2.17	8×5=40	534.2	0.7	1.0
4	30×12.46×7.5	827	14.0	31.1	2.50	2.07	12×6=72	535.7	1.0	1.0

Table 4 Design results of pile foundation by the JSHBS before the revision

Model No.	Footing Size $L_F \times L_B \times H_F$ (m×m×m)	Size of Pile: Diameter D (mm)	Thickness t_p (mm)	Length l_p(mm)	Interval x_T (m)	x_L (m)	Number of Piles	Bearing Capacity (t)	Natural Period (sec): Longitudinal Direction	Transverse Direction
1	20× 7.52×5.0	752	14.0	33.6	2.50	1.89	8×4=32	455.1	0.5	0.5
2	30× 9.45×7.5	756	14.0	31.1	2.50	1.88	12×5=60	475.5	0.5	0.5
3	20×10.55×5.0	844	14.0	33.6	2.50	2.11	8×5=40	510.7	0.7	1.0
4	30×12.75×7.5	850	14.0	31.1	2.50	2.13	12×6=72	514.4	1.0	1.0

Table 5 Design results of pile foundation by the JSHBS after the revision

Model No.	Footing Size $L_F \times L_B \times H_F$ (m×m×m)	Size of Pile: Diameter D (mm)	Thickness t_p (mm)	Length l_p(mm)	Interval x_T (m)	x_L (m)	Number of Piles	Bearing Capacity (t)	Natural Period (sec): Longitudinal Direction	Transverse Direction
1	20× 7.25×5.0	725	14.0	33.6	2.50	1.81	8×4=32	438.8	0.5	0.5
2	30× 9.06×7.5	725	14.0	31.1	2.50	1.81	12×5=60	438.8	0.5	0.5
3	20×10.26×5.0	821	14.0	33.6	2.50	2.05	8×5=40	496.9	0.7	1.0
4	30×12.30×7.5	820	14.0	31.1	2.50	2.05	12×6=72	496.3	1.0	1.0

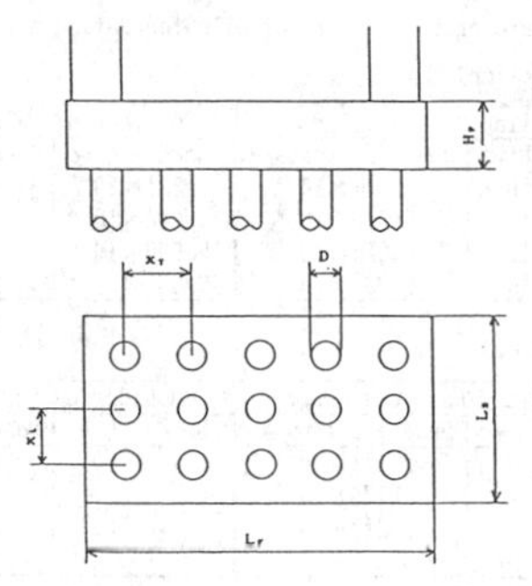
Fig. 6 Structural Size of Pile Foundation

of Highway 1990) after revision at 1990. In Figs.2 through 5, it is clearly seen that the design results by the OHBDC are a little larger than those of the JSHBS. This difference lies in the reason that practical safety factors of the JSHBS and the OHBDC under the combination D+E are 2.0 and about 2.8, respectively. Making a comparison between the design results before and after revision at 1990, we can see that the results after revision is a little smaller than the results before revision. The improvement of accuracy for estimating bearing capacity is one of reasons.

4. SAFETY ESTIMATION OF PILE FOUNDATION

4.1 *Definition of limit state function of pile foundation*

In this study, for simplicity, limit state function f(•) is defined as follows;

$$f(R_u, P_N) = R_u - P_N \qquad (6)$$

where R_u is bearing capacity of the pile in the vertical direction, and P_N is vertical external force acting on the pile top. When f(•)<0, structure is assumed to be failed. This means that the failure of pile is occurred when reaction force at the pile top exceeds bearing capacity of the pile.

4.2 *Modeling of vertical bearing capacity of pile*

For simplicity, considering only dominant uncertainties, the bearing capacity R_u in the vertical direction is formulated by Eq.(7). This equation is derived using the condition that R_u in the ultimate limit state is equal to ultimate bearing capacity obtained by the pile load test in Japan.

$$R_u = \alpha_R' \cdot R_n' \qquad (\alpha_R' = \alpha_R \cdot R_n/R_n') \qquad (7)$$

where R_n is ultimate bearing capacity obtained by the JSHBS, R_n' is ultimate bearing capacity obtained by the OHBDC, and α_R is random variable which reflects such uncertainties as error at design formulation including error at modeling for bearing mechanism and ground etc. This α_R is expressed by the ratio R_e/R_n in which R_e is ultimate bearing capacity estimated by the pile load test in Japan. α_R is modeled by log normal distribution with expected value 0.99 and standard deviation 0.48. Moreover, the pull-out bearing capacity is modeled by a random variable with log normal distribution with expected value 1.0 and standard deviation 0.5.

4.3 *Modeling of vertical load acting on pile top*

Dead, live and earthquake loads are considered as loads acting on pile foundation structure. Two combination cases for these three loads are specified in the OHBDC as follows; 1) dead load and live load (D+L), 2) dead load and earthquake load (D+E). Each actual load component is modeled based on extensive investigation (HDL Committee 1986) on actual conditions of various loads acting on urban expressway bridges of Hanshin (Osaka-Kobe) area in Japan. Outlines of the modeling are as follows;

(1) Dead Load: Only the own weight of structure is considered as dead load and is assumed to be deterministic.

(2) Live Load: The actual live load is modeled as the reaction force on the piers by using the Monte-Carlo simulation technique. The CDF of live load concerning maximum value during bridge lifetime period T (=50 years) is as follows;

$F_L(x)=\{F_L'(x)\}^n$

$$=\exp[-n\bullet\exp\{-0.1942(x-120.1)\}] \quad (8)$$

$(x>0, \text{ unit: tonf})$

where $F_L'(x)$ is the CDF of live load concerning maximum value during one month and n=600 is number of months during lifetime period T. $F_L'(x)$ is modeled by Weibull distribution. Mean value and standard deviation of Eq.(8) are 156.0(tonf) and 6.6(tonf), respectively. These characteristics are evaluated for one supports on the pier.

(3) Earthquake Load: Actual earthquake load is modeled by linear acceleration response spectrum, S_A. The CDF of the maximum S_A during lifetime period T is given by

$$F_{E50}(x)=\exp[-\nu T\{1-F_E(x)\}] \quad (9)$$

where $F_E(x)$ is the CDF of S_A evaluated on the assumption that the occurrence of earthquake is Poisson distributed and that its average return period is considered to be greater than 2 years. And ν is occurrence rate (=0.5/year) of the earthquake of $F_E(x)$. $F_E(x)$ is provided by the CDF of the Weibull distribution as follows;

$$TT=0.5(sec): FE(x)=1-\exp[-\{x41.28)/34.24\}0.913] \quad (x>41.28, \text{ Unit:gal})$$
$$TT=0.7(sec): FE(x)=1-\exp[-\{x25.88)/26.12\}0.879] \quad (x>25.88, \text{ Unit:gal}) \quad (10)$$
$$TT=1.0(sec): FE(x)=1-\exp[-\{x17.91)/18.05\}0.850] \quad (x>17.91, \text{ Unit:gal})$$

where TT indicates the natural frequency of the structure.

4.4 *Numerical example*

The numerical calculations are carried out using the IFM (Iterative Fast Monte-Carlo Procedure) (Bucher 1988).

(1) Comparison with Design Based on the JSHBS: Tables 6 and 7 show failure probability P_f of design results shown in section 3.4. In these tables, it is shown that P_f of design results calculated according to the OHBDC is much smaller than those of design results by the JSHBS. This is that the safety margin of pile design by the OHBDC is larger than that of the JSHBS as mentioned in section 3.4.

Table 6 Failure probability P_f of pile foundation by the OHBDC

(Longitudinal Direction)

Model No.	Neglecting Pull-Out Force		Considering Pull-Out Force	
	Dead + Live	Dead +Earthquake	Dead + Live	Dead +Earthquake
1	2.558×10^{-4}	8.547×10^{-3}	1.371×10^{-3}	4.342×10^{-3}
2	1.588×10^{-4}	6.010×10^{-3}	1.246×10^{-3}	3.492×10^{-3}
3	1.967×10^{-4}	8.625×10^{-5}	5.609×10^{-4}	2.881×10^{-4}
4	8.330×10^{-4}	9.232×10^{-5}	9.331×10^{-4}	2.182×10^{-4}

(Transverse Direction)

Model No.	Neglecting Pull-Out Force		Considering Pull-Out Force	
	Dead + Live	Dead +Earthquake	Dead + Live	Dead +Earthquake
1	-	2.111×10^{-3}	-	6.075×10^{-4}
2	-	1.176×10^{-3}	-	3.868×10^{-4}
3	-	8.217×10^{-5}	-	1.596×10^{-4}
4	-	4.258×10^{-6}	-	7.947×10^{-6}

Table 7 Failure probability P_f of pile foundation by the JSHBS

(Longitudinal Direction)

Model No.	Before Revision		After Revision	
	Dead + Live	Dead +Earthquake	Dead + Live	Dead +Earthquake
1	2.448×10^{-3}	1.231×10^{-2}	2.731×10^{-3}	1.126×10^{-2}
2	4.036×10^{-3}	1.793×10^{-2}	4.385×10^{-3}	1.616×10^{-2}
3	1.065×10^{-3}	2.953×10^{-3}	1.134×10^{-3}	2.741×10^{-3}
4	2.813×10^{-3}	6.612×10^{-3}	2.943×10^{-3}	6.284×10^{-3}

(Transverse Direction)

Model No.	Before Revision		After Revision	
	Dead + Live	Dead +Earthquake	Dead + Live	Dead +Earthquake
1	-	4.227×10^{-3}	-	4.084×10^{-3}
2	-	6.542×10^{-3}	-	6.144×10^{-3}
3	-	2.168×10^{-3}	-	2.114×10^{-3}
4	-	3.715×10^{-3}	-	3.663×10^{-3}

(2) Discussions on Limit State Design Method: Fig.7 shows the change of P_f for four models. Failure probabilities of Models 1 and 2 are calculated for the same external forces such as live and earthquake loads without dead load. Failure probabilities of Models 3 and 4 are also calculated for the same live and earthquake loads without dead load. Therefore, it is shown that

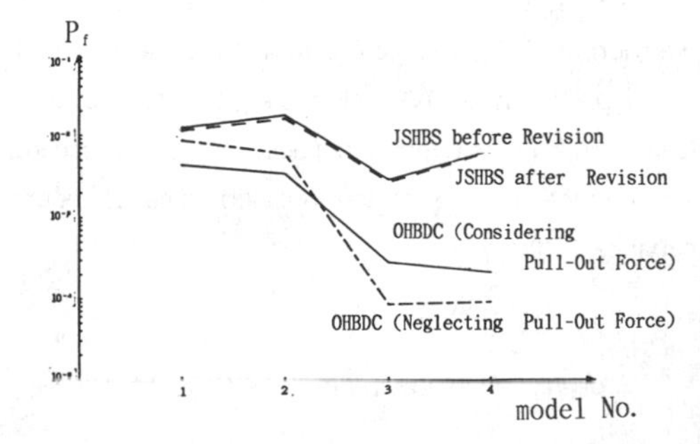

Fig.7 Failure probability P_f versus model No.

design results by the OHBDC have almost consistent safety for Models 1 and 2, and for Models 3 and 4, regardless different type of structures. But, it is also shown that the change of safety for pile foundation designed by the OHBDC for different external forces is larger than change of safety of that designed by the JSHBS.

5. CONCLUDING REMARKS

The main results obtained in this study are as follows;

(1) The reliability levels of pile foundations designed by the OHBDC and the JSHBS are evaluated, and then it is shown that the OHBDC gives rather higher level of reliability as compared with that of the JSHBS.

(2) The OHBDC insures the consistent level of reliability for different type of pile foundations subjected to same earthquake load, on the other hand, the JSHBS does not insures it. This result clarifies the effectiveness of reliability-based LSDM.

(3) Discussions on reliability-based limit state design of pile foundations of highway bridges in Japan are presented.

REFERENCES

Bucher,C.G. 1988. Adaptive Sampling - An Iterative Fast Monte-Carlo Procedure. *Structural Safety:* Vol.5, pp.119-126.

HDL Committee 1986. *Report on Investigation of Design Load Systems on Hanshin Expressway Bridges:* Hanshin Expressway Public Corporation. (in Japanese)

Japanese Society of Highway 1980. *Japanese Steel Highway Bridge Specification and its comments.* Vols. I ,IV and V. (in Japanese)

Japanese Society of Highway 1990. *Japanese Steel Highway Bridge Specification and its comments.* Vols. I ,IV and V. (in Japanese)

Ministry of Transportation and Communications 1983. *Ontario Highway Bridge Design Code and Commentary:* Ontario.

Shiraki,W, Matsuho,S and Takaoka,N. 1987. Reliability Analysis of Pile Foundation Supporting Rigid-Frame Steel Piers of Highway Bridge. *Proc. of JCOSSAR'87:* Vol.1. pp.207-212. (in Japanese)

Shiraki,W, Matsuho,S and Takaoka,N. 1989. Probabilistic Evaluation of Load Factors for Steel Rigid-Frame Piers on Urban Expressway Network. *Proc.of ICOSSAR'89:* pp.1987-1994.

Structural Safety & Reliability, Schuëller, Shinozuka & Yao (eds) © 1994 Balkema, Rotterdam, ISBN 90 5410 357 4

Quantification of slope instability risk and cost parameters for geotechnical applications in a highway project and in a regional study

T. Shuk
Universidad de los Andes, Bogotá, Colombia

A.J. González
Universidad Nacional and Análisis Geotécnicos Colombianos Ltda, Bogotá, Colombia

ABSTRACT: This paper briefly describes some of the interesting aspects and results of the application of slope stability, failure probability and economic decision analyses for two practical cases. In the first one, the Bogotá-Villavicencio highway, these analyses resulted in total costs functions which consisted of the sum of the initial (where applicable) and maintenance costs plus the direct and indirect expected failure costs, and which allowed the determination of the best decision among the three planning alternatives which were analyzed. The purpose of the analyses for the second practical case, - the Utica disaster area -, was to obtain an answer to the question of whether the town should be relocated or not, after it suffered damage due to an avalanche triggered by the accidental damming of the Negra Creek, and the subsequent failure of this dam. The massive geotechnical quantification required for these analyses could not have been carried out without the new "Natural Slope Methodology" (NSM).

1 INTRODUCTION

The topics presented in this paper are arranged according to a sequence in which a very brief description of the Natural Slope Methodology (NSM) is presented first. The next two headings correspond to the two practical application cases of the Bogotá-Villavicencio Highway and the Utica Disaster Area, each of which includes a short description of the study area, the purpose of the study, the highlights of its analytical process, and of its most pertinent final results. The closing text corresponds to the pertinent conclusions and acknowledgements.

2 THE NATURAL SLOPE METHODOLOGY (NSM)

A summarized version of this methodology and its pertinent applications is presented in a companion paper in this Volume (T. Shuk; "Key elements of the Natural Slope Methodology (NSM) and of its applications for slope stability, failure probability and economic decision analyses"). In short, by means of NSM one can obtain, - with a highly attractive cost and time effectiveness -, the estimates required for these analyses simply on the basis of adequate topographic maps with general geologic formation (or member) boundaries, and, - contrary to the prevailing conventional slope stability methodologies -, without carrying out borings or laboratory tests.

The nature of this paper is totally descriptive. For a better understanding of some NSM terms such as family, population, method of universes, etc., and of the underlying analytical bases, the reader is urged to read first the previously mentioned companion paper, or to refer to it whenever necessary. *

3 THE BOGOTA-VILLAVICENCIO HIGHWAY RELIABILITY STUDY

The 107 km Bogotá-Villavicencio highway traverses rugged mountainous topography, and is the main access to the "Llanos" (flatland) region of the eastern part of Colombia, which extends to the border with Venezuela. This region, which has an area of approximately 250,000 km^2, covers nearly one-fourth of the country's area. The town of Villavicencio, with a population of approximately 350,000 inhabitants, is the main transportation and distribution hub of the "Llanos", and is undergoing an accelerated growth process, spurred by recent major oil field discoveries.

Notwithstanding its economic importance, the Bogotá-Villavicencio highway has defi-

*) Note by the editors: for the accompaning paper please contact the author directly.

cient characteristics and is subject to frequent traffic interruptions, especially during the rainy seasons, due to landslides, avalanches, and other types of mass movements. These interruptions range from several hours, or a few days, up to the duration of the last major interruption (Sept, 1991), which lasted almost one month. The Villavicencio Chamber of Commerce estimated that the economic loss due to this interruption was of the order of US$38,000,000; other less reliable estimates go up to figures as high as US$60,000,000.

3.1 Purpose of the study

The purpose of this study was the comparative economic analysis of three different decisions regarding the future of this highway:

Decision 1: To do nothing about the presently existing highway; except, of course, to continue with preventive measures, and the routine maintenance operations (and the extraordinary ones, whenever necessary).

Decision 2: To build a so-called "pair-way" which is in an advanced design stage, and which involves the building of a two-lane highway on the opposite side of the Negro River to the one where the existing two-lane highway now traverses, with three bridges (one at each end of the Negro River stretch, and one bridge at an intermediate location) connecting the two branches of "pair-way", plus two short tunnels, some layout realignments and alignment changes, and others; the Negro River stretch, which is approximately 40 km long, is one of the most unstable ones of this highway. The basic purpose of the "pair-way" is not that of using fully its 4 lane traffic capacity, but to detour traffic from the existing highway to the opposite side of the Negro River whenever it becomes obstructed due to mass movements in this stretch, and then to return the traffic to the existing highway.

Decision 3: To design and build an Italian type viaduct-tunnel ("autostrada") expressway.

3.2 Highlights of the analytical process

The geotechnical risk data base for this study consisted of the measurement of 249 natural slope families in 18 different lithologies, with a gradual younger to older age-wise west to east transition from quaternary fluvioglacial, colluvial, and terrace deposits, to sedimentary rocks of Cretaceous and Jurassic age, and finally, sedimentary and metamorphic rocks of Paleo-

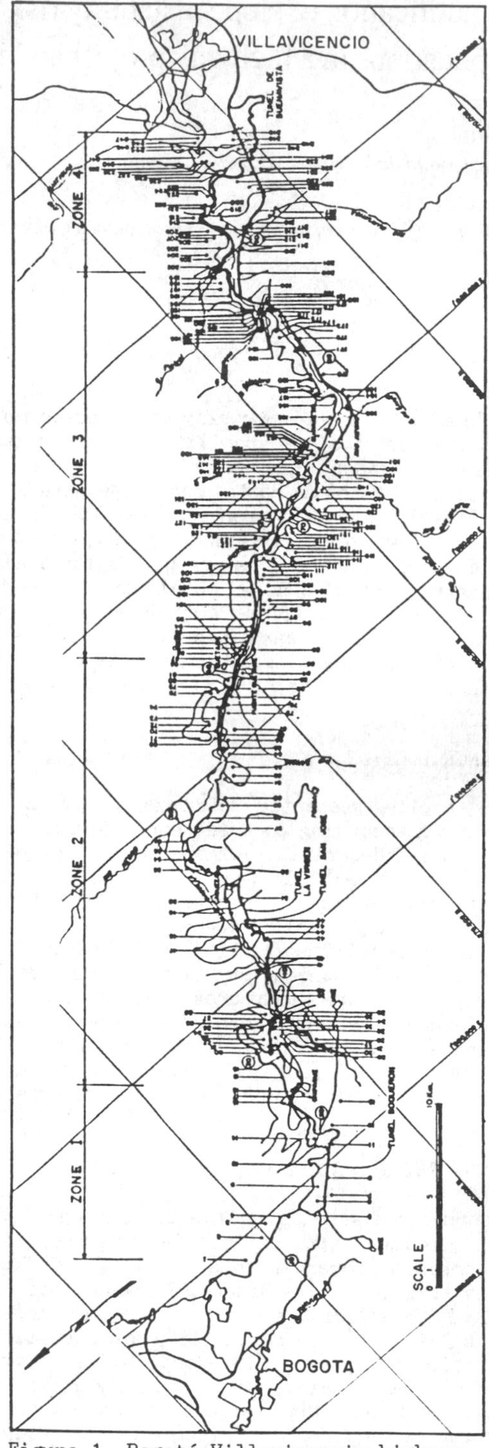

Figure 1. Bogotá-Villavicencio highway. Natural slope families

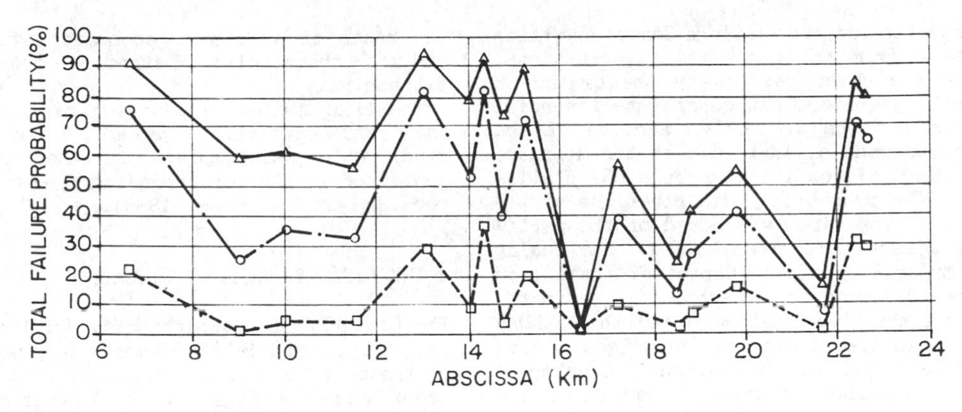

Figure 2. Total (relative) failure probability profile, Zone 1, Bogotá-Villavicencio highway

zoic age. Figure 1 gives an idea of the density of natural slope measurements.

In addition to the very rugged topography, these other uncommon features, which had to be involved in the analysis, posed some very special problems: the high rainfall, - from 1000 to 1500 mm/year in the area neighboring Bogotá, up to 3000 to 4000 mm/year as the alignment approaches the Llanos foothills next to Villavicencio-, and its high seismic susceptibility, which is configured by a system of large geologic faults, some of them intersecting the highway, and which are subsidiary to the major seismogenic source of the Llanos Frontal Fault; this fault runs parallel to the Llanos foothills and is intersected by the highway very close to Villavicencio.

Due to the fact that the design of the pair-way had not been completed at the time of this study, and hence, no design slopes were available, the method of "universes", - which results in relative safety factor and failure probability values -, had to be used.

The 1:10,000 scale topographical maps used for the NSM measurements were published in 1965, and were based on aerial photographs taken during the same year. Part of the results of this study consisted of total (relative) failure probability versus highway abscissa graphs, such as the one illustrated in Figure 2 for the three Safety Factor time periods described previously.

One of the major highlights of this study was the fact that if NSM would have existed at that time, it would have been possible with these graphs to pinpoint three major disasters which started occurring in 1974, one of which caused a large number of deaths.

Apart from the many other interesting features of the analysis, - and which, due to space limitations, cannot be included -, there are two aspects which require presentation, both of them related to the two components of the expected (probabilistic) cost member of the total expected cost function, and both of which were estimated for two economic lifetimes (25 and 40 years) of each one of the three analyzed decision:

Expected Direct Failure Costs: these are costs to be covered by the owning or operating entity and are directly related to the damage caused by mass movements; they involve the costs of removing mass movement debris, the complete reconstruction of the damaged length of highway including its side-slopes, and others; in the Bogotá-Villavicencio case they did not include property loss (destruction of homes, vehicles, etc.). In the Colombian cost assessment practice, these expected direct failure costs have three dimensional bases: Volume - V[m^3] for the removal of the debris and other items, Area - A[m^2] among others, for some slope reconstruction, and slope failure protective and preventive measures; and Length - L[km] for the complete reconstruction of the damaged stretch of highway. The probability distributions for these three dimensional bases (V,A and L) were estimated by the combination of one theoretical and two empirical methods; these last ones were based on correlations between the four functional NSM parameters (a, b, L_{LD}, H_{LD}) and a sizeable amount of information on these dimensional bases obtained in Colombia, a part of which has been published (Bernal and Solano, 1989; Shuk, 1989). For presentation purposes, all were finally converted to a single dimensional base (Length - L).

Expected Indirect Failure Costs: are those arising from the highway's interruptions; these are never covered by the owning or operating entity, but constitute financial losses for the productive sectors of the economic zone of influence of the highway, and hence affect the country's GNP statistics. The probability distributions of the interruption times was based on the pertinent direct cost items, and on the owning entity's logistical capacity (based on historical records and subjective estimates), for the time required to put the highway back into operation; the interruption time cost basis was the Villavicencio Chamber of Commerce estimate mentioned previously. One of the most interesting features of this analysis was the approximate solution obtained for the probabilities of the simultaneous interruptions of the two branches of the pair-way on the two sides of the Negro River stretch. The same as for the expected direct costs, the indirect ones were converted to a single dimensional base (Length - L).

3.3 Results

Both the direct and indirect expected failure cost components were estimated for 4 typical zones of the highway, where variations due to geotechnical, geological, meteorological and seismicity conditions were evidenced by the results. Due to space limitations, only the overall results for an economic lifetime of 40 years are shown in Table 1.

Table 1. Total cost in million US dollars per kilometer

	Decision*		
	Nº.1	Nº.2	Nº.3
Construction	0	3,065	28,000
Maintenance	184	286	572
Failure:Direct	306	602	0
Failure:Indirect	9,200	2,200	0

*Nº.1: Leave highway in its present state.
Nº.2: Build pair-way (see text).
Nº.3: Build italian "auto-strada" type expressway.

Based on the results of Table 1, the construction of the pair-way (Decision 2) was the optimum decision at the time of the preparation of the report. This optimum decision coincides with the one estimated on the basis of a 25 year economic lifetime, and with the results of standard cost-benefit analysis.

The details of the methodological, analytical and estimation procedures of this study, which was completed in 6 months, are contained in its corresponding report (Universidad de los Andes, 1992).

4 THE DISASTER AREA OF UTICA

The town of Utica (Department of Cundinamarca) is located 119 km northwest from Bogotá, at the deltaic type confluence of the Negra Creek and the Negro River (as shown on Figure 3.), at an altitude of 497 m.a.s.l; in 1988 the municipality had approximately 4400 inhabitants, of which nearly 60% dwelled within the town's limits, whose extension is around 50 hectares. The town's mean annual temperature is 26°C and its climate is relatively dry. The Negra Creek's watershed has an extension of 70.2 km², with altitudes ranging between 497 and 2065 m.a.s.l., and at these latter ones the temperature decreases to 18°C. The mean annual rainfall ranges between 1350 mm at Utica and a maximum of 2000 mm at an intermediate elevation (1700-1800 m.a.s.l.) of the watershed, and it evidences two major annual rainy seasons (March to May, and September to November). The Negra Creek and its four tributaries all display a high hydraulic gradient and are of a torrential nature.

On the night of November 17/1988, during a very intense rainy season which affected a great part of Colombia, a debris flow avalanche from the Negra Creek affected a part of the town, covering several streets and causing large economic losses and some casualties, flooded several buildings, and damaged the railway bridge and cemetery. This avalanche originated from a temporary dam formed on one of the tributaries of the Negra Creek and its subsequent failure. Approximately three to four days after, a big landslide of 1.5 million m³ developed on the left side of the Negro River 5 km upstream from Utica (Santa Bárbara landslide area, Figure 3), which buried 600 m of the railroad track and temporarily dammed the river; after the fortunately gradual failure of this small dam, it heavily damaged cultivated land and dwellings within the municipality, but only barely touched the town of Utica itself; no human casualties were reported.

The lithology of the Negra Creek Valley corresponds to that of the Villeta Group, which consists of shales, siltstones and some limestones in the western hillslopes,

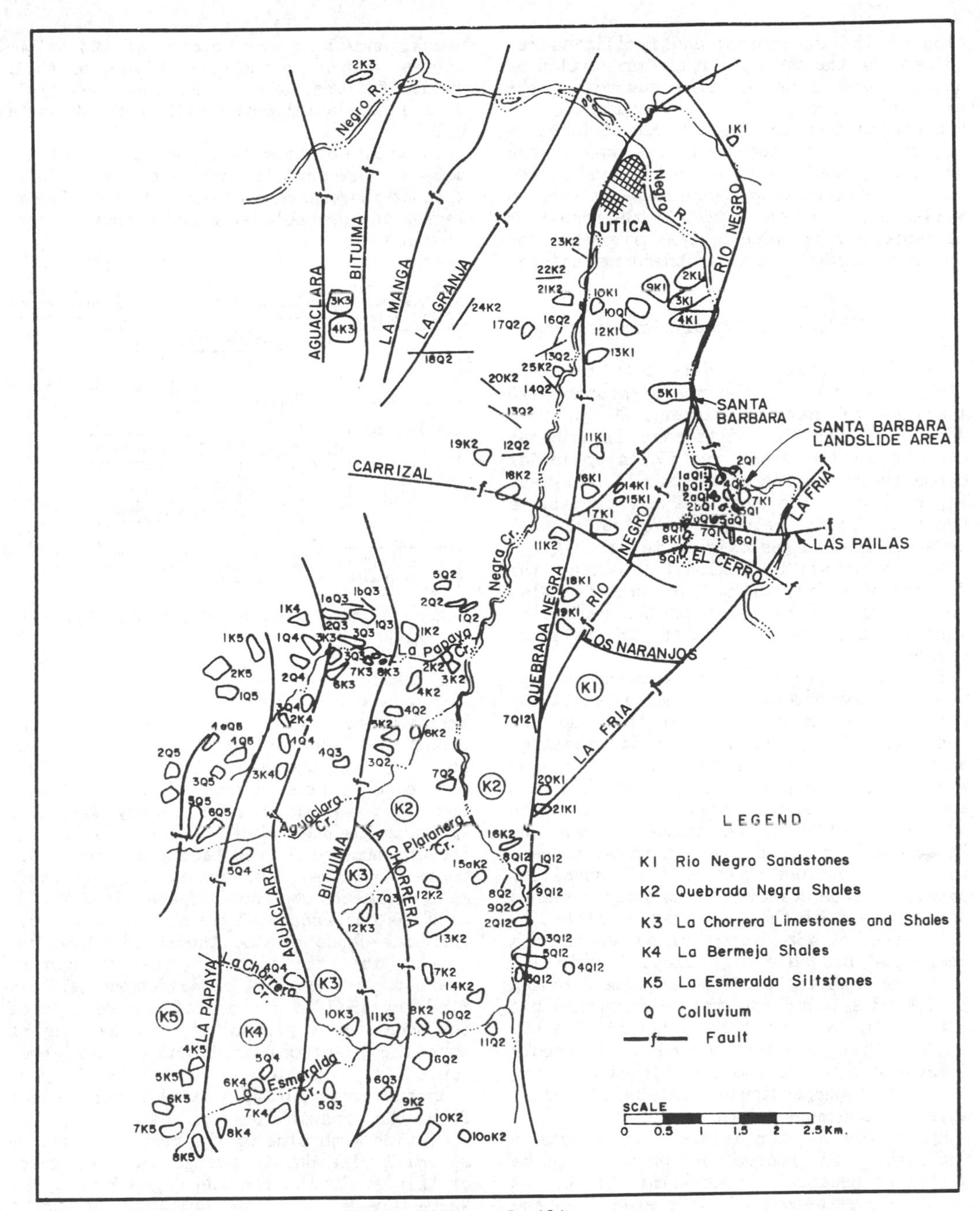

Figure 3. Utica disaster area. Natural slope families

and of sandstones, siltstones and some shales in the eastern flank. All the rocks are intensely faulted and most of them are covered by abundant colluvial deposits susceptible to mass movements.

4.1 Purpose of the study

The purpose of this study was to establish an economic comparison of the costs of relocating the town of Utica and its inhabitants, versus the costs of not relocating the town, which involved those of construction of the structures and facilities required for the protection and prevention of damages and loss of life caused by the expected (probabilistic) occurrence of avalanches from the Negra Creek of increasing extents (in terms of the town's area which they would cover), up to, and including, the simultaneous occurrence of such an avalanche and of one caused by the formation of temporary dam (due to mass movements) on the Negro River, and its subsequent failure.

4.2 Highlights of the analytical process

The geotechnical risk data base for this study consisted of the measurement of 126 families of natural slopes, 69 in rock masses of the five different lithologies present in the area, and 57 in colluvial deposits overlaying these same lithologies. Figure 3 shows the locations and approximate boundings of the measured families, as well as some notable geologic features.

As is usual in regional studies, the factors of safety and failure probabilities were estimated by means of the method of "universes", which results in relative values.

For another type of mass movement susceptibility evaluation, a general stability zoning map was prepared on the basis of conventional geotechnical unit ranking, slope gradients, morphodynamic units, drainage density, and climate, vegetation and erosion factors; all these converged to the establishment of 5 stability categories. The locations and boundaries of these conventionally obtained geotechnical stability category zones coincided with those obtained on the basis of the quantitative stability parameters of NSM, except in one small area where NSM showed a high slope instability category, while the conventional zoning indicated a lower one; this discrepancy was settled in favor of NSM, because it became evident that this area covered an intensely fractured zone due to the intersection of two major perpendicular geologic faults, with its consequent harmful effects.

Apart from the many interesting aspects of the analytical process, and which cannot be described because of space limitations, the outstanding feature of this study was the rather complex and cumbersome process leading to the evaluation of avalanche risk. This evaluation was based on the estimation of the factors and probabilities (and their interrelations) which are shown in the Mass Flow Hazard Event Sequence Diagram of Figure 4, and included, among others, the pioneering probabilistic analysis of the height of temporary dams that could form on the Negra Creek, and/or on one or more of its tributaries, and/or on the Negro River, of their sudden failure, and of the area that these mass flows would cover within the town of Utica.

It would not have been possible to assess some of the aspects of these evaluations, without the elements provided by NSM directly for the probabilistic and economic decision analyses.

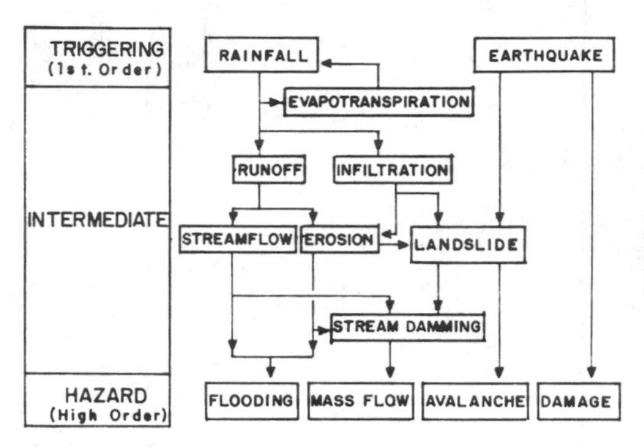

Figure 4. Mass flow hazard event sequence

4.3 Results

From a conceptual standpoint, an interesting result of the probability interrelation analysis, in the sense that it coincides with a subjective "a-priori" expectation, - and which also applies for the Bogotá-Villavicencio highway study -, is the following: the maximum total probability of the occurrence of landslides, avalanches and other mass movements does not necessarily coincide with the occurrence of the maximum magnitude (or exceedence of the threshold value required for failure) of the triggering events; in fact, no coincidences of this kind appeared in any one of the results of the many total probability evaluations of both the practical application cases presented in this paper.

Another interesting partial result arises from the possibility of evaluating the evolution with time of the maximum monetary material risk due to avalanches; the graph of this evolution for the Negro River, the Negra Creek, and the combined or joint action of both waterways is shown on Figure 5. The results of this figure indicate that the upper limit of the justifiable investment for corrective, protective and preventive structures, facilities and measures within a long-term consideration (approximately 39 years), was around US$ 14,000,000.

A probabilistic opportunity cost-benefit cost analysis was carried out for each of

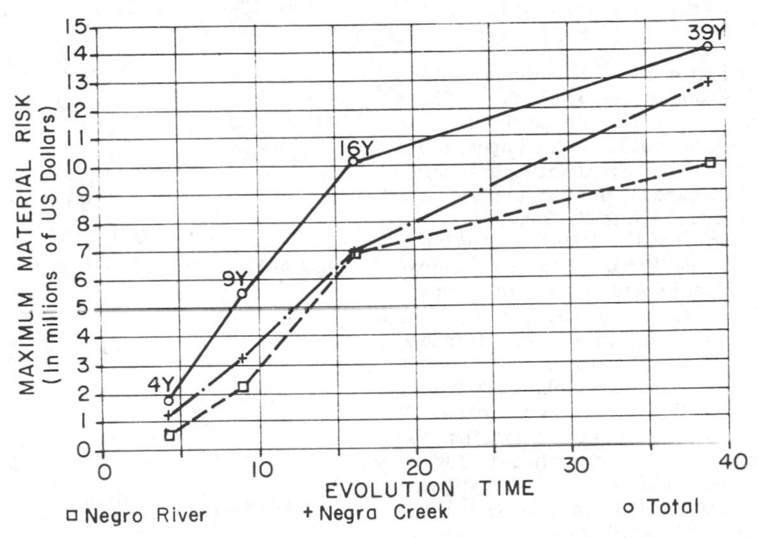

Figure 5. Utica disaster area. Evolution of avalanche risk with time

the many structures, facilities and other protective and preventive measures proposed for the alternative of not relocating the town of Utica; the results of this analysis are too lengthy for their presentation herein. The final result of this study was that the total sum of those cost items which showed a positive economic opportunity benefit, - of the order of US$ 7,000,000 (which was much less than the upper justifiable investment limit of the previous paragraph) -, came out much cheaper than the cost of relocating the town and its inhabitants (approximately US$ 20,000,000). The consequent recommendation against the relocation of the town, and for the construction of these positive economic benefit structures, facilities, and other preventive and protective measures, was fully accepted by the corresponding authorities.

The details of the methodological, analytical and estimation procedures of this study, which was completed in 7 months, are included in its corresponding report (Análisis Geotécnicos Colombianos Ltda., 1991). A very summarized version in English of this study is available (González, 1992).

5 CONCLUSIONS

The short descriptions of the two practical cases presented in this paper show that it is now possible, - within practical cost and time limits -, to obtain answers based on a totally quantitative analytical process, to problems arising from geotechnical slope instability factors in both linear projects and regional studies; previously these had to be mostly to nearly totally based on qualitative criteria, categories and classifications. This development could not have taken place without the massive quantification capabilities of the Natural Slope Methodology (NSM), which is a necessity not only for the geothecnical data base, but also for the estimates of other factors which pertain directly to the failure probability and economic decision analyses; these capabilities also promote the opening of new analytical avenues which could not have been even attempted before.

ACKNOWLEDGEMENTS

The authors are grateful to the following Colombian entities for the opportunity of carrying out the studies mentioned in this paper, and for the permission to publish their results: National Planning Department, National Fund for Development Studies, Government of the Department of Cundinamarca, Regional Emergency Committee of Cundinamarca, the Institute of Investigations on Geosciences, Mining and Chemistry, and Universidad de los Andes. Their gratitude is expressed also to Clara Inés Bernal and Zsolt Gácsi, for their permanent and creative collaboration towards both the NSM research effort, and the presentation and edition of this paper.

REFERENCES

Análisis Geotécnicos Colombianos Ltda. (1991) - "Estudios de Amenaza y Obras Alternativas de Protección a Utica - a Nivel de Prefactibilidad". Unpublished report for the Government of the Department of Cundinamarca, and for other state entities. Volumes 1 and 2. Bogotá.

Bernal,C.I. & Solano,J. (1989) - "Análisis Probabilístico de Estabilidad de Laderas Naturales". Thesis submitted in partial fulfillment of the requirements for the degree of Civil Engineer, Universidad Nacional de Colombia, Bogotá.

González,A.J.(1992) - "Avalanche Risk Evaluation at Utica - Colombia". Proceedings of the First International Symposium on Remote Sensors and Geographical Information Systems (GIS), single volume, p.356-378, "Agustín Codazzi" Geographical Institute. Bogotá, April.

Shuk,T.(1989) - "Aproximación a Dos Aspectos del Costo de Falla de Taludes". Proceedings of 1st Southamerican Symposium on Landslides, Vol.1, p.726-743. Paipa (Colombia), August.

Universidad de los Andes (1992) - "Estudio Preliminar de la Confiabilidad de la Vía Bogotá-Villavicencio". Center of Studies and Investigations of the Faculty of Engineering. Unpublished report for the National Fund for Development Studies. 2 Volumes. Bogotá, November.

Structural Safety & Reliability, Schuëller, Shinozuka & Yao (eds) © 1994 Balkema, Rotterdam, ISBN 90 5410 357 4

Stochastic evaluation of differential settlements of reclaimed lands by means of Pasternak Model

H.Tanahashi
Osaka Structural Engineering Office of Nikken Sekkei Ltd, Osaka & Nikken Sekkei Nakase Geotechnical Institute, Kawasaki, Japan

N.Uchida
Osaka Structural Engineering Office of Nikken Sekkei Ltd, Japan

M.Fukui
Nikken Soil Research Inc., Osaka, Japan

ABSTRACT: A classical Winkler Model is often used in soil-structure interaction analyses, where the shear rigidity of soils is ignored. However, the shear rigidity of soils has some effects of reducing differential settlements actually. In order to evaluate the reduction effects, use of a Pasternak Model as an improved and refined foundation model is studied in this paper. The reduction effects are evaluated by means of the scale of fluctuation. It is shown analytically that the scale of fluctuation has a lower limit. The results are discussed comparing with actual settlements measured.

1 INTRODUCTION

Prediction of differential settlements of structures due to non-homogeneity of soil properties and loads is essential in evaluating serviceability and safety of structures on reclaimed lands. In order to predict differential settlements rationally, it is neccesary to make stochastic evaluation of differential settlements based on soil-structure interaction.

It is well known that the rigidity of structures reduces their differential settlements and at the same time it causes additional stresses on structures through soil-structure interaction. A classical Winkler Model is often used in soil-structure interaction analyses, where the shear rigidity of soils is ignored.

However, the shear rigidity of soils has some effects of reducing differential settlements actually. In order to evaluate this reduction effects produced by the shear rigidity of soils, the authors proposed to use a Pasternak Model (Kerr 1964) as an improved and refined foundation model (Tanahashi et al. 1991).

Reclaimed lands are modelled as a Pasternak Model and the reduction effects are evaluated by means of the scale of fluctuation. The results are discussed comparing with actual settlements measured.

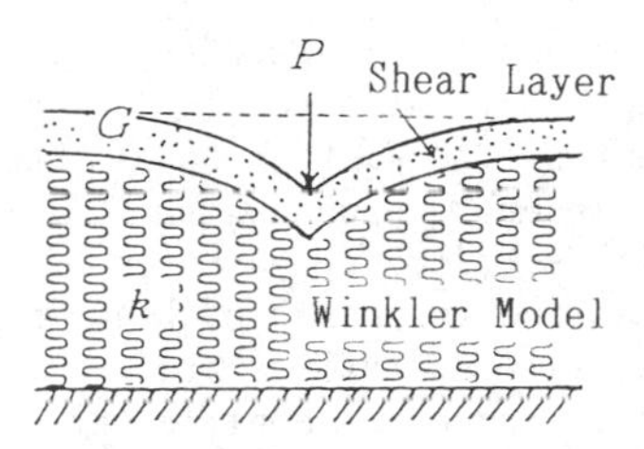

Fig.1 Pasternak Model

2 PASTERNAK MODEL

The Pasternak Model is regarded as a shear layer on a Winkler Model (Fig.1). The authors selected the Pasternak Model as a simple and appropriate model by which the reduction effects of soils can be evaluated. The load-displacement relation of this model is expressed by the following equation under the condition of plane strain (Kerr 1964),

$$p = ky - Gd^2y/dx^2 \quad \cdots\cdots (1)$$

where, y: vertical displacement of the surface,
p: vertical load,
k: coefficient of subgrade reaction,
G: shear rigidity of the shear layer in the vertical direction.

The solution of Eq.(1) due to a vertical concentrated load P at the origin of x coordinate is represented as,

$$y = \frac{P\gamma_s}{2k} e^{-\gamma_s|x|} \quad \cdots\cdots (2)$$

where $\gamma_s = \sqrt{k/G}$ is the characteristic value of the Pasternak Model.

This model has two independent parameters. The first parameter k is the same as that of the Winkler Model. The second parameter G produces the differential settlement reduction effects.

In general the responses of beams due to differential settlements are formulated as reduction coefficients of differential settlements which are represented in a form of frequency (wave number) response functions (Tanahashi et al.1991).

In the case of a Pasternak Model the reduction coefficients are also defined as the ratio of the amplitude of a cosine settlement curve of a soil with shear rigidity to that with no shear rigidity in the vertical direction.

Therefore the reduction coefficient of differential settlements $\alpha_s(\omega)$ of the Pasternak Model is expressed as follows (Tanahashi et al.1991).

$$\alpha_s(\omega) = \frac{1}{1 + \omega^2/\gamma_s^2} \quad \cdots\cdots (3)$$

where, $\omega = 2\pi/L$: wave number,
L: wave length of a settlement curve.

The reduction coefficients are exact when the coefficient of subgrade reaction is constant and they are approximate when the coefficient of subgrade reaction varies.

3 MODELLING OF RECLAIMED LANDS

The parameter γ_s can be evaluated on the basis of Vlasov's general variational method (Vlasov et al. 1960). According to his method γ_s depends upon elastic moduli, Poisson's ratios and depths of the layers and function $\psi(z)$ that determines the rate of decrease of the vertical displacements with depth (Fig.2).

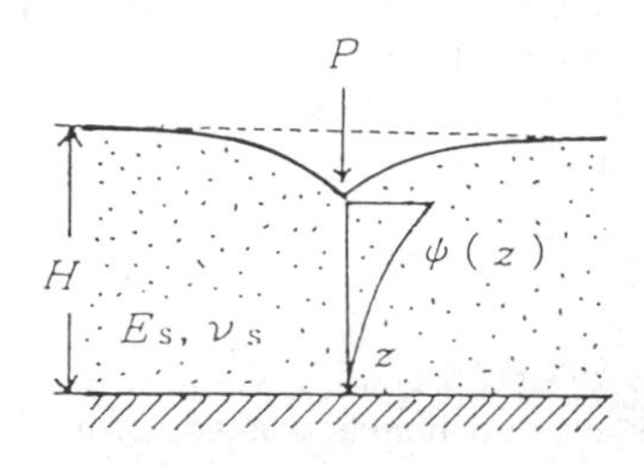

Fig.2 Function $\psi(z)$ of Vlasov Model

$$\gamma_s = \sqrt{k/2t} \quad \cdots\cdots (4)$$

where,

$$\left.\begin{aligned} k &= \int_0^H \frac{(1-\nu_s)E_s}{(1+\nu_s)(1-2\nu_s)} \left(\frac{d\psi}{dz}\right)^2 dz \\ 2t &= \int_0^H \frac{E_s}{2(1+\nu_s)} \psi^2 dz \end{aligned}\right\} \quad (5)$$

$2t$: shear rigidity, which is the same as G,
E_s : elastic modulus of a soil,
ν_s : Poisson's ratio of a soil,
H : depth of a soil layer.

In the case of a typical reclaimed land which consists of a sandy reclaimed layer and an alluvial clay layer beneath it, it is reasonable that the soil is modelled as a Pasternak Model with two layers; the sandy layer which is very rigid for vertical displacements and the clay layer which has an elastic modulus which yields an elastic settlement equivalent to the consolidation settlement of the clay layer.

Therefore the function $\psi(z)$ is assumed in Fig.3 and the parameters of the Pasternak Model are obtained as follows.

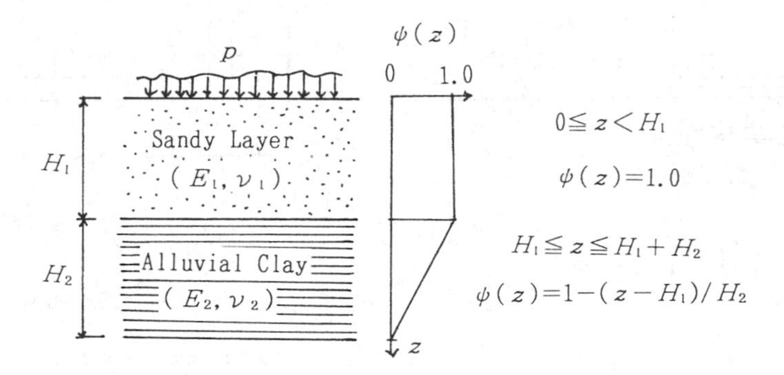

Fig.3 Function $\psi(z)$ in Reclaimed Lands

The elastic modulus of the clay layer is determined using the coefficient of volume compressibility m_V as (Yamaguchi 1984),

$$\frac{(1-\nu_2)E_2}{(1+\nu_2)(1-2\nu_2)} = \frac{1}{m_V} \quad \cdots\cdots (6)$$

Thus

$$k = 1/(m_V H_2) \quad \cdots\cdots (7)$$

The shear rigidity of the sandy layer is determined by E_1 and ν_1. The shear rigidity of the clay layer is obtained from $E_2 = E_{50}$ of one-dimensional compression test and undrained Poisson's ratio $\nu_2 = \nu_U = 0.5$.

$$2t = \frac{E_1 H_1}{2(1+\nu_1)} + \frac{E_{50} H_2}{6(1+\nu_U)} \quad \cdots\cdots (8)$$

Therefore γ_s is obtained as,

$$\gamma_s = \sqrt{\frac{2(1+\nu_1)}{(1+\xi/3)\, m_V E_1 H_1 H_2}} \quad \cdots\cdots (9)$$

where,

$$\xi = (1+\nu_1) E_{50} H_2 / \{(1+\nu_U) E_1 H_1\}$$

represents the ratio of shear rigidity of the clay layer to that of the sandy layer.

4 PARAMETERS IN KOBE PORT ISLAND

Let us try to evaluate γ_s of a reclaimed land, i.e., Kobe Port Island, Kobe Japan, which was filled with sand of weathered granite on the soft seabed of the alluvial clay layer. The soil profile and soil properties are shown in Fig.4.

The elastic modulus of the sandy layer is determined using average blow counts N=10 of Standard penetration test as,

$$E_1 = 280N\ \text{tf/m}^2 = 2800\ \text{tf/m}^2 \fallingdotseq 28\text{MPa}$$

The Poisson's ratio $\nu_1 = 0.3$ is taken.

The average elastic modulus of the clay layer is as,

$$E_{50} = 500\ \text{tf/m}^2 \fallingdotseq 5\text{MPa}$$

As for the coefficient of volume compressibility of the clay layer, the mean value is taken corresponding to the effective stress 0.2～0.3MPa of the

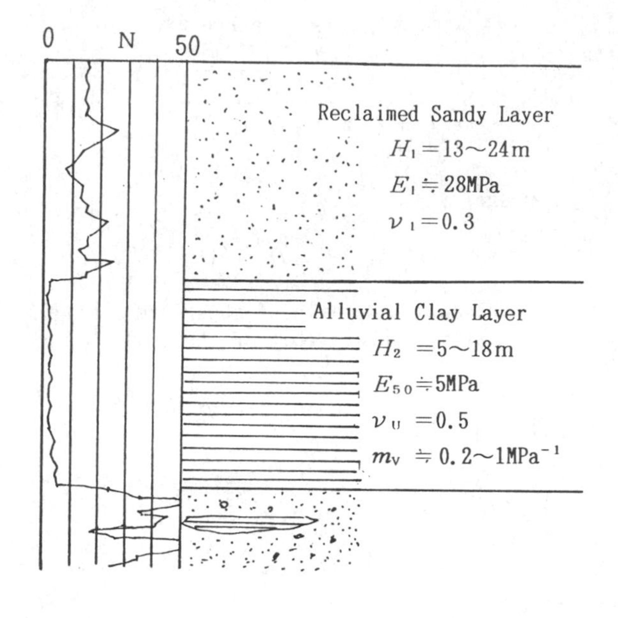

Fig.4 Soil Profile and Soil Properties of Kobe Port Island

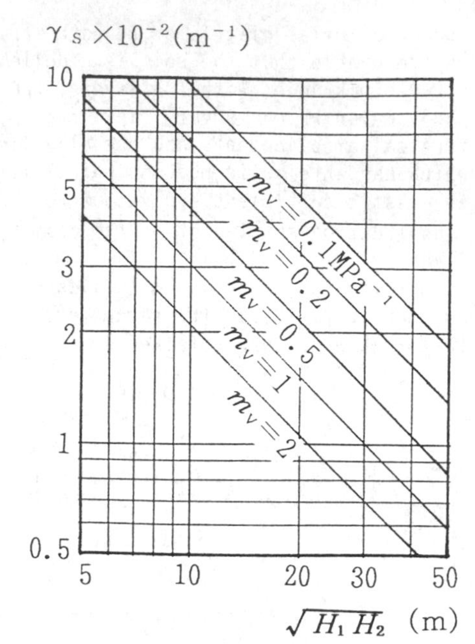

Fig.5 Diagram of Parameter γ_s

middle of the layer as,

$$m_v=0.002\sim0.01\text{m}^2/\text{tf}≒0.2\sim1\text{MPa}^{-1}$$

Since $H_1=13\sim24$m and $H_2=5\sim18$m,

$\sqrt{1+\xi/3}≒1$ and Eq.(9) yields,

$$\gamma_s ≒1.61/\sqrt{m_v E_1 H_1 H_2} \quad \cdots\cdots (10)$$

This means the shear rigidity of the clay layer is ignored because it is negligible in comparison with that of the sandy layer.

Eq.(10) is represented in Fig.5, where γ_s is in inverse proportion to $\sqrt{H_1 H_2}$ when $E_1=28$MPa and m_v is constant.

$\sqrt{H_1 H_2}$ varies from 11m to 21m, then γ_s varies in a range of 0.014m^{-1}～0.061 m^{-1}.

5 CHANGES IN AUTOCORRELATION FUNCTIONS

Eq.(3) shows that the reduction effects depend on the wave number and the characteristic value and that shorter the wave lengths become, larger the effects are.

These effects cause some changes in the power spectral density function and the autocorrelation function concerning differential settlements of the ground surface.

The autocorrelation function of reduced differential settlements $\bar{R}(\Lambda)$ is represented as,

$$\bar{R}(\Lambda)=\int_{-\infty}^{\infty}\{\alpha_s(\Omega)\}^2 S(\Omega)\cos\Omega\Lambda\, d\Omega \quad \cdots\cdots (11)$$

$$S(\Omega)=(1/2\pi)\int_{-\infty}^{\infty}R(\Lambda)\cos\Omega\Lambda\, d\Lambda \quad \cdots\cdots (12)$$

where, $\Omega=\omega/\gamma_s$: non-dimensional wave number,
$\Lambda=\gamma_s\lambda$: non-dimensional distance separation,
λ : distance separation.

$S(\Omega)$ is the power spectral density function and $R(\Lambda)$ is the autocorrelation function, each concerning differential settlements of ground surfaces when the shear rigidity of soils is ignored.

$\alpha_s(\Omega)$ is the non-dimensional reduction

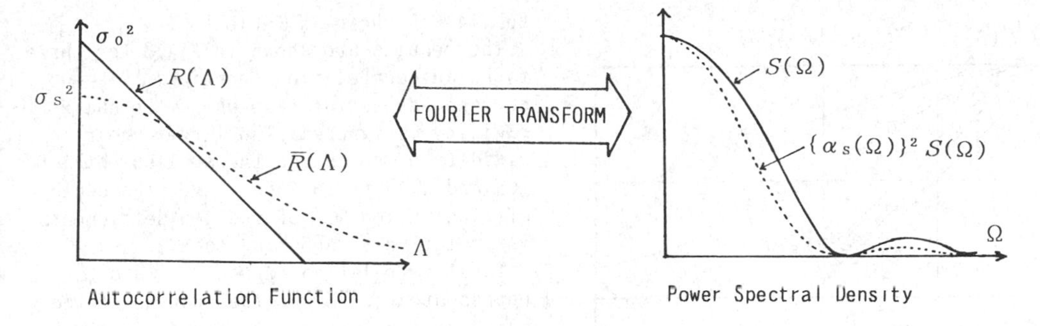

Fig.6 Relation between Autocorrelation Function and Power Spectral Density

coefficient of differential settlements which is expressed as,

$$\alpha_s(\Omega) = 1/(1+\Omega^2) \quad \cdots\cdots (13)$$

The autocorrelation functions of reduced settlements $\overline{R}(\Lambda)$ are obtained by the Fourier transform based on the Wiener-Khintchine relationship shown in Eqs.(11) and (12), which is shown conceptually in Fig.6.

The reduction effects are evaluated by means of the scale of fluctuation which is defined as follows (Fig.7).

$$L_s = (2/\sigma_0{}^2)\int_0^{L_0} R(\Lambda)\, d\Lambda \quad \cdots\cdots (14)$$

where, $L_s = \gamma_s l_s$: non-dimensional scale of fluctuation,
l_s : scale of fluctuation,
σ_0 : standard deviation of differential settlements,
$L_0 = \gamma_s l_0$: non-dimensional minimum distance separation which meets $R(L_0) = 0$ including infinity.

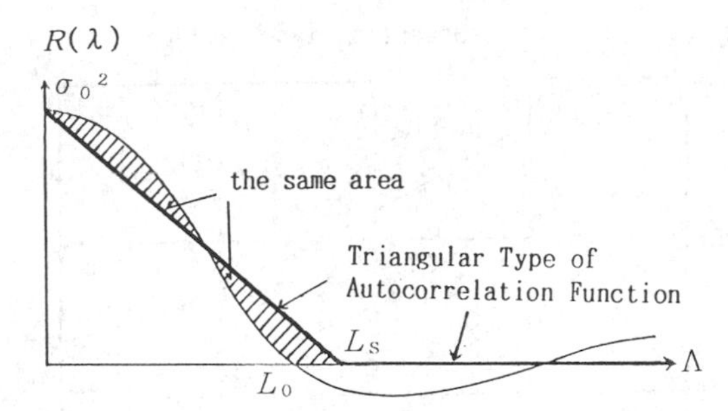

Fig.7 Definition of Scale of Fluctuation

The authors proposed this scale of fluctuation by modifying Vanmarcke's definition (Vanmarcke 1983) for a main parameter which determines the autocorrelation characteristics (Tanahashi et al. 1991).

Table 1 Autocorrelation Functions

TYPE	Autocorrelation Function
Triangular Type	$R(\Lambda) = \begin{cases} \sigma_0{}^2(1-\lvert\Lambda\rvert/L_s) & : \lvert\Lambda\rvert \leq L_s \\ 0 & : \lvert\Lambda\rvert > L_s \end{cases}$
Exponential Type	$R(\Lambda) = \sigma_0{}^2 \exp(-2\lvert\Lambda\rvert/L_s)$
Low-pass White Noise Type	$R(\Lambda) = \sigma_0{}^2 \dfrac{\sin(\nu\pi\Lambda/L_s)}{\nu\pi\Lambda/L_s}$ $: \nu = \sum_{n=1}^{\infty} \dfrac{2(-1)^{n+1}\pi^{2n-2}}{(2n-1)\cdot(2n-1)!} \fallingdotseq 1.1788$

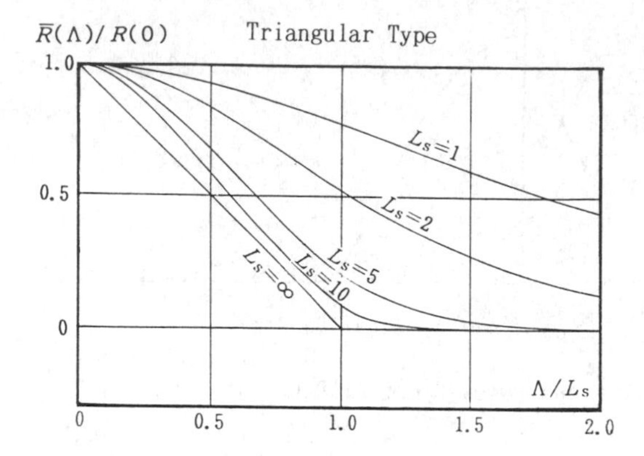

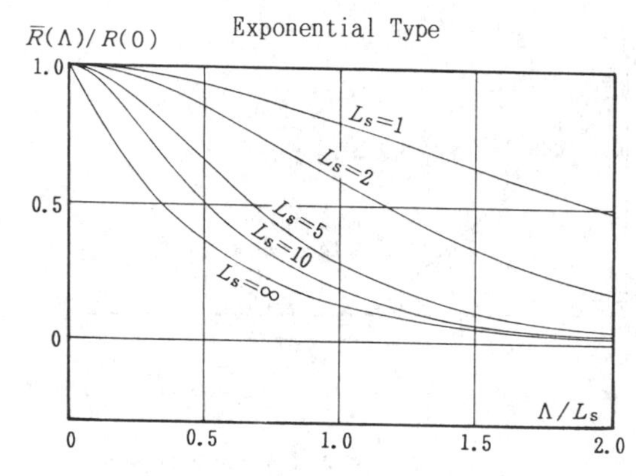

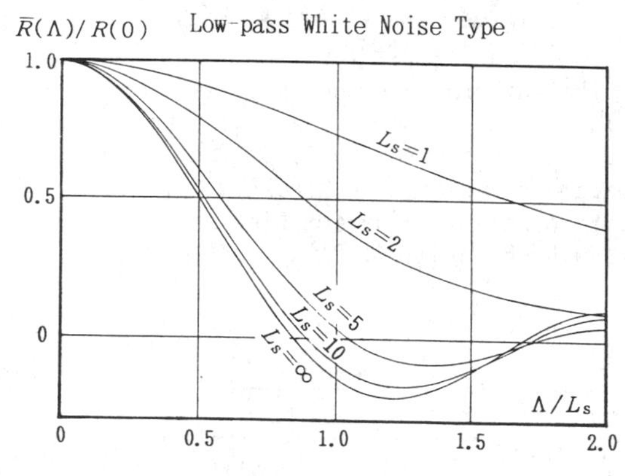

Fig.8 Changes in Autocorrelation Function

Here the power spectral density functions of reduced settlements are analysed numerically concerning three typical types of autocorrelation functions which satisfy $S(0)\neq 0$, such as a triangular type, an exponential type and a low-pass white noise type. They are normalized by L_s in order to satisfy Eq.(14) as shown in Table 1.

The results are shown in Fig.8 for three types autocorrelation functions. $L_s=\infty$ corresponds to the case where the shear rigidity is ignored. The larger shear rigidity of soils is, the smaller both of γ_s and L_s are and the longer the scale of fluctuation $\overline{L}_s$ of reduced settlements is, relatively comparing to L_s.

Thus the relation between L_s and $\overline{L}_s$ is represented in Fig.9. The changes of the scale of fluctuation are found clearly; when L_s increases to infinity, $\overline{L}_s$ equals L_s; when L_s decreases to zero, $\overline{L}_s$ decreases to 4 asymptoticly.

This result is confirmed analytically as follows.

$$\overline{L}_s=(2/\sigma_s{}^2)\int_0^{L_0}\overline{R}(\Lambda)\,d\Lambda$$
$$=(2/\sigma_s{}^2)\{\int_0^{\infty}\overline{R}(\Lambda)\,d\Lambda-\int_{L_0}^{\infty}\overline{R}(\Lambda)\,d\Lambda\}$$
$$=(2/\sigma_s{}^2)\{\pi\,S(0)-\int_{L_0}^{\infty}\overline{R}(\Lambda)\,d\}$$

where σ_s is the standard deviation of reduced differential settlements.

Since,

$$\lim_{Ls\to 0}\sigma_s{}^2=\lim_{Ls\to 0}\int_{-\infty}^{\infty}\{\alpha_s(\Omega)\}^2 S(\Omega)\,d\Omega$$
$$=\lim_{Ls\to 0}S(0)\int_{-\infty}^{\infty}\{\alpha_s(\Omega)\}^2\,d\Omega$$
$$=\lim_{Ls\to 0}S(0)\int_{-\infty}^{\infty}1/(1+\Omega^2)^2\,d\Omega$$
$$=\pi\,S(0)/2$$

and if L_s decreases to 0, L_0 increases to infinity,

$$\lim_{Ls\to 0}\int_{L_0}^{\infty}\overline{R}(\Lambda)\,d\Lambda\ =\ 0$$

Therefore under the condition of $S(0)\neq 0$,

$$\lim_{Ls\to 0}\overline{L}_s\ =\ 4$$

When $L_s\to\infty$, that is, the shear rigidity of soils is zero, clearly,

$$\lim_{Ls\to\infty}\overline{L}_s\ =\ L_s$$

According to this relation $\overline{L}_s$ has the minimum value 4 and therefore $\overline{l}_s$ has a lower limit $4/\gamma_s$.

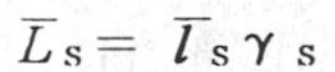

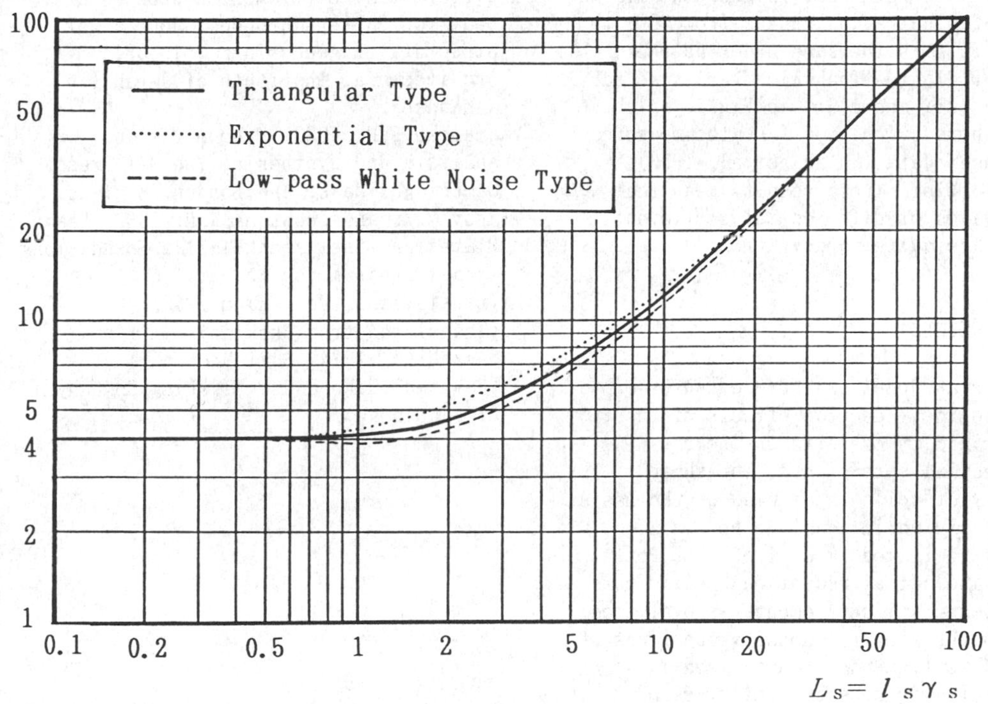

Fig.9 Relation between L_s and $\overline{L}_s$

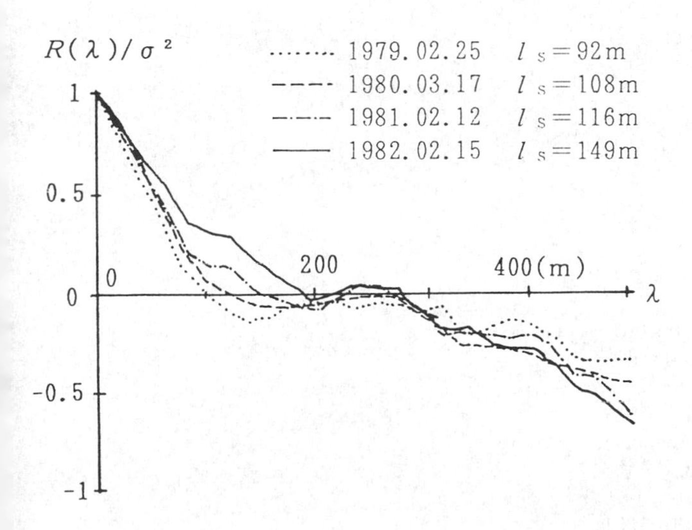

Fig.10 Autocorrelation Functions of Settlements and Scales of Fluctuation in Kobe Port Island

Since γ_s ranges from $0.014m^{-1}$ to $0.061m^{-1}$ in Kobe Port Island, the minimum scale of fluctuation ranges from 66m to 286m. This is the same order as the scale of fluctuation which varies so far between 92m and 149m in Fig.10 which the authors calculated from the measured settlement data (Kutara et al. 1983).

In addition, these results mean that short waves of differential settlements take place quite rarely.

6 CONCLUSIONS

A Pasternak Model can be used in order to evaluate the reduction effects of the shear rigidity of soils on their differential settlements. The shear rigidity of soils is assumed on the basis of Vlasov's variational method in a typical reclaimed land, i.e. Kobe Port Island and actual reduction coefficients of differential settlements is evaluated.

The power spectral density functions of reduced settlements are analysed numerically in some typical types of autocorrelation functions. The reduction effects are evaluated by means of the scale of fluctuation which the authors proposed by modifying Vanmarcke's definition.

The results show the scale of fluctuation has a lower limit, which is the same order as the measured scale of fluctuation.

Moreover, these results mean that short waves of differential settlements take place quite rarely.

REFERENCES

Kerr, A. D. 1964. Elastic and Viscoelastic Foundation Models, Journal of Applied Mechanics, Vol.31: 491～498.

Kutara, K., Gomadou, M. and Takeuchi, T. 1983. Differential Settlement on Soft Grounds. Civil Engineering Reports. 25-12. (in Japanese).

Tanahashi, H., Uchida, N. and Fukui, M. 1991. Probability-based Prediction for Differential Settlements on Structures. Proc. JCOSSAR'91: A-34 (in Japanese).

Tanahashi, H., Uchida, N. and Fukui, M. 1992. Differential Settlement Responses of Beams on Pasternak Model. Journal of Structural Engeenering. Vol.38B: Architectural Institute of Japan. (in Japanese).

Vanmarcke, E. H. 1983. Random Fields; Analysis and Synthesis: The MIT Press, Cambridge, Mass. and London, England.

Vlasov, V. Z. and Leont'ev, N. N. 1960. Beams, Plates and Shells on Elastic Foundations Israel Program for Scientific Translations. Jerusalem, 1966. (Translated from Russian).

Yamaguchi, H. 1984. Soil Mechanics (Revised): Gihoudou Publishing Company. (in Japanese).

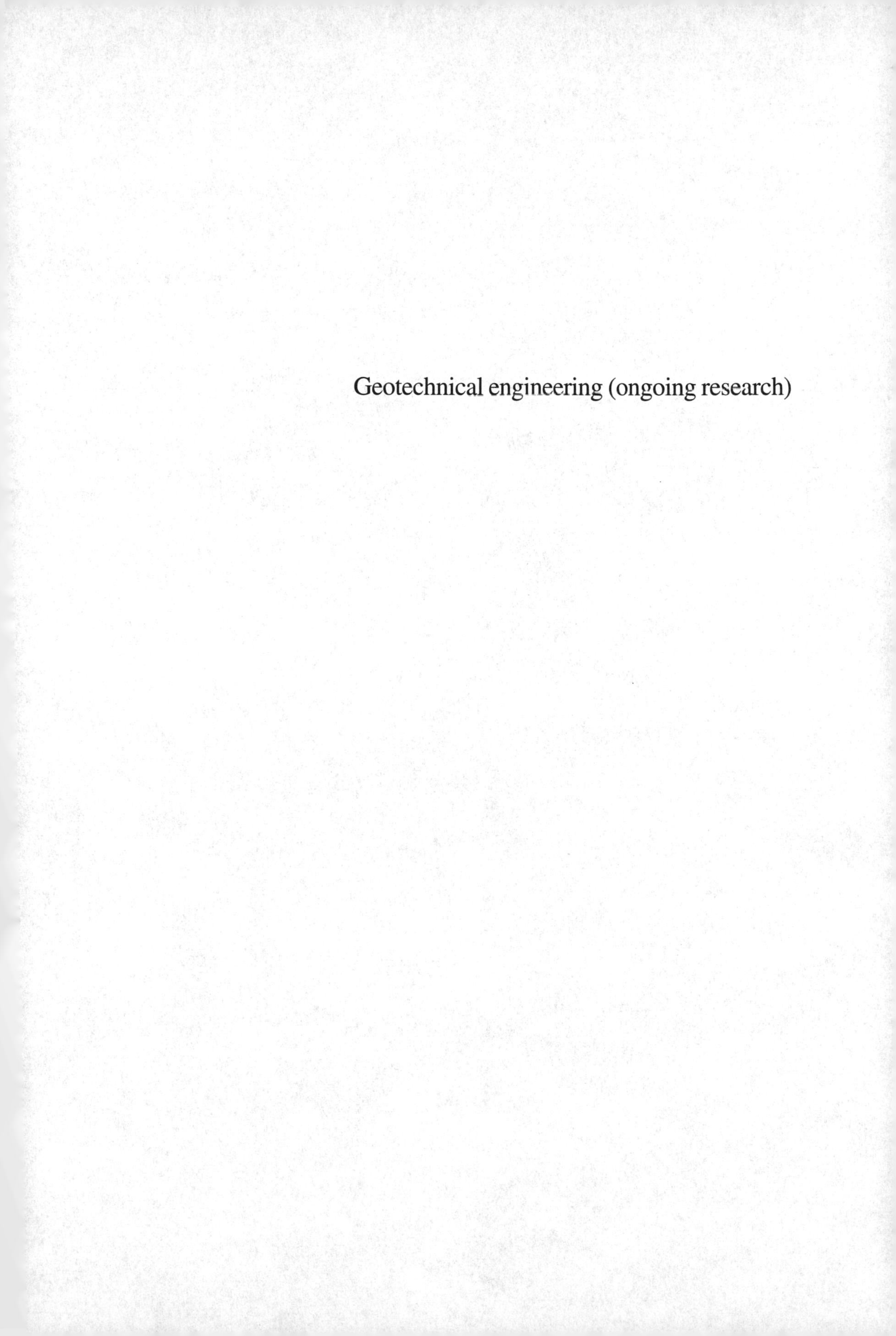

Geotechnical engineering (ongoing research)

Structural Safety & Reliability, Schuëller, Shinozuka & Yao (eds) © 1994 Balkema, Rotterdam, ISBN 90 5410 357 4

Entropy as a tool for statistical analysis of drilling data

M. Bourget
Scétauroute, Saint Quentin en Yvelines, France

F. Crémoux
LERGGA, University of Bordeaux I, France

Abstract: A stationary Gaussian model seems to be suited for modelling the interaction between a drilling bit and an homogenous geological setting. According to that, entropy is a mathematical tool which is powerful for dividing a recording versus depth of drilling parameters into homogeneous zones.

1. INTRODUCTION

Analogical drilling data have been used for a long time by civil engineers. The recent acquisition in numerical form should enable use of signal processing techniques. The erratic character of drilling data requires statistical processing. Nevertheless, common statistical treatments are only able to handle the case of homogeneous medium. Their inability to take into account heterogeneneities has not been overcome. The layered structure is the most commonly case encountered in civil engineering surveys which investigate domains ranging from the sub-surface down to tens of meters deep.

To handle this situation, a new approach based on the entropy has been developed at the LERGGA. The governing idea consists in breaking up drilling data into a set of homogeneous sub-domains.

2. MATHEMATICAL DEVELOPMENT

Consider X(z), a mean square differentiable random signal (z denotes the depth). The absolute value of the derivative, $|X'(z)|$, is assumed stationary of second order and is characterised by:

- a constant mean: $m_{|X'|}(z)=E[|X'|]=m_{|X'|}$
- a constant standard deviation $\sigma_{|X'|}(z)$:
  $\sigma^2_{|X'|}(z)=E[(|X'|-m_{|X'|})^2]=\sigma^2_{|X'|}$
- a covariance invariant by translation:

$$K_{|X'|}(z',z)=E[(|X'(z')|-m_{|X'|})(|X'(z)|-m_{|X'|})]$$
$$=K_{|X'|}(z,z')=K_{|X'|}(\tau) \text{ where } \tau=|z'-z|$$

The entropy H(z) of the signal X(z) is expressed by:

$$H(z)=\int_0^z |X'(u)|du, \text{ thus } E[H(z)]=E\left[\int_0^z |X'(u)|du\right]$$
$$=\int_0^z E[|X'(u)|]du=m_H(z)=zm_{|X'|}.$$

Therefore E[H(z)] is a linear function of the depth and the slope is proportional to the average of the absolute value of the signal first derivative.

Consider the deviation from linearity: $\beta(z)=H(z)-m_H(z)$. One gets:

$$E[(\beta(z)-\beta(z'))^2]=E[(H(z)-m_H(z)-H(z')-m_H(z'))^2]$$
$$=E[(\int_0^z(|X'(u)|-m_{|X'|})du-\int_0^{z'}(|X'(u')|-m_{|X'|})du')^2]$$
$$=E[\int_{z'}^{z}\int_{z'}^{z}(|X'(u')|-m_{|X'|})(|X'(u)|-m_{|X'|})du'du]$$
$$=\int_{z'}^{z}\int_{z'}^{z}K_{|X'|}(u',u)du'du.$$

Taking into account that $K_{|x'|}(u',u)$ is a function of $\tau=|u'-u|$, the last integral becomes:

$$2\int_{z-z'}^{0}(z'-z+\tau)K_{|x'|}(\tau)d\tau=2\ (T+\tau)K_{|x'|}(\tau)d\tau=I(T).$$

I(T) is invariant by translation with respect to z and I(-T)=I(T). There hence, $E[(\beta(z)-\beta(z'))^2]$ is invariant by translation with respect to z.

This enables us to conclude that, when $|X'(z)|$ is stationary of order 2, the entropy of X(z) is characterised by a linear mean and the deviations from the linearity are stationary of order 2.

The relation between stationarity of both signal and absolute value of its derivative is investigated hereafter:

Resume the case of a mean square differentiable random signal X(z) stationary of order 2. It meets:

- a constant mean: $m_x(z)=m_x$,
- a constant standard deviation: $\sigma_x(z)=\sigma_x$,
- a covariance invariant by translation: $K_x(z,z')=K_x(z',z)=K_x(\tau)$ where $\tau=|z'-z|$.

Thus, its derivative admits:

- a zero mean: $m_{x'}(z)=m_{x'}=0$,
- a covariance invariant by translation $K_{x'}(z,z')=K_{x'}(z',z)=K_{x'}(\tau)=-\frac{\partial^2}{\partial\tau^2}K_x(\tau)$
- a constant standard deviation $\sigma_{x'}$: $\sigma_{x'}^2=K_{x'}(0)=\text{constant}$.

Notice that:

$E[(|X'(z)|-m_{|x'|}(z))^2]=E[(X'(z))^2]-m_{|x'|}(z)^2$,

it comes: $\sigma_{|x'|}^2(z)=\sigma_{x'}^2-m_{|x'|}^2(z)$.

Introducing the means of positive and negative values of the derivative (respectively m^+ and m^-) and assuming the probability density of X'(z) to be symmetric yields to:

$m^++m^-=0$,

$\text{Probability}(|X'|=|x'|)=2\text{Probability}(X'=x')$ and $m_{|x'|}(z)=m^+=\text{constant}$, thus $\sigma_{|x'|}(z)$ is constant.

It can be shown that the covariance of the absolute derivative of X(z) is invariant by translation when the cross probability density $P(X'(z)=x'_1,X'(z')=x'_2)$ is symmetric with respect to x'_1 and x'_2.

Hence, if the random signal X(z) is mean square differentiable, stationary of order 2 with a symmetric cross probability density, its entropy admits a linear mean and deviations from linearity stationary of order 2.

Consider now the example of a Gaussian signal, stationary and mean square differentiable. Its probability density is given by:

$$P(X=x)=\frac{1}{\sqrt{2\pi}\sigma_X}\exp\left(-\frac{1}{2}\left(\frac{x-m_X}{\sigma_X}\right)^2\right)$$

In the Gaussian case, stationarity of second order amounts to stationarity in the broad sense. The probability density of the signal derivative is symmetric, has a zero mean and a constant standard deviation $\sigma_{x'}$. The probability density of the absolute value of the derivative of the signal is characterised by:

- its average $m_{|x'|}=\sigma_{x'}(2/\pi)^{1/2}$,
- its standard deviation $\sigma_{|x'|}=\sigma_{x'}(1-2/\pi)^{1/2}$,
- its probability density:

$$P(X=x)=\frac{2}{\sqrt{2\pi}\sigma_{X'}}\exp\left(-\frac{1}{2}\left(\frac{|x'|}{\sigma_{X'}}\right)^2\right).$$

The slope of the entropy is accordingly:

$$m_{|x'|}=\sigma_{x'}(2/\pi)^{1/2}.$$

The relation between $\sigma_{x'}$ and σ_x depends on the covariance of the signal $K_x(\tau)$. Take the more specific case: $K_x(\tau)=\sigma_x^2\exp(-\tau^2/a^2)$, the Gaussian model, which stands for a variogram characterising a mean square differentiable random variable with a correlation length equal to $a\sqrt{3}$. We find that:

$$\frac{\partial^2}{\partial\tau^2}K_X(\tau)=-\frac{2\sigma_X^2}{a^2}\left(1-\frac{2\tau^2}{a^2}\right)\exp\left(-\frac{\tau^2}{a^2}\right).$$

Hence, $\sigma_{X'}^2=-\frac{\partial^2}{\partial\tau^2}K_X(\tau)\Big|_{\tau=0}=\frac{2\sigma_X^2}{a^2}$ and the slope p of the entropy becomes: $p=\frac{2\sigma_X}{a\sqrt{\pi}}$.

This means that, on regards this particular case, the slope of the entropy is proportional to the standard deviation and inversely proportional to the correlation length. Using the characteristic length put forth by J. M.

Duchamps in the form $DC=4\sigma_x/p$, it follows that DC is proportional to the correlation length.

3. APPLICATION TO DRILLING DATA.

Assume that the investigated medium amounts to a set of homogeneous layers. Homogeneity is achieved if drilling response parameters (rate of penetration, torque, fluid injection pressure) obtained in response to constant excitation parameters (weight on bit, rotating speed, rate of inflow) come down to random signals exhibiting a stationarity of order 2, in each strata. The underlying idea is that, when the response parameters are random signals, stationary of order 2, the geo-technical characteristics of the medium involved in the drilling mechanism are also stationary.

Under this assumption, interpretation of drilling data aims at identifying, locating and characterising the different homogeneous layers encountered while drilling. The study of several sets of real data has underlined that the derivative of a response parameter exhibits amplitude and wave-length fluctuations oscillating around the naught value. Its average is equal to zero at both global and local scales.

The figures presented in this paper deal with the rate of penetration, the most representative response parameter.

Figure 1 shows the recording versus depth of drilling parameters: ROP (the rate of penetration), W (the weight on bit), DROP (the derivative of the ROP) for one drill. The associated geological log is also depicted.

The entropy drawn all along the drill does not highlight the different strata (figure 2). The Fourier spectrum of the ROP (figure 3) shows a high frequency content of pure random type. It acts as a noise perturbing the data and obscuring the transitions between strata.

Having filtered such additional noise, the entropy (figure 4) permits to isolate several layers.

Figure 5 illustrates Gaussian approximations of the rate of penetration and its derivative in the limestone layer.

Figure 6 presents the variogram of the ROP within the limestone layer computed directly from the data and a second one modelled using a Gaussian variogram, the slope of the entropy and the standard deviation within the limestone layer.

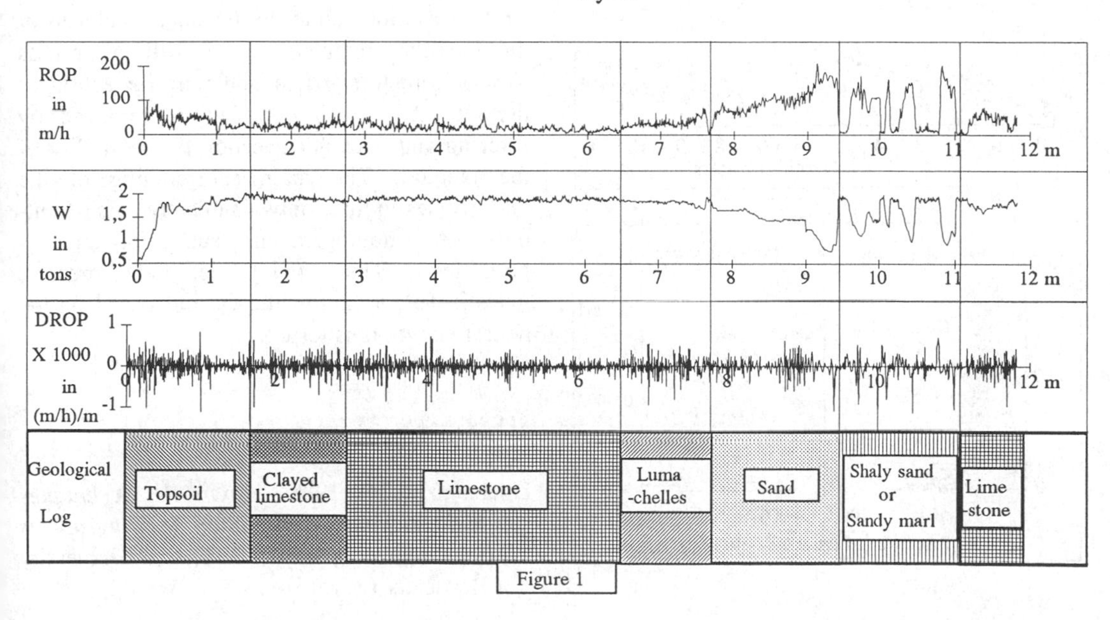

Figure 1

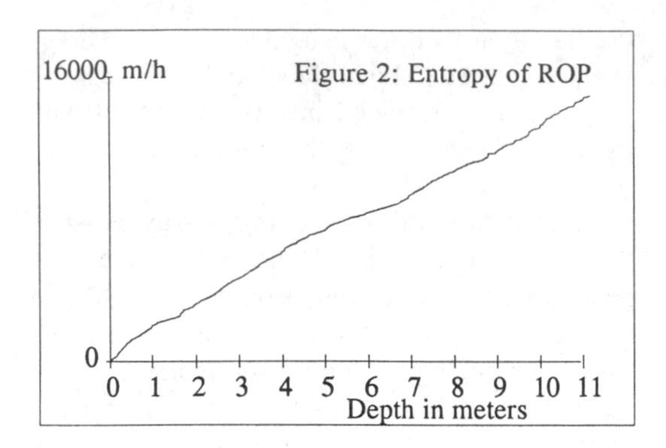

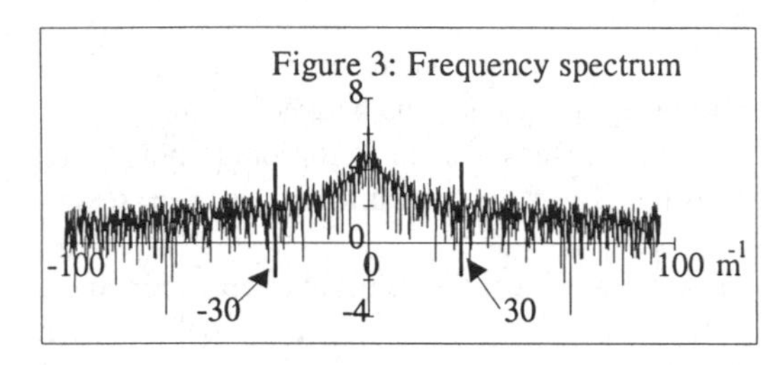

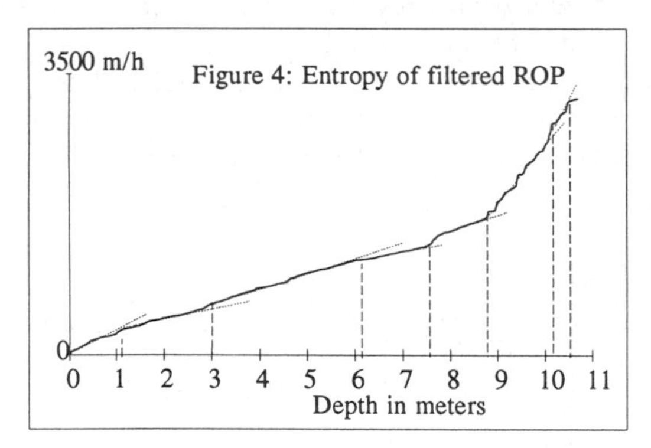

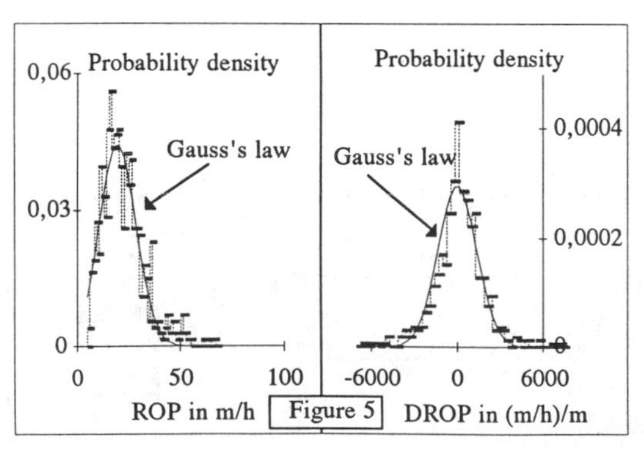

Figure 5

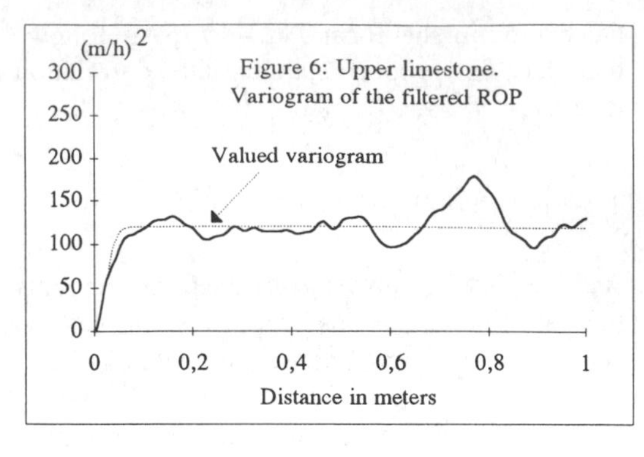

4. CONCLUSION

The entropy turns out to be a reliable tool for processing the drilling data. Studies of real data have confirmed that a drilling response associated with homogeneous zones can be modelled by a stationary Gaussian mean square differentiable random signal. The method developed at the LERGGA is tractable given its simplicity as opposed to more common techniques (for example: the Fisher algorithm). Use of entropy in real time, i.e. while drilling, would be very helpful for locating a particular formation for sampling. The high frequency noise which obscures the transitions between strata is ascribed either to the small thickness of the medium involved in the drilling process (few millimeters) or to non uniform action of the bit. This noisy part can be removed by filtering and does not question the reliability of the method. The main shortcoming of the entropy lies in that, only a subjective visual cut-out into homogeneous sub domains is permitted. This point is, at present, investigated: an automatic cut-out based on the neural networks is tested.

REFERENCES

Duchamp, J.M. 1988. *Apport des techniques statistiques pour l'exploitation des diagraphies instantanées en génie civil*, Thesis, University of Bordeaux I.

Structural Safety & Reliability, Schuëller, Shinozuka & Yao (eds) © 1994 Balkema, Rotterdam, ISBN 90 5410 357 4

Uncertainties in the dynamic response of soils in the San Francisco Bay region

Stephanie A. King & Anne S. Kiremidjian
The John A. Blume Earthquake Engineering Center, Department of Civil Engineering, Stanford University, Calif., USA

ABSTRACT: For purposes of seismic microzonation it is important to assess the sensitivities of forecasted surface ground motions. In order to study the effect of uncertainties in various soil parameters on the amplification of earthquake ground motion, several case studies are performed for soils in the San Francisco Bay region. In particular, the variability in the number of soil layers and the shear wave velocity, damping and unit weight of each soil layer are considered in the analysis. The results from the case study for Treasure Island, California are presented in this paper. It is found that the post-strain site period is dependent on the overall depth of the soil deposit but not on the variations in soil properties within the soil deposit. Peak ground acceleration and spectral accelerations are most sensitive to the number of layers considered in the soil deposit. These parameters increase considerably as the number of soil layers increase when the overall soil depth is kept constant. The forecasted ground motions appear to be rather insensitive to the choice of probability distribution for the various soil parameters. In contrast the post-strain site period is highly dependent on the probability distribution selected to represent the variability in soil parameters. Thus, when developing a predominant period map it is important to obtain reliable information on the depth and the average shear wave velocity of the soil deposits in the region. For ground motion forecasting, it is important to obtain more accurate information on the variation of soil properties with depth.

1 INTRODUCTION

Difficulties with predicting site ground motions arise in part due to the lack of information on the local soil properties and in part due to our inbability to properly model dynamic soil behavior. When site specific ground motions are desired, it may be feasible to obtain detailed soil information in order to develop more accurate amplification models of the ground motions at the site. For regional seismic zonation purposes, however, it is difficult to gather extensive soil parameter data and ground motion amplifications are obtained as averages over large areas. Thus, it is important to assess the variability and uncertainty associated with predicted ground motions over these areas.

In this paper, the effect of uncertainties in soil properties on the dynamic characteristics of the soil are studied. For this purpose, the depth, shear wave velocity, damping and unit weight of each soil layer, and the number of soil layers are treated as random variables. The study investigates the effect of these parameters on the variability of the post-strain site period, the peak surface ground motion, the peak spectral amplification and the period at which the peak spectral amplification occurs. The method of analysis and the result from a case study are described in the following sections.

2 UNCERTAINTY ANALYSIS

In order to study the effect of soil properties on ground motion parameters, a simple bi-linear soil dynamics model for ground motion amplification is considered (Idriss and Seed, 1968; Seed and Idriss, 1969). The soil profile for the Treasure Island site in the San Francisco Bay area is used as a test case. The computer program SHAKE (Schnabel, et al., 1972) is utilized to simulate the ground motions at the ground surface. The recording from the 1989 Loma Prieta earthquake at nearby Yerba Buena Island is used as the input motion at the bedrock level at Treasure Island.

The soil is assumed to be composed of 8, 5, 3 or 2 layers, respectively for each test. In each case the overall depth of the soil is kept to a total of 285 feet. The parameters for the eight layer case corresponds to the actual data available for the site. For the smaller number of soil layers, parameters were lumped together for soil layers with similar properties. The purpose of this modification is to determine the degree of error introduced by assuming a simpler soil profile. For most regions, knowledge of local soil is limited to gross geologic information leading to a one or at most two-layer soil characterization. It is expected that the analysis presented in this paper will provide some insight to the degree of error introduced by soil profile simplification.

The following parameters are assumed to be uncertain: soil depth, h_i, shear wave velocity, v_i, damping, γ_i, and unit weight, w_i,. Each of these are assumed to be uniformly, normally or lognormally distributed. In addition, various combinations of these distributions are also considered. The mean for each parameter corresponds to the measured mean. Test cases are performed for two values of the coefficient of variation of 20% and 35%. Table 1 lists the average value of the parameters for the eight soil layer case.

Table 1. Mean values of the soil parameters with depth

Layer	Layer Dpth ft	Shear Wave Vel. ft/sec	Damping	Unit Wght	Soil Type	Soil Factor
1	30	550	0.05	0.11	2	0.50
2	15	600	0.05	0.11	2	0.70
3	30	500	0.05	0.11	1	0.40
4	25	700	0.05	0.11	1	0.60
5	40	1100	0.05	0.125	2	1.25
6	60	1200	0.05	0.130	1	2.75
7	85	1400	0.05	0.130	1	3.25
8		3500	0.05	0.150		

Figure 1 shows the variation of post-strain site period as a function of the number of soil layers at the test site. The post-strain site period increases only from two to three layers and appears to reach a plateau after five layers. The amplification is approximately about 1.22. The standard deviation of post-strain period does not show a particular pattern with increasing number of soil layers.

The variation of post-strain period with different distributions is also shown in Figure 1. The post-strain site period changes only by 1% to 2% when different distributions are considered for each parameter. An increase of approximately 8% to 15% is observed when the coefficient of variation is increased from 20% to 35% for each random variable. Thus, the variability in the coefficient of variation is found to have the strongest effect on the predicted value of the post-strain site period.

The effect of uncertainty in the various parameters on the predicted surface peak ground acceleration is shown in Figure 2. The value of peak acceleration appears to increase with increasing number of layers even though the total depth is kept constant. That increase appears to be as much as a factor of two. In that figure, the dependence of predicted peak acceleration values on the choice of distributions for the random soil parameters is also shown. For a fixed number of layers, the variation in predicted peak ground acceleration is small and in all cases is less than 10%. Thus, the simple bi-linear model for ground motion amplification appears to be insensitive to the distributions chosen for the shear wave velocity, damping and unit weight. Furthermore, the predicted mean peak ground accelerations vary only slightly when the coefficient of variation in each soil parameter increase from 20% to 35%. This insensitivity appears to be due primarily to the model used for estimating the peak ground acceleration.

The standard deviations of predicted peak ground motion acceleration values were also found to increase with increasing number of layers. For example, the standard deviations increased from 0.005g to 0.015 g when the soil parameters are assumed to be normally distributed with coefficient of variation of 35%. Similarly, the choice of probability distribution appears to have had little effect on the predicted standard deviations with the exception of the case of uniformly distributed parameters. For this latter case, the predicted standard deviation is a factor of almost two larger than any of the other standard deviations.

The sensitivity analysis for the maximum

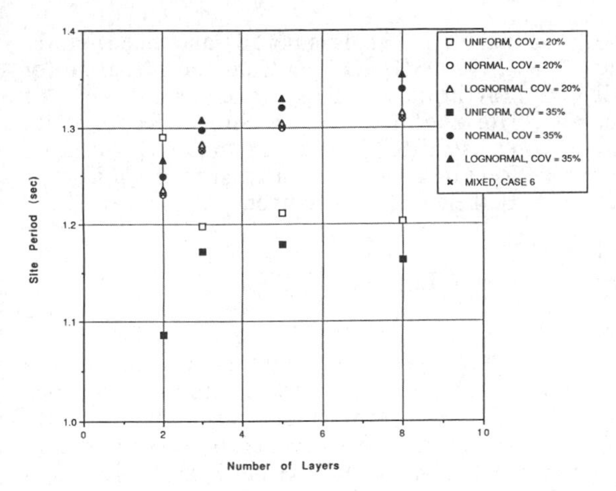

Fig. 1 Mean Post-Strain Site Period as a Function of the Number of Layers

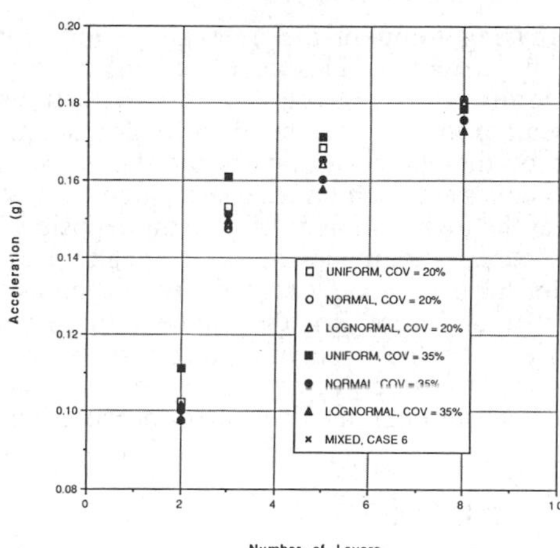

Fig. 2 Mean Surface Ground Acceleration as a Function of the Number of Layers

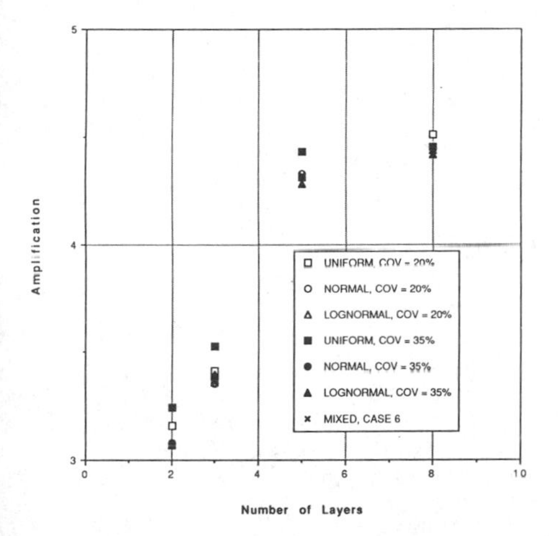

Fig. 3 Mean Peak Spectral Amplification as a Function of the Number of Layers

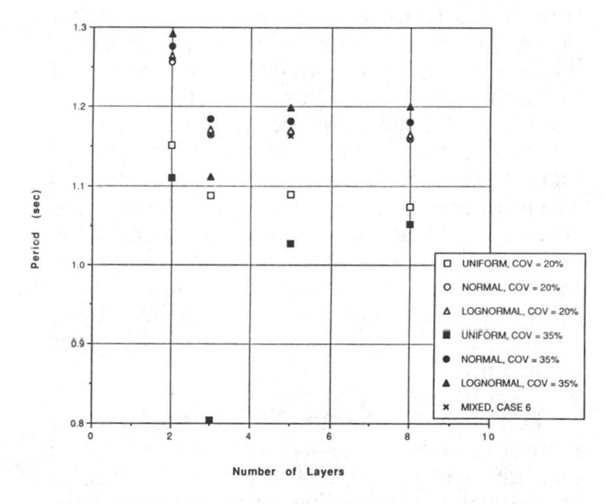

Fig. 4 Mean Period at Peak Spectral Accel. as a Function of the Number of Layers

spectral amplification is shown in Figure 3. The observations for the maximum spectral amplification are similar to these for the peak ground acceleration. The peak amplification values increase dramatically between three layers and five layers, from 3.5 to 4.3. The average of the predicted peak amplification value, however, do not seem to depend on the type of distribution used for the soil parameters. In contrast, predicted standard deviations are highly sensitive to the choice of the distribution and double in value when the coefficients of variation of each soil parameter is increased from 20% to 35% reflecting a cumulative uncertainty effect of the different parameters.

The period at which the maximum spectral acceleration occurs is very sensitive to the choice of distribution for each parameter and even more so to the degree of uncertainty in each variable as represented by the coefficient of variation. Figure 4 shows the variations in predicted period of maximum spectral acceleration. From that figure, the period of the maximum spectral acceleration value does not show much sensitivity to the number of layers used in the analysis, although a slight decreasing trend in the average period value can be detected. This is expected since the predominant period of the soil deposit will depend primarily on the overall depth rather than on the number of layers and the variations in thickness of each layer. These parameters will affect the higher periods of the soil deposit. The predicted standard deviations of peak spectral period also vary only slightly with increasing number of layers, but they appear to be very sensitive to the type of distribution and the amount of uncertainty assigned to each parameter through the coefficient of variation.

3 CONCLUSIONS

The sensitivity of a simple bi-linear model for predicting earthquake ground motion amplification was studied in this paper through a series of simulations. Peak ground acceleration and response spectral acceleration amplification were found to be very sensitive to the number of layers in the model and to the degree of uncertainty in each soil parameter expressed through their respective coefficients of variation. The choice of probability distribution for representing the uncertainty in ground motion parameters does not appear to have a strong influence on the mean peak ground and spectral amplification values. In contrast, the period at the maximum spectral amplification as well as the post-strain site period appear to be most sensitive to the choice of distribution and almost insensitive to the number of layers, as expected.

In all cases, the amplification of ground motion is observed to vary between 1.5 and a maximum of about 5 for the peak spectral amplification values. The larger amplifications are observed at the period range of 0.6 sec to about 1.7 sec. This range, however, reflects only the soil depth considered in this sensitivity study.

REFERENCES

Idriss, I.M. and Seed, H. B. (1968). Seismic Response of Horizontal Soil Layers.", *Journal of Soil Mechanics and Foundations Divisions, ASCE*, Vol. SM4, pp. 1003-1031.

Seed, H. B. and Idriss, I. M., (1969). Influence of Soil Conditions on the Ground Motions During Earthquakes, *Journal of Soil Mechanics and Foundations Divisions, ASCE*, Vol. SM 1, pp. 99-137.

Schnabel, P. B., Lysmer, J. and Seed, H. B. (1972). SHAKE : A Computer Program for Earthquake Response Analysis of Horizontal Layered Sites., Report No. UCB/EERC-72/12, Earthquake Engineering Research Center, University of California, Berkeley, CA, December, 102p.

ACKNOWLEDGMENTS

The authors would like to thank Dr. Roger Borcherdt for the numerous instructive discussions on soil amplification. Suggestions by Steve Weinterstein are also greatly appreciated. This research was partially supported by the NSF Grant EID-9024032 and by the John A. Blume Earthquake Engineering Center.

Structural Safety & Reliability, Schuëller, Shinozuka & Yao (eds) © 1994 Balkema, Rotterdam, ISBN 90 5410 357 4

Settlement prediction by using one-dimensional consolidation analysis with special reference to variability of soil parameters and observational procedure

S. Nishimura, H. Fujii & K. Shimada
Okayama University, Japan

ABSTRACT: The coefficients of volume compressibility, m_v , and permeability, k , are necessary in the calculation of settlement of soft ground. It is well known that these parameters have great variability. First, this paper discusses one-dimensional consolidation analysis, considering the variability of these parameters. The finite element method combined with the Monte Carlo simulation method (MFEM) is used to consider variability and nonlinearity of m_v and k . This approach is applied for the settlement prediction of a soft ground with the embankment construction herein. The second topic of this paper is the observational procedure for settlement prediction. Asaoka's method(Asaoka,1978) is used herein. It can evaluate mean , standard deviation and 95% confidence interval of future settlement by using observation result. The prediction results for some different identification periods are compared with one another.

1 NUMERICAL METHOD

In this research, FEM coupling method is used to calculate settlement and pore pressure. It is formulated as follows:

$$\begin{pmatrix} K & C \\ C^t & -H\Delta t \end{pmatrix}\begin{pmatrix} U_{t+\Delta t} \\ P_{t+\Delta t} \end{pmatrix}=\begin{pmatrix} K & C \\ C^t & 0 \end{pmatrix}\begin{pmatrix} U_t \\ P_t \end{pmatrix}+\begin{pmatrix} \Delta F_t \\ 0 \end{pmatrix} \quad (1)$$

in which t = time, Δt = time increment, U_t = displacement vector at time = t, P_t = pore pressure vector at time = t, ΔF_t = applied load vector at time = t, K = stiffness matrix, C = nodal displacement - volume change transformation matrix and H = permeability matrix. This formula gives $U_{t+\Delta t}$ and $P_{t+\Delta t}$. K matrix involves Young's modulus E and Poisson's ratio ν and H matrix dose permeability k. E and k are the function of the vertical effective stress. In the case of one-dimensional consolidation, Young's modulus is rclated to the coefficient of volume compressibility m_v by the next equation ,when the ground consolidates one-dimensionally under the K_0 condition.

$$E=\frac{(1-2\nu)(1+\nu)}{m_v(1-\nu)} \quad (2)$$

m_v and k are decided to be log- normal probabilistic variables in this research and the Monte Carlo simulation method is applied for the normal variables $\log_{10}m_v$ and $\log_{10}k$. The Monte Carlo method which is used herein was formulated by Shinozuka(1972).

2 STATISTICAL MODELS OF PARAMETERS

Our numerical approach is used to predict the settlement of the soft ground under the embankments, which were constructed in Kasaoka-Bay reclaimed land (Okayama Pref. in Japan). Figs.1-2 show the distributions of m_v and k . They are the oedometer test results of the undisturbed samples at the vertical effective stress $p=p_o+\Delta p/2$, in which p_o = the initial effective overburden stress and Δp = 39.2kN/m² (≒the total applied load) . These parameters are decided to be log-normal variable herein and separated into three or four layers. The statistical model of $M_v(Z)=\log_{10}m_v$ is formulated for each layer as follows:

$$M_v(Z) = \mu_{mv}(Z) + \sigma_{mv}(Z)\cdot u_{mv}(Z) \quad (3)$$

$$\mu_{mv}(Z) = A_{mv}+B_{mv}\cdot Z \quad (4)$$

in which $\mu_{mv}(Z)$= mean function of $M_v(Z)$, A_{mv} and B_{mv} = constants of regression line of $M_v(Z)$, $\sigma_{mv}(Z)$ = standard deviation function of $M_v(Z)$ and $u_{mv}(Z)$ = N(0,1) type probabilistic variable. A_{mv}, B_{mv} , $\sigma_{mv}(Z)$ are defined for each layer herein. These are not constant in the consolidation process; therefore, they are given as the functions of Δp in Fig.3. A_k, B_k, $\sigma_k(Z)$ and $u_k(Z)$ are the parameters as to log-permeability $\log_{10}k$ and defined like as A_{mv}, B_{mv}, $\sigma_{mv}(Z)$. A_k, B_k, $\sigma_k(Z)$ are given in Fig.4. The random variables $u_{mv}(Z)$ and $u_k(Z)$ can be assumed to have no correlation with each other herein, because the calculated coefficient of correlation is -0.1~-0.3 in the consolidation process. These variables are assumed to have the next exponential type auto-correlation:

$$r(\Delta Z) = exp(-\Delta Z/\delta) \quad (5)$$

in which ΔZ = distance between two points along the vertical direction and δ = auto-correlation distance. δ is 5.2m for m_v and 1.4m for k. These values are the means in the consolidation process. The parameters $u_{mv}(Z)$ and $u_k(Z)$ change slightly in the consolidation process, but they are assumed to be constant herein for simplification. This assumption is conservative, because it presents a greater variability of settlement than the strict one.

3 ANALYTICAL MODEL

Fig.5 shows the finite element model of the analysis. Two embankments (No.6C,No.7H) are analyzed in this research. Water is drained from the sand drain and the analytical model is axis-symmetric. Poisson's ratios are 0.385(0.0m-0.6m depth), 0.379(0.6m-3.0m depth), 0.373(3.0m-9.0m depth) and 0.391(9.0m-12.0m depth), but they are not important parameters in the

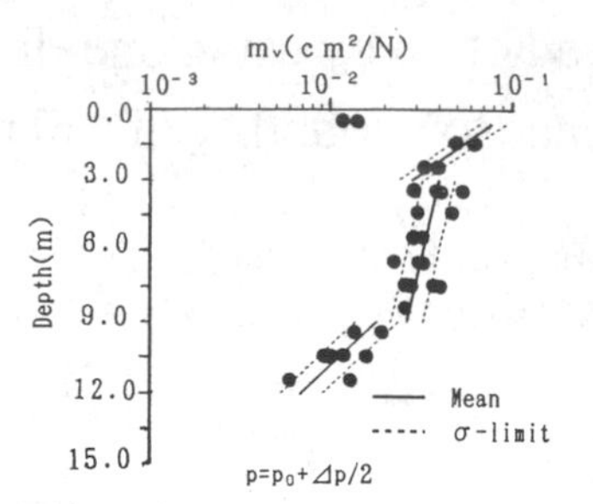

Fig.1 Distribution of coefficient of volume compressibility m_v

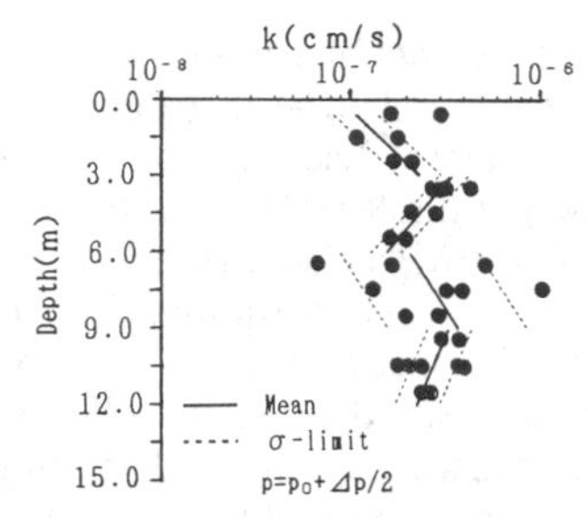

Fig.2 Distribution of coefficient of permeability k

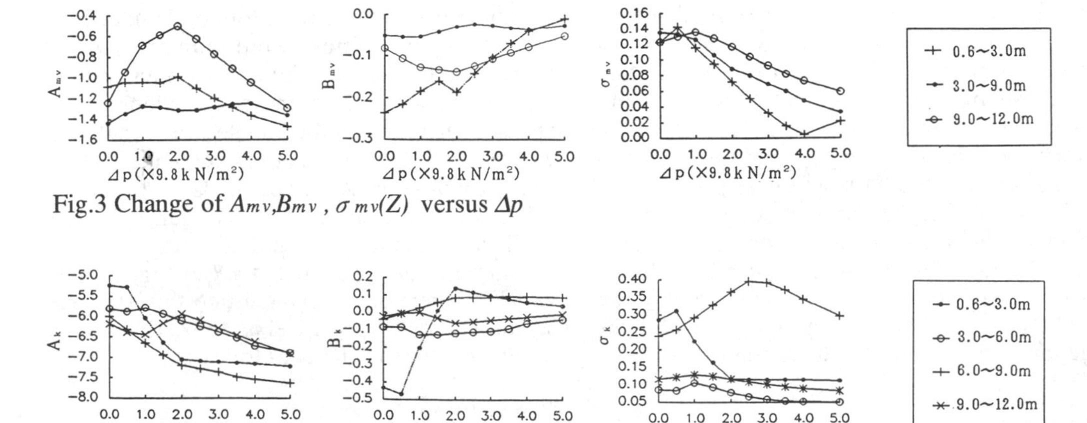

Fig.3 Change of A_{mv},B_{mv} , $\sigma_{mv}(Z)$ versus Δp

Fig.4 Change of A_k, B_k, $\sigma_k(Z)$ versus Δp

case of a one-dimensional consolidation problem and are treated as deterministic parameters. Young's modulus of the sand drain is assumed to be 980kN/m^2 and the coefficient of permeability was decided to be 21.6m/d from a laboratory test. The loading processes are presented in Fig.6.

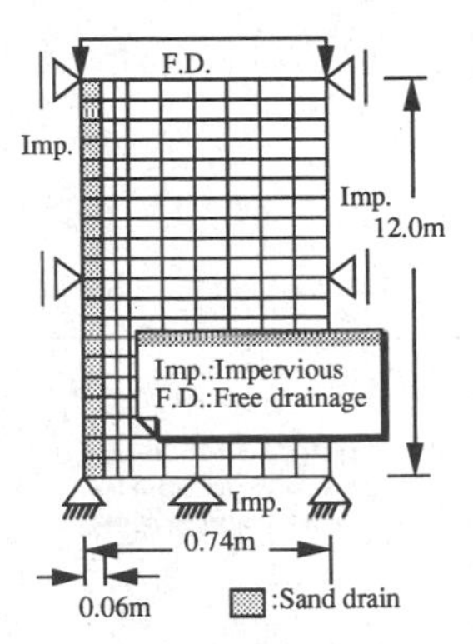

Fig.5 Analytical model

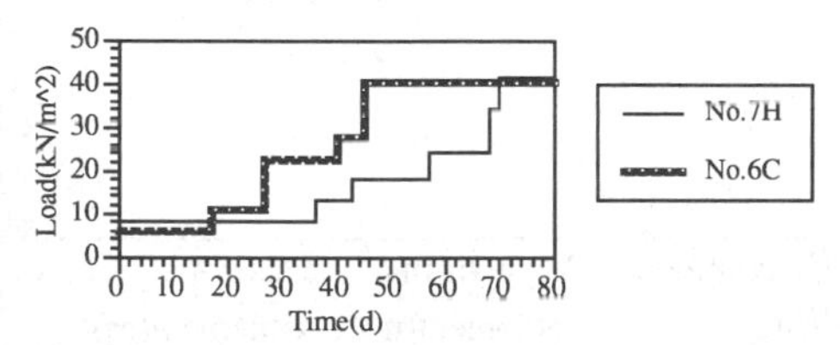

Fig.6 Loading process

4 RESULT OF MFEM ANALYSIS

Figs.7(a),(b) show the time-settlement curve of the No.6C and No.7H points. In those figures, "+" stands for the observed settlement. The solid and broken lines signify the means and σ,2σ-limits.

The means and the standard deviations of the analytical settlement increase gradually and the coefficients of variation are almost constant (0.11-0.14) in the consolidation process. The final settlement is given in Table1. The observed final settlements are involved in the σ-limit calculated by MFEM. Strictly speaking, the observed final settlement shown here means the predicted value by using Asaoka's method and all observed settlements.

Table 1 Final settlement

Site	Mean(m)	S.D.(m)	Observed(m)
No.6C	1.38	0.19	1.36
N0.7H	1.41	0.20	1.55

S.D.:Standard deviation

5 OBSERVATIONAL PROCEDURE

Generally, it is difficult to predict settlement exactly by using only laboratory test results. Therefore, the observational procedure is available to predict future settlement.

In this rescarch , Asaoka's method is used as an observational prediction method. This is based on the following auto-regressive model:

$$S_i = \alpha_0 + \alpha_1 S_{i-1} + \sigma_{ei} \quad (6)$$

in which S_i = settlement at time=t_i, S_{i-1}= settlement at time=t_{i-1} =t_i-Δt (Δt = 20 d), α_0, α_1 = constants of auto-regression and σ_{ei} = the Gaussian random variable.

$$E[\sigma_{ei},\sigma_{ej}] = \begin{Bmatrix} \sigma_e^2 & i=j \\ 0 & i \neq j \end{Bmatrix} \quad (7)$$

Figs.8(a),(b) show the results of observational procedures. The calculations are conducted for three identification periods. The solid lines stand for the means of the analytical settlements. These figures show that a 100-days observation period is necessary for exact prediction. The short-term observation predicts much smaller settlement than the observed one. Table 2 shows the prediction results of the settlements for the days when the last observations were conducted, and the analytical means of the final settlements.

6 CONCLUSIONS

1. MFEM is applied to a non-linear consolidation problem in this study. This method can show the variability of settlement and pore pressure of soft ground.
2. MFEM presents 0.11-0.14 constant coefficients of variation for the settlement in the consolidation process. (Fig.7)
3. The observed final settlements are involved in the σ-limit calculated by MFEM. (Table 1)

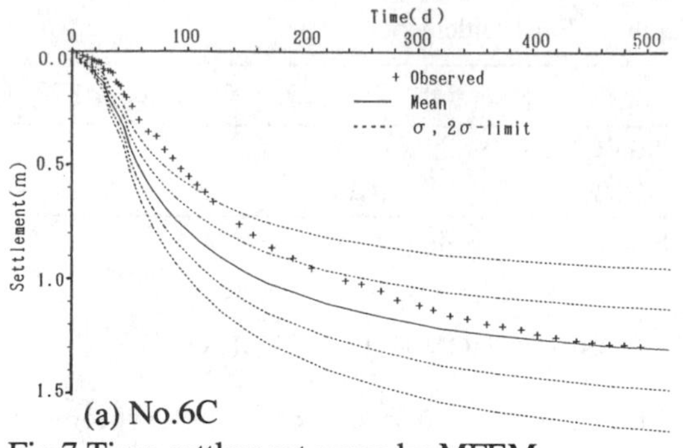

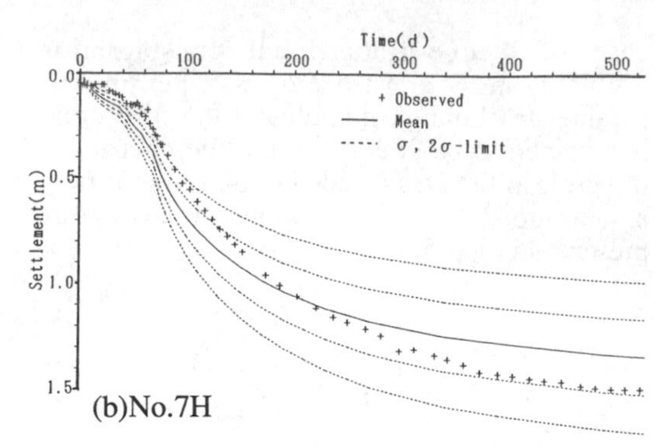

Fig.7 Time-settlement curve by MFEM

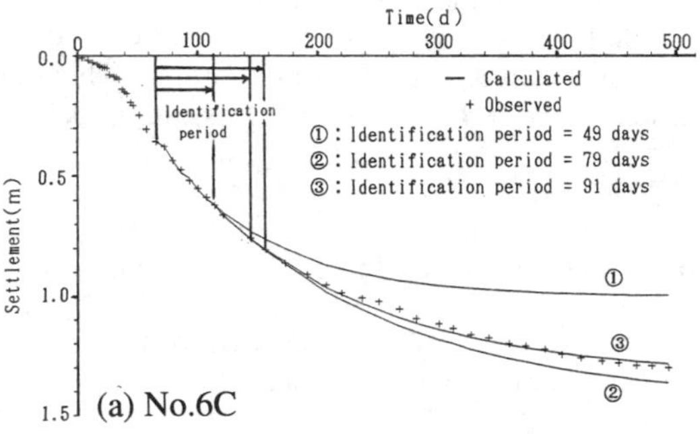

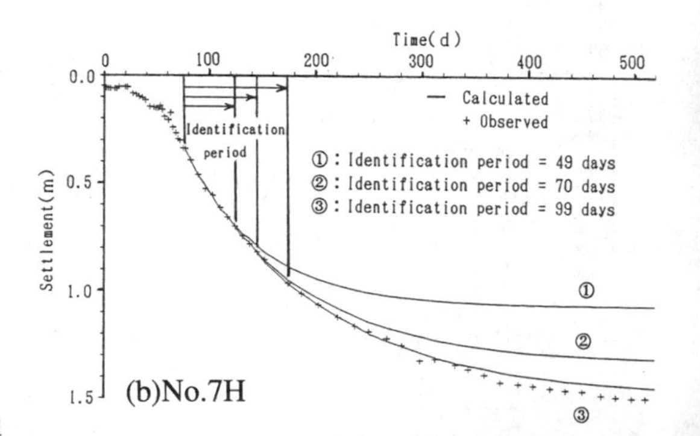

Fig.8 Time-settlement curve by Asaoka's method

Table 2 Results of observational procedure

Observation point	Identification period(d)	Mean(m)*	Standard deviation(m)*	95% confidence level(m)*	Mean of final settlement(m)	Observed settlement(m)*
	49	0.997	0.008	0.958-1.036	1.002	
No.6C	79	1.364	0.013	1.348-1.381	1.445	1.302
	91	1.284	0.011	1.278-1.291	1.336	
	49	1.065	0.012	1.027-1.102	1.066	
No.7H	70	1.312	0.013	1.297-1.328	1.326	1.495
	99	1.446	0.013	1.441-1.451	1.479	

*For the time=day524(No.6C),day486(No.7H)

4. In the case of this research, an observation period of at least 100 days is necessary to predict the future settlement exactly by the observational procedure. (Fig.8)

5. The standard deviations of the settlements determined by the observational procedure are very small and independent of the identification periods.(Table 2)

6. The shorter the observation period is, the greater is the 95% confidence interval. (Table 2)

REFERENCES

Asaoka,A.1978. Observational procedure of settlement prediction: *Soils and foundations*.18.No.4: 87-101.

Shinozuka,M. and Jan,C-M.1972.Digital simulation of random processes and its application. Journal of *Sound and Vibration*.25.No.1:111-128.

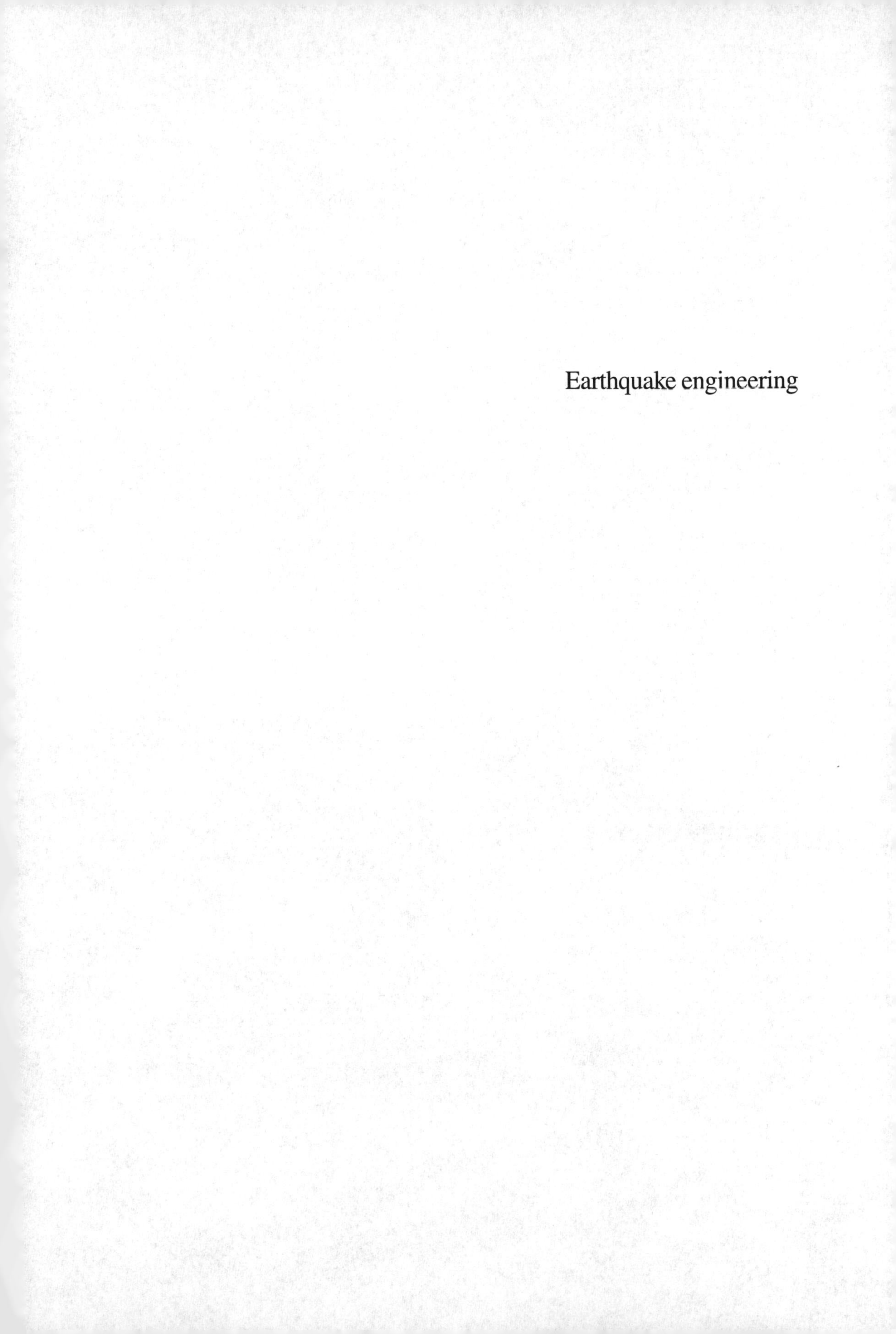

Earthquake engineering

Structural Safety & Reliability, Schuëller, Shinozuka & Yao (eds) © 1994 Balkema, Rotterdam, ISBN 90 5410 357 4

A simplified estimation method for first excursion probability of structures subjected to earthquake excitations

Shigeru Aoki
Tokyo Metropolitan Technical College, Japan

ABSTRACT: It has been pointed out that safety assessment of structures subjected to earthquake excitations should be evaluated in probabilistic manner. There are many failure modes which are considered in such structures. First excursion failure is one of the most important failure modes which should be considered. On the other hand, concept of average response spectrum is widely accepted to actural aseismic design codes. In this paper, considering above mentioned situation, a simplified estimation method for first excursion probability using average response spectrum is presented. It is found that this method can be applied to secondary system and nonlinear system with hysteretic restoring force-deformation relation.

1 INTRODUCTION

Important mechanical systems should be designed so as to survive even if they are subjected to destructive earthquake excitations. It has been pointed out that probabilistic evaluation is important for safety assessment of mechanical systems (Sundararajan 1986). Author has been studied about first excursion probability of mechanical systems (Aoki & Suzuki 1985), (Aoki 1989). Studies about first excursion probability are mainly concerning with introducing theoretical estimation method of probability (Ang 1988). In generally, complicated calculation and long calculation time are needed for probability estimation. Thus, development of a simplified probability estimation method is one of important problems for seismic safety assessment of mechanical systems. On the other hand, concept of average response spectrum is widely accepted to aseismic design codes of actual structures (Clough and Penzien 1975).

In this paper, considering above mentioned points, a simplified estimation method for first excursion probability using average response spectrum is proposed. First, by simulation method using a simple analytical model, it is shown that first excursion probability can be conventionally estimated when tolerance level is normalized by average response spectrum. Next, obtained results are examined by a theoretical method. Next, this method is applied to secondary system installed on primary system. Finally, application of this method to system with hysteretic restoring force-deformation relation is examined.

2 NOMENCLATURE

- P_f : first excuraion probability
- B_D : tolerance level
- x : absolute displacement of mass
- y : absolute displacement of input
- z : displacement of mass relative to input point, x-y
- h : damping ratio
- ω_n : natural circular frequency
- T_n : natural period, $2\pi/\omega_n$
- Z : expected value of the maximum response of $|z|$
- σ_g : standard deviation of input acceleration
- σ_z : standard deviation of z
- x_s : absolute displacement of secondary system
- x_p : absolute displacement of primary system
- z_s : displacement of secondary system relative to primary system, $x_s - x_p$
- γ : mass ratio of secondary system to primary system
- h_s : damping ratio of secondary system
- h_p : damping ratio of primary system

T_s : natural period of secondary system
T_p : natural period of primary system
Z_s : expected value of the maximum response of $|z_s|$
σ_{Z_s} : standard deviation of z_s

3 FIRST EXCURSION PROBABILITY ESTIMATION METHOD

In this paper, mechanical system is simulated by a single-degree-of-freedom system as shown in Fig.1 for simplicity. It is assumed that failure of mechanical system occurs when absolute value of z first crosses B_D. As input acceleration excitations, stationary white noise and nonstationary white noise of which envelope function is given as Fig.2 are used. 100 time histories for each white noise are generated for estimation of P_f. Attention is focused on P_f at the time when T second passes from beginning of input. T is selected as 30s.

4 ESTIMATION RESULTS OF FIRST EXCURSION PROBABILITY

Fig.3 shows P_f for stationary white noise input. These figures show relation between P_f and μ which is defined as the following equation:

$$\mu = B_D / \sigma_g \tag{1}$$

where σ_g is standard deviation of input acceleration which is generated by computer simulation. Since σ_g is related to the maximum input acceleration, Eq.(1) means B_D is devided by the value which corresponds to the maximum value of input. Fig.3(a) shows P_f for different values of h in the case where T_n is 0.3s. Fig.3(b) shows P_f for different values of T_n in the case of h=0.05. From Fig.3(a), P_f decreases as damping ratio increases. From Fig.3(b), P_f increases as natural period increases. In these figures, variation of P_f is large when h and T_n are changed. In order to suppress this variation, tolerance level is normalized by Z as follows.

$$\lambda = B_D / Z \tag{2}$$

Results are shown in Fig.4. These figures correspond to Fig.3. Comparing Fig.4 with Fig.3, variation of P_f for the same value of λ is very small. From this result, P_f is approximately represented by one curve independent of h and T_n. In Eq.(2), Z corresponds to average response spectrum. Concept of average response spectrum is widely accepted to aseismic design. Therefore, P_f for arbitrary value of h and

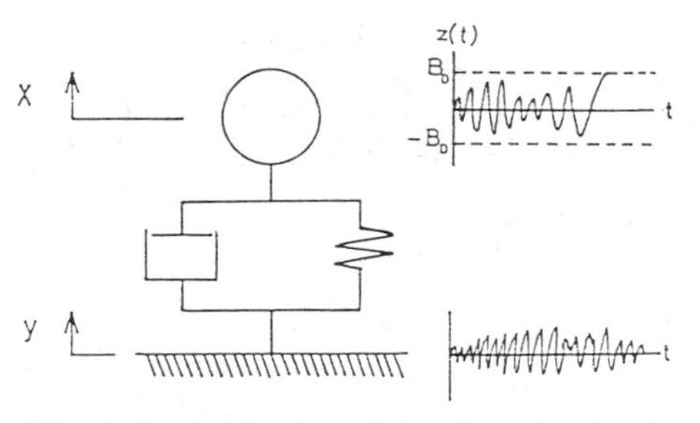

Fig.1 Analytical model of mechanical system

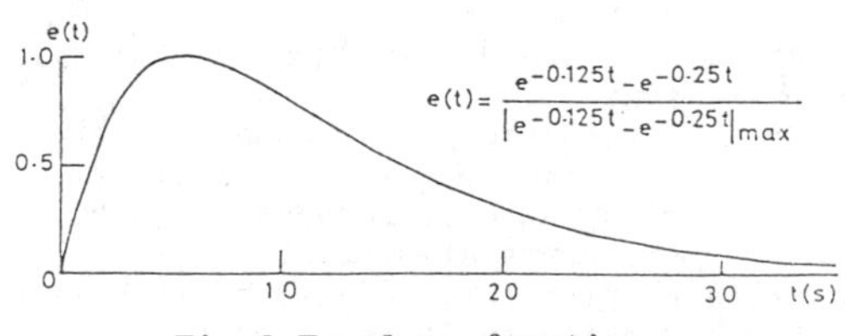

Fig.2 Envelope function

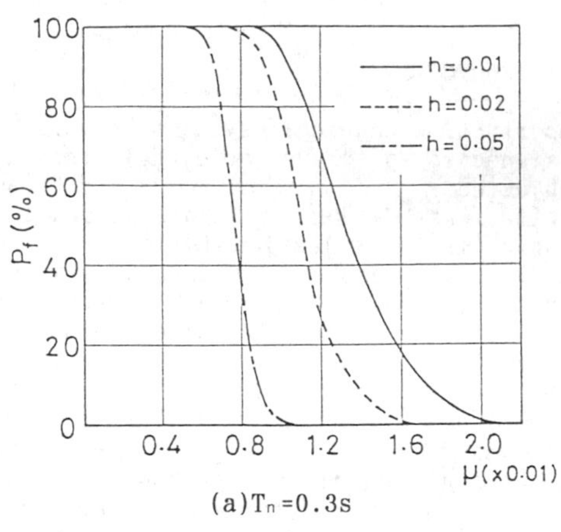

(a)T_n=0.3s

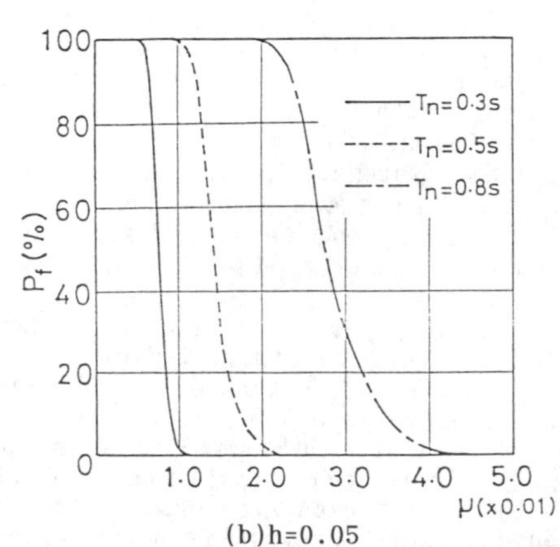

(b)h=0.05

Fig.3 First excursion probability for stationary white noise input

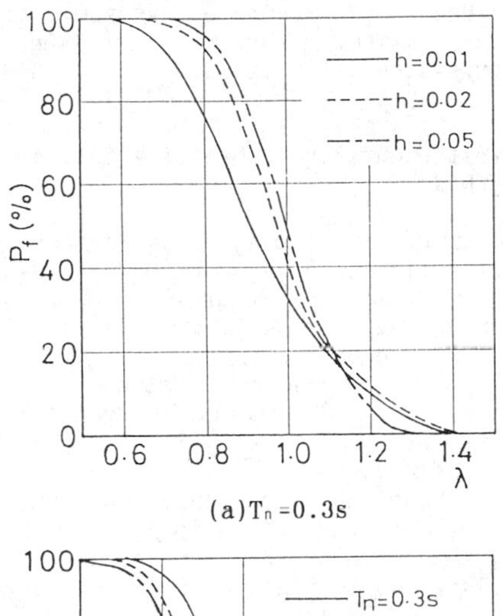

(a)T_n=0.3s

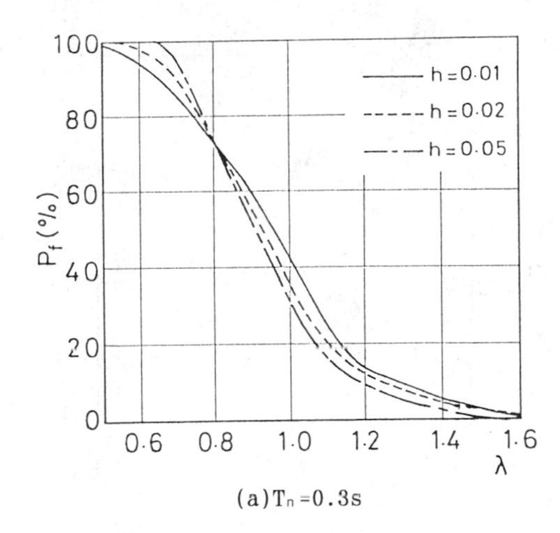

(a)T_n=0.3s

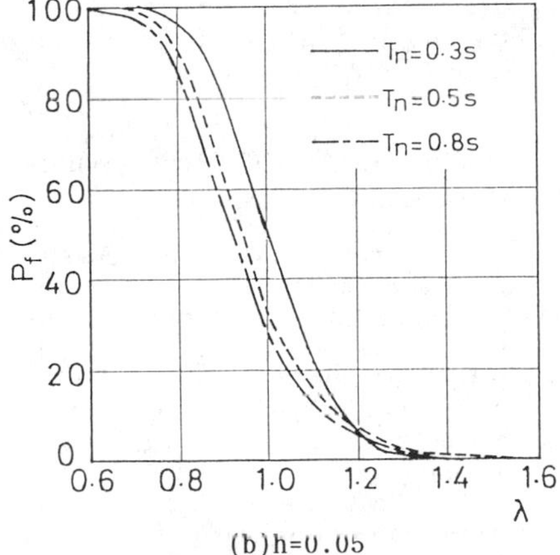

(b)h=0.05

Fig.4 First excursion probability for stationary white noise input

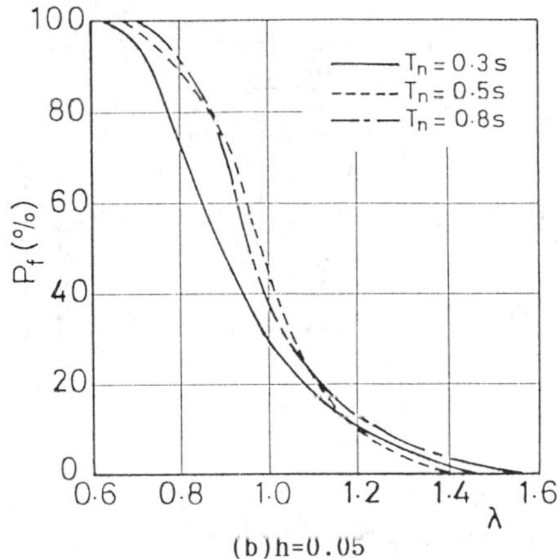

(b)h=0.05

Fig.5 First excursion probability for nonstationary white noise input

T_n is obtained when one curve of P_f is given for a certain seismic motion.

Next, P_f for nonstationary white noise input is estiamted in order to examine effect of nonstationary characteristic of input. Results are shown in Fig.5. Variation of P_f for the same value of λ is very small. Therefore, in the case of nonstaitonary white noise input, P_f is also conventionally estimated independent of h and T_n when tolerance level is normalized as Eq.(2).

5 EXAMINATION BY THEORETICAL METHOD

Obtained results in Chapter 4 are examined by a theoretical method for the case of stationary white noise input. When system is subjected to stationary random input, P_f at time T is obtained by the following equation:

$$P_f(T)=1-\exp(-2\nu T) \tag{3}$$

It is assumed that instants at which z(t) crosses B_D are statistically independent. ν is given as follows:

$$\nu=\frac{1}{2\pi}\omega_n \exp\left(-\frac{B_D{}^2}{2\sigma_z{}^2}\right) \tag{4}$$

σ_z is given as next equation.

$$\sigma_z=\sqrt{\frac{\omega_n{}^3 S_0}{2h}} \tag{5}$$

Where S_0 is power spectral density of input. When value of B_D for P_f=0.5 is used as expected value of the maximum response, denominator of Eq.(2), P_f is obtained by using λ as follows:

$$P_f(T)=1-\exp\left\{-\frac{2T}{T_n}\left(-\frac{T_n}{2T}\log 0.5\right)^{\lambda^2}\right\} \tag{6}$$

In Fig.6, P_f obtained by Eq.(6) are shown. Fig.6(a) shows P_f for different values of h in the case of T_n=0.3s. Fig.6(b) shows P_f for different values of T_n in the case of h=0.05. These figures

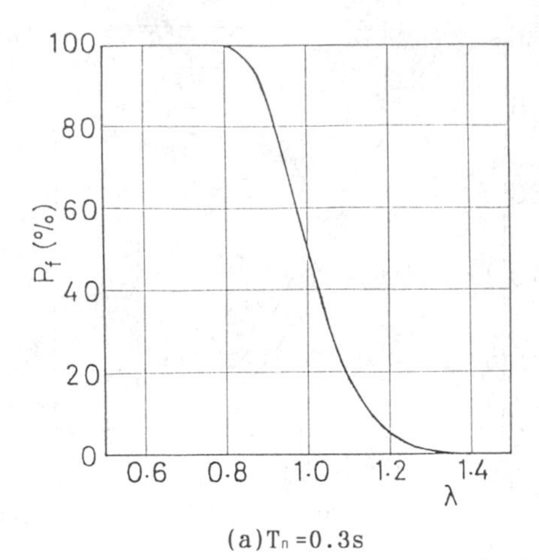

(a)T_n=0.3s

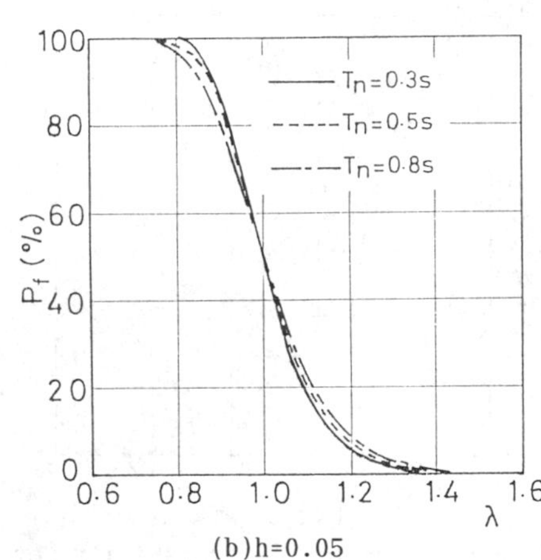

(b)h=0.05

Fig.6 First excursion probability obtained by theoretical method

correspond to Fig.4. Since h is not included in Eq.(6), there is no variation of P_f due to change of h. When T_n changes, variation of P_f for the same value of λ is very small. Gradient of curve in Fig.4 and that in Fig.6 are different. This reason is that assumption is not strictly appropriate that instants at which z(t) crosses B_D are statistically independent (Crandall & Mark 1963). However, there is a simple relation between P_f obtained by theoretical method and that by simulation method (Aoki & Suzuki 1985). Therefore, it is theoretically demonstrated that variation of P_f for the same value of λ is very small when B_D is normalized as Eq.(2).

6 APPLICATION TO SECONDARY SYSTEM

Application of obtained results for single-degree-of-freedom system to secondary system which is installed on primary system is examined.

6.1 First excursion probability estimation method

As an analytical model, a two-degree-of-freedom system as shown in Fig.7 is used. In this model, secondary system and primary system are simulated by single-degree-of-freedom system respectively. For response of secondary system, z_s is obtained. B_D is normalized by Z_s as follows.

$$\lambda_s = B_D / Z_s \tag{7}$$

In this case, when response spectrum is given, Z_s is conventionally estimated by method proposed by the author (Suzuki & Aoki 1981). h_p is fixed as 0.05. Values of γ, h_s, T_s and T_p are changed.

6.2 Estimation results of first excursion probability

In Fig.8, obtained results are shwon for the case of stationary white noise input. Fig.8(a) shows P_f for different values of γ. Fig.8(b) shows P_f for different values of h_s. Fig.8(c) shows P_f for different values of natural period in the case where T_s coincides with T_p. Fig.8(d) shows P_f for different values of natural period ratio T_s/T_p. From these figures, it is obvious that variation of P_f for the same value of λ_s is very small. Therefore, P_f can be conventionally estimated independent of mass ratio, damping ratio and natural period when tolerance level is normalized as Eq.(7).

Obtained results in Fig.8 are examined by the theoretical method used in Chapter 5. Instead of σ_z in Eq.(4), σ_{z_s} is

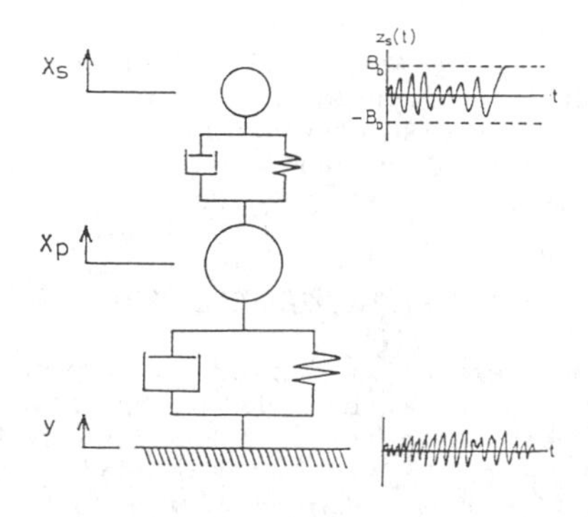

Fig.7 Analytical model of secondary system

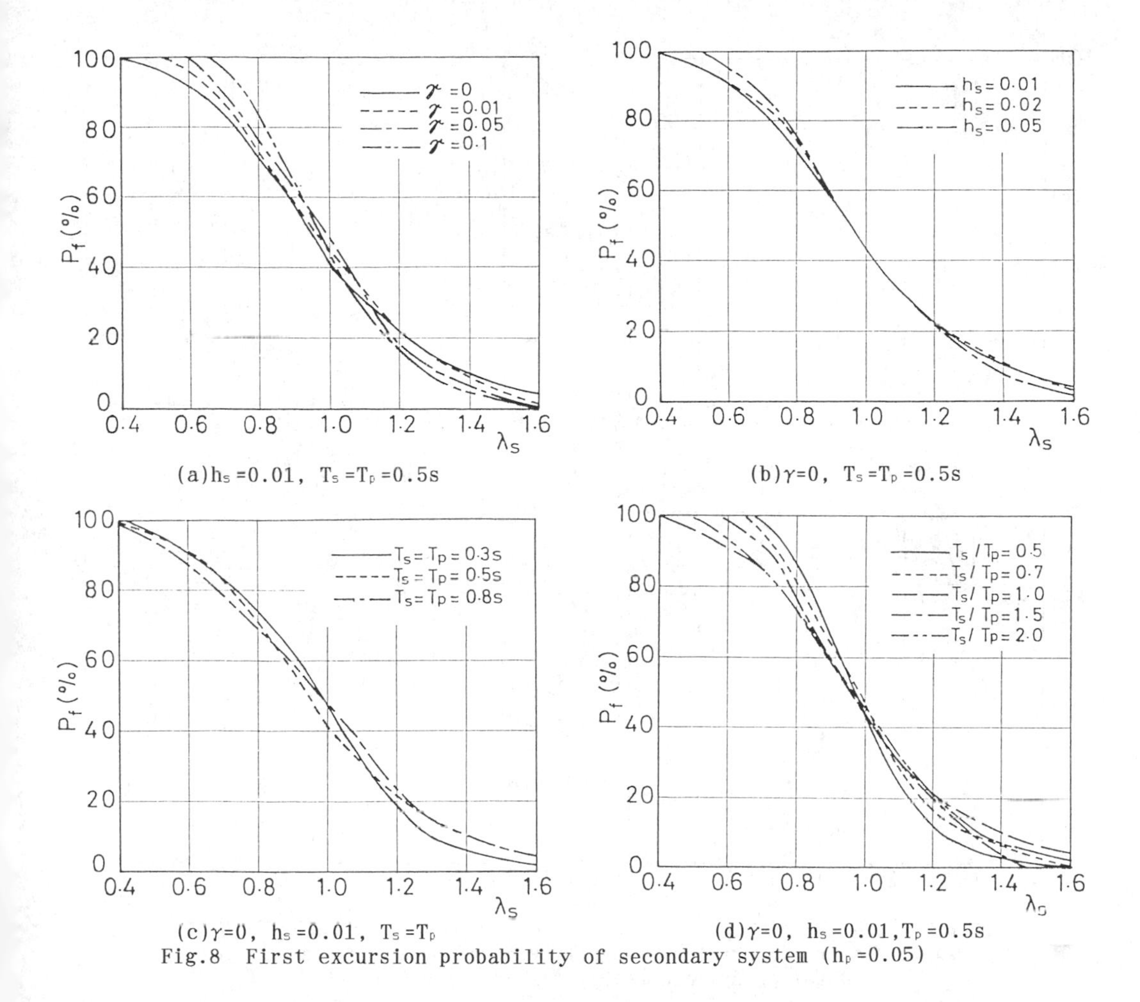

(a) $h_s = 0.01$, $T_s = T_p = 0.5s$

(b) $\gamma = 0$, $T_s = T_p = 0.5s$

(c) $\gamma = 0$, $h_s = 0.01$, $T_s = T_p$

(d) $\gamma = 0$, $h_s = 0.01$, $T_p = 0.5s$

Fig.8 First excursion probability of secondary system ($h_p = 0.05$)

required. σ_{Z_s} is obtained by the following equation.

$$\sigma_{Z_s} = \sqrt{\int_{-\infty}^{\infty} |H_s(\omega)|^2 S_{\ddot{g}}\, d\omega} \qquad (8)$$

Where $H_s(\omega)$ is frequency response function of z_s to input acceleration.

P_f is obtained by substituting σ_{Z_s} obtined by Eq.(8) into σ_z in Eq.(4). Obtained results are shown in Fig.9. These figures correspond to Fig.8. From Fig.9, it is found that variation of P_f is very small when tolerance level is normalized as Eq.(7). Therefore, characteristics obtained by Fig.8 are demonstrated by the theoretical method.

7 APPLICATION TO NONLINEAR SYSTEM

Application of results for linear system to single-degree-of-freedom system with hysteretic restoring force-deformation relation is examined. As hysteretic characteristic, bilinear type as shown in Fig.10 is used. One yielding force F_1 is determined as follows.

$$F_1 = \alpha \omega_n^2 Z \qquad (9)$$

Where Z is expected value of the maximum response for linear system. The other yielding force F_2 is determined as follows.

$$F_2 = \beta F_1 \qquad (10)$$

Tolerance level is normalized by expected value of the maximum response of nonlinear system Z_n as follows.

$$\lambda_n = B_D / Z_n \qquad (11)$$

Fig.11 shows relation between P_f and λ_n for different value of β in the case of h=0.01, T_n=0.5s and α=0.7. Variation of P_f is very small for the same value of λ_n. Same characteristic can be seen when values of other parameters are changed. Therefore, P_f can be estimated independent of damping ratio, natural period and

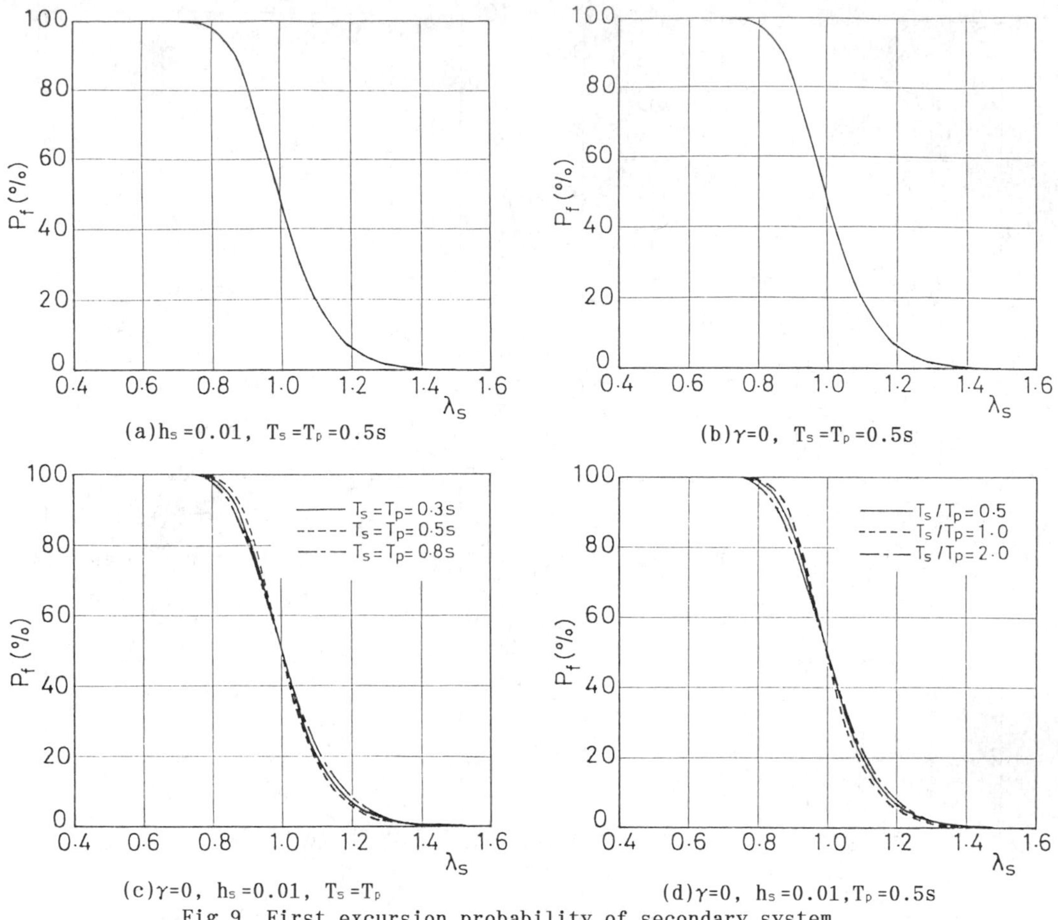

(a) $h_s = 0.01$, $T_s = T_p = 0.5s$

(b) $\gamma = 0$, $T_s = T_p = 0.5s$

(c) $\gamma = 0$, $h_s = 0.01$, $T_s = T_p$

(d) $\gamma = 0$, $h_s = 0.01$, $T_p = 0.5s$

Fig.9 First excursion probability of secondary system obtained by theoretical method ($h_p = 0.05$)

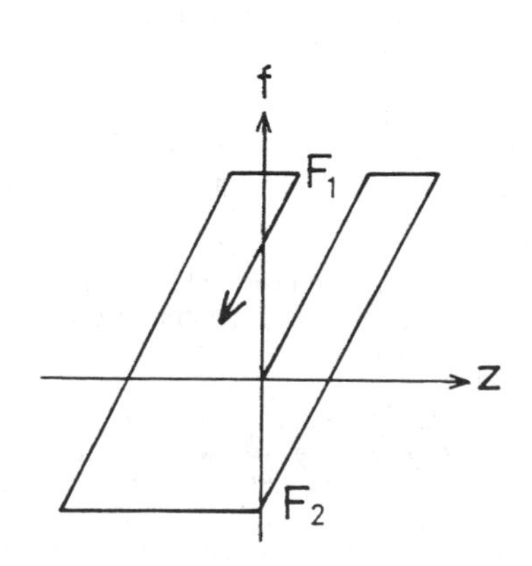

Fig.10 Hysteretic restoring force-deformation relation

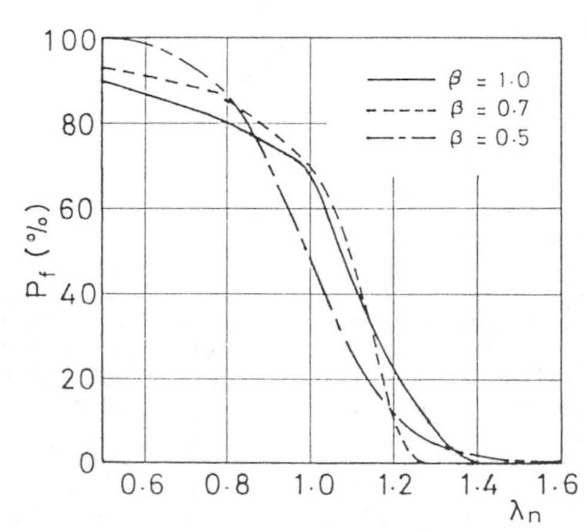

Fig.11 First excursion probability of secondary system ($h = 0.05$, $T_n = 0.3s$, $\alpha = 0.7$)

yielding force when tolerance level is normalized as Eq.(11).

8 CONCLUSIONS

Obtained results are summarized as follows.

(1) When tolerance level is normalized by expected value of the maximum response, average response spectrum, first excursion probabiblity can be conventionally estimated independent of damping ratio and natural period.

(2) Obtained results are demonstrated by a theoretical method.

(3) For secondary system, first excursion probability can be estimated independent of mass ratio of secondary system to primary system, damping raito and natural period of both systems when tolerance level is normalized as Eq.(7).

(4) For system with hysteretic restoring force-deformation relation, first excursion probability can be estimated independent of damping ratio, natural period and yielding force when tolerance level is normalized as Eq.(11).

REFERENCES

Ang,A.H S. 1988. Probabilistic seismic safety and damage assessment of structures. Proceedings of Ninth World Conference on Earthquake Engineering. 8: 717-728.

Aoki,S. & K.Suzuki 1985. First excursion probability estimation method of mechanical appendage system subjected to nonstationary earthquake excitations. Proceedings of 4th International Conference on Structural Safety and Reliability. 1: 201-210.

Aoki,S. 1989. Reduction of first excursion probability of nonlinear secondary system. Proceedings of 5th International Conference on Structural Safety and Reliability. 2: 1233-1240.

Clough,R.W. & J.Penzien 1975. Dynamics of structures. McGraw-Hill.

Crandall,S.H. & W.D.Mark 1963. Random vibration in mechanical systems. Academic Press.

Sundararajan,C. 1986. Probabilistic assessment of pressure vessel and piping reliability. Transactions of American Society of Mechanical Engineers. Journal of pressure Vessel Technology. 108 February: 1-13.

Suzuki,K. and S.Aoki 1981. Conventional estimating method of earthquake response of mechancial appendage system installed in the nuclear structural facilities. Transactions of 6th International Conference on Structural Mechanics in Reactor Technology. K10/3.

Structural Safety & Reliability, Schuëller, Shinozuka & Yao (eds) © 1994 Balkema, Rotterdam, ISBN 90 5410 357 4

Structural design based on the optimum seismic reliability and response distribution

Koichiro Asano & Hiroyuki Harada
Department of Architecture, Kansai University, Suita, Osaka, Japan

ABSTRACT : An analytical approach to structural design parameters is developped on the basis of the optimum seismic reliability and inter-story response distribution. An earthquake-like stationary random excitation being assumed, probabilistic earthquake response is estimated as a solution of a simple simultaneous algebraic equation. Based on this probabilistic earthquake response, the optimum parameters are determined by selecting seismic structural reliability and inter-story response distribution index as objective functions.

1 INTRODUCTION

In general the anti-seismic safety of structural systems has been examined on the basis of analysis of structural behavior subjected to actually recorded strong earthquake motions. Structural systems have been checked to ascertain whether the earthquake response of their elements is less than the safety threshold level corresponding to these elements, but optimality has not hitherto been investigated. Since one objective of anti-seismic safety design of structural systems lies essentially in offering aseismic safety to the system, a structural design should in turn preserve this objective.

This paper deals with an optimum design of structural systems, i.e. the prediction of the optimum dynamic parameters of the system by selecting seismic reliability in addition to response distribution index as objective functions. For this objective, we adopt a stationary earthquake-like non-white random process with a prescribed maximum amplitude, predominant period and spectral shaping factor for the earthquake acceleration functions. A stationary process is used here rather than a non-stationary one, because of its suitability for the parametric surveying method.

The analytical procedures developed here are, (i) derivation of the probabilistic second order moment response and its derivatives with respect to design parameters as a solution of the algebraic simultaneous equation; and (ii) calculation of the derivatives of the objective functions and determination of design parameters optimizing the objective functions, derived using the Newton-Raphson method[1]. The presented approach is partly dependent upon such hypotheses as the stationarity of earthquake-like random excitation, the Poisson probability distribution function (p.d.f.) of the anti-seismic probabilities of the respective stories, and their independence of the safety probability, all of which remain to be pursued in the future.

2 ANALYTICAL METHOD

2.1 FUNDAMENTAL EQUATION OF MOTION OF STRUCTURAL SYSTEM

The fundamental equation of motion of lumped mass structural systems may be written in a matrix form as

$$[\tilde{m}]\{\ddot{u}\} + [\tilde{c}]\{\dot{u}\} + [\tilde{k}]\{\phi\} = -[\tilde{m}]\{1\}f \quad (1)$$

$$: f = -2h_g\omega_g\dot{z} - \omega_g{}^2 z \; ; \; \ddot{z} + 2h_g\omega_g\dot{z} + \omega_g{}^2 z = -\ddot{w} \quad (2)$$

where $[\tilde{m}]$, $[\tilde{c}]$ and $[\tilde{k}]$ are respectively the matrices associated with mass, viscous damping and stiffness, $\{u\}$ and $\{\phi\}$ are respectively the inter-story displacement response and restoring force vectors, f is the non-white random input process to the structural system, z is the relative displacement response of a surface layer with viscous damping h_g and predominant period ω_g, which is subjected to white random base motion $\ddot{w}$. In Eq.(1) $\{\phi\} = \{u\}$ for the linear system, and for the bi-linear hysteretic system[2)]

$$\{\phi\} = [r]\{u\} + [r']\{g_1(\dot{u}, y)\} \quad (3)$$

$$\{\dot{y}\} = \{g_2(\dot{u}, y)\} \quad (4)$$

in which

$$g_{1_i} = y_i\{u(y_i+\delta_i)-u(y_i-\delta_i)\} + \delta_i\{u(y_i-\delta_i)u(\dot{u}_i)-u(-y_i-\delta_i)u(-\dot{u}_i)\} \quad (5)$$

$$g_{2_i} = \dot{u}_i\{u(y_i+\delta_i)-u(y_i+\delta_i)+u(y_i-\delta_i)u(-\dot{u}_i) + u(-y_i-\delta_i)u(\dot{u}_i)\} \quad (6)$$

and u(·) is a unit step function , $\{y\}$ is the displacement vector associated with a coulomb slip element which is one of the distributed elements for the composition of a bi-linear hysteretic system, $[r]$ and $[r']$ are respectively the diagonal matrices with the elements r_i and $1-r_i$, in which r_i is the dimensionless plastic branch slope of a bi-linear hysteretic characteristics, and non-linear function vectors $\{g_1\}$ and $\{g_2\}$ can be replaced with linear ones using the stochastic equivalent linearization technique[3] as

$$\{g_1\} \cong [C_1]\{\dot{u}\}+[C_2]\{y\} \quad (7)$$

$$\{g_2\} \cong [C_3]\{\dot{u}\}+[C_4]\{y\} \quad (8)$$

in which $[C_1]\sim[C_4]$ are the diagonal matrices whose elements are

$$C_{1_i} = \frac{\delta_i}{\sqrt{2\pi}\sigma_{\dot{u}_i}}\,\mathrm{erfc}\left(\frac{\delta_i}{\sigma_{y_i}\sqrt{2(1-\rho_i^2)}}\right) \quad (9)$$

$$C_{2_i} = \mathrm{erf}\left(\frac{\delta_i}{\sqrt{2}\sigma_{y_i}}\right) - \frac{\delta_i}{\sqrt{2\pi}\sigma_{y_i}}\exp\left(-\frac{\delta_i^2}{2\sigma_{y_i}^2}\right) \times \mathrm{erfc}\left(\frac{\rho_i\delta_i}{\sigma_{y_i}\sqrt{2(1-\rho_i^2)}}\right) \quad (10)$$

$$C_{3_i} \cong \frac{1}{2}\left\{1+\mathrm{erf}\left(\frac{\delta_i}{\sqrt{2}\sigma_{y_i}}\right)\right\} - \frac{1}{\pi}\exp\left(-\frac{\delta_i^2}{2\sigma_{y_i}^2}\right)\frac{\rho_i}{\sqrt{1-\rho_i^2}} \times \left\{1-\frac{\rho_i^2}{3(1-\rho_i^2)}\left(\frac{\delta_i^2}{2\sigma_{y_i}^2}+1\right)\right\} \quad (11)$$

$$C_{4_i} = -\frac{\rho_i\delta_i\sigma_{\dot{u}_i}}{\sqrt{2\pi}\sigma_{y_i}^2}\exp\left(-\frac{\delta_i^2}{2\sigma_{y_i}^2}\right) \times \left\{1+\mathrm{erf}\left(\frac{\rho_i\delta_i}{\sigma_{y_i}\sqrt{2(1-\rho_i^2)}}\right)\right\} \quad (12)$$

where δ_i is the elastic-limit deformation of a bi-linear hysteretic characteristic, $\sigma_{\dot{u}_i}$, σ_{y_i} are respectively the root mean square values of $\dot{u}_i$ and y_i , ρ_i is the coefficient of correlation of $\dot{u}_i$ and y_i , and erf(·) and erfc(·) are respectively the error function and complementary error function.

2.2 MOMENT RESPONSE AND ITS DERIVATIVE WITH RESPECT TO DESIGN PARAMETERS

The fundamental equation of the motion of structural systems (1) can be rewritten as

$$\dot{u}_j = \sum_{i=1}^{\tilde{n}} a_{ji}u_i + b_j f \quad (13)$$

in which $\tilde{n}$ is the maximum number of state variables u_j necessary for specifying the motion of the system, j is a variable number, a_{ji} is specified by the stiffness and viscous damping coefficient of the elastic system plus the elastic-limit deformation, the dimensionless plastic branch slope and the equivalent linearization coefficients of the bi-linear hysteretic characteristics, the predominant period ω_g, and the shaping factor h_g of the earthquake excitation, while b_j is the coefficient associated with the excitation intensity. Probabilistic non-stationary second order moment response $m_{ij} = E(u_iu_j)$ is estimated by the differential equations,

$$\dot{m}_{ij} = \sum_{l=1}^{\tilde{n}}(a_{il}m_{lj}+a_{jl}m_{li}) \quad (14)$$

$$: m_{\tilde{n}\tilde{n}} = (\sigma_f^2 - 4h_g\omega_g^3 m_{\tilde{n}\tilde{n}-1} - \omega_g^4 m_{\tilde{n}-1\tilde{n}-1})/4h_g^2\omega_g^2 \quad (15)$$

in which $i=1\sim\tilde{n}$, $j=i\sim\tilde{n}-1$, σ_f^2 is the mean square of the excitation, and $m_{\tilde{n}\tilde{n}}$, appearing on the right hand side of Eq.(14), has to be substituted using Eq.(15) for the numerical calculation. For the stationary case , $\dot{m}_{ij}=0$ and from Eq.(14) is obtained

$$\sum_{l=1}^{\tilde{n}}(a_{il}m_{lj}+a_{jl}m_{li}) = 0 \quad (16)$$

Substitution of Eq.(15) into Eq.(16) leads to the following simultaneous algebraic equation for $M_J = m_{ij}$,

$$[A]\{M\} = \{B\}\sigma_f^2 \Rightarrow \sum_{J=1}^{N} A_{IJ}M_J = B_I\sigma_f^2 \quad (17)$$

in which the elements of the coefficient matrix [A] are expressed in terms of those of the matrix $[a]$, and the elements of {B} are expressed in terms of ω_g and h_g. By selecting the variables $\{X_k\}(k=1\sim\mu)$ that specify structural elements as design parameters, the first and second order derivatives with respect to these variables are derived from Eq.(17) as

$$\left\{\frac{\partial M}{\partial X_k}\right\} = -[A]^{-1}\left[\frac{\partial A}{\partial X_k}\right]\{M\} \quad (18)$$

$$\left\{\frac{\partial^2 M}{\partial X_k \partial X_{k'}}\right\} = -[A]^{-1}\left(\left[\frac{\partial^2 A}{\partial X_k \partial X_{k'}}\right]\{M\} + \left[\frac{\partial A}{\partial X_k}\right]\left\{\frac{\partial M}{\partial X_{k'}}\right\} + \left[\frac{\partial A}{\partial X_{k'}}\right]\left\{\frac{\partial M}{\partial X_k}\right\}\right) \quad (19)$$

2.3 DESIGN PARAMETERS BASED ON THE OPTIMUM SEISMIC RELIABILITY AND RESPONSE DISTRIBUTION

Here seismic reliability $R(t)$ is selected as an objective function in determination of the optimum parameters for structural elements; provided that the p.d.f. of the structural maximum response process is Poissonian, it is expressed approximately by

$$R(t) \cong \exp\left(-\sum_{j=1}^{n} p_j(t)\right) \quad (20)$$

$$: p_j(t) = \frac{t\sqrt{m_{\dot{u}_j\dot{u}_j}}}{\pi\sqrt{m_{u_ju_j}}}\exp\left(-\frac{\tilde{u}_j^2}{2m_{u_ju_j}}\right) \quad (21)$$

in which p_j and $\tilde{u}_j$ are respectively the probability and the threshold level of safety for a given storey j and t is duration of the input process.

The derivative of $R(t)$ with respect to the variable X_k gained from Eq.(20) is

$$\frac{\partial R(t)}{\partial X_k} = \left(-\sum_{j=1}^{n}\frac{\partial p_j(t)}{\partial X_k}\right)\exp\left(-\sum_{j=1}^{n} p_j(t)\right) \quad (22)$$

$$: \frac{\partial p_j(t)}{\partial X_k} = \frac{t}{2\pi}\exp\left(-\frac{\tilde{u}_j^2}{2m_{u_ju_j}}\right)\sqrt{\frac{m_{\dot{u}_j\dot{u}_j}}{m_{u_ju_j}}} \times \left[\left(\frac{\tilde{u}_j^2}{m_{u_ju_j}} - 1\right)\frac{\partial m_{u_ju_j}}{\partial X_k}/m_{u_ju_j} + \frac{\partial m_{\dot{u}_j\dot{u}_j}}{\partial X_k}/m_{\dot{u}_j\dot{u}_j}\right] \quad (23)$$

and the optimization of $R(t)$ to X_k is tantamount to solving the following algebraic simultaneous equation to X_k

$$\{{}_1J_k\} = \left\{\sum_{j=1}^{n}\frac{\partial p_j(t)}{\partial X_k}\right\}R(t) = \{0\} \quad (24)$$

Given that $m_{u_ju_j} = m_{11}$ and $m_{\dot{u}_j\dot{u}_j} = m_{22}$, the derivatives of ${}_1J_k$ with respect to $X_{k'}$ are gained by

$$\frac{\partial_1 J_k}{\partial X_{k'}} = \frac{\partial^2 R(t)}{\partial X_k \partial X_{k'}} = \exp\left(-\sum_{j=1}^{n} p_j(t)\right) \times \left(\sum_{j=1}^{n}\frac{\partial p_{j(t)}}{\partial X_k}\sum_{j=1}^{n}\frac{\partial p_{j(t)}}{\partial X_{k'}} - \sum_{j=1}^{n}\frac{\partial^2 p_{j(t)}}{\partial X_k \partial X_{k'}}\right) \quad (25)$$

in which

$$\frac{\partial^2 p_j(t)}{\partial X_k \partial X_{k'}} = \frac{t}{2\pi}\sqrt{\frac{m_{22}}{m_{11}}}\exp\left(-\frac{\tilde{u}_j^2}{2m_{11}}\right)\left[\frac{1}{2}\left\{\left(\frac{\tilde{u}_j^2}{m_{11}} - 1\right) \times \frac{\partial m_{11}}{\partial X_{k'}}/m_{11} + \frac{\partial m_{22}}{\partial X_{k'}}/m_{22}\right\} \times \left\{\left(\frac{\tilde{u}_j^2}{m_{11}} - 1\right)\frac{\partial m_{11}}{\partial X_k}/m_{11} + \frac{\partial m_{22}}{\partial X_k}/m_{22}\right\} + \left(1 - \frac{\tilde{u}_j^2}{m_{11}}\right)\frac{\partial m_{11}}{\partial X_k}\frac{\partial m_{11}}{\partial X_{k'}}/{m_{11}}^2 - \frac{\partial m_{22}}{\partial X_k}\frac{\partial m_{22}}{\partial X_{k'}}/{m_{22}}^2 + \left(\frac{\tilde{u}_j^2}{m_{11}} - 1\right) \times \frac{\partial^2 m_{11}}{\partial X_k \partial X_{k'}}/m_{11} + \frac{\partial^2 m_{22}}{\partial X_k \partial X_{k'}}/m_{22}\right] \quad (26)$$

For the comparative discussion of optimum design parameters, here the square value of coefficient of variation of story drift moment response ${C_v}^2$ is selected as an objective function in determination of the optimum structural parameters, which is expressed by

$${C_v}^2 = \frac{E({m_{jj}}^2)}{E(m_{jj})^2} - 1 \quad (27)$$

in which $E(m_{jj})$ and $E({m_{jj}}^2)$ are respectively the mean and square mean values of the story drift moment response $m_{jj} = m_{u_ju_j}$. The derivative of ${C_v}^2$ with respect to the X_k obtained from Eq.(27) is

$$\frac{\partial {C_v}^2}{\partial X_k} = \frac{1}{E(m_{jj})^2}\frac{\partial E({m_{jj}}^2)}{\partial X_k} - \frac{2E({m_{jj}}^2)}{E(m_{jj})^3}\frac{\partial E(m_{jj})}{\partial X_k} \quad (28)$$

in which

$$\frac{\partial E(m_{jj})}{\partial X_k} = \sum_{j=1}^{n}\frac{\partial m_{jj}}{\partial X_k}/n$$

$$\frac{\partial E({m_{jj}}^2)}{\partial X_k} = 2\sum_{j=1}^{n}\left(m_{jj}\frac{\partial m_{jj}}{\partial X_k}\right)/n \quad (29)$$

and the optimization of ${C_v}^2$ to X_k is tantamount to solving the following algebraic simultaneous equation to X_k

$$\{{}_2J_k\} = \left\{\frac{\partial {C_v}^2}{\partial X_k}\right\} = \{0\} \quad (30)$$

The derivatives of ${}_2J_k$ with respect to $X_{k'}$ are gained by

$$\begin{aligned}\frac{\partial_2 J_k}{\partial X_{k'}} &= -\frac{2}{E(m_{jj})^3}\frac{\partial E(m_{jj})}{\partial X_{k'}}\frac{\partial E(m_{jj}{}^2)}{\partial X_k} \\ &+ \frac{1}{E(m_{jj})^2}\frac{\partial^2 E(m_{jj}{}^2)}{\partial X_k \partial X_{k'}} \\ &- \frac{2}{E(m_{jj})^3}\frac{\partial E(m_{jj}{}^2)}{\partial X_{k'}}\frac{\partial E(m_{jj})}{\partial X_k} \\ &+ \frac{6E(m_{jj}{}^2)}{E(m_{jj})^4}\frac{\partial E(m_{jj})}{\partial X_{k'}}\frac{\partial E(m_{jj})}{\partial X_k} \\ &- \frac{2E(m_{jj}{}^2)}{E(m_{jj})^3}\frac{\partial^2 E(m_{jj})}{\partial X_k \partial X_{k'}} \qquad (31)\end{aligned}$$

in which

$$\frac{\partial^2 E(m_{jj})}{\partial X_k \partial X_{k'}} = \sum_{j=1}^{n} \frac{\partial^2 m_{jj}}{\partial X_k \partial X_{k'}} / n$$

$$\frac{\partial^2 E(m_{jj}{}^2)}{\partial X_k \partial X_{k'}} = \frac{2}{n}\sum_{j=1}^{n}\left(\frac{\partial m_{jj}}{\partial X_{k'}}\frac{\partial m_{jj}}{\partial X_k} + m_{jj}\frac{\partial^2 m_{jj}}{\partial X_k \partial X_{k'}}\right) \qquad (32)$$

The application of the Newton-Raphson method is effective in the solution of Eqs.(24) and (30). The increment amount of the design variable $\{\Delta X_k\}$ with the i-step approximate solution $\{X_k\}$ is given by solving the simultaneous equation,

$$\left[\frac{\partial_v J_k}{\partial X_k}\right]\{\Delta X_k\} = -\{{}_v J_k\} \quad ; \quad v = 1,2 \qquad (33)$$

where $[\partial_v J_k/\partial X_k]$ and $\{{}_v J_k\}$ are calculated using the i-step solution of $\{X_k\}$.

3 NUMERICAL EXAMPLES

3.1 ELASTIC STRUCTURAL SYSTEMS

To examine the validity of the presented analytical approach, the optimum distribution of dynamic elastic structural properties is discussed.

Let the given elastic structural system have six-degree-of-freedom, its mass distribution $\{m\}$ be uniform $\{\overline{m}\}$ ($\overline{m}$: a standardized amount of mass) and its stiffness distribution be given as[4)]

$$\{k_i\} = \bar{k}\left\{1 - \lambda\left(\frac{i-1}{n-1}\right)^{\nu}\right\} \qquad (34)$$

where $\overline{k}$ is a standardized stiffness , λ and ν are the indicies governing the distribution of $\{k_i\}$, and the damping ratio is assumed to be proportional to the stiffness, assuming three percent of the critical damping as the fundamental mode of vibration. Further, let the level of the envelope function of σ_f , the shaping factor of the power spectra h_g and the predominant angular frequency ω_g of the excitation be respectively 50~100(gal) , 0.3 ~ 0.5 and 3.0~20.0(rad/sec) , while the standardized mass and stiffness are determined so that the fundamental period of structural systems T_1 is 0.5~2.0(sec) . The stiffness matrix of the structural system with six-degree-of-freedom is calculated from Eq.(34) as in Eq.(35).

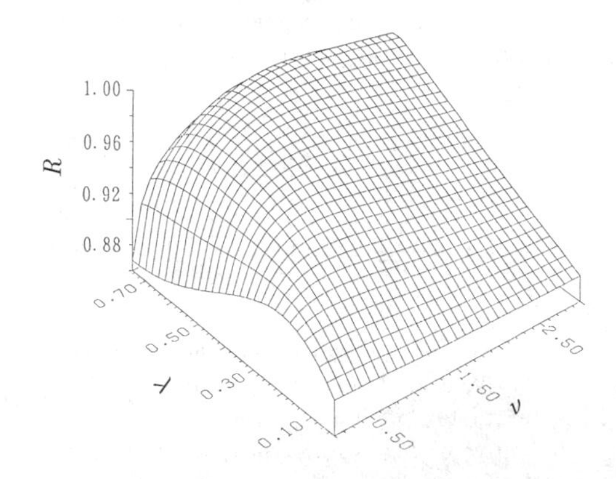

Fig.1 Reliability Perspective as a Function of λ and ν : 6 d.o.f. elastic system

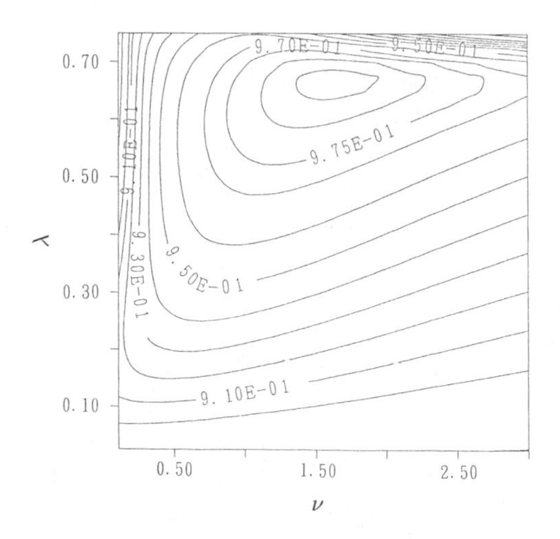

Fig.2 Reliability Contour lines as a Function of λ and ν : 6 d.o.f. elastic system

$$
[k'] = \begin{bmatrix}
2-\lambda/5^{\nu} & -(1-\lambda/5^{\nu}) & 0 & 0 & 0 & 0 \\
-(1-\lambda/5^{\nu}) & 2-\lambda(1+2^{\nu})/5^{\nu} & -(1-\lambda/(2/5)^{\nu}) & 0 & 0 & 0 \\
0 & -(1-\lambda/(2/5)^{\nu}) & 2-\lambda(2^{\nu}+3^{\nu})/5^{\nu} & -(1-\lambda(3/5)^{\nu}) & 0 & 0 \\
0 & 0 & -(1-\lambda(3/5)^{\nu}) & 2-\lambda(3^{\nu}+4^{\nu})/5^{\nu} & -(1-\lambda(4/5)^{\nu}) & 0 \\
0 & 0 & 0 & -(1-\lambda(4/5)^{\nu}) & 2-\lambda(4^{\nu}+5^{\nu})/5^{\nu} & -(1-\lambda) \\
0 & 0 & 0 & 0 & -(1-\lambda) & 1-\lambda
\end{bmatrix} \tag{35}
$$

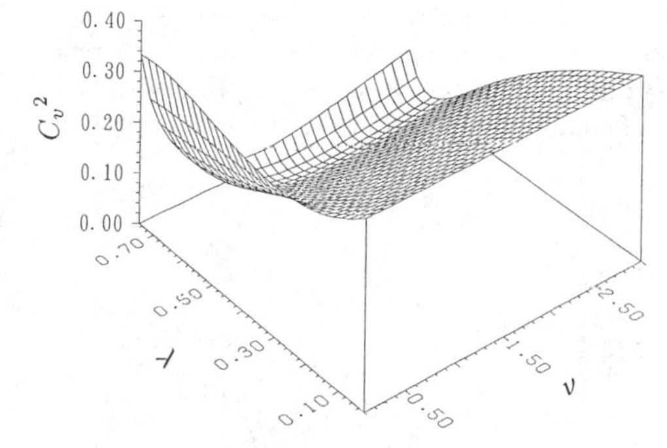

Fig.3 Uniformity Perspective as a Function of λ and ν : 6 d.o.f. elastic system

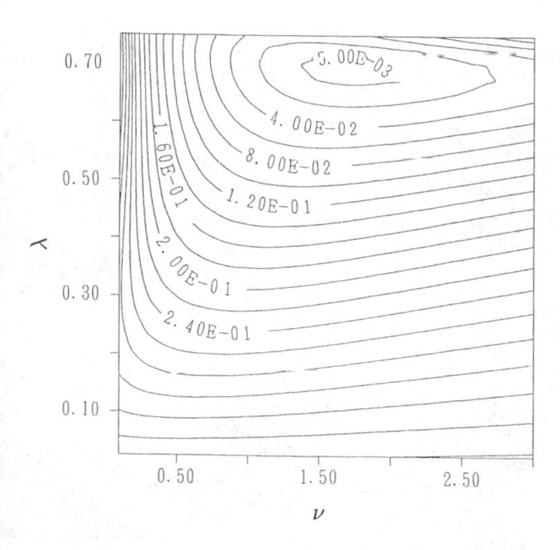

Fig.4 Uniformity Contour lines as a Function of λ and ν : 6 d.o.f. elastic system

and the elements of the matrix $[a]$ in Eq.(1) with damping proportional to its stiffness are given as

$$
[a] = \begin{bmatrix} [a_{11}] & [a_{12}] & [a_{13}] \\ [a_{21}] & [a_{22}] & [a_{23}] \\ [a_{31}] & [a_{32}] & [a_{33}] \end{bmatrix} \quad :
$$

$$
[a_{11}] = [a_{13}] = [a_{31}] = [a_{32}] = [0] \ , \ \ [a_{12}] = [E] \ ,
$$

$$
[a_{21}] = \overline{\Omega}^2[\tilde{k}] \ : \ \overline{\Omega}^2 = \overline{k}/\overline{m} \ ,
$$

$$
[a_{22}] = \eta[a_{21}] \ : \ \eta = 2h_1/\omega_1 \ ,
$$

$$
[a_{23}] = \begin{bmatrix} {\omega_g}^2 & 2h_g\omega_g \\ 0 & 0 \\ \vdots & \vdots \\ 0 & 0 \end{bmatrix} \ ,
$$

$$
[a_{33}] = \begin{bmatrix} 0 & 1 \\ -{\omega_g}^2 & -2h_g\omega_g \end{bmatrix} \tag{36}
$$

Numerical calculations of seismic reliability were made based on Eqs.(17), (20) and (21), using the parameters described immediately above. The results are shown in Figs.1 and 2, where $T_1 = 1.8(sec)$, the elastic modal critical damping ratio = 0.03, the safety level in terms of the storey drift angle = 1/200, $\sigma_f = 160/3(gal)$, $\omega_g = 20(rad/sec)$, $h_g = 0.5$ and the duration of stationary excitation= 10 (sec). In Fig.1, seismic reliability is plotted as a function of λ and ν , the latter ranging respectively over 0.025~0.75 and 0.1~3.0 ; and the contour line corresponding to this is plotted in Fig.2.

Figs.1 and 2 indicate that the seismic reliability is extremely sensitive to λ , but not so sensitive to ν. Parametric survey calculations were also made for the given parameters, the optimum combination of which gives for the maximum seismic reliability of the system with $\lambda = 0.675$ and $\nu = 1.700$, while the optimum parameters based on the presented approach are $\lambda = 0.669$ and $\nu = 1.645$. The mutual consonance of these indices is quite satisfactory.

Numerical caluculations of the response distribution index ${C_v}^2$ — the square value of coefficient of variation were also made based on Eqs.(17) and (27). The results are shown in Figs.3 and 4 , which respectively show the perspective and the contour lines of the index ${C_v}^2$ for the same exitation parameter used for the seismic reliability caluculation in Figs.1 and 2.

Figs.3 and 4 indicate that the distribution index (uniformity) is also sensitive to λ , but not so sensitive to ν as in Figs.1 and 2. Parametric survey calculations give the optimum distribution index for the combination values $\lambda = 0.700$ and $\nu = 1.700$, while the optimum parameters based on the presented approach are $\lambda = 0.693$ and $\nu = 1.714$.

These results indicate that the strict values of optimum parameters are dependent upon the selection of the objective function but that the mutual consonance of the optimum parameters obtained from the completely different basis is quite good from the engineering viewpoint. Therefore, the response distribution uniforming basis may lead to the optimum seismic safety design of structural systems as far as the uniform seismic safety levels are chosen to the respective inter-story displacement response.

3.2 ELASTO-PLASTIC STRUCTURAL SYSTEMS

Here is discussed the optimum distribution of dynamic elasto-plastic structural properties by taking the bi-linear hysteretic characteristics $\{\phi\}$ in Eq.(3) as an example. The parameters considered here are the dimensionless plastic slope r_i and the elastic-limit deformation δ_i , the distribution of which is given as

$$\{\delta_i\} = \overline{\delta}\left\{1 - \alpha\left(\frac{i-1}{n-1}\right)^{\beta}\right\} \qquad (37)$$

where $\overline{\delta}$ is a standardized elastic-limit deformation, and α and β are the similar indices defining the distribution of $\{\delta_i\}$ as in Eq.(34). The elasto-plastic structural system has six-degree-of-freedom, the optimum elastic stiffness distribution derived in section 3.1, the uniform mass distribution, the elastic modal critical damping ratio = 0.03, the elastic fundamental period = 1.2(sec), and the safety threshold level in terms of the ductility factor = 2.0. The excitation parameters $\sigma_f = 365/3(gal)$, $\omega_g = 20(rad/sec)$ and $h_g = 0.5$ are used for numerical calculations.

The elements of the coefficient matrix of $[A]$ corresponding to those in Eq.(17) are expressed for this case in terms of the elastic stiffness and the associated damping plus the equivalent linearization coefficients for bi-linear hysteretic characteristics of the respective story. Similar numerical calculations were made on the basis of Eqs.(17), (20) and (21).

In Fig.5, the seismic reliability of bi-linear hysteretic structural systems with $r_i = 0.1$ is plotted as a function of α and β utilizing a parametric survey calculation method, and the corresponding contour line is plotted in Fig.6. These figures indicate that the seismic reliability is highly sensitive to α but completely insensitive to β, and that the reliability attenuates very rapidly with increase of the index α from zero to 0.5.

Numerical calculations of the response distribution index ${C_v}^2$ were also made for this case. The perspective and the corresponding contour line are plotted as a function of α and β in Figs.7 and 8 respectively.

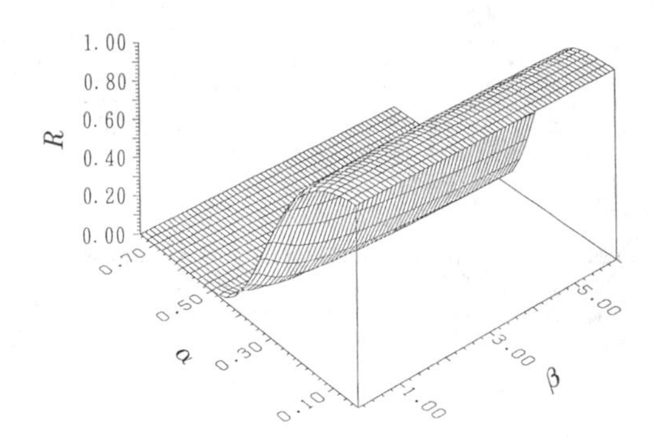

Fig.5 Reliability Perspective as a Function of α and β : 6 d.o.f. bi-linear system, $r_i = 0.1$

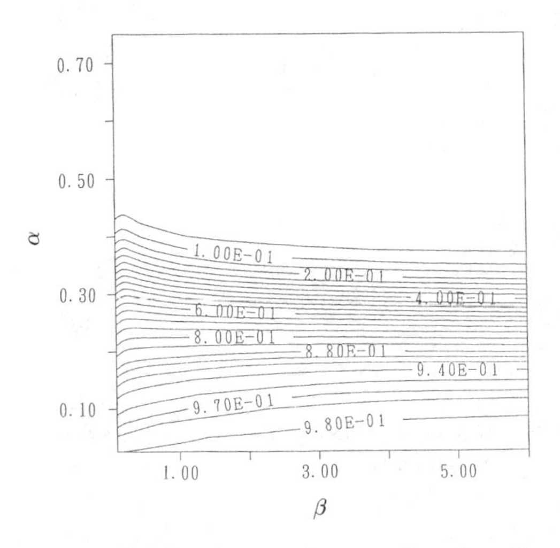

Fig.6 Reliability Contour lines as a Function of α and β : 6 d.o.f. bi-linear system, $r_i = 0.1$

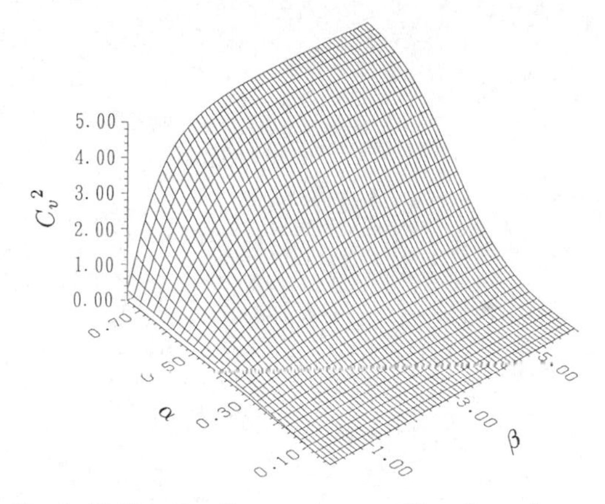

Fig.7 Uniformity Perspective as a Function of α and β : 6 d.o.f. bi-linear system, $r_i = 0.1$

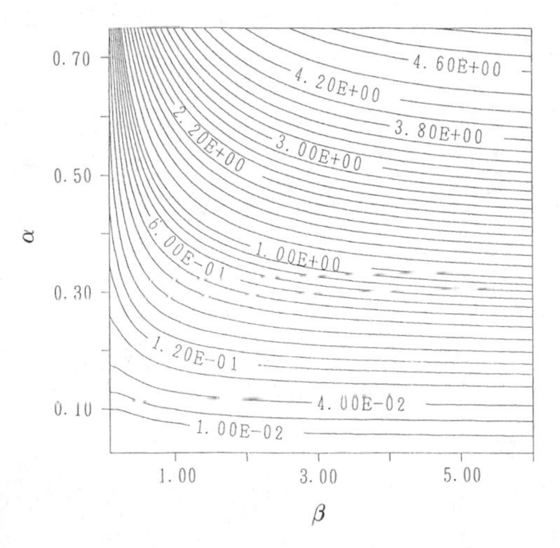

Fig.8 Uniformity Contour lines as a Function of α and β : 6 d.o.f. bi-linear system, $r_i = 0.1$

These results are different from those in Figs.5 and 6 in detail, but the closer examination of these results indicate that the parameters optimizing both the reliability and response distribution are given by

$$\{\delta_i\} - \{\bar{\delta}\} \tag{38}$$

, which means the elastic deformation should be chosen equal to the respective stories.

4 CONCLUSIONS

An analytical approach has been developed for elastic and elasto-plastic structural systems based on the optimality of two objective functions — seismic reliability and inter-story response distribution index. The validity of the approach was demonstrated by examining the perspective and contour lines of the objective function as a function of design parameters for six-degree-of-freedom elastic structural systems. The optimum parameters were determined and compared for elastic systems using the Newton–Raphson method, and the optimum ones for bi-linear hysteretic systems were also determined using a parametric survey caluculation.

ACKNOWLEDGEMENT

The authours would like to express their sincere gratitude to Mr. M. Kanagawa, graduate student of Kansai university, for his numerical calculations and preparation of this paper.

REFERENCES

1) H.Ozaki and H.Hagiwara , 1963 , Electrical Mathematics *III* , Ohm–publisher Co. , pp.80~81.
2) K.Asano and W.D.Iwan , 1984 , An Alternative Approach to the Random Response of Bilinear Hysteretic Systems , Earthquake Engineering & Structural Dynamics ,Vol.12,No.2,pp 229~236.
3) T.S.Atalik and S.Utku , 1976 , Stochastic Linearization of Multi–Degree–of–Freedom Non–Linear Systems , Earthquake Engineering & Structural Dynamics , Vol.14 , pp.441~420.
4) T.Kobori , R.Minai and M.Kawano , 1970 , An Approach to Optimization of Structural Earthquake Response , Annual Report of Disaster Prevention Reseach Institute , Kyoto University , No.13 A , pp.303~321.

Structural Safety & Reliability, Schuëller, Shinozuka & Yao (eds) © 1994 Balkema, Rotterdam, ISBN 90 5410 357 4

Generation and use of secondary uniform risk spectra

R. L. Bruce, P. K. Prakash & C. P. Rogers
ABB Impell Ltd, Warrington, UK

ABSTRACT: Current practice in the seismic design of structures for the nuclear industry involves the use of broadbanded, piecewise linear design response spectra. In an attempt to avoid the inherent conservatism in these spectra there is a move within the industry towards the use of Uniform Risk Spectra (URS). Unfortunately the benefit gained by using URS as input motion can be swamped by the code requirements for analysing a range of soil conditions together with enveloping and broadening the resultant secondary spectra. This paper investigates the use of structural reliability theory using Monte Carlo simulation in order to develop secondary level URS which fully account for uncertainties in the free field spectrum and uncertainties in soil and structural properties, and coherence between spectrum values at different frequencies. The techniques developed in this paper open the way to full probabilistic analysis of plant and structures subject to seismic loads. An outline of how this can be achieved is presented.

1 INTRODUCTION

Seismic design of nuclear related structures has developed over recent decades to a stage where a series of conservatisms are now inherent in the whole design process. The boundaries between disciplines and between different teams of engineers addressing each stage of the process has precluded the development of a rational and consistent approach. The geophysicist, the structural analysts and the M & E and vessel designers have all adopted techniques which have their own varying degrees of conservatism. The interface between disciplines is usually the handing over of a design spectrum, be it free field, site modified, secondary or tertiary floor spectra. There is seldom however any indication of the level of conservatism contained within each of these spectra.

As a consequence several prestigious projects of recent years have been faced with tertiary structural components having predicted responses of up to 50g. This situation has not arisen out the misapplication of accepted techniques but rather due to conservatisms compounding at each stage of the design. Further, the pressures of project timescales tend to dictate that these results are accepted and the design proceeds against these highly conservative accelerations. There is usually insufficient time during a major project to start rationalising and optimising design techniques.

The disjointed deterministic approach to seismic design is clearly not condusive to rationalisation. The answer lies in a probabilistic approach to seismic design. The uncertainties at each stage can be described by their probability distribution instead of conservative upper or lower bound valves. The correlations between parameters can be modelled where appropriate and the parameter uncertainty can be carried through to give a true measure of the probability distribution of these parameters which feed into the next stage of design.

In principal this idea can be extended to the whole design process so that the probability of the stress exceeding yield in any part of a vessel can be reliably determined.

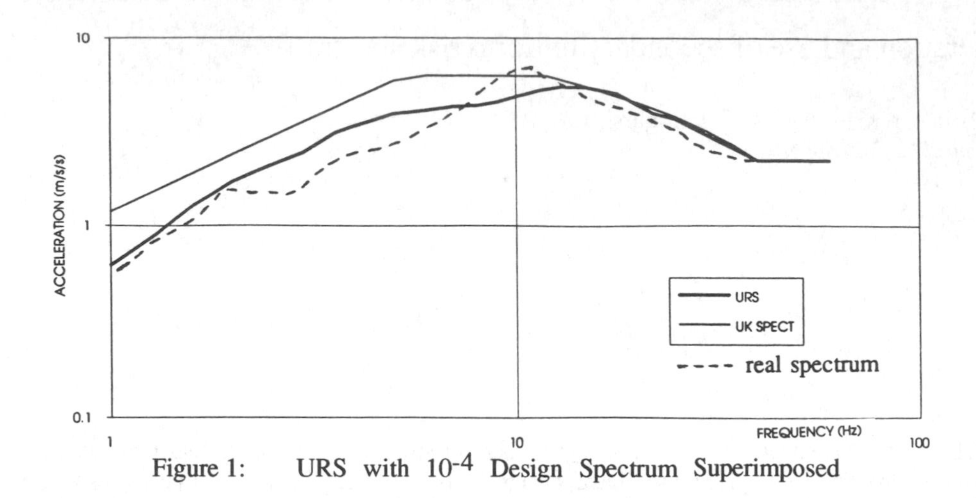

Figure 1: URS with 10^{-4} Design Spectrum Superimposed

The natural uncertainty of the occurrence of earthquakes, the uncertainty of the free field spectrum to be expected, the uncertain knowledge about soil properties, structural mass, stiffness and damping and vessel properties can all be treated probabilistically and a reliable measure of the probability of failure can be derived.

1.1 *Uniform Risk Spectra*

Some progress has been made in the direction of probabilistic design with the advent of uniform risk spectra (URS). Instead of the convention of piecewise linear, broad banded design spectra as in Fig. 1, some practitioners are starting to look at the use of URS.

The broadbanded design spectrum is conservative in two ways. First of all it is specified at the mean plus one standard deviation level ($m+\sigma$), i.e. the 84 percentile level. Secondly, real earthquakes rarely, if ever, consist of spectral energy at the level of the design spectrum across all frequencies.

It is clear from inspection of real earthquake spectra that real events are predominantly of one narrow frequency band. At its predominant frequency a real earthquake spectrum will most likely exceed the URS (corresponding to the URS level at the pga frequency) but at other frequencies it will be lower than the URS.

One way of describing the above observed characteristics, and one which allows us to simulate real earthquake spectra is as follows. The URS can be considered as a mean spectrum and the variations of real spectra (with the same pga) about this mean curve can be described by their density functions. In addition the correlation coefficient of the deviation of any one spectrum about this mean, from one frequency (f_i) to another (f_j) can be determined.

The above considerations have been employed to permit earthquake spectra to be simulated and employed in Monte Carlo Simulation of structural response in order to generate **secondary** URS, (SURS).

2 METHODOLOGY

The objective is to generate secondary uniform risk spectra given an input URS, its frequency correlation and information about the random nature of the governing structural parameters (masses, stiffnesses etc). This is achieved using Monte Carlo Simulation with Importance Sampling (MCSIS). Monte Carlo Simulation (MCS) is the technique whereby random processes are simulated using the random number generation capabilities of a computer. It is shown below how synthetic input spectra can be simulated such that they exhibit all the necessary characteristics of real input spectra. These can be used to analyse a structure in which the dynamic properties are similarly simulated using the user defined probabilistic information about the distribution of those parameters.

The above procedure is repeated a large

number of times and the output response spectra are analysed to produce secondary uniform risk spectra. The steps involved are discussed in further details below. They are:-

1. Description of stochastic properties of real earthquakes.
2. Generation of synthetic stochastic input motions.
3. Generation of random dynamic properties of the structure
4. Response analysis
5. Analysis of response time histories to produce SURS

2.1 *Description of Stochastic Properties of Real Earthquakes*

As can be seen in fig 1 any one real spectrum rarely contains the same broad frequency spread of energy as the design spectrum. Consequently there is significant conservatism in use of design spectra.

The uniform risk spectrum (URS) was conceived partly as an attempt to remove this conservatism. The URS is defined as a series of spectral ordinates of equal exceedance probabilities. For example figure 1 shows the 10^{-4} URS compred with the 10^{-4} UK design spectrum.

The URS however contains no information about the variable frequency content of real spectra. Of two spectra with similar pga return periods one could be predominantly high frequency with little energy in the lower frequency bands and the other could be the converse, i.e. predominantly low frequency energy. It is only over a long term average that real spectra will have the exceedance probabilities reflected by the URS. No one real spectrum is ever likely to be the shape of a single URS curve.

This tendency of real spectra to be predominantly high or low frequency (or even intermediate frequencies) can be exemplified by studying the correlation of real spectra across different frequencies.

In order to do this a collection of real spectra were compiled and normalised to the same pga. The covariance between two frequencies f_i and f_j, was then defined as σ_{ij} and given by:-

$$\sigma_{ij} = E\left[\left(S(f_i) - \overline{S}(f_i)\right)\left(S(f_j) - \overline{S}(f_j)\right)\right]$$

or (1)

$$\sigma_{ij} = E\left[s_i s_j\right]$$

or in matrix form

$$C = SS^T$$

where E [ ] is the expected valuc (avcraging) operator

$S(f_i)$ is the response spectral acceleration at frequency f_i

$\overline{S}(f_i)$ is the mean value of $S(f_i)$ across all spectra analysed.

The covariance σ_{ij} can be normalised by dividing by the standard deviation at each frequency thus giving the correlation coefficient ρ_{ij}:

$$\rho_{ij} = \frac{\sigma_{ij}}{\sigma_{ii}\sigma_{jj}} \qquad (2)$$

Figure 2 shows the result of this analysis for all possible combinations and f_i and f_j. As expected the correlation is 1.0 when $f_i = f_j$ and falls off as f_i and f_j become more widely separated. The interesting point to note about figure 2 is the clear negative correlation exhibited between low frequencies and high frequencies. This corroborates the earlier statement that real spectra tend to be dominated by energy in just a small frequency band.

2.2 *Generation of Synthetic Stochastic Input Motions*

Having derived the covariance's of spectral values at different frequencies we can use this information to help simulate synthetic spectra.

The steps are as follows:

i) simulate a vector of uncorrected uniformly distributed variables **u** where $u_i \in U[0,1]$

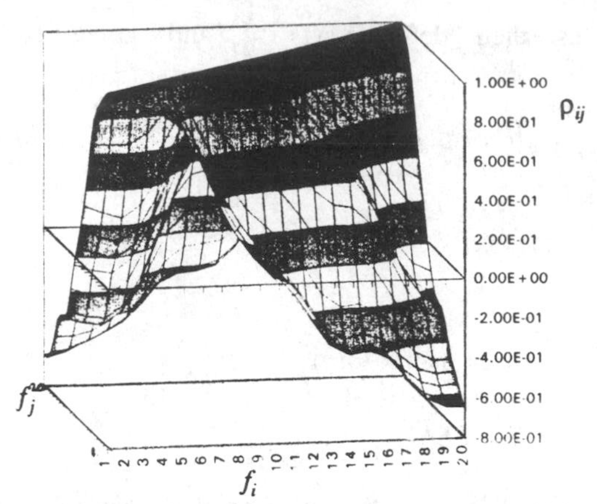

Figure 2: Correlation of Real Spectra Across Different Frequencies

ii) transform **u** to a vector of uncorrelated normally distributed variables with zero mean and unit variance, **x** where $x_i \in N[0,1]$

iii) transform **x** to **s** such that the covariances between the s_i values are the same as the target covariance matrix **C**. s_i represents the deviation of the spectral ordinate from the mean spectrum i.e. $s_i = S(f_i) - \overline{S}(f_i)$.

iv) transform the s_i spectral values to the observed log-normal distribution, thus obtaining a synthetic response spectrum.

v) simulate a time history of ground acceleration, $\ddot{x}_g$, from the synthetic response spectrum, using for example the method given in Khan (1987).

The key step in the above procedure (step iii) is transforming between the uncorrelated unit vector **x** and the vector **s** which must exhibit the same covariance matrix **C** as the real earthquake data. If **L** represents a transformation matrix from **x** to **s**;

$$\text{i.e.} \quad \mathbf{s} = \mathbf{L}\,\mathbf{x} \tag{3}$$

Then;

$$E[\mathbf{s}\mathbf{s}^T] = E[\mathbf{L}\mathbf{x}\mathbf{x}^T\mathbf{L}^T] = \mathbf{C} \tag{4}$$

$$= \mathbf{L}E[\mathbf{x}\mathbf{x}^T]\mathbf{L}^T = \mathbf{C}$$

However, by definition

$$E[\mathbf{x}\mathbf{x}^T] = \mathbf{I}$$

thus

$$\mathbf{L}\mathbf{L}^T = \mathbf{C} \tag{5}$$

Since **C** is known from analysis of real data, then **L** can be found by any suitable decomposition routine. This then allows step (iii) to be performed and this can be followed by steps (iv), and (v) thus giving an artificial ground acceleration time history with the following properties:

- pga suitably sampled to fit the actual hazard curve for the site.
- a correlation between frequencies which matches that observed in real data.
- average exceedance probabilities at each frequency commensurate with the specified URS.

2.3 *Generation of Random Dynamic Properties of the Structure*

The response of any part of a structure is a function of frequency ω_j, damping ratio β_j, participation factor Γ_j, mode shape ϕ_{ij} and the input $\ddot{x}_g$.

Section 2.2 has already described how $\ddot{x}_g$ can be simulated. We shall now discuss the random simulation of ω_j, β_j, Γ_j and ϕ_{ij}.

The damping β_j is often assumed to be frequency independent and so can easily be simulated by sampling from a suitable probability distribution, e.g. log-normal.

Frequency, participation factor and mode shape however are functions of many variables within the structure such as stiffness and mass of key members. Simulation of ω_j, Γ_j and ϕ_{ij} can thus be based on user defined probability distributions of such basic variables as stiffness and mass. The problem however is that this implies that an eigenvalue analysis is required for every simulation, and this will be of the order of several hundred in a typical MCS study.

To circumvent this problem a Taylor series xpansion of the dynamic parameters in terms of he basic variables is used, e.g.

$$\Gamma' = \overline{\Gamma} + \Delta^T \partial \Gamma + \frac{1}{2!}(\Delta^T \partial)^2 \Gamma + \dots \qquad (6)$$

where: Γ = vector of modal participation factors

Γ' = random value of Γ

$\overline{\Gamma}$ = Γ evaluated for the mean values of the basic variables

∂ = vector of partial derivative operators w.r.t. basic variables

Δ = vector of random variations of basic variables

The random variations of the basic parameters mass, stiffness etc.) within Δ can be simulated within each MCS loop based on the user specified .d.f. for that variable.

The other dynamic parameters ω_j and ϕ_{ij} can oth be simulated as above using an analogous quation to equation 6.

.4 *Response Analysis*

he acceleration time history of any point on a ructure relative to the ground, due to seismic nput is given by:

$$\ddot{x}_i = \sum_{i=1}^{N} \phi_{ij} \ddot{y}_j \qquad (7)$$

here $\ddot{x}_i$ = response in dof i

$\ddot{y}_j$ = modal response in mode j

ϕ_{ij} = mode shape in dof i for mode j

The sum is either taken over a sufficient umber of modes to represent all of the dynamic ass, or provided all resonant modes are included, a static correction term, proportional to the ground acceleration $\ddot{x}_g$, can be added.

The modal responses, $\ddot{y}_j$, are found by solving the individual modal equations:

$$\ddot{y}_j + 2\beta_j \omega_j \dot{y}_j + \omega_j^2 y_j = \Gamma_j \ddot{x}_g \qquad (8)$$

where β_j = modal damping

ω_j = modal frequency

Γ_j = participation factor

$\ddot{x}_g$ = ground acceleration time history

It is important to note that the computational effort required here is not excessive. For each simulation loop it is necessary to solve equation 8 for just a few dozen modes at most. By including a static correction term where necessary a reasonably exact response analysis can be achieved. Since the modal equations are each SDOF systems their numerical integration is almost trivial. A standard Newmark-β algorithm has been used in this work.

Having determined the response acceleration time history for each simulation loop this can be stored, for possible input into a secondary structure, and also used to produce secondary level response spectra.

2.5 *Analysis of Response Time Histories to Produce SURS*

At its simplest level Monte Carlo Simulation involves sampling each random variable in the problem from its parent distribution. In the case of the pga however, the distribution of which is shown in Fig. 3, such sampling would result in a pga of zero for roughly 95% of samples and the rest would be heavily biased towards the lower acceleration values. To avoid this problem the technique of importance sampling is adopted, as described below:

$P(S_a(f)>T)$ = Probability that the secondary response spectrum at frequency f will exceed threshold T

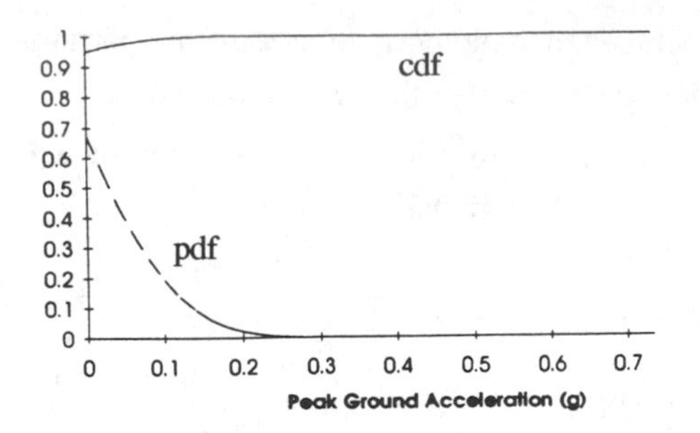

Figure 3: Cdf and Pdf of Peak Ground Acceleration

$$P(S_a(f)>T) = \int_{\mathbf{x}} I(\mathbf{x}) \frac{f(\mathbf{x})}{h(\mathbf{x})} h(\mathbf{x}).d\mathbf{x} \quad (9)$$

where $I(\mathbf{x})$ is an indicator function equal to 1 if $S_a(f)>T$ for parameter sample set $\mathbf{x}$ and zero otherwise

$f(\mathbf{x})$ is the actual joint distribution of the parameter set (including pga)

$h(\mathbf{x})$ is the sampling distribution of the parameter set

Thus the pga can be sampled from, say, a uniform distribution over the range of interest (0-0.8g) and the exceedance probabilities can be calculated using equation 9.

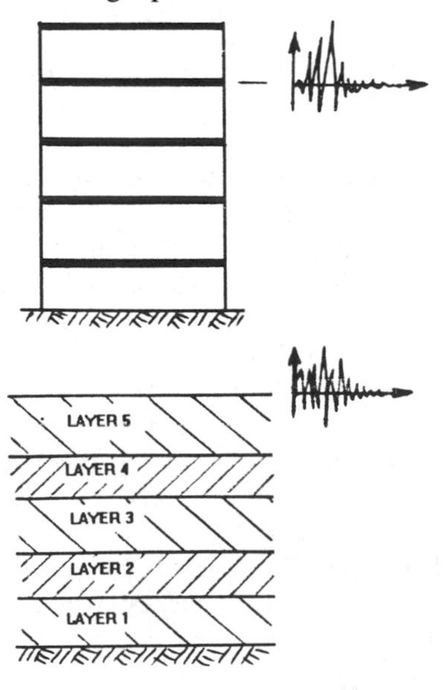

Figure 4: Soil and Structure Models Used

3 RESULTS

Figure 4 shows two test models, one of a soil column and one of a simple five storey portal structure. The free field URS was used as input at bedrock level (deconvolution from the surface was not included in this study but would of course normally be performed).

A total of 100 input spectra were simulated using pga values sampled from a uniform distribution. These were then used to simulate input time histories using the method of Khan (1987). The shear modulus of the soil was sampled from a log-normal distribution which was chosen to fit the following fractiles:

10 % fractile	$G' = {}^2/_3 G_{mean}$
50 % fractile	$G' = G_{mean}$
90 % fractile	$G' = 1.5 G_{mean}$

where G_{mean} is the best estimate soil modulus.

This choice of fractiles was based on the conventional approach to seismic design, as specified in ASCE (1986), whereby the analysis is performed for G_{mean}, ${}^2/_3G_{mean}$ and $1.5G_{mean}$ and the results enveloped. Figure 5 shows the log-normal distribution adopted.

Figure 6 shows the secondary URS at the soil surface following application of the MCSIS

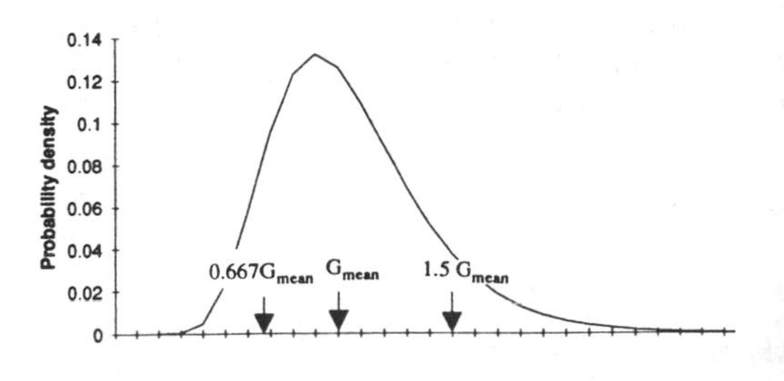

Figure 5: Log-Normal Distribution of Soil Modulus

procedure outlined in this paper. Figure 7 shows the 10^{-4} URS compared with the secondary spectrum generated using the conventional approach.

The time histories generated during the simulation at the soil surface, in addition to being used to generate the secondary URS , were also used as input to a MCSIS analysis of the simplified structure.

Figure 8 shows the **tertiary** URS and figure 9 compares the 10^{-4} tertiary URS with the tertiary spectrum generated using the conventional approach, (the enveloped soil surface secondary spectrum was used as input to the structure and the response spectrum was broadened by ±15% in accordance with ASCE (1986))

It is clear from these results, especially figures 7 and 9 that there are significant conservatisms in the conventional approach which have been avoided in this work.

4 APPLICATION OF SURS TO DESIGN

Having produced secondary or tertiary URS it becomes straightforward to apply these to the probabilistic design/analysis of plant items. This can proceed in a number of ways.

4.1 *Direct use of SURS*

The probability that some plant response R_i (e.g. a stress) exceeds some allowable value R_a can be expressed in terms of the modal contributions to R_i

$$R_i = \left(\sum_{j=1}^{N} R_{ij}^{\,2} \right)^{\frac{1}{2}} \tag{10}$$

$$P(R_i > R_a) = P\left(\left(\sum_{j=1}^{N} R_{ij}^{\,2} \right)^{\frac{1}{2}} > R_a \right) \tag{11}$$

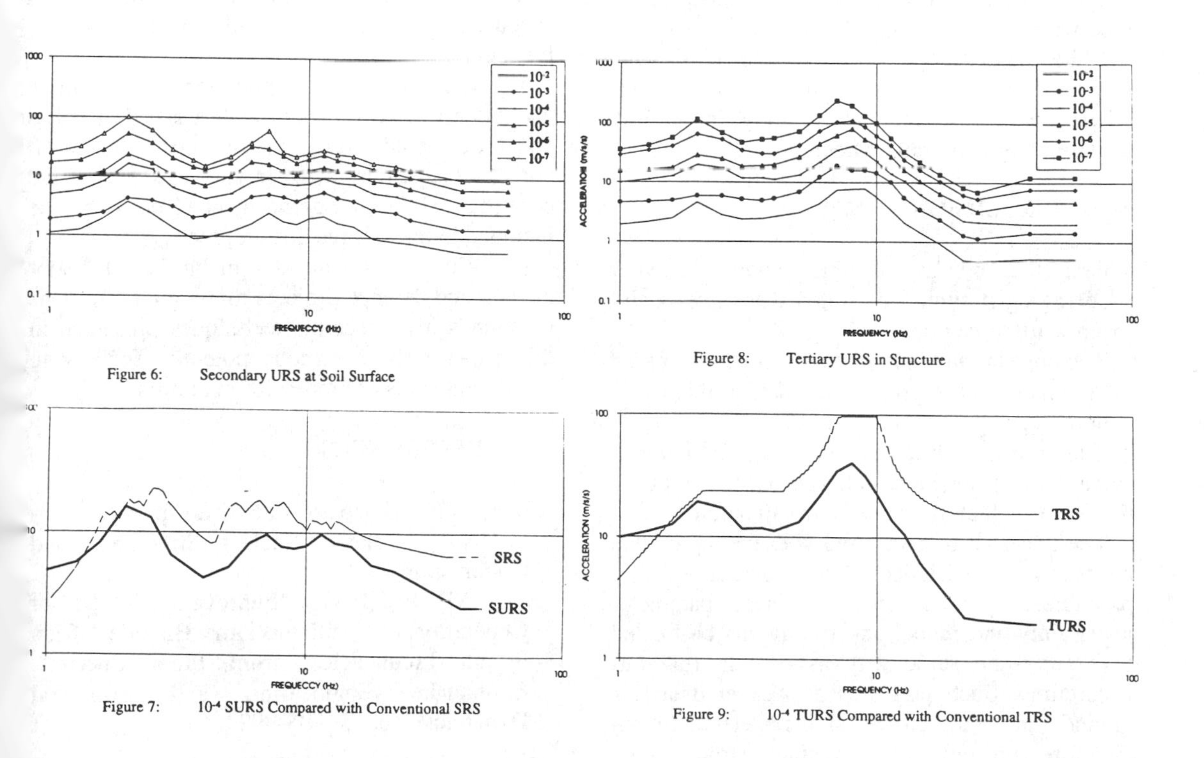

Figure 6: Secondary URS at Soil Surface

Figure 8: Tertiary URS in Structure

Figure 7: 10^{-4} SURS Compared with Conventional SRS

Figure 9: 10^{-4} TURS Compared with Conventional TRS

Equations 10 and 11 can of course be further developed to include the effects of modal correlations.

Since the exceedence probabilities of the individual modal responses, R_{ij}, are by definition, given by the SURS curves, then evaluation of equation (11) becomes straightforward.

4.2 *Use of Secondary Response Time Histories*

Since the secondary response time histories have been produced anyway, in order to produce the SURS, then it is a simple matter to use these as direct input to an MCSIS analysis of the plant item. The modal properties of the component can be simulated in much the same way as for the soil or the structure and then the pga sampling distribution h(**x**) can be introduced and equation (9) rewritten as :

$$P(R_i > R_a) = \int_{\mathbf{x}} I(\mathbf{x}) \frac{f(\mathbf{x})}{h(\mathbf{x})} h(\mathbf{x}).d\mathbf{x} \qquad (12)$$

This time however I (**x**) is 1 if $R_i > R_a$ and 0 otherwise.

Either way, $P(R_i > R_a)$ is generated and this can be compared with some suitable acceptance criteria or fed into a PRA study in a fully consistent and rational manner.

5 DISCUSSION

Whilst the results of this study are very encouraging it should be noted that there is still much work to be done.

The correlation study of real spectra was based on an inadequate supply of real data in this work. a more detailed study is required.

The choice of fractiles for fitting distributions needs further consideration. The sensitivity of the results to such choices needs investigation.

The uncertainty associated with the input URS needs to be included. One advantage of a probabilistic approach is that parameter uncertainty and model uncertainty can be treated in exactly the same way as random (natural) uncertainty. Such problems as lack of data (for ground motion or structural parameters) do not therefore prevent these methods from being applied.

Finally the effect of non-linearities such as ductility or gaps should be investigated, although with the time history approach adopted in this work there is no reason why such refinements should not easily be included.

6 CONCLUSIONS

The following conclusions can be drawn from this study :

- Realistic earthquake spectra, accurately representing the probabilistic characteristics of real spectra, including correlation across different frequencies, can be simulated using the techniques presented in this paper.
- The above synthesised spectra and the use of Taylor series expansions of stochastic structural properties can be combined with Monte Carlo Simulation with Importance Sampling to produce **secondary** (and even **tertiary**) Uniform Risk Spectra.
- Secondary and tertiary URS, so produced, show a significant reduction in conservatism when compared with the conventional design spectrum approach (i.e. with enveloping and broadening)
- The method of application of secondary and tertiary URS for plant and vessel design has been discussed. It is clear that significant conservatisms, introduced in various stages of the conventional design approach, need no longer be included in the seismic design process.
- With the seismic design burden on major nuclear and defence projects running into millions of pounds it is clear the techniques presented in this paper have enormous potential for saving significant sums of money for operators.

7 REFERENCES

ASCE, 1986. Standard 4-86 - "Seismic Analysis of Safety Related Nuclear Structures (and Commentary).

Khan, M. R. 1987. "Improved Method of Generation of Artificial Time Histories, Rich in all Frequencies, from Floor Spectra". Earthquake Engineering and Structural Dynamics. Vol 15, 985-992.

Structural Safety & Reliability, Schuëller, Shinozuka & Yao (eds) © 1994 Balkema, Rotterdam, ISBN 90 5410 357 4

A damage model for reinforced concrete buildings: Further study with the 1985 Mexico City earthquake

D. De Leon & A. H-S. Ang
University of California, Irvine, Calif., USA

ABSTRACT: The Park-Ang (1985) damage model for reinforced concrete structures is calibrated using a sample of 8 reinforced concrete buildings located in the soft soil of Mexico city and that were damaged during the 1985 earthquake. A nonlinear hysteretic random vibration analysis is performed for each building to estimate the statistics of the response and the median damage index. The estimated global damage indices are consistent with observed damages. A damage index (i.e. repairable) of D = 0.5 appears to be the limit of the tolerable damage . Fragility curves against collapse, (D$\geq$ 1.0) can be developed as illustrated for a specific building.

1 INTRODUCTION

Experience has shown that reinforced concrete structures can develop high inelastic excursions and sustain a large number of load reversals under strong earthquakes. In these cases, energy dissipation through hysteresis plays an important role in the response, as shown by a progressive degradation of stiffness and strength which leads the building, eventually, to its collapse. Seismic damage assessment has been studied from different viewpoints, e.g. Aoyama (1981), Meyer and Roufaiel (1989), Powell and Alahabadi (1988), Reinhorn ,et al. (1989), among others. A probabilistic approach to predict damages in reinforced concrete structures, has been proposed by Park and Ang (1985). Damage prediction appears to be useful for the evaluation of existing buildings as well as design of new structures. Practical application of avalaible models, however, requires calibration with observed damages of actual structures. The wide variety of damages of R/C buildings caused by the 1985 Mexico city earthquake provides a valuable source of information to calibrate the available models and to extend their capabilities. A further calibration and extension of the Park-Ang model (1985) through the evaluation of 8 R/C buildings that sustained different levels of damage during the 1985 Mexico city earthquake, is presented. According to this model the damage index for a structural component is expressed in terms of the expected responses, namely the maximum interstory displacement δ_{max} and the dissipated hysteretic energy E_h as follows:

$$D = E[\delta_{max}]/\delta_u + \beta_0 E[E_h]/(Q_y\delta_u) \quad (1)$$

where: D = damage index, E[ ] = expected value, δ_u = ultimate deformation, Q_y = yielding force, β_0 = constant.

2 STRUCTURAL PARAMETERS AND SEISMIC LOAD MODEL

A shear beam model is assumed for the multistory building and an elasto- plastic skeleton is taken as the force-displacement relationship which includes both the hysteretic and degradating characteristics (Wen (1980) and Baber (1980)). The equilibrium equation is:

$$M\ddot{x} + C\dot{x} + Kx + (1-\alpha)Kz = -M\ddot{x}_g \quad (2)$$

where: $\ddot{x}, \dot{x}$ and x = interstory acceleration, velocity and displacement vectors; M, C and K = mass, viscous damping and initial stiffness matrices,respectively, z = vector of hysteretic displacement, α = ratio of post-yielding to pre-yielding stiffness, $\ddot{x}_g$ = vector representation for ground acceleration and the hysteretic-degradating model is:

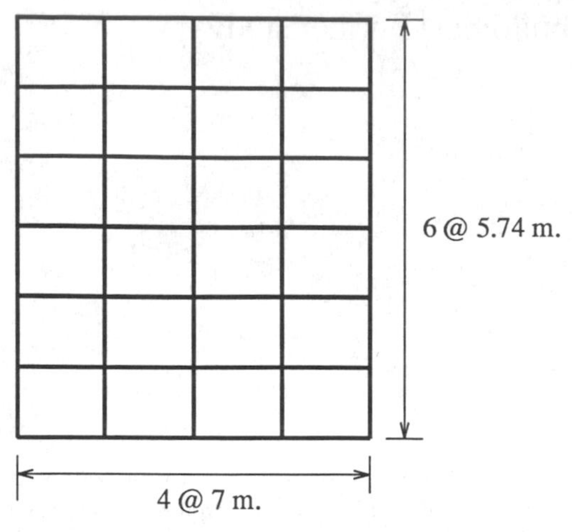

Fig. 1 Plan view for building LR 07
Direction of analysis ⟶

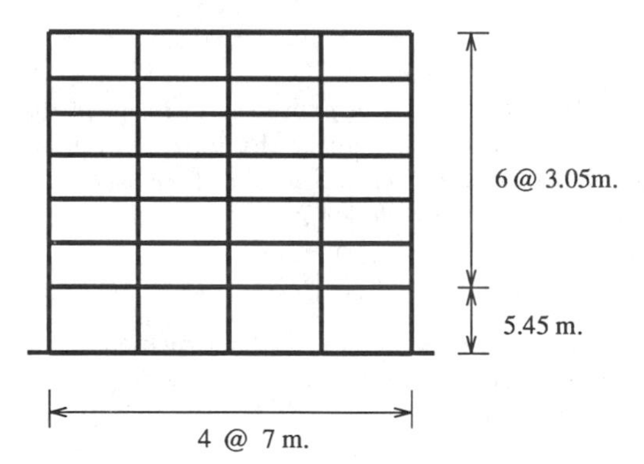

Fig. 2 Elevation for building LR 07

$$\dot{z} = [A\dot{x} - \nu(\beta|\dot{x}||z|z + \gamma\dot{x}|z|^2)]/\eta \tag{3}$$

where: A, ν and η = degradation parameters which govern the amplitude of the loop following to Sues et al. (1983):

$$\beta = -3\gamma \tag{4}$$

$$\beta + \gamma = \text{A}\ [(1\text{-}\alpha)\text{K}/Q_y]^2 \tag{5}$$

$$\text{A}(E_h) = 1.0 \ \text{-}\delta\ \text{A}\ E_h \tag{6}$$

$$\nu(E_h) = 1.0\ +\delta\nu E_h \tag{7}$$

$$\eta(E_h) = 1.0\ +\delta\eta E_h \tag{8}$$

and where: $\delta A,\ \delta\nu\, and\, \delta\eta$ are the degradation rate

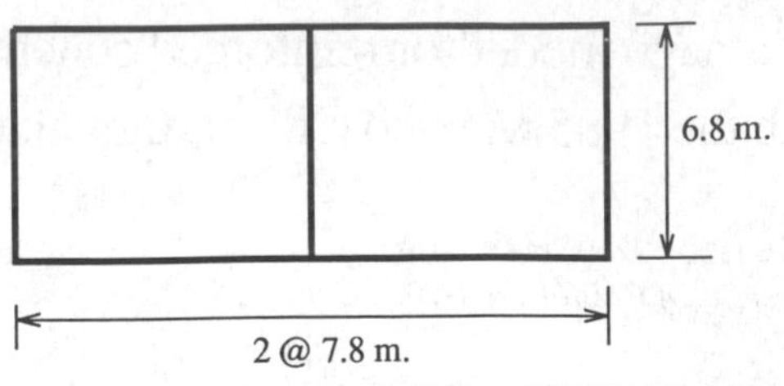

Fig. 3 PLan view for building HUATABAMPO
Direction of analysis ⟶

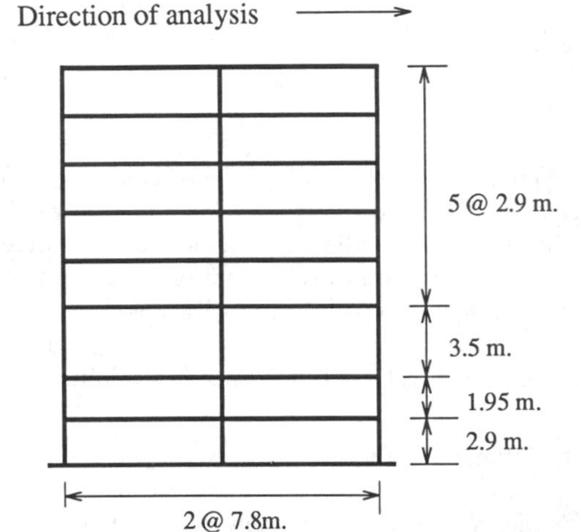

Fig. 4 Elevation for building HUATABAMPO

parameters, and $E_h = (1-\alpha)\text{K}\int_{t_0}^{t_f} z\dot{x}dt$, t_0, t_f = initial and final times for the excitation. Structural properties are carefully determined from the respective design plans and related studies (Meli (1986)). Interstory properties, e.g. stiffness, shear strength and potential energy, are calculated in terms of those of the elements by linear superposition. Ultimate deformation is obtained as the deformation up to which the interstory potential energy is the sum of those corresponding to the elements. And the element ultimate deformation is represented (Park, et al. (1985)) through: $\delta_u = \mu\delta_y$ where: δ_y = yielding deformation $=\delta_f + \delta_b + \delta_s + \delta_e$, $\delta_f, \delta_b, \delta_s, \delta_e$ = flexure, bond-slippage, inelastic shear and elastic shear deformations.

$$\mu = (\frac{\epsilon_p}{\epsilon_0})^{(0.218\rho_w - 2.15)}\ e^{(0.654\rho_w + 0.38)} \tag{9}$$

ϵ_p is the maximum principal normalized by ϵ_0, and ρ_w is the confinement ratio. The buildings range from 5 to 10 stories and show a structural system of frames, or "waffle" slabs over columns or frames with shear and infill walls. Typical plan and elevation are shown in figs. 1 to 4.

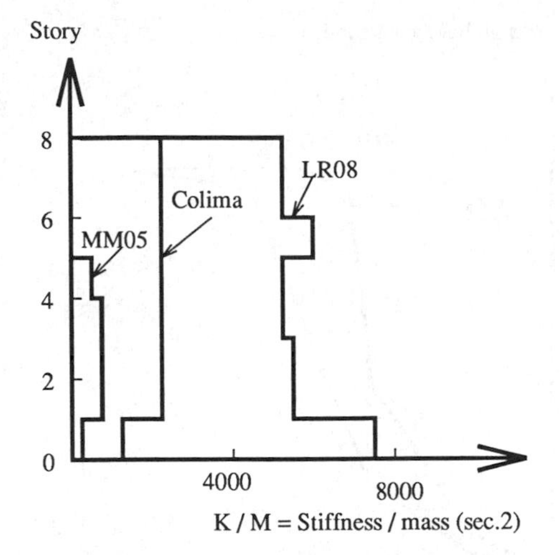

Fig. 5 Distribution of K/M with heigth

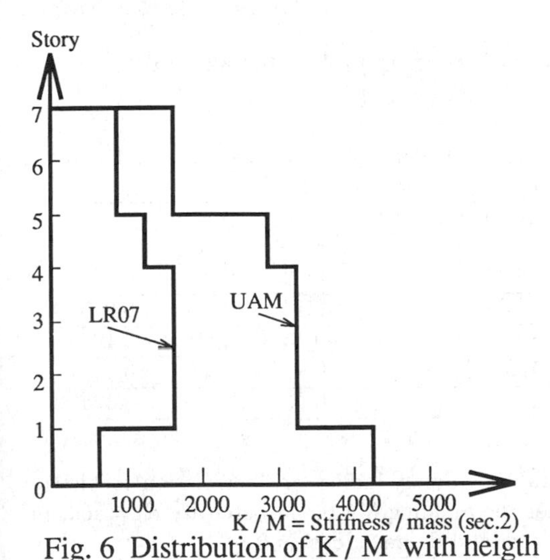

Fig. 6 Distribution of K / M with heigth

Some special characteristics of the buildings are: "weak" first story, "weak" intermediate story, stiff intermediate story, unfavourable distribution of infill walls with heigth, or thin columns in one direction, among others. The excitation is considered as a stationary random process with a power spectral density modulated by a deterministic envelope (Amin and Ang (1968)), and the peak ground acceleration, dominant frequency and duration of the strong motion are taken as recorded during the earthquake in the EW direction of the SCT (Secretaria de Comunicaciones y Transportes) station on soft ground.

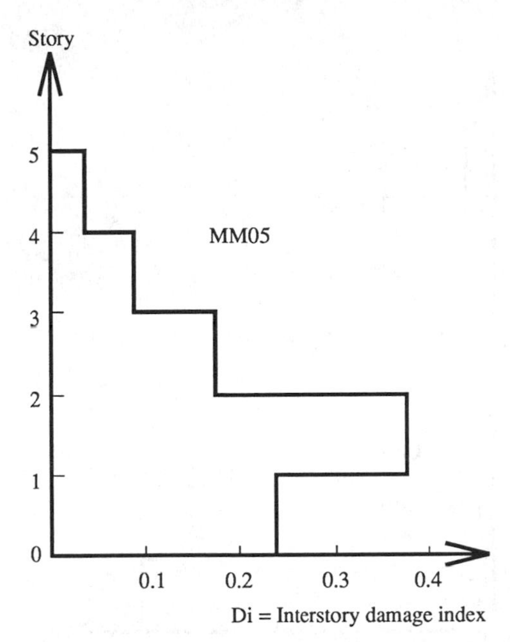

Fig. 7 Distribution of Di with heigth

3 RESULTS AND DISCUSSION

Structural properties for typical buildings are shown in tables 1 and 2 and figs. 5 and 6.

Table 1. Structural properties and response of the building Colima

Story no.	stiffness (tn/cm.)	strength (tn.)	maximum displacement(cm)
1	270.29	397.	24.4
2	498.58	437.0	1.07
3	498.58	437.0	0.79
4	498.58	437.0	0.58
5	498.58	437.0	0.43
6	498.58	437.0	0.30
7	498.58	437.0	0.20
8	498.58	437.0	0.10

Table 2. Structural properties and responses of the building MM05-03

Story no.	stiffness (tn/cm.)	strength (tn.)	maximum displacement(cm)
1	84.79	670.1	11.4
2	268.11	528.0	2.36
3	268.11	647.8	1.32
4	268.11	608.4	0.84
5	187.68	524.8	0.56

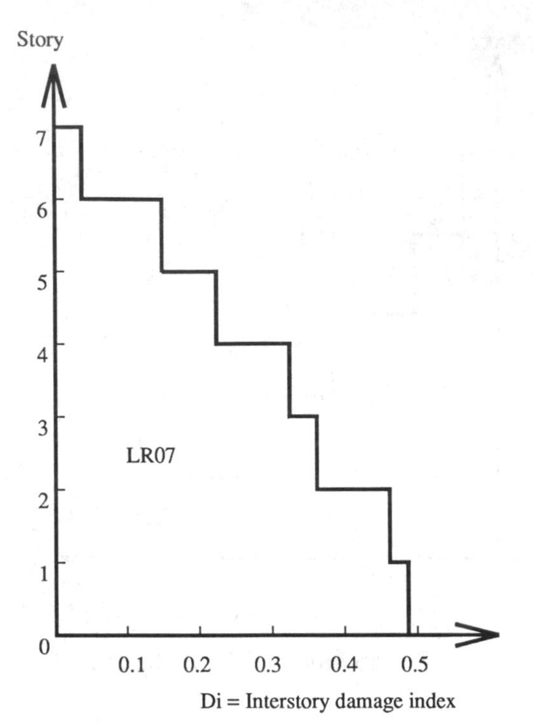

Fig. 8 Distribution of Di with heigth

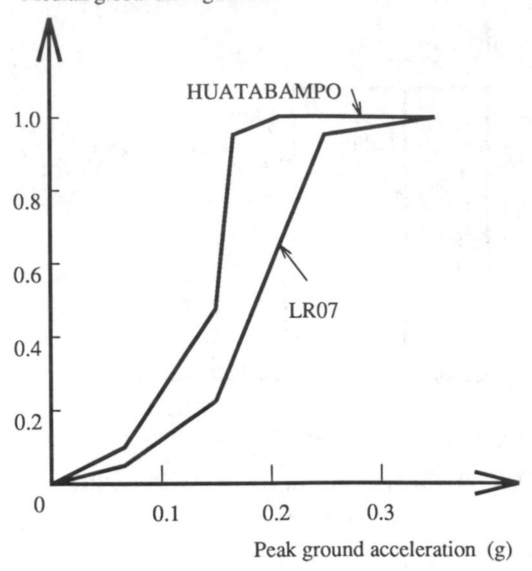

Fig. 9 Median global damage index

Interstory damage indices for 2 typical buildings appear in figs. 7 and 8.

The median global damage index per building is estimated in terms of the probabilities that this index is exceeded at the stories as components. Fig. 9 shows the variation of this index with the intensity for 2 typical buildings and table 3 shows it for a 3rd. one.

Safety indices, correlation coefficients and 2nd. order probability bounds are calculated for every story and damage level, assuming a lognormal distribution for the global damage index (Kim (1992), table 4).

The expected maximum responses obtained through the random iteration of eq.(2) are compared to those simulated by Monte Carlo techniques; on this basis, the results of Random Vibration analysis were more conservative (10% larger). Another comparison is made between the damage indices estimated in this paper with those reported by other authors (Meli (1986)), and the reported levels of damage observed during the earthquake for the respective buildings. For 1 of the 2 collapsed buildings, the predicted damage index for the 1st story (D = 1.25) agrees with the reported demolition of the entire building, whereas for the other the predicted D = 1.1 (4th story collapsed) appears also to agree with the reported demolition from the 4th to the 9th stories. Failures can be associated with a too flexible 1st story ($k1/k2 = .54$) for

Table 3. Median global damage index D for several intensities (for building MM05-03)

Spectral intensity cm^2/sec^3	peak ground acceleration (g)	D
46.90	0.10	0.020
79.35	0.13	0.050
105.5	0.15	0.100
140.9	0.17	0.239
187.7	0.20	0.424

Table 4. Safety indices and correlation coefficients for the building MM05-03, intensity $S_0 = 46.9 cm^2/sec^3$ and damage level = 0.8

Story :	1	2	3	4	5
Safety Index	2.47	2.61	2.80	3.72	3.92
Correla-	-.642	-.258	-.360	-.250	-.150
tion coe-	-.005	-.001	-.000	-.000	-.000
fficient	0.793	0.966	0.931	0.975	0.988

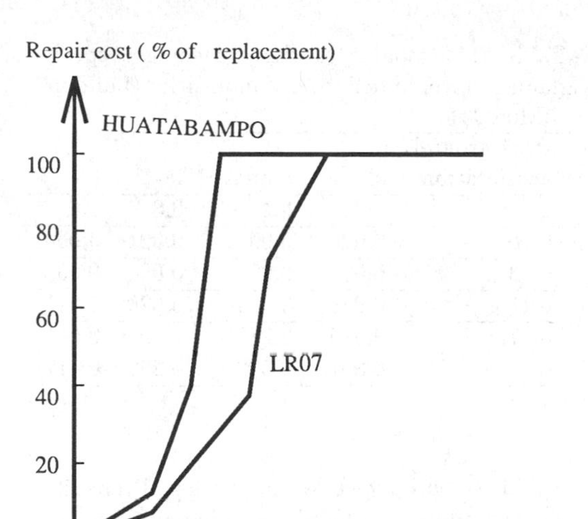

Fig. 10 Repair cost as a function of PGA

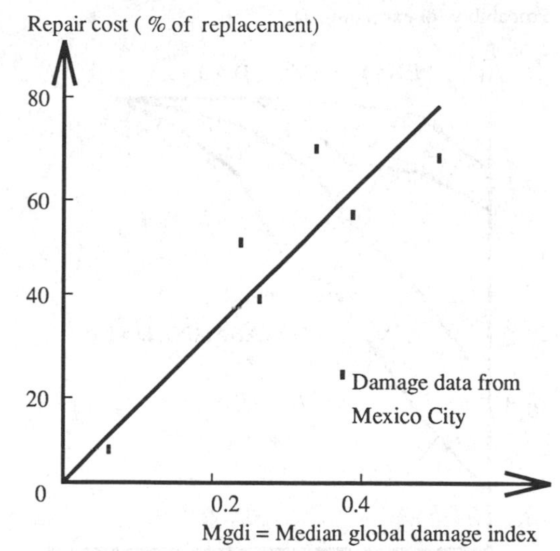

Fig. 11 Damage/repair cost as a function of Mgdi

Table 5. Gross estimation of replacement and repair costs (in million Pesos)

Building	replacement (1)	repair	% of (1)
UAM	1050.	21.	2.0
MM05-03	2175.	711.4	32.7
LR08-06	2225.	1083.0	49.0
LR10-07	20400.	11628.0	57.0
MM10-02	6480.	4283.0	66.1
LR07-04	5530.	3669.0	66.3

the 1st building, and with a too flexible 4th story (k4/k1 = .3, k4/k2 = .11, k4/k3 = .63) for the 2nd building. Five buildings (LR07, MM10, LR10, LR08 and MM05) had severe estimated damages. On the basis of grossly estimated costs of reposition and repairment (from Guerrero, 1990) the damage for the buildings LR07 and MM10 can be considered as non-repairable (table 5):

Median Global Damage is associated to repair cost (as a % of the replacement) as it is shown in fig. 11.

Variation of the repair cost with the intensity (fig. 10) can be calculated in terms of this association and the variation of the global index with the intensity showed before.

Comparing the estimated with the observed damages for these 5 cases (table 6) a reasonable agreement is found in both the magnitude and distribution of the damages in the stories.

Table 6. Comparison between estimated and observed damages

Building	story	estimated	observed *
LR07-04	1	0.49	severe
	2 - 3	0.47, 0.35	strong
	4	0.34	moderate
MM10-02	1 - 2	0.46, 0.37	severe
	3 - 4	0.13, 0.13	severe
	5 - 6	0.04, 0.02	severe
LR10-07	5 , 9	0.29, 0.22	moderate
	6 , 8	0.47, 0.32	strong
	7	0.27	severe
LR08-06	1	0.37	strong
	2 - 4	0.29, 0.24, 0.20	slight
	5 , 6	0.10, 0.02	slight
	7 , 8	0.31, 0.11	strong
MM05-03	1 , 3	0.25, 0.17	moderate
	2	0.38	severe

**According to Meli* (1986)

Any difference between the observed and estimated damage can be explained as follows: a).- Severe damage due to strong plastification at both ends of the beams from the 1st to the 6th stories, building MM10, is not reproduced analytically because of the shear beam model assumed in this paper for

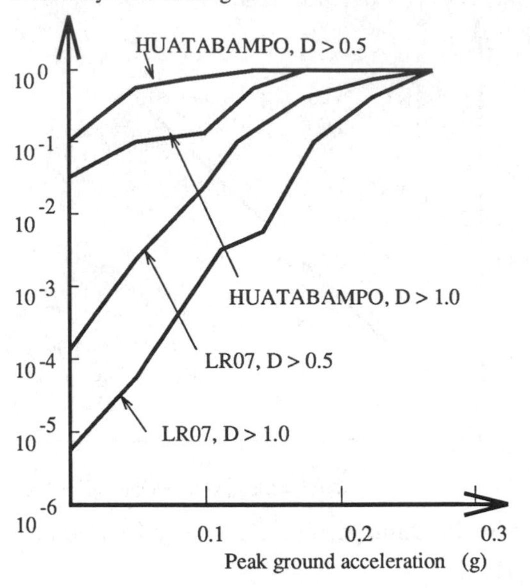

Fig. 12 Fragility curves for 2 buildings

the response analysis. b).- Moderate damage in the columns located at one end of a wall or a large opening in the 7th and 8th stories, building LR08, is underestimated given that the structural model does not consider walls or openings. c).- Non observed damage in the 1st story, building MM05, is estimated as moderate probably because of the wall and stiff box foundation not represented in the structural model. Strong damages in the set of 5 buildings can be associated with some design details which make them more vulnerable, independently of the exceptional characteristics of the earthquake, namely:

weak first or intermediate story, including ratios of 1st story heigth/other stories heigth = 2 for the buildings LR07 and MM05 and = 1.4 for the MM10 ,

different use for different stories in the same building, LR08, with a number of interior walls changing from 0 (1st story, parking use) to a maximum (2nd to 8th stories, housing use) , leading to a large ratio of story mass/1st story mass ,

thin columns in one direction (70X30cms.) with a few ties, as in building LR07, or quite thin ties as in building MM05-03,

too different mean strength index and high variability for the concrete used in different stories, leading to some "weak" zones in the building (e.g. LR08). Slight damage observed for the building

Table 7. Probabilities for exceeding a specified damage level under several intensities (building MM05-03)

Peak ground acceleration	<--	damage	level	-->
g	0.2	0.4	0.8	1.0
0.10	0.023	0.005	0.001	0.0001
0.13	0.094	0.024	0.004	0.002
0.15	0.248	0.085	0.020	0.011
0.17	0.575	0.288	0.095	0.061
0.20	0.851	0.532	0.290	0.117

UAM is reproduced by the model. Probabilities for exceeding a global damage index are obtained for several intensities (table 7):

Fragility curves, for both severe damage ($D > 0.5$) and collapse ($D > 1.0$) and for 2 buildings, are developed and shown in fig. 12.

4 CONCLUSIONS

The following conclusions can be summarized:

a).- Expected maximum response obtained with the random vibration approach is more conservative (by 10%) than the calculated by Monte Carlo simulation,

b).- A significant agreement is found between the estimated and observed damages,

c).- A damage of $D = 0.5$ appears to be limiting or tolerable damage, beyond which a building would be demolished, otherwise ($D < 0.5$) would be repairable,

d).- The model is useful for predicting the potential seismic damage of given structures, and for formulating recommendations for design or retrofit in terms of a prescribed level of tolerable damage.

5 ACKNOWLEDGMENTS

The authors thank to Dr. Roberto Meli from UNAM , Mexico and to M. in E. Jesus Iglesias and his group from UAM ,Mexico for the information given to make possible this study. The computational support from Dr. Jose Pires from UCI is also acknowledged and appreciated.

6 REFERENCES

Amin, M. and Ang,A. H-S. 1968. "Nonstationary stochastic model of earthquake motions".*Journal of Engeneering Mechanics. ASCE* .Vol.94, pp. 671-691.

Aoyama,H. 1981. "A method for the evaluation of the seismic capacity of existing reinforced concrete buildings in Japan".*Bulletin New Zealand Natural Society for Earthquake Engineering* . Vol. 14, no. 3, pp. 105-130.

Baber,T.T. and Wen,Y.K. 1908. "Stochastic equivalent linearization for hysteretic, degradating and multistory structures". *Civil Engineering Studies* SRS no. 471. Department of Civil Engineering. University of Illinois; Urbana, Illinois.

Guerrero, J.J. 1990. "Vulnerabilidad sismica de edificios de concreto reforzado en la Delegacion Cuauhtemoc". Master thesis (in Spanish), DEPFI, UNAM, Mexico, D. F.

Kim,W. J. 1992. "Damage assessment of existing bridge structures" Ph. D. dissertation. University of California, Irvine.

Meli,R. et al. 1986. "Evaluacion de los efectos de los sismos de septiembre en las estructuras de la Ciudad de Mexico. Parte II". *Internal report (in Spanish), Institute of Engineering* ., UNAM, Mexico, D. F.

Meli,R. 1986. "Evaluation of performance of concrete buildings damaged by the sep.19, 1985 Mexico earthquake".*International Conference on: The Mexico earthquakes 1985. Factors involved and lessons learned. ASCE* . Mexico, D.F.

Meyer,C. ; Roufaiel,M.S.L. 1989. "Reliability of damaged reinforced concrete frames". *VIII WCEE* . San Francisco, CA, pp. 535-542.

Park, Y. J. , Ang, A. H-S. and Wen, Y. K. 1985. "Mechanistic seismic damage model for reinforced concrete".*Journal of Structural Engineering. ASCE* . Vol. 111, no.4, pp. 722-739.

Powell,G.H. and Allahabadi,R. 1988. "Seismic damage prediction by deterministic methods: concepts and procedures". *Earthquake Engineering and Structural Dynamics* , Vol. 16, pp.719-734.

Reinhorn,A.M. ;Mander,J.B. ;Bracci,J. and Kunnath,S.K. 1989, "Simulation of seismic damage of reinforced concrete structures in Eastern US". V ICOSSAR , San Francisco, CA, pp. 407-414.

Sues,R.H. ;Wen Y.K. and Ang,A.H-S. 1983. "Stochastic seismic performance. Evaluation of buildings". *Civil Engineering Studies* SRS no. 506. Department of Civil Engineering. University of Illinois; Urbana, Illinois.

Wen,Y.K. 1980. "Equivalent linearization for hysteretic systems under random excitation". *Journal of Applied Mechanics. Transactions of the ASME* . Vol. 47.

Structural Safety & Reliability, Schuëller, Shinozuka & Yao (eds) © 1994 Balkema, Rotterdam, ISBN 90 5410 357 4

Modeling of seismic hazards for dynamic reliability analysis

S. Fukushima
Tokyo Electric Power Services Co. Ltd, Japan

M. Mizutani
Tokyo, Japan

H. Katukura & Y. Akao
Shimizu Corporation, Tokyo, Japan

ABSTRACT : In order to assure the safety of structures, it is neccssary to consider the uncertainty of each factor, such as loading or capacity, since damage to structures is caused by very extreme events. Generally, uncertainty consists of random uncertainty and modeling uncertainty. It is important to mitigate the modeling uncertainty to improve the accuracy of reliability analysis. In this paper, first of all, variabilities of indices of input ground motion are investigated. Trough these investigations, as an index of seismic hazard, we have shown the efficiency of employing the total energy of acceleration which has less variability than the peak acceleration. Also indicated is the approach of reliability analysis based on the total energy.

1 INTRODUCTION

In many cases, damage to structures is caused by very extreme events. Therefore, it is important to consider the uncertainty of each factor, such as loading or capacity, to assure the safety of structures. The abovementioned uncertainty consists of random uncertainty and modeling uncertainty. The former is a substantial or physical variability. On the other hand, the latter is a variability due to the shortage of knowledge and data, and will be expected to decrease by accumulating such information.

Authors have been involved in the study on dynamic reliability analysis (*DRA*), in which both uncertainties can be treated properly. Process of *DRA* developed by authors involves the following three steps,
1) seismic hazard analysis,
2) fragility analysis, and
3) calculation of probability of failure.

In many *DRA*s, input ground acceleration is modeled as a random process having a specified spectral shape and random phases. Structural responses can also be obtained by means of transfer functions including both uncertainties. Therefore, for *DRA* developed in the frequency domain, peak value, such as peak acceleration defined in the time domain, may not be a suitable index of input ground motion.

Figure 1 shows the relationships between the earthquake record and indices of input ground motion. It is apparent that peak acceleration is not an accurate index, since it must be combined with the uncertain factors, such as duration time and peak factor, to define the input ground motion suitable for the following response analysis. Even though peak acceleration is considered to have an important meaning in the actual design stage, it is necessary to employ more suitable index for the implementation of accurate *DRA*. In addition, when obtaining the index mentioned above, it may also be expected that modeling uncertainty in the *DRA* can be mitigated by considering the structure of variability.

In this study, variabilities of indices of input ground motion are examined. Through this examination, an index suitable for DRA is proposed, followed by the construction of *DRA* process based on the index newly proposed.

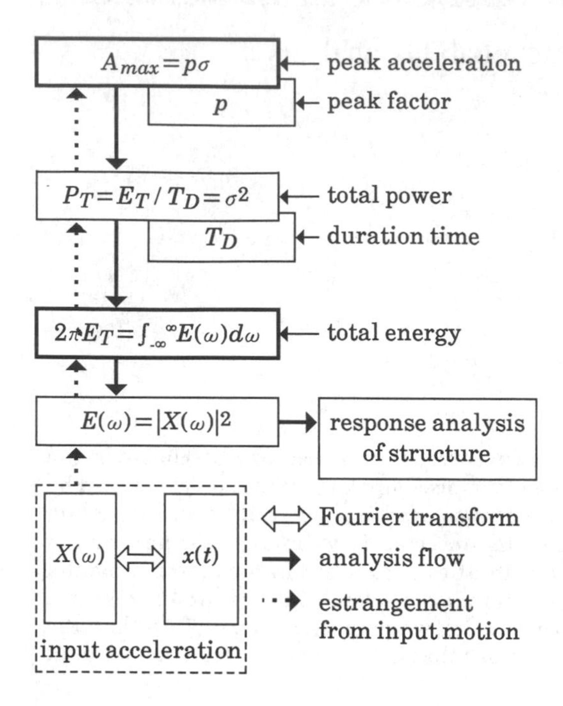

Fig.1. Indices of input ground motion

2 RELATIONSHIPS AMONG INDICES OF INPUT GROUND MOTION

Figure 1 shows the relationships among the indices of input ground motion, where A_{max}, P_T, E_T, T_D, p denote the peak acceleration, total power, total energy, duration time, and peak factor, respectively. It is shown that the A_{max}, P_T, and E_T of input ground motion are related to each other through T_D and p.

It should be noted that E_T is not a physical energy of input ground motion as defined in Fig. 1. Table 1 shows the several definitions of A_{max}^2 and E_T, for reference. Using A_{max}^2 instead of A_{max} is to make the comparison with E_T easier.

3 MAGNITUDE DEPENDENCY OF FAULT PARAMETERS AND DURATION TIME

In order to obtain the basic relationships between the properties of ground motion and seismic magnitude, magnitude dependencies of fault parameters and duration time are arranged and are shown in Table 2, where L, D̄, S, M_0, T_D, and M denote the fault length,

Table 1. Definitions of peak acceleration and total energy

Index	Definitions
A_{max}^2	A_{max}^2 $A_{max}^2 = p^2\sigma^2$ $A_{max}^2 = p^2E_T/T_D$
E_T	$E_T = T_DA_{max}^2/p^2$ $E_T = T_D\sigma^2$ E_T

average displacement, fault area, seismic moment, duration time, and magnitude, respectively. The variability for magnitude is expressed by the standard deviation which is the value with ±.

Table 2. Magnitude dependencies of fault parameters and duration time

Regression equation
$\log_{10} L = 0.5(M\pm0.498) - 1.882$
$\log_{10} \bar{D} = 0.5(M\pm0.494) - 1.368$
$\log_{10} S = 1.0(M\pm0.407) - 4.082$
$\log_{10} M_0 = 1.5(M\pm0.372) + 16.19$
$\log_{10} T_D = 0.31M - 0.774$

4 VARIABILITY OF INDICES OF INPUT GROUND MOTION

Magnitude dependencies of indices of input ground motion are investigated by using ground motion records. The observation station, Tomioka in eastern Japan, which lies on a hard rock with thin alluvium, is selected for this study. Instead of employing the usual method which takes the whole duration of a ground motion into account, we introduced the segmentation method by which a direct S-wave segment of the record is identified. In this research, ground motions from 58 records are analyzed using the segmentation method described above.

Through the investigation using ground motion records, the prominent differences of the magnitude dependency among the indices can be obtained. The magnitude dependency of an index I is described by the magnitude

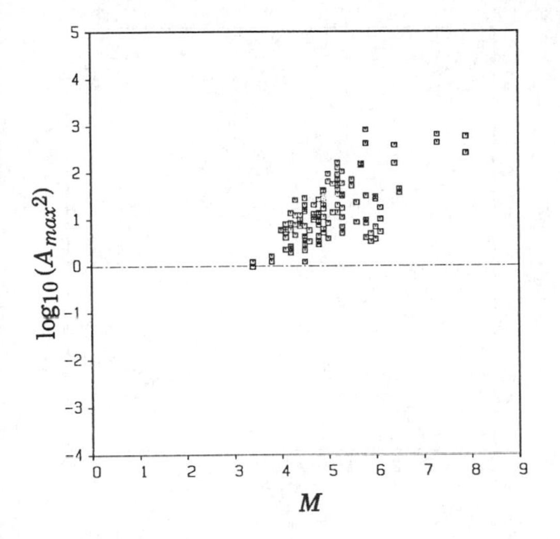

Fig.2. Variability of A_{max}^2

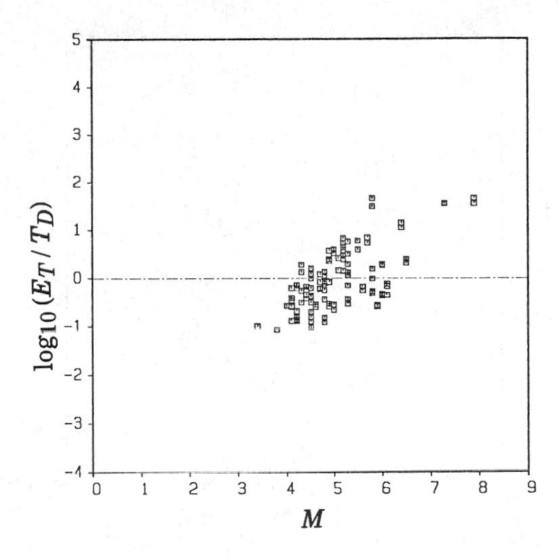

Fig.3. Variability of P_T

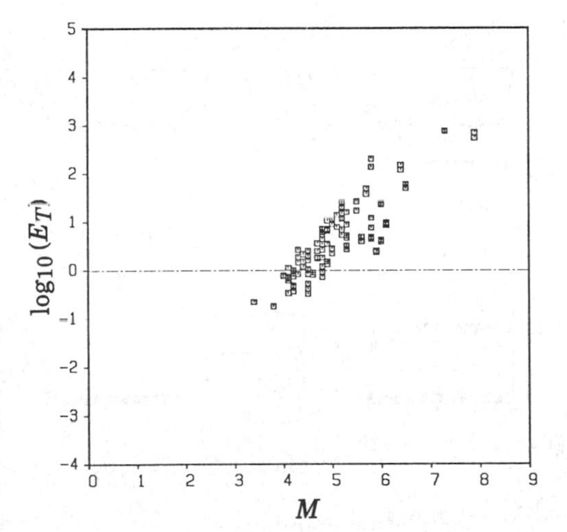

Fig.4. Variability of E_T

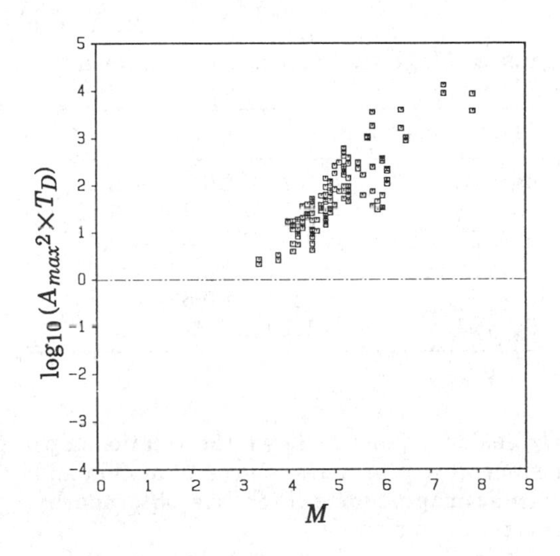

Fig.5. Variability of $A_{max}^2 \times T_D$

coefficient a, which is utilized in the following equation,

$$\log_{10} I = aM + c \qquad (1)$$

where M denotes magnitude and c is a constant. Of course, in order to derive the above equation for various records, the index I should be modified based on some attenuation relation. In this paper, $X^{-1.73}$ is adopted for this purpose, where X is a focal distance of each record. The magnitude coefficient a is obtained by the regression analysis for the records, in which the variance indicates the level of fitting.

In this study, three indices of ground motion, A_{max}, P_T, and E_T are investigated. Figures 2, 3, and 4 show the variability of each index, respectively. Figure 5 shows the variability of the product of A_{max}^2 and T_D. In Figs. 6 and 7, the variability of T_D and p^2 are shown, respectively. It must be noted that A_{max} and p are investigated in the form of square to make the comparison with P_T and

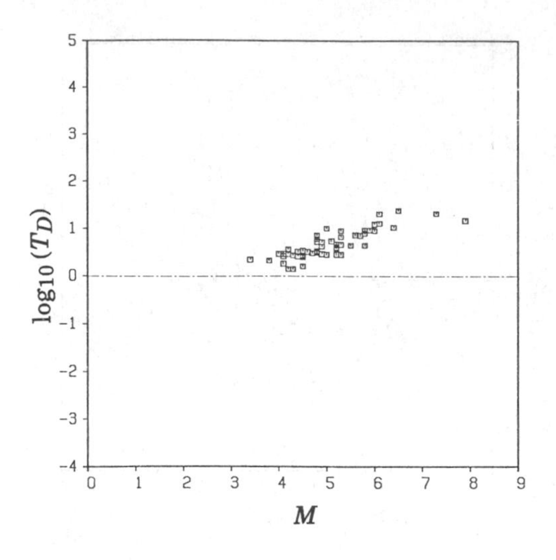

Fig.6. Variability of T_D

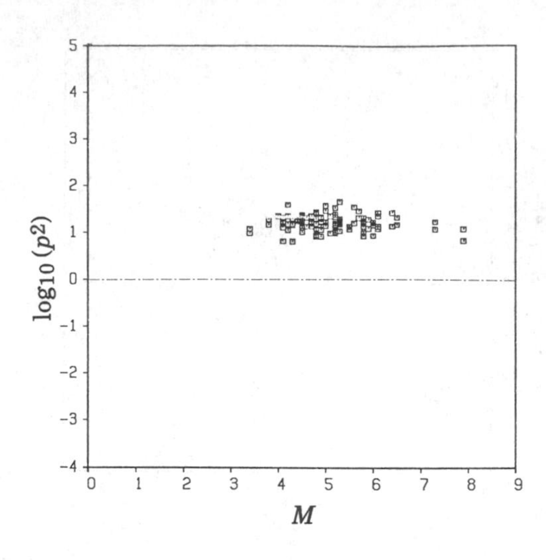

Fig.7. Variability of p^2

Table 3. Magnitude dependencies of seismic indices

Regression equation	
$\log_{10} A_{max}^2$	$= 0.7\,(M \pm 0.72) - 2.40$
$\log_{10} P_T$	$= 0.7\,(M \pm 0.70) - 3.59$
$\log_{10} E_T$	$= 1.0\,(M \pm 0.41) - 4.24$
$\log_{10} (T_D A_{max}^2)$	$= 1.0\,(M \pm 0.44) - 3.25$
$\log_{10} T_D$	$= 0.3\,(M \pm 0.52) - 0.85$
$\log_{10} p^2$	$= 1.20 \pm 0.16$

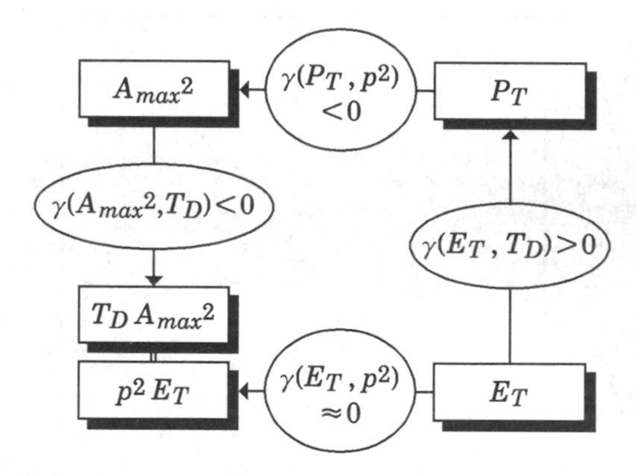

Fig.8. Correlation of indices

E_T easier. Table 3 shows the relationships between the properties of ground motion and seismic magnitude, which are obtained by regression analysis.

From the examination of the regression equations, the following results are derived.

1) Variability of A_{max}^2 is greater than that of E_T. From the comparison of Table 2 and Table 3, it is shown that variability of E_T is close to that of fault area. Also shown is that variability of T_D is close to that of fault length. Variability of A_{max}^2 is much greater than those. Although these relationships can hardly have a physical meaning, they will be effective as measures of variability of indices.
2) P_T derived from E_T and T_D has the similar variability as A_{max}^2. This indicates that the variability of P_T is also greater than that of E_T. Therefore, it can be concluded that E_T is the index with the least variability. It must be noticed that the variability of A_{max}^2 can be mitigated by multiplying T_D as shown in Fig. 5.
3) Magnitudes of the variability of A_{max}^2, P_T, and E_T support the relationships indicated in Fig. 1. That is, the more uncertain factors exist, the greater the variability is.

   Figure 8 shows the correlations among the indices.

In this study, the variability is given to seismic magnitude M as indicated in Table 3. This implies that for the ground motion with specified magnitude, the extent of M to be examined will differ by selecting the index. That is, by selecting E_T instead of A_{max}^2, it can be possible to consider the smaller extent of M, expecting the mitigation of the uncertainties in the seismic hazard analysis.

5 PROCESS OF DYNAMIC RELIABILITY ANALYSIS BASED ON TOTAL ENERGY

Figure 1 indicates that greater variability will be included in the reliability analysis by selecting A_{max} as an index, which contains more uncertain factors, such as peak factor and duration time. This hypothesis is supported by the examination using ground motion records shown before. In order to improve the accuracy of reliability analysis, it is necessary to mitigate the variability due to the modeling uncertainty. Therefore, it can be concluded that the index of ground motion should be described by E_T, which has less uncertain factors.

The process of *DRA* based on E_T is shown below.

1) Estimation of attenuation relation on E_T
   The following attenuation relation may be obtained with the variability.

$$\log_{10} E_T = 1.0\, M - 1.73 \log_{10} X + c \qquad (2)$$

2) Estimation of seismic hazard curves of E_T using data of M and X
3) Estimation of attenuation relation on spectral shape $S(\omega)$
   This yields the energy spectrum $E(\omega)$ based on the following equation.

$$E(\omega) = E_T \times S(\omega) \qquad (3)$$

4) Calculation of the energy spectrum of the response $E'(\omega)$
   $E'(\omega)$ is obtained from $E(\omega)$ and $H(\omega)$ based on the following equation,

$$E'(\omega) = |H(\omega)|^2 \times E(\omega) \qquad (4)$$

   where $H(\omega)$ is a transfer function of the structure.
5) Estimation of the duration time of response T'_D
   T_D is considered to be the parameter which is predicted more determinately with the consideration of fault model reflecting the seismological knowledge. T'_D will be estimated from T_D and property of the structure, using random vibration theory. The envelope function can also be used in order to distribute E_T.
6) Calculation of the power spectrum of the response $P'(\omega)$
   $P'(\omega)$ is obtained from $E'(\omega)$ and T'_D based on the following equation,

$$P'(\omega) = E'(\omega) / T'_D \qquad (5)$$

7) Calculation of the conditional probability of failure from $P'(\omega)$, based on random vibration theory.
8) Calculation of the annual probability of failure by convolving seismic hazard curve and the conditional probability of failure

As shown above, the process of *DRA* based on E_T has a similar structure as that based on A_{max}. The major differences between both process are the steps 1), 2), and 3), yielding the estimation of input motion without T_D and p, which increase the variability. It can also be said that the process of *DRA* is now improved as the rational structure with less variability by selecting E_T as an index.

6 CONCLUSIONS

Total energy is proposed as an index of input ground motion for the implementation of more rational reliability analysis based on the approach in the frequency domain, and is examined. As mentioned before, accuracy of reliability analysis is improved by mitigating the modeling uncertainties. In this study, two indices, peak acceleration and total energy are compared from the view point of variability. Followings are the results obtained in this study.

1) From the ground motion records it is confirmed that, total energy which has less variability for seismic magnitude than the square of peak acceleration, is more suitable as an index of ground motion. Variability of total power is somewhat smaller than that of the square of peak acceleration. As long as variabilities of duration time and peak factor exist, total energy possesses the least variability of the three. Therefore, it

is effective to describe the input motion by the total energy in order to carry out more accurate reliability analysis.

2) Above magnitudes of the variabilities of indices support the relationships of the variability shown in Fig. 1. Although, power spectrum is employed in the stationary random vibration theory, it can be said that energy spectrum is more advantageous to develop the theory, from the variability point of view.
3) As shown in Table 3, regression equations derived by using peak acceleration has greater variabilities than those derived by using total energy. This implies that by employing total energy, it is possible to relate the seismic magnitude to the index of input motion, regardless of the type of index.
4) Process of the dynamic reliability analysis based on the total energy as an index of input ground motion, is newly proposed. This process is improved as the rational structure with less variability.

REFERENCES

Hisada, T. and Ando, H., : Relation between duration of earthquake ground motion and the magnitude, Kajima institute of construction technology, 19-1, 1976

Mizutani, M., et al. : A study on the use of seismic data in the seismic hazard estimation of PRA, 10th SMiRT, Vol. P, pp.181-186, 1989

Mizutani, M., Kai, Y. and Matsubara, M. : A study on the applicability of the response estimation of our PRA procedure, 11th SMiRT, Vol. M, pp.181-186, 1991

Sato, R., et al. : Research on seismic fault model, Part 2 on evaluation of fault parameters[II], Investigation and research on earthquake insurance 12, Fire and marine insurance rating association of Japan, 1985 (in Japanese)

Shinozuka, M., Mizutani, M., Takeda, M. and Takemura, M. : A seismic PRA procedure in Japan and its application to a building performance safety estimation, Part 1, 2 and 3, ICOSSAR'89, Proc. I, pp.621-644, 1989

Structural Safety & Reliability, Schuëller, Shinozuka & Yao (eds) © 1994 Balkema, Rotterdam, ISBN 90 5410 357 4

Reliability analysis of frame shear wall type of tall buildings subjected to earthquakes by the failure mode approach

He Guangqian
China Academy of Building Research, Beijing, People's Republic of China

ABSTRACT: Based upon the study of very large number of elasto-plastic time history dynamic analysis of frame shear wall type of tall buildings under various earthquake ground motion excitations, failure of the building structures generally occurs at the so called weak stories. By the statistical characteristics from the results of such analyses, a reliability evaluation of the building structures could be formulated which is susceptible to further development and simplification to supplement the reliability analysis of the current aseismic design code now enforced in the People's Republic of China.

1 INTRODUCTION

As the frame shear wall type of current tall buildings has been much used in China, very large number of elasto-plastic time history dynamic analyses of such building structures has been carried out under various seismic recorded ground motions as well as through artificial earthquakes of the relevant localities since the mammoth Tangshan earthquake in China in 1976. In certain cases, the random earthquake ground motion has also been simulated as a stationary filtered non-white noise and its spectral density is given by

$$S_\Lambda(\omega) = A_g \cdot S_o[1 + (\omega^2/\omega_h^2)] \quad (1)$$

where

$$A_g = [1+4\xi_g^2(\omega^2/\omega_g^2)]/[(1-\omega^2/\omega_g^2)^2+4\xi_g^2\omega^2/\omega_g^2]$$

in which ω_g and ξ_g are respectively the predominant frequency and damping ratio of the site soil, ω_h is a parameter and is equal to 8 rad|s, S_o is the strength factor which, together with the stationary duration T can be determined according to the design standard of earthquake actions and the site soil classifications.

2 SOME DEFORMATION CHARACTERISTICS OF FRAME SHEAR WALL TYPE TALL BUILDINGS SUBJECTED TO STRONG EARTHQUAKES

By investigating a large number of frame shear wall buildings, it has been found that the distribution of yield strengths, either in shear or in combined shear and bending is, except in a number of typical floors, quite non-uniform along the different stories in the vertical direction of the building. It will become more accentuated by comparing the relative strength ratio coefficient $\xi(i)$ of a certain story i with those of the adjacent stories. For frame shear wall structures, there are two strength coefficients, one being the shear strength coefficient $\xi_y(i)$ and the other being the combined shear and bending strength coefficient $\xi_m(i)$. They can be defined as follows:

$$\xi_v(i) = V_y(i)/V_e(i) \quad (2)$$

$$\xi_m(i) = M_y(i)/M_e(i) \quad (3)$$

in which $\xi_v(i)$, $\xi_m(i)$ are respectively the strength ratio coefficient in shear and that in combined bending and shear in between the upper and lower floor of the i th story; $V_y(i)$, $M_y(i)$ are respectively the yielding shearing force and yielding due to combined bending and shear of the i th story while $V_e(i)$, $M_e(i)$ are respectively those as above through a pseudo elastic analysis, i.e. by considering the structure to be always in the elastic stage no matter how great be the action of the earthquake excitations.

In a non-uniform structure along the various stories of the building in the vertical direction, $\xi(i)$ varies from story to story and the smaller the value of ξ that a story possesses, the weaker will be its capability against earthquake actions. If the value of $\xi(i)$ is much smaller than those of the upper and lower adjacent stories, then that story is a weak story. This conclusion which is the outcome of the very many computational results of the elasto-plastic time history dynamic analyses can be interpreted as follows:

In a non-uniform structure, $\xi(i)$ varies from story to story but the smaller the value of $\xi(i)$ that a story possesses, the weaker will be its capability against earthquake actions. Under a strong earthquake ground motion, the energy therefrom will be generally transmitted to the upper stories from the bottom of the building. As the weak stories are generally susceptible to absorb large amount of energy by yielding to large story drifts, the remaining energy transmitted to their upper stories would be much lessened , thus resulting in much smaller story drifts above. This phenomenon, called the Plastic Deformation Concentration at the weak stories, is strongly related to the failure modes of the system. As a matter of fact, the variation of the story drifts along the height of the building will become much more pronounced in the inelastic stage than that in the elastic stage as shown in the accompanying figure 1.

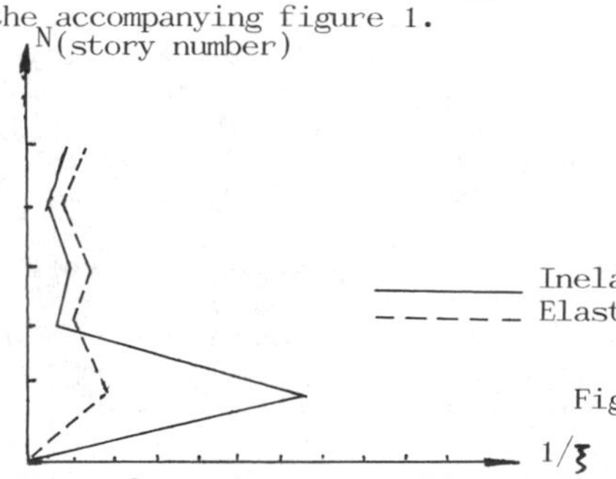

N(story number)

Inelastic
Elastic

$1/\xi$

Figure 1

Very often, in a non-uniform building of the frame shear wall type, there are more than one weak story. Furthermore, as are revealed by the elasto-plastic time history dynamic analyses, it could happen that for a certain earthquake input, the weak story at the lower part of the building might not yield to collapse during a certain phase of the vibration so that another weak story in the upper story of the same building will absorb more energy from below thereby causing a bigger story drift than any of the stories below as shown in the accompanying figure 2.

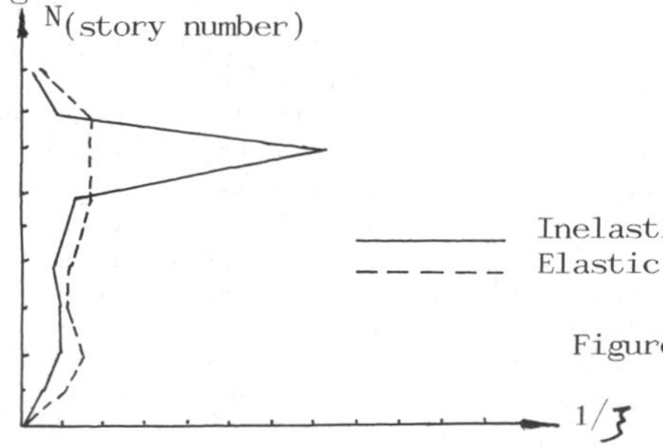

N(story number)

Inelastic
Elastic

$1/\xi$

Figure 2

This can be considered as another failure mode at a different weak story of the building.

However, in other cases, large story drift might occur simultaneously at each of the weak stories leading to only one of them or several of them to collapse during a same earthquake ground motion excitation. In general, different earthquake record inputs often alter the sequential occurrence of the plastic deformation concentrations at the weak stories so that each of them should be considered as a potential failure mode. There were cases where two or more weak stories collapsed simultaneously in a building structure under a same earthquake excitation. For instance, the 13 storied reinforced concrete frame structure of a distillation tower at Tanggu district of Tianjin area collapsed at the 6th and 11th story simultaneously under a single earthquake excitation. On the other hand, if the frame shear wall type tall building is very uniform in the yielding strength for all the stories, which is almost next to impossible in actuality, the problem of reliability analysis should be studied through other lines of approach and will not be included in this paper.

3 STATISTICAL CHARACTERISTICS OF ULTIMATE DEFORMATION OF FRAME SHEAR WALL ELEMENTS

There are many factors influencing the ultimate deformation capability of frame shear elements or structures, some of which are quite indeterministic in nature such as the parameters of earthquake action inputs, the mathematical modelling of the frame shear wall structures and the correlation in between etc. Model tests also present many difficulties such as the innumerable combinations of the magnitudes of the axial compression ratios of frame columns, the reinforcements in shear walls as well as the amount of ties in columns and the amount of stirrups in beams, the ratio of the distance of the concentrated load nearest the support to the effective depth of the beam (sometimes abbreviated as the shear span ratio in certain papers) and the strengths of materials etc. Some of the design parameters are even correlated with each other. Paper 3 as given in the reference has made some studies in this respect and arrives at at a preliminary conclusion that the coefficient of variation of the maximum deformation response of the weak story in a frame shear wall type of tall buildings is 0.57 - 0.60 and can be taken as 0.59 for simplicity and that its probability distribution curve is of the Extreme Value Distribution Type I.

From the tests of many shear walls with

edge members, it has been found that the average maximum angle of the story drift is 0.015, the standard deviation is 0.053 and that the coefficient of variation is 0.35. Through K-S tests, the probability distribution curve of the ultimate deformation capacity of shear walls is of the type of Lognormal Distribution.

Another interesting statistics shows that for shear type structures, the maximum top displacements by the elasto-plastic time history dynamic analysis designated as $\Delta u_p(N)$ bear a certain relationship to those $\Delta u_e(N)$ by the same time history dynamic analysis of the same structures by supposing that the latter are fully elastic no matter how big are the stresses or strains., i.e.

$$\Delta u_p(N) = \eta \Delta u_e(N) \tag{4}$$

in which $\eta = \alpha[(T_o - T)/\xi] + 0.75$

the details of which being given in Paper 4.

But for the case of frame shear wall type of structures, a relatively more stable relationship has been found and this can be given by

$$\Delta u_p(N) = 0.9 \Delta_e(N) \tag{5}$$

By use of the statistical characteristics given above, a simplified method of evaluating the failure probability of a frame shear wall type of tall buildings can be approached.

4 RELIABILITY ANALYSIS OF FRAME SHEAR WALL TYPE OF TALL BUILDINGS BY FAILURE MODE APPROACH

Through the statistical characteristics from the elasto-plastic time history dynamic analyses, we can evaluate the failure probability of frame shear wall type of tall buildings by the following formula:

$$P_{fi} = \iint_{R-S<0} f_R(r)\cdot f_s(S/I)drds = \int_0^\infty f_s(S/I)\cdot F_R(S)ds \tag{6}$$

in which f_R, F_R are respectively the probability density function and the probability distribution function of the ultimate deformation R;

$f_s(S/I)$ is the probability density function of the maximum deformation response of the weak story of the aforesaid structure under earthquake intensity I.

For a certain length of time in the coming years, the occurrence and magnitudes of the future earthquakes are all random factors. But through earthquake hazard analysis, there is the possibility of giving the probabilities of the occurrences of different magnitudes of future earthquakes of a given locality. Thus the global failure probability of the frame shear wall type of tall building can be figured by the following formula:

$$P_f = \sum_{i=1}^{n} P_{fi}\cdot P(I_i) \tag{7}$$

in which P_{fi} is the failure probability of the frame shear wall type of tall building under the actions of an earthquake of intensity i;

$P(I_i)$ is the probability of the occurrence of earthquake of intensity i.

For instance, for a 10 storied reinforced concrete frame shear wall building with its structural parameters as listed in table I and the probability distribution function of the occurrences of the earthquakes of the locality as listed in Table II, we can proceed to evaluate the global failure probability of the structure as follows:

Table I: Yield bending moments of shear walls and yield shearing strengths of columns of the structure

Story Number	A	B	C
10	4810.0	86.0	75.0
9	5010.0	101.0	84.0
8	5150.0	122.0	108.0
7	5310.0	141.0	123.0
6	5460.0	162.0	142.0
5	5610.0	198.0	168.0
4	5760.0	236.0	182.0
3	5910.0	273.0	201.0
2	6020.0	283.0	264.0
1	6230.0	264.0	258.0

in which A represents M_y of shear walls in kN-m;

B represents V_y of interior columns in kN;

C represents V_y of side columns in kN.

Table II: Probability distribution function of earthquake intensities of a locality in the coming 50 years

Intensities	5.5^o	6.0^o	6.5^o	7.0^o
$P(I\leq I_i)$	0.100	0.271	0.496	0.700
Intensities	7.5^o	8.0^o	8.5^o	9.0^o
$P(I\leq I_i)$	0.841	0.9235	0.9658	0.9856
Intensities	9.5^o	10.0^o	10.5^o	11.0^o
$P(I\leq I_i)$	0.9943	0.9978	0.9992	0.9997

After the elasto-plastic time history dynamic analysis, it has been found that the structure is in the full elastic stage under earthquake of intensity 7 but that it enters into the elasto-plastic stage at the earthquake of intensity 8 with $\xi_m = 0.72$. Under earthquake intensity 10, the value of $\xi_m = 0.18$ and for for earthquakes of intensity greater than 10.5 or of 11, the structure collapses. Thus we have to calculate the failure probabilities of the structure at the bottom story for earthquakes of intensities 8, 9 and 10, the results of which being listed in Table III.

Table III: Failure probability of a 10 storied frame shear wall building at the bottom story under earthquake actions of different intensities

A	B	B	D	E	F
8	0.0028	0.59	0.015	0.35	0.00254
9	0.0071	0.59	0.015	0.35	0.11118
10	0.0246	0.59	0.015	0.35	0.73515

in which A represents earthquake intensity;
B represents the average maximum bottom story drift angle;
C represents coefficient of variation;
D represents average of maximum deformation drift angle;
E represents coefficient of variation of maximum deformation drift angle;
F represents the failure probability.

As mentioned above, the structure is still in the full elastic stage up to earthquakes of intensity 7 which means that the failure probability at that stage is zero. Starting from intensity 8, the probability of its occurrence is, through earthquake hazard analysis, 0.1248, that for intensity 9 is 0.0285, that for intensity 10 is 0.0049 and that for earthquake of intensity 11 and above is 0.0007. Then by formula (7), we have

$$P_f = \sum_{i=1}^{n} P_{fi} \cdot P(I_i)$$
$$= 0.00254 \times 0.1248 + 0.11118 \times 0.0285 + 0.73515 \times 0.0049 + 1.0 \times 0.00007$$
$$= 0.0078$$

which is the global failure probability of the 10 storied frame shear wall building.

In the most general case, there could be more than one weak story in a frame shear wall building. Furthermore, we are quite unsure of the time history of the actual ground motions of the coming earthquakes in spite of the fact that we might have already in possession of several recorded earthquakes in the past for the same locality. This does explain why we need to investigate all the different possible pontential failure modes M_i of the structure.

It is evident that collapse of the frame shear wall structure will occur if any of the failure modes does appear. This can be formulated as:

$$E = \bigcup_{i=1}^{n} M_i \qquad (8)$$

The collapse probability will then be given by:

$$P(E) = P(\bigcup_{i=1}^{n} M_i) \qquad (9)$$

while the reliability of frame shear wall type of tall buildings can be given by:

$$R = 1 - P(\bigcup_{i=1}^{n} M_i) \qquad (10)$$

5 CONCLUSION

In the aseismic design of tall building structures, the reliability analysis has been much needed in the advanced studies of such important engineering projects as it also bears an important influence to the global optimization of the structural design. This paper is only a step toward the goal and further development and simplication should still be carried out for practical design applications.

ACKNOWLEDGMENT

The author would like to express his thanks to Dr. Gao Xiaowang for his help in the preparation of this paper.

REFERENCES

1. Chen, G.H. 1984. Simplified calculation of elastoplastic story drift responses of multistory shear type structures subjected to earthquakes. Journal of Building Structures, No.2, Vol.5.
2. Research report 1987. Study of shear wall type tall building structures. China Academy of Building Research.
3. Gao, X.W. et al. 1989.Assessment of seismic damages and reliability of reinforced concrete buildings by the method of structural reliability. Research report, China Academy of Building Research
4. He, G.Q. et al. 1981. On elasto-plastic deformations of multi-storeyed shear type structures due to earthquake ground motions. US/PRC workshop on seismic analysis and design of reinforced concrete structures. 43-55.
5. Ang A.H-S. 1991. Computational damage analysis and reliability assessment. Proc. Sino-U.S.Joint Symposium/Workshop.

Structural Safety & Reliability, Schuëller, Shinozuka & Yao (eds) © 1994 Balkema, Rotterdam, ISBN 90 5410 357 4

Realistic earthquake damage assessment for high-rise reinforced concrete buildings by nonlinear simulation

U. Hanskötter, W.B. Krätzig & K. Meskouris
Institute for Statics and Dynamics, Ruhr-University Bochum, Germany

ABSTRACT: In the first part of this paper the problem of the classification of probable ground motions at a site according to their damage potential is discussed while the second part deals with nonlinear analyses of high-rise reinforced concrete buildings stiffened by moment-resisting frames as well as by shear walls. Last but not least nonlinear computer simulations of such buildings are presented.

1. INTRODUCTION

Any realistic earthquake damage assessment for high-rise reinforced concrete buildings by nonlinear simulation depends on the description of the damage potential of an earthquake and on the nonlinear modelling of the structure. The description of the seismic risk by a single parameter (e.g. the expected peak ground acceleration, PGA) is not sufficient because the observed damages of the buildings do not correlate well with measured PGA values. Other descriptors which are closely connected with the energy of the motion, such as energy spectra and strong motion duration values are much better in this respect.

2. DESCRIPTION OF THE DAMAGE POTENTIAL

The standard description of site seismic risk involves a smooth broad-band response spectrum which can be scaled by the local PGA value. The following example illustrates a possible pitfall, leading to drastic reduction of the safety level in the structure thus designed.

We consider two measured accelerograms (Montenegro 1979 earthquake, measured in Petrovac and Mexico City 1985 earthquake, SCT record) and scale them to the same PGA, namely 0.2g. Linear response spectra for a damping value of 5% critical for these scaled motions exhibit the same spectral ordinate for a period of about 0.7s. We conclude that structures with first natural period about 0.7s will react similarly to both motions, as long as linear behavior subsists. To illustrate, Fig.1 shows time histories of the horizontal displacement of the first floor of a building for both motions, calculated for linear system behavior. This picture changes drastically, however, when we consider the true nonlinear behavior of reinforced concrete: Fig.2 depicts results of a nonlinear time-history analysis. In the case of the Mexico City excitation, its longer duration which leads to a large number of load reversals in the structure induces structural collapse, while the relatively short Petrovac motion causes elastoplastic drift values which are almost the same as in the linear case.

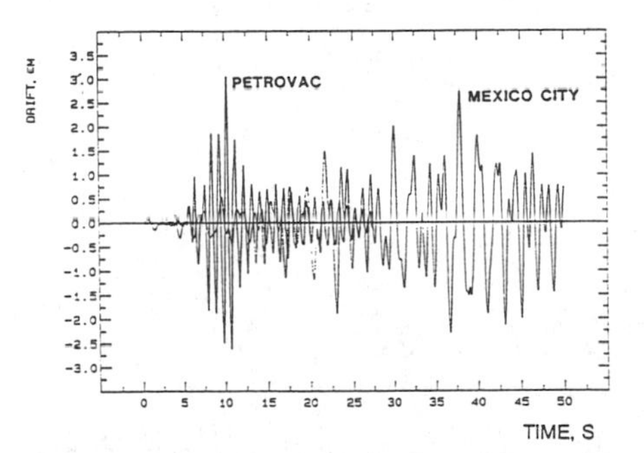

Fig.1: Ground floor drift for two motions, linear model

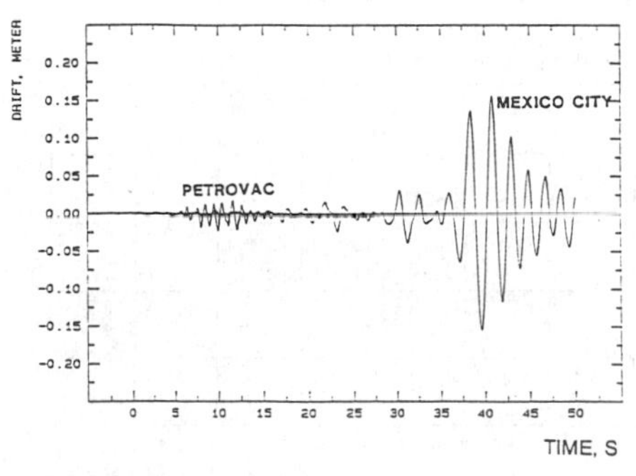

Fig.2: Ground floor drift for two motions, nonlinear model

Obviously, such effects cannot be described by linear models. We conclude that the design process by response spectra should be supplemented by a nonlinear "verification stage" which ascertains that even for the most unfavorable probable earthquake events, with correspondingly large return periods, structural behavior and degree of damage are as expected. This conclusion is of course valid only for sites with high seismic risk, whose probable motions are characterized by high energy content and long duration, implying high numbers of load reversals in the structure. The classification shown in Fig.3 of probable site motions can serve as orientation aid. For sites with S- or M-type motions, linear or quasilinear design models are quite acceptable, while the design of structures on sites with probable L-type motions should include a nonlinear verification stage. Of course, the table Fig.3 can be regionally adjusted and amplified by introducing additional parameters such as ARIAS intensities, spectral intensities and, prominently, energy spectra [1].

S-type *(High frequency, low energy)*

-Strong motion duration $< 10s$

-Central period $T_0 < 1.0s$

$a/v > 1.0g/m/s$

L-type *(Low frequency, high energy)*

-Strong motion duration $> 15s$

-Central period $T_0 > 1.2s$

$a/v < 0.8g/m/s$

M-type

Intermediate values

Fig.3: Classification of ground motion records

In order to illustrate the ability of energy spectra to highlight high-risk motions, we show in Fig.4 energy spectra of four motions: SCT record in Mexico City, Petrovac record (PET), IGN record of the 1986 San Salvador event and Tolmezzo /Friuli 1976 (TOL).

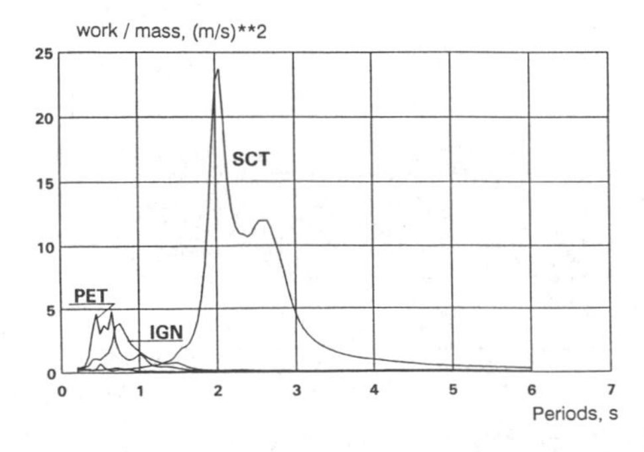

Fig.4: Energy response spectra (SCT, PET, IGN, TOL)

If measured records are scarce or not available for the site in question, standard geophysical models of the vicinity may be used for the generation of artificial accelerograms. On the other hand, strong motion durations may also be predicted by such models and should be used in addition to site response spectra as important parameters of its seismic risk.

Now to the mentioned "verification stage" of the design process. A first possibility lies in conducting a static "ultimate load analysis" for a full loading cycle [2]. This relatively simple procedure, somewhat similar to the U.S. "pushover tests", yields quite meaningful information on the nonlinear behavior of the system. However, for buildings and structures whose continuous function during a strong earthquake is of vital importance (e.g. lifelines, hospitals, communication centers), detailed dynamic time-history analyses are advocated, which allow for monitoring the pertinent damage evolution patterns and provide in-depth information on weak spots. Details on the corresponding nonlinear models for the special case of high-rise reinforced concrete buildings are given in the following.

3. NONLINEAR ANALYSES OF HIGH-RISE REINFORCED CONCRETE BUILDINGS

Three-dimensional models of high-rise reinforced concrete buildings (e.g. as in Fig.5), which are stiffened by moment-resisting frames as well as by shear walls are considered. In contrast to buildings stiffened exclusively by moment-resisting frames, where story drifts exhibit large variations, shear walls act as "stiff backbones" ensuring a much more uniform

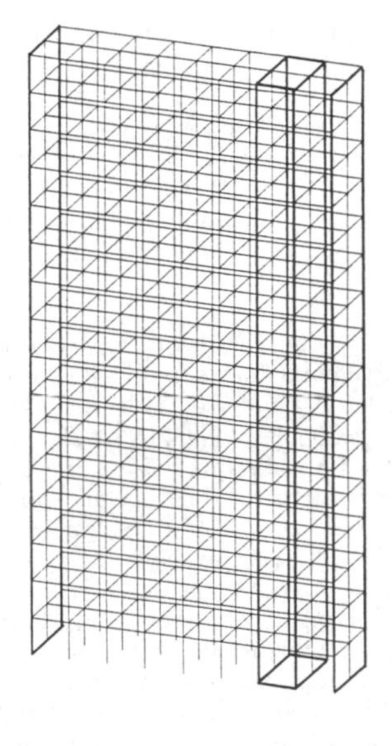

Fig.5: 16-story building

drift distribution and also a drastic limitation of the single story drifts. Nonlinear dynamic analyses of such structures are quite complicated and time-consuming, starting with the description of the highly nonlinear material "reinforced concrete". Therein, nonlinear influences must be considered for its single components (concrete and steel) as well as for their joint behavior (bond slip effects). For modelling the concrete component of the shear-walls we use the material law after DARWIN/PECKNOLD [3]. For the idealization of the concrete component of the beams the material law after MEYER [4] is employed, in the enhanced form given by GARSTKA [5]. The steel component is modelled by the uniaxial stress-strain law after PARK/KENT [6], which includes BAUSCHINGER phenomena and cyclic hardening. Finally, tension stiffening effects are considered according to GILBERT/WARNER [7] by increasing the steel stiffness.

For the finite element discretization of the reinforced concrete buildings nonlinearly formulated beam and slab elements are needed. The beam elements are idealized by composite macro-elements, whose end regions consist of coupling elements or layered cracked-beam elements [8]. The coupling elements describe, on the substructural level, the behavior of the beam's (or beam-column's) yielding portions, considering also the interaction between normal forces and bending moments. The underlying nonlinear constitutive law is usually of the half-empirical TAKEDA type. In these layered cracked-beam elements the cross section is divided into concrete and steel layers and the overall element matrices result from numerical integration over the cross-section and the element length.

For reinforced concrete shear walls there also exist several discretization options. A crude approximation models the shear wall with just one nonlinear spring element per story, coupling neighboring displacement degrees of freedom while another element type makes use of a truss analogy to compute approximative constitutive laws. The central drawback of both elements in a nonlinear context is their inability to predict "unexpected" failure modes, which were not explicitly considered beforehand. This argument is partially valid also for the "Multiple vertical line" element family, of which the "Three vertical line" element by KABEYASAWA is probably the best known.

We advocate the modelling of the shear walls by nonlinear layered finite elements with discrete concrete and steel layers. In addition to the standard 8-DOF element we also use refined 12-DOF elements after [9] which by including the four rotational DOF allow a much better modelling of bending deformations. As an example, Fig.6 shows the results of convergence tests of both element types.

The central problem of every nonlinear time-history analysis of moderately large buildings lies in the (possibly prohibitively) large working load. Even for

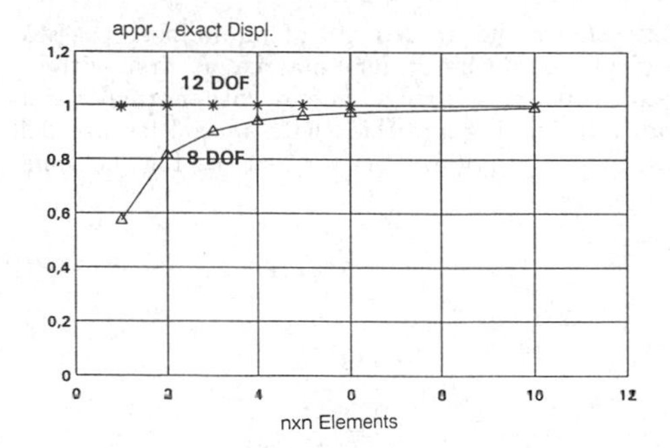

Fig.6: Convergence tests of slab elements

modern computer workstations, CPU time is often measured in days rather than hours, so every possibility of economizing time and reducing the overall effort must be utilized.

The following techniques have been implemented in the FEMAS program system [10]:

• In the 3D-model of the building floor slabs are assumed to be rigid in their own plane, so that only the two horizontal diplacements and the rotation of the vertical axis remain as independent degrees of freedom per floor. This implies a kinematic transformation of all degrees of freedom of a floor as a function of the three master degrees of freedom.

• In order to reduce the necessary storage space the data are not stored element-wise but rather following a special element-group- storing scheme. Corresponding generation algorithms ensure correct access to any special member data set in the course of the computation of the element matrices.

• The nonlinear constitutive laws have been coded using graph theory and knowledge-based techniques, leading to "lean" subroutines without superfluous loops and jumps [11].

• A special band-width optimizer has been developed for the present case of kinematically coupled degrees of freedom. By a static condensation of the non-active degrees of freedom it has been possible to further decrease the required storage space.

• The total computing time has been further reduced by utilizing special techniques for incorporating the changes due to nonlinear effects in the course of the computation.

A complete description of these strategies for optimizing the FE-concept for seismic damage analyses is given in [12].

As a typical example we consider the 16-story building of Fig.5 investigated for the Kalamata earthquake of the year 1986.

Fig.7 shows the procedures of minimizing the storage places of the system matrix, at first without band-width optimizing, second with normal band-width optimizing and third with special band-width optimizing. In Fig.8 the CPU-times for the time-

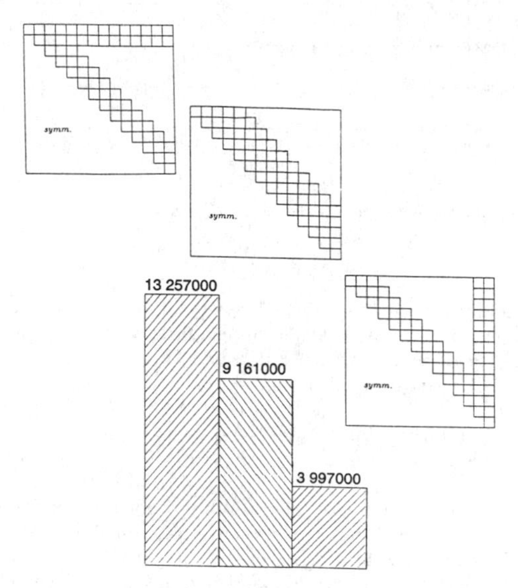

Fig.7: Minimizing the computer storage of the tangential system stiffness matrix

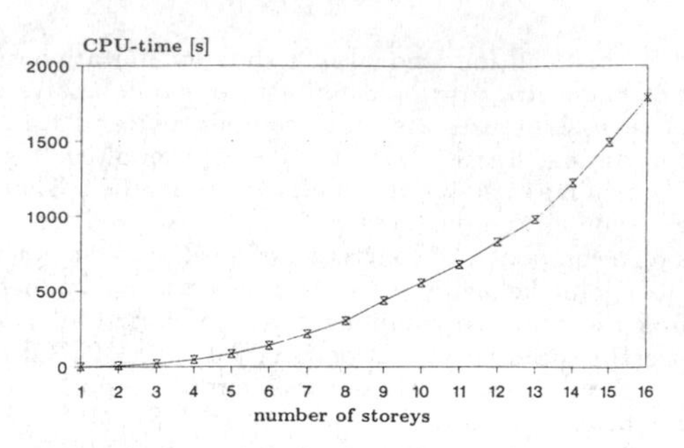

Fig.8: Minimizing the CPU-time for triangulation of the tangential system matrix

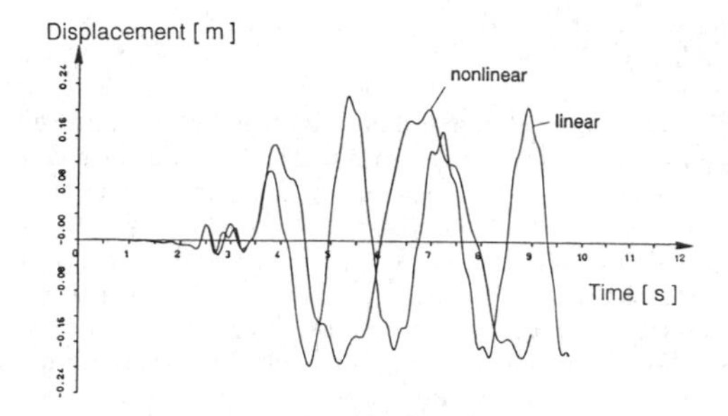

Fig.9: Linear and nonlinear histories of the horizontal displacement of the 16th floor

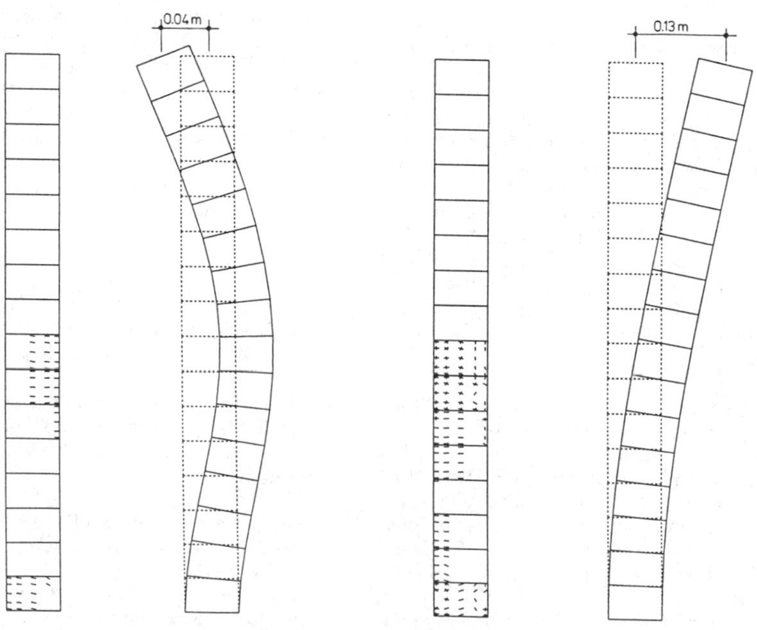

Fig.10: Crack patterns and deformed configurations for time instances t=2.8s and t=4.0s

wasting triangulation of the tangential system matrix for this example is presented. In case of the condensation of non-active DOFs the CPU-time can be reduced drastically.
Fig.9 shows time histories of the horizontal displacement for linear and nonlinear behavior. As it can be seen, the uncracked system vibrates initially in its second mode. Ground mode components becoming dominant at a later stage. To illustrate the damage of the outer single shear wall, its crack patterns have been depicted in Fig.10 as computed for the time instances t=2.8s and t=4.0s. The corresponding deformed configurations are also reproduced.

4. RESULTS AND CONCLUSIONS

Any realistic earthquake damage assessment for high-rise reinforced concrete buildings depends on the description of the damage potential of an earthquake and on the quality of the nonlinear model of the structure itself. Since the damage potential of an earthquake depends heavily on its duration and energy, standard linear-elastic design concepts can lead to insufficient safety margins if nonlinear effects are disregarded for high-energy, long-duration events. In view of the ever- increasing availability of cheap computing power and advanced software for conducting nonlinear dynamic time-history analyses of structures, the wider use of these techniques for design verification is advocated. Some pertinent techniques for reducing the overall effort in the special case of wall-stiffened reinforced concrete buildings were presented, as they were implemented in the FEMAS program environment.

ACKNOWLEDGMENTS

The presented results stem from the special research project B1 of the SFB 151 at Bochum. The continuous financial support of the "Deutsche Forschungsgemeinschaft (DFG)" is gratefully ack- nowledged.

REFERENCES

[1] Bertero, V.V, Uang, C.-M.: Issues and future directions in the use of an energy approach for seismic-resistant design of structures. In: P. Fajfar, H. Krawinkler (Hrsg.), Nonlinear Seismic Analysis and Design of Reinforced Concrete Buildings, Elsevier Applied Science, London/New York, 1992, p. 3.

[2] Krätzig, W.B., Meskouris, K.: Nachweiskonzepte für erdbeben- sichere Ingenieurtragwerke. Tagungsheft "Baustatik-Baupraxis 4", Hannover 1990.

[3] Darwin, D. and Pecknold, D.A., Nonlinear Biaxial Law for Concrete, J. Eng. Mech. Division, ASCE, Vol. 105 (1979), p. 623.

[4] Meyer, I.F.: Ein werkstoffgerechtes Schadens- und Stababschnittsmodell für Stahlbeton unter zyklisch nichtlinearer Beanspruchung, Techn.-wiss. Mitteilungen 88-4, Institut für Konstruktiven Ingenieurbau, Ruhr-Univ. Bochum, 1988

[5] Garstka, B.: Untersuchung zum Trag- und Schädigungsverhalten stabförmiger Stahlbetonbauteile mit Berücksichtigung des Schubeinflusses bei zyklischer nichtlinearer Beanspruchung, Techn.-wiss. Mitteilungen 93-2, Institut für Konstruktiven Ingenieurbau, Ruhr-Univ. Bochum, 1993

[6] Park, R., Kent, D.C., Sampson, R.A., Reinforced Concrete Members with Cyclic Loading, J. Struct. Division, ASCE, Vol. 98 (1972), p. 1341.

[7] Gilbert, R.I., Warner, R.F., Tension Stiffening in Reinforced Concrete Slabs, J. Struct. Div., ASCE, Vol.105 (1978), p. 1885.

[8] Meyer, I.F., Krätzig, W.B., Stangenberg, F., Meskouris, K.: Damage Prediction in reinforced concrete frames under seismic actions. European Earthquake Engineering 3, 1988, p. 9.

[9] Kwan, A.K.H.: Analysis of Buildings Using Strain-Based Element with Rotational DOFs. J. Struct. Engineering, ASCE, Vol. 118 (1992), p. 1191.

[10] Beem, H. et al., FEMAS, Finite Element Moduln Allgemeiner Strukturen, Benutzerhandbuch Version 90.1, Institut für Statik und Dynamik, Ruhr-Universität Bochum 1991.

[11] Hanskötter, U. et al., A knowledge-based approach to the numerical treatment of constitutive laws for cyclic loading. Structural Dynamics, Proceedings EURODYN 1990, Balkema-Verlag, Vol.1, p.465.

[12] Hanskötter, U.: Strategien zur Minimierung des numerischen Aufwands von Schädigungsanalysen seismisch erregter, räumlicher Hochbaukonstruktionen mit gemischten Aussteifungssystemen aus Stahlbeton, Thesis under work, Bochum, 1993.

Structural Safety & Reliability, Schuëller, Shinozuka & Yao (eds) © 1994 Balkema, Rotterdam, ISBN 90 5410 357 4

A stochastic wave model of earthquake ground motion

Takanori Harada
Department of Civil Engineering, Miyazaki University, Japan

ABSTRACT: The spatial variation of earthquake ground motions is an important factor that should be carefully considered in the seismic design of buried lifelines such as tunnels and pipelines. The consideration of the spatial variation of ground motions may also have significant effects on the earthquake response of structures with spatially extended foundations or multiple supports. The temporal and spatial variability of earthquake ground motions has been inferred experimentally using data from closely spaced seismograph arrays. In this paper, the earthquake ground motion varying in time-space domain is described analytically by the stochastic wave filtered by a surface soil layer of irregular thickness resting on half space. The SH wave is transmitted to the soil layer from the half space. The analytic expressions of the earthquake responses of irregular ground are based on the perturbation method, and their accuracy is examined by comparison with the responses computed by the direct boundary element method for many rough surface models consisting of topography with sinusoidal shape. Finally, an analytic expression for the frequency wavenumber spectrum of the stochastic wave at irregular ground surface is then derived.

1 INTRODUCTION

The spatial variation of earthquake ground motions is an important factor that should be carefully considered in the seismic design of buried lifelines such as tunnels and pipelines. The consideration of the spatial variation of ground motion may also have significant effects on the seismic response of structures with spatially extended foundations or multiple supports. In fact, for buried lifelines, the seismic deformation method was developed (Public Works Research Institute, 1977) and is now in practical use in Japan. For the seismic design of the Akashi Kaikyo Bridge foundations, a modified response spectrum was used taking into account the spatial variation of ground motions around the foundations (Kashima et al., 1984; Kawaguchi et al., 1987).

The temporal and spatial variability of earthquake ground motions has been inferred experimentally using data from closely spaced seismograph arrays. The SMART-1 array, for example, located at Lotung in the NE corner of Taiwan has provided valuable data for the analysis of ground motions in time-space domain. Numerous studies using the SMART-1 array data have been reported (Loh et al., 1982; Harada, 1984; Harada and Shinozuka, 1986; Harichandran, 1988; Abrahamson, 1985 and 1991). It is common in these studies that the accelerograms from each seismic event are described as samples from space-time stochastic processes or stochastic waves and eventually the spatial coherence functions or the frequency wavenumber spectra are estimated.

In this paper, the earthquake ground motion varying in time-space domain is described by the stochastic wave filtered by a single surface soil layer of irregular thickness, due to the irregular shape of the free-surface, resting on half space. The SH wave is transmitted to the soil layer from the half space. The analytic expressions of the earthquake responses of irregular ground are presented on the basis of the perturbation method, and their accuracy is examined by comparison with the responses computed by the direct boundary element method for many rough surface models consisting of topography with sinusoidal shape. Finally, an analytic expression for the frequency wavenumber spectrum of the stochastic wave at irregular ground surface is then derived.

2 FUNDAMENTALS OF SPACE-TIME CORRELATION STRUCTURE

In the stationary-homogeneous approach, it is well known that a stochastic wave $u_0(\mathbf{x}, t)$ with zero mean value is characterized by the autocorrelation function defined as:

$$R_{u_0 u_0}(\boldsymbol{\xi}, \tau) = E\left[u_0(\mathbf{x}+\boldsymbol{\xi}, t+\tau)u_0(\mathbf{x}, t)\right] \quad (1)$$

in which, $E[\cdot]$ denotes the expectation operator, $\mathbf{x}$ is a vector of spatial coordinates (x_1, x_2) and t the time variable. Since the stochastic wave is assumed to be stationary in time and homogeneous in space, the autocorrelation function possesses the symmetric nature with respect to the origin:

$$R_{u_0 u_0}(\boldsymbol{\xi}, \tau) = R_{u_0 u_0}(-\boldsymbol{\xi}, -\tau) \quad (2)$$

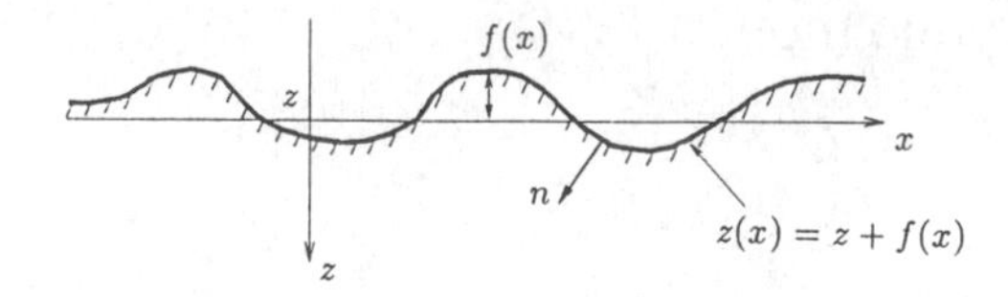

Fig. 1. An irregular interface between two homogeneous media

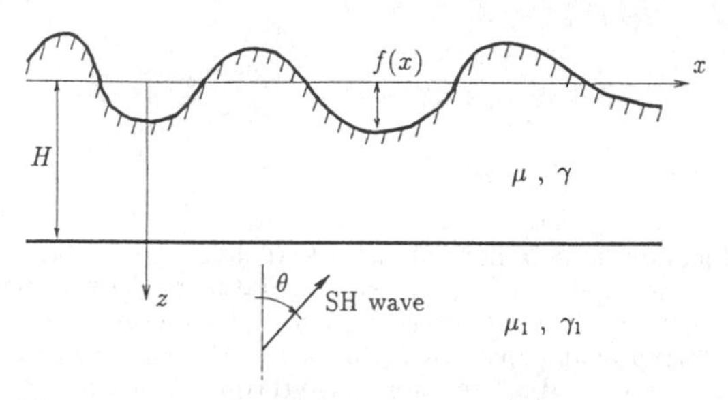

Fig. 2. A single soil layer with irregular free-surface resting on half space

Assuming that the Fourier transform of the autocorrelation function exists, the frequency wavenumber spectrum of $u_0(\mathbf{x},t)$ is defined as:

$$S_{u_0u_0}(\boldsymbol{\kappa},\omega)=\frac{1}{(2\pi)^3}\iiint R_{u_0u_0}e^{-i(\boldsymbol{\kappa}\cdot\boldsymbol{\xi}-\omega\tau)}d\xi_1d\xi_2d\tau \tag{3a}$$

and its inverse transform is given by:

$$R_{u_0u_0}(\boldsymbol{\xi},\tau)=\iiint S_{u_0u_0}(\boldsymbol{\kappa},\omega)e^{i(\boldsymbol{\kappa}\cdot\boldsymbol{\xi}-\omega\tau)}d\kappa_1d\kappa_2d\omega \tag{3b}$$

The above two equations represent the Wiener Khintchine relations in three dimensions, where $\boldsymbol{\kappa}$ is a vector of the wavenumbers (κ_1,κ_2) and ω the frequency.
It can be shown from Eqs. (2) and (3a) that the frequency wavenumber spectrum is symmetric:

$$S_{u_0u_0}(\boldsymbol{\kappa},\omega)=S_{u_0u_0}(-\boldsymbol{\kappa},-\omega) \tag{4}$$

It is also shown according to Bochner (1956) that the frequency wavenumber spectrum is real and nonnegative:

$$S_{u_0u_0}(\boldsymbol{\kappa},\omega)\geq 0 \tag{5}$$

It should be noted here that the autocorrelation function or the frequency wavenumber spectrum provide all information concerning the correlational characteristics of the stochastic wave $u_0(\mathbf{x},t)$. In fact other correlation functions can be defined using either the autocorrelation function or the frequency wavenumber spectrum. For example, the frequency-dependent autocorrelation function $C_{u_0u_0}(\boldsymbol{\xi},\omega)$ is defined as:

$$C_{u_0u_0}(\boldsymbol{\xi},\omega)=\frac{1}{2\pi}\int R_{u_0u_0}(\boldsymbol{\xi},\tau)e^{i\omega\tau}d\tau$$
$$=\iint S_{u_0u_0}(\boldsymbol{\kappa},\omega)e^{i\boldsymbol{\kappa}\cdot\boldsymbol{\xi}}d\kappa_1d\kappa_2 \tag{6}$$

The frequency-dependent autocorrelation function is usually used to represent the correlational characteristics of stochastic wave because the correlation is in general decreased with separation distance and frequency. The frequency-dependent autocorrelation function represents directly such a correlation.

3 SH WAVE FILTERED BY A SINGLE LAYER WITH IRREGULAR FREE SURFACE RESTING ON HALF SPACE

Response of a single layer with irregular surface resting on half space subjected to an SH wave as shown in Fig. 2 is considered in this section. The method is based on the perturbation approach (Kennet, 1972). In subsection 3.1, the perturbation method by Kennet is summarized, and then the response of the ground with irregular surface in Fig. 2 is derived in subsection 3.2.

3.1 *Basic Equations for Ground with Irregular Interface*

For harmonic excitation with frequency ω, the displacement wave field $v(x,z)\exp(-i\omega t)$ corresponding to SH wave propagations in a two-dimensional homogeneous isotropic medium is known to be expressed as the matrix wave equation in the form (Kennet, 1972):

$$\frac{\partial}{\partial z}\mathbf{B}(x,z)=\mathbf{A}_0(z)\mathbf{B}(x,z) \tag{7}$$

where $\mathbf{B}$ is the displacement-stress vector and $\mathbf{A}_0$ is the operator matrix defined as:

$$\mathbf{B}_{SH}=col\ [v\ ,\ \tau_{yz}] \tag{8}$$

$$\mathbf{A}_{0_{SH}}=\begin{bmatrix}0 & \frac{1}{\mu}\\ -\mu\partial_{xx}-\rho\omega^2 & 0\end{bmatrix} \tag{9a}$$

with

$$\partial_x=\partial/\partial x,\quad \partial_{xx}=\partial^2/\partial xx \tag{9b}$$

where μ is the complex shear modulus, and ρ the density of the medium.

Welded boundary conditions at the irregular interface (see Fig. 1) require continuity of displacement and traction at each point on the interface. Therefore, a new displacement-stress vector $\mathbf{b}$, being measured with respect to the local tangent plane at each point on the interface, has to be introduced. The new displacement-stress vector takes the form (Kennet, 1972):

$$\mathbf{b}(x,f)=(\mathbf{I}+\frac{\partial f}{\partial x}\mathbf{Q}_0)\ \mathbf{B}(x,f) \tag{10}$$

where $\mathbf{b}$ and $\mathbf{B}$ are evaluated along the irregular interface located at the depth $z(x)$ defined by:

$$z(x)=z+f(x) \tag{11}$$

with z being the average depth of the interface and $f(x)$ being the lateral fluctuation of the interface (see Fig. 1). In Eq. (10), $\mathbf{I}$ is the unit matrix and $\mathbf{Q}_0$ is given by:

$$\mathbf{Q}_{0SH} = \begin{bmatrix} 0 & 0 \\ -\mu\partial_x & 0 \end{bmatrix} \qquad (12)$$

The irregular interface boundary condition can be expressed in the form:

$$\mathbf{b}_1(x,f) = \mathbf{b}_2(x,f) \qquad (13)$$

where subscripts indicate the respective media.

In order to obtain the approximation of Eq. (10), the scattered wave field $\mathbf{B}(x,f)$ (displacement-stress vector in the medium with irregular interface) is approximately related to the background field $\mathbf{B}(x,z)$ (displacement-stress vector in the medium with horizontal plane interface) by using a Taylor's expansion around the average interface depth $z(x) = z$, and then substituting into Eq. (10), and omitting terms of order higher than f and $\partial f/\partial x$, one obtains:

$$\mathbf{b}(x,f) = \{\mathbf{I} + f\mathbf{A}_0 + \frac{\partial f}{\partial x}\mathbf{Q}_0\}\mathbf{B}(x,z) \qquad (14)$$

Furthermore, introducing the new notations, $\mathbf{b}^0$ and $\mathbf{B}^0$, which represent the background wave fields, and denoting the first-order approximations of $\mathbf{b}$ and $\mathbf{B}$ by $\mathbf{b}^I$ and $\mathbf{B}^I$, respectively, one obtains the first-order approximation of Eq. (14) as:

$$\mathbf{b}^I(x,f) = \mathbf{B}^I(x,z) + \{f\mathbf{A}_0 + \frac{\partial f}{\partial x}\mathbf{Q}_0\}\ \mathbf{B}^0(x,z) \qquad (15)$$

with the following condition because $\mathbf{b}$ equals to $\mathbf{B}$ at the interface $z(x) = z$ in the case of horizontal plane interface ($f = 0$):

$$\mathbf{b}^0(x,z) = \mathbf{B}^0(x,z) \qquad (16)$$

The Fourier transform of Eq. (15) with respect to x, using the result that the Fourier transform of a product is the convolution of the Fourier transform, yields:

$$\mathbf{b}^I(\kappa,f) = \mathbf{B}^I(\kappa,z) + \int f(\kappa-\kappa')\mathbf{J}(\kappa,\kappa')\mathbf{B}^0(\kappa',z)d\kappa' \qquad (17)$$

where

$$\mathbf{J}(\kappa,\kappa') = \mathbf{A}_0(\kappa') + i(\kappa-\kappa')\mathbf{Q}_0(\kappa') \qquad (18)$$

with

$$\mathbf{J}_{SH} = \begin{bmatrix} 0 & \dfrac{1}{\rho C_s^2} \\ \rho C_s^2\kappa\kappa' - \rho\omega^2 & 0 \end{bmatrix} \qquad (19)$$

In Eq. (19), C_s is the complex S-wave velocity given by:

$$C_s = C_s^0(1 - iD_s) \qquad (20)$$

with C_s^0 being the elastic S-wave velocity and D_s being the ratio of the linear hysteretic damping for S-wave.

The irregular interface boundary condition given by Eq. (13) can be written for the first-order approximations in frequency wavenumber domain as:

$$\mathbf{B}_1^I(\kappa,z) = \mathbf{B}_2^I(\kappa,z) + \mathbf{S}(\kappa,z) \qquad (21a)$$

where

$$\mathbf{S}(\kappa,z) = \int_{-\infty}^{\infty} f(\kappa-\kappa')\mathbf{L}_{21}(\kappa,\kappa')\mathbf{B}_2^0(\kappa',z)d\kappa' \qquad (21b)$$

$$\mathbf{L}_{21} = \mathbf{J}_2 - \mathbf{J}_1 = -\mathbf{L}_{12} \qquad (21c)$$

Equation (21a) indicates that the presence of irregular interface results in discontinuity in the scattered wave field $\mathbf{B}(x,z)$ at $z(x) = z$. This discontinuity acts like a seismic source $\mathbf{S}$ which can be evaluated directly from the background wave field.

3.2 *Response of Ground with Irregular Free Surface*

Response of a single soil layer with irregular free surface resting on half space subjected to incident SH wave as shown in Fig. 2 is considered. For a free surface the traction has to vanish, so that for an irregular free surface $z(x) = f(x)$, $(z = 0)$, the scattered SH wave field takes the form:

$$\mathbf{b}(\kappa,f) = \mathbf{B}(\kappa,f) = col\ [\mathbf{U}(\kappa,f),\mathbf{0}] \qquad (22)$$

Then, the first-order approximation of the interface boundary condition given by Eq. (17) can be expressed as:

$$\mathbf{B}^I(\kappa,f) = \mathbf{B}^I(\kappa,0) + \int_{-\infty}^{\infty} f(\kappa-\kappa')\mathbf{J}(\kappa,\kappa')\mathbf{B}^0(\kappa',0)d\kappa' \qquad (23)$$

Making use of the propagator matrix $\mathbf{P}(\kappa,z,z_0)$ which satisfies (Kennet, 1972)

$$\frac{\partial}{\partial z}\mathbf{P}(\kappa,z,z_0) = \mathbf{A}_0(\kappa,z)\mathbf{P}(\kappa,z,z_0) \qquad (24a)$$

$$\mathbf{P}^{-1}(\kappa,z,z_0) = \mathbf{P}(\kappa,z_0,z) \qquad (24b)$$

the displacement-stress vector at the bedrock, $z(x) = H$, can be transferred to that at the free surface, $z(x) = 0$, such as:

$$\mathbf{B}(\kappa,0) = \mathbf{P}(\kappa,0,H)\mathbf{B}(\kappa,H) \qquad (25)$$

For an SH wave the propagator matrix is given by:

$$\mathbf{P}_{SH}(\kappa,z,z_0) = \begin{bmatrix} \cos\gamma(z-z_0) & \dfrac{\sin\gamma(z-z_0)}{\mu\gamma} \\ -\mu\gamma\sin\gamma(z-z_0) & \cos\gamma(z-z_0) \end{bmatrix} \qquad (26a)$$

where γ is the vertical (z axis) wavenumber given by:

$$\gamma = \sqrt{(\frac{\omega}{C_s})^2 - \kappa^2}\quad ;\quad Im(\gamma) \geq 0 \qquad (26b)$$

In order for the physical condition of a wave propagating in the positive direction of the vertical axis (z axis) to be zero at infinity, then the imaginary part of the vertical wavenumber $Im(\gamma)$ in Eq. (26b) must be positive.

It is assumed now that an input earthquake motion is an inclined SH wave with the amplitude v_{in} and the horizontal wavenumber κ_0. Then, the input motion (represented by displacement-stress vector) at the bedrock is expressed in the form:

$$\mathbf{B}(\kappa', H) = \mathbf{B}(\kappa_0, H)\ \delta(\kappa' - \kappa_0) \tag{27a}$$

where δ is the Delta function, and κ_0 is given by:

$$\kappa_0 = \frac{\omega \sin\theta}{C_s}. \tag{27b}$$

in which θ is the angle of the incident SH wave, measured clockwise from the vertical axis as shown in Fig. 2. For vertical incidence, $\theta = 0°$, then $\kappa_0 = 0$ from Eq. (27b). The input motion $\mathbf{B}$ at the bedrock is related to the incident SH wave such that:

$$\mathbf{B}(\kappa, H) = \mathbf{R}(\kappa)\ \mathbf{v_0}(\kappa, H) \tag{28a}$$

where $\mathbf{v_0}$ is the displacement vector in half space and $\mathbf{R}$ is a transform matrix from the displacement vector to the displacement-stress vector which are defined as:

$$\mathbf{v_0} = col\ [v_{in}\ ,\ v_{out}] \tag{28b}$$

$$\mathbf{R} = \begin{bmatrix} 1 & 1 \\ i\mu_1\gamma_1 & -i\mu_1\gamma_1 \end{bmatrix} \tag{28c}$$

in which v_{out} is the displacement amplitude of the outgoing SH wave in half space.

Substituting Eqs. (25) and (27) into (23), one obtains:

$$\mathbf{B}^I(\kappa, f) = \mathbf{P}(\kappa, 0, H)\mathbf{B}^I(\kappa, H)\delta(\kappa - \kappa_0)$$

$$+f(\kappa - \kappa_0)\mathbf{J}(\kappa, \kappa_0)\mathbf{P}(\kappa_0, 0, H)\mathbf{B}^0(\kappa_0, H) \tag{29}$$

By considering the boundary conditions in which a) traction vanishes at the free surface and b) the incident SH wave is specified at the bedrock, these, together with the relationship of Eq. (28a), then Eq. (29) can be more explicitly expressed in the partitioned form:

$$\begin{bmatrix} \mathbf{U}^I(\kappa, f) \\ \mathbf{0} \end{bmatrix} = \begin{bmatrix} \mathbf{F}_{11} & \mathbf{F}_{12} \\ \mathbf{F}_{21} & \mathbf{F}_{22} \end{bmatrix} \begin{bmatrix} v_{in}(\kappa, H) \\ v^I_{out}(\kappa, H) \end{bmatrix} \delta(\kappa - \kappa_0)$$

$$+f(\kappa - \kappa_0) \begin{bmatrix} \mathbf{J}_{11} & \mathbf{J}_{12} \\ \mathbf{J}_{21} & \mathbf{J}_{22} \end{bmatrix} \begin{bmatrix} \mathbf{F}^0_{11} & \mathbf{F}^0_{12} \\ \mathbf{F}^0_{21} & \mathbf{F}^0_{22} \end{bmatrix} \begin{bmatrix} v_{in}(\kappa_0, H) \\ v^0_{out}(\kappa_0, H) \end{bmatrix} \tag{30}$$

where $\mathbf{F}_{ij}$ represents the (ij)th component of the matrix $\mathbf{PR}$, and $\mathbf{F}^0_{ij} = \mathbf{F}_{ij}(\kappa_0, 0, H)$. In Eq. (30), the relation $v^I_{in}(\kappa, H) = v_{in}(\kappa, H)$ is used because the incident motion displacement is specified at the bedrock. From Eq. (30), one can obtain the scattered wave field $\mathbf{U}^I(\kappa, f) = v^I(\kappa, f)$ at irregular free surface such as:

$$v^I(\kappa, f) = [\frac{p}{\cos\gamma H}\delta(\kappa - \kappa_0)$$

$$+f(\kappa - \kappa_0)q\gamma_0\frac{\sin\gamma H}{\cos\gamma_0 H \cdot \cos\gamma H}(\frac{\kappa\kappa_0 - \kappa_0^2}{\gamma\gamma_0} - \frac{\gamma_0}{\gamma})]2v_{in} \tag{31a}$$

where

$$p = \frac{1}{1 + i\frac{\mu\gamma}{\mu_1\gamma_1}\frac{\sin\gamma H}{\cos\gamma H}} \tag{31b}$$

$$q = p_0\frac{1 - i\frac{\mu\gamma}{\mu_1\gamma_1}\frac{\cos\gamma H}{\sin\gamma H}}{1 + i\frac{\mu\gamma}{\mu_1\gamma_1}\frac{\sin\gamma H}{\cos\gamma H}} \tag{31c}$$

In Eq. (31a), the first term of the right-hand side represents the response displacement at free surface with a horizontal plane surface, and the second term the scattered wave displacement due to irregular free surface.

4 ACCURACY OF THE PERTURBATION METHOD

Since the analytic expression in Eq. (31) is based on the perturbation method, it is necessary to examine the accuracy and to reveal the practical limits imposed on Eq. (31). For this purpose, the response displacements at free surface are compared with those derived by the direct boundary element method. Four ground models are used in this comparison, each having the free surface topography of sinusoidal shapes with different heights and wave lengths. The accuracy of the direct boundary element method used in this comparison is described in subsection 4.1, and then, the accuracy of the perturbation method is examined in subsection 4.2.

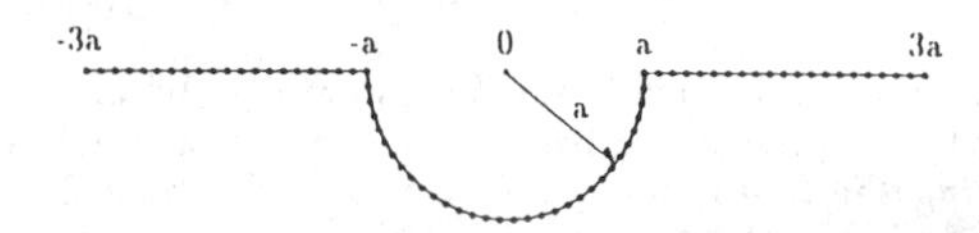

Fig. 3(a). Distribution of boundary elements along the surface of a semicylindrical canyon (71 constant elements are used)

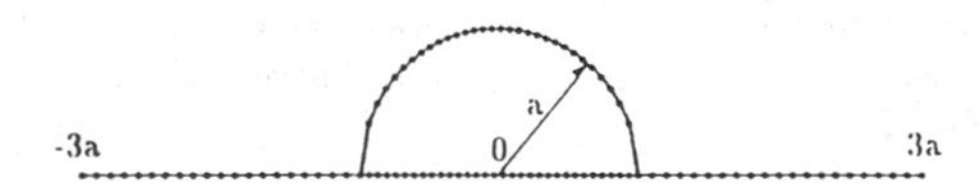

Fig. 3(b). Distribution of boundary elements along the surface of a semicylindrical hill (71 constant elements are used)

4.1 *Accuracy of the Direct Boundary Element Method*

The boundary element method used in this study is based on the formulation (Wong and Jennings, 1975). The frequency responses of the two-dimensional semi-cylindrical canyon and hill as shown in Fig. 3 are computed by the boundary element method, and are compared with the exact solutions (Trifunac, 1973; Men and Yuan, 1992). Fig. 3 shows the distribution of boundary elements along the surface of a canyon and hill used in this study. A total of seventy-one constant elements are used for calculation. Fig. 4 shows an example on the frequency response distribution along the surface in and around the canyon by using discrete points (boundary element method), together with the exact solutions (Trifunac, 1973) using solid curves. Only the case for the dimensionless

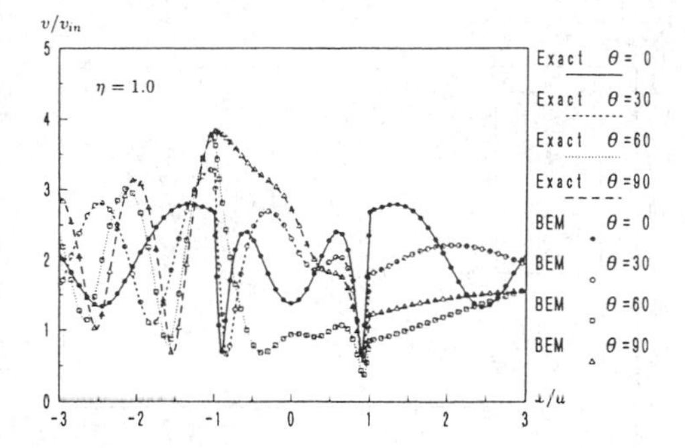

Fig. 4. Comparisons in the frequency responses of semicylindrical canyon computed by boundary element method (discrete points) and the exact solution (lines) by Trifunac (1973)

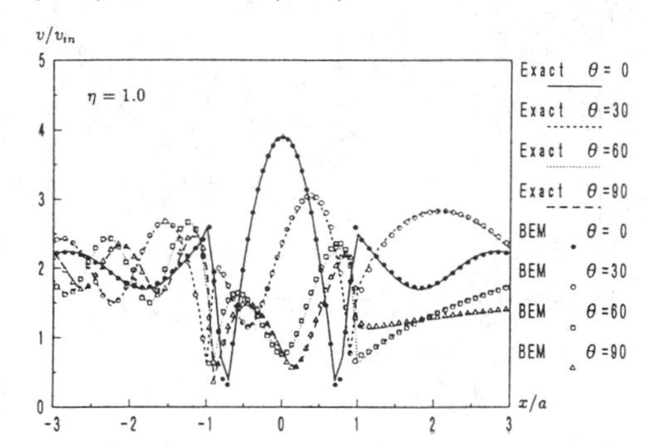

Fig. 5. Comparisons in the frequency responses of semicylindrical hill computed by boundary element method (discrete points) and the exact solution (lines) by Yuan and Men (1992)

frequency $\omega a/\pi C_s = 1.0$ is shown, where one half of the S-wave length is equal to the radius a of the canyon. Note that this is the dimensionless frequency η in Trifunac's paper. The horizontal axis is the dimensionless space coordinate x/a, and the vertical axis represents the amplitude of frequency response relative to the incident wave amplitude v_{in}. In Fig. 4 the comparisons are shown for four different incidence angles θ (0°, 30°, 60° and 90°). For the same conditions in the comparisons using a semicylindrical canyon in Fig. 4, the comparisons using a semicylindrical hill are also performed. The results of comparison between the boundary element method and the exact solution (Men and Yuan, 1992) are shown in Fig. 5. Excellent agreements are observed from Figs. 4 and 5 in the results by both methods, indicating the accuracy of the boundary element method used in this study.

4.2 *Accuracy of the Perturbation Method*

The analytic expression in Eq. (31) for the response of the irregular free surface to the incident SH wave may be applied to the response of a semicylindrical canyon and hill used in subsection 4.1 for examining the accuracy of the boundary element method. However the direct comparisons between the perturbation result in Eq. (31) and the exact solutions for a semicylindrical canyon and hill are not possible. The numerical Fourier transform is necessary for comparisons because the perturbation result in Eq. (31) is expressed in the frequency-wavenumber domain ($\omega - \kappa$), while the results of the boundary element method and the exact solutions for a canyon and hill are expressed in the frequency-space domain ($\omega - x$). In order to perform the direct comparison between the perturbation method and the boundary element method without using the numerical Fourier transform, the frequency responses to incident SH wave are compared for the half space ground models with irregular free surface consisting of sinusoidal shapes having four different heights and wave lengths. The four half space ground models are shown in Fig. 6, and the free surface heights and wave lengths in each model are indicated in Table 1.

Table 1.

Model	Height (m)	Wave Length (m)
A	10	320
B	10	160
C	10	80
D	20	80

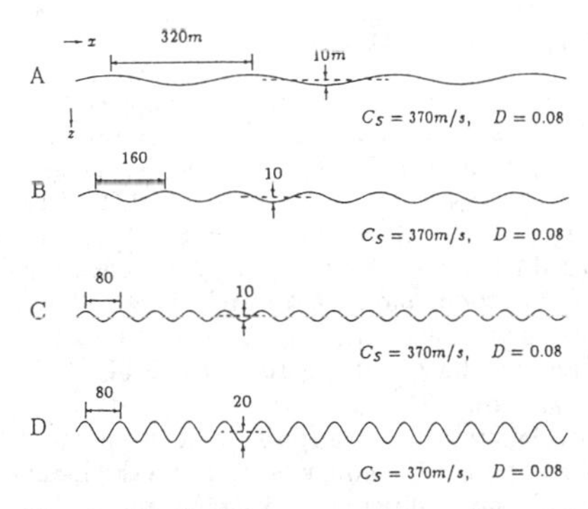

Fig. 6. The half space ground models with four different sinusoidal topographies used in the examination of accuracy of the perturbation method

For the half space ground model with irregular free surface consisting of sinusoidal shapes, the analytic expression of Fourier transform is possible, and then the response displacement in frequency-space domain with $\kappa_0 = 0$ can be obtained as:

$$v^I(x, f(x), \omega) = \int_{-\infty}^{\infty} v^I(\kappa, f(\kappa)) e^{i\kappa x} d\kappa$$

$$= [1 + f_0 \frac{(\frac{\omega}{C_s})^2}{\sqrt{(\kappa^*)^2 - (\frac{\omega}{C_s})^2}} \cos \kappa^* x] 2 v_{in} \qquad (32)$$

where f_0 is the height of sinusoidal topography of free surface, and $\kappa^*=2\pi/L$ is the horizontal wavenumber of sinusoidal topography of free surface with L being the periodicity length (wave length) of sinusoidal topography. In the case of a plane free surface, $f_0 = 0$, and then the frequency response displacement is constantly $2v_{in}$ in every position x. In the case of sinusoidal free surface the spatial variation of the frequency response displacement is sinusoidal ($\cos \kappa^* x$) which is the same as the spatial variation of free surface topography since the perturbation approach in this study truncates the higher order terms. In Eq. (32), for the frequency that satisfies the condition $\kappa^* = \omega/C_s$, the response displacement approaches infinity. However this phenomena is caused by neglecting the higher order scattering waves. To remove this unrealistic higher response to the frequency $\omega = \kappa^* C_s$, material damping is taken into consideration in Eq. (32), even though the half space ground model used in the computation by the boundary element method has no material damping. In the comparison the material damping ratio is assumed to be $D = 0.08$ in both methods.

In the perturbation method the analytic expression is obtained by neglecting terms of order higher than irregular height f and slope $\partial f/\partial x$. Therefore, the height and the slope are the key parameters for the bounds of accuracy of the perturbation method. In this comparison, the parameters α and β are introduced which correspond to the relative height and the approximate slope defined as:

$$\alpha = \frac{f_{max}}{S_{wavelength}} = \frac{f_{max}\omega}{2\pi C_s}, \quad \beta = \frac{4 f_{max}}{L} \tag{33}$$

where f_{max} is the maximum height of irregular surface, C_s is the S-wave velocity, and L is the minimum periodicity length of irregular surface. In the models used in the comparisons, β = 0.125, 0.25, 0.50, and 1.00, since f_{max} = 10, 20 m, and L = 80, 160, and 320 m are used (Table 1). By changing the excitation frequency in the ground model with each slope parameter β, ten cases with different height parameter α are computed, i.e, α= 0.02, 0.06, 0.10, 0.12, 0.14, 0.16, 0.18, 0.20, 0.22, and 0.24.

In order to facilitate the comparison of the perturbation method and the boundary element method, the following two L_2 norm differences are computed:

$$ABS_{L_2 Error} = 100 \sqrt{\frac{\sum_{x=1}^{n}(|v_B| - |v_P|)^2}{\sum_{x=1}^{n}|v_B|^2}} \tag{34a}$$

$$RE_{L_2 Error} = 100 \sqrt{\frac{\sum_{x=1}^{n}[Re(v_B) - Re(v_P)]^2}{\sum_{x=1}^{n}Re(v_B)^2}} \tag{34b}$$

where v_B is the frequency response displacement by the boundary element method, v_P is the frequency response displacement by the perturbation method, and Re represents the real part of the complex variable.

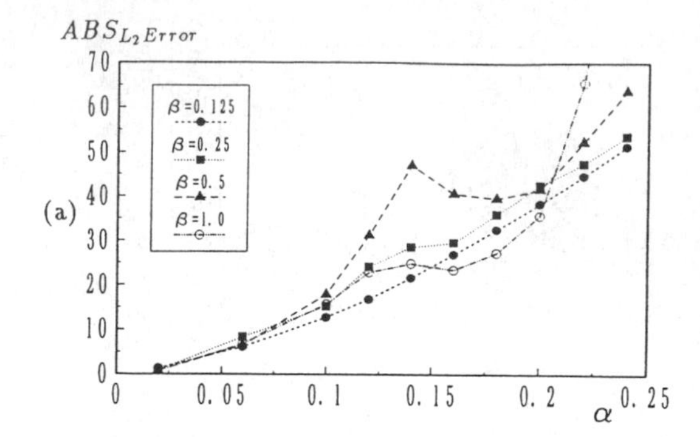

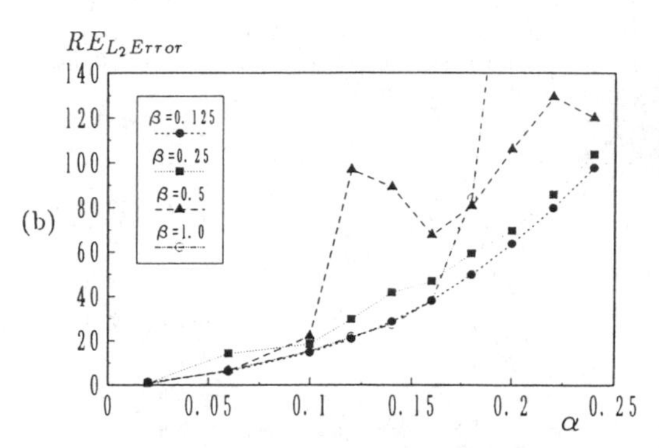

Fig. 7. L_2 norm comparison of the perturbation method and the boundary element method: (a) $ABS_{L_2 Error}$; (b) $RE_{L_2 Error}$

The two L_2 norm differences are plotted in Fig. 7 against ten height parameters α for the four slope parameters β. L_2 norm differences increase rapidly when increasing the height parameter and the slope parameter. In this particular example it may be acceptable for the values of $\alpha \leq 0.10$ and $\beta \leq 0.25$. Therefore, the perturbation solution may be acceptable for the following ranges:

$$f_0 \leq \alpha\lambda, \quad f_0 \leq \frac{\beta}{4}L, \quad \alpha = 0.1, \quad \beta = 0.25 \tag{35}$$

where f_0 is the height of sinusoidal topography of free surface, λ is the S-wave length, and L is the periodicity length (wave length) of sinusoidal free surface topography. Fig. 8 shows two examples of results of comparison for the two height parameter α = 0.06, 0.10, with constant slope parameter $\beta = 0.25$. It is observed from Fig. 8 that the differences in both methods are small.

5 FREQUENCY WAVENUMBER SPECTRUM OF STOCHASTIC WAVE AT THE IRREGULAR FREE SURFACE OF A SINGLE LAYER DUE TO INCIDENT SH WAVE

By assuming that the incident SH wave motion is the

(a)
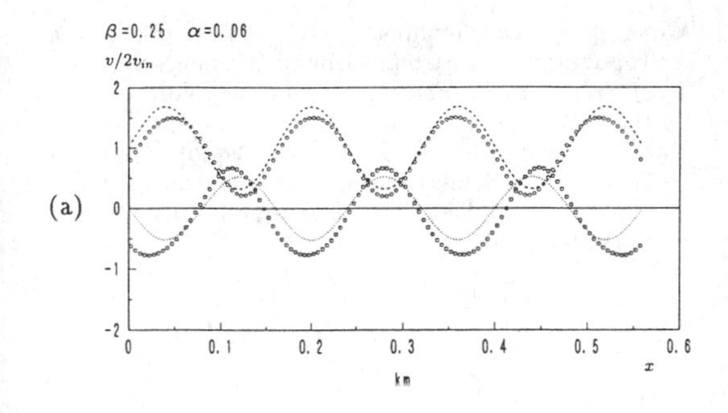

(b)
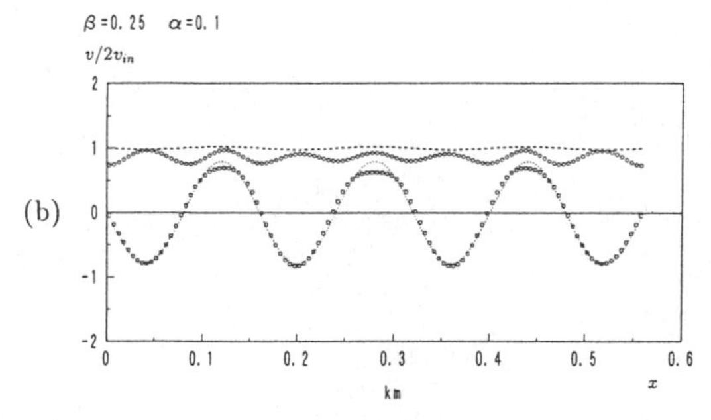

Fig. 8. Examples of the frequency responses along the sinusoidal surface computed by the perturbation method (dotted lines) and the boundary element method (discrete points): (a) $\alpha = 0.06$; (b) $\alpha = 0.1$

stationary stochastic process and the fluctuation of irregular free surface is the stochastic field with zero mean, eventually the response displacement wave field may be a stationary and homogeneous stochastic wave. Then, its frequency wavenumber spectrum is obtained as:

$$S_{vv}(\kappa, \omega) = |v(\kappa, \omega)|^2 \tag{36}$$

Substitution of Eq. (31a) into Eq. (36) yields:

$$S_{vv}(\kappa, \omega) = [|A|^2\delta(\kappa - \kappa_0) + |B|^2 S_{ff}(\kappa - \kappa_0)]S_{v_{in}v_{in}}(\omega) \tag{37a}$$

where S_{ff} is the wavenumber spectrum of the fluctuation of irregular free surface $f(x)$. The quantities A and B appearing in Eq. (37) are given by:

$$A = \frac{2p}{\cos \gamma H} \tag{37b}$$

$$B = \frac{2q\gamma_0 \sin \gamma H}{\cos \gamma H \cos \gamma_0 H}\left(\frac{\kappa\kappa_0 - \kappa_0^2}{\gamma\gamma_0} - \frac{\gamma_0}{\gamma}\right) \tag{37c}$$

where p and q are given in Eqs. (31b) and (31c), and γ_0 corresponds to the value of γ when $\kappa = \kappa_0$. The validity ranges of Eq. (37) may be estimated from Eq. (35) by changing f_0 and L to the RMS (root mean square) height and the mean wave length of stochastic field $f(x)$ which are obtained from the wavenumber spectrum moments of stochastic field $f(x)$ such as:

$$f_{RMS} = \sqrt{\int_{-\infty}^{\infty} S_{ff}(\kappa)d\kappa} \tag{38a}$$

$$L_{MEAN} = 2\pi\sqrt{\frac{\int_{-\infty}^{\infty} S_{ff}(\kappa)d\kappa}{\int_{-\infty}^{\infty} \kappa^2 S_{ff}(\kappa)d\kappa}} \tag{38b}$$

where f_{RMS} is the RMS height of $f(x)$, and L_{MEAN} is the mean wave length of $f(x)$.

6 CONCLUSIONS

In this paper, a closed form analytic expression of the frequency wavenumber spectrum is established. The corresponding earthquake ground motion is produced by the earthquake response of a single soil layer of irregular thickness resting on half space. The earthquake motion, assumed by a single plane SH wave, is transmitted to the soil layer from the half space. The analytic expressions of the earthquake responses of irregular ground are based on the perturbation method, and their accuracy is examined by comparison with the responses computed by the direct boundary element method for many rough surface models consisting of topography with sinusoidal shape. The error in the perturbation method is acceptable for the height of sinusoidal topography of less than about 10 percent of the shortest S-wave length and the slope of less than about 0.25. Finally, an analytic expression for the frequency wavenumber spectrum of the stochastic wave at irregular ground surface is then derived. Although the present paper considers the inhomogeneity of local soil layer caused by the lateral variation of the thickness of soil layer only, the effect of the inhomogeneity caused by the spatial variation of soil properties on the ground motions can be also studied using the perturbation method presented in this paper (Harada et al., 1990). The sample stochastic wave form can be efficiently simulated on the basis of the spectral representation method where the simulated wave consists of the superposition of a number of plane waves having amplitudes consistent with the frequency wavenumber spectrum (Shinozuka et al., 1987).

7 ACKNOWLEDGEMENT

The author wishes to thank Miyazaki University graduate students Satoshi Teramoto and Wilson Gorges for their kind assistance in carrying out, respectively, the numerical computations and editing necessary for the completion of the manuscript.

8 REFERENCES

Abrahamson, N. A. Estimation of Seismic Wave Coherency and Rupture Velocity Using The SMART-1 Strong-Motion Array Recordings. UCB/EERC, 85/02, Earthquake Engineering Research Center, University of California, Berkeley, 1985.

Abrahamson, N. A., Schneider, J. F., and Stepp,

J. C. Empirical Spatial Coherency Functions for Application to Soil-Structure Interaction Analysis. *Earthquake Spectra, Professional Journal of the Earthquake Engineering Research Institute,* Vol. 7, No. 1, pp. 1-27, 1991.

Bochner, S. Lectures on Fourier Integrals, *Annals Mathematical Studies,* No.42, Princeton University Press, 1956.

Bouchon, M. Discrete Wavenumber Representation of Elastic Wave Fields in Three-Space Dimensions. *Journal of Geophysical Research,* Vol. 84, pp. 3609-3614, 1979.

Harada, T. Probabilistic Modeling of Spatial Variation of Strong Earthquake Ground Displacements. *Proceedings of 8th World Conference on Earthquake Engineering,* Prentice Hall, pp. 605-612, 1984.

Harada, T., and Fugasa, T. Characteristics of Seismic Responses of 3-Dimensional Ground with Stochastic Soil Properties. Memoirs of the Faculty of Engineering, Miyazaki University, No. 36, September, 1990.

Harada, T., and Shinozuka, M. Ground Deformation Spectra. *Proceedings of 3rd U. S. National Conference on Earthquake Engineering,* Earthquake Engineering Research Institute, pp. 2191-2202, 1986.

Harichandran, R. Local Spatial Variation of Earthquake Ground Motion. *Earthquake Engineering and Soil Dynamics - Recent Advances in Ground Motion Evaluation,* edited by Lawrence Vom Thun, J., Geotechnical Special Publication, No. 20, ASCE, pp. 203-217, 1988.

Kashima, N., Kawashima, K., Harada, T., Isoyama, R., and Masuda, S. Soil-Structure Interaction and Its Implication for Seismic Design of Structures. *Proceedings of 9th World Conference on Earthquake Engineering,* Prentice Hall, pp. 605-612, 1984.

Kawaguchi, K., Masuda, S., Isoyama, R., and Saeki, M. Aseismic Design of Akashi Kaikyo Bridge Foundations. *Proceedings of New Zealand-Japan Workshop on Base Isolation of Highway Bridges,* Technology Research Center for National Land Development, pp. 52-63, 1987.

Kennett, B. L. N. Seismic Wave Scattering by Obstacles on Interfaces. *Geophysical Journal of the Royal Astronomical Society,* Vol. 28, pp. 249-266, 1972.

Loh, C. H., Penzien, J., and Tsai, Y. B. Engineering Analysis of SMART-1 Accelerograms. *Earthquake Engineering and Structural Dynamics,* Vol. 10, pp. 579-591, 1982.

Public Works Research Institute. A Proposal for Earthquake Resistant Design Method. Technical Memorandum of PWRI, Ministry of Construction, No. 1185, March 1977 (in Japanese).

Shinozuka, M., Deodatis, G., and Harada, T. Digital Simulation of Seismic Ground Motion. *Stochastic Approaches in Earthquake Engineering,* edited by Lin, Y. K. et al., Springer-Verlag, pp. 252-298, 1987.

Trifunac, M. D. Scattering of Plane SH Waves by a Semicylindrical Canyon, *Earthquake Engineering and Structural Dynamics,* Vol. 1, pp. 267-281, 1973.

Wong, H. L. and Jennings, P. C. Effect of Canyon Topography on Strong Ground Motion, *Bulletins of Seismological Society of America,* Vol. 65, pp. 1239-1257, 1975.

Yuan, X. and Men, F. L. Scattering of Plane SH Waves by a Semicylindrical Hill, *Earthquake Engineering and Structural Dynamics,* Vol. 21, pp. 1091-1098, 1992.

Structural Safety & Reliability, Schuëller, Shinozuka & Yao (eds) © 1994 Balkema, Rotterdam, ISBN 90 5410 357 4

Seismic risk analysis based on nonstationary Poisson occurrence model

Li-Ling Hong
Department of Civil Engineering, National Cheng Kung University, Tainan, Taiwan

Shyue-Wen Guo
Sinotech Engineering Consultants, Inc., Taipei, Taiwan

ABSTRACT: Compared to a stationary Poisson process, the nonstationary Poisson process characterized with a U-shape mean occurrence rate function of the recurrence time is more flexible in modelling the occurrence of clustering earthquakes. The parameters of such a U-shape function are determined by plotting the data points of the recurrence time on an exponential probability paper. The seismic risk analysis is performed with a little modification to the case of a stationary Poisson occurrence model, and then the hazard curves induced by using different Poisson occurrence models are compared for a simple case.

1 INTRODUCTION

In modelling the earthquake occurrence, the stationary Poisson process has been used extensively (Cornell 1968, Der Kiureghian & Ang 1977, Anagnos & Kiremidjian 1988) because of its simplicity and easy application. Although the stationary Poisson occurrence model can be accepted in some seismic regions, the assumption of constant mean occurrence rate is obviously invalid where earthquakes occur in sequence of clusters. Each typical cluster of earthquakes includes a few foreshocks, one mainshock, and several aftershocks, and any two successive clusters are separated by a relatively long waiting period. Thus, the mean occurrence rates within and between clusters are quite different. Kameda and Ozaki (1979) proposed a double Poisson renewal model with two distinct mean occurrence rates to simulate the earthquake occurrences during the waiting period and within the incoming cluster, whereas a continuous mean occurrence rate function, such as the case of Weibull interarrival times (Savy *et al.* 1980), is used for the same reason.

Instead of monotonically increasing functions, a U-shape function representing the variation of the instantaneous mean occurrence rate is presented in this study. Shortly after an earthquake occurs, the mean occurrence rate would be high because the next earthquake will follow soon with great possibility as a mainshock or aftershock. As the time to the last earthquake passes, the hazard of incoming earthquakes is reduced because it is more and more believed that this current cluster of earthquakes will end. The mean occurrence rate then reaches a minimum during the waiting period. However, it would increase eventually in the beginning of the following cluster of earthquakes. Therefore, the U-shape function with three different parts, namely decreasing, constant, and increasing functions of the recurrence time, represents the varying mean occurrence rate within the current cluster, between clusters, and a little before the next cluster, respectively.

The other non-Poissonian models, such as the slip- and time-predictable models (Kiremidjian & Anagnos 1984, Anagnos & Kiremidjian 1984), generally consider the dependence between the magnitude and the interarrival time of earthquakes. Their assumptions may be more reasonable in the geophysical sense, but their applications are limited due to the difficulty in either the parameter estimation or the seismic risk evaluation. On the contrary, a memoryless nonstationary Poisson model takes the advantage in evaluating the seismic hazard, whereas the parameter estimation is slightly laborious compared to that in a stationary Poisson model.

2 NONSTATIONARY POISSON PROCESS

A nonstationary Poisson process is a counting process based on the assumption of independent increments and at most one occurrence at any instant of time. The number of occurrences in the time interval $[0, t]$, $N(t)$, is a random variable with a probability mass function as

$$P[N(t)=n]=\frac{[\int_0^t \nu(\tau)d\tau]^n}{n!}\exp[-\int_0^t \nu(\tau)d\tau] \qquad n=0,1,2,... \quad (1)$$

While in the time interval $[t_1, t_2]$, the probability mass function becomes

$$P[N(t_2)-N(t_1)=n]=\frac{[\int_{t_1}^{t_2} \nu(\tau)d\tau]^n}{n!}\exp[-\int_{t_1}^{t_2} \nu(\tau)d\tau] \qquad n=0,1,2,... \quad (2)$$

where $\nu(\tau)$ is the instantaneous mean occurrence rate.

The recurrence time T of an event following a nonstationary Poisson process has the distribution as

$$F_T(t)=1-\exp[-\int_0^t \nu(\tau)d\tau] \quad t\ge 0 \quad (3)$$

Therefore, the mean occurrence rate function can be derived from the cumulative distribution function of the recurrence time, *i.e.*,

$$\nu(t)=\frac{d}{dt}[\int_0^t \nu(\tau)d\tau]=\frac{d}{dt}\{-\ln[1-F_T(t)]\} \quad t\ge 0 \quad (4)$$

It must be emphasized that t in Eq. (4) is the time elapsed since the last occurrence. The differentiation in Eq. (4) gives

$$\nu(t)=\frac{f_T(t)}{1-F_T(t)}=h(t) \quad t\ge 0 \quad (5)$$

where $f_T(t)$ and $h(t)$ are the probability density function and the hazard function of the recurrence time T, respectively.

Eq. (5) indicates that the mean occurrence rate function of a nonstationary Poisson process is exactly the hazard function of the recurrence time. Once $\nu(t)$ or $h(t)$ is well known, the nonstationary Poisson process is completely defined. Based on Eq. (4), the next section will present a method to obtain $\nu(t)$ in which earthquakes occur in sequence of clusters.

Two special cases of nonstationary Poisson processes are notable. The first one is that $\nu(t)$ is a constant corresponding to a stationary Poisson process, where the recurrence time is an exponential variate. The other one is the process having

$$\nu(t)=\frac{k}{w}(\frac{t}{w})^{k-1} \quad t\ge 0$$

and

$$F_T(t)=1-\exp[-(\frac{t}{w})^k] \quad t\ge 0$$

i.e., its recurrence time is Weibull distributed with parameters k and w. When k = 1 in the Weibull distribution, the Poisson process becomes stationary.

3 U-SHAPE HAZARD FUNCTION

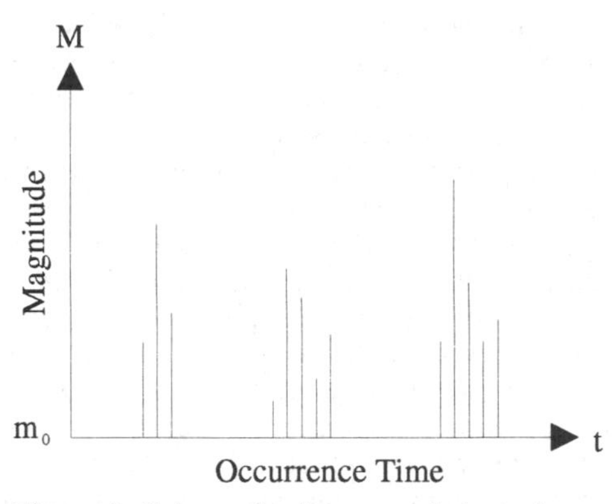

Figure 1. Schematic diagram of clustering earthquakes

To predict the failure probability of a piece of equipment under operation, the U-shape hazard function is always used. In the beginning of operation and after long usage the higher hazards are due to poor manufacture and aging, respectively. Observing earthquake sequences also find its application in modelling the recurrence times of clustering earthquake events. Fig. 1 shows the schematic representation of earthquake clusters, in which the recurrence times of earthquakes are characterized with either relatively short or relatively long values. Therefore, the ideal hazard function is U-shaped, as shown in Fig. 2. The hazard is the lowest in the waiting period between clusters, whereas it is higher immediately after an earthquake occurs and when the time elapsed since the last earthquake occurrence is longer than a representative waiting time.

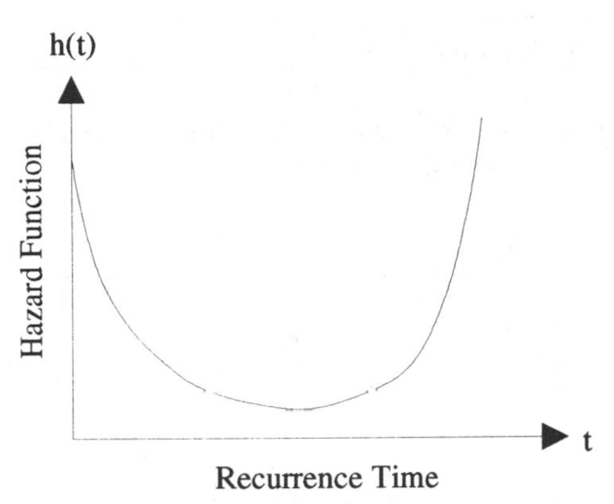

Figure 2. U-shape hazard function

Theoretically, three different formulae should be decided to simulate the decreasing, constant, and increasing behavior of the U-shape function; but, unfortunately, the recorded earthquakes are not so sufficient as to determine the two points of continuity and the associated parameters in each formula. Among many feasible continuous functions, a hyperbolic cosine function is chosen here, *i.e.*,

$$h(t) = \nu(t) = ab\cosh(bt - c) \tag{6}$$

where a, b and c are three parameters estimated through a nonlinear regression to the data of recurrence times plotted on an exponential probability paper by using the following equation:

$$-\ln[1 - F_T(t)] = a[\sinh(bt - c) + \sinh c] \tag{7}$$

Eq. (7) is the integration of Eq (6). When Eq. (7) is plotted in an exponential probability paper, a and b are the scale parameters, and c is the location parameter in the recurrence time axis.

4 CASE STUDY

The seismic risk can be evaluated as follows:

$$P(Y > y) = 1 - \exp[-\sum_{i=1}^{n_s} p_i \int_{T_i}^{T_i+t} \nu_i(\tau)d\tau] \tag{8}$$

where y is the required intensity, say the peak ground acceleration, at a given site; n_s is the number of potential seismic zones nearby, in which the numbers of earthquake occurrences are assumed to be independent; p_i is the probability of $Y > y$ when an earthquake occurs in Zone i; T_i is the time elapsed since the last earthquake occurrence in Zone i; and $\nu_i(\tau)$ is the mean occurrence rate function in Zone i. The procedures to calculate p_i are different in the Point Source Model (Cornell 1968) and Fault-Rupture Model (Der Kiureghian & Ang 1977).

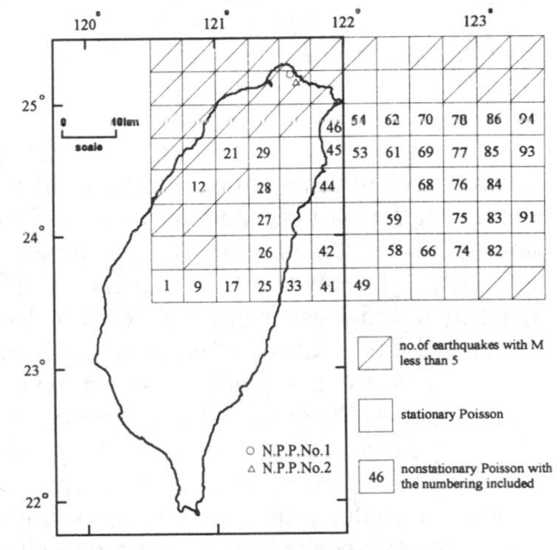

Figure 3. Seismogenic zoning scheme

In order to compare the seismic hazards induced by using different nonstationary Poisson occurrence models, the seismic risk analysis is performed for the site of Taiwan Power Company's Nuclear Power Plant No. 2 located at Kuosheng in the North of Taiwan. Since no strong information to identify the fault directions and types of seismic sources, all the seismic regions nearby are modelled as Type III areal sources in the Fault-Rupture Model. Furthermore, as shown in Fig. 3, the source area is divided into many small grids, each of 0.25° latitude by 0.25° longitude, and all the future earthquakes are assumed to occur at the centers of the grid squares, from which the epicenter distances are taken.

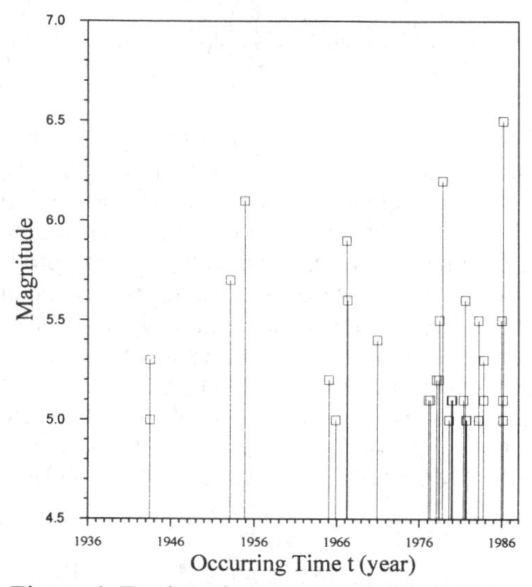

Figure 4. Earthquake sequence in Zone 46

Also shown in Fig. 3, there are many grids where the occurrences of earthquakes fail to pass the test of the assumed stationary Poisson model, and among which Zone 46 is the one nearest to the site. Twenty-eight earthquakes with the magnitude ≥ 5.0 occur in Zone 46 from 1936 to 1988, and the sequence of earthquakes is shown in Fig. 4. At the first glance of Fig. 5, where the data points of the earthquake recurrence time in Zone 46 are plotted, the assumption that the recurrence time is Weibull distributed can be reasonably accepted. In view of Fig. 6, however, the use of a U-shape hazard function to simulate the recurrence of earthquakes in Zone 46 is better than that of the Weibull distribution. In Fig. 6, the regression curve of the Weibull distribution is obtained in a Weibull probability paper, *i.e.*, Fig. 5, but the regression curve of the U-shape hazard function is decided according to the data points on an exponential probability paper by using Eq. (7).

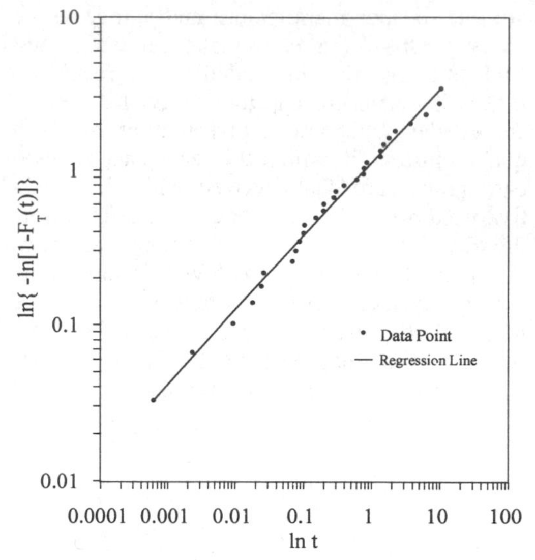

Figure 5. Recurrence times in Zone 46 plotted on Weibull probability paper

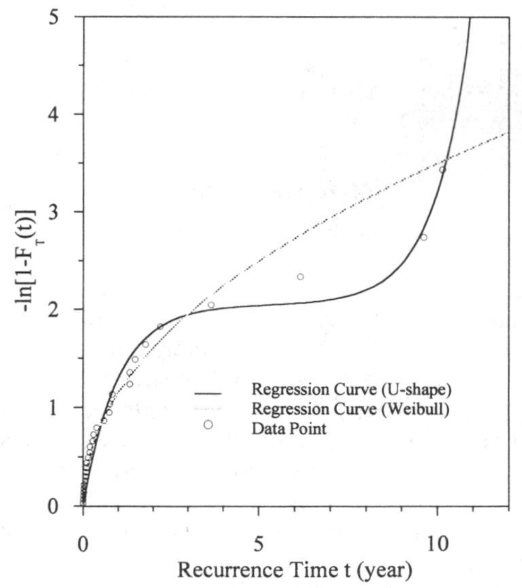

Figure 6. Recurrence times in Zone 46 plotted on exponential probability paper

A modified Kanai's attenuation equation is used here, *i.e.*,

$$y = 0.0305 \exp(0.4144 m)(d_{\min}^2 + 0.8512 h^2 + 10)^{-0.2883} \tag{9}$$

with $\sigma_{\ln Y}$ = 0.691. In Eq. (9), y is the peak ground acceleration in the unit of g, m is the local magnitude, $d_{\min}$ is the shortest horizontal distance in km from a site to the fault line, h is the focal depth also in km. Because the epicentral distance and the focal depth play the different roles on the seismic wave propagation, they are separated in the attenuation equation. Such a separation reduces the value of $\sigma_{\ln Y}$ slightly (Hong *et al.* 1992). Moreover, the use of the shortest horizontal distance in Eq. (9), instead of the epicentral distance, is consistent with the assumption of the Fault-Rupture Model. In the nonlinear regression analysis to obtain Eq. (9), a special scheme is used to determine the shortest horizontal distance (Hong *et al.* 1992) if the fault direction of the recorded earthquake is unknown. Finally, the risk calculated from Eq. (8) should be corrected when the uncertainty associated with Eq. (9) is considered (Der Kiureghian & Ang 1977).

As a simple case for comparison, the seismic risk curves induced solely by the future earthquakes in Zone 46 are evaluated at the site of Nuclear Power Plant No. 2 by applying three different stationary and nonstationary Poisson occurrence models, *i.e.*, the recurrence time of an exponential distribution, a Weibull distribution, and a U-shape hazard function, respectively. All

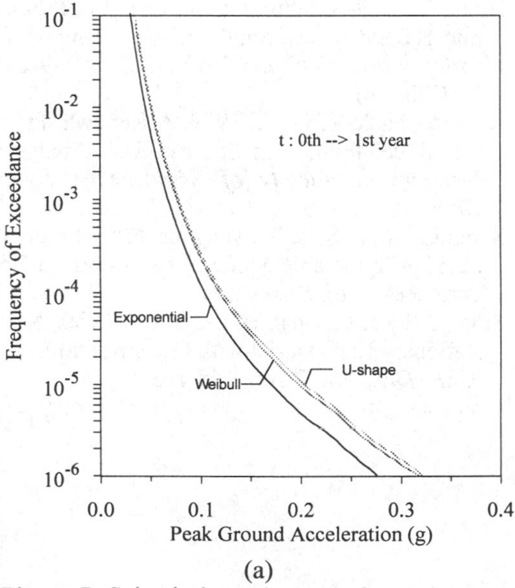

(a)

Figure 7. Seismic hazard curves for various occurrence models

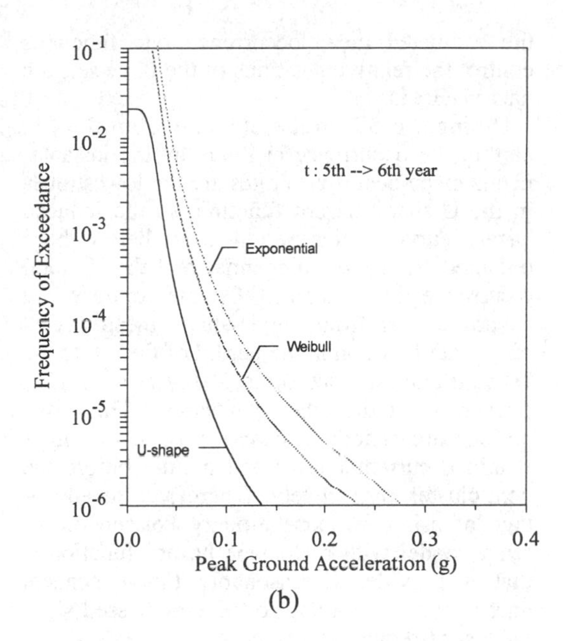

(b)

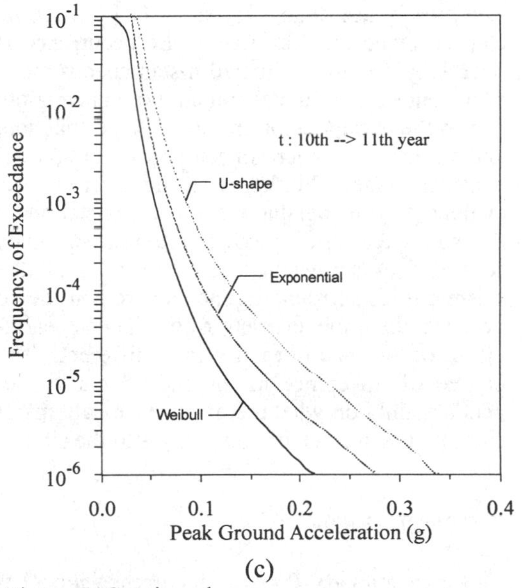

(c)

Figure 7. Continued

the other required seismic parameters are in Eastern International Engineers, Inc. (1984). Figs. 7(a)-(c) show the results in the same period of 1 year but differed in the time elapsed to the last earthquake occurring in Zone 46, *i.e.*, 0, 5 and 10 years, respectively. It can be shown in Figs. 7(a)-(c) and demonstrated by Eq. (8) that the estimated mean occurrence rate functions control the relative positions of the three seismic hazard curves.

During the 5th to the 6th year from the last earthquake occurrence in Zone 46, the instantaneous mean occurrence rates are the lowest ones in the U-shape hazard function, so the induced hazard curve in this period is far below those obtained from the exponential and the Weibull recurrence time variates. On the contrary, the hazard curves from the U-shape mean occurrence rate function in the period of the 0th to the 1st year and the 10th to the 11th year are higher than those of the other two cases. These two periods are exactly corresponding to the times within a current cluster and a little before the next cluster, respectively. Therefore, the advantage of using the nonstationary Poisson occurrence model with a U-shape hazard function is that it provides a reasonably time-dependent hazard curve according to the time passed since the last earthquake occurrence.

The maximum recorded recurrence time in Zone 46 is less than 11 years. Once the time elapsed since the last earthquake occurrence is over 11 years, the estimated instantaneous mean occurrence rate will have much more uncertainty due to the extrapolation in the U-shape function, and so does the other nonstationary Poisson occurrence model . Furthermore, the hazard curves evaluated with the stationary and nonstationary Poisson occurrence models are not so much separated as shown in Figs. 7(a)-(c) if all the seismic grids surrounding the site are considered because the time elapsed from the last earthquake occurrence in each grid is different. The degree of difference in the hazard curves depends mainly on what the occurrence patterns of earthquakes are in the nearest grids to the site.

5 CONCLUSION

The nonstationary Poisson occurrence model is comprehensive in modelling the occurrence of earthquakes and applicable in calculating the seismic risk. A specific U-shape hazard function of the recurrence time is provided to estimate the instantaneous mean occurrence rate because of its suitability and flexibility in simulating the occurrences of clustering earthquakes.

6 ACKNOWLEDGMENT

This study was supported by the National Science Foundation, R.O.C. under Grant NSC 80-0414-P006-07B. The authors are grateful to the Institute of Earth Sciences, Academia Sinica, R.O.C. for providing the earthquake data recorded in Taiwan.

7 REFERENCES

Anagnos, T. & A.S. Kiremidjian 1984. Stochastic Time-Predictable Model for Earthquake Occurrences. *Bull. Seism. Soc. Am.* 74: 2593-2611.

Anagnos, T. & A.S. Kiremidjian 1988. A Review of Earthquake Occurrence Models for Seismic Hazard Analysis. *Prob. Eng. Mech.* 3: 3-11.

Cornell, C. 1968. Engineering Seismic Risk Analysis. *Bull. Seism. Soc. Am.* 58: 1583-1606.

Der Kiureghian, A. & A.H-S. Ang 1977. A Fault-Rupture Model for Seismic Risk Analysis. *Bull. Seism. Soc. Am.* 67: 1173-1194.

Eastern International Engineers, Inc. 1984. Seismic Hazard Analysis of Taiwan Power Company's Nuclear Power Plant No. 2 at Kuosheng. Final Report submitted to A.E.C., R.O.C.

Hong, L-L., S-M. Wu & S-W. Guo 1992. Seismic Hazard in the Southern-West Taiwan. *J. Chinese Inst. Civil and Hydr. Eng.* 4: 249-258 (in Chinese).

Kameda, H. & Y. Ozaki 1979. A Renewal Process Model for Use in Seismic Risk Analysis. *Memoirs of Faculty of Engineering, Kyoto Univ.* XLI: 11-35.

Kiremidjian, A.S. & T. Anagnos 1984. Stochastic Slip-Predictable Model for Earthquake Occurrences. *Bull. Seism. Soc. Am.* 74: 739-755.

Savy, J.B., H.C. Shah & D. Boore 1980. Nonstationary Risk Model with Geophysical Input. *J. Str. Div., ASCE* 106: 145-163.

Structural Safety & Reliability, Schuëller, Shinozuka & Yao (eds) © 1994 Balkema, Rotterdam, ISBN 90 5410 357 4

Stochastic interpolation of earthquake wave propagation

Masaru Hoshiya & Osamu Maruyama
Department of Civil Engineering, Musashi Institute of Technology, Tokyo, Japan

ABSTRACT: An analytical method is proposed that interpolates motions between measured points such that predicted ones coincide with measured waves at the points of the installation. The major contributions of this paper makes use of simple auto-regressive modeling in the time domain which expresses the very property of spatially and temporally deviating phenomena such as an earthquake wave propagation, and the development of an effective method in terms of computer time.

1 INTRODUCTION

In recent years, earthquake ground motions are simultaneously measured at several spatial points through a network of densely installed accelerometers in many countries.

The analyses of obtained data have led to modeling of stochastic characteristics of such spatially and temporally deviating phenomena. However, in order to design line-like structures such as pipelines, we need input waves along the structures. To establish urban disaster prevention program, prediction of more detailed ground motions is also necessary at points other than measured ones. In any case, analytical basis must be developed that provides a tool to accomplish the objective.

In order to visualize the real aspect, a conditional simulation method has been devised and available (Matheron(1973), Ripley(1981) and etc.) that allows one to realize a sample field at other locations which satisfies with the properties of a stochastic field and is compatible with measured values at sample locations. Vanmarcke and Fenton(1991) studied this method to simulate a local field of earthquake ground motions in a Fourier series, they applied the Kriging method on the Fourier coefficients of motions at unrecorded locations, through which compatible accelerograms conditional on recorded motions are produced. Kameda and Morikawa(1992) provided an analytical framework that evaluates the probability distribution of Fourier coefficients for conditioned stochastic processes on which a sample field is simulated in the field of ground motions.

This paper employed the following equation which is basically same as a conditional simulation(Hoshiya 1992,1993)

$$W^*(X,t)=\hat{W}(X,t)+[W(X,t)-\hat{W}(X,t)] \quad (1)$$

where $W^*(X,t)$ =simulated process at unrecorded spatial coordinate X and time t. $\hat{W}(X,t)$=Kriging estimate at X and t.

It is pointed out that since the sample value $\underline{W}(X,t)$ of $W(X,t)$ is not available, the error value $W(X,t) - \hat{W}(X,t)$ remains unknown. However, if the stochastic properties of $W(X,t)-\hat{W}(X,t)$ are clarified, this error may be simulated. Thus, equation(1) may provides a mean for the purpose. In addition, it should be emphasized that with the advantage of a linear interpolation applied on a gaussian field, this method will give a precise solution as in Kameda and Morikawa(1992).

2 THEORY

It is considered that a temporal and spatial stochastic field $W(X,t)$ is homogeneous in time t and stationary in space X with zero mean. The field is assumed to be gaussian with known covariance of

$$C(d, \tau)=E[W(X,t)W(Y,s)] \quad (2)$$

where X and Y are space vectors and t and s are time.
Since a stochastic field is assumed to be homogeneous and stationary, we have $d=Y-X$ and $\tau=s-t$.

Consider that data $\underline{W}(Xi,k)$ are observed at discrete locations $i=1,2,\cdots,N$ and discrete time $t=k\Delta$; $(k=1,2,\cdots,NT)$ where Δ is a sampling interval.

In order to interpolate $W(Xr,k)$ at unrecorded locations $r=1\sim J$ which are consistent with the stochastic property $C(d,\tau)$ and are compatible with recorded data, the following equations are used simply extended from Hoshiya(1992,1993).

$$W^*(Xr,k)=\hat{W}(Xr,k)+\varepsilon(Xr,k) \tag{3}$$

$$\hat{W}(Xr,k)=\sum_{i=1}^{N}\sum_{j=-M}^{M}\lambda_{ij}(Xr)W(Xi,k+j) \tag{4}$$

and $$\varepsilon(Xr,k)=W(Xr,k)-\hat{W}(Xr,k) \tag{5}$$

It is indicated that $W^*(Xr,k)$ is the estimate of $W(Xr,k)$, and in a sample field, $\underline{W}^*(Xr,k)=\hat{\underline{W}}(Xr,k)+\underline{\varepsilon}(Xr,k)$. In equation (3), $\hat{W}(Xr,k)$ is a linear interpolation of $W(Xr,k)$, $i=1,2,\cdots,N$. $\varepsilon(Xr,k)$ is the error function. Equation (4) indicates that in time direction $W(Xr,k)$ is interdependent only for the range of $k-M$ to $k+M$ centering around time k, since correlation between $W(Xr,k)$ and $W(Xr,\iota)$ is generally very weak for many physical phenomena if time lag $(k-l)\Delta$ exceeds the range. It is clear that $\hat{W}(Xr,k)$ is the unbiased estimate since,
$E[\hat{W}(Xr,k)]=E[W(Xr,k)]=0$ and that
$E[\varepsilon(Xr,k)]=0$ and $E[W^*(Xr,k)]=0$.

The unknown coefficients $\lambda_{ij}(Xr)$ may be determined such that the mean square of the error function should be minimum. Then, it leads to

$$C(drm,-n)=\sum_{i=1}^{N}\sum_{j=-M}^{M}\lambda_{ij}(Xr)C(dim,j+n), \tag{6}$$

where $m=1\sim N$, $n=-M\sim M$, $drm=Xm-Xr$ and $dim=Xm-Xi$.

It is noted that Xm, $m=1\sim N$ and Xi, $i=1\sim N$ are recorded locations whereas Xr is an unrecorded location.
Since equation(6) is a simultaneous algebraic equation, $\lambda_{ij}(Xr)$ may be solved unless the field is perfectly correlated.
When equation (6) holds, the variance of $\varepsilon(Xr,k)$ is obtained as follows.

$$\sigma^2\varepsilon(Xr,k)=C(0,0)-\sum_{i=1}^{N}\sum_{j=-M}^{M}\lambda_{ij}(Xr)C(dij,-j) \tag{7}$$

When equation(6) holds, the following properties may be proved as to $\varepsilon(Xr,k)$.

$$E[\hat{W}(Xr,k)\varepsilon(Xr,k)]=0 \tag{8}$$

$$E[W(Xi,k+\iota)\varepsilon(Xr,k)]=0 \quad i=1\sim N,\ \iota=-M\sim M \tag{9}$$

$$\begin{aligned}E[\varepsilon(Xr,k)\varepsilon(Xr,\iota)]&=C(0,\iota-k)\\&-\sum_{i=1}^{N}\sum_{j=-M}^{M}\lambda_{ij}(Xr)C(dri,\iota-k-j)\\&\sum_{i=1}^{N}\sum_{j=-M}^{M}\lambda_{ij}(Xr)C(dir,\iota+j-k)\\&-\sum_{i=N}^{M}\sum_{j=-M}^{M}\sum_{i'=1}^{N}\sum_{j'=-M}^{M}\lambda_{ij}(Xr)\lambda_{i'j'}(Xr)C(di'i,\iota-k+j-j')\end{aligned} \tag{10}$$

Equations (8) and (9) provide orthogonality relationships and equation (10), the autocorrelation function of $\varepsilon(Xr,k)$ in time direction.

3 CONDITIONAL MEAN AND CONDITIONAL VARIANCE

Because of a linear interpolation, the conditional mean of $W^*(Xr,k)$, given a sample field at Xi, $i=1,2,\cdots,N$ is simply obtained from equations (3) and (4) as follows.

$$\begin{aligned}&\mu\,W^*(Xr,k)/\text{cond.}\\&=E[W^*(Xr,k)/\underline{W}(Xi,j),i=1\sim N,j=1\sim NT]\\&\doteqdot E[W^*(Xr,k)/\underline{W}(Xi,k+j),i=1\sim N,j=-M\sim M]\\&\quad=\sum_{i=1}^{N}\sum_{j=-M}^{M}\lambda_{ij}(Xr)\underline{W}(Xi,k+j)\\&+E[\varepsilon(Xr,k)/\underline{W}(Xi,k+j),i=1\sim N,j=-M\sim M]\\&\quad=\sum_{i=1}^{N}\sum_{j=-M}^{M}\lambda_{ij}(Xr)\underline{W}(Xi,k+j)\end{aligned} \tag{11}$$

Where, NT is total number of observation at discrete at times.

Similarly the conditional variance of $W^*(Xr,k)$ is given by

$$\sigma^2W^*(Xr,k)/\text{cond.}=E[\varepsilon(Xr,k)^2]=\sum_{i=1}^{N}\sum_{j=-M}^{M}\lambda_{ij}(Xr)C(dir,-j) \tag{12}$$

The following comment is valid. If equation (4) were a nonlinear function of $W(Xi,k+j)$, the output $W(Xr,k)$ would be nongaussian, which is contradictory to the prerequisite of a gaussian field. Therefore, equation (4) should be linear, and in fact conditional mean of equation (11) and conditional variance of (12) are not approximate and they construct the exact conditional gaussian probability density function.

4 CONDITINAL SIMULATION OF SAMPLE FIELD

To generate a conditional sample field, both $\hat{\underline{W}}(Xr,k)$ (or $\lambda_{ij}(Xr)$) and $\underline{\varepsilon}(Xr,k)$ must be evaluated. While unknown coefficients

$\lambda ij(Xr,k)$ may be easily obtained from equation (6), the determination of $\underline{\varepsilon}(Xr,k)$ is rather difficult because of the following reason. Although $\varepsilon(Xr,k)$ is independent of $\widehat{W}(Xr,k)$ and $W(Xi,k+j)$,j=-M ~ M from equation (8) and (9), it is found that

$$E[\widehat{W}(Xr,k+\iota)\varepsilon(Xr,k)]\neq 0 \quad \iota=-1\sim -M \tag{13}$$

Therefore, $\varepsilon(Xr,k)$ should be simulated so as to satisfy both the correlations in equations (10) and (13), and it becomes a quite complicated way, since we must employ the Cholesky decomposion and/or other methods. to avoid this, equation (3) is converted into

$$W^*(Xr,k) = \sum_{i=1}^{N}\sum_{j=-M}^{M}\lambda'ij(Xr)W(Xi,k+j) + \sum_{\iota=-M}^{-1}\lambda'r\iota(Xr)W(Xr,k+\iota)+\delta(Xr,k) \tag{14}$$

Equation (14) indicates that the error function $\varepsilon(Xr,k)$ in equation (3) is decomposited into an autoregressive type model in which the error function $\delta(Xr,k)$ is a gaussian white noise. It is recognized that the second and third terms in equation (14) correspond to $\varepsilon(Xr,k)$.

Equation (14) involves unknown values $W(Xr,k+\iota)$,(ι=-M~-1) , however, sample estimates $W^*(Xr,k+\iota)$ are treated as known values since $\underline{W}^*(Xr,k+\iota)$ may be obtained by a sequential manner of simulation in the time direction as in the following paragraphs.

The unknown coefficients $\lambda'ij(Xr)$ and $\lambda'r\iota(Xr)$ may be determined such that the mean square of the error function should be minimum. Then, it leads to

$$\mathbf{C}\ \mathbf{Xr} = \mathbf{C}\ \boldsymbol{\lambda}'(Xr) \tag{15}$$

Where,

$\boldsymbol{\lambda}'^{T}(Xr)=[\lambda'r\text{-}M(Xr)\ \lambda'r\text{-}M+1(Xr) \cdots \lambda'r\text{-}1(Xr) \vdots \lambda'1\text{-}M(Xr) \cdots \lambda'10(Xr) \cdots \lambda'1M(Xr) \vdots \lambda'2\text{-}M(Xr) \cdots \lambda'20(Xr) \cdots \lambda'2M(Xr) \vdots \cdots \lambda'N\text{-}M(Xr) \cdots \lambda'N0(Xr) \cdots \lambda'NM(Xr)]$

$\mathbf{W}_K^{T} = [\ W(Xr,k\text{-}m)\ W(Xr,k\text{-}M+1) \cdots W(Xr,k\text{-}1) \vdots W(X1,k\text{-}M) \cdots W(X1,k) \cdots W(X1,k\text{-}M) \vdots W(X2,k\text{-}M) \cdots W(X2,k) \cdots W(X2,k\text{-}M) \cdots W(XN,k\text{-}M) \cdots W(XN,k) \cdots W(XN,k\text{-}M)\ ]$

$\mathbf{C}\ \mathbf{Xr} = E[\ \mathbf{W}_K W(Xr,k)]$

$\mathbf{C}\ = E[\ \mathbf{W}_K \mathbf{W}_K^{T}]$

When equation (15) holds, the variance of $\delta(Xr,k)$ is given as follows.

$$\sigma^2\ \delta(Xr,k) = C(0,0)-\boldsymbol{\lambda}'^{T}(Xr)\ \mathbf{C}\ \mathbf{Xr} \tag{16}$$

And $\delta(Xr,k)$ has following stochastic properties.

$$E[W(Xi,k+j)\delta(Xr,k)]=0, i=1\sim N, j=-M\sim M \tag{17}$$

$$E[W(Xr,k+\iota)\delta(Xr,k)]=0,\ \iota=-M\sim 1 \tag{18}$$

Equations (17) and (18) indicate that $\delta(Xr,k)$ is uncorrelated with $W(Xi,k+j)$as well as $W(Xr,k+\iota)$. Therefore, a sample field may be obtained with the following steps. The first step is to calculate the unknown coefficients $\lambda'ij(Xr)$ and $\lambda'r\iota(Xr)$ based on equation (15). The second step is to simulate a $\underline{\delta}(Xr,k)$ with $E[\delta(Xr,k)]=0$ and with equation (16), without consideration of any correlation, since it is proved that $E[\delta(Xr,k)\delta(Xr,\iota)]=0,\ k\neq\iota$. After $\underline{W}(Xr,k)$ is obtained, the time k will be shifted to k+1 and similarly $\underline{W}(Xr,k+1)$ will be simulated. When $\underline{W}(Xr,k)$,k=1 ~ NT is completed, $\underline{W}(Xr,k)$ may be included into the observation data. By doing so, a step by step expansion will be carried out for multiple point. (Refer to Hoshiya,1992,1993).

If Xr coincides with one of the recorded locations Xm, the solutions of equations(15) are $\lambda'm0(Xm)=1$; i=m,j=0 otherwise $\lambda'ij(Xm)=0$ and equation (16) becomes

$$\sigma^2\ \delta_{(Xr,k)} = 0 \tag{19}$$

Therefore,

$$\underline{\delta}\ (Xr,k) = 0 \tag{20}$$

Consequently, we have from equation (14).

$$\underline{W}^*(Xm,k)=\underline{W}(Xm,k) \tag{21}$$

This means that the estimate $\underline{W}^*(Xm,k)$ coincides with $\underline{W}(Xm,k)$ at a recorded location. On the other hands, if Xr is remote from any of the recorded locations, $\lambda'ij(Xr)$ becomes zero in equation (14). Hence, $W^*(Xr,k)=\varepsilon(Xr,k)$ and $\sigma^2\varepsilon(Xr,k) = C(0,0)$. Clearly, $W^*(Xr,k)$ becomes a unconditional field.

5 NUMERICAL EXAMPLE

Covariances that are required in the conditional simulation are obtained by the Fourier transformation of a cross spectral density function as follows.

$$C(\mathbf{d},\tau) = \int_{-\infty}^{\infty} S(\mathbf{d},f)\cdot\exp(i2\pi f\tau)df \tag{22}$$

In general, a cross spectral density function for propagating ground motions may be expressed as

$$S(\mathbf{d},f) = S(f)\ \gamma\ (\mathbf{d},f) \tag{23}$$

where s(f) is a power spectral density function which is common to a homogeneous and stationary field, and $\gamma(\mathbf{d},f)$ is a coherence spectrum. Goto and Kameda(1969) proposed the following model.

$$ {}_{A}S(f)=(2\pi f)^2 {}_{V}S(f)=(2\pi f)^4 {}_{D}S(f) = (64/6\pi f_g^5)f^4\cdot\exp\{-4|f|/f_g\} \quad (24) $$

Where suffices A, V and D respectively stand for acceleration, velocity and displacement.

The following model, (TYPE1; Kawakami(1983) , TYPE2; Harichandran and Vanmarcke(1986)) are employed as a coherency spectrum.

$$ \text{TYPE1.} \quad \gamma(d,f) = \exp(\alpha|fd|c) *\exp(-i2\pi fdc) \quad (25) $$

$$ \text{TYPE2.} \quad \gamma(d,f) = |\gamma(d,f)| *\exp(-i2\pi fe) \quad (26) $$

$$ |\gamma(d,f)| = A\cdot\exp\{-2|d|(1-A+\alpha' A)/\alpha'\theta(f)\} + (1-A)\cdot\exp\{-2|d|(1-A+\alpha' A)/\theta(f)\} $$

$$ \theta(f) = K\{1+(f/f_0)^b\}^{-1/2} $$

$$ e = \vec{c}\cdot d/|\vec{c}|^2 $$

Further details in model is entrusted to the above references, where data in Table 1 are used. A example of a homogeneous field is numerically demonstrated.

Table 1. Input Data

Input Data	Values
f_g in eq.(24)	1.0(Hz)* 2.0(Hz)
$\lvert\vec{c}\rvert$	1000(m/sec)*
α in eq.(25)	2.0
α' A K in eq.(26) f_0 b	0.147 0.736 5210(m) 1.09 2.78
Δt M in eq.(3)and(14)	0.1(sec)* 40*

* used for Type1.

5.1 EXAMPLE 1 (1-D homogeneous field)

Figure (1) shows displacement sample fields of a one dimensional wave propagation of velocity of 1km/s conditionally simulated according to equation (14), where TYPE1 model is used. Figures (2) and (3) show respectively conditional means and conditional variances. They are obtained both by the theoretical formula and simulation (100 samples). They coincide perfectly in Figure 2 and almost identically in Figure 3.

5.2 EXAMPLE 2 (2-D homogeneous field)

Simulated sample waves are shown in Figure 4. They satisfy the property of the 2-D random field.

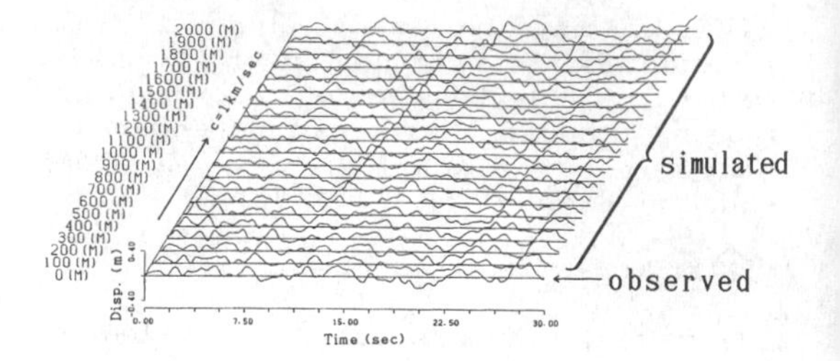

(a) Observation Given at One Point

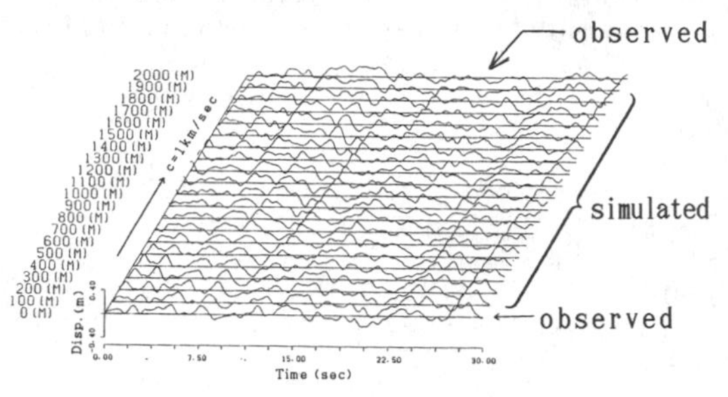

(b) Observation Given at Two Points

Figure 1. Conditional Sample Fields

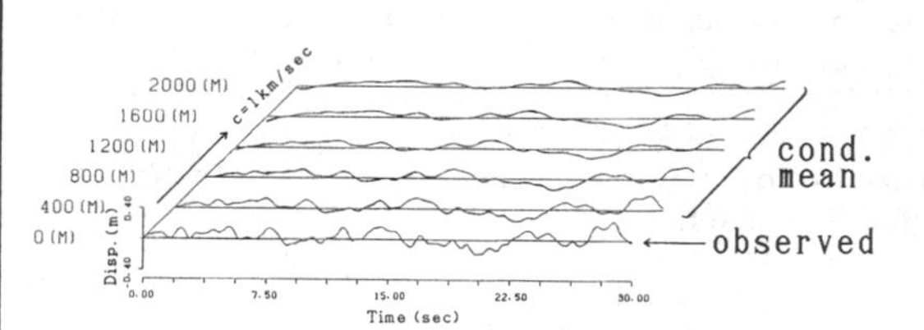

(a) Observation Given at One Point

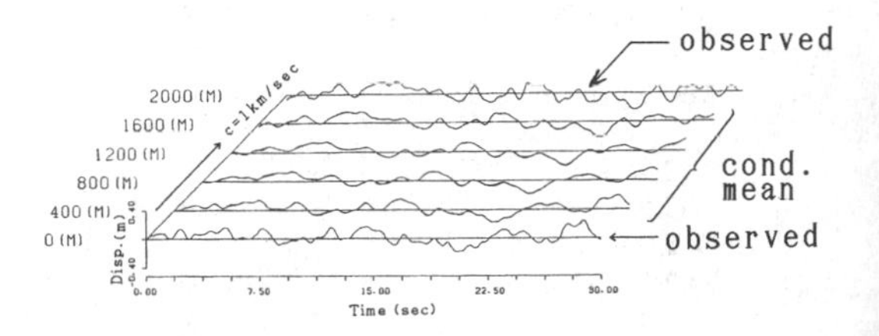

(b) Observation Given at Two Points

Figure 2. Conditional Means (Analytical solution and simulation coincide perfectly)

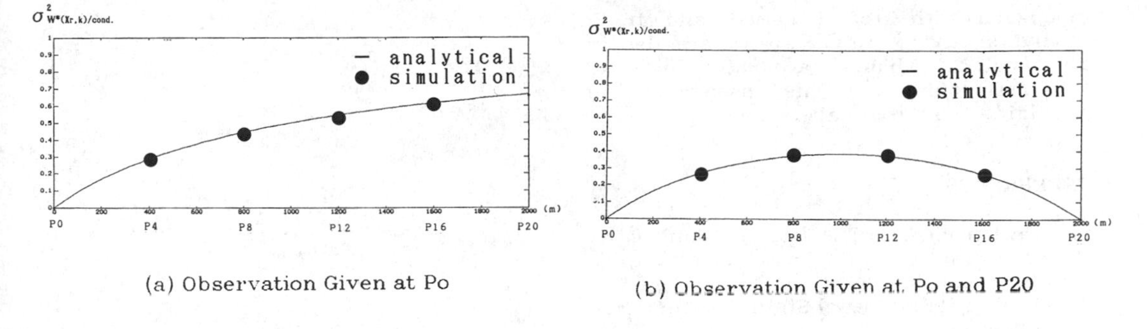

(a) Observation Given at Po

(b) Observation Given at Po and P20

Figure 3 Conditional Variances

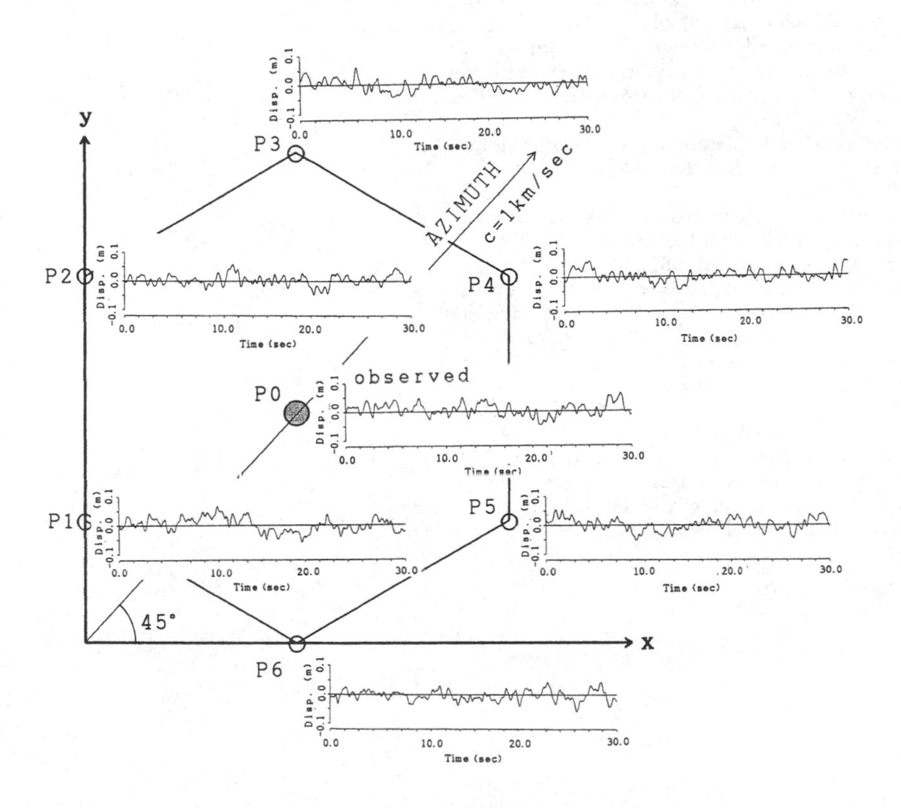

Figure 4. Simulated Displacements Conditional on Observation at Po

6 CONCLUDING REMARKS

This paper proposed a stochastic interpolation method of conditional simulation on a temporal and spatial stochastic Gaussian field. It has been pointed out that this method may be very promising for the study of wave propagation using such as array data of earthquake ground motions.

The theoretical framework of this paper is developed cooperatively by the both writers.

ACKNOWLEDGMENTS

In performing this study, informative

discussion with Prof. H. Kameda and Mr. H. Morikawa (Kyoto Univ. Japan) are deeply appreciated. Thanks are also due to Mr.K.Yamaguchi for his assistance in providing the numerical examples.

REFERRECES

1) G.Matheron(1073);The Intrinsic Random Function and Their applications, Adv. Appl. Prob.5, pp.439 ~ 468.
2) D.Ripley(1981);Spatial Statistics, John Wiley and Sons Inc.
3) E.H.Vanmarcke and G.A.Fenton(1991); Conditional Simulation of Local Fields of Earthquake Ground Motion,Jour. of Structural Safety,10, pp247 ~ 264.
4) H.Kameda and H.Morikawa (dated March 1992); Conditioned Stochastic Processes for Conditional Random Field,Submitted to ASCE, Jour,of E.M.Div.
5) M.Hoshiya(1993);Simulation of a Conditional Stochastic Field, JSCE No. 459, I-22, pp.113-118,
6) M.Hoshiya(1993);Conditional Simulation of a Stochastic Field,Submitted to ICOSSAR'93.
7) H.Goto and H.Kameda(1969);Statistical Inference of the Future Earthquake Ground Motion,Proc.4WCEE,Chile,Vol.1,A-1,pp.39 ~ 54.
8) H.kawakami(1983);Effect of Deformation of Seismic Waves on Estimated Value of Ground Relative Displacement or Strain, JSCE, No.337, pp37 ~ 46,(in japanese).
9) R.S.Harichandran and E.H.Vanmarcke (1986); Stochastic Variation of Earthquake Ground Motion in Space and Time,Jour.of E.M.Div.,ASCE,Vol.112,No.2,pp.154 ~ 174.

Structural Safety & Reliability, Schuëller, Shinozuka & Yao (eds) © 1994 Balkema, Rotterdam, ISBN 90 5410 357 4

Seismic hazard analysis using source models for interplate earthquakes

Takashi Inoue
Technical Research Institute, Hazama Corporation, Tsukuba, Japan

Jun Kanda
Department of Architecture, University of Tokyo, Japan

ABSTRACT: The seismic hazard of the Tokyo and Osaka areas is evaluated by thc Probabilistic Model Method and Cumulative Distribution Method using historical earthquake data for the last 300 years together with Kanai's attenuation relationship. The hazard curve obtained by the Probabilistic Model Method agrees well with results from the Cumulative Distribution Method. Seismic source areas of interplate earthquakes are separately modeled. The seismic hazard of Tokyo is greatly affected by the extent of the source area. Seismic hazard based on nonstationary earthquake occurrences at interplate faults is also studied.

1 INTRODUCTION

Seismic hazard at a site can be evaluated by both the Probabilistic Model Method (PMM) and Cumulative Distribution Method (CDM). PMM uses probabilistic models based on the Gutenberg-Richter formula and Poisson earthquake occurrence (Cornell 1968). CDM statistically obtains the cumulative distribution of the annual maximum seismic intensity at a site directly using historical earthquake data and an attenuation equation (Kawasumi 1951). An extreme value distribution is often fitted to the cumulative distribution.

In PMM, the parameters for earthquake occurrence models can be determined using earthquake catalogs and active fault data. However, appropriateness of the models employed must be examined. CDM is straightforward as it directly uses earthquake catalogs, but the quality of earthquake data is not uniform in time and space, and the number of data is not sufficient to appraise the maximum level. The results from these methods are compared.

As it is difficult to identify the existence of active faults in regions with thick soil deposits such as Tokyo, source-areas are often used for PMM in Japan. On the other hand, investigations have been carried out on the activity and location of interplate faults at subduction zones. To take into account the effects of interplate faults which cause severe damage to the east coast of Japan, the seismic source for the interplate faults is separately modeled from other source-areas. Non-stationarity of earthquake occurrences is also taken into account using a simple model.

2 METHOD OF SEISMIC HAZARD ANALYSIS

Earthquake catalogs for 300 years (1686-1985) are used. A focal depth of 30km is adopted for all events. Kanai's attenuation equation for the peak velocity of bedrock motion v_0 (cm/sec) is used (Kanai et al. 1966):

$$v_0=10^{0.61M-(1.66+\frac{3.60}{x})\log x-(0.631+\frac{1.83}{x})} \tag{1}$$

where M and x denote earthquake magnitude and hypocentral distance in kilometers respectively. Variation of the attenuation is not considered.

For the PMM, the program EQRISK (McGuire 1976) was used. Probability of exceeding an intensity y at a site is given by

$$P(Y>y)=\int_M\int_x P(Y>y|m,x)f_M(m)f_x(x)dmdx \tag{2}$$

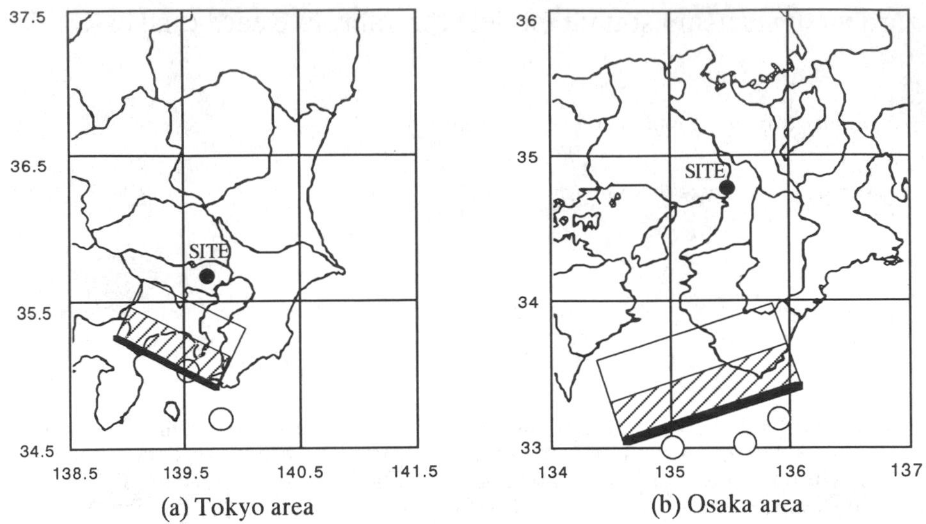

(a) Tokyo area

(b) Osaka area

Fig. 1 Sites and seismic source areas

where $f_M(m)$ and $f_x(x)$ are probability density functions of magnitude m and distance x. If a distribution of magnitude is assumed to conform to the Gutenberg-Richter formula $\log N(M) = a - bM$ (where $N(M)$ is the number of earthquakes having magnitude greater than M, and a and b are constants), $f_M(m)$ is given by

$$f_M(m) = \frac{\beta \exp\{-\beta(m - m_0)\}}{1 - \exp\{-\beta(m_1 - m_0)\}} \tag{3}$$

where m_0 and m_1 are lower-bound and upper-bound magnitudes respectively. Assuming that earthquakes occur as Poisson arrivals, the probability of exceeding an intensity y in Δt years is given by

$$P(Y_{\Delta t} > y) = 1 - \exp\{-\nu \Delta t P(Y > y)\} \tag{4}$$

where ν is the annual rate of occurrence of events having magnitude greater than m_0.

For the CDM, the peak velocity by Eq. (1) at a site is averaged for a circular area with a radius of 40 km (Kanda and Dan 1987). The non-exceedance probability, F_n, for the nth annual maximum velocity is postulated according to the Hazen method, i.e.,

$$F_n = 1 - \frac{n - 0.5}{T_s} \tag{5}$$

where T_S is the observation period (T_S = 300 years).

3 SEISMIC HAZARD ANALYSIS BY ONE DEGREE SQUARE SOURCE AREAS

Sites and seismic source areas are shown in Fig. 1. Parameters of the Gutenberg-Richter equation were obtained for 9 source areas of one-degree longitude-latitude squares around the sites using 300-year earthquake catalogs. The data for $M \geq 7.5$, $M \geq 6.0$, and $M \geq 5.0$ are used for the year range 1686-1884 (by Usami), 1885-1925 (by Utsu), and 1926-1985 (by JMA) respectively (Katayama 1979). The value of b, the annual occurrence rate of events having magnitudes greater than 5.0, ν, and the maximum magnitude, m_1, for each source area are shown in Fig. 2. The maximum magnitude of a region is assessed using long-term data over 1307 years.

Seismic hazard curves obtained by PMM and CDM (1 to 50th peak velocity) are shown in Fig. 3 where the annual maximum velocity is plotted versus the reduced variate of non-exceedance probability i.e. $y = -\ln(-\ln F)$. Although the result by PMM gives a slightly larger velocity than that by CDM, the results by the two methods agree well. One of the reasons that PMM gives larger velocity for small y values (high probability) appears to be the lack of small magnitude earthquakes in the historical data. For large y values (low probability), the annual maximum velocity at a ,

Latitude	138.5°–139.5°	139.5°–140.5°	140.5°–141.5°
37.5°–36.5°	0.82 0.21 7.5	1.04 0.35 7.5	1.22 3.49 7.5
36.5°–35.5°	0.80 0.64 7.5	1.06 2.48 7.5	1.44 6.77 7.5
35.5°–34.5°	0.91 (0.74) 1.59 (1.14) 7.9 (7.4)	0.84 (0.86) 1.37 (1.34) 8.1 (7.3)	0.78 0.66 7.5

Legend
b ν m_1

(a) Tokyo area

Latitude	134°–135°	135°–136°	136°–137°
36°–35°	0.77 0.38 7.2	0.82 0.68 7.4	0.73 0.22 8.0
35°–34°	0.94 0.12 7.0	0.99 0.48 7.5	0.95 0.64 7.6
34°–33°	0.73 (0.94) 0.31 (0.50) 8.4 (7.0)	0.65 (0.89) 0.54 (0.81) 8.4 (7.5)	0.76 0.55 7.9

(b) Osaka area

Fig. 2 Parameters of one-degree square source areas
Values in ( ) are those without large interplate events

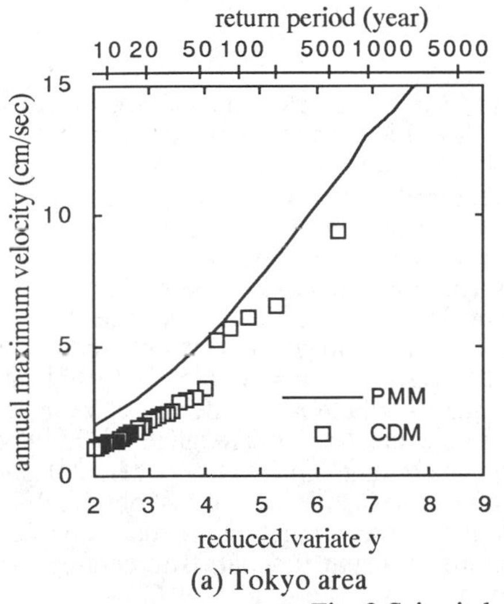

(a) Tokyo area

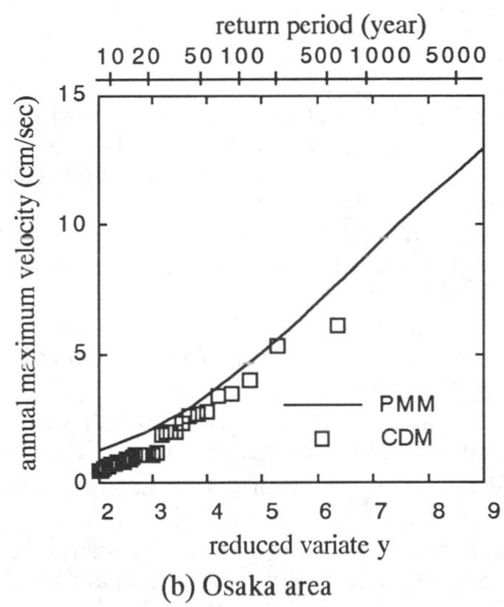

(b) Osaka area

Fig. 3 Seismic hazard curves by PMM and CDM

site is considerably higher in the result by PMM than that by CDM because the PMM assumes the maximum magnitude earthquake of an area source could occur anywhere in the source area.

4 SEPARATE MODELING FOR INTERPLATE EARTHQUAKES

For the evaluation of the seismic hazard at sites along the Pacific coast of Japan, modeling of large interplate earthquakes that repeatedly occur at subduction zones is important. The interarrival time of interplate earthquakes is generally shorter than that of intraplate earthquakes, and this may be estimated from historical earthquakes. Sources of large interplate earthquakes at the Sagami and Nankai Troughs are separately modeled using proposed fault locations and earthquake recurrence data.

The parameters of 1 degree square source areas without large events at the Troughs (M=8.1 in 1703 and M=7.9 in 1923 at the Sagami Trough; M=8.4 in 1707, M=8.4 in 1854,

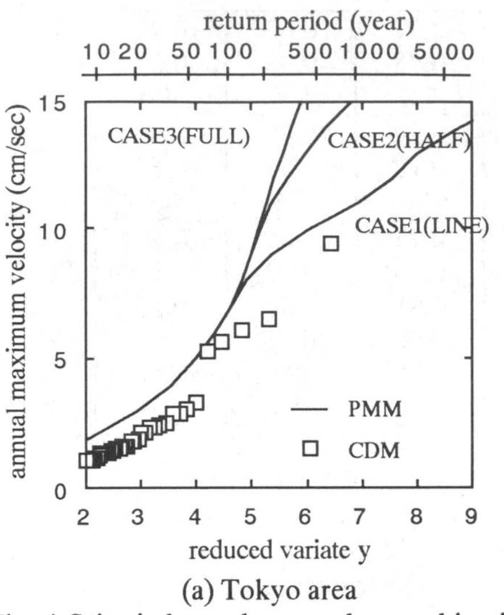

(a) Tokyo area

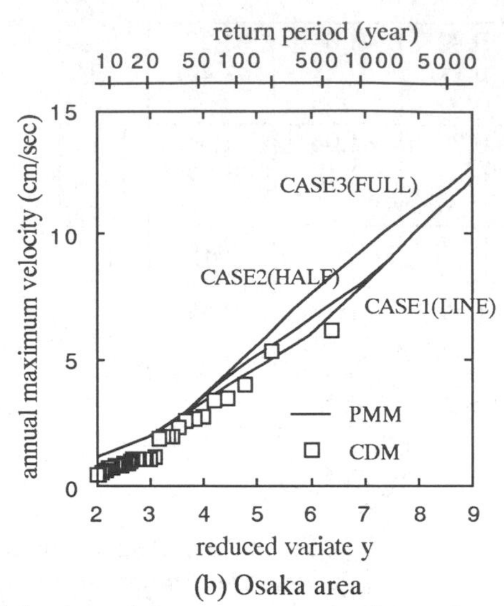

(b) Osaka area

Fig. 4 Seismic hazard curves by combination of the interplate source area and one-degree square source areas

and M=8.0 in 1946 at the Nankai Trough) are shown in the parenthesis of Fig. 2. It is difficult to define the source area of interplate earthquakes only from historical data. Therefore, three configurations of the source model are used in the analysis based on proposed fault models.

Case 1: A line source at the location of the upper edge of a fault plane. The depth of the line source is 30km.

Case 2: An area source of the upper half of an inclined fault plane.

Case 3: An area source of a full inclined fault plane.

The fault models for the Kanto earthquake (1923) and Nankai earthquake (1854) are used for the Sagami and Nankai Troughs respectively. The parameters of the fault models are shown in Table 1 (Matu'ura et al. 1923, Aida 1981). The epicenters of the large events and the source models (Case 1 is shown as a bold

Table 1. Fault Parameters

Parameters	Kanto	Nankai
Fault length (km)	95	150
Fault width (km)	54	70
Depth of upper edge (km)	1.9	10
Dip Angle (°)	25	10

line and Case 2 as a hatched area) are shown in Fig. 1. Case 1 is close to the reported epicenters of the large events in the earthquake catalogs.

It is assumed that only large magnitude events repeatedly occur at the subduction zones, while the magnitude distribution of smaller events conform to the Gutenberg-Richter law (Wesnousky et al. 1984). At the Sagami Trough, it is assumed that events occur at Poisson arrivals of ν = 0.005 and that the magnitude distribution is uniform between 7.9 and 8.1. At the Nankai Trough, ν = 0.00855 and the magnitude range is between 8.0 and 8.4. The seismic hazard curves obtained by PMM using a combination of the interplate source area and the 9 square source areas are shown in Fig. 4. The results by CDM are also shown in the figure.

The result by CDM is similar to that by PMM in Case 1 where the line source is located close to epicenters of large interplate events. For Case 2 and 3 of Tokyo in which a half and a full area of the suggested fault plane are used as a source area of large interplate events, velocity for y > 5 (return period > 150 years) is considerably larger than that for Case 1 and CDM. The seismic hazard of Tokyo is greatly affected by the location of the source of large interplate events. On the other hand, the seismic hazard of Osaka, where the interplate fault is far from the site, is not affected greatly by the location of the fault.

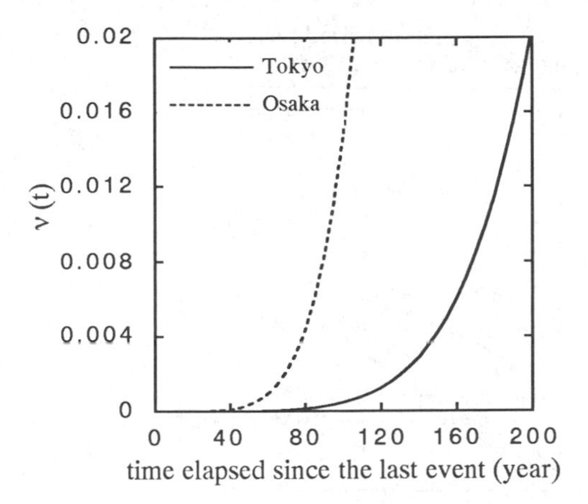

Fig. 5 Hazard rate for the Sagami and Nankai troughs as a function of the time since the last event

5 NON-STATIONARY OCCURRENCE MODEL

Nonstationary earthquake occurrences should be considered when the time period of interest is shorter than the return period of the major events. The distribution of interarrival time $F_T(t)$ can be modeled by the Weibull distribution as follows.

$$F_T(t) = 1 - \exp(-\alpha t^{\gamma}) \tag{6}$$

For the Nankai Trough, Utsu obtained $\alpha = 2.98 \times 10^{-14}$ and $\gamma = 6.44$ based on six large events ($M \geq 7.9$) from 1361 to 1946 (Utsu 1984). The mean interval time is 117 years and the standard deviation 21 years. For the Sagami Trough, only two large events are recorded. Wesnousky et al. (1984) assumed normally distributed interarrival time having a mean of 200 years and standard deviation of 20 years using fault displacement data. The standard deviation of 20 years considerably limits the time of occurrence. In this study, a Weibull distribution with 200-years mean interval time is used. Assuming the C.O.V. is the same as the Nankai Trough, a standard deviation of 36 years was adopted, and the corresponding parameters are $\alpha = 7.0 \times 10^{-16}$ and $\gamma = 6.5$. The hazard rate for the Weibull distribution is given by

$$\nu(t) = \alpha \gamma t^{\gamma - 1} \tag{7}$$

The hazard rates of the Sagami and Nankai Troughs are shown in Fig. 5.

Seismic hazard curves at time t = 50, 100, 150, 200 years since the last large event (in 1923 at Sagami and in 1946 at Nankai) are shown in Fig. 6. Case 2 is used for the interplate source model. Dashed lines are the results by Poisson occurrence. Seismic hazard of the sites is evaluated by the combination of large interplate earthquakes of nonstationary Poisson occurrences and other earthquakes of Poisson occurrences. The results by Poisson occurrence (dashed lines) agree with t = 150 years at the Tokyo area and t = 90 years at the Osaka area. As the time after the last event increases, the probability of occurrence increases and the hazard curves approach the curve given if the interplate earthquake certainly occurs.

6. CONCLUSIONS

The results obtained by PMM using one-degree square source areas agrees well with that by CDM. In the high probability range (small y), CDM gives smaller annual peak velocities than PMM. As the upper limit peak velocity by PMM depends on the maximum magnitude of the source areas, PMM gives greater peak velocity than that by CDM in a low probability range (large y). In the case of separate modeling of interplate earthquakes, the shape of hazard curves at Tokyo changes according to the region of the interplate source area. At Osaka, where the interplate source area is far, the change is small. By the non-stationary occurrence modeling of interplate earthquakes, the seismic hazard of the sites increases as the time since the last event increases, and the seismic hazard curves approach the curve that is given if an interplate event certainly occurs.

CDM is a purely statistical method to define the design earthquake level. On the other hand, many researches are carried out to deterministically predict earthquake ground motion by interplate fault rupture models. The method proposed here lies between these two methods and enables parametric studies on the interplate fault. The accuracy of the analysis can be improved by further investigation on size, location, and frequency of interplate events and the attenuation relationship near the epicenter.

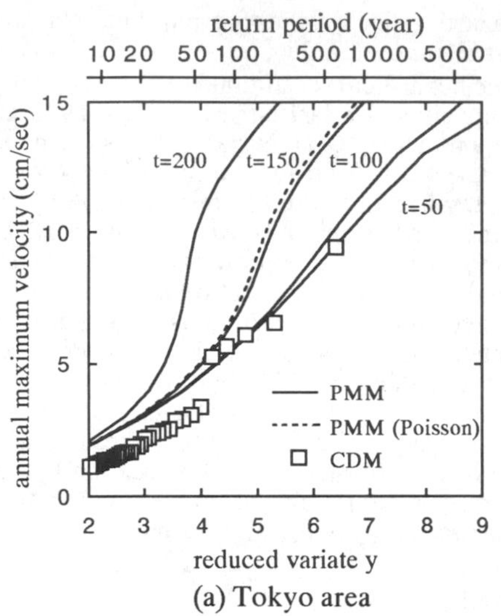

(a) Tokyo area

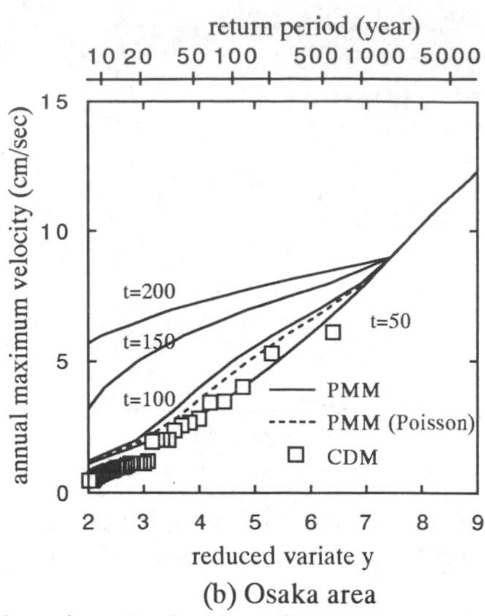

(b) Osaka area

Fig. 6 Seismic hazard curves for 50, 100, 150, 200 yr since the last large interplate event

REFERENCES

Aida, I. 1981. Numerical experiments for the tsunamis generated off the coast of the Nankaido district, Bull. Earthq. Res. Inst.,Vol.56: 713-730.

Cornell, C.A. 1968. Engineering seismic risk analysis. Bull. Seism. Soc. Am., Vol.58, No.5,: 1583-1606.

Kawasumi H. 1951. Measures of earthquake danger and expectancy of maximum intensity throughout Japan as inferred from the seismic activity in historical times. Bull. Earthq. Res. Inst., Vol.29: 469-482.

Kanai, K., Hirano, K. and Yoshizawa, S. 1966. Observation of strong earthquake motions in Matsushiro area, Part 1, Bull. Earthq. Res. Inst. Vol.44: 1269-1296.

Kanda, J. and Dan, K. 1987. Distribution of seismic hazard in Japan based on an empirical extreme value distribution, Structural Safety, Vol. 4: 229-239.

Katayama, T. 1979. Seismic risk analysis in terms of acceleration response spectra, Proc. 2nd U.S. National Conference on Earthquake Engineering: 117-126.

McGuire, R.K. 1976. FORTRAN Computer program for seismic risk analysis, U.S. Geological Survey Open-File Report 76-67.

Matsu'ura M. et al. 1980. Statical and dynamical study on faulting mechanism of the 1923 Kanto earthquake, J. Phys. Earth, Vol.28: 119-143.

Utsu, T. 1984. Estimation of parameters for recurrence models of earthquakes, Bull. Earthq. Res. Inst., Vol.59: 53-66.

Wesnousky, S.G. et al. 1984. Integration of geological and seismological data for the analysis of seismic hazard: a case study of Japan, Bull. Seism. Soc. Am., Vol.74, No.2: 687-708.

Structural Safety & Reliability, Schuëller, Shinozuka & Yao (eds) © 1994 Balkema, Rotterdam, ISBN 90 5410 357 4

Probabilistic evaluation of earthquake response spectrum and its application to response analysis

H. Ishida
Kajima Technical Research Institute, Kajima Corporation, Japan

ABSTRACT : This paper proposes a method of determining an earthquake motion input to a structural system in terms of a probabilistic model of the acceleration response spectrum. The input model is composed of the response spectral components with discrete periods whose logarithms are jointly normal. The joint distribution of the logarithmic response spectral components is determined from the expected values and standard deviations of all the components and the correlation coefficients of all couples of the components. For the purpose of evaluating the effect of the above correlation coefficients, formulae incorporating them have been derived to obtain statistics of the response of a multi-degree-of-freedom system subjected to the input model. A simple case study has been performed based on observed earthquake motion records. A probabilistic model of the response spectrum has been set up first by means of a multiple regression analysis. The response of a three-degree-of-freedom system subjected to the input has been evaluated and the effect of the correlation of the logarithmic response spectral components is examined. As a result, it has become clear that the correlation is important in evaluating the response of a structural system.

1 INTRODUCTION

Some seismic probabilistic risk assessments or reliability analyses for structural systems require the earthquake input in terms of a probabilistic model of the response spectrum. In such probabilistic models, a response spectral component with a discrete period is regarded as a random variable. However, the correlation of a couple of the response spectral components with different periods has not been sufficiently examined. The so-called uniform hazard response spectra, which represent uniform probabilities of exceedance for the response spectral values over the entire period range of interest, apparently do not include any information about the correlation.

This paper proposes a method of forming a probabilistic model for the response spectrum incorporating the above correlation and stresses its importance with regard to its effect on the response of a structural system. For this purpose, a simple method of evaluating the response of a multi-degree-of-freedom system subjected to the proposed input model will be developed. While the peak response value of the multi-degree-of-freedom system can be evaluated by means of SRSS or CQC, this paper regards the peak response value as an random variable. The effect of the correlation on the response will be examined in terms of the probability of exceedance for the response. It should be noted that response nonlinearity is not considered here. A case study will be carried out to obtain a probabilistic model of the response spectrum through detailed examination of earthquake motion records observed around Tokyo Bay in Japan and examine how the correlation affects the response of a structural system.

2 PROBABILISTIC INPUT MODEL

Consider a ground motion caused by an earthquake with a magnitude m and observed at a site $x$km from the source. Although its acceleration response spectrum is the essential subject of examination, the natural logarithm of the spectrum is mainly used for simpler formulation. The logarithm of the spectrum is denoted by

$$Z(T_i,m,x) = \log S(T_i,m,x), \qquad (1)$$

where T_i, i=0,1,...,N-1, are periods and $S(T_i,m,x)$ are components of the acceleration response spectrum with a period T_i and a damping factor of 5%. The variables $S(T_i,m,x)$ and $Z(T_i,m,x)$ are considered to be random. The random variable will be identified by Gothic type. If the random variable $Z(T_i,m,x)$ for each period T_i has a normal distribution, it can be expressed as

$$Z(T_i,m,x) = \bar{z}(T_i,m,x) + e(T_i), \tag{2}$$

where $\bar{z}(T_i,m,x)$ is the expected value of $Z(T_i,m,x)$ and $e(T_i)$ is a normal random variable with the expected value of 0 ;

$$\mathrm{E}[e(T_i)] = 0. \tag{3}$$

Hence, the covariance of $e(T_i)$ and $e(T_j)$ is given by

$$\mathrm{E}[e(T_i)e(T_j)] = r_{ij}\beta_i\beta_j, \tag{4}$$

where β_i and β_j are, respectively, the standard deviations of $e(T_i)$ and $e(T_j)$, and r_{ij} is the correlation coefficient of $e(T_i)$ and $e(T_j)$. It is clear that the values of β_i and r_{ij} are the same as those of the random variables $Z(T_i,m,x)$. If random variables $Z(T_i,m,x)$ are assumed to be jointly normal, their joint density function is given by

$$f_Z(z) = \frac{1}{(2\pi)^{N/2}|C|^{1/2}} \exp\left[-\frac{1}{2}(z-\bar{z})^{\mathrm{T}}C^{-1}(z-\bar{z})\right], \tag{5}$$

where z and $\bar{z}$ are vectors consisting of outcomes of $Z(T_i,m,x)$ and its expected values, respectively, and C is a matrix whose (i,j) component is the covariance given by Equation (4).

Now, consider the logarithms of the spectral components under the condition that the peak ground acceleration (PGA) equals a specific value a,

$$Z(T_i,m,x,a) = \log S(T_i,m,x,a). \tag{6}$$

Statistics of $Z(T_i,m,x,a)$ will be used to prescribe the probabilistic input to a structural system in combination with the hazard curve, or the annual probabilities of exceedance for the PGA range of interest. Using this type of input, the annual probabilities of exceedance associated with the response of the structural system can be obtained. Assuming that the PGA equals the response spectral component with the shortest period considered in the analysis, 0.02sec for example, the random variables $Z(T_i,m,x,a)$ become jointly normal, even under the condition that the PGA equals a specific value a. Thus, if the PGA is denoted by A as a random variable, their conditional density function is given by

$$f_Z(z|A=a) = \frac{1}{(2\pi)^{N/2}|D|^{1/2}} \exp\left[-\frac{1}{2}(z-\bar{\zeta})^{\mathrm{T}}D^{-1}(z-\bar{\zeta})\right], \tag{7}$$

where $\bar{\zeta}$ is a vector whose i-th component $\bar{\zeta}_i$ given by

$$\bar{\zeta}_i = \bar{z}(T_i,m,x) - r_{0i}\frac{\beta_i}{\beta_0}(\bar{z}(T_0,m,x) - \log a) \tag{8}$$

and D is a covariance matrix whose (i,j) component is obtained as (Johnson and Wichern 1988)

$$d_{ij} = (r_{ij} - r_{0i}r_{0j})\beta_i\beta_j. \tag{9}$$

The variable with the subscript 0 means that it is associated with the PGA, which is approximately given by the acceleration response spectral component with the shortest period T_0 in the period range for analysis.

3 RESPONSE OF A STRUCTURAL SYSTEM

3.1 *Probability distribution of response acceleration*

Consider a multi-degree-of-freedom system subjected to an input motion with ground acceleration α. The relative acceleration of the k-th mass, $\ddot{y}_k$, satisfies the inequality,

$$|\ddot{y}_k + \alpha|_{\mathrm{peak}} \leq \sum_n |{}_n\phi\, {}_nu_k|\, S({}_nT, {}_nh), \tag{10}$$

where $S({}_nT, {}_nh)$ is the acceleration response spectral component with the n-th natural period ${}_nT$ and modal damping factor ${}_nh$, ${}_n\phi$ is the n-th participation factor, and ${}_nu_k$ is the k-th component of the n-th eigenvector.

Replacing $|\ddot{y}_k + \alpha|_{\mathrm{peak}}$ and $S({}_nT, {}_nh)$ in Inequality (10) with random variables Y and ${}_nS$, respectively, and rewriting $|{}_n\phi\,{}_nu_k|$ as ${}_nv$, Inequality (10) yields a probabilistic equation,

$$Y = U\sum_n {}_nv\, {}_nS, \tag{11}$$

where U is a random variable which can take a value from 0 to 1 and could depend on the soil condition, the input motion and the structural system. Assuming that U is independent of ${}_nS$ for all n, and ${}_nS$ are jointly normal, the expected values of Y and Y^2 are given by

$$\mathrm{E}[Y] = \mathrm{E}[U]\sum_n {}_nv \exp\left[{}_n\bar{z} + \frac{1}{2}{}_n\beta^2\right] \tag{12}$$

and

$$\mathrm{E}[Y^2] = \mathrm{E}[U^2]\sum_m\sum_n {}_mv\,{}_nv \times\exp\left[{}_m\bar{z} + {}_n\bar{z} + \frac{1}{2}\left({}_m\beta^2 + 2\,{}_{mn}r\,{}_m\beta\,{}_n\beta + {}_n\beta^2\right)\right], \tag{13}$$

respectively, where ${}_n\bar{z}$ and ${}_n\beta$ are the expected value and standard deviation of the logarithm of ${}_nS$, respectively, and ${}_{mn}r$ is the correlation coefficient of the logarithms of ${}_mS$ and ${}_nS$.

If the random variable Y is log-normal, the median $\hat{y}$ and logarithmic standard deviation β_Y of Y are obtained from

$$\hat{y} = \frac{\mathrm{E}^2[Y]}{\mathrm{E}^{1/2}[Y^2]} \tag{14}$$

and

$$\beta_Y^2 = \log \frac{E[Y^2]}{E^2[Y]}, \tag{15}$$

respectively. Hence, the probability density function of Y is determined as

$$f_Y(y) = \frac{1}{\beta_Y\, y \sqrt{2\pi}} \exp\left[-\frac{1}{2}\left(\frac{\log y - \log \hat{y}}{\beta_Y} \right)^2 \right]. \tag{16}$$

Using this density function, the probability of Y exceeding a specific value y is given by

$$P\{ Y > y \} = \int_y^\infty f_Y(y')\, dy'. \tag{17}$$

This probability is considered to be the probability of exceedance for the response acceleration when a specific earthquake occurs.

3.2 *Annual probability of exceedance*

If seismic sources for a specific site are appropriately modeled, and proper attenuation equation of the PGA is used, the annual mean number, $\nu(a)$, of the PGA exceeding a specific value a at the site can be obtained. Moreover, assuming that the time series of earthquake occurrences is a Poisson process, the annual probability of exceedance for the PGA is given by

$$P_{yr}(a) = 1 - e^{-\nu(a)}. \tag{18}$$

As the probability density function, $f_{yr}(a)$, of the annual maximal PGA is calculated easily from Equation (18), the annual probability of exceedance for the response acceleration of a multi-degree-of-freedom system is given by

$$P_{yr}(y) = \int_0^\infty P\{ Y > y \mid A = a \} f_{yr}(a)\, da. \tag{19}$$

The conditional probability $P\{ Y > y \mid A = a \}$ on the right side of Equation (19) is obtained from the the density function (16) and whose parameters are determined under the condition of $A = a$.

4 CASE STUDY

4.1 *Evaluation of earthquake motions on an outcrop*

First of all, a data set will be constructed for a regression analysis of the acceleration response spectra in consideration of inhomogeneous geological conditions for earthquake motion records. Secondly, the regression analysis will be performed and on the basis of the results a probabilistic model for the acceleration response spectrum will be set up. Lastly, a probabilistic model for the input to a structural system will be proposed.

This analysis uses one hundred four horizontal acceleration records from twenty one earthquakes observed at four observation stations surrounding Tokyo Bay in Japan. These records have been selected only when the magnitude is equal to or greater than 4.5 on the Japan Meteorological Agency scale and the hypocentral distance is less than about 200km. Figure 1 shows the epicentral distribution of the earthquakes and the locations of the observation stations. Stations KOT and KSR utilize a vertical instrument array installed on the surface of a soil deposit and in base rock. Station KOT has a 41.5m-thick surface layer with an S-wave velocity less than about 400m/sec, and KSR has a corresponding 14.5m-thick layer. Stations KMT and SYK have a seismometer installed in rock under a surface layer thin enough to be neglected in terms of the earthquake motion frequency components of interest.

The observed records have already been filtered to remove components strongly affected by noise coming mainly from the observation system. The effective period range for the data is 0.1sec to 5sec.

Since input motions to structural systems are usually taken as acting on the outcrop of base rock in Japan, earthquake inputs are estimated in this way in the present paper. Therefore, for KOT and KSR earthquake motions observed in the rock at depth have to be transformed to motions on the outcrop. In these two stations, some geotechnical boring logs were performed before earthquake observation and as a result soil profiles have been obtained. On the basis of the soil profiles and the one-dimensional wave propagation theory, earthquake motions on the outcrop are obtained from the observed records

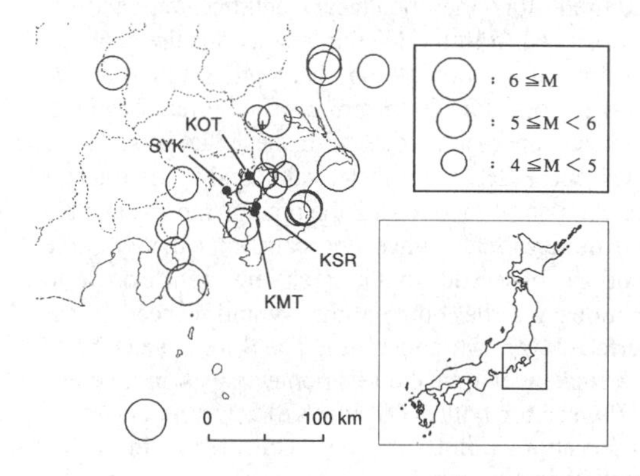

Fig.1 Epicentral distribution of selected earthquakes

(Ikeura et al. 1989). However, the logging results show a geologic column only along one vertical line in the soil and inevitably include errors. Therefore, logging results sometimes contradict observation results. Moreover, in situ investigation of soil damping has seldom been performed. If an incorrect subsurface structure model is used, the transformed motions obtained from it are of no use. This is why such soil parameters should be optimized before the transformation. In this paper, the damping factors of soil are assumed to depend on frequency f as

$$h = h_0 f^{-\eta}, \tag{20}$$

where h_0 and η are constants. This formulation makes an optimized model simpler and better than that without frequency-dependency of damping factors. Constants h_0, η and S-wave velocities of layers are the parameters to be optimized.

To set up the objective function for optimization, the amplitude of the transfer function from the observation point in the rock to that on the surface is obtained for each event from

$$|H_{A/B}(f)| = \sqrt{\frac{S_{AA}(f)}{S_{BB}(f)}}\,, \tag{21}$$

where $S_{AA}(f)$ and $S_{BB}(f)$ are the energy spectra of earthquake motions on the surface and in the rock, respectively. The geometrical mean amplitude of the transfer function, $|\overline{H}_{A/B}(f_k)|$, is used to form the objective function as

$$F(p) = \sum_k \frac{1}{\sigma^2(f_k)}\left(|\overline{H}_{A/B}(f_k)| - |\hat{H}_{A/B}(f_k, p)|\right)^2, \tag{22}$$

where f_k is the k-th frequency where the spectral ratio is given, p is the vector whose components are soil parameters to be optimized, $|\hat{H}_{A/B}(f_k, p)|$ is the amplitude of the transfer function based on the one-dimensional propagation theory, and $\sigma^2(f_k)$ is the variance of Equation (21) (Ishida 1991).

The initial model for the subsurface structure is required for the nonlinear optimization and is constructed from the logging results and the geologic column of a borehole so that optimization can be worked out for all parameters. Figures 2 and 3 compare optimal models with the initial models for KOT and KSR in parameter value and in the transfer function amplitude. Using the optimal models and the one-dimensional wave propagation theory, the motions observed in the rock are transformed to motions on the outcrop that would appear if the surface layers were removed. The S-wave velocity of the rock in which the seismometer was installed is 470m/sec for both KOT and KSR. Layers under the observation point are not considered in either observation station.

4.2 *Probabilistic modeling of response spectrum*

The multiple regression analysis has been performed using the transformed earthquake motions for KOT and KSR and observed records at KMT and SYK. The regression equation is expressed as

$$\log \hat{s}(T_i, m, x) = a(T_i)\,m - b(T_i)\,x - \log x + c_j(T_i), \tag{23}$$

where T_i is the i-th period, $a(T_i)$ and $b(T_i)$ are regression coefficients, and $c_j(T_i)$ is another regression coefficient for the j-th observation station (Takemura et al. 1987). If $\log \hat{s}(T_i, m, x)$ is regarded as $\bar{z}(T_i,m,x)$ in Equation (2), the residual of this regression equation can be regarded as a random variable $e(T_i)$ in Equation (2). Therefore, the standard deviations and correlation coefficients of the logarithmic response spectral components $Z(T_i,m,x)$ are obtained from Equation (4).

The regression coefficients are shown in Figure 4, and the standard deviations in Figure 5. It is seen

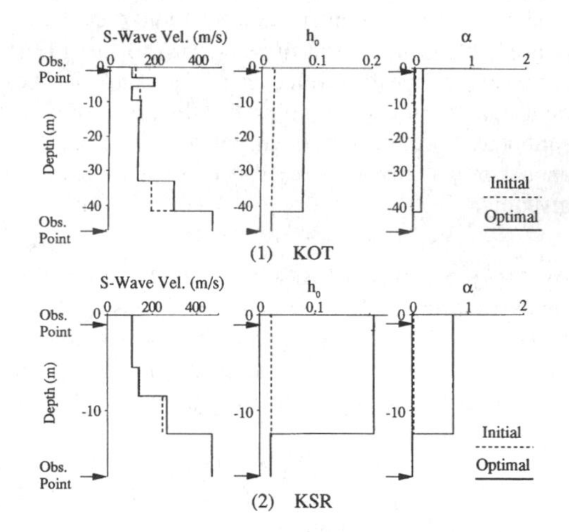

Fig.2 Initial and optimal models of soil profile

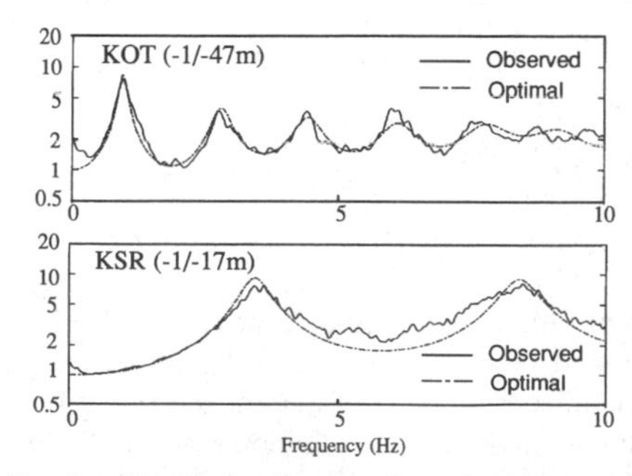

Fig.3 Amplitudes of transfer functions from observed records and optimal models

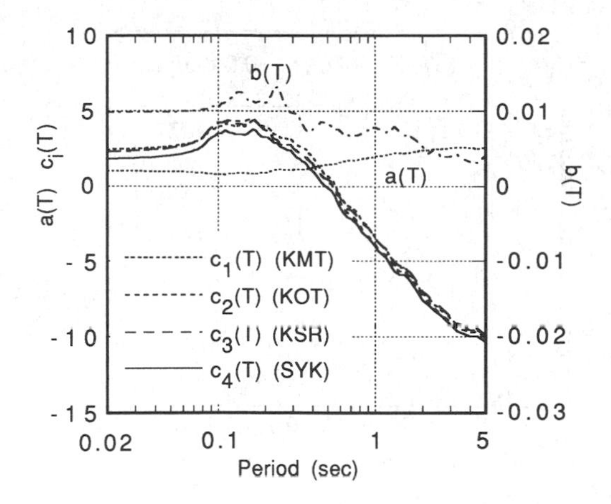

Fig.4 Regression coefficients of acceleration response spectrum

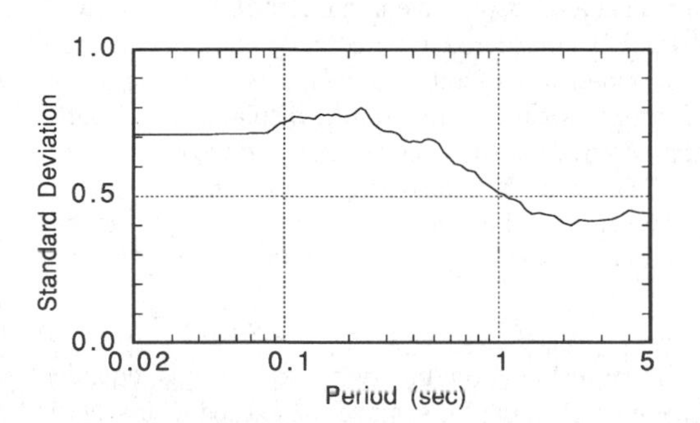

Fig.5 Standard deviation of acceleration response spectrum

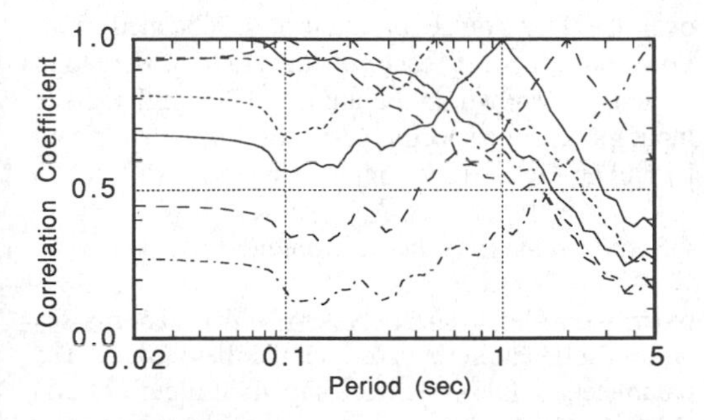

Fig.6 Correlation coefficient of seven logarithmic acceleration response spectral components over range of 0.02 to 5sec

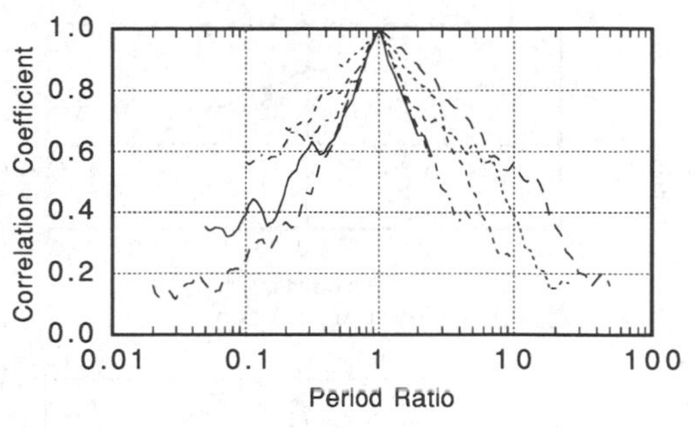

Fig.7 Relation of correlation coefficient to ratio of two periods

from Figure 5 that the standard deviation is pretty large in the short period range, but becomes smaller with increase in period. Figure 6 shows the correlation coefficients of seven components with 0.02, 0.1, 0.2, 0.5, 1.0, 2.0 and 5.0sec periods over the range of 0.02sec to 5sec. For example, the solid line taking a value of 1 at 1sec shows the correlation coefficient of the 1-second component and each component over the period range. It is seen from this figure that the correlation coefficient decreases almost monotonously in the effective period range of 0.1sec to 5sec with the distance between the two periods, but approaches asymptotically to a certain value if one of the two periods becomes shorter than 0.1sec. This is considered to be due to filtering. On closer examination, a slight asymmetry can be detected in the correlation coefficient curves even in the effective period range. This is considered to be attributed to the asymmetry of the response amplification curve of the one-mass system about the natural period.

Figure 7 is obtained from Figure 6 by changing the horizontal axis to the ratio of two periods on the logarithmic scale in the effective period range. It is seen from this figure that the correlation coefficient is related almost linearly to the ratio of the two periods in this range. Referring to Figures 5, 6 and 7, the standard deviation and correlation coefficient for the logarithmic response spectral components are modeled as shown in Figure 8.

Here, consider an active fault which generates earthquakes with a magnitude of 7 and is located 50km from a site. The median of the acceleration response spectrum for each of the four observation stations is obtained from Equation (23), as shown in Figure 9. When an earthquake occurs on the fault, the probabilistic model of the earthquake input is determined from Figures 8 and 9.

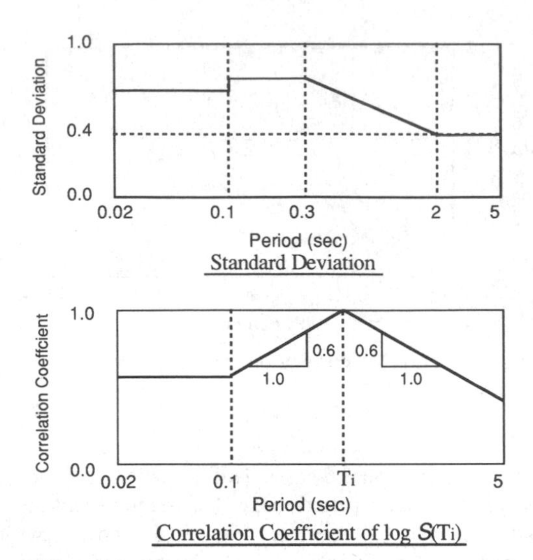

Fig.8 Model for standard deviation and correlation coefficient of logarithmic acceleration response spectral components

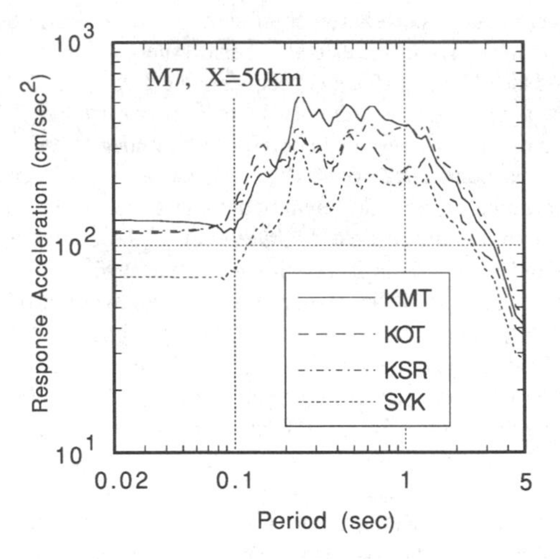

Fig.9 Acceleration response spectra for four observation stations

When evaluating the annual probabilities of exceedance for the response of a structural system, the seismic hazard curve of the PGA has to be included in the earthquake input model. The mean annual number of earthquake occurrences for the fault can be estimated from

$$\nu_0 = \frac{0.01}{10^{0.6M - 4.0}}, \tag{24}$$

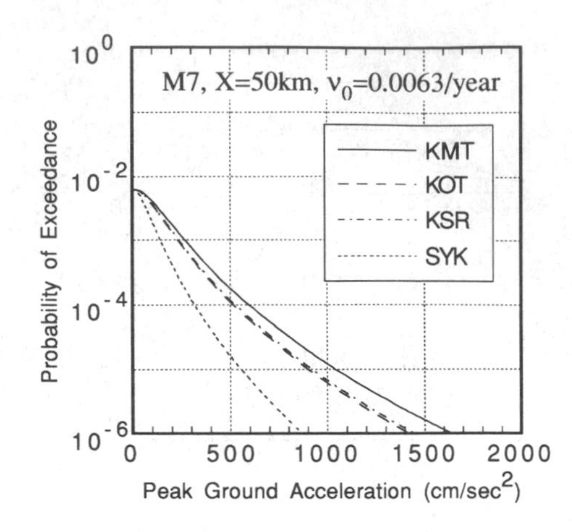

Fig.10 Hazard curves for four observation stations

where the numerator on the right side means the mean annual displacement in meter of the active fault, which is considered to be near the upperbound value for inland active faults in Japan, and the denominator is an empirical equation for the displacement in meter caused by an earthquake with magnitude M (Matsuda 1975). Given M equal to 7, ν_0 is obtained as 0.0063 per year. If there are no sources around the site except this fault, $\nu(a)$ in Equation (18) is expressed as

$$\nu(a) = \nu_0 \, P\{A>a \mid m=7, x=50km\}. \tag{25}$$

The probability on the right side of this equation means that of the PGA at the site exceeding a specific value a when the fault moves. It can be obtained from the attenuation equation and the logarithmic standard deviation for the acceleration response spectral component with a period of 0.02sec. The seismic hazard curve, or the annual probability of exceedance over the PGA range of interest, is obtained from Equation (18) for each of the four observation stations as shown in Figure 10. The probabilistic input model for a specific period of time is completed by adding Figure 10 to Figures 8 and 9.

4.3 *Application to response evaluation*

Now, consider a shear system with three masses whose foundation is fixed to the base rock. The parameters of this system are shown in Figure 11 and Table 1. The first step toward response evaluation is to estimate the expected values of the random variable U and its square. Using again the earthquake motion records used in the regression analysis, they have been obtained as

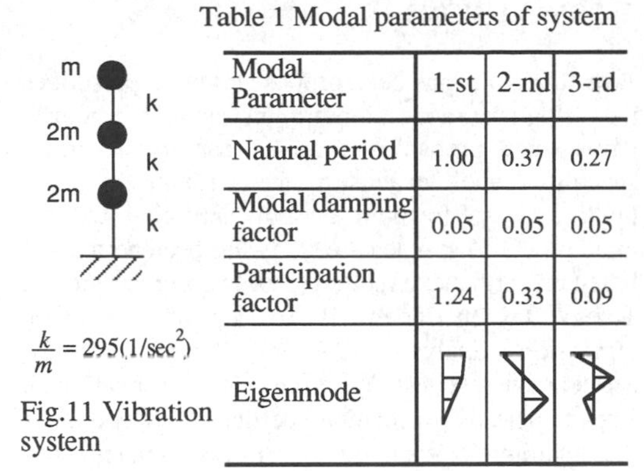

Table 1 Modal parameters of system

Modal Parameter	1-st	2-nd	3-rd
Natural period	1.00	0.37	0.27
Modal damping factor	0.05	0.05	0.05
Participation factor	1.24	0.33	0.09
Eigenmode			

Fig.11 Vibration system

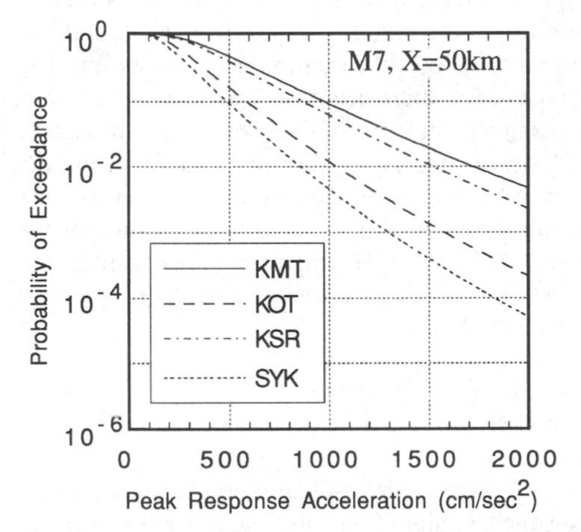

Fig.12 Probability of exceedance for response acceleration for a specific earthquake

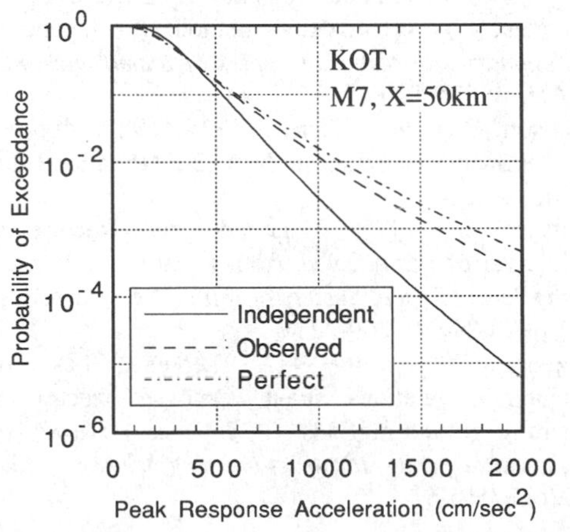

Fig.13 Effect of correlation on probability of exceedance for a specific earthquake

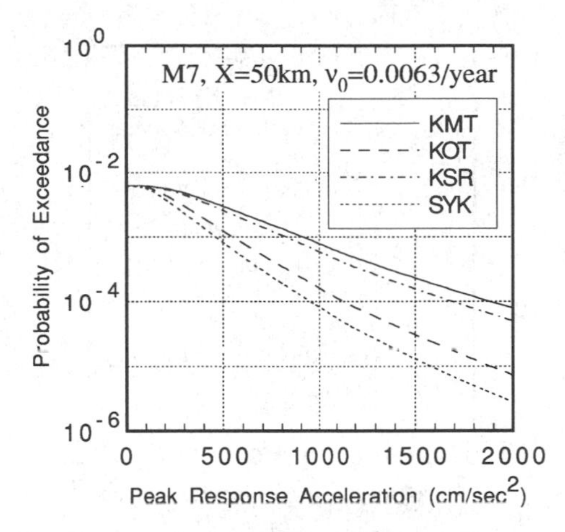

Fig.14 Annual probability of exceedance for response acceleration

$$\mathrm{E}[U] = 0.716, \quad \mathrm{E}\left[U^2\right] = 0.520 \cdot \tag{26}$$

When an earthquake with a magnitude of 7 occurs 50km from a site, the probabilities of exceedance for the response acceleration on the top mass are evaluated as shown in Figure 12. The probability of the response acceleration exceeding 1000m/sec^2 for KMT is found to be about 0.1, which is about 20 times larger than that for SYK. This difference is attributed to that between the medians shown in Figure 9.

To understand the effect of correlation between two spectral components on the response evaluation, two extreme cases are examined. One is the Perfect Case, where all correlation coefficients between two logarithmic response spectral components with different periods are 1, and the other is the Independent Case where they are 0. Figure 13 compares the results of the two cases with the result shown in Figure 12 for KOT. It is seen from this figure that the probability of exceedance becomes greatest in the Perfect Case and smallest in the Independent Case of the three cases.

When considering an input including the seismic hazard curves of Figure 10, the annual probability of exceedance for the response acceleration on the top mass of the three-degree-of-freedom system is obtained from Equation (19), which is shown in Figure 14. For KOT, the probability of exceedance for the response acceleration for a year, a decade and a centenary are shown in Figure 15, and the probability of exceedance for a centenary is compared with those of the Perfect Case and the Independent Case in Figure 16.

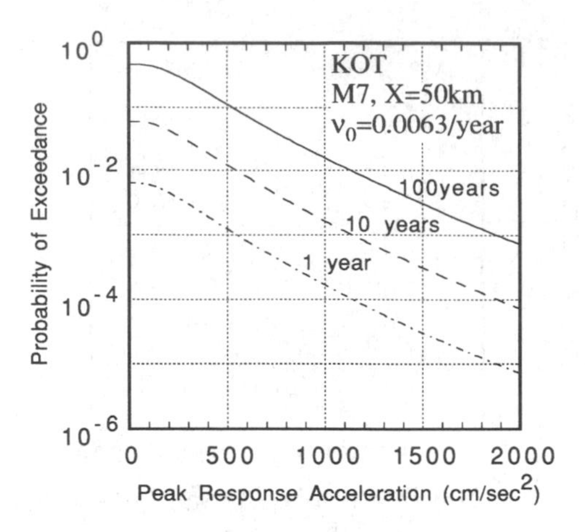

Fig.15 Effect of period on probability of exceedance

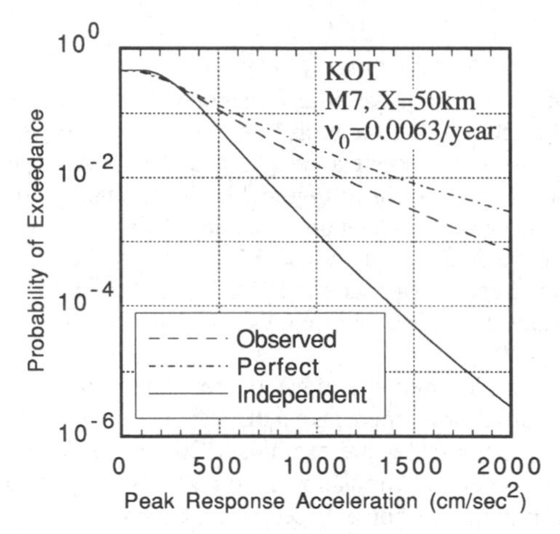

Fig.16 Effect of correlation on probability of exceedance per centenary

It is seen from these figures that the annual probability of exceedance for the response acceleration is approximately the same as the probability of exceedance for the specific earthquake multiplied by the probability of the earthquake occurrence per year, that the probability of exceedance is approximately proportional to the length of period, and that the effect of the correlation of the logarithmic response spectral components on the probability of exceedance for the response acceleration for a specific period tends to be similar to that of Figure 13.

5 CONCLUSIONS

A methodology has been proposed, which prescribes the earthquake motion input to a structural system in terms of a probabilistic model for the response spectrum and evaluates the response of a multi-degree-of-freedom system subjected to this input model. A simple case study has been performed based on earthquake motion records observed around Tokyo Bay in Japan. By means of a multiple regression analysis, the distribution of the acceleration response spectrum has been examined. In particular, the correlation coefficient of a couple of the logarithmic response spectral components has been found to be related almost linearly to the logarithm of the ratio of the couple of periods around a ratio of 1. As a result, a probabilistic model for the acceleration response spectrum has been set up. The response of a three-degree-of-freedom system has been examined in terms of the effect of the correlation coefficient on the probability of exceedance for the response. It has been found that the larger the correlation coefficients, the larger the response of the system. Hence, it is important to consider the correlation coefficients when evaluating the response of a structural system.

REFERENCES

Ikeura, T. et al. 1989. A study on characteristics of seismic motions in the rock with vertical instrument array (Part XV). (in Japanese) *Summaries of technical papers of annual meeting*, AIJ, B: 751-752.

Ishida, H. 1991. A study on subsoil structure model obtained by optimization method. (in Japanese) *Summaries of technical papers of annual meeting*, AIJ, B: 505-506.

Johnson, R.A. and Wichern, D.W. 1988. *Applied multivariate statistical analysis*. 2/E. Prentice-Hall, Inc.

Matsuda, T. 1975. Magnitude and Recurrence Interval of Earthquakes from a Fault. *Journal of the Seismological Society of Japan*, Second series, Vol.28, No.2: 269-283

Takemura, M. et al. 1987. Theoretical basis of empirical relations about response spectra of strong ground-motions. (in Japanese) *Journal of structural and construction engineering*, AIJ, Vol.375: 1-9.

Structural Safety & Reliability, Schuëller, Shinozuka & Yao (eds) © 1994 Balkema, Rotterdam, ISBN 90 5410 357 4

Scenario earthquakes vs probabilistic seismic hazard

Y. Ishikawa
Institute of Technology, Shimizu Corporation, Tokyo, Japan

H. Kameda
Urban Earthquake Hazard Research Center, Disaster Prevention Research Institute, Kyoto University, Japan

ABSTRACT : While both the probabilistic seismic hazard analysis and scenario earthquakes have been widely used for engineering seismic hazard assessment, there have been no logical links between them. This paper proposes a method for the determination of the scenario earthquakes in the context of the probabilistic seismic hazard analysis. It consists of the probabilistic re-evaluation of the magnitudes and the epicentral locations for individual source-areas in terms of their hazard-consistent magnitudes and distances. Case studies are presented for some specific regions to demonstrate its usefulness for engineering application.

1 INTRODUCTION

In the seismic hazard assessment, two typical methods are generally used in regard to the future earthquake occurrences. One is the probabilistic seismic hazard analysis (PSHA) and the other is the use of deterministic scenario earthquakes (SE). PSHA is useful as it is capable of determining the ground motion intensity parameter corresponding to a single risk index such as annual probability of exceedance, and has been used as a major tool for the site ground motion estimation. However, the physical characteristics of earthquakes such as their magnitudes and epicentral locations are not clear in the seismic hazard results. It is convenient to clarify the physical characteristics of earthquakes in case of the estimation of not only ground motion intensity but also other ground motion parameters, such as strong motion duration.

SE, on the other hand, has been used to estimate regional ground motion distributions for urban earthquake hazard mitigation planning, assessment of seismic performance of wide-spread lifeline networks, etc. In such a case, the physical characteristics of SE are unique and deterministic, and hence, not only ground motion intensity parameter but also other parameters can be simultaneously estimated all over the region. SE are normally determined from geological and seismo-tectonic considerations, and the relationship between SE and a risk index is not clear in most cases. However, as long as SE is used for engineering application, information should be provided regarding the chance of its occurrence. It is therefore desirable to determine SE on a rigorous probabilistic basis.

This study discusses a possibility of establishing a link between SE and PSHA, and proposes a method in which the methodology of PSHA is used in the determination of SE. The method is developed on the basis of the concept of the hazard-consistent magnitude and distance that have been defined as conditional means by Ishikawa and Kameda (1988). It was presented by Ishikawa and Kameda (1991). This paper deals with the results of some methodological refinement and discussion on engineering application.

2 OBJECTIVE OF SEISMIC HAZARD ASSESSMENT : SITE MOTIONS AND REGIONAL DISTRIBUTIONS

Seismic hazard assessment is reviewed from its two aspects ; *i.e.* site ground motions and regional ground motion distributions. In the site ground motion estimation, it is primarily required to determine the design earthquake load. For this purpose, PSHA is useful to be able to determine the ground motion intensity parameter as a function of a single risk index. The hazard curve which is a plot of the ground motion intensity against the risk index is commonly used as a major output of PSHA. The uniform hazard spectrum which shows a relation between the value of maximum response and the structural period given the risk level is also used in order to consider the frequency content of the earthquake ground motions. Although both of them are widely applied to evaluate the site ground motions, they tend to eliminate information on the physical characteristics of earthquakes such as their magnitude and epicentral locations. This aspect can be regarded as an advantage of PSHA, as it greatly simplifies the issue of design seismic load evaluation. With recent increasing demand for dynamic seismic design of structures, it is often the case that the ground motion duration and other parameters in addition to the intensity should be determined. This again requires the physical characteristics of the earthquakes we are dealing with be made clear. It is therefore desirable to use information on typical magnitude and epicentral location under the specific risk level.

On the other hand, in view of the urban earthquake hazard mitigation, it is necessary to estimate the ground motion distributions in the urban region caused by candidate scenario earthquakes. When PSHA is performed independently for individual locations in the region (*i.e.* "seismic hazard maps"), the conditional probability distributions of magnitude and distance under the same risk level are different from location to location, and thus it is impossible to represent the attenuation distribution of the seismic intensity caused by a specific earthquake. Hence, SE whose physical characteristics (magnitude and distance) are specified deterministically has an advantage over PSHA in the estimation of the regional ground motion distributions. By using SE, not only ground motion intensity but also other parameters can be simultaneously estimated all over the region. However, in most cases, the relationship between a certain SE and a risk index is not clear quantitatively, because the risk index is influenced by many earthquakes whose occurrence rates, magnitudes and epicentral locations are different.

From the foregoing discussion, it should be pointed out that there are no logical links between PSHA and SE, while both PSHA and SE have many advantages in the seismic hazard assessment. It is therefore quite useful and practical in the seismic hazard assessment if SE can be determined corresponding to the single risk index which has been used in PSHA.

3 A LINK BETWEEN PROBABILISTIC SEISMIC HAZARD AND SCENARIO EARTHQUAKES

The possibility of combining the methodology of PSHA with the determination of SE is discussed. This is achieved by applying the hazard-consistent magnitude and distance which have been proposed by Ishikawa and Kameda (1988). They have been defined as the conditional mean values of magnitude and distance under the condition that the ground motion intensity exceeds a specific value corresponding to a prescribed risk index.

Definition of the hazard-consistent magnitude and distance is as follows. The seismic region surrounding the site is divided into several areas (so called "source-area"). Within each source-area, parameters characterizing its seismic activities, such as the occurrence rate, b-value of earthquake recurrences, upper bound magnitude, etc. are assumed to be uniform. Earthquakes are assumed to occur randomly and independently according to the Poisson process in each source-area. Let p_0 denote the annual probability that the ground motion intensity Y exceeds a value of $y(p_0)$ $(Y \geq y(p_0))$ at the site ; *i.e.*,

$$p_0 = 1 - \exp\left[-\sum_k w_k(p_0)\right] \qquad (1)$$

where $w_k(p_0)$ is the annual occurrence rate of events such that $Y \geq y(p_0)$ caused by earthquakes in the source-area k, and can be represented by using the probability distribution of magnitude, probability distribution of distance and the conditional probability of $Y \geq y(p_0)$ given the earthquake magnitude and distance.

Once the value of $y(p_0)$ is determined corresponding to a target probability level p_0, the following discussion is made under the condition that $Y \geq y(p_0)$. Hazard-consistent magnitude for the source-area k, which is denoted by $\overline{M}_k(p_0)$, is defined as the conditional mean value of magnitude given that $Y \geq y(p_0)$ under earthquakes occurring in the source area k. Hazard-consistent distance for the source-area k $\overline{\Delta}_k(p_0)$ is defined likewise. Details of the formulation of $\overline{M}_k(p_0)$ and $\overline{\Delta}_k(p_0)$ are presented in Ishikawa and Kameda (1988).

This study proposes a method for determining the probability-based SE. It consists of the probabilistic re-evaluation of the magnitudes and epicentral distances in terms of $\overline{M}_k(p_0)$ and $\overline{\Delta}_k(p_0)$ corresponding to the representative earthquakes for important source-areas with large contribution factors. The contribution factor for the source-area k $c_k(p_0)$ is defined as the conditional probability that given the ground motion intensity Y exceeding $y(p_0)$ at the site in an earthquake, the earthquake has occurred in the source-area k ; *i.e.*,

$$c_k(p_0) = w_k(p_0) / \sum_k w_k(p_0) \qquad (2)$$

In many cases, definition of SE requires the determination of not only their epicentral distances but also their exact locations. In such cases, the exact location of SE can be determined by using the hazard-consistent azimuth for the source-area k $\overline{\Theta}_k(p_0)$ in addition to $\overline{\Delta}_k(p_0)$. $\overline{\Theta}_k(p_0)$ is defined in the same manner as $\overline{M}_k(p_0)$ and $\overline{\Delta}_k(p_0)$.

In this way, the probability-based SE can be determined corresponding to a target probability level p_0. It can be variable depending on the value of p_0 and also the frequency region of interest.

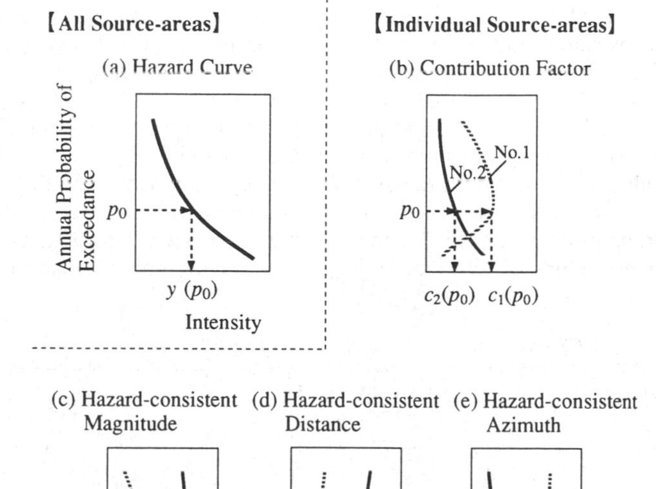

Fig. 1 Determination of probability-based scenario earthquakes

4 DETERMINATION OF PROBABILITY-BASED SCENARIO EARTHQUAKES

A method for the probability-based determination of SE is developed as an application of the hazard-consistent magnitude and distance. The procedure of this method, illustrated in Fig.1, is described as follows.

1. Divide the seismic region surrounding the site into several areas (so-called "source-areas") taking into account the regional seismicity. SE are to be determined for each of these source-areas.

2. Perform conventional PSHA for the site and generate the hazard curve. Determine the ground motion intensity $y(p_0)$, peak ground motion, spectral amplitude, or whatever else, corresponding to a target probability level p_0 (Fig.1 (a)). Note that in PSHA it is necessary to choose the appropriate ground motion intensity parameter for the structural characteristics such as the peak ground acceleration, spectral response at the specific frequency, etc.

3. Determine the contribution factor for the source-area k $c_k(p_0)$ by using eq.(2). Fig.1 (b) shows a typical relation between $c_k(p_0)$ and p_0.

4. Identify important source-areas that have large values of the contribution factor $c_k(p_0)$ for the prescribed value of p_0. They are regarded as those having dominant influence on the site under the probability level p_0.

5. Define the probability-based SE for the selected source-areas. This is achieved by determining the hazard-consistent magnitude $\overline{M_k}(p_0)$, the hazard-consistent distance $\overline{\Delta_k}(p_0)$ and the hazard-consistent azimuth $\overline{\Theta_k}(p_0)$ for the selected source-area k. The procedure is illustrated in Fig.1 (c) ~(e).

In the case of Fig.1, two probability-based SE are selected. One is an earthquake whose magnitude is $\overline{M_1}(p_0)$, epicentral distance is $\overline{\Delta_1}(p_0)$ and azimuth is $\overline{\Theta_1}(p_0)$. The other is an earthquake of $\overline{M_2}(p_0)$, $\overline{\Delta_2}(p_0)$ and $\overline{\Theta_2}(p_0)$.

5 INTERPRETATION OF PROBABILITY-BASED SCENARIO EARTHQUAKES IN VIEW OF SOURCE-AREA CHARACTERISTICS

5.1 *Case studies*

(1) *Earthquake occurrence model*

Usefulness of this method for engineering application is described with some case studies. Two sites of Tokyo and Osaka in Japan are analyzed. Fig.2 shows the source-areas models for Tokyo and Osaka that were identified according to the characteristics of earthquake occurrence. For Tokyo, three inter-plate earthquakes of Kanto, Tokai and Sanriku are considered separately from other earthquakes occurring in the 16 source-areas. Nankai earthquake is considered as a inter-plate earthquake for Osaka. The characteristic earthquake model is used for these inter-plate

(a) Tokyo

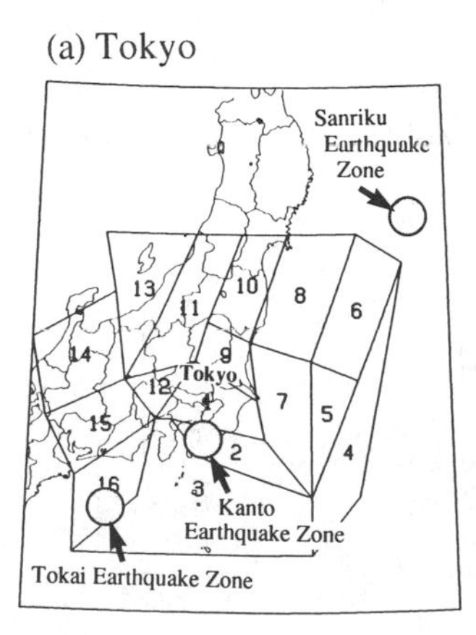

(b) Osaka

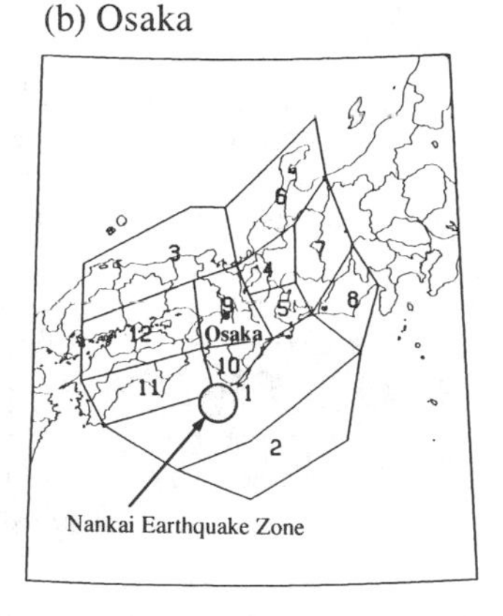

Fig. 2 Source-areas model

earthquakes and the b-value model is used for other earthquakes in order to calculate the relative frequency of magnitude. Both the inter-plate earthquakes and other earthquakes are assumed to occur according to the stationary Poisson process in the time and space domain. Seismic parameters for all source-areas such as the occurrence rate, upper bound of magnitude etc. are obtained based on the historical earthquake data in Japan.

(2) *Ground motion model*

Three ground motion intensity parameters, the peak acceleration, the peak velocity and the peak displacement are dealt with. Attenuation equations by Kameda and Sugito (1985) for the peak acceleration and velocity and by Shino and Katayama (1988) for the peak displacement are used. Attenuation uncertainties are assumed as the logarithmic normal distribution with coefficient of variation of 0.425 for acceleration, 0.448 for velocity and 0.533 for displacement.

(3) *Criteria for the selection of probability-based scenario earthquakes*

In this study, the probability-based SE are determined for the source-areas whose contribution factors are larger than 0.1 ($c_k(p_0) \geq 10\%$) against the target probability level p_0. Characteristics of the probability-based SE corresponding to three probability levels of p_0 = 0.02, 0.01, 0.005 are mainly discussed.

5.2 *Relation between probability-based scenario earthquakes and target probability level*

Fig.3 shows the PSHA results for Tokyo by using the peak displacement as the ground motion intensity. From Fig.3 (a) which is the hazard curve, the peak displacements $y(p_0)$ are estimated as 7.9cm, 10.0cm and 12.5cm

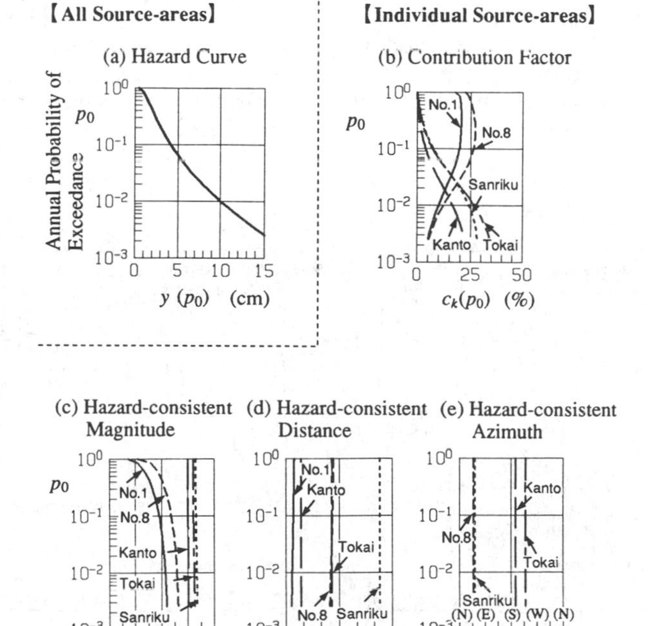

Fig. 3 Seismic hazard results for Tokyo by using the peak displacement

corresponding to p_0 of 0.02, 0.01 and 0.005 respectively. In this case, the contribution factors of five source-areas, No.1, No.8, Kanto, Tokai and Sanriku, are relatively large as shown in Fig.3 (b).

In these results, five probability-based SE's have been determined against p_0 =0.02 as such that $c_k(p_0) \geq 10\%$, while only three SE's have been defined under the same criterion when p_0=0.005. Fig.4 shows the magnitudes, epicentral distances and epicentral locations of these SE with their contribution factors.

The influence of p_0 on the probability-based SE can be explained from two aspects. One is the difference in the selected source-areas that satisfy $c_k(p_0) \geq 10\%$. As p_0 decreases, the selected source-areas are generally limited to those with large magnitudes and short distances. In this example, only three inter-plate earthquakes of Kanto, Tokai and Sanriku types are determined against p_0=0.005 as compared with the case of p_0=0.02 in which five probability-based SE's are determined.

The other influence of p_0 is to vary the values of the magnitude $\overline{M_k}(p_0)$ and the epicentral distance $\overline{\Delta_k}(p_0)$ of the probability-based SE. In general, as p_0 decreases, the magnitude of SE tend to increase and the epicentral distance tend to decrease. The azimuth $\overline{\Theta_k}(p_0)$ of SE are scarcely influenced by p_0 in most cases.

5.3 *Relation between probability-based scenario earthquakes and ground motion intensity*

Fig.5 shows the probability-based SE by using the peak acceleration as compared with them by using the peak displacement corresponding to p_0=0.01 for Tokyo. By using the peak displacement, five probability-based SE's are determined as shown in Fig.5(b). Especially, contribution factors of three inter-plate earthquakes are relatively large. On the other hand, by using the peak acceleration, only two SE's are determined. One is an earthquake occurring in the source-area No.1 and the other is the Kanto-type inter-plate earthquake. The contribution factor of the former earthquake is remarkably large in this example.

Even if the target probability level p_0 is the same, the probability-based SE's are different depending on the selection of the ground motion intensity parameter. By using the peak acceleration, the contribution factors for the source-areas located at close distance from the site are relatively large as compared with them by using the peak displacement. This reflects the fact that the acceleration represents the short-period component of earthquake ground motions with rapid attenuation with distance, while the displacement corresponds to long-period region with slower attenuation.

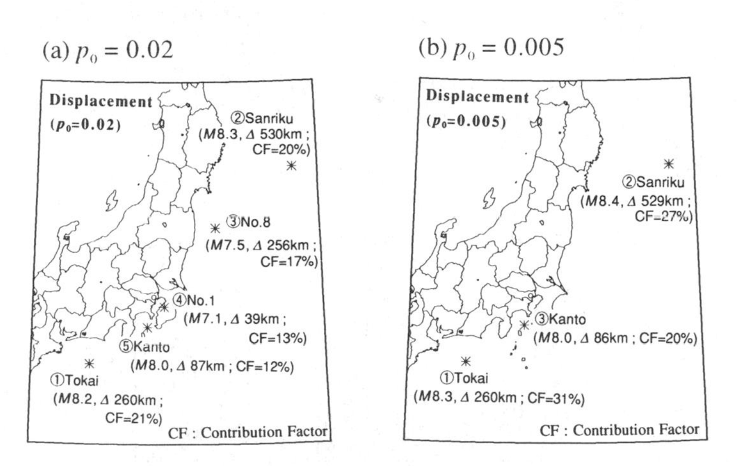

Fig. 4 Probability-based scenario earthquakes for Tokyo

5.4 *Calibration between ground motion intensity based on probability-based SE and $y(p_0)$*

One typical way of using the probability-based SE for a source-area k is to use their hazard-consistent magnitude $\overline{M_k}(p_0)$ and hazard-consistent distance $\overline{\Delta_k}(p_0)$ directly in the attenuation equation ; *i.e.*,

$$y_k'(p_0) = \widehat{y}\ \{\overline{M_k}(p_0)\ ,\ \overline{\Delta_k}(p_0)\} \qquad (3)$$

in which $\widehat{y}$ stands for the median value of attenuation equation. However, $y_k'(p_0)$ thus obtained does not coincide with $y(p_0)$.
In a probabilistic sense, the calibration can be made between $y_k'(p_0)$ and the conditional mean $\overline{y_k}(p_0) = E\ [Y\ |Y \geq y(p_0)$; source-area k]. As $\overline{y_k}(p_0)$ incorporates attenuation uncertainty while $y_k'(p_0)$ does not, the calibration may be formulated as ;

$$\overline{y_k}(p_0) = \exp\{\rho_{1k}(p_0)\cdot\zeta\}\ \cdot\ y_k'(p_0) \qquad (4)$$

in which $\rho_{1k}(p_0)$ is the "calibration factor for conditional mean", and ζ is the scale of attenuation uncertainty given in terms of the logarithmic standard deviation.

In engineering applications, the threshold value $y\ (p_0)$ is employed as the reference value of the ground motion intensity. In this case, $y\ (p_0)$ and $y_k'(p_0)$ are calibrated in the following manner ;

$$y\ (p_0) = \exp\{\rho_{2k}(p_0)\cdot\zeta\}\ \cdot\ y_k'(p_0) \qquad (5)$$

(a) Peak Acceleration

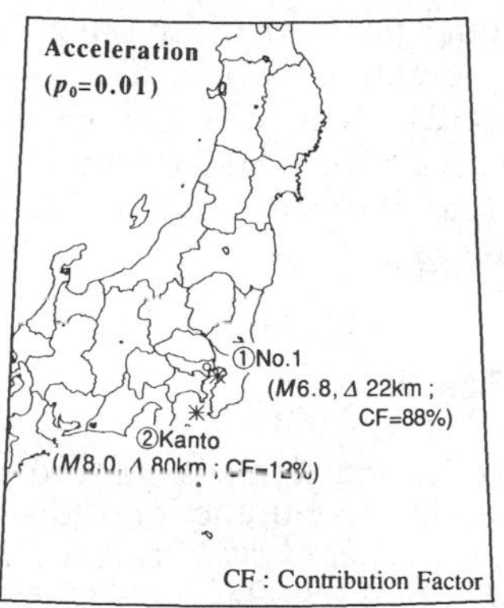

(b) Peak Displacement

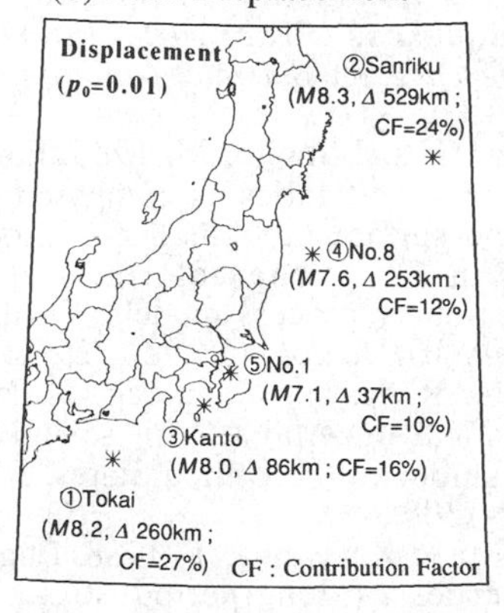

Fig. 5 Probability-based scenario earthquakes for Tokyo

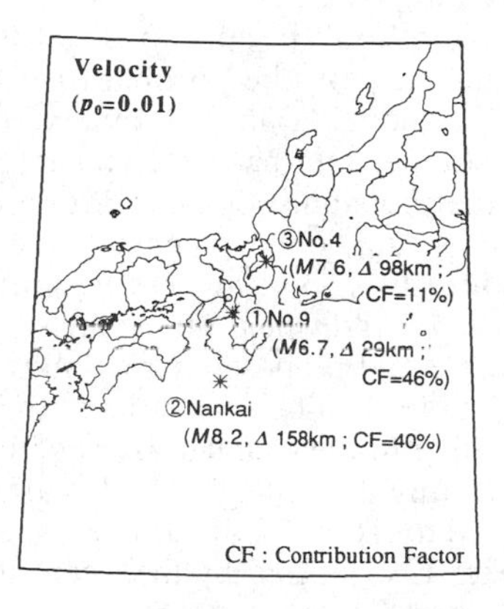

Fig. 6 Probability-based scenario earthquakes for Osaka

Table 1 Calibration Factors for Osaka (p_0=0.01)

source-area	$y_k'(p_0)$ (cm/s)	$\overline{y_k}(p_0)$ (cm/s)	$\rho_{1k}(p_0)$	$y(p_0)$ (cm/s)	$\rho_{2k}(p_0)$
No.9	12.4	20.1	1.13	13.0	0.11
Nankai	13.0	19.2	0.91	13.0	0.0
No.4	11.2	19.1	1.25	13.0	0.35

in which $\rho_{2k}(p_0)$ is the "calibration factor for threshold intensity".

Fig.6 shows the probability-based SE corresponding to p_0=0.01 for Osaka by using the peak velocity. Two intra-plate earthquakes occurring in the source-areas No.4 and No.9, and the Nankai-type inter-plate earthquake are determined. Table 1 shows the values of the ground motion intensity parameters and the calibration factors discussed above based on the results for the Osaka region.

6 CONCLUSIONS

The main results of this study may be summarized as follows.

1. Discussion is made on a link between the probabilistic seismic hazard analysis and the scenario earthquakes.
2. A method is proposed for the determination of the probability-based scenario earthquakes in the context of the probabilistic seismic hazard analysis on the basis of the probabilistic re-evaluation of the magnitudes and the epicentral locations corresponding to the representative earthquakes for individual source-areas.
3. The method proposed herein allows one to define more than one scenario earthquakes corresponding to the characteristics of source-areas affecting the site. In such cases, the relative importance of those scenario earthquakes is evaluated by their contribution factors.
4. Case studies are presented for some specific regions in Japan and its usefulness for engineering application is demonstrated.

ACKNOWLEDGMENTS

Following the presentation of Ishikawa and Kameda (1991), there were chances of correspondence and direct discussion between C. Allin Cornell and the authors on the methodologies developed herein. Prof. Cornell's interest in this subject was related to a U.S. trend to incorporate a semi-probabilistic method in the determination of controlling earthquakes for seismic design of nuclear power plants. It was indicated by Prof. Cornell that the discussion going on in the U.S. uses the concept of "effective magnitude" $\bar{M}$ and "effective distance" $\bar{D}$ defined by McGuire and Shedlock (1981). Their definition is mathematically identical with "hazard-consistent magnitude $\overline{M_k}(p_0)$ and "hazard-consistent distance" $\overline{\Delta_k}(p_0)$ used in this study. While the authors' $\overline{M_k}(p_0)$ and $\overline{\Delta_k}(p_0)$ have been proposed under a definite intention to establish a link between probabilistic seismic hazard and deterministic scenario earthquakes through zone-by-zone evaluation, McGuire and Shedlock's $\bar{M}$ and $\bar{D}$ were defined as part of a set of parameters for discussing statistical uncertainties in the seismic hazard evaluation. This is a major reason for the authors not to have recognized McGuire and Shedlock's definition before publishing Ishikawa and Kameda (1988, 1991). The authors, however, would like to express their respect to McGuire and Shedlock's work that has motivated the aforementioned trend in the U.S.

Including this and many other interesting points, the authors would like to gratefully acknowledge that discussion with Prof. Cornell was very informative and has stimulated the authors to deepen the methodology.

REFERENCES

Ishikawa, Y. and Kameda, H. 1988. Hazard-consistent magnitude and distance for extended seismic risk analysis. *Proc. 9th WCEE, Vol.II* : 89-94.

Ishikawa, Y. and Kameda, H. 1991. Probability-based determination of specific scenario earthquakes. *Proc. 4th International Conference on Seismic Zonation, Vol.II* : 3-10.

Kameda, H. and Sugito, M. 1985. Earthquake motion uncertainties as compared between ground surface motions and bedrock input motion --- characterization using evolutionary process models. *Transactions of 8th SMIRT, Vol. M1K1/3* : 297-302.

McGuire, R. K. and Shedlock, K. M. 1981. Statistical uncertainties in seismic hazard evaluations in the United States. *BSSA 71* : 1287-1308.

Shino, I. and Katayama, T. 1988. Engineering properties of long-period strong motion evaluated from displacement seismograph records. *Proc. 9th WCEE, Vol.II* : 251-256.

Structural Safety & Reliability, Schuëller, Shinozuka & Yao (eds) © 1994 Balkema, Rotterdam, ISBN 90 5410 357 4

Seismic safety evaluation of existing buildings in Japan

Jun Kanda & Ryoji Iwasaki
University of Tokyo, Japan
Hajime Kobayashi
Kajima Technical Research Institute, Tokyo, Japan
Bruce R. Ellingwood
Johns Hopkins University, Baltimore, Md., USA

ABSTRACT: The seismic safety of recent six existing buildings in Japan is evaluated in terms of a probabilistic second-moment seismic safety measure. These buildings are analyzed using lumped-mass models and the seismic hazard is estimated from a probability-based seismic hazard map. The distribution of the safety measure with story suggests that significant damage may be concentrated in only a few stories in some buildings. The seismic safety of existing buildings can be approximated by the proposed second-moment seismic margin index. The feasibility of the proposed index is demonstrated.

1. Introduction

It has been a long-term issue to evaluate the performance and reliability of existing buildings against seismic hazards in a reasonably objective manner. Especially in Japan, it has been necessary to extend the existing reliability-based design procedures to include cases in which inelastic structural response to strong ground motion must be taken into account because earthquake-resistant design is of primary concern. Reflecting the worldwide trend that probability-based design procedures are being introduced into codes of practice, LRFD design procedures were proposed by the Architectural Institute of Japan (AIJ) and these procedures recently have been published as the AIJ Limit States Design Standard for Steel Construction (1990)[1].

Usually the design and construction of buildings predates the introduction of probabilistic procedures. Especially when the perceived hazard is unusually severe, and particularly for reactor buildings, up-to-date probabilistic safety assessments may be performed to evaluate an existing building. When public health and safety are at issue, however, objective and quantitative data should be factored as much as possible into the safety assessment of an existing building.

In this study, therefore, we propose for this purpose a relatively simple seismic safety evaluation method based on second-moment reliability principles. The methodology is illustrated by performing seismic hazard evaluations, inelastic response analyses, and relative seismic safety assessments for six existing steel and reinforced concrete buildings.

2. Summary of Analysis[2)]

The analytical procedure is illustrated in Fig. 1. First of all, the mean of the 50-year maximum ground acceleration at a building site, which is assumed to define the seismic hazard, is estimated. The relationship between structural response, in terms of the equivalent elastic response, and the peak acceleration of the input ground motion is established. Finally, the seismic margin index β for each story in the building is developed, which leads to the final safety evaluation for the building.

2.1 Seismic Hazard Evaluation

The mean of the 50-year maximum bedrock velocity was obtained according to a seismic

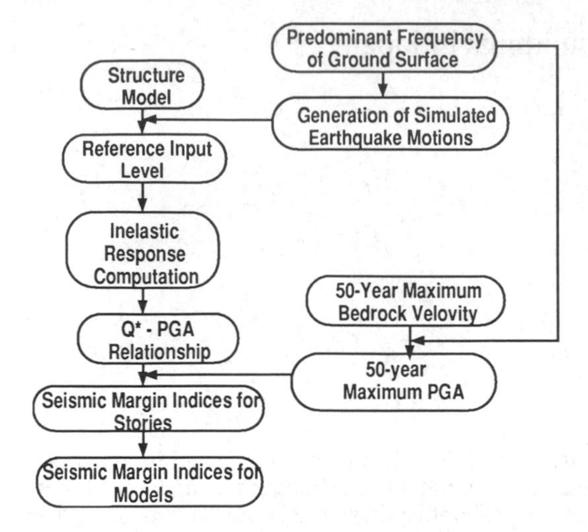

Fig. 1 Summary of Analytical Procedure

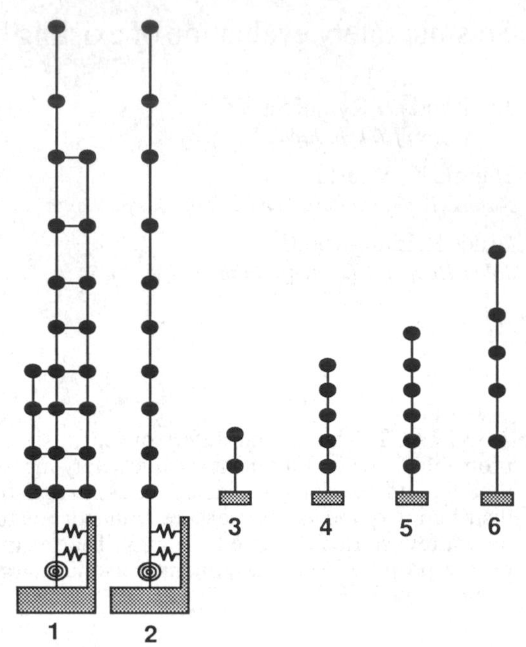

Fig. 3 Building Models

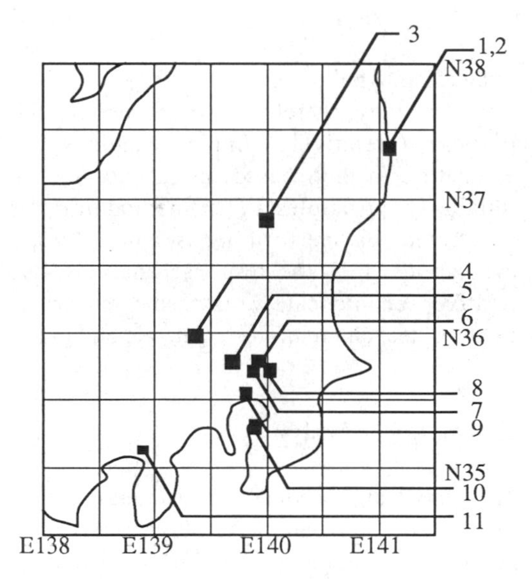

Fig. 2 Location of Buildings

Table 1 Outline of Buildings

Model#	Stories	Type	Occupancy	Height (m)	C_B	S/D*1
1	9	RC	NPP*2	70.0	0.658	D
2	9	RC	NPP	70.0	0.671	D
3	2	RC	Warehouse	8.59	0.200	S
4	5	S	Office	19.6	0.230	S
5	6	S	Office	23.6	0.262	S
6	5	S	TPP*3	36.5	0.593	D

*1 This indicates the information used to determine the restoring force characteristics of each story was based on whether static analysis (S) or dynamic analysis (D) was employed.
*2 NPP = Nuclear Power Plant Building
*3 TPP = Thermal Power Plant Building

hazard map proposed by Kanda and Dan[5]. The mean of the 50-year maximum ground acceleration, which represents the intensity of an input ground motion, was obtained by multiplying the bedrock velocity by the magnification ratio of the surface layer at the building site[4].

As for the coefficient of variation (c.o.v.) of P.G.A. was taken as 60 percent in this study[4]. The probability distribution of P.G.A. was assumed to be log-normal according to the central limit theorem, since the P.G.A. is modeled as the product of random variables.

2.2 Modeling

Fig. 2 shows the location of the buildings in the central Honshu region of Japan. The buildings were modeled as stick-type, lumped-mass models with rigid-floors, except for Models 1 and 2 where a slightly more detailed model was employed, as shown in Fig. 3. These are of a reactor building and has sway-rocking springs at the bottom to take into account soil-structure interaction.

The hysteretic restoring force characteristics of each story were defined by the Origin-Oriented (O.O.) rule for reinforced concrete structures (RC) or steel-reinforced concrete structures (SRC) and by a Normal Bilinear (N.B.) rule for steel structures (S). The O.O. rule allows for degradation in concrete stiffness due to progressive cracking during reversals of load. The damping coefficients were assumed to be 5 percent for RC and SRC structures and 2 percent for steel structures, which are customarily used in Japan. Additional information about the buildings is summarized in Table 1. In Table 1, C_B represents the base shear coefficient, which is defined as the ratio of the yield shear force of the first story to the weight of the building.

2.3 Relationship between Equivalent Elastic Response and P.G.A.

In order to evaluate the inelastic response of a structure, the equivalent elastic response, Q^*, with equal strain energy is defined by a transformation of the maximum inelastic restoring force characteristic. The relationship between equivalent elastic response and P.G.A. is deduced from inelastic response computations using simulated ground motions that are suitable for an individual site. Simulated ground motions are generated for a given earthquake magnitude and given Fourier amplitude spectrum[5]. These values were chosen to represent the earthquake magnitudes because both the wave form and the response differs depending on the magnitude of the input motions.

2.4 Definition of Ultimate Limit State

The ultimate limit state must be identified for each structural type in order to have a basis for computing a reliability measure. The shear strain, the deflection and the cumulative deformation were used as measures for the ultimate limit state of a reactor building, and RC, SRC, and S buildings, respectively. Values for these limit state measures are determined as follows;

RC bearing wall of reactor building : $\gamma=5.0x10^{-3}$ (shear strain)

RC moment resisting frame : $\gamma=1/50$ (interstory deflection angle)

SRC moment resisting frame : $\gamma=1/30$ (interstory deflection angle)

S moment resisting frame : $\eta=14$ (cumulative plastic deformation ratio)

The value $\eta=14$ was obtained by defining the ductility factor $\mu=8$ and introducing the relationship $\eta=2(\mu-1)$[6]. The equivalent elastic resistance, R, is obtained in the same manner as Q^* is obtained.

2.5 Seismic Margin Index β

Making the simplifying assumption that the probability distribution of both equivalent elastic response, Q^*, and resistance, R, are log-normal, a seismic margin index β based on second-moment reliability can be obtained for each story as follows[4];

$$\beta=\frac{\ln m_R-\frac{\zeta_R^2}{2}+\frac{\ln m_R}{\ln m_a}\left(\frac{\zeta_a^2}{2}+\ln\frac{a_y}{\mu_a}\right)}{\sqrt{\zeta_R^2+\left(\frac{\ln m_R}{\ln m_a}\zeta_a\right)^2}} \qquad (1)$$

in which $m_R = R/Q_y^*$, $m_a = a_R/a_y$, and ζ_Q* and ζ_R are logarithmic standard deviation of Q^* and R. Q_y^* means the yield shear force response in terms of equivalent elastic response caused by a_y; a_R is the P.G.A. corresponding to R. μ_a stands for the 50-year maximum P.G.A.

A seismic margin index β for each story is calculated according to eq.(1). The probability of failure, P_f, can be obtained from the standard normal cumulative distribution function, $\Phi(.)$, as $P_f=\Phi(-\beta)$.

Assuming that the failure probability of a structure is approximated by the largest failure probability of any story in the building, the seismic margin index of a structure is obtained as the smallest β of any story. This β is associated with a given earthquake magnitude, either $M=5.5$ or $M=8.5$, and should be regarded as 'conditional' β (β_5 or β_8). The 'unconditional' seismic margin

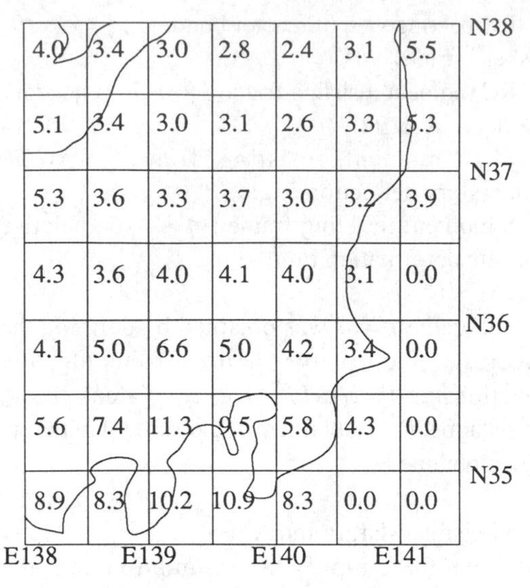

Fig. 4 Seismic Hazard Map

Table 2 Predominant Period of Surface Layer and Seismic Hazard at Each Site

Model No.	T_G (sec)	μ_v (cm/s)	μ_a (cm/s^2)
1,2	0.400	4.162	130.76
3	0.247	4.963	155.92
4	1.011	6.596	208.36
5	1.710	5.016	206.08
6	0.922	8.893	279.39

Table 3 Comparison of Seismic Hazard Evaluation using Different Existing Seismic Hazard Maps in terms of the 50-year Maximum P.G.A

Model No.	1	2	3	4	5	6
1,2	131	119	112	119	127	380
3	156	356	149	296	127	476
4	208	356	150	296	127	476
5	206	356	149	296	166	622
6	279	356	186	356	159	476

(The numbers in the top row corresponds to the seismic hazard maps, i.e., Dan and Kanda (1), Kawasumi (2), Kanai (3), Goto (4), Ozaki (5), and Makino (6), respectively. This table is based on the following publication: AIJ, 'Seismic Loading - state of the art and future developments', p.86 (1987) (in Japanese))

index β for a structure is calculated from β_5, and β_8 by weighing the occurrence frequency N_1, and N_2 as follows;

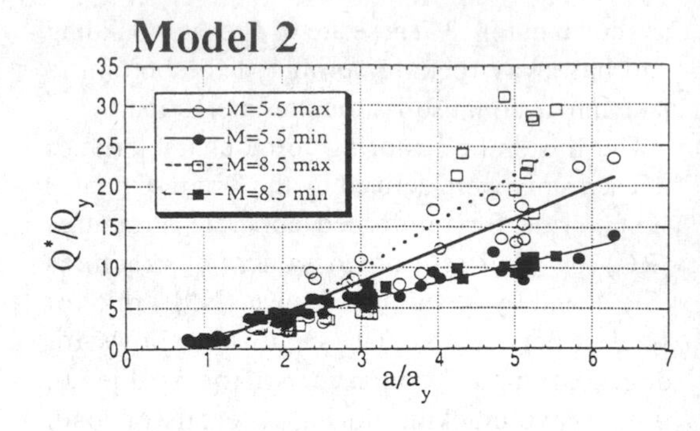

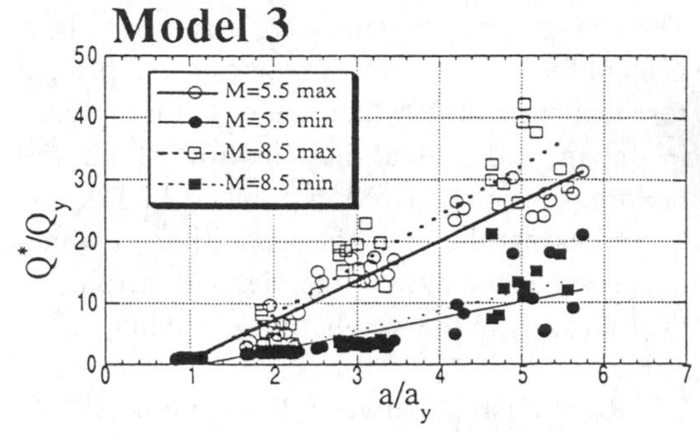

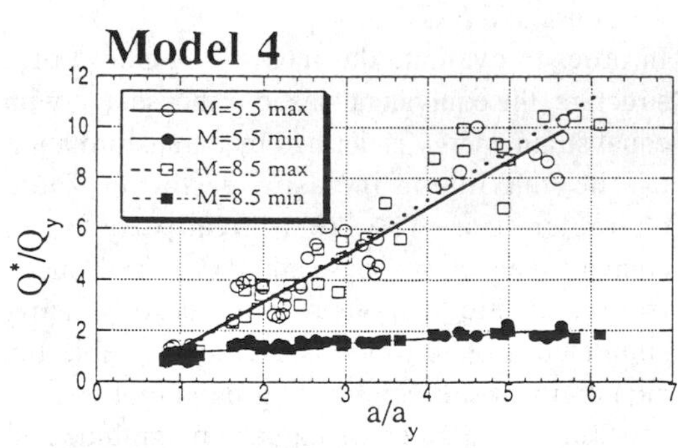

Fig. 5 Relationship of Equivalent Elastic Response v.s. P.G.A.

$$\beta = -\Phi^{-1}\{\frac{1}{N_1+N_2}[N_1\Phi(-\beta_5) + N_2\Phi(-\beta_8)]\} \quad (2)$$

Lumping the seismic hazard into two magnitudes simplifies the safety assessment and was found to have a negligible impact on the results.[4)]

3. Result of Analysis and Considerations

3.1 Seismic Hazard Evaluation

Fig. 4 shows the seismic hazard in terms of bedrock velocity (cm/s) for the central Honshu region of Japan. Table 2 summarizes the predominant period of the surface layer, the mean of the 50-year maximum bedrock velocity and the ground acceleration. There have been proposed several different approaches to evaluate seismic hazard and also some seismicity maps have been available based on those methods. Table 3 shows the results of seismic hazard evaluation due to different seismicity maps including results obtained in this study in terms of P.G.A. Although some fairly large discrepancies can be found among the estimated values for P.G.A. at a site, the values used in this study seem to be around the average and regarded as appropriate.

3.2 Relationship between Equivalent Elastic Response and P.G.A.

The relationship between Q^* and P.G.A. is shown in Fig. 5 for the stories in building models 2, 3 and 4 where the inclination of the regression line is the largest and the smallest. As seen from the results for the models for which the hysteretic rule is O.O. (Models 2 and 3; see Table 1), the effect of duration on magnitudes on inelastic response is considerable. There are several models where the difference in inclinations for individual stories is considerable; at the same time, there are also some models where this difference is not significant (e.g., Model 2). These differences may be caused by the concentration of damage in a particular story.

3.3 Variability of Seismic Hazard and Structural Response

The coefficient of variation (c.o.v.) of the equivalent elastic response, Q^*, around the regression line approximating the relationship between Q^* and P.G.A., as shown in Fig. 5, is 20% to 30%. Meanwhile, c.o.v. of the seismic hazard in terms of P.G.A. was assumed to be 60%[3]. Thus the variability of seismic hazard is far greater than that of structural response, it has been shown that c.o.v. of P.G.A. affects more than c.o.v. estimation of Q^*. Although some other previous studies dealt with the variability of seismic hazard as greater values like 75%[8], the value 60% assumed in this study appears to be more reasonable because this assumption was made taking into account the variability of various factors such as the variability of bedrock velocity, which is assumed to be around 40%, and

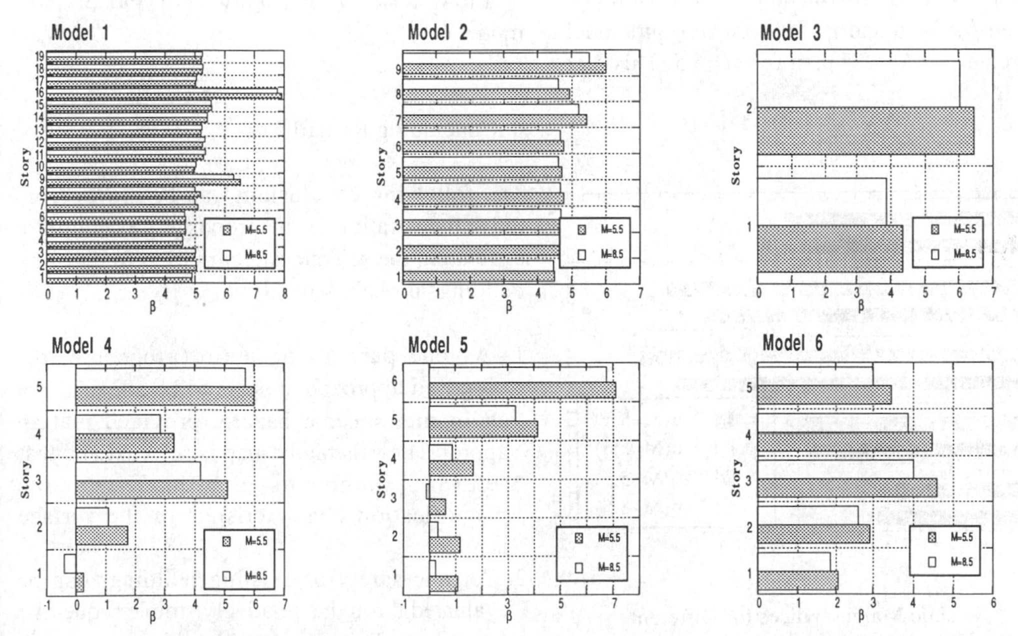

Fig. 6 Seismic Margin Indices for Stories

uncertainty in ground amplification. Since it seems scarcely possible to give any definite values as c.o.v. of seismic hazard, it may have to be left to a sound engineering judgment at this stage.

3.4 Seismic Margin Index

Figures 6(a)-6(f) show the seismic margin indices for all stories of the six building models. For models 1, 2, and 3, the differences between β for all stories are generally small. In contrast, differences are significant for models 4, 5, and 6; this may be explained by the concentration of damage in a particular story as was expected from the Q^*-P.G.A. relationship. β for the bottom story is not always the smallest because of the relatively low elastic limit in shear strength in some of the higher stories.

Interestingly, the difference in β when $M=5.5$ and $M=8.5$ usually is not large. This suggests that a discrete earthquake magnitude modeling may be a reasonable simplification. The seismic margin indices computed on the basis of maximum ductility factor, μ, and in terms of cumulative plastic response ratio, η, are compared in Fig. 7 for the steel structures. Obviously, the difference between β values for different magnitudes is more significant when cumulative response is used; for instance, the difference between βs for μ and η of the 1st story in model 6 is greater for $M=8.5$ than for $M=5.5$. This fact can be attributed to the effect of duration of ground motion.

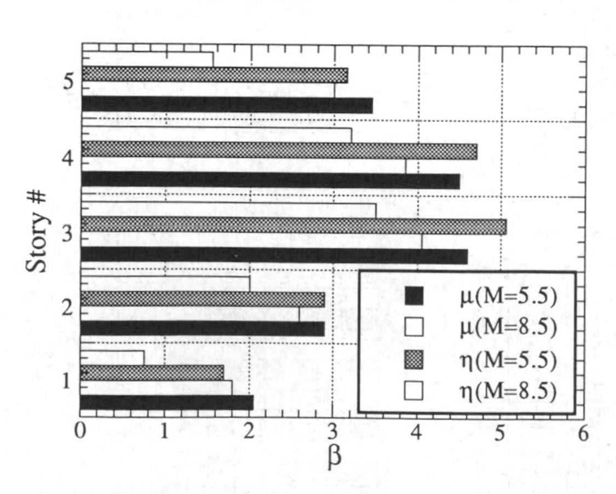

Fig. 7 Seismic Margin Indices for Different Response Measures

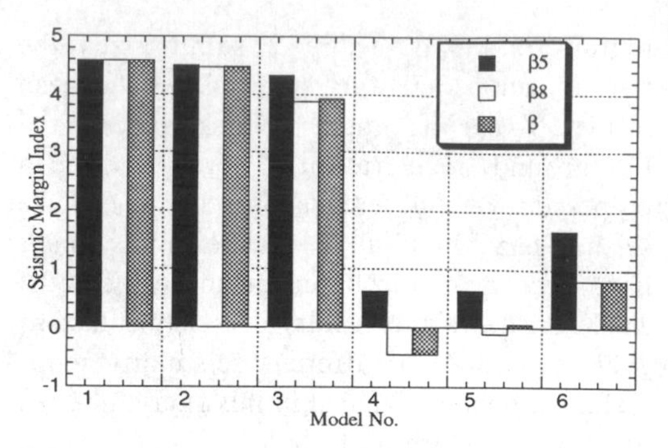

Fig. 8 Seismic Margin Indices for Models

A comparison of seismic margin indices for all building models is shown in Fig. 8. Maximum response was used for every model in this comparison. The seismic margin indices for the buildings which can be regarded as important from the viewpoint of occupancy types, e.g., reactor building, are relatively high.

Since the yield and ultimate resistance for model 4 have been evaluated based on the allowable stress, β for that model is very low. A high seismic hazard level and significant damage concentration in a few stories lead to a low β for model 5 and model 6.

4. Concluding Remarks

The following conclusions can be drawn from the demonstration of the proposed procedure for evaluating the seismic safety margin for existing buildings probabilistically:

1. As one part of the entire procedure, the proposed approach appears to be adequate for estimating seismic hazard, provided that an appropriate seismicity map is selected and that relevant judgments are made on the magnification characteristics of the surface layer.
2. Seismic safety of existing buildings can be evaluated through a relatively simple sequential procedure. There is considerable variation in

the seismic margin among buildings. Highly important buildings should have a relatively higher seismic margin.

3. The probability-based seismic margin index β focuses on damage concentration effects, which would considerably influence the seismic safety of a building. It has also been shown that the estimation of structural resistance plays an important role in evaluating seismic safety of a building.
4. The seismic margin index was originally determined based on the maximum total or interstory deflection. Results based on cumulative deflection response, however, show that cumulative response can also be a basis for a probability-based seismic margin index.

5. Recommendations for Further Study

The sequential analytical procedure requires further study since various assumptions have been introduced in order to simplify the procedure. Among the items requiring further research are the following:

1. Further examination of the c.o.v. of seismic hazard is desirable in order to validate the probabilistic seismic safety measures.
2. The proposed procedure can be extended for a building with a more complicated shape or numerous failure modes by system-failure analysis and examining the variability and correlation between strengths of individual stories.
3. Objective evaluation of the ultimate resistance is urgently needed for more rational seismic safety evaluation. An analytical procedure to predict the statistics of this resistance at the design stage would be desirable.

Acknowledgment

Tokyo Electric Power Company is gratefully appreciated for supporting this study.

References

(1) AIJ. 1990. Limit-State Design Standard for Steel Construction, AIJ
(2) J. Kanda et al. 1987. Stochastic Evaluation of Inelastic Seismic Response of a Simplified Reactor Building Model, Transactions of the 9th SMiRT, Vol.K1: 403-408
(3) J. Kanda, and K. Dan. 1987. Distribution of seismic hazard in Japan based on an empirical extreme value distribution , Structural Safety 4(3): .50-56.
(4) J. Kanda et al. 1990. Probability-Based Seismic Margin Evaluation of Existing Buildings, Proc. 8th Japan Earthquake Engineering Symposium: 1575-1580.
(5) H. Tajimi. 1960. A Statistical Method of Determining the Maximum Response of a Building Structure During an Earthquake, Proc. 2WCEE: 781-797.
(6) B. Kato et al. 1988. Evaluation of Structural Characteristic Coefficient for Steel Structures, Research Report, Grant No.61460173, Ministry of Education, Japan: 113.
(7) J. Sakamoto. 1988. Load Factors for Use in Probability-based Limit States Design, Transactions of AIJ, No. 393: 71-80.
(8) AIJ. 1987. Seismic Loading - state of the art and future developments, AIJ: 87

Structural Safety & Reliability, Schuëller, Shinozuka & Yao (eds) © 1994 Balkema, Rotterdam, ISBN 90 5410 357 4

Uncertainty analysis in seismic hazard evaluations

Chin-Hsiung Loh & Wen-Yu Jean
Department of Civil Engineering, National Taiwan University, Taiwan

ABSTRACT: The purpose of this paper is to determine the uncertainties in seismic hazard evaluation resulting from uncertainties in models and parameters used as input to the analysis. With the aim of determining uncertainties, the following two subjects are discussed: (1) effect of dispersion on peak ground acceleration to the result of seismic hazard analysis, (2) effect of modal parameters of spectral acceleration probability distribution to the influence on uniform hazard response spectra. An uniform hazard response spectrum was generated and discussed by using different methods of parameter estimation on the distribution of spectral accumulation.

1. INTRODUCTION

Seismic hazard analysis (SHA) is concerned with estimating the probability of occurrence of seismic hazards, principally ground shaking, during a given time period. It is common to state the probability as a return period (R_T) in year for a seismic event exceeding a stated intensity [Cornell, 1968; DerKiureghian & Ang, 1977]. There are considerable uncertainties and simplifications associated with the derivation of seismic hazard from earthquake and this should be reflected in the design rules. The major sources of uncertainty include: probabilistic modelling, selection of source zones, attenuation law et al. Among them the most important input into seismic hazard analysis is the attenuation law which describes the dependence of earthquake shaking on source type, source-site distance and site conditions. Typically this is of the (empirical) form:

$$\ln Y = a + b\, M_L + c\, \ln \left[R + d\, \exp(e M_L)\right] + \varepsilon \quad (1)$$

where Y represents peak ground motion acceleration (PGA), M_L is the Richter magnitude, R is the hypocentral distance, and ε is a zero-mean random error term. The coefficients a, b, c, d, and e typically are estimated from empirical data. The difference between the observed and the predicted values of the natural logarithm of PGA at specific value of magnitude and distance is the dispersion between prediction model and data that must be taken into account in seismic hazard analysis.
In the development of response spectra for use in designing seismic resistant structures, it has been common practice to first establish sets of spectra normalized to a peak ground acceleration (PGA) equal to 1 g and then to scale them down to specified PGA levels in their design applications [Seed, *et al.*, 1976; Newmark, *et al.*, 1968]. It should be recognized that the resulting spectra do not represent the same probability of exceedance over the full frequency range of interest. Because of this deficiency, the present trend in engineering practice is to develop sets of design response spectra which do represent uniform probability of exceedance over the entire frequency of interest. These so-called "uniform-hazard response spectra" are now being established [Penzien, *et al.*, 1992] by generating sets of seismic hazard curves. Uncertainties involve in generating the uniform-hazard response spectra are carefully discussed.

2. EFFECTS OF PGA DISPERSION ON SEISMIC RISK

To consider the effect of the PGA uncertainty on the calculated seismic risks, the actual intensity Y_a is defined

$$Y_a = N\, Y \quad (2)$$

where N is the correction for error in the PGA attenuation equation. The required probability, including the effect of this uncertainty becomes:

$$P[Y_a > y] = P[NY > y] = \int_0^{\infty} P[NY > y | N = \nu]\, f_N(\nu)\, d\nu \quad (3)$$

where $f_N(\nu)$ is the probability density function of N. The lognormal distribution was used to represent the dispersions of the PGA values about their mean level. Set $Z = \ln N$, then Z will follow the normal distribution. Through transformation, Eq. (3) can be rearranged in the following form

$$P[Y_a > y] = a \int_{ZL}^{ZU} P[Y > ye^{-z}]$$
$$\left\{ \frac{1}{\sigma\sqrt{2\pi}} \exp\left[-\frac{1}{2}\left(\frac{z-\mu}{\sigma}\right)^2 \right] \right\} dz \qquad (4)$$

in which μ denotes the mean-value of Z and σ^2 denotes the variance of Z, ZU and ZL specify the upper and lower bound of integration to truncate the normal distribution to avoid integration from negative infinity to positive infinity. A correction factor, "a", is introduced that makes the area under the normal probability density function between these integration cut-off limits to one. Now define a new random variable w as given by

$$w = (z - \mu)/\sigma \qquad (5)$$

The probability density function $p(w)$ for variable w will be of the symmetrical Guassian (normal) form shown in Fig. 1. Suppose, for example, the tails of this function are cut off at $w = +2$ and $w = -2$. Since the area under $p(w)$ between these cut-off limits equals 0.9544, that protion of the function in the range $-2 < w < 2$ could simply be scaled up by the factor 1/0.9544 yielding a new function $f(w)$ which does satisfy the unit-area requirement within these cut-off limits. While this rather standard truncation procedure can be used, it results in the probability density function $f(w)$ having discontinuities in both its oridnate and slope at each cut-off point. Such discontinuities are, of course, unrealistic from a physical point of view.

To remove the above-described discontinuities on uncertainty, a modified truncation procedure is used herein which makes use of the modified probability density function

$$g(w) = \begin{cases} \frac{1}{\sqrt{2\pi}} \exp\left[-\frac{1}{2}w^2 \right] + A\cos\alpha w + B; \\ \qquad\qquad -\beta < w < \beta \\ 0 \qquad\qquad |w| > \beta \end{cases} \qquad (6)$$

having unknown constants A, B, and α which must be evaluated so as to satisfy the realistic boundary conditions and the unit-area condition

$$g(w = \beta) = g(w = -\beta) = 0 \qquad (7a)$$

$$\left[\frac{dg(w)}{dw}\right]_{w=+\beta} = \left[\frac{dg(w)}{dw}\right]_{w=-\beta} = 0 \qquad (7b)$$

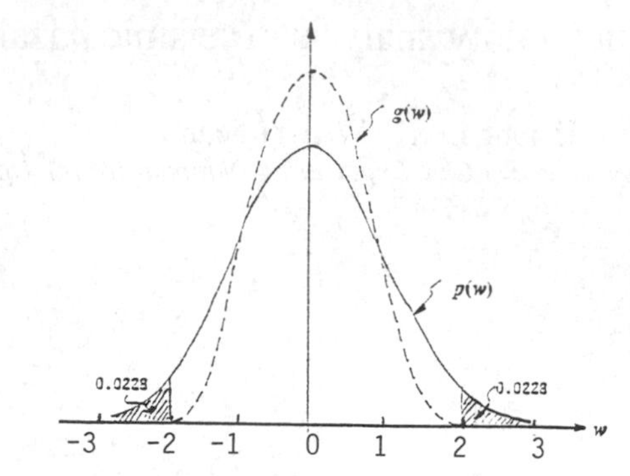

Fig. 1 Normal probability density function $p(w)$ and its truncated function $g(w)$.

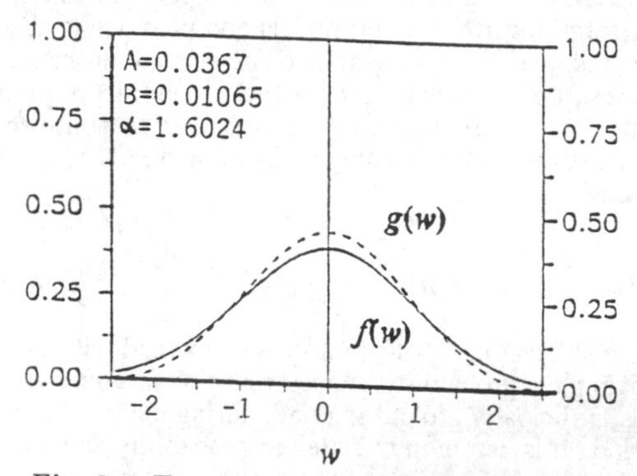

Fig. 2 Truncated probability density function $f(w)$ and $g(w)$ for $\beta = \pm 2.5$.

$$\int_{-\beta}^{\beta} g(w)\, dw = 1 \qquad (7c)$$

The numerical values of these constants will depend unon the cut-off limits $\pm\beta$. It is only the upper cut-off limit which has a significant effect on the pertinent numerical results presented later; however, the lower cut-off limit is applied to keep symmetry of the modified probability density function, thus allowing a more efficient analytical solution.

2.1 Seismic Hazard Curves for PGA

Using nearly 200 recorded strong motion accelerograms for rock sites in Taiwan, it was found that for $z = \ln \mathrm{PGA}$, the dispersion on z was totally confined to the region between cut-offs $\beta = \pm 2.5$; thus, these β values have been used in generating seismic hazard curves for PGA. Figure 2 shows the truncated functions $f(w)$ and $g(w)$ for $\beta = \pm 2.5$. To illustrate the effects of truncating the lognormal distribution in the process of generating the seismic hazard curve for PGA, a set of such hazard curves

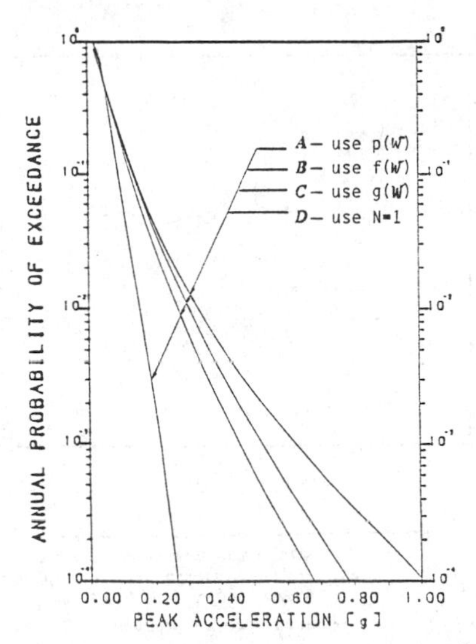

Fig. 3 Seismic hazard curves on peak ground acceleration (PGA).

are shown in Fig. 3 which were generated for a particular rock site in Taiwan. Before going into detail description on the meaning of each hazard curve in Fig. 3, reasonable bounded values of PGA dispersion as well as its distribution function of dispersion must be determined. Generally, the scatter of the PGA values about their predicted values can be estimated through the analysis of normalized weighted residual (NWR_i). It is defined as

$$NWR_i = \frac{(\ln PGA_i - \overline{\ln PGA_i}) - MWR}{\sigma} \quad (8)$$

where $MWR = \frac{1}{n}\sum_n^{i=1}(\ln PGA_i - \overline{\ln PGA_i})$. A normal probability plot comparing the observed distribution of residuals for the ground motion model [Campbell's form modified by Taiwan data] with that based on a normal distribution and the modified distribution $g(w)$ are shown in Figs. 4(a) and 4(b). It is clear that both distribution functions of PGA dispersion can be used in hazard calculation. Since no data fall outside the normal score of ± 2.0. It will suggest to use $w = \pm 2.5$ as the truncation values.

In Fig. 3, Curve D was generated totally ignoring the dispersion of PGA values about the mean (or median) attenuation relations; Curve A was generated considering full dispersion of PGA values in the lognormal form without applying cut-offs to the tails of the distribution; Curve B was generated considering dispersion of PGA values in the lognormal form but applying the standard truncation procedure to the function $p(w)$, with cut-offs at $w = \pm 2.5$, yielding the function $f(w)$; and, Curve C was generated considering dispersion of PGA values in the lognormal form but applying the modified truncation procedure to $p(w)$, with cut-offs at $w = \pm 2.5$ ($\beta = \pm 2.5$), yielding the function $g(w)$. Clearly, the dispersion of PGA values must be considered in a consistent probabilistic approach when generating the seismic hazard curve. As pointed out, Taiwan data indicate that the cut-off points can be set at the $\pm 2.5\sigma$ levels, i.e., $\beta = \pm 2.5$. Of the two procedures described, the modified procedure is preferred; thus, Curve C in Fig. 3 has been used in generating the uniform hazard response spectra.

3. ANNUAL PROBABILITY OF EXCEEDING MAXIMUM CREDIBLE RESPONSE SPECTRAL VALUE

In the equivalent static method on seismic design, the base shear is given by the expression

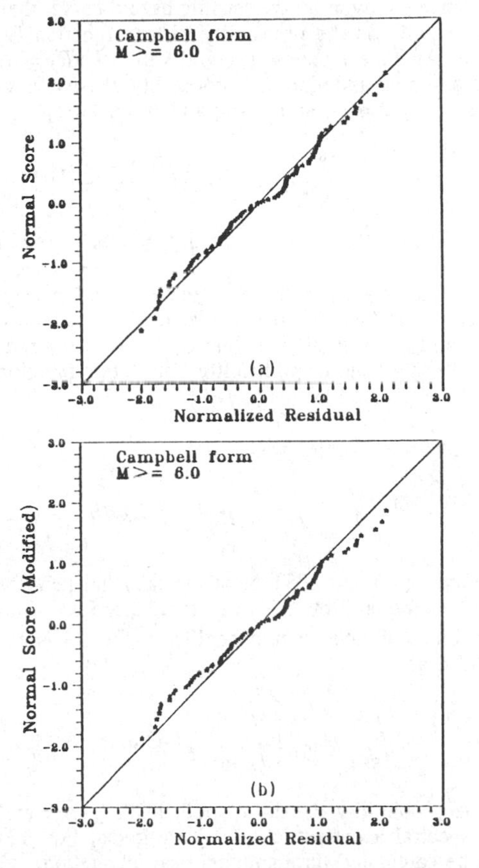

Fig. 4 A normal probability plot comparing the observed distribution of residuals with that of (a) normal distribution, (b) modified probability distribution, $g(w)$.

$$V = C_d(T_1)\, W_t \tag{9}$$

where $C_d(T_1)$ is the acceleration ordinate of a response spectrum whose value depends on the fundamental period of the structur T_1 and the geotechnical characteristics of the site. The value of W_t is taken as the total seismic weight of the structure above the base level. $C_d(T_1)$ can also be defined as the maximum credible response spectral value and expressed in the following form

$$C_d(T) = A_{mc}\, S_a(T) \tag{10}$$

where A_{mc} is the maximum credible effective peak ground acceleration taken from the seismic hazard curve at specified annual probability of exceedance. $S_a(T)$ is the normalized (PGA $= 1\ g$) design response spectral acceleration value at the (mean $+1\sigma$) level. Since A_{mc} having an annual probability of exceedance as given by the seismic hazard curve, then define $p(A)$ as the probability density function for A_{mc}. And the response spectral value $S_a(T) = S$ was assumed as lognormal probability density function, $p(S)$, expressed as [Ang and Tang, 1975]

$$p_S(S) = \frac{1}{\sqrt{2\pi}\zeta S} \exp\left[-\frac{1}{2}\left(\frac{\ln S - \lambda}{\zeta}\right)^2\right] \quad ; \qquad 0 < S < \infty \tag{11}$$

λ and ζ are scale and shape parameters, respectively. Since C_d is the multiple fo two random variables, A_{mc} and S_a, it is obvious that $C_d(T)$ is also a random variable and its probability distribution function can be expressed as:

$$P(C) = \text{Prob}[C \le C_R] = \int_{S=0}^{\infty} \int_{A=0}^{G_R/S} p(A,S)\, dA\, dS \tag{12}$$

Assume $p(A)$ and $p(S)$ are statistically independent then the annual probability of exceedance for andom variable C greater than a specified value C_R is expressed as:

$$P(C) = \int_{S=0}^{\infty} p(S) \left[\int_{A=0}^{C_R/S} p(A)\, dA\right] dS \tag{13}$$

If the probability density of both PGA ($p(A)$) and the spectral acceleration ($p(S)$) are given, Eq. (13) can be calculated through numerical integration.

4. COEFFICIENT OF VARIATION AND DISTRIBUTION OF $S_{an}(T, 0.05)$

Having generated the set of 5%-damped normalized acceleration response spectra for the strong motion accelerograms recorded on rock sites in Taiwan, the coefficient of variation (standard deviation divided by mean value) of $S_{an}(T, 0.05)$ was evaluated as a function of period T. The result is shown by the solid curve in Fig. 5a. For using in the generation of results described subsequently, this solid curve was smoothed somewhat as indicated by the dashed curve. To test the validity of this smoothed coefficient-of-variation function, it was multiplied by the mean 5%-damped normalized acceleration response spectrum published by H. B. Seed for a rock site to obtain the corresponding one-standard-deviation function $\sigma(T)$. Adding this function to Seed's mean spectrum shown in Fig. 5b gives the dashed curve in this same figure which matches very well with Seed's mean $+1\ \sigma$ spectrum; thus, showing the coefficient-of-variation function for Taiwan data (Fig. 5a) can be applied to the corresponding data obtained elsewhere. Furthermore, let S denote the random normalized acceleration response spectral value $S_{an}(T, \xi)$ for a specified period T and damping ratio ξ. Having obtained a set of mean normalized acceleration response spectrum curves,

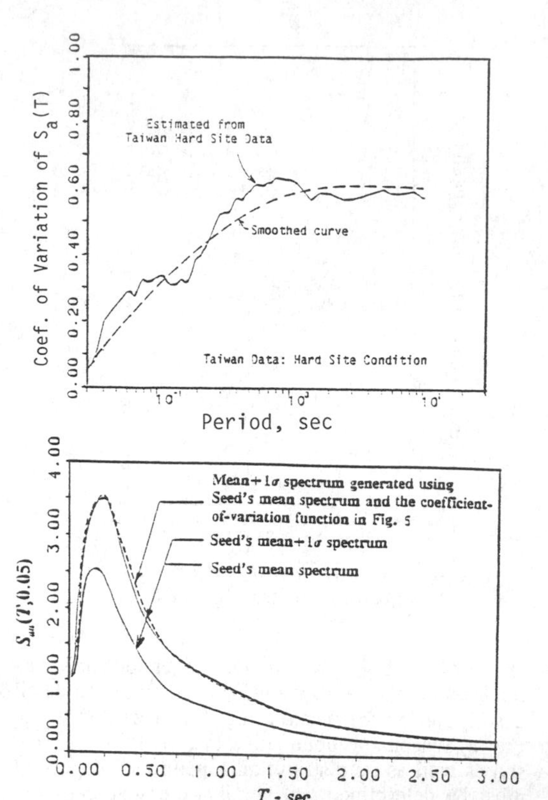

Fig. 5 (a) Coefficient of variation of random variable $S_{an}(T, 0.05)$, (b) mean and mean $+1\sigma$ normalized acceleration response spectra.

denoted by $\mu = \mu(T, \xi)$, and the corresponding set of variance curves, denoted by $\sigma^2 = \sigma^2(T, \xi)$, the probability density function for random variable S can be expressed using the lognormal form. To estimate the modal prameters from data,three methods are used: method of momnet, maximum likelihood method and nonlinear regression analysis. Figures 6a and 6b show the comparison of cumulative distribution of S_a with respect to data distribution at period of 0.5 sec and 1.0 sec, respectively, by using the results from these three estimation methods.

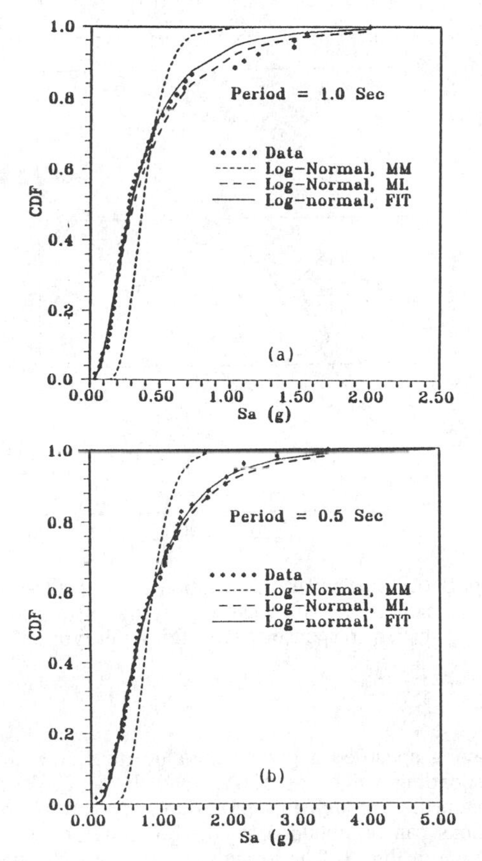

Fig. 6 Comparison on the distribution of $S_a(T)$ by using three different methods of estimation with the data, (a) for $T = 1.0$ sec, (b) for $T = 0.5$ sec.

5. SENSITIVITY ANALYSIS ON UNIFORM HAZARD RESPONSE SPECTRUM

To generate the uniform-hazard 5%-damped acceleration response sepctra, use will be made from the hazard curve on PGA (curve C) in Fig. 3 to represent the probability of exceedance function $P(A)$. Since the coefficient-of-variation function $c(T)$ shown in Fig. 5a for Taiwan data is equally applicable to Seed's data, as demonstrated in Fig. 5b, it can be used directly in parameter estimation to obtain the parameters of lognormal distribution. Having the lognormal probability density function for random variable $S_{an}(T, 0.05)$, $p[S_{an}(T, 0.05)]$, it was truncated using the modified procedure for $\beta = \pm 3.0$. The truncated version of this function together with the hazard curve for PGA (curve C in Fig. 3), are substituted into Eq. (13) and the double integration is carried out numerically for discrete values of T to obtained the corresponding desired probability distribution functions $P[S_a(T, 0.05)]$ conssitent with $p(S_a) = p(C)$. Having the probability distribution functions $P[S_a(T, 0.05)]$ for discrete values of T, the corresponding annual probability of exceedance functions $Q[S_a(T, 0.05)]$ are obtained using the relation

$$Q[S_a(T, 0.05)] = 1 - P[S_a(T, 0.05)] \qquad (14)$$

Setting a fixed numerical value for $Q[S_a(T, 0.05)]$, say, 0.001, over the entire period range of interest, the corresponding acceleration response spectrum $S_a(T, 0.05)$ is obtained directly from $Q[S_a(T, 0.05)]$. Since mean return period measured in years is the reciprocal of annual probability of exceedance, the above example probability of exceedance equal to 0.001 corresponds to a mean return period equal to 1000 years. Other uniform-hazard acceleration response spectra can similarly be obtained for other

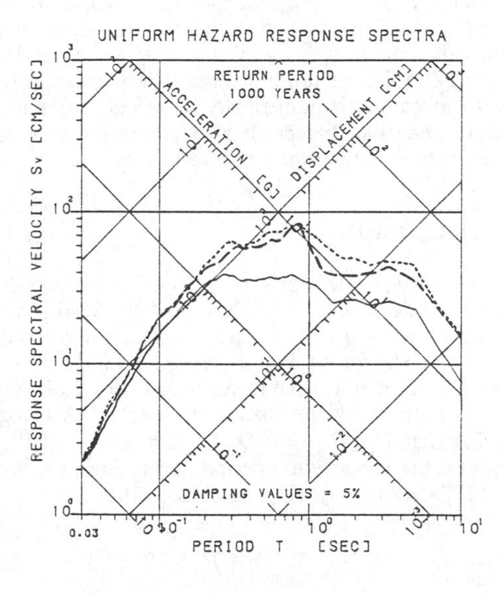

Fig. 7 Uniform-hazard response spectra developed by using three different methods of parameter estimation on $S_a(T)$ distribution.

discrete values of annual probability of exceedance. As discussed in Fig. 6, there are three different ways of parameter estimation on the distribution function $S_a(T)$. The distribution functions of $S_a(T)$ show a little difference. The influence of this difference to the calculation of uniform-hazard response spectra can be observed from Fig. 7. Generally, the result from maximum likelihood method on modal parameters can give a more conservative uniform hazard response spectrum than other methods. If different coefficient of variation of spectral accelerations for different soil types are used (4 types of soil COV values were obtained from Seed's spectrum), as shown in Fig. 8a, site-dependent uniform hazard response spectra can be generated. Figure 8b shows the uniform hazard response spectra with 4 types of soil conditions.

6. CONCLUSIONS

The effect of PGA dispersion on seismic hazard analysis was studied first. Both truncated and modified probability density function at $\beta = \pm 2.5$, $f(w)$ and $g(w)$, were suggested for the calculation of hazard curve of PGA. In generating the uniform hazard response spectrum, the probability density function of PGA as well as spectral acceleration $p(S_a)$ were used. In carrying out the double integration (Eq. 13), the upper limit to the first integral was changed from infinity to the upper-limit finite value set by the truncation procedure applied to $p(S)$. Effect of coefficient of variation of S_a for different site conditions as well as modal parameters of density function $p(S_a)$ to the influence of response spectra were investigated. The uncertainties analysis in hazard calculation, both on PGA hazard curves and on the developing of uniform hazard response spectra, were investigated numerically. Not only the uncertainty analysis, this paper also presents a new approach to generate unfiorm hazard response spectra.

7. DISCUSSION

As an example, let curve C in Fig. 3 represents the seismic hazard curve on PGA and let Seed's 5%-damped mean $+1\sigma$ standard acceleration response spectrum shown in Fig. 5 represents $S_d(T, 0.05)$. Let $R_A(A)$ denotes the mean return period of PGA $= A$ as given by the hazard curve on PGA recognizing that $R_A(R) = 1/Q(A)$, and let $R_d(A, T, \xi)$ denotes the mean return period measured in years of $S_a(T, \xi) = A \times S_d(T, \xi)$ as obtained by

$$R_d(A, T, \xi) = 1 \big/ \{1 - P\left[S_a(T, \xi)\right]\} \qquad (15)$$

in which $P[S_a(T, \xi)]$ is given by Equation (13).

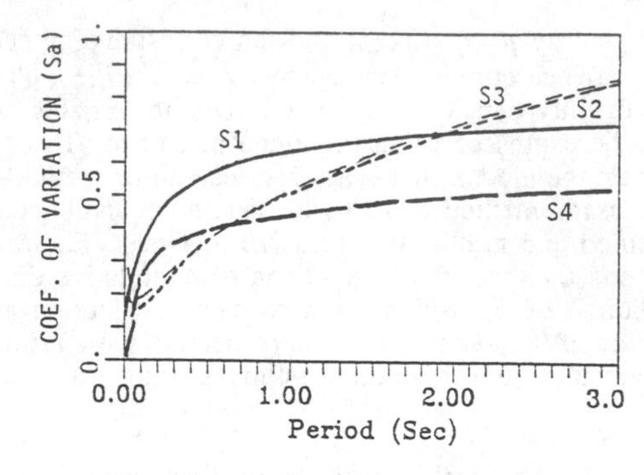

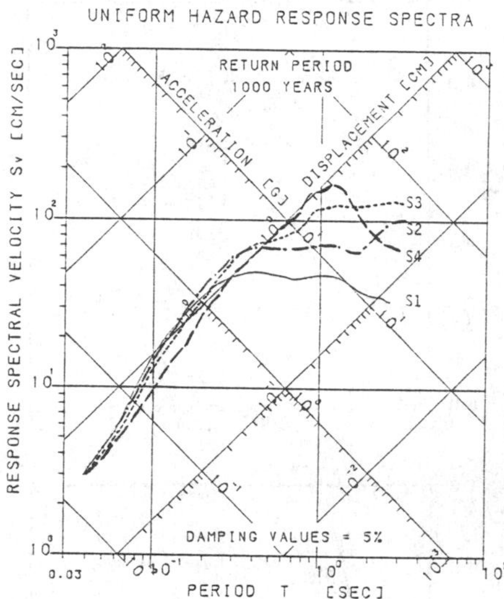

Fig. 8 (a) Coefficient of variation of four different soil types, (b) sensitivity study on uniform-hazard response spectra for 4 soil types.

Having specified a particular value of A, the corresponding values of $R_A(A)$ and $R_d(A, T, \xi)$ are obtained. Through this procedure, R_d and R_a relations can be obtained for discrete value of T, as shown in Fig. 9. The irregular form of the function R_d versus T is due to the difference between hazard curves of $Q[S_a(T, 0.05)]$ and $P(A) \cdot S_{an}$. To explain the use of R_d versus R_A curves in Fig. 9, choose $R_A = 475$ years (which corresponds to $A = 0.4\ g$ in curve C), which corresponds to $R_d = 1000$ years (if $\beta = \pm\ 2.0$ is used in the truncation of spectral acceleration distribution) and $R_d = 740$ years (if $\beta = \pm\ 3.0$ is used in the truncation of spectral acceleration) for $T = 1.0\ sec$ respectively. This shows that when scaling the mean plus one standard deviation normalized response spectra by some value A for design purposes, the return periods on the scaled

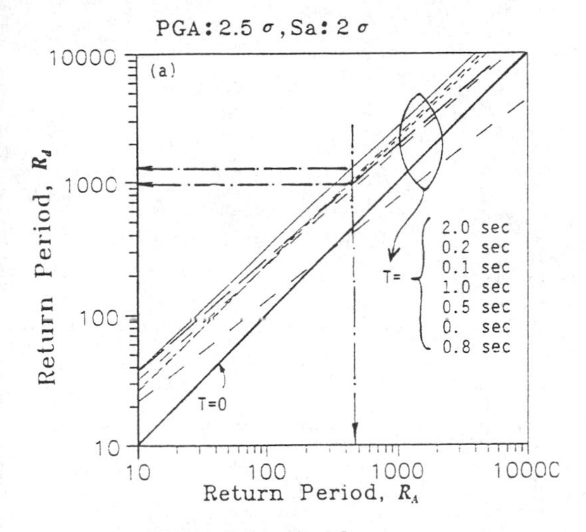

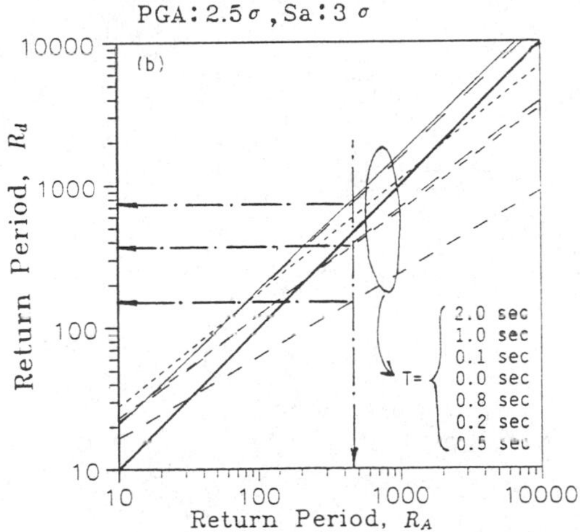

Fig.9: Return period vs for discrete values of period T; (a) for $\beta = \pm 2.$ as truncational bound in Sa distribution, (b) for = 3.0 as truncational bound in Sa distribution.

spectral values are generally greater than the return period on PGA value (only few exceptions, such as $T = 0.5\ sec$ for $\beta = \pm\ 3.0$). It is concluded that if the uniform-hazard response spectra developed by using $\beta = \pm\ 2.0$ as truncation was used for design purpose, then the present practical design procedure (normalized response spectra by their associated PGA level) is quite conservative. However, if the uniform-hazard response spectra developed by using $\beta = \pm\ 3.0$ as truncation was used for design, then the present practical design procedure will be a little nonconservative for some structural period.

8. ACKNOWLEDGEMENTS

This study was supported by a grant from the National Science Council, ROC (No. NSC-82-0414-P-002-016-BY).

REFERENCES

Ang, A. H-S. and Tang, W. H., ''Probability Concepts in Engineering Planning and Design,'' Basic Principles, Vol. I, 1975.

Cornell, C. A., ''Engineering Seismic Risk Analysis,'' Bull. Seism. Soc. Amer. Vol. 58, No. 5, 1583--1606, 1968.

Der Kiureghian, A. and Ang, A. H., ''A Fault-Rupture Model for Seismic Risk Analysis,'' Bull. Seism. Soc. Amer. Vol. 67, No. 4, 1173--1194, 1977.

Newmark, N. M., Blume, J. A., and Kapur, K. K., ''Seismic Design Spectra for Nuclear Power Plants,'' Jour. Power Division, ASCe, Vol. 99, No. P02, 1973.

Loh, C. H., Jean, W. Y., and Penzien J., ''Uniform-Hazard Response Spectra — An Alternative Method,'' will appear in Earthquake Engineering and Structural Dynamics, 1993.

Seed, H. B., Ugas, J., and Lysmer, J., ''Site-Dependent Spectra for Earthquake Resistant Design,'' Bull. of the Seismological Society of America, Vol. 66, No. 1, 1976.

Structural Safety & Reliability, Schuëller, Shinozuka & Yao (eds) © 1994 Balkema, Rotterdam, ISBN 90 5410 357 4

Structural response spectra to different frequency bandwidth earthquakes

Dan Lungu, S. Demetriu & L. Muscalu
Faculty of Civil Engineering, Bucharest, Romania

Ovidiu Coman
National Commission for Nuclear Activity Control, Bucharest, Romania

Tiberiu Cornea
EBASCO Services Inc., N.Y., USA

ABSTRACT: In the analysis of the frequency content of strong ground motions it was found that related concepts to the power spectral density as unitless measure, ε (Cartwright & Longuet-Higgins) and f_{10}, f_{50} and f_{90} frequencies can be recommended for the classification of the width of the frequency band. A very strong correlation between these measures and the control frequencies of the response spectra (Newmark) was found. The concepts are applied to earthquake records of extreme frequency content and high peak acceleration from Romania (1990, 1986, 1977), Turkey (1992), Armenia (1988), Mexico (1985), Iran and USA.

1. FREQUENCY BANDWIDTH OF SEISMIC PROCESSES ACCORDING TO ε CRITERION

The strong ground motions recorded in the last 15 years in different areas of the world, suggest to differentiate the narrow frequency band motions with large predominant periods, recorded in soft soil conditions, in respect with the wide frequency band motions, recorded on rock or hard soil conditions.

For the analysis of the frequency content of the seismic motion, the Power Spectral Density (PSD) and its related concepts are used associated with the structure response spectra.

The stochastic diagnosis of the frequency band width of the ground acceleration process was based on three indicators: ε measure, median frequency f_{50} and frequency bandwidth $\Delta f = f_{90} - f_{10}$.

f_{10}, f_{50} and f_{90} (Kennedy & Shinozuka) are frequencies below which 10%, 50% and 90% of the total cumulative power occurs.

The dimensionless indicator, ε (Cartwright & Longuet - Higgins) is defined as function of the spectral moments of the PSD:

$$\varepsilon = \sqrt{1 - \lambda_2^2 / \lambda_0 \lambda_4}$$

where λ_i is i-th moment of the spectral density function $S(\omega)$,

$$\lambda_i = \int_{-\infty}^{+\infty} \omega^i \, S(\omega) \, d\omega$$

The ε values are characterised with the limiting cases (Clough & Penzien):

White noise process	2/3
Narrow frequency band process approaching the single harmonic process	0
Superposition of a single harmonic process with a band limited process	1.

Strong motion duration over which PSD is evaluated is defined as $D = T_{0.9} - T_{0.1}$ where $T_{0.9}$ and $T_{0.1}$ are the times at which 90% and 10% of the total cumulative energy of the acceleration process are reached.

Alternative definitions of the duration of the stationary power of the accelerogram segment as $T_{0.95}-T_{0.05}$, $T_{0.85}-T_{0.15}$, $T_{0.75}-T_{0.05}$ could be commented.

The earthquake records in Table 1 are characterised by extreme frequency content and by high peak accelerations, Fig. 1 and Fig. 3.

The results in Table 1 are bordered by the narrowest frequency band records of Mexico City, SCT, 1985 (ε =0.99) and by the broadest frequency band Iranian accelerograms (ε =0.64). As a rule, ε is greater than 0.9 for narrow frequency band seismic motions and around 0.8–0.85 for wide frequency band motions.

The predominant frequency of the narrow frequency band motion is the abscissa of the highest peak of PSD.

Table 1. Classification of frequency content of strong ground motions according to ε criterion.

Earthquake record	Comp	Peak accel. cm/s^2	0.8E Cum. power cm^2/s^4	Measures of frequency bandwith ε	f_{10} Hz	f_{50} Hz	f_{90} Hz	Response spectra control freq. ξ=0.05 f_A Hz	f_B Hz	Frequency content
USA, El Centro May 18, 1940	VERT	261.9	$2.72 \cdot 10^3$	0.53	3.2	8.90	12.4	0.22	9.80	Broadest
Romania, Vrancea, Focsani St. Aug 31, 1986	VERT	122.7	$8.75 \cdot 10^2$	0.58	2.8	7.27	9.9	0.17	4.55	
Turkey, Erzinkan, Meteo St. March 13, 1992	VERT	244.4	$3.91 \cdot 10^3$	0.76	1.5	5.27	11.5	0.24	1.91	Broad
Romania, Vrancea, Bucharest - INCERC, March 4, 1977	VERT	105.9	$3.51 \cdot 10^2$	0.82	0.5	2.57	8.3	0.51	1.34	
Iran, Naghan April 6, 1977	L	709.5	$1.16 \cdot 10^5$	0.64	2.5	7.78	14.3	0.79	7.49	Broadest
Iran, Tabas Sept 16, 1978	N16W	915.4	$5.95 \cdot 10^4$	0.81	1.1	3.81	7.4	0.21	1.77	
USA, El Centro May 18, 1940	NS	318.5	$5.84 \cdot 10^3$	0.85	0.7	2.38	6.5	0.42	1.53	Broad
Romania, Vrancea, Focsani St. Aug 31, 1986	EW	297.1	$5.76 \cdot 10^3$	0.85	1.1	2.25	4.6	0.59	2.03	
Turkey, Erzican, Meteo St. March 13, 1992	EW	491.9	$1.61 \cdot 10^4$	0.86	0.9	2.51	5.1	0.40	1.80	Transition
Romania, Vrancea, Focsani St. Aug 31, 1986	NS	273.3	$5.80 \cdot 10^3$	0.88	0.6	2.13	4.5	0.53	1.40	
Armenia, Spitak, Gukasian St. Dec 7, 1988	EW	186.6	$1.31 \cdot 10^3$	0.91	0.8	2.06	4.9	0.37	1.36	
Turkey, Erzincan, Meteo St. March 13, 1992	NS	389.6	$1.63 \cdot 10^4$	0.95	1.0	1.13	4.6	0.44	0.68	Narrow
Romania, Vrancea, Bucharest - INCERC, March 4, 1977	NS	194.9	$4.21 \cdot 10^3$	0.97	0.4	0.69	2.00	0.53	0.75	Narrow. Predominant frequency=0.625 Hz
Mexico, Mexico City SCT St. Sept 19, 1985	EW	180.1	$6.29 \cdot 10^3$	0.99	0.31[1)]	0.44	0.56[2)]	0.44	0.51	Narrowest. Predominant frequency=0.49 Hz

1) f_5 2) f_{95}

The predominant period of the ground motion, (i.e. the reverse of the predominant frequency) can be determined from the periodicity of the autocorrelation function, Fig.5.

The predominant period of the NS and EW components recorded in SCT station during the Mexico City 1985 earthquake is: T_g = 2.05 s.

The predominant period of NS component recorded in Bucharest-INCERC station during 1977 Vrancea earthquake is: T_g = 1.60 s.

In Nuclear Engineering, the frequency band width descriptor can be applied to the stationary process of the artificial time-history, whose PSD must be not less than a target PSD compatible with NRC Regulatory Guide 1.60 Response Spectrum:

PSD 1 $\quad 0.2 < f < 34$ Hz

$$G(f)=6900\,\frac{in^2}{s^3}\cdot\frac{1+4\xi_g^2\left(f/f_g\right)^2}{\left[1+\left(f/f_g\right)^2\right]^2+4\xi_g^2\left(f/f_g\right)^2}$$

PSD 2 $\quad 0.4 < f < 15$ Hz

$$G(f)=20000\,\frac{in^2}{s^3}\cdot f^{-2.1} \le 3500\,\frac{in^2}{s^3}$$

The coefficients are scaled for one sided PSD and for 1.0 g peak ground acceleration.

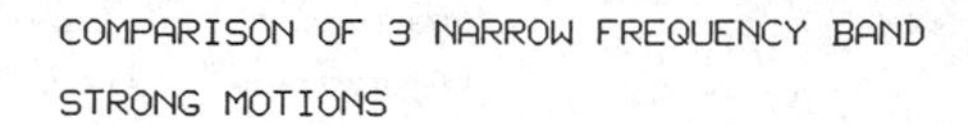
COMPARISON OF 3 NARROW FREQUENCY BAND
STRONG MOTIONS

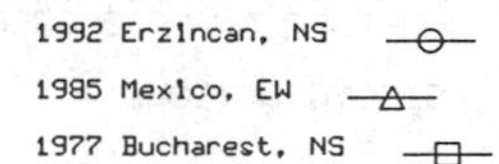
1992 Erzincan, NS
1985 Mexico, EW
1977 Bucharest, NS

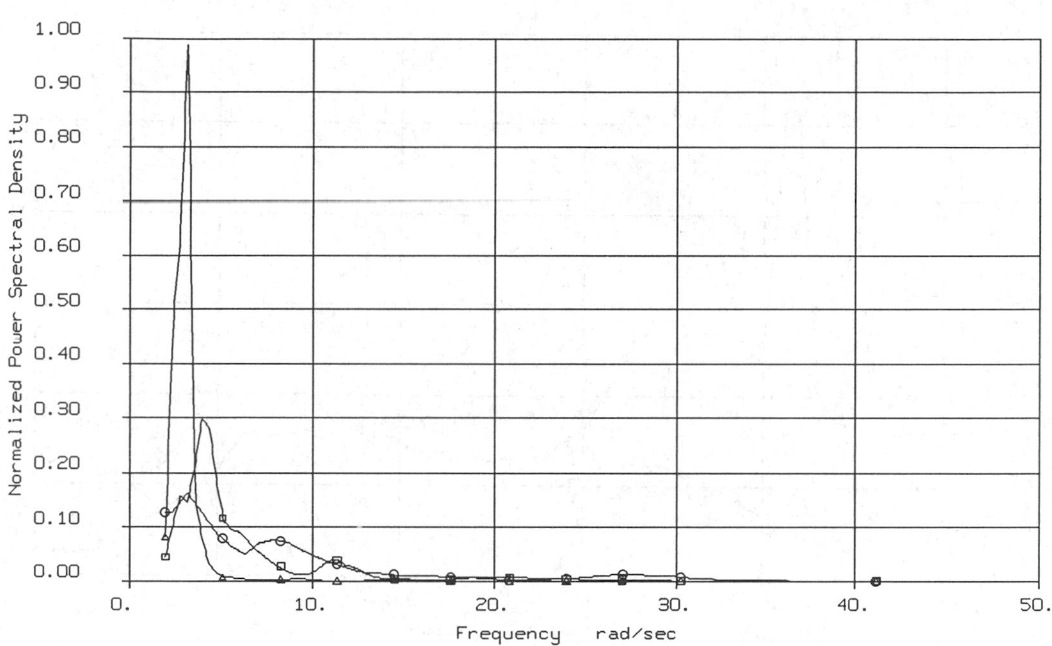
Normalized Power Spectral Density
1.00
0.90
0.80
0.70
0.60
0.50
0.40
0.30
0.20
0.10
0.00
0.
10.
20.
30.
40.
50.
Frequency rad/sec

Fig. 1

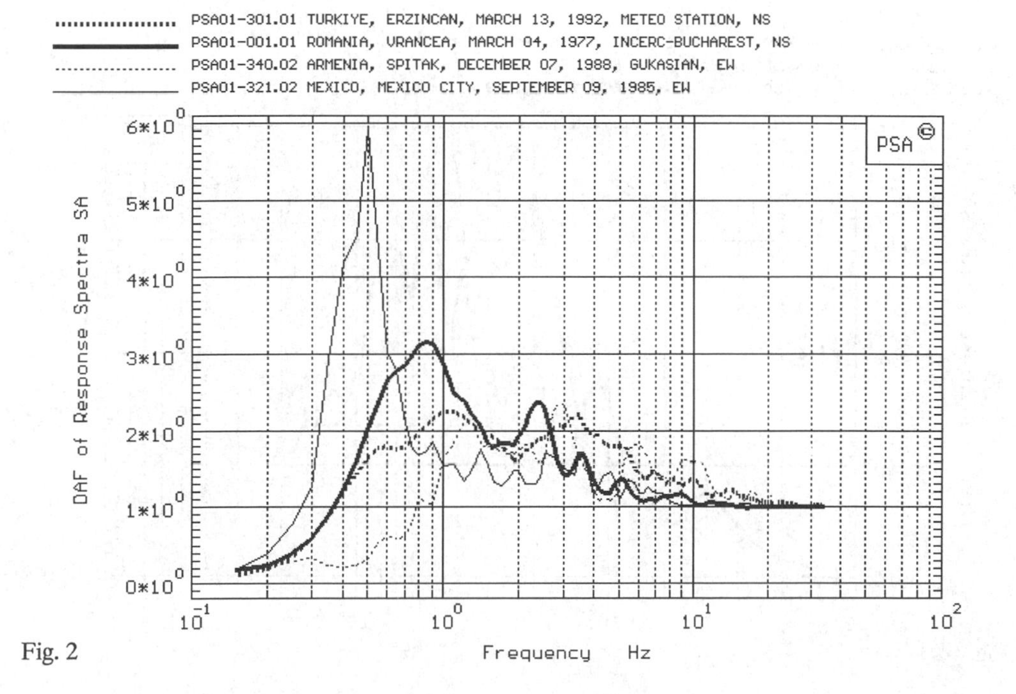
Dynamic Amplification Factor of SA
Damping : 0.05
PSA01-301.01 TURKIYE, ERZINCAN, MARCH 13, 1992, METEO STATION, NS
PSA01-001.01 ROMANIA, VRANCEA, MARCH 04, 1977, INCERC-BUCHAREST, NS
PSA01-340.02 ARMENIA, SPITAK, DECEMBER 07, 1988, GUKASIAN, EW
PSA01-321.02 MEXICO, MEXICO CITY, SEPTEMBER 09, 1985, EW
PSA©
DAF of Response Spectra SA
Frequency Hz

Fig. 2

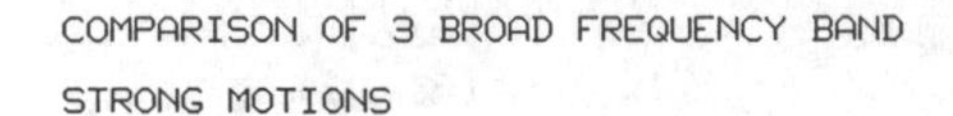

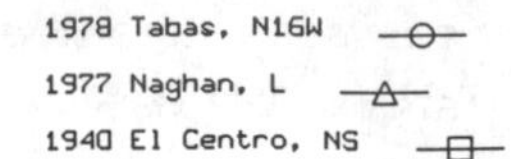

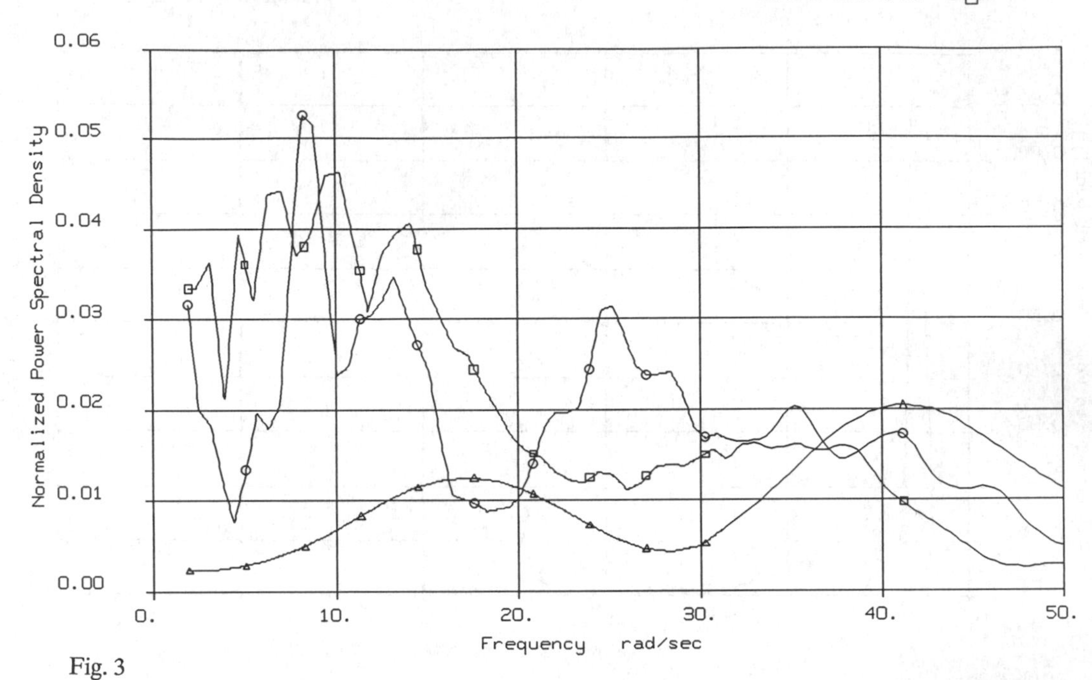

Fig. 3

Dynamic Amplification Factor of SA

Damping : 0.05

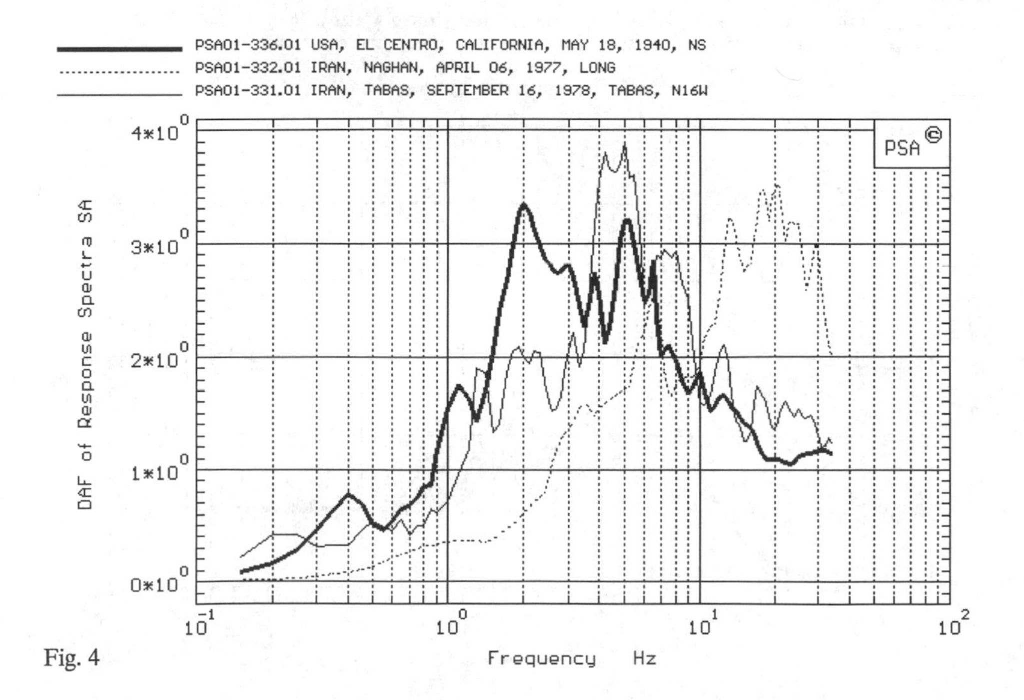

Fig. 4

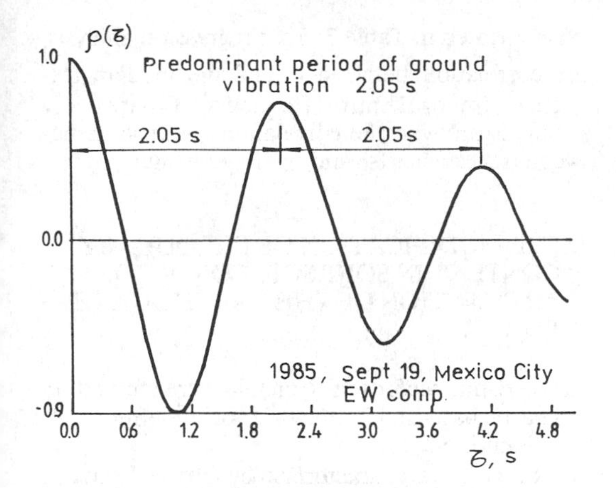

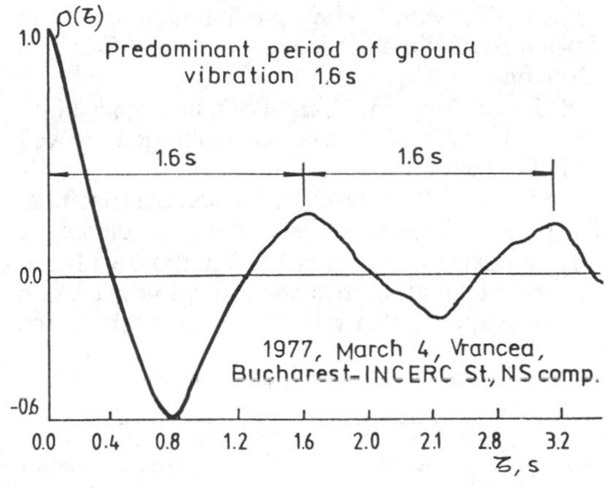

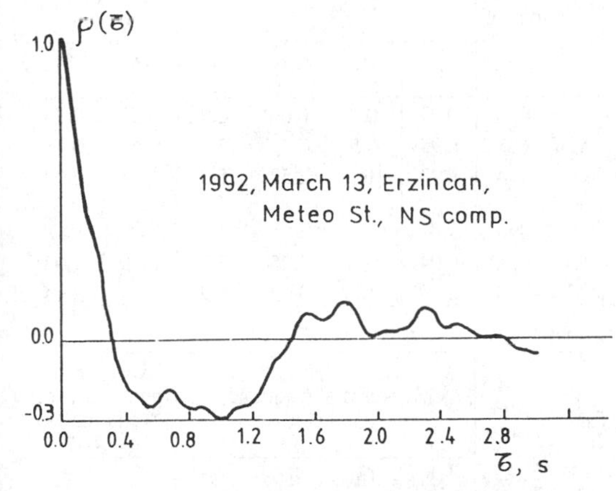

Fig. 5 Normalised autocorrelation function.

Since the recommended values for the narrow frequency band (of low predominant period) PSD 1are ω_g=10.66 rad/s and ξ_g =0.9793, the $\mathcal{E}$ values for PSD 1 and PSD 2 indicate processes with different frequency bandwidth.

The results for Kanai-Tajimi PSD 1 are given in Table 2 and for PSD 2, $\mathcal{E}$ = 0.831.

Table 2. The $\mathcal{E}$ descriptor of Kanai-Tajimi PSD

	$\omega_g = 2\pi$	4π	8π rad/s
ξ_g =0.3	$\mathcal{E}$ =0.995	0.99	0.97
0.6	0.986	0.97	0.94
0.9	0.975	0.95	0.91

2. STRUCTURE RESPONSE TO EARTHQUAKE RECORDS OF DIFFERENT FREQUENCY BANDWIDTH

The classical measure of the frequency content of the ground acceleration is the response of the simple oscilator to the specified accelerogram.

Acceleration response spectra, SA or structure Dynamic Amplification Factors (with respect to the Peak Ground Acceleration)

$$\mathrm{DAF} = \frac{\mathrm{SA}}{\mathrm{PGA}}$$

for the records classified in Table 1 according to the width of the frequency band, are given in Fig. 2 and Fig. 4. It is emphasized:

1. For the ciclostationary process Mexico SCT, 1985, NS and EW components, with the narrowest frequency band ever met,
$$\max \mathrm{DAF} \cong 6$$
2. For the narrow frequency band Bucharest - INCERC record in 1977 Vrancea earthquake,
$$\max \mathrm{DAF} = 3.15$$

The pair of DAF for the Mexican records are the highest values ever obtained from the earthquake accelerograms. The max DAF = 3.15 is also the effect of the narrowing of the frequency bandwidth in the case of a large predominant period (1.6 sec).

Related measures of the structure maximum response to the ground motions are the control frequencies f_B and f_A defined by Newmark in the tripartite log plot of response spectra:

$$f_B = \frac{1}{2\pi}\frac{\max \mathrm{SA}}{\max \mathrm{SV}} \qquad f_A = \frac{1}{2\pi}\frac{\max \mathrm{SD}}{\max \mathrm{SV}}$$

SD, SV and SA are, respectively, spectral relative displacement and velocity and spectral absolute acceleration of the SDOF structure.

The control frequencies of SEAOC 1990 normalised spectrum shape are given as function of site (i.e. soil conditions): type 1 f_B=2.56, type 2 f_B=1.71 and type 3 f_B=1/0.915 s=1.09 Hz.

The deterministic f_B and f_A frequencies are listed in Table 1 next to the stochastic indicators ε, f_{10}, f_{50}, f_{90}.

Table 3. Estimated correlation coefficients between the measures of frequency bandwidth

14 strong motions : Table 1

	ε	f_{50}	Δf	f_B
ε	1	– 0.98	–0.79	–0.90
f_{50}		1	0.83	0.91
Δf			1	0.69
f_B				1

22 ground motions: 1986 Vrancea earthquake

	ε	f_{50}	Δf	f_B
ε	1	–0.94	–0.49	–0.88
f_{50}		1	0.56	0.84
Δf			1	0.47
f_B				1

As indicated in Table 3, from regression analysis, the correlation between ε and the median frequency, f_{50} or control frequency, f_B are very strong. Moreover, the correlation coefficients between frequencies f_{50} and f_B are very high.

3. THE MODIFICATION OF FREQUENCY CONTENT IN SOFT SOIL CONDITIONS AS FUNCTION OF THE EARTHQUAKE POWER

The seismic records of Romania were studied on stochastic basis and 2 categories of site-effects were identified:

(i) Site effects, characterized by narrow frequency band PSD with 1.4-1.6 s predominant period, in Bucharest and in other locations in the SE of Romania

(ii) Broad frequency band PSD, in Vrancea epicentral area, in Moldavia and Dobrogea, as well as in Bucharest.

The first site effect correspond to soft soil conditions and the second one to stiff soil and rock conditions.

To the narrow frequency band motions with 1.4-1.6 predominant periods correspond acceleration response spectra, SA with important peaks in the

Table 4. Modification of narrow band frequency content in soft soil condition of Bucarest, as function of the earthquake power.

Station location	Vrancea earthquakes				Comp	Peak accel	0.8 Cum. power	Measures of frequency bandwidth				Response spectra control frequencies ξ=0.05	
	Year	M	h Km	Δ Km		cm/s^2	cm^2/s^4	ε	f_{10} Hz	f_{50} Hz	f_{90} Hz	f_A Hz	f_B Hz
Bucharest - INCERC	1977	7.2	109	100	NS	194.9	4212	0.97	0.4	0.69	2.0	0.53	0.75
	1986	7.0	133	123	NS	89.2	628	0.95	0.5	0.74	3.8	0.61	0.76
	1990	6.7	90	161	NS	76.6	194	0.78	1.1	2.57	5.2		2.00
Bucharest - Magurele	1977						No record						
	1986	7.0	133	135	NS	135.1	1209	0.94	0.5	1.25	3.7	0.69	1.02
	1990	6.7	90	173	NS	89.6	273	0.79	1.2	3.63	8.9	0.39	3.44

Table 5. Broad band frequency content in Vrancea and Moldavia.

Station location	Vrancea earthquakes				Comp	Peak accel	0.8 Cum. power	Measures of frequency bandwidth				Response spectra control frequencies ξ=0.05	
	Year	M	h Km	Δ Km		cm/s^2	cm^2/s^4	ε	f_{10} Hz	f_{50} Hz	f_{90} Hz	f_A Hz	f_B Hz
Vrancioaia in epicentral area	1986	7.0	133	42	NS	82.3	682	0.87	0.8	1.82	3.5	0.40	1.70
					EW	140.8	1612	0.85	1.2	2.42	4.2	0.36	2.19
	1990	6.7	90	6	NS	119.6	774	0.80	0.6	2.51	4.5	0.52	1.52
					EW	157.3	2196	0.70	1.5	3.26	4.6	0.49	3.12
Iasi in Moldavia	1986	7.0	133	201	NS	66.9	327	0.87	1.0	2.38	6.6	0.60	2.04
					EW	99.6	365	0.86	1.1	2.00	6.1	0.79	2.11
	1990	6.7	90	166	NS	95.8	574	0.72	1.1	4.01	9.3	1.06	3.11
					EW	106.5	560	0.70	1.2	5.89	8.8	0.74	4.70

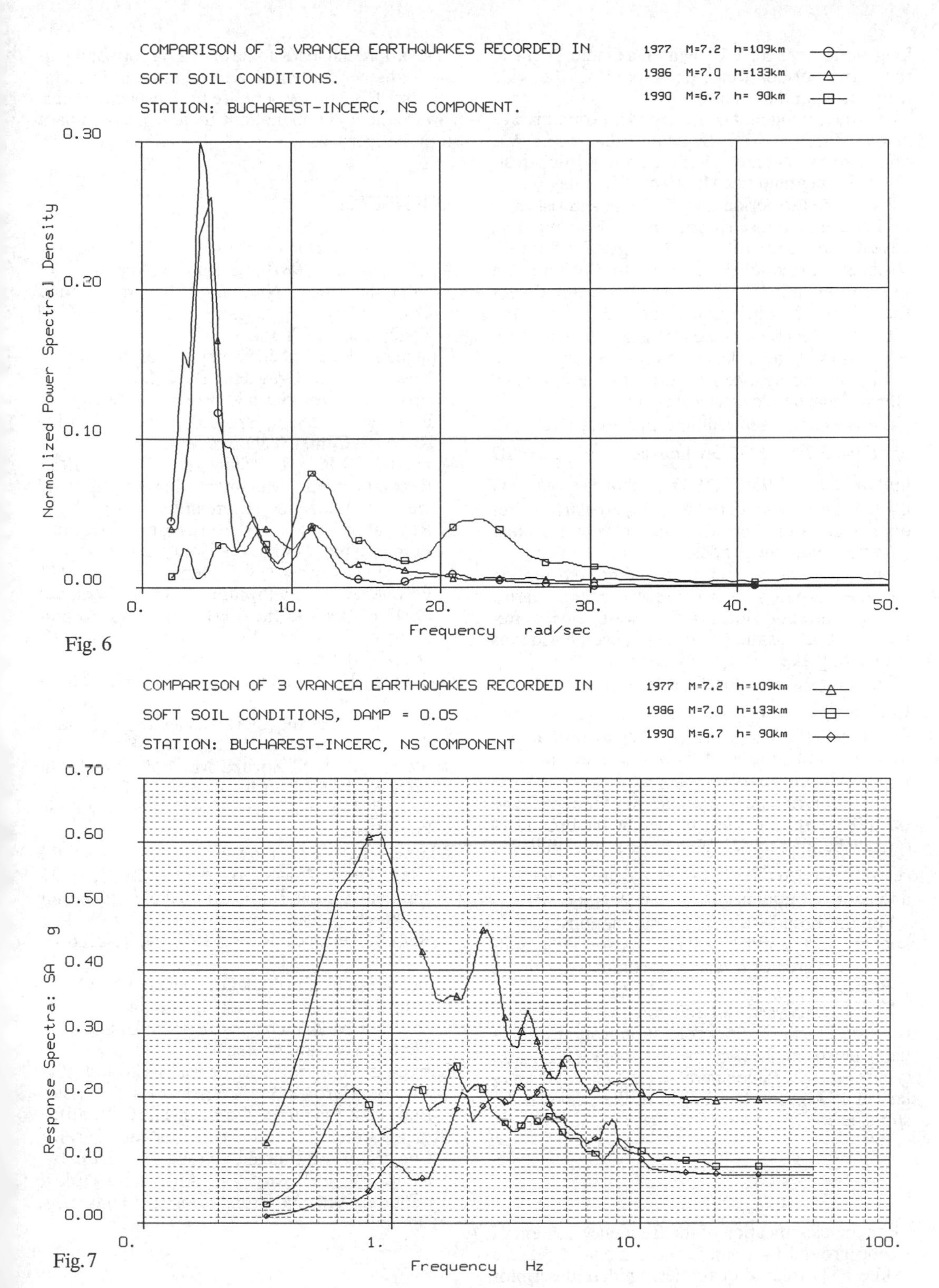
COMPARISON OF 3 VRANCEA EARTHQUAKES RECORDED IN
SOFT SOIL CONDITIONS.
STATION: BUCHAREST-INCERC, NS COMPONENT.
1977 M=7.2 h=109km
1986 M=7.0 h=133km
1990 M=6.7 h= 90km
Normalized Power Spectral Density
0.30
0.20
0.10
0.00
0.
10.
20.
30.
40.
50.
Frequency rad/sec

Fig. 6

COMPARISON OF 3 VRANCEA EARTHQUAKES RECORDED IN
SOFT SOIL CONDITIONS, DAMP = 0.05
STATION: BUCHAREST-INCERC, NS COMPONENT
1977 M=7.2 h=109km
1986 M=7.0 h=133km
1990 M=6.7 h= 90km
Response Spectra: SA g
0.70
0.60
0.50
0.40
0.30
0.20
0.10
0.00
0.
1.
10.
100.
Frequency Hz

Fig.7

long period range. SA having maximum peaks in the short periods range correspond to the wide frequency band motions.

The first strong motion recorded in Romania was in the March 4, 1977. By chance this record was on the soft soil condition of the Station Bucharest-INCERC characterized by 16 m of soft clay.

Predominant period and PSD characteristics as well as the response spectra, in this soil condition, have been comparatively investigated for the 3 Vrancea earthquakes recorded in Bucharest in 1977, 1986 and 1990, Table 4 (M - magnitude, h - focal depth, Δ - epicentral distance).

It was concluded that the frequency content of the seismic motion recorded on soft soil is strongly affected by the earthquake power, i.e. by the combined effect of magnitude and focal depth, Fig. 6.

The NS component, station Bucharest-INCERC, has ε =0.97 for 1977 earthquake (M=7.2, h=109 km) and ε =0.95 for 1986 earthquake (M=7.0, h=133 km), both corresponding to narrow frequency band motions with, respectively, 1.6 s and 1.4 s predominant periods.

The predominant period of the ground motion has the tendency to become larger as earthquake intensity increase.The width of the frequency band has the tendency to become wider as the earthquake power decrease.

The same NS component, station Bucharest-INCERC, has ε =0.78 for 1990 earthquake (M=6.7, h=90 km), corresponding to a broad frequency band motion. This is the effect of both magnitude and source mechanism.

The different source mechanism of the 1990 and the 1986 earthquaks was proved by the broad frequency band records, Table 5. As a rule, the 1990 records have smaller ε values and greater median and control frequencies than the 1986 records.

For the broad frequency band motions, the response spectra of the 1990 records are slighty shifted toward greater frequencies compared with the 1986 results.

For the analysed soft soil conditions in Bucharest, the height of the acceleration response spectrum peaks in the long period range (>1s) are: greater (in 1977), comparable (in 1986) and smaller (in 1990) than the height of the peaks in the short period range (<0.5 s), Fig. 7.

4. CONCLUSIONS

For the classification of the frequency content of strong ground motions, is recommended the use of the PSD related quantities: unitless descriptor ε. and median frequency f_{50}.

The approach leads to new criteria for zoning of the frequency content of ground motions in countries like Romania, where are both broad and narrow frequency band motions of large predominant period recorded.

REFERENCES

ASCE 4-86, 1987. Seismic Analysis of Safety-Related Nuclear Structures and Commentary. American Society of Civil Engineers, New York.

Clough R, Penzien J. Dynamics of Structures. New York: Mc. Graw Hill Book Co.

Iranian Code for Seismic Resistant Design of Buildings, 1988. Building and Housing Research Center, Teheran.

Kennedy R.P. & Shinozuka M., 1989. Recommended Minimum Power Spectral Density Functions Compatible with NRC Regulatory Guide 1.60 Response Spectrum, Prepared for Brookhaven National Laboratory.

Lungu D., Cornea T., Demetriu S. 1992. Frequency Bandwidth of Vrancea Earthquakes and the 1991 Edition of Seismic Code of Romania. Proc. 10WCEE, Vol 10: 5633-5638. Rotterdam: Balkema.

O'Connor M.J., Elingwood B.R., 1992. Site - Dependent Models of Earthquake Ground Motion. Earthquake Engineering and Structural Dynamics. Vol 21: 573-589.

Schueler G.I. & Shinozuka M., 1987. Stochastic Methods in Structural Dynamics. Dordrecht, Boston, Lanchaster: Martinus Nijhoff Publishers.

Whitman R.V. Ed., 1992. Proceedings from the Site Effects Workshop, Oct. 24-25, 1991. National Center for Earthquake Engineering Research, State University of New York at Buffalo. Technical Report NCEER 92-0006.

Armenia Spitak Earthquake. Digitized by Okada Lab., I.I.S. Univ. of Tokyo, Japan (Courtesy Dr. M.G. Melkumian, Earthquake Engineering Center, Yerevan, Armenia).

Corrected Accelerograms of the 1986 and 1990 Vrancea Earthquakes, Center of Earth Physics, Bucharest-Magurele (Courtesy Dr. C. Radu).

Strong Ground Motion Records of Turkiye. Copyright: Earthquake Research Departament. PK 763 Kizilay - Ankara - Turkiye (Courtesy prof. M. Erdik, Bogazici University, Istanbul).

Structural Safety & Reliability, Schuëller, Shinozuka & Yao (eds) © 1994 Balkema, Rotterdam, ISBN 90 5410 357 4

Conditional random fields with an application to earthquake ground motion

H. Morikawa
Kyoto University, Japan

H. Kameda
Urban Earthquake Hazard Research Center, Disaster Prevention Research Institute, Kyoto University, Japan

ABSTRACT: Analytical solutions are presented for conditional random fields that are conditioned by a set of deterministic time functions using exact probabilistic theory. On this basis, the first-passage solution and the method of numerical simulation for conditional random fields are presented. The methodology developed herein is applied to stochastic earthquake waves propagating through a ground with random soil properties.

1 INTRODUCTION

In this paper, conditional random fields are formulated and analytical solutions for their probabilistic properties are derived. On this basis, the theory is applied to earthquake wave fields. This study will perform a key part in stochastic interpolation of earthquake ground motion and the development of regional seismic monitoring systems for urban earthquake hazards reduction(Kameda 1991). Furthermore, the method developed in this study can be applied to other engineering problems of space-time random fields such as wind engineering(Morikawa and Maruyama 1993).

Some pioneering studies of the subject dealt with herein have been performed by Kawakami and Ono(1992), and Vanmarcke and Fenton (1991), which are important accomplishments. It is noted, however, that they are confined to the technique of numerical simulation of interpolating motions. On the other hand, this work is an attempt to present a variety of probabilistic information, using the theory of conditional random fields on a purely probabilistic basis(Kameda and Morikawa 1992, 1993). They include not only methods of numerical simulation but also further steps such as the expectation and uncertainty evaluation of conditioned processes, and the first-passage solutions. In this paper, these theoretical developments are summarized, and in addition the method is applied to stochastic earthquake waves propagating through a ground with random soil properties.

It is noted that stochastic interpolation in the context of random field theory has been presented by Ditlevsen(1991) where no explicit discussion is made regarding the treatment of time-dependent phenomena. The present work, which was performed under engineering motivations mentioned above can be regarded as a counterpart that deals with stochastic interpolation of time functions.

2 THEORETICAL PROPERTIES OF CONDITIONAL RANDOM FIELDS

2.1 *Formulation*

In this study, the conditional random field is represented in terms of a multi-variate stochastic process that is conditioned by a set of deterministic time functions. This way was employed to facilitate analytical treatment of probabilistic conditioning. It is also assumed that the stochastic processes before they are conditioned by deterministic time functions are zero-mean stationary Gaussian processes.

Then the multi-variate conditioned stochastic process may be represented by $U_j(t \mid u_i(t); i = 1,2,\ldots,m); j = m+1,\ldots,n$ at locations $\boldsymbol{x}_j$. Here $u_i(t)$ stands for the deterministic time function specified at the location $\boldsymbol{x}_i$. For simplicity, $U_j(t \mid u_i(t); i = 1,2,\ldots,m)$ may be represented by $U_j(t \mid \mathit{condition})$. The word *condition* will be used in the same context for other cases; i.e., to mean that any quantity under discussion is conditioned by $u_i(t); i = 1,2,\ldots,m$.

Generally, stochastic processes $U_i(t); i = 1,2,\ldots,n$ can be expanded in a Fourier series as

$$U_i(t) = \sum_k (A_{ik}\cos\omega_k t + B_{ik}\sin\omega_k t) \quad (i = 1,2,\ldots,n), \tag{1}$$

in which A_{ik} and B_{ik} = Fourier coefficients at frequency ω_k. The coefficients A_{ik} and B_{ik} ($i = 1,2,\ldots,n$; k = fixed) are mutually independent and zero-mean Gaussian random variables, and their covariance matrix $\boldsymbol{V}_k$ is represented by means of $S_i(\omega_k)$ and $S_{ij}(\omega_k)$(Kameda and Morikawa 1992), where $S_i(\omega_k)$ = power spectral density function associated with $U_i(t)$, and $S_{ij}(\omega_k)$ = cross spectral density function between $U_i(t)$ and $U_j(t)$.

The deterministic time functions $u_i(t)$; $i = 1,2,\ldots,m$ are treated as realized values in the random field. Then these deterministic time functions can be represented in a way similar to Eq.(1) as

$$u_i(t) = \sum_k (\tilde{a}_{ik}\cos\omega_k t + \tilde{b}_{ik}\sin\omega_k t) \quad (i = 1,2,\ldots,m), \tag{2}$$

where $\tilde{a}_{ik}$ and $\tilde{b}_{ik}$ denote realized values of Fourier coefficients corresponding to A_{ik} and B_{ik} at frequency ω_k.

2.2 *Analytical Reduction of Conditioned Stochastic Processes*

The $2n$-dimensional joint Gaussian probability density function(p. d. f.) is derived for the Fourier coefficients A_{ik} and B_{ik} ($i = 1,2,\ldots,n$) at frequency ω_k, by means of their covariance matrix $\boldsymbol{V}_k$. Then one can obtain the $2(n-m)$-dimensional conditional joint p. d. f. for Fourier coefficients A_{ik} and B_{ik} ($i = m+1,\ldots,n$) conditioned by the realized values $\tilde{a}_{ik}$ and $\tilde{b}_{ik}$ ($i = 1,2,\ldots,m$)(Kameda and Morikawa 1992, 1993).

(a) *Fourier coefficients*

It is analytically derived that the conditioned Fourier coefficients $A_{\bar{j}k}$ and $B_{\bar{j}k}$ are mutually independent Gaussian variables. Then their mean values and variances reduce to

$$\begin{cases} \langle A_{\bar{j}k} \mid \mathit{condition}\rangle = -\dfrac{\sum_{i=1}^{m}\left({}^{(\bar{j})}q_{i\bar{j}k}\tilde{a}_{ik} - {}^{(\bar{j})}r_{i\bar{j}k}\tilde{b}_{ik}\right)}{{}^{(\bar{j})}p_{\bar{j}k}} \\ \langle B_{\bar{j}k} \mid \mathit{condition}\rangle = -\dfrac{\sum_{i=1}^{m}\left({}^{(\bar{j})}r_{i\bar{j}k}\tilde{a}_{ik} + {}^{(\bar{j})}q_{i\bar{j}k}\tilde{b}_{ik}\right)}{{}^{(\bar{j})}p_{\bar{j}k}}, \end{cases} \tag{3}$$

$$\sigma^2_{A_{\bar{j}k}|condition} = \sigma^2_{B_{\bar{j}k}|condition} = \frac{1}{{}^{(\bar{j})}p_{\bar{j}k}}, \tag{4}$$

where $\langle\ \cdot\ \rangle$ denotes the expectation and the coefficients ${}^{(h)}p_{ik}$, ${}^{(h)}q_{ijk}$, ${}^{(h)}r_{ijk}$ are derived as the following recurrence formulas;

$$\begin{cases} {}^{(h-1)}p_{ik} = {}^{(h)}p_{ik} - \dfrac{{}^{(h)}q^2_{ihk} + {}^{(h)}r^2_{ihk}}{{}^{(h)}p_{hk}} \\ {}^{(h-1)}q_{ijk} = {}^{(h)}q_{ijk} - \dfrac{{}^{(h)}q_{ihk}{}^{(h)}q_{jhk} + {}^{(h)}r_{ihk}{}^{(h)}r_{jhk}}{{}^{(h)}p_{hk}} \\ {}^{(h-1)}r_{ijk} = {}^{(h)}r_{ijk} + \dfrac{{}^{(h)}q_{ihk}{}^{(h)}r_{jhk} - {}^{(h)}r_{ihk}{}^{(h)}q_{jhk}}{{}^{(h)}p_{hk}} \end{cases} \quad (m+1 \le h \le n). \tag{5}$$

Initial step of Eq.(5) is given by ${}^{(n)}p_{ik} = \Lambda_{2i-1\,2i-1\,k}$, ${}^{(n)}q_{ijk} = \Lambda_{2i-1\,2j-1\,k}$, and ${}^{(n)}r_{ijk} = \Lambda_{2i-1\,2j\,k}$, where $\{\Lambda_{ijk}\} = \boldsymbol{\Lambda}_k = \boldsymbol{V}_k^{-1}$.

As the coherency involved in the covariance matrix decreases, $A_{\bar{j}k}$ and $B_{\bar{j}k}$ tend to zero-mean Gaussian variate. Correspondingly, it can be proved(Kameda and Morikawa 1992) that the Fourier amplitudes tend to Rayleigh variates, and the Fourier phase angle, to uniform variates.

(b) *Mean values and variances*

It is noted that the conditioned stochastic process $U_{\bar{j}}(t \mid \mathit{condition})$ is no longer of zero-mean. Its time-varying mean value is represented by

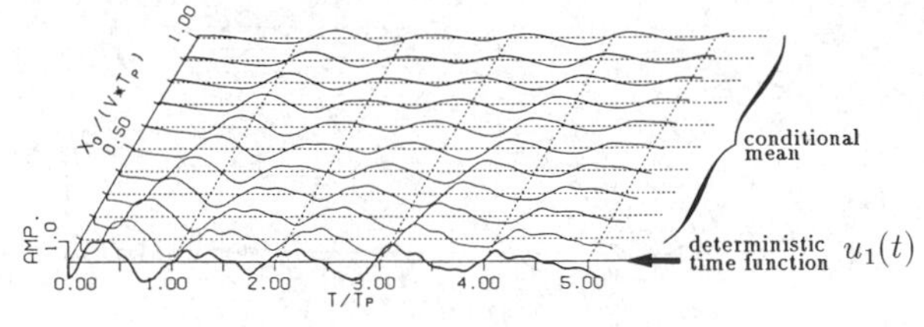

(a) Single deterministic time function

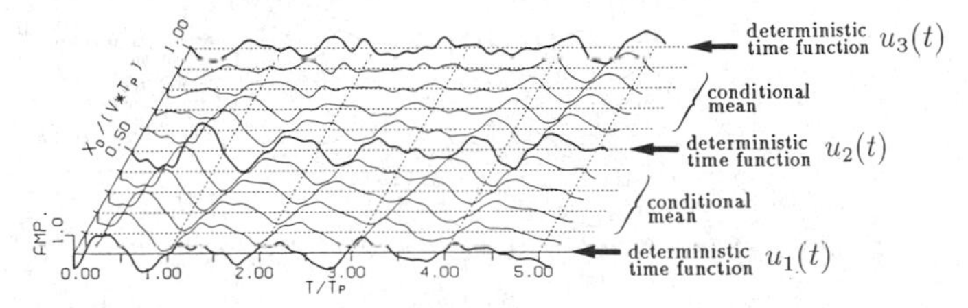

(b) Three closely spaced deterministic time functions

Fig. 1 Time-varying mean values conditioned by one or three deterministic time functions.

$$\mu_{U_{\bar{j}}|condition}(t) = \sum_k \Big\{\langle A_{\bar{j}k} \mid condition\rangle \cos\omega_k t + \langle B_{\bar{j}k} \mid condition\rangle \sin\omega_k t\Big\}. \tag{6}$$

As a numerical example for a typical case*, **Fig. 1** shows the conditioned mean values of $U_{\bar{j}}(t \mid condition)$ presented in Eq.(6), conditioned by one or three deterministic time functions. Observe that the conditional mean values are time-varying, and also that they tend to zero as the distance from the locations of the deterministic time functions increases. From Eq.(4), the variance of $U_{\bar{j}}(t \mid condition)$ is derived in the following form which is independent of the realized values $\tilde{a}_{ik}$ and $\tilde{b}_{ik}$; $i = 1, 2, \ldots, m$;

$$\begin{aligned}\sigma^2_{U_{\bar{j}}|condition} &= \sum_k \Big\{\sigma^2_{A_{\bar{j}k}|condition} \cdot \cos^2\omega_k t + \sigma^2_{B_{\bar{j}k}|condition} \cdot \sin^2\omega_k t\Big\} \\ &= \sum_k \frac{1}{^{(\bar{j})}p_{\bar{j}k}}.\end{aligned} \tag{7}$$

Furthermore, the conditional mean value and variance for time derivative $\dot{U}_{\bar{j}}(t \mid condition)$ of conditioned stochastic processes; namely, $\mu_{\dot{U}_{\bar{j}}|condition}(t) = \dot{\mu}_{U_{\bar{j}}|condition}(t)$ and $\sigma^2_{\dot{U}_{\bar{j}}|condition}$ can be readily derived from Eqs.(6) and (7)(Kameda and Morikawa 1992, 1993). They will be used later in the analysis of the first-passage problem.

It is interesting to note that the conditioned stochastic process can be rewritten as

$$U_{\bar{j}}(t \mid condition) = \mu_{U_{\bar{j}}}(t \mid condition) + U^*_{\bar{j}}(t \mid condition), \tag{8}$$

in which $U^*_{\bar{j}}(t \mid condition)$ is represented by

$$U^*_{\bar{j}}(t \mid condition) = \sum_k \Big(A^*_{\bar{j}k}|condition \cdot \cos\omega_k t + B^*_{\bar{j}k}|condition \cdot \sin\omega_k t\Big), \tag{9}$$

where $A^*_{\bar{j}k} \mid condition$ and $B^*_{\bar{j}k} \mid condition$ are random components of $A_{\bar{j}k} \mid condition$ and $B_{\bar{j}k} \mid condition$, respectively; namely, they are mutually independent Gaussian variables with zero-mean and variance shown in Eq.(4).

(c) *Power spectrum of random component*

Whereas power spectrum of conditioned stochastic processes $U_{\bar{j}}(t \mid condition)$ is equal to given $S_{\bar{j}}(\omega_k)$ if there were no loss or absorption of energy, that of random component $U^*_{\bar{j}}(t \mid condition)$ is given by

$$S_{U^*_{\bar{j}}|condition}(\omega_k)d\omega = \frac{1}{2} \cdot \frac{1}{^{(\bar{j})}p_{\bar{j}k}}. \tag{10}$$

*See Kameda and Morikawa(1992, 1993) for details of assumption about this numerical example.

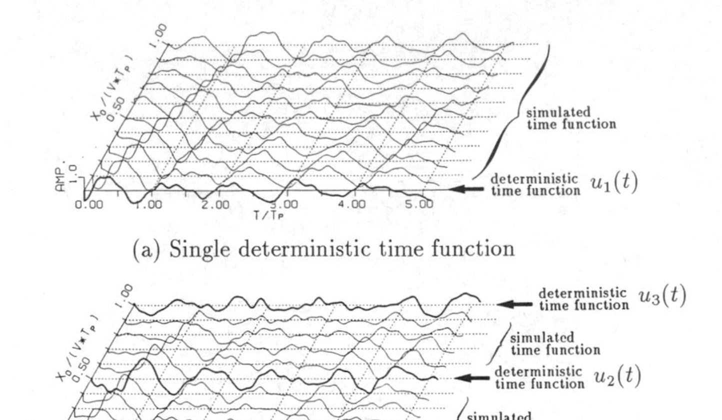

(a) Single deterministic time function

(b) Three closely spaced deterministic time functions

Fig. 2 Simulated time functions conditioned by one or three deterministic time functions.

3 FIRST-PASSAGE SOLUTION

In engineering applications of conditional random fields developed in this study, it will be important to deal with their time maxima. This involves the first-passage problem of the non zero-mean conditioned processes. This problem is converted to a first-passage of a zero-mean process across a time-varying threshold.

For this purpose, the threshold-value crossing rate of the conditioned processes should first be discussed. The upward(+) or downward(−) crossing rate $\nu^{\pm}_{U_{\bar{\jmath}}|condition}(\zeta;t)$ at time t and threshold ζ has been analytically derived(Shinozuka and Yao 1967). Applying its results to the present problem leads to the following:

$$\begin{aligned}\nu^{\pm}_{U_{\bar{\jmath}}|condition}(\zeta;t) &= \frac{1}{2\pi}\frac{\sigma_{\dot{U}_{\bar{\jmath}}|condition}}{\sigma_{U_{\bar{\jmath}}|condition}}\exp\left[-\frac{1}{2}\left(\frac{\zeta-\mu_{U_{\bar{\jmath}}|condition}(t)}{\sigma_{U_{\bar{\jmath}}|condition}}\right)^2\right] \\ &\cdot\left\{\exp\left[-\frac{1}{2}\cdot\left(\frac{\dot{\mu}_{U_{\bar{\jmath}}|condition}(t)}{\sigma_{\dot{U}_{\bar{\jmath}}|condition}}\right)^2\right]\right. \\ &\pm\sqrt{\frac{\pi}{2}}\frac{\dot{\mu}_{U_{\bar{\jmath}}|condition}(t)}{\sigma_{\dot{U}_{\bar{\jmath}}|condition}}\left\{ \left. 1+\operatorname{erf}\left(\pm\frac{1}{\sqrt{2}}\frac{\dot{\mu}_{U_{\bar{\jmath}}|condition}(t)}{\sigma_{\dot{U}_{\bar{\jmath}}|condition}}\right)\right\}\right\},\end{aligned} \tag{11}$$

where erf(·) denotes the error function. The crossing rate at two sided barriers, namely upward crossing of $\zeta(\geq 0)$ or downward crossing of $-\zeta$ is represented by $\nu^{a}_{U_{\bar{\jmath}}|condition}(\zeta;t) = \nu^{+}_{U_{\bar{\jmath}}|condition}(\zeta;t) + \nu^{-}_{U_{\bar{\jmath}}|condition}(-\zeta;t)$.

With this result and with an assumption of independent threshold-value crossings, the first-passage solution for two-sided barriers is given by the cumulative distribution function of the time-maximum Z in time duration $[0,\tau]$; setting $a_0(\zeta) = \text{Prob}\left[\,|U_{\bar{\jmath}}(0 \mid condition)| \leq \zeta\right]$,

$$\begin{aligned}&F_{Z|condition}(\zeta;\tau) \\ &= a_0(\zeta)\cdot\exp\left[-\int_0^{\tau}\nu^{a}_{U_{\bar{\jmath}}|condition}(\zeta;t)\,dt\right].\end{aligned} \tag{12}$$

4 NUMERICAL SIMULATION OF CONDITIONAL RANDOM FIELDS

In order to simulate the conditional random fields numerically, it is necessary to simulate A_{jk} and B_{jk} $(j = m+1,\ldots,n)$ which satisfy the $2(n-m)$-dimensional conditional probability density functions stated at the beginning of 2.2. Since marginal p.d.f. of $A_{\ell k}$ and $B_{\ell k}$ conditioned by A_{jk} and B_{jk} $(j = 1,2,\ldots,\ell-1;\ ^{\forall}\ell)$ are analytically derived, $A_{\ell k}$ and $B_{\ell k}$ are recursively simulated, by means of the realized values $\tilde{a}_{jk}$ and

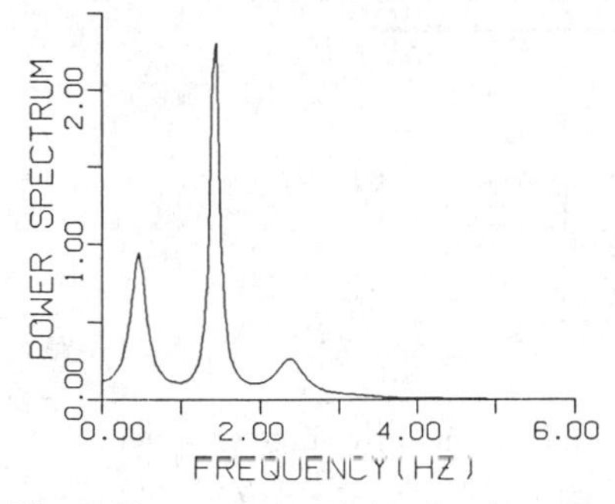

Fig. 3 Power spectral density function at free surface.

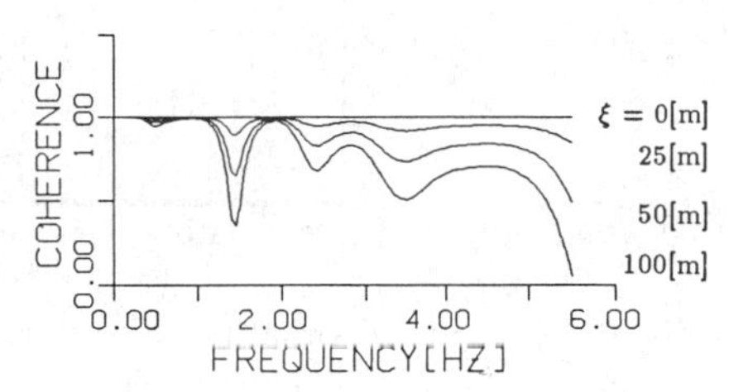

Fig. 4 Coherence function at free surface. (ξ =distance between two locations)

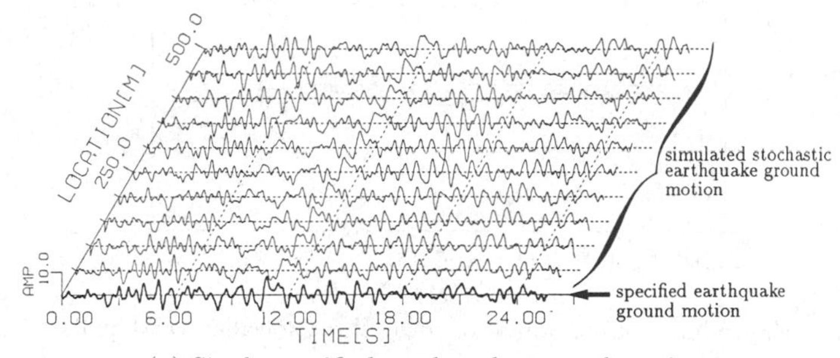

(a) Single specified earthquake ground motion

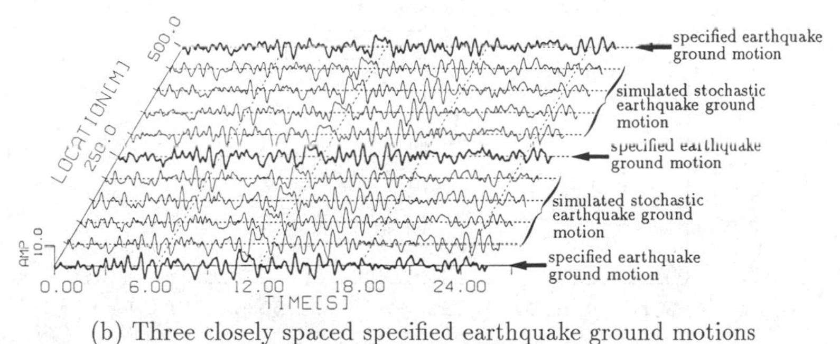

(b) Three closely spaced specified earthquake ground motions

Fig. 5 Simulated stochastic earthquake ground motions conditioned by one or three specified earthquake ground motions. ($c = 1000$[m/s], $\sigma_{ff} = 5\%$)

$\tilde{b}_{jk}$ ($j = 1, 2, \ldots, \ell - 1$; $m + 1 \leq \ell \leq n$) already obtained(Kameda and Morikawa 1993).

Numerical examples of simulated time functions are shown in **Fig. 2**. The parameter values employed in **Fig. 1** has also been used as conditions for this figure.

5 APPLICATION TO STOCHASTIC INTERPOLATION OF EARTHQUAKE GROUND MOTIONS

5.1 *Harada's Spectrum Based on Stochastic Ground*

This chapter is devoted to apply the theory of conditional random fields to earthquake wave field. Some numerical examples are presented for

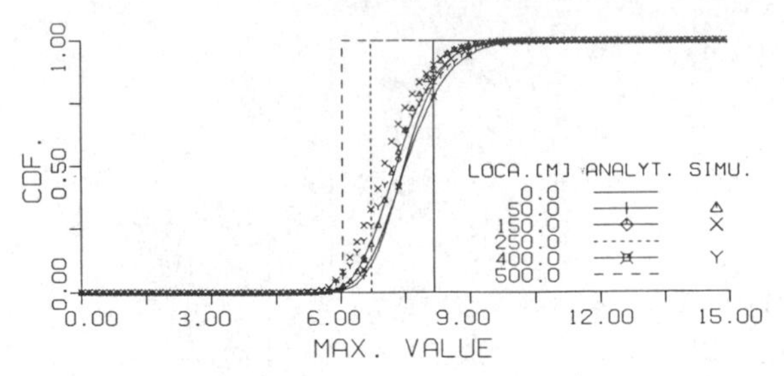

Fig. 6 Cumulative distribution of peak ground motion conditioned by three specified earthquake ground motions. ($c = 1000$[m/s], time $\tau = 25.0$[s], $\sigma_{ff} = 5\%$)

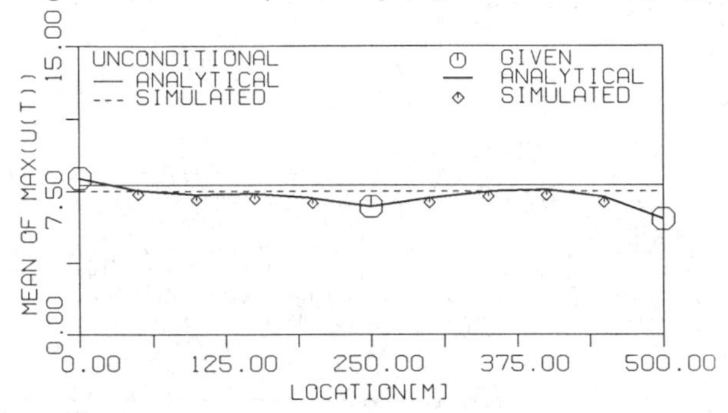

Fig. 7 Mean peak ground motions conditioned by three specified earthquake ground motions. ($c = 1000$[m/s], time $\tau = 25.0$[s], $\sigma_{ff} = 5\%$)

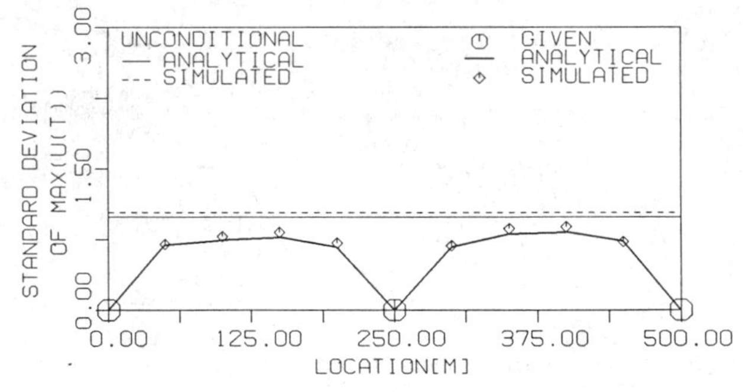

Fig. 8 Standard deviation of peak ground motions conditioned by three specified earthquake ground motions. ($c = 1000$[m/s], time $\tau = 25.0$[s], $\sigma_{ff} = 5\%$)

probabilistic estimation of earthquake wave field including the recorded ground motions, introducing the spectrum derived by Harada and Shinozuka(1988) for shear waves horizontally propagating 2-dimensional homogeneous ground with random soil properties. In the following numerical examples, the power spectrum shown in **Fig. 3** and the coherence functions shown in **Fig. 4** are used.

5.2 *Numerical Simulation*

Fig. 5 shows results of numerical simulation for stochastic earthquake ground motion including one or three specified deterministic motions.

5.3 *Conditional Probability Distribution of Peak Ground Motion*

By means of Eq.(12), numerical results for the peak ground motions have been obtained as shown in **Figs. 6–8**. These figures are a case with three specified earthquake ground motions. This case corresponds to **Fig. 5**(b). As Eq.(12) is based on an approximation assuming independent threshold-value crossings, Monte Carlo simulation has been carried out to evaluate its accuracy. The simulated results in **Figs. 6–8** are based on 1000 samples generated.

Fig. 6 shows the cumulative distribution function $F_{Z|condition}(\zeta;\tau)$ for a duration $\tau = 25$[s] and various values of the distance. In this figure, $F_{Z|condition}(\zeta;\tau)$ becomes a step function at three distances 0, 250 and 500[m] where earthquake ground motions are deterministic.

Figs. 7 and **8** show the mean values and the standard deviations of peak ground motion. In these figures, the horizontal lines stand for the unconditional cases. Along with the analytical results, simulated results based on 10000 samples are also plotted in these figures. In **Fig. 7**, the ordinates of the points indicated by circles are equal to the peak values of the specified ground motions. The conditioned mean $\langle\zeta;\tau \mid condition\rangle$ tends to the unconditional means as the location gets apart from these deterministic motions. **Fig. 8** showing the standard deviation $\sigma_{Z|condition}$ of the peak ground motion gives a significant engineering information. It vanishes at locations of specified ground motions, while it increases with distance, approaching the unconditional standard deviation.

6 CONCLUSIONS

The conclusions derived from this study may be summarized as follows.

1. Conditional random fields were formulated in terms of multi-variate stochastic processes conditioned by a set of deterministic time functions specified at discrete locations. Their probabilistic properties were analytically obtained.
2. The first-passage solution for conditioned stochastic processes was obtained.
3. Numerical simulation of conditional random fields was performed.
4. The methodology developed herein was applied to stochastic earthquake waves propagating through a ground with random soil properties, and numerical examples of stochastic estimation are presented for earthquake wave fields including specified earthquake ground motions.

ACKNOWLEDGMENTS

Throughout this study, informative discussion with Profs. J. Akamatsu, DPRI, Kyoto Univ. and T. Harada of Miyazaki Univ. are greatly appreciated. Hoshiya and Maruyama(1993) recently developed a theory of conditional random field by means of the time domain analysis. It is gratefully acknowledged that the authors enjoyed a series of in-depth discussion conducted with them.

REFERENCES

Ditlevsen, O. 1991. Random field interpolation between point by point measured properties. *Computational Stochastic Mechanics*, edited by P. D. Spanos and C. A. Brebbia, Proc. 1st International Conference on Computational Stochastic Mechanics: 801-812. Corfu, Greece.

Harada, T. and Shinozuka, M. 1988. Stochastic analysis of seismic ground motions in space and time. *Proc. 9WCEE* Vol. II: 825-830. Tokyo-Kyoto.

Hoshiya, M. and Maruyama, O. 1993. Stochastic Interpolation of Earthquake Wave Propagation. *ICOSSAR '93*. Innsbruck.

Kameda, H. 1991. Upgrading urban seismic safety and reliability — Proposal of regional seismic monitoring systems —. *Frontier R & D for Constructed Facilities*, edited by A. H-S. Ang and C-K. Choi and N. Shiraishi, Proc.

U.S./Korea/Japan Trilateral Seminar: 218-232. Honolulu, Hawaii.

Kameda, H. and Morikawa, H. 1992. An interpolating stochastic process for simulation of conditional random fields. *Probabilistic Engineering Mechanics* 7: 243-254.

Kameda, H. and Morikawa, H. 1993. Conditioned stochastic processes for conditional random fields. accepted for publication in the *Journal of Engineering Mechanics*, ASCE. (submitted March 1992)

Kawakami, H. and Ono, M. 1992. Simulation of space-time variation of earthquake ground motion using a recorded time history. *Structural Engineering/Earthquake Engineering*, Proc. JSCE No.441/I-18: 167-175. (in Japanese)

Morikawa, H. and Maruyama, T. 1993. Numerical simulation of time variations of wind speed on inlet boundary condition using theory of conditional random fields. *Proc. 8th Numerical Simulation of Turbulence Symposium*: 41-48. Institute of Industrial Science, University of Tokyo. (in Japanese)

Shinozuka, M. and Yao, J. T. P. 1967. On the two-sided time-dependent barrier problem. *Journal of Sound and Vibration* 6: 98-104.

Vanmarcke, E. and Fenton, G. 1991. Conditioned simulation of local fields of earthquake ground motion. *Structural Safety* 10: 247-264.

Structural Safety & Reliability, Schuëller, Shinozuka & Yao (eds) © 1994 Balkema, Rotterdam, ISBN 90 5410 357 4

Nonstationary response and reliability of linear systems under seismic loadings

T. Nakayama
Hiroshima Institute of Technology, Japan

S. Komatsu
Osaka Sangyo University, Japan

H. Fujiwara
Onoda Cement Co., Ltd, Chiba, Japan

N. Sumida
Japan Highway Public Corporation, Tokyo, Japan

ABSTRACT: This paper presents an estimating procedure for first-passage probabilities of both single- and multi-degree-of-freedom systems subjected to seismic excitation which can be defined as a nonseparable process, considering the statistical dispersion of a safe barrier. First, a technique of computing evolutionary spectrum of response based on the input-output relationship in frequency domain is introduced by using the complex demodulation method. The modulating function of excitation in the technique is given in complex form and enables precise computation of evolutionary spectrum of response. Second, statistical expressions for the nonstationary response of those systems are derived on the preceding basis. Finally, an approximate first-passage probability of a linear system under seismic excitation is evaluated, taking the scattering of a safe barrier into account.

1 INTRODUCTION

Approximations to a barrier crossing probability in a stochastic process has been used somewhat effectively so far to develop the methodology for evaluating the safety and reliability of a structure subjected to seismic excitation(Spanos 1980, 1984). Nevertheless, it has often been assumed to be a stationary process, even though its statistical properties are nonstationary in nature. Furthermore, in most instances, a certain barrier specifying a safety margin of a structure is represented by a deterministic quantity without including its inherent randomness.

As to the former problem, there are appreciable amounts of investigations in which seismic excitations expressed by nonstationary stochastic processes are adopted. However, in many cases, a nonstationary process is expressed as the product of a stationary process by a deterministic function of time which has an appropriate form based on statistical analyses of real accelerograms and is referred to as intensity function;the process is usually called "separable" stochastic process. It is therefore desirable to develop more sophisticated computational technique for realistically evaluating the nonstationary characteristic of both frequency and amplitude.

Next, as to the latter problem, it is preferable that a safe barrier is considered as a random variable with a certain probability distribution function such as normal or lognormal distribution.

We already proposed individual methods including these improvements;the first is for a nonstationary response analysis in frequency domain using a "nonseparable" stochastic process(Komatsu et al. 1986, Nakayama et al. 1986) without the restriction that "modulating function" is a real positive one(Hammond 1968);another is for a simplified mathematical evaluation of a first-passage probability performed with due regard to the randomness of a safe barrier(Komatsu and Nakayama 1978).

In this paper, it is shown that the procedure to approximately evaluate the first-passage probabilities of both single- and multi-degree-of-freedom systems under seismic excitation can be developed by using these two methods;the former method which can reflect nonstationary characteristics of amplitude and frequency is used to compute the variance and the covariance of response of systems;the latter method is used to evaluate the first-passage probabilities, taking the randomness of safe barriers into account.

The effectiveness of the procedure is demonstrated through numerical examples about a single-degree-of-freedom system.

2 FORMULATION OF RESPONSE ANALYSIS FOR SINGLE-DEGREE-OF-FREEDOM SYSTEM

2.1 *Relationship between excitation and response*

Hammond(1968) and Shinozuka(1970) developed the input-output relationship of a nonstationary process $x(t)$ and $y(t)$ as follows:

$$f_y(t,\omega) = f_x(t,\omega)\frac{|H(\omega)|^2\,|G(t,\omega)|^2}{|A(t,\omega)|^2} \tag{1}$$

where $f_x(t,\omega)$ and $f_y(t,\omega)$ are evolutionary spectra(Priestley 1965) of input and output, respectively; $H(\omega)$ is the frequency response function; $A(t,\omega)$ is a modulating function of $x(t)$;$G(t,\omega)$, as being a modulating function of $y(t)$, is given by

$$G(t,\omega) = \int_{-\infty}^{\infty} h(\tau)A(t-\tau,\omega)\frac{e^{-i\omega\tau}}{H(\omega)}d\tau \tag{2}$$

in which $h(\tau)$ is the unit impulse response function .
According to the definition of the evolutionary spectrum, $f_x(t,\omega)$ is written as

$$f_x(t,\omega) = |A(t,\omega)|^2 f_x(\omega) \tag{3}$$

where $f_x(\omega)$ is the stationary spectrum of input.

In the above formulations, modulating functions seem to take complex forms in a general way, but there is no concrete descriptions on the definition of the modulating function in papers(Hammond 1968, Priestley 1965, Shinozuka 1970). Particularly, Hammond said in his paper, "$A(t,\omega)$ which is not calculated explicitly in the estimation of the evolutionary spectral density and consequently the equations would appear to be of academic interest only." However, he assumed that $A(t,\omega)$ was merely a square root of $|A(t,\omega)|^2$ in equation (3) as he said, "$A(t,\omega)$ empirically derived."

Here, we now define the modulating function $A(t,\omega)$ to be complex and show in the following section that the modulating function $A(t,\omega)$ with complex form enables precise computation of equation (2).

2.2 *Evaluation of modulating function*

The nonstationary excitation $x(t)$ is generally expressed by

$$x(t) = \int_{-\infty}^{\infty} e^{i\omega t}dF(t,\omega) \tag{4}$$

where $dF(t,\omega)$ is defined by a modulating function $A(t,\omega)$ and an orthogonal process $X(\omega)$ as follows:

$$dF(t,\omega) = A(t,\omega)dX(\omega) \tag{5}$$

Let $dF(t,\omega)$ be

$$dF(t,\omega) = |dF(t,\omega)|e^{i\phi(t,\omega)} \tag{6}$$

Then, equation (4) is rewritten as

$$x(t) = \int_{-\infty}^{\infty} |dF(t,\omega)|e^{i(\omega t+\phi(t,\omega))} \tag{7}$$

where $\phi(t,\omega)$ is a time-varying phase angle dependent on the real and imaginary parts, $dF_R(t,\omega)$ and $dF_I(t,\omega)$ of $dF(t,\omega)$ and given as follows:

$$\phi(t,\omega) = \arctan\left\{\frac{dF_I(t,\omega)}{dF_R(t,\omega)}\right\} \tag{8}$$

Since $x(t)$ is a real function and the right-hand side of equation (7) can approximately be expressed by the sum of all ω, $x(t)$ may be represented by the following equation:

$$x(t) = 2\sum_{i=1}^{N}|dF(t,\omega_i)|\cos\{\omega_i t + \phi(t,\omega_i)\} \tag{9}$$

Then, $x_i(t)$ can be obtained by the use of a narrow band-pass filter as follows:

$$x_i(t) = 2|dF(t,\omega_i)|\cos\{\omega_i t + \phi(t,\omega_i)\} \tag{10}$$

Multiplying both sides of equation (10) by $\cos\omega_i t$ yields

$$x_i(t)\cos\omega_i t = |dF(t,\omega_i)|\left[\cos\phi(t,\omega_i) + \cos\{2\omega_i t + \phi(t,\omega_i)\}\right] \tag{11}$$

Repeating a similar procedure, the following equation is obtained:

$$x_i(t)\sin\omega_i t = |dF(t,\omega_i)|[-\sin\phi(t,\omega_i) + \sin\{2\omega_i t + \phi(t,\omega_i)\}] \quad (12)$$

By eliminating the circular frequency component $2\omega_i$ in equations (11) and (12) with a low-pass filter F, we obtain

$$\left.\begin{aligned} F\{x_i(t)\cos\omega_i t\} &= |dF(t,\omega_i)|\cos\phi(t,\omega_i) \\ F\{x_i(t)\sin\omega_i t\} &= -|dF(t,\omega_i)|\sin\phi(t,\omega_i) \end{aligned}\right\} \quad (13)$$

Then it is shown that the time-varying phase angle at ω_i can be expressed by

$$\phi(t,\omega_i) = \arctan\left[\frac{-F\{x_i(t)\sin\omega_i t\}}{F\{x_i(t)\cos\omega_i t\}}\right] \quad (14)$$

Comparing equation (8) with equation (14) leads to

$$\left.\begin{aligned} dF_R(t,\omega_i) &= F\{x_i(t)\cos\omega_i t\} \\ dF_I(t,\omega_i) &= -F\{x_i(t)\sin\omega_i t\} \end{aligned}\right\} \quad (15)$$

Consequently, it is found that the real and imaginary parts of $dF(t,\omega_i)$ can be obtained by filtering $x(t)$ twice. Since both sides of equation (5) can be divided into real and imaginary parts, the following expression is obtained:

$$\{A_R(t,\omega_i) + iA_I(t,\omega_i)\}\{dX_R(\omega_i) + iX_I(\omega_i)\} = dF_R(t,\omega_i) + idF_I(t,\omega_i) \quad (16)$$

where $dX_R(\omega_i)$ and $dX_I(\omega_i)$ are the real and imaginary parts of $dX(\omega_i)$, respectively.

We now have from equation (16)

$$\begin{bmatrix} dX_R(\omega_i) & -dX_I(\omega_i) \\ dX_I(\omega_i) & dX_R(\omega_i) \end{bmatrix}\begin{Bmatrix} A_R(t,\omega_i) \\ A_I(t,\omega_i) \end{Bmatrix} = \begin{Bmatrix} dF_R(t,\omega_i) \\ dF_I(t,\omega_i) \end{Bmatrix} \quad (17)$$

As a result, the real and imaginary parts of the modulating function can be obtained by resolving the following equation:

$$\begin{Bmatrix} A_R(t,\omega_i) \\ A_I(t,\omega_i) \end{Bmatrix} = \begin{bmatrix} dX_R(\omega_i) & -dX_I(\omega_i) \\ dX_I(\omega_i) & dX_R(\omega_i) \end{bmatrix}^{-1}\begin{Bmatrix} dF_R(t,\omega_i) \\ dF_I(t,\omega_i) \end{Bmatrix} \quad (18)$$

$dX_R(\omega_i)$ and $dX_I(\omega_i)$ in equation (18) can be given as

$$\left.\begin{aligned} dX_R(\omega_i) &= (2\pi/T)F_{x,R}(\omega_i) \\ dX_I(\omega_i) &= (2\pi/T)F_{x,I}(\omega_i) \end{aligned}\right\} \quad (19)$$

where $F_{x,R}(\omega_i)$ and $F_{x,I}(\omega_i)$ are the real and imaginary parts of Fourier coefficients of $x(t)$, respectively.

2.3 *Numerical examples*

Significant matters in the results reported in paper(Nakayama et al. 1986) are briefly mentioned here.

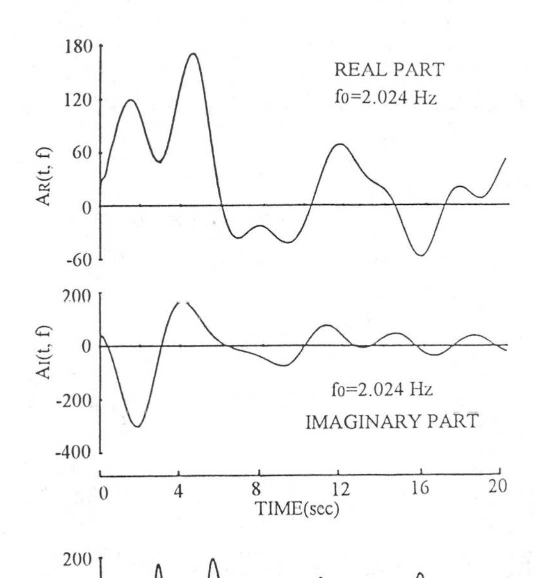

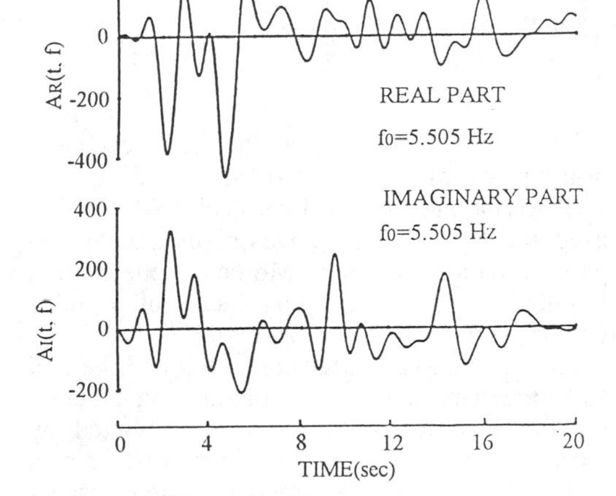

Fig.1 Modulating function

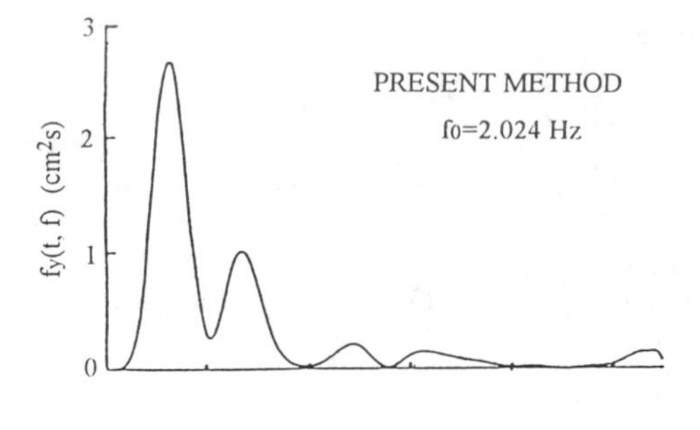

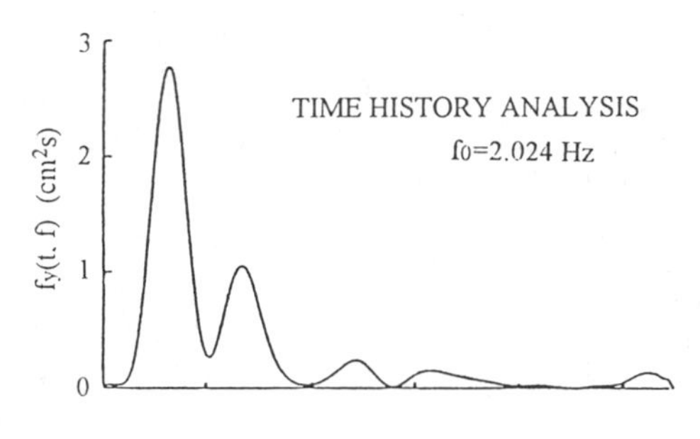

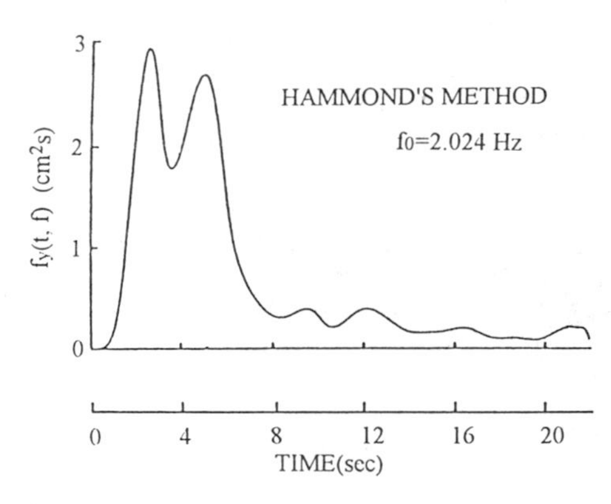

Fig.2 Comparison of evolutionary spectra computed by three methods

In the numerical analysis, the digital acceleration record of N-S component of EL Centro 1940 earthquake sampled for each 0.01 sec was used and Ormsby's filter was applied as a low-pass or band-pass filter. Moreover, the natural frequency and the damping factor of a linear system were taken 2.0 Hz and 0.05, respectively.

The evolutionary spectrum $f_x(t,\omega)$ and the real and imaginary parts of the modulating function for the seismic excitation were computed by means of methods presented in paper(Komatsu et al. 1986) and in the previous section, respectively. The evolutionary spectrum $f_y(t,\omega)$ of the response in frequency domain is computed by using equations in section 2.1.

As examples of results, the real and imaginary parts of the modulating functions corresponding to central frequencies f_0=2.0 Hz and 5.5 Hz in Ormsby's filter evaluated by the present method are shown in Figure 1. From the figure, it is found that the fluctuations of modulating functions with time become more remarkable with the increase of the central frequency.

Figure 2 shows the evolutionary spectra of response at central frequency 2.0 Hz, which is close to the resonance point. These results have been obtained by the present method, Time History Analysis method, and Hammond's method. The definitions of the latter two methods are as follows:

1. In Time History Analysis method, the evolutionary spectrum of response is obtained by the evolutionary spectral analysis(Komatsu et al. 1986) of time series calculated by Newmark's β method;the result by this method can be "bench mark" to judge the precision of the present method.

2. In Hammond's method, $|A(t-\tau,\omega)|$ in the equation (2) is used, instead of $A(t-\tau,\omega)$.

It is seen in Figure 2 that the result based upon Hammond's m11ethod shows the significant difference from "bench mark" in the vicinity of time t=5.0 sec. On the other hand, the result given by the present method shows good agreement with "bench mark" over the entire time domain.

3 FIRST-PASSAGE PROBABILITY OF SINGLE-DEGREE-OF-FREEDOM SYSTEM SUBJECTED TO NONSTATIONARY EXCITATION

3.1 *Variance and covariance function of response*

It is assumed that the response $y(t)$ of a single-degree-of-freedom system under nonstationary excitation can be given by the following equation.

$$y(t) = \int_{-\infty}^{\infty} G(t,\omega) e^{i\omega t} dY(\omega) \qquad (20)$$

where $\boldsymbol{Y}(\omega)$ is an orthogonal process and $G(t,\omega)$ is given by equation (2).

Then, using the expected values $\boldsymbol{E}[|dY(\omega)|^2]$ of

$|dY(\omega)|^2$, the autocovariance function $R_y(t,s)$ of $y(t)$ is represented as follows:

$$R_y(t,s) = \int_{-\infty}^{\infty} G(t,\omega)G^*(s,\omega)e^{i\omega(t-s)}\boldsymbol{E}\left[|d\boldsymbol{Y}(\omega)|^2\right] \tag{21}$$

where symbol * denotes a conju
gate complex number.

With the substitution $s=t$ in equation (21), the autocorrelation function $R_y(t,t)$ of $y(t)$ is obtained by

$$R_y(t,t) = 2\int_0^{\infty} |G(t,\omega)|^2 \boldsymbol{E}\left[|d\boldsymbol{Y}(\omega)|^2\right] \tag{22}$$

If the one-sided evolutionary spectrum $f_y(t,\omega)$ of $y(t)$ is expressed as

$$f_y(t,\omega)d\omega = 2|G(t,\omega)|^2 \boldsymbol{E}\left[|d\boldsymbol{Y}(\omega)|^2\right] \tag{23}$$

the variance function $\sigma_y^2(t)$ of $y(t)$ can be derived by the use of equation (1) as follows:

$$\sigma_y^2(t) = \int_0^{\infty} f_x(t,\omega)\frac{|H(\omega)|^2 |G(t,\omega)|^2}{|A(t,\omega)|^2}d\omega \tag{24}$$

Similarly, the variance function $\sigma_{\dot{y}}^2(t)$ of velocity and the covariance function $\sigma_{y\dot{y}}^2(t)$ of displacement and velocity are derived as follows:

$$\sigma_{\dot{y}}^2(t) = \int_0^{\infty} f_x(t,\omega)\frac{|H(\omega)|^2 |Q(t,\omega)|^2}{|A(t,\omega)|^2}d\omega \tag{25}$$

$$\sigma_{y\dot{y}}^2(t) = \int_0^{\infty} f_x(t,\omega)|H(\omega)|^2 \left[G_R(t,\omega)\dot{G}_R(t,\omega) + G_I(t,\omega)\dot{G}_I(t,\omega)\right] / |A(t,\omega)|^2 d\omega \tag{26}$$

where subscripts R and I denote real and imaginary parts, respectively;dot represents differentiation with respect to time;$Q(t,\omega)$ is expressed as

$$Q(t,\omega) = \dot{G}(t,\omega) + i\omega G(t,\omega) \tag{27}$$

Then the correlation coefficient $\rho_{y\dot{y}}(t)$ between displacement and velocity is calculated by

$$\rho_{y\dot{y}}(t) = \frac{\sigma_{y\dot{y}}^2(t)}{\sqrt{\sigma_y^2(t)}\sqrt{\sigma_{\dot{y}}^2(t)}} \tag{28}$$

3.2 *Procedure of evaluation of first-passage probability*

If the evolutionary spectrum and the modulating function of excitation are given by the spectral analysis by means of the complex demodulation method, we can obtain the variances of displacement and velocity, and their covariance according to equations (24), (25) and (26), respectively.

Then, based upon the assumption that excursions of the response over a certain barrier occur independently, the first-passage probability $P_f(t_d)$ can be approximately evaluated(Komatsu and Nakayama 1978);for example, in the case that a barrier is represented with a normally distributed stochastic variable.

$$P_f(t_d) = 1 - \text{Prob}\left[|y(0)| \le B\right]\exp\left\{-2\int_0^{td} \nu_y^+(t)dt\right\} \tag{29}$$

where:

$$\nu_y^+(t) = \frac{1}{2\pi}\frac{\sigma_{\dot{y}(t)}}{\sigma_{y(t)}}\left[\frac{1-\rho_{y\dot{y}}^2(t)}{\mu(t)}\exp\left\{-\frac{\overline{m}^2(t)}{2\mu^2(t)}\right\} + \frac{\rho_{y\dot{y}}(t)}{\sqrt{\pi}\,\overline{m}(t)\delta_B} \cdot \int_{-\infty}^{\infty} Z\exp\left\{-\frac{(1+\overline{m}^2(t)\delta_B^2)Z^2 - 2\overline{m}(t)Z + \overline{m}^2(t)}{2\overline{m}^2(t)\delta_B}\right\} \cdot erf\left\{-\frac{\rho_{y\dot{y}}(t)Z}{\sqrt{2(1-\rho_{y\dot{y}}^2(t))}}\right\}dZ\right]$$

$\mu(t) = \sqrt{1-\rho_{y\dot{y}}^2(t) + \overline{m}^2(t)\delta_B^2}$, $\overline{m}(t) = \overline{B}/\sigma_y(t)$,
$Z = B/\sigma_y(t)$, $erf(r) = \int_r^{\infty} \exp(-s^2)ds$,
$\overline{B}$:mean value of barrier; δ_B:coefficient of variation of barrier.

3.3 *Numerical examples*

As an example, the first-passage probability of a simple structure idealized in a single-degree-of-freedom system was computed with the procedure developed in the previous section.

Though it is desirable that we can use the evolutionary spectrum and the modulating function representing nonstationary characteristics of earthquake ground motion under various conditions, such physical quantities have not yet been obtained.

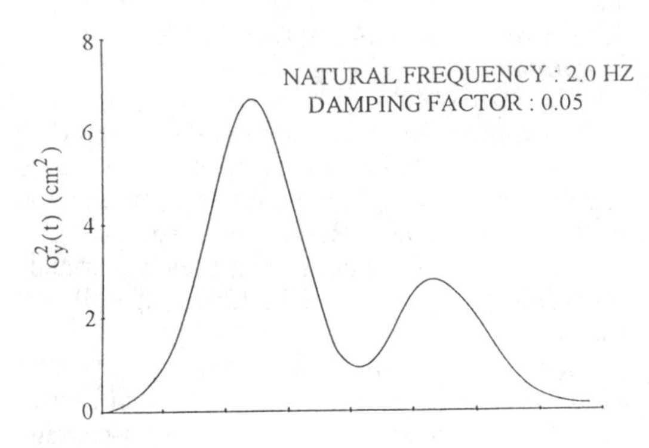

(a) Variance of displacement

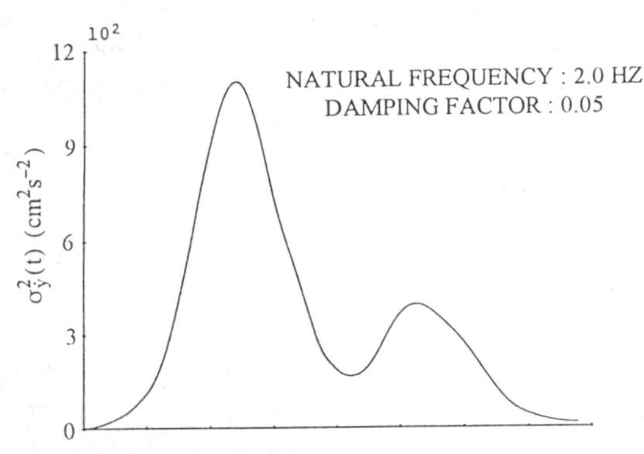

(b) Variance of velocity

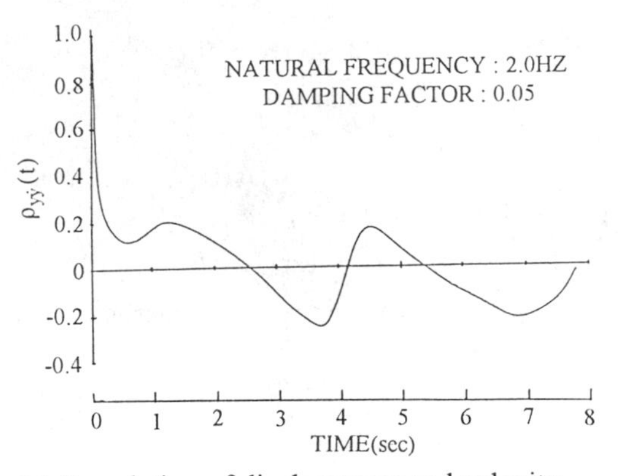

(c) Correlation of displacement and velocity

Fig.3 Time variation of variance and correlation of response

In this paper, El Centro earthquake record was adopted as an example. The natural frequency and the damping factor of the system were assumed to be 2.0 Hz and 0.05, respectively.

Figure 3 (a), (b), and (c) show the variances of displacement and velocity, and their correlation coefficient, respectively. We can find the variances of displacement and velocity become maximum at the time t=2.3 sec.

Table 1 denotes the first-passage probabilities of the system obtained under several conditions of safe barrier.

Table 1 First-passage probability

$\overline{m}^*$ \ δ_B	0.05	0.10	0.15
4.0	$1.2\text{x}10^{-3}$	$2.8\text{x}10^{-3}$	$8.6\text{x}10^{-3}$
4.5	$1.6\text{x}10^{-4}$	$6.1\text{x}10^{-4}$	$2.9\text{x}10^{-3}$
5.0	$1.8\text{x}10^{-5}$	$1.2\text{x}10^{-4}$	$1.0\text{x}10^{-3}$

The coefficients of variation of the barrier were assumed to be 0.05, 0.10 and 0.15. In the table, $\overline{m}^*$ is given by dividing the mean value of barrier by the maximum value of standard deviation of the response. As can be seen from this table, it is necessary to take much account of the statistical scatter of the barrier to precisely estimate the first-passage probability.

4 FIRST-PASSAGE PROBABILITY OF MULTI-DEGREE-OF-FREEDOM SYSTEM SUBJECTED TO NONSTATIONARY EXCITATION

In this section, we show the procedure for the evaluation of a first-passage probability of a multi-degree-of-freedom system subjected to nonstationary excitation.

On the basis of the modal analysis technique, we can approximately calculate the first-passage probability of the system by the same procedure as single-degree-of-freedom system.

It is assumed that the rth normal coordinate $\alpha_r(t)$ may be expressed by the modulating function $G_r(t,\omega)$ as follows:

$$\alpha_r(t) = \int_{-\infty}^{\infty} G_r(t,\omega) e^{i\omega t} d\mathbf{Y}_r(\omega) \tag{30}$$

where $G_r(t,\omega)$ is given by

$$G_r(t,\omega) = \int_{-\infty}^{\infty} h_r(\tau) A_r(t-\tau,\omega) \frac{e^{-i\omega\tau}}{H_r(\omega)} d\tau \tag{31}$$

where $A_r(t,\omega)$ is the modulating function of $\alpha_r(t)$; $h_r(\tau)$ and $H_r(\omega)$ are the rth unit response function and the rth frequency response function, respectively.

Then, the displacement of the system is given

by the following equation.

$$y(z,t) = \int_{-\infty}^{\infty} \sum_{r=1}^{\infty} \varphi_r(z) G_r(t,\omega) e^{i\omega t} d\boldsymbol{Y}_r(\omega) \qquad (32)$$

where $\boldsymbol{Y}_r(\omega)$ and $\phi_r(z)$ denote an orthogonal process and the rth natural mode, respectively.

As we can neglect the coupling effect between the modes in the case where the natural frequencies are essentially separate, the variance function $\sigma_{y_z}{}^2(t)$ of $y(z,t)$ is obtained as

$$\sigma_{y_z}^2(t) = 2\int_0^{\infty} \sum_{r=1}^{\infty} \varphi_r^2(z) |G_r(t,\omega)|^2 \boldsymbol{E}\left[|d\boldsymbol{Y}_r(\omega)|^2\right] \qquad (33)$$

Introducing the evolutionary spectrum $f_{Pr}(t,\omega)$ of the generalized force $P_r(t)$ which is related to both the participation factor β_r for the rth mode and the evolutionary spectrum $f_x(t,\omega)$ of excitation, we obtain

$$\sigma_{y_z}^2(t) = \int_0^{\infty} \sum_{r=1}^{\infty} \varphi_r^2(z) fp_r(t,\omega) \frac{|H_r(\omega)|^2 |G_r(t,\omega)|^2}{|A_r(t,\omega)|^2} d\omega \qquad (34)$$

$$\sigma_{y_z}^2(t) = \int_0^{\infty} \frac{f_x(t,\omega)}{|A(t,\omega)|^2} \cdot \sum_{r=1}^{\infty} \beta_r^2 \varphi_r^2(z) |H_r(\omega)|^2 |G_{r,0}(t,\omega)|^2 d\omega \qquad (35)$$

where

$$G_{r,0}(t,\omega) = \int_{-\infty}^{\infty} h_r(\tau) A(t-\tau,\omega) \frac{e^{-i\omega\tau}}{H_r(\omega)} d\tau \qquad (36)$$

Similarly, the variance function $\sigma_{\dot{y}_z}{}^2(t)$ of velocity $\dot{y}(z,t)$ is derived as follows:

$$\sigma_{\dot{y}_z}^2(t) = \int_0^{\infty} \frac{f_x(t,\omega)}{|A(t,\omega)|^2} \cdot \sum_{r=1}^{\infty} \beta_r^2 \varphi_r^2(z) |H_r(\omega)|^2 |Q_{r,0}(t,\omega)|^2 d\omega \qquad (37)$$

where

$$Q_{r,0}(t,\omega) = \dot{G}_{r,0}(t,\omega) + i\omega G_{r,0}(t,\omega) \qquad (38)$$

Furthermore, under the condition that the effect of coupling between the modes is neglected, the covariance function $\sigma_{y_z \dot{y}_z}{}^2(t)$ of displacement $y(z,t)$ and velocity $\dot{y}(z,t)$ can be obtained as

$$\sigma_{y_z\dot{y}_z}^2(t) = \int_0^{\infty} \frac{f_x(t,\omega)}{|A(t,\omega)|^2} \sum_{r=1}^{\infty} \beta_r^2 \varphi_r^2(z) |H_r(\omega)|^2 \cdot \left[G_{r,0,R}(t,\omega)\dot{G}_{r,0,R}(t,\omega) + G_{r,0,I}(t,\omega)\dot{G}_{r,0,I}(t,\omega)\right] d\omega \qquad (39)$$

Therefore, using equations (35), (37) and (39), a first-passage probability in case of a multi-degree-of-freedom system can be evaluated in the same manner as a single-degree-of-freedom system.

5 CONCLUSIONS

The real and imaginary parts of the modulating function relative to seismic excitation have been given the complex demodulation method. Furthermore, it has been shown that the evolutionary spectrum of response can be computed by such modulating function precisely in comparison with the results by Time History Analysis method as described in section 2.3.

The variance and the covariance of the response of both single- and multi-degree-of-freedom systems under seismic excitation have been formulated. Based on the formulation of these statistics, the procedure to approximately compute the first-passage probabilities of linear systems has been developed, taking the scattering of safe barriers into account. In addition, it has been suggested that the consideration of the randomness of a specified barrier is needed in precise analysis of the structural safety and reliability.

REFERENCES

Hammond, J.K. 1968. On the response of single and multidegree of freedom systems of non-stationary random excitations. *Journal of Sound and Vibration*, Vol.7: 393-416.

Komatsu, S. and Nakayama, T. 1978. First-passage failure probabilities of structures with scattered material strength under nonstationary random excitation. *Proc. Japan Society of Civil Engineers*, No.278: 25-38 (in Japanese).

Komatsu, S.,et al. 1986. Nonstationary spectrum analysis of earthquake ground motion by means of complex demodulation method. *Proc. Japan Society of Civil Engineers*, No.368:311-318 (in Japanese).

Nakayama, T.,et al. 1986. Non-stationary response analysis of dynamic structural systems. *Proc. Japan Society of Civil Engineers*, No.374: 541-548 (in Japanese).
Priestley, M.B. 1965. Evolutionary spectra and nonstationary processes. *Journal of Royal Statistical Society*, Vol.27: 204-237.
Shinozuka, M. 1970. Random processes with evolutionary power. *Proc. American Society of Civil Engineers*, Vol.96, EM:543-545.
Spanos, P-T.D. and Lutes, L.D. 1980. Probability of response to evolutionary process. Proc. *American Society of Civil Engineers*, Vol.106, EM:213-224.
Spanos, P-T.D. and Solomos, G.P. 1984. Barrier crossing due to transient excitation. *Proc. American Society of Civil Engineers*, Vol.110, EM:20-36.

Structural Safety & Reliability, Schuëller, Shinozuka & Yao (eds) © 1994 Balkema, Rotterdam, ISBN 90 5410 357 4

Importance index of damaged lifeline-network components

Toshihiko Okumura
Ohsaki Research Institute, Shimizu Corporation, Tokyo, Japan

ABSTRACT : This paper proposes a simplified method to evaluate the importance of a damaged link of a lifeline network where the importance of a link is defined as the degree of the network's functional serviceability which cannot be restored without repairing the link. When a damaged network does not have a redundancy, the network can be effectively restored by using the proposed importance index. Although the proposed method does not exactly evaluate the importance if the network contains loops, examples indicate that the calculated importance index still provides an effective order of restoration when the functional serviceability of the network is measured either in terms of connectivity or the total water head.

1. INTRODUCTION

Damage to lifeline systems has attracted earthquake engineers' attention since the 1971 San Fernando earthquake. This is related to the fact that people's activity relies more and more on the lifeline systems such as electric power, water, gas, transportation, and telecommunication systems. Thus, not only the physical damage to structures but also the losses of urban system's serviceability reduce the social and economic activities in urban areas. Therefore, it is essential to prepare and to take measures to cope with damage caused by forthcoming earthquakes. Structural upgrading of components, adding redundancy and buckups to systems, and preparing restoration materials are some of the examples to be completed before an earthquake. The procedure for reliability upgrading of lifeline systems proposed by Moghtaderi-Zadeh and Der Kiureghian (1983) falls under this category. After an earthquake, location and degree of damage should be firstly identified. The system should be under the emergency operation in order to minimize the extension of the damage to other parts of the system. Then, the damaged system should be restored to minimize the total inconvenience. Because lifeline systems contain various subsurface structures and because they are distributed over a wide area with different ground conditions, damage to the existing lifeline systems cannot be prevented. Therefore, post-earthquake action is very important and if it works well, the loss can be reduced remarkably.

Among several studies on the post-earthquake restoration strategy for lifeline systems, Nojima and Kameda (1992) proposed a restoration process of network connectivity given that the location of all the damaged components is known.

This study deals with the similar subject to Nojima and Kamdea's, and proposes a new index which provides an effective order of restoration of damaged links of a network. However, the method proposed in this study can directly treat a network with loops, and can restore the network's total water head although the restoration process in such cases is not necessarily optimal.

In the following sections, the importance index is defined in conjunction with the functional importance of a damaged link. Then, some numerical examples are presented to demonstrate the efficiency of the use of the index in the restoration of the damaged network.

2. DEFINITION OF IMPORTANCE

This study deals with a network which consists of some links and nodes. Only damage to links is considered here. When some links of a network sustain damage, each damaged link contributes to the lowering of the functional serviceability of the entire network. The degree of contribution, however, may be different from link to link. Therefore, it is essential to quantify the degree of contribution of each damaged link because a link with greater contribution should have a priority in the restoration process in order to recover the function of the network most effectively.

In the network shown in Figure 1, links 1, 2, 3, and 4 are damaged. Let us assume that damage to a link results in the loss of connection between the tied nodes, and that the functional serviceability of the network is measured in terms of the total number of nodes which are connected to the supply node. Then, obviously, three nodes (directly two nodes) lose their connection to the supply node due to the damage to link 1. Similarly, the damage to link 2 and link 4 result in the loss of connection of one and two nodes, respectively. On the other hand, the damage to link 3 does not affect the function of the network since the link has a detour. From another viewpoint, the three nodes in the downstream side of link 1 cannot be restored without fixing link 1, and thus, the restoration of link 1 is a necessary condition for the recovery of the connectivity of these three nodes.

In order to restore the network shown in Figure 1 most effectively, the order of links to fix is either 1→4→2→3 or 4→1→2→3. When the importance of a damaged link is defined as "the number of nodes which cannot be restored without fixing the link," it can be used as an index to determine the order of restoration. In the example above, the importance of the links 1 to 4 are 3, 1, 0 and 2, respectively, and lead to the order of restoration of 1→4→2→3, which is one of the most effective orders.

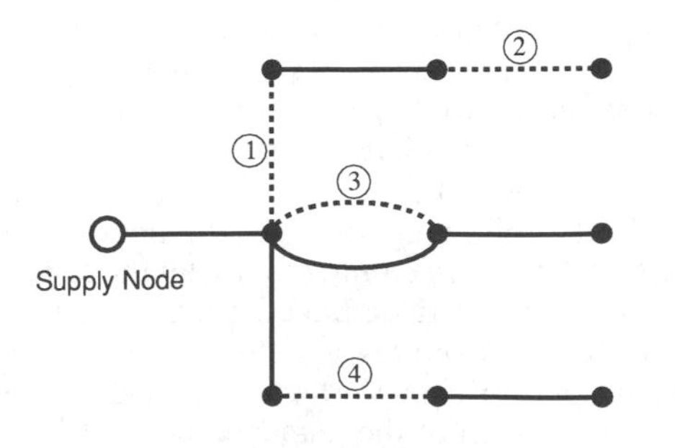

Figure 1 Example of Damaged Network

Since the connectivity is not the only functional measure of a network, it is convenient to redefine the importance of a damaged link in more general form, i.e., "the degree of the network's functional serviceability which cannot be restored without fixing the link."

3. EVALUATION OF IMPORTANCE INDEX

When the connectivity is used as the functional serviceability measure of a network, the importance defined in the previous section can be approximately evaluated by employing the analysis method of flow in a pipe network. In general, the following two equations are necessary to calculate flow in a pipe network:

$$\sum_j Q_{ij} + p_i = 0 \qquad \cdots (1)$$

$$Q_{ij} = \frac{1}{R_{ij}} \left| E_i - E_j \right|^a \qquad \cdots (2)$$

where j represents the nodes directly connected to node i, Q_{ij} is the flow rate in pipe ij, p_i is the amount of external demand at node i, E_i is the water head at node i, and R_{ij} and a are the coefficients. The value taken by the coefficient a is around 0.5 for water. Thus, the equation (2) is a non-linear equation.

Now, for simplicity, let us assume a unit demand p_i=1 at each demand node of the network shown in Figure 1. This assumption is appropriate when all the demand nodes have the same weight. When a link with a very large resistance coefficient $R_{ij}=R^D$ replaces the damaged links and very small resistance coefficient $R_{ij}=R^I$ is assigned to the intact links, the flow rate in each link will take the value shown in Figure 2. When there exists a bypassing route for a damaged link, the flow avoids the damaged link and takes the detour in which virtually no head loss is produced. On the other hand, if there is not any detour, the flow exists in a damaged link to satisfy the demand at the nodes even though it results in great head loss. Since the purpose of this study is not to calculate the actual flow in the network, the coefficient a in Equation (2) can be set to 1 to make the equation linear. In this case, the amount of head loss in a damaged link is proportional to the amount of flow in the link, i.e., the number of nodes in the downstream side of the damaged link because a unit demand is assigned to each node. The calculated head loss, therefore, can be used as an index of the importance because the greater the head loss, the greater the number of nodes whose connection to the supply node is interrupted by the damage to the link.

The calculated head loss corresponds to the connectivity-based importance of the damaged link when the network does not contain any loop or when at most one of the links which form a loop sustains damage. However, when some links constitute a loop and more than one of these links are damaged, the importance is divided among these damaged links. Thus, in such a case, the calculated value does not correspond to the importance defined in the previous section. Hereafter, the head loss evaluated by using the linearized flow equation is called "importance index" in order to distinguish it from the "importance." In the calculation of the importance index, weight for nodes and links can be considered by varying the value of p_i and R_{ij}^D. It is obvious from the above explanation that the calculation of the importance indices of all links requires to solve an $N \times N$ linear system (when head equations are used) or $L \times L$ linear system (when flow rate equations are used) only once, where N is the number of demand nodes and L is the number of links.

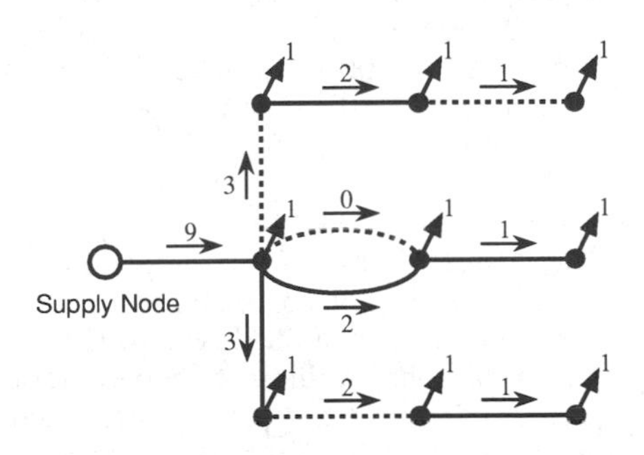

Figure 2 Application of Flow Analysis to Evaluation of Importance Index

4. RESTORATION STRATEGY BASED ON IMPORTANCE INDEX

In this section, damaged networks are restored in the order of the importance index proposed in the previous section. The restoration process is compared with those by other restoration strategies.

4.1 *Functional measure of a network*

There are a number of measures to quantify the remaining functional serviceability of a damaged network. In this study, the following two different measures are used. One is the connectivity, which is the total number or the total weighted number of the nodes connected to the supply nodes. The other is the total water head of all the connected nodes, which is obtained by performing a flow analysis of the network. This measure is used under the assumption that the network considered here is a water delivery system.

4.2 *Restoration strategies*

The restoration strategies compared in this study are as follows:

Strategy I : repair links in the order of the proposed importance index;

Strategy C : proceed step by step and at each step repair the link which most improves the connectivity of the network;

Strategy H : proceed step by step and at each step repair the link which most improves the total head of the network.

More concretely, the order of restoration in strategies C and H is determined step by step in the following manner. Let n be a total number of the damaged links of the network. Then, there are n candidates to choose from for this restoration step. Select an arbitrary link from these n links and evaluate the amount of network's function recovered by repairing the link. Repeat this procedure n times and select the link which improves the network most effectively as the link to be fixed in the step being considered. Therefore, the strategies C and H are based on the local optimization sequences which do not assure the global optimization (Nojima and Kameda, 1992). The calculation effort required to determine the order of restoration by strategies C and H is

$n+(n-1)+\cdots+2=[n(n+1)/2]-1$ times of connectivity or flow analysis.

4.3 *Application to network without loop*

First, the importance indices are calculated for a damaged network which does not contain any loop. In this case, the importance index of a link corresponds to the importance defined in Section 2, i.e., the (weighted) number of nodes which cannot be restored without repairing the link.

Figure 3 shows a network which consists of 29 links, one supply node, and 29 demand nodes. The number on each demand node together with the node number is a randomly assigned weight ranging from 0 to 2. Out of 29 links, 13 links shown by the dotted lines are damaged resulting in the loss of connection of 24 nodes from the supply node. These damaged links are repaired by employing the three strategies mentioned earlier. In these strategies, only one link is repaired at each restoration step. In strategy I, the resistance coefficients R^I and R^D are assumed to be 0.01 and 200.0, respectively, in the calculation of the importance indices. In strategy H, the following values are used in the flow analysis: the inner diameter, length, and Hazen-Williams roughness coefficient of each link are assumed to be 0.5m, 250m, and 140, respectively; each demand node is assumed to have an external demand of ($0.1\times$*weight*) m^3/sec and to have the same elevation; a fixed water head of 50m is assigned to the supply node.

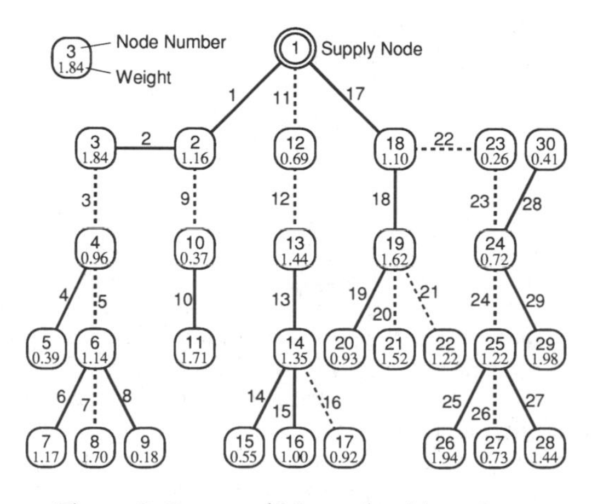

Figure 3 Damaged Network without Loop

The restoration sequences determined by strategies I, C, and H are as follows:

I : 22 → 23 → 11 → 3 → 24 → 12 → 5 → 9 → 7 → 20 → 21 → 16 → 2

C : 9 → 20 → 3 → 5 → 7 → 21 → 11 → 12 → 16 → 22 → 23 → 24 → 2

H : 3 → 5 → 9 → 11 → 12 → 22 → 23 → 24 → 16 → 21 → 20 → 7 → 26

Figure 4 (a) compares how the network is restored in terms of the restoration ratio based on the weighted number of restored nodes after each step. The restoration ratio is defined by the following equation.

$$\text{restoration ratio} = \frac{\text{restored function}}{\text{initial degree of damaged function}} \quad \cdots (3)$$

Although the difference among the three strategies is relatively small during the first five steps, strategy I which uses the proposed importance index restores the network most effectively after the sixth step. On the other hand, the restoration ratio by strategy C is lower than those by strategies I and H despite the fact that strategy C is based on the local optimization of the connectivity. One of the reasons for this is that the importance index contains the information of all the nodes while strategy C accounts for the information of only one step of restoration. For example, the first link to be fixed in strategy I is link 22 which only recovers the connectivity of node 23 (weight = 0.26) but is necessary to fix in order to restore nodes 23 to 30 (total weight = 8.70). Therefore, by strategy C, the links that can recover two nodes such as link 9 are fixed prior to link 22 to leave nodes 23 to 30 disconnected till a later stage of the restoration.

Next, 100 samples of damaged networks are generated by randomly choosing thirteen links to be damaged in each sample, and they are restored by using strategies I, C, and H. Figure 4 (b) compares the mean values of the restoration curves of the 100 samples by these strategies. As seen in Figure 4 (a), strategy I restores the network most effectively except during the early stage of the restoration. However, unlike Figure 4 (a), strategy C gives a better result than strategy H on average.

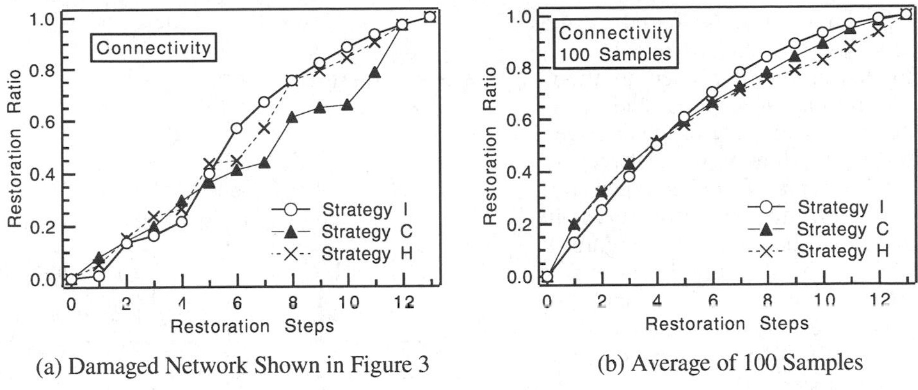

(a) Damaged Network Shown in Figure 3 (b) Average of 100 Samples

Figure 4 Restoration Curves in Terms of Connectivity for Network without Loop

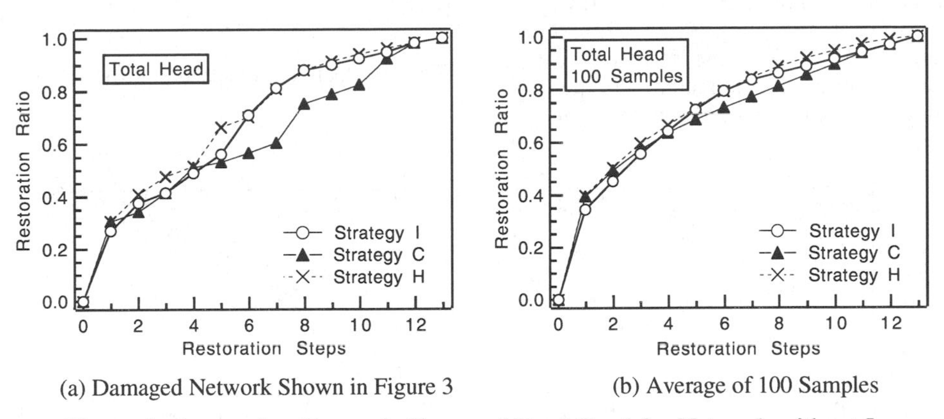

(a) Damaged Network Shown in Figure 3 (b) Average of 100 Samples

Figure 5 Restoration Cruves in Terms of Total Head for Network without Loop

Figure 5 (a) compares the restoration curves in terms of the total water head when the damaged network shown in Figure 3 is restored by the three strategies. In this case, strategy H which is based on the local optimization of the water head gives the best result throughout the steps, and the importance index-based strategy I follows closely. The restoration ratio by strategy C is lower than those by other strategies from the sixth to tenth step. Figure 5 (b) which depicts the mean restoration curves of 100 samples exhibits the same tendency as Figure 5 (a).

From these examples, as long as a network without any loop is concerned, the order of restoration determined by referring to the importance index provides the excellent results when the functional serviceability of a network is measured either in terms of the connectivity or total water head. This is probably because the importance index is based on the connectivity of a tree-shaped network and because the index is calculated by the simplified flow analysis of a network.

4.4 *Application to network with loops*

Figure 6 depicts a network generated by adding four links (30 to 33) to the network shown in Figure 3 to form several loops. This network is restored by strategies I, C, and H, and the restoration processes are compared.

Figure 7 (a) shows the restoration curves in terms of the connectivity. The connectivity-based strategy C provides the best result during the entire restoration procedure, then, strategies I and H follow in this order. The same tendency is observed in Figure 7 (b) which depicts the average restoration curves of 100 samples of damaged network. When a network contains some loops, a disconnected demand node can be reconnected to a supply node when one of the parallel routes is fixed. However, as mentioned earlier, the importance is shared by the damaged links in parallel when it is evaluated by using the proposed method. Thus, the importance index underestimates the priority of some of the damaged links if the function of the network is measured in terms of the connectivity. On the contrary, the restoration curves in terms of the total head shown in Figures 8 (a) and (b) exhibit the completely different results from Figures 7 (a) and (b); strategy H gives the best results while strategy C exhibits low restoration ratio after the fourth step.

Figure 9 compares the networks after four steps of restoration by strategies I, C, and H. By strategy C, links 9, 16, 31, and 33 have been fixed at the fourth step leaving 10 nodes disconnected. In the fifth step, link 24 is repaired to recover

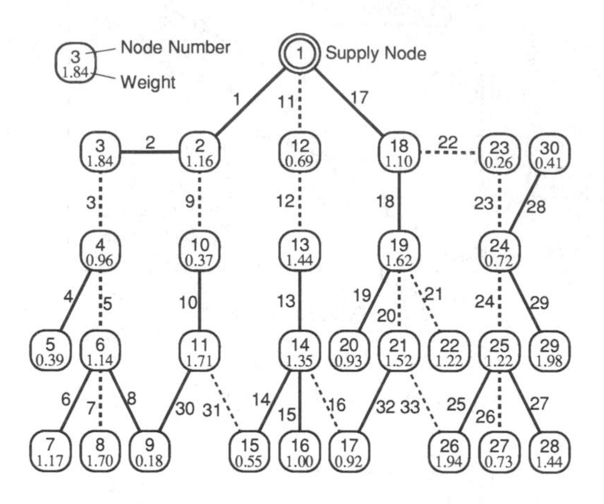

Figure 6 Damaged Network with Loops

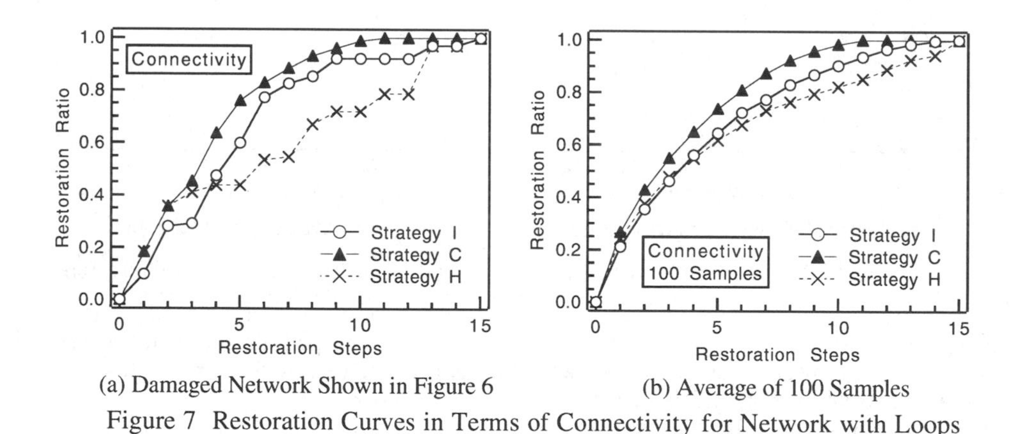

(a) Damaged Network Shown in Figure 6 (b) Average of 100 Samples

Figure 7 Restoration Curves in Terms of Connectivity for Network with Loops

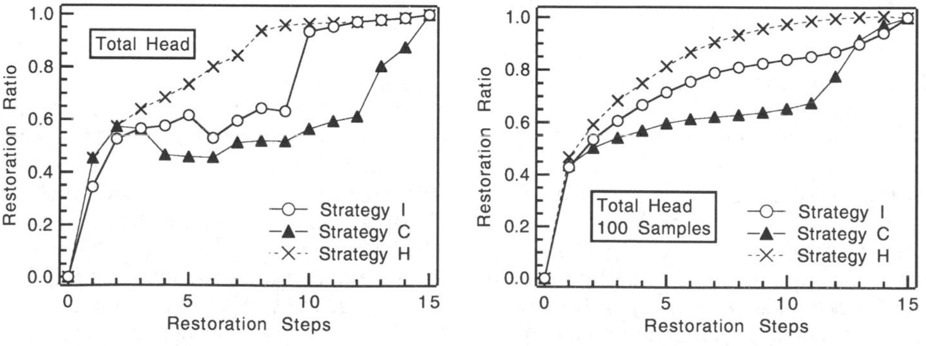

(a) Damaged Network Shown in Figure 6 (b) Average of 100 Samples

Figure 8 Restoration Curves in Terms of Total Head for Network with Loops

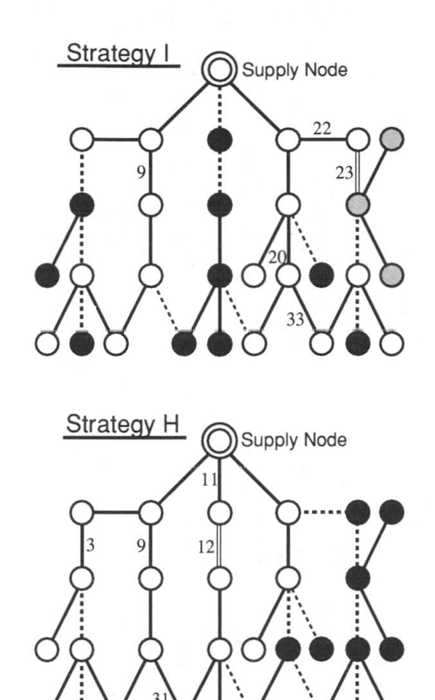

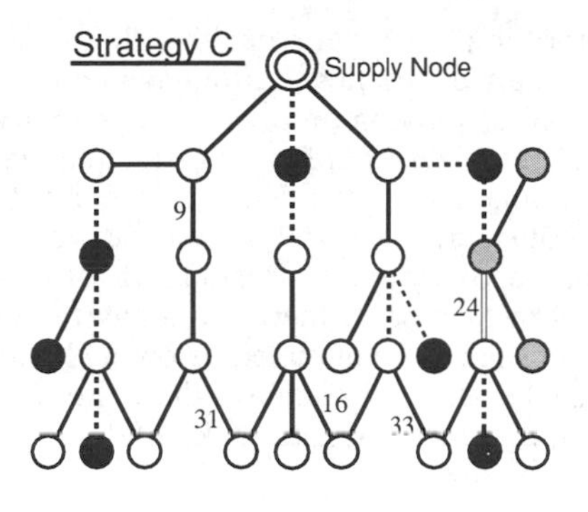

○ : Connected Node
◍ : Restored in the 5th Step
● : Disconnected Node
═ : To Be Fixed in the 5th Step

Figure 9 Comparison of Networks after 4 Steps of Restoration

the three nodes which are shaded in the figure. This strategy tends to create a long supply route in the network in order to increase the number of connected nodes, but such a route is hydraulically unfavorable. Therefore, strategy C is the best among the three if the function of the network is measured in terms of the connectivity but the worst if the total water head is the function. In particular, as seen in Figure 8 (a), the total water head may decrease in spite of the increase in number of the connected nodes. On the other hand, twelve nodes are disconnected after the four restoration steps, and the next link to be fixed is link 12 which does not improve the connectivity but improves the hydraulic performance of the network. Thus, strategy H does not improve the connectivity as effectively as strategy C, but provides the best order of restoration when the total water head is the functional measure of the network.

When the restoration of a network with loops is concerned, the restoration strategy should be elaborated taking into account which function of the network is the most important. However, considering that the optimization of the hydraulic properties such as the total head requires enormous amount of calculation, the proposed importance index will be a practical index for a realistic large network.

5. CONCLUSIONS

The major conclusions of this study can be summarized as follows:

1. The functional importance of a damaged link of a network is defined as "the degree of the network's functional serviceability which cannot be restored without repairing the link." A method to evaluate the importance is then proposed based on the linearized flow analysis method.
2. The restoration strategy based on the calculated importance index gives an excellent result if a network does not contain any loop. When a network contains several loops, the importance index does not exactly correspond to the importance of the link, but the result is still comparable to the results by the strategies based on the local optimization sequences.
3. Taking the calculation effort into account, the proposed importance index will be a practical index for realistic large networks.

The restoration procedure dealt with in this study is a part of the post-earthquake actions. In the actual restoration process, the possibility of assignment of repair crews, construction machines and materials to the damaged site should also be considered as well as the network's functional recovery rate. Besides, as briefly mentioned in Section 1, there are many things to be established and completed before and after an earthquake. Effort to achieve these pre- and post-earthquake plans would make lifeline systems more reliable.

REFERENCES

Moghtaderi-Zadeh, M., and A. Der Kiureghian (1983) : Reliability Upgrading of Lifeline Networks for Post-Earthquake serviceability," *Earthquake Engineering and Structural Dynamics*, Vol. 11, pp. 557-566.

Nojima, H., and H. Kameda (1992) : "Optimal Strategy by Use of Tree Structure for Post-Earthquake Restoration of Lifeline Systems," *Proceedings of the Tenth World Conference on Earthquake Engineering*, Madrid, Spain, Vol. 9, pp. 5541-5546.

Structural Safety & Reliability, Schuëller, Shinozuka & Yao (eds) © 1994 Balkema, Rotterdam, ISBN 90 5410 357 4

Ground motion incoherence, its modelling and effects on structures

O. Ramadan & M. Novak
Faculty of Engineering Science, University of Western Ontario, London, Ont., Canada

ABSTRACT: Spatially variable seismic ground motions are mathematically characterized in frequency as well as time domains and their effects on the lateral response of long gravity dams are theoretically investigated. Published data for spatial coherency models are critically reviewed and unified in a general model. A novel simulation technique is outlined for digitally generated, spatially multi-dimensional ground motions which satisfies both the target auto-spectrum and coherency function. The paper provides a simple yet sufficiently accurate approach for describing spatially incoherent ground motions and demonstrates the important effects of this incoherence on the seismic response of extensive structures.

INTRODUCTION

The seismic design of structures is usually conducted with the ground motion prescribed in terms of a power spectral density function, a response spectrum or a single realization. Such ground motion models are sufficient for the analysis of structures which have small dimensions in plan but may not be adequate for the design of extensive structures because the latter can suffer not only from the ground motion variation in time but also its variation in space. Spatial variations of seismic ground motion are clearly indicated in the data recorded at large-scale seismographic arrays during recent earthquakes [e.g. 2 to 5]. To describe the random field of the ground motion allowing for randomness in both space and time, complex-valued cross-spectra between any two stations along the foundation base are needed. This task can be considerably simplified by assuming that the random field of the ground motion is homogeneous. The analysis of data recorded during past earthquakes justifies this assumption [3]. Then, the cross-spectra can be defined as the product of the local, invariant power spectrum and the spatial coherency function; an exponential term is added to account for the effects of travelling waves.

A number of models for the spatial coherency function were suggested. They are based on experience with other random fields [6], simplified theoretical studies on wave scattering in random media [13] or signal processing of actual data [e.g. 2, 3, 5]. These models, while intended to describe the same natural phenomena, differ considerably and the parameters involved in one model are unrelated to those of another one. The published cohere ncy models are evaluated in this paper and it is found that the simple, joint coherency model suggested by Novak and Hindy [6] fourteen years ago is capable of describing many seismic events just by altering the numerical values of its two parameters.

To describe the space-time random field of the ground motion in the time domain, simulation techniques which generate motions satisfying both the power spectrum and coherency function (or the cross-spectral matrix) are needed. The two available procedures [11,12] require the decomposition of the spectral matrix at each frequency and are computationally quite demanding as the order of this matrix increases. As an alternative, a novel, computationally efficient simulation technique which requires only the summation of trigonometric functions is outlined and is implemented in response analysis of a dam-reservoir-foundation system to demonstrate the effects of ground motion incoherence.

FREQUENCY DOMAIN CHARACTERIZATION OF GROUND MOTIONS

The homogeneous space-time random field of the seismic ground motion can be characterized by the cross-spectral density function between any two stations, expressed as

$$S(\vec{r},\omega) = S_\ell(\omega)\ R(\vec{r},\omega)\exp(i\omega\tau_r) \tag{1}$$

in which, S_ℓ is the local, invariant power spectrum, $R(\vec{r},\omega)$ is the spatial coherency function and ω is the circular frequency; $\vec{r}$ is the separation vector between the two stations, τ_r is the time lag between them due to surface travelling waves and $i = \sqrt{-1}$. Equation 1 allows for the direction dependency of the coherency function, a feature necessary in the cases in which the epicentral distances are not much larger than the size of the structure [10]. For most engineering applications, however, the coherency function may be assumed isotropic (i.e. direction-independent). In this case, the coherency becomes dependent on the true separation $r = |\vec{r}|$ rather than the separation vector.

Spatial Coherency Function

The spatial coherency functions suggested for seismic ground motions may be grouped in three categories as follows: (i) - the coherency which jointly describes the dependence on both separation and frequency in terms of a single dimensionless parameter, $(\omega r/V)$, i.e. [6]

$$R(r,\omega) = \exp[-c(\frac{\omega r}{V})^\gamma] \tag{2}$$

in which c, γ and V are a decay parameter, a power factor and a suitable wave velocity; (ii) - exponential models involving a frequency dependent decay parameter, c_ω, in the form

$$R(r,\omega) = \exp(-c_\omega r) \tag{3}$$

where the decay parameter is assumed to be linear or quadratic in frequency [e.g. 5]; and (iii) - more complex models including more than three parameters [2,3].

To investigate the validity of the dimensionless parameter in Eq. 2, the coherency functions are evaluated using the SMART-1 array data for three different cases, i.e. radial component of event 20 [3], tangential component of event 20 [4], and horizontal component of event 40 [2]). The data reported for individual frequencies or separations are replotted versus $(\omega r/V)$ in Fig. 1. Also shown in this figure is the best fit of Eq. 1 to these data. The individual coherencies for the different frequencies collapse onto one curve or very close to it. This suggests that the joint coherency model defined in terms of the dimensionless frequency $(\omega r/V)$ is adequate for those events, although it may not match all data reported. The joint coherency model with $\gamma=2$ is also supported by theoretical studies on wave scattering in random media [13]. Other advantages of the joint coherency model are its simplicity and the use of two parameters only. The low number of parameters simplifies the establishment of these parameters from actual data, the model application in seismic response analysis of structures and also the correlation of these parameters to physical parameters such as focal depth, epicentral distance, foundation parameters and earthquake intensity. More details on the coherency function are given in [10].

SIMULATION OF SPATIALLY INCOHERENT GROUND MOTIONS

Seismic ground motions satisfying Eq. 1 can be digitally generated for use in time domain response analysis of large structures. Considering one spatial coordinate, s, and assuming upward propagating shear waves (i.e. no surface travelling waves), these motions can be generated by

$$u_s(t) = \sum_{k=0}^{N}\sum_{i=1}^{M} a_k\{\sqrt{S_d(\omega_{ik})\Delta\omega}\cos[\omega_{ik}t+\phi_{ik}+\pi ks/L_i] + \sqrt{S_d(\omega'_{ik})\Delta\omega}\ \cos[\omega'_{ik}t+\phi'_{ik}-\pi ks/L_i] \tag{4}$$

In Eq. 4, S_d is the displacement power spectrum, ϕ_{ik} and ϕ_{ik}' are two statistically

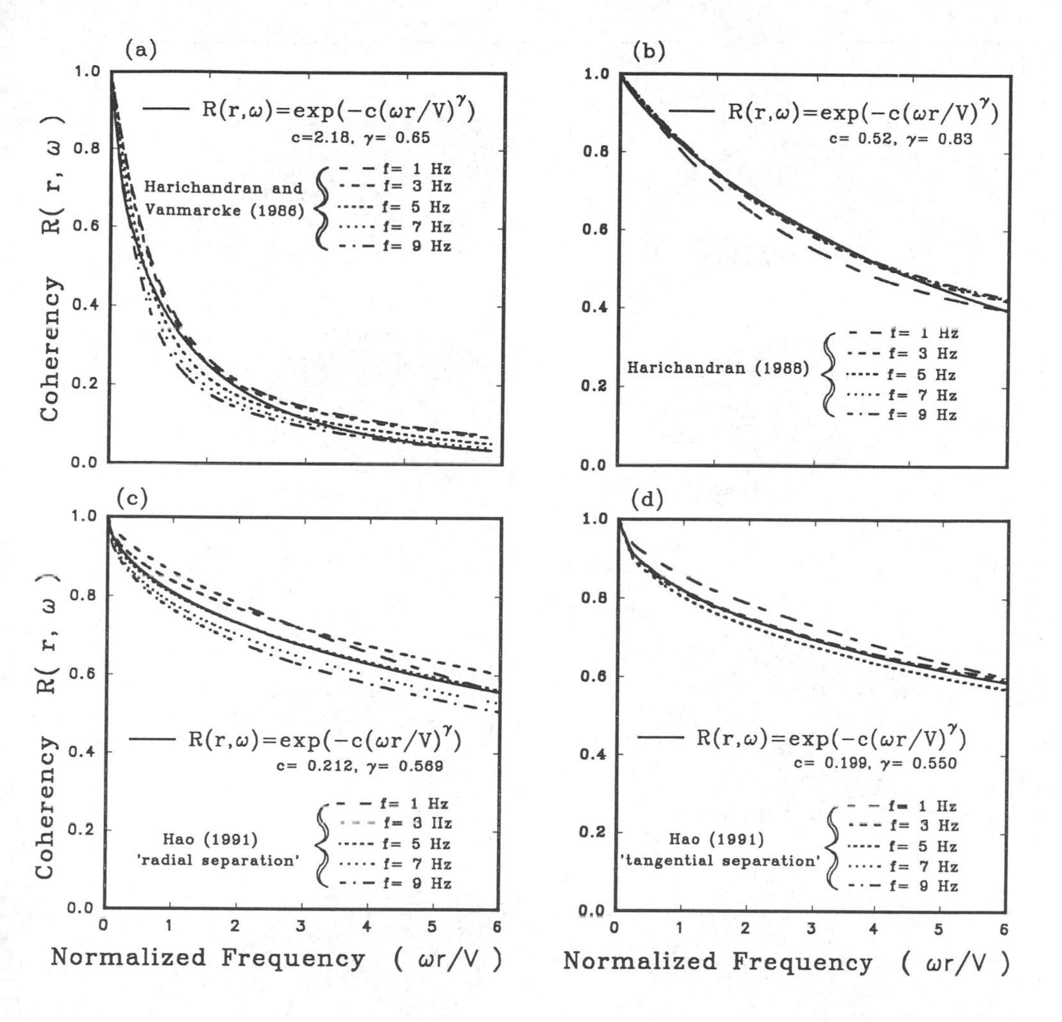

Figure 1 Coherency versus dimensionless frequency for the SMART-1 array data: (a) radial component of event 20; (b) tangential component of event 20; (c) event 40 (radial separation); and (d) event 40 (tangential separation).

independent random variables (phase shifts) uniformly distributed between 0 and 2π, and a_k and L_i are amplitude modifiers and correlation lengths dependent on the coherency function; finally,

$$\omega_{ik} = i(\Delta\omega)-k(\Delta\omega)/N, \quad \omega_{ik}' = i(\Delta\omega)-(k-0.5)\Delta\omega/N$$

where $\Delta\omega = \omega_{max}/M$ and ω_{max} is the cut-off frequency.

It was shown in [7] that the motion $u_s(t)$ by Eq. 4 satisfies the power spectrum for large M independent of N, and can match any coherency function by suitably specifying a_k and L_i. If the joint coherency function by Eq. 2 is assumed with $\gamma=1$, these parameters become

$$Li = \alpha V/[c(i-0.5)\Delta\omega]$$
$$a_o^2 = \frac{1}{\alpha}(1-e^{-\alpha}), \text{ and}$$
$$a_k^2 = 2[1-(-1)^k e^{-\alpha}]/(\alpha[1+(\pi k/\alpha)^2])$$

The coherency match depends on the values of both N and α but for $N \geq 5$ and $\alpha \geq 3$, the coherency function of the simulated motions compares very well with the target one. Effects

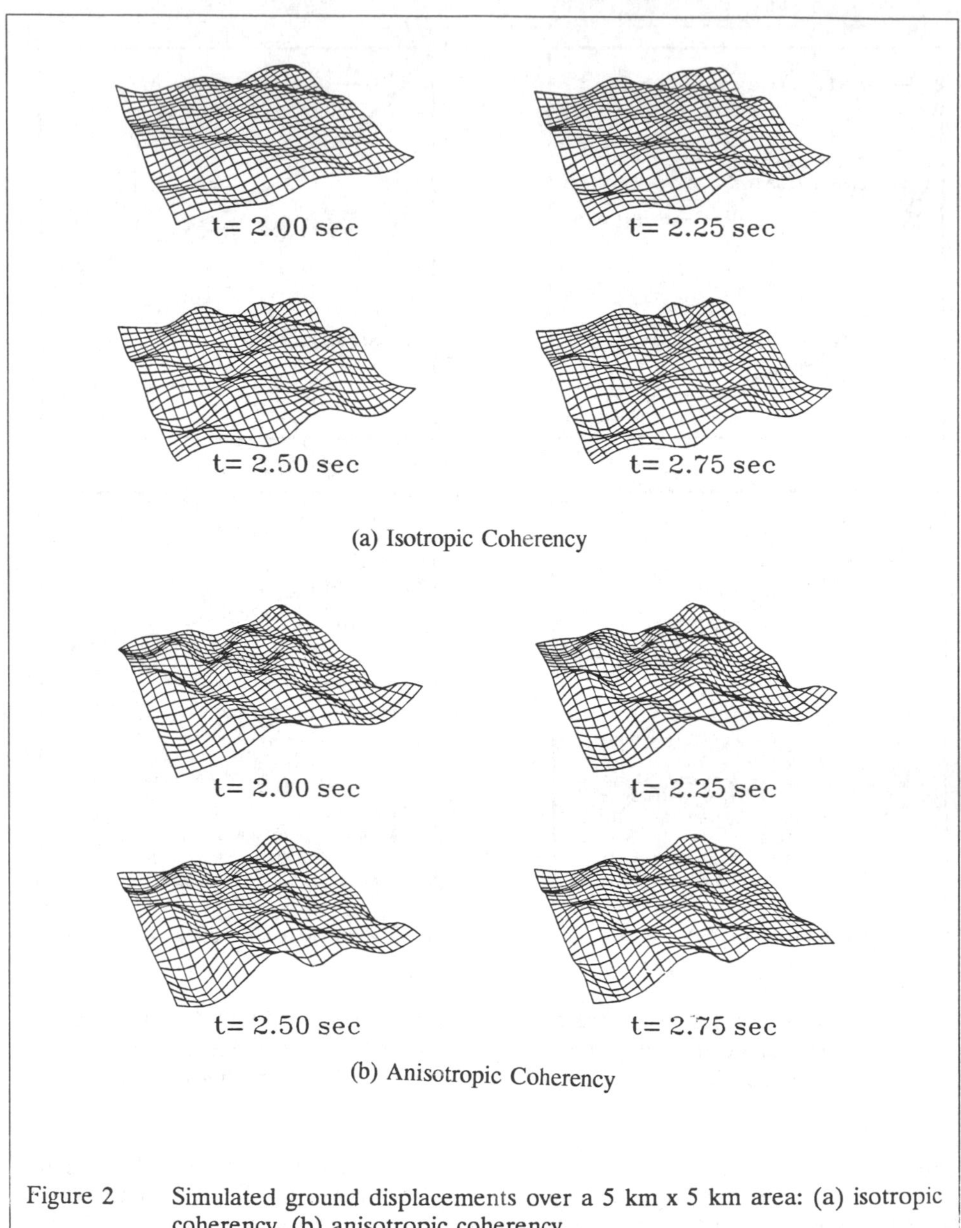

Figure 2 Simulated ground displacements over a 5 km x 5 km area: (a) isotropic coherency, (b) anisotropic coherency.

of surface travelling waves can be incorporated in Eq. 4 by replacing the time t by $(t-\tau_s)$ where τ_s is the time lag at station s with respect to the reference point, s=0.

Equation 4 can be extended to include two spatial coordinates (s_1,s_2) by replacing the term $\{\pi ks/L_i\}$ by $\{\pi ks_1/L_{i1}+(-1)^k \pi ks_2/L_{i2}\}$, where L_{i1} and L_{i2} correspond to the coherencies in directions s_1 and s_2, respectively, to allow for aniso-

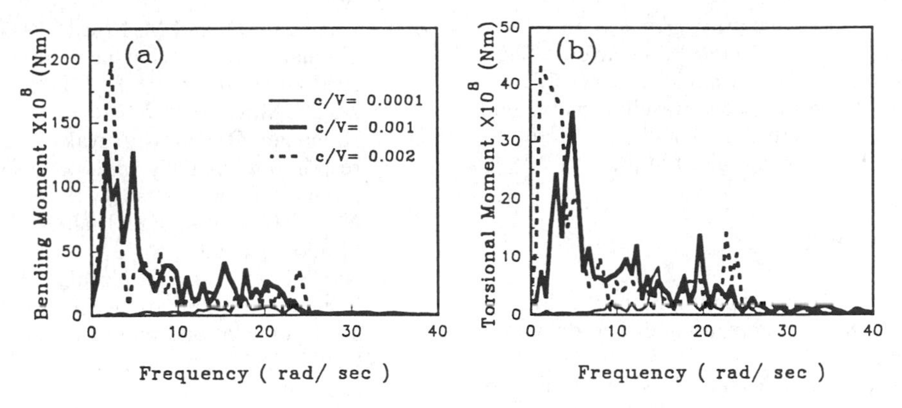

Figure 3 Dam response to motions with different degrees of spatial incoherence.

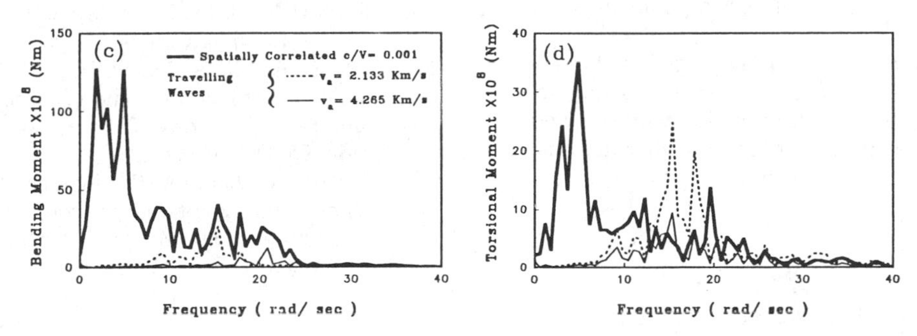

Figure 4 Dam response to spatially incoherent ground motions and to fully coherent travelling waves.

tropic coherencies. Figure 2 shows the simulated ground displacement over a 5 km x 5 km area for both isotropic and anisotropic coherency structures. The power spectrum assumed is of the modified Kanai-Tajimi type [1] with $\omega_g=10$ $\omega_f=5\pi$ and $\zeta_g=\zeta_f=0.60$, while the coherency is given by Eq. 2 with $\gamma=1$, $(c/V)_1=(c/V)_2=0.001$ s/m for the isotropic case, and $\gamma=1$, $(c/V)_1=2(c/V)_2=0.001$ for the anisotropic case. The spatial variability of the ground motion is clearly manifested in Fig. 2. For simulation in three-dimensional domains, the technique can be extended in a similar fashion [9].

DAM RESPONSE TO INCOHERENT GROUND MOTIONS

Seismic response of an 853 m long concrete gravity dam to incoherent ground motions varying along the dam longitudinal axis was investigated in detail. Dam interaction with both its foundation and water in the reservoir was accounted for. The dam midpoint response to motions with the power spectrum specified earlier but different c/V ratios is displayed in Fig. 3. The figure shows that as the degree of spatial incoherence increases (higher c/V values), the dam internal stresses increase. For a maximum acceleration of 0.2 g and a degree of spatial incoherence similar to that of event 20 at the SMART-1 array (c/V=0.001), the critical bending and shear stresses are 8.76 and 1.46 MPa for continuous monolithic dams but reduce to 1.66 and 1.91 MPa for segmented dams, respectively. These stresses are significant and remain totally undetected when the ground motions are assumed to be fully correlated. Finally, Fig. 4 compares the dam mid-point response to spatially incoherent motions (c/V= 0.001) with its response to fully correlated surface travelling waves featuring the same time domain variations and wave propagation velocities of 2.133 and 4.265 km/s, respectively. The effects of the incoherent motions markedly differ from those of surface travelling waves and much exceed the latter in magnitude. More details on this study are given in [8].

CONCLUSIONS

Spatially incoherent ground motions are described in both the frequency and time domains in a simple, consistent way and a novel simulation technique is presented. It is found that the joint coherency function expressed in terms of dimensionless frequency is capable of matching the data from different events by merely altering the values of its two parameters.

Response analysis of large gravity dams indicates that the ground motion spatial incoherence can produce significant stresses and that these stresses are much higher than those produced by surface travelling waves.

REFERENCES

1. Clough, R.W. and J. Penzien. 1975. Dynamics of Structures. New York: McGraw-Hill.
2. Hong Hao. 1991. Response of multiply supported rigid plate to spatially correlated seismic excitations. Earthq. Eng. Struct. Dyn., Vol. 20, pp. 821-835.
3. Harichandran, R.S. and E.H. Vanmarcke. 1986. Stochastic variation of earthquake ground motion in space and time. J. Eng. Mech., ASCE, Feb., pp. 154-174.
4. Harichandran, R.S. 1988. Local spatial variation of earthquake ground motion. Earthq. Engrg. and Soil Dyn. II - Recent Advances in Ground Motion Evaluation, Proceedings, J.L. Von Thun (Ed.), ASCE, pp. 203-217.
5. Loh, C.H. and S.G. Lin. 1990. Directionality and simulation in spatial variation of seismic waves. Eng. Struct., Vol. 12, pp. 134-143.
6. Novak, M. and A. Hindy. 1979. Seismic response of buried pipelines. 3rd Canadian Conf. on Earthq. Engrg., Montreal, 1, pp. 177-203.
7. Ramadan, O. and M. Novak. 1993. Simulation of spatially incoherent random ground motions. J. Eng. Mech., ASCE, May, pp. 997-1016.
8. Ramadan, O. and M. Novak. 1992. Dam response to spatially variable seismic ground motions Parts I & II. Report No. GEOT-14-92, Fac. Eng. Sc., U.W.O., London, Canada.
9. Ramadan, O. and M. Novak. 1993. Simulation of multi-dimensional, anisotropic ground motions. Report No. GEOT-15-92, Fac. Eng. Sc., UWO, London, Canada.
10. Ramadan, O. and M. Novak. 1993. Coherency functions for spatially correlated seismic ground motions. Report No. GEOT-09-93, Fac. Eng. Sc., UWO, London, Canada.
11. Shinozuka, M. and C-M. Jan. 1972. Digital simulation of random processes and its applications. J. Sound and Vibration, 25(1), pp. 111-128.
12. Spanos, P.D. and M.P. Mignolet. 1990. Simulation of stationary random processes: two-stage MA to SARMA approach. J. Eng. Mech., ASCE, Vol. 116, No. 3, pp. 620-641.
13. Uscinski, B.J. 1977. The Elements of Wave Propagation in Random Media. New York: McGraw-Hill.

Structural Safety & Reliability, Schuëller, Shinozuka & Yao (eds) © 1994 Balkema, Rotterdam, ISBN 90 5410 357 4

On the frequency dependence of the strong ground motion duration

R.J. Scherer
Institute of Structural Concrete and Construction Material Technology, University Karlsruhe, Germany

ABSTRACT

The frequency-dependent strong ground motion duration used in this analysis is defined by the exceedance durations of the bandpass-filtered accelerograms. For the filtering the same procedure was applied as for the estimation of evolutionary power spectra (Scherer, 1993). The analysis was carried out for a subset of 6 stations with about 40 horizontal strong motion recordings from the 1976 Friuli aftershock sequence. The frequency dependent strong motion durations obtained show quite stable shape functions systematically depending a) on the exceedance level defined by the ratio of the maximum bandpass-filtered acceleration, b) on the local soil conditions, and c) on the absolute seismic intensity at the site. For smaller exceedance levels below about 25 % of the maximum bandpass-filtered acceleration, the frequency-dependent shape functions of the strong motion duration are of exponential type for soft soil sites, whereas for rock sites they tend to a linear type with a weaker frequency dependence.

1 INTRODUCTION

Strong motion acceleration records show a limited duration with a build up phase - mostly short, a more or less stable high amplitude phase and a long die down phase. Such a simple shape function is mostly true for small to moderate earthquakes (M<6.5) where the rupture happens approximately as a homogenous stochastic process. For more complex earthquakes the build up phase and more often the tail of the die down phase are ambiguous in the sense that a certain duration cannot be estimated easily.

The strong motion duration should represent that part of a strong motion record which strongly excites a structure and which may cause damage or structural failure. Since the most important part of the transfer function of the structural system is at the structural resonance frequency, the frequency should be one of the most important parameters when describing a strong motion duration.

The methods available to estimate the strong motion duration can be categorized in four classes:

a) *exceedance duration,* the total time above a certain threshold
b) *interval duration (or bracket duration)*, the time between the first up crossing of a threshold and the last down crossing of the same threshold,
c) *shape duration*, the time between the build up and the die down of the acceleration amplitudes and therefore defined by the shape of the envelope independent of a threshold,
d) *wave duration*, the time between the first arrival of a certain wave type and the end of a certain, not necessarily the same, wave type.

In the past, several studies were performed to obtain an empirical formula for the strong motion duration by regression analysis. Most of them assumed a frequency-independent duration. The more important ones were carried out by Gutenberg & Richter (1956), Esteva & Rosenblueth (1964), Housner (1965), Husid et al. (1965), Bolt (1974), Trifunac & Brady (1975), Trifunac & Westermo (1978), Dobry et al. (1978), McCann & Shah (1979), Vanmarke & Lai (1980) and Caillot & Bard (1992). Only four of them take the frequency dependence into account.

The regression results of these studies demonstrated that the strong motion duration has an unambiguous frequency dependence. However, none of them gave a physically based explanation of this behaviour. Consequently, no special attention was be paid in the regression analysis on the frequency dependence. It was been obtained by chance. The regression parameters adopted were the standard ones used in the regression analysis of seismic quantities such as magnitude and source distance. The sample space was only roughly classified according to the soil condition (rock, hard soil, soft soil) and therefore too general. Accordingly, no deep insight into the physical background of the frequency dependence was possible. The obtained frequency dependence was weak, representing only the overall trend.

2 METHOD

The necessary frequency separation before computing the strong motion duration can be done by bandpass filtering, i.e. by applying a series of Butherworth filters as carried out by Caillot & Bard (1992). However, the shortcoming is the time leakage due to the filter inertness.

Combining the deterministic filter techniques with stochastic considerations a filter procedure was developed by Scherer et al. (1982) which strongly reduced the time leakage without reducing or destroying the frequency resolution. It is based on the Evolutionary Power Spectrum which was established by Priestley (1981) in order to describe nonstationary stochastic processes. The estimation procedure for the Evolutionary Power Spectrum has its roots in the bandpass multiple filter technique (e.g. Arnold 1970). As filter elements viscously damped harmonic oscillators are chosen and the filter inertness is approximately considered and removed from the filtered signal. The quality of the filter procedure was demonstrated by Scherer et al. (1982) and several successful applications for the analysis of strong motion records according to source and site effects were subsequently carried out (e.g. Scherer & Schuëller 1988, Scherer 1993).

The key quantity in this analysis, the two-sided Evolutionary Power Spectrum is defined as

$$S(f,t)\,df = E\{[A(t,f)\,dZ(f)]^2\} \tag{1}$$

where dZ is the differential of the orthogonal random process Z(f), E{} is the expectation operator and A(t,f) the amplitude modulation function. The mean envelope of the bandpass filtered random process is expressed, at f with bandwidth Δf, as

$$E\{env[a(t,f,\Delta f)]\} = \sqrt{2}\cdot\left[\int_{f-\Delta f/2}^{f+\Delta f/2} 2\cdot S(f,t)\,df\right]^{1/2} \tag{2}$$

Further it is assumed that Z(f) is a unit stationary shot noise process, i.e. dZ(f) = 1,0 and S(t,f) = $A^2(t,f)$, which is done for convenience.

For the analysis of strong motion records a filter bandwidth of Δf = 0,25 Hz is used, which results in the simple relationship that

$$E\{env[a(t,f,\Delta f)]\} = A(t,f) \tag{3}$$

In the following the exceedance duration is applied. From the amplitude modulation function A(t,f) the frequency-dependent strong motion duration can be derived, which is defined as

$$SMD(a_e,f,\Delta f) = \int_0^T H(a_e)\cdot dt \tag{4}$$

where $H(a_e)$ is the Heavyside Function. It is defined as

$$H(a_e) = \begin{cases} 0 & \text{for} \quad A(t,f) < a_e \\ 1 & \text{for} \quad A(t,f) \geq a_e \end{cases} \tag{5}$$

a_e is the threshold and T is the total duration of the strong motion record.

3 SELECTED STATIONS AND EVENTS

The developed method is applied in a first step to some well selected near source stations which recorded the Friuli Earthquake Sequence 1976 in order to have the influencing parameters under control. The stations and events are selected according to the following criteria:

1. each station had to be a statistically significant representative of one local soil and topology condition
2. the seismic events had not to be too complex but nevertheless moderate
3. several moderate aftershocks had to be recorded at the same station
4. the total seismic process (P-, S- and surface wave) had to be recorded.

The available stations and events are summarized in Console & Rovelli (1981), Rovelli et al. (1991) and Scherer et al. (1985).

Based on these requirements 6 stations (tab.1) and 4 aftershocks (tab.2) were selected.

Tab. 1 Selected stations

Station	Soil condition
Buia	soft soil in the lower Tagliamento valley
Forgaria	medium soil at the slop of the Tagliamento
San Rocco	valley
Tarcento	weathered outcropping rock located nearby Forg.
Breginj	thin alluvial deposit (5-15 m) over rock (Flysch)
Somplago	thin alluvial deposit (≈ 10 m) over rock (limest.)
	compact hard rock about 300 m below surface

4 STRONG MOTION DURATION SHAPE FUNCTION

For the six stations and the four events the Evolutionary Power Spectra are computed using a half-power bandwidth of $\Delta f = 0{,}25$ Hz and a time window for the statistical average of $\Delta t \approx 4\pi f$. They are visualized as contour plots, where the contour levels are chosen in a logarithmic scale ($\Delta a_e = 5$ db) and with equal values for all stations except for Somplago, where the contour levels are chosen 1/10 of the contours of the other stations. As an example the EW components at Forgaria from the second and third event are shown in fig. 1. The Evolutionary Spectra clearly show the strong nonstationary stochastic character of the seismic acceleration process, which is discussed elsewhere (e.g. Scherer 1993). In the Evolutionary Spectra the mapping of two different physical phenomena can be observed. One is the soil resonance phenomenon, which is more evidently reflected by the higher contours. It is of upmost importance for the seismic risk of structures because it represents the high excitation amplitudes. The other phenomenon is the low pass filter effect due to damping and to multiple reflections and scattering of the waves. It is mapped by the L-shape of the lower contours.

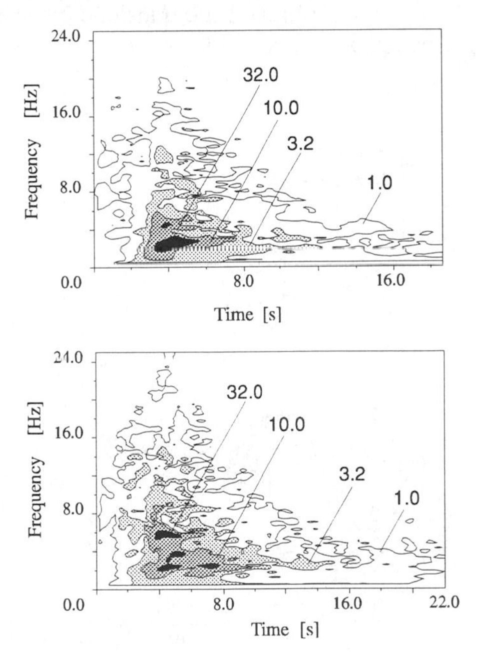

Fig. 1 Amplitude modulation function of the Evolutionary Power Spectra of the EW-component recorded at Forgaria on 09/11/76, 16:35 (top) and on 09/15/76, 3:15 (bottom). The contour units are $cm/s^{(3/2)}$

From the contour plots of the Evolutionary Spectra the shape of the exceedance duration can be already imagined. In fig. 2 the two EW-components obtained during the second and third event at Forgaria are shown. For an easier comparison the same values are chosen as exceedance level as for the Evolutionary Spectra. Both phenomena already discussed for the Evolutionary Spectra are well reflected by the frequency dependent strong motion durations. For high exceedance levels the soil resonance phenomenon stamps the strong motion duration, which is strongly frequency dependent. For low exceedance levels the multiple reflection, refraction and scattering effect forms the frequency dependent strong motion duration. The shape of the function is in-between a straight line and an exponential function. The fluctuation superposed on this trend is comparably small and independent from the frequency. The corresponding variance is approximately constant. The two strong motion duration contours in-between the lowest and highest contour represent the transition between these two phenomena.

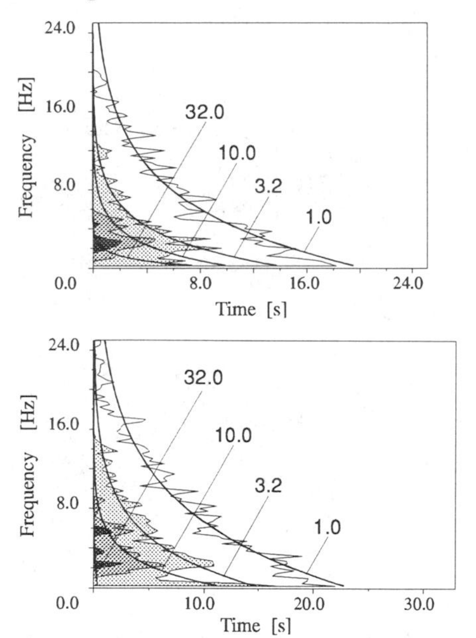

Fig. 2 Exceedance durations for the same records shown in fig. 1 and the according regression curves. The threshold units are cm/s2

The frequency dependent strong motion durations for the six selected stations show significant stable shapes for each station, regardless of the different events and the different source distances. Therefore, the shape of the duration functions is a characteristic of the site or a site effect. The shapes of the frequency dependent strong motion durations are always stamped by the above mentioned physical phenomena - the soil resonance phenomenon mainly caused by direct P- and S-waves, and the low pass filter and statistical averaging phenomenon from the multiple reflection, refraction and scattering of the non-direct seismic waves and surface waves. The shape of the duration functions for high exceedance threshold is a very locally focused site effect, whereas the lower exceedance duration shape function is determined by local topography and soil layers in the broader environment of the station.

The shape functions of the strong motion duration for the six different stations are shown qualitatively in fig. 3. In the scattering domain the shape can be described by an exponential function

$$d(f) = c_1 e^{c_2 f} \tag{6}$$

The most pronounced exponential shape is obtained for Buia (soft soil), a medium exponential shape for Forgaria (medium soil), a flat exponential shape for San Rocco (weathered outcropping rock) an a very flat exponential shape nearly approximating a straight line for Somplago Turbin (hard crystalline rock). The two stations Tarcento and Breginj, with a thin deposit over hard rock, most probably non-weathered, show also exponential shapes in the very high frequency ranges but these exponential shapes are cut off in the medium frequency ranges by straight lines.

$$d(f) = c_3 + c_4 f \tag{7}$$

These lines are often constant or show a positive or negative inclination. The shape functions at Tarcento and Breginj seems to be a combination of the shape functions obtained at Forgaria (medium soil) and those obtained at Somplago (hard rock).

As a consequence of the independence of the shape functions on PGA or magnitude and source distance respectively, strong motion durations for linearly scaled exceedance thresholds are calculated for Forgaria. In fig. 4 the duration functions are shown for 75%, 50%, 25% and 5% thresholds related to the maximum of the modulation functions. These contour plots clearly reveal that the local resonance phenomenon determines the shape of the 75% and 50% exceedance functions, whereas the 5% exceedance function is determined by the scattering phenomenon. The 25% exceedance function is in the transient region, sometimes dominated by the resonance phenomenon, sometimes by the scattering phenomenon.

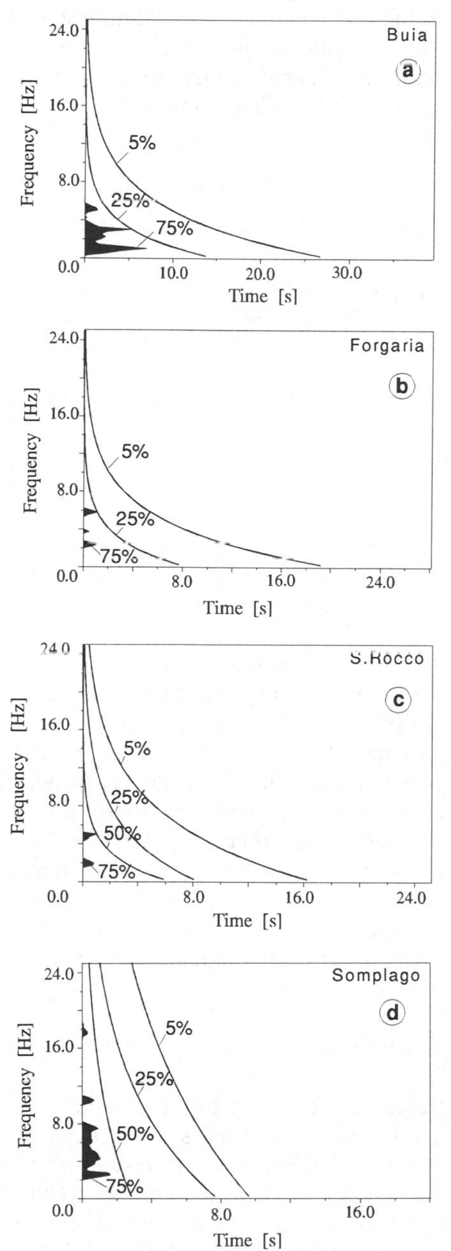

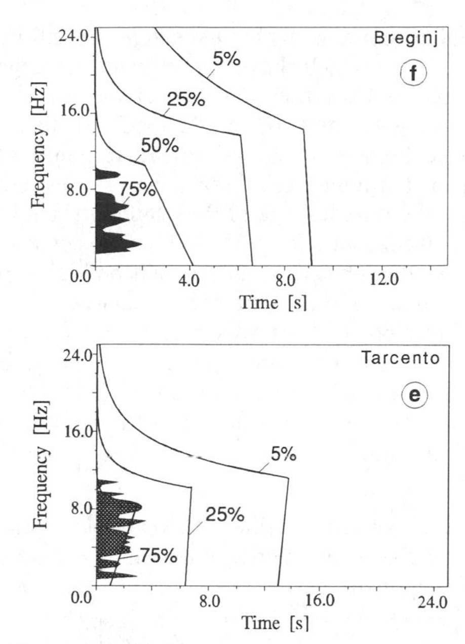

Fig. 3 Qualitative exceedance durations for different local soil and topography conditions (soft to rock: a - d, thin soft layer over rock: e, f)

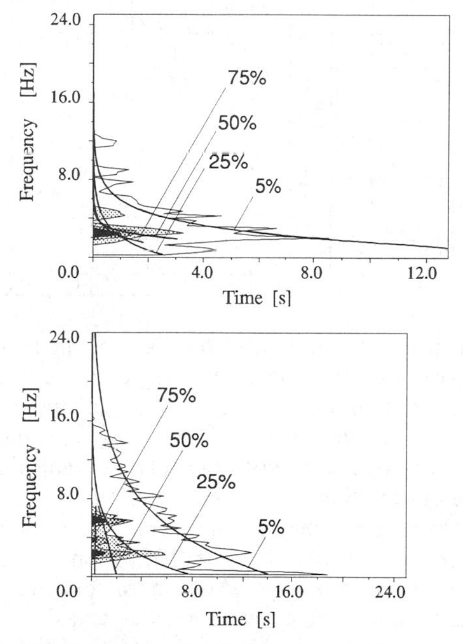

Fig. 4 Exceedance durations and the according exponential and linear regression curves for the records as shown in fig. 1. The thresholds represent 75, 50, 25 and 5 % of the maximum amplitude of the modulation function.

Similar contour plots were represented by Perez (1973) which shows the velocity envelope spectrum and the time duration of the velocity response envelope of a damped oscillator. Because he was interested in the response of damped harmonic oscillators, he a) did not correct the time leakage of the oscillators and b) chose a constant damping and not a constant filter bandwidth. His results would be biased if they were used in the scope of this analysis.

Regression analysis was carried out for the duration functions in order to test the hypothesis that the shape of the duration functions are local site effects, i.e. statistical characteristic values for each station.

Tab. 2 Parameters of the LN-Regression Analysis for the strong motion duration at Forgaria, EW-component

Event Date Time m_b PGA A_{max}	1 11.09.1976 10:31 5.5 113 17.8			2 11.09.1976 16:35 5.5 230 88.7			3 15.09.1976 03:15 6.0 214 51.4			4 15.09.1976 09:21 5.9 329 105.0		
a_e %	c_1 s	c_2 cm/s²	ρ %	c_1 s	c_2 cm/s²	ρ %	c_1 s	c_2 cm/s²	ρ %	c_1 s	c_2 cm/s²	ρ %
1	-	-	-	-0.15	3.01	-93	-	-	-	-0.13	3.31	-95
2	-	-	-	-	-	-	-0.13	3.16	-94	-	-	-
3	-0.13	2.88	-96	-0.27	2.71	-80	-0.17	3.13	-92	-0.17	2.63	-91
4	-	-	-	-0.31	2.70	-77	-	-	-	-	-	-
5	-0.14	2.76	-95	-0.45	2.71	-74	-0.17	2.68	-93	-0.19	2.31	-86
6	-0.16	2.79	-94	-	-	-	-0.22	2.74	-90	-	-	-
10	-0.22	2.57	-86	-0.56	2.80	-79	-0.19	2.18	-81	-0.37	2.45	-70
11	-	-	-	-0.42	2.40	-79	-	-	-	-	-	-
18	-0.18	2.07	-80	-	-	-	-	-	-	-	-	-
19	-	-	-	-	-	-	-0.40	2.51	-68	-	-	-
25	-0.22	1.70	-47	-0.50	1.06	-19	-0.34	2.13	-74	-0.37	1.32	-52
30	-	-	-	-	-	-	-	-	-	-0.59	0.53	-37
36	-	-	-	-1.33	2.33	-75	-	-	-	-	-	-
50	-0.72	1.75	-53	-1.45	3.87	-72	-0.25	0.40	-17	1.28	-0.24	45
56	-0.82	1.50	-57	-	-	-	-	-	-	-	-	-
62	-	-	-	-	-	-	-0.27	-0.87	-12	-	-	-
75	-0.04	0.35	-45	-0.92	1.99	-64	-0.46	-2.76	-28	-1.16	3.14	-34

In the following the results obtained for Forgaria are represented. The regression curves are plotted in fig. 2 and 4 onto the duration curves. For the regression analysis frequency windows were applied to exclude the vanishing values in the high frequency range and those in the low frequency range for the higher exceedance thresholds. The regression coefficients do not always show a simple linear dependence. However, they reveal an unambiguous trend. The exponential coefficient increases with increasing threshold, whereas the factorial coefficient decreases. This trend always happens for the different events and, more over, the order of the coefficient often seems to be about constant for the same threshold (in % of the maximum amplitude of the modulation function) but for different events. The regression parameters are summarized in tab. 2 for the constant thresholds (32, 10, 3.2, 1.0 cm/s²) as well as for the thresholds in % of the maximum modulation function (75, 50, 25, 10, 5, 3 %). The regression coefficient usually takes values between 0.7 and 0.9 for the scattering domain. In the transition domain the regression coefficient drops down, indicating that the exponential shape function is no longer valid.

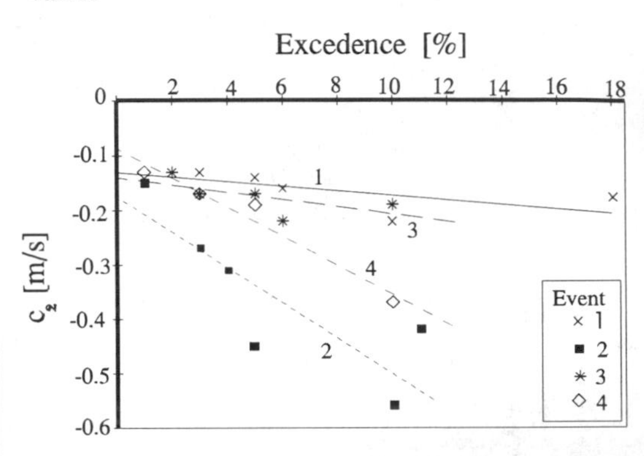

Fig. 5 Regression parameters for the scattering duration obtained at Forgaria (EW-component) drawn over the threshold level

In fig. 5 the regression parameter c_2 is plotted over the threshold. Only those values are shown, for which the correlation coefficient is above 0.70, because it is found that such a correlation coefficient is a confidential value for testing the hypothesis of an exponential shape for the duration function, i.e. that the duration is the scattering duration. For the resonance duration no simple regression curve can be suggested. A straight line, band limited in the frequency domain, may be used for a pragmatic description.

5 DISCUSSION

The frequency dependence of the strong motion duration has not been discussed consequently in the literature before, but the results obtained may serve for a comparison. A frequency separation of the strong motion duration was also carried out by Bold (1974), Trifunac &

Brady (1975), Trifunac & Westermo (1970) and Caillot & Bard (1992).

Bolt analysed several stations of the San Fernando earthquake 1971, which is a complex, multi-rupture earthquake with M_S = 6.5. For a threshold of a_e = 0.05 g he plotted the interval duration of these stations over the frequency, but he did not come up with a functional dependence. He only gave a hypothetical upper limit as a plotted curve, which represents the shape function in the transient zone, as it may be expected for a threshold of 50 cm/s2 of a M_S = 6.5 earthquake for near source strong motion records.

Trifunac & Brady (1975) analysed the interval duration (5 - 95 %) of the signal energy (Arias Intensity) of the San Fernando 1971 strong motion records. The frequency dependence was approximately included in their analysis, because they evaluated the strong motion duration for the accelerations, velocities and displacements. The frequency independent duration from the acceleration is the duration for the moderate frequency range (about 1 - 5 Hz), the duration from the velocity is that for the low frequency range (about 0.5 - 1.0 Hz) and the duration from the displacement is that for the very low frequency range (about less then 0.5 Hz). They choose very low threshold levels and therefore their duration values should reflect the scattering duration. The trend of their duration values increase from acceleration through velocity to displacement according to the proposed trend of the scattering duration.

Trifunac & Westermo (1978) extended their former work (see above) by using six band-pass filters (0.2, 0.5, 1.1, 2.7, 7.0 and 18 Hz) for each analysed quantity (acceleration, velocity, displacement). In comparison to the former work they selected the exceedance duration of the signal energy using a threshold of 10% of the bandpass filtered signal energy, i.e. this duration should be a scattering duration. Establishing frequency dependent duration from their given regression parameters exponential shape functions result, which supports the findings of this paper.

Caillot & Bard (1992) used the same definition of duration as Trifunac & Brady (1978), i.e. the 5 - 95 % of the signal energy, but they applied this interval duration definition to band-pass filtered strong motion records using a 40 dB/oct Butterworth filter with filter center frequencies of $0.6 \cdot 2^{n/2}$, $n = 0-9$. They analysed 260 Italian strong motion records from the ENEA-ENEL databank. Because of the low exceedance level chosen, they analysed the scattering duration. They draw the evaluated durations over log frequency. The resulting curves are almost linear, i.e. the duration is an exponential function of frequency, as proposed in this paper.

They classified the stations only in soft soil, medium soil and rock stations. This is not detailed enough, as our study shows, and therefore they obtained only the overall trend for the scattering duration, which is a medium exponential function.

High exceedance levels, which are of upmost interest for small to moderate earthquakes were not investigated in the past. However, by analogy, it is evident from amplification studies that high exceedance durations must map the resonance bands.

6 CONCLUSIONS

This study shows that the frequency dependent strong motion duration is formed by two physical phenomena, the local resonance phenomenon and the local damping and multiple reflection and refraction (scattering) phenomenon. The first forms the duration function for high exceedance values, the second for low exceedance values. For moderate earthquakes, i.e. with non-complex rupture processes shape function for the two phenomena are deduced. It is shown that the regression parameters of the shape function are extremely stable for each station, but the shape function vary with stations, i.e. with the soil and topological condition of the local site. Therefore the shape functions of the frequency dependent exceedance duration is a site characteristic.

For a complex rupture process, e.g. strong earthquakes with M_S > 6.5, the strong motion duration should be a superposition of the durations of several non-complex rupture processes, where the rupture evolution and the radiation pattern determines the superposition. Therefore a simple generalizable shape function cannot be forecasted.

In this study the shape of the duration function and its physical reason was of main interest and not the absolute values of the strong motion duration. On the basis of the findings of this study it may be easy to classify the available strong motion records properly and perform reliable regression analysis for determining the frequency dependent strong motion duration.

7 ACKNOWLEDGEMENT

The author may thank Mrs. Cengiz, Mr. Katranuschkov and Mr. Michels for carrying out the extensive computations. The partial support of this study by the local state of Baden-Württemberg is grateful acknowledged.

8 REFERENCES

Arias A. 1970 A Measure of Earthquake Intensity. In Seismic Design for Nuclear Power Plants. Robert J. Hansen (ed), MIT Press

Arnold C.R. 1970. Spectral estimation for Transient Waveforms. IEEE Trans. on Audio and Electroacoustics, Vol. AU-18, No. 3, p. 248-257.

Bolt B.A. 1974. Duration of Strong Ground Motion. Proceedings 5. WCEE, Vol.1, Rome.

Caillot V. & Bard P.Y. 1992. Band Limited Duration and Spectral Energy: Empirical Dependence on Frequency, Magnitude, Hypocentral Distance and Site Conditions. Proceedings 10. WCEE, Vol. 2, Madrid.

Console R. & Rovelli A. 1981. Attenuation Parameters for Friuli Region from Strong Motion Acceleration Spectra. Bull. Seism. Soc. Am., Vol. 71, No. 6, p. 1981-1991.

Dobry R., Idriss I.M. & Ng E. 1978. Duration Characteristics of Horizontal Components of Strong-Motion Earthquake Records. Bull. Seism. Soc. Am., Vol.68, No.5, p. 1487-1520.

Housner G.W. 1975. Measures of Severity of Earthquake Ground Shaking. Proc. U.S. Nat. Conf. Eartquake Eng., Ann Arbor, Michigan.

McCann M.W., Jr. & Shah H.C. 1979. Determining Strong Motion Duration of Earthquakes. Bull. Seism. Soc. Am., Vol. 69, No. 4, p. 1253-1265.

Perez V. 1974. Peak Ground Accelerations and their Effect on the Velocity Response Envelope Spectrum as a Function of Time, San Fernando Earthquake, February 9, 1971. Proceedings 5. WCEE, Vol.1, Rome.

Priestley M.B. 1981. Spectral Analysis and Time Series. Vol.2, Academic Press, London.

Rovelli A., Cocco M., Console R., Alessandrini B. & Mazza S. 1991. Ground motion waveforms and source spectral scaling from close-distance accelerograms in a compressional regime area, (Friuli, North-Eastern Italy). Bull. Seism. Soc. Am., Vol.81, p. 57-80.

Scherer R.J., Riera J.D. & Schuëller G.I. 1982. Estimation of the Time-Dependent Frequency Content of Earthquake Accelerations. J. Nuclear Eng. and Design, Vol. 71, p. 301-310.

Scherer R.J. & Schuëller G.I. 1985. Records and Power Spectra of Corrected and Integrated Strong Motion Earthquake Data. Vol.I, Friuli Earthquake Sequence of 1976, Univ. Innsbruck/Techn. Univ. München.

Scherer R.J. & Schuëller G.I. 1988. Quantification of Local Site and Regional Effects on the Nonstationarity of Earthquake Accelerations. 9th World Conference on Earthquake Engineering, Tokyo-Kyoto.

Scherer R.J. 1993. Advancement in Strong Motion Analysis by the Use of Evolutionary Power Spectra. New Horizons in Strong Motion, Special Issue of Tectonophysics, Vol. 218, No. 1-3, p. 83-91.

Trifunac M.D. & Brady A.G. 1975. A Study of the Duration of Strong Earthquake Ground Motion. Bull. Seism. Soc. Am., Vol. 65, No. 3, p. 581-626.

Trifunac M.D. & Westermo B.D. 1978. Dependence of the Duration of Strong Earthquake Ground Motion on Magnitude, Epicentral Distance, Geological Conditions at the Recording Station and Frequency of Motion. Publication No. 59, Inst. Earthquake Engineering and Engineering Seismology, University "Kiril and Metodij", Skopje, Yugoslavia.

Vanmarcke E.H. & Lai S.-S.P. 1980. Strong-Motion Duration and RMS Amplitude of Earthquake Records. Bull. Seism. Soc. Am., Vol. 70, No. 4, p. 1293-1307.

Structural Safety & Reliability, Schuëller, Shinozuka & Yao (eds) © 1994 Balkema, Rotterdam, ISBN 90 5410 357 4

Ground motion parameter for damage estimation of high and low-rise buildings

Hitoshi Seya, Yasuhiko Abe, Mitsuo Ishiguro & Hideo Nanba
Takenaka Corporation, Tokyo, Japan

ABSTRACT: The shape of seismic waves is a major factor of uncertainty in relation to the safety evaluation of structures. In order to make clear how ground motion parameters (such as peak acceleration and peak velocity, which define seismic intensity for fragility)depend on the structure type, a 24-story steel building and a 6-story reinforced concrete building are analysed. Ten observations made from earthquake records and ten simulated ground motion experiments concerning the possibility of earthquakes in the Tokyo area are utilized as the variables of seismic waves. Correlations between a nonlinear response and the seismic intensity are examined. These indicate the peak velocity is a better descriptor of damage than that of peak acceleration.

1 INTRODUCTION

Among many uncertain factors involved in fragility analyses, the uncertainty of the seismic wave is known to be one of the most significant factors. In Probabilistic Seismic Safety Assessments (PSA) of nuclear power plants, fragilities were usually performed in terms of the peak ground acceleration. However, in order to extend the PSA method to tall buildings, the peak ground velocity would be a better parameter for measuring ground motion intensity, mainly because tall buildings are associated with a longer period.

In this paper, the sensitivity of two ground motion parameters, the peak ground acceleration and the peak ground velocity, against the fragility of two different types of structures were examined. In order to ascertain that the appropriate parameter of seismic intensity depends on structure type, a 24-story steel building and a 6-story reinforced concrete building were analyzed. By performing the nonlinear analysis, which is conventionally used in designing the response characteristics of each structure, two parameters were examined.

2 EARTHQUAKE RECORDS

The acceleration time histories used in this study consists of ten observed records, which have been used as a standard for the designs, and another ten synthetic ground motion records were simulated on the basis of two past earthquakes in the Tokyo area. The response velocity spectrum of observed records in Table 1 are shown in Fig.1. As these records are prepared for a general building design without adequate seismic information of the site, the response spectrum covers a broad period, from 0.3 - 5.0 seconds. These observed records have been conventionally used in designing tall buildings in Japan. The synthetic ground motion records are based on two past hazardous earthquakes, namely the Kanto earthquake in 1923 and the Ansei earthquake in 1855. However, since we could not get the observed records for both earthquakes, we synthesized time histories based on the magnitude of the earthquakes and the epicentral distances from the sites. The target and the five simulated response spectrums for the Kanto and Ansei earthquakes are shown in Figs. 2 and 3. When these five time histories were generated, phase angles randomly varied. The amplification of ground motion through soil layers, as shown in Table 2, is significant at this site. Therefore applying a one-dimensional shear wave propagation analysis, the time histories obtained at the earthquake bedrock, where the shear wave

Table 1 Observed Earthquake List

No.	Name	Year	Velocity cm/sec	Acceleration cm/sec^2
1	EL CENTRO NS	1940	33.45	341.70
2	EL CENTRO EW	1940	36.92	210.14
3	TAFT NS	1952	15.72	152.70
4	TAFT EW	1952	17.71	175.95
5	TOKYO-101 NS	1956	7.63	74.00
6	SENDAI - 501 NS	1963	3.46	57.50
7	OSAKA-205 EW	1968	5.08	25.00
8	HACHINOHE NS	1968	34.08	225.00
9	HACHINOHE EW	1962	35.81	182.90
10	TH030-1FL EW	1978	27.57	202.57

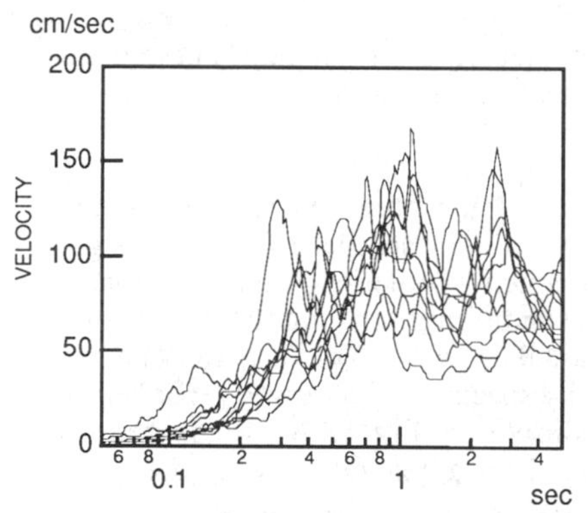

Fig. 1 Response Velocity Spectrum of Observed Earthquake (h=0.05, Vmax=50cm/s)

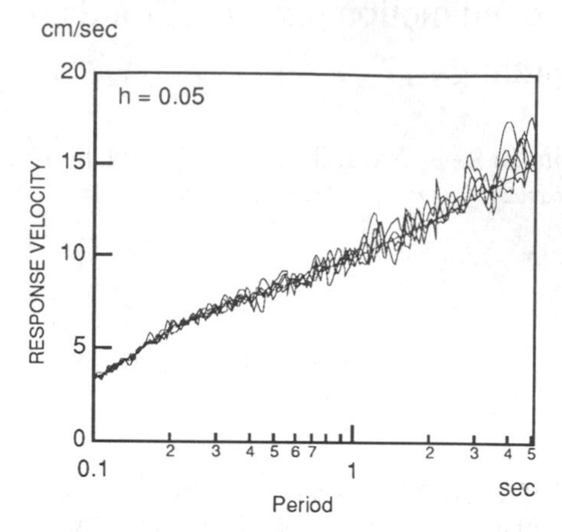

Fig.2 Target and simulated Response Spectrum for Kanto Earthquake (1923)

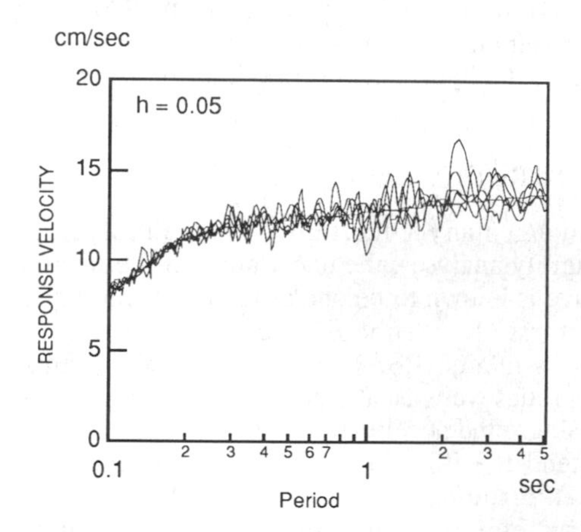

Fig.3 Target and simulated Response Spectrum for Ansei Earthquake (1855)

velocity is more than 3000 m/sec, are developed into the foundation level of the structures.
Figures 4 and 5 show each synthetic time history sample at the earthquake bedrock for the Kanto earthquake and Ansei earthquake. Figure 6 shows the response velocity spectrum for the synthetic records thus obtained. And it indicates that the soil layers reflect on the peaks at a period of three seconds. Figure 7 shows the relationship between the peak velocity, Vmax, versus the peak acceleration, Amax, for the ten observed records in Table 1 and the ten simulated records at the structure foundation level. The ratio of Amax to Vmax tends to have the value of 8.

Table 2 Soil Layer Scheme

GL - (m)	Vs (m/sec)
0 ~ 22.0	150
22.0 ~ 25.9	440
25.9 ~ 33.2	400
33.2 ~ 250	580
250 ~ 1500	700
1500 ~ 2300	1500
2300 ~	3000

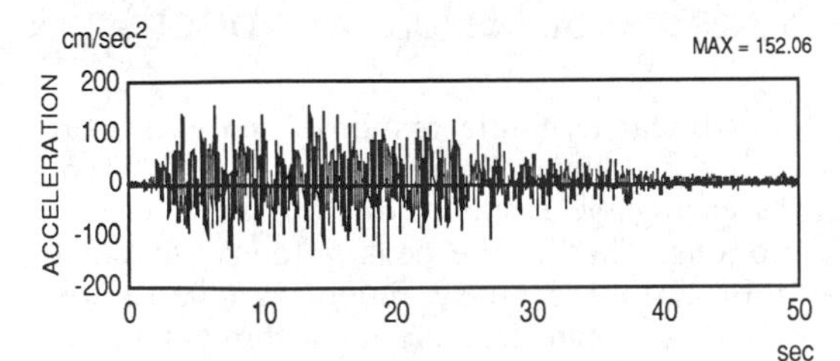

Fig.4 Example of Simulated Wave for KANTO (1923)

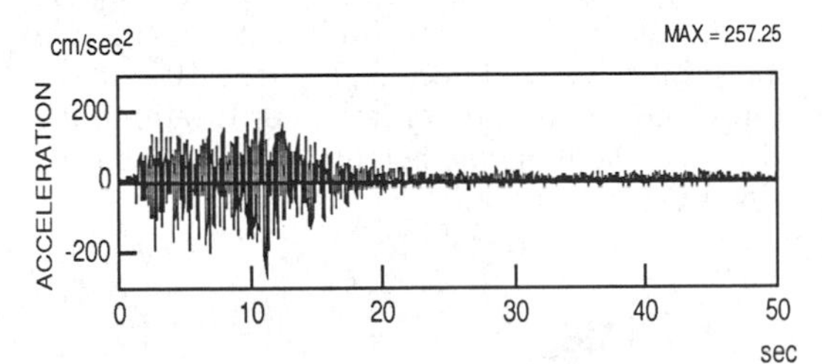

Fig.5 Example of Simulated Wave for Ansei Earthquake (1855)

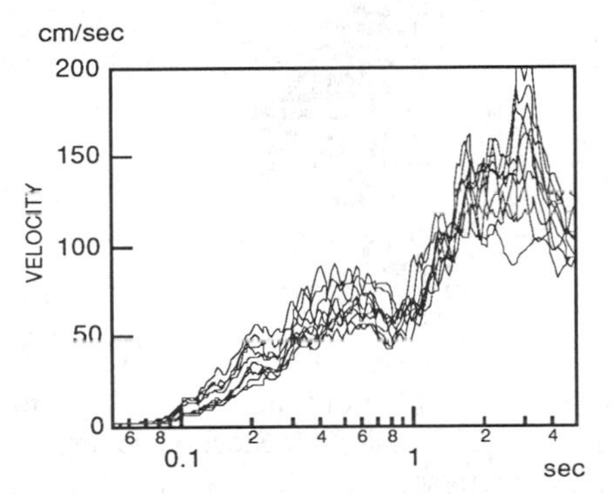

Fig.6 Response Velocity Spectun of Simurated Earthquake (h=0.05,Vmax=50cm/s)

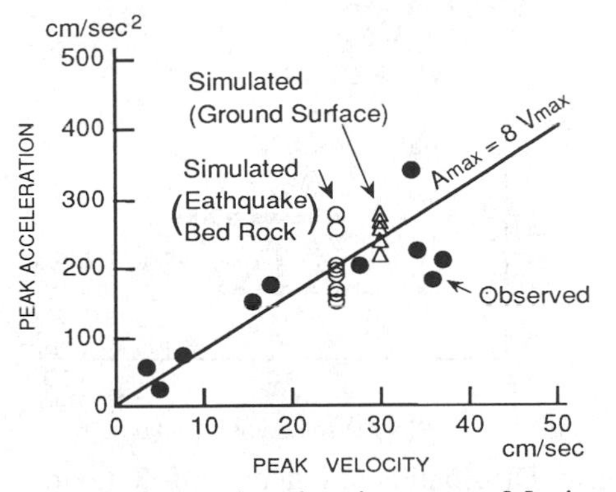

Fig.7 Maximum Acceleration versus Maximum Verocity of Earthquake Records

3 HIGH AND LOW-RISE BUILDING MODEL

The structures used in this study are a 24-story office building and a 6-story office building located in Tokyo. The tall building is a steel rigid frame structure with the natural period of 2.4 seconds. The low building is a reinforced concrete frame structure with the natural period of 0.6 seconds. The foundation of the 24-story building rests on hard soil, where Vs equals 440 m/sec in Table 2. The foundation of the 6-story building rest on the ground surface,where Vs equals 150 m/sec, and are supported by piles. These buildings were assumed to be properly designed according to the Japanese building design codes. That is, for a seismic force the tall building is designed by applying nonlinear dynamic response analysis with the typical observed records in Table 1 recommended for the design. On the other hand, the seismic design for the low building is based on a standard, where the seismic loads are statistically defined. The plans for these buildings are shown in Figs.8-9 with the direction of the analyzed seismic fore.

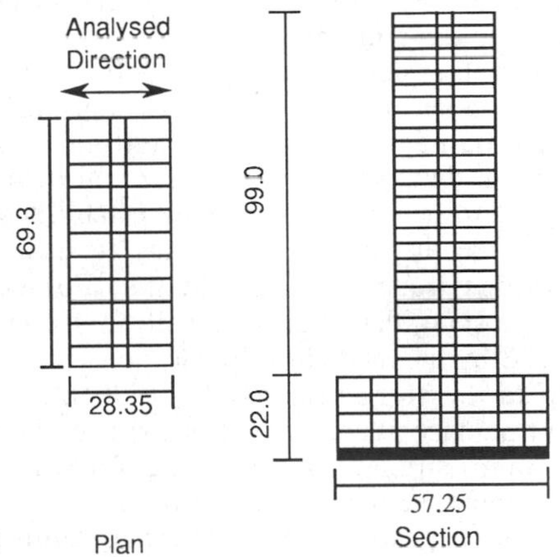

Fig.8 Shape of 24 Story Steel Building

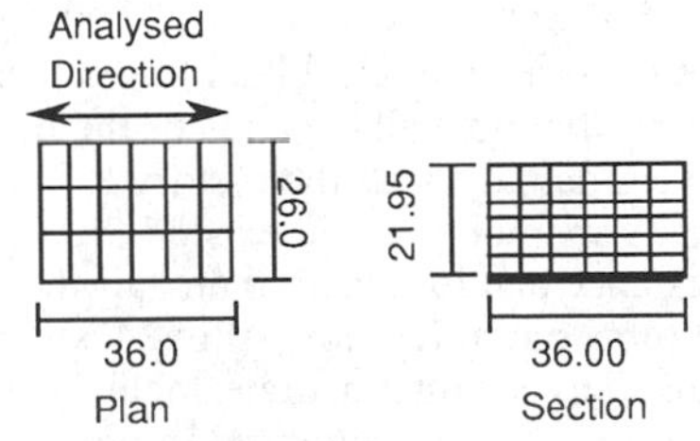

Fig.9 Shape of 6 Story RC Building

Table 3 Ratio of Lateral Drift to Story Height for Each Structural Damage Category

	Slight Damage	Moderate	Severe
24 -story Steel	$\frac{1}{125}$	$\frac{1}{75}$	$\frac{1}{50}$
6 - story RC	$\frac{1}{200}$	$\frac{1}{125}$	$\frac{1}{100}$

The structures were represented by the multi degree of freedom stick models fixed at the base. Non linearity was introduced in the structural model by a tri-linear hysteretic relationship between the restoring shear force and the lateral drift.

4 DAMAGE CATEGORIES

Structural damage is defined in terms of the inelastic lateral drifts which generally correspond to significant strength degradation of major structural elements. The ratio of lateral drift to the story height during an earthquake is also used as a measure of structural damage and is used as the index of design criteria for both static and dynamic procedures. In this analysis the overall structural damage is defined by the largest value among all the peak story drifts. Three damage states representing slight structural damage, moderate structural damage and severe structural damage are established to categorize the degree of damage incurred in buildings during an earthquake. For each damage state, the median capacity is shown in Table 3 by the medium value of the lateral drift ratio to the story height (AIJ (1990)). At the slight damage state, the structures reach the maximum strength of the lateral force in the floor where the maximum drift ratio to the story height is supposed to marks the values defined in Table 3. At the moderate damage state, the 6-story RC structure makes hinges at almost all the ends of the beams and the 24-story steel structure has a slight local buckling effect on the members at almost all the ends of the beams. At the severe damage state the 6-story RC structure has many members which exceed their capacity of resistance and lose more than half their lateral resistance capacity, and the 24-story steel structure has an intermediate local buckling at their ends. At the severe damage state the structures are very rarely usable again.

5 NUMERICAL RESULT AND DISCUSSION

Non linear dynamic response analyses were carried out all the way up to structural failure by increasing the intensity of the twenty ground motions. That is, the peak velocity increased from 25 to 125 cm/sec intervals. The sample statistics, mean, and the logarithmic standard deviation, of the calculated responses were evaluated. The median intensity of the ground motion, either for the peak acceleration or the peak velocity, for each limit state of the two buildings were obtained. Figures 10 and 11 show the distribution of lateral drift ratio to the story height along the height when Vmax equals 75 cm/sec.

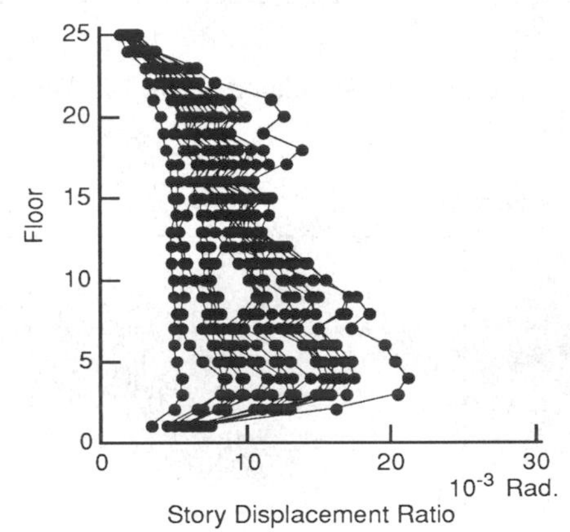

Fig.10 Distribution of Lateral Drift Ratio to Story Height for 24-story Steel Structure (Vmax=75cm/sec)

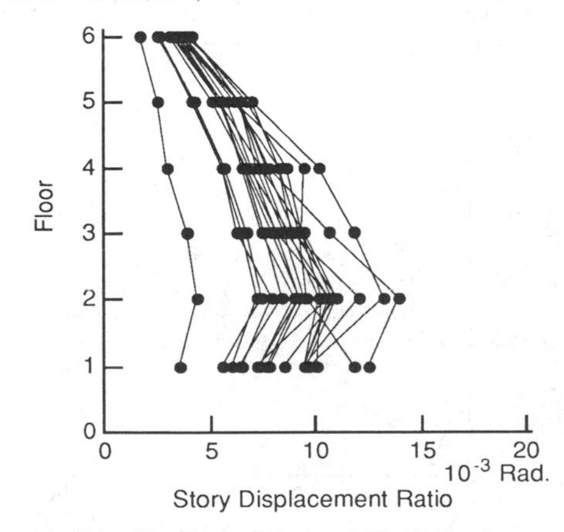

Fig.11 Distribution of Lateral Drift Ratio to Story Height for 6-story RC Structure (Vmax=75cm/sec)

The maximum story drifts for the 24-story building have two peaks over the height of the structure on account of higher modes contributing to the response; however, the maximum drifts occurred predominantly at the lower floor for the 6-story building. The correlations between the structural damage and the intensity of the ground motion parameter were quantified and compared. The plot of the peak input values versus the maximum ratio story drifts using logarithmic axes are shown in Figs.12 and 13 for each structure and indicate a proportional relationship with a rather uniform deviation (Kobayashi (1973)). Therefore the medium value of the peak acceleration, Amax, and the peak velocity, Vmax, can be related to the maximum lateral drift ratio to the story height by performing a linear regression analysis. For the 24-story steel structure the center lines in Fig.12-a,b are given by

$$A_{max} = 12.8\ \phi^{1.52} \qquad (1)$$

$$V_{max} = 4.34\ \phi^{1.11} \qquad (2)$$

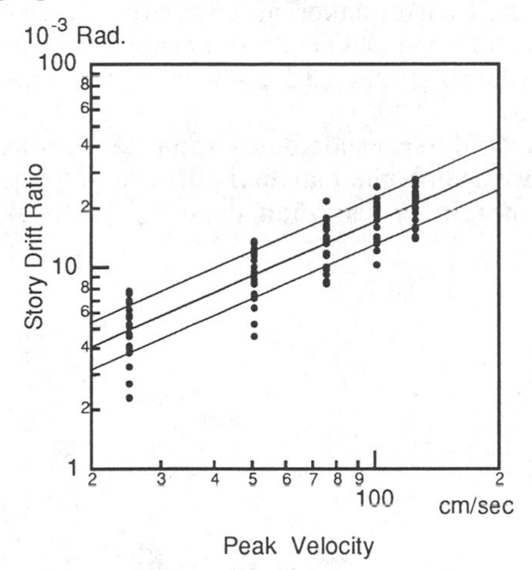

Fig.12-a Maximum Drift Ratio to Story Height versus Peak Velocity for 24-story Steel Structure

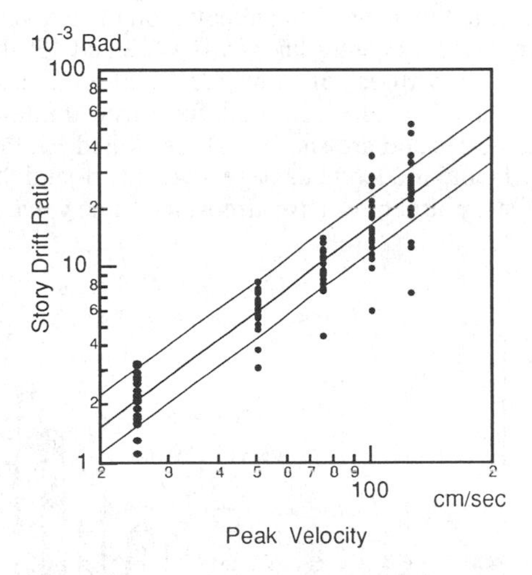

Fig.13-a Maximum Drift Ratio to Story Height versus Peak Velocity for 6-story RC Structure

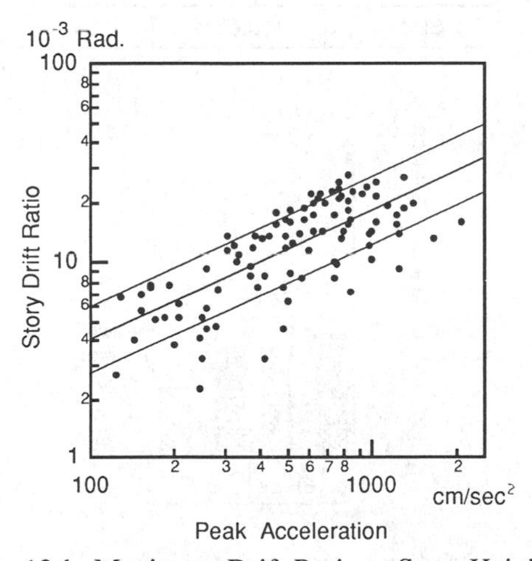

Fig.12-b Maximum Drift Ratio to Story Height versus Peak Acceleration for 24-story Steel Structure

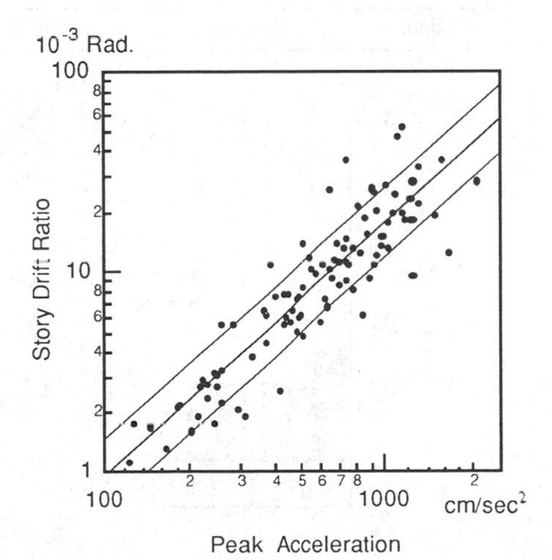

Fig.13-b Maximum Drift Ratio to Story Height versus Peak Acceleration for 6-story RC Structure

where the unit of Amax is cm/sec^2, the unit of Vmax is cm/sec ; ϕ is lateral drift ratio to the story height and the unit of 10^{-3} rad.

For the six story reinforced concrete structure the following equations can indicate center lines in Fig.13, a-b.

$$\mathrm{Amax} = 105 \ \phi^{0.79} \quad (3)$$
$$\mathrm{Vmax} = 15.2 \ \phi^{0.68} \quad (4)$$

The powers of ϕ for the steel structure are about double of those for the reinforced concrete structure. Substituting the damage drift ratio indicated in Table 3 to the equation (1) to (4) as expressed by center lines in Figs.12 and 13, the median values of the peak velocity and acceleration are obtained for three damage categories and are shown in Figs.14 and 15. For slight and moderate damage state, the 6- and the 24-story structures have almost the same value is for the peak velocity and acceleration. Conserning severe damage, the 6-story structure occurs with 20 percent increase of peak velocity and acceleration from moderate damage values, compared to the 24-story structure which occurs at about double the moderate damage values.

Fig.15 shows that the 24-story structure has large difference of the peak acceleration between the observed earthquakes and simulated earthquakes in the moderate and severe damage states. This difference can be presented by the fact that the observed earthquakes have less amplitude for longer period compared to simulated earthquakes as compared betwen Figs.1 and 6 and that the 24-story structure shifts its fundamental period longer under the more damage.

The logarithmic standard deviation of the peak velocity against the maximum drift ratio to the story height is less than those of the peak

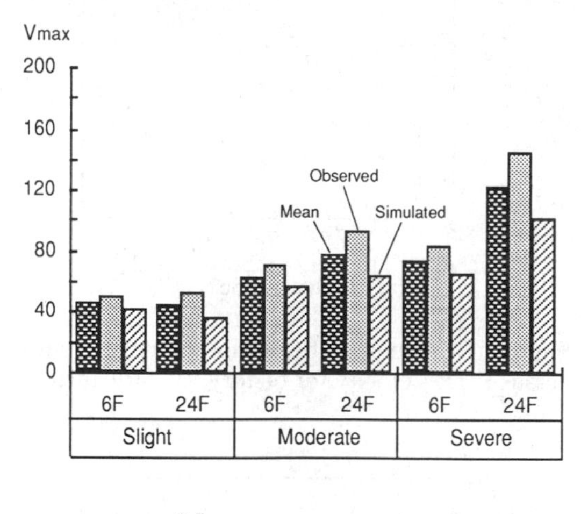

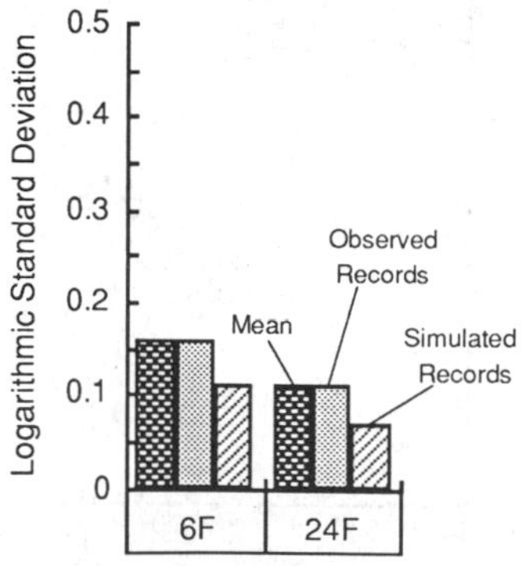

Fig.14 Median Peak Velocity and Logarithmic Standard Deviation of Response for Each Damage Category

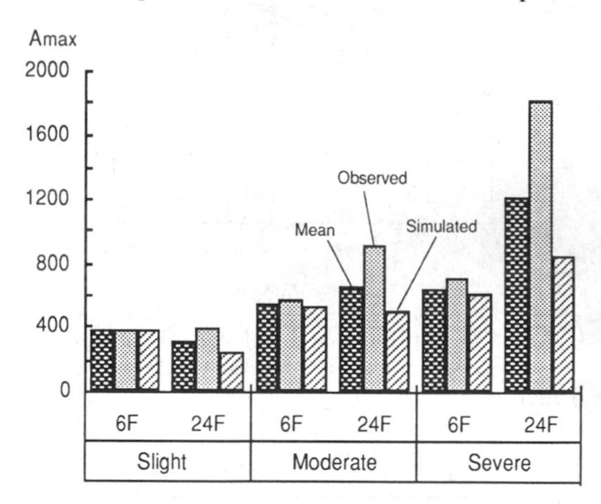

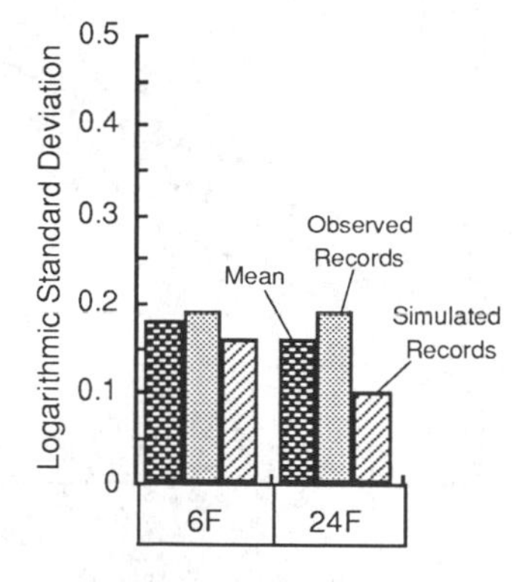

Fig.15 Median Peak Acceleration and Logarithmic Standard Deviation of Response for Each Damage Category

acceleration particularly for the 24-story structure. This indicate that the peak velocity can be better parameter than the peak acceleration for these two structures.

6 CONCLUSION

Suitability of the ground motion parameters (such as the peak ground acceleration and the peak ground velocity which define seismic intensity for fragility), is analysed by the samples of the 24-story steel building and the 6-story reinforced concrete building, which are properly designed against the earthquakes in Japan. It is concluded that the peak ground velocity is substantially better description of damage than that of the peak acceleration.

REFERENCES

Architectural Institute of Japan, 1990. Ultimate Strength and Deformation Capacity of Buildings in Seismic Design.

Kobayashi, H. and Nagahashi, S. 1973. Evaluation of Earthquake Effects on The Deformation of Multi-Storied Buildings by Ground Motion Amplitudes, Transactions of Architectural Institute of Japan, Vol.210.

Structural Safety & Reliability, Schuëller, Shinozuka & Yao (eds) © 1994 Balkema, Rotterdam, ISBN 90 5410 357 4

Sine-square modification to Kanai-Tajimi earthquake ground motion spectrum

Masanobu Shinozuka, Ruichong Zhang & George Deodatis
Princeton University, N.J., USA

ABSTRACT: A new formula is proposed for the frequency spectrum of horizontal earthquake ground acceleration at the far-field. The formula takes into account the effect of wave propagation in the earth medium (described by the classic Kanai-Tajimi spectrum) and the effect of a shear dislocation seismic source (described by a ramp-type slip function). Numerical computations are carried out to compare the proposed formula with theoretical results and with those given by the Kanai-Tajimi and Clough-Penzien formulas.

1 INTRODUCTION

Since structural damage caused by earthquakes is closely related to the frequency spectrum of ground motion, it is obvious that a simple, closed-form expression of the frequency spectrum is of paramount importance to designers in the area of earthquake engineering. In order to accurately capture the underlying physical process, such a spectrum must take into account both the seismic source mechanism which generates the seismic waves and the earth medium in which these waves propagate. For a comprehensive survey, readers are referred to Aki and Richards (1980) and Shinozuka and Deodatis (1988). Since theoretical solutions are complicated and can be particularly time-consuming when calculated numerically (e.g. Deodatis et al. 1990a, 1990b and Zhang et al. 1991a, 1991b), empirical formulas are proposed to approximate the theoretical solutions (e.g. Hanks and McGuire, 1981, Boore and Atkinson, 1987, and Joyner and Boore, 1988). The most widely used one is perhaps the Kanai-Tajimi spectrum (Kanai, 1957 and Tajimi, 1960), which characterizes the seismic wave propagation in the earth medium. Since the Kanai-Tajimi acceleration spectrum is not equal to zero at zero frequency (which is inconsistent with earthquake records), several modifications have been proposed to overcome this drawback. One of the most important is the Clough-Penzien spectrum (1975), which incorporates a high-passing filter to the Kanai-Tajimi spectrum. Mathematically, such a high-passing filter may have many representations, which, however, may not have a clear physical meaning. In this paper, a sine-square modification to the Kanai-Tajimi earthquake ground motion spectrum is proposed, which takes into account both the shear dislocation seismic source and the wave propagation in the earth medium. Numerical computations are carried out to compare the proposed formula with theoretical results and with those given by the Kanai-Tajimi and Clough-Penzien spectra.

2 EARTHQUAKE GROUND MOTION SPECTRUM

The theoretical solution for the frequency spectrum of horizontal ground acceleration due to a point shear dislocation source idealized as a double couple can be represented in the following form (Deodatis et al. 1990a, 1990b)

$$S_a(\omega) = |\omega^2 M_0 \tilde{m}(\omega)|^2 \int_{-\infty}^{\infty}\int_{-\infty}^{\infty} |G(k_x,k_y,\omega)|^2 dk_x dk_y \quad (1)$$

where M_0 is the seismic moment, $G(k_x,k_y,\omega)$ is the frequency-wave number domain acceleration response, due to a double couple impulse, (G describes the wave propagation in the earth medium), and $\tilde{m}(\omega)$ is the Fourier transform of slip function at seismic source ($\tilde{m}$ describes the seismic source). When the slip is assumed as a ramp function, which is usually adopted in both seismology and engineering, we may obtain

$$\omega^2 \tilde{m}(\omega) = \frac{1}{T}(e^{i\omega T} - 1) \quad (2)$$

or

$$|\omega^2 \overline{m}(\omega)|^2 = \frac{4}{T^2}\sin^2(\omega T/2) \quad (3)$$

with T being the rise time of the ramp function. It can be shown (Theoharis, 1991) that the same expression for $\tilde{m}(\omega)$ would appear in the expression for $S_a(\omega)$ if the source was idealized as a Haskell-type one. However, the model that a shear dislocation occurs at a seismic fault with one deterministic slip function is of course unrealistic since the earthquake occurrence is a random process. Therefore, it is reasonable to assume that the earthquake is a result of a series of point sources with random final slip and random rise time. The random final slip may be taken into account in the seismic moment. The effects of random rise time may be obtained by taking ensemble average of equation (2), i.e.

$$\omega^2\overline{\tilde{m}(\omega)} = E\left[\frac{1}{T}(e^{i\omega T} - 1)\right] \quad (4)$$

where overline stands for the average value and E denotes ensemble average. The probabilistic distributed function of rise time may be assumed as Beta distribution, i.e.

$$p_T(t) = \begin{cases} \dfrac{t^{\mu}(t_f - t)^{\lambda}}{t_f^{\mu+\lambda+1} B(\mu+1,\lambda+1)}, & 0<t<t_f \\ 0, & t_r<0 \text{ or } t>t_f \end{cases} \quad (5)$$

where t_f is the maximum rise time under consideration and Beta function B(x,y) is defined as

$$B(x,y) = \int_0^1 t^{x-1}(1-t)^{y-1} dt \quad (6)$$

For simplicity, both μ and λ are chosen to be one. The probabilistic density function is plotted in Fig. 1 with t_f=2 sec. As shown in Fig. 1, $t_f/2$ is the dominant rise time. Substituting equation (5) into equation (4), we may obtain

$$|\omega^2\overline{\tilde{m}(\omega)}|^2 = 1 + \frac{8}{(\omega t_f)^3}\left[\frac{\omega t_f}{2}\cos(\omega t_f) - \sin(\omega t_f)\right] + \frac{16}{(\omega t_f)^4}\sin^2\left(\frac{\omega t_f}{2}\right) \quad (7)$$

A numerical computation of $\omega^4\overline{\tilde{m}(\omega)}^2$ is presented in Fig. 2, which indicates that $\omega^4\overline{\tilde{m}(\omega)}^2$ is a kind of high passing filter with a turning frequency around $\omega=\pi/T$. The expression for the ground motion spectrum of equation (1) has usually a complicated form because of $G(k_x,k_y,\omega)$ and expression of equation (7). To have a simple form for the ground motion spectrum, which may be easily applied in practice, we propose the following empirical formula, which stems from the theoretical representation of equation (1), i.e.

$$S_a(\omega) = S_{K-T}(\omega)\, m(\omega) \quad (8)$$

where

$$S_{K-T}(\omega) = S_0 \frac{\omega_g^4 + 4\zeta_g^2\omega_g^2\omega^2}{(\omega_g^2-\omega^2)^2 + 4\zeta_g^2\omega_g^2\omega^2} \quad (9)$$

and

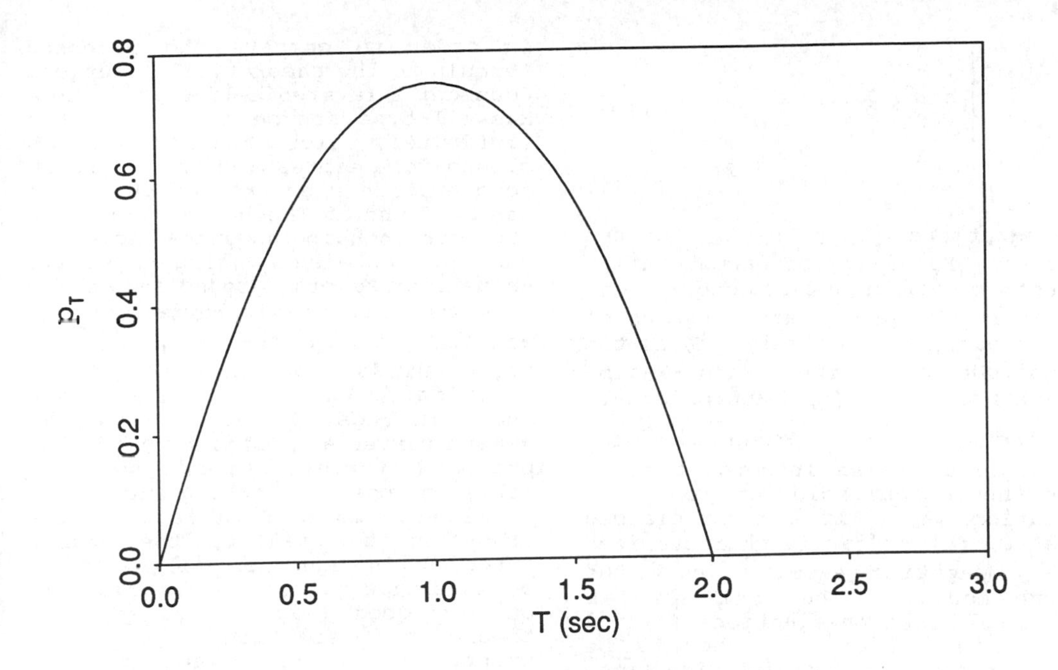

Fig. 1 Beta distribution with $\mu=\lambda=1$

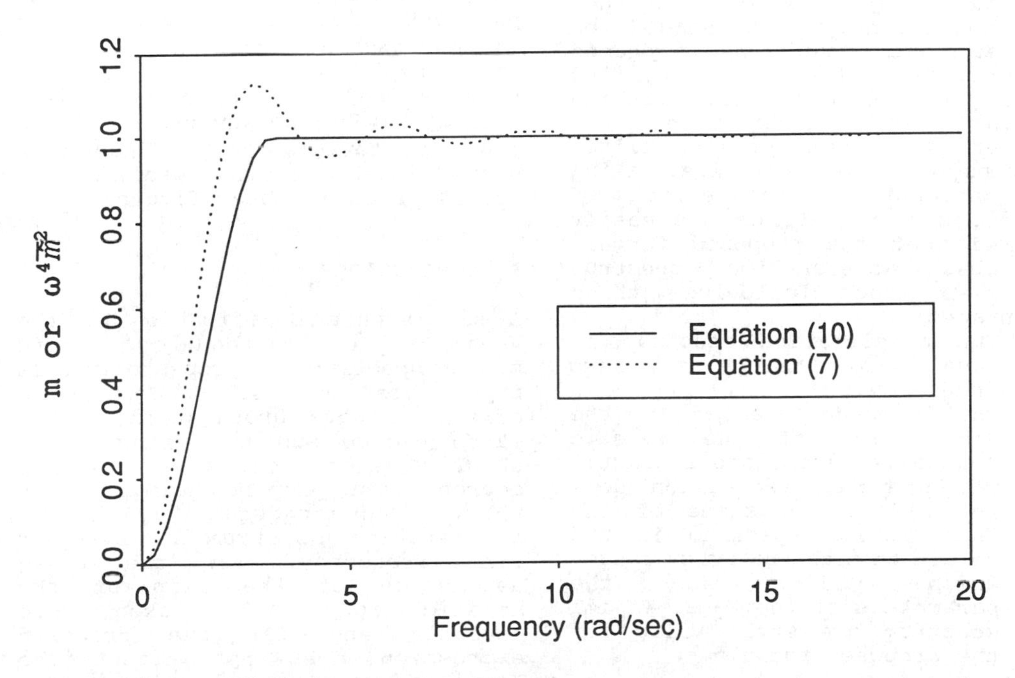

Fig. 2 Comparison of statistical and proposed factors associated with slip function (equations (7) and (10)

$$m(\omega) = \begin{cases} \sin^2(\frac{\omega T}{2}) , & \omega < \frac{\pi}{T} \\ 1 , & \omega \geq \frac{\pi}{T} \end{cases} \quad (10)$$

In equations (8)-(10), S_{K-T} is the classic Kanai-Tajimi acceleration spectrum with ω_g and ζ_g denoting the dominant frequency and damping of the ground, respectively. S_0 is the magnitude of the Kanai-Tajimi spectrum. $S_{K-T}(\omega)$ approximately replaces the product of M_0^2 multiplied by the integral term in the theoretical solution given in equation (1). It can be claimed that $S_{K-T}(\omega)$ primarily characterizes the propagating seismic waves in the earth medium. The term $m(\omega)$ is obtained based on equations (3) and (7), in which T should be interpreted as a dominant rise time. The comparison of the proposed simple form in equation (10) and original one in equation (7) is plotted in Fig. 2, which shows they are consistent. It should be pointed out that the proposed formula is actually a modified Kanai-Tajimi spectrum using a high passing filter of equation (10). However, this high passing filter has a physical meanings, i.e. taking into account of the effect of seismic source. It can be easily checked that the proposed formula for the acceleration spectrum possesses the following three advantages:

(a) The acceleration spectrum is equal to zero at zero frequency;
(b) The time-dependence of the dislocation of the seismic source is taken into account;
(c) The proposed expression does not affect the shape of the Kanai-Tajimi spectrum in the middle to high frequency range, which implies that the parameters of the Kanai-Tajimi spectrum are still valid for the proposed formula.

3 NUMERICAL EXAMPLE

A numerical example is carried out in order to compare the proposed formula to the theoretical solution. Consider a layered half-space with a Haskell-type source located in the (infinitely thick) bottom layer. A closed-form expression of the ground acceleration spectrum for this case can be found in Theoharis (1991) and the corresponding numerical solution can be obtained using specific values for several ground and source parameters. Consequently, parameters ω_g, ζ_g and S_0 in equation (4) can be selected based on this theoretical solution. Numerical results are shown in Figs. 3 and 4, where the smooth curves are obtained using the proposed formula (equations (8)-(10)) and the oscillating curves are obtained by using a 2-D Fast Fourier Transform to evaluate the double integral in equation (1). It is obvious that the proposed formula is quite good in fitting the theoretical solution. The proposed ground acceleration spectrum is also compared with the Kanai-Tajimi and Clough-Penzien spectra in Figs. 5 and 6. For the numerical computation, the following parameters are used: ω_g=5 and 15 rad/sec and ω_1=1 and 2 rad/sec (used in Clough-Penzien's formula) for Figs. 5 and 6, respectively, T=1 sec, ζ_g= 0.5. The advantages of the proposed formula (which have been stated in the previous section) are clearly seen in these figures.

4 CONCLUSIONS

A sine-square modification of the Kanai-Tajimi earthquake ground motion spectrum was proposed in this paper, taking into account the effect of wave propagation in the earth medium and the effect of a shear dislocation type seismic source. The proposed spectrum has three advantages: (1) the acceleration spectrum is zero at zero frequency; (2) the time-dependence of the slip of the seismic source is taken into account; and (3) the proposed expression does not affect the Kanai-Tajimi spectrum in the middle to high frequency range. Numerical results show that the proposed formula is very good in fitting the theoretical solution.

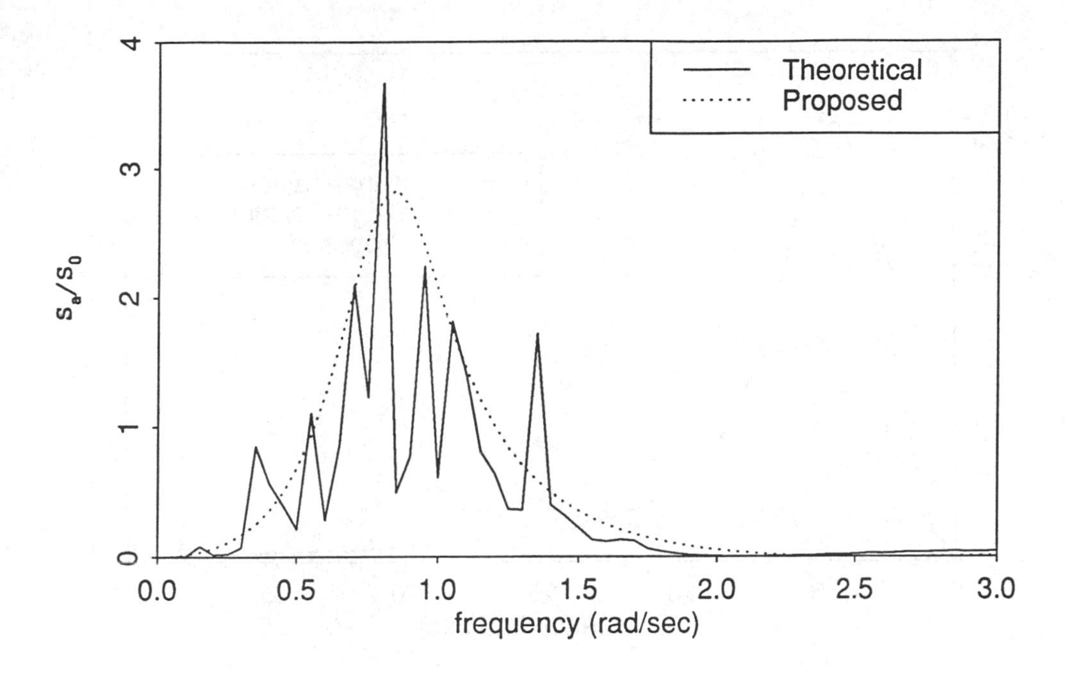

Fig. 3 Comparison of proposed formula with the theoretical solution in a uniform earth medium

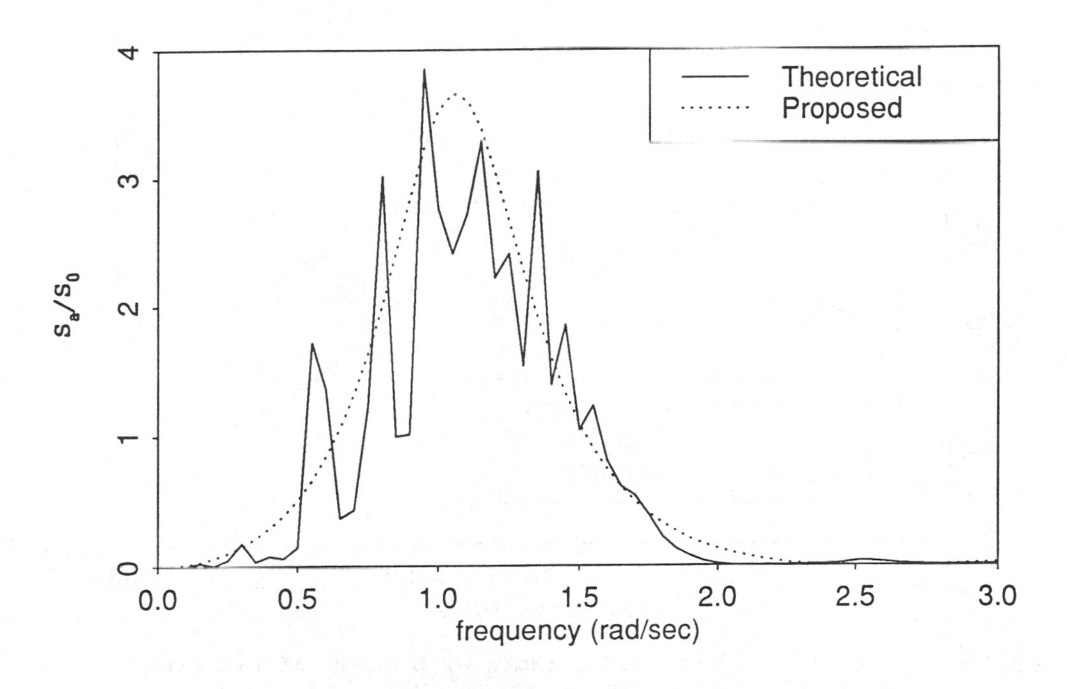

Fig. 4 Comparison of proposed formula with the theoretical solution in a two-layer medium

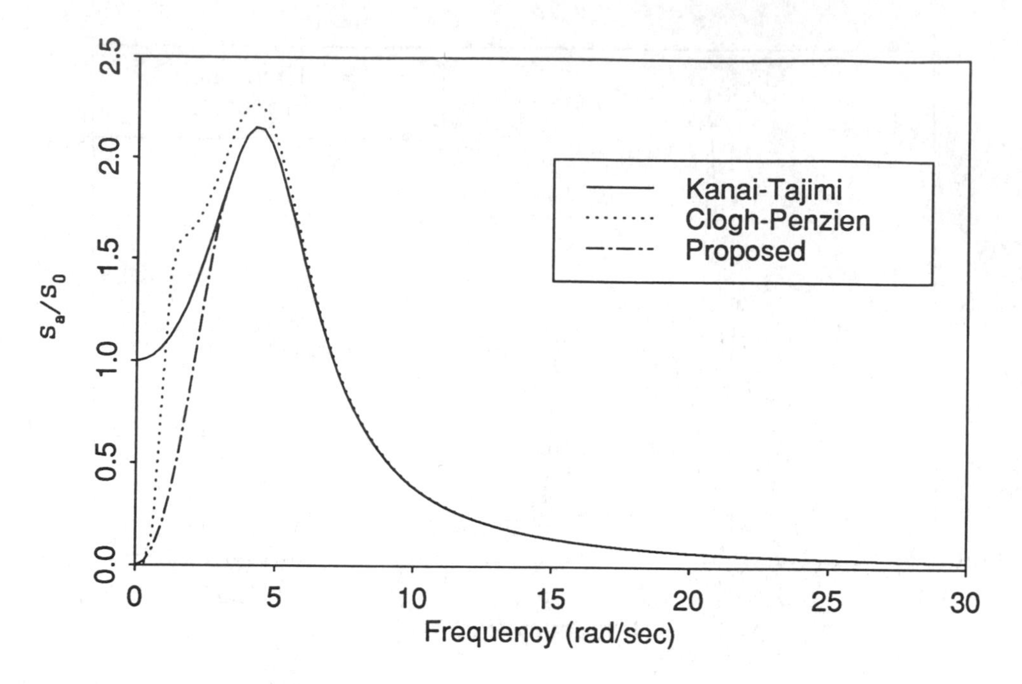

Fig. 5 Comparison of proposed formula with those of Kanai-Tajimi and Clough-Penzien (ω_g=5 rad/sec, ζ_g=0.5)

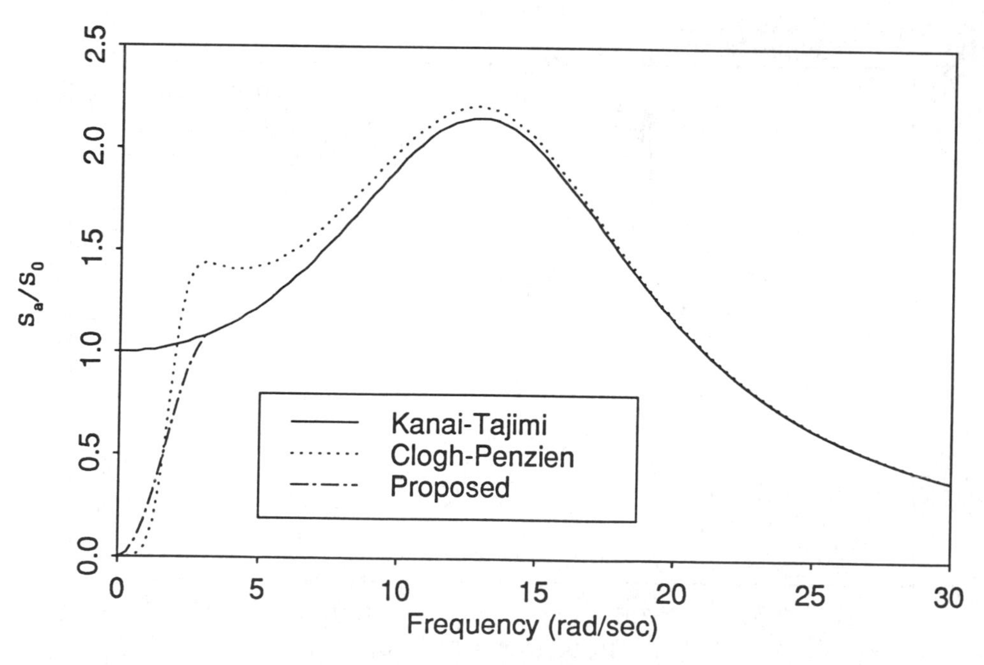

Fig. 6 Comparison of proposed formula with those of Kanai-Tajimi and Clough-Penzien (ω_g=15 rad/sec, ζ_g=0.5)

ACKNOWLEDGMENTS

This work was supported by Contract No. NCEER 92-3302A under the auspices of the National Center for Earthquake Engineering Research.

REFERENCES

Aki, K. and Richards, P.G. 1980. Quantitative Seismology ---- Theory and Methods, Vol. 1 and 2, W.H. Freeman and Company.

Boore, D.M. and Atkinson, G.M. 1987. Stochastic prediction of ground motion and spectral response parameters at hard-rock sites in eastern north America, Bull. Seis. Soc. Am. 77, 2, 440-467.

Clough, R.W. and Penzien, J. 1975. Dynamics of Structures. McGraw-Hill Inc.

Deodatis G., Shinozuka M., and Papageorgiou A. 1990a. Stochastic wave representation of seismic ground motion. I: F-K spectra, J. Engrg. Mech., 116, 11, 2363-2379.

Deodatis G., Shinozuka M., and Papageorgiou A. 1990b. Stochastic wave representation of seismic ground motion. II: simulation, J. Engrg. Mech., 11, 11, 2381-2399.

Hanks, T.C. and McGuire, R.K. 1981. The character of high-frequency strong motion, Bull. Seis. Soc. Am., 71, 6, 2071-2095.

Kanai, K. 1957. Semi-empirical formula for the seismic characteristics of the ground, Bull. Earthq. Res. Inst., 35, 309-325.

Shinozuka, M., and Deodatis, G. 1988. Stochastic process models for earthquake ground motion, Prob. Engrg. Mechanics, 3, 3, 114-123.

Tajimi, H. 1960. A statistical method of determining the maximum response of a building structure during an earthquake, Proc. 2nd World Conference of Earthquake Engineering, Tokyo and Kyoto, Vol. II, pp. 781-798.

Theoharis, A.P. 1991. Wave representation of seismic ground motion, Dissertation of Department of Civil Engineering and Operations Research, Princeton University.

Joyner, W.B. and Boore, D.M. 1988. Measurement, characterization, and prediction of strong ground motion, Earthquake Engineering & Soil Dynamics II (Lawrence Von Thun, ed.), 43-102.

Zhang, R.C., Yong, Y. and Lin, Y.K. 1991a. Earthquake ground motion modeling I: Deterministic point source, J. Eng. Mech., 117, 9, 2114-2132.

Zhang, R.C., Yong, Y. and Lin, Y.K. 1991b. Earthquake ground motion modeling I: Stochastic line source, J. Eng. Mech., 117, 9, 2133-2148.

Structural Safety & Reliability, Schuëller, Shinozuka & Yao (eds) © 1994 Balkema, Rotterdam, ISBN 90 5410 357 4

On the load factors of seismic loads for the port and offshore structures by limit state design method

S.Shiraishi, S.Ueda & T.Uwabe
Port and Harbour Research Institute, Ministry of Transport, Yokosuka, Japan

ABSTRACT: The design code of concrete structures was revised in 1986 and 1991 by the Japan Society of Civil Engineers. In this code, standard values of load factors and material factors are shown. On the application of the code to port and offshore structures, safety factors such as the load factor must be examined taking into account the characteristics of loads and structures. Previously, two of authors reported the load factors of seismic loads for port structures in the proceedings of ICOSSAR'89. In this paper, the relation between seismic coefficients and acceleration records are revised and the load factors of seismic loads are calculated again and standard value of load factor of seismic loads for port and offshore structures is proposed.

1 INTRODUCTION

The design code of concrete structures was revised in 1986 and 1991 by the Japan Society of Civil Engineers (JSCE,1991). In this code, limit state design method is adopted instead of allowable stress design method and standard values of load factors and material factors are shown. On the application of the code to port and offshore structures, safety factors such as the load factor must be examined taking into account the characteristics of loads and structures.

Previously, two of authors reported the load factors of seismic loads for port structures (Shiraishi et al.,1987 and 1989). Those reports were discussed load factor of seismic load computed by means of the relation between the horizontal seismic coefficients and maximum horizontal acceleration proposed by Noda et al.(1975). However, seismic coefficient based on the Noda's relation gives large values than the seismic coefficient shown in technical standard for port and harbour facilities(1989).

Then, in this report, the relation between horizontal seismic coefficients and acceleration are revised and the load factors of seismic loads are calculated again(Shiraishi et al.,1991).

2 METHOD OF CALCULATION OF LOAD FACTOR

2.1 Relation between seismic coefficient and acceleration of ground

Gravity type and sheet pile type port structures are mainly designed by seismic coefficient method. Then, in this paper load factor is discussed taking into account the N-year maximum **mean values** (hereafter denoted **MEANs**) and **the coefficient of variation** (hereafter denoted **COVs**) of horizontal seismic coefficients.

Kitazawa et al. (1984) were

calculated the maximum base rock acceleration for 190 points along coasts of Japan against each earthquake occurred after 1885 to 1981. In the calculation, the maximum base rock acceleration at each point was calculated by means of the relation between the magnitude of earthquake and the epicentral distance. Also the upper twenty values of base rock accelerations were presented and the expected values in recurrence interval were calculated.

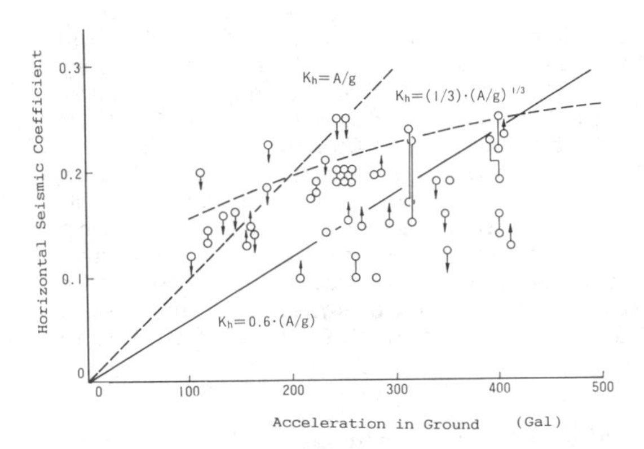

Figure 1 Relation between acceleration and horizontal seismic coefficient

Noda et al.(1975) were analyzed the stability of gravity type port structures by use of failure examples of past earthquakes and showed the relation between horizontal seismic coefficient and acceleration in ground as shown **Figure 1.** While horizontal seismic coefficients show about 40% variation to same acceleration, Noda et al. proposed the Eq.(1)(hereafter denoted Noda's Equation).

$$K_h = \left(\frac{1}{3}\right)\left(\frac{A}{g}\right)^{1/3} \qquad (A \geq 200 \text{ Gal})$$

$$K_h = \left(\frac{A}{g}\right) \qquad (A < 200 \text{ Gal}) \qquad (1)$$

in which K_h is maximum horizontal seismic coefficient, A is maximum acceleration in ground and g is gravity acceleration. Because of Noda's Equation shows upper limits of the relation between horizontal seismic coefficient and ground acceleration, horizontal seismic coefficients computed by Eq.(1) give large values than the seismic coefficient given in technical standards of port and harbour structures(1989). Then, in this paper horizontal seismic coefficients are computed by means of Eq.(2). This gives average value in the relation between horizontal seismic coefficient and ground acceleration.

$$K_h = 0.6\left(\frac{A}{g}\right) \qquad (2)$$

2.2 Method of N-year maximum horizontal seismic coefficient

In order to evaluate N-year maximum **MEANs** and **COVs** of loads, it is necessary to obtain data in long term, but it is usually difficult to obtain such kind of data. In this paper, the N-year maximum **MEANs** and **COVs** are calculated by means of the N-year maximum **probability distribution function**(hereafter denoted **PDF**) of loads which are approximately defined from the yearly maximum **PDFs** of loads. In this paper, following two types of asymptotic distribution forms are examined.

Gumbel distribution:

$$P(x)=\exp\left[-\exp\left\{-\frac{x-B_1}{A_1}\right\}\right] \qquad (3)$$

Weibull distribution:

$$P(x)=1-\exp\left[-\left\{\frac{x-B_2}{A_2}\right\}^k\right] \qquad (4)$$

in which $P(x)$ is **PDF** of loads and A_i, B_i are parameters where subscripts $i=1,2$ correspond to the Gumbel and Weibull distribution, respectively.

The yearly maximum **PDFs** are obtained fitting to one of two types of asymptotic distribution forms of extreme values. As for Weibull distribution, the parameters A_2, B_2 are fitted to each shape parameter k as 0.75, 0.85, 1.0, 1.1, 1.25, 1.5, 2.0, respectively (Petruaskas et al., 1970).

The N-year maximum **PDFs** $P_N(x)$ is approximately defined from the yearly maximum **PDF** $P(x)$ as follows.

$$P_N(x)=[P(x)]^N \tag{5}$$

Gumbel distribution :

$$P_N(x)=\exp[-\exp\{-\frac{x-B_1-A_1\ln N}{A_1}\}] \tag{6}$$

Weibull distribution :

$$P_N(x)=[1-\exp[-\{\frac{x-B_2}{A_2}\}^k]]^N \tag{7}$$

As for Gumbel distribution function, the N-year maximum **MEANs** x_{NM} and **COVs** V_N are written with functions as Eq.(8) and (9).

$$x_{NM} = (B_1+A_1)+A_1\ln N \tag{8}$$

$$V_N = \frac{\pi A_1}{\sqrt{6}\, x_{NM}} \tag{9}$$

As for the Weibull distribution function, the N-year maximum **MEANs** and **COVs** cannot be written with functions. Two of authors showed method by the numerical integration of **PDF**(Shiraishi et al. 1987).

2.3 Method of computation of load factors

Structural safety verified by Eq.(10) in limit state design method.

$$\frac{R(\frac{f_k}{r_m})}{r_b} > r_i \sum_{j=1}^{M} r_a S(r_{fj}F_{kj}) \tag{10}$$

in which f_k is characteristic value of material strength, r_m is material factor, $R(f_k/r_m)$ is computed value of proof strength, r_b is member factor, r_i is structure factor, r_a is structural analysis factor, r_{fj} is load factor for F_{kj}, F_{kj} is characteristic value of load, $S(r_{fj}F_{kj})$ is computed value of stress resultant and M is a number of combination of loads.

In this paper, we discuss the load factors taking into account the **COVs** of loads, then r_b, r_i and r_a are supposed as 1.0.

Safety index of structure is computed by Eq.(11).

$$\beta = \frac{\ln(\frac{R_m}{S_m})}{\sqrt{V_R^2+V_S^2}} \tag{11}$$

in which R is proof strength, S is stress resultant , R_m is a **MEAN** of R, S_m is a **MEAN** of S, V_R is a **COV** of R and V_S is a **COV** of S.

Then, load factor is computed by Eq.(12)

$$r_{fj}= (\frac{S_{jm}}{S_{fj}})\exp(\alpha_1\alpha_2\beta V_{sj}) \tag{12}$$

in which S_{jm} is a **MEAN** of load, S_{fj} is a characteristic value of load, α_1, α_2 are linearized coefficients(here α_1, α_2 =0.75 (Lind,1971)) and V_{sj} is a **COV** of the loads.

3 MEAN AND COEFFICIENT OF VARIATION OF N-YEAR MAXIMUM HORIZONTAL COEFFICIENT

3.1 Zoning

In this paper, load factors of seismic load are computed for five zones shown in **Figure 2.** These zones are same one used in the prediction method of sand liquefaction in port and harbour structures.

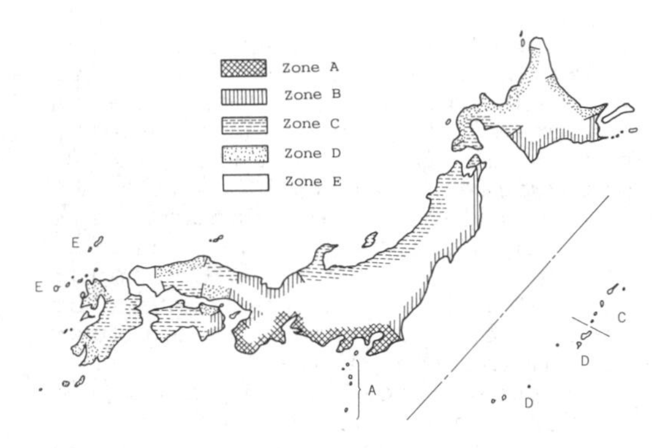

Figure 2 Zone of seismic coefficient

3.2 Relation of between N-year maximum and R-year recurrent value

Figure 3(a) shows the relation between 50-yr. maximum horizontal seismic coefficient and 50-yr. recurrent interval value for 190 points along coasts of Japan. The ratio of 50-yr. horizontal seismic coefficient to 50-yr. recurrent interval value is almost 1.2 to 1.3. Also **Figure 3(b)** shows the relation between 50-yr. maximum horizontal seismic coefficient and 75-yr. recurrent interval value. The ratio of 50-yr. horizontal seismic coefficient to 75-yr. recurrent interval value is almost 1.1. Goda(1988) showed the following approximately relation formula between the recurrent interval R-yr. and life time N-yr.

$$R = 1.76N + 0.5 \tag{13}$$

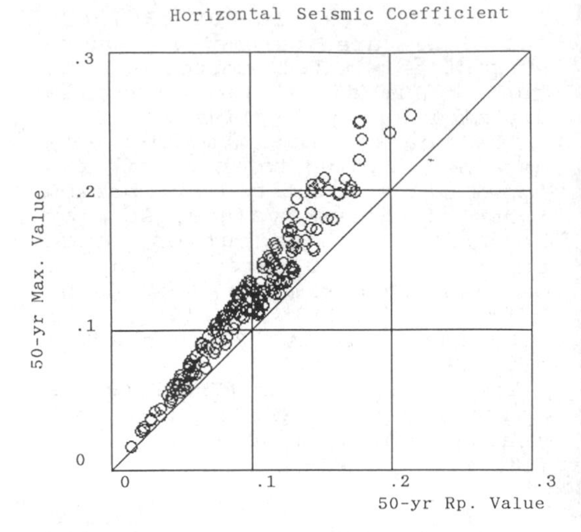

Figure 3(a) Comparison between 50-yr. maximum horizontal seismic coefficient and 50-yr. recurrent interval value

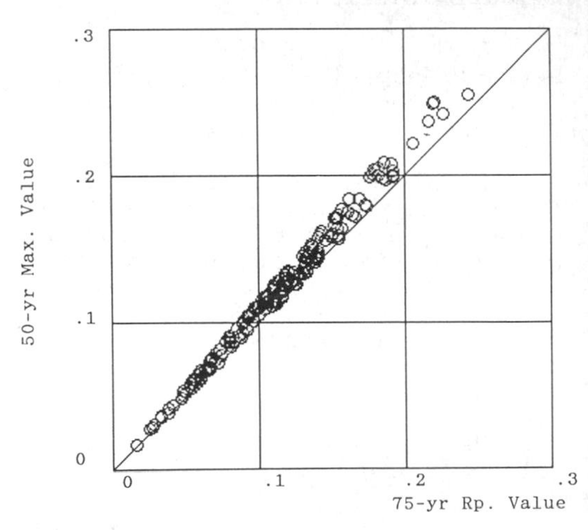

Figure 3(b) Comparison between 50-yr. maximum horizontal seismic coefficient and 75-yr. recurrent interval value

By means of above relation, N=42 years is corresponding to R=75 years. Technical standard of port and harbour structures in

Japan(1989) shows the seismic coefficient are determined based on the accelerations of 75 years recurrent intervals. Then, in this paper load factor of seismic load is discussed for N= 50yr.

3.3 MEANs and COVs of 50-yr. horizontal seismic coefficient

Figure 4 shows the relation between **COVs** and **MEANs** of 50-yr. maximum horizontal seismic coefficient. **COVs** of 50-year maximum horizontal seismic coefficient show 0.15 to 0.60.

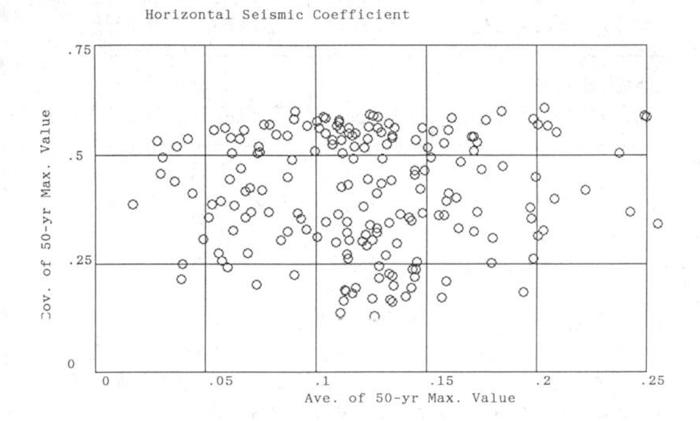

Figure 4 Covs of 50-yr. maximum horizontal seismic coefficient

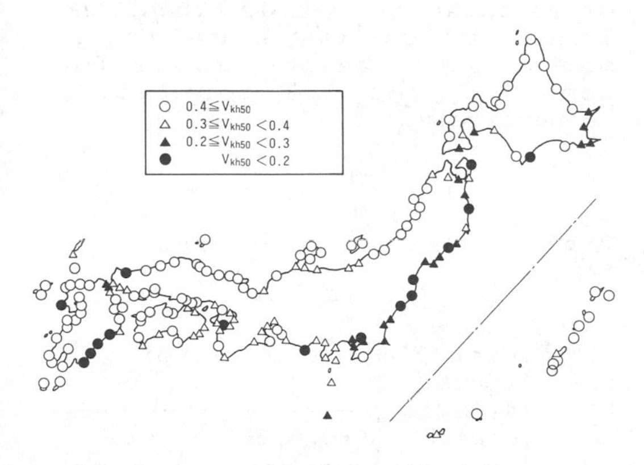

Figure 5 Regional distribution of COV of horizontal seismic coefficient

Figure 5 shows regional distribution of **COVs** of horizontal seismic coefficient. Open circles, open and closed triangles and closed circles show the locations where **COVs** of horizontal seismic coefficients are corresponding to greater than 0.4, in the range of 0.3 to 0.4, in the range of 0.2 to 0.3, and less than 0.2, respectively.

By comparison of **Figure 2** and **Figure 5** , the regional distribution of **COVs** shows nearly same characteristics of zones shown in **Figure 5**. Accordingly, for the point belonging to Zone A and Zone B, **Covs** of horizontal seismic coefficient show small value than the other zones. Then, we decided to compute load factors for five zones which are shown in **Figure 2**.

Figure 6 shows mean values of N-yr. maximum horizontal seismic coefficients which are computed as average of each zone. In this figure N are considered as 20, 30, 50, 100, respectively. **MEANs** of the 50-yr. maximum horizontal seismic coefficient are 0.172, 0.145, 0.124, 0.095, 0.058 for zone A to E, respectively.

Figure 7 shows **COVs** of N-yr. maximum horizontal seismic coefficient which are computed as average of each zone. The coefficient of variation of the 50-yr. maximum horizontal seismic coefficient are 0.35, 0.31, 0.45, 0.49, 0.44 for zone A to E, respectively. **COVs** for zone A and B are smaller than the values for zone C,D,E.

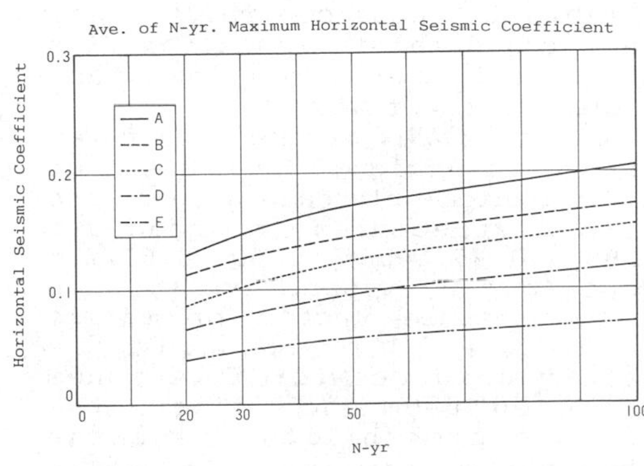

Figure 6 MEANs of Horizontal seismic coefficient

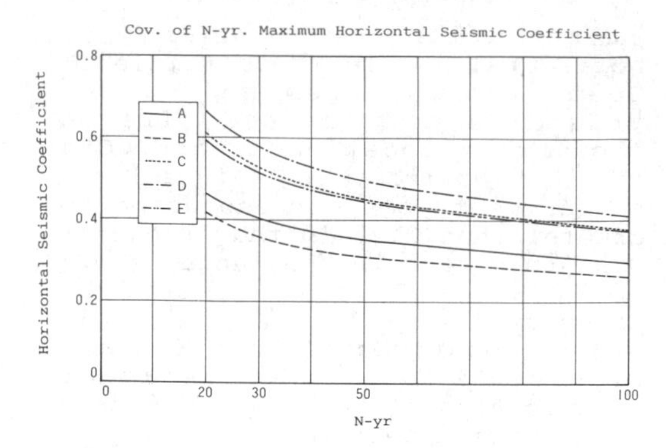

Figure 7 COVs of horizontal Seismic coefficient

4 LOAD FACTOR FOR HORIZONTAL SEISMIC COEFFICIENT

4.1 Safety index

In the calculation of load factors for seismic loads, target safety indices for port and harbour facilities are decided by means of verifying the safety of current structures. In current technical standard of port and harbour facilities in Japan, the design seismic coefficient shall be determined with consideration to the classification of region where structures is located, that of the subsoil condition and the degree of importance of structure. The coefficient of importance of structure are 1.5,1.2,1.0,0.5 for Class S,A,B,C, respectively. And factors for subsoil conditions are 0.8,1.0,1.2 for each class.

Table 1 shows central safety factor and safety index for five zone and for three important factor, respectively.

Safety index of the structures which are designed by means of the technical standard of port and harbor structures in Japan are 0.85, 0.95, 0.65, 0.60, 0.67 for zone A to E, respectively, when the importance factor for seismic load is 1.0. Although the safety indices for five zones differed from each other, the failure probability is large when the safety index is small. The design of structures is reasonable to get same failure probability of structure for each zone. So, load factors shall be decided to be the safety factor as 1.0 which is the upper value of the safety index of structures for zone A to E.

Table 1 Central safety factor and safety index

Zone	Cov. of seismic coef.	Safety Index		
		B	A	S
A	0.346	0.85	1.35	1.99
B	0.307	0.95	1.51	2.22
C	0.449	0.65	1.05	1.52
D	0.493	0.60	0.96	1.41
E	0.444	0.67	1.06	1.56
Central safety factor		1.35	1.61	2.02

4.2 Load factor

Table 2 shows the result of computation of load factors. In this table, the results which safety factor are assumed 0.6 and 0.8 are also shown. Load factors for seismic loads are 1.22, 1.19, 1.29, 1.32, 1.29 for each zone, when the safety index of structures is 1.0. Then, the load factor on limit state design method shall be set to 1.2 for zone A,B and 1.3 for C,D,E, respectively.

Table 2 Safety index and load factor

Zone	COV. of hori-zontal seismic coef.	Load factor		
		β=0.6	β=0.8	β=1.0
A	0.346	1.12	1.17	1.22
B	0.307	1.11	1.15	1.19
C	0.449	1.16	1.22	1.29
D	0.493	1.18	1.25	1.32
E	0.444	1.16	1.22	1.29

The design code of concrete structures(JSCE,1991) is proposed the load factor for standard value as 1.0 to 1.2. Although, the load factor obtained by our research for zone A and B is 1.2 which is same to upper value of JSCE proposal, the values for zone C,D,E are a little larger than the value presented JSCE.

5 CONCLUSION

Major results obtained in this research are as follows.

1) Mean of the 50-yr. maximum horizontal seismic coefficient are 0.172, 0.145, 0.124, 0.095, 0.058 for zone A to E, respectively. And the coefficient of variation of the 50-yr. maximum horizontal seismic coefficient are 0.35, 0.31, 0.45, 0.49, 0.44 for zone A to E, respectively.
2) Safety index of the structures which are designed by use of the technical standard of port and harbor structures in Japan are 0.85, 0.95, 0.65, 0.60, 0.67 for zone A to E, respectively, when the importance factor for seismic load is 1.0. Although the safety indices for five zones differed from each other, the failure probability is large when the safety index is small. The design of structures is reasonable to get same failure probability of structure for each zone. So, load factors shall be decided to be the safety factor as 1.0 which is the upper value of the safety index of structures for zone A to E.
3) Load factors for seismic loads are 1.22, 1.19, 1.29, 1.32, 1.29 for each zone, when the safety index of structures is 1.0. Then, the load factor on limit state design method set to 1.2 for zone A,B and 1.3 for C,D,E,respectively.

REFERENCES

Goda, Y. 1988. Numerical Investigation on Plotting Formulas and Confidence Intervals of Return Values in Extreme Statistics, *Report of the Port and Harbour Research Institute (PHRI), Vol.27, No.1*, pp.31-92

Japanese Port and Harbour Association 1989. *Technical Standards for Port and Harbour Facilities (Revised Version)*, pp.194-197

Japan Society of Civil Engineers(JSCE) 1991. *Design Code of Concrete Structures*, 206p.

Kitazawa,S. et al. 1984. Expected Values of Maximum Base Rock Accelerations along Coasts of Japan, *Tcchnical Note of the PHRI, No.486*, 137p.

Lind, N.C. 1971. Consistent Partial Safety Factors, *Proc. of ASCE, Vol.97, No.ST6*, pp.1651-1669

Noda,S. et al. 1975. Relation between Seismic Coefficient and Ground Acceleration for Gravity Quaywall, *Report of the PHRI, Vol.14, No.4*, pp.67-111

Petruaskas, C. and P.M. Aagaard 1970. Extrapolation of Historical Storm Data for Estimating Design Wave Heights, *Proc. of 2nd Offshore Technology Conference*, pp.I-409-428

Shiraishi,S. and S. Ueda 1987. Study on the Method of Verification of Structural Safety of Port and Offshore Structures, *Report of the PHRI, Vol.26, No.2*, pp.493-576

Shiraishi,S. and S. Ueda 1989. The N-year Maximum Mean Values and the Coefficients of Variation of Loads Subjected to Port and Offshore Structures, *Proc. of 5th ICOSSAR*, pp.1791-1798

Shiraishi,S., S.Ueda and T. Uwabe 1991. Study on the Load Factor of Seismic Loads on Limit State Design Method, *Technical Note of the PHRI, No.708*, 27p.

Structural Safety & Reliability, Schuëller, Shinozuka & Yao (eds) © 1994 Balkema, Rotterdam, ISBN 90 5410 357 4

Probabilistic estimation of design value of earthquake effects considering possibility of existence of unrecorded earthquakes

T. Sugiyama
Asian Institute of Technology, Bangkok, Thailand

ABSTRACT: In order to estimate the design value of seismic response acceleration spectrum for highway bridge considering the possibility of existence of unrecorded earthquakes, what scope of earthquake records should be used in statistical analysis has been discussed in this paper. The result shows that the use of all available data in statistical analysis is not necessarily reasonable and one subjective method to determine the scope of data has been proposed.

1. INTRODUCTION

Once it happens, large earthquake often inflicts big damages on structures. And in the region where earthquake occurs frequently -- e.g. Japan, United States, and so on --, the design of structure is often controlled by earthquake effects. Hence it is necessary to grasp the characteristics of earthquake occurrence probabilistically and incorporate these features into structural design codes. Although sufficient data observed for many years, at least 100 years, are required for the purpose described above, however, the number of available data about earthquakes is not necessarily enough at present. Therefore, it may be natural that the statistical analysis should be done by using the earthquake data which includes the latest record of earthquake. Fortunately information network system has been well developed and the utilization of data-base is popularized in these days. Then it has become easy to analyze the updated earthquake records statistically.

Under these situations, the initial purpose of a series of the author's studies on this topic was to develop the computer program through which design value of seismic response spectra for highway bridge could be obtained taking account of the following three factors; that is, regional differences with respect to the occurrence of earthquake, the kind of foundation of construction site, and dynamic characteristics of highway bridge structure. The target region of this study was Japan. After construction of data-file about earthquakes and development of this computer programs, we have been able to obtain easily the design value of seismic response spectra for highway bridge only by inputting longitude and latitude of construction site, damping coefficient and natural period of bridge structure, and kind of foundation of construction site (Sugiyama, et al 1991)

However, it was revealed that the obtained value of seismic response spectrum depended on how many earthquake records were used in statistical analysis. For example, in a certain region, the value of seismic response spectrum obtained from all available earthquake records is smaller than that from recent several decades's records. This serious problem must be due to the followings; precise records are left if earthquakes occurred after the middle of twentieth century; before this age, there may be the possibility of existence of unrecorded earthquakes, however. Taking these results into account, it is desirable to establish some kinds of criteria which prescribe the scope of data to be used in the determination of design value of seismic effect. Few works discuss about this problem. However, the earthquake occurrence rate and the magnitude of seismic ground motion are separately taken into account in the work (Goto and Kameda 1968). That is, the problem about the scope of data used in statistical analysis, in which both earthquake occurrence and magnitude are considered together, has not been studied and no one has put forward any idea regarding this subject yet.

The purpose of this paper is to discuss how the possibility of unrecorded earthquakes should be considered statistically when we determine the design value of earthquake effects. In other words, what scope of earthquake records should be used in the determination of design value of seismic effect is discussed. In this study, a few typical regions are picked up to discuss the above mentioned problem first, and next, an attempt to establish the general criteria is made. The records of earthquakes which occurred in and around Japan from A.D. 684 to A.D. 1990 are used and the seismic response spectra for highway bridge is chosen as the subject of study.

2. DATA-FILE AND CALCULATION OF SEISMIC RESPONSE SPECTRA

The earthquakes which occurred from A.D. 684 to A.D. 1952 are taken from the book (Usami 1987) and those which occurred between A.D. 1953 and A.D. 1990 are extracted from the reports (Japan Meteological Agency 1953-1990). The total number of earthquake records input into the data-file is 2753. Each earthquake record is composed of occurrence date, serial number, name, the longitude and latitude of epicenter, and magnitude of earthquake.

By using this data-file about earthquake records, the seismic response spectrum for highway bridge caused by each earthquake is obtained through the following steps (Sugiyama 1991).

(1) to calculate the distance from the epicenter to the construction site considering that the globe is sphere

(2) to obtain the response acceleration spectrum by the following equation (Public Works Research Institute 1983 and Kawashima et. al 1985).

$$S_A(\mathrm{T,h}) = \left(\frac{1.5}{40h+1} + 0.5\right) \cdot S_A(T,0.05)$$

where

$$S_A(T,0.05) = a(T,\mathrm{GCi}) \times 10^{b(T,\mathrm{GCi})M} \times (\Delta+30)^{-1.178}$$

T : natural period of bridge

h : damping constant of bridge

Δ : distance between epicenter to construction site

M_s : magnitude of earthquake

a and b : constants which depend on T and kind of foundation

(3) to pick up the earthquakes whose values of response acceleration spectra are equal to or greater than 100[gal] under the conditions that T = 0.1, h = 0.05 and kind of foundation is Type-III.
[because smaller earthquakes may not cause damage to bridges]

(4) to judge whether occurrence of earthquake may follow the Poisson process or not by using the following index

$\varepsilon \leq 0.5 \rightarrow$ Poisson Process

$\varepsilon > 0.5 \rightarrow$ non Poisson Process

where y_i : Gumbel's probability

$$\varepsilon = \lambda \times \sqrt{\frac{\sum^{N} \{t_i - \log_{10}(1-y_i)/\alpha\}^2}{N-2}} \qquad (Eq.1)$$

$$\alpha = \sum^{N} t_i \times \log_{10}(1-y_i) / \sum^{N} t_i^2$$

$$\lambda = -\alpha/\log_{10} e, \quad e = 2.71828$$

(5) to calculate the 10% exceedance probability value for life-time (= 50 years) maximum value as follows

[in case of Poisson Process]

$$F_z(x) = \sum_{k}^{\infty} \{F_s(x)\}^k \cdot \frac{(\nu T_s)^k}{k!} \exp(-\nu T_s) \qquad (Eq.2)$$

where $F_s(x)$: CDF of response acceleration spectrum (assumed to be Extreme type-I largest value in this study)

ν: mean occurrence rate of earthquake

[in case of not Poisson Process (Sugiyama et al 1978)]

$$F_z(x) = 1 - \{1 - F_s(x)\} \sum^{\infty} F_s(x)^{k-1} \times F_{\tau k}(T_c + T_s) \qquad (Eq.3)$$

where $F_{\tau k}(t) = \int_0^t F_{\tau k-1}(u) \cdot f_\tau(t-u)du$

$$F_{\tau 1} = \{F_\tau(t) - F_\tau(T_c)\}/\{1 - F_\tau(T_c)\}$$

$$f_\tau(t) = dF_\tau(t)/dt$$

where $F_T(t)$: CDF of time interval between two continuous earthquakes (assumed to be Weibull distribution, i.e.

$$F_T(t) = 1 - \exp\left(-\frac{Kt^{m+1}}{m+1}\right)$$

T_C: time period between the latest earthquake occurrence and construction time of bridge

3. RESULTS OF STATISTICAL ANALYSIS AND DISCUSSION

Fig. 1(a) and (b) show the earthquake records diagram at Tokyo and Kyoto. In Fig. 1, the ordinate and the abscissa represent response acceleration spectra (unit: gal) and time (unit: year of A.D.), respectively. From these figures, the following facts are recognized.

i) With respect to Tokyo, most of earthquake data were recorded after A.D. 1600.

ii) On the other hand, relatively large earthquakes occurred at approximately uniform rate between A.D. 684 and A.D. 1990 although relatively small earthquakes were recorded frequently after A.D. 1600 than before that age.

The values of seismic response acceleration spectra obtained from all of available earthquake records at Tokyo and Kyoto are shown in Fig. 2(a) and (b), respectively. In order to investigate how the calculated seismic response acceleration spectra are influenced by the scope of earthquake data used in statistical analysis, the values of response acceleration spectra calculated from the earthquake records left after A.D. 1600, 1800, 1900, and 1953 are also drawn in Fig. 2(a) and (b). The reason why the case of the earthquake records left after A.D. 1953 is taken up is as follows; after A.D. 1953, the earthquake records were observed precisely because of the improvement of seismograph and complete equipment of earthquake observation network.

From Fig.2, the following facts are recognized.

i) From the point of view of safety of designed structure, to use the earthquake records left after A.D. 1600 may be most desirable at Tokyo.

ii) On the contrary, to use all of available earthquake records is most desirable at Kyoto.

The above fact i) indicates that to use all of available data is not necessarily appropriate and that the existence of unrecorded earthquakes

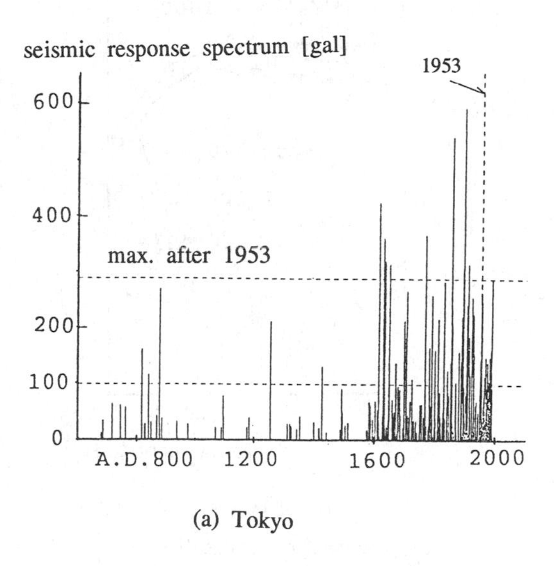

(a) Tokyo

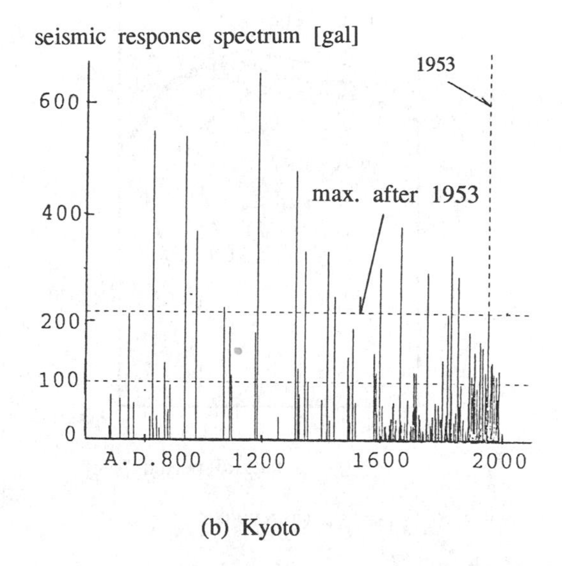

(b) Kyoto

Fig. 1 Earthquake occurance diagram

should be taken into account in the determination of design value of earthquake effect for highway bridge.

Fig. 3(a) shows the earthquake records diagram at Urakawa which is located in Hokkaido. As most of large earthquakes were recorded after A.D. 1953 at Urakawa, it may be expected that the use of earthquake records left after A.D. 1953 must be better. Fig. 3(b), which presents the calculated seismic response acceleration spectra, proves that this expectation is correct.

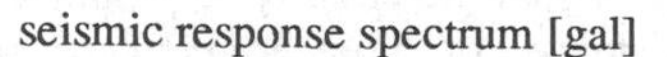

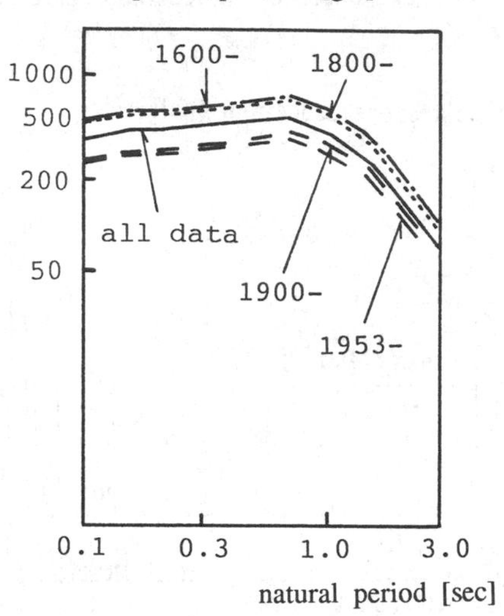

(a) Tokyo

seismic response spectrum [gal]

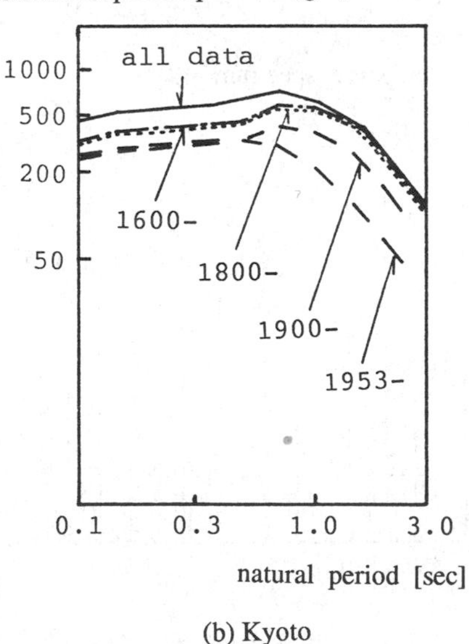

(b) Kyoto

Fig. 2 Seismic response acceleration spectrum vs. natural period

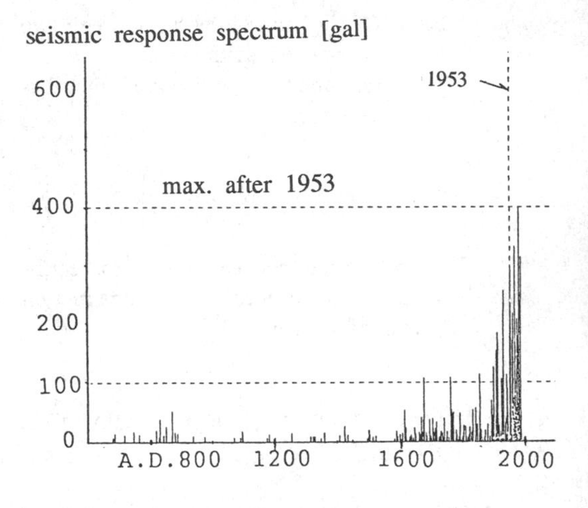

(a) Earthquake occurance diagram

seismic response spectrum [gal]

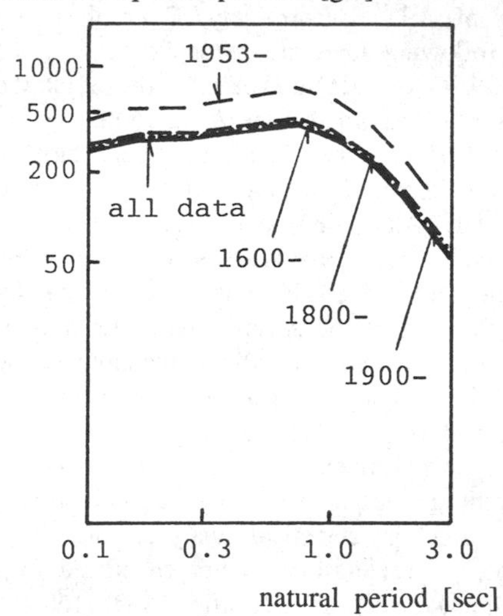

(b) Seismic response acceleration spectrum vs. natural period

Fig. 3 Earthquake occurance diagram and seismic response acceleration spectrum vs. natural frequency at Urakawa

These results indicate that it is desirable to establish some kinds of criteria which prescribe the scope of data to be used in the determination of design value of seismic effect.

As described before, it is said that the earthquake records were observed precisely after A.D. 1953. Therefore, it can be concluded that the use of only these records in the determination of design value is desirable if its value is estimated statistically with enough accuracy. However, when we determine the design value, not only the probabilistic characteristics of earthquake occurrence but also those of magnitude of seismic response spectra should be considered, as recognized easily from Eq. (2) or Eq. (3). Here, a few typical regions are picked up to discuss the above mentioned problem first, and next, an attempt to establish the general criteria is made. The discussions are made by paying attention to the following two facts,that is, mean occurrence rate of earthquake and conditional standard deviation defined by Eq. (1).

In Table 1, the relations between the scope of data used in statistical analysis and mean occurrence rate at Urakawa, Tokyo, and Kyoto are listed. From this table, the mean occurrence rate decreases as the increase of the scope of used data. This means that there may be the possibility of existence of unrecorded earthquakes as we go back to the older ages.

Table 2 shows the relations between the scope of data used in statistical analysis and the values of conditional standard deviation at Urakawa, Tokyo, and Kyoto. It is assumed in this paper that the earthquake occurrence is modeled by Poisson Process if the conditional standard deviation takes the value of less than 0.5 (Sugiyama et al 1991). The numbers of data used in statistical analysis for each scope are listed in Table 3.

In the case of Urakawa, the number of earthquake records left after A.D. 1953 seems to be enough (Sugiyama et al 1985) and the earthquake whose response acceleration spectrum takes the maximum value among all data was recorded after A.D. 1953. And it is recognized from Table 2 that the earthquake occurrence is well modeled by Poisson Process even if the records left after A.D. 1953 are used. Accordingly, in the region such as Urakawa, the use of only earthquake records left after A.D. 1953 is reasonable.

In the case of Tokyo, before A.D. 1953 there were many earthquake records whose response acceleration spectra are greater than the maximum value among records left after A.D. 1953. This indicates that the records left before A.D. 1953 should be used in statistical analysis. However, if these data are used, the mean occurrence rate obtained from these data becomes smaller, as shown in Table 1. Let's take notice of the conditional standard deviation here. It is recognized from Table 2 that the occurrence of earthquake does follow the Poisson Process in case of the use of records left after A.D. 1600, although not in case of the ones left after A.D. 1953. If all of the available records are used, the earthquake occurrence is not modeled by

Table 1 Mean occurance rate of earthquake

	all available data	after A.D. 1600	after A.D. 1800	after A.D. 1900	after A.D. 1953
Urakawa	0.0604	0.0604	0.1741	0.5114	1.8058
Tokyo	0.0183	0.1553	0.2628	0.3598	0.6249
Kyoto	0.0323	0.0648	0.1008	0.1570	0.1488

Table 2 Conditional standard deviation defined by Eq. (1)

	all available data	after A.D. 1600	after A.D. 1800	after A.D. 1900	after A.D. 1953
Urakawa	1.0745	1.0745	1.0356	0.7403	0.3436
Tokyo	1.0363	0.4922	0.4805	0.7129	0.5609
Kyoto	0.1944	0.2587	0.2188	0.2409	0.3216

Table 3 Number of data used in statistical analysis

	all available data	after A.D. 1600	after A.D. 1800	after A.D. 1900	after A.D. 1953
Urakawa	100	100	98	97	85
Tokyo	96	91	76	59	38
Kyoto	47	26	22	14	6

Poisson Process. Judging from these facts subjectively, it may be desirable to use the records left after A.D. 1600 at Tokyo.

In the case of Kyoto, earthquake records have been left on an average on time-axis. Furthermore, the occurrence of earthquake can be modeled by Poisson Process even if all of available records are used, although the value of mean occurrence rate decreases as the increase of the scope of used data. Taking account of the fact that the total number of available records is not so sufficient, too, it is better to use all of available records in Kyoto.

In the same manner as described above, the discussions about the scope of data used in statistical analysis were conducted for many regions in Japan. Based on the results obtained from these discussions, almost all regions are classified into three groups with respect to the scope of data used in statistical analysis. The first is the group such as Urakawa. The second is the one such as Tokyo. And the third is the one like Kyoto.

In every cases, it is important to satisfy the conditions that the conditional standard deviation takes the value of less than 0.5 approximately and the value of mean occurrence rate must be stable even if the oldest record included in the specified scope of data is not used.

4. CONCLUDING REMARKS

In order to estimate the design value of seismic response acceleration spectrum for highway bridge considering the possibility of existence of unrecorded earthquakes, what scope of earthquake records should be used in statistical analysis has been discussed in this paper.

The results show that the use of all available data in statistical analysis is not necessarily reasonable and one of the following ways have to be selected.

1) If enough number of earthquake records have been left after A.D. 1953 and the maximum value of response spectrum among these data is greater than that among the data recorded before A.D. 1953, the use of only the data left after A.D. 1953 is desirable.

2) If the earthquake records have been left on an average on time axis and the occurance of earthquake can be modeled by Poisson Process, it is better to use all of available records.

3) If before A.D. 1953 there exist many earthquake records whose response acceleration spectra are greater than the maximum value among records left after A.D. 1953, the records left before A.D. 1953 should be used in statistical analysis. And its scope of data is determined so that the occurrence of earthquake can be modeled by the Poisson Process by using such data. In this case, the value of mean occurrence rate must be stable even if the oldest record included in the specified scope of data is not used. In the regions classified into this group, it may be desirable to use the earthquake records left after A.D. 1600.

Though the target region of this study is Japan, the results obtained here may be applicable to other regions.

It is now being considered to develop a criterion to determine the scope of earthquake data not subjectively but quantitatively. The author intend to present its results in the near future.

REFERENCES:

Goto, H. and Kameda, H. November, 1968. Probabilistic Consideration on Maximum Seismic Acceleration. *Proc. of JSCE* Vol.159:pp.1-12 (in Japanese).

Japan Meteological Agency January, 1953 - December, 1990. *Weather Monthly Report* (in Japanese).

Kawashima, K., Aizawa, K. and Takahashi, K. September, 1985. Attenuation Law of Maximum Seismic Ground Motion and Seismic Response Spectrum, *Report of PWRI*, Vol. 166 (in Japanese).

Public Works Research Institute July, 1983. Methods to Estimate the Maximum Seismic Ground Motion and Seismic Response Spectrum. *Material of PWRI*, No. 2001 (in Japanese).

Sugiyama, T. et al. September, 1978. Consideration on Probabilistic Characteristics of Incidental Occurrence Loads. *the 33rd annual convention of JSCE*, I-128:248-249 (in Japanese).

Sugiyama, T., Fujino, Y. and Ito, M. March, 1985. Determination of Distribution Model and Characteristic Value from Statistical Data. *Jour. of Structural Engineering*. Vol. 31A:287-300 (in Japanese).

Sugiyama, T., Togashi, N. and Orii, K. March, 1991. Determination of Design Value of Seismic Response Spectra Considering Probabilistic Characteristics of Earthquake Occurrence. *Jour. of Structural Engineering* 37A: 607-616 (in Japanese).

Usami, T. March, 1987. *New Version of Summary of Earthquake Damage in Japan*. University of Tokyo Press (in Japanese).

Structural Safety & Reliability, Schuëller, Shinozuka & Yao (eds) © 1994 Balkema, Rotterdam, ISBN 90 5410 357 4

Reliability of elasto-plastic frame under seismic excitation

Y.P.Teo, S.T.Quek & T.Balendra
National University of Singapore, Singapore

ABSTRACT: A method is developed to evaluate the reliability of nonlinear plane frames with elasto-plastic nodes subjected to seismic excitation. The elastic capacity is controlled by the plastic potential function of the axial force and moment and the plastic flow rule is enforced. The ground excitation is modelled as a zero-mean uniformly-modulated Gaussian process defined by a seismological model. The governing equations are written as a set of nonlinear first-order differential equations. By taking small discrete time steps, the nonlinear terms are replaced by their expected value at each time step, ensuring stability, simplicity and economy in the solution process. The linearized equations are solved using complex modal decomposition technique giving the mean value and autocorrelation functions at each time-step. The reliability of the frame is estimated using the concept of crossing rate with the failure events assumed to constitute a Poisson process. A numerical example is solved and compared with simulation results showing reasonably good agreement.

1 INTRODUCTION

For realistic reliability analysis of frames under seismic excitation, the nonlinearity of the frame, the nonstationarity of the excitation and the uncertainty in the model parameters adopted in modelling the problem must be considered. The reliability of linear SDOF oscillator under nonstationarity seismic excitation incorporating the effects of uncertainty in the model parameters have previously been studied (Quek et al 1991; Wall & Bucher 1987). Under seismic excitation of high magnitude, it is likely that frame structures will undergo nonlinear behaviour. The method of equivalent linearization is often employed in solving such nonlinear problems (Pradlwarter & Schueller 1987; Roberts & Spanos 1990; Simulescu et al 1989). Poisson assumption is often invoked when performing the reliability computation and the crossing rate computed using Rice's formula. The response process and its derivative process are assumed to be uncorrelated which is untrue for nonstationary Gaussian process. In addition, when the interaction of the axial force and moment are considered in the elasto-plastic analysis, the enforcement of the associated flow rule complicates the problem and is sometimes ignored (Nielsen et al 1990). The purpose of this study is to present a practical solution for nonlinear structures with multiple degrees of freedom under nonstationary seismic excitation accounting for the correlation between the response and its derivative process as well as enforcing the associated flow rule in the constitutive relationship. The uncertainty in the model parameters are considered.

2. STRUCTURAL MODEL

Plane frames are modelled as assembly of discrete beam elements with a node at each end of each element. Each node has two translational and one rotational degrees of freedom. The nodes are assumed to be elasto-plastic following the associated plastic flow rule. The ith node of an element becomes plastic when the plastic potential function $\Phi_i(P_e) = 0$ and is moving outward, i.e. $\dot{\Phi}_i(P_e,\dot{P}_e) > 0$, where P_e is the vector of element forces at node i and the superscript . refers to the differential of the variable with respect to time. A typical axial force-moment interaction surface (Φ =0) used in this study for a tubular section is shown in Fig. 1.

The equilibrium condition for each element can be specified in terms of the vector of rate of element nodal force $\{\dot{P}_e\}$, vector of rate of element total nodal displacements $\{\dot{u}_e\}$ and plastic component of displacement rate $\{\dot{u}_e^p\}$

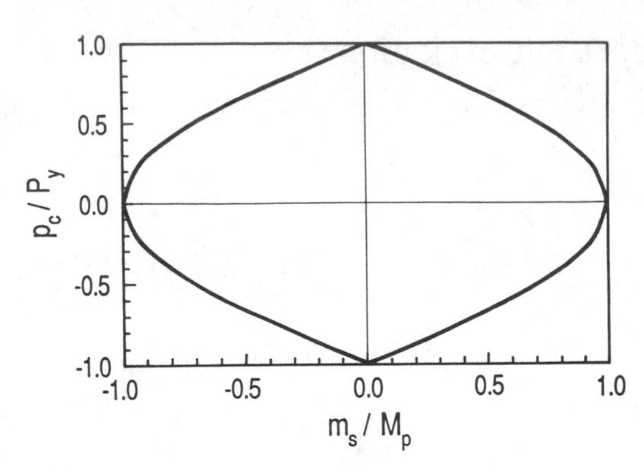

Fig. 1 Plastic potential function for tubular section

through the element elastic tangent stiffness matrix $[K_t^e]$

$$\{\dot{P}_e\} = [K_t^e](\{\dot{u}_e\}-\{\dot{u}_e^p\}) \tag{1}$$

where $\{\dot{u}_e^p\}=[\nabla\Phi]([\nabla\Phi]^T[K_t^e][\nabla\Phi])^{-1}[\nabla\Phi]^T[K_t^e]\{\dot{u}_e\}$, $[\nabla\Phi]$ is the vector of partial derivatives of Φ wrt $\{P_e\}$ and the superscript T denotes transpose. The equation of motion for frame structure with m elements is governed by

$$\begin{bmatrix} [M_{ff}][M_{fs}] \\ [M_{sf}][M_{ss}] \end{bmatrix} \begin{Bmatrix} \{\ddot{u}_f\} \\ \{\ddot{u}_s\} \end{Bmatrix} + \begin{bmatrix} [C_{ff}][C_{fs}] \\ [C_{sf}][C_{ss}] \end{bmatrix} \begin{Bmatrix} \{\dot{u}_f\} \\ \{\dot{u}_s\} \end{Bmatrix} +$$

$$\begin{bmatrix} [T_{gf}]^T \\ [T_{gs}]^T \end{bmatrix} \begin{Bmatrix} \{P_t^e\}_1 \\ \vdots \\ \{P_t^e\}_m \end{Bmatrix} = \begin{Bmatrix} \{Q_f\} \\ \{Q_s\} \end{Bmatrix} - \begin{bmatrix} [M_{ff}][M_{fs}] \\ [M_{sf}][M_{ss}] \end{bmatrix}$$

$$\begin{Bmatrix} \{I_{nf}\} \\ \{I_{ns}\} \end{Bmatrix} a_g(t) \tag{2}$$

where M denotes mass matrix, C damping matrix, T_g transformation matrix from element to global coordinates, Q static load vector, I_n the influence vector for the displacements from a unit ground motion, $a_g(t)$ the ground acceleration. The subscript s represents the known degrees of freedom (dof) and f the unrestrained dof. For the case where the known dof have zero values, the complete governing equation is (Teo 1992)

$$\frac{d}{dt}\{\dot{u}_f\} = -[M_{ff}]^{-1}[C_{ff}]\{\dot{u}_f\} - [M_{ff}]^{-1}[T_{gf}]^T\{\hat{P}_e\}$$

$$+ [M_{ff}]^{-1}\{Q_f\} - [M_{ff}]^{-1}\Big[[M_{ff}][M_{fs}]\Big]$$

$$\begin{Bmatrix} \{I_{nf}\} \\ \{I_{ns}\} \end{Bmatrix} a_g(t)$$

and
$$\frac{d}{dt}\{\hat{P}_e\} = [\hat{K}_t^e]\ \Big([T_{gf}]\{\dot{u}\}-\{\hat{\dot{u}}_e^p\}\Big) \tag{3}$$

where
$$[\hat{P}_e] = \begin{Bmatrix} \{P_e\}_1 \\ \vdots \\ \{P_e\}_m \end{Bmatrix},\quad [\hat{K}_t^e] = \begin{bmatrix} [K_t^e]_1 & \cdots & [0] \\ \vdots & \ddots & \vdots \\ [0] & \cdots & [K_t^e]_m \end{bmatrix}$$

and
$$[\hat{\dot{u}}_e^p] = \begin{Bmatrix} \{\dot{u}_e^p\}_1 \\ \vdots \\ \{\dot{u}_e^p\}_m \end{Bmatrix}$$

Equation 3 is a set of first-order differential equations where the nonlinearity comes from $\{\dot{u}_e^p\}$ which is a function of $\{P_e\}$, $\{\dot{P}_e\}$ and $\{\dot{u}\}$.

3. EXCITATION MODEL

The ground accelerations $a_g(t)$ is modelled as a Gaussian non-stationary uniformly-modulated stochastic process with zero mean

$$a_g(t) = \mathcal{I}(t)\tilde{a}_g(t) \tag{4}$$

$\mathcal{I}(t)$ is the deterministic time-modulating function given by $\mathcal{I}_o(e^{-bt}-e^{-\chi bt})$ and $\tilde{a}_g(t)$ is a zero-mean stationary random process having a two-sided power spectral density $S_{\tilde{a}\tilde{a}}(\omega) = \bar{A}^2(\omega)/(2\pi T_o)$. $\bar{A}(\omega)$ is the Fourier amplitude of the ground acceleration and T_o is the strong motion duration. In this study, the seismological model proposed by Boore(1983) is adopted to obtain $A(\omega)$ and the details are summarized in Teo (1992).

4. SOLUTION TECHNIQUE

The equivalent linearization technique is often used to solve nonlinear stochastic differential equation such as that of eq. 3 (Pradlwarter & Schueller 1987; Roberts & Spanos 1990; Simulescu et al 1989). For problems with slight nonlinearity, this method is superior; for other cases, the linearized coefficients can take on values that may result in numerical problems during the eigen-solution stage. This problem can be overcome in some cases by using very small time step which may incur large computing time. A practical compromise is to use small time-steps for such nonlinear problems and to replace the nonlinear terms by constants at each time step. By minimising the error between the nonlinear and approximate equations, the constants are merely the expected value of the nonlinear term evaluated at each time step. Solution for the expectation may be derived in closed form for some distribution functions (pdfs) whereas for complex cases, numerical integration using simulation techniques is a practical, efficient and simple alternative, especially when truncated joint pdfs are involved. For the ith time step, we have

$$\{\dot{u}_e^p(\mathbf{P}_e,\dot{\mathbf{P}}_e,\dot{u})\}_i = \{E[\dot{u}_e^p(\mathbf{P}_e,\dot{\mathbf{P}}_e,\dot{u})]\}_i \tag{5}$$

At each time step, the joint moments of $\mathbf{P}_e$, $\dot{\mathbf{P}}_e$ and $\dot{u}$ are approximated by values from the previous time step. Since the solutions are obtained using the linearized equations subject to Gaussian random excitations, the response variables are assumed to be Gaussian. The bending moment and axial forces are however confined by the yield surface described by $\Phi = 0$. Hence the pdfs for these two variables are taken as truncated joint Gaussian taking discrete delta functions at the yield surface. Resulting from the linearization process, a set of linear first-order differential equations are obtained in terms of the state variables $\{y(t)\}$ as

$$\frac{d}{dt}\{y(t)\} = [A]\{y(t)\} + \{F\} + \{G\}\mathcal{G}(t)\tilde{a}_g(t) \tag{6}$$

where $\{y\} = \langle\{\dot{u}\}\ \{u\}\ \{P_e\}_1 \cdots \{P_e\}_m\rangle^T$

$$[A] = \begin{bmatrix} -[M_{ff}]^{-1}[C] & [0] & -[M_{ff}]^{-1}[T_{gf}]_1^T & \cdots & -[M_{ff}]^{-1}[T_{gf}]_m^T \\ [I] & [0] & [0] & \cdots & [0] \\ [K_t^e]_1[T_{gf}]_1 & [0] & [0] & \cdots & [0] \\ \vdots & \vdots & \vdots & \ddots & \vdots \\ [K_t^e]_m[T_{gf}]_m & [0] & [0] & \cdots & [0] \end{bmatrix}, \quad \{F\} = \begin{Bmatrix} [M_{ff}]^{-1}\{Q_f\} \\ \{0\} \\ -[K_t^e]_1 E\{\dot{u}_e^p\}_1 \\ \vdots \\ -[K_t^e]_m E\{\dot{u}_e^p\}_m \end{Bmatrix}$$

$$\text{and } \{G\} = \begin{Bmatrix} -[M_{ff}]^{-1}\left[[M_{ff}][M_{fs}]\right]\begin{Bmatrix}\{I_{nf}\} \\ \{I_{ns}\}\end{Bmatrix} \\ \{0\} \\ \{0\} \\ \vdots \\ \{0\} \end{Bmatrix}$$

Equation 6 is decoupled using the complex eigenvalues and eigenvectors of the non-symmetric matrix [A]. The state variables within the kth and (k+1)th time-set can be decomposed as

$$\{y(t_k+\tau)\} = \sum_{i=1}^{\ell} \{\phi\}_i z_i^k(\tau) \tag{7}$$

where $0 \le \tau \le (t_{k+1}-t_k)$, $\{\phi\}_i$ is the ith right hand column eigenvector, $z_i^k(\tau)$ are the generalized variables and ℓ the number of dominant eigenvectors. Substitution eq. 7 into 6 and solving (for $i = 1,2,\ldots,\ell$) gives

$$z_i^k(\tau) = z_i^k(0)e^{\lambda_i\tau} + \frac{f_i^k}{\lambda_i}\left[e^{\lambda_i\tau}-1\right] + g_i\int_0^{\tau} \mathcal{G}(t_k+s)\tilde{a}_g(t_k+s)e^{\lambda_i(\tau-s)}\,ds \tag{8}$$

where $f_i^k = \langle\psi\rangle_i\{F^k\}$, $g_i = \langle\psi\rangle_i\{G\}$ and $\langle\psi\rangle_i$ is the ith left hand row eigenvector. Due to the nature of the approximation, since $\tilde{a}_g(t)$ is Gaussian, $z^k(\tau)$ and hence $y(t)$ are Gaussian processes.

5. STATISTICS OF RESPONSE

From eqs. 7-8, the mean response at the end of each time step is

$$E\{y(t_k+\Delta t)\} = \sum_{i=1}^{\ell} \{\phi\}_i E[z_i^k(0)]e^{\lambda_i\Delta t} +$$

$$\{\phi\}_i \frac{f_i^k}{\lambda_i}\left[e^{\lambda_i \Delta t} - 1\right] \tag{9}$$

where $E[z_i^k(0)] = E[z_i^{k-1}(t_k - t_{k-1})]$. The mean values at t=0 are taken as the static response of the system prior to being dynamically excited. The matrix of covariance functions at the end of the kth time step given by

$$[K_{yy}(t_k+\Delta t)] = \sum_{i=1}^{\ell} \{\phi\}_i \sum_{j=1}^{\ell} \{\phi\}_j^{T*} K_{z_i^k z_j^{k*}}(\Delta t) \tag{10}$$

where $K_{z_i^k z_j^{k*}}(\Delta t) = e^{(\lambda_i + \lambda_j^*)\Delta t} K_{z_i^k z_j^{k*}}(0) +$

$$e^{\lambda_i \Delta t} g_i g_j^* \int_o^{\Delta t}\int_o^{\Delta t} e^{\lambda_i(\Delta t - s)} e^{\lambda_j^*(\Delta t - v)}$$

$$\mathcal{I}(t_{k-1}+s)\mathcal{I}(t_k+v)R_{\tilde{a}_g \tilde{a}_g}(\Delta t+v-s)dvds +$$

$$e^{\lambda_j^* \Delta t} g_i g_j^* \int_o^{\Delta t}\int_o^{\Delta t} e^{\lambda_i(\Delta t - s)} e^{\lambda_j^*(\Delta t - v)}$$

$$\mathcal{I}(t_{k-1}+v)\mathcal{I}(t_k+s)R_{\tilde{a}_g \tilde{a}_g}(\Delta t+v-s)dvds +$$

$$g_i g_j^* \int_o^{\Delta t}\int_o^{\Delta t} e^{\lambda_i(\Delta t - s)} e^{\lambda_j^*(\Delta t - v)}$$

$$\mathcal{I}(t_k+s)\mathcal{I}(t_k+v)R_{\tilde{a}_g \tilde{a}_g}(v-s)dvds \tag{11}$$

Equation 11 is the result of the simplifying assumption that the ground excitation is wide band and its autocorrelation function decays drastically as the time difference increases such that only the correlation between the excitation and modal response at the previous time step contributes to the current time step computation. The original equation involves summation of double integrals which is computationally costly even when the assumption that assumption that the initial condition of the system is deterministic and uncorrelated with the excitation is made (Teo 1992)

6. RELIABILITY ESTIMATES

In addition to the mean and variance of the response, the performance of the system with respect to some prescribed requirements (e.g. serviceability or ultimate limit states) are of practical interest. Failure to perform is assumed to be the result of randomness in the excitation as well as uncertainty in the parameters of the idealized model used for analysis. For the latter, the two structural parameters considered are the two lowest modal damping ratios, ξ_1 and ξ_2, whereas the five uncertain parameters of the seismological model are γ_M which is a bias factor for the recorded earthquake magnitude, $\Delta\sigma$ the stress drop, κ_o the uncertainty in the attenuation of seismic wave with the source-to-site distance relationship, τ_o the uncertainty in the strong motion duration and ε the rise time fraction of the time-modulating function. The statistics of these parameters are obtained from earthquake data and summarized in Quek et al (1991). The reliability of the system is estimated through a two-step approach. The first step deals with the random ground excitation where the reliability conditioned on a prescribed set of model parameters is estimated based on the mean crossing rates accounting for the correlation between the response and its derivative process. The second step accounts for the random model parameters by unconditioning the conditioning reliability function using the AFOSM method (Teo 1992).

7. NUMERICAL EXAMPLE

A simple structure comprising two beam elements with a lump mass of 407700kg shown in Fig. 2 is subject to a horizontal static load of 1000 kN as well as earthquake excitation of local magnitude 7 and epicentral distance 10 km. The material properties for both elements are assumed identical and shown in Table 1. The statistical properties of the random variables in Table 2 are adopted in the computation.

By conditioning the model parameters on their statistical modal values, the conditional mean values and standard deviation of the response quantities are obtained. Comparison of results with simulation based on 2000 cycles show good agreement (Fig. 3) except for the standard deviations of the vertical displace ment and moment (Figs. 3b & d) where the nonlinear results cannot be captured satisfactorily. This is due mainly to the permanent deformation at some nodes of the structure not being fully captured as a result of approximating the nonlinear terms by their expected values at each time step. Using a performance function consisting of two parallel hyperplane

$$G(t_n) = \eta - \max_{i=1}^{n} |u_2(t_i)| \tag{12}$$

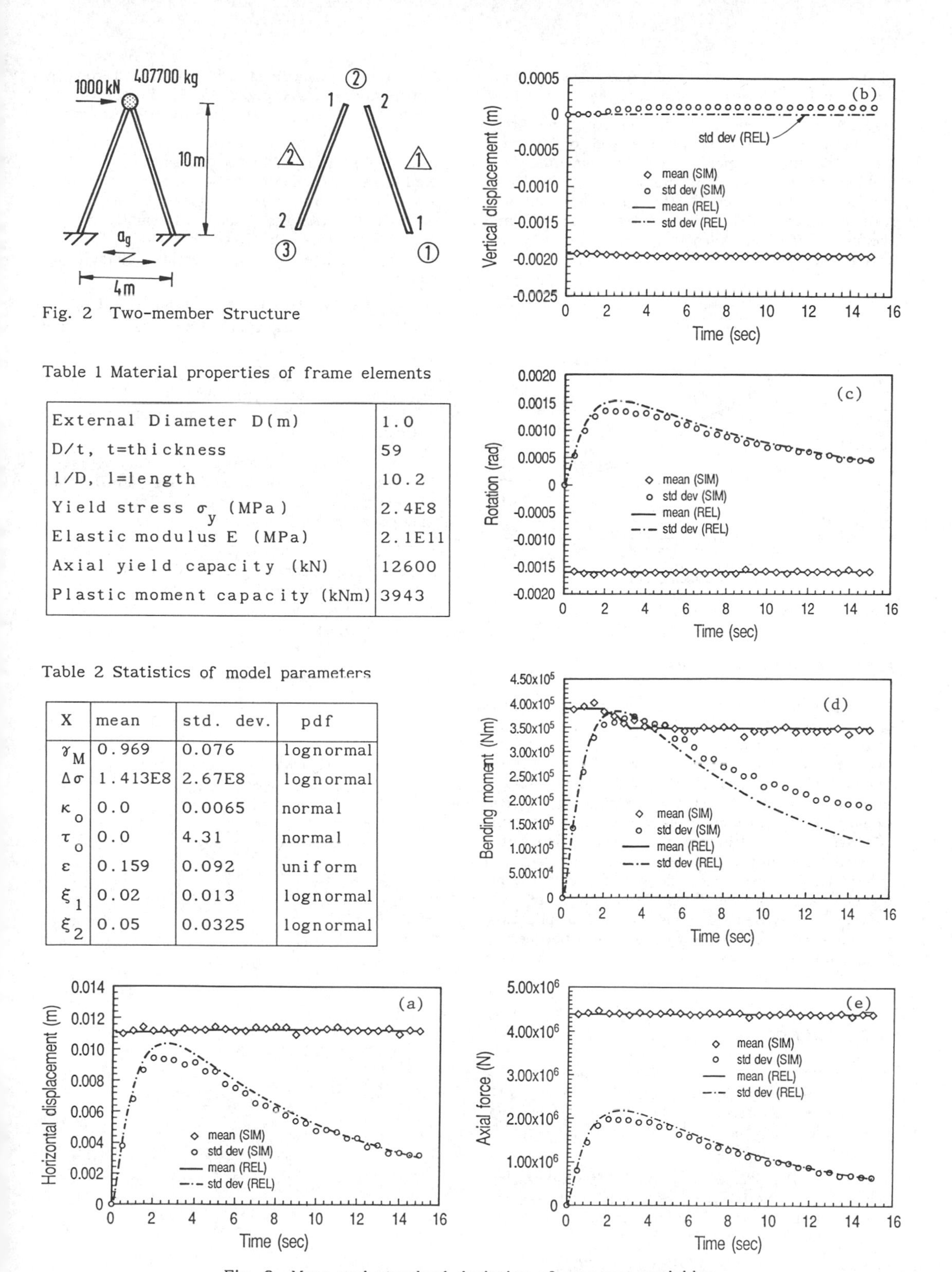

Fig. 2 Two-member Structure

Table 1 Material properties of frame elements

External Diameter D(m)	1.0
D/t, t=thickness	59
l/D, l=length	10.2
Yield stress σ_y (MPa)	2.4E8
Elastic modulus E (MPa)	2.1E11
Axial yield capacity (kN)	12600
Plastic moment capacity (kNm)	3943

Table 2 Statistics of model parameters

X	mean	std. dev.	pdf
γ_M	0.969	0.076	lognormal
$\Delta\sigma$	1.413E8	2.67E8	lognormal
κ_o	0.0	0.0065	normal
τ_o	0.0	4.31	normal
ε	0.159	0.092	uniform
ξ_1	0.02	0.013	lognormal
ξ_2	0.05	0.0325	lognormal

Fig. 3 Mean and standard deviation of response variables

where η is the prescribed displacement limit (or barrier level) and u_2 the horizontal displacement at node 2 at the ith time step, the failure probability at the end of 15 secs for different barrier levels is computed and plotted in Fig. 4. The disagreement for high barrier levels is due to the fact that for low failure probabilities, simulation results are accurate if sufficiently large number of cycles are used.

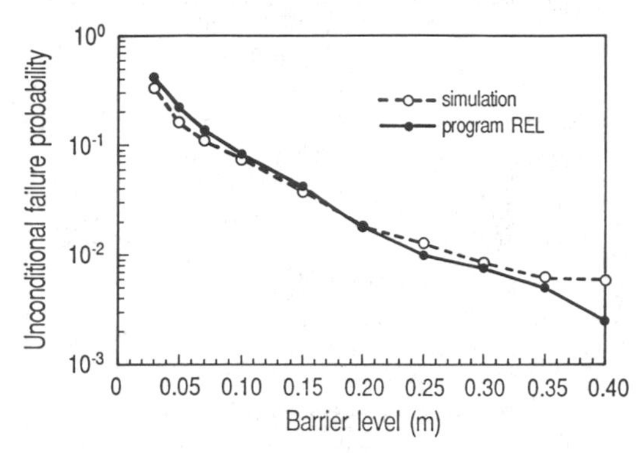

Fig. 4 Failure probability (at t = 15s)

8. CONCLUDING REMARKS

A simple solution for the reliability analysis of elasto-plastic plane frames under nonstationary seismic excitation is presented accounting for uncertainty in the model parameters. The method is economical, stable and simpler than methods such as the original statistical linearization for problems where time-stepping is involved. Although the results presented showed good agreement with simulation results, the method nevertheless does not recover completely all the nonlinear behaviour of the system. However, it does provide a practical approach to problems involving many degrees of freedom and where the associated flow rule need to be enforced.

9. REFERENCES

Boore, D. M. 1983. Stochastic simulation of high frequency ground motion based on seismological models of the radiated spectra. Bulletin of the Seismological Society of America. 73:1865-94.

Nielsen, S.R., Mork, K.J. & Thoft-Christensen, P. 1990. Reliability of hysteretic system subjected to white noise excitation. Structural Safety. 8:369-379.

Pradlwarter, H. J. & Schueller, G. I. 1987. Accuracy and limitations of the method of equivalent linearization for hysteretic multi-storey structures. In F. Ziegler & G. Schueller (eds.), IUTAM Symposium,Austria. 3-21.

Quek, S.T., Teo, Y.P. & Balendra, T. 1991. Reliability estimate of linear oscillator with uncertain input parameters. In I. Elishakoff & Y.K. Lin (eds.), Stochastic Structural Dynamics 2: New Practical Applications, Springer-Verlag Berlin, Heidelberg, 139-154.

Roberts, J.B. & Spanos, P.D. 1990. Random vibration and statistical linearization. John Wiley & Sons.

Simulescu, I., Mochio, T. & Shinozuka, M. 1989 Equivalent linearizaton method in nonlinear FEM. Journal of Engineering Mechanics. 115 (3):475-492.

Teo, Y.P. 1992. Reliability of elasto-plastic frames under seismic excitation. Ph.D. thesis, Dept. of Civil Engrg., National University of Singapore.

Wall, F.J. & Bucher, C.G. 1987. Sensitivity of expected exceedance rate of SDOF-system response to statistical uncertainties of loading and system parameters. Probabilistic Engineering Mechanics. 2(3):138-146.

Structural Safety & Reliability, Schuëller, Shinozuka & Yao (eds) © 1994 Balkema, Rotterdam, ISBN 90 5410 357 4

Reliability of current code procedures against seismic loads

Y.K.Wen, D.A.Foutch, D.Eliopoulos & C-Y.Yu
University of Illinois, Urbana, Ill., USA

ABSTRACT: The reliability of provisions for seismic loads in the recently procedures U.S. code procedures, i.e., UBC and SEAOC is evaluated. The emphasis is on accurate modeling of seismic source, ground motion, structural system, and inelastic response behavior. The study concentrates on low- to medium-rise steel buildings. Both time history and random vibration methods are used for the response analysis. Limit states considered include maximum story drift, damage due to low cycle fatigue to members and connections. The risks implied in the current procedure, for example those based on various R_w factors for different structural types, are calculated and their consistency examined.

1 INTRODUCTION

The commonly accepted philosophy in design of a building under seismic loads is to ensure that it will withstand a minor or moderate earthquake without structural damage and survive a severe one without collapse. In view of the large uncertainties normally associated with the seismic excitation and the structural resistance, however, this design philosophy can be put into practice only in terms of acceptable risks of limit states of buildings and structures. This has not yet been fully accounted for in current practice in building design, although the need for consideration of the uncertainty involved has long been recognized. The objective of this study, therefore, is to evaluate the reliability of buildings designed according to the recently proposed and adopted procedures; namely, the provisions recommended by the Structural Engineering Association of California (SEAOC) and Uniform Building Code (UBC). The results will provide basis for developing reliability-based design criteria. The progress is summarized in the following.

2 SITE RISK ANALYSIS AND MODELING OF GROUND MOTION

Two sites are considered in this study, both in Southern California.One of these is at Imperial Valley 5km from the Imperial Fault, and the other one is at Santa Monica Boulevard, 60km from the Mojave Segment of the Southern San Andreas fault. The potential future earthquakes that present a threat to the sites are characterized as either characteristic or non-characteristic. The former are major seismic events which occur along the major fault and with relatively better understood magnitude and recurrence time behavior (U.S.G.S. Working Group Report, 1988), therefore, treated as a renewal process. The latter are local events that their occurrences collectively can be treated as a Poisson process (Algermissen, et al., 1982; Cornell and Winterstein, 1988). The major source parameters of the characteristic earthquake are magnitude (M), epicenter distance to the site (R) and attenuation, whereas parameters for non-characteristic earthquakes are local (MMI) intensity, I, and duration. The duration t_D is defined as the significant duration associated with the strong phase of the ground motion (Trifunac and Brady, 1975). It depends on M and R for characteristic earthquakes and on I for non-characteristic earthquakes. These parameters define the ground motion which is modeled as a nonstationary random process as follows.

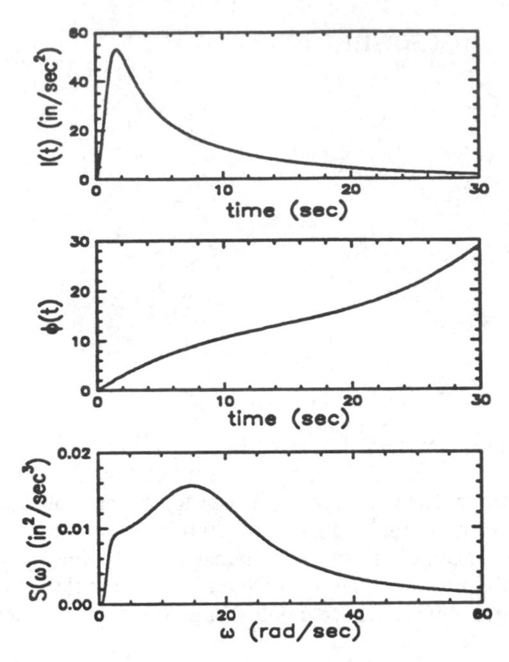

Figure 1 Identified functions of the ground motion model for the Los Angeles site due to a non-characteristic earthquake ($I=8.7$, $t_D = 10.66$ sec)

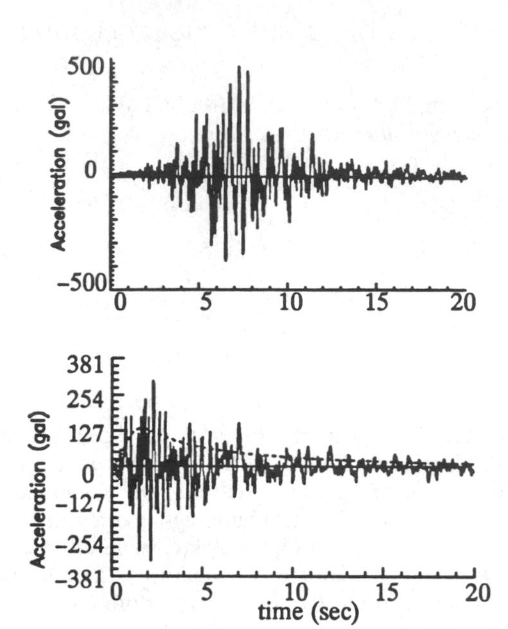

Figure 2 Simulated ground motion time histories: characteristic earthquake at the Imperial Valley site (top) and non-characteristic earthquake at the L. A. site (bottom).

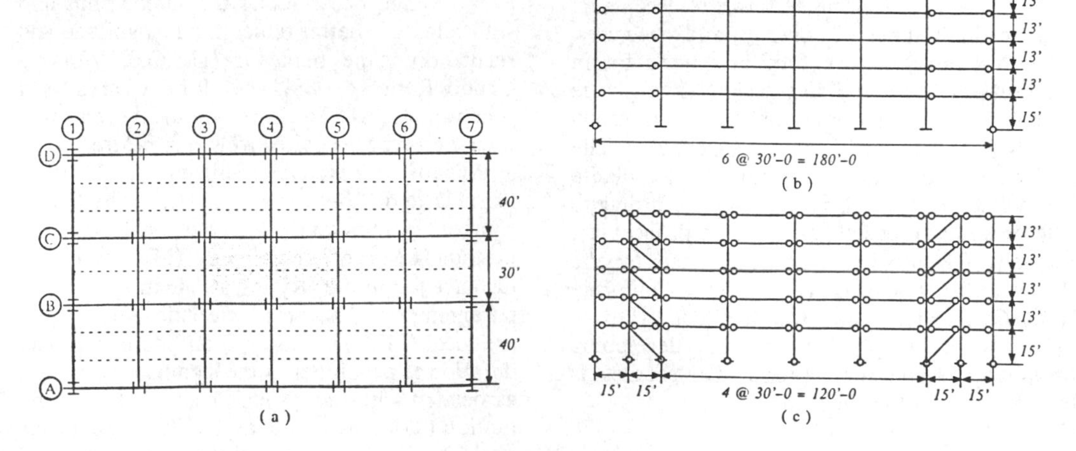

Figure 3 (a) Plan view of building. (b) Elevation of SMRSF and OMRSF. (c) Elevation of CBF

The model is based on that of Yeh and Wen (1990) whose intensity and frequency content vary with time.

$$a(t) = I(t)\zeta[\phi(t)] \quad [1]$$

in which $\zeta(\phi)$ is a zero mean, unit variance, stationary white noise filtered through a Clough-Penzien type linear filter. I(t) is the intensity envelope function and $\phi(t)$ is a frequency modulation function. While I(t) and $\phi(t)$ control the nonstationarity, the filter parameters determine the frequency content of the ground motion. It has been shown that the time dependent (instantaneous) power spectral density of $\zeta(t)$ at time t is

$$S_{aa}(t,\omega) = \frac{1}{\phi'(t)} S_{CP}\left[\frac{\omega}{\phi'(t)}\right] \quad [2]$$

in which S_{CP} = Clough and Penzien (CP) spectral density. This model allows straightforward identification of parameters from actual ground accelerograms, computer simulation of the ground motion for time history response analysis, and analytical solution of inelastic structure response by the method of random vibration (Wen, 1989).

For the Imperial Valley site, earthquake ground motion records of the characteristic earthquake are available. It is assumed that future earthquake ground motions can be inferred from the past records; therefore, model parameters are estimated directly from the records. For the Los Angeles site where no such records are available, a procedure (Eliopoulos and Wen, 1991) has been established to determine the model parameters as functions of those of the source, i.e., magnitude, epicentral distance, etc. based on empirical results given in Trifunac and Lee (1984). For characteristic earthquakes the empirical Fourier amplitude spectrum as a function of the source parameters is used in which a frequency (or period) dependent attenuation law and local geology (site) condition are considered. The Arias intensity can be evaluated and used as a scaling factor for the intensity and the frequency content (parameters in the C-P filter) can be determined from the Fourier amplitude spectrum. For non-characteristic earthquakes, a similar procedure is used to determine the ground motion parameters as a function of MMI and the site condition. Also, since the Imperial Valley site is very close to the fault, the important directivity effect of the rupture surface is considered in the ground motion model which is known to affect significantly the frequency content and duration of the ground motion.

Figure 1 shows identified intensity, frequency modulation, and spectral density functions of the ground motion model for the Los Angeles site due to characteristic earthquake. Note that I(t) depends on the random duration and attenuation. The energy and zero crossing functions calculated from the San Fernando earthquake of 1971 and Whittier Narrow earthquake of 1987 at the site are used to identify the shape of I(t) and $\phi(t)$ for the noncharacteristic earthquakes. Figure 2 shows sample ground motion time histories of characteristic earthquakes at the two sites and the noncharacteristic earthquake at the Los Angeles site generated by the model. Again, these are sample histories and there are large variations in intensity and duration from sample to sample.

Note that for the Imperial Valley site, the rupture propagation toward the site is assumed giving rise to long duration pulses which are known to be most damaging when the structure becomes inelastic.

3 BUILDING DESIGN AND RESPONSE AND DAMAGE ANALYSIS

Six low-rise steel building types are designed according UBC; namely, (i) ordinary moment-resisting space frame (OMRSF), (ii) special moment-resisting space frame (SMRSF), (iii) concentric braced frame (CBF), (iv) eccentric braced frame (EBF), (v) dual system with CBF, D/CBF, and (vi) dual system with EBF, D/EBF. The R_w value varies from 6 (OMRSF) to 12 (SMRSF). R_w is the response modification factor and is used to reduce the elastic design forces to account for inelastic behavior. One five story building using each of the above framing systems was designed for Zone 4 in accordance with the 1988 UBC. Structural engineers were consulted to ensure that the floor plan and the design loads and procedures would be consistent with those used in practice. A plan review of the building is shown in Figure 3(a). Lateral loads are carried by the perimeter frames. All beams-to-columns connections at interior joints are assumed to be pinned. An

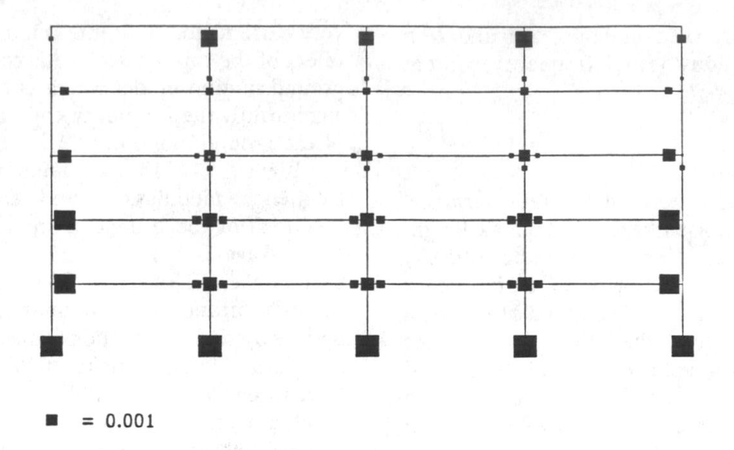

Figure 4 Damage index for the SMRSF frame at the Imperial Valley Site.

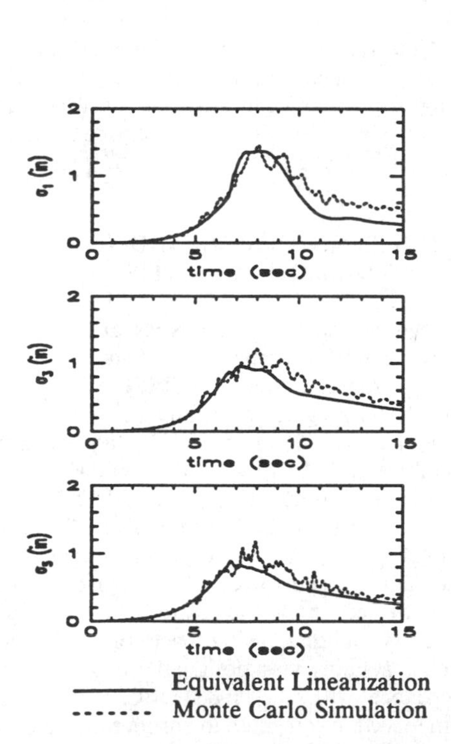

Figure 5 Comparison between nonstationary root mean square story drifts obtained by Monte Carlo simulations and by the statistical equivalent linearization method (Imperial Valley, 1979 earthquake)

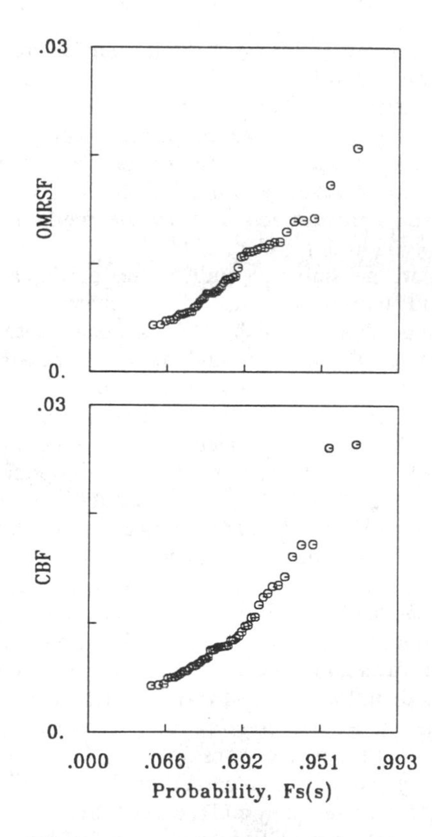

Figure 6 Maximum first-story drift for 50 Imperial Valley Accelerograms.

elevation view for the SMRSF and OMRSF frames as shown in Figure 3(b) and for the CBF frame in Figure 3(c).

Given the occurrence of an earthquake, the response of the building is calculated by both time history and random vibration methods. The foregoing ground motion model provides the ground excitation either in the form of time histories or nonstationary random processes in which the effect of the source parameters and their uncertainties have been properly accounted for. The responses of interests are: (1) story drifts, (2) damage to nonstructural elements, (3) energy dissipation demand, and (4) damage index. In the time history method, the well known finite element program DRAIN 2DX is modified and used and the statistics of the above responses are obtained for a large number of ground motions. An example of the damage index calculations is shown in Figure 4 for the SMRSF frame under a very severe sample ground excitation at the Imperial Valley site. A damage index of 1.0 at any location indicates failure of the connection by low-cycle fatigue. These results indicate that failures of well-constructed connections are not likely to be a problem during earthquakes.

In the random vibration analysis, the time domain approach for an inelastic system (Wen, 1989) is used. It gives response statistics of interest such as maximum interstory displacement and hysteretic energy dissipation. For the SMRSF frame a strong-column weak-beam (SCWB) model is developed which localizes inelastic behavior at the base and the floor level at the beams. It allows lateral displacement and floor rotation. The hysteretic restoring moments are described by the smooth differential equation model which allows solution by the equivalent linearization method. Comparisons between the results for different acceleration records by the SCWB and DRAIN 2DX indicate that the former reproduces well the inelastic response behavior of the SMR frame. Also, the accuracy of the random vibration results are verified by comparison with simulation (Fig. 5). Details are available in Eliopoulos and Wen (1991).

4 LIMIT STATE RISK EVALUATION

The random vibration analyses of the structural response provide the conditional statistics and probability of limit states being reached given the occurrence of the earthquake and that the seismic event and ground motion parameters are known. These parameters (e.g., attenuation coefficient, duration, etc.), however, are known to have large variabilities and, therefore, are treated as random variables. They may be correlated or functionally dependent (e.g., on magnitude and epicentral distance) and often play a significant, sometimes dominant, role in the overall risk evaluation. An extensive literature survey of information on these parameters and their uncertainties (e.g., USGS Report 1982, 1988; Joyner and Boore, 1988) has been carried out, and models developed based on the survey results. To include these parameters uncertainties into the risk analysis, a fast integration technique (Wen and Chen, 1986) based on the first order reliability method is used. In the time history/simulation approach, these parameters have been randomized according to their distributions and incorporated into the simulated ground motion time histories. Figure 6 shows a plot of maximum first story drift for OMRSF and CBF frames at the Imperial Valley site on type I extreme values probability papers given the occurrence of a characteristic earthquake.

At the Imperial Valley site, the major threat is characteristic earthquakes of magnitude 6.5. A 50% probability of fault rupture propagation toward or away from the site is assumed. The results are found to be not particularly sensitive to this assumption. At the Los Angeles site, both types of earthquakes contribute. The characteristic earthquakes, though of a larger magnitude of 7.5, contribute less than the noncharacteristic earthquakes primarily due to the distance (60km) from the site to the Mojave segment. The random vibration/fast integration method gives comparable results. The proposed method allows one to calculate the response statistic given the occurrence of an earthquake. The earthquake occurrence probability at the site is then incorporated to evaluate the risk of limit states being reached as a function of the length of the time window considered and the dormant period since the last characteristic earthquake. The last event occurred in 1979 at the Imperial Valley fault and in 1867 at the Mojave segment.

Table 1 shows the drift level (% of story height) being exceeded corresponding to various probability levels for the next 50 years at the two

Table 1. Interstory drift (% of story height) level corresponding to an exceedance probability in 50 years.

Site	Frame	Prob (%)	Story 1	2	3	4	5
Los Angeles	OMRSF	50	.19	.30	.32	.31	.35
		25	.30	.46	.49	.47	.53
		15	.39	.59	.63	.61	.71
		10	.50	.73	.78	.76	.91
		5	.76	1.09	1.15	1.14	1.40
	SMRSF	50	.22	.34	.38	.38	.41
		25	.34	.51	.57	.57	.62
		15	.46	.68	.75	.74	.84
		10	.60	.85	.94	.92	1.07
		5	.94	1.30	1.41	1.38	1.66
	CBF	50	.18	.22	.26	.31	.31
		25	.28	.33	.38	.47	.48
		15	.38	.44	.51	.63	.65
		10	.50	.56	.64	.82	.84
		5	.78	.86	.99	1.28	1.32
	D/CBF	50	.16	.25	.26	.28	.26
		25	.24	.37	.39	.42	.40
		15	.33	.50	.51	.56	.54
		10	.43	.64	.65	.72	.69
		5	.68	.98	.99	1.11	1.08
Imperial Valley	OMRSF	50	.64	.90	.96	1.03	1.20
		25	.89	1.15	1.22	1.28	1.42
		15	1.04	1.31	1.39	1.43	1.55
		10	1.16	1.43	1.51	1.54	1.65
		5	1.36	1.63	1.71	1.72	1.81
	SMRSF	50	.76	.97	1.07	1.10	1.40
		25	1.04	1.25	1.36	1.38	1.70
		15	1.21	1.43	1.55	1.56	1.89
		10	1.34	1.56	1.69	1.70	2.03
		5	1.55	1.78	1.93	1.92	2.26
	CBF	50	.59	.64	.77	.98	1.00
		25	.87	.90	1.01	1.26	1.23
		15	1.09	1.08	1.16	1.44	1.37
		10	1.26	1.21	1.27	1.57	1.47
		5	1.54	1.44	1.45	1.78	1.63
	D/CBF	50	.47	.75	.82	.88	.82
		25	.70	1.02	1.09	1.09	.98
		15	.88	1.21	1.26	1.23	1.08
		10	1.01	1.36	1.39	1.33	1.16
		5	1.23	1.59	1.60	1.49	1.28

sites for the four buildings. Note that although both sites are in Zone 4, at equal probability levels, response (hence damage) are much higher at the Imperial Valley site. On the other hand, the variability in the response is much larger at the Los Angeles site largely due to the large uncertainties in the intensity of noncharacteristic earthquakes and the attenuation equation for the characteristic earthquakes. Whereas at the Imperial Valley site, the uncertainty is much less due to the knowledge of the last two events (in 1940 and 1979) at the Imperial Fault. As expected, the brace frame and dual system give lower responses. The differences, however, are not as large as the different R_w values used in the design would suggest.

5 CONCLUSION AND FUTURE WORK

A method for evaluation of reliability is presented for steel building designed according to current code procedures in the U.S. The results indicate that the risks of different drift limit states show variation among the four different types of buildings which are consistent with the design philosophy and appear to be reasonable. There is, however, a greater discrepancy in the risks and response levels for the two sites considered, both in Zone 4 of the UBC designation. Additional uncertainties currently being investigated are those in the seismic risk and ground motion modeling as well as in the structural resistance, in particular, those due to nonstructural components (partition walls and cladding). Recent evidence indicates that cladding, etc., may increase the stiffness of the bare frame by a large factor and significantly increase its strength (e.g., Foutch, et al., 1986). Another area being investigated is calibration for more risk-consistent design. The R_w factors for different building types are being carefully examined to develop design procedures which provide a given amount of drift and damage control.

6 ACKNOWLEDGMENT

This research is supported by the National Science Foundation under grant NSF CES-88-22690 and BCS-91-06390. The support is gratefully acknowledged.

REFERENCES

Algermissen, S. T., O. M. Perkins, P. C. Theuhaus, S. L. Hanson, & B. L. Benden 1982. Probabilistic estimates of maximum acceleration and velocity in rock in the contiguous United States. USGS open-file report 82-1033, 1982.

Cornell, C. A. & S. R. Winterstein 1988. Temporal and magnitude dependence in earthquake recurrence models. *Bull. Seis. Soc. of Am.*, 78: 1522-1537.

Eliopoulos, D. & Y. K. Wen 1991. Method of seismic reliability evaluation for moment resisting steel frames. *Structural Research Series No. 562*, Univ. of IL.

Foutch, D. A., S. C. Goel, & C. W. Roeder 1987. Seismic testing of a full-scale steel building - Part 1. *J. of Struc. Engr., ASCE*, 113, 11:2111-2129.

Joyner, W. B. & D. M. Boore 1988. Measurement, characteristics, and prediction of strong ground motion. *Proc. Erth. Engr. and Soil Dyn., Geotech. Div., ASCE*, Utah.

Trifunac, M.O. & A. G. Brady 1975. A study of strong earthquake ground motion. *Bull. Seis. Soc. of Am.*, 65, 3: 581-626.

Trifunac, M. O. & V. W. Lee 1989. Empirical models for scaling Fourier amplitudes spectra of strong earthquake accelerations in terms of magnitude, source to station distance, site intensity and recording site conditions. *Soil Dyn. and Erth. Engr.*, 1, 3.

United States Department of the Interior, U.S. Geological Survey 1988. Probabilities of large earthquake occurring in California, on the San Andreas fault. Open-file report 88-398.

Wen, Y. K. 1989. Method of random vibration for inelastic structures. *Appl. Mech. Rev.*, 42, 2.

Wen, Y. K. & H. C. Chen. On fast Integration for Time Variant Structural Reliability. *J. of Prob. Engr. Mech.*, 2, 3: 156-162.

Yeh, C-H. & Y. K. Wen 1990. Modeling of nonstationary ground motion and analysis of inelastic structural response. *Struc. Safety*, 8: 281-298.

Structural Safety & Reliability, Schuëller, Shinozuka & Yao (eds) © 1994 Balkema, Rotterdam, ISBN 90 5410 357 4

Partial correlation consideration in quantification of seismic failure sequences of a fast breeder reactor

Akira Yamaguchi
O-arai Engineering Center, Power Reactor and Nuclear Fuel Development Corporation, Japan

ABSTRACT: Safety systems of a liquid metal cooled fast breeder reactor (FBR) are designed with emphasis on redundancy and diversity so that high reliability is achieved. However, they sustain a seismic load at the same time and common cause failures are a major point of concern. Considerable effort is required to identify the common root causes and to quantify the correlation coefficients. The purpose of this study is to investigate the influence of partial correlation consideration on the system failure frequencies of typical seismic event sequences of the FBR. It is concluded that sequences including a number of cut sets can be evaluated with the assumption of the extreme (full correlation) without deteriorating the results. On the other hand, the partial correlation should be taken into account for sequences where one or two cut sets dominate the annual frequency of failure. It is advantageous that one can concentrate the effort on the correlation analyses of a few critical components and systems.

1. INTRODUCTION

A seismic probabilistic safety analysis (SPSA) study (Aizawa 1991) is being conducted for a liquid metal cooled fast breeder reactor (FBR). Safety systems of the FBR are designed with emphasis on redundancy and diversity so that high reliability is achieved. However, the safety systems sustain a seismic load at the same time and common cause failures are a major point of concern. Therefore, it should be assured that the redundancy and diversity, which are successful design concept for internal initiators, be effective under seismic conditions as well.

In the SPSA study, the fault trees developed for the internal event analysis are used, which assures the same level of detail between the internal and external evaluations. Thousands of components are included in the original systems model developed for the internal PSA study. Since considerable effort is required to identify the common root causes and to quantify the correlation coefficients, it is time-consuming to consider the partial correlation for all the combinations of components and systems in the failure sequences. If the sensitivity of the correlation coefficient to the plant-level failure frequency is not significant, one can simplify the evaluation of the failure frequency by assuming the extremes, i.e., no or full correlation.

The sensitivity of the correlation coefficients to the annual frequency of failure sequence has been compared with that of seismic hazard curves. The uncertainty of the seismic hazard estimate is another important aspect and the uncertainty of the annual frequency of failure caused by the uncertainty of the hazard curves is considered to be quite large in general. The purpose of this study is to investigate the necessity of the partial correlation considerations on the system failure probability of typical seismic event sequences of the FBR.

2. FBR PLANT SYSTEM

2.1. Plant description

The FBR analyzed in this study is a loop-type plant. The schematics of the plant is shown in

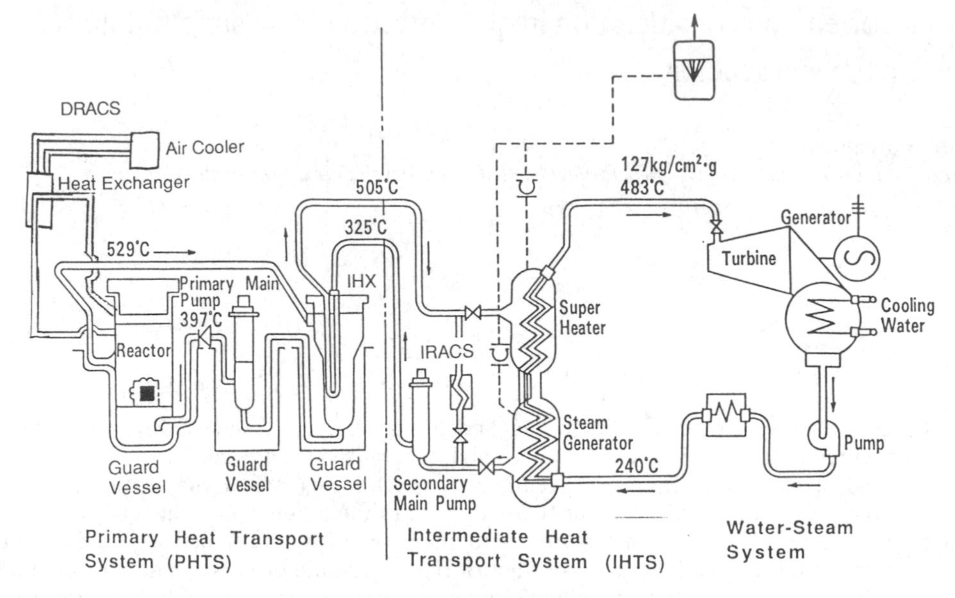

Fig. 1 Schematics of the FBR plant (only one loop is illustrated).

Figure 1. Although it has three heat transport loops (i.e., A, B and C loops), one of them is shown in Fig. 1. The plant system is made up of the reactor vessel, primary heat transport system (PHTS), intermediate heat transport system (IHTS), and a water and steam system (WSS). The PHTS and IHTS are interrelated through the intermediate heat exchanger (IHX). Heat produced in the reactor core is transported via the PHTS and IHTS to the WSS to generate electricity. Cooling of the reactor core during normal operation is accomplished by three heat transport system loops. In the decay heat removal mode following reactor shutdown, the intermediate reactor auxiliary cooling system (IRACS) shown in Fig. 1 is used instead of the WSS. The IRACS is arranged in parallel to the WSS. Sodium valves indicated in Fig. 1 switch over the coolant path when the reactor changes the operation mode from power generation with the WSS to the decay heat removal with the IRACS. The direct reactor auxiliary cooling system (DRACS) is another way of decay heat removal. The system has the primary and secondary loops, a heat exchanger in the midst of them, and an air cooler. The coolant in the DRACS is circulated directly from the reactor vessel without resorting to the PHTS and IHTS.

2.2. Seismic initiating events

Seismic initiating events to be evaluated have been selected based on the plant response and the influence of the failed structure/components on the safety systems. Nine initiating events were identified taking into consideration the plant behavior given a certain earthquake level (Nakai 1993).

Table 1 Seismic initiating events.

Abbr.	Descriptions
SI-1	Reactor trip with no coolant boundary failure assuming loss of off-site power
SI-2	PHTS leakage within guard vessel
SI-3	PHTS leakage outside of guard vessel
SI-4	DRACS leakage
SI-5	Reactor vessel failure
SI-6	Core support structure failure
SI-7	Double leakage in PHTS in two/more loops
SI-8	Inner containment concrete structure failure
SI-9	Reactor auxiliary building failure

The structural failures (SI-8 and SI-9) are supposed to lead to core damage directly because of multiple components will lose the support system. SI-5 and SI-6 result in core damage (loss-of-core-configuration: LOCC). SI-7 sequence causes loss-of-reactor-level (LORL) in the reactor vessel. Therefore, the

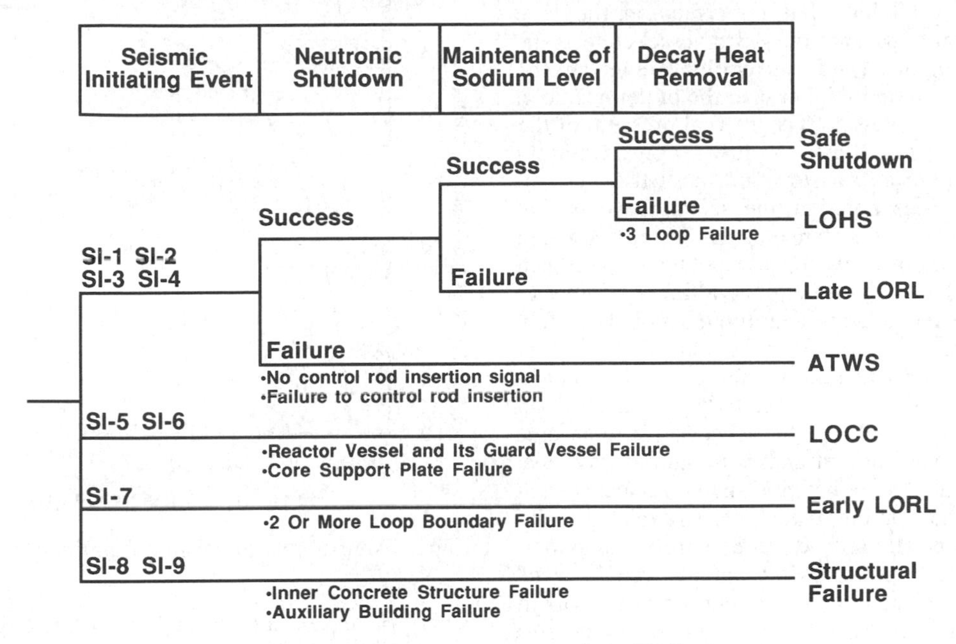

Fig. 2 Functional event tree of the FBR plant.

event trees need not be developed for sequences from SI-5 to SI-9 and detailed event trees have been developed for SI-1, SI-2, SI-3 and SI-4 sequences.

2.3. Safety functions of FBR

Essential safety functions of the FBR are 1) reactor power reduction to decay heat level, 2) decay heat removal following the neutronic shutdown, and 3) structural integrity that assures the coolant path from the core to the air coolers in the decay heat removal system. The functional event tree for the safe shutdown is shown in Fig. 2. The accident sequences are categorized as loss-of-heat-sink (LOHS), early or late LORL, anticipated transient without scram (ATWS), LORL and structural failure.

The reactor power reduction is achieved by the reactor protection system that is doubly redundant. In other words, the primary and secondary shutdown systems are installed. Double failure of the both systems results in failure of neutronic shutdown and ATWS sequences follow.

As mentioned above, the plant has three main heat transport loops and one additional DRACS loop for decay heat removal. An essential passive safety feature of the FBR is the capability of decay heat removal from the core by fully natural circulation. The IRACS has triple multiplicity and significant potential of natural circulation. Only one loop out of three is required for either forced or natural circulation. In the natural circulation decay heat removal, no active components such as a pump and a blower and no electricity are required and one can leave the whole system to take its own course. It suggests no active operations are required for the safe shutdown of the reactor once neutronic shutdown is achieved by the insertion of control rods. On the other hand, the DRACS depends its operation on the electrical power system because electro-magnetic pumps drive the coolant flow. It provides independence from the IRACS loops because the design and its location in the building are different.

For the maintenance of the sodium level, a guard vessel is provided for each of reactor vessel, primary pumps, and IHXs as shown in

Fig. 1. Therefore, coolant leakage at one place in a PHTS loop has no effect on the core coolability because the guard vessel collects the sodium and hold the coolant level in the primary system higher than the outlet nozzle of the reactor vessel. The internal pressure of the coolant boundary is low (approximately atmospheric pressure) in the FBR system. Hence it is noted in the seismic events that significant importance of the primary coolant boundary integrity that assures the coolant path from the reactor core to the ultimate heat sink.

The redundancy and diversity of the safety systems mentioned above contribute to the achievement of high reliability of the nuclear power plant system. The failure sequences can be terminated by the successful neutronic shutdown, sodium make-up and decay heat removal. Many systems and components such as pumps, piping, panels, etc. are related to the accomplishment of the safety functions. Therefore, it seems the annual frequency of failure is dominated by a number of cut sets in general. On the other hand, no safety systems appear in the sequences initiated by the coolant boundary failure in two or more loops (SI-7) or failure of neutronic shutdown (ATWS). Therefore, one has to consider only one cut set, i.e., multiple failure of coolant boundary structure for the SI-7 sequence or double failure of the primary and secondary shutdown systems for the ATWS sequences.

3. QUANTIFICATION OF SEISMIC FAILURE SEQUENCES

3.1. Methodology

Seismic fragility of structure or equipment is defined as a failure probability on condition that an earthquake takes place. The double logarithmic normal distribution model is widely used to describe the seismic fragility curves. Using the peak ground acceleration a as an intensity parameter of an earthquake, the seismic fragility is expressed as the following equation:

$$F(a,Q) = \Phi\left[\frac{ln\left(\frac{a}{A_m}\right) + \beta_U \Phi^{-1}(Q)}{\beta_R}\right] \quad (1)$$

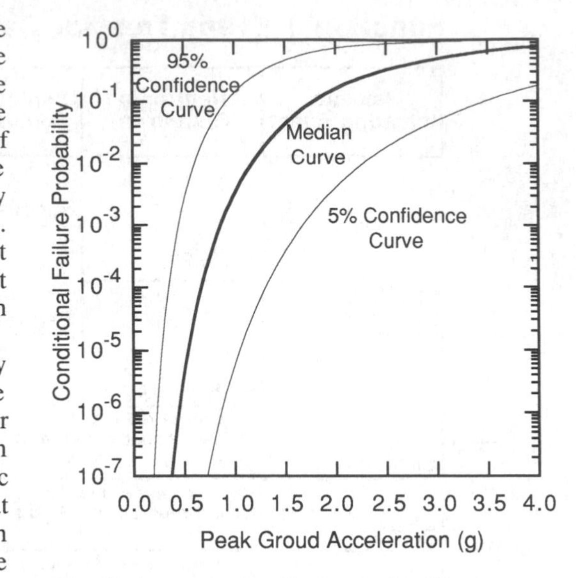

Fig. 3 An example of seismic fragility curves.

where β_R and β_U are variabilities of the failure probability associated with random uncertainty and modeling uncertainty, respectively. The variabilities are given as the lognormal standard deviation of the probability function. $\Phi(\bullet)$ is the cumulative normal distribution function and $\Phi(\bullet)^{-1}$ is its inverse function. Q is the degree of belief regarding the median capacity A_m. An example of the seismic fragility is shown in Fig. 3. β_R and β_U reflect the slope of the fragility curves and the width of the upper and lower bound curves, respectively. High confidence (95%) low probability (5%) of failure (HCLPF) is an index that describes the relative ruggedness of a component reflecting the median capacity as well as the random and modeling uncertainties. The HCLPF value is calculated as:

$$HCLPF = A_m \, exp\left[-1.645(\beta_R + \beta_U)\right] \quad (2)$$

Conditional occurrence probabilities of failure sequences are evaluated at various applied load levels. The system failure sequence is expressed as Boolean algebra of a number of constituent components. Let us consider the situation where the plant-level failure sequence is expressed as union of N second-order cut sets or third-order cut sets:

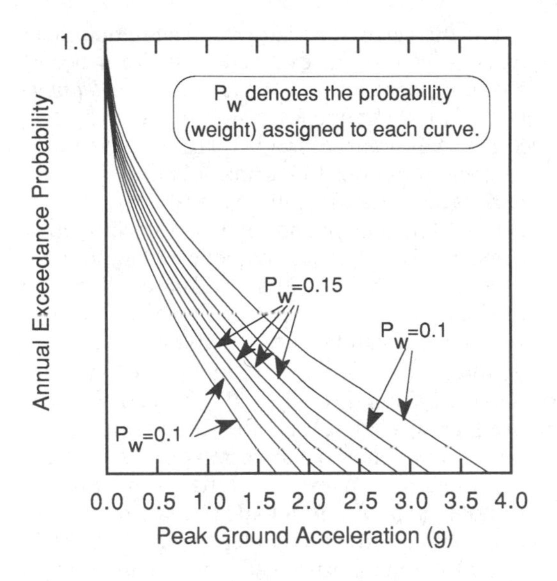

Fig. 4 An example of seismic hazard curves.

$$(Failure\ Sequence) = \bigcup_{i=1}^{N} (C_{1,i} \cap C_{2,i}) \tag{3}$$

$$(Failure\ Sequence) = \bigcup_{i=1}^{N} (C_{1,i} \cap C_{2,i} \cap C_{3,i}) \tag{4}$$

where '$\cap$' and '$\cup$' denote intersection and union, respectively. $C_{j,i}$ represents component failure. N is the number of cut sets involved in the failure sequence. The components are more or less dependent each other under seismic load.

Once the conditional failure probability of a sequence (plant-level fragility) is obtained, hazard curves are convoluted with the plant-level fragility curves to evaluate the annual occurrence frequency of the sequence. The hazard curve is defined as annual exceedance probability as a function of the peak ground acceleration. Typical hazard curves are show in Fig. 4. Being the hazard curves $H(a)$ given, annual frequency of failure P is evaluated as:

$$P = -\int_{0}^{\infty} \frac{dH(a)}{da} F(a)da \tag{5}$$

where $F(a)$ is the plant level seismic fragility as a function of the peak ground acceleration. The convolution of the hazard and fragility curves is executed using discrete probability distribution method.

The uncertainty of the annual frequency of failure is influenced from the plant-level fragility and the seismic hazard. The sensitivity of the correlation coefficients among components and systems has been investigated in this study. In addition, the hazard curve has a great deal of random and modeling uncertainties. Hence, a sensitivity study has been also performed as to the hazard cut-off value and the variability of the hazard curve. The sensitivity of the correlation assumption to the failure frequency is estimated and it is compared with the uncertainty range derived from the hazard curve that is one of the greatest contributors to the uncertainty of the annual frequencies of plant-level fragility.

Two approaches are possible to refine the annual frequency of seismically induced failure of the FBR system: 1) to reduce the uncertainty of the seismic hazard, and 2) to enhance the plant-level seismic capacity. For the second approach, enhancement of the individual equipment capacity or increase in the diversity of the safety systems might be effective, both of which require design modification more or less. The author takes notice on the diversity of the safety systems and consideration of partial correlation instead of full correlation among the constituent components.

In the preceding SPSA studies, the diversity was tend to be neglected because of the difficulty in the correlation coefficient estimate and failure sequence quantification. The evaluation of the degree of dependence and the quantification of correlation coefficient is not straightforward and engineering judgment is needed. Therefore, the most conservative assumption (full correlation) from the viewpoint of the failure frequency was made. Ravindra and Johnson (1991) discussed the necessity of considering partial correlation in the SPSA of nuclear power plants. On the basis of the method proposed by Reed et al. (1985) the author has developed a computer program that deals with the partial correlation and has evaluated the multiple failure mode in the heat transport system of an FBR (Yamaguchi 1991). The program evaluates the conditional probability of the arbitrary Boolean expression consisted of correlated components. Execution of the integral of Eq. (5) given the seismic hazard curves yields the annual frequency of failure. This program has been used for the calculation described in the following section.

3.2. Analytical results and discussions

Three equipment types (i.e., fragile, moderate and rugged) have been studied as shown in Tables 2. Median capacity A_m, random uncertainty β_R, modeling uncertainty β_U and HCLPF value (see Eq. (2)) are tabulated in Table 2. A_m and HCLPF are given in terms of peak ground acceleration. The uncertainties β_R and β_U are not changed for the three cases.

Table 2 Fragility parameters of components.

Equipment type	A_m (g)	β_R	β_U	HCLPF (g)
Fragile	1.0	0.3	0.4	0.32
Moderate	2.0	0.3	0.4	0.63
Rugged	3.0	0.3	0.4	0.95

Three seismic hazard cases shown in Table 3 are considered. The first seismic hazard, termed H_1, is a standard case that is based on a preliminary hazard analysis at the reference reactor site. In the standard case, no upper cut-off value of peak ground acceleration is utilized. Various earthquake attenuation relations are used in the preliminary analysis. As a result, the lognormal standard deviation (LNSD) is quite large; e.g., it lies in the range of 1.0 at 0.1 g, 2.5 at 1.0 g, and 5.0 at 5.0 g. The second seismic hazard H_2 employs the upper cut-off value at 1.0 g with the same LNSD as in the first one. In the third hazard case H_3, it is assumed that the LNSD is halved at every acceleration level and no cut-off value of the peak ground acceleration is used.

Table 3 Seismic hazard curves.

Hazard case	Cut-off value (g)	Variation of hazard curve
$H1$	No cut-off	Large (LNSD=1.0-5.0)
$H2$	Cut-off at 1.0 g	ditto.
$H3$	No cut-off	Halved(LNSD=0.5-2.5)

For the seismic hazard cases and equipment types shown in Tables 2 and 3, the influence of changing correlation coefficients on the plant-level failure frequency has been investigated. Since the seismic fragility of equipment is subject to double lognormal distribution, two correlation coefficients ρ_R and ρ_U can be defined for random and modeling uncertainties, respectively. Annual frequency of failure for five cases have been evaluated for the equipment types defined in Table 2. Table 4 shows the notation of the analytical cases. P_i denotes the annual frequency of failure obtained from Eq. (5). P_1, P_2 and P_3 have been compared to see the reduction of the failure frequency if the correlation assumption is refined by considering partial correlation. It is noted that P_3 is an ideal case where the components are fully independent. Different hazard curves are used for the calculations of P_1 , P_4 and P_5. One can see the reduction of the annual frequency of failure by improving the hazard curve estimate.

Table 4 Notation of the annual frequency of failure for each analytical cases.

Correlation \ Hazard case	H_1	H_2	H_3
Full correlation ($\rho_R = \rho_U = 1.0$)	P_1	P_4	P_5
Partial correlation ($\rho_R = \rho_U = 0.5$)	P_2	——	——
No correlation ($\rho_R = \rho_U = 0.0$)	P_3	——	——

A reduction factor R_i is defined as the relative change of annual frequency of failure resulted from modifying the correlation assumptions or hazard curves. In other words, P_i divided by P_1 (i = 2, 3,..5) is the reduction factor R_i. The reduction factors for fragile equipment case (A_m=1.0 g) are shown in Figure 5. They are evaluated for the second-order cut set system given in Eq. (3) as a function of the number of cut set N. A system with larger N corresponds to a more complicated system. From Fig. 5, it is seen that the reduction factor by taking into account the independence or partial correlation among the components gradually increases as the system becomes complicated. It exceeds 1.0 when N=4 for the partial correlation case. It

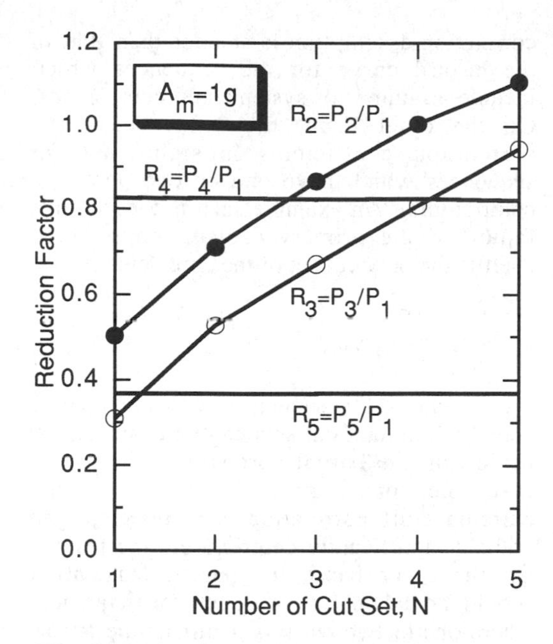

Fig. 5 Reduction factor as a function of number of cut sets (A_m=1g).

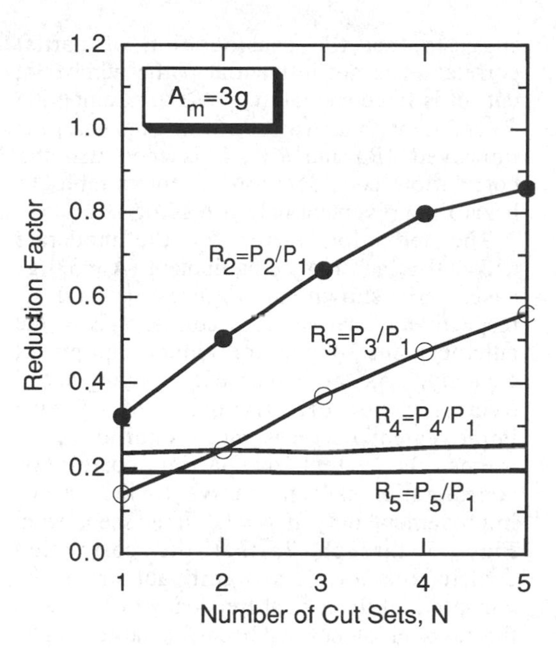

Fig. 7 Reduction factor as a function of number of cut sets (A_m=3g).

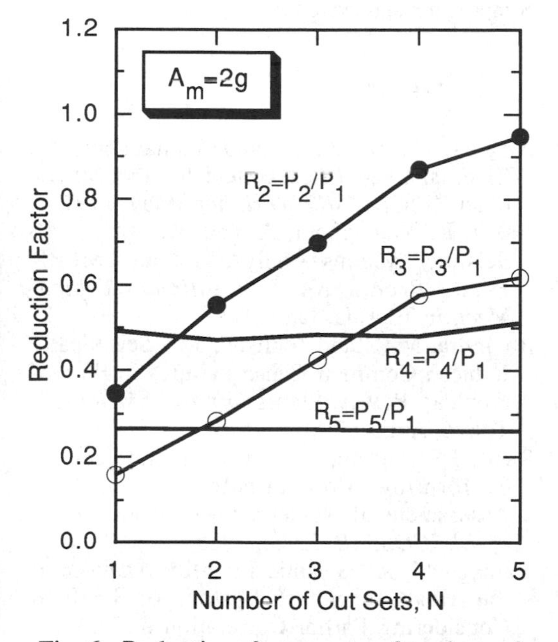

Fig. 6 Reduction factor as a function of number of cut sets (A_m=2g).

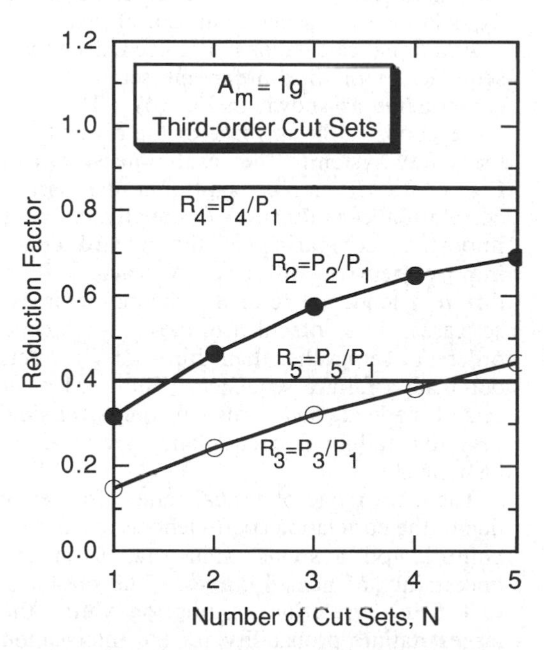

Fig. 8 Reduction factor as a function of number of cut sets (A_m=1g, third order cut set case).

suggests that the consideration of partial correlation is not influential if the number of cut set is large enough (e.g. $N \geq 3$). Comparing R_2 and R_3 with the cases in which the hazard is improved (R_4 and R_5), it is seen that the correlation consideration is comparable to hazard improvement only if $N \leq 2$ or 3.

The reduction factor for the moderate (A_m=2.0 g) and rugged equipment (A_m=3.0 g) cases are shown in Figures 6 and 7, respectively. These cases corresponds to the enhancement of the individual equipment capacity. As the equipment is strengthened overall, the effectiveness of hazard improvement becomes more significant. It seems the effect of partial correlation consideration is comparative to the hazard improvement only if N=1. It is seen, from Figs. 5 through 7, that the correlation coefficients are less significant from the viewpoint of the annual frequency of failure as the number of cut set becomes larger. The hazard cut-off has minor effect on the annual frequency of failure if a sequence consists of the fragile components which median capacity A_m is in the same range as the cut-off value.

Additional calculations have been made for a sequence with third-order cut set of fragile components as shown in Eq. (4). This case corresponds to the increase of the diversity of the safety system. The results are shown in Figure 8. It can be said that the partial correlation consideration is equally or more important comparing to the hazard curve improvement regardless of N value. It is different tendency form the second-order cut set cases. It is noted that in most cases second order cut sets rather than third-order cut sets dominate a failure sequence. Third-order cut sets of moderate and/or rugged equipment yield very low failure frequency and they need not be noticed.

The error range of annual failure frequency due to the correlation coefficients is smaller for complicated systems than that from the uncertainty of hazard curves. The reason is that full correlation assumption yields the largest failure probability for the intersection system and the smallest one for the union system. As the system becomes complicated, the conservatism of the intersection and unconservatism of the union compensate each other. The influence of the uncertainty in the correlation assumption is smaller than that of the hazard curve for the sequences which include a number of systems and components. On the other hand, the influence of the correlation coefficients is significant for sequences which have one or two dominant contributors. An example such is the multiple failure of the primary coolant loops which itself is the only cut set of the core damage.

4. CONCLUSIONS

It is concluded that sequences with a number of equally dominant cut sets can be evaluated by neglecting the partial correlation. In such a case, the conservative assumption of the extreme (full correlation) can be employed without significantly deteriorating the results. On the other hand, the partial correlation should be taken into account for sequences where one or two cut sets dominate the annual frequency of failure. It is advantageous that one can concentrate the effort on the correlation analyses of a few critical components and systems.

5. REFERENCES

Aizawa, K., Nakai, R. and Yamaguchi, A., External Events Assessment for an LMFBR Plant, *OECD/BMU Workshop*, May, 1991.

Nakai, R. Yamaguchi, A. and Morishita, M., Seismic Systems Analysis for an LMFBR Plant, *Proc. PSA International Topical Meeting*, Florida, Jan. 1993.

Ravindra, M.K. and Johnson, J.J., Seismically Induced Common Cause Failures in PSA of Nuclear Power Plants, *Proc. SMiRT-11*, Tokyo, August 1991.

Reed, J.W., et al., Analytical Techniques for Performing Probabilistic Seismic Risk Assessment of Nuclear Power Plants, *Proc. of 4th ICOSSAR*, Kobe, 1985.

Yamaguchi, A., Seismic Fragility Analysis of the Heat Transport System of LMFBR Considering Partial Correlation of Multiple Failure Modes, *Proc SMiRT-11*, Tokyo, August, 1991.

Structural Safety & Reliability, Schuëller, Shinozuka & Yao (eds) © 1994 Balkema, Rotterdam, ISBN 90 5410 357 4

Use of neural networks for earthquake damage estimation

Fumio Yamazaki, Gilbert L. Molas & Maliha Fatima
Institute of Industrial Science, The University of Tokyo, Japan

ABSTRACT: This paper proposes an application of neural networks to damage estimation of structures subjected to strong earthquake motions. Since it is not so easy to correlate strong motion parameters and resultant structural damage using simple mathematical equations, neural networks are conveniently employed to construct such a relationship. The peak ground acceleration, the peak ground displacement and the spectrum intensity from 79 actual earthquake records are considered as input parameters of a neural network and the damage observations near the recording sites are used as desirable outputs of the network. After iterations of supervised learning, the network converges and can be used for future estimations. Although the input parameters and learning data are still preliminary, the methodology may be useful for an early damage detection of structures due to earthquakes.

1 INTRODUCTION

When an earthquake strikes large cities, it is important to estimate the resultant damage soon after the earthquake. Especially, for city gas networks, a decision of whether or not to shut off the supply is urgent otherwise secondary disasters may follow structural damage. However, this decision must be made carefully. If we shut off the gas supply without severe structural damage, recovery may take time and inconvenience to customers may be more serious. In order to make an early but accurate estimation of damage to customers' buildings and pipelines, an extensive monitoring system of earthquake intensities was developed in Japan (Nakane et al., 1992). This system measures the peak ground acceleration (PGA) and the spectrum intensity (SI) of many points within a service area. The monitored PGA and SI values are transmitted to the headquarters of the gas company, and the damage estimation is conducted.

To estimate structural damage from these earthquake indices, a large number of studies exist. Obviously, if we specify structures and specify the input motion in terms of time history, sophisticated response analyses can be conducted. However, if we must estimate overall damage of many types of structures from the measured earthquake ground motion indices, a quick and robust method is necessary. The PGA is the most commonly used index to describe the severity of earthquake ground motion. However, it is well known that a large PGA value is not necessarily followed by severe structural damage. Katayama et al. (1988) demonstrated that the SI value has a better correlation with structural damage than the PGA. Some other indices of earthquake ground motion, e.g., the peak ground velocity (PGV), the peak ground displacement (PGD), the duration of strong motion, and the spectral characteristics of various definitions, can also be considered in such a damage estimation. Ando et al. (1990) demonstrated that PGA, PGV and SI, and PGD are related to the damage of short-period, intermediate-period, and long-period structures, respectively.

However, to correlate these input parameters with observed damage in a mathematical form is not an easy task since a large number of uncertainties are involved and the relationship must be highly nonlinear. A conventional way to construct such a relationship from observed data is to use the multiple regression analysis. But in such a case, we must assume some functional forms to relate input and output parameters. To avoid this, the use of neural networks is proposed in this paper for the earthquake damage estimation problem.

Among several new techniques of computer science, neural networks or parallel distributed processing (PDP) recently has drawn considerable attention in various fields of science and technology. Along with the development of theories and computational algorithms (see e.g., Aleksander and Morton, 1990), the technique has been applied to fields like automation, character recognition, electro-communication and noise filtering, image processing, industrial control problems, etc. Several recent studies can also be found where the neural networks are applied to problems in earthquake engineering, e.g., active vibration control of structures (Nekomoto et al., 1991), seismic hazard prediction (Wong et al., 1992).

The use of neural networks for the earthquake damage estimation problem has several other advantages: adding new data to the network is quite easy; once the network has been set up, the damage estimation for new inputs is very fast. However, since the estimation is highly dependent on learning data, we must prepare well-examined data sets. In this paper, however, the learning data set is not so complete. But we can still demonstrate the usefulness of the technique for earthquake damage estimation problems. Collection of more extensive data is, of course, very important to make the relationship more reliable.

2 DATA

In order to construct a relationship between earthquake ground motion and structural damage using neural networks, a data set comprising inputs (strong ground motion parameters) and outputs (structural damage) must be prepared. There are basically two methods for doing this: one is collecting actual earthquake records and damage data near the recording site; the other is performing earthquake response analyses for given inputs and models and obtaining the resultant damage (outputs). The former is more convincing because it uses actual damage data. But good recordings obtained near structural damage are few. With the latter, it is easier to prepare well-distributed data. But, since it is not based on actual observations, a lot of care should be taken in selecting structural models and input motions. In this paper, we use the former approach as a first attempt although available data are rather limited. The latter approach was also used and the results were presented by the authors elsewhere (Molas and Yamazaki, 1993).

To prepare the data set consisting of earthquake recordings and structural damage information, an extensive literature survey was conducted for past earthquakes (Japan Gas Association, 1991). Among these records, 73 records from 11 earthquakes were selected. Structural damage around these recording sites was rather well-known. The records of two recent earthquakes in Japan; namely, the 1993 Kushiro-Oki (Nagata et al., 1993) and Noto Peninsula-Oki earthquakes, are also used. The number of records and damage are summarized in Table 1. At these locations, damage extent is given in three levels: negligible, moderate, and severe. The definition of these categories is given in the paper by Iwata et al. (1992), where damage to wooden houses and underground pipelines is primarily used to judge the damage extent. This judgement was mostly done based on the literature survey. But for three recent earthquakes, the 1987 Chibaken-Toho-Oki and the 1993 Kushiro-Oki earthquakes in Japan, and the 1989 Loma Prieta earthquake in California, site investigation was also performed to judge the degree of damage.

As simplest indices of the ground motion severity, the PGA, PGV, PGD and SI of the 79 records were

Table 1. List of earthquake records and their damage classification

Damage Category	Negligible	Moderate	Severe	Total
Niigata (1964)	0	0	1	1
Matsushiro (1965-66)	20	2	1	23
Tokachi-Oki (1968)	0	0	3	3
Miyagiken-Oki (1978)	2	0	2	4
Nihonkai-Chubu (1983)	0	0	2	2
Chibaken-Toho-Oki (1987)	7	2	0	9
Izu Pen. Toho-Oki (1989)	2	2	0	4
Kushiro-Oki (1993)	2	2	1	5
Noto Pen. Oki (1993)	1	0	0	1
Imperial Valley (1940)	0	0	1	1
San Fernando (1971)	0	2	3	5
Mexico (1985)	0	0	2	2
Loma Prieta (1989)	4	2	13	19
Total	38	12	29	79

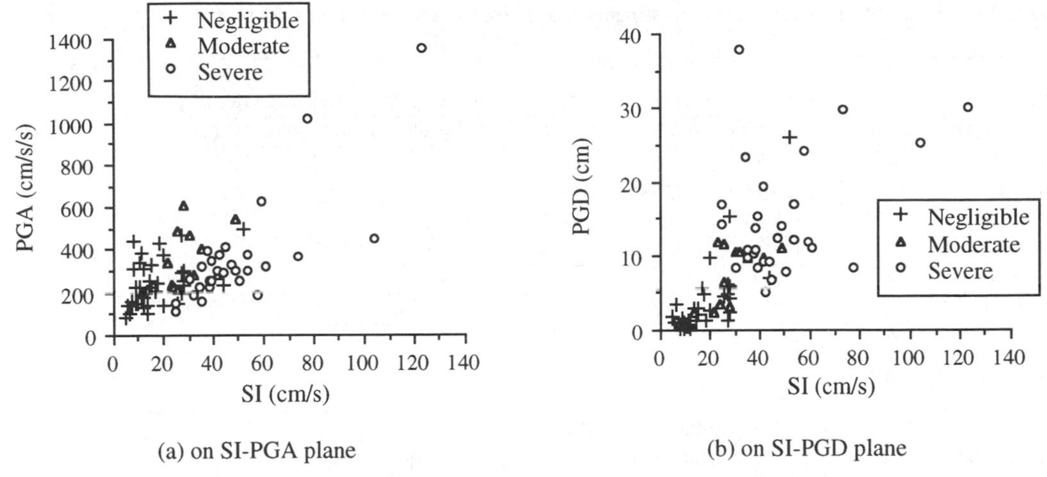

(a) on SI-PGA plane

(b) on SI-PGD plane

Figure 1. Distribution of peak values of the 79 earthquake records used in the analysis

calculated from two horizontal components of the acceleration time histories. Note that in this study, the SI value is defined as the average velocity response spectrum of 20 % damped single-degree-of-freedom systems with natural period between 0.1 s to 2.5 s as (Katayama et al., 1988)

$$SI = \frac{1}{2.4} \int_{0.1}^{2.5} S_V(T,h{=}0.2)\, dT \qquad (1)$$

For the peak values of the ground motion, the maximum of the resultants of the two horizontal components is used. SI values are computed by rotating the axis from the East-West to the North-South axis in 9 degree intervals and the maximum is used in the analysis. Figure 1 shows the relationship between (a) SI and PGA, and (b) SI and PGD together with the damage category for the 79 records. Due to the difference in the waveform and frequency contents, the data points look quite scattered in the SI-PGA plane. But severe damage is only seen for SI values larger than 25. In the SI-PGD plane, damage looks to be proportional to both SI and PGD although some exceptions are found. The relationship between PGV and SI is not shown here, but they were found to be very similar values. Thus, we decided to use only SI instead of PGV in neural network analyses since SI value (and PGA) is actually observed by a new type of seismometer of a Japanese gas company (Nakane et al., 1992).

Other parameters showing severity (or characteristics) of ground motion, duration and dominant frequency contents of motion should be taken into account in damage estimation. But in this first trial using neural networks, we just considered three parameters; PGA, PGD and SI.

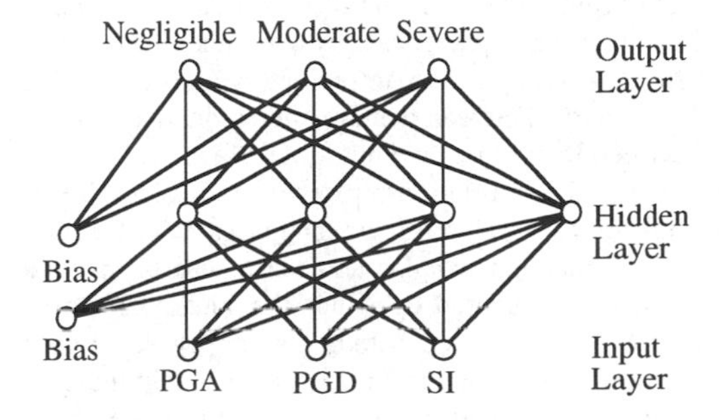

Figure 2. A neural network model for earthquake damage estimation

3 NEURAL NETWORK MODEL

A neural network is a collection of parallel processors connected together in a form of a directed graph. A network consists of neurons or Processing Elements (PEs) which are arranged in layers. In this study, a typical feed-forward network with supervised learning using the extended delta-bar-delta learning algorithm (Minai and Williams, 1990) is employed. Neural network analyses are run on Neural Works Professional II simulator (NeuralWare, Inc., 1991).

Figure 2 shows a three-layered network model used in this study. PGA, PGD and SI are considered as input parameters. To represent the three damage categories, three PEs are considered in the output layer. Each of the PEs represents one damage category. For a particular damage category, the corresponding PE is activated as 1 and the two other PEs are inhibited as 0.

Table 2 Mean output values and correctly estimated case ratios by recall tests for the 79 records

Input parameters	No. of PEs in hidden layer	Mean of network's estimation			Correctly estimated case (ratio)		
		negligible	moderate	severe	negligible	moderate	severe
PGA, PGD,	4	0.855	0.557	0.870	0.92	0.50	0.97
and SI	5	0.933	0.685	0.844	0.95	0.67	0.93
	6	0.944	0.502	0.931	0.95	0.50	0.97
PGA and SI	4	0.800	0.418	0.894	0.82	0.33	0.97
PGD and SI	4	0.836	0.439	0.822	0.92	0.33	0.97
PGA only	4	0.575	0.238	-0.002	0.55	0.33	0.10
SI only	4	0.763	0.249	0.797	0.74	0.00	0.86

An output vector of {1,0,0} represents negligible damage while moderate and severe damage categories are represented by {0,1,0} and {0,0,1}, respectively.

One hidden layer is considered between the input and output layers. The number of PEs in the hidden layer is usually determined by experience or trial and error. Several cases are tested in this study (see Table 2). The PEs between the input and hidden layers and between the hidden and output layers are fully interconnected. The relationship between connected PEs is represented by a hyperbolic tangent transfer function with connection weights and bias. The connection weights are updated to reduce the squares of errors between desired and actual outputs by supervised learning. The bias term is included to help the convergence of the weights in an expectable range. After thousands of iterations (learning), the network converges.

Similar neural network analyses are also performed for two input parameter cases (PGA and SI; and PGD and SI) and one input parameter cases (PGA only and SI only).

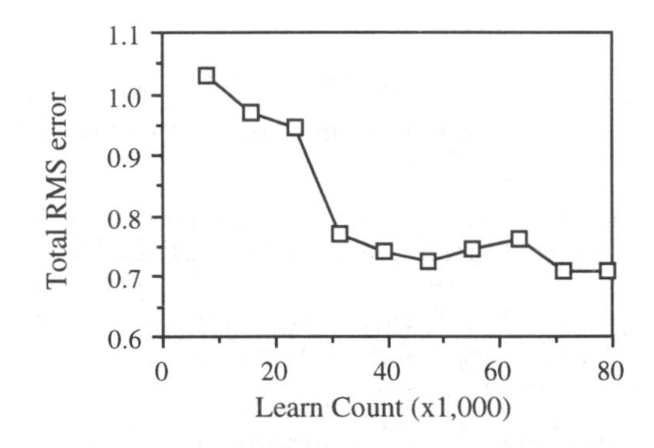

Figure 3. Convergence of neural network with 5 hidden layer PEs using PGA, PGD, and SI as input

4 RESULTS AND DISCUSSIONS

The results of neural network analysis are summarized in Table 2 for the number of learning counts =79,000 (1000 iterations of the 79 records of the training data set). In this study, the connection weights are updated after one complete pass of the training data set. Figure 3 shows the convergence of the neural network with 5 hidden layer PEs using the PGA, PGD and SI as input. The convergence is based on the sum of the root-mean-square (RMS) errors of each of the three output PEs for the whole training data set. After the supervised learning, the converged network suggests three output values for a given set of input parameter values. Although the ideal output for a severe damage case is {0,0,1}, the network gives, for example, {0.03, 0.32, 0.89}. The numbers in the three left columns in Table 2 are the mean values of the network estimation for the recall test with the 79 data. Figure 4 plots these numbers for different input parameter combinations.

It is observed that the estimated values corresponding to negligible damage and severe damage are high (mostly more than 0.8), except for the case that uses only PGA as input parameter. In fact, it can be seen that the PGA alone cannot adequately estimate the damage. The network that uses only SI gives a better estimation than the one using only PGA. This is consistent with the observations of other researchers (Katayama et al., 1988). The estimation using both the PGA and SI, however, gives a better estimation than PGA alone or SI alone.

The values for moderate damage are not so high (often less than 0.5). This fact indicates that moderate damage is most difficult to identify. Actually, the criteria for this category are most uncertain and the number of data used for the learning is smallest (only 12 as shown in Table 1). Considering uncertainties

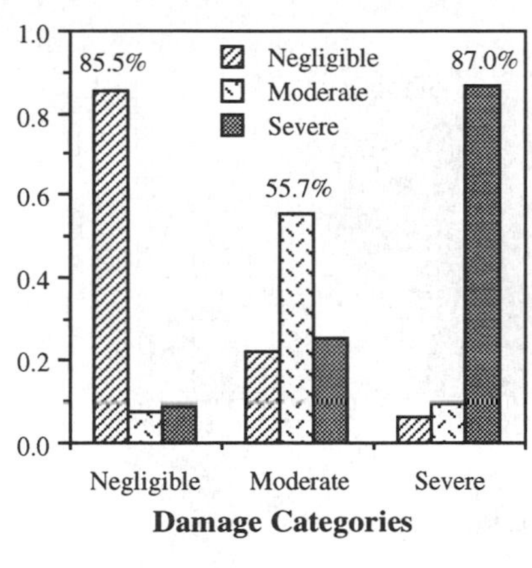

(a) PGA, PGD and SI input (4 hidden PEs)

1.0
0.8
0.6
0.4
0.2
0.0
-0.2
93.3%
68.5%
84.4%
Negligible
Moderate
Severe
Negligible
Moderate
Severe
Damage Categories

(b) PGA, PGD and SI (5 hidden PEs)

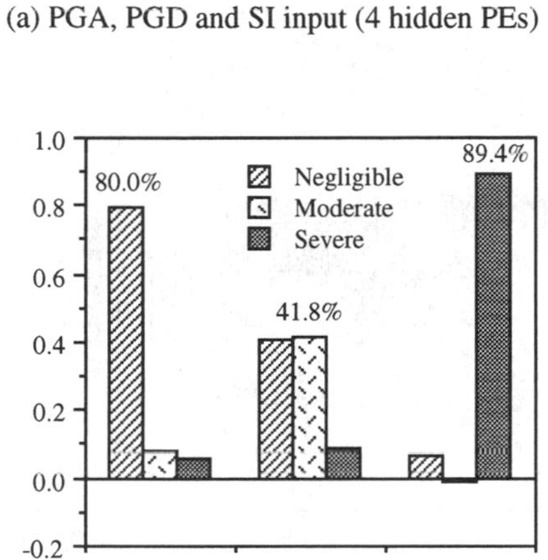

(c) PGA and SI input (4 hidden PEs)

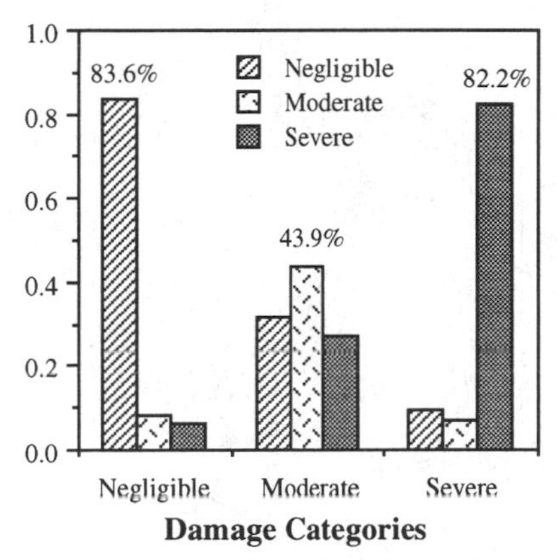

(d) PGD and SI input (4 hidden PEs)

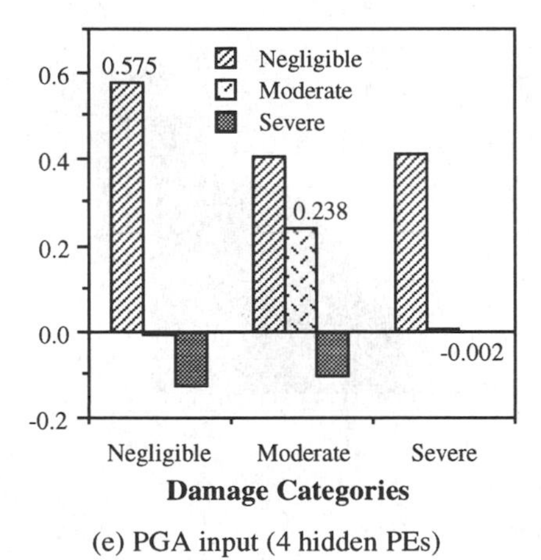

(e) PGA input (4 hidden PEs)

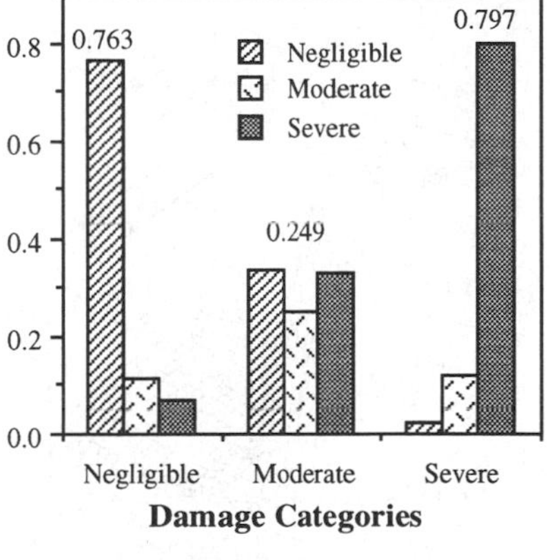

(f) SI input (4 hidden PEs)

Figure 4. Mean output values of networks by recall tests for the 79 records at 79,000 learning counts

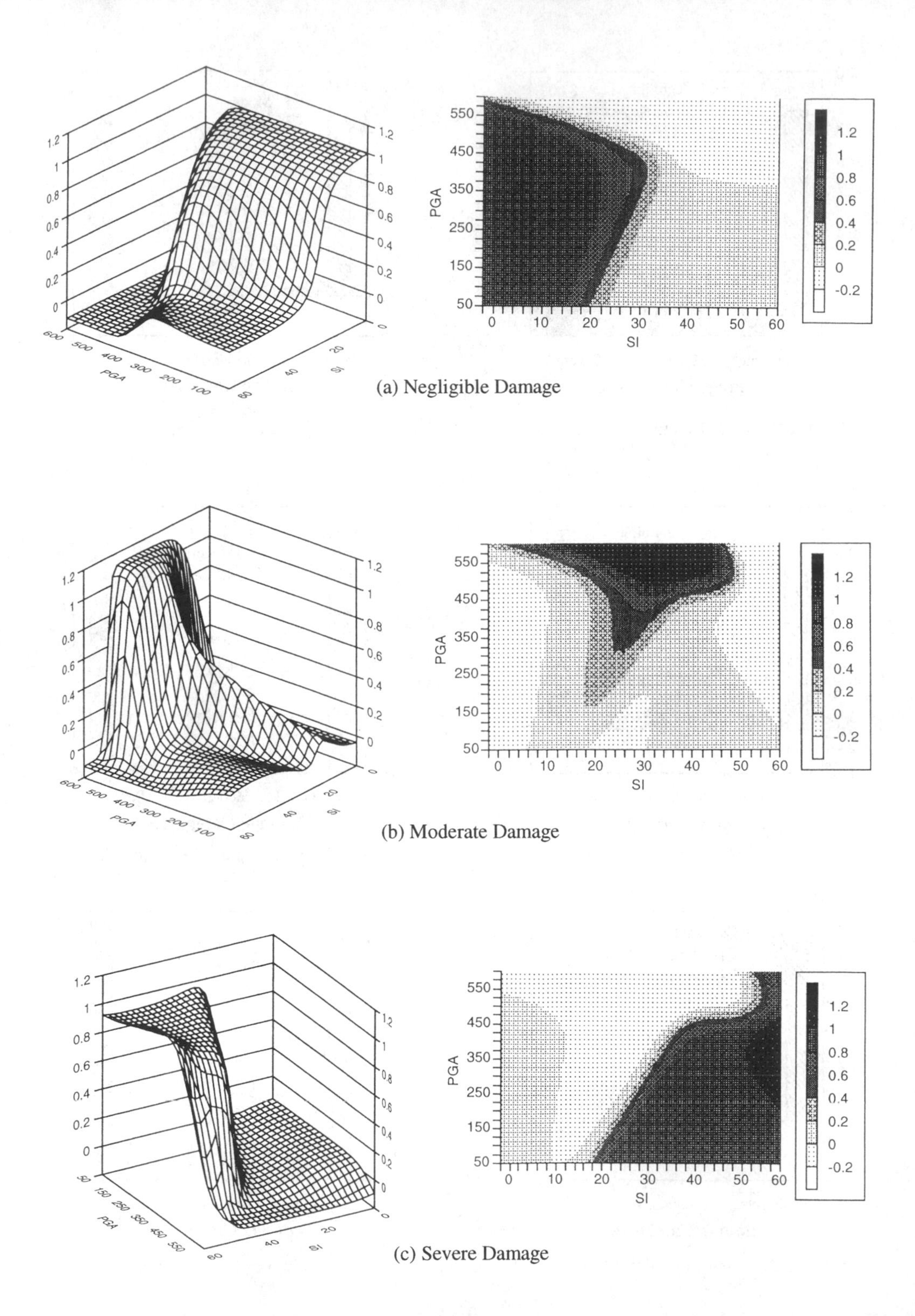

Figure 5. Output values by the converged network for various new inputs of PGA and SI

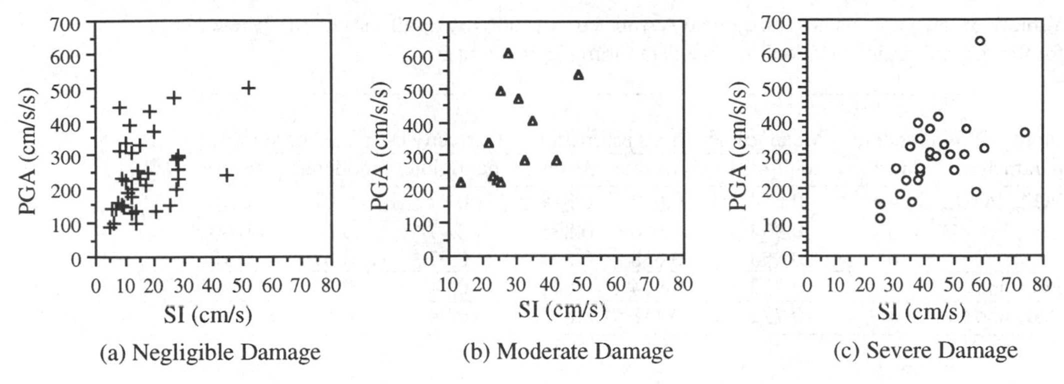

Figure 6. Distribution of training data using the PGA and SI as input

involved in the input-output relation, the estimation can be concluded to be fairly good.

The numbers in the three right columns in Table 2 indicate the ratio of correctly estimated cases, which means that if the maximum of the three output values is larger than 0.5 and it is the correct damage category, it is considered as a correct estimation. These ratios look also high.

Among three cases of input parameter combinations, the three parameter case (PGA, PGD and SI) gave the highest estimated values. However, the two parameter cases (PGA and SI; PGD and SI) also gave high values, indicating the feasibility of damage estimation based on observed SI and PGA by the new type of seismometer.

For the three input parameter case (PGA, PGD and SI), the appropriate number of elements in the hidden layer was sought. Among 4, 5 and 6 PEs' networks, the 5 and 6 hidden layer PEs gave the best estimation for negligible damage; the 5 hidden layer PEs gave the best estimation for moderate damage; and the 4 and 6 hidden layer PEs gave the best estimation for severe damage. The appropriate number of PEs in hidden layers seems to be problem (model)-dependent and data-dependent. Hence, a sensitivity analysis may be necessary for each problem.

The converged network is a kind of damage criterion in numerical sense. Figure 5 shows the estimated category values for various new input values of PGA and SI. The plot for negligible damage shows that this category is more dependent on SI than PGA: if SI is less than 20, the damage is negligible even when the PGA is high. The severe damage category seems to be affected by both PGA and SI although SI looks more influential. The range of SI and PGA for moderate damage lies between negligible and severe damage. But the estimated values are low, which indicates low confidence to assign this category, except for a small region where the training data distribution indicates moderate damage only.

By looking at the separate plots of the distribution of the damage categories with respect to the PGA and SI (Figure 6) and comparing with Figure 5, it can be seen that the regions for the neural networks damage category are very similar to the training data distribution. In the case of the regions where the damage data overlap, the network identified the damage as the one with the most number of cases in the training data. Since moderate damage has the least number of data in these regions, it will be considered by the network as "noise" and be disregarded.

In order to avoid this effect, the network was retrained using the moderate damage data twice. This means that the number of moderate damage data is increased by 12 and the number of training data becomes 91. Table 3 shows the results of the retrained neural networks. Compared with Table 2, the estimation for the moderate damage was improved, but this is accompanied by a decline in the estimation of negligible and severe damage. This demonstrates the dependence of the network on the training data used.

Although the examples shown here are highly data-dependent, the general tendency is close to the damage observations in previous studies (e.g., Katayama et al., 1988). It is also noted that since it is rather difficult to construct damage criterion of structures in terms of several strong motion parameters (e.g., PGA and SI), the method used in this paper may be conveniently applied to such cases. However, it must be emphasized that although a neural network can estimate damage extent, the estimation is learning data-dependent. Hence, to prepare a good data set for the learning is the most important thing in the practical use of neural networks in engineering problems.

Table 3. Mean output values and correctly estimated case ratios by recall tests for the 79 records for the network trained by doubling the data for moderate damage

Input parameters	No. of PEs in hidden layer	Mean of network's estimation			Correctly estimated case (ratio)		
		negligible	moderate	severe	negligible	moderate	severe
PGA, PGD,	4	0.814	0.612	0.819	0.92	0.67	0.90
and SI	5	0.934	0.686	0.835	0.97	0.67	0.90
	6	0.854	0.638	0.841	0.97	0.67	0.90
PGA and SI	4	0.712	0.604	0.841	0.74	0.75	0.93
PGD and SI	4	0.722	0.418	0.683	0.74	0.50	0.69

5 CONCLUSIONS

A use of neural networks for the damage estimation of structures subjected to strong earthquake motions was demonstrated. Since it is not so easy to correlate strong motion parameters and resultant structural damage using simple mathematical equations, neural networks were conveniently introduced to construct such a relationship based on given data. The peak ground acceleration, the peak ground displacement, and the spectrum intensity from 79 actual earthquake records were used as input parameters while the corresponding observed damage extent (e.g., negligible, moderate, and severe) was considered as a desired output. After iterations of supervised learning, the network converged. The recall tests using the learning data showed fairly good accuracy of estimation. Although the input parameters and data used are still preliminary, the method may be a useful tool for early damage estimation of structures based on observed strong motion parameters.

REFERENCES

Aleksander, I. and Morton, H. 1990. *An introduction to neural computing.* Chapman & Hall.

Ando, Y., Yamazaki, F., and Katayama, T. 1990. Damage estimation of structures based on indices of earthquake ground motion. *Proc. 8th Japan Earthq. Eng: Symposium:* Vol. 1, 715-720 (in Japanese).

Iwata, T., Nakane, H., Tazo, T., Shimizu, K., and Kataoka, S. 1992. Demonstrative evaluation of variables indicating severity of earthquake applicable to earthquakes sensors for control. *Proc. 10th World Conf. on Earthq. Eng.:* 381-385.

Japan Gas Association. 1991. *Research report on earthquake information for emergency response.* (in Japanese).

Katayama,T., Sato, N., and Saito, K. 1988. SI-sensor for the identification of destructive earthquake ground motion. *Proc. 9th World Conf. on Earthq. Eng.:* Vol . VII, 667-672.

Minai, A.A. and Williams, R.D. 1990. Acceleration of back-propagation through learning rate and momentum adaptation. *Int'l Joint Conf. on Neural Networks:* Vol. I, 676-679.

Molas, G.L. and Yamazaki, F. 1993. Quick structural damage estimation by neural networks. *Proc. 22nd JSCE Earthq. Eng. Symposium:* 587-590.

Nakane, H., Kodama, E., and Yokoi, I. 1992. New earthquake-induced ground motion severity sensing apparatus for reliable system shutdown. *Proc. 10th World Conf. on Earthq. Eng.:* 5559-5562.

Nagata, S., Molas, G.L., and Yamazaki, F. 1993. Ground acceleration records of 1993 Kushiro-Oki earthquake. *Bull. Earthquake Resistant Struc. Res. Center, IIS, Univ. of Tokyo:* 26, 19-29.

Nekomoto, Y., Fujita, K., and Tanaka, M. 1991. Simulation on the active vibration control of structures by neural networks. *Proc. Symposium on Application of Fuzzy and Neural Networks, JSME:* 93-96 (in Japanese).

NeuralWare, Inc. 1991. *Neural Computing: NeuralWorks Professional II/PLUS and NeuralWorks Explorer.*

Wong, F.S., Tung, A.T.Y., and Dong, W. 1992. Seismic hazard prediction using neural nets. *Proc. 10th World Conf. on Earthq. Eng.:* 339-343.

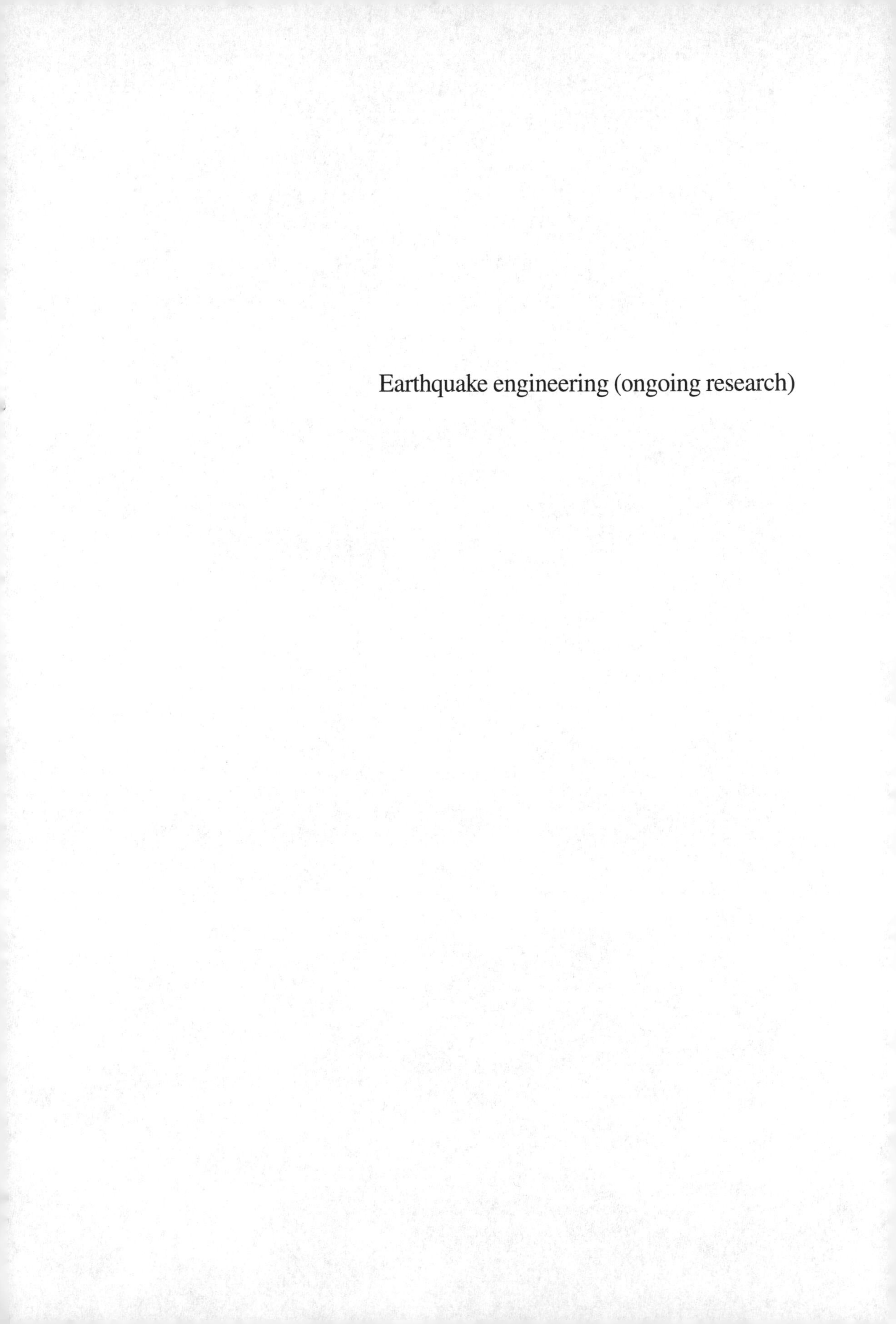

Earthquake engineering (ongoing research)

Structural Safety & Reliability, Schuëller, Shinozuka & Yao (eds) © 1994 Balkema, Rotterdam, ISBN 90 5410 357 4

Random vibration of structure with asymmetric hysteretic restoring force-deformation relation

Shigeru Aoki
Tokyo Metropolitan Technical College, Japan

ABSTRACT: This paper deals with asymmetric effect of yielding force on random vibration characteristics of structure with perfectly-elasto-plastic restoring force-deformation relation. By simulation method, effect of asymmetric characteristic on the maximum response, permanent deformation and absorbed energy by plastic deformation are examined. One yielding force is fixed and the other one is changed. The maximum response of acceleration and velocity are decreases as one yielding force decreases. The maximum response of displacement takes the minimum value when yielding forces are symmetric. Displacement response is significantly related to plastic deformation. Displacement response increases to the direction where one yielding force is smaller than the other one. Absorbed energy by plastic deformation increases as one yielding force decreases.

1 INTRODUCTION

Many studies about response of structures with hysteretic restoring force-deformation relation have been carried out (Zaiming, Katukura & Izumi 1991), (Ang 1988). In most studies, hysteretic characteristics are assumed to be symmetric. However, it is considered that there are some asymmetric characteristics in hysteretic restoring force-deformation relation of actual structures. It is important to evaluate effect of asymmetric characteristics on the response of structures.

In this paper, effect of asymmetric characteristic of hysteretic restoring force-deformation relation on response of structure subjected to random vibration such as earthquake excitaitons is examined. First, by an analysis using a simple analytical model, effect of asymmetric characteristic on the maximum response is evaluated. Second, permanent deformation which occurs in plastic region is obtained since asymmetric effect on displacement response is great. Third, absorbed energy by hysteresis loop is examined.

2 ANALYTICAL MODEL AND INPUT EXCITATIONS

Fig.1 shows an analytical model used in this study. A single-degree-of-freedom system is used for simplicity. It is assumed that hysteretic restoring force-deformation relation of this model is represented by perfectly-elasto-plastic model shown in Fig.2. Equation of motion with respect to displacement of mass relative to input point, z=x-y, is written as follows:

$$\ddot{z}+2\zeta\omega_n\dot{z}+f=-\ddot{y} \qquad (1)$$

where ζ is damping ratio, ω_n is natural circular frequency and f is restoring force. As input acceleration excitations $\ddot{y}$, stationary white noises are used in order to get basic response characteristics. 50 response time histories are obtained. Values of parameters are selected as ζ is 0.01 and natural period $T_n(=2\pi/\omega_n)$ is 1.0s. Upper yielding force of model shown in Fig.2 is determined as

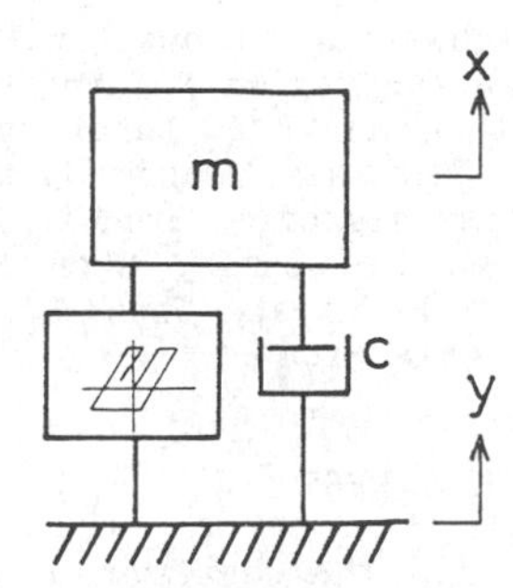

Fig.1 Analytical model used in this study

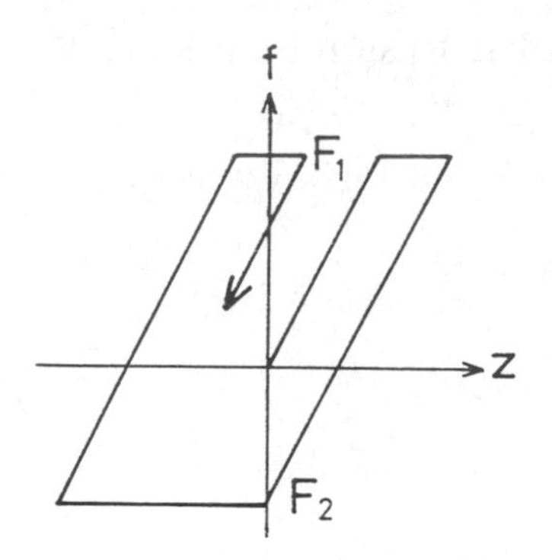

Fig.2 Perfectly-elasto-plastic restoring force-deformation relation

$$F_1 = \alpha \omega_n^2 Z_m \quad (2)$$

where α is a parameter which represents yielding effect. Z_m is expected value of the maximum displacement response of linear system which does not have hysteretic characteristic. Lower yielding force is determined as

$$F_2 = \beta F_1 \quad (3)$$

where β is a parameter which represents asymmetric effect.

3 MAXIMUM RESPONSE

In order to examine effect of asymmetric characteristic on the maximum response, expected values of the maximum response of absolute acceleration, relative velocity and relative displacement are obtained. In Table 1, results for α=0.7 are shown. Values in Table 1 are ratio of the maximum response to standard deviation of input excitations which is generated by computer simulation. Acceleration reponse and velocity response decrease as β decreases. On the other hand, displacement response takes the minumum value when β=1.0. It increases as β is greater than 1.0 or less than 1.0. Same characteristics can be seen for other values of α.

In Table 2, the number of the maximum values which occur in positive region N_+ and in negative region N_- are shown. When β is greater than 0.7, total numbers are less than 50. This reason is that some response time histories are within elastic limit. From this table, when β is less than 1.0, N_- is great. When β is greater than 1.0, N_+ is great.

4 DISPLACEMENT RESPONSE

In Table 1, the maximum displacement response takes the minimum value when β=1.0. This characteristic is examined from view point of permanent deformation. When hysteretic characteristic is taken into consideration, effect of permanent deformation due to plastic deformation on displacement response is great. In Fig.3, examples of displacement response time histories for β=0.5, β=1.0 and β=1.5 are shown. When β=0.5, resopnse increases to negative direction. When β=1.5, response increases to positive direction. When β=1.0, response vibrates around equilibrium position. In order to examine results of Fig.3 in detail, sum of plastic deformation in positive region Z_+ and that in negative region Z_- which are shown in Fig.4 are obtained. When β is less than 1.0, that is, lower yielding force is less than upper one, Z_- increases. When β is greater than 1.0, that is, upper yielding force is less than lower one, Z_+ increases. From these results, it is obvious that permanent deformation increases for the direction where one yielding force is less than the other one.

Table 1 Maximum response

β	Acc.	Vel. ($\times 10^{-1}$ s)	Disp. ($\times 10^{-2}$ s^2)
0.5	1.45	2.67	16.00
0.6	1.61	2.88	13.10
0.7	1.74	3.03	10.08
0.8	1.83	3.22	8.74
0.9	1.88	3.34	7.12
1.0	1.91	3.45	6.04
1.1	2.03	3.56	6.94
1.2	2.12	3.62	7.76
1.3	2.16	3.65	8.02
1.4	2.16	3.64	8.07
1.5	2.16	3.66	8.08

Table 2 Number of maximum response N_+ and N_-

β	Acc.		Vel.		Disp.	
	N_+	N_-	N_+	N_-	N_+	N_-
0.5	0	50	2	48	0	50
0.6	0	50	5	45	0	50
0.7	0	49	6	43	2	47
0.8	2	45	8	39	2	45
0.9	3	42	12	33	3	42
1.0	28	14	23	19	19	23
1.1	41	1	28	14	37	5
1.2	41	1	33	9	39	3
1.3	41	1	36	6	39	3
1.4	41	1	36	6	39	3
1.5	41	1	36	6	39	3

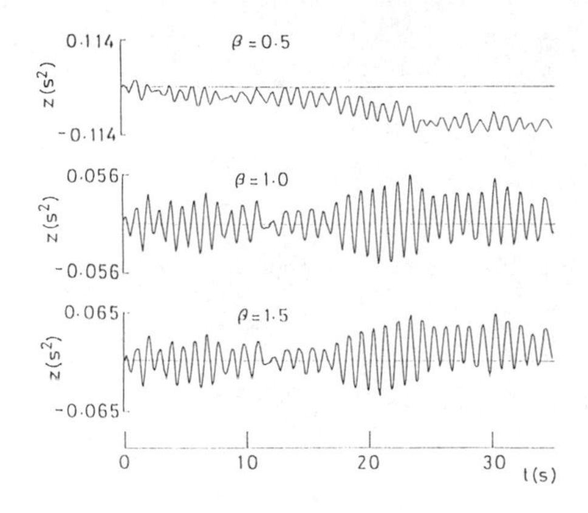

Fig.3 Time histories of displacement response

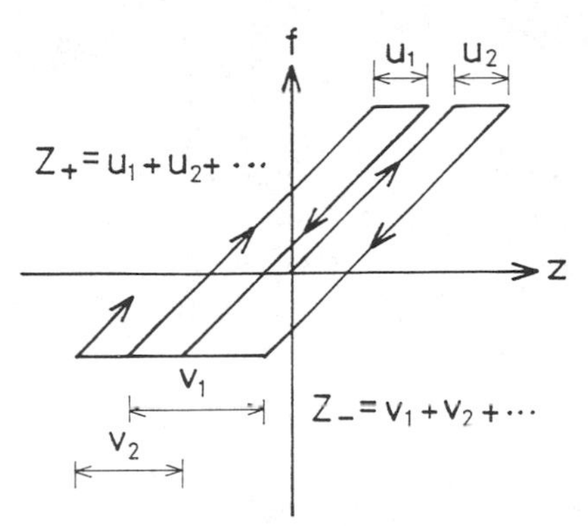

Fig.4 Sum of plastic deformation

Difference between Z_+ and Z_- is permanent deformation. Therefore, effect of permanent deformation on displacement response is great.

5 ABSORBED ENERGY BY PLASTIC DEFORMATION

Absorbed energy by plastic deformation of which effect on response characteristics is great is examined. Absorbed energy is obtained by the following equation.

$$E_n = F_1 Z_+ + F_2 Z_- \quad (4)$$

In Fig.6, expected value of E_n is shown. From this figure, E_n increases as β decreases. E_n is related to velocity response (Akiyama 1988). From Table 1 and Fig.6, the maximum velocity response decreases as E_n increases.

6 CONCLUSIONS

For structure with perfectly-elasto-plastic restoring force-deformation

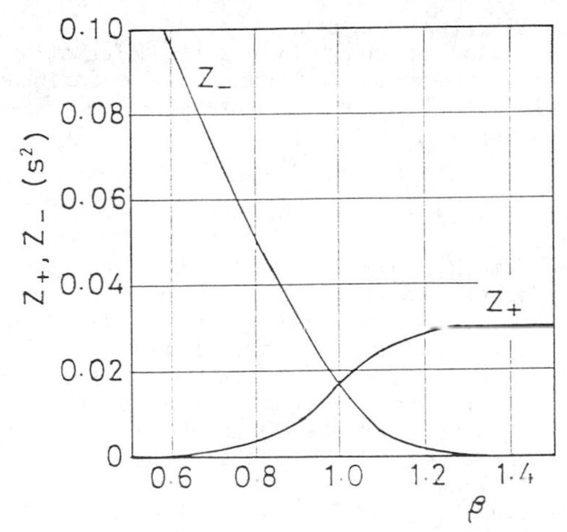

Fig.5 Sum of plastic deformation at positive region and negative region

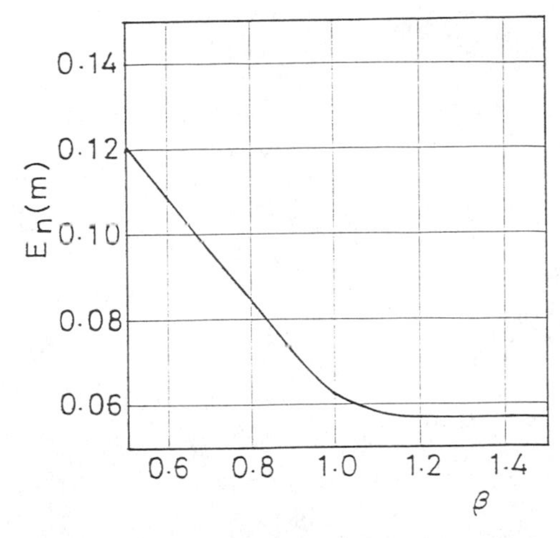

Fig.6 Absorbed energy by plastic deformation

relation, asymmetric effect of yielding force on random vibration characteristics is examined. Obtained results are summarized as follows.

(1) The maximum acceleration response and velocity response decreases as β which represents asymmetric effect defined by Eq.(3) decreases. The maximum response of displacement response takes the minumum value when β=1.0. When β is greater than 1.0 or less than 1.0, diaplacement response increases.

(2) Displacement response is significantly related to permanent deformation. Permanent deformation increases to the direction where one yielding force is

smaller than the other one.
(3) Absorbed energy by plastic deformation E_n increases as β decreases. The maximum velocity respone decreases as E_n increases.

REFERENCES

Akiyama,H. 1988. Earthquake resistant design based on the energy concept. Proceedings of Ninth World Conference on Earthquake Engineering. 5:905-910.

Ang,A.H-S. 1988. Probabilistic seismic safety and damage assessment of structures. Proceedings of Ninth World Conference on Earthquake Engineering. 8:717-728.

Zaiming,L., H.KATUKURA & M.IZUMI 1991. Synthesis and extension of one-dimensional nonlinear hysteretic models. Journal of Engineering Mechanics. ASCE, 117-1:100-109.

Structural Safety & Reliability, Schuëller, Shinozuka & Yao (eds) © 1994 Balkema, Rotterdam, ISBN 90 5410 357 4

A stochastic fault behavior model for spatially and temporally dependent earthquakes

Kimberly A. Lutz & Anne S. Kiremidjian
The John A. Blume Earthquake Engineering Center, Department of Civil Engineering, Stanford University, Calif., USA

ABSTRACT: Evidence suggests that large magnitude earthquakes on long faults exhibit both temporal and spatial dependence. Thus there is a need for a fault behavior model that incorporates these dependencies. The model presented in this paper is a generalized semi-Markov process, which is chosen for its ability to characterize the spatial dependence and represent random state transition times. The model assumes that a fault is divided into segments with well characterized behavior. It is further assumed that the stress accumulation rate is constant at all cells that comprise a characteristic fault segment. The model is applied to the northern section of the San Andreas fault. The probabilities of occurrences of events on this segment of the fault appear to be in close agreement with forecasts obtained by the Working Group on California Earthquake Probabilities (1990). Similarly, good agreement is obtained for the expected number of events as function of magnitudes.

1 INTRODUCTION

Previous earthquake occurrence models have assumed either spatial independence or temporal independence or both. Earthquake events can be assumed to be spatially independent if the co-seismic fault rupture occurs randomly along the length of the fault. A wide variety of earthquake occurrence models, including time- and slip-predictable models, assume spatial independence (e.g.,Kiremidjian and Anagnos, 1984). Temporal dependence is the property of the earthquake generating process that relates the time of occurrence of subsequent events to the time of occurrence of preceeding events. Poisson models assume temporal independence (as well as spatial independence) and require that only the rate of earthquakes and not with their clustering be specified. The assumption of spatial independence may be reasonable for characteristic type earthquakes, and the assumption of temporal independence may be reasonable for small magnitude earthquakes. However, for large magnitude events (approximately moment magnitude 6.5 and above) occurring infrequently along long faults, the evidence indicates that the assumptions of temporal and spatial independence are not valid.

It has been recognized that long faults do not rupture completely during a single earthquake. This has given rise to the concept of fault segmentation, which attempts to divide a long fault into segments, each of which is capable of rupturing independently. There are studies suggesting that physical controls in the fault zone define the ends of segments, and that these segments persist through many seismic cycles (Schwartz, 1988). Thus, it can be hypothesized that earthquakes do have a spatial correlation because the rupture zones depend upon the physical controls governing the segmentation rather than being uniformly distributed over the length of the fault.

These observations highlight the need for a fault behavior model that includes both spatial and temporal dependence. In this paper we present such a model based on generalized semi-Markov process theory. The model is applied to the northern portion of the San Andreas fault.

2 GENERALIZED SEMI-MARKOV PROCESSES

The fault behavior model is specified as a generalized semi-Markov process (GSMP). A

GSMP is comprised of one or more state variables that take on different values to describe the process as a function of time. A GSMP moves from state to state at random time intervals, which makes it particularly useful for modeling earthquakes. The theoretical development of GSMP is given in Haas and Shedler (1985) and Iglehart and Shedler (1983).

In a GSMP the probability of going from state i to state j depends not only on the states i and j, but also on the state transition mechanism. Similarly, the times between state transitions for such a process form a holding time distribution that again depends only upon the current state and the next state and on the mechanism causing the state transitions. Furthermore, a GSMP can have a complex state of state space comprised of many state variables. The basic physical quantity that the fault behavior model will track is the accumulated slip. Since it can be calculated from the slip rate and the elapsed time, accumulated slip is a convenient variable to describe the state of the fault. In addition, the slip released during an earthquake can be directly related to the moment release and thus the moment magnitude of the earthquake. Since the model is to have spatial dependence, it must keep track of the spatial distribution of accumulated slip. The use of a complex state space will make this possible by discretizing the fault into cells. The amount of slip accumulated on each one of these cells will then be represented by one state variable. The array of state variables defining the state of individual cells represents the state of accumulated slip on the entire fault.

3 FAULT BEHAVIOR MODEL

In order to develop the GSMP that will underlie the fault behavior model, a fault or a fault segment capable of completely rupturing in one earthquake is assumed to be composed of L subsegments with homogeneous properties (slip rate and earthquake interarrival time statistics). Each of the subsegments is further divided into cells of uniform length such that the entire fault is composed of N cells. Figure 1 shows a model of the northern San Andreas fault that was suggested by the second Working Group on California Earthquake Probabilities (1990); it has been discretized onto 30 km cells. The smallest earthquake simulated by the model corresponds to a rupture length of one cell. The fault behavior model traces through time the slip accumulated on each cell and the amount of slip release on each cell due to earthquake occurrences.

Define $\{X(t), t \geq 0\}$ to be a stochastic process where

$$X(t)=\{(A_1(t), B_1(t)), (A_2(t), B_2(t)), \ldots (A_N(t), B_N(t))\} \quad (1)$$

Each cell j has two state variables associated with it, $A_j(t)$ and $B_j(t)$. At time t, $A_j(t)=k$ if cell j is capable of rupturing k cells. This means that at time t, there is enough slip accumulated on cell j to cause a rupture with length of k cells. Since j is an index on the cell number, it can assume values from one to the maximum number of cells N. The number k refers to the number of cells that cell j can rupture, so it can assume values from zero (no rupture is possible) to N (the entire fault can rupture). An empirical relationship between average displacement from an earthquake and rupture length is used to relate the rupture length of a cell to the accumulated slip (Wells and Coppersmith, 1991). In this relationship it is assumed that the accumulated slip on a cell can be represented by the average displacement of an earthquake. At any time t, the value of Bj(t) is either 0 or 1. This state variable is needed to strictly formulate the model as a GSMP.

All the values that the process X(t) can assume form its state space S.

$$S = \{((a_1,b_1), (a_2,b_2),(a_N,b_N)) \;\varepsilon\; (\{0,1, \ldots N\} \times \{0,1\}^N\} \quad (2)$$

Equation 2 states that each of the a_j variables may take on values in the set {0,1, ... N} and that each of the b_j variables may take on values in the set {0,1}. (The lower case aj and bj are used, rather than the upper case $A_j(t)$ and $B_j(t)$ used above, to denote specific values of the state variables, rather than their values through time.) There are no restrictions on the permissible values of the state space as defined above.

In order to determine how and when the process X(t) moves between states, it is necessary to enumerate the event set E, which contains all the different events that can occur in this process. Each event is scheduled by simulating a random number from a distribution describing the time it takes for the event to occur. A clock with this amount of time is set for the event and as the time passes this clock will always show the amount of time remaining until the event occurs. The event that triggers the state transition will be the one with the shortest amount of time on its clock, i.e. the first to occur. The event set E for the fault behavior process is

$$E = \{e11, \ldots e_{N1}, e12, \ldots e_{N2}, e13, \ldots e_{N3}\} \tag{3}$$

For each cell j, the event e_{j1} is the event that cell j increments by one the number of cells that it is capable of rupturing. This occurs when enough time has passed that the cell has accumulated enough slip to rupture a length of fault that is longer by the length of one cell. For each cell j, the events e_{j2} and e_{j3} are events that cell j triggers and earthquake.

When the process X(t) is in a given state s, the events that can occur and trigger a transition onto the next state must be determined. This set of events E(s) is a subset of the entire event set E defined above. The event set mapping tells how to determine which events are in E(s).

For $s = \{((a_1,b_1), (a_2,b_2),(a_N,b_N))\} \in S$:

$e_{j1} \in E(s)$ if and only if $a_j \neq N$

$e_{j2} \in E(s)$ if and only if $b_j = 0$ (4)

$e_{j3} \in E(s)$ if and only if $b_j = 1$

Thus for each cell j, the events e_{j1} is scheduled unless there is already enough accumulated slip on cell j to rupture the entire fault. Event e_{j2} is scheduled when $b_j = 0$, and the event e_{j3} is scheduled when $b_j = 1$.

Since the events to be scheduled in the current state have been determined, it becomes necessary to specify how to set the clock for each event. For events e_{j1} the clock is set deterministically according to an empirical equation of average displacement versus surface rupture length. Dividing the average displacement by the slip rate yields the amount of time it takes for enough slip to accumulate to rupture one cell. In order to determine the additional amount of time it takes to accumulate enough slip to rupture m cells when currently m-1 cells can rupture, the time to rupture m and m-1 cells is determined in the same manner and then the time increment is applied.

To set the clock corresponding to event e_{j2}, it is necessary to determine if cell j triggered an earthquake that caused the current state transition. If so, then the clock for e_{j2} is set by simulating a random number from the distribution D_j, which is the distribution of times between cell j triggering an earthquake. This is not the same as the distribution of interarrival times for cell j rupturing during an earthquake because cell j can break due to another cell triggering rupture. The mean and standard deviation of D_j are determined by trial and error based on the estimated mean and standard deviation of the earthquake interarrival times. D_j must have only positive values so it is assumed to be lognormal, though other distributions, such as Weibull, could also be assumed.

Setting the clock corresponding to event e_{j2} when cell j did not trigger an earthquake is similar. When cell j breaks and $b_j = 0$, events e_{j3} is scheduled by simulating from D_j. If $b_j = 1$, event e_{j2} is scheduled. It becomes unnecessary to know if a given cell that ruptures triggered the transition if e_{j2} and e_{j3} are alternatively scheduled for all cells that break. The clock speeds for the process are specified as unity.

The final component that must be specified to completely describe the fault behavior process is the state transition mechanism. The state transition mechanism tells how the new state s' of the process X(t) is determined when the old state s and the trigger event set E* causing the state transition are known.

4 APPLICATION OF THE MODEL TO THE NORTHERN SAN ANDREAS FAULT

As an example of its application, the fault behavior model will be used to simulate the behavior of the northern section of the San Andreas fault, which ruptured as a whole in 1906. The fault depth will be assumed to be 20 km throughout. Results are presented from the 30 km long cells shown in Figure 1. Results will be obtained using slip rates of 15, 19, and 23 mm/yr. Figure 2 shows the number of events with a given moment magnitude (M_W) or greater per year. The largest magnitude earthquake generated for all six cases falls between $M_W = 8$ and $M_W = 8.2$, which is consistent with the assumed magnitude of the 1906 earthquake. Figure 3 shows the probabilities of each cell rupturing within the given time frames (50, 100, and 150 years) assuming 30 km cells and a slip rate of 19mm/yr. Note that the probability of an earthquake anywhere on the fault within 50 years is 0.02 when a slip rate of 19 mm/yr is

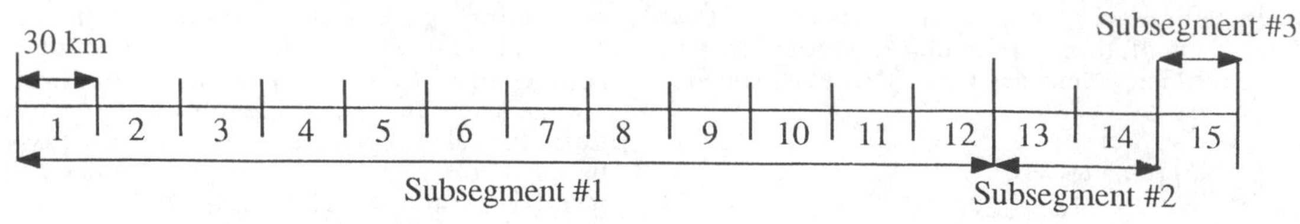

Figure 1. Model of the northern San Andreas fault with 30 km cells.

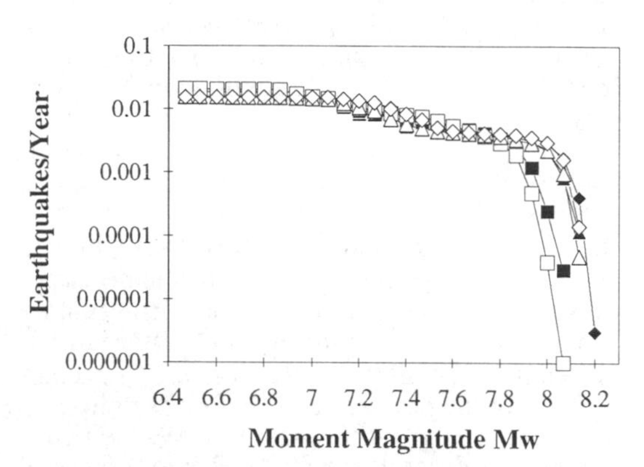

Fig. 2. Cumulative number of earthquakes per year as a function of M_W.

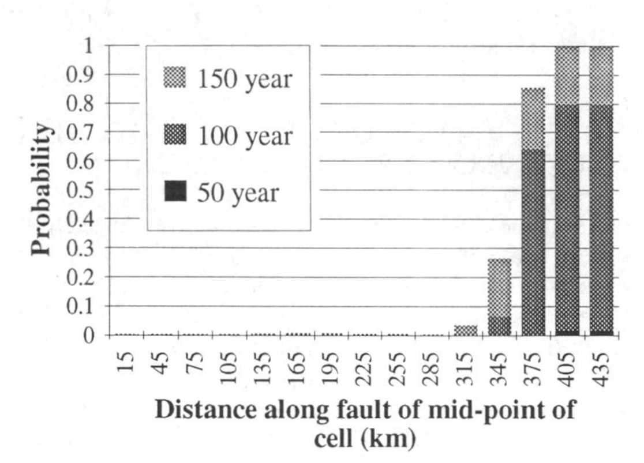

Fig. 3. Probability of rupturing 30 km cells on the north. San Andreas fault with a slip rate of 19mm/yr.

assumed. Similar results were obtained for slip rates of 15 mm/r and for cell of 20 km. The overall probabilities are similar for the different cell sizes, but with the 20 km cells, there is a more gradual transition form the low rupture probability for the cells at the northern end of the fault to the high rupture probability for the cells at the southern end.

ACKNOWLEDGMENTS

The authors would like to thank Professor Gerald Shedler who provided guidance on the mathematical aspects of GSMPs. We express our gratitude to Carl Stepp and John Schneider whose coments were valuable in confirming the rupture mechanism. This material is based upon work supported under a National Science Foundation Graduate Fellowship. Additional support was provided under EPRI Grant RP2356-91 and NSF Grant EID-9024032 .

REFERENCES

Haas, P.J. and G.S. Shedler 1985. Regenerative Simulation Methods for Local Area Computer Networks. *IBM J. Res. Develop.* Vol. 29. No. 2: 194-205.

Iglehart, D.L. and G.S. Shedler 1983. Simulation of Non-Markovian Systems. *IBM J. Res. Develop.* Vol. 27, No. 5: 472-480.

Kiremidjian, A.S. and T. Anagnos 1984. Stochastic Slip-Predictable Model for Earthquake Occurrences. *Bull. Seism. Soci. Am.* Vol. 74, No. 2: 739-755.

Schwartz, D.P. 1988. Geologic Characterization of Seismic Sources: Moving into the 1990s. *Earthquake Engineering and Soil Dynamics II - - Recent Advances in ground Motion Evaluation*: 1-42 Proceedings of the Specialty Conference on June 27-30, Park Citv. Utah.

Wells, D.L. and K.J. Coppersmith February 20, 1991. Personal communication.

Working Group on California Earthquake Probabilities 1990. Probabilities of Large Earthquakes in the San Francisco Bay Region, California. *U.S. Geological Survey Circular 1053.*

Structural Safety & Reliability, Schuëller, Shinozuka & Yao (eds) © 1994 Balkema, Rotterdam, ISBN 90 5410 357 4

Codes and reliability of reinforced concrete buildings in seismic zones

G. Sara' & R. Nudo
Dipartimento di Costruzioni, University of Florence, Italy

ABSTRACT: In this paper a procedure is proposed, aimed to check, with a specific reference to reinforced concrete buildings, efficacy of code tools to secure, in seismic zones, constructive products supplied with suitable standard reliability.

1 INTRODUCTION

Seismic risk of buildings depends on environment specific aggressiveness - seismic liability of site zone, related to local seismicity and mechanical and morphologic properties of placing site - and also on seismic vulnerability of constructive system due to architectonic-structural characteristics (Sara' (1987)).

The level of seismic vulnerability of a building, and then its reliability in a given environmental context, depends on realization procedures (relatively to design and execution phases) as influenced by construction rules provided by normative set.

This research is aimed to define a suitable procedure to evaluate reliability of code tools, that is efficacy in performing their natural role of warrantors for attainment of constructive products supplied with suitable standard reliability.

For this purpose, with reference to constructive types ordinarily adopted for R.C. buildings, a study has been carried out on actual capacity of italian and european present technical codes (D.M.LL.PP. (1986) and Eurocode 8 (1989)) to predict and prevent activation of potential sources of seismic vulnerability just as pointed out, most of all, by damages analysis on the occasion of earthquake real tests.

Definition of links among damage set, vulnerability sources, code prescriptions set, allows to check, with reference to specific seismic events, efficacy of code tools devoted to seismic prevention for buildings.

2 STRUCTURE AND CHARACTERISTICS OF SEISMIC CODES FOR R.C. BUILDINGS

In order to make easy analysis about contents of normative documents and comparison among them, prescriptions of examined codes have been assembled and arranged according to a common paradigm (table 1) (Sara' and Nudo (1992)). On the inside of the assumed paradigm, various topics, closely connected with traditional design phases, are recognizable: identification of design object (system and relating environment, points 1.1 and 1.2); indication of binding requirements on construction (placing site, building type, structural characteristics, points 2.1 and 2.2); indication of procedures to perform reliability verifications (point 2.3).

With a specific reference to prevention targets, instructions relating to design (constructive types, quality, dimensions, calculation requirements) may be summarily classified as follows:

a) prescriptions directed to check "demand" (internal forces and deformations, point 2.2.1) and related prediction (points 2.3.1 and 2.3.2.1);
b) prescriptions directed to check structural "capacity" (strength and ductility, point 2.2.2) and related prediction (point 2.3.2.2);
c) prescriptions directed to check building reliability through "demand-capacity balance" (point 2.3.2.2).

It is worthwhile to remark that binding prescriptions, aimed to check demand (point 2.2.1) and capacity (point 2.2.2), are primarily directed to make up for lacks of prediction calculation.

Table 1. Prescriptions of italian and european seismic codes.

INSTRUCTIONS		ITALIAN TECHNICAL CODE - 1986	EUROCODE 8 - 1988
1 Definition of construction general features			
1.1 Definition of placing site characteristics		Seismicity degree of the site zone	Max. ground acceleration of the site zone Type of soil profile
1.2 Definition of building general characteristics			Building configuration: characteristics and class of regularity Type of structural system: -type of earthquake-resistant structural system -class of ductility
2 Design of buildings			Design of regular buildings
2.1 Placing requirements			
2.1.1 Foundation soil requirements		Potentially unstable soils (sloping or subjected to liquefaction soils): check of implantation stability	Potentially unstable soils (sloping or subjected to liquefaction soils): check of implantation stability
2.1.2 Distance among buildings requirements		Street nearness: observance of min. roadway width Isolation: reference to municipal regulations Adjacency: observance of min. distance	Adjacency: observance of min. distance
2.2 Building requirements			
2.2.1 Building type requirements	2.2.1.1 Building on the whole		Height: reference to national regulations Building configuration: regular (if possible)
	2.2.1.2 Building components	Foundations: -foundations connection: connection by a grid of beams (otherwise verification of relative displacements effects) -foundation layer: soil zone scantly subjected to seasonal variations of water content Non-structural walls: -infilling and partition masonry walls: requirements about high and large panels -openings: requirements about strengthening	Foundations: code section not yet available (EC8 - part 5) Non-structural walls: requirements about making of infilling masonry walls
2.2.2 Materials and structural elements requirements	2.2.2.1 Materials		Concrete: requirements about min. strength Bars: -steel: requirements about mechanical characteristics (ultimate strain, ultimate and yielding stress) -critical regions of earthquake-resistant elements (high stresses zones): high bond longitudinal bars
	2.2.2.2 Frames		Beams and columns: -requirements about cross section dimensions -requirements about reinforcements (longitudinal reinforcements, transverse reinforcements in critical regions) Beam-column joints: -beam-column eccentricity: observance of max. eccentricity -requirements about reinforcements
	2.2.2.3 Single or coupled structural walls		Structural walls: -requirements about thicknesses -requirements about reinforcements Coupling beams: requirements about reinforcements
2.3 Reliability verifications			
2.3.1 Calculation of response to design seismic actions	2.3.1.1 Calculation of design seismic actions	Calculation of storey horizontal seismic forces (by simplified dynamic analysis; modal analysis obligatory for buildings with relatively high fundamental period): -direction of the seismic action: according to two orthogonal conventional directions -live load masses: reduced -distribution coefficients: linearly decreasing toward the base -intensity: depending on seismicity degree, elastoplastic normalized response spectrum and coefficients of foundation, behaviour and importance Calculation of vertical seismic forces: for overhangs, great span horizontal members, thrusting members exclusively	Calculation of storey horizontal seismic forces (by simplified dynamic analysis; modal analysis obligatory for buildings with relatively high fundamental period): -direction of the seismic action: according to two orthogonal conventional directions -live load masses: reduced -distribution coefficients: linearly decreasing toward the base -intensity: depending on max. ground acceleration, elastic normalized response spectrum and factors of site, behaviour and damping Calculation of vertical seismic forces: for overhangs, great span horizontal members, thrusting members, beams bearing columns exclusively
	2.3.1.2 Calculation of effects due to design seismic actions	Modelling of the structural system: spatial system of earthquake-resistant elastic elements Calculation of storey torsional moments: referring to conventional eccentricity of storey horizontal forces Definition of seismic action combinations: horizontal seismic action in a single direction at a time plus vertical seismic action Calculation of combined effects: internal forces and deformations	Modelling of the structural system: spatial system of earthquake-resistant elastic elements taking into account the influence of non-structural infilling masonry walls Calculation of storey torsional moments: referring to conventional eccentricity of storey horiz. forces Definition of seismic action combinations: by means of combination coefficients Calculation of combined effects: internal forces and deformations
2.3.2 Verifications	2.3.2.1 Calculation of response to overall design actions	Definition of design action combinations: dead loads + + live loads + seismic actions	Definition of design action combinations: by means of importance and combination factors Calculation of design internal forces for beams and columns: elastic internal forces with redistribution (when allowed) and adjustments based on consideration of limit equilibrium and 2nd order effects
	2.3.2.2 Definition of limit capacities and reliability verifications	Verifications: according to allowable stresses method	Verifications (according to limit states method): ultimate resistance, ductility (check of max. axial force), global stability (overturning and sliding), foundations (EC8 - part 5, not yet available), damage limitation (adjacent buildings displacements, interstorey drifts)

3 ANALYSIS OF DAMAGES, SOURCES OF VULNERABILITY AND COVERAGE OF CODES

Effectiveness of code tools towards reduction of building vulnerability degree is checked, on the occasion of seismic events, through evaluation of damage levels and modes.

The analysis of damages allows to trace crisis mechanisms and related constructive sources of activation. A subsequent vulnerability analysis permits to determine, with reference to specific damages, the mechanical cause of vulnerability (relating to the expected demand-capacity balance) and the origin of the same on the inside of building realization process (design and execution phases). Examination of code prescriptions concerning identified constructive sources of vulnerability permits to evaluate its preventing efficacy.

In order to evaluate effectiveness of present italian seismic code relating to R.C. buildings, almost unchanged since 1975, we refer to a survey of damages on R.C. earthquake-resistant buildings performed in S. Angelo dei Lombardi (Irpinia, Italy) (Aristodemo, Sara' and Vulcano (1985)) after the strong earthquake of november 1980 (Richter magnitude about 6.5), earthquake characterized by an intensity consistent with the seismic category of the zone.

Analyses of damages and vulnerability sources (Sara' (1987)) are summarily reported in table 2, supplemented by a conclusive judgement on actual covering capacity - with respect to the various sources of vulnerability - provided by prescriptions of present italian code; for the purpose of comparison, such evaluation has been then extended to the european seismic code.

To explain compilation criteria of table 2, it is helpful to observe what follows:

a) Damage location. "Primary" damages have been considered, namely those considered of initial activation. Damages in foundations, generally verifiable with difficulty, have not been reported. The analyzed elements have been classified as upright elements (frame columns, including nodes, and shear walls), transverse elements (frame beams, staircase climbing beams, coupling beams for shear walls), floors and slabs, non-structural walls (infilling masonry walls).
b) Crisis mechanism. Crisis mechanisms can be concomitant; the mechanism supposed as predominant relating to the examined damage has been reported.
c) Constructive source of crisis mechanism activation. The indicated sources, related to the building characteristics, can be directly or indirectly connected with the specific damage to which they are referred.
d) Cause of vulnerability. "Pathological" vulnerability has been considered, that is vulnerability derived by lack of design prediction. In the case of concomitant causes of vulnerability, the cause assumed as prevalent to determine a demand-capacity balance more unfavourable than the expected one, has been considered. The following causes have been recognized: causes referable to action excess, that is excess of absorbed energy due to increase of motion (torsion, resonance, soft soil effects) or because of scanty dissipation; causes referable to stress excess in structural elements and connected with a more severe stress distribution than the expected one; causes referable to resisting capacity shortage, connected with deficiency in strength and ductility of sections and/or materials.
e) Origin of vulnerability. Causes of missed achievement of expected demand-capacity balance have been ascribed to design or execution phases.
f) Coverage of code prescriptions. Seismic codes, integrated with R.C. building regulations, have been considered. Judgement on prevention capacity of code instructions has been graduated referring to three different levels of efficacy, also as regards explicitness and completeness.

4 CONCLUSIONS

The developed survey procedure, by means of examination of seismic code prescriptions in connection with recognized sources of vulnerability, permits to evaluate efficacy degree by which codes perform their role, that is to assure obtainment of suitable standard reliability levels for buildings.

In particular, it can be noticed that present italian code - in spite of some slight improvements with respect to code in force at the time of last italian strong earthquake (Irpinia, 1980) - is not sufficiently watchful against most of potential sources of vulnerability; particularly as regards influence, on the ductility demand, of architectonic-structural irregularities and execution of constructive details in critical regions of structural elements.

European code assures, in general, a better prevention coverage; however, it should be desirable a ductility control based, to a greater extent, on analytical verifications; this ought to allow a reduction of the numerous prescriptive restraints - limitations on structural type and dimen-

Table 2. Seismic damages and coverage of codes: analysis referred to damages in S. Angelo dei Lombardi, after Irpinia Earthquake of 1980.

ANALYSIS OF DAMAGES		ANALYSIS OF VULNERABILITY SOURCES			COVERAGE OF CODES	
damage location	crisis mechanism	constructive source of crisis mechanism activation	cause of vulnerability	origin of vulnerability	italian code	eurocode
upright structural elements	interstorey shear	lack of transverse connecting members	stress excess	design phase	unsatisfactory	unsatisfactory
		transverse connection by low beams	stress excess	design phase	satisfactory	satisfactory
		pilotis interstorey	stress excess	design phase	scanty	satisfactory
		foundations on different levels	stress excess	design phase	scanty	satisfactory
	interstorey torsional moment	asymmetry (2) or lack of proportion in plan	action excess	design phase	scanty	satisfactory
	shear-bending in upright structural elements	squat column (3)	stress excess	design phase	scanty	scanty
		shear wall (4)	stress excess	design phase	unsatisfactory	satisfactory
		cavities for conduits	strength shortage	execution phase	satisfactory	satisfactory
		exuberance of longitudinal reinforcement	strength shortage	design phase	unsatisfactory	satisfactory
		shortage of reinforcement spacing	strength shortage	design and/or execution phase	satisfactory	satisfactory
		shortage of stirrups	strength shortage	design and/or execution phase	unsatisfactory	satisfactory
		low grade concrete	strength shortage	execution phase	satisfactory	satisfactory
		homogeneity and continuity lack in concrete (5)	strength shortage	execution phase	unsatisfactory	unsatisfactory
	hammering	building adjacency	action excess	design phase	satisfactory	satisfactory
transverse structural elements	shear-bending in transverse structural elements	coupling beams for shear walls	stress excess	design phase	unsatisfactory	satisfactory
		bad anchorage of longitudinal bars	strength shortage	design and/or execution phase	satisfactory	satisfactory
floors and slabs	bending (1)	lack of beams in the floor (6)	stress excess	design phase	unsatisfactory	unsatisfactory
		staircase climbing slabs	stress excess	design phase	unsatisfactory	scanty
non-structural walls	shear	low compactness of infilling walls	strength shortage	execution phase	unsatisfactory	scanty
	jolting	infilling walls of bow-windows	action excess	design phase	unsatisfactory	scanty

Notes:
1) Reference to bending induced by lateral displacements of building.
2) Asymmetry due to plan shape and to irregular disposition of shear walls, infilling walls or staircase frames.
3) Particular reference to parts of columns defined by columns connection with staircase climbing beams.
4) Specific reference to the bases of shear walls.
5) Particular reference to components segregation and recasting.
6) Reference to beams directed as the floor joists.

sions - at the present justified by uncertainties and difficulties about effective modelling of the actual dynamic behaviour of R.C. structural systems beyond the elastic range.

ACKNOWLEDGMENTS

The research has been supported by Italian Scientific Council CNR-GNDT.

REFERENCES

Sara', G. 1987. Valutazione della vulnerabilita' sismica degli edifici. Specialization Course on "Criteri e metodologie d'intervento per il restauro statico del patrimonio edilizio". Ordine degli Ingegneri della Provincia di Catania. Catania.

D.M.LL.PP. 24 gennaio 1986. Norme tecniche relative alle costruzioni in zone sismiche. Gazzetta Ufficiale No. 108 of 12/05/1986.

Eurocode 8 (draft, may 1988). Strutture in zone sismiche. Progetto Commissione CEE. Rapporto EUR 12266 IT. Luxembourg, 1989.

Sara', G. and Nudo, R. 1992. Meaning and evolution of building codes: Eurocode 8 and reinforced concrete constructions in seismic zones. Proc. 10th WCEE: 5661-5666. Madrid.

Aristodemo, M., Sara', G. and Vulcano, A. 1985. Esperienze relative al comportamento sotto sisma delle costruzioni. In : Sara', G., Ingegneria antisismica, chapter XIII. Napoli, Liguori.

Structural Safety & Reliability, Schuëller, Shinozuka & Yao (eds) © *1994 Balkema, Rotterdam, ISBN 90 5410 357 4*

Lifeline response to multiple support seismic excitations

A. Zerva
Drexel University, Philadelphia, Pa., USA

ABSTRACT: This study analyzes the sensitivity of the response of buried pipelines and above-ground lifelines when different models are used to describe the spatial variability of the seismic ground motions at their supports. Differential displacements and differential response spectra are developed from spatial incoherence models commonly used in lifeline earthquake engineering, and the contribution of the models to the quasi-static and dynamic response of the structures is compared. It is shown that the spatial variability models induce different lifeline response, and that the differences correlate with the degree of exponential decay of the models with separation distance and frequency.

1 INTRODUCTION

The seismic response of lifelines, such as bridges and pipelines, can be significantly affected by the spatial variability of the seismic motions at their supports. The spatial variability is usually described as a function that decays exponentially with separation distance and frequency. Various expressions and degrees of exponential decay appear to fit data recorded at different sites or at the same site but for different earthquakes, and, as a result, there is a multitude of expressions that describe the spatial variability of the seismic motions.

Analyses of the effects of the spatial variability on the response of extended structures are commonly based on only one of the several existing descriptions for the spatial variability of the seismic motions; the results of such analyses, however, may strongly depend on the particular spatial variability model that was used. This study investigates the sensitivity of the response of buried pipelines and above-ground lifelines to the spatial variability expression chosen to model the incoherence of the seismic motions. Differential displacements and differential response spectra are developed from the spatial variability models and compared. The characteristics of the models that result in the highest quasi-static and dynamic response of the structures are identified.

2 DESCRIPTION OF SEISMIC GROUND MOTION

In stationary random vibration analyses of lifeline systems, the seismic ground motion is described by its cross spectral density between the motions of two stations at a distance ξ apart from each other as:

$$S(\xi,\omega) = S(\omega)\, \rho(\xi,\omega) \tag{1}$$

$S(\omega)$, with ω indicating frequency in rad/sec, is the power spectral density (PSD) of the motions, which is assumed to be the same at all locations on the ground surface. In this study, the PSD of the seismic motions is described by the Clough-Penzien (1975) spectrum:
-acceleration ($\ddot{u}$) PSD:

$$S_{\ddot{u}}(\omega)=S_o \frac{\omega_g^4+4\zeta_g^2\omega_g^2\omega^2}{(\omega_g^2-\omega^2)^2+4\zeta_g^2\omega_g^2\omega^2} \cdot \frac{\omega^4}{(\omega_f^2-\omega^2)^2+4\zeta_f^2\omega_f^2\omega^2} \tag{2a}$$

-displacement (u) PSD:

$$S_u(\omega) = S_{\ddot{u}}(\omega) / \omega^4 \tag{2b}$$

with ω_g=15.46rad/sec, ω_f=1.636rad/sec, ζ_g=0.623 ζ_f=0.619 (Hindy and Novak, 1980); these values correspond to firm soil conditions. It is assumed that S_o is equal to 1cm^2/sec^3.

$\rho(\xi,\omega)$ in Eq. 1 is the frequency dependent spatial correlation function. From the available descriptions for the spatial variability, the following two are compared herein, mainly, because they are commonly used in random vibration analyses of lifeline systems. These are: Harichandran and Vanmarcke's (1986) model-model I- developed from the analysis of the recorded data of Event 20 at the SMART-1 array:

$$\rho(\xi,\omega)=c \exp\{\frac{-2|\xi|(1-c+ac)}{a\theta(\omega)}\}+(1-c) \exp\{\frac{-2|\xi|(1-c+ac)}{\theta(\omega)}\}$$

$$\theta(\omega)=k\ [1+ (|\omega|/2\pi f_o)^b]^{-1/2} \tag{3}$$

with $c=0.736$; $a=0.147$; $k=5210m$; $f_o=1.09Hz$ and $b=2.78$; and Luco and Wong's (1986) model-model II-developed from the analysis of wave propagation through random media:

$$\rho(\xi,\omega)= \exp[-\alpha^2\omega^2\xi^2] \quad (4)$$

in which, α is the incoherence parameter that controls the exponential decay of the model. $\alpha=2\cdot10^{-4}$sec/m -case 1- and $\alpha=10^{-3}$sec/m -case 2- are used in the analysis. The variation of the models with frequency at separation distances $\xi=100$ and 300m is presented in Fig. 1. Zerva (1992a, b) has shown through simulations that spatial variability models such as models I and II represent the spatial incoherence of the seismic motions (change in the shape of the motions); the apparent propagation of the motions on the ground surface is not reproduced by these models. Spatial incoherence effects overshadow the apparent propagation effects in near source regions, where the values of the apparent propagation velocity are high (Zerva, 1992b); this analysis concentrates on the effects of spatial incoherence on the response of lifelines.

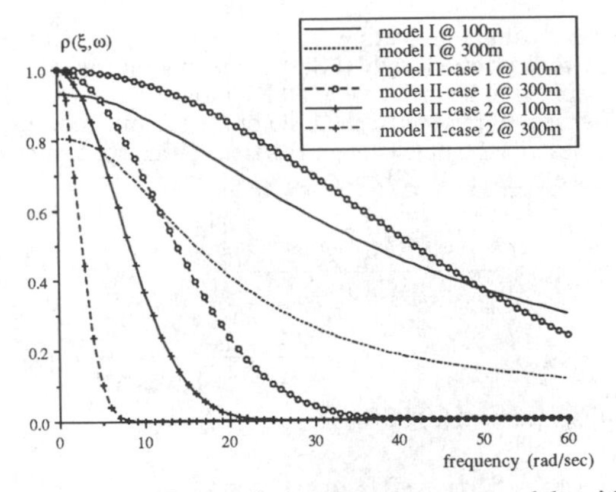

Figure 1. Variation of spatial incoherence models with separation distance and frequency

3 EFFECT OF CHOICE OF SPATIAL INCOHERENCE MODEL ON LIFELINE RESPONSE

Buried pipelines--The parameter that controls the seismic response of straight, buried pipelines is the seismic ground strain along their axis, which can be estimated from the ratio of the differential displacement between two stations divided by their separation distance. Root-mean-square (rms) differential displacements can be obtained from:

$$\sigma_{\Delta u}(\xi)= \left\{ \int_{-\infty}^{+\infty} 2S_u(\omega)[1-\rho(\xi,\omega)]\, d\omega \right\}^{-1/2} \quad (5)$$

The rms differential displacements based on models I and II (cases 1 and 2) for support separation distances ranging from 0 to 500m are presented in Fig. 2. The differential rms displacements of model II increase with increasing incoherence parameter, since, as α increases, the motions become less correlated (Fig. 1). Figure 2 indicates that the rms differential displacements of model II-case 1 are considerably lower than those of models I and II-case 2. These differences result from the different behaviour of the spatial variation functions (Fig. 1) in the low frequency range that controls the displacements (ω<5rad/sec from Eq. 2b). Model II-case 1 produces essentially full correlation in this frequency range (Eq. 4) resulting in low values for the differential displacements. On the other hand, model I results in partial correlations even at zero frequency for finite separation distances (Eq. 3) and yields high values for the differential displacements. Model II-case 2 produces differential displacements of comparable amplitude as those of model I, although the exponential decay of the spatial incoherence of the two models is significantly different (Fig. 1). The spatial incoherence of model II-case 2 is full at frequencies close to zero, but decays sharply at low frequencies and causes large differential displacements. It is noted that fully coherent motions ($\rho(\xi,\omega)=1$ for any separation distance and frequency) result in zero differential displacements (Eq. 5). It follows then that spatial incoherence models with partial correlation at low frequencies produce higher differential displacements and seismic ground strains, and, consequently, higher response in buried pipelines than models fully correlated in that frequency range.

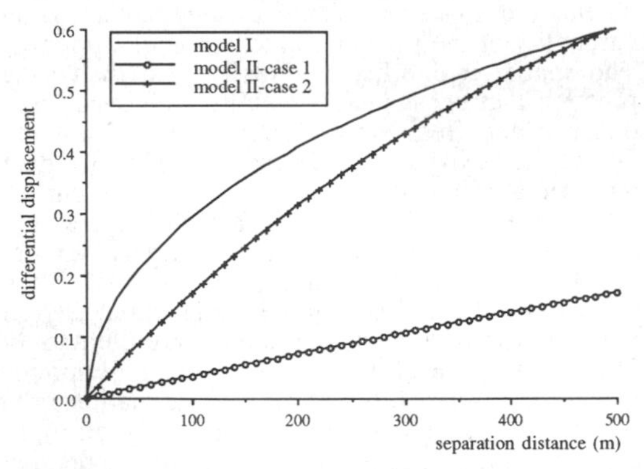

Figure 2. RMS differential ground displacement [cm]

Above-ground lifelines--The PSD of the seismic response at location y along the axis of an extended structure (e.g., a beam) on multiple supports subjected to spatially variable seismic ground motions is, generally, given by (Zerva, 1992c):

$$S(y,\omega)=\sum_{J=1}^{N}\sum_{L=1}^{N} r_J^{(n)}(y)\, r_L^{(n)}(y)\, S_{u_J u_L}(\omega)$$

$$+\sum_{J=1}^{N}\sum_{L=1}^{N}\sum_{k=1}^{K} [\Phi_k^{(n)}(y)\{r_J^{(n)}(y)R_{kL}H_k(\omega)S_{u_J\ddot{u}_L}(\omega) + r_L^{(n)}(y)R_{kJ}H_k^*(\omega)S_{\ddot{u}_J u_L}(\omega)\}]$$

$$+\sum_{J=1}^{N}\sum_{L=1}^{N}\sum_{k=1}^{K}\sum_{m=1}^{K} \Phi_k^{(n)}(y)\Phi_m^{(n)}(y)R_{kJ}R_{mL}\cdot$$

$$\cdot H_k^*(\omega) H_m(\omega) S_{\ddot{u}_J \ddot{u}_L}(\omega) \quad (6)$$

in which, $r_J^{(n)}(y)$ indicates the nth derivative with respect to y of the shape function for unit displacement at support J; $\Phi_k^{(n)}(y)$ is the nth derivative of the modeshape for mode k; R_{kJ} is the participation factor for mode k and the excitation at support J; N is the number of supports and K the number of modes required for the evaluation for the response quantity under consideration; $H_k(\omega)=[(\omega^2-\omega_k^2)+2i\zeta_k\omega_k\omega]^{-1}$ is the frequency transfer function for mode k, with ω_k and ζ_k indicating the frequency and the damping coefficient for mode k; and * indicates the complex conjugate.

$S_{u_J u_L}(\omega)=S_u(\omega)\rho(\xi_{JL},\omega)$; $S_{\ddot{u}_J \ddot{u}_L}(\omega)= S_{\ddot{u}}(\omega)\rho(\xi_{JL},\omega)$;

and $S_{\ddot{u}_J u_L}(\omega) = S_{\ddot{u}_J u_L}(\omega) = -S_{\ddot{u}}(\omega)\, \rho(\xi_{JL},\omega) / \omega^2$ indicate the cross spectral densities between the ground motions (displacements and accelerations) at supports J and L; the distance between the supports is ξ_{JL}. When n=0, Eq. 6 describes the PSD of total displacements. For n>0, the equation provides the PSDs of the derivatives of total displacements with respect to y, and is proportional to the PSDs of internal forces in the structure.

In Eq. 6 the double summation represents the quasi-static contribution, the quadruple summation the dynamic contribution and the triple summation the contribution of the cross correlation between the quasi-static and dynamic terms to the total response along the structure. Der Kiureghian and Neuenhofer (1991) indicated that the cross terms between the quasi-static and the dynamic response in Eq. 6 can be neglected if the natural frequencies of the structure are higher than approximately 0.5Hz. They also suggested that, for incoherent seismic motions, such as the ones described by models I and II, and for firm soil conditions, the cross terms between modes in the quadruple summation (Eq. 6) are small for well separated frequencies. Accordingly, the dominant terms in Eq. 6, that control the lifeline response, are the quasi-static ones (double summation in Eq. 6):

$$S_{QS}(y,\omega)= \sum_{J=1}^{N} \sum_{L=1}^{N} r_J^{(n)}(y)\, r_L^{(n)}(y) S_{u_J u_L}(\omega) \quad (7)$$

and from the dynamic terms (quadruple summation in Eq. 6), the ones that correspond to the excitation of individual modes:

$$S_D(y,\omega)=$$

$$\sum_{J=1}^{N} \sum_{L=1}^{N} \sum_{=1}^{K} [\Phi_k^{(n)}(y)]^2 R_{kJ} R_{kL} |H_k(\omega)|^2 S_{\ddot{u}_J \ddot{u}_L}(\omega) \quad (8)$$

It can be shown (Zerva, 1992c) that the rms quasi-static internal forces (n>0 in Eq. 7) in above-ground lifelines are equal to the weighted superposition of the rms differential displacements between the supports, with the weights being functions of the structural properties. Accordingly, the discussion on the effect of the exponential decay of the spatial incoherence models on the response of buried pipelines applies also to the contribution of the models to the quasi-static internal forces in above-ground lifelines.

The rms dominant dynamic response can be evaluated from Eq. 8 as:

$$\sigma_D^2(y) = \sum_{k=1}^{K} \frac{[\Phi_k^{(n)}(y)]^2}{\omega_k^4} \{ \sum_{J=1}^{N} R_{kJ}^2 A^2(\omega_k,\zeta_k)$$

$$+ \sum_{J=1}^{N-1} \sum_{L=J+1}^{N} R_{kJ} R_{kL} A^2(\omega_k,\zeta_k,\xi_{JL})\} \quad (9)$$

in which,

$$A^2(\omega_k,\zeta_k)=\omega_k^4 \int_{-\infty}^{+\infty} |H_k(\omega)|^2\, S_{\ddot{u}}(\omega)\, d\omega \quad (10)$$

and

$$A^2(\omega_k,\zeta_k,\xi_{JL})=2\omega_k^4 \int_{-\infty}^{+\infty} |H_k(\omega)|^2\, S_{\ddot{u}_J \ddot{u}_L}(\omega)\, d\omega \quad (11)$$

$A(\omega_k,\zeta_k)$ is the rms response of a single-degree-of-freedom oscillator with frequency ω_k and damping ζ_k subjected to the seismic motions described by the Clough-Penzien spectrum; it is proportional to the ordinates of the conventional response spectrum. $A(\omega_k,\zeta_k,\xi_{JL})$, termed differential response spectrum, is the rms contribution of the spatial incoherence between the accelerations of two supports J and L to the dynamic response of the structure filtered through a single-degree-of-freedom oscillator with frequency ω_k and damping ζ_k. The comparison of the effects of the spatial variability models on the dynamic response of lifelines reduces then to the comparison of their corresponding differential response spectra, which depend only on the seismic ground motion characteristics and given values for the modal frequency and the damping coefficient.

Figure 3 presents the differential response spectra obtained from equal support excitations and the spatial incoherence models for support separation distances of 100 and 300m. The modal frequency varies from 3.14(=π) to 60 rad/sec and the damping coefficient is equal to 5% of critical for all cases. Figure 3 indicates that equal support excitations produce the highest ordinates for the differential response spectra, which are independent of separation distance since $\rho(\xi,\omega)=1$. For the shorter separation distance of 100m, model II-case 1 produces higher amplitudes for the differential response spectra than model I; the differential response spectra of these two models tend to similar values as the modal frequency increases. For the longer separation distance (300m) model II-case 1 produces slightly higher amplitudes up to 12rad/sec than model I, but lower ones for higher natural frequencies. Model II-case 2 yields considerably lower amplitudes than the other models for both separation

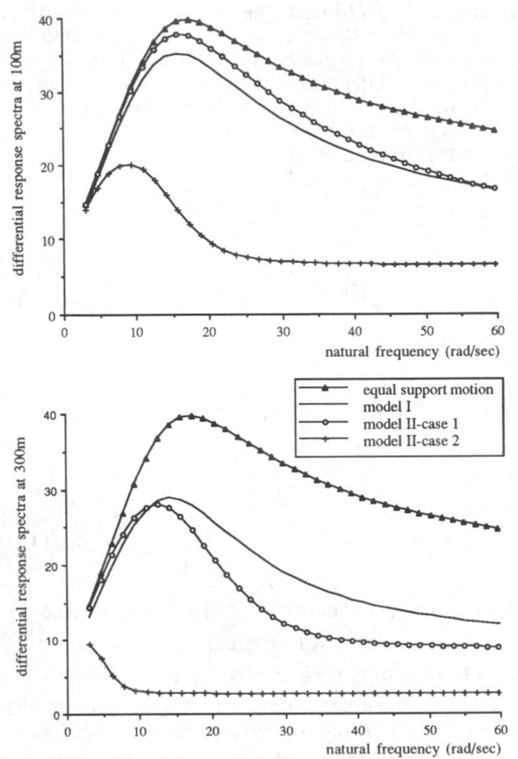

Figure 3. Differential response spectra [cm/sec^2]

distances. It is interesting to note that, although models I and II-case 2 produce a comparable quasi-static response, their contribution to the dynamic response of the structures is significantly different.

When the modal frequency falls within the range of the dominant frequencies of the seismic excitation, the major contribution to the response results from frequencies in the vicinity of the modal frequency under consideration. The seismic excitation for the evaluation of the differential response spectra in the case of equal support motions is twice the PSD of the ground acceleration. When the spatial incoherence is taken into account, the seismic excitation for the differential response spectra is twice the cross spectral density of the seismic motions (Eq. 11), which is the product of the PSD, common for all cases, and the spatial variability (Fig. 1). It follows then that the spatial variability model with higher correlation in the vicinity of the modal frequency yields higher ordinates for the differential response spectra. Indeed, the assumption of equal support motions, which yields the highest correlation (=1) for all frequencies and separation distances, produces the highest contribution to the dynamic response. For almost the entire frequency range analyzed when ξ=100m, and for ω<12rad/sec when ξ=300m, model II-case 1 produces higher correlations than model I (Fig. 1), and results in higher amplitudes for the differential response spectra (Fig. 3). For the range of higher frequencies at ξ=300m, the exponential decay of model II-case 1 is sharper than that of model I (Fig. 1), and the ordinates of its differential response spectrum become significantly lower than those of model II (Fig. 3). Model II-case 2, which decays sharply with separation distance and frequency, yields the lowest amplitudes for the spectra.

It should be emphasized at this point that the contribution of the spatial variability models to the quasi-static response of the structures through the differential displacements as well as the differential response spectra constitute only a part of the total response of the structures. The total response of a particular structure is the weighted superposition of these terms, the weights being functions of the structural characteristics (Eq. 6). Which spatial incoherence model induces the highest total response relies heavily on the structural configuration and properties.

4 CLOSURE

This study analyzes the variability that results in the seismic response of lifelines when the spatial variation of the seismic motions at their supports is described by different models. Differential displacements and differential response spectra are evaluated for this purpose. The analysis suggests that spatial incoherence models that yield partial correlations at low frequencies produce higher response in buried pipelines and higher contributions to the quasi-static internal forces in above-ground lifelines than models fully correlated in that frequency range. On the other hand, the comparison of the differential response spectra resulting from the models indicates that spatial variability models that decay slowly with separation distance and frequency contribute more to the dynamic response than models that exhibit a sharper exponential decay.

ACKNOWLEDGMENT

This study was supported by Grant No. BCS-9114895 from the National Science Foundation.

REFERENCES

Clough, R. W. & J. Penzien 1975. *Dynamics of Strucures*, McGraw-Hill, NY.

Der Kiureghian, A. & A. Neuenhofer 1991. A response spectrum method for multiple-support seismic excitations. University of California at Berkeley, Report No. UBC/EERC -91/08.

Harichandran, R. S. & E. H. Vanmarcke 1986. Stochastic variation of earthquake ground motion in space and time. *J. Eng. Mech.* 112: 154-174.

Hindy, A. & M. Novak 1980. Pipeline response to random ground motion.*J. Eng.Mech.* 106:339- 360.

Luco, J. E. & H. L. Wong 1986. Response of a rigid foundation to a spatially random ground motion *Earthq. Eng. Structural Dyn.* 14: 891-908.

Zerva, A. 1992a. Seismic ground motion simulations from a class of spatial variability models.*Earthq. Eng. Structural Dyn.* 21: 351-361.

Zerva, A. 1992b. Spatial incoherence effects on seismic ground strains. *Prob. Eng. Mech.* 7: 217-226.

Zerva, A. 1992c. Seismic loads predicted by spatial variability models. *Structural Safety* 11: 227-243.

Structural Safety & Reliability, Schuëller, Shinozuka & Yao (eds) © 1994 Balkema, Rotterdam, ISBN 90 5410 357 4

Probabilistic system performance measures for reinforced concrete masonry wall structures

George T. Zorapapel
Englekirk and Sabol Inc., Consulting Engineers, Los Angeles, Calif., USA
Gary C. Hart
Department of Civil Engineering, University of California, Los Angeles, Calif., USA
Dan M. Frangopol
Department of Civil Engineering, University of Colorado, Boulder, Colo., USA

ABSTRACT: This paper explores several probabilistic system performance measures for reinforced concrete masonry wall structures under lateral loads. In this context, the study defines both system ductility and system redundancy measures. Monte Carlo simulations provide useful properties of these system measures for both balanced and unbalanced reinforced concrete masonry wall systems.

1 INTRODUCTION

Masonry structures are commonly designed using reinforced concrete masonry walls to resist lateral loads. Two extreme types of performance are possible in a reinforced concrete masonry wall under lateral loads: either the wall can undergo large deformations while partially maintaining its strength (ductile behavior) or the wall can suddenly lose its strength (brittle behavior) and potentially collapse. The use of confinement reinforcement can improve the performance of masonry walls with respect to lateral loads. If concrete masonry is confined, it can generally sustain large strain levels and enhance the flexural ductility. It is only recently that studies have been performed to evaluate practical and efficient ways to confine concrete masonry (Carter 1990, Sajjad 1990). These studies concluded that the stress-strain relationship for confined concrete masonry is different from the stress-strain relationship of unconfined masonry. In fact, as demonstrated by Carter (1990), among others, stress-strain curves of confined masonry are expected to exhibit flatter declining slopes, higher residual stresses, and increased terminal strains.

An in-depth understanding of the performance of confined reinforced concrete masonry wall structures is necessary to develop more effective design and analysis methods. In this regard, the present study is an attempt to define probabilistic performance measures for reinforced concrete masonry wall structures, including system ductility and system redundancy. In order to quantify these performance measures, Monte-Carlo simulations are performed on *(a)* a concrete masonry wall with minimum vertical reinforcement and confinement of the vertical steel, and *(b)* structural systems composed of several flexural walls connected with rigid diaphragms. The reliability of a wall or a structural system is viewed as the conditional probability that ductility demand will be less than ductility capacity. Monte Carlo simulations provide useful properties of system ductility and system redundancy measures.

2 SYSTEM PERFORMANCE

The major difficulty in estimating the probability of failure of any element or structure subjected to earthquake loading is generally created by the extremely large uncertainty associated with the load. Even if the activity in some potential earthquake sources may be described probabilistically with a relatively good confidence, the uncertainty associated with the local soil amplification is generally very large.

In the current design practice, this shortcoming is overcome by providing structures with critical components that do not fail in a brittle manner but, rather, continue to deform inelastically. The amount of deformation that the structure can sustain from the first yield limit state to the ultimate limit state quantifies the ductility of the structure. Our objective, therefore, is to provide the subject structure with the ductility necessary to undergo the deformations required. Levels of *component ductility* and/or *redundancy* are now required in modern building codes but no conscious effort is made to produce acceptable levels of structural *system ductility* and/or *redundancy*.

In view of the above, the needs for quantifying system ductility and redundancy are evident. This study defines both system ductility and redundancy measures for reinforced concrete masonry wall structures. Properties of these measures are also investigated.

3 PROBABILISTIC SYSTEM DUCTILITY AND REDUNDANCY MEASURES

Under conditions of large uncertainty in the level of lateral loading, a flexural wall must behave in a ductile manner to withstand a major earthquake. In this study, the *ductility index of a wall* is defined as

$$DI_{wall} = (\bar{D}_{wall} - 1)/\sigma_{D_{wall}} \tag{1}$$

where $\bar{D}_{wall}$ and $\sigma_{D_{wall}}$ are the mean and standard deviation of the wall ductility, D_{wall}, which can be either the curvature ductility or the displacement ductility. In the case of masonry wall structures, a ductility index based on displacement was preferred to that based on curvature (Zorapapel 1991). The sensitivity of $\bar{D}_{wall}$ and $\sigma_{D_{wall}}$ to various input parameters (e.g., axial force, masonry stress and strain, confinement characteristics) was reported by Zorapapel (1991).

For a reinforced concrete masonry wall structure three limit states may be defined as follows: (a) *limit state 1:* all walls fail by reaching ultimate strain in tensioned steel prior to masonry crushing in compression; (b) *limit state 2:* at least one wall attains its yield strain in tension before the masonry crushes in compression and the structure reaches its ultimate displacement; and (c) *limit state 3:* all walls fail by masonry crushing in compression before the steel attains its yield strain in tension. Limit states 1 and 2 are of ductile type, and limit state 3 is brittle.

Besides the parameters that govern the ductility of a wall, there are additional factors that have an influence on the system performance (e.g., system ductility, system redundancy) such as: the size of the system (i.e., number of walls), the correlation among material parameters distributed among walls, and the distribution and correlation of axial load among walls. The effects of these factors on system performance for structures consisting of two to eight walls were obtained by Monte Carlo simulation (Zorapapel 1991).

For a reinforced concrete masonry wall structure, *system ductility* is defined as

$$D_{system} = \Delta_{ult,system}/\Delta_{yield,system} \tag{2}$$

where $\Delta_{ult,system}$ and $\Delta_{yield,system}$ are the ultimate and yield displacement of the structure. Fig. 1 presents the load-deflection response (the overturning moment at the base against the displacement at the top) for three statistically identical walls and for the structure consisting of these three walls. The yield displacement of the structure is defined consistently by the yield displacement of the bilinear model. Based on extensive Monte Carlo simulations, Figs. 2(a) and 2(b) present the change in mean and coefficient of variation of system ductility, respectively, versus number of walls for various levels of the axial load P.

The *system ductility index,* which is the measure of system robustness, is defined as a function of the statistics of D_{system} as follows

$$DI_{system} = (\bar{D}_{system} - 1)/\sigma_{D_{system}} \tag{3}$$

System ductility indices were computed for both balanced and unbalanced reinforced concrete masonry wall structures and are reported by Zorapapel (1991).

Another probabilistic system performance

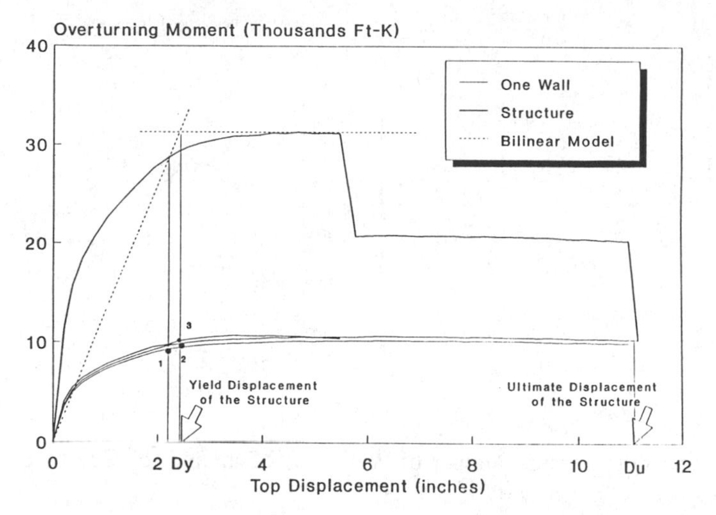

Fig. 1 Moment-Displacement Response of a Three-Wall Structure

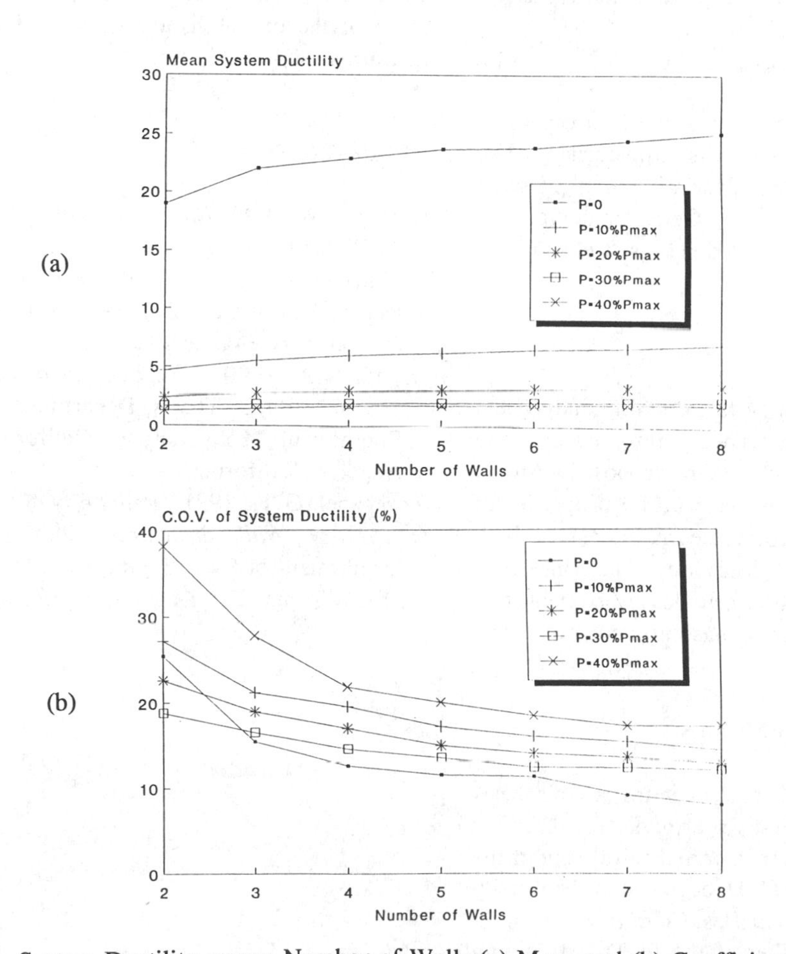

Fig. 2 System Ductility versus Number of Walls:(a) Mean and (b) Coefficient of Variation

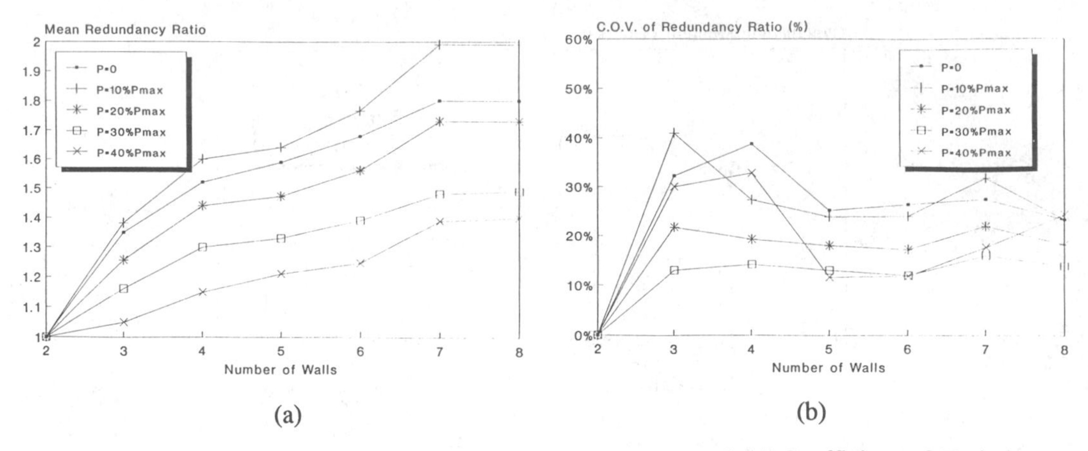

Fig. 3 Redundancy Ratio versus Number of Walls: (a) Mean and (b) Coefficient of Variation

measure is the *redundancy ratio* defined as

$$R_{system} = \Delta_{ult,system} / \Delta_{ult,first\ wall} \quad (4)$$

where $\Delta_{ult,first\ wall}$ is the displacement corresponding to the first wall reaching its ultimate state. The mean and coefficient of variation of R_{system} were also computed by Monte Carlo simulation and some results are presented in Figs. 3(a) and 3(b).

4 CONCLUSIONS

In this study, both system ductility and system redundancy measures for reinforced concrete masonry wall structures are proposed. Monte Carlo simulations indicate useful properties of these system performance measures for both balanced and unbalanced cases. These measures deserve more attention in the design and evaluation of reinforced concrete masonry wall structures.

ACKNOWLEDGEMENTS

Results presented in this paper are part of a thesis prepared by the first writer under the direction of the second writer in partial fulfillment of the requirements for a Ph.D. degree at the University of California, Los Angeles, California. The third writer, as a member of the Ph.D. Examining Committee, contributed to the development of some of the probabilistic system performance measures used in this study.

REFERENCES

Carter, E.W. 1990. *Effects of Confinement Steel on Flexural Behavior of Reinforced Masonry Shear Walls,* Master of Science Thesis, Department of Civil Engineering, University of Colorado, Boulder, Colorado.

Sajjad, N.A. 1990. *Confinement of Concrete Masonry,* Ph.D. Thesis, Department of Civil Engineering, University of California, Los Angeles, California.

Zorapapel, G.T., 1991. *Reliability of Concrete Masonry Wall Structures,* Ph.D Thesis, Department of Civil Engineering, University of California, Los Angeles, California.

Structural Safety & Reliability, Schuëller, Shinozuka & Yao (eds) © 1994 Balkema, Rotterdam, ISBN 90 5410 357 4

Author index